Higher MATHEMATICS for AQA GCSE

Tony Banks and David Alcorn

Causeway
Press

Pearson Education Limited
80 Strand
London
WC2R ORL
England

ISBN-13: 978-1-4058-3143-7
ISBN-10: 1-4058-3143-X

ARP impression 98

Exam questions
Past exam questions, provided by the *Assessment and Qualifications Alliance*, are denoted by the letters AQA. The answers to all questions are entirely the responsibility of the authors/publisher and have neither been provided nor approved by AQA.

Every effort has been made to locate the copyright owners of material used in this book. Any omissions brought to the notice of the publisher are regretted and will be credited in subsequent printings.

Page design
Billy Johnson

Reader
Barbara Alcorn

Artwork
David Alcorn

Cover design
Raven Design

Typesetting by Billy Johnson, San Francisco, California, USA

Printed by Ashford Colour Press Ltd

preface

Higher Mathematics for AQA GCSE has been written to meet the requirements of the National Curriculum Key Stage 4 Programme of Study and provides full coverage of the new **AQA Specifications for the Higher Tier of entry**.

The book is suitable for students preparing for assessment at the Mathematics Higher Tier of entry on either a 1-year or 2-year course or as a revision text.

In preparing the text, full account has been made of the requirements for students to be able to use and apply mathematics in written examination papers and be able to solve problems both with and without a calculator. Some chapters include ideas for investigational, practical and statistical tasks and give the student the opportunity to improve and practice their skills of using and applying mathematics.

The planning of topics within chapters and sections has been designed to provide efficient coverage of the specifications. Depending on how the book is to be used you can best decide on the order in which chapters are studied.

Chapters 1 - 9 Number
Chapters 10 - 22 Algebra
Chapters 23 - 34 Shape, Space and Measures
Chapters 35 - 40 Handling Data

Each chapter consists of fully worked examples with explanatory notes and commentary; carefully graded questions, a summary of what you need to know and a review exercise.
The review exercises provide the opportunity to consolidate topics introduced within the chapter and consist of exam-style questions, which reflect how the work is assessed, plus lots of past examination questions (marked AQA).

Further opportunities to consolidate skills acquired over a number of chapters are provided with section reviews, which have been organised into two parts for non-calculator and calculator practice.

As final preparation for the exams a further compilation of exam and exam-style questions, organised for non-calculator paper and calculator paper practice, has been included.

The book has been designed so that it can be used in conjunction with the companion book
Higher Mathematics for AQA GCSE - Student Support Book
Without Answers: ISBN 1-405834-92-7
With Answers: ISBN 1-405834-93-5

contents

CHAPTER 4 Percentages and Money

CHAPTER 5 Working with Number

CHAPTER 6 Standard Index Form

CHAPTER 7 Ratio and Proportion

CHAPTER 8 Speed and Other Compound Measures

CHAPTER 9 Extending the Number System

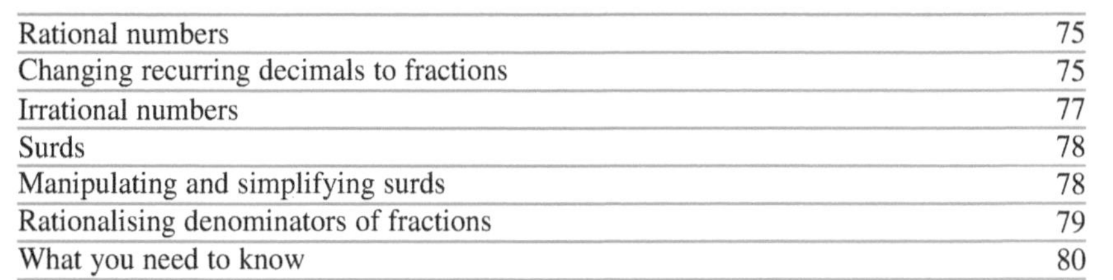

SR Section Review - Number

CHAPTER 10 Introduction to Algebra

CHAPTER 11 Solving Equations

CHAPTER 17 Further Graphs

CHAPTER 18 Direct and Inverse Proportion

CHAPTER 19 Quadratic Equations

CHAPTER 20 Simultaneous Equations

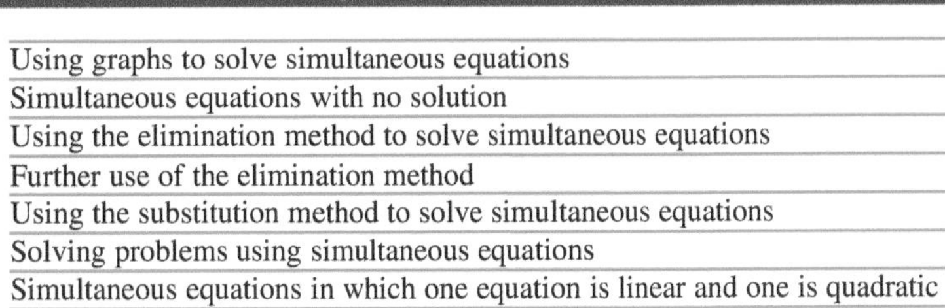

CHAPTER 21 Algebraic Methods

CHAPTER 22 Transforming Graphs

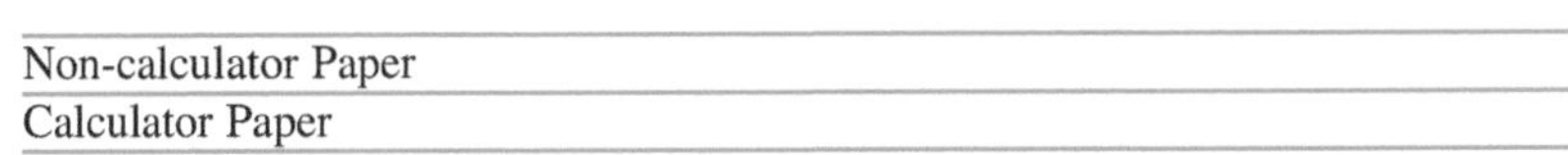

SR Section Review - Algebra

CHAPTER 23 Angles and 2-D Shapes

CHAPTER 24 Circle Properties

CHAPTER 25 Perimeters and Areas

CHAPTER 26 Maps, Loci and Constructions

CHAPTER 27 Transformations

CHAPTER 28 Pythagoras' Theorem

CHAPTER 29 Trigonometry

CHAPTER 30 3-D Shapes

CHAPTER 31 Understanding and Using Measures

CHAPTER 32 Congruent Triangles and Similar Figures

Whole Numbers

The numbers 0, 1, 2, 3, 4, 5, ... can be used to count objects.
Such numbers are called **whole numbers**.
Whole numbers that end in 0, 2, 4, 6 or 8 are called **even** numbers.
Whole numbers that end in 1, 3, 5, 7 or 9 are called **odd** numbers.

Place value

Our number system is made up of the digits 0, 1, 2, 3, 4, 5, 6, 7, 8 and 9.
The position a digit has in a number is called its **place value**.
In the number 5384 the digit 8 is worth 80, but in the number 4853 the digit 8 is worth 800.

Reading and writing numbers

$543 = 5 \times 100 + 4 \times 10 + 3 \times 1$
The number 543 is written or read as, "five hundred and forty-three."

For numbers bigger than one thousand:

- split the number into groups of three digits, starting from the units,
- combine the numbers of *millions* and *thousands* with the number less than 1000.

4567205 = 4 567 205 is written and read as,
"four *million* five hundred and sixty-seven *thousand* two hundred and five."

Non-calculator methods for addition and subtraction

Addition

Write the numbers in tidy columns according to place value.
Add together the numbers of units, tens, hundreds, etc.
If any of these answers are 10 or more, something is carried to the next column.

EXAMPLE

Work out 4567 + 835.

```
  4567    7 + 5 = 12                   which is 2 carry 1.
+  835    6 + 3 + carried 1 = 10,      which is 0 carry 1.
  ----    5 + 8 + carried 1 = 14,      which is 4 carry 1.
  5402    4 + carried 1 = 5.
  111
```

Subtraction

Write the numbers in columns according to place value.
Then, in turn, subtract the numbers of units, tens, hundreds, etc.
If the subtraction in a column cannot be done because the number being subtracted is greater, borrow 10 from the next column.

Addition is the opposite (inverse) operation to subtraction.
If $a - b = c$,
then $c + b = a$.

EXAMPLE

Work out 7238 − 642.

```
 6 11 1
 7 2 3 8      Units: 8 − 2 = 6
−   6 4 2      Tens: 3 − 4 cannot be done, so, borrow 10 from the 2 in the next column.
 -------             Now  10 + 3 − 4 = 9.
 6 5 9 6   Hundreds: 1 − 6 cannot be done, so, borrow 10 from the 7 in the next column.
                     Now  10 + 1 − 6 = 5.
           Thousands: 6 − 0 = 6.
```

You can use addition to check your subtraction. *Does 6596 + 642 = 7238?*

Exercise 1.1

Do not use a calculator for this exercise.

1. Write the following numbers in figures.
   (a) five hundred and forty-six
   (b) seventy thousand two hundred and nine
   (c) one million two hundred thousand and fifty-two

2. Write the following numbers in words.
   (a) 23590 (b) 6049 (c) 93145670 (d) 9080004

3. Which digit has the greater value, the 7 in 745 or the 9 in 892?

4. Write these numbers in ascending order (smallest to largest).
   3842, 5814, 3874, 3801, 4765

5. Write these numbers in descending order (largest to smallest).
   9434, 9646, 9951, 9653, 9648

6. By using each of the digits 5, 1, 6 and 2 make:
   (a) the largest four-digit **even** number,
   (b) the smallest four-digit **odd** number.
   In each case explain your method.

7. Work these out in your head.
   (a) 15 + 12 (b) 19 + 16 (c) 26 + 48 (d) 29 + 41 (e) 13 + 99
   (f) 26 − 9 (g) 87 − 38 (h) 204 − 99 (i) 1000 − 425 (j) 1003 − 999

8. Work these out by writing the numbers in columns.
   (a) 324 + 628 (b) 1273 + 729 (c) 3495 + 8708 (d) 431 + 865 + 245
   (e) 407 − 249 (f) 2345 − 1876 (g) 8045 − 1777 (h) 10 000 − 6723

9. Marcus drove 1754 miles on his holiday.
   His milometer reading was 38 956 at the start.
   What was his milometer reading at the end?

10. What is the total weight of these packages?

11. The last four attendances at a football stadium were:
    21 004, 19 750, 18 009, 22 267.
    What is the total attendance?

12. A batsman has scored 69 runs. How many more runs must he score to get a century?

13. A car park has spaces for 345 cars. On Tuesday 256 spaces are used.
    How many spaces are not used?

14. (a) Ricky buys a bunch of spring onions.
    He pays with 50p.
    How much change is he given?
    (b) Liz buys a cucumber and a lettuce.
    She pays with £1.
    How much change is she given?

Salad Specials

Cucumber	39p each
Lettuce	55p each
Spring onions	33p a bunch

15. The table shows the milometer readings for three cars at the start and end of a year.

	Car A	Car B	Car C
Start	2501	55667	48050
End	10980	67310	61909

Which car has done the most miles in the year?

Non-calculator methods for multiplication and division

Short multiplication

Short multiplication is when the multiplying number is less than 10, e.g. 165×7.
Multiply the units, tens, hundreds, etc. in turn.

$$\begin{array}{r} 165 \\ \times\ 7 \\ \hline 1155 \\ \hline {\scriptstyle 1\ 4\ 3} \end{array}$$

Units: $7 \times 5 = 35$, which is 5 carry 3.
Tens: $7 \times 6 = 42 +$ carried $3 = 45$, which is 5 carry 4.
Hundreds: $7 \times 1 = 7 +$ carried $4 = 11$, which is 1 carry 1.
There are no more digits to be multiplied by 7, the carried 1 becomes 1 thousand.

Long multiplication

Long multiplication is used when the multiplying number is greater than 10, e.g. 24×17.

Work out 24×17.

$$\begin{array}{r} 24 \\ \times\ 17 \\ \hline 168 \\ +\ 240 \\ \hline 408 \\ \hline \end{array}$$

$168 \leftarrow 24 \times 7 = 168$
$240 \leftarrow 24 \times 10 = 240$

A standard non-calculator method for doing long multiplication multiplies the number by:

- the units figure, then
- the tens figure, then
- the hundreds figure, and so on.

All these answers are added together.

Short division

The process of dividing a number by a number less than 10 is called **short division**.
Short division relies on knowledge of the Multiplication Tables.
What is $32 \div 8$, $42 \div 7$, $72 \div 9$, $54 \div 6$?

Work out $882 \div 7$.

$7\overline{)8\,{}^{1}8\,{}^{4}2}$
$\quad 1\,2\,6$

Starting from the left:
$8 \div 7 = 1$ remainder 1, which is 1 carry 1.
$18 \div 7 = 2$ remainder 4, which is 2 carry 4.
$42 \div 7 = 6$, with no remainder.
So, $882 \div 7 = 126$.

You can check your division by multiplying.
Does $126 \times 7 = 882$?

Multiplication is the opposite (inverse) operation to division.
If $a \div b = c$,
then $c \times b = a$.

Long division

Long division works in exactly the same way as short division, except that all the working out is written down.

Work out $952 \div 7$.

$$\begin{array}{r} 136 \\ 7\overline{)952} \\ 7 \\ \hline 25 \\ 21 \\ \hline 42 \\ 42 \\ \hline 0 \\ \hline \end{array}$$

$9 \div 7 = 1$ and a remainder.
What is the remainder?
$1 \times 7 = 7$ (write below the 9).
$9 - 7 = 2$ (which is the remainder).
Bring down the next figure (5) to make 25.
Repeat the above process.
$25 \div 7 = 3$ and a remainder.
$3 \times 7 = 21$, $25 - 21 = 4$ (remainder).
Bring down the next figure (2) to make 42 and repeat the process.
$42 \div 7 = 6$, but there is no remainder.
$6 \times 7 = 42$, $42 - 42 = 0$ (remainder).
There are no more figures to be brought down and there is no remainder.
So, $952 \div 7 = 136$.

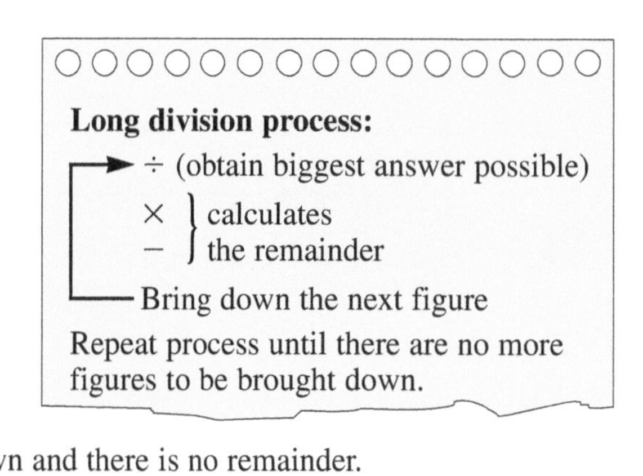

Long division process:

→ $\div$ (obtain biggest answer possible)
$\times$ } calculates
$-$ } the remainder
Bring down the next figure

Repeat process until there are no more figures to be brought down.

Exercise 1.2

Do not use a calculator for this exercise.

1. Work out. (a) 183 × 3 (b) 36 × 7 (c) 264 × 8 (d) 3179 × 5

2. Work out these divisions. Use multiplication to check your answers.
(a) 85 ÷ 5 (b) 816 ÷ 6 (c) 824 ÷ 4 (d) 9882 ÷ 9 (e) 80 560 ÷ 4

3. When John uses a store card he gets 4 points for every pound he spends. He spends £18.
How many points does he get?

4. A caretaker puts out 7 rows of chairs. There are 13 chairs in each row.
How many more chairs are needed for 120 chairs to be put out?

5. Sanjay is paid £7 per hour. One week he is paid £238.
How many hours did he work that week?

6. Work out. (a) 537 × 40 (b) 260 × 50 (c) 412 × 80 (d) 239 × 90

7. PLAYHOUSE THEATRE
PRESENT
CABARET
Tickets: £19 Programmes: £3

On the first night of the show, 70 tickets and 40 programmes are sold.
How much money is paid in total for the tickets and programmes?

8. There are 400 metres in one lap of a running track. How many metres are there in 25 laps?

9. A secretary buys one dozen boxes of staples. There are 5000 staples in each box.
How many staples does she buy?

10. Work out. (a) 7590 ÷ 30 (b) 7110 ÷ 90 (c) 21 480 ÷ 40 (d) 7560 ÷ 60

11. The total cost of a New Year's Eve party for 80 people is £2800.
What is the cost for each person?

12. A computer can print 50 000 characters per minute.
How many minutes would it take to print one million characters?

13. Work out. (a) 76 × 32 (b) 234 × 57 (c) 537 × 103 (d) 693 × 875

14. What is the total cost of 14 desks and 23 cabinets?

15. The organisers of a concert sell 1624 tickets at £27 each.
How much money is collected from the sale of tickets?

16. Work out. (a) 473 ÷ 11 (b) 7560 ÷ 15 (c) 9430 ÷ 23 (d) 20 928 ÷ 32

17. John collects £17 each from his friends to buy tickets for a football match.
He collects a total of £391. How many tickets does he buy?

18. Work these out and state the remainder each time.
(a) 410 ÷ 25 (b) 607 ÷ 24 (c) 800 ÷ 45 (d) 525 ÷ 37

19. A pint of milk costs 35 pence. How many pints of milk can be bought for £5?
How much change will there be?

20. Tins of beans are packed in boxes of 24. A supplier has 1000 tins of beans.
(a) How many full boxes is this? (b) How many tins are left over?

Order of operations in a calculation

What is $4 + 3 \times 5$? It is not sensible to have two possible answers.
It has been agreed that calculations are done obeying certain rules:

First	Brackets and Division line
Second	Divide and Multiply
Third	Addition and Subtraction

EXAMPLES

1. $4 + 3 \times 5 = 4 + 15 = 19$
2. $10 \div 2 + 3 = 5 + 3 = 8$
3. $10 \div (2 + 3) = 10 \div 5 = 2$
4. $(5 + 6) \times 3 + 4 = 11 \times 3 + 4 = 33 + 4 = 37$
5. $\frac{12}{11 - 8} - 3 = \frac{12}{3} - 3 = 4 - 3 = 1$

   This is the same as $12 \div (11 - 8) - 3$.

Exercise 1.3

Do not use a calculator for this exercise.

1. Work out the value of $17 - 3 \times (6 - 1)$.
2. Work these out.
   (a) $7 + 6 \times 5$ (b) $7 - (6 - 2)$ (c) $24 \div 6 + 5$
   (d) $7 \times 6 + 8 \times 2$ (e) $10 \div 5 + 8 \div 2$ (f) $(5 - 2) \times 7 + 9$
   (g) $60 \div (5 + 7)$ (h) $60 \div 5 + 7$ (i) $4 \times 3 + 2$
   (j) $4 \times (3 + 2)$ (k) $12 \times (20 - 2) \div 9$ (l) $36 \div (5 + 4)$
   (m) $4 \times 12 \div 8 - 6$ (n) $\frac{15}{18 - 3} + 4$ (o) $\frac{22 - 4}{9} + 12 \div 3$
3. Use brackets and the signs $+$, $-$, $\times$ and $\div$ to complete these sums.
   (a) $7 \quad 2 \times 3 = 15$ (b) $3 + 5 \quad 2 = 4$ (c) $(4 \quad 1) \times 7 \quad 2 = 25$
4. Using all the numbers 6, 3, 2 and 1 in this order, brackets and the signs $+$, $-$, $\times$ and $\div$, make all the numbers from 1 to 10.
   $6 - 3 \times 2 + 1 = 1$, $6 - 3 - 2 + 1 = 2$, and so on.
5. Claire is 16 cm taller than Rachel. Their heights add up to 312 cm.
   How tall is Rachel?
6. A box, which contains 48 matches, has a total weight of 207 g.
   If each match weighs 4 g, what is the weight of the empty box?
7. The admission charges to a zoo are £7 for a child and £12 for an adult.
   Zoe is organising a trip to the zoo for a group of people and worked out that the total cost would be £336.
   She collected £84 from the adults in the group.
   (a) How many children are in the group?
   (b) What is the total number of people in the group?

Negative numbers

The number line can be extended to include **negative whole numbers**.

−5 −4 −3 −2 −1 0 1 2 3 4 5 6

As you move from left to right along the number line the numbers become bigger.

As you move from right to left along the number line the numbers become smaller.

Negative whole numbers, zero and positive whole numbers are called **integers**.
−5 can be read as "minus five" or "negative five".
A number written without a sign before it is assumed to be positive. +5 has the same value as 5.
Many problems in mathematics involve calculations with negative numbers.

Adding and subtracting negative numbers

EXAMPLES

1. Work out $2 + (-6)$.

   $+-$ can be replaced with $-$.
   Start at 2 and move 6 to the left.
   $2 + (-6) = 2 - 6 = -4$

2. Work out $-2 - (-8)$.

   $--$ can be replaced with $+$.
   Start at −2 and move 8 to the right.
   $-2 - (-8) = -2 + 8 = 6$

To add or subtract negative numbers:
Replace double signs with the single sign.

+ + can be replaced by +
− − can be replaced by +
+ − can be replaced by −
− + can be replaced by −

Multiplying and dividing negative numbers

EXAMPLES

1. Work out $(-4) \times (-8)$.

   Signs: $- \times - = +$
   Numbers: $4 \times 8 = 32$
   So, $(-4) \times (-8) = 32$.

2. Work out $(+8) \div (-2)$.

   Signs: $+ \div - = -$
   Numbers: $8 \div 2 = 4$
   So, $(+8) \div (-2) = -4$.

Use these rules for multiplying and dividing negative numbers:

When multiplying:	When dividing:
$+ \times + = +$	$+ \div + = +$
$- \times - = +$	$- \div - = +$
$+ \times - = -$	$+ \div - = -$
$- \times + = -$	$- \div + = -$

Work logically:
Work out the sign first. Then work out the numbers.

Exercise 1.4

Do not use a calculator for this exercise.

1. List these temperatures from coldest to hottest. 23°C, −28°C, −3°C, 19°C, −13°C.
2. List these numbers from lowest to highest. 31, −78, 51, −39, −16, −9, 11.
3. At midnight, the temperature in York is 3°C below freezing and in Bath the temperature is 2°C above freezing.
   What is the difference in temperature between York and Bath?

4 Adrian is 5 kg heavier than Tim. Matt is 3 kg lighter than Tim.
What is the difference in weight between Matt and Adrian?

5 Work out.

(a) $-3 + (+5)$ (b) $5 + (-4)$ (c) $-2 + (-7)$
(d) $-1 + (+9)$ (e) $7 + (-3)$ (f) $15 + (-20)$
(g) $-11 + (+4)$ (h) $11 + (-4)$ (i) $8 + (-7)$
(j) $-8 + (-7)$ (k) $3 + (+3) + (-9)$ (l) $-7 + (-5) + 6$

6 Work out.

(a) $8 - (-5)$ (b) $-4 - (-10)$ (c) $10 - (+3)$
(d) $6 - (-1)$ (e) $-5 - (-10)$ (f) $-4 - (+8)$
(g) $-7 - (-6)$ (h) $7 - (-6)$ (i) $-2 - (+9)$
(j) $2 - (-9)$ (k) $5 - (+5) + 9$ (l) $-10 - (-6) + 4$

7 Work out.

(a) $10 + 5 - 8 + 6 - 7$ (b) $12 + 8 - 15 + 7 - 20$ (c) $30 - 20 + 12 - 50$
(d) $6 + 12 - 14 - 4$ (e) $37 - 23 - 24 - 25$ (f) $12 + 13 + 14 - 20$

8 Work out.

(a) $5 + (-4) - (-3) + 2 - (-1)$ (b) $5 - 4 + (-3) - (-2) + (-1)$
(c) $10 - (-11) + (-12) + 13 - (-14)$ (d) $-7 - (-7) + 6 + (-3) + (-9)$
(e) $12 + 8 - (-8) + 9 - (-1)$ (f) $15 - (-5) + 5 - (-10) + (-20)$
(g) $-5 + (-5) + (-5) + (-5) - (-5)$ (h) $5 - (-5) - (-5) - (-5) - (-5)$

9 What is the difference in temperature between
(a) Edinburgh and Rome,
(b) Edinburgh and New York,
(c) Moscow and New York,
(d) Moscow and Cairo?

City	Temperature
Edinburgh	−7°C
Moscow	−22°C
New York	−17°C
Rome	3°C
Cairo	15°C

10 The temperature inside an igloo is −5°C. The temperature outside the igloo is 17°C cooler.
What is the temperature outside the igloo?

11 Work out.

(a) $(-7) \times (+5)$ (b) $(-7) \times (-5)$ (c) $8 \times (-3)$
(d) $(-5) \times 9$ (e) $(-8) \times (-8)$ (f) $(-7) \times 6$
(g) $8 \times (-10)$ (h) $(-4) \times (-8)$ (i) $(+5) \times (-2) \times (+2)$
(j) $(+4) \times (-3) \times (-5)$ (k) $(-3) \times (-2) \times (-5)$ (l) $(-5) \times (+3) \times (-4)$
(m) $(-5) \times (+3) \times (+4)$ (n) $(-5) \times (-4) \times (-5)$ (o) $(-3) \times (+7) \times (+9)$

12 Work out.

(a) $(-8) \div (+2)$ (b) $(-8) \div (-2)$ (c) $(+20) \div (+4)$
(d) $(+20) \div (-4)$ (e) $(-20) \div (+4)$ (f) $(-20) \div (-4)$
(g) $(+18) \div (+3)$ (h) $(-18) \div (+3)$ (i) $(-24) \div (-6)$
(j) $(+24) \div (-3)$ (k) $(-30) \div (-5)$ (l) $(-30) \div (+6)$
(m) $\frac{(-3) \times (+8)}{(-6)}$ (n) $\frac{(-6) \times (-8)}{(+3) \times (-4)}$ (o) $\frac{(-2) \times (-3) \times (+4)}{(+3) \times (-8)}$

What you need to know

- You should be able to read and write numbers expressed in figures and words.
- Be able to recognise the place value of each digit in a number.
- Know the Multiplication Tables up to 10 × 10.
- Use non-calculator methods for addition, subtraction, multiplication and division.
- Know the order of operations in a calculation.
- Add (+), subtract (−), multiply (×) and divide (÷) with negative numbers.

Review Exercise 1

Do not use a calculator for this exercise.

1. By using each of the digits 4, 8, 7 and 5,
write the smallest 4-digit **odd** number you can.

2. Gaby stated: "The sum of three consecutive numbers is always an odd number."
By means of an example, show that this statement is not true.

3. The chart shows the distances in miles between some towns.

Liverpool			
109	Nottingham		
77	44	Sheffield	
102	87	61	York

Isaac drives from Liverpool to Nottingham, from Nottingham to York and then from York back to Liverpool. Calculate the total distance he drives.

4. A plank of wood is 396 cm in length. The plank is cut into two pieces.
One piece is 28 cm longer than the other.
How long is the shorter piece of wood?

5. (a) Work out the values of $(1 + 3) \times 5 + 5$
(b) Add brackets ( ) to make each statement correct.
You may use more than one pair of brackets in each statement.
(i) $1 + 3 \times 5 + 5 = 40$ (ii) $1 + 3 \times 5 + 5 = 31$

6. Use the information that $45 \times 230 = 10\,350$ to find the value of 44×230.

7. Work out.
(a) $-7 - (-11) + (-5)$ (b) $(-8) \times (-5)$ (c) $\frac{(-1) \times (-5) \times (+6)}{(-3)}$

8. One thousand chocolate biscuits are packed in boxes of 6.
(a) How many full boxes will there be?
(b) How many biscuits will be left over?

9. Angelique is taking 149 pupils to the theatre to see the Royal Ballet.
The cost of the theatre ticket is £23 per person. Work out the total cost of the tickets. AQA

10. Sue has collected £544 from her friends at work for theatre tickets. The tickets cost £17 each.
Work out the number of theatre tickets she can buy. AQA

11. A survey counted the number of visitors to a website on the Internet.
Altogether it was visited 30 million times. Each day it was visited 600 000 times.
Based on this information, for how many days did the survey last? AQA

12. Given that $530 \times 97 = 51\,410$, work out $\frac{514\,100}{53}$.

13. The temperature inside a fridge is 3°C. The temperature inside a freezer is −18°C.
(a) How much colder is it inside the freezer than inside the fridge?
(b) Calculate $\frac{9 \times (-18)}{5} + 32$ to find the temperature of the freezer in degrees Fahrenheit.
AQA

14. Ama and Bob take part in a quiz. The rules are:

For each correct answer +4 points. For each wrong answer −3 points.

After the first round Ama has +6 points and Bob has +13 points.
In the second round each person answers 5 more questions.
Bob scores −1 point in this round. This is added to his previous points.
What is the least number of questions Ama must answer correctly in the second round to have more points than Bob **in total**? AQA

CHAPTER 2 Decimals and Fractions

Numbers and quantities are not always whole numbers.
The number system you met in Chapter 1 can be extended to include **decimal numbers** and **fractions**.

Decimals

A **decimal point** is used to separate the whole number part from the decimal part of the number.

73.26 — This number is read as seventy-three point two six.

whole number, 73 — decimal part, 2 tenths + 6 hundredths (which is the same as 26 hundredths)

Many measurements are recorded using decimals, including money, time, distance, weight, volume, etc.
The metric and imperial measures you need to know are given in Chapter 31.

Non-calculator methods for addition and subtraction of decimals

Write the numbers in tidy columns according to place value.
This is easily done by keeping the decimal points in a vertical column.
Start the addition or subtraction from the right, just as you did for whole numbers.
Use the same methods for carrying and borrowing as well.

EXAMPLE

Work out 17.1 − 8.72.

```
  6 10 1
 1 7.1 0
−  8.7 2
 -------
   8.3 8
```

Check the answer by addition.

Useful tip:
Writing 17.1 as 17.10 can make the working easier.
This does not change the value of 17.1.

Exercise 2.1

Do not use a calculator for this exercise.

1. On these scales what numbers are shown by the arrows?
   (a) Scale from 5 to 6 with arrows A and B.
   (b) Scale from 0.5 to 0.6 with arrows C and D.
2. Write these decimals in ascending order. 3.567, 3.657, 3.576, 3.675, 3.652
3. Kevin is 0.15 m shorter than Sally. Sally is 1.7 m tall. How tall is Kevin?
4. Work these out in your head.
   (a) 2.5 + 8.4 (b) 0.7 + 0.95 (c) 0.36 + 0.54 (d) 6.47 + 4.53
   (e) 2.7 − 1.5 (f) 1.3 − 0.7 (g) 0.4 − 0.16 (h) 15.3 − 6.4
5. Work out.
   (a) 6.54 + 0.27 + 0.03 (b) 10 − 4.78 (c) 9.57 − 4.567
   (d) 2.22 + 0.78 (e) 9.13 − 7.89 (f) 5.564 + 0.017 + 10.2
6. Add these amounts of money: 31p, £0.25, 27p, £2.49.
   What is the change from £5?
7. Fred cuts three pieces of wood of length 0.95 m, 1.67 m and 2.5 m from a plank 10 m long.
   How much wood is left?

Non-calculator method for multiplying decimals

1. Ignore the decimal points and multiply the numbers using long multiplication.
2. Count the total number of decimal places in the numbers being multiplied together.
3. Place the decimal point so that the answer has the same total number of decimal places.

The result of multiplying two numbers is called the **product**.

EXAMPLES

1 Work out 0.2×0.4.

$$\begin{array}{r} 0.2 \\ \times\, 0.4 \\ \hline 0.0\,8 \\ \hline \end{array}$$

0.2 has 1 decimal place.
0.4 has 1 decimal place.
The answer has 2 decimal places.

$4 \times 2 = 8$
The answer must have 2 decimal places.
Noughts are used in the answer to locate the decimal point and to preserve place value.

2 Work out 4.25×0.18.

$$\begin{array}{r} 4.2\,5 \\ \times\, 0.1\,8 \\ \hline 3\,4\,0\,0 \\ 4\,2\,5\,0 \\ \hline 0.7\,6\,5\,0 \\ \hline \end{array}$$

3400 ← 425×8
4250 ← 425×10
0.7650 The answer must have 4 decimal places because 4.25 has 2 and 0.18 has 2.

$4.25 \times 0.18 = 0.7650$ This can be written as 0.765 which has the same value as 0.7650.

Non-calculator method for dividing decimals

1. Multiply the dividing number by a power of 10 (10, 100, 1000, …) so that it becomes a whole number.
2. Multiply the number to be divided by the same number.
3. If necessary the answer will have a decimal point in the same place.

The result of dividing one number by another is called the **quotient**.

EXAMPLES

1 Work out $4 \div 0.8$.

Multiply both numbers by 10.
$40 \div 8 = 5$

2 Work out $2 \div 0.25$.

Multiply both numbers by 100.
$200 \div 25 = 8$

3 Work out $9 \div 4$.

$$4\overline{)9.{}^{1}0{}^{2}0}$$
$$2.2\,5$$

Noughts are added until the division is finished.
Adding noughts does not change the value of the number. 9 has the same value as 9.00.
Continue dividing until either there is no remainder or the required accuracy is obtained.

Exercise 2.2

Do not use a calculator for this exercise.

1 A biro costs 25 pence. (a) How much will 10 cost? (b) How much will 100 cost?

2 (a) 100 calculators cost £795. How much does one cost?
(b) 1000 pencils cost £120. How much does one cost?
(c) 10 litres of petrol cost £8.69. How would the cost of 1 litre be advertised?

3 Write down pairs of calculations which give the same answer.

12.3×1000	$12.3 \div 100$	12.3×0.1	$12.3 \div 0.01$	12.3×10	$12.3 \div 0.1$
12.3×100	12.3×0.001	$12.3 \div 0.001$	$12.3 \div 10$	$12.3 \div 1000$	12.3×0.01

4 A puzzle costs £1.90. How much will 4 puzzles cost?

5 (a) I buy 5 kites which cost £2.99 each. What is the total cost?
(b) How much change will I get from £20?

6 Work these out in your head.
(a) 1.7×5 (b) 2.6×0.5 (c) 0.6×0.8 (d) 0.3×0.2
(e) $6 \div 0.4$ (f) $10 \div 2.5$ (g) $6 \div 0.12$ (h) $12 \div 0.5$

7 A cup of coffee costs £1.15. How much will I have to pay for 9 cups?

8 A pack of 7 tubes of oil paint costs £9.45. How much does each tube of oil paint cost?

9 Calculate these products.
(a) 0.25×0.3 (b) 2.5×3.5 (c) 8.7×1.9 (d) 4.1×0.25
(e) 0.9×4.32 (f) 13.4×0.7 (g) 0.06×0.72 (h) 0.35×0.08

10 Work out the cost of these vegetables.
(a) 0.6 kg of carrots at 35p per kilogram.
(b) 4.6 kg of potatoes at 40p per kilogram.
(c) 1.2 kg of cabbage at 65p per kilogram.

11 Work out.
(a) $8 \div 5$ (b) $10.5 \div 6$ (c) $9 \div 8$ (d) $2.46 \div 0.2$
(e) $0.146 \div 0.05$ (f) $100.1 \div 0.07$ (g) $0.05 \div 0.004$ (h) $0.3 \div 0.008$

12 (a) Multiply each of these numbers by 0.6. (i) 5 (ii) 2.5 (iii) 0.4
(b) What do you notice about the original numbers and each of your answers?

13 (a) Divide each of these numbers by 0.6. (i) 6 (ii) 3.6 (iii) 0.18
(b) What do you notice about the original numbers and each of your answers?

14 What is the cost of each of 4.5 metres of linen at £13.50 per metre?

15 (a) Work out the cost for each of these portions of cheese.
(i) 0.7 kg of Stilton.
(ii) 1.6 kg of Cheddar.
(iii) 0.8 kg of Sage Derby.
(iv) 0.45 kg of Cotherstone.
(b) Corrine buys:
0.25 kg of Stilton and 1.4 kg of Cheddar.
She pays with a £10 note.
How much change will she get?

16 8 bandages cost £14. How much does each bandage cost?

17 A steel bar is 12.73 metres long.
How many pieces, each 0.19 metres long, can be cut from it?

18 A jug holds 1.035 litres. A small glass holds 0.023 litres.
How many of the small glasses would be required to fill the jug?

Numerical calculations with a calculator

EXAMPLE

Calculate the value of $\frac{326.4}{(5 - 3.3)} - 500$

This is the same as $326.4 \div (5 - 3.3) - 500$.
So, the division line is the same as brackets.

To do the calculation enter the following sequence into your calculator.

3 2 6 . 4 ÷ (5 − 3 . 3) −
5 0 0 = (The answer is −308)

Exercise 2.3 Use a calculator in this exercise.

1. (a) Calculate the value of $\frac{2.76 + 3.2}{1.25}$
   (b) Write down the sequence of key presses that you made to work out (a).
2. (a) Calculate $\frac{43.7}{22.5 \times 0.92}$
   (b) Write down the sequence of key presses that you made to work out (a).
3. (a) Work out $\frac{5.76 \times 3.6}{6.4 - 4.7}$
   (b) Write down a sequence of calculator keys that could be used to answer (a).
4. Jill used the following key sequences to answer the questions below.
   A 5 . 4 2 ÷ 2 . 3 = + 5 . 7 5 =
   B 5 . 4 2 + 5 . 7 5 = ÷ 2 . 3 =
   C 5 . 4 2 + (5 . 7 5 ÷ 2 . 3) =
   D 5 . 4 2 ÷ (2 . 3 + 5 . 7 5) =
   (a) $5.42 + \frac{5.75}{2.3}$ (b) $\frac{5.42 + 5.75}{2.3}$ (c) $\frac{5.42}{2.3 + 5.75}$ (d) $\frac{5.42}{2.3} + 5.75$
   Which key sequences can be used to answer the questions correctly?
5. Calculate the following. Write down the full calculator display.
   (a) $\frac{4.326 \times 0.923}{5.623 \times 9.123}$ (b) $\frac{4.326 + 0.923}{5.623 + 9.123}$ (c) $\frac{5.19 \times 37.2}{16.7 + 0.78}$ (d) $\frac{5.2 \times 2.5 + 6.3}{10.34 - 4.7}$
   (e) $300 - \frac{5.7 \times 8.3}{4.6 - 6.7}$ (f) $19.3 \div (5.7 - 2.3 \div 0.4)$ (g) $(5 - (5.6 - 4.1) \times 2.3) \div (6.9 + 3.1)$

Fractions

$\frac{3}{8}$ is a **fraction**.

In a fraction: the top number is called the **numerator**,
the bottom number is called the **denominator**.

$\frac{3}{8}$ of this rectangle is shaded.

Equivalent fractions

Fractions which are equal are called **equivalent fractions**.

Each of the fractions $\frac{1}{4}$, $\frac{3}{12}$, $\frac{6}{24}$,
is the same fraction written in different ways.

These fractions are all equivalent to $\frac{1}{4}$.

To write an equivalent fraction:
Multiply the numerator and denominator by the **same** number.

For example. $\frac{1}{4} = \frac{1 \times 3}{4 \times 3} = \frac{3}{12}$

Simplifying fractions

Fractions can be **simplified** if both the numerator and denominator can be divided by the **same number**.

To write a fraction in its **simplest form** divide both the numerator and denominator by the largest number that divides into them both.

This is sometimes called **cancelling** a fraction.

Remember:
Multiplication and division are inverse (opposite) operations.
Equivalent fractions can also be made by dividing the numerator and denominator of a fraction by the same number.

EXAMPLE

Write the fraction $\frac{25}{30}$ in its simplest form.

The largest number that divides into both the numerator and denominator of $\frac{25}{30}$ is 5.

$\frac{25}{30} = \frac{25 \div 5}{30 \div 5} = \frac{5}{6}$ $\quad$ $\frac{25}{30} = \frac{5}{6}$ in its simplest form.

Types of fractions

Numbers like $2\frac{1}{2}$ and $1\frac{1}{4}$ are called **mixed numbers** because they are a mixture of whole numbers and fractions.
Mixed numbers can be written as **improper** or '**top heavy**' fractions.
These are fractions where the numerator is larger than the denominator.

EXAMPLES

1 Write $3\frac{4}{7}$ as an improper fraction.

$3\frac{4}{7} = \frac{(3 \times 7) + 4}{7} = \frac{21 + 4}{7} = \frac{25}{7}$

2 Write $\frac{32}{5}$ as a mixed number.

$32 \div 5 = 6$ remainder 2.

$\frac{32}{5} = 6\frac{2}{5}$

Finding fractions of quantities

EXAMPLES

1 Find $\frac{2}{5}$ of £65.

Divide £65 into 5 equal parts.
£65 ÷ 5 = £13.

Each of these parts is $\frac{1}{5}$ of £65.

Two of these parts is $\frac{2}{5}$ of £65.

So, $\frac{2}{5}$ of £65 = 2 × £13 = £26.

2 A coat costing £138 is reduced by $\frac{1}{3}$.
What is the reduced price of the coat?

Find $\frac{1}{3}$ of £138.

$\frac{1}{3}$ of £138 = £138 ÷ 3 = £46

So, reduced price = £138 − £46 = £92

Exercise 2.4

Do this exercise without using a calculator.

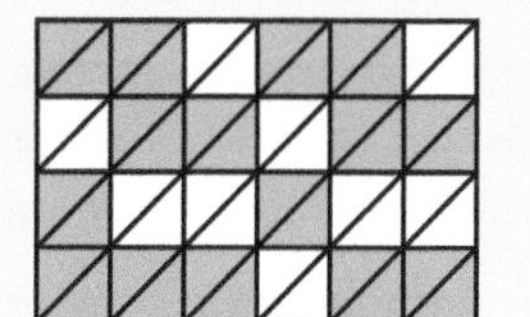

1. Write three equivalent fractions for the shaded part of this rectangle.
   What is the simplest form of the fraction for the shaded part?
2. Write these fractions in order, smallest first. $\frac{2}{3}$ $\quad$ $\frac{3}{5}$ $\quad$ $\frac{7}{10}$ $\quad$ $\frac{8}{15}$

3 Which of these fractions is the largest? $\frac{4}{5}$ $\frac{13}{20}$ $\frac{7}{10}$ $\frac{3}{4}$

4 Write these fractions in their simplest form. (a) $\frac{12}{15}$ (b) $\frac{18}{27}$ (c) $\frac{50}{75}$ (d) $\frac{16}{40}$

5 Change these fractions to mixed numbers: (a) $\frac{3}{2}$ (b) $\frac{17}{8}$ (c) $\frac{15}{4}$ (d) $\frac{23}{5}$

6 Change these numbers to improper fractions: (a) $2\frac{7}{10}$ (b) $1\frac{3}{5}$ (c) $5\frac{5}{6}$

7 Write 42 as a fraction of 70. Give your answer in its simplest form.

8 A box of 50 chocolates includes 30 soft-centred chocolates.
What fraction of the chocolates are soft-centred?

9 Calculate:

(a) $\frac{1}{4}$ of 12 (b) $\frac{1}{5}$ of 20 (c) $\frac{1}{10}$ of 30 (d) $\frac{1}{6}$ of 48 (e) $\frac{2}{5}$ of 20

(f) $\frac{3}{10}$ of 30 (g) $\frac{2}{7}$ of 42 (h) $\frac{5}{9}$ of 36 (i) $\frac{5}{6}$ of 48 (j) $\frac{3}{8}$ of 32

10 Richard has 30 marbles. He gives $\frac{1}{5}$ of them away.
(a) How many marbles does he give away?
(b) How many marbles has he got left?

11 Alfie collects £12.50 for charity. He gives $\frac{3}{5}$ of it to Oxfam.
How much does he give to other charities?

12 In a sale all prices are reduced by $\frac{3}{10}$.
What is the sale price of a microwave which was originally priced at £212?

13 A publisher offers a discount of $\frac{3}{20}$ for orders of more than 100 books.
How much would a shop pay for an order of 250 books costing £15 each?

14 Lauren and Amelia share a bar of chocolate. The chocolate bar has 24 squares.
Lauren eats $\frac{3}{8}$ of the bar. Amelia eats $\frac{5}{12}$ of the bar.
(a) How many squares has Lauren eaten?
(b) How many squares has Amelia eaten?
(c) What fraction of the bar is left?

Adding and subtracting fractions

Calculate $1\frac{3}{4} + \frac{5}{6}$

Change mixed numbers to improper ('top heavy') fractions. $1\frac{3}{4} = \frac{7}{4}$

The calculation then becomes $\frac{7}{4} + \frac{5}{6}$

Find the smallest number into which both 4 and 6 will divide.
4 divides into: 4, 8, 12, 16, …
6 divides into: 6, 12, 18, …
So, 12 is the smallest.

Change the original fractions to equivalent fractions using the smallest number as the new denominator. $\frac{7}{4} = \frac{21}{12}$ and $\frac{5}{6} = \frac{10}{12}$

Add the new numerators. Keep the new denominator the same. $\frac{21}{12} + \frac{10}{12} = \frac{21 + 10}{12} = \frac{31}{12}$

Write the answer in its simplest form. $\frac{31}{12} = 2\frac{7}{12}$

Fractions must have the **same denominator** before addition (or subtraction) can take place.

*What happens when you use a denominator that is **not** the smallest?*

EXAMPLES

1 Work out $\frac{5}{8} - \frac{7}{12}$.

24 is the smallest number into which both 8 and 12 divide.

$\frac{5}{8} = \frac{5 \times 3}{8 \times 3} = \frac{15}{24}$ and $\frac{7}{12} = \frac{7 \times 2}{12 \times 2} = \frac{14}{24}$

$\frac{5}{8} - \frac{7}{12} = \frac{15}{24} - \frac{14}{24} = \frac{1}{24}$

2 Calculate $3\frac{3}{10} - 1\frac{5}{6}$.

$3\frac{3}{10} = \frac{33}{10}$ and $1\frac{5}{6} = \frac{11}{6}$

$\frac{33}{10} = \frac{99}{30}$ and $\frac{11}{6} = \frac{55}{30}$

$3\frac{3}{10} - 1\frac{5}{6} = \frac{99}{30} - \frac{55}{30} = \frac{44}{30}$

$\frac{44}{30} = \frac{22}{15} = 1\frac{7}{15}$

Remember:
- Add the numerators only.
- When the answer is an improper fraction change it into a mixed number.

Exercise 2.5

Do not use a calculator in this exercise.

1 Work out:
(a) $\frac{1}{4} + \frac{1}{8}$ (b) $\frac{1}{3} + \frac{1}{4}$ (c) $\frac{1}{2} + \frac{1}{5}$ (d) $\frac{1}{3} + \frac{1}{5}$ (e) $\frac{1}{2} + \frac{1}{7}$

2 Work out:
(a) $\frac{1}{4} - \frac{1}{8}$ (b) $\frac{1}{3} - \frac{1}{4}$ (c) $\frac{1}{2} - \frac{1}{5}$ (d) $\frac{1}{3} - \frac{1}{5}$ (e) $\frac{1}{2} - \frac{1}{7}$

3 Work out:
(a) $\frac{1}{2} + \frac{3}{4}$ (b) $\frac{2}{3} + \frac{5}{6}$ (c) $\frac{3}{4} + \frac{4}{5}$ (d) $\frac{5}{7} + \frac{2}{3}$ (e) $\frac{3}{8} + \frac{5}{6}$

4 Calculate:
(a) $\frac{5}{8} - \frac{1}{2}$ (b) $\frac{13}{15} - \frac{1}{3}$ (c) $\frac{5}{6} - \frac{5}{24}$ (d) $\frac{7}{15} - \frac{2}{5}$ (e) $\frac{3}{4} - \frac{5}{12}$

5 Calculate:
(a) $2\frac{3}{4} + 1\frac{1}{2}$ (b) $1\frac{1}{2} + 2\frac{1}{3}$ (c) $1\frac{3}{4} + 2\frac{5}{8}$ (d) $2\frac{1}{4} + 3\frac{3}{5}$ (e) $4\frac{3}{5} + 1\frac{5}{6}$

6 Calculate:
(a) $2\frac{1}{2} - 1\frac{2}{5}$ (b) $1\frac{2}{3} - 1\frac{1}{4}$ (c) $3\frac{3}{4} - 2\frac{3}{8}$ (d) $5\frac{2}{5} - 2\frac{1}{10}$ (e) $4\frac{5}{12} - 2\frac{1}{6}$

7 Colin buys a bag of flour. He uses $\frac{1}{3}$ to bake a cake and $\frac{2}{5}$ to make a loaf.
What fraction of the bag of flour is left?

8 Edward, Marc, Dee and Lin share an apple pie. Edward has $\frac{1}{3}$, Marc has $\frac{1}{5}$ and Dee has $\frac{1}{4}$.
How much is left for Lin?

9 Both Lee and Mary have a packet of the same sweets.
Mary eats $\frac{1}{3}$ of her packet. Lee eats $\frac{3}{4}$ of his packet.
(a) Find the difference between the fraction Mary eats and the fraction Lee eats.

Lee gives his remaining sweets to Mary.
(b) What fraction of a packet does Mary now have?

10 Jon, Billy and Cathy are the only candidates in a school election.
Jon got $\frac{7}{20}$ of the votes. Billy got $\frac{2}{5}$ of the votes.
(a) What fraction of the votes did Cathy get?
(b) Which candidate won the election?

Multiplying fractions

Calculate $\frac{3}{8} \times \frac{1}{9}$

Simplify, where possible, by cancelling. $\frac{\cancel{3}^{1}}{8} \times \frac{1}{\cancel{9}_{3}}$

Multiply the numerators. Multiply the denominators. $\frac{1 \times 1}{8 \times 3} = \frac{1}{24}$

Write the answer in its simplest form. $\frac{3}{8} \times \frac{1}{9} = \frac{1}{24}$

To simplify:
Divide a numerator **and** a denominator by the **same number.**

In this case:
3 and 9 can be divided by 3.
$3 \div 3 = 1$ and $9 \div 3 = 3$.

EXAMPLES

1 Work out $\frac{3}{8} \times 12$

Note: $12 = \frac{12}{1}$

$\frac{3}{8} \times \frac{12}{1}$

Simplify by cancelling.

$= \frac{3}{\cancel{8}_{2}} \times \frac{\cancel{12}^{3}}{1}$

Multiply out.

$= \frac{3 \times 3}{2 \times 1} = \frac{9}{2}$

$= 4\frac{1}{2}$

2 Calculate $1\frac{1}{2} \times 1\frac{3}{5}$

Change mixed numbers to improper fractions.

$1\frac{1}{2} \times 1\frac{3}{5} = \frac{3}{2} \times \frac{8}{5}$

Simplify by cancelling.

$= \frac{3}{\cancel{2}_{1}} \times \frac{\cancel{8}^{4}}{5}$

Multiply out.

$= \frac{3 \times 4}{1 \times 5} = \frac{12}{5} = 2\frac{2}{5}$

Exercise 2.6

Do not use a calculator in this exercise.

1 Work out. Give your answers as mixed numbers.
(a) $\frac{1}{2} \times 7$ (b) $\frac{3}{5} \times 3$ (c) $\frac{5}{8} \times 10$ (d) $\frac{3}{4} \times 12$ (e) $\frac{7}{10} \times 15$

2 Work out:
(a) $\frac{1}{4} \times \frac{1}{5}$ (b) $\frac{3}{5} \times \frac{1}{6}$ (c) $\frac{1}{2} \times \frac{3}{4}$ (d) $\frac{1}{10} \times \frac{5}{8}$ (e) $\frac{3}{10} \times \frac{1}{6}$

3 Work out: **Hint** - Change mixed numbers to improper fractions first.
(a) $1\frac{1}{2} \times 5$ (b) $1\frac{1}{4} \times 6$ (c) $1\frac{1}{5} \times 4$ (d) $3\frac{2}{3} \times 2$ (e) $2\frac{3}{5} \times 3$
(f) $4\frac{1}{2} \times \frac{1}{2}$ (g) $3\frac{3}{4} \times \frac{1}{5}$ (h) $2\frac{2}{5} \times \frac{1}{3}$ (i) $2\frac{2}{3} \times \frac{1}{4}$ (j) $4\frac{3}{8} \times \frac{1}{7}$

4 Calculate:
(a) $\frac{2}{3} \times \frac{3}{4}$ (b) $\frac{3}{4} \times \frac{2}{5}$ (c) $\frac{2}{5} \times \frac{5}{6}$ (d) $\frac{2}{5} \times \frac{5}{7}$ (e) $\frac{3}{10} \times \frac{5}{6}$

5 Calculate:
(a) $1\frac{1}{2} \times \frac{3}{4}$ (b) $1\frac{1}{2} \times 1\frac{1}{3}$ (c) $2\frac{1}{4} \times 1\frac{2}{3}$ (d) $2\frac{3}{4} \times 1\frac{2}{5}$ (e) $1\frac{1}{2} \times 3\frac{1}{4}$
(f) $1\frac{3}{5} \times 1\frac{1}{6}$ (g) $6\frac{3}{10} \times 2\frac{2}{9}$ (h) $3\frac{3}{4} \times 3\frac{3}{5}$ (i) $1\frac{1}{4} \times 2\frac{4}{25}$ (j) $2\frac{5}{8} \times 3\frac{1}{3}$

6 Kylie has $\frac{2}{3}$ of a litre of orange. She drinks $\frac{2}{5}$ of the orange.
(a) What fraction of a litre does she drink?
(b) What fraction of a litre is left?

7 Tony eats $\frac{1}{5}$ of a bag of sweets.
He shares the remaining sweets equally among Bob, Jo and David.
(a) What fraction of the bag of sweets does Bob get?
(b) What is the smallest possible number of sweets in the bag?

Dividing fractions

The method normally used when one fraction is divided by another is to change the division to a multiplication. The fractions can then be multiplied in the usual way.

Calculate $2\frac{2}{15} \div 1\frac{3}{5}$

Change mixed numbers to improper ('top heavy') fractions. $2\frac{2}{15} = \frac{32}{15}$ and $1\frac{3}{5} = \frac{8}{5}$

Change the division to a multiplication. $\frac{32}{15} \div \frac{8}{5} = \frac{32}{15} \times \frac{5}{8}$

Simplify, where possible, by cancelling. $\frac{\cancel{32}^{4}}{\cancel{15}_{3}} \times \frac{\cancel{5}^{1}}{\cancel{8}_{1}}$

Multiply the numerators.
Multiply the denominators. $\frac{4 \times 1}{3 \times 1} = \frac{4}{3}$

Write the answer in its simplest form. $\frac{4}{3} = 1\frac{1}{3}$

EXAMPLES

1 Work out $\frac{2}{3} \div 5$.

$\frac{2}{3} \div 5$

$= \frac{2}{3} \times \frac{1}{5}$

$= \frac{2}{15}$

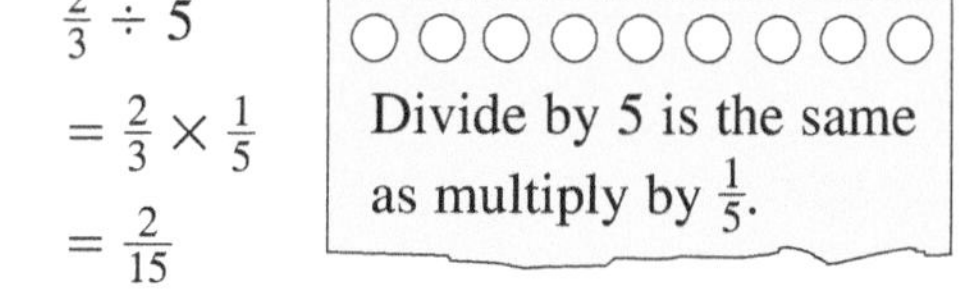

2 Calculate $1\frac{3}{5} \div \frac{4}{9}$.

$1\frac{3}{5} \div \frac{4}{9} = \frac{8}{5} \div \frac{4}{9}$

$= \frac{8}{5} \times \frac{9}{4}$

$= \frac{\cancel{8}^{2}}{5} \times \frac{9}{\cancel{4}_{1}}$

$= \frac{18}{5} = 3\frac{3}{5}$

Divide by $\frac{4}{9}$ is the same as multiply by $\frac{9}{4}$.

Exercise 2.7

Do not use a calculator in this exercise.

1 Work out. Give your answers in their simplest form.
(a) $\frac{1}{2} \div 5$ (b) $\frac{3}{4} \div 2$ (c) $\frac{2}{3} \div 2$ (d) $\frac{2}{5} \div 4$ (e) $\frac{6}{7} \div 3$

2 Work out: **Hint** - Change mixed numbers to improper fractions first.
(a) $1\frac{1}{2} \div 2$ (b) $1\frac{2}{3} \div 5$ (c) $1\frac{5}{7} \div 3$ (d) $3\frac{1}{9} \div 7$ (e) $2\frac{2}{3} \div 4$

3 Calculate:
(a) $\frac{1}{2} \div \frac{1}{4}$ (b) $\frac{1}{5} \div \frac{1}{2}$ (c) $\frac{7}{8} \div \frac{1}{3}$ (d) $\frac{2}{3} \div \frac{1}{5}$ (e) $\frac{3}{4} \div \frac{1}{8}$

4 Calculate:
(a) $\frac{2}{3} \div \frac{4}{5}$ (b) $\frac{3}{8} \div \frac{2}{3}$ (c) $\frac{3}{5} \div \frac{3}{4}$ (d) $\frac{2}{5} \div \frac{3}{10}$ (e) $\frac{3}{8} \div \frac{9}{16}$
(f) $\frac{7}{12} \div \frac{7}{18}$ (g) $\frac{4}{9} \div \frac{2}{3}$ (h) $\frac{7}{10} \div \frac{3}{5}$ (i) $\frac{9}{20} \div \frac{3}{10}$ (j) $\frac{21}{25} \div \frac{7}{15}$

5 Calculate:
(a) $1\frac{3}{4} \div \frac{1}{8}$ (b) $1\frac{1}{2} \div \frac{3}{10}$ (c) $1\frac{3}{5} \div \frac{4}{5}$ (d) $2\frac{1}{10} \div \frac{7}{20}$ (e) $3\frac{3}{4} \div \frac{5}{8}$
(f) $1\frac{1}{4} \div 1\frac{9}{16}$ (g) $3\frac{1}{5} \div 2\frac{2}{15}$ (h) $2\frac{1}{4} \div 1\frac{4}{5}$ (i) $2\frac{1}{7} \div 1\frac{7}{8}$ (j) $1\frac{2}{5} \div 1\frac{2}{3}$

6 Neil uses $\frac{1}{2}$ of a block of paté to make 5 sandwiches.
What fraction of the block of paté does he put on each sandwich?

7 Lauren uses $\frac{2}{3}$ of a bag of flour to make 6 muffins.
What fraction of the bag of flour is used for each muffin?

8 A shelf is $40\frac{3}{4}$ cm long. How many books of width $1\frac{3}{4}$ cm can stand on the shelf?

Problems involving fractions

Exercise 2.8 Do not use a calculator in this exercise.

1. Ben spends $\frac{5}{8}$ of his pocket money. He has £1.20 left.
   How much pocket money did Ben get?
2. A cyclist travels from A to B in two stages. Stage 1 is 28 km which is $\frac{2}{7}$ of the total journey.
   How long is stage 2?
3. Robert pays £9.60 for a CD in the sale. **Music Sale** - $\frac{1}{4}$ off all CDs
   How much did he save?
4. Sara's hourly wage is increased by $\frac{1}{10}$. Her new hourly wage is £6.60.
   What was her original hourly wage?
5. A young tree is $\frac{3}{8}$ taller in August than it was in May. In August it is 132 centimetres tall.
   How tall was it in May?

Fractions on a calculator

Fraction calculations can be done quickly using the fraction button on a calculator.

On most calculators the fraction button looks like this ... [$a^b/_c$]

EXAMPLE

Use a calculator to work out $4\frac{3}{5} \div 2\frac{1}{4}$.

This can be calculated with this calculator sequence.

[4] [$a^b/_c$] [3] [$a^b/_c$] [5] [÷] [2] [$a^b/_c$] [1] [$a^b/_c$] [4] [=]

This gives the answer $2\frac{2}{45}$.

Use a calculator to check your answers to some of the questions in Exercises 2.4 to 2.8.

Fractions and decimals

All fractions can be written as decimals and vice versa.

Changing decimals to fractions

EXAMPLES

Change the following decimals to fractions in their simplest form.

1. 0.02 $\quad 0.02 = \frac{2}{100} = \frac{2 \div 2}{100 \div 2} = \frac{1}{50}$
2. 0.36 $\quad 0.36 = \frac{36}{100} = \frac{36 \div 4}{100 \div 4} = \frac{9}{25}$

Changing fractions to decimals

EXAMPLES

Change the following fractions to decimals.

1. $\frac{1}{5} = 1 \div 5 = 0.2$
2. $\frac{11}{20} = 11 \div 20 = 0.55$

Remember:
$11 \div 20 = 11.00 \div 20$

$$20\overline{)1\,1.0\,0} = 0.5\,5$$

Recurring decimals

Some decimals have recurring digits. These are shown by:

a single dot above a single recurring digit,
a dot above the first and last digit of a set of recurring digits.

For example: $\frac{1}{3} = 0.3333333\ldots = 0.\dot{3}$

$\frac{123}{999} = 0.123123123\ldots = 0.\dot{1}2\dot{3}$

Exercise 2.9

Do not use a calculator for questions 1 to 3.

1 Change the following decimals to fractions in their simplest form.
(a) 0.12 (b) 0.6 (c) 0.32 (d) 0.175 (e) 0.45 (f) 0.65
(g) 0.22 (h) 0.202 (i) 0.28 (j) 0.555 (k) 0.625 (l) 0.84

2 Change the following fractions to decimals.
(a) $\frac{1}{4}$ (b) $\frac{1}{10}$ (c) $\frac{2}{5}$ (d) $\frac{3}{4}$ (e) $\frac{7}{10}$ (f) $\frac{4}{5}$

3 Change the following fractions to decimals.
(a) $\frac{3}{20}$ (b) $\frac{4}{25}$ (c) $\frac{7}{100}$ (d) $\frac{19}{20}$ (e) $\frac{9}{25}$ (f) $\frac{106}{200}$

4 Change these fractions to decimals.
(a) $\frac{1}{8}$ (b) $\frac{5}{8}$ (c) $\frac{9}{40}$ (d) $\frac{29}{40}$

5 Write these decimals using dots to represent recurring digits.
(a) 0.77777... (b) 0.11111... (c) 0.363636... (d) 0.828282...
(e) 0.135135... (f) 0.216216... (g) 0.166666... (h) 0.285714285714...

6 Write each of these fractions as recurring decimals, using dots to represent recurring digits.
(a) $\frac{8}{9}$ (b) $\frac{4}{9}$ (c) $\frac{17}{33}$ (d) $\frac{8}{11}$ (e) $\frac{1}{7}$
(f) $\frac{6}{7}$ (g) $\frac{1}{30}$ (h) $\frac{7}{15}$ (i) $\frac{5}{6}$ (j) $\frac{17}{22}$

What you need to know

- **Without using a calculator** you should be able to add, subtract, multiply and divide decimals.
- Carry out a variety of calculations involving decimals.
- Know that: when a number is **multiplied** by a number between 0 and 1
  the result will be **smaller** than the original number,
  when a number is **divided** by a number between 0 and 1
  the result will be **larger** than the original number.
- The top number of a fraction is called the **numerator**,
  the bottom number is called the **denominator**.
- $2\frac{1}{2}$ is an example of a **mixed number**. It is a mixture of whole numbers and fractions.
- $\frac{5}{2}$ is an **improper** (or '**top heavy**') fraction.
- Fractions must have the **same denominator** before **adding** or **subtracting**.
- Mixed numbers must be changed to **improper fractions** before **multiplying** or **dividing**.
- All fractions can be written as decimals.
  Some decimals have **recurring digits**. These are shown by:
  a single dot above a single recurring digit, e.g. $\frac{2}{3} = 0.6666\ldots = 0.\dot{6}$
  a dot above the first and last digit of a set of recurring digits, e.g. $\frac{5}{11} = 0.454545\ldots = 0.\dot{4}\dot{5}$

Review Exercise 2

Do not use a calculator for questions 1 to 12.

1 (a) Write 0.45 as a fraction. Give your answer in its simplest form.
(b) Write as a decimal. (i) $\frac{7}{25}$ (ii) $\frac{5}{9}$

2 Work out.
(a) $2.94 + 9.47$ (b) $10 - 5.67$ (c) $8 \div 0.4$ (d) 0.2×0.4

3 An examination in French is marked out of 80.
(a) Jean scored $\frac{4}{5}$ of the marks. How many marks did she score?
(b) Pete scored 35 marks. What fraction of the total did he score?

4 36 girls and 24 boys applied to go on a rock climbing course.
$\frac{2}{3}$ of the girls and $\frac{3}{4}$ of the boys went on the course.
What fraction of the 60 students who applied went on the course?
Write the fraction in its simplest form.

5 (a) Convert $\frac{1}{7}$ to a decimal, giving your answer correct to three decimal places.
(b) Place the following numbers in order of size, starting with the smallest.
11.14 $1\frac{1}{7}$ 1.14 1.41 1.014

AQA

6 Work out. (a) $\frac{1}{4} + \frac{2}{3}$ (b) $\frac{2}{5} - \frac{1}{8}$ (c) $\frac{3}{5} \times \frac{5}{7}$ (d) $\frac{5}{8} \div \frac{3}{4}$

7 At Wendy's party $\frac{1}{3}$ of the children have blue eyes and $\frac{2}{5}$ have brown eyes.
What fraction of the children do not have blue or brown eyes?

8 Given that $275 \times 41 = 11\,275$ work out the value of:
(a) 275×0.0041 (b) $112.75 \div 4.1$

9 One kilometre is approximately five eighths of a mile.
The distance by car between Southampton and Birmingham is 130 miles.
What is the distance in kilometres?

AQA

10 In a school $\frac{8}{15}$ of the pupils are girls. $\frac{3}{16}$ of the girls are left-handed.
What fraction of the pupils in the school are left-handed girls?

11 Calculate: (a) $2\frac{5}{7} + 1\frac{4}{9}$ (b) $4\frac{3}{10} - 1\frac{2}{3}$ (c) $2\frac{2}{3} \times 3\frac{3}{4}$ (d) $3\frac{3}{5} \div 2\frac{1}{10}$

12 The price of a coat is reduced by $\frac{2}{5}$ to £48. What was the original price of the coat?

13 Sonya uses her calculator to work out $\frac{2.34 + 1.76}{3.22} + 1.85$
Sonya presses keys as follows.
[2] [.] [3] [4] [+] [1] [.] [7] [6] [÷] [3] [.] [2] [2] [+] [1] [.] [8] [5] [=]
(a) Explain clearly what is wrong with Sonya's method.
(b) Calculate the correct answer.

14 Tony uses his calculator to work out $\frac{4.2 \times 86}{3.2 \times 0.47}$
What answer should he get?

AQA

15 Work out $\frac{4.7 \times 20.1}{5.6 - 1.8}$
Write down your full calculator display.

AQA

16 In Switzerland some goods can be bought using francs or euros.
£1 = 1.56 euros. £1 = 2.27 francs. How much is 100 francs in euros?

17 Tom spends his wages as follows: $\frac{3}{20}$ on tax, $\frac{1}{4}$ on rent and $\frac{1}{5}$ on fares. He has £80 left.
How much were his wages?

18 Is the value of $\frac{a + c}{b + d}$ always between the fractions $\frac{a}{b}$ and $\frac{c}{d}$? Explain your answer.

Approximation and Estimation

Approximation

In real-life it is not always necessary to use exact numbers. A number can be **rounded** to an **approximate** number. Numbers are rounded according to how accurately we wish to give details. For example, the distance to the Sun can be given as 93 million miles.

Can you think of other situations where approximations might be used?

Rounding to the nearest 10, 100, 1000

If there were 7487 people at a football match a newspaper report could say,

"7000 at the football match."

The number 7487 can be approximated as 7490, 7500 or 7000 depending on the degree of accuracy required.

7487 rounded to the nearest 10 is 7490.
7487 rounded to the nearest 100 is 7500.
7487 rounded to the nearest 1000 is 7000.

It is a convention to round a number which is in the middle to the higher number.
75 to the nearest 10 is 80.
450 to the nearest 100 is 500.
8500 to the nearest 1000 is 9000.

EXAMPLE

A lifeguard says, "There are 120 people in the pool today."
This figure is correct to the nearest 10.
What is the smallest and largest possible number of people in the pool?

The smallest whole number that rounds to 120 is 115.
The largest whole number that rounds to 120 is 124.
So, the smallest possible number of people is 115 and the largest possible number of people is 124.

Exercise 3.1

1. Write each of these numbers to the nearest 10. (a) 47 (b) 53 (c) 65
2. Round the number 7475
   (a) to the nearest 10, (b) to the nearest 100, (c) to the nearest 1000.
3. (a) Rearrange the cards: 5 4 6 3 to make the smallest possible number.
   (b) Round your number to the nearest 100.
4. Write down a number each time which fits these roundings.
   (a) It is 750 to the nearest 10 but 700 to the nearest 100.
   (b) It is 750 to the nearest 10 but 800 to the nearest 100.
   (c) It is 8500 to the nearest 100 but 8000 to the nearest 1000.
   (d) It is 8500 to the nearest 100 but 9000 to the nearest 1000.
5. Write down these figures to appropriate degrees of accuracy.
   (a) The class raised £49.67 for charity.
   (b) The population of a town is 24 055.
   (c) The land area of the country is 309 123 km^2.
   (d) The distance to London is 189 km.

6. 64 537 people signed a petition. A newspaper report stated: '65 000 people sign petition'
To what degree of accuracy is the number given in the newspaper report?

7. Pete says, "I had 40 birthday cards." The number is correct to the nearest 10.
(a) What is the smallest possible number of cards Pete had?
(b) What is the largest possible number of cards Pete had?

8. The number of people at a concert is 2000 to the nearest 100.
(a) What is the smallest possible number of people at the concert?
(b) What is the largest possible number of people at the concert?

9. **"43 000 spectators watch thrilling Test Match."**
The number reported in the newspaper was correct to the nearest thousand.
What is the smallest possible number of spectators?

10. Carl has 140 postcards in his collection. The number is given to the nearest ten.
What is the smallest and greatest number of postcards Carl could have in his collection?

11. "You require 2700 tiles to tile your swimming pool." This figure is correct to the nearest 100.
What is the greatest number of tiles needed?

Rounding in real-life problems

In a real-life problem a rounding must be used which gives a sensible answer.

EXAMPLES

1. A Year group in a school are going to Alton Towers.
There are 242 students and teachers going.
Each coach can carry 55 passengers. How many coaches should be ordered?
$242 \div 55 = 4.4$ This should be rounded up to 5.
4 coaches can only carry 220 passengers ($4 \times 55 = 220$).

2. Filing cabinets are to be placed along a wall. The available space is 460 cm.
Each cabinet is 80 cm wide. How many can be fitted in?
$460 \div 80 = 5.75$ This should be rounded down to 5.
Although the answer is nearly 6 the 6th cabinet would not fit in.

Exercise 3.2

Do not use a calculator for this exercise.

1. 49 students are waiting to go to the Sports Stadium.
A minibus can take 15 passengers at a time. How many trips are required?

2. A classroom wall is 700 cm long.
How many tables, each 120 cm long, could be fitted along the wall?

3. There are 210 students in a year group. They each need an exercise book.
The exercise books are sold in packs of 25. How many packs should be ordered?

4. Car parking spaces should be 2.5 m wide.
How many can be fitted into a car park which is 61 m wide?

5. How many 30p stamps can be bought for £5?

6. Kim needs 26 candles. The candles are sold in packs of 4. How many packs must she buy?

7. A sweet manufacturer puts 17 sweets in a bag.
How many bags can be made up if there are 500 sweets?

Rounding using decimal places

What is the cost of 1.75 metres of material costing £3.99 a metre?

$1.75 \times 3.99 = 6.9825$

The cost of the material is £6.9825 or 698.25p.
As you can only pay in pence, a sensible answer is £6.98, correct to two decimal places (nearest penny).
This means that there are only two decimal places after the decimal point.

> Often it is not necessary to use an exact answer. Sometimes it is impossible, or impractical, to use the exact answer.

To round a number to a given number of decimal places

When rounding a number to one, two or more decimal places:

1. Write the number using one more decimal place than asked for.
2. Look at the last decimal place and
   - if the figure is 5 or more round up,
   - if the figure is less than 5 round down.
3. When answering a problem remember to include any units and state the degree of approximation used.

EXAMPLES

1 Write 2.76435 to
(a) 2 decimal places, (b) 1 decimal place.

(a) Look at the third decimal place. **4**
This is less than 5, so, round down.
Answer 2.76

(b) Look at the second decimal place. **6**
This is 5 or more, so, round up.
Answer 2.8

2 Write 7.104 to 2 decimal places.
$7.104 = 7.10$ to 2 d.p.
The zero is written down because it shows the accuracy used, 2 decimal places.

> **Notation:**
> Often decimal place is shortened to d.p.

3 $5.98 = 6.0$ to 1 d.p.
Notice that the next tenth after 5.9 is 6.0.

Exercise 3.3

1 Write the number 3.9617 correct to
(a) 3 decimal places, (b) 2 decimal places, (c) 1 decimal place.

2 Write the number 567.654 correct to
(a) 2 decimal places, (b) 1 decimal place, (c) the nearest whole number.

3 The display on a calculator shows the result of $34 \div 7$.

4.857142857

What is the result correct to two decimal places?

4 

The scales show Gary's weight.
Write Gary's weight correct to one decimal place.

5 Copy and complete this table.

Number	2.367	0.964	0.965	15.2806	0.056	4.991	4.996
d.p.	1	2	2	3	2	2	2
Answer	2.4						

6 Carry out these calculations giving the answers correct to

(a) 1 d.p. (b) 2 d.p. (c) 3 d.p.

(i) 6.12×7.54 (ii) 89.1×0.67 (iii) 90.53×6.29
(iv) $98.6 \div 5.78$ (v) $67.2 \div 101.45$

7 In each of these short problems decide upon the most suitable accuracy for the answer.
Then calculate the answer.
Give a reason for your degree of accuracy.

(a) One gallon is 4.54596… litres.
How many litres is 9 gallons?
(b) What is the cost of 0.454 kg of cheese at £5.21 per kilogram?
(c) The total length of 7 equal sticks, lying end to end, is 250 cm.
How long is each stick?
(d) A packet of 6 bandages costs £7.99.
How much does one bandage cost?
(e) Petrol costs 91.4 pence a litre. I buy 15.6 litres.
How much will I have to pay?

Rounding using significant figures

Consider the calculation $600.02 \times 7500.97 = 4500732.0194$
To 1 d.p. it is 4500732.0, to 2 d.p. it is 4500732.02.
The answers to either 1 or 2 d.p. are very close to the actual answer and are almost as long.
There is little advantage in using either of these two roundings.
The point of a rounding is that it is a more convenient number to use.

Another kind of rounding uses **significant figures**.
The **most** significant figure in a number is the figure which has the greatest place value.

Consider the number 237.
The figure 2 has the greatest place value. It is worth 200.
So, 2 is the most significant figure.

In the number 0.00328, the figure 3 has the greatest place value.
So, 3 is the most significant figure.

Noughts which are used to locate the decimal point and preserve the place value of other figures are not significant.

To round a number to a given number of significant figures

When rounding a number to one, two or more significant figures:

1. Start from the most significant figure and count the required number of figures.
2. Look at the next figure to the right of this and
   - if the figure is 5 or more round up,
   - if the figure is less than 5 round down.
3. Add noughts, as necessary, to locate the decimal point and preserve the place value.
4. When answering a problem remember to include any units and state the degree of approximation used.

EXAMPLES

1 Write 4 500 732.0194 to 2 significant figures.

The figure after the first 2 significant figures **45** is 0.
This is less than 5, so, round down, leaving 45 unchanged.
Add noughts to 45 to locate the decimal point and preserve place value.
So, 4 500 732.0194 = 4 500 000 to 2 sig. fig.

Notation:
Often significant figure is shortened to sig. fig.

2 Write 0.000364907 to 1 significant figure.

The figure after the first significant figure 3 is 6.
This is 5 or more, so, round up, 3 becomes 4.
So, 0.000364907 = 0.0004 to 1 sig. fig.

Notice that the noughts before the 4 locate the decimal point and preserve place value.

Choosing a suitable degree of accuracy

In some calculations it would be wrong to use the complete answer from the calculator.
The result of a calculation involving measurement should not be given to a greater degree of accuracy than the measurements used in the calculation.

EXAMPLE

What is the area of a rectangle measuring 4.6 cm by 7.2 cm?

Note:
To find the area of a rectangle: multiply length by breadth.

$4.6 \times 7.2 = 33.12$
Since the measurements used in the calculation (4.6 cm and 7.2 cm) are given to 2 significant figures the answer should be as well.
33 cm^2 is a more suitable answer.

Exercise 3.4

1 Write these numbers correct to one significant figure.
(a) 17 (b) 523 (c) 350 (d) 1900 (e) 24.6
(f) 0.083 (g) 0.086 (h) 0.00948 (i) 0.0095

2 Copy and complete this table.

Number	456 000	454 000	7 981 234	7 981 234	1290	19 602
sig. fig.	2	2	3	2	2	1
Answer	460 000					

3 Copy and complete this table.

Number	0.000567	0.093748	0.093748	0.093748	0.010245	0.02994
sig. fig.	2	2	3	4	2	2
Answer						

4 This display shows the result of 3400 ÷ 7.
What is the result correct to two significant figures?

485.7142857

5 Carry out these calculations giving the answers correct to
(a) 1 sig. fig. (b) 2 sig. fig. (c) 3 sig. fig.
(i) 672×123 (ii) 6.72×12.3 (iii) 78.2×12.8
(iv) $7.19 \div 987.5$ (v) $124 \div 65300$

6 A rectangular field measures 18.6 m by 25.4 m.
Calculate the area of the field, giving your answer to a suitable degree of accuracy.

7 In each of these short problems decide upon the most suitable accuracy for the answer.
Then work out the answer, remembering to state the units.
Give a reason for your degree of accuracy.
(a) The area of a rectangle measuring 13.2 cm by 11.9 cm.
(b) The area of a football pitch measuring 99 m by 62 m.
(c) The total length of 13 tables placed end to end measures 16 m.
How long is each table?
(d) The area of carpet needed to cover a rectangular floor measuring 3.65 m by 4.35 m.

Estimation

It is always a good idea to find an **estimate** for any calculation.
An estimate is used to check that the answer to the actual calculation is of the right magnitude (size).
If the answer is very different to the estimate then a mistake has possibly been made.

Estimation is done by approximating every number in the calculation to one significant figure.
The calculation is then done using the approximated values.

EXAMPLES

1 Estimate 421×48.

Round 421 to one significant figure: 400
Round 48 to one significant figure: 50
$400 \times 50 = 20\,000$

Use long multiplication to calculate 421×48.
Comment on your answer.

2 Use estimation to show that $\frac{78.5 \times 0.51}{18.7}$ is close to 2.

Approximating: $78.5 = 80$ to 1 sig. fig.
$0.51 = 0.5$ to 1 sig. fig.
$18.7 = 20$ to 1 sig. fig.

$\frac{80 \times 0.5}{20} = \frac{40}{20} = 2$ (estimate)

Using a calculator $\frac{78.5 \times 0.51}{18.7} = \frac{40.035}{18.7} = 2.140909\ldots$

Is 2.140909 reasonably close to 2? Yes.

Remember:
When you are asked to estimate, write each number in the calculation to one significant figure.

Exercise 3.5

1 John estimated 43×47 to be about 2000. Explain how he did it.

2 Make estimates to these calculations by using approximations to one significant figure.
(a) (i) 39×21 (ii) 115×18 (iii) 797×53 (iv) 913×59
(b) (i) $76 \div 18$ (ii) $597 \div 29$ (iii) $889 \div 61$ (iv) $3897 \div 82$

3 Lilly ordered 39 prints of her holiday photographs. Each print cost 52 pence.
Use suitable approximations to **estimate** the total cost of the prints. Show your working.

4 (a) When estimating the answer to 29×48 the approximations 30 and 50 are used.
How can you tell that the estimation must be bigger than the actual answer?
(b) When estimating the answer to $182 \div 13$ the approximations 200 and 10 are used.
Will the estimate be bigger or smaller than the actual answer? Explain your answer.

5 Bernard plans to buy a conservatory costing £8328 and furniture costing £984.
(a) By using approximations, estimate the total amount Bernard plans to spend.
(b) Find the actual cost.

6 Kath uses her calculator to work out the value of 396×0.470.
The answer she gets is 18.612.
Use approximations to show that her answer is wrong.

7 (a) Calculate $\frac{49.7 + 10.6}{9.69 \times 3.04}$
(b) Do not use your calculator in this part of the question.
By using approximations show that your answer to (a) is about right.

8 Find estimates to these calculations by using approximations to 1 significant figure.
Then carry out the calculations with the original figures.
Compare your estimate to the actual answer.
(a) $\frac{7.9 \times 3.9}{4.8}$ (b) $\frac{400 \times 0.29}{6.2}$ (c) $\frac{81.7 \times 4.9}{1.9 \times 10.3}$ (d) $\frac{4.12 \times 49.7}{0.096}$

9 (a) Calculate $\frac{78.9}{0.037 \times 5.2}$
(b) Show how you can use approximations to check your answer is about right.

10 Niamh calculates $5\,967\,000 \div 0.029$. She gets an answer of $2\,057\,586\,207$.
Use approximations to check whether Niamh's answer is of the right magnitude.

Accuracy in measurement

No measurement is ever exact.
Measures which can lie within a range of possible values are called **continuous measures**.
The value of a continuous measure depends on the accuracy of whatever is making the measurement.

Jane is 160 cm tall to the nearest 10 cm.
What are the limits between which her actual height lies?

Height is a continuous measure.
When rounding to the nearest 10 cm:
The minimum value that rounds to 160 cm is 155 cm.
155 cm is the minimum height that Jane can be.

The maximum value that rounds to 160 cm is $164.\dot{9}$ cm.
$164.\dot{9}$ cm is the maximum height that Jane can be.
For ease the value $164.\dot{9}$ cm is normally called 165 cm.

So, Jane's actual height is any height from 155 cm to 165 cm.
This can be written as the inequality:
155 cm $\leqslant$ Jane's height $<$ 165 cm

155 cm is the **lower bound** of Jane's height.
165 cm is the **upper bound** of Jane's height.

All possible heights for Jane can be shown on a number line.

155 160 165

The hollow circle indicates that 165 is **not** included

If a **continuous measure**, c, is recorded to the nearest x, then:
Lower bound $= c - \frac{1}{2}x$ **Upper bound** $= c + \frac{1}{2}x$

The **limits** of the possible values of c can be written as $c \pm \frac{1}{2}x$

EXAMPLES

1. The length of a pencil is 17 cm to the nearest centimetre.
What are the limits between which the actual length of the pencil lies?
When rounding to the nearest centimetre: The smallest value that rounds to 17 cm is 16.5 cm.
The largest value that rounds to 17 cm is $17.4\dot{9}$ cm.
So, the actual length of the pencil lies between 16.5 cm and 17.5 cm.
$16.5\text{ cm} \leqslant \text{length of pencil} < 17.5\text{ cm}$

2. A concrete block weighs 1.8 kg, correct to the nearest tenth of a kilogram.
What is the minimum possible weight of the concrete block?
Minimum weight $= 1.8\text{ kg} - 0.05\text{ kg} = 1.75\text{ kg}$

3. The length of this page is 26 cm to the nearest centimetre.
What are the upper and lower bounds of the true length of this page?
Upper bound $= 26 + 0.5 = 26.5$ cm Lower bound $= 26 - 0.5 = 25.5$ cm

Exercise 3.6

1. A girl's height is 168 cm, correct to the nearest centimetre.
What is the minimum possible height of the girl?

2. The height of a building is 9 m, correct to the nearest metre.
Copy and complete the inequality: ...... $\leqslant$ height of building $<$ ......

3. A brick weighs 840 g, correct to the nearest 10 g.
What is the minimum and maximum possible weight of the brick?

4. An athlete completed a race in 11.6 seconds, correct to the nearest tenth of a second.
What is the minimum possible time the athlete could have taken?

5. A pane of glass weighs 9.4 kg, correct to one decimal place.
What is the minimum possible weight of the pane of glass?

6. The length of a table is 2.7 m, correct to the nearest tenth of a metre.
Write down the least and greatest possible length of the table.

7. A glass contains 24 ml of milk, correct to the nearest millilitre.
Find the minimum possible number of millilitres in four glasses.

8. Loaves of bread each weigh 0.8 kg, correct to the nearest 100 g.
Write down the minimum and maximum possible weight of ten loaves of bread.

9. At birth a baby is 39 cm in length, correct to the nearest centimetre, and has a mass of 3.4 kg, correct to one decimal place.
(a) What is the lower bound of the baby's length?
(b) What is the upper bound of the baby's mass?

10. A tea bag weighs 3.2 g, correct to the nearest tenth of a gram.
What is the lower bound of the weight of 60 tea bags?

11. Duncan drives to work and back 5 days a week.
The journey to work is 8.6 km, correct to one decimal place.
What is the lower bound of the distance Duncan drives to and from work each week?

12. A book weighs 0.53 kg, correct to 2 decimal places.
(a) What are the limits between which the true weight of the book lies?
(b) What is the upper bound of the weight of 10 copies of the book?

Calculations involving bounds

Calculation involving ...	For upper bound calculate ...	For lower bound calculate ...
Adding measures	Upper bound + Upper bound	Lower bound + Lower bound
Multiplying measures	Upper bound × Upper bound	Lower bound × Lower bound
Subtracting measures	Upper bound − Lower bound	Lower bound − Upper bound
Dividing measures	Upper bound ÷ Lower bound	Lower bound ÷ Upper bound

EXAMPLE

Two strips of wood have lengths of 124 cm and 159 cm, to the nearest centimetre.
(a) What is the lower bound of the total length of the strips of wood?
(b) What is the upper bound of the difference between the lengths of the strips of wood?

Shorter strip	Longer strip
Lower bound = 124 cm − 0.5 cm = 123.5 cm	Lower bound = 159 cm − 0.5 cm = 158.5 cm
Upper bound = 124 cm + 0.5 cm = 124.5 cm	Upper bound = 159 cm + 0.5 cm = 159.5 cm

(a) To find the lower bound of the total length use the lower bounds of both strip lengths.
Lower bound of the total length = 123.5 + 158.5 = 282 cm.

(b) The upper bound of the difference in lengths is found from:
Upper bound of the longer strip − lower bound of the shorter strip = 159.5 − 123.5 = 36 cm

Exercise 3.7

1. Fiona measures a corridor as being 126 paces in length.
The length of her pace is 90 cm, to the nearest 10 cm.
What is the difference between the maximum and minimum possible lengths of the corridor?

2. The distance between A and B is 12.2 km to the nearest 100 m.
The distance between B and C is 14.34 km to the nearest 10 m.
Trevor walks from A to B and then from B to C.
Calculate the upper and lower bounds of the total distance that Trevor walks.

3. A bag of potatoes weighs 5.0 kg to the nearest 100 g.
(a) Find the upper bound of the weight of 100 bags of potatoes.
1.5 kg of potatoes to the nearest 100 g are taken from a bag.
(b) What is the lower bound of the weight of potatoes left in the bag?

4. Books are packed into boxes for delivery.
The total weight of a box should **not exceed** 30 kg.
A book weighs 2.7 kg, to the nearest 0.1 kg.
What is the maximum number of books that can be packed in a box?

5. Petra buys 24 litres of petrol at 89 pence per litre.
The amount of petrol and the price of petrol are both given correct to 2 significant figures.
Find the difference between the minimum and maximum amount that Petra could have spent on petrol.

6. $m = 7.4$, correct to two significant figures.
$n = 3.68$, correct to three significant figures.
(a) (i) What is the upper bound of m? (ii) What is the lower bound of n?
(b) Calculate the lower and upper bounds of $m - n$.
(c) Calculate the least possible value of mn.
(d) Calculate the greatest possible value of $\frac{m}{n}$.

What you need to know

- A number can be rounded to an **approximate** number.
- How to approximate using **decimal places**.
  1. Write the number using one more decimal place than asked for.
  2. Look at the last decimal place and
     - if the figure is 5 or more round up,
     - if the figure is less than 5 round down.
- How to approximate using **significant figures**.
  1. Start from the most significant figure and count the required number of figures.
  2. Look at the next figure to the right of this and
     - if the figure is 5 or more round up,
     - if the figure is less than 5 round down.
  3. Add noughts, as necessary, to locate the decimal point and preserve the place value.
- When answering a problem, include any units and state the degree of approximation used.
- You should be able to choose a suitable degree of accuracy.
- Be able to use approximations to estimate that the actual answer to a calculation is of the right magnitude.
- Be able to recognise limitations on the accuracy of data and measurements.

  A **discrete measure** can only take a particular value and a **continuous measure** lies within a range of possible values which depends upon the degree of accuracy of the measurement.

> If a **continuous measure**, c, is recorded to the nearest x, then:
> **Lower bound** $= c - \frac{1}{2}x$ **Upper bound** $= c + \frac{1}{2}x$

Review Exercise 3

Do not use a calculator for questions 1 to 8.

1 "2 million listen to the Cup Final on the radio."
The number is given to the nearest million. What is the smallest possible number of listeners?

2 Find **estimates** to these calculations by using approximations to one significant figure.
(a) 86.5×1.9 (b) $2016 \div 49.8$

3 Aimee uses her calculator to multiply 18.7 by 0.96. Her answer is 19.752.
Without finding the exact value of 18.7×0.96, explain why her answer must be wrong.

4 Jonathan uses his calculator to work out the value of 42.2×0.027.
The answer he gets is 11.394. Use approximation to show that his answer is wrong. AQA

5 Alun has a part-time job. He is paid £28 each day he works. Last year he worked 148 days.
Estimate Alun's total pay for last year. Write down your calculation and answer. AQA

6 Pat needs 136 crackers for a Christmas party. The crackers are only sold in packs of 12.
How many packs must Pat buy?

7 The number of people at a beach party is 90, to the nearest 10.
What is the smallest and largest possible number of people at the party?

8 The display shows the result of $179 \div 7$.
What is the result correct to:
(a) two decimal places,
(b) one decimal place,
(c) one significant figure?

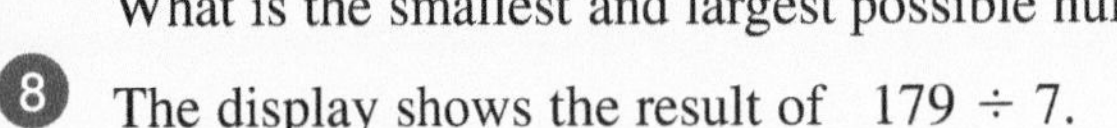

9 Calculate $72.5 \div 7.9$ (a) to 1 decimal place, (b) to 2 decimal places.

10 Calculate $107.9 \div 72.5$ (a) to 1 significant figure, (b) to 2 significant figures.

11 The number of visitors to a local museum was reported as 2600, to the nearest hundred.
(a) What is the smallest possible number of visitors?
(b) What is the largest possible number of visitors?

12 A bus is 18 m in length, correct to the nearest metre.
(a) What is the minimum possible length of the bus?
(b) Copy and complete this inequality: ...... $\leqslant$ length of bus $<$ ......

13 Find the value of $\frac{1000}{3} + 18\pi$, giving your answer correct to three significant figures.

14 (a) Calculate $\frac{89.6 \times 10.3}{19.7 + 9.8}$.
(b) Do not use your calculator in this part of the question.
By using approximations show that your answer to (a) is about right.
You **must** show all your working. AQA

15 Use your calculator to find the value of $3.94 + \frac{12.89}{7.15 + 3.28}$
Give your answer to an appropriate degree of accuracy. AQA

16 (a) A rectangular lawn measures 27 metres by 38 metres.
A firm charges £9.95 per square metre to turf the lawn.
Estimate the charge for turfing the lawn.
(b) Is your estimate too large or too small?
Give a reason for your answer.

17 John uses his calculator to work out $\frac{0.39 \times 85.2}{5.8}$. He gets an answer of 57.3.
Without using a calculator, use approximation to find whether John's answer is of the correct order of magnitude. You **must** show all your working. AQA

18 Work out $\frac{4.7 \times 20.1}{5.6 - 1.8}$
(a) Write down your full calculator display.
(b) Use estimation to check your answer. Show each step of your working. AQA

19 The floor of a lounge is a rectangle which measures 5.23 m by 3.62 m.
The floor is to be carpeted.
(a) Calculate the area of carpet needed.
Give your answer to an appropriate degree of accuracy.
(b) Explain why you chose this degree of accuracy.

20 A lorry weighs 1800 kg, correct to the nearest 100 kg.
What are the lower and upper bounds of the weight of the lorry?

21 A book weighs 3.2 kg, given to the nearest 100 g.
Find the minimum possible weight of 6 copies of the same book. AQA

22 Clive drove for 3 hours along a motorway, averaging 60 mph.
Both quantities are given to one significant figure.
(a) What is the minimum time for which Clive could have driven?
(b) What is the minimum distance Clive could have driven? AQA

23 $p = 600$, correct to the nearest 100.
$q = 50$, correct to the nearest 10.
(a) (i) What is the lower bound of p? (ii) What is the upper bound of q?
(b) Calculate (i) the upper bound of $p + q$, (ii) the lower bound of pq.
(c) Calculate the least possible value of $\frac{p}{q}$.

24 A field is 50 m in width and 110 m in length.
The width is given correct to the nearest 5 metres.
The length is given correct to the nearest 10 metres. Find the maximum area of the field. AQA

25 Pond A has a surface area of 150 m^2, to an accuracy of 2 significant figures.
Pond B has a surface area of 90 m^2, to an accuracy of 2 significant figures.
What is the lower bound of the difference between the surface areas of Ponds A and B?

CHAPTER 4 Percentages and Money

The meaning of a percentage

'Per cent' means 'out of 100'.
The symbol for per cent is %.
A percentage can be written as a fraction with denominator 100.

> 10% means 10 out of 100.
> 10% can be written as $\frac{10}{100}$.
> 10% is read as '10 per cent'.

Changing percentages to decimals and fractions

> To change a percentage to a decimal or a fraction: **divide by 100**

EXAMPLES

1 Write 38% as a fraction in its simplest form.

38% means '38 out of 100'.

This can be written as $\frac{38}{100}$.

$\frac{38}{100} = \frac{38 \div 2}{100 \div 2} = \frac{19}{50}$

$38\% = \frac{19}{50}$

2 Write 42.5% as a decimal.

$42.5\% = \frac{42.5}{100} = 42.5 \div 100 = 0.425$

> **Remember:**
> To write a fraction in its **simplest form** divide both the numerator and denominator of the fraction by the **largest** number that divides into them both.
>
> **To change a fraction to a decimal:**
> divide the numerator by the denominator.

Changing decimals and fractions to percentages

> To change a decimal or a fraction to a percentage: **multiply by 100**

EXAMPLES

1 Change 0.3 to a percentage.

$0.3 \times 100 = 30$
So, 0.3 as a percentage is 30%.

2 Change $\frac{11}{25}$ to a percentage.

$\frac{11}{25} \times 100 = 11 \times 100 \div 25$
$= 1100 \div 25 = 44\%$

3 Ben scored 17 out of 20 in a Maths test and 21 out of 25 in a History test.
Which is Ben's better mark?

Change each mark to a percentage.

Maths: $\frac{17}{20}$ $\quad \frac{17}{20} \times 100 = 17 \times 100 \div 20 = 85\%$

History: $\frac{21}{25}$ $\quad \frac{21}{25} \times 100 = 21 \times 100 \div 25 = 84\%$

So, Ben's better mark was his Maths mark of 85%.

> Fractions can be compared by first writing them as percentages.

Expressing one quantity as a percentage of another

This involves writing a fraction and then changing it to a percentage.
The units of any quantities in the numerator and denominator of the fraction must be the same.

EXAMPLES

1 Express 30p as a percentage of £2.

$\frac{30p}{£2} = \frac{30p}{200p} = \frac{30}{200}$

$\frac{30}{200} = \frac{30 \div 2}{200 \div 2} = \frac{15}{100} = 15\%$

This means that 30p is 15% of £2.

2 A newspaper contains 48 pages,
6 of which are Sports pages.
What percentage of the pages are Sports pages?

6 out of 48 pages are Sports pages.

$\frac{6}{48} = 6 \div 48 = 0.125$

$0.125 \times 100 = 12.5$

12.5% of the pages are Sports pages.

Exercise 4.1

Do not use a calculator for questions 1 to 12.

1 Change these percentages to fractions in their simplest form.
(a) 10% (b) 25% (c) 5% (d) 35% (e) 48% (f) 12.5%

2 Change these percentages to decimals.
(a) 20% (b) 15% (c) 1% (d) 72% (e) 87.5% (f) 150%

3 Change these fractions to percentages.
(a) $\frac{17}{50}$ (b) $\frac{12}{25}$ (c) $\frac{30}{200}$ (d) $\frac{4}{5}$ (e) $\frac{135}{500}$ (f) $\frac{13}{20}$ (g) $\frac{2}{3}$ (h) $\frac{2}{9}$

4 Change these decimals to percentages.
(a) 0.15 (b) 0.32 (c) 0.125 (d) 0.07 (e) 1.12 (f) 0.015

5 Write in order of size, lowest first:
(a) $\frac{1}{2}$ 60% $\frac{2}{5}$ 0.55 (b) 43% $\frac{9}{20}$ 0.42 $\frac{11}{25}$ (c) $\frac{23}{80}$ 28% $\frac{57}{200}$ 0.2805

6 What is (a) 30 as a percentage of 50, (b) 42 as a percentage of 200?

7 James saved £30 and then spent £9. What percentage of his savings did he spend?

8 A Youth Club has 200 members. 80 of the members are boys.
(a) What percentage of the members are boys?
(b) What percentage of the members are girls?

9 240 people took part in a survey. 30 of them were younger than 18.
What percentage were younger than 18?

10 A bar of chocolate has 32 squares. Jane eats 12 of the squares.
What percentage of the bar does she eat?

11 Billy earns £9 per hour. He gets a wage rise of 27 pence per hour.
What is his percentage wage rise?

12 Rohima achieved the following results.

Geography: 32 out of 40 English: 21 out of 25 Maths: 17 out of 20

In which subject did she do best?

13 In an ice hockey competition Team A won 8 out of the 11 games they played
whilst Team B won 5 of their 7 games.

Which team has the better record in the competition?

14 What is (a) 6 minutes as a percentage of 1 hour,
(b) 30 mm as a percentage of 5 cm,
(c) 150 g as a percentage of 1 kg?

15 A new car costs £13 500. The dealer gives a discount of £1282.50.
What is the percentage discount?

16 There are 600 pupils in Years 9 to 13 of a High school.
360 pupils are in Years 10 and 11
15% of pupils are in Years 12 and 13.
What percentage of pupils are in Year 9?

Finding a percentage of a quantity

EXAMPLE

Find 20% of £56.

Step 1 Divide by 100.
£56 ÷ 100 = £0.56

Step 2 Multiply by 20.
£0.56 × 20 = £11.20

So, 20% of £56 is £11.20.

To find 1% of a quantity divide the quantity by 100.
To find 20% of a quantity multiply 1% of the quantity by 20.
This is the same as the method you would use to find $\frac{20}{100}$ of a quantity.

Percentage change

EXAMPLES

1 A shirt normally priced at £24 is reduced by 15% in a sale.
How much does it cost in the sale?

Reduction in price = 15% of £24
15 ÷ 100 × 24 = 0.15 × 24 = 3.6
15% of £24 = £3.60
The shirt costs £24 − £3.60 = £20.40.

2 A packet of cereals weighs 440 g.
A special offer packet weighs 30% more.
What is the weight of the special offer packet?

Extra contents = 30% of 440 g
= 440 ÷ 100 × 30
= 132 g
440 + 132 = 572
The special offer packet weighs 572 g.

Exercise 4.2

Do not use a calculator for questions 1 to 10.

1 Find

(a) 20% of £80	(b) 75% of £20	(c) 30% of £220	(d) 15% of £350
(e) 5% of £500	(f) 9% of 300 kg	(g) 45% of £25	(h) 60% of 20 m

2 There are 450 seats in a theatre. 60% of the seats are in the stalls.
How many seats are in the stalls?

3 Tony invests £400 in a building society. He earns 5% interest per year.
How much interest does he get in one year?

4 Jenny gets a 15% discount on a theatre ticket. The normal cost is £18.
How much does she save?

5 A salesman earns a bonus of 3% of his weekly sales.
How much bonus does the salesman earn in a week when his sales are £1400.

6 Increase:
(a) £50 by 60% (b) £10 by 30% (c) £15 by 10% (d) £50 by 15%

7 Decrease:
(a) £600 by 15% (b) £55 by 90% (c) £42 by 20% (d) £63 by 35%

8 A mobile telephone company offers a 20% discount on calls made in March.
The normal cost of a peak time call is 50 pence per minute.
How much does a peak time call cost in March?

9 Abdul earns £200 per week. He gets a wage rise of 7.5%. What is his new weekly wage?

10 A packet of breakfast cereal contains 660 g. A special offer packet contains an extra 15%.
How many grams of breakfast cereal are in the special offer packet?

Questions 11 to 16. Where appropriate give your answers to 3 significant figures.

11 The price of a gold watch is £278. What does it cost with a 12% discount?

12 The price of a used car is £5200. What does it cost with a 9.5% discount?

13 Milk costs 35 pence a pint. How much does it cost after a 14% increase?

14 Petrol costs 89.9 pence a litre. What does it cost after a 2.4% decrease?

15 A car was valued at £13 500 when new. After one year it lost 22% of its value.
What was the value of the car after one year?

16 In 2001 house prices increased by 19.6%. In 2002 house prices increased by 17.4%.
A house was valued at £78 000 at the beginning of 2001.
What was the value of the house at the end of 2002?

Percentage increase and decrease

Sadik and Chandni took Maths tests in October and June.

Who has made the most improvement?

They have both improved by a score of 18% so by one measure they have both improved equally.
Another way of comparing their improvement is to use the idea of a percentage increase.

$$\text{Percentage increase} = \frac{\text{actual increase}}{\text{initial value}} \times 100\%$$

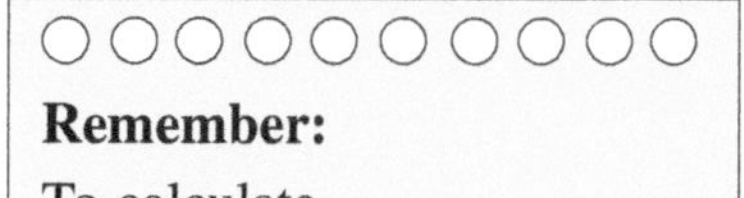
Remember:
To calculate
% increase or % decrease
always use the initial value.

Comparing percentage increases is the best way to decide whether Sadik or Chandni has made the most improvement.
Explain why.

For Sadik
% increase $= \frac{18}{54} \times 100\% = 33.3\%$

For Chandni
% increase $= \frac{18}{42} \times 100\% = 42.9\%$

Both calculations are correct to one decimal place.

A percentage decrease can be calculated in a similar way.

$$\text{Percentage decrease} = \frac{\text{actual decrease}}{\text{initial value}} \times 100\%$$

EXAMPLES

1 A shop buys pens for 15 pence and sells them for 21 pence.
What is their percentage profit?

Actual profit = 21 pence − 15 pence
= 6 pence

$$\% \text{ profit} = \frac{\text{actual profit}}{\text{initial value}} \times 100$$
$$= \frac{6}{15} \times 100 = 40\%$$

2 Pam buys a micro-scooter for £24.
She sells the micro-scooter for £15.
What is her percentage loss?

Actual loss = £24 − £15
= £9

$$\% \text{ loss} = \frac{\text{actual loss}}{\text{initial value}} \times 100$$
$$= \frac{9}{24} \times 100 = 37.5\%$$

Exercise 4.3

Do questions 1 to 5 without a calculator.

1. A shop buys calculators for £5 and sells them for £6. Find the percentage profit.
2. Geri's wages of £7.50 per hour are increased to £9.00 per hour.
Find the percentage increase in her earnings.
3. A man buys a boat for £25 000 and sells it for £18 000. Find his percentage loss.
4. In October, Hinn scored 50% in an English test. In January he improved to 66%.
In the same tests, Becky scored 40% and 56%.
Who has made the most improvement? Explain your answer.
5. The rent on Karen's flat increased from £80 to £90 per week.
(a) Find the percentage increase in her rent.
At the same time Karen's wages increased from £250 per week to £280 per week.
(b) Find the percentage increase in her wages.
Comment on your answers.
6. The value of car A when new was £13 000. The value of car B when new was £16 500.
After one year the value of car A is £11 200 and the value of car B is £13 500.
Calculate the percentage loss in the values of cars A and B after one year.
7. During 2004 the population of a village decreased from 323 to 260.
Find the percentage decrease in the population.
8. At the start of May a flower was 12.3 cm high. In May it grew by 14.5%.
At the start of July it was 16.7 cm high.
What was its percentage growth during June?
9. A rectangle has length 12 cm and width 8 cm.
The length is increased by 7.5% and the width decreased by 12.5%.
Find the change in area as a percentage of the original area of the rectangle.
10. In 2003 Miles bought £2000 worth of shares.
In 2004 the value of his shares decreased by 10%.
In 2005 the value of his shares increased by 20%.
By what percentage has the value of his shares changed from 2003 to 2005?
11. The price of a micro-scooter is reduced by 10%.
In a sale, the new price is reduced by a further 10%.
By what percentage has the original price of the micro-scooter been reduced in the sale?
12. The price of telephone calls is increased by 15%.
Companies are given a 10% discount off the new price.
What is the percentage increase in the price of calls for companies?

Reverse percentage problems

EXAMPLES

1 A shop sells videos with a 20% discount.
Petra buys a video and pays £10.
How much does the video normally cost?

Discount price is normal price less 20%.
So, 80% of normal price = £10.
So, 1% of normal price = £10 ÷ 80
= £0.125
So, normal price = £0.125 × 100
= £12.50

2 Tara gets a 5% wage rise.
Her new wage is £126 per week.
What was Tara's wage before her wage rise?

New wage = old wage + 5%
So, 105% of old wage = £126
1% of old wage is 126 ÷ 105 = £1.20
Old wage = 1.2 × 100 = £120

Exercise 4.4

Do questions 1 to 4 without a calculator.

1. A special bottle of pop contains 10% more than a normal bottle.
The special bottle contains 660 ml.
How much does the normal bottle contain?

2. Jim saves 15% of his monthly salary. Each month he saves £90.
What is his monthly salary?

3. May gets a 20% wage rise. Her new wage is £264 per week.
What was May's wage before her wage rise?

4. In a high jump event, Nick jumps 1.8 metres.
This is 5% lower than the best height he can jump.
What is the best height he can jump?

5. A house is valued at £350 000.
This is a 12% increase on the value of the house a year ago.
What was the value of the house a year ago?

6. 30 grams of a breakfast cereal provides 16.2 mg of vitamin C.
This is 24% of the recommended daily intake.
What daily intake of vitamin C is recommended?

7. Tom gets a 3% increase in his salary. His new salary is £1462.60 per month.
What was Tom's salary before his wage rise?

8. A one year old car is worth £6720. This is a decrease of 16% of its value from new.
What was the price of the new car?

9. Here is some data about the changes in the number of pupils in schools A and B.
School A's numbers increased by 4% to 442.
School B's numbers decreased by 6% to 423.
How many pupils were in schools A and B before the change in numbers?

10. John sells his computer to Dan and makes a 15% profit.
Dan then sells the computer to Ron for £391. Dan makes a 15% loss.
How much did John pay for the computer?
Explain why it is not £391.

11. Kim sells her bike to Sara.
Sara sells it to Tina for £121.50. Both Kim and Sara make a 10% loss.
How much did Kim pay for the bike?
Explain why it is not 20% more than £121.50.

Wages

Hourly pay

Many people are paid by the hour for their work. In most cases they receive a **basic hourly rate** for a fixed number of hours and an **overtime rate** for any extra hours worked.

EXAMPLE

A car-park attendant is paid £6.20 per hour for a basic 40-hour week.
Overtime is paid at time and a half.
One week an attendant works 48 hours.
How much does he earn?

Basic Pay: £6.20 × 40 = £248.00
Overtime: 1.5 × £6.20 × 8 = £74.40
Total pay = £322.40

Overtime paid at 'time and a half' means 1.5 × normal hourly rate. In this example, the hourly overtime rate is given by: 1.5 × £6.20

Common overtime rates are 'time and a quarter', 'double time', etc.

Income tax

The amount you earn for your work is called your **gross pay**.
Your employer will make deductions from your gross pay for income tax, National Insurance, etc.
Pay after all deductions have been made is called **net pay**.

The rates of tax and the bands (ranges of income) to which they apply vary.

The amount of **income tax** you pay will depend on how much you earn.
Everyone is allowed to earn some money which is not taxed, this is called a **tax allowance**.
Any remaining money is your **taxable income**.

EXAMPLE

George earns £6310 per year.
His tax allowance is £4895 per year and he pays tax at 10p in the £ on his taxable income.
How much income tax does George pay per year?

Taxable income: £6310 − £4895 = £1415
Income tax: £1415 × 0.10 = £141.50

George pays income tax of £141.50 per year.

An income tax rate of 10% is often expressed as '10p in the pound (£)'.

Exercise 4.5

1. A secretary earns £352.80 a week.
She is paid £9.80 per hour.
How many hours a week does she work?

2. A chef is paid £12.40 per hour for a basic 38-hour week. Overtime is paid at time and a half.
How much does the chef earn in a week in which she works 50 hours?

3. A hairdresser is paid £8.20 per hour for a basic 35-hour week.
One week she works two hours overtime at time and a half and $3\frac{1}{2}$ hours overtime at time and a quarter. How much is she paid that week?

4. A driver is paid £68.85 for $4\frac{1}{2}$ hours of overtime. Overtime is paid at time and a half.
What is his basic hourly rate of pay?

5 Lyn earns £6080 per year.
Her tax allowance is £4895 per year and she pays tax at 10p in the £ on her taxable income.
(a) What is her annual taxable income?
(b) How much income tax does she pay per year?

6 Kay has an annual salary of £23 980.
Her tax allowance is £4895 per year.
She pays tax at 10p in the £ on the first £2090 of her taxable income and 22p in the £ on the remainder.
How much income tax does she pay per year?

7 Les has an annual salary of £29 600. His tax allowance is £4895 per year.
He pays tax at 10p in the £ on the first £2090 of his taxable income and 22p in the £ on the remainder.
He is paid monthly. How much income tax does he pay per month?

8 Alf 's income is £25 800 per year.
He pays 9% of his gross income into a pension scheme on which he does not pay tax.
Alf also has a tax allowance of £4895 per year.
He pays tax at 10p in the £ on the first £2090 of his taxable income and 22p in the £ on the remainder.
Calculate how much income tax he pays per year.

9 Reg has an annual salary of £49 880. His tax allowance is £4895 per year.
He pays tax at 10p in the £ on the first £2090 of his taxable income, 22p in the £ on the next £30 310 and 40p in the £ on the remainder.
Calculate how much income tax he pays per year.

10 Alex has an annual salary of £41 240. Her tax allowance is £4895 per year.
She pays tax at 10p in the £ on the first £2090 of her taxable income, 22p in the £ on the next £30 310 and 40p in the £ on the remainder. She is paid monthly.
How much income tax does she pay per month?

Spending

Spending money is part of daily life.
Every day people have to deal with many different situations involving money.
Money is needed to buy fares for journeys, for purchases at shops, for hiring cars and equipment and for buying large items such as furniture.

When a large sum of money is needed to make a purchase, **credit** may be arranged.
This involves paying for the goods over a period of time by agreeing to make a number of weekly or monthly repayments.
It may also involve paying a **deposit**. The cost of credit may be more than paying cash.

A motor home costs £19 950.
It can be bought on credit by paying a deposit of £7000 and 36 monthly payments of £395.
How much more is paid for the motor home when it is bought on credit?

Deposit: £ 7 000
Payments: £395 × 36 = £14 220

Credit Price: £21 220

Difference: £21 220 − £19 950 = £1270
Credit price is £1270 more.

Best buys

When shopping we often have to make choices between products which are packed in various sizes and priced differently. If we want to buy the one which gives the better value for money we must compare prices using the same units.

EXAMPLE

Peanut butter is available in small or large jars, as shown.
Which size is the better value for money?

Compare the number of grams per penny for each size.
Small: $250 \div 58 = 4.31\ldots$ grams per penny.
Large: $454 \div 106 = 4.28\ldots$ grams per penny.

The small size gives more grams per penny and is better value.

Household bills

The cost of living includes many bills for services provided to our homes. Electricity, gas and telephone charges are all examples of **quarterly bills** which are sent out four times a year.
Some bills are made up of two parts:

A fixed (standing) charge, for providing the service.
A charge for the quantity of the service used (amount of electricity, duration of telephone calls, etc.)

Other household bills include taxes payable to the local council, water charges and the cost of the insurance of the house (structure) and its contents.

Exercise 4.6

1. Mr Jones pays £4.14 for 400 g of Brie and 250 g of Stilton. Stilton costs £7.60 per kilogram. How much per kilogram is Brie?

2. A washing machine costs £475. It can be bought on credit by paying a deposit of 10% of the cash price and 24 monthly payments of £19.50.
How much more is paid for the washing machine when it is bought on credit?

3. Jars of pickled onions are sold at the following prices: 460 g at 65p or 700 g at 98p.
Which size is better value for money?

4. Mrs Dear checks her water bill. She has used 58 cubic metres of water at 97.04 pence per cubic metre and there is a standing charge of £11. How much is her bill?

5. George insures his house valued at £284 000 and its contents valued at £27 500.
The annual premiums for the insurance are:

   Buildings: £1.35 per £1000 of cover, Contents: 56p per £100 of cover.

   Calculate the total cost of the insurance premium.

6. Mr Peters has an annual council tax of £1982.28. He pays the council tax in 10 instalments. The first instalment is £200.28 and the remaining amount is payable in 9 instalments of equal value. How much is the second instalment?

7. Mr Jones receives an electricity bill for £56.84.
The bill includes a quarterly charge of £10.40 and the cost per unit is 6.85 pence.
Calculate to the nearest whole number, the number of units he has used.

8. Sheila rents a flat and pays £69.44 to insure its contents.
Contents insurance costs 56p for each £100 insured.
For how much are the contents insured?

VAT

Some goods and services are subject to a tax called **value added tax**, or **VAT**, which is calculated as a percentage of the price or bill. Total amount payable = cost of item or service + VAT

For most purchases the rate of VAT is 17.5%. For gas and electricity the rate of VAT is 5%. Some goods are exempt from VAT.

EXAMPLES

1 A bill at a restaurant is £24 + VAT at 17.5%. What is the total bill?

VAT: £24 × 0.175 = £4.20

Total bill: £24 + £4.20 = £28.20

The total bill is £28.20.

Remember: $17.5\% = \frac{17.5}{100} = 0.175$

2 An electricity bill of £49.14 includes VAT at 5%. How much VAT is paid?

Cost = cost without VAT + 5% VAT.
£49.14 = 105% of cost without VAT.

1% of cost without VAT is given by £49.14 ÷ 105 = £0.468

VAT = 5% of cost without VAT, so, VAT = £0.468 × 5 = £2.34.

Exercise 4.7

Do not use a calculator for questions 1 and 2.

1. Naomi's gas bill is £120 plus VAT at 5%. How much VAT does she have to pay?
2. Joe receives an electricity bill for £70 plus VAT at 5%.
   (a) Calculate the amount of VAT charged.
   (b) What is the total bill?
3. A car service costs £90 plus VAT at 17.5%.
   (a) Calculate the amount of VAT charged.
   (b) What is the total cost of the service?
4. A bike costs £248 plus VAT at 17.5%. What is the total cost of the bike?
5. James receives a telephone bill for £66 plus VAT at 17.5%. How much is the total bill?
6. A loft conversion costs £23 000 plus VAT at 17.5%. What is the total cost?
7. George buys vertical blinds for his windows.
   He needs three blinds at £65 each and two blinds at £85 each.
   VAT at 17.5% is added to the cost of the blinds.
   How much do the blinds cost altogether?
8. A car is hired for two days and driven 90 miles.
   VAT at 17.5% is added to the hire charges.
   How much does it cost to hire the car altogether?

9. A computer costs £1233.75 including VAT at 17.5%.
   How much of the cost is VAT?
10. A gas bill of £67.20 includes VAT at 5%. How much VAT is paid?
11. VAT at 17.5% on a washing machine is £43.75.
    What is the price of the washing machine including VAT?
12. The bill for a new central heating boiler includes £436.10 VAT. VAT is charged at 17.5%.
    What is the total bill?

Savings

Money invested in a savings account or a bank or building society earns **interest**, which is usually paid once a year.

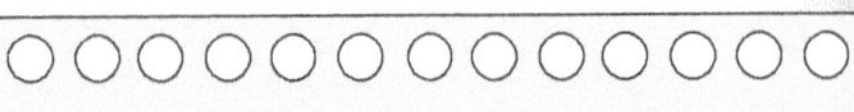

Banks and building societies advertise the **yearly rates** of interest payable.

For example, 6% per year.

Interest, usually calculated annually, can also be calculated for shorter periods of time.

Simple Interest

With **Simple Interest**, the interest is paid out each year and not added to your account.

> The amount of Simple Interest an investment earns can be calculated using:
>
> Simple Interest = Amount invested × Time in years × Rate of interest per year

Compound Interest

With **Compound Interest**, the interest earned each year is added to your account and also earns interest the following year.

For example, an investment of 5% per annum means that the amount invested earns £5 for every £100 invested for one year.

So, after the first year of the investment, every £100 invested becomes £100 + 5% of £100.
£100 + 5% of £100 = £100 + £5 = £105
So, after the second year of the investment, every £100 of the original investment becomes £105 + 5% of £105.
£105 + 5% of £105 = £105 + £5.25 = £110.25

This can also be calculated as: $100 \times (1.05)^2 = £110.25$
Explain why this works.

EXAMPLES

1 Find the Simple Interest paid on £600 invested for 6 months at 8% per year.

Simple Interest

$= 600 \times \frac{6}{12} \times \frac{8}{100}$

$= 600 \times 0.5 \times 0.08$

$= £24$

The Simple Interest paid is £24.

> **Note:** Interest rates are given 'per year'.
> The length of time for which an investment is made is also given in years.
> 6 months $= \frac{6}{12}$ years.
> *Explain why.*

2 Find the Compound Interest paid on £600 invested for 3 years at 6% per year.

1st year	Investment	= £600
	Interest: £600 × 0.06	= £ 36
	Value of investment after one year	= £636
2nd year	Investment	= £636
	Interest: £636 × 0.06	= £ 38.16
	Value of investment after two years	= £674.16
3rd year	Investment	= £674.16
	Interest: £674.16 × 0.06	= £ 40.45
	Value of investment after three years	= £714.61

Compound Interest = Final value − Original value
= £714.61 − £600 = £114.61

This could also be calculated as follows: $600 \times (1.06)^3 - 600 = £114.61$

Exercise 4.8

Do not use a calculator for questions 1 to 7.

1. Find the simple interest paid on £200 for 1 year at 5% per year.
2. Calculate the simple interest on £500 invested at 6% per year after:
   (a) 1 year, (b) 6 months.
3. Calculate the simple interest paid on an investment of £6000 at 7.5% per year after 6 months.
4. Find the simple interest on £800 invested for 9 months at 8% per year.
5. Calculate the simple interest on £10 000 invested for 3 months at 9% per year.
6. Jenny invests £200 at 10% per annum compound interest.
   What is the value of her investment after 2 years?
7. Which of the following investments earn more interest?
   (a) £200 for 3 years at 5% compound.
   (b) £300 for 2 years at 5% compound.
   Show your working.
8. £10 000 is to be invested for 3 years.
   Calculate the final value of the investment if the interest rate per annum is 6%.
   Give your answer to a suitable degree of accuracy.
9. Interest on a loan of £2000 is charged at the rate of 21% per annum.
   Interest is calculated on the outstanding loan at the **start** of each year.
   (a) How much is owed immediately the loan is taken out?
   Repayments are £600 per year.
   (b) How much is owed at the **start** of the third year of the loan?
10. A bouncy ball is dropped from the top of a skyscraper 256 m high.
    After each bounce it reaches a height 25% less than its previous height.
    What is the height of the ball after 4 bounces?
11. A man buys a new car for £13 000. The car loses value at the rate of 14% per annum.
    (a) (i) What is its value after 3 years?
    (ii) Express its value after 3 years as a percentage of its original value.
    (b) Repeat (a) for a new car originally valued at £20 000.
    (c) What do you notice about your answers to (a) and (b)?
12. How long does it take £100 to double in value at 9% per annum?

What you need to know

- 'Per cent' means 'out of 100'. The symbol for per cent is %. 10% can be written as $\frac{10}{100}$.
- To change a decimal or a fraction to a percentage - **multiply by 100**.
  For example: 0.12 as a percentage is $0.12 \times 100 = 12\%$.
  $\frac{3}{25}$ as a percentage is $\frac{3}{25} \times 100 = 3 \times 100 \div 25 = 12\%$.
- To change a percentage to a decimal or a fraction - **divide by 100**.
  For example: 18% as a decimal is $18 \div 100 = 0.18$.
  18% as a fraction is $\frac{18}{100}$ which in its simplest form is $\frac{9}{50}$.
- Percentage increase $= \frac{\text{actual increase}}{\text{initial value}} \times 100\%$ Percentage decrease $= \frac{\text{actual decrease}}{\text{initial value}} \times 100\%$
- **Hourly pay** is paid at a **basic rate** for a fixed number of hours.
  Overtime pay is usually paid at a higher rate such as time and a half,
  which means each hour's work is worth 1.5 times the basic rate.
- Everyone is allowed to earn some money which is not taxed. This is called a **tax allowance**.
- Tax is only paid on income earned in excess of the tax allowance. This is called **taxable income**.

Percentages and Money

- **Value added tax**, or **VAT**, is a tax on some goods and services and is added to the bill.
- Gas, electricity and telephone bills are paid **quarterly**.
  Some bills consist of a standing charge plus a charge for the amount used.
- When considering a **best buy**, compare quantities by using the same units.
  For example, find which product gives more grams per penny.
- Money invested in a savings account at a bank or building society earns **interest**, which is usually paid once a year.
  With **Simple Interest**, the interest is paid out each year and not added to your account.

  $$\text{Simple Interest} = \text{Amount invested} \times \text{Time in years} \times \text{Rate of interest per year}$$

  With **Compound Interest**, the interest earned each year is added to your account and also earns interest the following year.

Review Exercise 4

Do not use a calculator for questions 1 to 4.

1. A bag contains 60 beads.
   (a) Emily uses 30% of the beads to make a necklace. How many beads does she use?
   (b) Laura uses 12 beads to make a bracelet. What percentage of the beads does she use?

2. Joe earns £650 in May. In June he earns 20% more. How much does he earn in June? AQA

3. A roll of carpet is 20 m long. Beryl buys 18 m of carpet from the roll.
   What percentage of the roll did she buy?

4. An estate agent makes the following charge for the sale of a house.

Sale price of house	Up to £50 000	Over £50 000
Charge by estate agent	3% of the sale price	3% of the first 50 000 plus 2% on the remainder

   Calculate the charge made by the estate agent for a house sold for £229 000. AQA

5. David buys 0.6 kg of grapes and 0.5 kg of apples. He pays £1.36 altogether.
   The grapes cost £1.45 per kilogram. How much per kilogram are apples? AQA

6. An electricity bill is £73.28 plus VAT at 5%. Calculate the VAT charged. AQA

7. A PC costs £499 + VAT at $17\frac{1}{2}$%. What is the total cost of the PC?

8. Josie decides to buy a motorcycle on credit.
   The credit terms are:

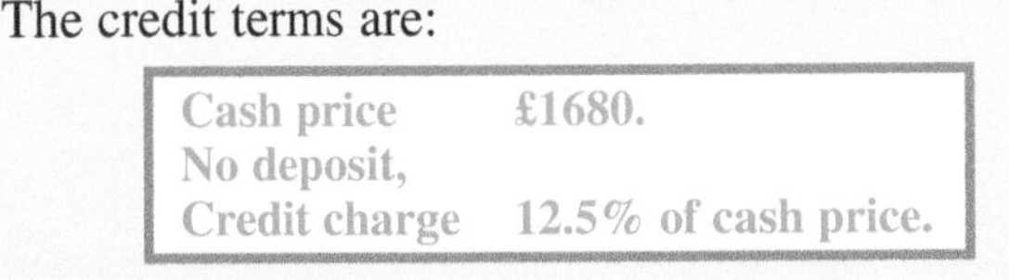

Cash price £1680.
No deposit,
Credit charge 12.5% of cash price.

   The total cost is the credit charge plus the cash price.
   She pays by 12 equal monthly payments.
   How much does she pay each month? AQA

9. Jam is sold in two sizes.
   A large pot of jam costs 88p and weighs 822 g.
   A small pot of jam costs 47p and weighs 454 g.
   Which pot of jam is better value for money?
   You must show all your working. AQA

10. In a sale a dress costs £32.40. The original price was reduced by 10%.
    What was the original price? AQA

11 Francis is paid £7.60 per hour for a basic 35-hour week.
One week Francis also works overtime at time and a half. His total pay that week was £311.60.
How many hours overtime did he work that week?

12 (a) A year ago Martin was 1.60 m tall. He is now 4% taller.
Calculate his height now.
(b) Martin now weighs 58 kg. A year ago he weighed 51 kg.
Calculate the percentage increase in his weight.
Give your answer to an appropriate degree of accuracy. AQA

13 Jason bought an old bicycle for £36. He repaired it and resold it for £52.
What was his percentage profit? AQA

14 Brian buys 300 CDs for £1400. He sells $\frac{3}{4}$ of them at £9 each.
He then reduces the price of the remaining CDs by 30%.
When he has sold 290 CDs, he gives the last 10 to a charity.
(a) How much money does he receive from selling the CDs?
(b) Find the percentage profit which Brian made on these CDs. AQA

15 The number of salmon in the River Dribble is expected to increase by about 15% every year up to the year 2010.
At the beginning of 2006 there were approximately 6000 salmon in the River Dribble.
(a) How many salmon are expected to be in the River Dribble at the beginning of 2007?
(b) The River Authority stated that by the year 2010 there will be over 10 000 salmon in the River Dribble. Do you agree with this? Show all your working.

16 In a sale, all the prices are reduced by 20%. I bought a coat for £68 in the sale.
What was the price of the coat before the sale? AQA

17 Hannah invests £360 in a building society account at 4.8% per year.
Find the simple interest paid on her investment after 4 months.

18 Shane invests £2000 at 7% per annum compound interest.
Calculate the value of his investment after 3 years.

19 Guy's house is in council Band G. He pays £1381.80 per year in council tax.
Properties in Band G pay 84% of the council tax for properties in Band H.
Calculate the council tax for properties in Band H.

20 Dipak's income is £48 564 per year. Dipak has a tax allowance of £4895 per year and he also pays 15% of his income into a pension scheme on which he does not pay tax.
He pays tax at 10p in the £ on the first £2090 of his taxable income, 22p in the £ on the next £30 310 and 40p in the £ on the remainder.
Calculate how much income tax he pays per year.

21 Ros needs to insure her car. The full premium is £848.
Ros gets a 12.5% discount for agreeing to be the only driver.
She then gets a further 5% discount off the reduced price for agreeing to pay the first £250 of any claim.
How much does Ros pay for her insurance?

22 In 1982 it was estimated that there were only 20 000 Minke whales left in the world.
The hunting of Minke whales was banned in 1982.
After 1982 the population increased by 45% each year.
(a) How many Minke whales were there in 1983 (1 year after the ban)?
(b) How many Minke whales were there in 1985 (3 years after the ban)?
Give your answer to a suitable degree of accuracy.
(c) It was agreed that when the Minke whale population reached 250 000 some hunting of Minke whales would be allowed again. In what year did this happen? AQA

CHAPTER 5

Working with Number

Multiples

Numbers in the 4 times table are called **multiples** of 4.
Multiples of a number are found by multiplying the number by 1, 2, 3, 4, ...
For example, the multiples of 6 are: 6, 12, 18, 24, ...
The 8th **multiple** of 7 is $8 \times 7 = 56$.

Factors

Pairs of **whole numbers** which have a product of 6 are 1×6 and 2×3.
1, 2, 3, and 6 are called **factors** of 6.
The **factors** of a number can be found from the multiplication facts that give the number.
For example: $1 \times 12 = 2 \times 6 = 3 \times 4 = 12$.
The factors of 12 are 1, 2, 3, 4, 6, 12.

Common factors

The factors of 20 are: 1, 2, 4, 5, 10, 20.
The factors of 50 are: 1, 2, 5, 10, 25, 50.
1, 2, 5 and 10 are factors of both 20 **and** 50.
They are called the **common factors** of 20 and 50.

Prime numbers

Numbers like 7 are called **prime numbers**.
A prime number has exactly **two** factors, 1 and the number itself.
The first few prime numbers are: 2, 3, 5, 7, 11, 13, ...
The number 1 is not a prime number because it has only one factor.

Exercise 5.1

Do not use a calculator for this exercise.

1. (a) Write down a multiple of 7 between 30 and 40.
   (b) Write down a multiple of 8 between 40 and 50.
2. Cameron states that the sum of four consecutive numbers is always a multiple of 4.
   Give an example to show that this statement is not true.
3. 3, 4, 5, 9, 14, 20, 27 and 35.
   Which of the above numbers are:
   (a) multiples of 7, (b) factors of 20, (c) prime numbers?
4. (a) What multiple of 6 is the third multiple of 4?
   (b) What multiple of 8 is the fourth multiple of 4?
   (c) What multiple of 20 is the tenth multiple of 10?
   (d) What multiple of 24 is the fourth multiple of 18?
5. Find all the factors of:
   (a) 16 (b) 28 (c) 36 (d) 45 (e) 48 (f) 50
6. Find the common factors of:
   (a) 10 and 15, (b) 12 and 20, (c) 16 and 18, (d) 24 and 36, (e) 12, 18 and 36.
7. (a) Find all the prime numbers between 30 and 40.
   (b) Is 49 a prime number? Give a reason for your answer.

Powers

Products of the same number, like

$$3 \times 3, \quad 5 \times 5 \times 5, \quad 10 \times 10 \times 10 \times 10 \times 10,$$

can be written in a shorthand form using **powers**.

For example:

$3 \times 3 = 3^2$	This is read as	'3 to the power of 2'.	3^2 has the value 9.
$5 \times 5 \times 5 = 5^3$	This is read as	'5 to the power of 3'.	5^3 has the value 125.
$10 \times 10 \times 10 \times 10 \times 10 = 10^5$	This is read as	'10 to the power of 5'.	10^5 has the value 100 000.

Index form

Numbers written in shorthand form like 3^2, 5^3 and 10^5 are said to be in **index form**.
This is sometimes called **power** form.

An expression of the form $a \times a \times a \times a \times a$ can be written in index form as a^5.
a^5 is read as 'a to the **power** 5'. a is the **base** of the expression. 5 is the **index** or **power**.

Prime factors

The factors of 18 are 1, 2, 3, 6, 9 and 18.
Two of these factors, 2 and 3, are prime numbers.
The **prime factors** of 18 are 2 and 3.

Those factors of a number which are prime numbers are called **prime factors**.

Products of prime factors

All numbers can be written as the product of their prime factors.
For example: $6 = 2 \times 3$ $\quad 20 = 2 \times 2 \times 5$ $\quad 168 = 2 \times 2 \times 2 \times 3 \times 7$

Powers can be used to write numbers as the product of their prime factors in a shorter form.
For example: $20 = 2^2 \times 5$ $\quad 168 = 2^3 \times 3 \times 7$

A **factor tree** can be used to help write numbers as the product of their prime factors.

For example, this factor tree shows that:

$40 = 2 \times 20$

$40 = 2 \times 2 \times 10$

$40 = 2 \times 2 \times 2 \times 5$

$40 = 2^3 \times 5$

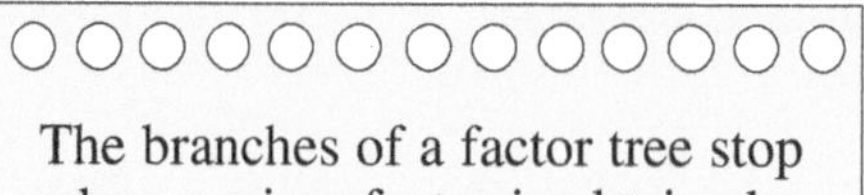

So, 40 written as the product of its prime factors is $2^3 \times 5$.

Least common multiples

The **Least Common Multiple** of two numbers is the smallest number that is a multiple of both.

For example, to find the least common multiple of 36 and 48:

Write each number as a product of its prime factors.

$$36 = 2^2 \times 3^2 \qquad 48 = 2^4 \times 3$$

Circle the prime factor with the **highest** power for **each** prime factor.

$$36 = 2^2 \times \textcircled{3^2} \qquad 48 = \textcircled{2^4} \times 3$$

Multiply these numbers together. $\quad 2^4 \times 3^2 = 16 \times 9 = 144$

The **least common multiple (LCM)** of 36 and 48 is 144.

Highest common factors

The **Highest Common Factor** of two numbers is the largest number that is a factor of both.

For example, to find the highest common factor of 36 and 48:

Write each number as a product of its prime factors.

$$36 = 2^2 \times 3^2 \qquad 48 = 2^4 \times 3$$

Circle the prime factor with the **lowest** power for prime factors **common** to both numbers.

$$36 = \textcircled{2^2} \times 3^2 \qquad 48 = 2^4 \times \textcircled{3}$$

Multiply these numbers together. $2^2 \times 3 = 4 \times 3 = 12$

The **highest common factor (HCF)** of 36 and 48 is 12.

Exercise 5.2

Do not use a calculator.

1. Write these products of primes in index form.
   (a) $2 \times 2 \times 3 \times 3$ (b) $2 \times 3 \times 3 \times 3 \times 5$ (c) $2 \times 3 \times 5 \times 5$
   (d) $2 \times 2 \times 2 \times 3 \times 5 \times 5$ (e) $5 \times 3 \times 3 \times 3 \times 5 \times 5$ (f) $7 \times 7 \times 7 \times 7$

2. Find the prime factors of:
   (a) 12 (b) 20 (c) 28 (d) 45 (e) 66 (f) 108

3. Write the following numbers as products of their prime factors.
   (a) 12 (b) 20 (c) 28 (d) 45 (e) 66 (f) 108

4. Write 128 as a product of its prime factors.

5. Write 1000 as a product of its prime factors.

6. A number written in terms of its prime factors is $2^3 \times 5^5$.
   (a) Calculate the number.
   (b) Write, in terms of its prime factors, a number which is 5 times the size.

7. Write each of the numbers 24, 40, 56 and 88 as the product of their prime factors.
   What do you notice? Find the next three numbers.

8. Find the least common multiple of:
   (a) 8 and 12 (b) 5 and 32 (c) 10 and 20 (d) 15 and 18
   (e) 30 and 45 (f) 4, 6 and 8 (g) 5, 8 and 10 (h) 45, 90 and 105

9. Find the highest common factor of:
   (a) 12 and 66 (b) 8 and 24 (c) 16 and 18 (d) 20 and 36
   (e) 33 and 88 (f) 16, 20 and 28 (g) 15, 39 and 45 (h) 45, 90 and 105

10. (a) Write 24 as a product of prime factors.
    (b) Write 54 as a product of prime factors.
    (c) What is the highest common factor of 24 and 54?
    (d) What is the least common multiple of 24 and 54?

11. (a) Find the value of x when $2^2 \times 3^x = 108$.
    (b) Write 162 as a product of prime factors.
    (c) What is the highest common factor of 108 and 162?
    (d) What is the least common multiple of 108 and 162?

12. The bell at St. Gabriel's church rings every 6 minutes.
    At St. Paul's, the bell rings every 9 minutes.
    Both bells ring together at 9.00 am.
    When is the next time both bells ring together?

Square numbers

Whole numbers raised to the power 2 are called **square numbers**.

For example: $3^2 = 3 \times 3 = 9$ 3^2 is read as '3 squared'. 9 is a square number.

Numbers that are not whole numbers can also be squared.
To **square a number** multiply it by itself.

For example: $1.6^2 = 1.6 \times 1.6 = 2.56$
2.56 is **not** a square number. *Why not?*

Squaring on a calculator

1.6^2 is read as '1.6 squared'.

To calculate 1.6^2 use this sequence of buttons:

[1] [.] [6] [x^2] [=]

Cube numbers

Whole numbers raised to the power 3 are called **cube numbers**.

For example: $3^3 = 3 \times 3 \times 3 = 27$ 3^3 is read as '3 cubed'. 27 is a cube number.

Numbers that are not whole numbers can also be cubed.

For example: $1.6^3 = 1.6 \times 1.6 \times 1.6 = 4.096$
4.096 is **not** a cube number. *Why not?*

Using a calculator

Powers

The **squares** and **cubes** of numbers can also be calculated using the [x^y] button on a calculator.

The [x^y] button can be used to calculate the value of a number x raised to the power of y.

For example, to calculate the value of 2.6^3, enter the following sequence into your calculator:

[2] [.] [6] [x^y] [3] [=] This gives $2.6^3 = 17.576$

Reciprocals

The **reciprocal** of a number is the value obtained when the number is divided into 1.

For example, the reciprocal of 5 is $\frac{1}{5}$ (or 0.2).

On a calculator: [5] [$\frac{1}{x}$]

The reciprocal of a number x is $\frac{1}{x}$.

A number times its reciprocal equals 1.

For example, the reciprocal of 2 is $\frac{1}{2}$, and $2 \times \frac{1}{2} = 1$.

To find the reciprocal of a number on a calculator use the [$\frac{1}{x}$] button.
0 (zero) has no reciprocal.

Square roots

The opposite of squaring a number is called finding the **square root**.

For example:
The square root of 16 is 4 because $4^2 = 16$.

$\sqrt{\ }$ This special symbol stands for the square root.

For example: $\sqrt{9} = 3$ $\sqrt{2.56} = 1.6$

Note: $-3 \times -3 = 9$ and $-1.6 \times -1.6 = 2.56$
So, the square root of a number can be **positive** or **negative**.
In most cases we only use the positive square root.

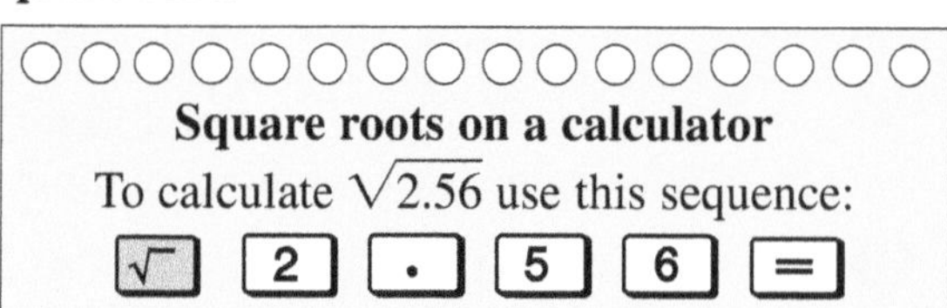

Square roots on a calculator

To calculate $\sqrt{2.56}$ use this sequence:

[$\sqrt{\ }$] [2] [.] [5] [6] [=]

Cube roots

The opposite of cubing a number is called finding the **cube root**.

For example: The cube root of 27 is 3 because $3^3 = 27$.

Exercise 5.3

Do not use a calculator for questions 1 to 6.

1. What is (a) the square of 7, (b) the cube of 5, (c) the reciprocal of 2?
2. Jake says, "The square root of 55 lies between 7 and 8." Is he right? Explain your answer.
3. Consider the numbers: 8 16 27 36 64 100
   Which of these numbers is both a square number **and** a cube number?
4. Connie says that $2^2 + 3^2 = 5^2$. Is she right? Explain your answer.
5. (a) Calculate the value of: (i) $(-3)^2$ (ii) $(-2)^3$ (iii) $(-4)^2$ (iv) $(-5)^3$
   (b) What do you notice about the signs of your answers?
6. Find the value of: (a) $\sqrt{64}$ (b) $\sqrt{49}$ (c) $\sqrt{3^2 + 4^2}$ (d) $\sqrt{13^2 - 12^2}$
7. Find the cube root of (a) 8, (b) 64, (c) 125.
8. Use the x^y button on your calculator to find the value of:
   (a) 2.5^2 (b) 0.8^2 (c) 15^3 (d) 2.4^3 (e) 7^5 (f) 0.5^4 (g) 6.5^6
9. Use the $\frac{1}{x}$ button on your calculator to find the reciprocals of:
   (a) 4 (b) 20 (c) 25 (d) 0.5 (e) 0.25 (f) 0.4 (g) 0.16
10. Show by means of an example, that a number times its reciprocal is equal to 1.
11. Use your calculator to find $\sqrt{96}$. Give your answer to one decimal place.
12. Calculate. (a) $\sqrt{5.3}$ (b) $\sqrt{300}$ (c) $\sqrt{4.8^2 - 2.7^2}$
    Give your answers correct to one decimal place.
13. Calculate the value of:
    (a) $3^3 \times 10^3$ (b) $10^5 \div 5^3$ (c) $2.6 + \frac{1}{2.6}$ (d) 2.2^3
    (e) $8.5^2 - 1.3^2$ (f) $\frac{5}{(0.4)^2}$ (g) $(1.9 + 2.2)^2 \times 1.5$ (h) $0.8^2 \times \frac{1}{0.5}$
    (i) $5^3 \times (4^5 - 4^2) + 6^3$ (j) $(8^7 - 8^5) \div (4^7 - 4^5)$
14. (a) Show that $16^3 = 4^6$.
    (b) Hence, or otherwise, solve the equation $2^x = 4^6$.
15. Use trial and improvement to solve $5^x = 10$. Give your answer to 2 d.p.

Multiplying and dividing numbers with powers

$3 \times 3 = 3^2$ and $5 \times 5 \times 5 = 5^3$ Both 3^2 and 5^3 are examples of numbers with powers.
These examples introduce methods for multiplying and dividing powers of the same number.

1 Calculate the value of $6^5 \times 6^4$ in power form.

$6^5 = 6 \times 6 \times 6 \times 6 \times 6$ and $6^4 = 6 \times 6 \times 6 \times 6$

$6^5 \times 6^4 = (6 \times 6 \times 6 \times 6 \times 6) \times (6 \times 6 \times 6 \times 6)$

$= 6 \times 6 \times 6 \times 6 \times 6 \times 6 \times 6 \times 6 \times 6$

This gives: $6^5 \times 6^4 = 6^9$

2 Calculate the value of $6^7 \div 6^4$ in power form.

$6^7 \div 6^4 = \frac{6^7}{6^4} = \frac{6 \times 6 \times 6 \times \cancel{6} \times \cancel{6} \times \cancel{6} \times \cancel{6}}{\cancel{6} \times \cancel{6} \times \cancel{6} \times \cancel{6}}$

$= 6 \times 6 \times 6 = 6^3$

This gives: $6^7 \div 6^4 = 6^3$

Can you see a quick way of working out the index (power) for each example?

Rules of indices

When **multiplying**: powers of the same base are **added**. In general: $a^m \times a^n = a^{m+n}$

When **dividing**: powers of the same base are **subtracted**. In general: $a^m \div a^n = a^{m-n}$

Raising a power to a power

$(a^m)^n = a^{mn}$

Two special results

$a^1 = a \quad a^0 = 1$

EXAMPLES Simplify. Leave your answers in index form.

(a) $2^9 \times 2^4 = 2^{9+4} = 2^{13}$
Add indices

(b) $2^9 \div 2^4 = 2^{9-4} = 2^5$
Subtract indices

(c) $(4^9)^3 = 4^{9 \times 3} = 4^{27}$
Multiply indices

Exercise 5.4

Do not use a calculator for this exercise.

1 Simplify. Leave your answers in index form.
(a) $2^3 \times 2^2$ (b) $3^5 \times 3^2$ (c) $5^6 \times 5^2$ (d) $7^3 \times 7$ (e) $9^5 \times 9^0$

2 Simplify. Leave your answers in index form.
(a) $2^3 \div 2^2$ (b) $3^5 \div 3^2$ (c) $5^6 \div 5^2$ (d) $7^3 \div 7$ (e) $9^5 \div 9^0$

3 Simplify. Leave your answers in index form.
(a) $3 \times 3^2 \times 3^3$ (b) $\frac{10 \times 10^3}{10^2}$ (c) $\frac{4^3 \times 4^3}{4}$ (d) $\frac{5^5 \times 5^2}{5^4}$
(e) $\frac{2 \times 2^5}{2^3}$ (f) $\frac{5 \times 5^2}{5^3}$ (g) $\frac{7^3 \times 7^2}{7^7}$ (h) $\frac{3^5 \times 3^0}{3^2}$

4 Simplify half of 2^{16}.

5 Simplify. Leave your answers in index form.
(a) $(2^3)^2$ (b) $(3^5)^2$ (c) $(5^2)^3$ (d) $(7^3)^3$ (e) $(9^0)^5$

6 Find x in each of the following.
(a) $2^3 \times 2^5 = 2^x$ (b) $3^5 \times 3^2 = 3^x$ (c) $7^2 \times 7^8 \times 7 = 7^x$ (d) $3^5 \div 3^2 = 3^x$
(e) $8^7 \div 8^6 = 8^x$ (f) $6^5 \div 6 = 6^x$ (g) $(2^3)^2 = 2^x$ (h) $(5^4)^5 = 5^x$
(i) $\frac{2^3 \times (2^2)^5}{2^8 \times 2^3} = 2^x$ (j) $\frac{(3^3 \times 3^2)^3}{3^7} = 3^x$ (k) $\frac{5^x \times 5^3 \times 5^4}{(5 \times 5^3)^3} = 5$ (l) $\frac{(3^x \times 3^2)^3}{(3 \times 3^5)^2} = 3^3$

Negative powers and reciprocals

Using patterns of powers

This list shows the powers of two extended to include negative indices.

$2^3 = 2 \times 2 \times 2 = 8$
$2^2 = 2 \times 2 = 4$
$2^1 = 2 = 2$
$2^0 = 1 = 1$
$2^{-1} = \frac{1}{2} = 0.5 = \frac{1}{2^1}$
$2^{-2} = \frac{1}{4} = 0.25 = \frac{1}{2^2}$
$2^{-3} = \frac{1}{8} = 0.125 = \frac{1}{2^3}$

Using the rules of indices

$\frac{1}{10} = 1 \div 10$
$\frac{1}{10} = 10^0 \div 10^1$
$\frac{1}{10} = 10^{0-1}$
$\frac{1}{10^1} = 10^{-1}$

$1 = 10^0$
$10 = 10^1$

$a^m \div a^n = a^{m-n}$

$\frac{1}{32} = 1 \div 32$
$\frac{1}{32} = 2^0 \div 2^5$
$\frac{1}{32} = 2^{0-5}$
$\frac{1}{2^5} = 2^{-5}$

$1 = 2^0$
$32 = 2^5$

All of the examples illustrate this general rule for negative indices.

$$a^{-m} = \frac{1}{a^m}$$

$\frac{1}{a^m}$ and, hence, a^{-m}, is called the **reciprocal** of a^m.

EXAMPLES

1 Simplify $10^{-4} \div 10^{-2}$.

$$10^{-4} \div 10^{-2} = 10^{-4 - -2} = 10^{-4+2} = 10^{-2}$$

2 Simplify $3^4 \times 2^3 \times 3^{-5} \times 2^5$.

$$3^4 \times 2^3 \times 3^{-5} \times 2^5 = 3^4 \times 3^{-5} \times 2^3 \times 2^5 = 3^{4+-5} \times 2^{3+5} = 3^{-1} \times 2^8$$

3 Find the reciprocal of $\left(\frac{1}{6}\right)^{-2}$.

$\left(\frac{1}{6}\right)^{-2} = \frac{1}{6^{-2}} = 6^2$.

The reciprocal of $\left(\frac{1}{6}\right)^{-2}$ is 6^{-2}.

This is equivalent to $\frac{1}{36}$.

4 If $2^x \div 2^5 = \frac{1}{8}$ find the value of x.

$\frac{1}{8} = \frac{1}{2^3} = 2^{-3}$

$2^{x-5} = 2^{-3}$

So, $x - 5 = -3$

This gives $x = 2$.

Remember: When multiplying and dividing powers with different bases each base must be dealt with separately.

Exercise 5.5

Do not use a calculator for this exercise.

1 Simplify. Leave your answers in index form.

(a) $9^2 \times 9^{-2}$ (b) $2^{-3} \times 2$ (c) $5^5 \times 5^{-7}$ (d) $8^{-2} \times 8^{-3}$
(e) $2^{-3} \div 2$ (f) $5^5 \div 5^{-7}$ (g) $11^{-2} \div 11^3$ (h) $7^{-4} \div 7^{-3}$

2 Simplify. Leave your answers in index form.

(a) $8^{-3} \times 8^5$ (b) $7^2 \div 7^7$ (c) $2.5^{-2} \div 2.5^{-1}$
(d) $4^3 \times 4^2 \times 4^{-5}$ (e) $10^{-3} \div 10^{-2}$ (f) $6^{-3} \times 6^4 \div 6^5$
(g) $0.1^{-7} \div 0.1^5$ (h) $5^{-7} \div (5^2 \times 5^6)$ (i) $4^2 \div (4^{-1} \times 4^{-2})$
(j) $4^{-3} \times 4^5 \times 8^5 \times 8^2$ (k) $4^{-1} \times 5^5 \times 5^{-7} \times 4^2$ (l) $2^{-5} \times 5^3 \times 2^3 \times 5^2$
(m) $\frac{3^5 \times 3^{-2}}{3^2}$ (n) $\frac{5^{-3} \times 5^4}{5^{-2}}$ (o) $\frac{2 \times 2^{-3} \times 2^{-1}}{2^2 \times 2}$

3 Express with positive indices.

(a) 3^{-2} (b) 2^{-3} (c) 3×3^{-2} (d) $5^{-2} \times 5^{-1}$ (e) $\frac{1}{3^{-2}}$
(f) $\frac{5}{5^{-1}}$ (g) $2 \div 2^{-3}$ (h) $3^{-3} \div 3^{-1}$ (i) $(5^{-3})^2$ (j) $(3^{-2})^{-3}$

4 Write down the value of:

(a) 5^0 (b) 3^{-1} (c) 2^{-3} (d) $\frac{1}{3^2}$ (e) $\frac{1}{2^{-3}}$ (f) $\frac{3}{3^{-2}}$

5 Calculate each of the following.

(a) $10 + 10^0 + 10^{-1}$ (b) $2 + 2^{-1} + 2^{-2} + 2^{-3}$ (c) $5^0 + 5^{-1} + 5^{-2}$
(d) $3^{-1} + 2^{-1}$ (e) $5^{-2} + 2^{-2}$ (f) $5^{-2} \times 2^{-2}$
(g) $5^{-2} \div 2^{-2}$ (h) $5 \times 6^{-1} + 2 \times 5^{-1}$

6 Find x in each of the following.

(a) $3^x = \frac{1}{81}$ (b) $5^x = \frac{1}{25}$ (c) $\left(\frac{1}{4}\right)^x = 16$ (d) $\left(\frac{1}{6}\right)^x = 216$
(e) $2^x = 0.5$ (f) $5^x = 0.04$ (g) $3 \times 10^x = 0.003$ (h) $2 \times 5^x = 0.4$

7 Calculate $\frac{4^5}{4^{-2}}$ giving your answer in the form 2^n.

Powers and roots

This symbol $\sqrt{\ }$ stands for root.
$\sqrt{16}$ means the square root of 16.
$\sqrt[3]{27}$ means the cube root of 27.
$\sqrt[6]{64}$ means the sixth root of 64.

The inverse (opposite) of raising to a power is finding a **root**.
The inverse of squaring is finding the **square root**.
The inverse of cubing is finding the **cube root**.
The inverse of raising to the power 5 is finding the **fifth root**.
For example:

The square root of 9 is 3 because $3^2 = 9$

The cube root of 64 is 4 because $4^3 = 64$

The fifth root of 32 is 2 because $2^5 = 32$

$\sqrt{6.25} = 2.5$ because $2.5^2 = 6.25$

$\sqrt[3]{1.728} = 1.2$ because $1.2^3 = 1.728$

$\sqrt[6]{15625} = 5$ because $5^6 = 15625$

The connection between powers and roots

If $a = b^n$, $\sqrt[n]{a} = \sqrt[n]{b^n} = b$

Using the rules of powers

$(b^n)^{\frac{1}{n}} = b^{n \times \frac{1}{n}} = b$

So, $(b^n)^{\frac{1}{n}}$ is the same as $\sqrt[n]{b^n}$.

So, $a^{\frac{1}{n}}$ is the same as $\sqrt[n]{a}$.

The inverse of "raising to the power n" is finding the nth root.
In general, finding the nth root of a number, a, can be written as:

$$\sqrt[n]{a} \quad \text{or} \quad a^{\frac{1}{n}}.$$

EXAMPLES

1 Calculate $81^{\frac{1}{4}}$.

$81^{\frac{1}{4}} = \sqrt[4]{81} = 3$

Because $3^4 = 81$.

2 Calculate $25^{-\frac{1}{2}}$.

$25^{-\frac{1}{2}}$ is the **reciprocal** of $25^{\frac{1}{2}}$.

$25^{-\frac{1}{2}} = \frac{1}{25^{\frac{1}{2}}} = \frac{1}{\sqrt{25}} = \frac{1}{5} = 0.2$

3 Calculate $512^{-\frac{1}{9}}$.

Use this key sequence: [5] [1] [2] [$x^{1/y}$] [9] [+/-] [=]

This gives $512^{-\frac{1}{9}} = 0.5$

Roots on a calculator:
Most calculators have a square root button and some also have a cube root button.
On most calculators the yth root of a number can be calculated using the [$x^{1/y}$] button.

Harder fractional powers

This section deals with evaluating expressions of the form $a^{\frac{m}{n}}$ and $a^{-\frac{m}{n}}$.

In general, using the rules of indices:

$$a^{\frac{m}{n}} = \left(a^{\frac{1}{n}}\right)^m = \left(\sqrt[n]{a}\right)^m$$

$$a^{-\frac{m}{n}} = \frac{1}{a^{\frac{m}{n}}} = \frac{1}{\left(\sqrt[n]{a}\right)^m}$$

To find the value of $a^{\frac{m}{n}}$:

1. Find the nth root of a.
2. Raise the nth root of a to the power m.

To find the value of $a^{-\frac{m}{n}}$:

Carry out the first two steps as before.

3. Write down the reciprocal of $a^{\frac{m}{n}}$.

EXAMPLES

1 Find the value of $32^{\frac{4}{5}}$.

Find the 5th root of 32.

$32^{\frac{1}{5}} = 2$

Raise to the power of 4.

$2^4 = 16$

$32^{\frac{4}{5}} = 16$

2 Find the value of $27^{-\frac{4}{3}}$.

$27^{-\frac{4}{3}}$ is the reciprocal of $27^{\frac{4}{3}}$.

So, first work out $27^{\frac{4}{3}}$.

$27^{\frac{4}{3}} = \left(27^{\frac{1}{3}}\right)^4 = 3^4 = 81$

The reciprocal of 81 is $\frac{1}{81}$.

So, $27^{-\frac{4}{3}} = \frac{1}{81}$.

Exercise 5.6

Do not use a calculator for questions 1 to 9.

1 Find the value of each of the following.
(a) $\sqrt{400}$ (b) $\sqrt[3]{27}$ (c) $\sqrt[3]{1000}$ (d) $\sqrt[4]{16}$ (e) $\sqrt[3]{64}$ (f) $\sqrt{6.25}$

2 Find the value of each of the following.
(a) $64^{\frac{1}{2}}$ (b) $8^{\frac{1}{3}}$ (c) $81^{\frac{1}{4}}$ (d) $32^{\frac{1}{5}}$ (e) $625^{\frac{1}{4}}$ (f) $36^{0.5}$

3 Find the value of each of the following.
(a) $100^{-\frac{1}{2}}$ (b) $49^{-0.5}$ (c) $16^{-\frac{1}{4}}$ (d) $125^{-\frac{1}{3}}$ (e) $256^{-\frac{1}{4}}$ (f) $243^{-\frac{1}{5}}$

4 Find the value of each of the following.
(a) $1000^{\frac{2}{3}}$ (b) $9^{\frac{3}{2}}$ (c) $16^{\frac{3}{4}}$ (d) $32^{\frac{2}{5}}$ (e) $4^{\frac{5}{2}}$
(f) $9^{2.5}$ (g) $125^{\frac{2}{3}}$ (h) $16^{\frac{5}{4}}$ (i) $243^{\frac{4}{5}}$ (j) $36^{1.5}$

5 Find the value of each of the following.
(a) $1000^{-\frac{2}{3}}$ (b) $16^{-\frac{3}{2}}$ (c) $8^{-\frac{2}{3}}$ (d) $32^{-\frac{3}{5}}$ (e) $4^{-\frac{3}{2}}$
(f) $100^{-\frac{5}{2}}$ (g) $25^{-\frac{3}{2}}$ (h) $16^{-\frac{3}{4}}$ (i) $128^{-\frac{5}{7}}$ (j) $125^{-\frac{2}{3}}$

6 (a) Evaluate. (i) $16^{0.5} \times 2^{-3}$ (ii) $49^{-0.5} \times 81^{0.25}$
Give your answers as fractions.
(b) Work out. (i) $(\sqrt{5})^4$ (ii) $27^{\frac{2}{3}}$

7 Find x in each of the following.
(a) $\sqrt[x]{27} = 3$ (b) $\sqrt[x]{16} = 2$ (c) $\sqrt[x]{32} = 2$ (d) $25^x = 5$ (e) $27^x = 9$
(f) $16^x = 2$ (g) $16^x = 8$ (h) $x^{\frac{3}{4}} = 64$ (i) $x^{\frac{2}{3}} = 25$ (j) $25^x = 125$

8 Calculate each of the following.
(a) $2^{-1} + \left(\frac{1}{16}\right)^{\frac{1}{2}}$ (b) $\left(\frac{1}{32}\right)^{-\frac{1}{5}} \times 8^{-\frac{2}{3}}$ (c) $27^{\frac{2}{3}} \times 3^{-1}$ (d) $49^{-\frac{1}{2}} + 16^{-\frac{3}{4}}$ (e) $9^{\frac{3}{2}} \div 8^{\frac{2}{3}}$

9 Use the rules of powers to show that:
(a) $0.25^{-\frac{7}{2}} = 16^{\frac{7}{4}}$ (b) $32^{\frac{3}{5}} \times 4^{-\frac{3}{2}} = 1$ (c) $25^{\frac{1}{2}} \times 36^{-\frac{1}{2}} = 8^{-\frac{1}{3}} + 3^{-1}$

10 Use your calculator to find the value of each of the following, writing the full display.
(a) $\sqrt{\frac{59.6}{(0.4)^2}}$ (b) $\left(\frac{3.51}{\sqrt{0.28}}\right)^3$ (c) $\left(\frac{2.96 \times 8.7}{5.4 + 13.9}\right)^4$
(d) $\sqrt[3]{\frac{47.6}{8.51 - 6.79}}$ (e) $\frac{9.3^2}{6.2 + \sqrt[3]{59.7}}$ (f) $\frac{69.7}{2.9^2} - \frac{3.7}{\sqrt{5.4}}$

What you need to know

- **Multiples** of a number are found by multiplying the number by 1, 2, 3, 4, …
  For example: the multiples of 8 are: $1 \times 8 = 8$, $2 \times 8 = 16$, $3 \times 8 = 24$, $4 \times 8 = 32$, …
- **Factors** of a number are found by listing all the products that give the number.
  For example: $1 \times 6 = 6$ and $2 \times 3 = 6$. So, the factors of 6 are: 1, 2, 3 and 6.
- The **common factors** of two numbers are the numbers which are factors of **both**.
- A **prime number** is a number with only two factors, 1 and the number itself.
  The first few prime numbers are: 2, 3, 5, 7, 11, 13, 17, 19, 23, 29, 31, …
- The **prime factors** of a number are those factors of the number which are prime numbers.
- The **Least Common Multiple** of two numbers is the smallest number that is a multiple of both.
- The **Highest Common Factor** of two numbers is the largest number that is a factor of both.
- An expression such as $3 \times 3 \times 3 \times 3 \times 3$ can be written in a shorthand way as 3^5.
  This is read as '3 to the power 5'. The number 3 is the **base** of the expression. 5 is the **power**.
- Powers can be used to help write any number as the product of prime factors.
- Numbers raised to the power 2 are **squared**. Whole numbers squared are called **square numbers**.
  Squares can be calculated using the x^2 button on a calculator.
  The opposite of squaring a number is called finding the **square root**.
  Square roots can be calculated using the $\sqrt{}$ button on a calculator.
  The square root of a number can be positive or negative.
  For example, the square root of 4 can be +2 or −2.
  The square root of a number can be written in index form. For example: $\sqrt{9} = 9^{\frac{1}{2}}$.
- Numbers raised to the power 3 are **cubed**. Whole numbers cubed are called **cube numbers**.
  The opposite of cubing a number is called finding the **cube root**.
  Cube roots can be calculated using the $\sqrt[3]{}$ button on a calculator.
  The cube root of a number can be written in index form. For example: $\sqrt[3]{27} = 27^{\frac{1}{3}}$.
- **Powers**
  The squares and the cubes of numbers can be worked out on a calculator by using the x^y button.
  The x^y button can be used to calculate the value of a number x raised to the power of y.
- **Roots** can be calculated using the $x^{1/y}$ button.
- **Reciprocals**
  The reciprocal of a number is the value obtained when the number is divided into 1.
  The reciprocal of a number can be found on a calculator by using the $\frac{1}{x}$ button.
  A number times its reciprocal equals 1. Zero has no reciprocal.
  The reciprocal of a number can be shown using an index of −1. For example: $5^{-1} = \frac{1}{5}$.
- **The rules of indices**

Multiplying powers with the same base	$a^m \times a^n = a^{m+n}$	
Dividing powers with the same base	$a^m \div a^n = a^{m-n}$	
Raising a power to a power	$(a^m)^n = a^{mn}$	
Raising any number to the power zero	$a^0 = 1$	(also $a^1 = a$)
Negative powers and reciprocals	$a^{-m} = \frac{1}{a^m}$	a^{-m} is the reciprocal of a^m
Fractional powers and roots	$a^{\frac{1}{n}} = \sqrt[n]{a}$	and $a^{\frac{m}{n}} = \left(a^{\frac{1}{n}}\right)^m = \left(\sqrt[n]{a}\right)^m$

Review Exercise 5

Do not use a calculator for questions 1 to 14.

1

3	9	20	25	29	75	92	100

Which of the numbers in the box are:
(a) square numbers, (b) factors of 100, (c) prime numbers? AQA

2 (a) State which of these numbers are multiples of 4.
2, 3, 8, 15, 22, 25, 29, 39.
(b) Jane tries to find a cube number in this list.
She says, "39 is a cube number because $13^3 = 39$." Explain why Jane is wrong. AQA

3 (a) What is the value of $2^3 - \sqrt{25}$?
(b) Work out the value of 10^4.

4 (a) Write down the multiple of 6 that is larger than 105, but smaller than 111.
(b) Write down all the factors of 111.
(c) Write down the largest prime number that is smaller than 107.
(d) Write down an even prime number. AQA

5 Work out (a) the cube of 5, (b) 2^6. AQA

6 (a) Find the value of p when $2^p \times 3 = 48$.
(b) Write 72 as a product of prime factors.
(c) What is the highest common factor of 48 and 72?
(d) What is the least common multiple of 48 and 72? AQA

7 A blue light flashes every 18 seconds and a green light flashes every 30 seconds.
The two lights flash at the same time.
After how many seconds will the lights next flash at the same time?

8 For each of the following equations, write down the value of n.
(a) $2^n = 32$ (b) $n^3 = 125$ (c) $8^n = 8$ AQA

9 Find x in each of the following.
(a) $5^3 \times 5^2 = 5^x$ (b) $2^5 \div 2^4 = 2^x$ (c) $(7^3)^2 = 7^x$ (d) $\frac{5^5 \times 5^3}{5^4} = 5^x$

10 (a) Show that $8^4 = 2^{12}$.
(b) Hence, or otherwise, solve the equation $4^x = 8^4$. AQA

11 (a) Write $2^{-2} \times 4$ as a single power of 2.
(b) Write $2^3 \div \frac{1}{8}$ as a single power of 2. AQA

12 Evaluate: (a) $16^{\frac{1}{4}}$ (b) 2^{-4} (c) $27^{-\frac{1}{3}}$ AQA

13 (a) Find the value of w in $2^w = 1$. (b) Find the value of x in $2^x = \frac{1}{4}$.
(c) Find the value of y in $32^y = 2$. (d) Find the value of z in $16^{z+3} = 2^z$. AQA

14 Find the value of each of the following. (a) $27^{\frac{2}{3}}$ (b) $25^{-\frac{3}{2}}$ (c) $81^{-\frac{3}{4}}$

15 Calculate: $2.6^3 \times \sqrt{4.3 + 2.8}$ Give your answer correct to one decimal place.

16 Find the reciprocal of 6. Give your answer correct to 3 d.p.

17 Computer magazines often use the fact that 2^{10} is approximately equal to 10^3.
To how many significant figures is this true? AQA

18 Calculate the value of each of these expressions.
(a) $2^5 \times 3^8 + 5^0$ (b) $2.5^5 \times (2.5^4 + 2.5^{-2})$ AQA

19 Calculate the value of: (a) $(\sqrt{7})^6$ (b) $7^{-\frac{3}{4}}$ (c) $\sqrt[4]{\frac{5.7}{(0.3)^2}}$ (d) $\left(\frac{2.4}{\sqrt{1.5}}\right)^{-5}$
Give your answers correct to 3 significant figures.

Working with Number

Standard Index Form

Standard index form is a shorthand way of writing very large and very small numbers.
Standard index form is often called **standard form** or **scientific notation**.

Large numbers and your calculator

Scientists who study the planets and the stars work with very large numbers.

Approximate distances from the Sun to some planets are:

Earth 149 000 000 km   Mars 228 000 000 km   Pluto 5 898 000 000 km

A calculator displays very large numbers in **standard form**.
To represent large numbers in standard form you need to use powers of 10.

Number	1 000 000	100 000	10 000	1000	100	10
Power of 10	10^6	10^5	10^4	10^3	10^2	10^1

For example: $2\ 600\ 000 = 2.6 \times 1\ 000\ 000 = 2.6 \times 10^6$
Therefore $2\ 600\ 000 = 2.6 \times 10^6$ in **standard form**.

Calculator displays

Work out $3\ 000\ 000 \times 25\ 000\ 000$ *on your calculator.*
Write down the display.

Most calculators will show the answer as: [7.5   13]

In **standard form** the answer should be written as 7.5×10^{13}.

A number written in standard form has **two** parts.
The first part must be a number between 1 and 10.
The second part is a power of 10.
The two parts are connected by a multiplication sign.

EXAMPLES

1 (a) Write 370 000 in standard form.
$370\ 000 = 3.7 \times 100\ 000 = 3.7 \times 10^5$

(b) Write 5.6×10^7 as an ordinary number.
$5.6 \times 10^7 = 5.6 \times 10\ 000\ 000 = 56\ 000\ 000$

2 Write the calculator display [7.3   05]

(a) in standard form,   (b) as an ordinary number.

(a) [7.3   05] means 7.3×10^5 (b) $7.3 \times 10^5 = 7.3 \times 100\ 000 = 730\ 000$

Exercise 6.1

1 (a) Write 600 000 in standard form.
(b) Write 9×10^6 as an ordinary number.

2 Copy the table and fill in all the different forms of each number.

	Ordinary number	Power of 10	Standard form
	300 000	$3 \times 100\,000$	3×10^5
(a)	75 000	$7.5 \times 10\,000$	
(b)		$8 \times 100\,000\,000$	
(c)			3.5×10^{13}
(d)	62 300 000 000 000		

3 Write each of these numbers in standard form.
(a) 300 000 000 000 (b) 80 000 000 (c) 700 000 000 (d) 2 000 000 000
(e) 42 000 000 (f) 21 000 000 000 (g) 3 700 000 000 (h) 630

4 Change each of these numbers to an ordinary number.
(a) 6×10^5 (b) 2×10^3 (c) 5×10^7 (d) 9×10^8
(e) 3.7×10^9 (f) 2.8×10^1 (g) 9.9×10^{10} (h) 7.1×10^4

5 Write these calculator displays:
(a) in standard form,
(b) as an ordinary number.

(i) [4.5 *03*] (ii) [7.8 *07*] (iii) [5.3 *05*] (iv) [3.25 *04*]

6 The table shows the estimated population for some countries in 2020.

Country	Estimated population in 2020
Albania	5.3×10^6
Brazil	2.5×10^8
Fiji	9.0×10^5
Greece	1.2×10^7

(a) (i) Which of these countries is estimated to have the largest population in 2020?
(ii) Write the estimated population as an ordinary number.
(b) In 2020, China is estimated to have a population of 1288 million people.
Write 1288 million in standard form.

Small numbers and your calculator

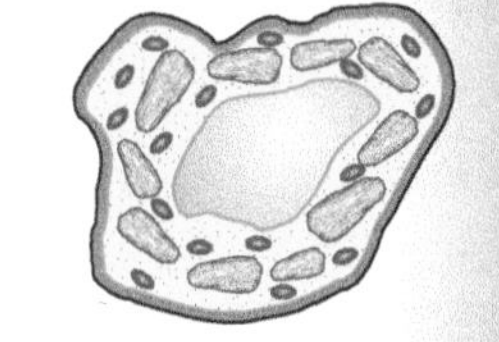

Scientists who study microbiology work with numbers that are very small.
The smallest living cells are bacteria which have a diameter of about 0.000 025 cm.
Blood cells have a diameter of about 0.000 75 cm.

A calculator displays small numbers in **standard form**.
It does this in the same sort of way that it does for large numbers.
To represent very small numbers in standard form you need to use powers of 10 for numbers less than 1.

Number	1000	100	10	1	0.1	0.01	0.001	0.000 1
Power of 10	10^3	10^2	10^1	10^0	10^{-1}	10^{-2}	10^{-3}	10^{-4}

For example: $0.000\,037 = 3.7 \times 0.000\,01 = 3.7 \times 10^{-5}$
Therefore $0.000\,037 = 3.7 \times 10^{-5}$ in **standard form**.

Calculator displays

Work out 0.000 007 × 0.000 9 *on your calculator.*
Write down the display.

Most calculators will show the answer as: `6.3 −09`

In standard form the answer should be written as 6.3×10^{-9}.

In **standard form** a number is written as: a number between 1 and 10 × a power of 10

A **large** number has a **positive power**. E.g. $160\,000\,000 = 1.6 \times 10^{8}$

A **small** number has a **negative power**. E.g. $0.000\,000\,06 = 6 \times 10^{-8}$

EXAMPLES

1 (a) Write 0.000 73 in standard form.
$0.000\,73 = 7.3 \times 0.000\,1 = 7.3 \times 10^{-4}$

(b) Write 2.9×10^{-6} as an ordinary number.
$2.9 \times 10^{-6} = 2.9 \times 0.000\,001 = 0.000\,002\,9$

2 Write the calculator display: `1.5 −03`

(a) in standard form, (b) as an ordinary number.

(a) `1.5 −03` means 1.5×10^{-3} (b) $1.5 \times 10^{-3} = 1.5 \times 0.001 = 0.001\,5$

Exercise 6.2

1 (a) Write 0.000 35 in standard form.
(b) Write 2.5×10^{-3} as an ordinary number.

2 Copy and complete the following table.

	Ordinary number	Power of 10	Standard form
	0.000 03	3 × 0.000 01	3×10^{-5}
(a)	0.007 5	7.5 × 0.001	
(b)	0.000 008 75		
(c)			3.5×10^{-9}
(d)	0.000 000 000 006 23		

3 Write each of these numbers in standard form.
(a) 0.007 (b) 0.04 (c) 0.000 000 005
(d) 0.000 8 (e) 0.000 000 002 3 (f) 0.000 000 045
(g) 0.023 4 (h) 0.000 000 002 34 (i) 0.006 7

4 Change each of these numbers to an ordinary number:
(a) 3.5×10^{-1} (b) 5×10^{-4} (c) 7.2×10^{-5} (d) 6.1×10^{-3}
(e) 1.17×10^{-10} (f) 8.135×10^{-7} (g) 6.462×10^{-2} (h) 4.001×10^{-9}

5 Write these calculator displays:
(a) in standard form, (b) as an ordinary number.

(i) `3.4 −03` (ii) `5.65 −05` (iii) `7.2 −04` (iv) `9.13 −01`

Using large and small numbers with a calculator

Calculations with large and small numbers can be done on a calculator by:

- changing the numbers to standard form,
- entering the numbers into the calculator using the Exp button.

If your calculator works in a different way to the example shown refer to the instruction booklet supplied with the calculator or ask someone for help.

EXAMPLE

Calculate the value of 62 500 000 000 × 0.000 000 003
Give your answer both as an ordinary number and in standard form.

$62\ 500\ 000\ 000 \times 0.000\ 000\ 003 = (6.25 \times 10^{10}) \times (3 \times 10^{-9})$

To do the calculation enter the following sequence into your calculator.

6 . 2 5 Exp 1 0 × 3 Exp 9 +/– =

Giving:
62 500 000 000 × 0.000 000 003
= 187.5 (ordinary number)
= 1.875×10^2 (standard form)

Some calculators display this result as: 187.5
Other calculators give this display: 1.875 *02*

Remember:
The calculator display **must** be changed to either **standard form** or an **ordinary number**.

Exercise 6.3

Use your calculator to work out each of the following questions.

Write each of the answers:
(i) as on your calculator display, (ii) in standard form, (iii) as an ordinary number.

1
(a) 300 000 × 200 000 000 (b) 120 000 × 80 000 000
(c) 15 000 × 700 000 000 (d) 65 000 × 2 000 000 000
(e) 480 000 × 500 000 000 (f) 50 000 × 50 000 000

2
(a) 0.000 03 × 0.000 000 2 (b) 0.000 045 × 0.000 003
(c) 0.000 75 × 0.000 000 04 (d) 0.002 3 × 0.000 000 05
(e) 0.053 × 0.000 000 08 (f) 0.000 006 4 × 0.000 015 2

Solving problems involving large and small numbers

Problems can involve numbers given in standard form.

EXAMPLE

The following figures refer to the population of China and the USA in 1993.

China 1.01×10^9 USA 2.32×10^8

By how much did the population of China exceed that of the USA in 1993?

The greater the power …
… the bigger the number.

1.01×10^9 is greater than 2.32×10^8.
You need to work out $1.01 \times 10^9 - 2.32 \times 10^8$
To do the calculation enter the following sequence into your calculator.

1 . 0 1 Exp 9 – 2 . 3 2 Exp 8 =

Giving: $1.01 \times 10^9 - 2.32 \times 10^8 = 778\ 000\ 000 = 7.78 \times 10^8$

Exercise 6.4

Use the **Exp** button on your calculator to answer these questions.

1 Give the answers to the following calculations as ordinary numbers.

(a) $(5.25 \times 10^9) \times (5 \times 10^{-5})$ (b) $(5.25 \times 10^9) \div (5 \times 10^{-5})$
(c) $(8.5 \times 10^6)^2$ (d) $(5 \times 10^{-3})^3$
(e) $(7.2 \times 10^5) \div (2.4 \times 10^{-5})$ (f) $(9.5 \times 10^6) \div (1.9 \times 10^{-7})^2$

2 Give the answers to the following calculations in standard form.

(a) 33 500 000 000 × 2 800 000 000 (b) 0.000 000 000 2 × 80 000 000 000
(c) $15\,000\,000\,000\,000^2$ (d) $0.000\,000\,000\,000\,5^3$
(e) 48 000 000 000 ÷ 0.000 000 000 2 (f) 25 000 000 000 ÷ 500 000 000 000

3 (a) In 1992 about 1 400 000 000 steel cans and about 688 000 000 aluminium cans were recycled.
What was the total number of cans that were recycled in 1992?
Give your answer in standard form.

(b) Alpha Centauri is about 40 350 000 000 000 km from the Sun.
Alpha Cygni is about 15 300 000 000 000 000 km from the Sun.
How much further is it from the Sun to Alpha Cygni than from the Sun to Alpha Centauri?
Give your answer in standard form.

4 Here are the diameters of some planets.

Saturn 1.2×10^5 km Jupiter 1.42×10^5 km Pluto 2.3×10^3 km

(a) List the planets in order of size starting with the smallest.
(b) What is the difference between the diameters of the largest and smallest planets?
Give your answer in standard form and as an ordinary number.

5 Here are the areas of some of the world's largest deserts.

The Sahara desert in North Africa	8.6×10^6 km²
The Gobi desert in Mongolia and North East China	1.166×10^6 km²
The Patagonian desert in Argentina	6.73×10^5 km²

(a) What is the total area of the Sahara and Patagonian deserts?
(b) What is the difference in area between the Gobi and the Patagonian deserts?
Give your answer in standard form.

6 Calculate $5.42 \times 10^6 \times 4.65 \times 10^5$
giving your answer in standard form correct to 3 significant figures.

7 Calculate $1.7 \times 10^3 \div 7.6 \times 10^7$
giving your answer in standard form correct to 2 significant figures.

8 The area of the surface of the Earth is about 5.095×10^9 square miles.
Approximately 29.2% of this is land.
Use these figures to estimate the area of land surface on Earth.

9 James wins a lottery prize of £1.764×10^6. He pays £5.29×10^5 for a house.
What percentage of his prize did he spend on the house?
Give your answer to a suitable degree of accuracy.

10 The mass of an oxygen atom is 2.7×10^{-23} grams.
The mass of an electron at rest is approximately 30 000 times smaller than this.
Estimate the mass of an electron at rest.

11 The modern human appeared on the Earth about 3.5×10^4 years ago.
The Earth has been in existence for something like 1.3×10^5 times as long as this.

(a) Estimate the age of the Earth.

Reptiles appeared on the Earth about 2.3×10^8 years ago.

(b) How many times longer than the modern human have reptiles been alive?
Give your answer in standard form.

Standard form calculations without a calculator

In some standard form problems the calculations can be handled without using a calculator.

EXAMPLES

1 Calculate the value of $(3 \times 10^2) + (4 \times 10^3)$.
Give your answer in standard form.

$3 \times 10^2 = 300 \qquad 4 \times 10^3 = 4000$

$$(3 \times 10^2) + (4 \times 10^3) = 300 + 4000$$
$$= 4300$$
$$= 4.3 \times 10^3$$

When adding or subtracting numbers in standard form without a calculator change to ordinary numbers first.

2 Calculate the value of ab where $a = 8 \times 10^3$ and $b = 4 \times 10^5$.

$$ab = (8 \times 10^3) \times (4 \times 10^5)$$
$$= 8 \times 4 \times 10^3 \times 10^5$$
$$= 32 \times 10^8$$
$$= 3.2 \times 10 \times 10^8$$
$$= 3.2 \times 10^9$$

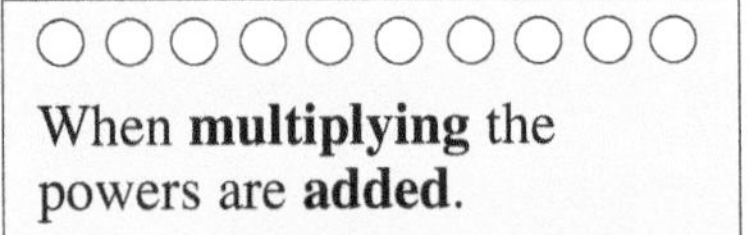

When **multiplying** the powers are **added**.

$10^3 \times 10^5 = 10^{3+5} = 10^8$

$10 \times 10^8 = 10^{1+8} = 10^9$

3 Calculate the value of x^2 where $x = 7 \times 10^{-8}$.

$$x^2 = (7 \times 10^{-8})^2$$
$$= 49 \times 10^{-16}$$
$$= 4.9 \times 10 \times 10^{-16}$$
$$= 4.9 \times 10^{-15}$$

Remember:
$(7 \times 10^{-8})^2 = 7^2 \times (10^{-8})^2$
$10 \times 10^{-16} = 10^{1-16} = 10^{-15}$

4 Calculate the value of $(1.2 \times 10^3) \div (4 \times 10^{-8})$.

$$(1.2 \times 10^3) \div (4 \times 10^{-8}) = (1.2 \div 4) \times (10^3 \div 10^{-8})$$
$$= 0.3 \times 10^{11}$$
$$= 3 \times 10^{-1} \times 10^{11}$$
$$= 3 \times 10^{10}$$

When **dividing** the powers are **subtracted**.
$10^3 \div 10^{-8} = 10^{3--8} = 10^{11}$

Exercise 6.5

Do not use a calculator. Give your answers in standard form.

1 For each of the following calculate the value of $p + q$.
(a) $p = 5 \times 10^3$ and $q = 2 \times 10^2$
(b) $p = 4 \times 10^5$ and $q = 8 \times 10^6$
(c) $p = 3.08 \times 10^4$ and $q = 9.2 \times 10^3$
(d) $p = 4.25 \times 10^4$ and $q = 7.5 \times 10^3$

2 For each of the following calculate the value of $p - q$.
(a) $p = 3 \times 10^3$ and $q = 2 \times 10^2$
(b) $p = 9.05 \times 10^5$ and $q = 5 \times 10^3$
(c) $p = 3.05 \times 10^7$ and $q = 5 \times 10^5$
(d) $p = 9.545 \times 10^8$ and $q = 4.5 \times 10^6$

3 For each of the following calculate the value of $p \times q$.
(a) $p = 4 \times 10^3$ and $q = 2 \times 10^4$
(b) $p = 2 \times 10^4$ and $q = 3 \times 10^3$
(c) $p = 4 \times 10^5$ and $q = 6 \times 10^2$
(d) $p = 9 \times 10^9$ and $q = 3 \times 10^5$

4 For each of the following calculate the value of $p \div q$.
(a) $p = 6 \times 10^5$ and $q = 2 \times 10^2$
(b) $p = 9 \times 10^5$ and $q = 3 \times 10^2$
(c) $p = 2.5 \times 10^5$ and $q = 5 \times 10^3$
(d) $p = 4 \times 10^8$ and $q = 2 \times 10^{-3}$
(e) $p = 1.2 \times 10^3$ and $q = 3 \times 10^{-3}$
(f) $p = 1.5 \times 10^{-5}$ and $q = 5 \times 10^{-3}$

5 $x = 3 \times 10^4$ and $y = 5 \times 10^{-5}$. Work out the value of each of these expressions.
(a) xy (b) x^3 (c) x^2y (d) y^3 (e) $\frac{x}{y}$

Standard Index Form

6 Find the value of each of the following.

(a) $(5 \times 10^6) \times (3 \times 10^4)$ (b) $(8 \times 10^{-5}) \times (3 \times 10^7)$
(c) $(7 \times 10^{-5}) \times (6 \times 10^{-4})$ (d) $(8 \times 10^3)^2$
(e) $(4 \times 10^5) \div (8 \times 10^3)$ (f) $(1.8 \times 10^3) \div (6 \times 10^{-7})$
(g) $(5 \times 10^{-7}) \div (8 \times 10^{-2})$ (h) $(2 \times 10^4) \div (8 \times 10^{-1})$

What you need to know

- **Standard index form**, or **standard form**, is a shorthand way of writing very large and very small numbers.
- In **standard form** a number is written as: $a \times 10^n$ where $1 \leqslant a < 10$ and n is an integer.
  For large numbers, greater than 10, n is **positive**. E.g. $15\,000\,000 = 1.5 \times 10^7$
  For small positive numbers, less than 1, n is **negative**. E.g. $0.000\,06 = 6 \times 10^{-5}$

Review Exercise 6

Try to do questions 1 and 2 without using your calculator.

1 In the box are six numbers written in standard form.

8.3×10^4	3.9×10^5	6.7×10^{-3}	9.245×10^{-1}	8.36×10^3	4.15×10^{-2}

(a) (i) Write down the largest number. (ii) Write your answer as an ordinary number.
(b) (i) Write down the smallest number. (ii) Write your answer as an ordinary number.
AQA

2 The number p written in standard form is 8×10^5.
The number q written in standard form is 5×10^{-2}.
(a) Calculate $p \times q$. Give your answer in standard form.
(b) Calculate $p \div q$. Give your answer in standard form. AQA

3 (a) Write 34 500 000 000 in standard form.
(b) Write 0.000 000 543 in standard form.
(c) Work out $\frac{7.2 \times 10^5}{6.4 \times 10^3}$. Give your answer in standard form. AQA

4 $p = 1.65 \times 10^7$, $q = 4.82 \times 10^6$ and $r = 6.17 \times 10^{-2}$.
Calculate the value of the following. Give your answers in standard form.
(a) $2p + 3q$ (b) $p \div r$ AQA

5 In 2005, a company paid out a total of £1.14×10^{10} in wages to 0.63 million employees.
Calculate the average annual wage per employee.
Give your answer in standard form correct to three significant figures.

6 The volume of water on Earth is approximately 1.436×10^9 km^3.
About 94% of this is contained in the Earth's oceans.
Use these figures to estimate the volume of water in the Earth's oceans.

7 Between 1980 and 1990 the population of the UK increased from 5.7×10^7 to 5.9×10^7.
Find the percentage increase in the population of the UK between 1980 and 1990.

8 Last year the population of the United Kingdom was approximately 5.9×10^7.
(a) An average of £680 per person was spent on food last year in the UK.
What was the total amount spent on food last year in the UK?
Give your answer in standard form.
(b) Last year there were 1.4×10^7 car drivers in the UK.
They spent a total of £1.5×10^{10} on their cars.
What was the average amount spent by each car driver?
Give your answer to a suitable degree of accuracy. AQA

CHAPTER 7 Ratio and Proportion

Equivalent ratios

Ratios are used only to **compare** quantities. They do not give information about actual values.

For example.
A necklace is made using red beads and white beads in the ratio **3 : 4**.
This gives no information about the actual numbers of beads in the necklace.
The ratio **3 : 4** means that for every 3 red beads in the necklace there are 4 white beads.
The **possible** numbers of beads in the necklace are shown in the table.

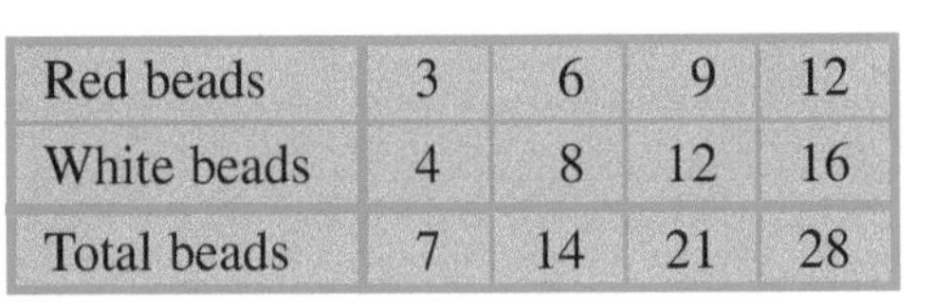

Red beads	3	6	9	12
White beads	4	8	12	16
Total beads	7	14	21	28

For the beads in this necklace:
The number of red beads is a **multiple** of 3.
The number of white beads is a **multiple** of 4.
The total number of beads is a **multiple** of 7.

Make similar tables when the ratio of red beads to white beads in the necklace is:
(a) 4 : 5 (b) 2 : 3 (c) 3 : 1

The ratios 3 : 4, 6 : 8, 9 : 12, ...
are different forms of the **same** ratio.
They are called **equivalent** ratios.
They can be found by multiplying or dividing each part of the ratio by the **same** number.

Simplifying ratios

To simplify a ratio divide both of the numbers in the ratio by the **same** number.
A ratio with whole numbers which cannot be simplified is in its **simplest form**.

EXAMPLES

1 The ratio of boys to girls in a school is 3 : 4.
There are 72 boys. How many girls are there?

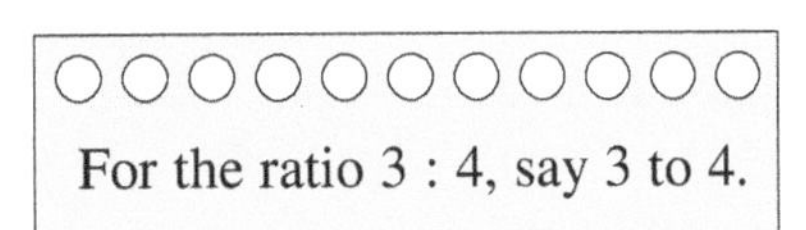

For the ratio 3 : 4, say 3 to 4.

$72 \div 3 = 24$
To find a ratio equivalent to 3 : 4 where the first number in the ratio is 72, multiply each number in the ratio by 24.
$3 \times 24 : 4 \times 24 = 72 : 96$
The number of girls = 96.

2 Write the ratio 2 cm : 50 mm in its simplest form.

2 cm : 50 mm = 20 mm : 50 mm = 20 : 50
Divide both parts of the ratio by 10.
$20 \div 10 : 50 \div 10 = 2 : 5$
The ratio 2 cm : 50 mm in its simplest form is 2 : 5.

In its simplest form a ratio contains **only** whole numbers. There are **no units**.
In order to simplify the ratio both quantities in the ratio must be in the **same units**.

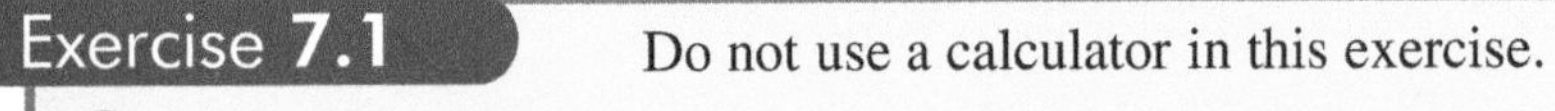

Exercise 7.1

Do not use a calculator in this exercise.

1 Give the simplest form of each of these ratios.
(a) 3 : 6 (b) 9 : 27 (c) 9 : 12 (d) 10 : 25 (e) 30 : 40
(f) 22 : 55 (g) 9 : 21 (h) 18 : 8 (i) 36 : 81 (j) 35 : 15

2. Each of these pairs of ratios are equivalent.
   (a) 3 : 4 and 9 : n. (b) 2 : 7 and 8 : n. (c) 8 : n and 2 : 25. (d) 25 : n and 5 : 4.
   In each case calculate the value of n.

3. The heights of two friends are in the ratio 7 : 9. The shorter of the friends is 154 cm tall.
   What is the height of the taller of the friends?

4. Sugar and flour are mixed in the ratio 2 : 3.
   How much sugar is used with 600 g of flour?

5. The ratio of boys to girls in a school is 4 : 5. There are 80 girls.
   How many boys are there?

6. I earn £300 per week.

   The amounts Jenny and James earn
   is in the ratio of their ages.
   Jenny is 20 years old. How old is James?

7. A necklace contains 30 black beads and 45 gold beads.
   What is the simplest form of the ratio of black beads to gold beads on the necklace?

8. On Monday a hairdresser uses 800 ml of shampoo and 320 ml of conditioner.
   Write in its simplest form the ratio of shampoo : conditioner used on Monday.

9. Denise draws a plan of her classroom. On her plan Denise uses 2 cm to represent 5 m.
   Write the scale as a ratio in its simplest form.

10. On a map a pond is 3.5 cm long. The pond is actually 52.5 m long.
    Write the scale as a ratio in its simplest form.

11. Write each of these ratios in its simplest form.
    (a) £2 : 50p (b) 20p : £2.50 (c) £2.20 : 40p (d) 6 m : 240 cm
    (e) 2 kg : 500 g (f) 1 kg : 425 g (g) 90 cm : 2 m (h) 5 km : 200 m

12. Sam spends 90p a week on comics. Tom spends £4 a week on comics.
    Write the ratio of the amounts Tom and Sam spend on comics in its simplest form.

13. An alloy is made of tin and zinc. 40% of the alloy is tin.
    What is the ratio of tin : zinc in its simplest form?

14. A box contains blue biros and red biros. $\frac{1}{3}$ of the biros are blue.
    What is the ratio of blue biros to red biros in the box?

15. A necklace is made from 40 beads. $\frac{2}{5}$ of the beads are white. The rest of the beads are red.
    Find the ratio of the number of red beads to the number of white beads in its simplest form.

Sharing in a given ratio

EXAMPLE

James and Sally share £20 in the ratio 3 : 2.
How much do they each get?

Add the numbers in the ratio. $3 + 2 = 5$
For every £5 shared: James gets £3, Sally gets £2.
$20 \div 5 = 4$ There are 4 shares of £5 in £20.
James gets £3 × 4 = £12. Sally gets £2 × 4 = £8.
So, James gets £12 and Sally gets £8.

Exercise 7.2

Do not use a calculator for questions 1 to 6.

1. (a) Share 9 in the ratio 2 : 1. (b) Share 20 in the ratio 3 : 1.
   (c) Share 35 in the ratio 1 : 4. (d) Share 100 in the ratio 9 : 1.

2. A bag contains 12 toffees. Bruce and Emily eat them at the ratio 1 : 2.
   How many toffees does Emily eat?

3. A box contains gold coins and silver coins. The ratio of gold coins to silver coins is 1 : 9.
   There are 20 coins in the box. How many silver coins are in the box?

4. Sunny and Chandni share £48 in the ratio 3 : 1. How much do they each get?

5. (a) Share £35 in the ratio 2 : 3. (b) Share £56 in the ratio 4 : 3.
   (c) Share £5.50 in the ratio 7 : 4. (d) Share £4.80 in the ratio 3 : 5.

6. A necklace contains 72 beads. The ratio of red beads to blue beads is 5 : 3.
   How many red beads are on the necklace?

7. £480 is shared in the ratio 7 : 3.
   What is the difference between the larger share and the smaller share?

8. A bag contains red beads and black beads in the ratio 1 : 3. What fraction of the beads are red?

9. A box contains red biros and black biros in the ratio 1 : 4.
   What percentage of the biros are black?

10. In a school the ratio of the number of boys to the number of girls is 3 : 5.
    What fraction of the pupils in the school are girls?

11. The ratio of non-fiction books to fiction books in a library is 2 : 3.
    Find the percentage of fiction books in the library.

12. John is 12 years old and Sara is 13 years old. They share some money in the ratio of their ages.
    What percentage of the money does John get?

13. In the UK there are 240 939 km^2 of land.
    The ratio of agricultural land to non-agricultural land is approximately 7 : 3.
    Estimate the area of land used for agriculture.

14. At the start of a game Jenny and Tim have 40 counters each.
    At the end of the game the number of counters that Jenny and Tim each have is in the ratio 5 : 3.
    (a) How many counters do Jenny and Tim have at the end of the game?
    (b) How many counters did Jenny win from Tim in the game?

15. On a necklace, for every 10 black beads there are 4 red beads.
    (a) What is the ratio of black beads to red beads in its simplest form?
    (b) If the necklace has 15 black beads how many red beads are there?
    (c) If the necklace has a total of 77 beads how many black beads are there?
    (d) Why can't the necklace have a total of 32 beads?

16. The lengths of the sides of a triangle are in the ratio 4 : 6 : 9.
    The total length of the sides is 38 cm. Calculate the length of each side.

17. To make concrete a builder mixes gravel, sand and cement in the ratio 4 : 2 : 1.
    The builder wants 350 kg of concrete. How much gravel does the builder need?

18. The angles of a triangle are in the ratio 2 : 3 : 4. Calculate each angle.

19. A bag contains some red, green and black sweets. 30% of the sweets are red.
    The ratio of the numbers of green sweets to black sweets is 5 to 9.
    What percentage of the total number of sweets are black?

Proportion

Some situations involve comparing **different** quantities.

Direct proportion

When a motorist buys fuel, the more he buys the greater the cost.
In this situation the quantities can change but the ratio between the quantities stays the same.
As one quantity increases, so does the other.
The quantities are in **direct proportion**.

Inverse proportion

By using more people to deliver a batch of leaflets, the time taken to complete the task is reduced.
In this situation, as one quantity increases the other quantity decreases.
The quantities are in **inverse proportion**.

EXAMPLES

1 4 cakes cost £1.20.
Find the cost of 7 cakes.

4 cakes cost £1.20
1 cake costs £1.20 ÷ 4 = 30p
7 cakes cost 30p × 7 = £2.10
So, 7 cakes cost £2.10.

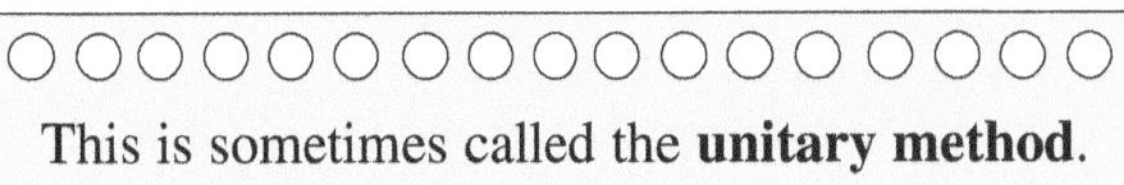

This is sometimes called the **unitary method**.
(a) **Divide** by 4 to find the cost of **1** cake.
(b) **Multiply** by 7 to find the cost of 7 cakes.

2 3 people take 8 hours to deliver some leaflets.
How long would it take 4 people?

3 people take 8 hours.
1 person takes 8 hours × 3 = 24 hours
4 people take 24 hours ÷ 4 = 6 hours
So, 4 people would take 6 hours.

This assumes that time is **inversely proportional** to the number of people.
(a) **Multiply** by 3 to find how long **1** person would take.
(b) **Divide** by 4 to find how long 4 people would take.

Exercise 7.3

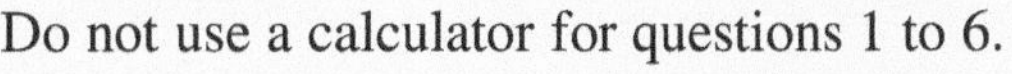

Do not use a calculator for questions 1 to 6.

1 5 candles cost 80 pence.
(a) What is the cost of 1 candle?
(b) What is the cost of 8 candles?

2 Alistair pays £1.90 for 2 cups of tea.
How much would he pay for 3 cups of tea?

3 Lauren can buy 3 toffee apples at 60 pence each.
How many could she buy if they cost 45 pence each?

4 The amount a spring stretches is proportional to the weight hung on the spring.
A weight of 5 kg stretches the spring by 60 cm.
(a) How much does a weight of 10 kg stretch the spring?
(b) What weight makes the spring stretch 24 cm?

5 Jean pays £168 for 10 square metres of carpet.
How much would 12 square metres of carpet cost?

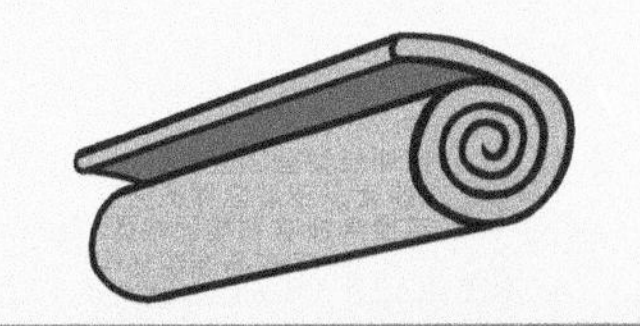

6. Three people can deliver 1000 leaflets in 30 minutes.
   How long would it take 5 people to deliver 1000 leaflets?

7. These ingredients are needed to make 20 scones.

500 g flour	100 g dried fruit	250 g butter	water to mix

   (a) How much dried fruit is needed to make 50 scones?
   (b) How much butter is needed to make 12 scones?

8. Aimee pays £1.14 for 3 kg of potatoes.
   How much would 7.5 kg of potatoes cost?

9. A farmer estimates it will take 12 workers, 10 days to harvest his crop.
   How many workers are needed to harvest the crop in 8 days?

10. 9 metres of stair carpet cost £83.70. How much does 9.6 metres cost?

11. It costs £15.60 each for 7 people to hire a minibus.
    How much would each person pay if 12 people hired the minibus?

12. A car travels 6 miles in 9 minutes. If the car travels at the same speed:
    (a) how long will it take to travel 8 miles,
    (b) how far will it travel in 24 minutes?

13. 5 litres of paint cover an area of $30\ m^2$.
    (a) What area will 2 litres of paint cover?
    (b) How much paint is needed to cover $72\ m^2$?

14. Meg can make 30 monthly payments of £148.80 to repay a loan.
    How many payments are needed if she can pay £186 a month?

15. Mary phones her uncle in New York.
    Phone calls to New York are charged at the rate of £2.20 for a 5-minute call.
    (a) How much would a 7-minute call to New York cost?
    (b) Mary's call cost £5.28. How long was her call?

16. A school is organising three trips to the zoo.

    (a) How much does Tuesday's trip cost?
    (b) How many students are going to the zoo on Wednesday?

17. At 75 kilometres per hour a journey takes 40 minutes.
    How long would the journey take at 60 kilometres per hour?

18. A 42-litre paddling pool is filled at the rate of 12 litres of water every 5 minutes.
    How long will it take to fill the pool?

19. A piece of beef weighs 1.5 kg and costs £5.22.
    How much would a piece of beef weighing 2.4 kg cost?
    Give your answer to a suitable degree of accuracy.

20. The lottery is won by 5 people and they each get £1 685 240.
    How much would they each have got if only two people had won?

What you need to know

- The ratio 3 : 2 is read '3 to 2'.
- A ratio is used only to **compare** quantities.
  A ratio does not give information about the exact values of quantities being compared.
- In its **simplest form**, a ratio contains whole numbers which have no common factor other than 1.
  All quantities in a ratio must have the **same units** before the ratio can be simplified.
  For example, £2.50 : 50p = 250p : 50p = 5 : 1.
- When two different quantities are always in the **same ratio** the two quantities are in **direct proportion**. For example, the amount and cost of fuel bought by a motorist.
- When, as one quantity increases the other decreases, the two quantities are in **inverse proportion**. For example, the number of people and the time taken to deliver a batch of leaflets.

Review Exercise 7

Do not use a calculator for questions 1 to 5.

1. A packet contains 5 white balloons and 15 red balloons.
   What is the ratio of white balloons to red balloons in its simplest form?

2. A sewing box contains pins and needles in the ratio 4 : 1. There are 36 pins in the box.
   How many needles are in the box?

3. Max shares £420 with a friend in the ratio 5 : 3. How much does each receive? AQA

4. What is the difference between the larger and smaller shares when £39 200 is shared in the ratio 7 : 9?

5. 5 litres of petrol costs £4.70. How much would 18 litres of petrol cost?

6. At 60 mph Matt takes $1\frac{1}{4}$ hours to complete a journey.
   How much longer would he take at 45 mph?

7. Fiona is given £24 for her birthday. She spends 5 times as much as she saves.
   How much does she save? AQA

8. On a map the distance between two houses is 19 mm.
   The actual distance between the houses is 3.8 km. What is the scale of the map?

9. Jack is building a brick wall for a garage. He decides to use a mortar mix given by:
   1 part cement, 1 part lime, 6 parts sand.
   (a) What fraction of the mix will be sand?
   (b) What percentage of the mix will be cement?
   (c) He uses 1200 cm³ of sand for the mix. What volume of lime will he need? AQA

10. (a) Phil has 50 birds; some are blue, the rest are yellow.
    The ratio of blue birds to yellow birds is 3 : 7.
    How many yellow birds are there?
    (b) Phil sells some of his birds and buys some others.
    The new ratio of blue birds to yellow birds is 5 : 2. There are 16 yellow birds.
    How many blue birds are there? AQA

11. Jane, Lilin and Huw share a bag of 40 sweets. Jane has $\frac{2}{5}$ of the sweets.
    Lilin and Huw share the remaining sweets in the ratio 7 : 5.
    What percentage of the bag of 40 sweets does Lilin have? AQA

12. Box *A* contains black counters and white counters in the ratio 3 : 5.
    (a) What percentage of the counters are black?

    Box *B* contains black counters and white counters in the ratio 7 : 5.
    Box *A* and Box *B* each contain the same number of counters.
    (b) What is the smallest possible number of counters in each box? AQA

CHAPTER 8

Speed and Other Compound Measures

Speed

Speed is a measurement of how fast something is travelling.
It involves two other measures, **distance** and **time**.
Speed can be worked out using this formula.

$$\text{Speed} = \frac{\text{Distance}}{\text{Time}}$$

Speed can be thought of as the **distance** travelled in **one unit of time** (1 hour, 1 second, ...)

Speed can be measured in:

kilometres per hour (km/h), metres per second (m/s), miles per hour (mph), and so on.

Average speed

When the speed of an object is **constant** it means that the object doesn't slow down or go faster.
However, in many situations, speed is not constant.
For example:

A sprinter needs time to start from the starting blocks and is well into the race before running at top speed.
A plane changes speed as it takes off and lands.

In situations like this the idea of **average speed** can be used.
The formula for average speed is:

$$\text{Average speed} = \frac{\text{Total distance travelled}}{\text{Total time taken}}$$

The formula linking speed, distance and time can be rearranged and remembered as:

(average) **speed** = (total) **distance** ÷ (total) **time**
(total) **distance** = (average) **speed** × (total) **time**
(total) **time** = (total) **distance** ÷ (average) **speed**

$S = D \div T$
$D = S \times T$
$T = D \div S$

EXAMPLES

1 Robert drives a distance of 260 km. His journey takes 5 hours.
What is his average speed on the journey?

$$\text{Speed} = \frac{\text{Distance}}{\text{Time}} = \frac{260}{5} = 260 \div 5 = 52\text{ km/h}$$

2 Lisa drives at an average speed of 80 kilometres per hour on a journey that takes 3 hours.
What distance has she travelled?

Distance = Speed × Time = $80 \times 3 = 240$ km So, in 3 hours she travels 240 km.

3 Lucy cycles at an average speed of 7 km/h on a journey of 28 km.
How long does she take?

Time = Distance ÷ Speed = $28 \div 7 = 4$ So, her journey takes 4 hours.

4 A cheetah takes 4 seconds to travel 100 m.
What is the speed of the cheetah?

$$\text{Speed} = \frac{\text{Distance}}{\text{Time}} = \frac{100}{4} = 100 \div 4 = 25\text{ m/s}$$

Exercise 8.1

Do not use a calculator.

1. John cycles 16 miles in 2 hours. What is his average speed in miles per hour?
2. Sue runs 21 km in 3 hours. What is her average speed in kilometres per hour?
3. Joe swims 100 m in 4 minutes. What is his average speed in metres per minute?
4. Calculate the average speed for each of the following journeys in kilometres per hour.

	Total distance travelled	Total time taken
(a)	60 km	3 hours
(b)	100 km	2 hours
(c)	10 km	$2\frac{1}{2}$ hours

5. Jackie runs 15 km in $1\frac{1}{2}$ hours. What is her average speed in kilometres per hour?
6. Beverley walks for 2 hours at an average speed of 4 km/h.
How many kilometres does she walk?
7. Howard cycles for 5 hours at an average speed of 6 miles per hour. How far does he cycle?
8. Aubrey runs at 6 km/h for $\frac{1}{2}$ hour. How far does he run?
9. Calculate the total distance travelled on each of the following journeys.

	Total time taken	Average speed
(a)	3 hours	50 km/h
(b)	2 hours	43 km/h
(c)	$\frac{1}{2}$ hour	80 km/h

10. Ahmed drives 30 miles at an average speed of 60 miles per hour.
How long does the journey take?
11. Lauren cycles 100 m at 5 metres per second. How long does she take?
12. A coach travels 75 km at an average speed of 50 km/h. How long does the journey take?
13. Calculate the total time taken on each of the following journeys.

	Total distance travelled	Average speed
(a)	30 km	10 km/h
(b)	80 km	40 km/h
(c)	210 km	60 km/h

14. Aimee cycles 27 km at 12 km/h. How long does she take?
15. Penny cycles to work at 18 km/h. She takes 20 minutes. How far does she cycle to work?
16. A train travels 100 km at an average speed of 80 km/h. How long does the journey take?
17. Bristol is 40 miles from Gloucester.
(a) How long does it take to cycle from Bristol to Gloucester at 16 miles per hour?
(b) How long does it take to drive from Bristol to Gloucester at 32 miles per hour?
18. A car travels 120 km in 2 hours.
(a) What is its average speed in kilometres per hour?
(b) How many hours would the car take to travel 120 km if it had gone twice as fast?
19. Liam drives for 60 km at an average speed of 40 km/h. He starts his journey at 9.50 am.
At what time does his journey end?

Further problems involving speed

EXAMPLE

The Scottish Pullman travels from London to York, a distance of 302.8 km in 1 hour 45 minutes.
It then travels from York to Edinburgh, a distance of 334.7 km in 2 hours 30 minutes.
Calculate the average speed of the train between London and Edinburgh.

Total distance travelled = 302.8 + 334.7 = 637.5 km
Total time taken = 1 hr 45 mins + 2 hr 30 mins = 4 hr 15 mins = 4.25 hours
Average speed = $\frac{637.5}{4.25}$ = 150 km/h

Exercise 8.2

1. On the first part of a journey a car travels 140 km in 3 hours.
   On the second part of the journey the car travels 160 km in 2 hours.
   (a) What is the total distance travelled on the journey?
   (b) What is the total time taken on the journey?
   (c) What is the average speed of the car over the whole journey?

2. Lisa runs two laps of a 400 m running track.
   The first lap takes 70 seconds. The second lap takes 90 seconds.
   What is her average speed over the two laps?

3. Jenny sets out on a journey at 10.20 am.
   She completes her journey at 1.05 pm. She travels a total distance of 27.5 km.
   Calculate her average speed in kilometres per hour.

4. Harry drives for 40 km at an average speed of 60 km/h. He starts his journey at 9.50 am.
   At what time does his journey end?

5. Sally cycles 38 km at an average speed of 23 km/h. She starts her journey at 9.30 am.
   At what time does she finish? Give your answer to the nearest minute.

6. Chandni runs from Newcastle to Whitley Bay and then from Whitley Bay to Blyth.

Newcastle to Whitley Bay	**Time taken:** 1 hr 20 min	**Distance:** 20 km
Whitley Bay to Blyth	**Average speed:** 0.2 km/min	**Distance:** 12 km

   (a) Calculate Chandni's average speed over the whole journey.
   (b) Chandni left Newcastle at 10.50 am.
   At what time did she arrive in Blyth?

7. Angela, Ben and Cathy drive from London to Glasgow.
   Angela takes 12 hours 30 minutes driving at an average speed of 64 km/h.
   Ben drives at an average speed of 100 km/h.
   (a) How long does Ben take?

   Cathy takes 7 hours 12 minutes.
   (b) What is Cathy's average speed?

8. Ron runs 400 m in 1 minute 23.2 seconds.
   Calculate his average speed in (a) metres per second, (b) kilometres per hour.

9. A cheetah runs at a speed of 90 km/h for 6 seconds.
   How many metres does the cheetah run?

10. The distance from the Sun to the Earth is about 150 million kilometres.
    It takes light from the Sun about 500 seconds to reach the Earth.
    Calculate the speed of light in metres per second.

Other compound measures

Density

Density is a compound measure because it involves two other measures, **mass** and **volume**.
The formula for density is:

$$\text{Density} = \frac{\text{Mass}}{\text{Volume}}$$

The formula linking density, mass and volume can be rearranged and remembered as:

$$\text{Volume} = \frac{\text{Mass}}{\text{Density}}$$

$$\text{Mass} = \text{Density} \times \text{Volume}$$

For example, if a metal has a density of 2500 kg/m³ then 1 m³ of the metal weighs 2500 kg.

EXAMPLES

1. A block of metal has mass 500 g and volume 400 cm³.
   Calculate the density of the metal.
   Density $= \frac{500}{400} = 1.25$ g/cm³

2. The density of a certain metal is 3.5 g/cm³. A block of the metal has volume 1000 cm³.
   Calculate the mass of the block.
   Mass $= 3.5 \times 1000 = 3500$ g.

Population density

Population density is a measure of how populated an area is. The formula for population density is:

$$\text{Population density} = \frac{\text{Population}}{\text{Area}}$$

EXAMPLE

Cumbria has a population of 489 700 and an area of 6824 km².
Surrey has a population of 1 036 000 and an area of 1677 km².
Which county has the greater population density?

The population densities are:

Cumbria $\frac{489\,700}{6824} = 71.8$ people/km². Surrey $\frac{1\,036\,000}{1677} = 617.8$ people/km².

Surrey has the greater population density.

Exercise 8.3

1. A metal bar has a mass of 960 g and a volume of 120 cm³.
   Find the density of the metal in the bar.

2. A block of copper has a mass of 2160 g. The block measures 4 cm by 6 cm by 10 cm.
   What is the density of copper?

3. A paperweight is made of glass. It has a volume of 72 cm³ and a mass of 180 g.
   What is the density of the glass?

4. A silver necklace has a mass of 300 g. The density of silver is 10.5 g/cm³.
   What is the volume of the silver?

5. A can in the shape of a cuboid is full of oil.
   It measures 30 cm by 15 cm by 20 cm. The density of oil is 0.8 g/cm³.
   What is the mass of the oil?

6 A copper plaque has a mass of 2.4 kg.
The density of copper is 3 g/cm³.
Calculate the volume of the plaque.

7 A bag of sugar has a mass of 1 kg.
The average density of the sugar in the bag is 0.5 g/cm³.
Find the volume of sugar in the bag.

8 A block of concrete has dimensions 15 cm by 25 cm by 40 cm.
The block has a mass of 12 kg.
What is the density of the concrete?

9 Metal A has density 3 g/cm³ and metal B has density 2 g/cm³.
600 g of metal A and 300 g of metal B are melted down and mixed to make an alloy which is cast into a block.
(a) Calculate the volume of the block.
(b) Calculate the density of the alloy.

10 The population of Northern Ireland is 1 595 000.
The area of Northern Ireland is 13 483 km².
Calculate the population density of Northern Ireland.

11 The table shows the total population, land area and the population densities for some countries in Europe.

	Country	Area km²	Population	Population density people/km²
(a)	Belgium	?	9 970 000	326.6
(b)	France	543 960	56 700 000	?
(c)	UK	244 090	?	235.2

Calculate the missing figures in the table.

12 Brass is an alloy made from zinc and copper.
The ratio of the volume of zinc to the volume of copper in the alloy is 1 : 3.
The density of zinc is 2.5 g/cm³. The density of copper is 3 g/cm³.
A brass ornament has a volume of 400 cm³.
Calculate the mass of the ornament.

What you need to know

- **Speed** is a compound measure because it involves **two** other measures.
- **Speed** is a measure of how fast something is travelling. It involves the measures **distance** and **time**.

  $$\text{Speed} = \frac{\text{Distance}}{\text{Time}}$$

- In situations where speed is not constant, **average speed** is used.

  $$\text{Average speed} = \frac{\text{Total distance travelled}}{\text{Total time taken}}$$

- The formula linking speed, distance and time can be rearranged and remembered as:
  (average) **speed** = (total) **distance** ÷ (total) **time**
  (total) **distance** = (average) **speed** × (total) **time**
  (total) **time** = (total) **distance** ÷ (average) **speed**
- Two other commonly used compound measures are **density** and **population density**.
- **Density** is a compound measure which involves the measures **mass** and **volume**.

  $$\text{Density} = \frac{\text{Mass}}{\text{Volume}}$$

- **Population density** is a measure of how populated an area is.

  $$\text{Population density} = \frac{\text{Population}}{\text{Area}}$$

Review Exercise 8

Do not use a calculator for questions 1 to 6.

1 The chart shows the distances in kilometres between some towns.
Mrs Hill drove from Manchester to Southampton.
She completed the journey in 4 hours.
What was her average speed for the journey in kilometres per hour?

London			
326	Manchester		
270	60	Sheffield	
129	376	334	Southampton

2 Lorraine has a jet-ski ride. She travels at an average speed of 28 km/h for half an hour.
How far does she travel?

3 A train travels 150 miles in 2 hours 30 minutes. Find its average speed in miles per hour. AQA

4 A lorry travels from Middlesbrough to York in $1\frac{1}{4}$ hours. The distance is 50 miles.
Find the average speed of the lorry in miles per hour.

5 Kath has to drive 18 km to work. Calculate her average speed for the journey when she leaves home at 0750 and gets to work at 0814.

6 Matt drove 120 miles at an average speed of 45 mph.
Calculate the time Matt took for the journey. Give your answer in hours and minutes.

7 Nick started his journey at 0945. He travelled 175 km at an average speed of 50 km/h.
At what time did his journey end?

8 Kelly cycles 36 kilometres in 4 hours 30 minutes.
Calculate her average speed in kilometres per hour.

9 Flik lives 2.4 km from school.
How many minutes does she take to walk to school if her average walking speed is 4 km/h?

10 An aeroplane flies from New York to Los Angeles, a distance of 2475 miles, at an average speed of 427 miles per hour.
How long does the flight take in hours and minutes? AQA

11 Jane cycles from *A* to *B* and then from *B* to *C*.
Details of each stage of her journey are given below.

A to *B*	Distance 55 km.	Average speed 22 km/h.
B to *C*	Time taken 1 hour 30 minutes.	Average speed 30 km/h.

Calculate Jane's average speed over the whole of her journey from *A* to *C*. AQA

12 Arnold runs 400 m in 1 minute. What is his average speed in km/h?

13 A tiger runs at a speed of 50 km/h for 9 seconds. How many metres does the tiger run? AQA

14 A glass ornament has a mass of 350 g and a volume of 120 cm^3.
Calculate the density of the glass.

15 A gold bracelet has a mass of 84 g. The density of gold is 19.3 g/cm^3.
What is the volume of the bracelet?

16 The area of Singapore is 4290 km^2. The population density of Singapore is 676 $people/km^2$.
Calculate the population of Singapore, correct to two significant figures.

17 Light travels at 1.86284×10^5 miles per second.
The planet Jupiter is 483.6 million miles from the Sun.
Using suitable approximations, estimate the number of **minutes** light takes to travel from the Sun to Jupiter. AQA

Extending the Number System

So far, we have looked at integers, fractions and decimals.
In this chapter we extend the number system to include **irrational numbers**.

All real numbers are either **rational** or **irrational**.

Rational numbers

Numbers which can be written in the form $\frac{a}{b}$, where a and b are integers ($b \neq 0$) are **rational**.
Examples of rational numbers are:

$2 \qquad -5 \qquad \frac{2}{5} \qquad 0.\dot{6} \qquad 3.47 \qquad 1\frac{3}{4}$

$\frac{a}{b}$ is a **proper fraction** if $a < b$.

$\frac{a}{b}$ is an **improper fraction** (top heavy) if $a > b$.

All fractions can be written as decimals.

$\frac{3}{4}$ can be thought of as $3 \div 4$ and is equal to 0.75.

Some decimals have recurring digits.
These are shown by:
a single dot above a single recurring digit,
a dot above the first and last digit of a set of recurring digits.

For example:

$$\frac{1}{3} = 0.3333333\ldots = 0.\dot{3}$$

$$\frac{41}{70} = 0.5857142857142\ldots = 0.5\dot{8}5714\dot{2}$$

$$\frac{123}{999} = 0.123123123\ldots = 0.\dot{1}2\dot{3}$$

$$\frac{3}{11} = 0.27272727\ldots = 0.\dot{2}\dot{7}$$

Changing recurring decimals to fractions

Recurring decimals can be converted to fractions.

Some are well known, such as $0.\dot{3} = \frac{1}{3}$ and $0.\dot{6} = \frac{2}{3}$.

To convert a recurring decimal to a fraction:
Let x = the recurring decimal.
Multiply both sides by the power of 10 that corresponds to the number of digits in the recurring pattern.

E.g. by $10^1 = 10$ if only 1 digit recurs,
by $10^2 = 100$ if 2 digits recur,
by $10^3 = 1000$ if 3 digits recur,
and so on.

Subtract the original equation from the new equation.

Solve the resulting equation for x.

Make sure that the answer is a fraction in its simplest form.

EXAMPLE

Write down the fraction, in its simplest form, which is equal to these recurring decimals.

(a) $0.\dot{4}$ (b) $0.\dot{5}\dot{7}$ (c) $0.1\dot{2}3\dot{4}$

(a) $x = 0.444...$
Only 1 digit recurs, so, multiply both sides by 10.
$10x = 4.444...$
Subtract the original equation from the new equation.
$9x = 4$
Divide both sides by 9.
$x = \frac{4}{9}$
$0.\dot{4} = \frac{4}{9}$

(b) $x = 0.5757...$
2 digits recur, so, multiply both sides by 100.
$100x = 57.5757...$
Subtract the original equation from the new equation.
$99x = 57$
Divide both sides by 99.
$x = \frac{57}{99} = \frac{19}{33}$ (in its simplest form)
$0.\dot{5}\dot{7} = \frac{19}{33}$

(c) $x = 0.1234234...$
3 digits recur, so, multiply both sides by 1000.
$1000x = 123.4234234...$
Subtract the original equation from the new equation.
$999x = 123.3$
Divide both sides by 999.
$x = \frac{123.3}{999}$
A fraction should consist of whole numbers.
So, multiply numerator and denominator by 10.
$x = \frac{1233}{9990} = \frac{137}{1110}$ (in its simplest form)
$0.1\dot{2}3\dot{4} = \frac{137}{1110}$

Exercise 9.1

1. Convert these recurring decimals to fractions in their simplest form.
   (a) $0.\dot{7}$ (b) $0.\dot{1}$ (c) $0.\dot{8}$ (d) $0.\dot{3}\dot{6}$ (e) $0.\dot{8}\dot{2}$
   (f) $0.\dot{1}3\dot{5}$ (g) $0.\dot{2}1\dot{6}$ (h) $0.\dot{4}2\dot{5}$ (i) $0.\dot{1}6\dot{2}$ (j) $0.\dot{2}8571\dot{4}$

2. Write each of these numbers as a fraction in its simplest form.
   (a) $0.1\dot{6}$ (b) $0.0\dot{3}$ (c) $0.6\dot{1}$ (d) $0.4\dot{6}$ (e) $0.2\dot{3}$
   (f) $0.8\dot{3}$ (g) $0.1\dot{8}\dot{2}$ (h) $0.1\dot{3}\dot{6}$ (i) $0.86\dot{1}$ (j) $0.77\dot{2}\dot{7}$

3. $0.\dot{1}17647058823529\dot{4}$
   When the number above is written as a fraction the numerator and denominator are both less than 20. What is the fraction?

4. Look at the decimal expansions for the fractions $\frac{1}{a}$, where $1 < a < 21$ (a is a whole number).
   For what values of a does the decimal not recur?
   Find a condition so that a fraction with numerator 1 does not recur.

Irrational numbers

Numbers are either **rational** or **irrational**.
An irrational number **cannot** be written as a fraction.
Irrational numbers include:
square roots of non-square numbers,
cube roots of non-cube numbers.
Examples of irrational numbers are:

$\sqrt{2}$ $\sqrt[3]{7}$ π $\sqrt{13}$

> Real numbers are either rational or irrational numbers.
> You will meet imaginary numbers, such as $\sqrt{-1}$, if you study Maths beyond GCSE.

EXAMPLES

1 Show that 0.425 and $0.\dot{2}$ are rational numbers.

0.425 is a **terminating decimal**.

$0.425 = \frac{425}{1000} = \frac{17}{40}$

$0.\dot{2}$ is a **recurring decimal**.

$x = 0.222...$
$10x = 2.222...$
$9x = 2$
$x = \frac{2}{9}$
$0.\dot{2} = \frac{2}{9}$

> All terminating and recurring decimals can be written in the form $\frac{a}{b}$ so they are all rational numbers.

2 State whether each of the following are rational or irrational numbers.
Where a number is rational write it as a fraction in its simplest form.

0.6 π $\sqrt[3]{7}$ $\sqrt{36}$

$0.6 = \frac{6}{10} = \frac{3}{5}$ rational — A rational number expressed in the form $\frac{a}{b}$ is in its simplest form if a and b have no common factor.

$\pi = 3.141592654...$ irrational — π is a non-recurring decimal and has no exact value.

$\sqrt[3]{7} = 1.91293118...$ irrational — $\sqrt[3]{7}$ is a non-recurring decimal and has no exact value.

$\sqrt{36} = 6$ rational — Note: $\sqrt{36}$ means the positive square root of 36.

Exercise 9.2

Do not use a calculator.

1 These numbers are rational.
Write each number as a fraction in its simplest form.

(a) 0.5 (b) 0.45 (c) 0.3 (d) 0.625 (e) $2\frac{1}{4}$

(f) -0.25 (g) $0.\dot{2}\dot{7}$ (h) $\sqrt{49}$ (i) $\frac{\sqrt{0.49}}{2}$ (j) $0.\dot{2}2\dot{5}$

(k) $\frac{\sqrt{64}}{\sqrt{16}}$ (l) $-\frac{\sqrt{0.16}}{2}$ (m) $\sqrt[3]{64}$ (n) $\frac{\sqrt[3]{27}}{4}$ (o) $0.1\dot{2}$

2 Which of these numbers are rational?

(a) 0.25 (b) $0.\dot{5}$ (c) $\sqrt{5}$ (d) $\sqrt{25}$ (e) $\frac{1.8}{3}$

(f) $-\sqrt{0.25}$ (g) $\frac{\pi}{2}$ (h) $2.\dot{1}$ (i) $\sqrt[3]{8}$ (j) $\sqrt{1\frac{7}{9}}$

Express each of the rational numbers in the form $\frac{a}{b}$, where a and b are integers.

Surds

Roots of rational numbers which **cannot** be expressed as rational numbers are called **surds**.
A surd is an irrational number.
These are examples of surds:

$\sqrt{2}$ $\sqrt{0.37}$ $\sqrt[3]{10}$ $3 + \sqrt{2}$ $\sqrt{7}$

$\sqrt{a}$ means the positive square root of a.

Numbers like $\sqrt{64}$, $\sqrt{0.25}$, $\sqrt[3]{27}$ are not surds because the root of each number is rational.

$(\sqrt{64} = 8,\ \sqrt{0.25} = 0.5,\ \sqrt[3]{27} = 3.)$

Manipulating and simplifying surds

Rules for surds.

$\sqrt{ab} = \sqrt{a} \times \sqrt{b}$ $m\sqrt{a} + n\sqrt{a} = (m + n)\sqrt{a}$ $\sqrt{\frac{a}{b}} = \frac{\sqrt{a}}{\sqrt{b}}$

To simplify surds look for factors that are square numbers.

EXAMPLE

Simplify the following leaving the answers in surd form.

(a) $\sqrt{32}$ (b) $\sqrt{8} + \sqrt{18}$ (c) $\sqrt{108} - \sqrt{75}$ (d) $\sqrt{\frac{72}{20}}$

(a) $\sqrt{32} = \sqrt{16} \times \sqrt{2} = 4\sqrt{2}$

(b) $\sqrt{8} + \sqrt{18} = \sqrt{4} \times \sqrt{2} + \sqrt{9} \times \sqrt{2} = 2\sqrt{2} + 3\sqrt{2} = 5\sqrt{2}$

(c) $\sqrt{108} - \sqrt{75} = \sqrt{36} \times \sqrt{3} - \sqrt{25} \times \sqrt{3} = 6\sqrt{3} - 5\sqrt{3} = \sqrt{3}$

(d) $\sqrt{\frac{72}{20}} = \frac{\sqrt{72}}{\sqrt{20}} = \frac{\sqrt{36}\sqrt{2}}{\sqrt{4}\sqrt{5}} = \frac{6\sqrt{2}}{2\sqrt{5}} = \frac{3\sqrt{2}}{\sqrt{5}}$

Exercise 9.3

1. Which of the following are surds?
   (a) $\sqrt{2}$ (b) $\sqrt{4}$ (c) $\sqrt[3]{8}$ (d) $\sqrt{0.4}$ (e) $\sqrt{0.09}$

2. Write the following surds in their simplest form.
   (a) $\sqrt{12}$ (b) $\sqrt{27}$ (c) $\sqrt{45}$ (d) $\sqrt{48}$ (e) $\sqrt{32}$
   (f) $\sqrt{50}$ (g) $\sqrt{54}$ (h) $\sqrt{24}$ (i) $\sqrt{98}$ (j) $\sqrt{80}$
   (k) $\sqrt{63}$ (l) $\sqrt{200}$ (m) $\sqrt{128}$ (n) $\sqrt{112}$ (o) $\sqrt{175}$

3. Simplify.
   (a) $\sqrt{2} + \sqrt{2}$ (b) $2\sqrt{5} - \sqrt{5}$ (c) $5\sqrt{3} + 2\sqrt{3}$
   (d) $\sqrt{18} + \sqrt{8}$ (e) $\sqrt{50} - \sqrt{32}$ (f) $\sqrt{45} + \sqrt{80}$
   (g) $\sqrt{75} - \sqrt{12}$ (h) $\sqrt{300} - \sqrt{48}$ (i) $\sqrt{50} + \sqrt{18} - \sqrt{8}$
   (j) $3\sqrt{20} + 2\sqrt{45}$ (k) $2\sqrt{48} + 3\sqrt{12}$ (l) $3\sqrt{45} - 2\sqrt{20}$

4. Simplify.
   (a) $\sqrt{200} - 2\sqrt{18} + \sqrt{72}$ (b) $\sqrt{300} + \sqrt{48} - 3\sqrt{27}$ (c) $3\sqrt{18} - 2\sqrt{8} + \sqrt{2}$

5. Simplify the following.
   (a) $\sqrt{\frac{9}{4}}$ (b) $\sqrt{\frac{25}{16}}$ (c) $\sqrt{\frac{18}{8}}$ (d) $\sqrt{\frac{24}{9}}$ (e) $\sqrt{\frac{6}{4}}$ (f) $\sqrt{\frac{12}{15}}$

6. Simplify the following.
   (a) $\sqrt{3} \times \sqrt{3}$ (b) $\sqrt{3} \times 2\sqrt{3}$ (c) $2\sqrt{5} \times 3\sqrt{5}$ (d) $\sqrt{2} \times \sqrt{8}$
   (e) $\sqrt{12} \times \sqrt{3}$ (f) $\sqrt{5} \times \sqrt{10}$ (g) $2\sqrt{6} \times \sqrt{3}$ (h) $2\sqrt{5} \times \sqrt{10}$
   (i) $\sqrt{8} \times \sqrt{18}$ (j) $3\sqrt{2} \times 2\sqrt{3}$ (k) $4\sqrt{3} \times 2\sqrt{2}$ (l) $\sqrt{27} \times \sqrt{32}$

7. Remove the brackets and simplify the following.
   (a) $\sqrt{2}(\sqrt{2} + 1)$ (b) $\sqrt{3}(\sqrt{6} - \sqrt{3})$ (c) $\sqrt{5}(\sqrt{10} - \sqrt{5})$

Rationalising denominators of fractions

When the denominator of a fraction is a surd it is usual to remove the surd from the denominator. This process is called **rationalising the denominator**.

For fractions of the form $\frac{a}{\sqrt{b}}$, multiply both the numerator (top) and the denominator (bottom) of the fraction by $\sqrt{b}$ and then simplify where possible.

EXAMPLE

Rationalise the denominator and simplify where possible: (a) $\frac{1}{\sqrt{2}}$, (b) $\frac{3\sqrt{2}}{\sqrt{6}}$.

(a) $\frac{1}{\sqrt{2}} = \frac{1}{\sqrt{2}} \times \frac{\sqrt{2}}{\sqrt{2}} = \frac{\sqrt{2}}{2}$

(b) $\frac{3\sqrt{2}}{\sqrt{6}} = \frac{3\sqrt{2} \times \sqrt{6}}{\sqrt{6} \times \sqrt{6}} = \frac{3\sqrt{2}\,\sqrt{6}}{6} = \frac{3\sqrt{2}\,\sqrt{2}\,\sqrt{3}}{6} = \frac{6\sqrt{3}}{6} = \sqrt{3}$

Exercise 9.4

1. Rationalise the denominator in each of the following and then simplify the fraction.
   (a) $\frac{1}{\sqrt{3}}$ (b) $\frac{1}{\sqrt{5}}$ (c) $\frac{1}{\sqrt{7}}$ (d) $\frac{2}{\sqrt{2}}$ (e) $\frac{5}{\sqrt{5}}$ (f) $\frac{4}{\sqrt{2}}$ (g) $\frac{6}{\sqrt{3}}$
   (h) $\frac{14}{\sqrt{7}}$ (i) $\frac{3}{\sqrt{6}}$ (j) $\frac{15}{\sqrt{5}}$ (k) $\frac{9}{\sqrt{3}}$ (l) $\frac{5}{\sqrt{15}}$ (m) $\frac{18}{\sqrt{6}}$ (n) $\frac{35}{\sqrt{5}}$

2. Express each of the following in its simplest form with a rational denominator.
   (a) $\frac{6}{\sqrt{8}}$ (b) $\frac{6}{\sqrt{24}}$ (c) $\frac{8}{\sqrt{32}}$ (d) $\frac{9}{\sqrt{18}}$ (e) $\frac{\sqrt{3}}{\sqrt{6}}$ (f) $\frac{\sqrt{12}}{\sqrt{3}}$
   (g) $\frac{\sqrt{18}}{\sqrt{2}}$ (h) $\frac{\sqrt{5}}{\sqrt{20}}$ (i) $\frac{3\sqrt{5}}{\sqrt{15}}$ (j) $\frac{4\sqrt{6}}{\sqrt{12}}$ (k) $\frac{2\sqrt{8}}{\sqrt{32}}$ (l) $\frac{5\sqrt{7}}{\sqrt{35}}$
   (m) $\frac{\sqrt{3}\,\sqrt{5}}{\sqrt{30}}$ (n) $\frac{\sqrt{2}\,\sqrt{3}}{\sqrt{18}}$ (o) $\frac{5}{2\sqrt{3}}$ (p) $\frac{\sqrt{8}}{2\sqrt{5}}$ (q) $\sqrt{\frac{9}{10}}$ (r) $\sqrt{\frac{4}{3}}$

What you need to know

- **Rational numbers** can be written in the form $\frac{a}{b}$, where a and b are integers ($b \neq 0$).
  Examples of rational numbers are: 2, -5, $\frac{2}{5}$, 0.6, 3.47, $1\frac{3}{4}$.
- All fractions can be written as decimals.
  For example, $\frac{1}{3} = 0.3333333... = 0.\dot{3}$, $\frac{123}{999} = 0.123123123... = 0.\dot{1}2\dot{3}$
- **Irrational numbers** cannot be written as fractions.
  Irrational numbers include: square roots of non-square numbers,
  cube roots of non-cube numbers.
  Examples of irrational numbers are: $\sqrt{2}$, $\sqrt[3]{7}$, π, $\sqrt{13}$.
- All **terminating** and **recurring decimals** are rational numbers.
- A **surd** is the root of a rational number which is not rational.
  A surd is an irrational number.
- Rules for manipulating and simplifying surds:
  $\sqrt{ab} = \sqrt{a} \times \sqrt{b}$ $\quad m\sqrt{a} + n\sqrt{a} = (m+n)\sqrt{a}$ $\quad \sqrt{\frac{a}{b}} = \frac{\sqrt{a}}{\sqrt{b}}$
- To **rationalise** the denominator of a fraction of the form $\frac{a}{\sqrt{b}}$ multiply both the numerator (top) and the denominator (bottom) of the fraction by $\sqrt{b}$.

Review Exercise 9

Do not use a calculator.

1 (a) Change $\frac{5}{9}$ into a decimal.
(b) Find the fraction which is equal to $0.2\dot{5}$. Give your answer in its simplest form.

2 Express $0.\dot{4}\dot{8}$ as a fraction in its simplest terms. AQA

3 Write down the fraction, in its simplest form, which is equal to these recurring decimals.
(a) $0.\dot{4}\dot{5}$ (b) $0.6\dot{4}\dot{5}$ (c) $0.\dot{3}2\dot{5}$ AQA

4 (a) Write $\sqrt{6} \times \sqrt{3}$ in the form $a\sqrt{b}$ where a and b are prime numbers.
(b) Simplify fully $5\sqrt{2} + \sqrt{8}$. AQA

5 Simplify.
(a) $3\sqrt{5} - \sqrt{5}$ (b) $\sqrt{\frac{36}{25}}$ (c) $\sqrt{2} \times 3\sqrt{2}$

6 Simplify the following leaving the answers in surd form.
(a) $\sqrt{12}$ (b) $2\sqrt{3} + 3\sqrt{3}$ (c) $\sqrt{\frac{15}{12}}$

7 (a) Simplify $\frac{10}{\sqrt{5}}$ by rationalising the denominator. Give your answer in its simplest form.
(b) Show that: $\frac{\sqrt{125} - \sqrt{45}}{\sqrt{125} + \sqrt{45}} = \frac{1}{4}$ AQA

8 Prove that: $0.4\dot{7} = \frac{43}{90}$ AQA

9 Rationalise the denominator and simplify fully $\frac{18}{\sqrt{2}}$ AQA

10 Show that: $\sqrt{12}(\sqrt{75} - \sqrt{48}) = 6$ AQA

11 (a) Simplify $\sqrt{8} + \sqrt{50}$
(b) Hence, simplify $(\sqrt{8} + \sqrt{50})(\sqrt{24} + \sqrt{54})$. Give your answer in its simplest form.
AQA

Number Non-calculator Paper

Do not use a calculator for this exercise.

1. Find the value of: (a) 10^5 (b) $\sqrt{49}$ (c) $8^2 \div 2^3$ (d) 0.1×0.9 (e) $\frac{9}{20}$ as a decimal

2. Every day, Ron drives a bus between Cambridge and Oxford. His bus can carry 40 passengers.
On Monday, $\frac{5}{8}$ of the seats for passengers are used.
(a) How many passengers are on the bus on Monday?
On Tuesday, 30 passengers are on the bus.
(b) What percentage of the seats are used? AQA

3. (a) Find $\frac{3}{5}$ of 60.
(b) Work out. (i) $\frac{2}{3} + \frac{3}{4}$ (ii) $\frac{1}{2} - \frac{2}{5}$ (iii) $\frac{3}{4} \times \frac{2}{5}$
(c) Write down a decimal that lies between $\frac{1}{4}$ and $\frac{1}{3}$.

4. A box contains red candles and white candles in the ratio 1 : 4. The box contains 30 candles.
How many red candles are there?

5. Raspberry jam is sold in two sizes.
Which size is the better value for money?
You **must** show all your working. AQA

Large 500 g — £2.69
Standard 250 g — £1.39

6. Debbie wants to calculate $\frac{70.24}{9.8 - 3.08}$.
(a) Write each of the numbers in Debbie's calculation to the nearest whole number.
(b) Hence, find an estimate of the answer for Debbie. AQA

7. A lorry travels 100 miles in 2 hours 30 minutes. Find its average speed in miles per hour.

8. There are 400 people at a fair.
Of these 400 people, $\frac{1}{4}$ are over 60 and $\frac{1}{5}$ are under 10.
(a) How many people are **not** over 60 and are **not** under 10?
(b) Of the 400 people, 240 are female. What percentage are female? AQA

9. The cash price of a computer is £960. The computer can be bought on credit by paying a deposit of 10% of the cash price and 12 monthly payments of £79.50.
How much **more** is paid when the computer is bought on credit? AQA

10. Ted pays £3.24 for 200 g of jelly babies and 300 g of toffees. 100 g of jelly babies cost 48p.
What is the cost of 100 g of toffees?

11. Which is bigger, 2^6 or 3^4? Show all your working. AQA

12. Parking 400 spaces
The number of spaces is given to the nearest ten.
What is the minimum and maximum possible number of spaces?

13. Use the calculation $487 \times 3.53 = 1719.11$ to find the value of:
(a) 487×0.0353 (b) 48700×0.00353 AQA

14. Work out. (a) $\frac{3}{4}$ of 5.6 (b) 0.2×0.4 (c) $3\frac{1}{4} - 1\frac{2}{5}$

15. Red tulips and yellow tulips are used for a flower display. The display uses 40 tulips.
The ratio of red tulips to yellow tulips is 3 : 5. How many tulips are red?

16 Kim invests £500 for two years, at 4% per annum compound interest.
Amir says that the interest will be £40.
Is Amir correct? Explain clearly how you obtained your answer. AQA

17 A sports pitch has a length of 75 metres, correct to the nearest metre.
Write down the least and the greatest possible length of this pitch. AQA

18 The volume of a metal prism is $30\,cm^3$. The mass of the prism is 210 g.
What is the density of the metal in g/cm^3?

19
(a) Express 72 as a product of its prime factors.
(b) Find the highest common factor of 48 and 72.

20 Andi tells Cato:
"When driving your car, if you increase your speed by 10% and then decrease your speed by 10% you will be travelling at a slower speed than you were to begin with."
Is he correct? Explain your answer.

21
(a) Use approximations to estimate the value of $\frac{9.67^2}{0.398}$. You **must** show all your working.
(b) (i) p and q are prime numbers. Find the values of p and q when $p^3 \times q = 24$.
(ii) Write 18 as a product of prime factors.
(iii) What is the least common multiple of 24 and 18? AQA

22 Are these statements true or false?
A: The product of two prime numbers is always a prime number.
B: The sum of two consecutive numbers is always odd.
Explain each answer.

23 Carol is paid £6.30 per hour for a basic 30-hour week. Overtime is paid at time and a third.
Last week Carol's total pay was £218.40.
How many hours and minutes overtime did Carol work last week? AQA

24 On a motorway there are three lanes, an inside lane, a middle lane and an outside lane.
One day, at midday, the speed of the traffic on these three lanes was in the ratio of 3 : 4 : 5.
The speed on the outside lane was 70 miles per hour.
Calculate the speed on the inside lane. AQA

25
(a) Find an approximate value of $\frac{584 \times 4.91}{0.198}$. You **must** show all your working.
(b) Work out $2\frac{3}{5} + 1\frac{2}{3}$.
(c) Find the value of 11^0.
(d) Work out $\frac{4 \times 10^8}{5 \times 10^{-6}}$. Give your answer in standard form. AQA

26 In a sale prices are reduced by 10%. The sale price of a tennis racket is £36.
What was the original price of the tennis racket?

27 Tim runs 100 m in 10 seconds.
What is his average speed? Give your answer in kilometres per hour. AQA

28 Work out the value of $4^1 - 4^0 + 4^{-1}$.

29
(a) Find the compound interest on £3000 invested for 2 years at 7%.
(b) Peggy has a bag of potatoes. She has used 20% of the potatoes.
The bag now contains 2 kg of potatoes.
How many kilograms of potatoes were in the bag to start with? AQA

30 Sam and Tom share 100 counters in the ratio 2 : 3.
They use the counters to keep score in a game.
At the end of the game Sam and Tom have counters in the ratio 7 : 13.
Calculate the change in the number of Sam's counters. AQA

31 (a) Work out $4 \times 10^7 \times 6 \times 10^{-5}$, giving your answer in standard form.
(b) Work out $\frac{3.2 \times 10^8}{4 \times 10^4}$, giving your answer in standard form.
(c) Work out $(4 \times 10^5)^2$, giving your answer in standard form. AQA

32 A running track is 100 m correct to the nearest metre.
Another running track is 400 m correct to the nearest metre.
Lauren runs four times around the 100 m track.
Aimee runs once around the 400 m track.
What is the upper bound of the difference between the distance Lauren runs and the distance Aimee runs?

33 (a) Write $0.\dot{3}\dot{4}$ as a fraction.
(b) Hence, or otherwise, write $0.10\dot{3}\dot{4}$ as a fraction in its simplest form. AQA

34 Simplify the following expressions, leaving your answer in surd form.
(a) $\sqrt{8} + \sqrt{18}$ (b) $\sqrt{32} \times \sqrt{27}$ AQA

35 (a) Calculate $125^{\frac{1}{3}}$.
(b) Calculate $49^{-\frac{1}{2}}$, giving your answer as a fraction.
(c) Calculate $\frac{3^{10}}{9^6}$, giving your answer as a power of 3.
(d) Calculate $128^{\frac{2}{7}}$. AQA

36 Liz estimated the cost of the material to make herself a dress.
When Liz goes to a shop to buy the material, the cost per metre is 40% more than she had estimated.
Liz is informed by the shopkeeper that she had miscalculated the amount of material and she really needs 20% more material than she had estimated.
How much more than her original estimate does the dress material cost?
Give your answer as a percentage of Liz's original estimate. AQA

37 Find the exact value of $64^{\frac{1}{3}} \times 196^{-\frac{1}{2}}$.
Give your answer as a fraction in its simplest form. AQA

38 The expression $n(n + 1)(n + 2)$ represents the product of three consecutive numbers.
(a) Explain why at least one of the numbers must be even.
(b) Prove that the value of the product must be a multiple of 6. AQA

39 Light travels at 1.86282×10^5 miles per second.
The Earth is 92.5 million miles from the Sun.
Using suitable approximations, estimate the number of **minutes** light takes to travel from the Sun to the Earth.

40 (a) Rationalise and simplify $\frac{1}{\sqrt{8}}$
(b) By simplifying $\sqrt{12} + \sqrt{108}$, write $\frac{\sqrt{12} + \sqrt{108}}{\sqrt{8}}$
in the form $a\sqrt{b}$ where a and b are integers. AQA

41 (a) Calculate $81^{-\frac{1}{2}}$, giving your answer as a fraction.
(b) Calculate $\frac{2^3}{2^{-2}}$, leaving your answer in the form 2^p.
(c) Write $\frac{\sqrt{3}}{\sqrt{2}}$ in the form $\frac{\sqrt{a}}{b}$, where a and b are whole numbers. AQA

42 Write $0.0\dot{6}\dot{3}$ as a fraction in its simplest form. AQA

Section Review

Number Calculator Paper

You may use a calculator for this exercise.

1 Dave drives 15 miles to work. The journey takes 20 minutes.
What is Dave's average speed in miles per hour? AQA

2 Carl buys 1.2 kg of potatoes and 0.4 kg of carrots. He pays 98p in total.
The potatoes cost 70p per kilogram. What is the cost of 1 kg of carrots? AQA

3 Use your calculator to find: (a) 2.3^2 (b) $(285 - 198) \div 2.9$ (c) $\sqrt{38}$, correct to 1 d.p.

4 (a) The weights and prices of two tins of pineapple are shown.
Which tin of pineapple gives more grams per penny?
You **must** show all your working.

YUM PINEAPPLE 227 g 27 pence

CORE PINEAPPLE 432 g 52 pence

(b) A greengrocer sold 16 cauliflowers and 24 cabbages.
The total cost was £18.16.
The cost of a cauliflower was 55p.
What was the cost of a cabbage? AQA

5 (a) John is 8.4 kg heavier than Alan. The sum of their weights is 133.2 kg.
How heavy is Alan?
(b) Before starting a diet Derek weighed 80 kg. He now weighs 8 kg less.
Calculate his weight loss as a percentage of his previous weight. AQA

6 A suitcase costs £56 plus VAT at $17\frac{1}{2}$%. What is the total price of the suitcase?

7 A sponge cake for eight people needs 120 g of sugar.
John makes a sponge cake for five people.
(a) Calculate the weight of sugar he needs.

John cuts his cake into five equal slices. He gives three slices to his mother.
(b) What percentage of the cake does John have left? AQA

8 Steff buys 5.1 kg of nails for £3.06. How much would she pay for 3.2 kg of the same nails? AQA

9 Zoe chooses three consecutive numbers.
Show that the sum of the first and the last of these three numbers is always an even number. AQA

10 (a) Use your calculator to find the value of $\frac{19.8^2}{7.19 + 2.73}$
(b) Show how you can use approximation to check your answer.

11 Wyn has 600 yo-yo's to sell. He sells $\frac{3}{5}$ of them at £2 each.
He then reduces the price of the remaining yo-yo's by 40%.
In the end he is left with 24 damaged yo-yo's which he cannot sell.
Wyn paid £500 for the yo-yo's.
Find the percentage profit which he made on the sale of these yo-yo's.

12 (a) Calculate the exact value of 5^9.
(b) Find the reciprocal of 7. Give your answer correct to 3 decimal places.
(c) Calculate $\frac{5.8 \times \sqrt{21.48}}{\sqrt{12}}$. Give your answer to one decimal place.

13 P is a prime number. Q is an odd number. State whether each of the following is always odd or always even or could be either odd or even.
(a) $P(Q + 1)$ (b) $Q - P$ AQA

14 The table shows the amount of foreign currency you can buy with £1.

EXCHANGE RATES - *£1 will buy:*	France 1.56 euros	USA 1.37 dollars

An American has 400 000 dollars to spend on a holiday villa in the south of France.
She sees a villa for sale for 450 000 euros.
Does she have enough money to buy it? You **must** show all your working.

15 Greg drove 115 miles at an average speed of 43 mph.
Calculate the time Greg took for the journey. Give your answer in hours and minutes. AQA

16 Three musicians received £100 between them for playing in a concert.
They divided their pay in the ratio of the number of minutes for which each played.
Angela played for 8 minutes, Fran played for 14 minutes and Dan played for 18 minutes.
How much did each receive? AQA

17 Jim invests £10 000 at 5.2% per year compound interest.
What is the percentage increase in the value of his investment after 3 years?

18 (a) Calculate $\dfrac{7.9 \times 8.2}{8.2 - 7.9}$ correct to two significant figures.
(b) Calculate $(4.9 \times 10^3)^2$. AQA

19 $P = 2^3 \times 3^2 \times 5 \qquad Q = 2 \times 3^3 \times 7$
(a) What number is the Highest Common Factor of P and Q?
(b) What number is the Least Common Multiple of P and Q?

20 Evaluate 0.9^7, giving your answer in standard form.

21 (a) (i) Chloe has 24 CDs and some cassettes. The ratio of CDs to cassettes is 3 : 5.
How many cassettes does Chloe have?
(ii) Chloe sells 10 of her cassettes.
What is the new ratio of CDs to cassettes? Give your answer in the form $1 : n$.
(b) Calculate 5% of 1.4×10^8. Give your answer in standard form.
(c) In a sale, prices are reduced by 20%. A toaster cost £8.80 in the sale.
What was the price of the toaster before the sale? AQA

22 Ahmed sells silk ties at a price which is 37% more than he paid for them.
The selling price of one tie is £36.99.
How much did Ahmed pay for this tie? AQA

23 (a) Calculate
(i) $\dfrac{34.3}{16.91 + 17.39}$ (ii) $\sqrt{3.1 - \frac{1}{3.1}}$ (iii) $(1.1 + 2.2 + 3.3)^2 - (1.1^2 + 2.2^2 + 3.3^2)$
(b) Show how you would make a quick estimation of the answer to part (a)(iii). AQA

24 The volume of a container is 1170 cu ft. The weight of its contents is 59 500 lb.
Find the density of the contents. AQA

25 In Portugal, Brian spends €2.80 on ice-cream.
This price includes VAT which is 12% in Portugal.
Find the amount of VAT which Brian paid. AQA

26 Americans eat 1.6×10^{10} quarts of popcorn a year.
The population of America is 276 million people.
How many quarts of popcorn on average does an American eat per year? AQA

27 Mark's height is 203 cm and Eileen's height is 185 cm.
Both heights are given to the nearest cm.
Find the maximum possible difference between the two heights. AQA

28 An Internet auction site has two identical cars for sale. Both cars are priced at £10 000.
The price of each car is to be reduced each week until they are sold.
The first car is reduced by 10% each week. The second car is reduced by £800 each week.
Assuming that no-one buys the cars, after how many weeks will the second car be cheaper than the first? AQA

29 (a) Calculate the value of 2×5^9.

(b) (i) Calculate $\frac{28.3 + 0.512}{(18.9 - 2.75)^2}$.

(ii) Paul gives his answer to (i) correct to 5 significant figures.
Give one reason why this is **not** an appropriate degree of accuracy.

(c) Calculate the value of $(2.34 \times 10^{-2})^{\frac{1}{2}}$. AQA

30 A plane flies from London to Naples at an average speed of 497 mph.
The distance is 1413 miles.
(a) Find the time taken. Give your answer in hours and minutes.

On one flight, the plane takes 2 hours 40 minutes for the journey from London to Naples.
(b) What is the average speed? AQA

31 A lorry moving ore in a gold mine carries 3.19×10^4 kilograms of ore.
The ore contains 1.89×10^2 grams of gold.
What percentage of the ore is gold? Give your answer in standard form. AQA

32 Express each of the following as a fraction in its simplest form.

(a) 4^{-3} (b) $\left(\frac{27}{125}\right)^{\frac{1}{3}}$ (c) $256^{-\frac{1}{2}}$ AQA

33 Cashchem sells sun oil in special offer bottles which contain an extra 25% free.
In a sale, Cashchem discounts all stock by 30%.
What percentage is the reduction of cost of 100 ml of sun oil when bought in the special offer bottles in the sale? AQA

34 (a) The numbers in this calculation are given to 3 significant figures.

Find the least possible value of $\frac{12.3}{15.6 - 7.20}$

(b) The maximum safe load of a lift is 1500 kg, to the nearest 50 kg.
The lift is loaded with boxes weighing 141 kg and 150 kg, both weights given to the nearest kilogram.
Can the lift safely carry 3 boxes weighing 141 kg each and 7 boxes weighing 150 kg each? You **must** show all your working. AQA

35 Prove that, for any positive integer, a, $a^3 + a$ is always even. AQA

36 (a) Simplify, leaving your answer where appropriate in surd form.

(i) $\sqrt{27} + \sqrt{147}$ (ii) $5\sqrt{3} + \sqrt{108}$ (iii) $\sqrt{18} \times \sqrt{12}$

(b) Simplify $\sqrt{\frac{2}{45}}$, leaving your answer with a rational denominator.

37 Tom and Chris run a 100 m race. Tom's time is 13.6 seconds. Chris's time is 14.1 seconds.
Their times are recorded to the nearest one tenth of a second.
(a) What is the lower bound in the difference in the times of Tom and Chris?

The 100 metre distance is measured to the nearest ten centimetres.
(b) What is the upper bound of Tom's average speed in the race? AQA

38 (a) If $p = \sqrt{a}$, where a is an integer, prove that it is always possible to find a number q, $p \neq q$, so that pq is also an integer.

(b) Write $(\sqrt{108} - \sqrt{2})^2$ in the form $p - q\sqrt{6}$ where p and q are integers. AQA

CHAPTER 10 Introduction to Algebra

Algebra is sometimes called the language of Mathematics.
Algebra uses letters in place of numbers.

A class of children line up. We cannot see how many children there are altogether because of a tree.
We can say there are n children in the line.
The letter n is used in place of an unknown number.

Three more children join the line.
There are now $n + 3$ children in the line.

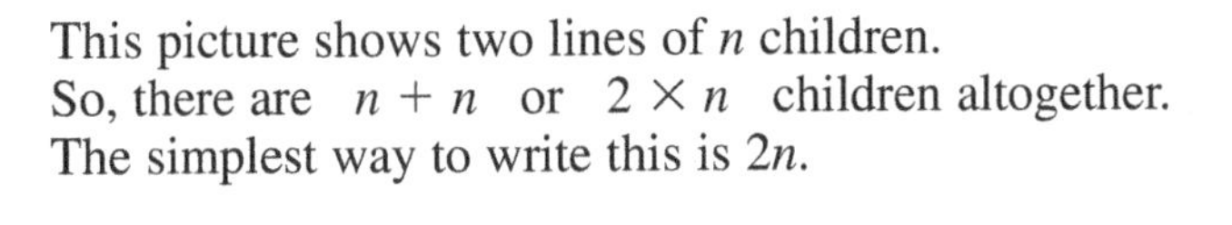

This picture shows two lines of n children.
So, there are $n + n$ or $2 \times n$ children altogether.
The simplest way to write this is $2n$.

Both $n + 3$ and $2n$ are examples of **algebraic expressions**.

Exercise 10.1

Write algebraic expressions for each of the following questions.

1. I have m marbles in a bag. I put in another 6 marbles.
   How many marbles are now in the bag?
2. I have m marbles. I lose 12 marbles.
   How many marbles do I have left?
3. I have 8 bags of marbles. Each bag contains m marbles.
   How many marbles do I have altogether?
4. There are p pencils in a pencil case. I take one pencil out.
   How many pencils are left in the pencil case?
5. There are p pencils in a pencil case. I put in another 5 pencils.
   How many pencils are now in the pencil case?
6. I have 25 pencil cases. There are p pencils in each pencil case.
   How many pencils do I have altogether?
7. I have 6 key rings. There are k keys on each key ring.
   How many keys do I have altogether?
8. What is the cost of b biscuits costing 5 pence each?
9. Three cakes cost a total of c pence. What is the cost of one cake?
10. Five kilograms of apples cost a pence. What is the cost of one kilogram of apples?
11. A group of 36 students are split into g groups. How many students are in each group?

Expressions and terms

Consider this situation:

$2n$ students start a typing course.
3 of the students leave the course.
How many students remain on the course?

$2n - 3$ students remain.
$2n - 3$ is an **algebraic expression**, or simply an **expression**.

An expression is just an answer made up of letters and numbers.
$+2n$ and -3 are **terms** of the expression.

Note:
A term includes the sign, + or −.
$2n$ has the same value as $+2n$.

Simplifying expressions

Adding and subtracting terms

You can add and subtract terms with the same letter.
This is sometimes called **simplifying an expression**.

$a + a = 2a$ $\quad$ $6a - 2a = 4a$
$5k + 3k = 8k$ $\quad$ $2d - 3d = -d$
$3p + 5 + p - 1 = 4p + 4$ $\quad$ $4x - 4x = 0$

$6 + a$ cannot be simplified.
$5p - 2q$ cannot be simplified.
$x^2 + x$ cannot be simplified.
$ab + ba = 2ab$

Note that:
A simpler way to write $1d$ is just d.

$-1d$ can be written as $-d$.

$0d$ is the same as 0.

Just as with ordinary numbers, you can add terms in any order.
$a - 2a + 5a = a + 5a - 2a = 4a$

EXAMPLE

Write down an expression for the perimeter of this shape.
Give your answer in its simplest form.

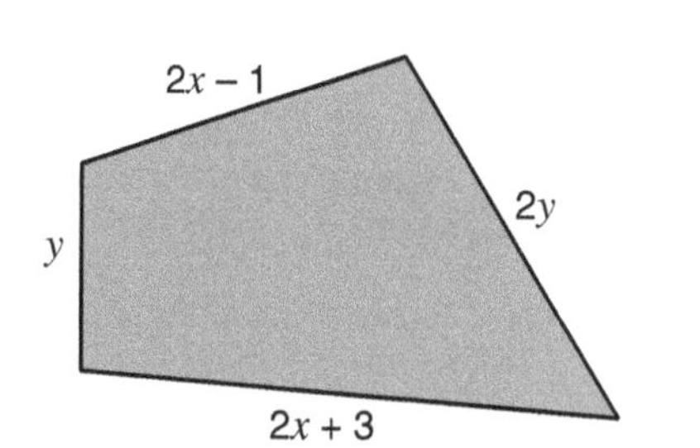

Perimeter is the total distance round the outside of the shape.
$y + 2x - 1 + 2y + 2x + 3$
Simplify this expression to get: $4x + 3y + 2$
The perimeter of the shape is $4x + 3y + 2$.

Exercise 10.2

1. Simplify these expressions.
   (a) $n + n + n$ (b) $2y + 3y$ (c) $5g + g + 4g$ (d) $5y - y$
   (e) $6m - m + 3m$ (f) $-2x - 3x + x$ (g) $2a - 5a - 12a + a$

2. Which of these expressions cannot be simplified? Give a reason for each of your answers.
   (a) $v + v$ (b) $v + 4$ (c) $2v + v + 4$ (d) $v + w$

3. Simplify where possible.
   (a) $5x + 3x + y$ (b) $w + 3v - v$ (c) $2a + b - 3b$
   (d) $-a + b + 2a$ (e) $5 - 9k + 4k$ (f) $2a - a + 3$
   (g) $3a - 5a + 2b + b$ (h) $3x - x + 5y - 2y$ (i) $2a - b + 3b - a$
   (j) $-f + g - f - g$ (k) $2v - w - 3w - v$ (l) $7 - 2t - 9 - 3t$

4. Simplify.
   (a) $xy + yx$ (b) $3pq - qp$ (c) $5ab - 2ba - a + b$
   (d) $3x^2 - x^2$ (e) $5y^2 + 4y^2 - y$ (f) $a^2 + 5a^2 - 2a^2 + a$
   (g) $d^2 - 2g^2 - g^2 + d^2$ (h) $3t^2 + t + 2t^2 - 2t$ (i) $3m^2 - 4m^2 + 7m - m$

5 Write down an expression for the perimeter of each shape.
Give each answer in its simplest form.

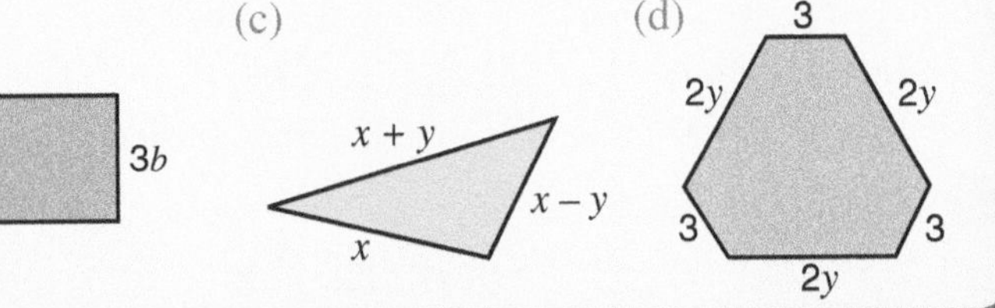

Multiplying terms

Multiply the numbers and then letters in alphabetical order.

$a \times b = ab$ $\quad 3x \times y = 3xy$ $\quad 3a \times 2b = 6ab$

$a \times a = a^2$ $\quad 5c \times 4c = 20c^2$ $\quad b \times b \times b = b^3$

$4a^2b \times 3abc^2 = 12a^3b^2c^2$ $\quad (4 \times 3 = 12,\ a^2 \times a = a^3,\ b \times b = b^2,\ c^2)$

Dividing terms

Divide the numbers and then letters in alphabetical order.

$8a \div 2 = 4a$ $\quad 6y \div 2y = 3$ $\quad 9x \div x = 9$

$\frac{4p^2q}{12pq^2} = \frac{p}{3q}$ $\quad \frac{4}{12} = \frac{1}{3},\ p^2 \div p = p,\ q \div q^2 = q^{-1} = \frac{1}{q}$

$\frac{4p^2q}{12pq^2}$ can be written as $4p^2q \div 12pq^2$

Multiplying and dividing algebraic expressions with powers

a^2 and b^3 are examples of **algebraic expressions with powers**.
The rules for multiplying and dividing powers of the same number can be used to simplify algebraic expressions involving powers.

When multiplying or dividing expressions that include both numbers and powers:
multiply or **divide** the **numbers**, **add** or **subtract** the **powers**.
Deal with the powers of different bases **separately**.

EXAMPLES

1 Simplify.

(a) $x^3 \times x^8 = x^{3+8} = x^{11}$

(b) $2x^2 \times 5x^7 = (2 \times 5) \times (x^2 \times x^7) = 10 \times x^{2+7} = 10x^9$

(c) $x^3y^2 \times xy^4 = (x^3 \times x) \times (y^2 \times y^4) = x^{3+1} \times y^{2+4} = x^4y^6$

(d) $(x^3y^2)^2 = x^{3 \times 2} \times y^{2 \times 2} = x^6y^4$

(e) $(3x^4)^2 = (3 \times 3) \times (x^{4 \times 2}) = 9x^8$

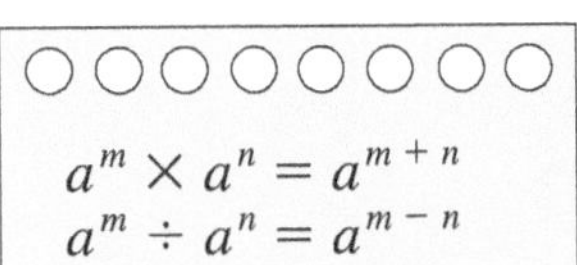

$a^m \times a^n = a^{m+n}$
$a^m \div a^n = a^{m-n}$
$(a^m)^n = a^{m \times n}$

2 Simplify.

(a) $x^8 \div x^5 = x^{8-5} = x^3$

(b) $6y^6 \div 2y^4 = (6 \div 2) \times (y^6 \div y^4) = 3 \times y^{6-4} = 3y^2$

Exercise 10.3

1 Write these expressions in a simpler form.

(a) $3 \times a$ $\quad$ (b) $2 \times 4 \times c$ $\quad$ (c) $3 \times 3 \times d$ $\quad$ (d) $f \times 8$

(e) $3 \times 2p$ $\quad$ (f) $3q \times 5$ $\quad$ (g) $r \times r$ $\quad$ (h) $2g \times g$

(i) $2g \times 3g$ $\quad$ (j) $3t \times 4t$ $\quad$ (k) $5u \times 3u$ $\quad$ (l) $3d \times 3d$

2 Simplify.

(a) $3 \times (-y)$ (b) $y \times (-5)$
(c) $(-2) \times (-y)$ (d) $3 \times (-2y)$
(e) $t \times (-t)$ (f) $2t \times (-t)$
(g) $(-2t) \times 5t$ (h) $(-2t) \times (-5t)$

Remember:
$2 \times (-x) = -2x$
$(-2) \times (-x) = 2x$

3 Simplify.

(a) $10a \div 2$ (b) $12x \div x$ (c) $20y \div y$ (d) $8y \div 4y$
(e) $18p \div p$ (f) $18p \div 6$ (g) $18p \div 6p$ (h) $18k \div 2k$
(i) $28g \div 7g$ (j) $10m \div 2m$ (k) $20t \div 5t$ (l) $27x \div 3x$

4 Simplify.

(a) $6y \div (-3)$ (b) $(-6y) \div 2$
(c) $(-5m) \div m$ (d) $(-5m) \div (-5)$
(e) $3a \div (-a)$ (f) $(-10d) \div 5d$
(g) $6g \div (-2g)$ (h) $(-3k) \div (-3k)$

Remember:
$2x \div (-2) = -x$
$(-2x) \div 2 = -x$
$(-2x) \div (-2) = x$

5 Simplify.

(a) $a \times b$ (b) $y \times y$ (c) $2 \times p \times q$ (d) $2 \times a \times a$
(e) $3 \times g \times 4 \times h$ (f) $a \times b \times c$ (g) $m \times m \times m$ (h) $2 \times d \times d \times d$
(i) $2x \times 3x \times x$ (j) $5m \times m \times 2n$ (k) $3a \times b \times c$ (l) $2p \times 3q \times 3r$

6 Simplify.

(a) $y^2 \times y$ (b) $t^3 \times t^2$ (c) $g^7 \times g^3$ (d) $m \times m^3 \times m^2$
(e) $3d^2 \times 2d^3$ (f) $4x^2 \times 2x^3$ (g) $2t \times t^2 \times 3t^2$ (h) $2r \times 3r^2 \times 4r^3$
(i) $mn \times m^2n$ (j) $a^2b \times ba^2$ (k) $3rs^3 \times 2r^2s$ (l) $2x^2y \times 5x^3y^2$

7 Simplify.

(a) $(t^2)^3$ (b) $(y^3)^2$ (c) $(g^3)^3$ (d) $(x^4)^2$
(e) $(3a)^2$ (f) $(2h)^3$ (g) $2 \times (m^3)^2$ (h) $(2m^3)^2$
(i) $3 \times (d^3)^2$ (j) $(3d^3)^2$ (k) $(3a^2)^3$ (l) $(2k^3)^3$
(m) $(xy)^2$ (n) $(mn^2)^3$ (o) $(2st)^3$ (p) $(3p^2q)^2$

8 Simplify.

(a) $y^3 \div y$ (b) $a^4 \div a^3$ (c) $x^5 \div x^5$ (d) $g \div g^2$
(e) $6b^3 \div b$ (f) $10m^3 \div 2m^2$ (g) $16t^3 \div 4t^2$ (h) $9h^2 \div 3h^3$
(i) $xy^2 \div xy$ (j) $m^3n \div mn^2$ (k) $2p^3q \div 3pq^2$ (l) $6r^2s^3 \div 2rs$

9 Simplify and express with positive indices.

(a) $t^5 \times t^{-3}$ (b) $x^3 \times x \times x^{-2}$ (c) $2ab^{-4} \times 6a^{-1}b$
(d) $5p^2q^{-1} \times 3p^{-1}q^3$ (e) $x^7 \div x^{-3}$ (f) $a^{-3} \div a^{-2}$
(g) $6x^3 \div 2x^{-1}$ (h) $8t^{-3}w^2 \div 4t^2w^{-2}$ (i) $(a^{-2})^3$
(j) $(y^{-4})^{-1}$ (k) $5(m^3)^{-3}$ (l) $(3n^{-2})^{-2}$

Remember:
$a^{-m} = \frac{1}{a^m}$

10 Simplify.

(a) $\frac{t^3}{t^2}$ (b) $\frac{g^2}{g^3}$ (c) $\frac{m^2 \times m}{m}$ (d) $\frac{y^2 \times y^3}{y^4}$

(e) $\frac{y \times y^3}{y^2}$ (f) $\frac{m^2 \times m^3}{m^6}$ (g) $\frac{2t^3 \times t}{t^2}$ (h) $\frac{6g^2 \times g}{2g^3}$

11 Simplify the following by cancelling.

(a) $\frac{2uv}{u}$ (b) $\frac{12p^2}{3p}$ (c) $\frac{a^2bc^3}{2ab}$ (d) $\frac{20mn}{5m^2n}$

(e) $\frac{0.8w^3}{0.05w^5}$ (f) $\frac{3ab}{2ac} \times \frac{4bc}{6b^2}$ (g) $\frac{8a^2b \times 5ab^2}{10a^3b^3}$ (h) $\frac{8ab^2c^3 \times 6a^2bc^{-5}}{12a^2b^3c}$

Brackets

Some expressions contain brackets. $2(a + b)$ means $2 \times (a + b)$.

You can multiply out brackets in an expression either by using a diagram or by expanding.

To multiply out $2(x + 3)$ using the **diagram method**:

$2(x + 3)$ means $2 \times (x + 3)$.
This can be shown using a rectangle.
The areas of the two parts are $2x$ and 6.
The total area is $2x + 6$.
$2(x + 3) = 2x + 6$

To multiply out $2(x + 3)$ by **expanding**:

$2(x + 3) = 2 \times x + 2 \times 3$
$= 2x + 6$

EXAMPLES

1 Multiply out the bracket $3(4a + 5)$.

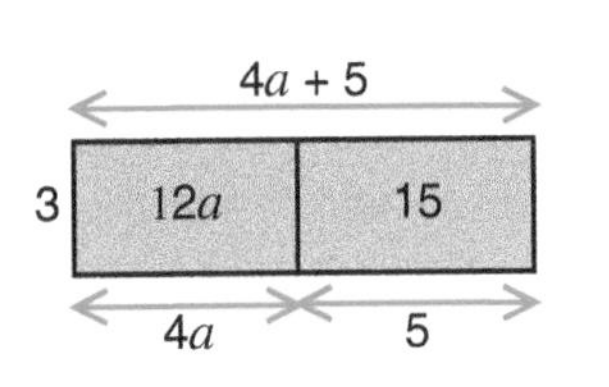

Diagram method

$3(4a + 5) = 12a + 15$

Expanding

$3(4a + 5) = 3 \times 4a + 3 \times 5$
$= 12a + 15$

2 Expand $x(x - 5)$.

$x \times x = x^2$ and $x \times -5 = -5x$

$x(x - 5) = x^2 - 5x$

Exercise 10.4

1 Use the diagrams to multiply out the brackets.

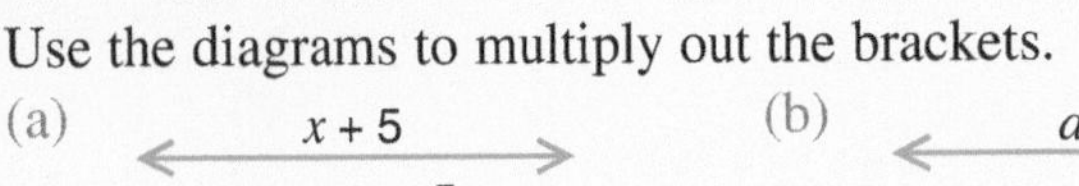

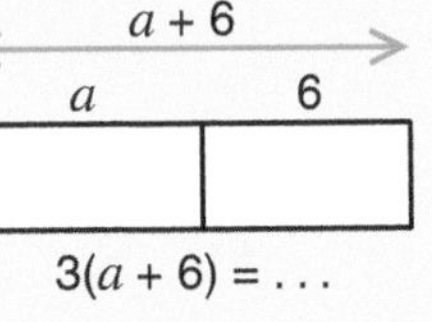

(a) $2(x + 5) = \ldots$ (b) $3(a + 6) = \ldots$ (c) $4(y + 3) = \ldots$

2 Draw your own diagrams to multiply out these brackets.

(a) $3(x + 2)$ (b) $2(y + 5)$ (c) $2(2x + 1)$ (d) $3(p + q)$

3 Use the diagrams to multiply out the brackets.

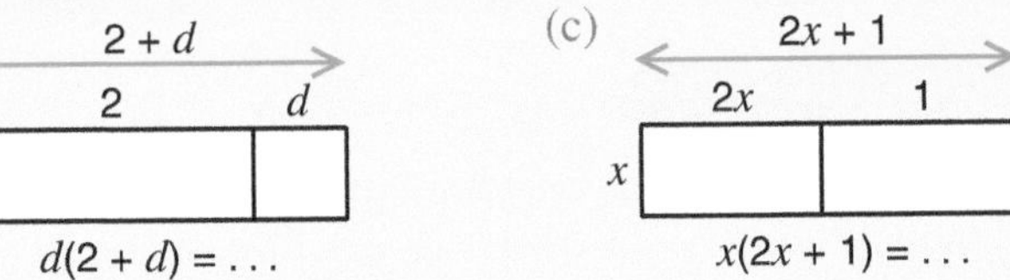

(a) $a(a + 1) = \ldots$ (b) $d(2 + d) = \ldots$ (c) $x(2x + 1) = \ldots$

4 Multiply out the brackets by expanding.

(a) $2(x + 4)$ (b) $3(t - 2)$ (c) $2(3a + b)$ (d) $5(3 - 2d)$
(e) $x(x + 3)$ (f) $t(t - 3)$ (g) $g(2g + 3)$ (h) $m(2 - 3m)$

5 Multiply out the brackets by expanding.

(a) $2p(p + 3)$ (b) $3d(2 - 3d)$ (c) $m(m - n)$ (d) $5x(x + 2y)$

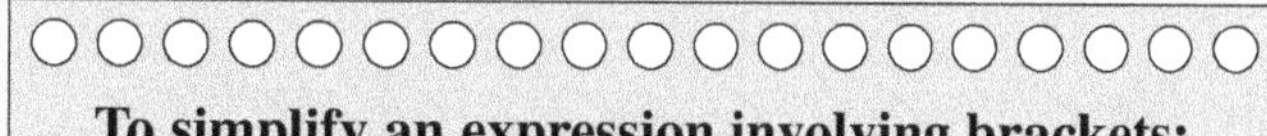

To simplify an expression involving brackets:

- Remove the brackets.
- Simplify by collecting like terms together.

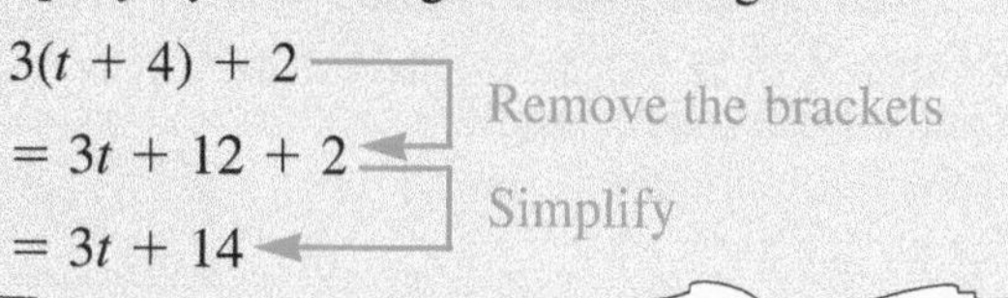

6 Multiply out the brackets and simplify.

(a) $2(x + 1) + 3$ (b) $3(a + 2) + 5$ (c) $6(w - 4) + 7$
(d) $4 + 2(p + 3)$ (e) $3 + 3(q - 1)$ (f) $1 + 3(2 - t)$
(g) $4(z + 2) + z$ (h) $5(t + 3) + 3t$ (i) $3(c - 2) - c$
(j) $2a + 3(a - 3)$ (k) $y + 2(y - 5)$ (l) $5x + 3(2 - x)$
(m) $4(2a + 5) + 3$ (n) $-2x + 4(3x - 3)$ (o) $3(p - 5) - p + 4$
(p) $3a + 2(a + b)$ (q) $3(x + y) - 2y$ (r) $2(p - q) - 3q$
(s) $2x + x(3 - x)$ (t) $a(a - 3) + a$ (u) $y(2 - y) + y^2$

7 Remove the brackets and simplify.

(a) $2(x + 1) + 3(x + 2)$ (b) $3(a + 1) + 2(a + 5)$ (c) $4(y + 2) + 5(y + 3)$
(d) $2(3a + 1) + 3(a + 1)$ (e) $3(2t + 5) + 5(4t + 3)$ (f) $3(z + 5) + 2(z - 1)$
(g) $7(q - 2) + 5(q + 6)$ (h) $5(x + 3) + 6(x - 3)$ (i) $8(2e - 1) + 4(e - 2)$
(j) $2(5d + 4) + 2(d - 1)$ (k) $m(m - 2) + m(2m - 1)$ (l) $a(3a + 2) + 2a(a - 3)$

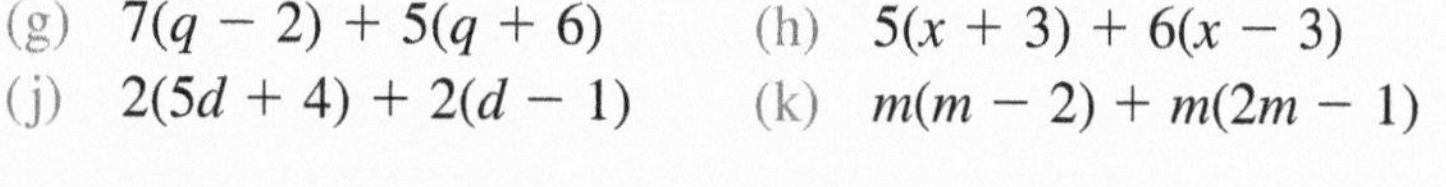

8 Multiply out the brackets and simplify.

(a) $-3(x + 2)$ (b) $-3(x - 2)$
(c) $-2(y - 5)$ (d) $-2(3 - x)$
(e) $-3(5 - y)$ (f) $-4(1 + a)$
(g) $5 - 2(a + 1)$ (h) $5d - 3(d - 2)$
(i) $4b - 2(3 + b)$ (j) $-3(2p + 3)$
(k) $5m - 2(3 + 2m)$ (l) $2(3d - 1) - d + 3$
(m) $-a(a - 2)$ (n) $2d - d(1 + d)$
(o) $x^2 - x(1 - x) + x$ (p) $-3g(2g + 3)$
(q) $t^2 - 2t(3 - 3t)$ (r) $2m - 2m(m - 3)$

Remember:

$(-2) \times (+3) = -6$
$(-2) \times (-3) = +6$

so, $-2(x + 3) = -2x - 6$
and $-2(x - 3) = -2x + 6$

9 Expand the brackets and simplify.

(a) $2(a + 1) - (3 - a)$ (b) $3(y - 1) - 2(y + 1)$ (c) $2(1 + m) - 3(1 - m)$
(d) $5(x - 2) - 2(x - 3)$ (e) $2(5 + d) - 3(3 + d)$ (f) $4(t - 1) - 3(t + 2)$
(g) $3(2m + 1) - 2(m - 3)$ (h) $4(2x - 3) - 3(3x + 2)$ (i) $2(4 + 3a) - 3(4a - 5)$

Factorising

Factorising is the opposite operation to removing brackets.
For example: to remove brackets $2(x + 5) = 2x + 10$

To factorise $3x + 6$ we can see that $3x$ and 6 have a **common factor** of 3 so,

$$3x + 6 = 3(x + 2)$$

A **common factor** can also be a **letter**.
Both y^2 and $5y$ can be divided by y.
To factorise $y^2 - 5y$ we take y as the common factor,
so, $y^2 - 5y = y(y - 5)$

Common factors:
The **factors** of a number are all the numbers that will divide exactly into the number.
Factors of 6 are 1, 2, 3 and 6.

A **common factor** is a factor which will divide into two or more terms.

In some instances the terms will have both a number and a letter as common factors and **both** must be taken out.

You can check that you have factorised an expression correctly by multiplying out the brackets.

EXAMPLES

1. Factorise $4x - 6$.

   Each term has a factor of 2.
   So, the common factor is 2.
   $4x - 6 = 2(2x - 3)$

2. Factorise $2x^2 - 6x$.

   Each term has a factor of 2.
   Each term has a factor of x.
   So, the common factor is $2x$.
   $2x^2 - 6x = 2x(x - 3)$

3. Factorise completely $2ab^2 - 6b^3$.

   $2ab^2 - 6b^3 = 2b^2(a - 3b)$

Exercise 10.5

1. Copy and complete.
   (a) $2x + 2y = 2(\ldots + \ldots)$ (b) $3a - 6b = 3(\ldots - \ldots)$ (c) $6m + 8n = 2(\ldots + \ldots)$
   (d) $x^2 - 2x = x(\quad)$ (e) $ab + a = a(\quad)$ (f) $2x - xy = x(\quad)$
   (g) $2ab - 4a = 2a(\quad)$ (h) $4x^2 + 6x = 2x(\quad)$ (i) $dg - dg^2 = dg(\quad)$

2. Factorise.
   (a) $2a + 2b$ (b) $5x - 5y$ (c) $3d + 6e$ (d) $4m - 2n$
   (e) $6a + 9b$ (f) $6a - 8b$ (g) $8t + 12$ (h) $5a - 10$
   (i) $4d - 2$ (j) $3 - 9g$ (k) $5 - 20m$ (l) $4k + 4$

3. Factorise.
   (a) $xy - xz$ (b) $fg + gh$ (c) $ab - 2b$ (d) $3q + pq$
   (e) $a + ab$ (f) $gh - g$ (g) $a^2 + 3a$ (h) $5t - t^2$
   (i) $d - d^2$ (j) $m^2 + m$ (k) $5r^2 - 3r$ (l) $3x^2 + 2x$

4. Factorise completely.
   (a) $3y + 6 - 9x$ (b) $t^3 - t^2 + t$ (c) $2d^2 + 4d$ (d) $3m - 6mn$
   (e) $2fg + 4g^2$ (f) $4pq - 8q$ (g) $6y - 15y^2$ (h) $6x^2 + 4xy$
   (i) $6n^2 - 2n$ (j) $4ab + 6b$ (k) $\frac{1}{2}a - \frac{1}{2}a^2$ (l) $20x + 4xy$
   (m) $a^3 + a^5 + a^2$ (n) $2\pi r + \pi r^2$ (o) $20a^2b + 12ab^2$ (p) $3pq - 9p^2q$

What you need to know

You should be able to:

- Write simple algebraic expressions.
- Simplify expressions by collecting like terms together.
  e.g. $2d + 3d = 5d$ and $3x + 2 - x + 4 = 2x + 6$
- Multiply simple expressions together. e.g. $2a \times a = 2a^2$ and $y \times y \times y = y^3$
- Recall and use these properties of powers:
  Powers of the same base are **added** when terms are **multiplied**.
  Powers of the same base are **subtracted** when terms are **divided**.
  Powers are **multiplied** when a power is raised to a power.

  $a^m \times a^n = a^{m+n}$
  $a^m \div a^n = a^{m-n}$
  $(a^m)^n = a^{m \times n}$
- Multiply out brackets. e.g. $2(x - 5) = 2x - 10$ and $x(x - 5) = x^2 - 5x$
- Factorise expressions. e.g. $3x - 6 = 3(x - 2)$ and $x^2 + 5x = x(x + 5)$

Review Exercise 10

1 A lollipop costs t pence. Write an expression for the cost of 6 lollipops.

2 Tom is x years old. Naomi is 3 years older than Tom.
How old is Naomi in terms of x?

3 (a) Aimee is n years old. Her brother Ben is two years younger.
Write down Ben's age in terms of n.
(b) Aimee's mother is three times as old as Aimee.
Write down her mother's age in terms of n.
(c) Aimee's father is four years older than her mother.
Write down her father's age in terms of n.
(d) Write an expression for the combined ages of all four members of the family.
Simplify your answer.
AQA

4 Cakes cost 25 pence each.

(a) Barry buys 6 cakes for his friends.
How much do the cakes cost altogether?
(b) Jane buys n cakes.
Write down an expression for the cost of n cakes.
(c) Yan buys y cakes. The total cost is x pence.
Which one of these equations shows this?
(i) $xy = 25$ (ii) $x = 25y$ (iii) $y = 25x$
AQA

5 Simplify. (a) $ab + 2ba$ (b) $a^2 - a + 3a$ (c) $3(x - 2) - x$ (d) $3x + 2(x + 1)$

6 In the triangle ABC, the side AB has length x units.
AC is twice the length of AB.
BC is three units shorter than AC.
(a) Write expressions in terms of x, for
(i) AC,
(ii) BC.
(b) Write an expression for the perimeter of the triangle, in terms of x.
Give your answer in its simplest form.
AQA

7 Multiply out and simplify where possible. (a) $y(y - 4)$ (b) $4(3y + 1) - 5y$

8 Multiply out (a) $3(x - 2y)$, (b) $2x(x + 3)$, (c) $x(x^2 - 3x)$.

9 Simplify. (a) $p \times p \times p$ (b) $2a \times 3b \times 4c$ (c) $x^3 \div x^3$
AQA

10 (a) Simplify. $4x + 3x + 7y - 2x + 3y$
(b) Expand and simplify. $x(2x - 3) + 4(x^2 + 1)$
AQA

11 Multiply out the brackets and simplify. (a) $x^2 - x(1 - x)$ (b) $4 - 3(x + 1)$

12 (a) Factorise. (i) $6x - 15$ (ii) $y^2 + 7y$
(b) Multiply out and simplify. $3(y + 2) - 2(y - 3)$

13 Simplify. (a) $t^3 \times t^5$ (b) $p^6 \div p^2$ (c) $\frac{a^3 \times a^2}{a}$
AQA

14 Simplify. (a) $3y^3 \times 2y^2$ (b) $8t^6 \div 4t^3$ (c) $(2a)^3$ (d) $a^2b^3 \times \frac{a^3}{b}$

15 (a) Factorise fully. (i) $6xy - 3y^2$ (ii) $4m^2 + 6m$
(b) Simplify. (i) $5y^3 \times 2y^2$ (ii) $6x^6 \div 2x^2$

16 Simplify. $\frac{4a^3c \times 3ab^3}{6a^2b}$
AQA

Solving Equations

Activity

Can you solve these puzzles?

- **Nueve** is a Spanish number.
  If you add 1 to **nueve** you get 10.
  What is **nueve**?
- What number must be put in each shape to make the statements correct?

  □ + 3 = 8 ⬡ × 3 = 30 2 × ○ − 3 = 7

These are all examples of **equations.**
Equations like these can be solved using a method known as **inspection**.
Instead of words or boxes, equations are usually written using letters for the unknown numbers.
Solving an equation means finding the numerical value of the letter which fits the equation.

EXAMPLES

Solve these equations by inspection.

1 $x - 2 = 6$

$x = 8$

Reason: $8 - 2 = 6$

2 $2y = 10$

$y = 5$

Reason: $2 \times 5 = 10$

Remember:
A letter or a symbol stands for an unknown number.
$2y$ means $2 \times y$.

Exercise 11.1

1 What number must be put in the box to make each of these statements true?

(a) □ + 4 = 7 (b) 15 − □ = 11 (c) 13 = □ + 4 (d) 11 = □ − 5

2 Solve these equations by inspection.

(a) $x + 2 = 6$ (b) $a + 7 = 10$ (c) $y - 4 = 4$
(d) $6 + t = 12$ (e) $h - 15 = 7$ (f) $d + 4 = 5$
(g) $z - 5 = 25$ (h) $p + 7 = 7$ (i) $c + 1 = 100$

3 What number must be put in the box to make each of these statements true?

(a) 3 × □ = 15 (b) □ × 4 = 20 (c) □ ÷ 2 = 9 (d) 7 = □ ÷ 3

4 Solve these equations by inspection.

(a) $3a = 12$ (b) $5e = 30$ (c) $8 = 2p$ (d) $4 = 8y$
(e) $\frac{d}{2} = 5$ (f) $\frac{t}{3} = 3$ (g) $\frac{m}{7} = 4$ (h) $\frac{x}{5} = 20$

5 What number must be put in the box to make each of these statements true?

(a) 2 × □ + 3 = 5 (b) □ × 3 + 5 = 17 (c) 3 + □ × 2 = 11
(d) 5 × □ − 1 = 9 (e) 4 × □ − 5 = 7 (f) □ × 3 − 6 = 9

Solving equations by working backwards

I think of a number and then subtract 3.
The answer is 5.
What is the number I thought of?

Imagine that x is the number I thought of.
The steps of the problem can be shown in a diagram.

x → subtract 3 → Answer 5

Now work backwards, doing the opposite calculation.

8 ← add 3 ← 5

The number I thought of is 8.

Remember:

Forwards	**Backwards**
add	subtract
subtract	add
multiply	divide
divide	multiply

EXAMPLE

I think of a number, multiply it by 3 and add 4.
The answer is 19.
What is my number?

x → multiply by 3 → add 4 → 19

5 ← divide by 3 ← 15 ← subtract 4 ← 19

The number I thought of is 5.

Exercise 11.2

Solve these equations by working backwards.

1. I think of a number and then multiply it by 2. The answer is 10. What is my number?
2. Jan thinks of a number and then subtracts 5. Her answer is 9. What is her number?
3. Lou thinks of a number. He multiplies it by 2 and then subtracts 5. The answer is 7.
   What is his number?
4. I think of a number, subtract 5 and then multiply by 2. The answer is 12.
   What is my number?
5. I think of a number, add 4 then multiply by 3. The answer is 24.
   What is my number?
6. Steve thinks of a number. He multiplies it by 5 and then adds 2. The answer is 17.
   What is his number?
7. I think of a number, multiply it by 3 and then subtract 5. The answer is 7.
   What is my number?
8. Solve this puzzle.

   Begin with x. Double it and then add 3. The result is equal to 17. What is the value of x?
9. Kathryn thinks of a number. She adds 3 and then doubles the result.
   (a) What number does Kathryn start with to get an answer of 10?
   (b) Kathryn starts with x. What is her answer in terms of x?
10. Sarah thinks of a number. She subtracts 2 and multiplies by 3.
   (a) What number does Sarah start with to get an answer of 21?
   (b) Sarah starts with x. What is her answer in terms of x?

The balance method

It is not always easy to solve equations by inspection.
To solve harder equations a better method has to be used.
Here is a method that works a bit like a balance.

These scales are balanced.

You can add the same amount to both sides and they still balance.

You can subtract the same amount from both sides and they still balance.

You can double (or halve) the amount on both sides and they still balance.

Equations work in the same way.
If you do the same to both sides of an equation, it is still true.

EXAMPLES

1 Solve $d - 13 = 5$.

$d - 13 = 5$
Add 13 to both sides.
$d = 18$

2 Solve $x + 7 = 16$.

$x + 7 = 16$
Subtract 7 from both sides.
$x = 9$

3 Solve $4a = 20$.

$4a = 20$
Divide both sides by 4.
$a = 5$

The aim is to find out what number the letter stands for, by ending up with **one letter** on one side of the equation and a **number** on the other side.

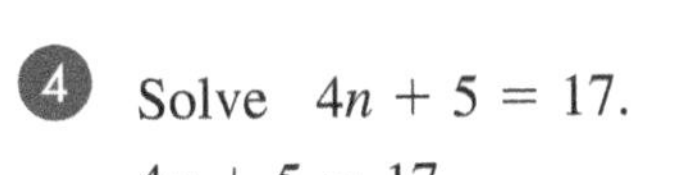

4 Solve $4n + 5 = 17$.

$4n + 5 = 17$
Subtract 5 from both sides.
$4n = 12$
Divide both sides by 4.
$n = 3$

Look at the examples carefully.
The steps taken to solve the equations are explained.
Notice that, doing the same to both sides means:
adding the **same number** to both sides.
subtracting the **same number** from both sides.
dividing both sides by the **same number**.
multiplying both sides by the **same number**.

Exercise 11.3

1. Use the balance method to solve these equations.

(a) $y + 4 = 7$	(b) $n - 7 = 9$	(c) $9 + x = 11$
(d) $y - 12 = 7$	(e) $14 + b = 21$	(f) $x - 9 = 20$
(g) $7 + m = 11$	(h) $k - 2 = 3$	(i) $5 + y = 12$

2. Use the balance method to solve these equations.

(a) $3c = 12$	(b) $5a = 20$	(c) $4f = 12$
(d) $\frac{d}{3} = 10$	(e) $\frac{e}{2} = 7$	(f) $\frac{m}{4} = 5$

3. Solve these equations.

(a) $2p + 1 = 9$	(b) $4t - 1 = 11$	(c) $3h - 7 = 14$
(d) $3 + 4b = 11$	(e) $5d - 8 = 42$	(f) $2x + 3 = 15$
(g) $2 + 3c = 17$	(h) $3n - 1 = 8$	(i) $4x + 3 = 11$

More equations

All the equations you have solved so far have had whole number solutions, but the solutions to equations can include negative numbers and fractions.

EXAMPLES

1. Solve $-4a = 20$.

$-4a = 20$
Divide both sides by -4.
$a = -5$

2. Solve $6m - 1 = 2$.

$6m - 1 = 2$
Add 1 to both sides.
$6m = 3$
Divide both sides by 6.
$m = \frac{1}{2}$

Exercise 11.4

1. Solve these equations.

(a) $4k = 2$	(b) $2a = -6$	(c) $-3d = 12$
(d) $-8n = 4$	(e) $t + 3 = -2$	(f) $n - 3 = -2$
(g) $2m + 1 = 4$	(h) $3x - 2 = 5$	(i) $2y + 5 = 4$

2. Solve these equations.

(a) $5x = -10$	(b) $2y + 7 = 1$	(c) $4t + 10 = 2$
(d) $5 - a = 7$	(e) $2 - d = 5$	(f) $3 - 2g = 9$
(g) $4t = 2$	(h) $2x = 15$	(i) $5d = 7$
(j) $4a - 5 = 1$	(k) $3 + 5g = 4$	(l) $2b - 5 = 4$

3. Solve.

(a) $x - 1 = -3$	(b) $3 + 2n = 2$	(c) $2 - a = 3$
(d) $4 - 3y = 13$	(e) $2x - 1 = -3$	(f) $3 - 5d = 18$
(g) $4x + 1 = -5$	(h) $-2 - 3x = 10$	(i) $2 - 4x = 8$
(j) $5 - 4n = -1$	(k) $\frac{1}{2}z + 2 = 7$	(l) $-2 = 5m + 13$
(m) $-3 = 17 - 5n$	(n) $-6p - 1 = 8$	(o) $12 + \frac{v}{3} = 15$

Equations with brackets

Equations can include brackets. Before using the balance method any brackets must be removed by multiplying out. This is called **expanding**.

Remember: $2(x + 3)$ means $2 \times (x + 3)$
$2(x + 3) = 2 \times x + 2 \times 3 = 2x + 6$

Once the brackets have been removed the balance method can be used as before.

EXAMPLES

1 Solve $3(x + 2) = 12$.

$3(x + 2) = 12$
Expand the brackets.
$3x + 6 = 12$
$3x = 6$
$x = 2$

2 Solve $5(3y - 7) + 15 = 25$.

$5(3y - 7) + 15 = 25$
Expand the brackets.
$15y - 35 + 15 = 25$
$15y = 45$
$y = 3$

Exercise 11.5

1 Solve.

(a) $2(x + 3) = 12$ (b) $4(a + 1) = 12$ (c) $5(t + 4) = 30$
(d) $3(p - 2) = 9$ (e) $6(c - 2) = 24$ (f) $2(x - 1) = 4$
(g) $6(d - 3) = 36$ (h) $7(2 + e) = 49$ (i) $5(f + 2) = 30$

2 Solve.

(a) $3(2w + 1) = 15$ (b) $2(4s + 5) = 34$ (c) $8(3t - 5) = 32$
(d) $5(3y + 2) = 25$ (e) $4(7 - 2x) = 4$ (f) $5(3y - 10) = 25$

3 Solve these equations. The solution will not always be a whole number.

(a) $3(p + 2) = 3$ (b) $2(3 - d) = 10$ (c) $2(1 - 3g) = 14$
(d) $2(x - 5) = 7$ (e) $5(y + 1) = 7$ (f) $2(1 + 3t) = 5$
(g) $2(2t - 1) = 5$ (h) $3(2a - 3) = 6$ (i) $5(m - 2) = 3$

4 Solve these equations.

(a) $2(x + 3) - 5 = 9$ (b) $3(a - 1) + 2 = 5$ (c) $5(3 - 2m) - 7 = 13$
(d) $5 + 2(y - 3) = 4$ (e) $2w + 3(1 + w) = -12$ (f) $e + 5(2e - 4) = 2$
(g) $3a - 7(1 - a) = 8$ (h) $2 - 3(t + 5) = 20$ (i) $5x - 2(3 - x) = 1$

Equations with letters on both sides

In some questions letters appear on both sides of the equation.

EXAMPLES

1 Solve $3x + 1 = x + 7$.

$3x + 1 = x + 7$
Subtract 1 from both sides.
$3x = x + 6$
Subtract x from both sides.
$2x = 6$
Divide both sides by 2.
$x = 3$

2 Solve $4(3 + 2x) = 5(x + 2)$.

$4(3 + 2x) = 5(x + 2)$
$12 + 8x = 5x + 10$
$8x = 5x - 2$
$3x = -2$
$x = -\frac{2}{3}$

Exercise 11.6

1 Solve the following equations.

(a) $3x = 20 - x$ (b) $5q = 12 - q$ (c) $2t = 15 - 3t$
(d) $5e - 9 = 2e$ (e) $3g - 8 = g$ (f) $y + 3 = 5 - y$
(g) $4x + 1 = x + 7$ (h) $7k + 3 = 3k + 7$ (i) $3a - 1 = a + 7$
(j) $3p - 1 = 2p + 5$ (k) $6m - 1 = m + 9$ (l) $3d - 5 = 5 + d$
(m) $2y + 1 = y + 6$ (n) $3 + 5u = 2u + 12$ (o) $4q + 3 = q + 3$

2 Solve.

(a) $3d = 32 - d$ (b) $3q = 12 - q$ (c) $3c + 2 = 10 - c$
(d) $4t + 2 = 17 - t$ (e) $4w + 1 = 13 - 2w$ (f) $2e - 3 = 12 - 3e$
(g) $2g + 5 = 25 - 2g$ (h) $2z - 6 = 14 - 3z$ (i) $5m + 2 = 20 + 2m$
(j) $5a - 4 = 3a + 6$ (k) $3 + 4x = 15 + x$ (l) $6y - 11 = y + 4$

3 Solve these equations. The solution will not always be a whole number.

(a) $3m + 8 = m$ (b) $2 - 4t = 12 + t$ (c) $5p - 3 = 3p - 7$
(d) $5x - 7 = 3x$ (e) $3 + 5a = a + 5$ (f) $2b + 7 = 11 - 3b$
(g) $4 - 4y = y$ (h) $7 + 3d = 10 - d$ (i) $f - 6 = 3f + 1$

4 Solve.

(a) $2(x + 3) - 5 = 9$ (b) $2(a - 1) + a = 3$ (c) $4(3 - 2m) - 7 = 2m$
(d) $3(a + 4) = 2 + a$ (e) $3(y - 5) = y - 4$ (f) $4(n + 2) = 2n + 5$
(g) $4d + 3 = 2(d - 3)$ (h) $7k + 2 = 5(k - 4)$ (i) $2(4t + 5) = t - 18$
(j) $5q - 2(q + 1) = 4$ (k) $x = 8 - 2(x + 3)$ (l) $4 - 3(a - 2) = a$

5 Solve.

(a) $2(3h - 4) = 3(h + 1) - 5$ (b) $2(3 - 2x) = 2(6 - x)$
(c) $2(3w - 1) + 4w = 28$ (d) $2(y + 4) + 3(2y - 5) = 5$
(e) $3(2v + 3) = 5 - 4(3 - v)$ (f) $5c - 2(4c - 9) = 5 + 5(2 - c)$
(g) $5(x + 2) + 2(2x - 1) = 7(x - 4)$ (h) $3(x - 4) = 5(2x - 3) - 2(3x - 5)$

Equations with fractions

You have already met some equations with fractions.
This section deals with harder equations with fractions.

For example: $\frac{3}{4}x = \frac{2}{5}$

With equations like this, it is easier to get rid of the fractions first.
To do this multiply both sides of the equation by the least common multiple of the denominators of the fractions.

Solve $\frac{3}{4}x = \frac{2}{5}$.

The multiples of 4 are: 4, 8, 12, 16, **20**, …
The multiples of 5 are: 5, 10, 15, **20**, …

The least common multiple of 4 and 5 is 20.
So, the first step is to multiply both sides of the equation by 20.

$$\frac{3}{4}x \times 20 = \frac{2}{5} \times 20$$

This is the same as: $x \times \frac{3}{4} \times 20 = \frac{2}{5} \times 20$

$$15x = 8$$

Divide both sides by 15.

$$x = \frac{8}{15}$$

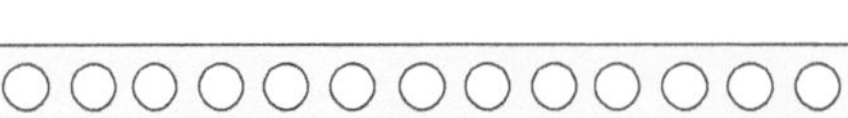

Remember:

$\frac{3}{4} \times 20$ is the same as $\frac{3}{4}$ of 20.

To find $\frac{3}{4}$ of 20:

$20 \div 4 = 5$ gives $\frac{1}{4}$ of 20.

$5 \times 3 = 15$ gives $\frac{3}{4}$ of 20.

So, $\frac{3}{4} \times 20 = 15$.

EXAMPLES

1 Solve $\frac{x}{2} + \frac{2x}{3} = 7$.

$$\frac{x}{2} + \frac{2x}{3} = 7$$

Multiply both sides by 6.

$$6 \times \frac{x}{2} + 6 \times \frac{2x}{3} = 6 \times 7$$
$$3x + 4x = 42$$
$$7x = 42$$
$$x = 6$$

2 Solve $\frac{x-1}{3} = \frac{x+1}{4}$.

$$\frac{x-1}{3} = \frac{x+1}{4}$$

Multiply both sides by 12.

$$4(x-1) = 3(x+1)$$
$$4x - 4 = 3x + 3$$
$$4x = 3x + 7$$
$$x = 7$$

Exercise 11.7

1 Solve these equations.

(a) $\frac{2}{3}x = 4$ (b) $\frac{2d}{5} = -4$ (c) $\frac{a}{8} = \frac{3}{4}$ (d) $\frac{2m}{8} = -\frac{3}{6}$ (e) $\frac{3t}{4} = \frac{1}{3}$

2 Solve.

(a) $\frac{h+1}{4} = 3$ (b) $\frac{2x-1}{3} = 5$ (c) $\frac{3a+4}{5} = -1$

(d) $\frac{7-d}{4} = \frac{5}{2}$ (e) $\frac{2a+1}{2} = \frac{3}{5}$ (f) $\frac{2-3h}{3} = -\frac{5}{6}$

(g) $\frac{x+2}{5} = \frac{3-x}{4}$ (h) $\frac{a-1}{2} = \frac{a+1}{3}$ (i) $\frac{2x-1}{6} = \frac{2-x}{3}$

3 Solve.

(a) $\frac{x}{2} + \frac{x}{4} = 1$ (b) $\frac{x}{2} - \frac{x}{3} = 2$ (c) $\frac{x}{4} + \frac{3x}{8} = -1$

(d) $\frac{2x}{3} - \frac{x}{6} = -2$ (e) $\frac{x+1}{2} + \frac{x-1}{3} = 1$ (f) $\frac{x+2}{3} - \frac{x+1}{4} = 2$

(g) $\frac{11-x}{4} = 2 - x$ (h) $\frac{x+2}{2} + \frac{x-1}{5} = \frac{1}{10}$ (i) $\frac{2x-3}{6} + \frac{x+2}{3} = \frac{5}{2}$

Using equations to solve problems

So far, you have been given equations and asked to solve them.
The next step is to **form an equation** first using the information given in a problem.
The equation can then be solved in the usual way.

EXAMPLE

The triangle has sides of length: x cm, $2x$ cm and 7 cm.

(a) Write an expression, in terms of x, for the perimeter of the triangle. Give your answer in its simplest form.

(b) The triangle has a perimeter of 19 cm. By forming an equation find the value of x.

x cm
7 cm
2x cm

(a) The perimeter of the triangle is: $x + 2x + 7$ cm
In its simplest form, the perimeter is: $(3x + 7)$ cm

(b) The perimeter of the triangle is 19 cm, so, $3x + 7 = 19$

$$3x = 12$$
$$x = 4$$

Exercise 11.8

1 The weights of three packages are shown.

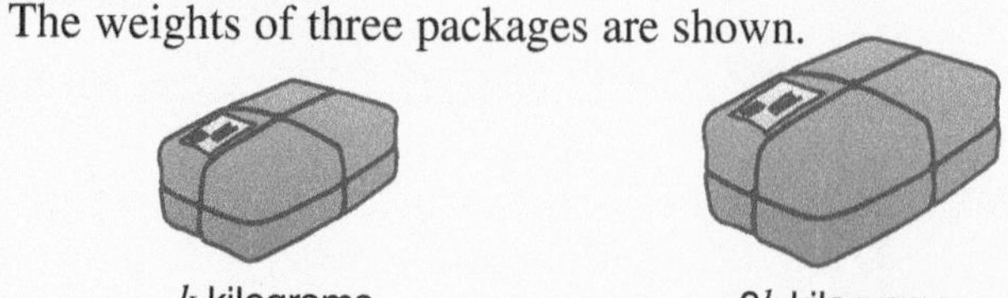

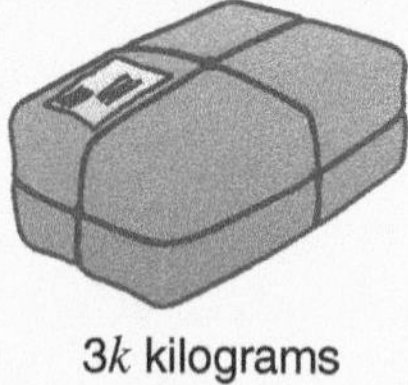

(a) Write an expression, in terms of k, for the total weight of the packages.
(b) The packages weigh 15 kilograms altogether.
By forming an equation find the weight of the lightest package.

2 A bag contains x yellow balls, $(2x + 1)$ red balls and $(3x + 2)$ blue balls.
(a) Write an expression, in terms of x, for the total number of balls in the bag.
(b) The bag contains 45 balls. How many yellow balls are in the bag?

3 Dominic is 7 years younger than Marcie.
(a) Dominic is n years old.
Write an expression, in terms of n, for Marcie's age.
(b) The sum of their ages is 43 years.
By forming an equation find the ages of Dominic and Marcie.

4 The diagram shows the lengths of three rods.

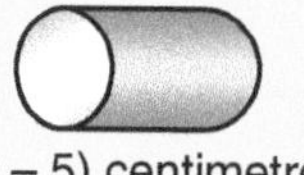

(a) Write an expression, in terms of y, for the total length of the rods.
(b) The total length of the rods is 30 centimetres.
What is the length of the longest rod?

5 Grace is given a weekly allowance of £p.
Aimee is given £4 a week **more** than Grace.
Lydia is given £3 a week **less** than Grace.
(a) Write an expression, in terms of p, for the amount given to
(i) Aimee, (ii) Lydia, (iii) all three girls.
(b) The three girls are given a total of £25 a week altogether.
By forming an equation find the weekly allowance given to each girl.

6 The cost of a pencil is x pence. The cost of a pen is 10 pence more than a pencil.
(a) Write an expression, in terms of x, for the cost of a pen.
(b) Write an expression, in terms of x, for the total cost of a pencil and two pens.
(c) The total cost of a pencil and two pens is 65 pence.
Form an equation in x and solve it to find the cost of a pencil.

7 A chocolate biscuit costs x pence.
A cream biscuit costs 4 pence less than a chocolate biscuit.
Jamie pays 77 pence for 3 chocolate biscuits and 2 cream biscuits.
By forming an equation find the cost of a cream biscuit.

8 (a) Write down a simplified expression, in terms of x, for the perimeter of the triangle.

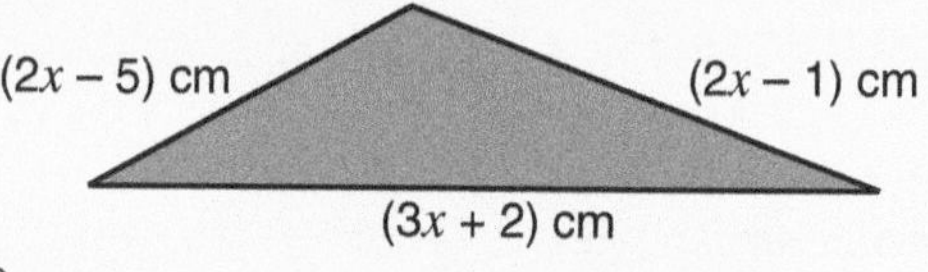

(b) The perimeter is 59 cm.
Write down an equation and solve it to find the value of x.
(c) Use your answer to find the length of each side of the triangle.

9 The length of a rectangle is x cm. The width of the rectangle is 4 cm less than the length.
(a) Write an expression, in terms of x, for the width of the rectangle.
(b) Write an expression for the perimeter of the rectangle in terms of x.
(c) The perimeter of the rectangle is 20 cm.
By forming an equation find the value of x.

10 Geoffrey knows that the sum of the angles of this shape adds up to 540°.
(a) Write down an equation in x.
(b) Use your equation to find the size of the largest angle.

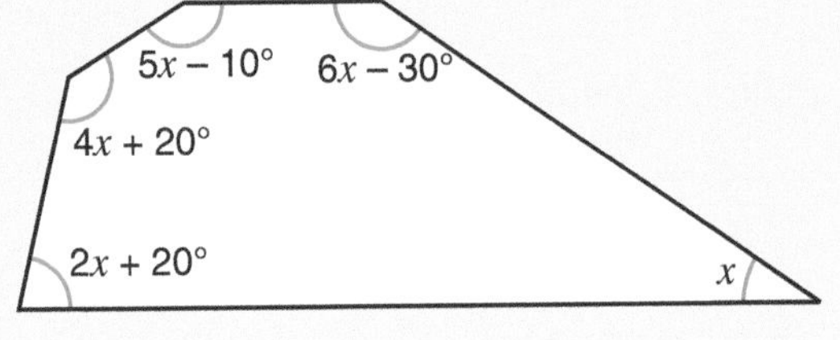

11 A drink costs x pence.
A cake costs 7 pence more than a drink.
The total cost of two drinks and a cake is 97 pence.
Form an equation in x and solve it to find the cost of a cake.

12 The areas of these rectangles are equal.

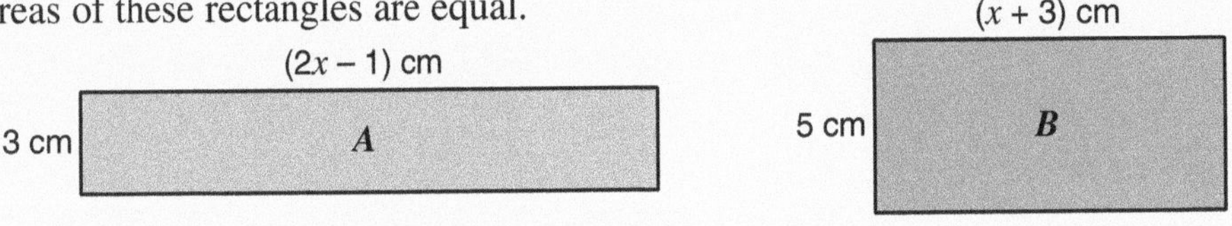

By forming and solving an equation in x, find the area of A.

13

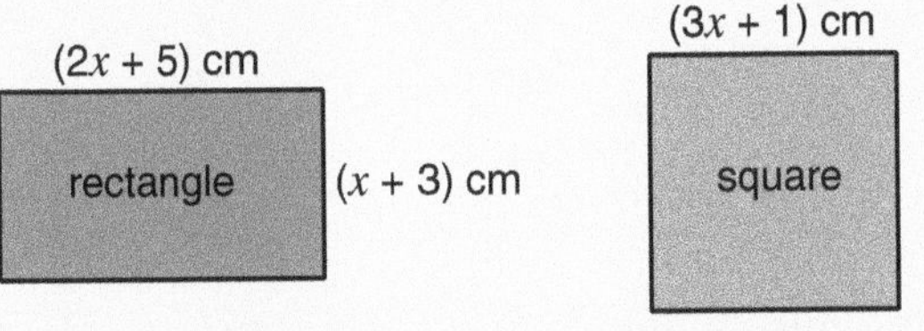

(a) Write down an expression, in terms of x, for the perimeter of the rectangle.
(b) The perimeter of the rectangle is equal to the perimeter of the square.
Form an equation and find the value of x.
(c) What is the perimeter of the rectangle, in centimetres?

14 John has x CDs in his collection.
Sarah has five times as many CDs as John.
John and Sarah each collect another 12 CDs. Now Sarah has twice as many CDs as John.
(a) Write down an expression, in terms of x, for the number of CDs Sarah has now.
(b) Form an equation and solve it to find how many CDs Sarah has now.

What you need to know

- The solution of an equation is the value of the unknown letter that fits the equation.

You should be able to:

- Solve simple equations by inspection. e.g. $x + 2 = 5$, $x - 3 = 7$, $2x = 10$, $\frac{x}{4} = 12$
- Solve simple equations by inspection by working backwards.
- Use the balance method to solve equations which are difficult to solve by inspection.
- Solve equations with brackets. e.g. $4(3 + 2x) = 36$
- Solve equations with unknowns on both sides of the equals sign. e.g. $3x + 1 = x + 7$
- Solve equations with fractions. e.g. $\frac{1}{4}x + 1 = 7$, $\frac{x+3}{4} = 2$, $\frac{x-1}{3} + \frac{x+1}{2} = \frac{1}{6}$
- Write, or form, equations using the information given in a problem.
- Use equations to solve problems.

Review Exercise 11

1 Jacob uses this rule.

"Start with a number, divide it by 2 and then add 3. Write down the result."

(a) What is the result when Jacob starts with 8?

(b) What number did Jacob start with when the result is 5? AQA

2 Solve. (a) $\frac{x}{5} = 4$ (b) $3x + 5 = 17$ (c) $5n - 3 = 7$

3 Solve these equations.

(a) $y + 3 = 5$ (b) $2t + 8 = 2$ (c) $4g = 2$ (d) $5x - 1 = 2$

4 Solve these equations. (a) $5x - 12 = 23$ (b) $3x + 4 = 5x + 8$ AQA

5 Solve these equations.

(a) $7x = 56$ (b) $5x + 7 = 15$ (c) $8x - 1 = 4x + 19$ AQA

6 Solve the following equations. (a) $4x + 6 = 11$ (b) $2(5 + 2x) = 16$ AQA

7 Solve the equations.

(a) $3x - 14 = 4$ (b) $5x + 7 = x + 9$ (c) $\frac{x - 7}{8} = 2$ AQA

8 Solve the equations. (a) $7x - 13 = 5(x - 3)$ (b) $\frac{10 - x}{4} = x - 1$ AQA

9 Solve these equations.

(a) $3 - 4q = 11$ (b) $4(2t - 3) + 4t = 6$ (c) $\frac{2x + 9}{5} = 6$

10 Hassan is twice as old as Ali. Their ages add up to 39 years.

How old is: (a) Ali, (b) Hassan? AQA

11 The diagram shows two cans of oil.

The cans hold a total of 3 litres of oil.

By forming an equation find the amount of oil in the larger can.

12 A drink costs x pence.

A packet of crisps costs 15 pence **less** than a drink.

(a) Write an expression, in terms of x, for the cost of a packet of crisps.

(b) A drink and two packets of crisps cost 96 pence.
By forming an equation find the cost of a drink.

13 The length of the sides of a triangle are $(x + 1)$ cm, $(x + 3)$ cm and $(x - 2)$ cm, as shown.

(a) Write an expression, in terms of x,
for the perimeter of the triangle.
Give your answer in its simplest form.

The perimeter is 23 cm.

(b) Write down an equation in x
and use it to find the value of x. AQA

14 Solve this equation. $3(3x + 2) - 2(x - 3) = x + 3$ AQA

15 Solve the equations. (a) $7x - 13 = 5(x - 3)$ (b) $\frac{10 - x}{4} = x - 1$ AQA

16 Solve the equation. $\frac{(x - 2)}{3} - \frac{(x - 3)}{2} = 1$

17 Solve the equation. $\frac{3x + 1}{2} - \frac{2x + 5}{3} = 1$ AQA

CHAPTER 12 Formulae

Most people at some time make use of **formulae** to carry out routine calculations.
A **formula** represents a rule written using numbers, letters and mathematical signs.
When using a formula you will need to **substitute** your own values for the letters in order to carry out your calculation.

Substitution

EXAMPLES

1 Find the value of $a - 3$, when $a = 4$.

$a - 3$
$= 4 - 3$
$= 1$

2 Find the value of $3m$, when $m = -4$.

$3m$
$= 3 \times -4$
$= -12$

3 Find the value of $pq - 3$, when $p = 5$ and $q = -2$.

$pq - 3$
$= 5 \times (-2) - 3$
$= -10 - 3$
$= -13$

Exercise 12.1

Do not use a calculator.

1. $t = 5$. Find the value of (a) $3 - t$ (b) $2t$ (c) $3t - 4$ (d) $t \times t$
2. $m = -3$. Find the value of (a) $m + 2$ (b) $m - 1$ (c) $4m$ (d) $2m + 9$
3. $a = 6$ and $b = 3$. Find the value of
   (a) $3a + 2b$ (b) $b - a$ (c) $\frac{a}{b}$ (d) $a \times b$ (e) $2(a + b)$
4. $x = 15$ and $y = 6$. Find the value of
   (a) $x + 2y$ (b) $x - 3y$ (c) $\frac{x}{y}$ (d) xy (e) $6(x - y)$
5. $p = -10$ and $q = 5$. Find the value of
   (a) $p + q$ (b) $p - 2q$ (c) $\frac{p}{q}$ (d) $2pq$ (e) $\frac{p - q}{p + q}$
6. $x = 15$ and $y = -6$. Find the value of
   (a) $x + 2y$ (b) $y - x$ (c) $\frac{x}{y}$ (d) $\frac{xy}{10}$ (e) $6(x + y)$
7. $a = -6$ and $b = -3$. Find the value of
   (a) $3a + 2b$ (b) $b - a$ (c) $\frac{a}{b}$ (d) $a^2 + b$ (e) $\frac{ab}{2a - b}$
8. Find the value of $pq + r$ when $p = 3$, $q = -5$ and $r = -2$.

Writing expressions and formulae

A lollipop costs 15 pence.
How much will n lollipops cost?
Write a formula for the cost, C, in pence, of n lollipops.

Each lollipop costs 15 pence.

So, n lollipops cost $15 \times n$ pence $= 15n$ pence.

> $15n$ is an **algebraic expression.**

If the cost of n lollipops is C pence, then $C = 15n$.

> $C = 15n$ is a **formula.**

Formulae can be used in lots of situations.

The grid shows the numbers from 1 to 50.
An **L** shape has been drawn on the grid.
It is called $\mathbf{L}_{14}$ because the lowest number is 14.

1	2	3	4	5	6	7	8	9	10
11	12	13	14	15	16	17	18	19	20
21	22	23	24	25	26	27	28	29	30
31	32	33	34	35	36	37	38	39	40
41	42	43	44	45	46	47	48	49	50

What is the sum of the numbers in $\mathbf{L}_{14}$*?*

The **L** shape can be moved to different parts of the grid.
We can find the sum of the numbers for each shape.

A formula for the sum of the numbers, $\mathbf{S}_n$,
can be written in terms of n for shape $\mathbf{L}_n$.

$\mathbf{S}_n = n + (n + 10) + (n + 20) + (n + 21)$

$\mathbf{S}_n = 4n + 51$

n	
$n + 10$	
$n + 20$	$n + 21$

An **expression** is just an answer using letters and numbers.

A **formula** is an algebraic rule.
It always has an equals sign.

EXAMPLES

1 A hedge is l metres long. A fence is 50 metres longer than the hedge.
Write an **expression**, in terms of l, for the length of the fence.

The fence is $(l + 50)$ metres long.

2 Boxes of matches each contain 48 matches.
Write down a **formula** for the number of matches, m, in n boxes.

$m = 48 \times n$ This could be written as $m = 48n$.

Exercise 12.2

1 I am a years old.
(a) How old will I be in 1 years time?
(b) How old was I four years ago?
(c) How old will I be in n years time?

2 Egg boxes hold 12 eggs each.
How many eggs are there in e boxes?

3 A child is making a tower with toy bricks.
He has b bricks in his tower.
Write an expression for the number of bricks in the tower after he takes 3 bricks from the top.

4 Paul is h cm tall.
Sue is 12 cm taller than Paul.
Write down an expression for Sue's height in terms of h.

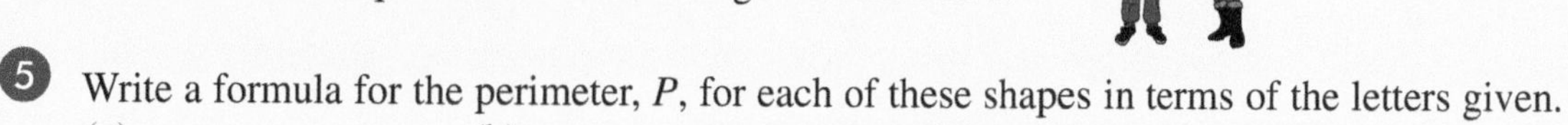

5 Write a formula for the perimeter, P, for each of these shapes in terms of the letters given.

(a) (b)

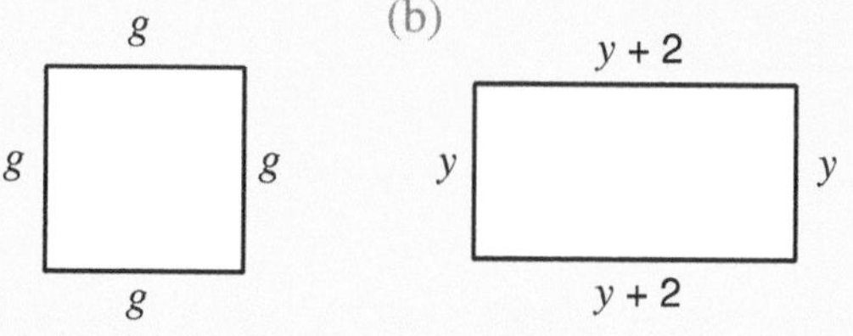

(c) (d)

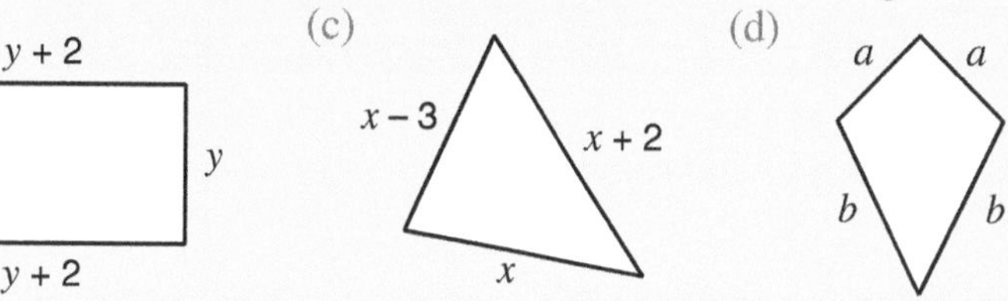

6 A caravan costs £25 per day to hire.
Write a formula for the cost, C, in £s, to hire the caravan for d days.

7 A packet of biscuits costs y pence.
Write down a formula for the cost, P pence, of another packet which costs
(a) two pence less than the first packet,
(b) twice the cost of the first packet.

8 The cost of hiring a ladder is given by:

£12 per day, plus a delivery charge of £8

(a) Sam hired a ladder for 6 days. How much did he pay?
(b) Fred hired a ladder for x days. Write down a formula for the total cost, £C, in terms of x.

9 The grid shows the numbers from 1 to 50.
A **T** shape has been drawn on the grid.
It is called $\mathbf{T}_{23}$ because the lowest number is 23.

1	2	3	4	5	6	7	8	9	10
11	12	13	14	15	16	17	18	19	20
21	22	23	24	25	26	27	28	29	30
31	32	33	34	35	36	37	38	39	40
41	42	43	44	45	46	47	48	49	50

Calculate the sum of the numbers in:
(a) $\mathbf{T}_{16}$ (b) $\mathbf{T}_{28}$ (c) $\mathbf{T}_{2}$

(d) The diagram on the right shows $\mathbf{T}_n$.
Copy and complete the **T** shape in terms of n.

n		

(e) Write a formula for the sum of the numbers, $\mathbf{S}_n$, in terms of n, for shape $\mathbf{T}_n$.
Write your answer in its simplest form.

Using formulae

The formula for the perimeter of a rectangle is $P = 2L + 2W$.
By **substituting** values for the length, L, and the width, W, you can calculate the value of P.

For example. To find the perimeter of a rectangle 5 cm in length and 3 cm in width,
substitute $L = 5$ and $W = 3$ into $P = 2L + 2W$.

$P = 2 \times 5 + 2 \times 3$
$= 10 + 6$
$= 16$

The perimeter of the rectangle is 16 cm.

EXAMPLES

1 Here is a formula for the area of a rectangle.

Area = length × width

Use the formula to find the area of a rectangle 8 cm in length and 3 cm in width.
Area = length × width
$= 8 \times 3$
$= 24\,\text{cm}^2$

2 $G = 4t - 1$.
Find the value of G when $t = \frac{1}{2}$.

$G = 4t - 1$
$= 4 \times \frac{1}{2} - 1$
$= 2 - 1$
$= 1$

3 $H = 3(4x - y)$
Find the value of H when $x = 5$ and $y = 7$.

$H = 3(4x - y)$
$= 3(4 \times 5 - 7)$
$= 3(20 - 7)$
$= 3(13)$
$= 39$

4 $W = x^2 + 2$
Find the value of W when $x = 3$.

$W = x^2 + 2$
$= 3 \times 3 + 2$
$= 9 + 2$
$= 11$

Exercise 12.3

Do not use a calculator for questions 1 to 10.

1. The number of points scored by a soccer team can be worked out using this formula.

   Points scored = 3 × games won + games drawn

   A team has won 5 games and drawn 2 games. How many points have they scored?

2. Here is a formula for the perimeter of a rectangle.

   Perimeter = 2 × (length + width)

   A rectangle is 9 cm in length and 4 cm in width.
   Use the formula to work out the perimeter of this rectangle.

3. $T = 5a - 3$. Find the value of T when $a = 20$.

4. $M = 4n + 1$. (a) Work out the value of M when $n = -2$.
   (b) Work out the value of n when $M = 19$.

5. $H = 3g - 5$. (a) Find the value of H when $g = 2.5$.
   (b) Find the value of g when $H = 13$.

6. $F = 5(v + 6)$. What is the value of F when $v = 9$?

7. $V = 2(7 + 2x)$. What is the value of V when $x = -3$?

8. $P = 3(5 - 2d)$. What is the value of P when $d = 0.5$?

9. $C = 8(p + q)$. What is the value of C when $p = -2$ and $q = -3$?

10. $S = ax + 4$. What is the value of S when $a = 5$ and $x = 0.4$?

11. $T = a(x + 4)$. What is the value of T when $a = -3$ and $x = \frac{1}{2}$?

12. $K = ab + c$. Work out the value of K when $a = 5$, $b = 3$ and $c = -2$.

13. $L = xy - z$. Work out the value of L when $x = -4$, $y = -2$ and $z = 3$.

14. The number of matches, M, needed to make a pattern of P pentagons is given by the formula: $M = 4P + 1$.
    Find the number of matches needed to make 8 pentagons.

15. The distance, d metres, travelled by a lawn mower in t minutes is given as: $d = 24t$.
    Find the distance travelled by the lawn mower in 4 minutes.

16. Convert 30° Centigrade to Fahrenheit using the formula: $F = C \times 1.8 + 32$

17. $T = 45W + 30$ is used to calculate the time in minutes needed to cook a joint of beef weighing W kilograms.
    How many minutes are needed to cook a joint of beef weighing 2.4 kg?

18. Convert 77 degrees Fahrenheit to Centigrade using the formula: $C = (F - 32) \div 1.8$

19. The voltage, V volts, in a circuit with resistance, R ohms, and current, I amps, is given by the formula: $V = IR$.
    Find the voltage in a circuit when $I = 12$ and $R = 20$.

20. A simple formula for the motion of a car is $F = ma + R$.
    Find F when $m = 500$, $a = 0.2$ and $R = 4000$.

21. The cost, £C, of n units of gas is calculated using the formula $C = 0.08n + 8.5$.
    Calculate the cost of 458 units of gas.

22. The formula $v = u + at$ gives the speed v of a particle, t seconds after it starts with speed u.
    Calculate v when $u = 7.8$, $a = -10$ and $t = \frac{3}{4}$.

Substitution into formulae with powers and roots

Exercise 12.4

Do not use a calculator for questions 1 to 12.

1. $S = a^2 - 5$. Find the value of S when $a = -3$.
2. $R = p^2 + 2p$. Find the value of R when (a) $p = 3$, (b) $p = -3$.
3. $K = m^2 - 5m$. Work out the value of K when $m = -4$.
4. $S = 2a^2$. Find the value of S when (a) $a = 3$, (b) $a = -3$.
5. $S = (2a)^2$. Find the value of S when (a) $a = 3$, (b) $a = -3$.
6. $T = 3a^2 - 9$. Find the value of T when $a = -4$.
7. $S = \frac{1}{2}p^2$. Find the value of S when (a) $p = 8$, (b) $p = -8$.
8. $S = \left(\frac{1}{2}p\right)^2$. Find the value of S when (a) $p = 8$, (b) $p = -8$.
9. $A = x^3$. Find the value of A when (a) $x = 2$, (b) $x = -3$.
10. $T = \sqrt{\frac{a}{b}}$ Work out the value of T when $a = 9$ and $b = 16$.
11. $S = \sqrt[3]{pq}$ Work out the value of S when
    (a) $p = 2$ and $q = 32$, (b) $p = 2\frac{1}{4}$ and $q = 1\frac{1}{2}$, (c) $p = 54$ and $q = -0.5$.
12. $L = \sqrt{m^2 + n^2}$ Work out the value of L when
    (a) $m = 6$ and $n = 8$, (b) $m = 0.3$ and $n = 0.4$.
13. The formula $F = \frac{mv^2}{r}$ describes the motion of a cyclist rounding a corner.
    Find F when $m = 80$, $v = 6$ and $r = 20$.
14. Use the formula $R = \left(\frac{t}{s}\right)^2$ to calculate the value of R when $t = -0.6$ and $s = \frac{2}{5}$.
15. Use the formula $v = \sqrt{u^2 + 2as}$ to calculate the value of v when
    (a) $u = 2.4$, $a = 3.2$, $s = 5.25$, (b) $u = 9.1$, $a = -4.7$, $s = 3.04$.
    Give your answers correct to one decimal place.

Writing and using formulae

EXAMPLE

(a) Nick has a birthday party for 10 people.
How much does it cost?

(b) Tony has a birthday party for x people.
Write a formula for the cost £T, in terms of x.

(c) Jean pays £140 for her birthday party.
How many people went to the party?

(a) Nick's party costs: £20 + 10 × £8
= £100

(b) Cost for x people in £ $= x \times 8 = 8x$
Total cost in £ $= 20 + 8x$
Total cost is £T
So, formula is $T = 20 + 8x$

(c) Using the formula $T = 20 + 8x$.
Jean's party costs £140, so, $T = 140$.
$140 = 20 + 8x$
$8x = 120$
$x = 15$

15 people went to Jean's party.

Exercise 12.5

1 The cost of a taxi journey is:

£3 plus £2 for each kilometre travelled

(a) Alex travels 5 km by taxi.
How much does it cost?

(b) A taxi journey of k kilometres costs £C.
Write a formula for the cost, C, in terms of k.

(c) Adrian paid £7 for a taxi journey.
Use your formula to find the number of kilometres he travelled.

2 A rule to find the cooking time, C minutes, of a chicken which weighs k kilograms, is:

multiply the weight of the chicken by 40 and then add 20

(a) Find the cooking time for a chicken which weighs 3 kg.

(b) Write a formula for C in terms of k.

(c) Use your formula to find the weight of a chicken which has a cooking time of 100 minutes.

3 An approximate rule for changing temperatures in degrees Celsius, C, to temperatures in degrees Fahrenheit, F, is given by the rule:

double C and add on 30

(a) Find the value of F when $C = 6$.

(b) Write down a formula for F in terms of C.

(c) Use your formula to find the value of C when $F = 58$.

4 A teacher uses this rule to work out the number of exercise books he needs for Year 11 students.

3 books per student, plus 50 extra books

(a) This year there are 120 students in Year 11.
How many books are needed?

(b) Using b for the number of books and n for the number of students, write down the teacher's rule for b in terms of n.

(c) For the next Year 11, he will need 470 books.
How many students will be in Year 11 next year?

Rearranging formulae

Sometimes it is easier to use a formula if you **rearrange** it first.

EXAMPLES

1 $k = \frac{8m}{5}$

Rearrange the formula to give m in terms of k.

$k = \frac{8m}{5}$

Multiply both sides by 5.

$5k = 8m$

Divide both sides by 8.

$\frac{5k}{8} = m$

We say we have **rearranged the formula** $k = \frac{8m}{5}$ to make m the **subject** of the formula.

2 $A = 3r^2$

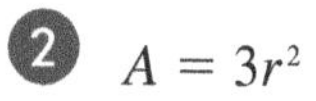

Make r the subject of the formula.

$A = 3r^2$

Divide both sides by 3.

$\frac{A}{3} = r^2$

Take the square root of both sides.

$\pm\sqrt{\frac{A}{3}} = r$

So, $r = \pm\sqrt{\frac{A}{3}}$

Exercise 12.6

1. Make m the subject of these formulae.
   (a) $a = m + 5$ (b) $a = x + m$ (c) $a = m - 2$ (d) $a = m - b$
2. Make x the subject of these formulae.
   (a) $y = 4x$ (b) $y = ax$ (c) $y = \frac{x}{2}$ (d) $y = \frac{x}{a}$ (e) $y = \frac{3x}{5}$
3. Make p the subject of these formulae.
   (a) $y = 2p + 6$ (b) $t = 5p + q$ (c) $m = 3p - 2$ (d) $r = 4p - q$
4. The cost, £C, of hiring a car for n days is given by $C = 35 + 24n$.
   Make n the subject of the formula.
5. $V = IR$. Rearrange the formula to give R in terms of V and I.
6. Make c the subject of these formulae.
   (a) $y = c^2$ (b) $y = \sqrt{c}$ (c) $y = dc^2$ (d) $y = \frac{\sqrt{c}}{3}$
   (e) $y = c^2 + x$ (f) $y = x + \sqrt{c}$ (g) $y = x + \frac{c^2}{d}$ (h) $y = \frac{\sqrt{c}}{a} - x$
7. Make a the subject of these formulae.
   (a) $b = a + c^2$ (b) $a^2 = b$ (c) $p = ma + d$ (d) $ma^2 = F$
8. Make x the subject of these formulae. (a) $3(x - a) = x$ (b) $2(a + x) = 3(b - x)$
9. Make a the subject of each of these formulae.
   (a) $3a - x = a + 2x$ (b) $a - b = ax$ (c) $a - 2 = ax + b$
10. $R = \left(\frac{t}{s}\right)^2$. Rearrange the formula to give t in terms of R and s.
11. $\frac{p + q}{3} = p - q$. Rearrange the formula to give q in terms of p.
12. Make a the subject of each of these formulae.
    (a) $a + 2 = x(3 + a)$ (b) $y = \frac{a - 3}{5 - a}$ (c) $x(a - 1) = b(a + 2)$
    (d) $y(a - 1) = 3(2 - a)$ (e) $\sqrt{\frac{a + x}{a - x}} = 2$ (f) $\sqrt{\frac{a}{x - a}} = 2x$

Solve problems by rearranging formulae

EXAMPLE

A cuboid has length 8 cm and breadth 5 cm.
The volume of the cuboid is 140 cm³.
Calculate the height of the cuboid.

h cm
5 cm
8 cm

The formula for the volume of a cuboid is $V = lbh$,

so, $h = \frac{V}{lb}$ (by dividing both sides of $V = lbh$ by lb)

Substitute $V = 140$, $l = 8$ and $b = 5$ in $h = \frac{V}{lb}$.

$h = \frac{140}{8 \times 5} = 3.5$

The height of the cuboid is 3.5 cm.

The values to be substituted may include whole numbers, negative numbers, decimals or fractions.

Exercise 12.7

Do not use a calculator for questions 1 to 5.

1 The perimeter of a square is $P = 4d$.
(a) Rearrange the formula to give d in terms of P.
(b) Find d when $P = 2.8\,\text{cm}$.

2 The area of a rectangle is $A = lb$.
(a) Rearrange the formula to give l in terms of A and b.
(b) Find l when $A = 27\,\text{cm}^2$ and $b = 4.5\,\text{cm}$.

3 The speed of a car is $S = \frac{D}{T}$.
(a) (i) Change the subject to D.
(ii) Find D when $S = 48\,\text{km/h}$ and $T = 2$ hours.
(b) (i) Change the subject to T.
(ii) Find T when $S = 36\,\text{km/h}$ and $D = 90\,\text{km}$.

4 The perimeter of a rectangle is $P = 2(l + b)$.
(a) Change the subject to b.
(b) Find b when $P = 18\,\text{cm}$ and $l = 4.8\,\text{cm}$.

5 $y = mx + c$
(a) Rearrange the formula to give x in terms of y, m and c.
(b) Calculate x when $y = 0.6$, $m = -0.4$ and $c = 1.8$.

6 $A = \frac{bh}{2}$
(a) Rearrange the formula to give b in terms of A and h.
(b) Calculate b when $A = 9.6$ and $h = 3$.

7 $p^2 = q^2 + r^2$
(a) Rearrange the formula to give q in terms of p and r.
(b) Calculate q when $p = 7.3$ and $r = 2.7$.

8 $V = \frac{4}{3}\pi r^3$
(a) Rearrange the formula to give r in terms of V and π.
(b) Calculate r when $V = 1.6 \times 10^5$ and $\pi = 3.14$.
Give your answer to a suitable degree of accuracy.

9 $\frac{x}{a} = \frac{b}{x}$
(a) Make x the subject of the formula.
(b) Calculate x when $a = 4.6 \times 10^3$ and $b = 3.5 \times 10^5$.
Give your answer in standard form correct to two significant figures.

10 $m = \frac{1 - n}{1 + n}$
(a) Make n the subject of the formula.
(b) Calculate n when $m = 8.5 \times 10^{-2}$.
Give your answer in standard form to a suitable degree of accuracy.

What you need to know

- A **formula** is an algebraic rule written using numbers, letters and mathematical signs.

You should be able to:

- Write algebraic expressions and formulae.
- Substitute positive and negative numbers in expressions and formulae.
- Use simple formulae to solve problems.
- Rearrange formulae to make another letter (variable) the subject.

Review Exercise 12

1. $V = a + bc$. Find the value of V when $a = -5$, $b = 3$ and $c = 4$.

2. $S = pq + r$. Find the value of S when $p = -3$, $q = 4$ and $r = -2$.

3. $P = 3(m + n)$. Find the value of P when $m = 0.5$ and $n = 2$.

4. What is the value of $t^3 - t$ when $t = 2$?

5. $T = m^2 - 7m$. Work out the value of T when $m = -5$.

6. Segville High School has a disco for Year 7 each year.
A teacher works out how many cans of drink to buy, using this rule:

two cans for each ticket sold, plus 20 spare cans

(a) This year, 160 tickets have been sold. How many cans will he buy?
(b) Using N for the number of cans and T for the number of tickets, write down the teacher's formula for N in terms of T.
(c) Last year, he bought 300 cans. How many tickets were sold last year? AQA

7. The cost of hiring a coach is £60 plus £3 for every mile travelled.
(a) How much will it cost for a journey of 72 miles?
(b) Write down an expression for the cost in £ of a journey of M miles.
(c) A journey costs £186.
  (i) Use your answer to part (b) to form an equation using this information.
  (ii) How many miles did the coach travel on this journey? AQA

8. A formula to estimate the number of rolls of wallpaper, R, for a room is $R = \frac{ph}{5}$ where p is the perimeter of the room in metres and h is the height of the room in metres.
The perimeter of Carol's bedroom is 15.5 m and it is 2.25 m high.
How many rolls of wallpaper will she have to buy? AQA

9. A formula is given as $t = 7p - 50$. Rearrange the formula to make p the subject. AQA

10. You are given the formula $v = u + at$.
(a) Work out the value of v when $u = 20$, $a = -6$ and $t = \frac{9}{5}$.
(b) Rearrange the formula to give t in terms of v, u and a. AQA

11. You are given the formula $y = 3x^2$.
(a) Calculate y when $x = -3$.
(b) Find a value of x when $y = 147$.
(c) Rearrange the formula to give x in terms of y. AQA

12. A cuboid has a square base of side x cm.
The volume of the cuboid is $V\text{cm}^3$ and the height is h cm.
Write down an expression for x in terms of V and h.

h cm
x cm
x cm

13. You are given the formula $t = \sqrt{\frac{r}{s}}$.
Rearrange the formula to give r in terms of s and t. AQA

14. The braking distance, D metres, of a car is shown by the formula $D = \frac{v^2}{252f}$ where v is its speed in kilometres per hour and f is the friction between the tyres and the road surface.
(a) Calculate D when $v = 80$ and $f = 0.25$.
(b) After an accident, it was shown from the skid marks that $D = 50$ and $f = \frac{3}{8}$.
Use the formula to calculate v. AQA

15. Make a the subject of the formula $c(a - 4) = b(4 - 2a)$ AQA

16. $m = \frac{cab}{a - b}$. Express b in terms of a, c and m. AQA

CHAPTER 13

Sequences

Continuing a sequence

A **sequence** is a list of numbers made according to some rule.
For example:

5, 9, 13, 17, 21, ...

The first term is 5.
To find the next term in the sequence, add 4 to the last term.
The next term in this sequence is 21 + 4 = 25.
What are the next three terms in the sequence?

The numbers in a sequence are called **terms**.
The start number is the **first term**, the next is the second term, and so on.

To continue a sequence:
1. Work out the rule to get from one term to the next.
2. Apply the same rule to find further terms in the sequence.

EXAMPLES

Find the next three terms in each of these sequences.

1 5, 8, 11, 14, 17, ...

To find the next term in the sequence, add 3 to the last term.
17 + 3 = 20, 20 + 3 = 23, 23 + 3 = 26.
The next three terms in the sequence are: 20, 23, 26.

2 2, 4, 8, 16, ...

To find the next term in the sequence, multiply the last term by 2.
$16 \times 2 = 32$, $32 \times 2 = 64$, $64 \times 2 = 128$.
The next three terms in the sequence are: 32, 64, 128.

3 1, 1, 2, 3, 5, 8, ...

To find the next term in the sequence, add the last two terms.
5 + 8 = 13, 8 + 13 = 21, 13 + 21 = 34.
The next three terms in the sequence are: 13, 21, 34.
This is a special sequence called the **Fibonacci sequence**.

Exercise 13.1

1 Find the next three terms in these sequences.

(a) 1, 5, 9, 13, ...
(b) 6, 8, 10, 12, ...
(c) 28, 25, 22, 19, ...
(d) 3, 8, 13, 18, 23, ...
(e) 3, 6, 12, 24, ...
(f) $\frac{1}{4}$, $\frac{1}{2}$, $\frac{3}{4}$, 1, $1\frac{1}{4}$, ...
(g) 32, 16, 8, 4, ...
(h) 0.5, 0.6, 0.7, 0.8, ...
(i) 10, 8, 6, 4, ...
(j) 80, 40, 20, 10, ...
(k) 1, 3, 6, 10, 15, ...
(l) 1, 3, 4, 7, 11, 18, ...

2 Find the missing terms from these sequences.

(a) 2, 4, 6, __, 10, 12, __, 16, ...
(b) 2, 6, __, 14, 18, __, 26, ...
(c) 1, 2, 4, __, 16, __, 64, ...
(d) 28, 22, __, 10, 4, __, ...
(e) 1, 4, 9, __, 25, __, 49, ...
(f) 1, 2, 3, 5, __, 13, __, 34, ...
(g) __, 8, 14, __, __, 32, 38, ...

3 Write down the rule, in words, used to get from one term to the next for each sequence.
Then use the rule to find the next two terms.

(a) 2, 9, 16, 23, 30, ...
(b) 3, 5, 7, 9, 11, ...
(c) 1, 5, 9, 13, 17, ...
(d) 31, 26, 21, 16, ...
(e) 64, 32, 16, 8, 4, ...
(f) 1, 3, 9, 27, ...
(g) −2, −4, −6, −8, ...
(h) 10, 7, 4, 1, −2, ...

4 A sequence begins 1, 4, 7, 10, ...
(a) What is the 10th number in this sequence?
(b) Explain how you found your answer.

5 A number sequence begins 1, 2, 4, ...
David says that the next number is 8.
Tony says that the next number is 7.
(a) Explain why they could both be correct.
(b) Find the 10th number in David's sequence.
(c) Find the 10th number in Tony's sequence.

6 Here is part of a number sequence: 3, 9, 15, 21, ...
Is the number 50 in this sequence?
Explain your answer.

Using rules

Sometimes you will be given a rule and asked to use it to find the terms of a sequence.

For example:
A sequence begins: 1, 4, 13, ...
The rule for the sequence is:

Multiply the last number by 3, then add 1

The next term in the sequence is given by:

$$13 \times 3 + 1 = 39 + 1 = 40$$

The following term is given by:

$$40 \times 3 + 1 = 120 + 1 = 121$$

So, the sequence can be extended to: 1, 4, 13, 40, 121, ...

Use the rule to find the next two terms in the sequence.

The **same rule** can be used to make different sequences.

For example:
Another sequence begins: 2, 7, 22, ...
Using the same rule, the next term is given by:

$$22 \times 3 + 1 = 66 + 1 = 67$$

The following term is given by:

$$67 \times 3 + 1 = 201 + 1 = 202$$

So, the sequence can be extended to: 2, 7, 22, 67, 202, ...

Use the rule to find the next two terms in the sequence.

EXAMPLE This rule is used to find each number in a sequence from the number before it.

Subtract 3 and then multiply by 4

Starting with 5 we get the following sequence: 5, 8, 20, 68, ...

(a) Write down the next number in the sequence.
(b) Using the same rule, but a different starting number, the second number is 16.
Find the starting number.

(a) $(68 - 3) \times 4 = 65 \times 4 = 260$
Notice that, following the rule, 3 is subtracted first and the result is then multiplied by 4.
The next number in the sequence is 260.

(b) Imagine the first number is x.

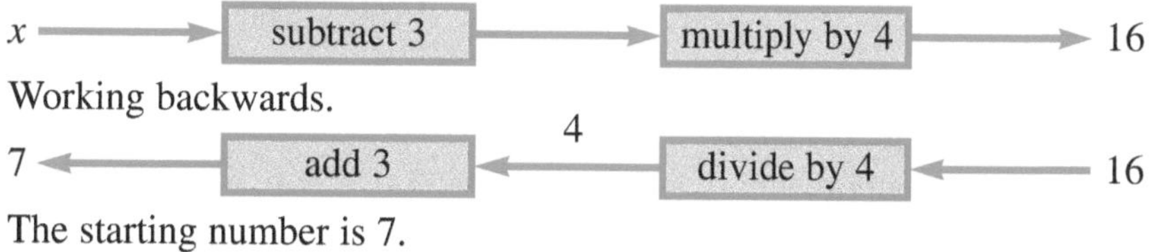

The starting number is 7.

Exercise 13.2

1 This rule is used to find each term of a sequence from the one before:

Subtract 3 then divide by 2

(a) The first term is 45.
(i) What is the second term? (ii) What is the **fourth** term?
(b) Using the same rule, but a different starting number, the second term is 17.
What is the starting number for the sequence?

2 This rule is used to find each term of a sequence from the one before:

Add 5 then multiply by 3

(a) The first term is 7.
(i) What is the second term? (ii) What is the **third** term?
(b) Using the same rule, but a different starting number, the second term is 45.
What is the starting number for the sequence?

3 A sequence is formed from this rule:

Add together the last two terms to find the next term

Part of the sequence is ... 5, 9, 14, 23, ...
(a) Write down the next two terms after 23 in the sequence.
(b) Write down the two terms that come before 5 in the sequence.

4 A sequence begins: 1, -3, ...
The sequence is continued using the rule:

Add the previous two numbers and then multiply by 3

Use the rule to find the next two numbers in the sequence.

5 A sequence begins: 4, 7, 13, 25, ...
The next number in the sequence can be found using the rule:

"Multiply the last term by 2 then subtract 1."

(a) Write down the next **two** terms in the sequence.
(b) The 11th term in the sequence is 3073.
Use this information to find the 10th term in the sequence.

Number sequences

A number sequence which increases (or decreases) by the same amount from one term to the next is called a **linear sequence**.
For example, terms in the sequence: 6, 11, 16, 21, ... increase by 5 from one term to the next.
We say that the sequence has a **common difference** of 5.

By comparing a sequence with multiples of the counting numbers: 1, 2, 3, 4, ...
we can write a rule to find the nth term of the sequence.

Sequence: 6 11 16 21 ...
Multiples of 5: 5 10 15 20 ...

To get the nth term add one to the multiples of 5.
So, the nth term is $5n + 1$.

Compare the sequence with multiples of the common difference.
In this case the common difference is 5, so, compare the sequence with multiples of 5.

A table can be used to find the nth term of a sequence.

The sequence 2, 8, 14, 20, ... has a common difference of 6.
The common difference is used to complete the table for terms 1, 2, 3, 4 and n.

Term	Term × common difference	Sequence	Difference
1	$1 \times 6 = 6$	2	$2 - 6 = -4$
2	$2 \times 6 = 12$	8	$8 - 12 = -4$
3	$3 \times 6 = 18$	14	$14 - 18 = -4$
4	$4 \times 6 = 24$	20	$20 - 24 = -4$
n	$n \times 6 = 6n$	$6n - 4$	

Term:
1 represents the first term,
2 the second term, and so on.
n represents the nth term.
Common difference:
2nd term − 1st term
$= 8 - 2 = 6$.
Differences:
Check that each pair of entries gives the same result.

The nth term of the sequence 2, 8, 14, 20, ... is $6n - 4$.

The rule for the nth term can be used to find the value of any term in the sequence.
To find the fifth term, substitute $n = 5$ into $6n - 4$.
$6 \times 5 - 4 = 30 - 4 = 26$.
The fifth term is 26.

EXAMPLE

(a) Find the nth term in the sequence 31, 28, 25, 22, ...
(b) Find the value of the 10th term in the sequence.

(a) The common difference is -3.

Term	Term × common difference	Sequence	Difference
1	$1 \times (-3) = -3$	31	$31 - (-3) = 34$
2	$2 \times (-3) = -6$	28	$28 - (-6) = 34$
3	$3 \times (-3) = -9$	25	$25 - (-9) = 34$
4	$4 \times (-3) = -12$	22	$22 - (-12) = 34$
n	$n \times (-3) = -3n$	$-3n + 34$	

The nth term is $-3n + 34$.
This can also be written as $34 - 3n$.

(b) The nth term is $34 - 3n$.
Substitute $n = 10$.
$34 - 3 \times 10 = 34 - 30 = 4$
The 10th term is 4.

Exercise 13.3

1 Find the common differences of the following sequences.

(a) 3, 6, 9, 12, … (b) 2, 5, 8, 11, … (c) 7, 13, 19, 25, …
(d) 12, 20, 28, 36, … (e) 20, 18, 16, 14, … (f) 7, 3, −1, −5, …

2 (a) The multiples of 3 are 3, 6, 9, 12, …
What is the nth multiple of 3?
(b) What is the nth multiple of 8?

3 A sequence of numbers starts: 4, 7, 10, 13, …
(a) What is the common difference?
(b) Copy and complete this table.

Term	Term × common difference	Sequence	Difference
1	1 × … =	4	4 − … =
2	2 × … =	7	7 − … =
3	3 × … =	10	10 − … =
4	4 × … =	13	13 − … =
n	n × … = …n	…n + …	

(c) Write down the nth term of the sequence.
(d) What is the value of the 8th term of the sequence?

4 A sequence of numbers starts: 9, 11, 13, 15, …
(a) What is the common difference?
(b) Copy and complete this table for the first four terms of the sequence.

Term	Term × common difference	Sequence	Difference
1	2	9	7
2	4	11	…
		13	

(c) Write down the nth term of the sequence.
(d) What is the value of the 20th term of the sequence?

5 A sequence of numbers starts: 20, 16, 12, 8, …
(a) What is the common difference?
(b) Copy and complete this table for the first four terms of the sequence.

Term	Term × common difference	Sequence	Difference
1	…	20	…
		16	

(c) Write down the nth term of the sequence.

6 Find the nth term of the following sequences.

(a) 1, 4, 7, 10, … (b) 19, 16, 13, 10, … (c) 5, 9, 13, 17, …
(d) 4, 8, 12, 16, … (e) 1, 3, 5, 7, … (f) 7, 11, 15, 19, …
(g) 6, 4, 2, 0, … (h) 5, 8, 11, 14, … (i) 3, 8, 13, 18, …
(j) 40, 35, 30, 25, … (k) 0, 1, 2, 3, … (l) −1, 1, 3, 5, …

7 Write down the first three terms of a sequence where the nth term is $n^2 + 3$.
Explain how you found your answer.

Sequences of numbers from shape patterns

Activity

These patterns are made using squares.

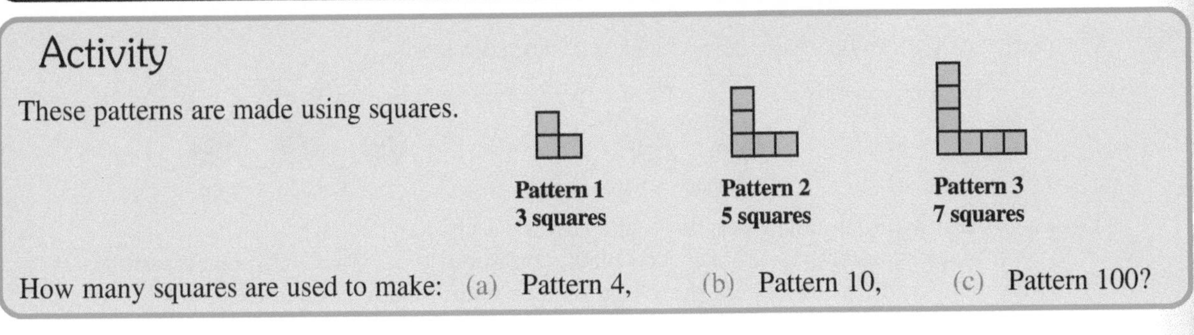

How many squares are used to make: (a) Pattern 4, (b) Pattern 10, (c) Pattern 100?

The number of squares used to make each pattern forms a **sequence**.

Pattern 4 is made using 9 squares.
You could have answered this: by drawing Pattern 4 or,
by continuing the sequence of numbers 3, 5, 7, ...
It is possible to do the same for Pattern 10, though it would involve a lot of work, but it would be unreasonable to use either method for Pattern 100.
Instead we can investigate how each pattern is made.

2 × 1 + 1 = 3 squares **2 × 2 + 1 = 5 squares** **2 × 3 + 1 = 7 squares**

Each pattern is made using a **rule**.
The rule can be **described in words**.
To find the number of squares used to make a pattern use the rule:
"Double the pattern number and add 1."

Pattern number	Rule	Number of squares
4	$2 \times 4 + 1$	9
10	$2 \times 10 + 1$	21
100	$2 \times 100 + 1$	201

The same rule can be **written using symbols**.
We can then answer a very important question: How many squares are used to make Pattern n?

Pattern n will have $2 \times n + 1$ squares.
This can be written as $2n + 1$ squares.

Special sequences of numbers

Square numbers

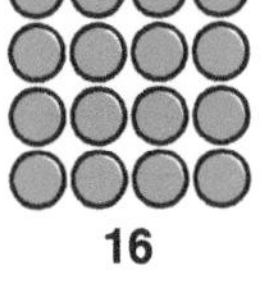
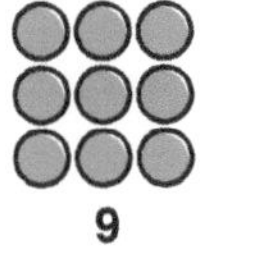
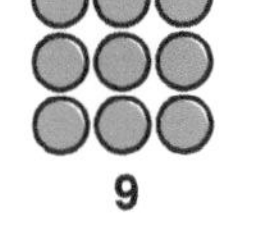
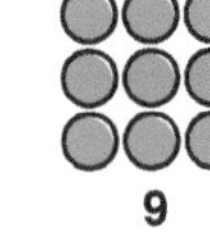
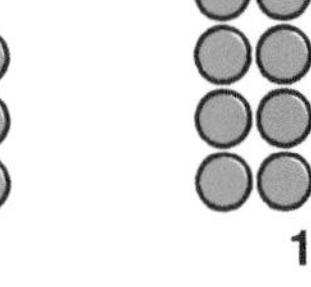

1 4 9 16

The sequence starts: 1, 4, 9, 16, ...
The numbers in this sequence are called **square numbers**.

Triangular numbers

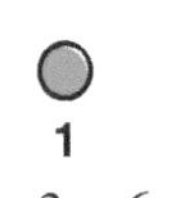
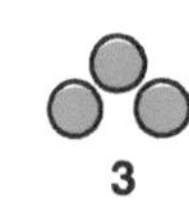

1 3 6 10

The sequence starts: 1, 3, 6, 10, ...
The numbers in this sequence are called **triangular numbers**.

Exercise 13.4

1 A sequence of patterns is made using black and white counters.

1 black, 3 white **2 black, 6 white** **3 black, 9 white**

(a) How many white counters are there in a pattern with
(i) 5 black counters, (ii) 10 black counters, (iii) 100 black counters?
(b) How many white counters are there in a pattern with n black counters?

2 Linking cubes of side 1 cm are used to make rods.
This rod is made using 4 linking cubes.

The surface area of the rod is 18 square centimetres.
(4 squares on each of the long sides plus one square at each end.)
(a) What is the surface area of a rod made using 5 linking cubes?
(b) What is the surface area of a rod made using 10 linking cubes?
(c) What is the surface area of a rod made using n linking cubes?

3 A sequence of patterns is made using sticks.

Pattern 1 **Pattern 2** **Pattern 3** **Pattern 4**

(a) How many sticks are used to make Pattern 5?
(b) Pattern n uses T sticks. Write a formula for T in terms of n.
(c) Use your formula to find the number of sticks used to make Pattern 10.
(d) One pattern uses 77 sticks. What is the pattern number?

4 These patterns are made using matches.

Pattern 1 **Pattern 2** **Pattern 3**

(a) How many matches are used to make Pattern 5?
(b) Which pattern uses 37 matches?
(c) Find a formula for the number of matches, m, in Pattern p.

5 These are the first four **square numbers**. 1, 4, 9, 16, ...
(a) What is the next square number?
(b) Copy and complete this list for the first 5 square numbers.

$$1^2 = 1, \quad 2^2 = 4, \quad 3^2 = 9, \quad \ldots = 16, \quad 5^2 = \ldots$$

(c) Write an expression for the nth square number.
(d) Find the 15th square number.

6 (a) Find the next term in each of these sequences. Explain how you found your answers.
(i) 2, 5, 10, 17, 26, ... (ii) 0, 3, 8, 15, 24, ...
(iii) 4, 7, 12, 19, 28, ... (iv) 2, 8, 18, 32, 50, ...
(v) 2, 6, 12, 20, 30, ...
(b) By comparing each of these sequences to the sequence of square numbers write an expression for the nth term of each sequence.

7 A sequence of shapes is made using small squares.

Shape 1

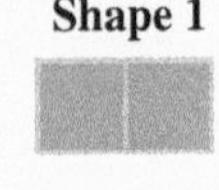

Shape 2

Shape 3

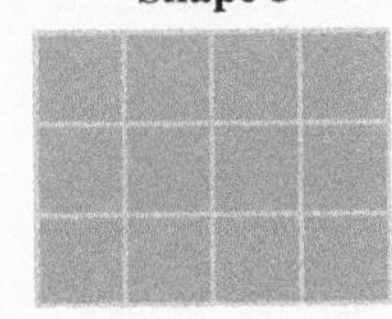

Shape 4

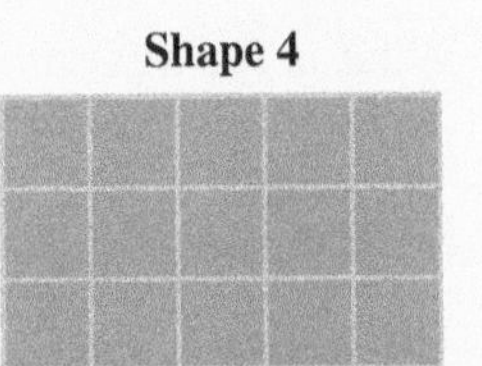

(a) Shape n uses T small squares.
Write a formula for T in terms of n.

(b) Use your formula to find the number of small squares in Shape 10.

8 These are the first four **triangular numbers**: 1, 3, 6, 10, ...

(a) What is the next triangular number?

(b) Copy and complete this list for the first 5 triangular numbers.

$\frac{1 \times 2}{2} = 1, \quad \frac{2 \times 3}{2} = 3, \quad \frac{3 \times 4}{2} = 6, \quad \frac{\quad}{2} = 10, \quad \ldots = \ldots$

(c) Write an expression for the nth triangular number.

(d) Find the 10th triangular number.

9 These are the first four **powers of 2**: 2, 4, 8, 16, ...

(a) What is the next power of 2?

(b) Copy and complete this list for the first 5 powers of 2.

$2^1 = 2, \quad 2^2 = 4, \quad 2^3 = 8, \quad \ldots = 16, \quad \ldots = \ldots$

(c) Write an expression for the nth power of 2.

(d) Find the 10th power of 2.

(e) What power of 2 is equal to 256?

10 (a) Write down the first 5 powers of 10.

(b) Write an expression for the nth power of 10.

(c) What power of 10 is equal to one million?

11 (a) Write an expression for the nth term of this sequence: 3, 9, 27, 81, ...

(b) Use your expression to find the 10th term of the sequence.

What you need to know

- A **sequence** is a list of numbers made according to some rule.
  The numbers in a sequence are called **terms**.
- **To continue a sequence:** 1. Work out the rule to get from one term to the next.
  2. Apply the same rule to find further terms in the sequence.
- A number sequence which increases (or decreases) by the same amount from one term to the next is called a **linear sequence**.
  The sequence: 2, 8, 14, 20, 26, ... has a **common difference** of 6.
- Special sequences **Square numbers:** 1, 4, 9, 16, 25, ...
  Triangular numbers: 1, 3, 6, 10, 15, ...
- Patterns of shapes can be drawn to represent a number sequence.
  For example, this pattern represents the sequence 3, 5, 7, ...

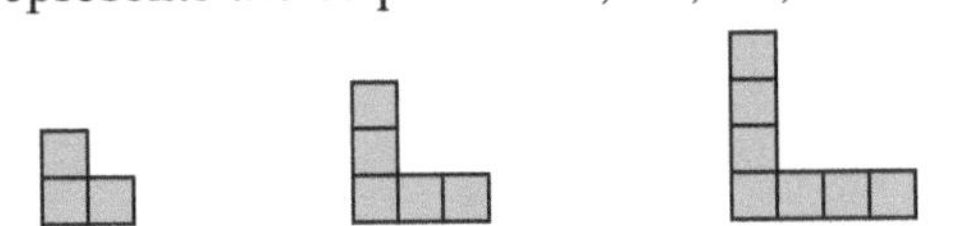

You should be able to:

- Draw patterns of shapes which represent number sequences.
- Continue a given number sequence.
- Find an expression for the nth term of a sequence.

Review Exercise 13

1 What is the next number in each of these sequences?

(a) 3, 6, 11, 18, 27, ... (b) 8, 4, 2, 1, $\frac{1}{2}$, ... AQA

2 Look at this number sequence: 16, 25, 36, 49, 64, ...

(a) Write down the next term in this sequence.

(b) Will the number 196 be in this sequence? Give a reason for your answer. AQA

3 The first two terms of a sequence are: 4, 8.

(a) Using the rule:

Add the two previous numbers and divide by two

write down the third and fourth terms of the sequence.

(b) If the sequence had begun 8, 4 instead of 4, 8 would the third and fourth terms be the same as those in part (a)? Give a reason for your answer. AQA

4 (a) Write down the 10th term of the sequence which begins: 3, 7, 11, 15, ...

(b) Write down an expression for the nth term of this sequence.

(c) Show that 1997 cannot be a term in this sequence.

(d) Calculate the number of terms in the sequence: 3, 7, 11, 15, ..., 399 AQA

5 (a) (i) What is the next number in this sequence? 1, 5, 9, 13, ...

(ii) One number in this sequence is x.
Write, in terms of x, the next number in the sequence.

(b) The nth term of a different sequence is $n^2 - 2$.
Find the 5th term in this sequence.

6 A sequence of numbers begins: 2, 5, 8, 11, ...

(a) Write down the next number in this sequence.

(b) Work out the 20th number in this sequence.

(c) Find the nth term of this sequence. AQA

7 (a) Write down the next two terms in the sequence: 3, 5, 9, 15, 23, ...

(b) (i) Write down the next term in the sequence: 3, 5, 9, 17, 33, ..., ...

(ii) Explain how you got your answer.

(c) Write down the nth term for the sequence: 3, 5, 7, 9, 11, ... AQA

8 Write down the nth term for each of the following sequences.

(a) 3, 6, 9, 12, ... (b) 1, 4, 7, 10, ... AQA

9 A sequence of patterns is formed using equilateral triangles.

Pattern 1 Pattern 2 Pattern 3

(a) How many triangles are in Pattern 10?

(b) Explain why a pattern in this sequence cannot have 40 triangles.

(c) Write an expression, in terms of p, for the number of triangles in Pattern p. AQA

10 A sequence of patterns is shown.

Pattern 1 Pattern 2 Pattern 3

Write an expression, in terms of n, for the number of white squares in the nth pattern of the sequence. AQA

CHAPTER 14 Straight Line Graphs

Coordinates

Coordinates are used to describe the position of a point.

Two lines are drawn at right angles to each other.
The horizontal line is called the ***x* axis**.
The vertical line is called the ***y* axis**.
The plural of axis is **axes**.
The two axes cross at the point called the **origin**.

On the diagram, the coordinates of point A are (3, 2).
To find point A: start at the origin and go right 3 squares then up 2 squares.
The coordinates of point B are $(-2, 3)$.

What are the coordinates of the points C and D?

Linear functions

Look at these coordinates (0, 1), (1, 2), (2, 3), (3, 4).
Can you see any number patterns?

The same coordinates can be shown in a **table**.

x	0	1	2	3
y	1	2	3	4

The diagram shows the coordinates plotted on a **graph**.
The points all lie on a **straight line**.

A **rule** connects the x coordinate with the y coordinate.
All points on the line obey the rule $y = x + 1$.

$y = x + 1$ is an example of a **linear function**.
The graph of a linear function is a **straight line**.

Special graphs

This diagram shows the graphs: $x = 4 \quad y = 1$
$x = -2 \quad y = -5$

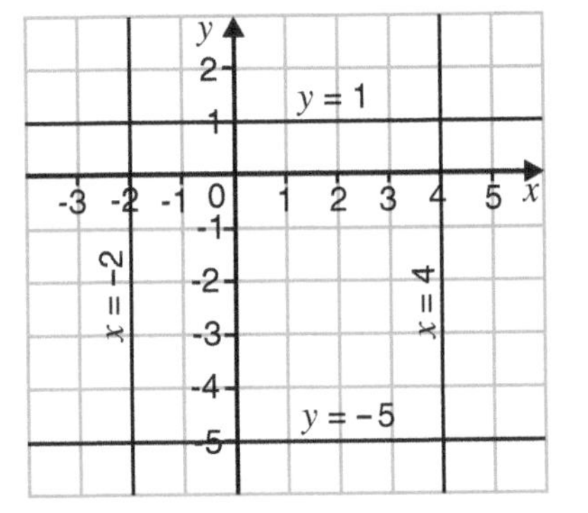

Notice that:
The graph of $x = 4$ is a **vertical** line.
All points on the line have x coordinate 4.

The graph of $y = 1$ is a **horizontal** line.
All points on the line have y coordinate 1.

$x = 0$ is the y axis.
$y = 0$ is the x axis.

Drawing a graph of a linear function

To draw a linear graph:
- Find at least two corresponding values of x and y.
- Plot the points.
- Join the points with a straight line.

EXAMPLES

1 (a) Complete the table of values for $y = 2x - 3$.

x	0	1	2	3
y			1	

(b) Draw the graph of the equation $y = 2x - 3$.

(a)

x	0	1	2	3
y	-3	-1	1	3

When $x = 0$, $y = 2 \times 0 - 3 = -3$.
When $x = 1$, $y = 2 \times 1 - 3 = -1$.
When $x = 3$, $y = 2 \times 3 - 3 = 3$.

(b) Plot the points $(0, -3)$, $(1, -1)$, $(2, 1)$, and $(3, 3)$.

The straight line which passes through these points is the graph of the equation $y = 2x - 3$.

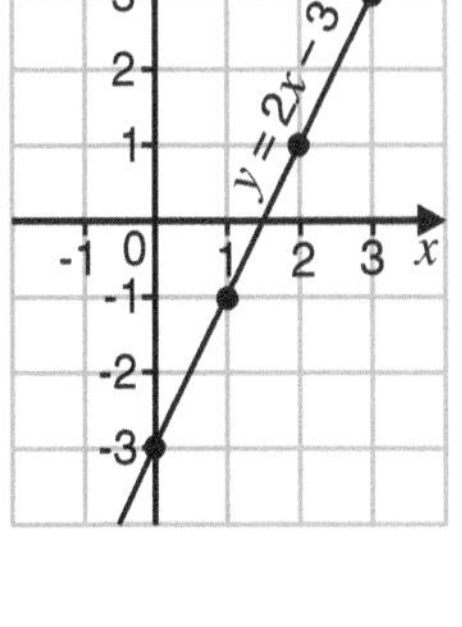
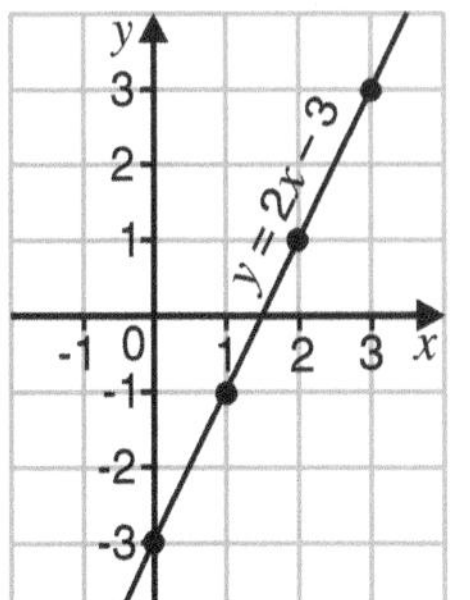

2 (a) Draw the graph of the equation $y = 4 - 2x$.
(b) Use your graph to find the value of y when $x = 2.5$.

(a) If values for x are not given in the question you must choose at least two of your own.

When $x = -1$, $y = 4 - 2 \times (-1) = 4 + 2 = 6$.
When $x = 0$, $y = 4 - 2 \times 0 = 4 - 0 = 4$.
When $x = 3$, $y = 4 - 2 \times 3 = 4 - 6 = -2$.
Plot the points $(-1, 6)$, $(0, 4)$ and $(3, -2)$.

The straight line which passes through these points is the graph of the equation $y = 4 - 2x$.

(b) Using the graph:
From 2.5 on the x axis, go down to meet the line $y = 4 - 2x$.
Then go left to meet the y axis at -1.
When $x = 2.5$, $y = -1$.

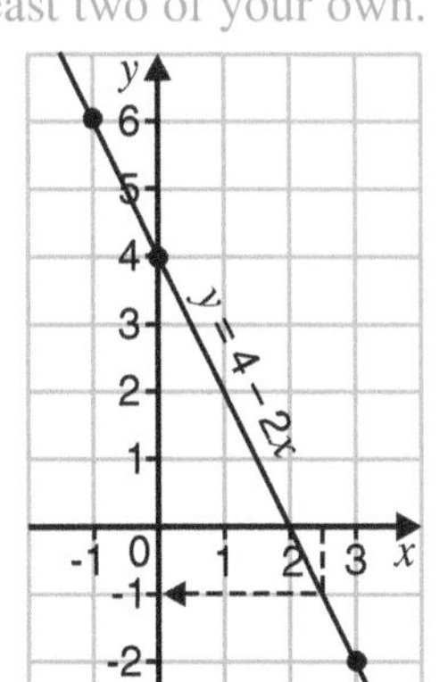

Exercise 14.1

1 The diagram shows the positions of points P, Q and R.
Write down the coordinates of these points.

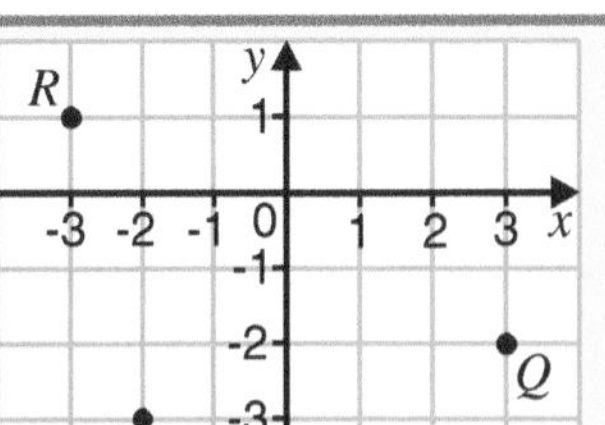

2 Draw x and y axes from -5 to 5.
(a) Plot the points $S(3, -2)$ and $T(-5, 4)$. Join ST.
(b) Write down the coordinates of the midpoint of ST.

3 Draw x and y axes from -2 to 4.
(a) Plot the points $A(3, 2)$, $B(3, -1)$ and $C(-1, -1)$.
(b) Points A, B and C are three corners of a rectangle.
Point D is the fourth corner of the rectangle. Plot point D on your diagram.
(c) What are the coordinates of point D?

4

(1) (2) (3) (4) (5)

(a) Write down the equations of the lines labelled on this graph.
(b) What are the coordinates of the point where lines (1) and (3) cross?
(c) What are the coordinates of the point where lines (2) and (4) cross?

5 Draw x and y axes from -4 to 4.
On your diagram, draw and label the graphs of these equations.
(a) $x = 3$ (b) $y = 2$ (c) $x = -2$ (d) $y = -1$

6 (a) Copy and complete a table of values, like the one shown, for each of these equations.
(i) $y = x + 2$
(ii) $y = 2x$
(iii) $y = -x$
(iv) $y = 2 - x$

x	0	1	2	3
y				

(b) Draw graphs for each of the equations in part (a).

7 Draw tables of values and use them to draw graphs of:
(a) $y = x - 1$ (b) $y = 2x + 1$ (c) $y = 3 - x$ (d) $y = 6 - 2x$
Draw and label the x axis from 0 to 4 and the y axis from -2 to 10.

8 (a) Copy and complete this table and use it to draw the straight line graph of $y = 4 - x$.

x	-2	-1	0	1	2
y		5			2

Draw and label the x axis from -3 to 3 and the y axis from -1 to 6.
(b) Use your graph to find the value of:
(i) y when $x = 1.5$, (ii) y when $x = -0.5$.

9 (a) Draw the graph of $y = 4x + 1$ for values of x from -2 to 2.
(b) Use your graph to find the value of:
(i) y when $x = -1.5$, (ii) x when $y = 3$.

10 (a) Draw the graph of $y = 2x - 1$ for values of x from -2 to 3.
(b) Use your graph to find the value of x when $y = 0$.

11 (a) Draw the graph of $y = 5 - 2x$ for values of x from -2 to 3.
(b) Use your graph to find the value of x when $y = 8$.

12 (a) Draw the graphs of $y = 2x - 3$ and $y = 3 - x$ on the same axes.
(b) Write down the coordinates of the point where the lines cross.

Activity

Draw these graphs **on the same diagram**:

$y = 2x + 2$ $y = 2x + 1$ $y = 2x$ $y = 2x - 1$

Draw and label the x axis from 0 to 3 and the y axis from -1 to 8.
What do they all have in common?
What is different?

Gradient and intercept

The gradient of a straight line graph is found by drawing a right-angled triangle.

$$\text{Gradient} = \frac{\text{distance up}}{\text{distance along}}$$

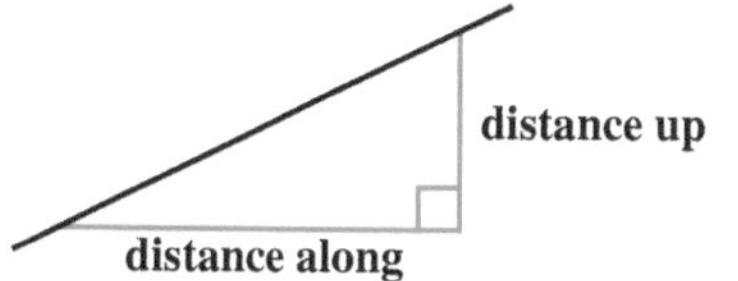

The gradient of a line can be positive, zero or negative.

Positive gradients **Zero gradients**

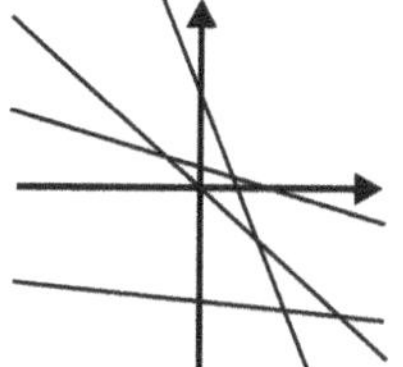

Negative gradients

The graphs of:
$y = 2x + 2$, $y = 2x + 1$, $y = 2x$, $y = 2x - 1$, go 2 squares up for every 1 square along.
The graphs are all parallel and have a gradient of 2.

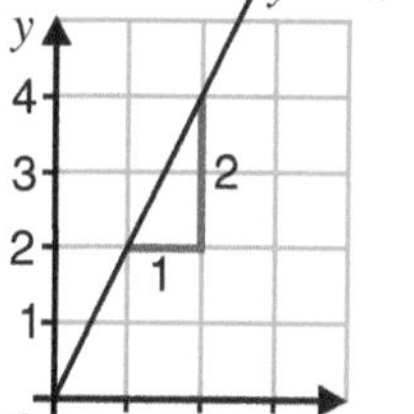

Lines that are **parallel** have the same **slope** or **gradient**.

The point where a graph crosses the y axis is called the ***y*-intercept**.

The y-intercept of the graph $y = 2x + 2$ is 2.
The y-intercept of the graph $y = 2x + 1$ is 1.

What are the y-intercepts of the graphs $y = 2x$ *and* $y = 2x - 1$?

In general, the equation of any straight line can be written in the form

$$y = mx + c$$

where m is the **gradient** of the line and c is the ***y*-intercept**.

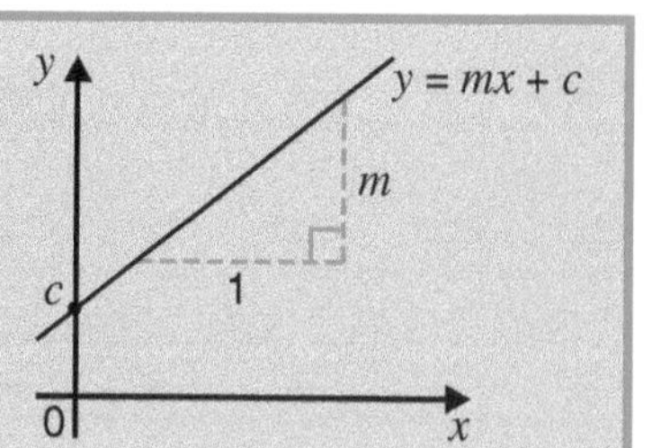

EXAMPLES

1 Write down the gradient and y-intercept for each of the following graphs.

(a) $y = 3x + 5$
(b) $y = 4x - 1$
(c) $y = 6 - x$

(a) Gradient $= 3$, y-intercept $= 5$.
(b) Gradient $= 4$, y-intercept $= -1$.
(c) Gradient $= -1$, y-intercept $= 6$.

2 Write down the equation of the straight line which has gradient -7 and cuts the y axis at the point (0, 4).

The general form for the equation of a straight line is $y = mx + c$.
The gradient, $m = -7$, and the y-intercept, $c = 4$.
Substitute these values into the general equation.
The equation of the line is $y = -7x + 4$.
This can be written as $y = 4 - 7x$.

EXAMPLE

3 Find the equation of the line shown on this graph.

First, work out the gradient of the line.
Draw a right-angled triangle.

$$\text{Gradient} = \frac{\text{distance up}}{\text{distance along}}$$

$$= \frac{6}{3}$$

$$= 2$$

The graph crosses the y axis at the point $(0, -3)$, so the y-intercept is -3.
The equation of the line is $y = 2x - 3$.

Exercise 14.2

1 (a) Draw these graphs **on the same diagram**:

(i) $y = x + 2$ (ii) $y = x + 1$ (iii) $y = x$ (iv) $y = x - 1$

Draw and label the x axis from 0 to 3 and the y axis from -1 to 5.

(b) What do they all have in common? What is different?

2 (a) Write down the gradient and y-intercept of $y = 3x - 1$.

(b) Draw the graph of $y = 3x - 1$ to check your answer.

3 Which of the following graphs are parallel?

$y = 3x$ $y = x + 2$ $y = 2x + 3$ $y = 3x + 2$

4 Copy and complete this table.

Graph	gradient	y-intercept
$y = 3x + 5$	3	
$y = 2x - 3$		
$y = 4 - 2x$		4
$y = \frac{1}{2}x + 3$		
$y = 2x$		
$y = 3$		

5 (a) Write down the equation of the straight line which has gradient 5 and crosses the y axis at the point $(0, -4)$.

(b) Write down the equation of the straight line which has gradient $-\frac{1}{2}$ and cuts the y axis at the point $(0, 6)$.

6 Match the following equations to their graphs.

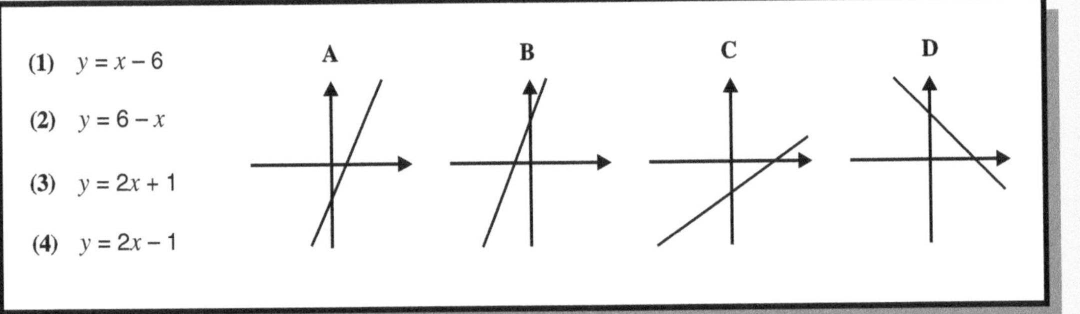

7 Find the equations of the lines shown on the following graphs.

(a)

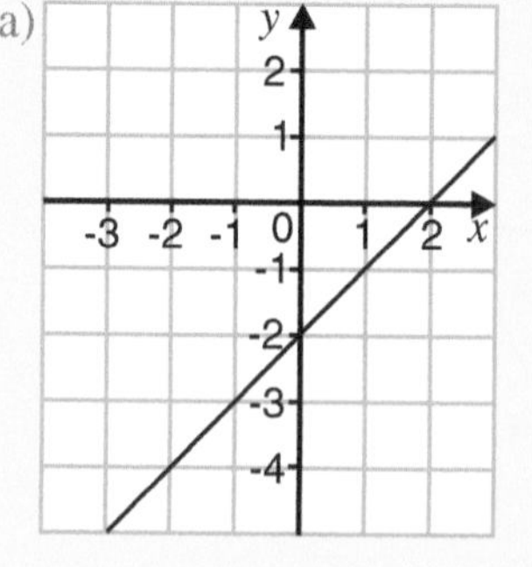

(b)

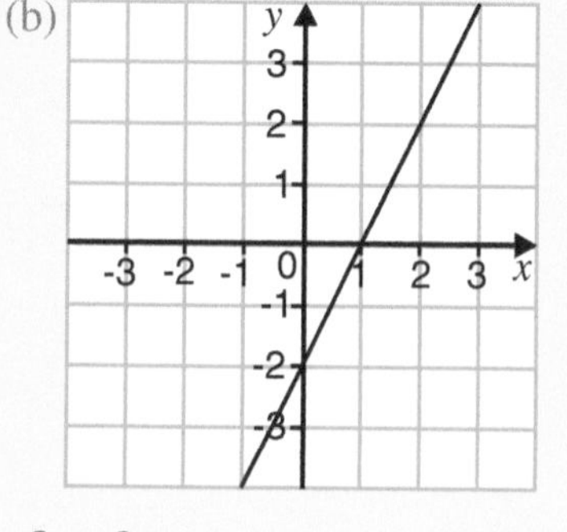

(c)

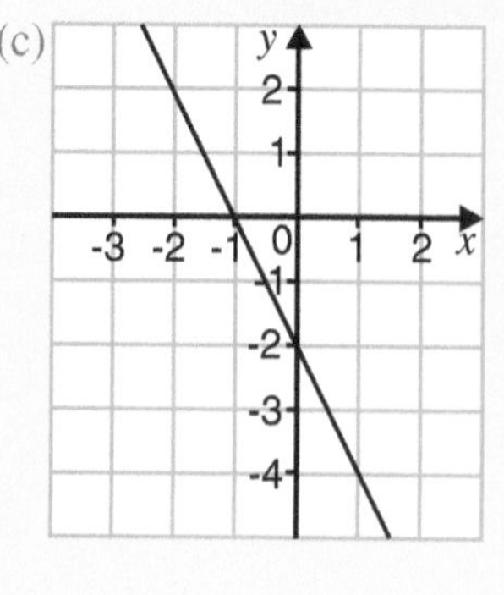

8 (a) Draw x and y axes from -8 to 8.
(b) Plot the points $A(-2, -6)$ and $B(8, 4)$.
(c) Find the gradient of the line which passes through the points A and B.
(d) Write down the coordinates of the point where the line crosses the y axis.
(e) Find the equation of the line which passes through the points A and B.

9 (a) Draw x and y axes from -8 to 8.
(b) Plot the points $P(-2, 3)$ and $Q(3, -7)$.
(c) Find the equation of the line which passes through the points P and Q.

10 What can you say about the slope of a line if the gradient is (a) 5, (b) -5, (c) 0?

11 A line, with a gradient of 3, passes through the origin. What is the equation of the line?

12 A plumber charges a fixed call-out charge and an hourly rate.
The graph shows the charges made for jobs up to 4 hours.

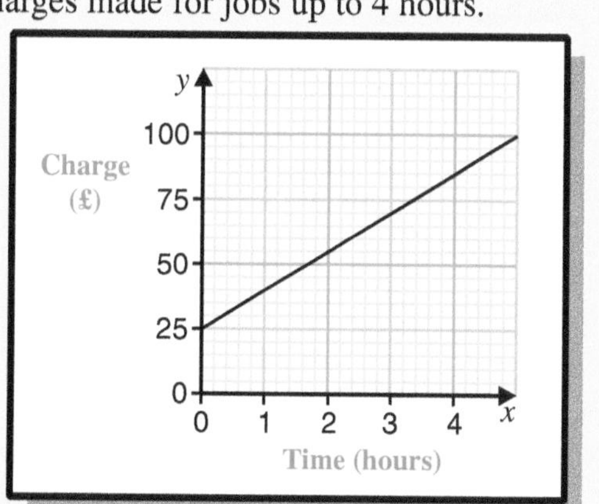

(a) What is the fixed call-out charge?
(b) What is the hourly rate?
(c) Write down the equation of the line in the form $y = mx + c$.
(d) Calculate the total charge for a job which takes 8 hours.

13 The graph shows the taxi fare for journeys up to 3 km.

(a) What is the fixed charge?

(b) What is the charge per kilometre?

(c) Write down the equation of the line in the form $f = md + c$.

(d) Calculate the taxi fare for a journey of 5 km.

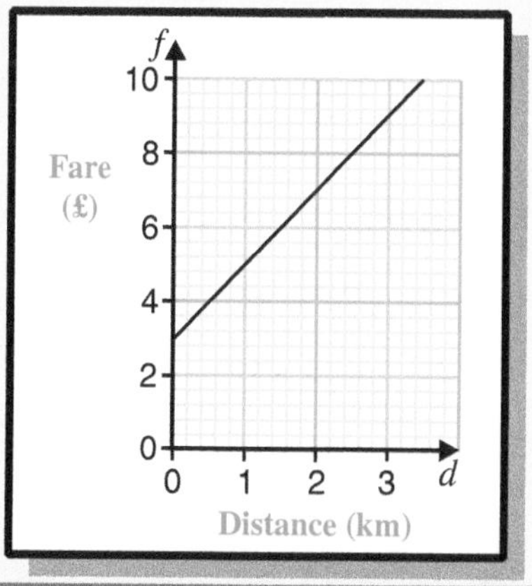

Drawing graphs of other linear equations

EXAMPLE

Draw the graph of the line given by the equation $x + 2y = 6$.

> By substituting $x = 0$ into the equation we can find the coordinates of the point where the line crosses the y axis.

$x + 2y = 6$
Substitute $x = 0$.
$0 + 2y = 6$
$2y = 6$
$y = 3$

The line crosses the y axis at the point (0, 3).

> By substituting $y = 0$ into the equation we can find the coordinates of the point where the line crosses the x axis.

$x + 2y = 6$
Substitute $y = 0$.
$x + 2 \times 0 = 6$
$x = 6$

The line crosses the x axis at the point (6, 0).

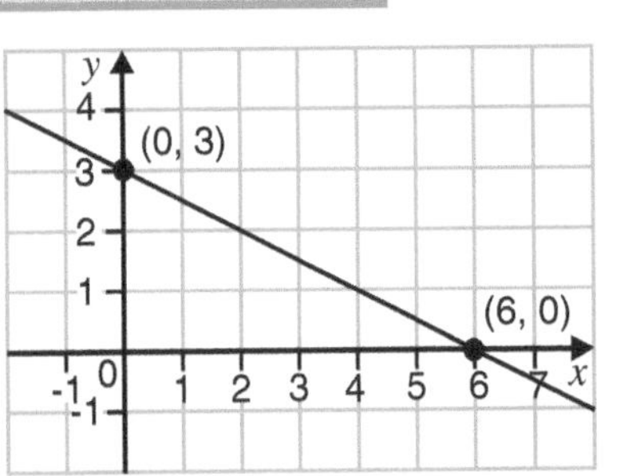

To draw the graph of $x + 2y = 6$:
1. Plot the points (0, 3) and (6, 0).
2. Using a ruler, draw a straight line which passes through the two points.

Exercise 14.3

1 A straight line has equation $x + y = 7$.
(a) By substituting $x = 0$ find the coordinates of the point where the line crosses the y axis.
(b) By substituting $y = 0$ find the coordinates of the point where the line crosses the x axis.
(c) Draw the graph of the line $x + y = 7$.

2 (a) Draw these graphs on the same diagram.
(i) $x + y = 2$ (ii) $x + y = 3$ (iii) $x + y = 5$
(b) What do they all have in common?

3 A straight line has equation $3x + y = 6$.
(a) By substituting $x = 0$ find the coordinates of the point where the line crosses the y axis.
(b) By substituting $y = 0$ find the coordinates of the point where the line crosses the x axis.
(c) Draw the graph of the line $3x + y = 6$.

4 Draw the graphs of lines with the following equations.
(a) $x + 2y = 6$ (b) $2y = 4 - x$ (c) $2y = x + 4$

5 A straight line has equation $3y + 5x = 15$.
(a) By substituting $x = 0$ find the coordinates of the point where the line crosses the y axis.
(b) By substituting $y = 0$ find the coordinates of the point where the line crosses the x axis.
(c) Draw the graph of the line $3y + 5x = 15$.

6 Draw the graphs of lines with the following equations, marking clearly the coordinates of the points where the lines cross the axes.
(a) $5y + 4x = 20$ (b) $4x - y = 4$ (c) $3y + 2x = 12$

Using graphs to solve linear equations

EXAMPLE

(a) Complete the tables for $y = x + 1$ and $y = 8 - x$.

x	1	2	3
$y = x + 1$			

x	1	2	3
$y = 8 - x$			

(b) Draw the graphs of $y = x + 1$ and $y = 8 - x$ on the same diagram.
(c) Use your graphs to solve the equation $8 - x = x + 1$.

(a)

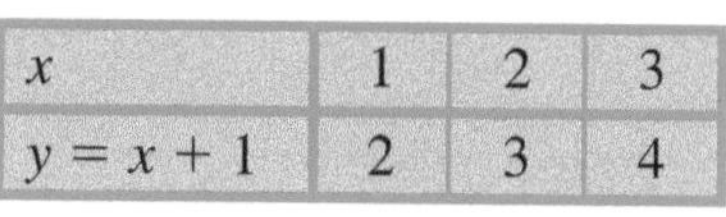

x	1	2	3
$y = x + 1$	2	3	4

x	1	2	3
$y = 8 - x$	7	6	5

(b)

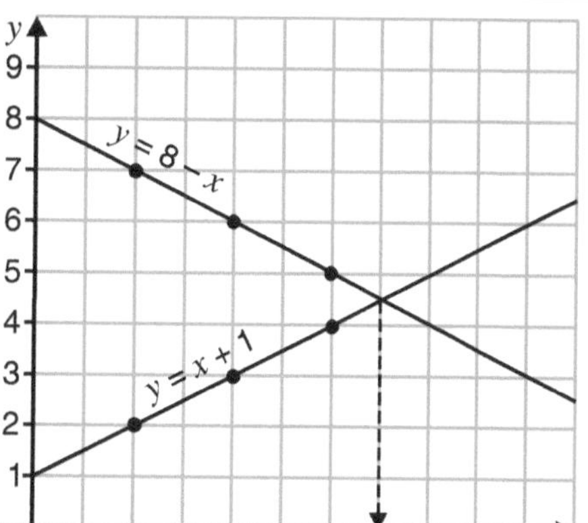

(c) The x value of the point where the two graphs cross gives the solution of the equation $8 - x = x + 1$.
Reading from the graph: $x = 3.5$

Check the graphical solution of the equation by solving $8 - x = x + 1$ *algebraically.*

Exercise 14.4

1 (a) Copy and complete the tables for $y = x + 2$ and $y = 5 - x$.

x	1	2	3
$y = x + 2$			

x	1	2	3
$y = 5 - x$			

(b) Draw the graphs of $y = x + 2$ and $y = 5 - x$ on the same diagram.
(c) Write down the coordinates of the point where the two lines cross.
(d) Use your graphs to solve $x + 2 = 5 - x$.

2 (a) Draw the graphs of $y = 3x + 1$ and $y = x + 6$ on the same diagram.
(b) Write down the coordinates of the point where the two lines cross.
(c) Use your graphs to solve the equation $3x + 1 = x + 6$.

3 (a) Draw the graphs of $y = x$ and $y = 3x - 1$.
(b) Use your graphs to solve the equation $x = 3x - 1$.

4 By drawing the graphs of $y = 2x$ and $y = 3 - x$, solve the equation $2x = 3 - x$.

5 (a) Draw the graph of $y = 3 + 2x$.
(b) What graph should be drawn to solve the equation $3 + 2x = 9$?
(c) Draw the graph and use it to solve the equation $3 + 2x = 9$.

Perpendicular lines

The diagram shows two lines, AB and CD.
The lines are **perpendicular**, at right angles, to each other.

Line AB has:	Line CD has:
gradient: 2	gradient: $-\frac{1}{2}$
y-intercept: -3	y-intercept: 2
equation: $y = 2x - 3$	equation: $y = -\frac{1}{2}x + 2$

Notice that:
Gradient of line AB × gradient of line $CD = -1$

In general, if two lines are perpendicular to each other, the product of their gradients $= -1$.

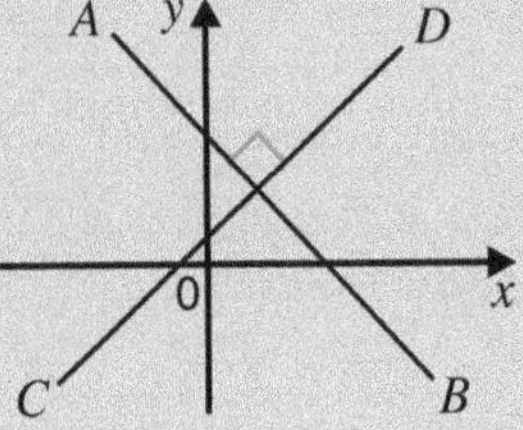

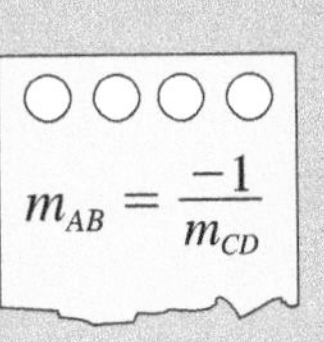

This can be written as:
$m_{AB} \times m_{CD} = -1$ where m_{AB} is the gradient of the line AB, and m_{CD} is the gradient of the line CD.

EXAMPLE

Write down the gradient of the line which is perpendicular to the line with equation $y = \frac{1}{5}x + 4$.
Gradient of the line $y = \frac{1}{5}x + 4$ is $\frac{1}{5}$.
Gradient of the line perpendicular to this line is $-1 \div \left(\frac{1}{5}\right) = -5$.

Exercise 14.5

1. Which of these lines are perpendicular to the line $y = x$?
   A $y = \frac{1}{2}x$ **B** $y = -x$ **C** $x + y = 3$ **D** $y = x - 1$
2. Write down the gradient of a line which is perpendicular to:
   (a) $y = 3x + 1$ (b) $y = \frac{1}{4}x - 3$ (c) $y = 4 - 2x$
3. (a) Find the equation of the line perpendicular to $y = x$ which passes through the point
   (i) (0, 0), (ii) (0, 2).
   (b) Find the equation of the line perpendicular to $y = \frac{1}{2}x$, which passes through the point
   (i) (0, 1), (ii) (1, 0).
   (c) Find the equation of the line through (2, 2) which is perpendicular to $y = 2x$.
4. Given the points $P(7, 4)$, $Q(9, -2)$, $R(-5, -20)$ and $S(-3, 4)$, show by means of gradients that PR is perpendicular to QS.
5. L is the point $(-2, -1)$, $M(0, 5)$ and $N(4, -2)$.
   (a) Find the equation of the line through N which is parallel to LM.
   (b) Find the equation of the line through N which is perpendicular to LM.
6. $ABCD$ is a rectangle, with A at $(-3, 1)$ and B at $(1, -1)$.
   Find the equations of the two sides AB and BC of the rectangle.

Rearranging equations

The general equation for a straight line graph is $y = mx + c$.
When an equation is in this form the gradient and y-intercept are given by the values of m and c.

The equation for a straight line can also be written in the form $px + qy = r$.
To find the gradient and y-intercept of this line we must first **rearrange** the equation.

For example, the graph of a straight line is given as $2y - 6x = 5$.
To rearrange: $2y - 6x = 5$

Add $6x$ to both sides.

$2y = 6x + 5$

Divide both sides by 2.

$y = 3x + 2.5$

Hence, the line has gradient 3 and y-intercept 2.5.

EXAMPLES

1 The graph of a straight line is given by the equation $4y - 3x = 8$.
(a) Write this equation in the form $y = mx + c$.
(b) Write down the gradient and the y-intercept of the line.

(a) $4y - 3x = 8$

Add $3x$ to both sides.

$4y = 3x + 8$

Divide both sides by 4.

$y = \frac{3}{4}x + 2$

(b) The line has gradient $\frac{3}{4}$ and y-intercept 2.

2 The equation of a straight line is $6x + 3y = 2$.
Write down the equation of another line which is parallel to this line.

Write the equation in the form $y = mx + c$.

$6x + 3y = 2$

Subtract $6x$ from both sides.

$3y = -6x + 2$

Divide both sides by 3.

$y = -2x + \frac{2}{3}$

The gradient of the line is -2.
To write an equation of a parallel line keep the same gradient and change the value of the y-intercept.
For example: $y = -2x + 5$

3 Find the equation of the line through (2, 3) which is perpendicular to the line $y = 2x - 3$.

The equation of the line will be of the form $y = mx + c$.
The gradient of the line $y = 2x - 3$ is 2.

$m \times 2 = -1$ $(m_{AB} \times m_{CD} = -1)$

$m = -\frac{1}{2}$

The line passes through the point (2, 3) and has gradient $-\frac{1}{2}$.

Substitute $x = 2$, $y = 3$ and $m = -\frac{1}{2}$ into $y = mx + c$.

$3 = -\frac{1}{2} \times 2 + c$

$3 = -1 + c$

$c = 4$

The equation of the line is $y = -\frac{1}{2}x + 4$.

Exercise 14.6

1. The graph of a straight line is given by the equation $2y - 3x = 6$.
   Write this equation in the form $y = mx + c$.

2. Write the equations of the following lines in the form $y = mx + c$.
   (a) $2y + x = 4$ (b) $5y + 4x = 20$ (c) $4 - 3y = 2x$ (d) $2x - 7y = 14$

3. Find the gradients of these lines.
   (a) $2y = x$ (b) $x - y = 0$ (c) $2x - y = 0$
   (d) $3y + x = 0$ (e) $4y - 3x = 0$ (f) $2x + 5y = 0$

4. These lines cross the y axis at the point $(0, a)$. Find the value of a for each line.
   (a) $y = x$ (b) $2y = x + 1$ (c) $3y + 6 = x$
   (d) $2y - 5 = x$ (e) $4y - 3x = 8$ (f) $x + 5y = 2$

5. Write down an equation which is parallel to each of these lines.
   (a) $y = x$ (b) $y = 2x - 3$ (c) $2y = x - 4$

6. The equation of a straight line is $2y + 3x = 4$.
   Write down the equation of another line which is parallel to this line.

7. Which of these pairs of lines are perpendicular to each other?
   (a) $y = 4x$, $y = \frac{1}{4}x$. (b) $3y = x - 6$, $y = 1 - 3x$. (c) $5y - 2x = 3$, $2y = 5x$.

8. The diagram shows a sketch of the line $2y = x + 4$.
   (a) Find the coordinates of points A and B.
   (b) What is the gradient of the line?

   Copy the diagram.
   (c) (i) Draw the sketch of another line that has the same gradient as $2y = x + 4$.
   (ii) What is the equation of the line you have drawn?

9. The diagram shows a sketch of the line $2y = 6 - x$.
   (a) Find the coordinates of points P and Q.
   (b) What is the gradient of the line?

   Copy the diagram.
   (c) (i) Draw the sketch of another line that has the same gradient as $2y = 6 - x$.
   (ii) What is the equation of the line you have drawn?

10. In each part, find the equation of the line, through P, which is
    (a) parallel to the given line,
    (b) perpendicular to the given line.
    (i) $P(0, 2)$, $y = x + 1$.
    (ii) $P(1, 3)$, $y = 2x - 1$.
    (iii) $P(-1, 5)$, $3y = x + 6$.
    (iv) $P(-4, -1)$, $4y + x = -8$.

11. In the diagram, the lines AB and CD are perpendicular to each other and intersect at the point $(1, 4)$.
    The line AB crosses the x axis at $(3, 0)$.

    Calculate the coordinates of the points P and Q.

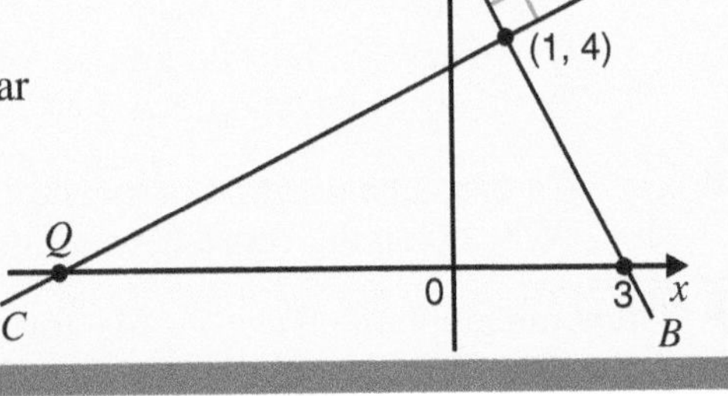

What you need to know

- **Coordinates** are used to describe the position of a point on a graph.
- The x axis is the line $y = 0$. The y axis is the line $x = 0$.
- The graph of a **linear function** is a **straight line**.
  The general equation of a linear function is $y = mx + c$, where m is the **gradient** of the line and c is the **y-intercept**.
- The **gradient** of a line can be found by drawing a right-angled triangle.

  $$\text{Gradient} = \frac{\text{distance up}}{\text{distance along}}$$

  Gradient can be positive, zero or negative.

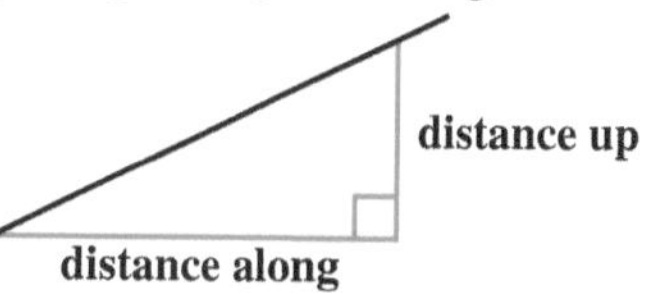

- If two lines are **perpendicular** to each other, the product of their gradients is -1.

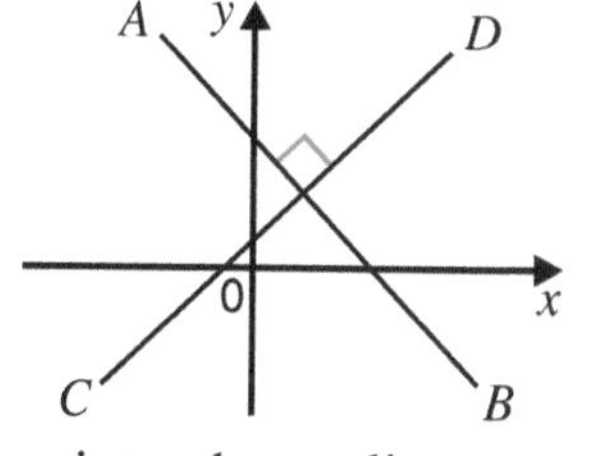

$$m_{AB} \times m_{CD} = -1$$

$$m_{AB} = \frac{-1}{m_{CD}}$$

- The points where a line crosses the axes can be found:
  by reading the coordinates from a graph,
  by substituting $x = 0$ and $y = 0$ into the equation of the line.
- Equations of the form $px + qy = r$ can be **rearranged** to the form $y = mx + c$.

You should be able to:

- Substitute values into given functions to generate points.
- Plot graphs of **linear functions**.
- Use graphs of linear functions to solve equations.
- Find the equation for a given line.

Review Exercise 14

1 (a) Write down the equations of the lines labelled on this graph.
(b) What are the coordinates of the point where lines (1) and (2) cross?

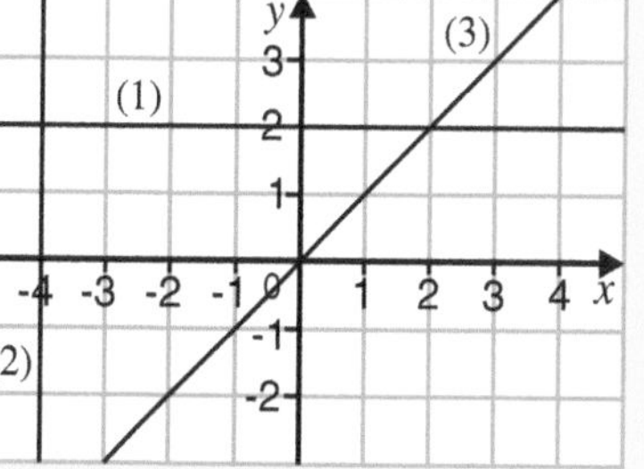

2 (a) Copy and complete the table of values for the graph of $x + y = 4$.

x	0	1	2	3	4
y	4			1	

(b) Draw the graph of $x + y = 4$.
(c) P is a point on the line $x + y = 4$.
David says, "The x coordinate of P is one greater than the y coordinate of P."
Write down the coordinates of P.

AQA

3 (a) On the same diagram draw and label the lines: $y = 2x$ and $x + y = 6$
(b) Write down the coordinates of the point where the lines cross.

4 Draw the graph of $2y + x = 4$ for values of x from -2 to 4.

5 In an experiment, weights are added to a spring and the length of the spring is measured.
The graph shows the results.

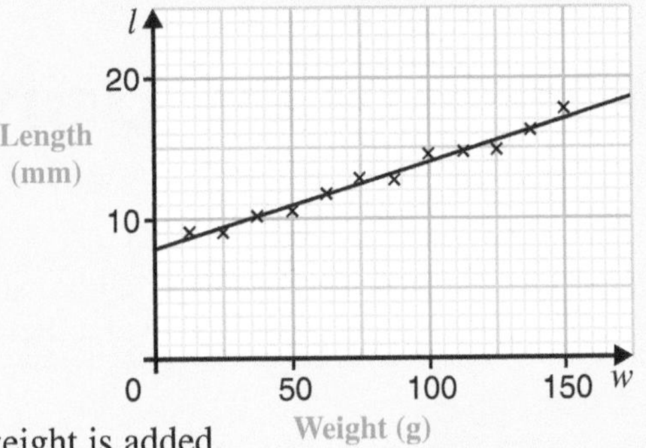

A line of best fit has been drawn.

(a) Estimate the length of the spring when no weight is added.

(b) Calculate the gradient of the line.

(c) Write down the equation of the line in the form $l = mw + c$.

(d) Use your equation to estimate the length of the spring for a weight of 300 g.

6 The graph shows the line AB. Work out the equation of the line AB.

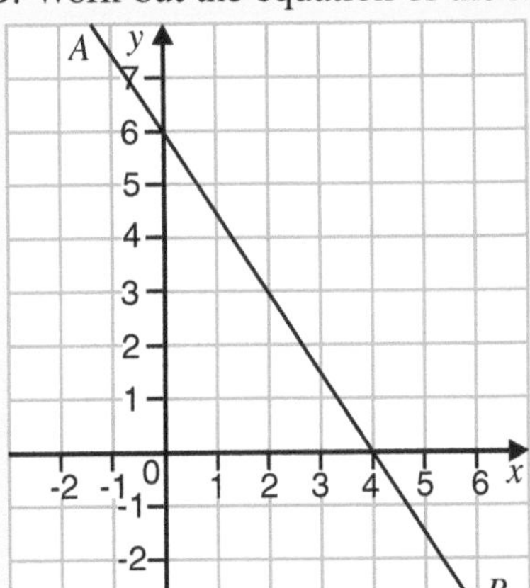

AQA

7 (a) The graph of a straight line is given by the equation $y - 3x = 2$.
Write down the equation of a line parallel to this line which passes through the point (0, 5).

(b) The graph of a different line is given by the equation $3y - 2x = -6$.

(i) What is the gradient of this line?

(ii) Find the coordinates of the point where this line crosses the y axis.

8 P is the point $(-1, -9)$ and Q is the point (2, 3).
Find the equation of the line segment PQ, in the form $y = mx + c$.

9 Show that the lines $2y = x + 2$ and $y = 6 - 2x$ are perpendicular to each other.

10 Find the equation of the straight line passing through the point (0, 5) which is perpendicular to the line $y = \frac{2}{3}x + 3$

AQA

11 The diagram shows a sketch of a straight line which passes through A (0, 4) and B (6, 0).

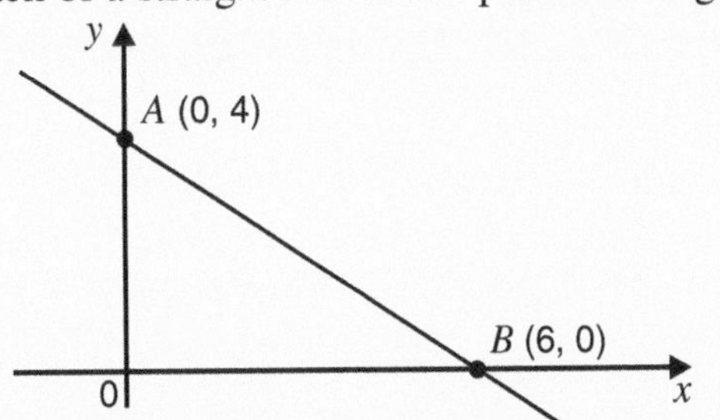

(a) Find the equation of the line which passes through A and B.

(b) Find the equation of the line which passes through (0, 2) and is parallel to AB.

(c) Find the equation of the line which passes through the midpoint of AB and is perpendicular to AB.

Using Graphs

Graphs are sometimes drawn to show real-life situations.
In most cases a quantity is measured over a period of time.

EXAMPLE

Craig drew a graph to show the amount of fuel in the family car as they travelled to their holiday destination. He also made some notes:

Part of Graph	Event
A	Leave home.
A to B	Motorway.
B to C	Car breaks down.
C to D	On our way again.
D to E	Stop for lunch.
E to F	Fill tank with fuel.
F to G	Country roads.
G	Arrive, at last!

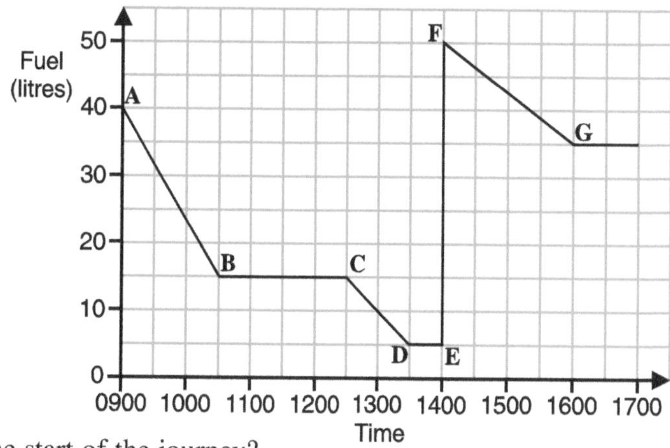

(a) How much fuel was in the tank at the start of the journey?
(b) At what time did the car break down?
(c) How long did the family stop for lunch?
(d) How much fuel was put into the tank at the garage?

(a) 40 litres (b) 1030 (c) $\frac{1}{2}$ hour (d) 45 litres

Use the graph to work out how many litres of fuel were used for the journey.

Notice that in this example and others involving graphs against time:

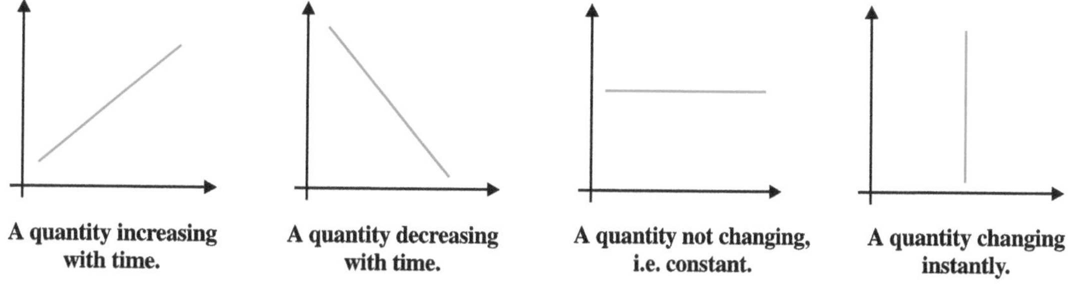

A quantity increasing with time. **A quantity decreasing with time.** **A quantity not changing, i.e. constant.** **A quantity changing instantly.**

Exercise 15.1

1. A climber pulls a rucksack up a vertical cliff face using a rope.
Which of the graphs below could represent the motion of the rucksack against time?

A Height / Time
B Height / Time
C Height / Time
D Height / Time

2 Match the graphs to the situations. In each graph, the horizontal axis represents time and the vertical axis represents speed.

1 A runner starts from rest and begins to pick up speed.

2 A runner keeps having to stop and start.

3 A runner gets tired and has to slow down.

4 A runner sees the finish, builds up speed and sprints to the line.

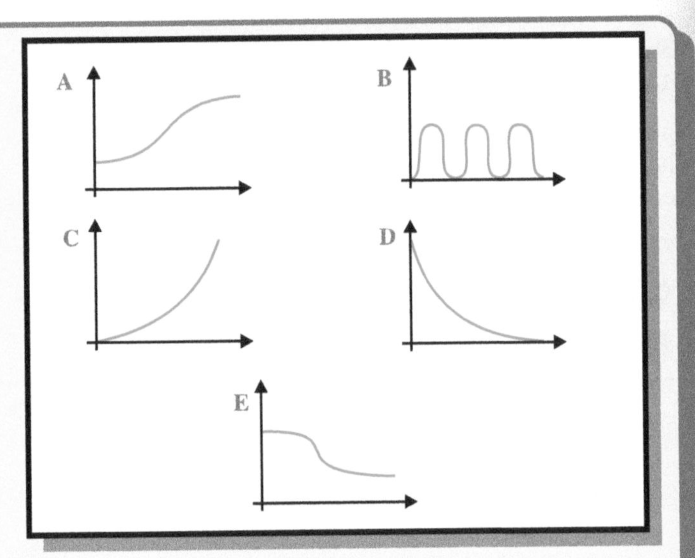

3 Cans of drink can be bought from a vending machine in the school canteen.
At 0900 one day the machine is three-quarters full.
During break, from 1045 to 1100, drinks are bought from the machine at a steady rate.
By the end of the break the machine is one-quarter full. At 1200 the machine is filled.
The lunch break is from 1230 to 1330. Someone complains at 1315 that the machine is empty.
The machine is filled at 1400, ready for the afternoon break from 1445 to 1500.
At 1600 the machine is three-quarters full.
Sketch a graph to show the number of drinks in the machine from 0900 to 1600.

4 Two companies each use a formula to calculate the charge made for hiring out scaffolding.

Company A uses the formula $c = 50 + 10d$,
Company B uses the formula $c = 15d + 20$,
where c is the total charge, in pounds,
d is the length of the hire period, in days.

(a) Draw the horizontal axis for d from 0 to 10 and the vertical axis for c from 0 to 180.
(b) Draw the graph of $c = 50 + 10d$.
(c) Draw the graph of $c = 15d + 20$.

Use your graph to answer the following.

(d) From which company is it cheaper to hire scaffolding for 2 days?
(e) From which company is it cheaper to hire scaffolding for 8 days?
(f) For what number of days do both companies make the same charge?

5 Water is poured into some containers at a constant rate.
Copy the axes given and sketch the graph of the depth of the water against time for each container as it is filled.

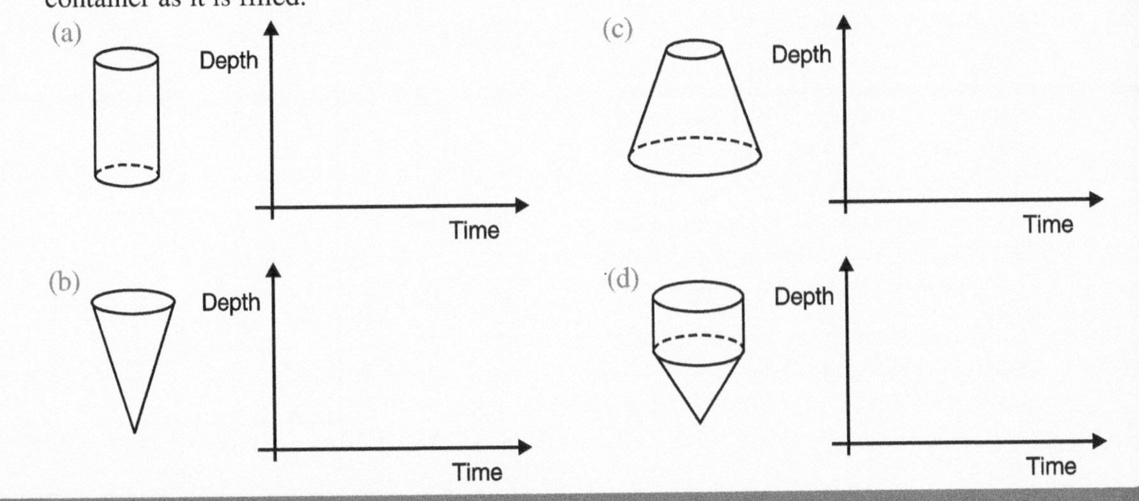

Distance-time graphs

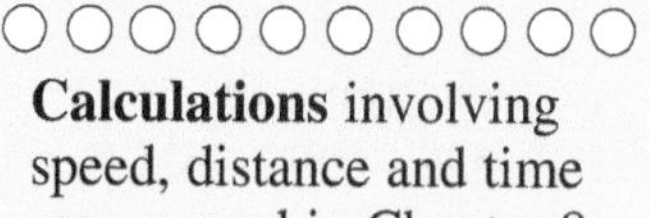

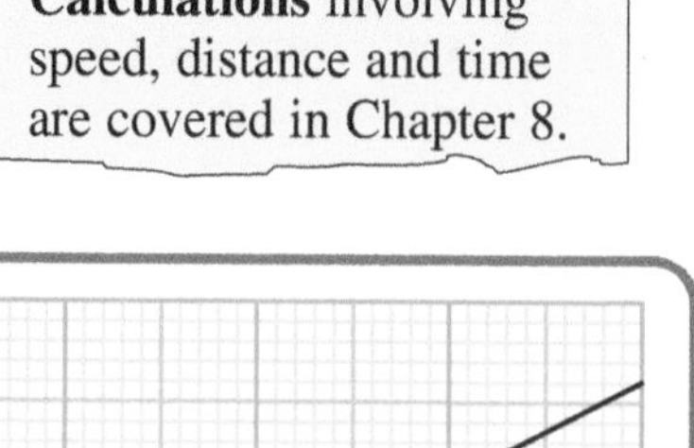

Calculations involving speed, distance and time are covered in Chapter 8.

Distance-time graphs are used to illustrate journeys.

Speed is given by the gradient, or slope, of the line.
The faster the speed the steeper the gradient.
Zero gradient (horizontal line) means zero speed (not moving).

EXAMPLES

1 The graph shows a bus journey.
(a) How many times does the bus stop?
(b) On which part of the journey does the bus travel fastest?

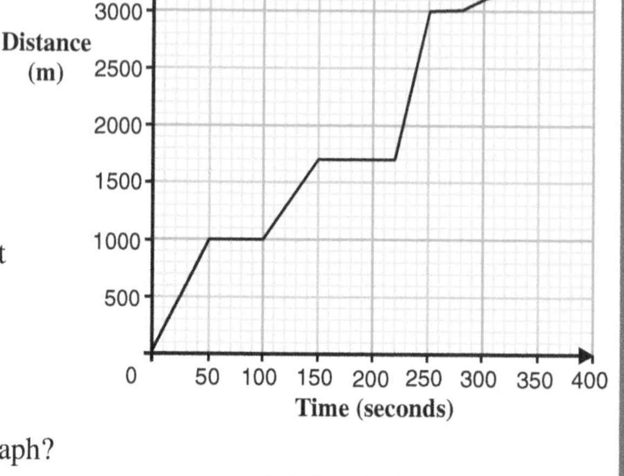

(a) At zero speed the distance-time graph is horizontal.
So, the bus stops 3 times.
(b) The bus travels fastest when the gradient of the distance-time graph is steepest.
So, the bus travels fastest between the second and third stops.

2 What speed is shown by this distance-time graph?
Give your answer in metres per second.

A distance of 30 metres is travelled in a time of 6 seconds.
Using Speed = Distance ÷ Time
Speed = 30 ÷ 6
= 5 metres per second

Remember:

Speed can be calculated from the gradient of a distance-time graph.

When the distance-time graph is **linear** the **speed is constant.**

When the distance-time graph is **horizontal** the **speed is zero.**

$$\text{Average speed} = \frac{\text{total distance}}{\text{total time}}$$

Exercise 15.2

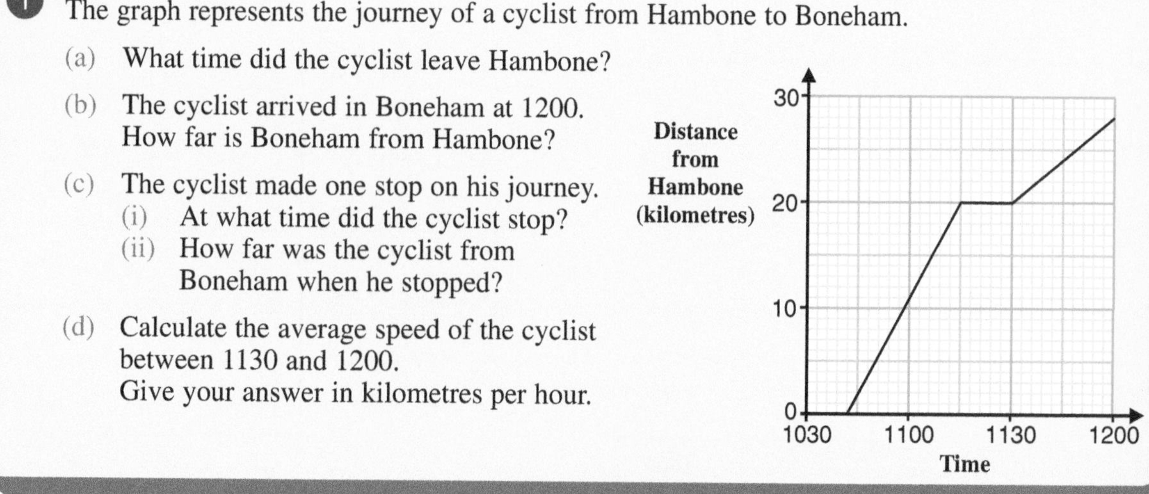

1 The graph represents the journey of a cyclist from Hambone to Boneham.
(a) What time did the cyclist leave Hambone?
(b) The cyclist arrived in Boneham at 1200.
How far is Boneham from Hambone?
(c) The cyclist made one stop on his journey.
(i) At what time did the cyclist stop?
(ii) How far was the cyclist from Boneham when he stopped?
(d) Calculate the average speed of the cyclist between 1130 and 1200.
Give your answer in kilometres per hour.

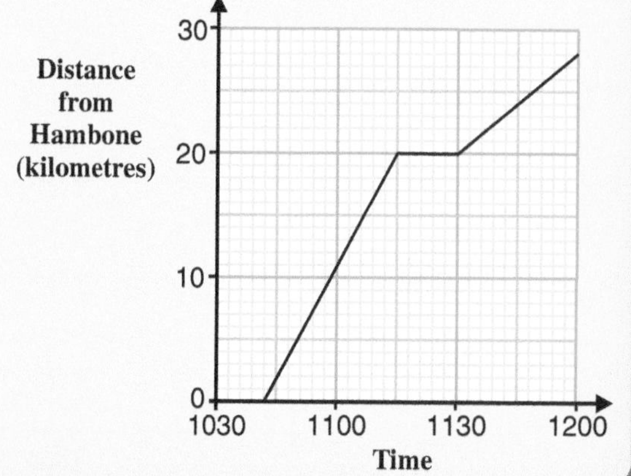

2 (a) The graph represents the journey of a car. What is the speed of the car in kilometres per hour?

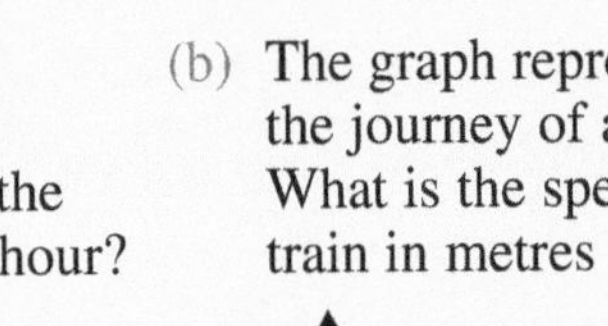

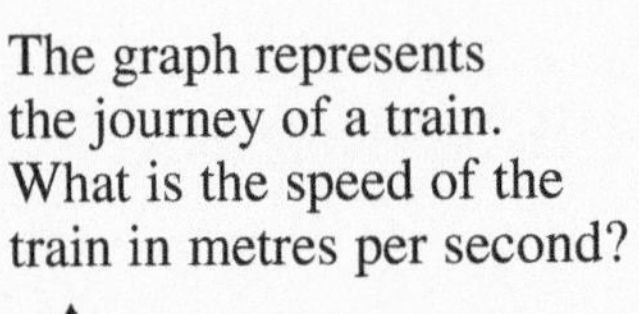

(b) The graph represents the journey of a train. What is the speed of the train in metres per second?

(c) The graph represents the speed of a cyclist. What is the speed of the cyclist in miles per hour?

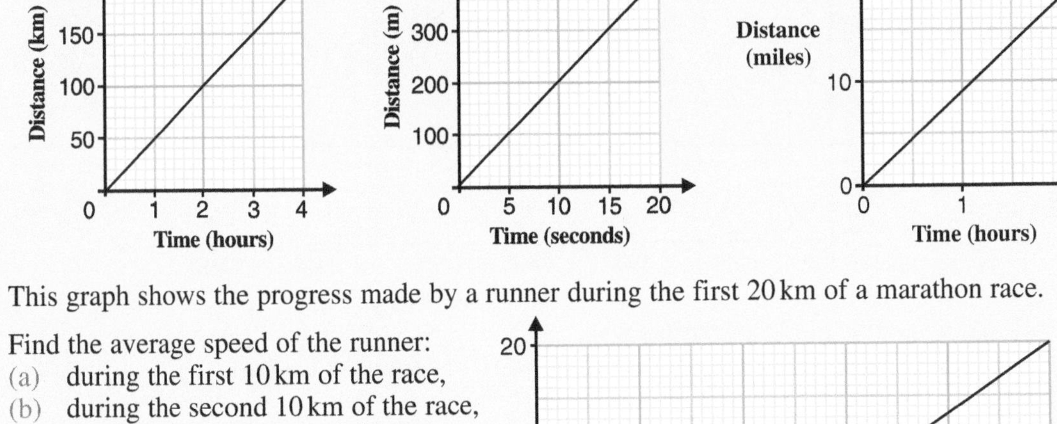

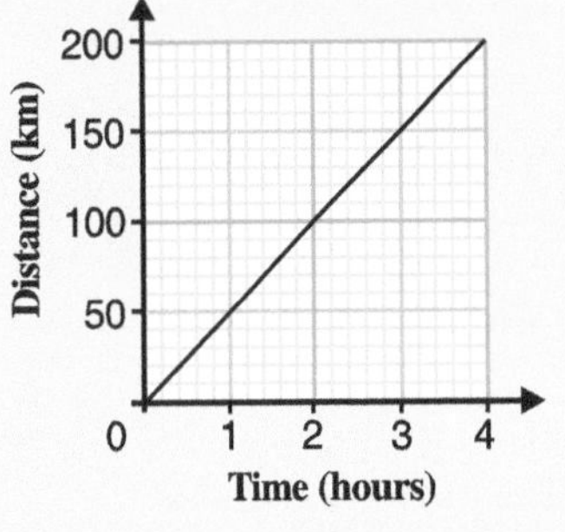

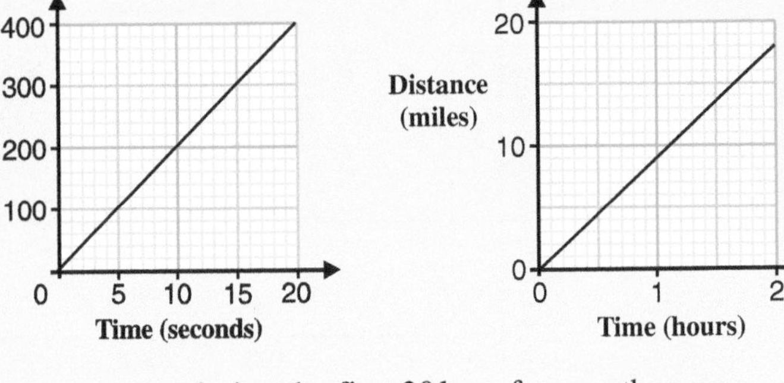

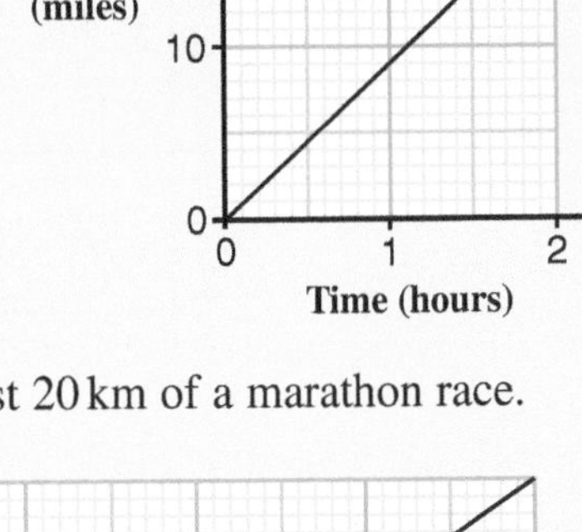

3 This graph shows the progress made by a runner during the first 20 km of a marathon race.

Find the average speed of the runner:

(a) during the first 10 km of the race,
(b) during the second 10 km of the race,
(c) during the first 20 km of the race.

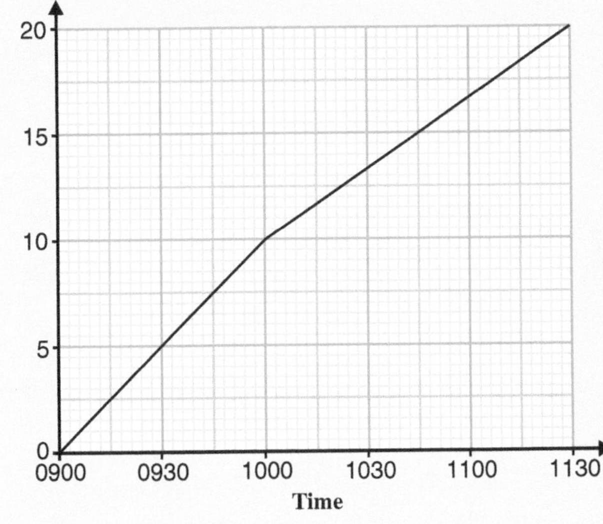

4 The graph represents a bus journey from Poole.

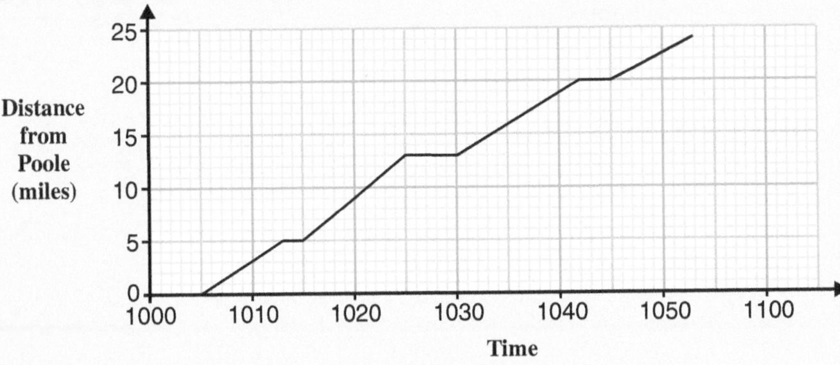

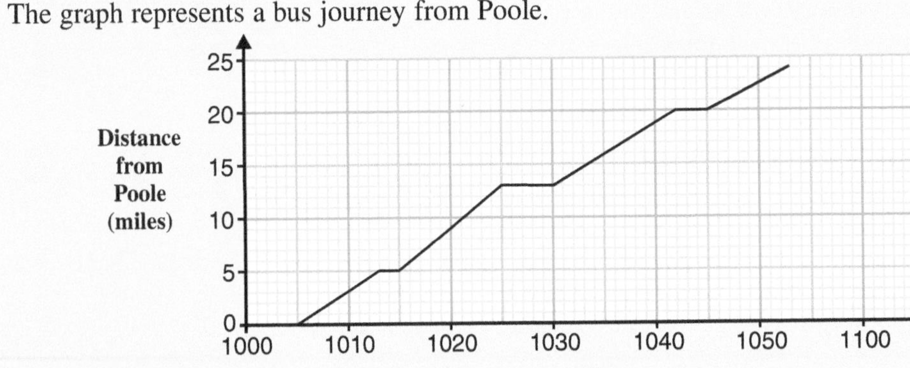

Calculate, in miles per hour, the average speed of the bus:

(a) between 1015 and 1025, (b) for the whole journey.

5 Selby is 20 miles from York.

(a) Kathryn leaves Selby at 1030 and drives to York.
She travels at an average speed of 20 miles per hour.
Draw a distance-time graph to represent her journey.

(b) At 1030 Matt leaves York and drives to Selby.
He travels at an average speed of 30 miles per hour.
(i) On the same diagram draw a distance-time graph to represent his journey.
(ii) At what time does Matt arrive in Selby?

6 Dan walks around Bolam Lake. He starts and finishes his walk at the car park beside the lake. The distance-time graph shows his journey.

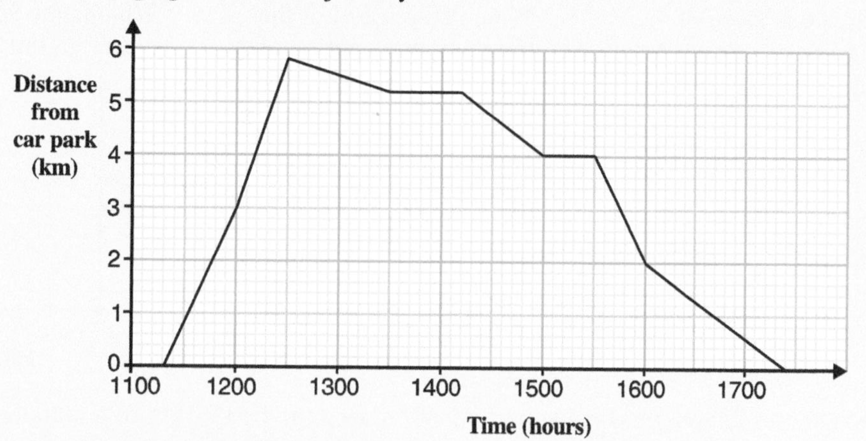

(a) At what time did Dan start his walk?
(b) At what time did Dan reach the furthest distance from the car park?
(c) How many times did Dan stop during his walk?
(d) Dan stopped for lunch at 1330.
For how many minutes did he stop for lunch?
(e) Between what times did Dan walk the fastest?
(f) At what speed did Dan walk between 1530 and 1600?

7 The graph represents the journey of a cyclist from Bournemouth to the New Forest.

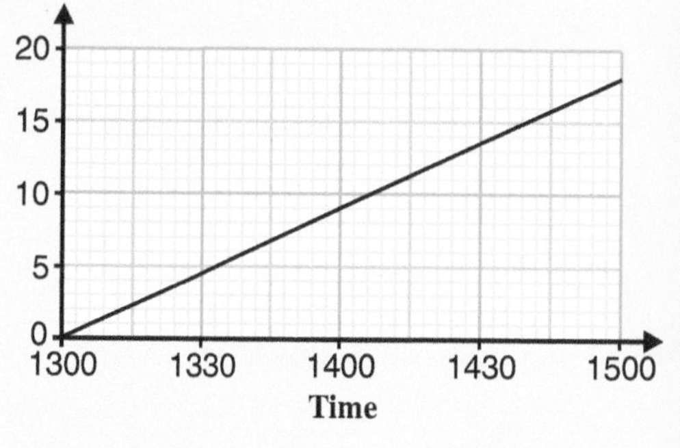

(a) What is the average speed of the cyclist in miles per hour?
(b) Another cyclist is travelling from the New Forest to Bournemouth at an average speed of 12 miles per hour. At 1300 the cyclist is 15 miles from Bournemouth.
(i) Copy the graph and on the same diagram draw a graph to show the journey of the cyclist to Bournemouth.
(ii) At what time does the cyclist arrive in Bournemouth?

8 Billy has a 10 km journey to school.
He leaves home at 8.05 am and takes 10 minutes to walk 1 km to Jane's house.
Billy gets a lift with Jane's mum at 8.20 am. Billy arrives at school at 8.45 am.
(a) Using 1 cm for 5 minutes on the time axis and 1 cm for 1 km on the distance axis draw a distance-time graph for Billy's journey.
(b) Find Billy's average speed for his whole journey in km/hour.

9 Emby and Ashwood are two towns 14 km apart.
Kim leaves Emby at 1000 and walks at a steady pace of 6 km/hour towards Ashwood.
Ray leaves Ashwood at 1040 and cycles towards Emby.
Ray travels at a steady speed of 21 km/hour for the first 20 minutes and then at a slower steady speed for the rest of his journey.
Ray arrives at Emby at 1130.
(a) On the same diagram and using a scale of 1 cm for 10 minutes on the time axis and 1 cm for 1 km on the distance axis, draw a distance-time graph to show both journeys.
(b) Use your diagram to find:
(i) the time that Kim arrives at Ashwood,
(ii) the time that Kim and Ray pass each other,
(iii) Ray's speed for the slower part of his journey.

Speed-time graphs

Acceleration is the rate of change of speed.
Acceleration is given by the gradient, or slope, of the line.
The greater the acceleration, the steeper the gradient.
Zero gradient (horizontal line) means zero acceleration (moving at a constant speed).
Acceleration = speed ÷ time
Acceleration is measured in metres per second squared (m/s^2).

Acceleration can be calculated from the gradient of a speed-time graph.

When the speed-time graph is **linear** the **acceleration is constant**.

When the speed-time graph is **horizontal** the **speed is constant** and the **acceleration is zero**.

The **area** enclosed by a speed-time graph represents the **distance** travelled.

EXAMPLE

The graph shows Jane's speed plotted against time during a run.

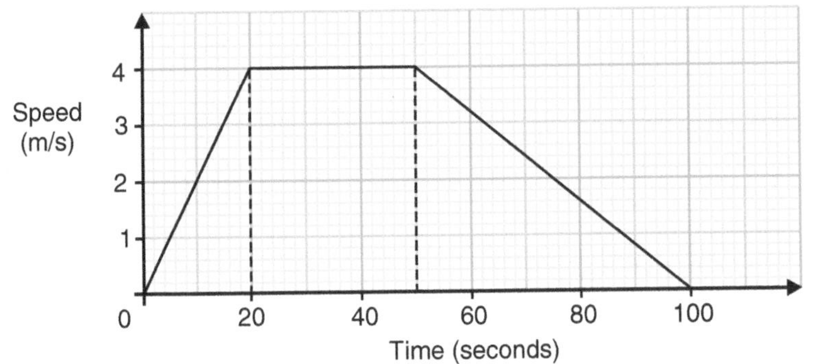

(a) Give a brief description of Jane's run.
(b) Calculate Jane's acceleration during the first 20 seconds of her run.
(c) Calculate Jane's deceleration during the last 50 seconds of her run.
(d) How far did Jane run?

(a) Jane accelerates at a constant rate for the first 20 seconds of her run.
She then runs at a constant speed of 4 m/s for 30 seconds before slowing at a constant rate to a stop after 100 seconds.

(b) Acceleration = speed ÷ time
= 4 ÷ 20
= $0.2\,m/s^2$

From the graph:
Positive gradient = getting faster, **accelerating**.
Negative gradient = slowing down, **decelerating**.

(c) Deceleration = speed ÷ time
= 4 ÷ 50
= $0.08\,m/s^2$

(d) The area under the graph represents the distance travelled, in metres.
Distance travelled = $\frac{1}{2}(30 + 100) \times 4$
= 260 m

Show that the distances Jane travels whilst accelerating, running at a constant speed and decelerating, are 40 m, 120 m and 100 m respectively.

Exercise 15.3

Try to do questions 1 to 3 without using your calculator.

1 A car accelerated steadily from rest to 12 m/s in 15 seconds.
What was the acceleration of the car in m/s^2?

2 Look at the points on this speed-time graph.

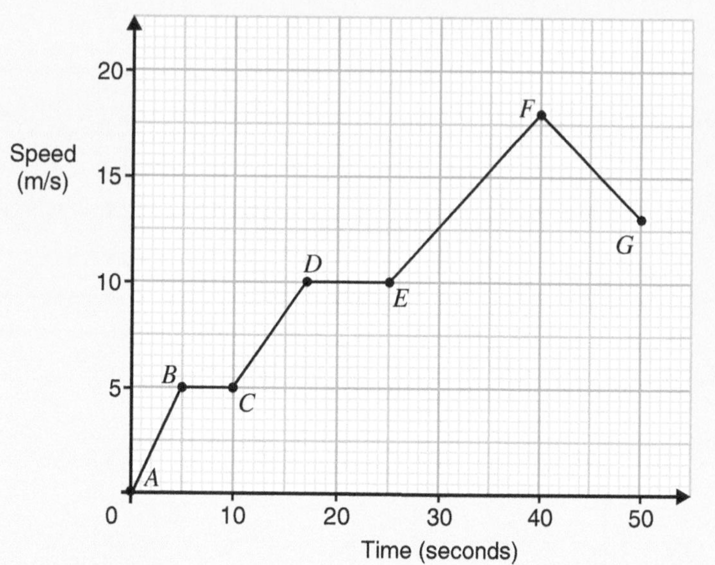

(a) Between which points is:
  (i) the speed constant,
  (ii) the acceleration greatest,
  (iii) the acceleration zero,
  (iv) the acceleration negative?

(b) Calculate the acceleration between points:
  (i) *E* and *F*,
  (ii) *F* and *G*.

3 The speed-time graph of an underground train travelling between two stations is shown.

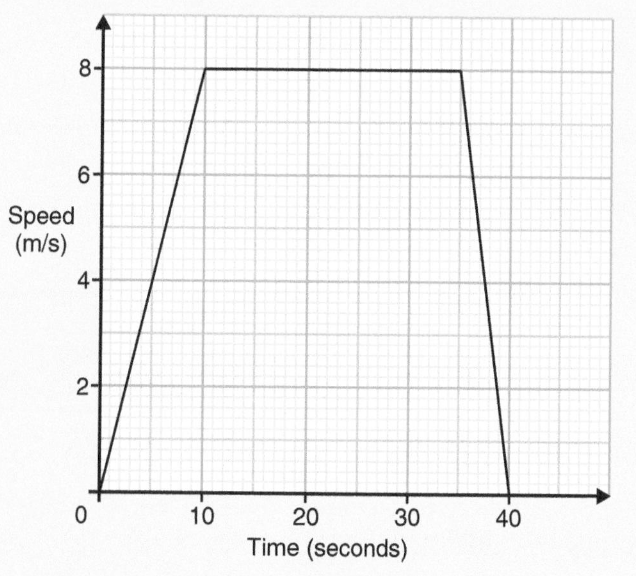

(a) What was the maximum speed of the train?
(b) Calculate the acceleration of the train at the beginning of the journey.
(c) Calculate the deceleration of the train at the end of the journey.
(d) Calculate the distance between the stations, in metres.

4 The diagram shows a speed-time graph for a train travelling between station A and station B.

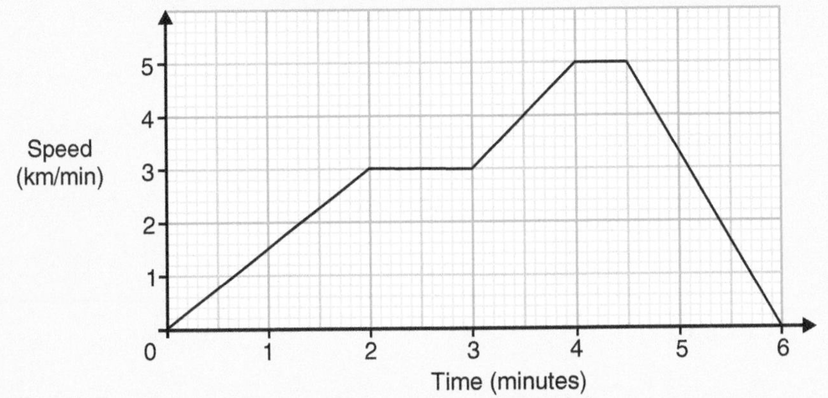

(a) For how long did the train accelerate at a constant rate at the start of its journey?
(b) What was the total time during the journey that the train was **not** accelerating?
(c) Show that the train was slowing down at a rate of approximately $0.9\,\text{m/s}^2$ as it approached station B.

5 A car takes 12 seconds to accelerate uniformly from rest to 24 m/s.
It then travels at 24 m/s for 30 seconds before slowing down steadily to rest.
The whole journey takes 50 seconds.
(a) Draw the speed-time graph of the car.
(b) How far does the car travel?

6 The speed-time graph shows the journey of a van between two road junctions.

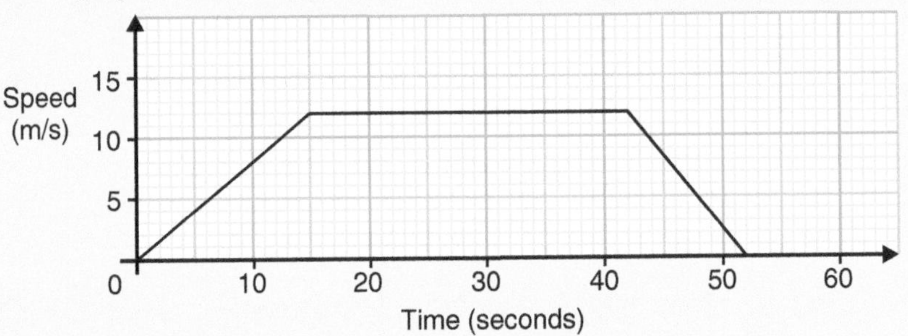

(a) Calculate the acceleration of the van at the beginning of the journey.
(b) Calculate the time it takes the van to travel half the distance between the junctions.

What you need to know

- A **gradient** measures the **rate of change** of one quantity with respect to another.
  A **positive** gradient represents a **rate of increase**.
  A **negative** gradient represents a **rate of decrease**.
- The gradient of a **distance-time graph** gives the speed.
  Speed is the rate of change of distance with respect to time.
  When the distance-time graph is **linear** the **speed is constant**.
  When the distance-time graph is **horizontal** the **speed is zero**.
- The gradient of a **speed-time graph** gives the acceleration.
  Acceleration is the rate of change of speed with respect to time.
  When the speed-time graph is **linear** the **acceleration is constant**.
  When the speed-time graph is **horizontal** the **speed is constant** and the **acceleration is zero**.
- The **area** enclosed by the graph on a speed-time graph represents the **distance** travelled.

You should be able to:

- Draw and interpret graphs which represent real-life situations.

Review Exercise 15

1 The graph illustrates a 1000 metre race between Nina and Polly.

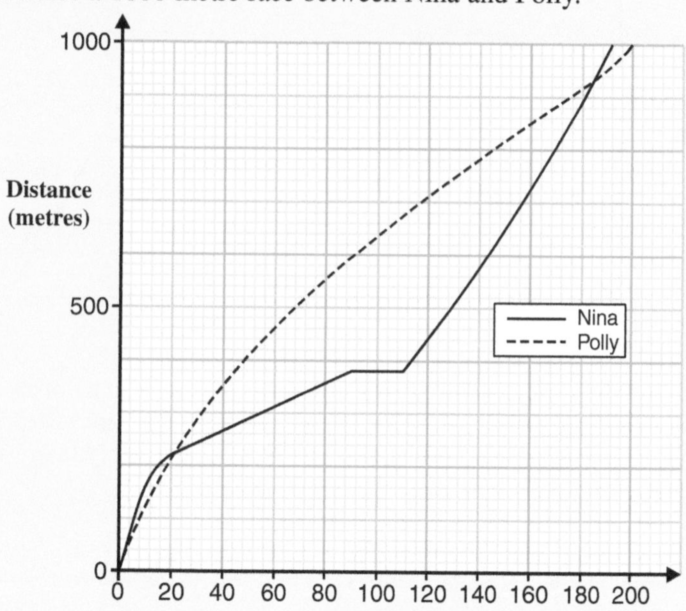

(a) Who was in the lead 10 seconds after the start of the race?
(b) Describe what happened 20 seconds after the start of the race.
(c) Describe what happened to Nina 90 seconds after the start of the race.
(d) Who won the race?

AQA

2 Decide which graph matches each relationship.

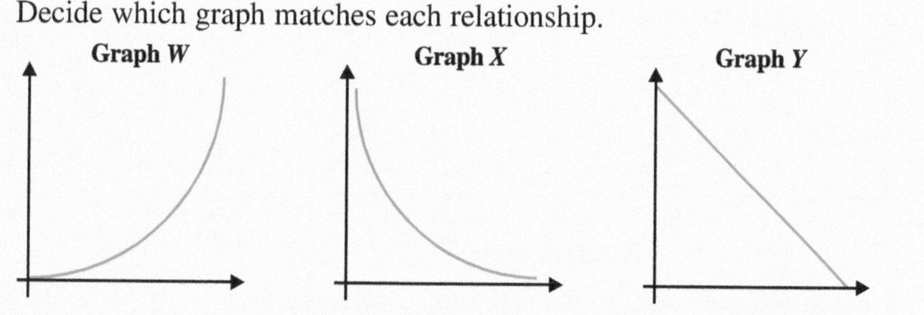

Relationships: *A*: The area of a circle plotted against its radius.
B: The circumference of a circle plotted against its radius.
C: The length of a rectangle of area $24\,\text{cm}^2$ plotted against its width.

AQA

3 Water flows into some containers at a constant rate.

(a) Sketch the graphs of the depths of the water against time.

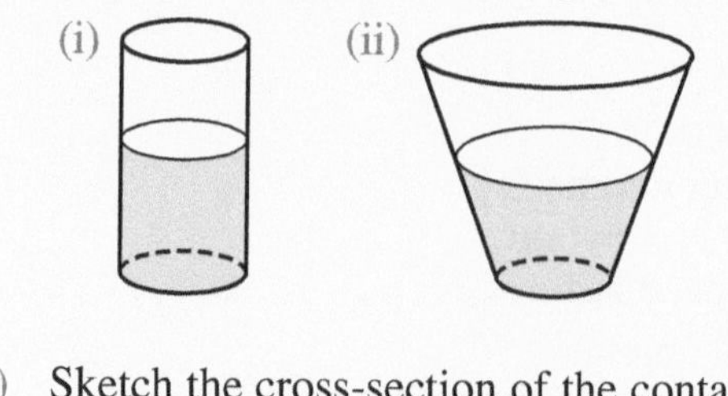

(b) Sketch the cross-section of the container that generated this graph.

AQA

4. Jenny cycles to school.
The graph shows her journey from home to school.
On the way she stops to talk to her friends.

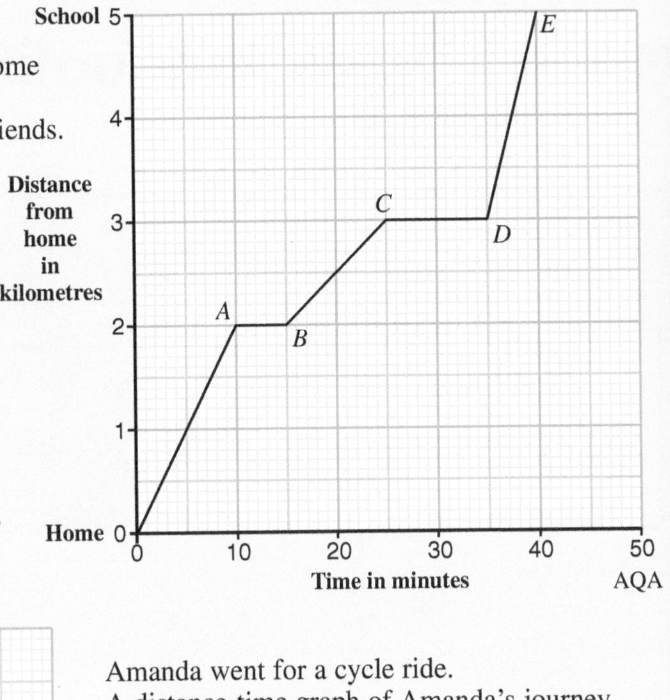

(a) How many times does she stop to talk to her friends?
(b) (i) How far does she travel in the first stage from home to A?
(ii) What is her average speed over the first ten minutes? Give your answer in kilometres per hour.
(c) On which stage of the journey is her average speed the fastest?

AQA

5. Amanda went for a cycle ride.
A distance-time graph of Amanda's journey is shown.

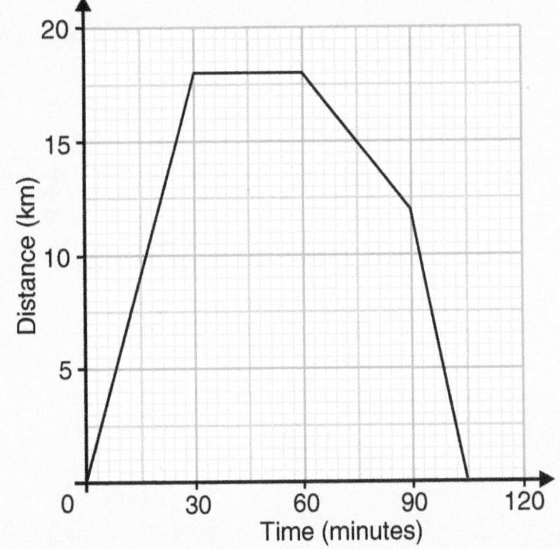

(a) How far did Amanda cycle before she stopped for a rest?
(b) Calculate Amanda's average speed for the whole cycle ride.

AQA

6. The distance from Ashby to Banborough is 16 km.
The distance from Banborough to Calby is 8 km.
John leaves Ashby at noon.
He walks towards Banborough for 1 hour at an average speed of 6 km/h and then rests for 20 minutes.
John then runs the remaining distance to Banborough at an average speed of 10 km/h.
At Banborough John talks to a friend for 20 minutes and borrows his bicycle.
He then cycles to Calby at an average speed of 16 km/h.
John leaves Calby at 1600. He cycles home over the same route arriving at 1650.
Using a scale of 1 cm for 20 minutes and 1 cm for 2 km draw a distance-time graph for John's journey.

7. This graph shows the speed of a van plotted against time for the first 50 seconds of a journey.

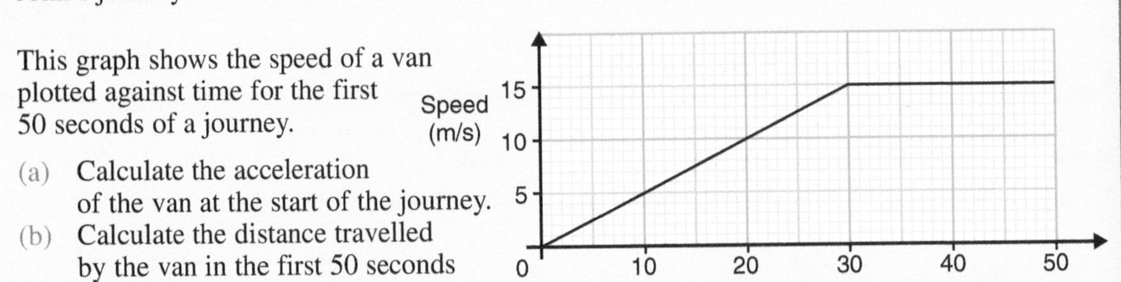

(a) Calculate the acceleration of the van at the start of the journey.
(b) Calculate the distance travelled by the van in the first 50 seconds of the journey.

Inequalities

Activity

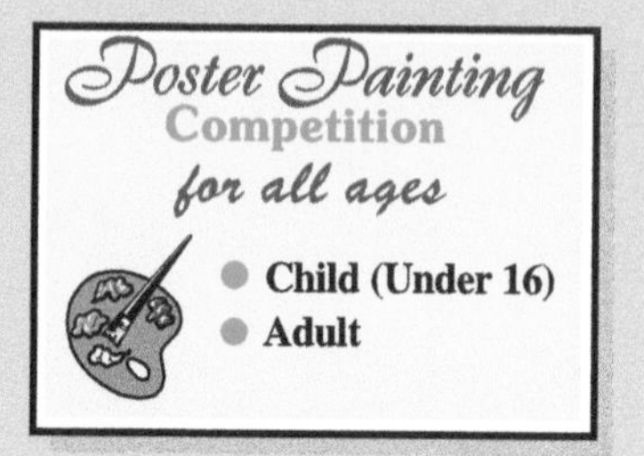

For all children who enter the competition we can say that
Age $<$ 16 years

For anyone riding the Big Dipper we can say that
Height $\geqslant$ 1.2 m

For all items sold in the store we can say that
Cost $\leqslant$ £1

These are examples of inequalities.
Can you think of other situations where inequalities are used?

Inequalities

An **inequality** is a mathematical statement, such as $x > 1$, $a \leqslant 2$ or $-3 \leqslant n < 2$.

In the following, x is an integer.

Sign	Meaning	Example	Possible values of x
$<$	is less than	$x < 4$	3, 2, 1, 0, −1, −2, −3, …
$\leqslant$	is less than or equal to	$x \leqslant 4$	4, 3, 2, 1, 0, −1, −2, −3, …
$>$	is greater than	$x > 6$	7, 8, 9, 10, …
$\geqslant$	is greater than or equal to	$x \geqslant 2$	2, 3, 4, 5, …

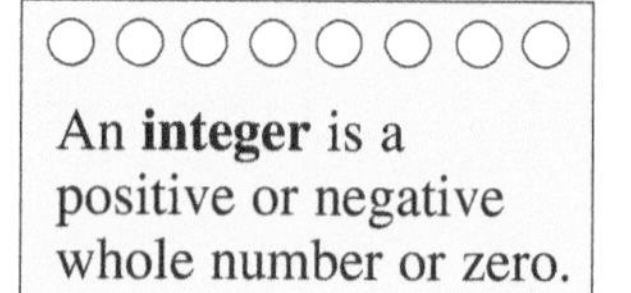

An **integer** is a positive or negative whole number or zero.

Explain the difference between the meanings of the signs $<$ and $\leqslant$.
Explain the difference between the meanings of the signs $>$ and $\geqslant$.

Number lines

Inequalities can be shown on a **number line**.

-6 -5 -4 -3 -2 -1 0 1 2 3 4 5 6

As you move to the right, numbers get bigger.
As you move to the left, numbers get smaller.

The number line below shows the inequality $-2 < x \leqslant 3$.

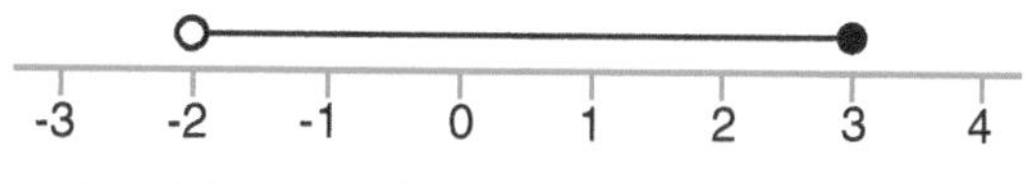

The circle at 3 is **filled** to show that 3 is **included**.
The circle at -2 is **not filled** to show that -2 is **not included**.

EXAMPLES

1 Draw number lines to show these inequalities:

(a) $x < 1.5$ (b) $x \geqslant -2$ (c) $x \leqslant 4$ and $x > -1$

(a) $x < 1.5$

(b) $x \geqslant -2$

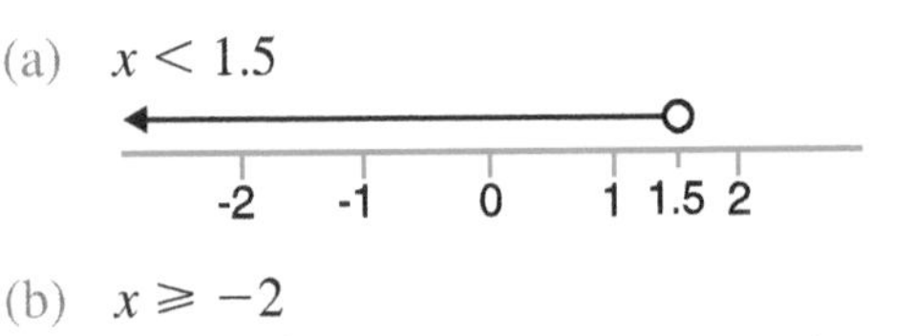

The circle is:
filled if the inequality is **included**,
not filled if the inequality is **not included**.

(c) $x \leqslant 4$ and $x > -1$.

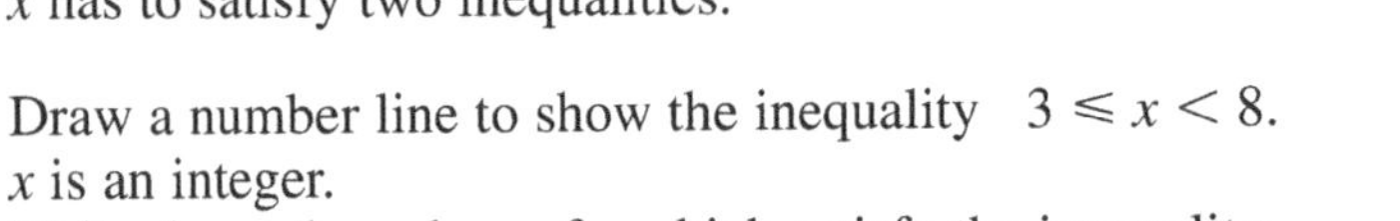

x has to satisfy two inequalities.

2 (a) Draw a number line to show the inequality $3 \leqslant x < 8$.
(b) x is an integer.
Write down the values of x which satisfy the inequality.

$3 \leqslant x < 8$
is a shorthand method of writing
$3 \leqslant x$ **and** $x < 8$.

(a)

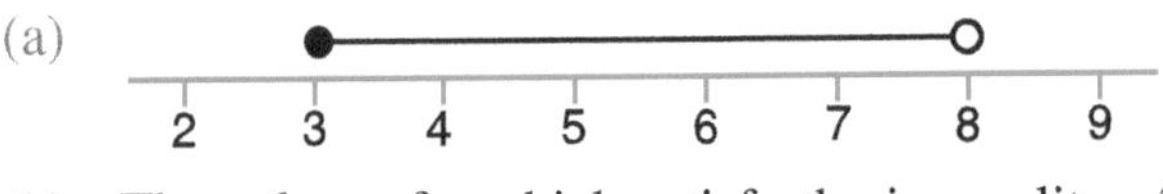

(b) The values of x which satisfy the inequality $3 \leqslant x < 8$ are: 3, 4, 5, 6, 7.

Exercise 16.1

1 Write down the following mathematical statements and say whether each is true or false.

(a) $4 < 7$ (b) $3 > -3$ (c) $4 \geqslant 4$ (d) $-2 > -1$
(e) $-8 \leqslant -8$ (f) $1.5 \geqslant 2.1$ (g) $3 \times 5 \leqslant 7 \times 2$ (h) $-4 \times (-2) > -4 - 4$

2 Write down an integer which could replace the letter.

(a) $x < 6$ (b) $a \geqslant -2$ (c) $c + 2 < 8$ (d) $2d \leqslant 14$
(e) $f - 3 > 7$ (f) $-2 < h < 0$ (g) $t \leqslant 5$ **and** $t > 4$ (h) $r \geqslant -6$ **and** $r < -1$

3 In this question x is an integer. Write down all the values of x which satisfy these inequalities.

(a) $1 < x < 5$ (b) $-2 < x \leqslant 3$ (c) $-4 \leqslant x \leqslant -1$ (d) $-1 \leqslant x < 3$

4 Write down a mathematical statement, using inequalities, for each of these diagrams.

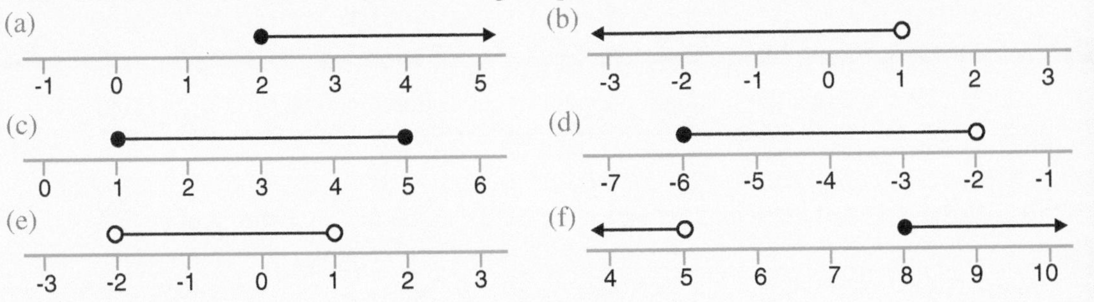

5 Draw number lines to show the following inequalities.
For each part, draw and label a number line from -5 to 5.

(a) $x > 2$ (b) $x \leqslant 1$ (c) $x \geqslant -4$ (d) $x < -2$
(e) $-2 \leqslant x < 3$ (f) $1 < x \leqslant 4$ (g) $-3 < x < 0$ (h) $x < -3$ and $x \geqslant 1$

Solving inequalities

Solve means to find the values of x which make the inequality true. The aim is to end up with **one letter** on one side of the inequality and a **number** on the other side of the inequality.

Solving inequalities is similar to solving equations.

EXAMPLES

1 Solve the inequality $5x - 3 < 27$ and show the solution on a number line.

$5x - 3 < 27$
Add 3 to both sides.
$5x < 30$
Divide both sides by 5.
$x < 6$

The solution is shown on a number line as:

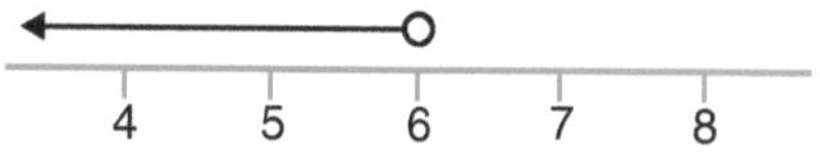

This means that the inequality is **true** for all values of x which are less than 6.

2 Solve the inequality $7a \geqslant a + 9$.

$7a \geqslant a + 9$
Subtract a from both sides.
$6a \geqslant 9$
Divide both sides by 6.
$a \geqslant 1.5$

This means that the inequality is true for all values of a which are equal to 1.5, or greater.

Substitute $a = 1.5$, $a = 2$ *and* $a = 1$ *into the original inequality.*
What do you notice?

Exercise 16.2

1 Solve each of the following inequalities and show the solution on a number line.

(a) $3n > 6$ (b) $2x < -4$ (c) $a + 1 < 5$ (d) $a - 3 < 1$
(e) $2d - 5 \leqslant 1$ (f) $t + 2 < -1$ (g) $5 + 2g > 1$ (h) $4 + 3y \geqslant 4$

2 Solve the following inequalities.
Show your working clearly.

(a) $a + 3 < 7$ (b) $5 + x \geqslant 3$ (c) $y + 2 < -1$ (d) $3c > 15$
(e) $2d < -6$ (f) $b - 3 \geqslant -2$ (g) $-2 + b \leqslant -1$ (h) $2c + 5 \leqslant 11$
(i) $3d - 4 > 8$ (j) $4 + 3f < -2$ (k) $8g - 1 \leqslant 3$ (l) $5h < h + 8$
(m) $3x < x - 6$ (n) $6j \geqslant 2j + 10$ (o) $7k > 3k - 16$ (p) $6m - 7 \leqslant m$

3 (a) (i) Solve the inequality $3x + 5 \geqslant 2$.
(ii) Write down the inequality shown by the following diagram.

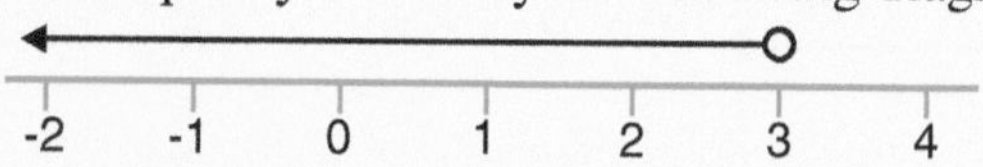

(b) Write down all the integers that satisfy both inequalities shown in part (a).

4 Solve the following inequalities.

(a) $\frac{1}{2}a \geqslant 3$ (b) $\frac{1}{3}b < 5$ (c) $4m + 2 > 2m - 11$
(d) $7n - 3 \leqslant 13 - n$ (e) $3p - 2 > 6 + 2p$ (f) $4q + 5 > 12 - 3q$
(g) $6r + 1 \geqslant 4r - 2$ (h) $2t - 10 > t + 3$ (i) $2(u - 5) \leqslant 8$
(j) $3(4v + 1) < -15$ (k) $\frac{1}{2}w + 3 > 7$ (l) $\frac{1}{2}(5x - 1) < 3$

Multiplying (or dividing) an inequality by a negative number

Activity

$-2 < 3$

Multiply both sides by -1.
$-2 \times (-1) = 2$ and $3 \times (-1) = -3$

$2 > -3$

To keep the statement true we have to reverse the inequality sign.

Multiply both sides of these inequalities by -1.

1. $3 > 2$
2. $3 > -2$
3. $-3 < -1$
4. $5 \geqslant 4$
5. $-4 \leqslant 5$
6. $-4 \geqslant -5$

The same rules for equations can be applied to inequalities, with one exception:
When you **multiply** (or **divide**) both sides of an inequality by a negative number the inequality is reversed.

EXAMPLES

1. Solve $-3x < 6$.

   Divide both sides by -3.
   Because we are dividing by a negative number the inequality is reversed.
   $x > -2$

2. Solve $3a - 2 \geqslant 5a - 9$.

   Subtract $5a$ from both sides.
   $-2a - 2 \geqslant -9$
   Add 2 to both sides.
   $-2a \geqslant -7$
   Divide both sides by -2.
   $a \leqslant 3.5$

Exercise 16.3

Solve the following inequalities. Show your working clearly.

1. $-4a > 8$
2. $-5b \leqslant -15$
3. $-3c \geqslant 12$
4. $3 - 2d < 5$
5. $14 - 3e \leqslant 4e$
6. $-5f > 4f - 9$
7. $4g < 7g + 12$
8. $5 - 3h \leqslant h - 3$
9. $-5 - j \geqslant 12j - 18$
10. $3 - 5k < 2(3 + 2k)$
11. $3(m - 2) > 5m$
12. $3(2n - 1) < 8n + 5$
13. $3p \geqslant 5 - 6p$
14. $2(q - 3) < 5 + 7q$
15. $n - 5 > 2(n - 7)$

Double inequalities

EXAMPLE

Find the values of x such that $-3 < x - 2 \leqslant 1$ and show the solution on a number line.

$-3 < x - 2 \leqslant 1$
Add 2 to each part of the inequality.
$-1 < x \leqslant 3$

The solution is shown on a number line as:

-2 -1 0 1 2 3 4

Inequalities involving integers

EXAMPLE

Find the integer values of n for which $-1 \leqslant 2n + 3 < 7$.

$-1 \leqslant 2n + 3 < 7$
Subtract 3 from each part.
$-4 \leqslant 2n < 4$
Divide each part by 2.
$-2 \leqslant n < 2$

Integer values which satisfy the inequality $-1 \leqslant 2n + 3 < 7$ are: $-2, -1, 0, 1$.

Exercise 16.4

1. Solve each of the following inequalities and show the solution on a number line.
   (a) $5 < x + 4 \leqslant 9$ (b) $-3 \leqslant x - 2 < 7$ (c) $2 < 9 + x \leqslant 13$

2. Find the values of x such that:
   (a) $2 < 2x \leqslant 6$ (b) $-6 \leqslant 3x < 12$ (c) $5 < 2x - 1 < 8$
   (d) $-2 \leqslant 3x - 1 \leqslant 11$ (e) $12 < 5x + 2 \leqslant 27$ (f) $-9 \leqslant 4x + 3 < 27$
   (g) $-16 < 7x - 2 < 12$ (h) $-1 \leqslant 3x - 10 < 8$ (i) $-4 \leqslant 5 + 2x \leqslant 3$

3. Find the integer values of n for which:
   (a) $3 < n - 2 < 7$ (b) $-2 < n + 1 \leqslant 5$ (c) $-2 < 2n \leqslant 4$
   (d) $5 \leqslant 2n - 3 < 13$ (e) $0 < 2n - 8 < 3$ (f) $5 < 4n + 1 \leqslant 13$
   (g) $-4 \leqslant 5n + 6 < 11$ (h) $-4 < 3n + 2 \leqslant 11$ (i) $-5 \leqslant \frac{1}{2}n - 3 \leqslant 0$
   (j) $-12 < 5 - n \leqslant -3$ (k) $-3 \leqslant 4 - 2n \leqslant 12$ (l) $-5 < 3(n + 5) < 0$

Graphs of inequalities

Activity

Line A has equation $y = 2$.

Describe the y coordinates of points **on** line A.
Describe the y coordinates of points **below** line A.
Describe the y coordinates of points **above** line A.

Above the line is the region $y > 2$.
Below the line is the region $y < 2$.

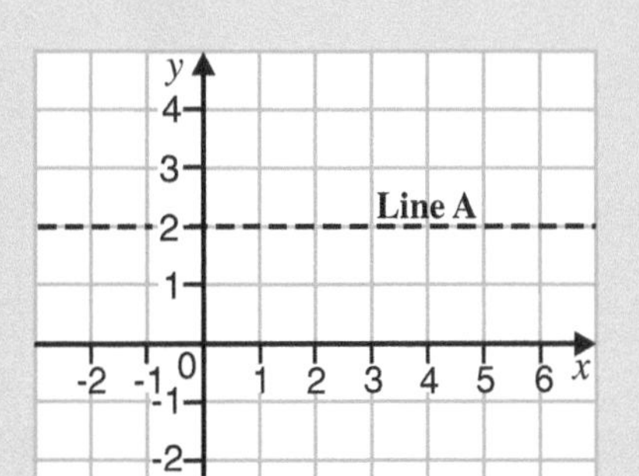

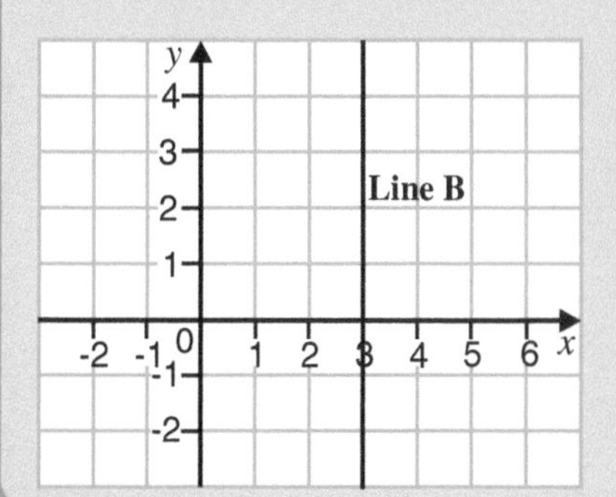

Line B has equation $x = 3$.

Use an inequality to describe the region to the right of line B.
Use an inequality to describe the region to the left of line B.

Regions

A line divides the graph into two **regions**.
The region $x \leqslant 2$ is to the **left** of the line $x = 2$, including the line itself.
The region $x \geqslant 2$ is to the **right** of the line $x = 2$, including the line itself.

The region $y < -3$ is **below** the line $y = -3$.
The region $y > -3$ is **above** the line $y = -3$.

A **solid line** is used when the points on the line are included.
A **broken line** is used when the points on the line are not included.

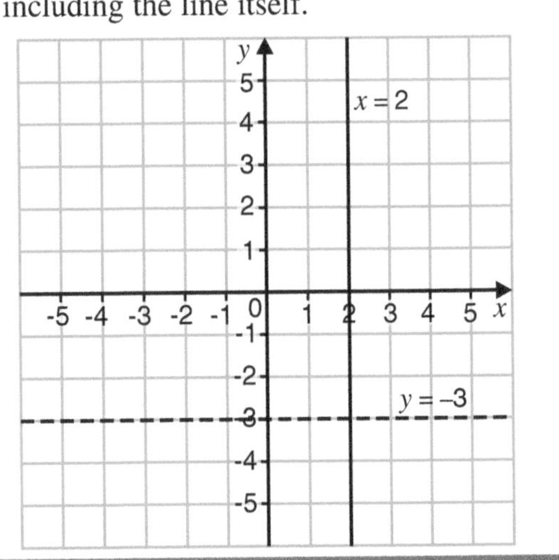

EXAMPLE

On a sketch, show the region where the inequalities $x > -2$ and $y \leqslant 1$ are both true.
Label the region R.

To do this:

1 Draw the x and y axes.
2 Draw and label the line $x = -2$, using a broken line.
3 Show the region $x > -2$ by shading out the **unwanted** region.
4 Draw and label the line $y = 1$ using a solid line.
5 Show the region $y \leqslant 1$ by shading out the **unwanted** region.
6 Label the region, R, where $x > -2$ **and** $y \leqslant 1$.
Note that the region R extends for ever to the right and downwards.

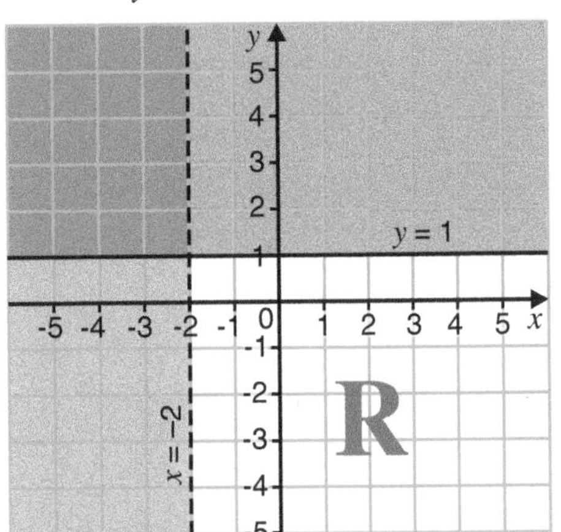

Shading regions on a graph:
Unless you are told to do otherwise, always shade out the **unwanted** region.

Exercise 16.5

For each question, draw and label axes for x and y from -5 to 5.
Shade the unwanted regions. Leave the given region unshaded and label it R.

1. $x > 2$ and $y \geqslant 1$
2. $x \geqslant -3$ and $y < 4$
3. $x \geqslant -2$, $x < 1$ and $y > 2$
4. $x < 5$, $x \geqslant -1$ and $y \leqslant 3$
5. $x \leqslant 2$, $x > -2$, $y < 3$ and $y \geqslant 1$
6. $x \geqslant -2$, $x \leqslant 0$, $y \geqslant 1$ and $y \leqslant 4$
7. $x \leqslant 1.5$, $y > -1$ and $y \leqslant 2.5$
8. $x > -2.5$, $x \leqslant 1.5$ and $y \leqslant 2.5$

Inequalities involving sloping lines

EXAMPLE

Show the region where the inequality $2x + 3y < 12$ is true.

First draw the line with equation $2x + 3y = 12$.
When $x = 0$, $y = 4$. Plot the point (0, 4).
When $y = 0$, $x = 6$. Plot the point (6, 0).

Next, draw the line through the two points.
Use a broken line because the inequality sign is $<$.

Point	Coordinates	Is $2x + 3y < 12$?
Above	(3, 4)	No
Below	(2, 1)	Yes

The inequality $2x + 3y < 12$ is true **below** the line, so we shade the unwanted region, above the line.

Exercise 16.6

1. Draw graphs to show the following. Leave unshaded the regions where the inequalities are true.
   (a) $y \geqslant x$ (b) $y < 2x$ (c) $y < x + 1$ (d) $y \geqslant 2x - 1$
   (e) $x + y \leqslant 3$ (f) $x + y > 2$ (g) $2x + y \geqslant 4$ (h) $x + 4y < 6$

2. Draw graphs to show the following.
   Label with the letter R the region defined by the inequalities.
   (a) $y < 3x$, $x < 2$ and $y > 0$ (b) $2x + 5y \leqslant 10$, $x \geqslant 0$ and $y \geqslant 0$
   (c) $3x + 4y \geqslant 12$, $x < 3$ and $y < 4$ (d) $x + 3y < 6$, $x > -1$ and $y > 1$
   (e) $2x + 3y \geqslant 6$, $1 < x \leqslant 3$ and $y < 4$

3. Match each of these inequalities to its **unshaded** region.
   A $2y < x + 2$ **B** $y > x$ **C** $2y < 4 - x$ **D** $y < 4 - x$

4. The shaded area can be described by three inequalities, one for each side of the shape.
   One of the inequalities is $x < 3$.
   Write down the other two inequalities.

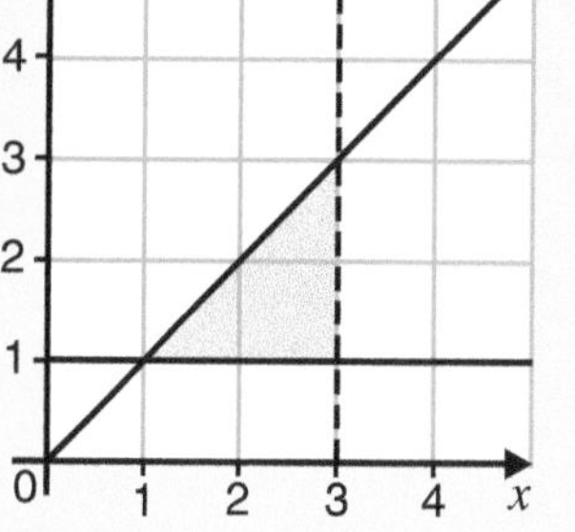

5. A point P has whole number coordinates and lies in the region $y < 3$, $1 < x < 4$ and $2y > x$.
   (a) Draw a diagram and show the region which satisfies these inequalities.
   (b) Write down the coordinates of all the possible positions of P.

6 Use inequalities to describe the shaded region in these diagrams.

(a)

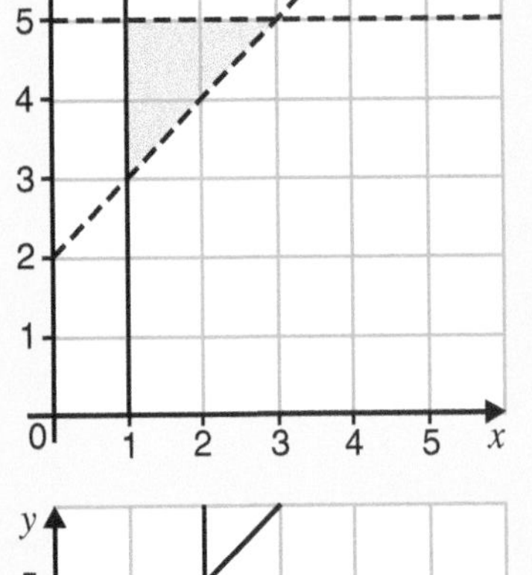

(b)

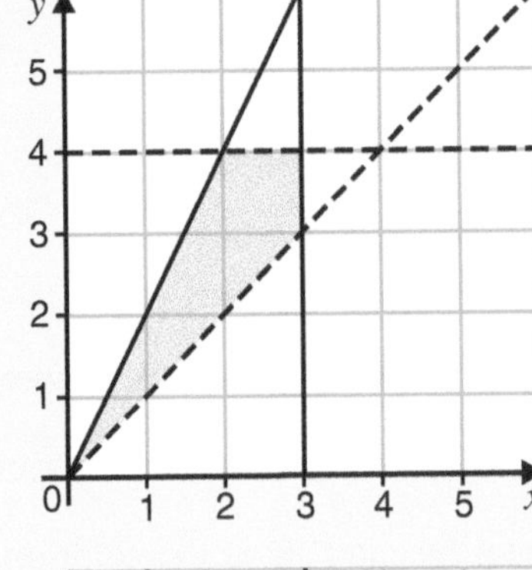

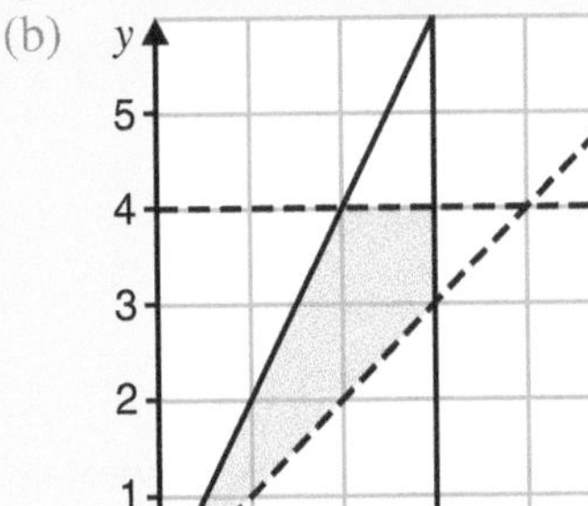

(c)

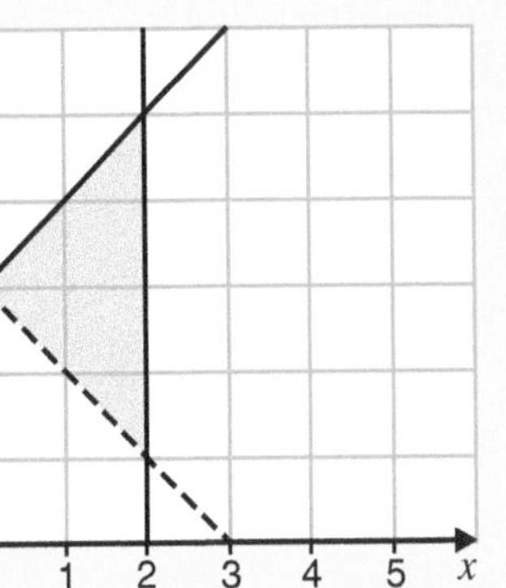

(d)

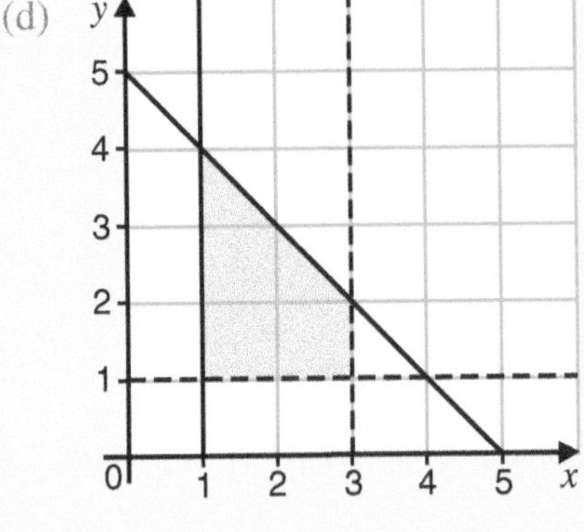

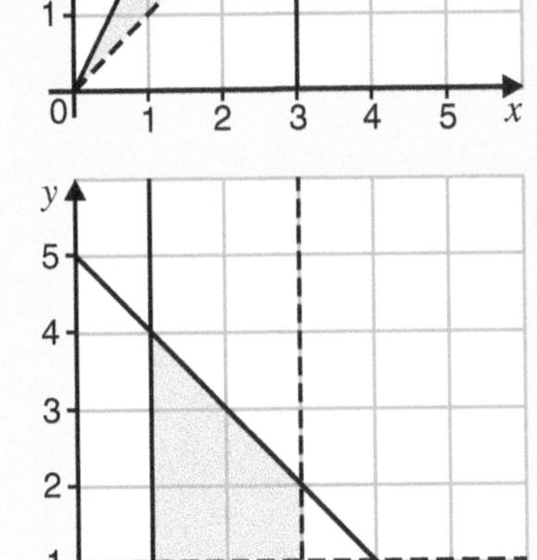

7 Draw and label axes for x and y from 0 to 8.

(a) On your diagram, draw and label the lines $y = x + 2$ and $y = 6 - x$.

(b) Write down the point, with whole number coordinates, which satisfies the inequalities: $y < x + 2$, $y < 6 - x$ and $y > 2$.

What you need to know

- Inequalities can be described using words or numbers and symbols.

Sign	Meaning
$<$	is less than
$\leqslant$	is less than or equal to

Sign	Meaning
$>$	is greater than
$\geqslant$	is greater than or equal to

- Inequalities can be shown on a **number line**.

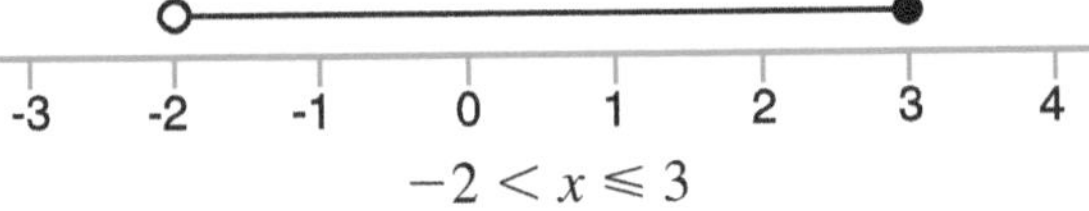

$$-2 < x \leqslant 3$$

The circle is:

filled if the inequality is **included** (i.e. $\leqslant$ or $\geqslant$),

not filled if the inequality is **not included** (i.e. $<$ or $>$).

- Solving inequalities.

Solve means find the values of x which make the inequality true.

The same rules for equations can be applied to inequalities, with one exception:

When you **multiply** (or **divide**) both sides of an inequality by a negative number the inequality is reversed.

For example, if $-3x < 6$ then $x > -2$.

- Inequalities can be shown on a graph. Replace the inequality sign by '=' and draw the line.

For $>$ and $<$ the line is **broken**. For $\geqslant$ and $\leqslant$ the line is **solid**.

Test a point on each side of the line to see whether its coordinates satisfy the inequality.

Shade the **unwanted** region.

Review Exercise 16

1 Draw number lines to show each of these inequalities.
(a) $x \leqslant -1$ (b) $x > 3$ (c) $4 < x \leqslant 9$ (d) $x < -2$ **and** $x > 5$

2 Solve each of these inequalities and show the solution on a number line.
(a) $2x < 6$ (b) $3x \geqslant -15$ (c) $x + 1 \geqslant 5$ (d) $7x - 3 \leqslant 18$

3 Solve the inequalities. (a) $5x + 1 \geqslant 11$ (b) $3x + 5 < 2$

4 List the values of x, where x is an integer number, such that $-3 \leqslant x < 5$. AQA

5 List the values of n, where n is an integer, such that $3 \leqslant n + 4 < 6$. AQA

6 Solve these inequalities and show the solution on a number line.
(a) $x - 3 < 1$ (b) $-2 < 2x \leqslant 4$ (c) $-1 \leqslant 3x + 2 < 5$ (d) $-1 < 2x + 5 < 3$

7 Find the integer values of n such that:
(a) $-4 < 2n \leqslant 8$ (b) $-3 \leqslant 3n + 6 < 12$ (c) $-4 \leqslant 5n + 6 \leqslant 1$

8 (a) List all the possible values of x, where x is an integer, such that $-4 \leqslant x < 2$.
(b) Solve $4x - 5 < -3$. AQA

9 Solve the inequality $3x - 7 > x - 3$. AQA

10 Solve the inequality $5(a - 3) > 3a - 5$. AQA

11 Solve the following inequalities.
(a) $2 > x - 4$ (b) $2(x + 3) > 3(2 - x)$ AQA

12 On separate diagrams, draw and label x and y axes from 0 to 6 and shade the regions where:
(a) $x \geqslant 2$, (b) $y \leqslant 4$, (c) $x + y \leqslant 4$.

13 (a) Solve the inequality $3x + 4 \leqslant 7$.
(b) (i) Solve the inequality $3x + 11 \geqslant 4 - 2x$.
(ii) If x is an integer what is the smallest possible value of x? AQA

14 Match each of these inequalities to its **unshaded** region.
A $2x + y < 4$ **B** $2x - y > 4$ **C** $y > 4 - 2x$ **D** $x + 2y < 4$

AQA

15 Draw and label axes for x from -3 to 8 and for y from 0 to 7.
(a) On your diagram draw and label the lines
$y = 3$ and $x + y = 5$.
(b) Show clearly on your diagram the single region that is satisfied by all of these inequalities.
$x \geqslant 0$, $y \geqslant 3$ and $x + y \leqslant 5$.
Label this region R.

16 Show, on a graph, the region where the following inequalities are true.
(a) $x \leqslant 2$, $y > -3$ and $y < 1$.
(b) $3x + 4y \leqslant 12$, $3x + y > 3$ and $y > -1$.

Further Graphs

Quadratic functions

Look at these coordinates: $(-3, 9)$, $(-2, 4)$, $(-1, 1)$, $(0, 0)$, $(1, 1)$, $(2, 4)$, $(3, 9)$.
Can you see any number patterns?

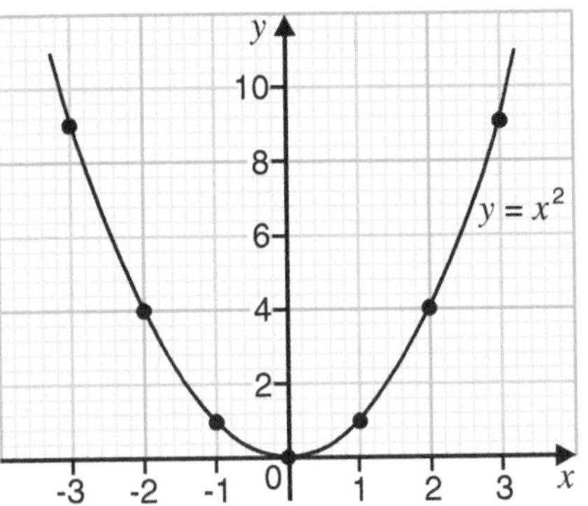

The diagram shows the coordinates plotted on a **graph**.
The points all lie on a **smooth curve**.

A **rule** connects the x coordinate with the y coordinate.
All points on the line obey the rule $y = x^2$.

$y = x^2$ is an example of a **quadratic function**.

The graph of a **quadratic function** is always a smooth curve and is called a **parabola**.
The general equation of a quadratic function is $y = ax^2 + bx + c$, where a cannot be equal to zero.

The graph of a quadratic function is symmetrical and has a **maximum** or a **minimum** value.

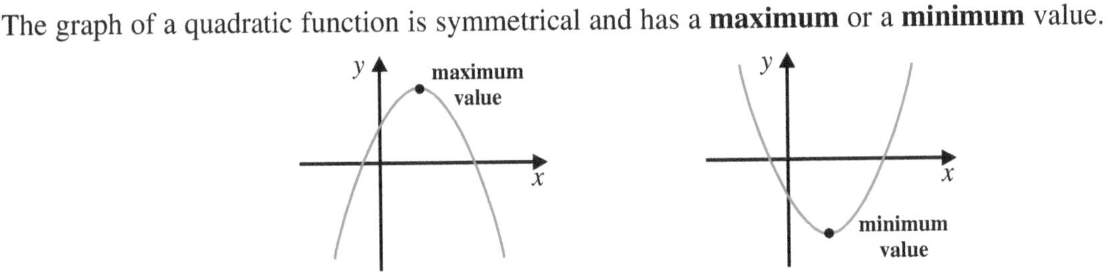

Drawing a graph of a quadratic function

To draw a quadratic graph:
- Make a table of values connecting x and y.
- Plot the points.
- Join the points with a smooth curve.

EXAMPLE

Draw the graph of $y = x^2 - 3$ for values of x from -3 to 3.

First make a table of values for $y = x^2 - 3$.

x	-3	-2	-1	0	1	2	3
y	6	1	-2	-3	-2	1	6

Plot these points.
The curve which passes through these points is the graph of the equation $y = x^2 - 3$.

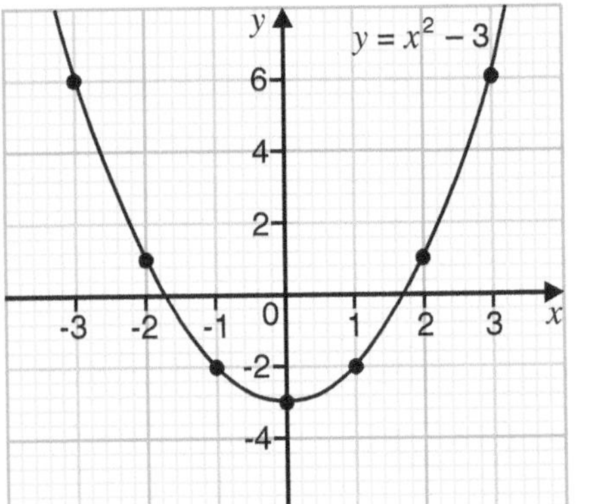

- Quadratic graphs are always symmetrical.
- Join plotted points using smooth curves and not a series of straight lines.

Exercise 17.1

1 (a) Copy and complete this table of values for $y = x^2 - 2$.

x	-3	-2	-1	0	1	2	3
y		2			-1		7

(b) Draw the graph of $y = x^2 - 2$.
Label the x axis from -3 to 3 and the y axis from -3 to 8.

(c) Use your graph to find the value of y when $x = -1.5$.

2 (a) Draw the graph of $y = x^2 + 1$ for values of x from -2 to 4.

(b) Use your graph to find the value of y when $x = 2.5$.

(c) Use your graph to find the values of x when $y = 4$.

(d) Write down the coordinates of the point at which the graph has a minimum value.

3 (a) Copy and complete this table of values for $y = 6 - x^2$.

x	-3	-2	-1	0	1	2	3
y		2		6			-3

(b) Draw the graph of $y = 6 - x^2$ for values of x from -3 to 3.

(c) Write down the coordinates of the points where the graph of $y = 6 - x^2$ crosses the x axis.

(d) Find the coordinates of the point at which the graph has a maximum value.

4 Draw the graph of $y = 2x^2$ for values of x from -2 to 2.

5 Draw the graph of $y = x^2 - x + 2$ for values of x from -2 to 3.

Using graphs to solve quadratic equations

The diagram shows the graph of $y = x^2 - 4$.

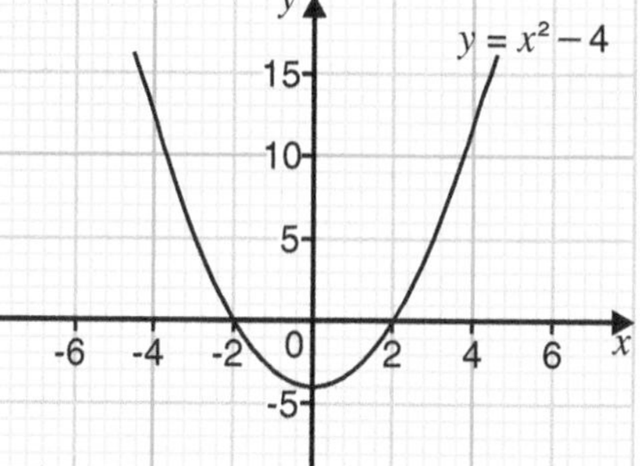

The values of x where the graphs of quadratic functions cross (or touch) the x axis give the **solutions to quadratic equations**.

At the point where the graph $y = x^2 - 4$ crosses the x axis the value of $y = 0$.

$$x^2 - 4 = 0$$

The solutions of this quadratic equation can be read from the graph: $x = -2$ and $x = 2$

The graph $y = x^2 - 4$ could be used to solve other quadratic equations.

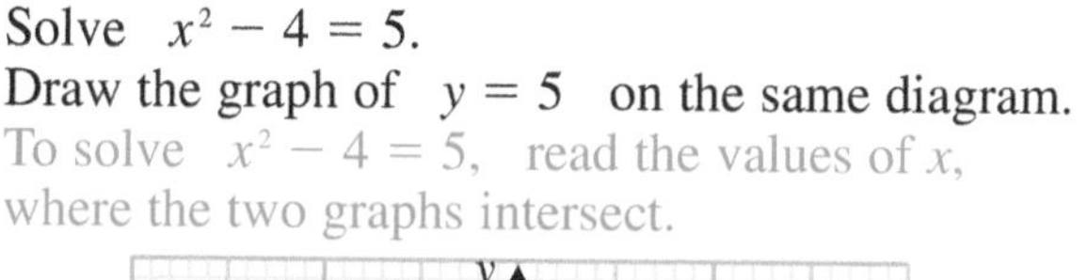

Solve $x^2 - 4 = 5$.
Draw the graph of $y = 5$ on the same diagram.
To solve $x^2 - 4 = 5$, read the values of x, where the two graphs intersect.

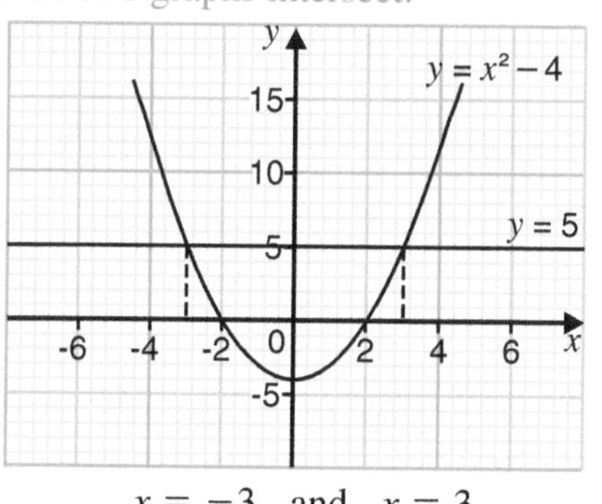

$x = -3$ and $x = 3$

Solve $x^2 - 4 = 3x$.
Draw the graph of $y = 3x$ on the same diagram.
To solve $x^2 - 4 = 3x$, read the values of x where the two graphs intersect.

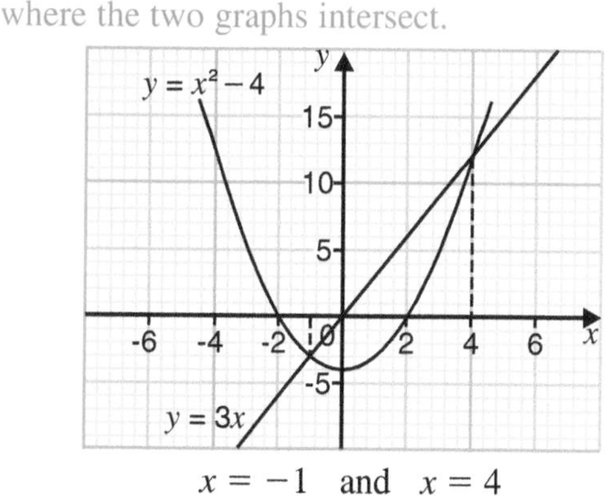

$x = -1$ and $x = 4$

EXAMPLES

1 (a) Draw the graph of $y = x^2 + 2x - 3$ for values of x from -4 to 2.

(b) Use your graph to find the solutions of the equation $x^2 + 2x - 3 = 0$.

(a) First make a table of values for $y = x^2 + 2x - 3$.

x	-4	-3	-2	-1	0	1	2
y	5	0	-3	-4	-3	0	5

Plot these points.
The curve which passes through these points is the graph of the equation $y = x^2 + 2x - 3$.

(b) To solve this equation, read the values of x where the graph of $y = x^2 + 2x - 3$ crosses the x axis. $x = -3$ and $x = 1$

2 You are given the graph of $y = 2x - x^2$.
By drawing a suitable linear graph, on the same axes, find the solutions of $x^2 - 3x + 1 = 0$.

To find the linear graph:
Rearrange $x^2 - 3x + 1 = 0$ to give $1 - x = 2x - x^2$.
Hence, the linear graph is $y = 1 - x$.

Draw the graph of $y = 1 - x$ on the same diagram.

To solve $x^2 - 3x + 1 = 0$, read the values of x where the two graphs intersect.

So, $x = 0.4$ and $x = 2.6$, correct to 1 d.p.

Exercise 17.2

1 The graph of $y = x^2 + 3x - 2$ is shown.
Use the graph to solve the equation $x^2 + 3x - 2 = 0$.
Give your answers correct to 1 d.p.

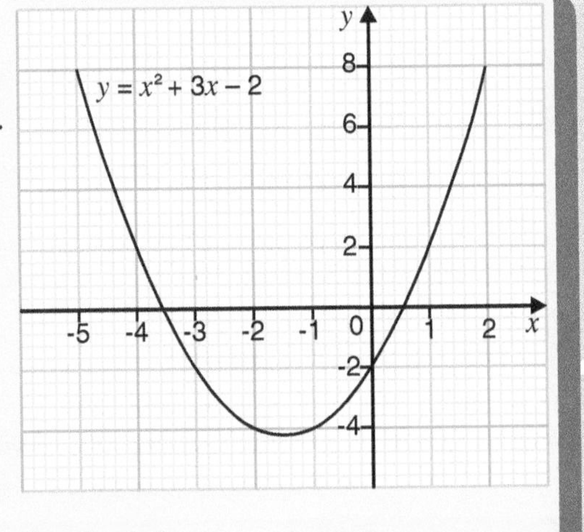

2 (a) Draw the graph of $y = x^2 + x$ for values of x from -3 to 2.

(b) Use your graph to solve the equation $x^2 + x = 0$.

(c) Find the coordinates of the point at which the graph has a minimum value.

3 (a) Copy and complete this table of values for $y = x^2 - 2x + 1$.

x	-2	-1	0	1	2	3
y						

(b) Draw the graph of $y = x^2 - 2x + 1$.

(c) Use your graph to solve the equation $x^2 - 2x + 1 = 0$.

4 (a) Draw the graph of $y = 10 - x^2$ for values of x from -4 to 4.

(b) Use your graph to solve the equation $10 - x^2 = 0$.

(c) Find the coordinates of the point at which the graph has a maximum value.

5 Draw suitable graphs to solve the following equations.
(a) $x^2 - 10 = 0$ (b) $5 - x^2 = 0$ (c) $y = x^2 - 3x + 2$ (d) $12 - 2x^2 = 0$

6 (a) Copy and complete this table of values for $y = 2x^2 - 3x + 1$.

x	−1	0	1	2	3
y		1			10

(b) Draw the graph of $y = 2x^2 - 3x + 1$.
(c) Hence, solve the equation $2x^2 - 3x + 1 = 0$.
(d) The graph of $y = 2x^2 - 3x + 1$ can also be used to solve the equation $2x^2 - 3x + 1 = 3$.
(i) What other graph would you need to draw on the same axes to solve $2x^2 - 3x + 1 = 3$?
(ii) Draw your graph in (d)(i) and hence, solve $2x^2 - 3x + 1 = 3$.

7 (a) Draw the graph of $y = x^2 - x$ for values of x from −2 to 3.
(b) Use your graph to solve the equation $x^2 - x = 0$.
(c) By drawing a suitable linear graph on the same diagram:
(i) solve the equation $x^2 - x = 4$,
(ii) solve the equation $x^2 - x = 2x$.

8 (a) On the same diagram draw the graphs $y = x^2$ and $y = x + 1$.
(b) Use your graphs to solve the equation $x^2 - x - 1 = 0$.

9 (a) Draw the graph of $y = x^2 + 2x - 2$ for values of x from −5 to 3.
(b) Use your graph to solve the equations: (i) $x^2 + 2x - 2 = 0$, (ii) $x^2 + 2x = 10$.

10 (a) Draw the graph of $y = x^2 - 4x + 3$ for values of x between −1 and 4.
(b) Use your graph to solve the equation $x^2 - 4x + 3 = 2$.
(c) By drawing an appropriate linear graph on the same axes solutions to the equation $x^2 - 3x = 0$ can be found.
(i) What is the equation of the linear graph?
(ii) By drawing the graph given in (c)(i) find the solutions of $x^2 - 3x = 0$.

11 (a) Draw the graph of $y = x^2 + x - 1$ for values of x from −2 to 4.
(b) By drawing a suitable linear graph on the same axes solve the equation $x^2 - 2x - 3 = 0$.

12 (a) Draw the graph of $y = x^2 - 4x + 1$ for values of x between −1 and 4.
(b) Use your graph to solve the equation $x^2 - 4x + 1 = 0$.
(c) By drawing an appropriate linear graph on the same axes find solutions to the equation $x^2 - 5x + 4 = 0$.

Cubic graphs

The general form of a **cubic function** is $y = ax^3 + bx^2 + cx + d$, where a cannot be equal to zero. As for quadratic graphs, the solutions of cubic equations can be found in a similar way.

Exercise 17.3

1 Draw the graphs of $y = x^3$, $y = x^3 - 2$ and $y = x^3 + x$ for values of x from −2 to 2.

2 (a) Copy and complete this table of values for $y = x^3 - 6x$.

x	−3	−2	−1	0	1	2	3
y	−9		5	0		−4	

(b) Draw the graph of $y = x^3 - 6x$ for $-3 \leqslant x \leqslant 3$.
(c) Use your graph to solve the equations (i) $x^3 - 6x = -6$, (ii) $x^3 - 6x = 2$.

The graph of the reciprocal function

The graph of the **reciprocal function** is of the form $y = \frac{a}{x}$ where x cannot be equal to zero.

The diagram shows the graph of $y = \frac{1}{x}$.

The graph of $y = \frac{a}{x}$ does not cross either the x axis or the y axis.

Explain this by investigating the values of y for very small and very large values of x.

Describe any symmetry you can see.

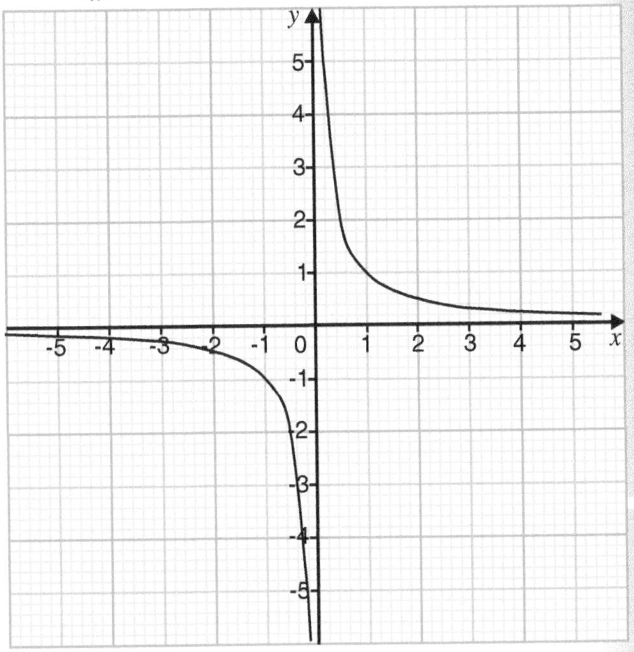

The graph of the exponential function

The graph of the **exponential function** is of the form $y = a^x$.

The diagram shows the graph of $y = 2^x$.

The table shows values of y, given to 2 d.p.

x	-3	-2	-1	0	1	2	3
y	0.13	0.25	0.5	1	2	4	8

Exponent is an alternative term for **index**.

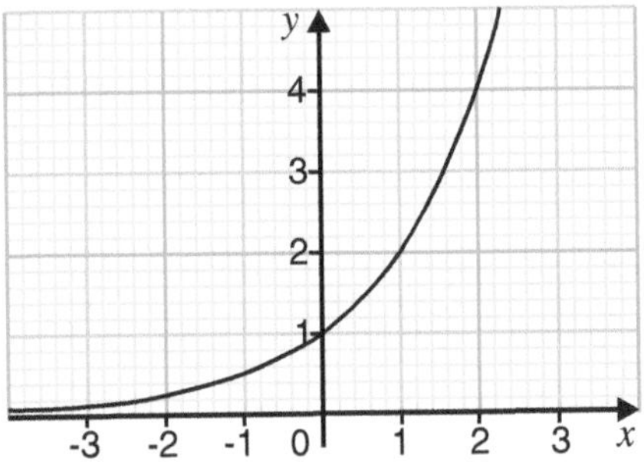

In general, the graph of $y = a^x$:
- does not touch the x axis,
- crosses the y axis at the point (0, 1).

Exercise 17.4

1. (a) Draw the graph of $y = \frac{1}{x}$ for values of x from -5 to 5.
   Label the y axis from -10 to 10.
   (b) On the same axes, draw the graph of $y = 4 - x$.
   (c) Use your graph to find the values of x for which $\frac{1}{x} = 4 - x$.
2. (a) Draw the graph of $y = \frac{3}{x}$ for values of x from -6 to 6.
   (b) On the same axes, draw the graph of $y = \frac{-3}{x}$.
   (c) Describe any patterns you can see.
3. (a) Draw the graph of $y = 2^x$ for values of x from -5 to 5.
   (b) On the same axes, draw the graph of $y = \left(\frac{1}{2}\right)^x$.
   (c) Compare and comment on the graphs drawn in parts (a) and (b).

4 (a) Copy and complete this table of values for $y = 3^x$.
Give values of y correct to 2 decimal places, where necessary.

x	-3	-2	-1	0	1	2	3
y							

(b) Draw axes marked from -3 to 3 for x and from 0 to 30 for y.
Draw the graph of $y = 3^x$ on your axes.

(c) Use your graph to find the value of
(i) y when $x = 1.5$, (ii) x when $y = 20$.

The graph of a circle

The graph of a **circle** is of the form $x^2 + y^2 = r^2$.

The diagram shows the graph of $x^2 + y^2 = 16$.
The graph shows a circle with centre (0, 0) and radius 4 units.

In general, the graph of a circle is of the form:
$$x^2 + y^2 = r^2$$
where r is the radius of the circle, with centre (0, 0).

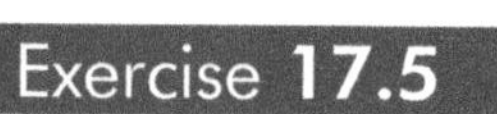

Exercise 17.5

1. The graph of a circle is given by $x^2 + y^2 = 9$.
   (a) What is the radius of the circle?
   (b) Write down the coordinates of the points where the graph of $x^2 + y^2 = 9$ cuts the x and y axes.
   (c) Draw the graph of $x^2 + y^2 = 9$.

2. (a) Draw the graph of the circle with centre (0, 0) and radius 5 units.
   (b) Write down the equation of the circle.
   (c) On the same axes draw the graph of the line $y = 4$.
   (d) Find the coordinates of the points of intersection of the circle with the line.

3. (a) On axes for x and y marked from -5 to 5, draw the graph of $x^2 + y^2 = 16$.
   (b) On the same axes, draw the graph of $y = 2x + 1$.
   (c) Use your graph to find the values of x where the graph of $x^2 + y^2 = 16$ cuts the graph of $y = 2x + 1$.

4. (a) Draw the graph of the circle $x^2 + y^2 = 1$.
   (b) On the same axes, draw the graph of $y = 2 - 2x$.
   (c) Find the coordinates of the points where the graphs intersect.

5. (a) Draw the graph of $x^2 + y^2 = 4$.
   (b) On the same axes, draw the graph of $y = 1 - \frac{1}{2}x$.
   (c) The graphs of $x^2 + y^2 = 4$ and $y = 1 - \frac{1}{2}x$ both cut the x axis at the point (2, 0).
   Use your graph to find the value of x at the second point where the graphs intersect.

Sketching graphs

You will need to be able to recognise and sketch **all** of these graphs.

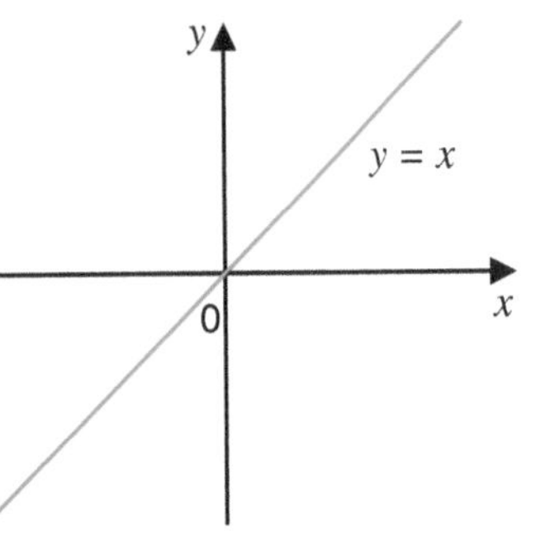

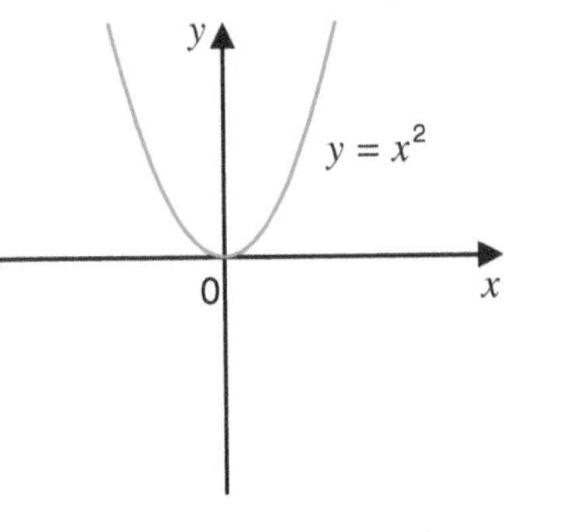

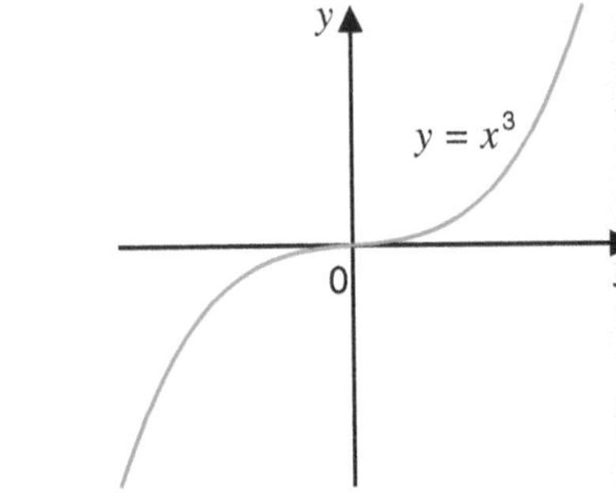

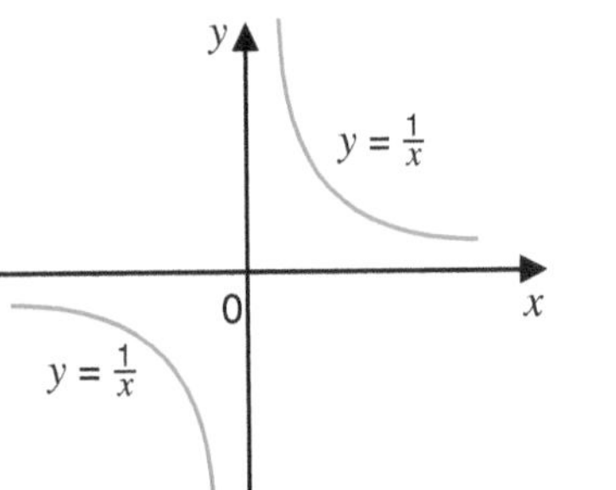

y
$y = 2^x$
1
0
x

y
5
$x^2 + y^2 = 25$
-5
0
5
x
-5

Exercise 17.6

1 Match the following equations to their graphs.

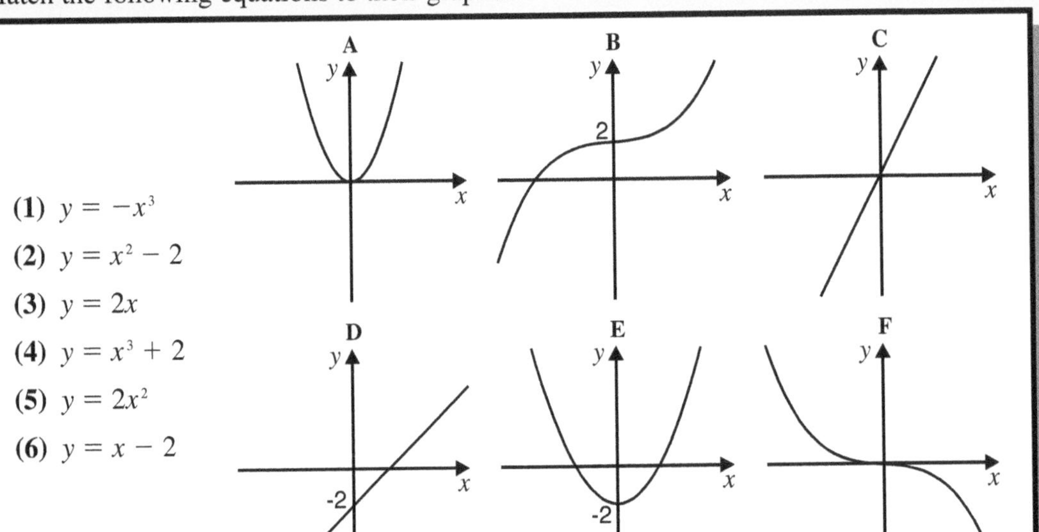

2 Sketch the graphs of these equations.

(a) $y = x^2 + 3$ (b) $y = x^3 - 3$ (c) $y = -x$ (d) $y = x - 3$

3 The sketches of some graphs are shown. Write down the equation of each graph.

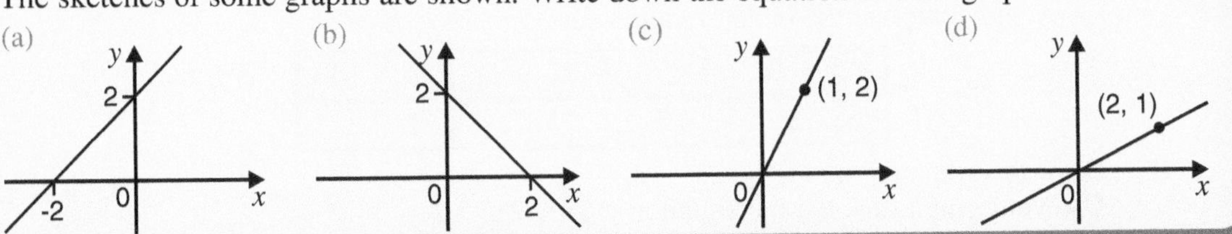

What you need to know

- The graph of a **quadratic function** is always a smooth curve and is called a **parabola**.
- The general form of a **quadratic function** is $y = ax^2 + bx + c$, where a cannot be zero.

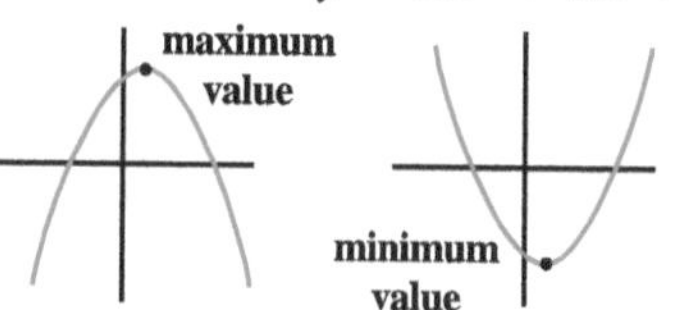

The graph of a quadratic function is symmetrical and has a **maximum** or **minimum** value.

- The general form of a **cubic function** is $y = ax^3 + bx^2 + cx + d$, where a cannot be zero.
- The graph of the **reciprocal function** is of the form $y = \frac{a}{x}$ where x cannot be equal to zero. E.g.

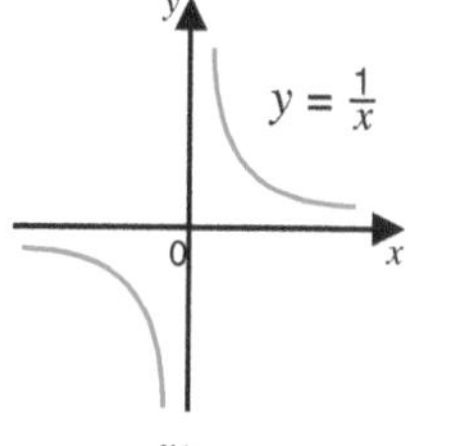

- The graph of the **exponential function** is of the form $y = a^x$. E.g.

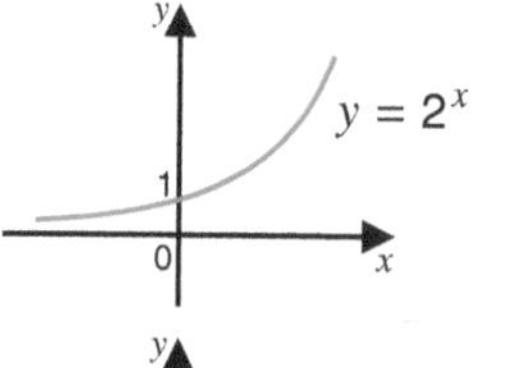

- The graph of a **circle**, centre (0, 0), is of the form $x^2 + y^2 = r^2$ where r is the radius of the circle. E.g.

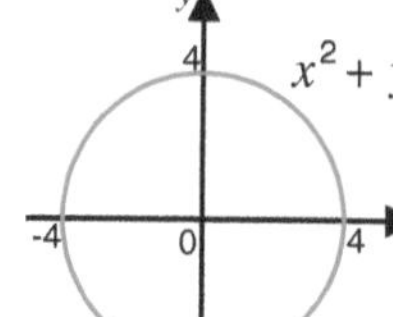

You should be able to:

- **Substitute** values into given functions to generate points.
- Plot the graphs of **quadratic** and other functions.
- Use graphs of functions to solve equations.
  This may include drawing another graph and looking for points of intersection.
- Sketch the graphs of functions.

Review Exercise 17

1 The graph of $y = x^2 - 4x$ is shown.

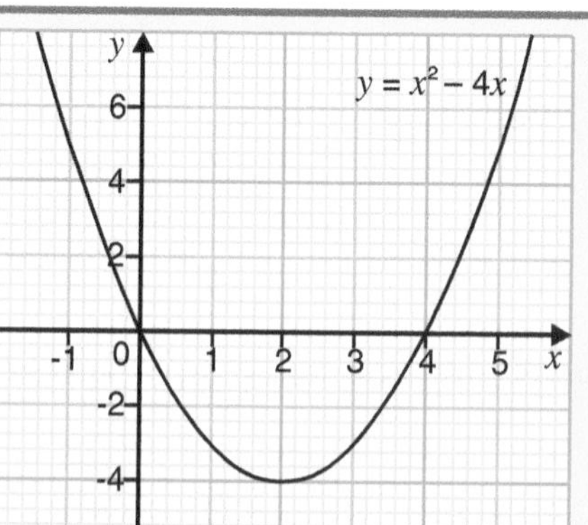

Use the graph to:

(a) solve the equation $x^2 - 4x = 0$,

(b) find the coordinates of the point at which the graph has a minimum value.

2 (a) Copy and complete the table of values for $y = x^2 - 2x - 2$.

x	−2	−1	0	1	2	3	4
y	6		−2			1	

(b) Draw the graph of $y = x^2 - 2x - 2$.

(c) Use your graph to solve the equation $x^2 - 2x - 2 = 0$.

AQA

3 (a) Complete the table of values for $y = 2x^2 - 4x - 1$.

x	-2	-1	-0	1	2	3
y	15		-1		-1	5

(b) Draw the graph of $y = 2x^2 - 4x - 1$ for values of x from -2 to 3.

(c) An approximate solution of the equation $2x^2 - 4x - 1 = 0$ is $x = 2.2$.

(i) Explain how you can find this from the graph.

(ii) Use your graph to write down another solution of this equation. AQA

4 Match each of these equations with its graph.

A $y = x^3$

B $y = 3^x$

C $y = 3x$

D $y = \frac{3}{x}$

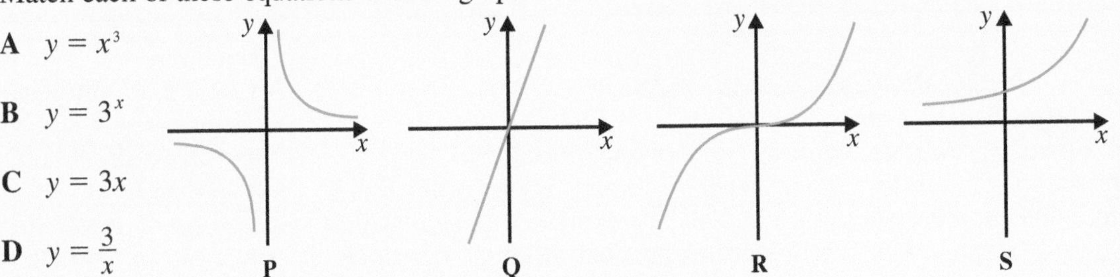

5 (a) Copy and complete the table of values for $y = (0.8)^x$.

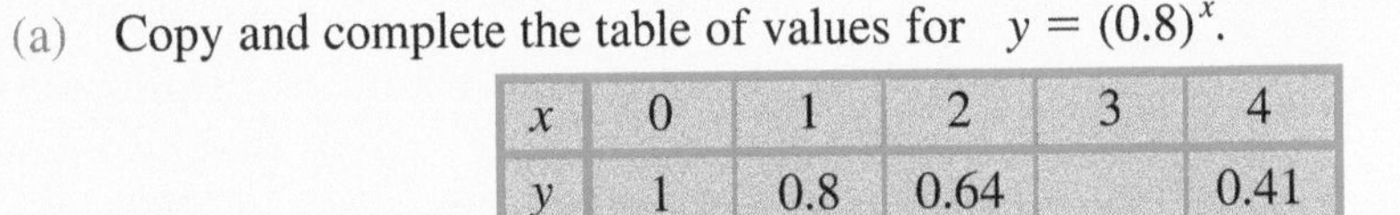

x	0	1	2	3	4
y	1	0.8	0.64		0.41

(b) Draw the graph of $y = (0.8)^x$ for values of x from 0 to 4.

(c) Use your graph to solve the equation $(0.8)^x = 0.76$. AQA

6 (a) Draw the graph of $y = \frac{2}{x}$ for $-4 \leqslant x \leqslant 4$.

(b) On the same axes, draw the graph of $y = x + 1$ for $-4 \leqslant x \leqslant 4$.

(c) Use your graph to solve the equation $\frac{2}{x} = x + 1$.

7 (a) Copy and complete the table of values for $y = 2x^2 - x - 3$.

x	-1.5	-1	-0.5	0	0.5	1	1.5	2
y	3		-2		-3		0	

(b) Draw the graph of $y = 2x^2 - x - 3$ for values of x from -1.5 to 2.

(c) Use your graph to solve the equation $2x^2 = x + 5$. AQA

8 (a) Draw the graph of $y = x^2 - 2x - 4$ for values of x from -3 to 5.

(b) On the same axes, draw the line with equation $y = 2x + 1$.

(c) Hence, solve the equation $x^2 - 2x - 4 = 2x + 1$.

(d) Show that $x^2 - 2x - 4 = 2x + 1$ can be simplified to $x^2 - 4x - 5 = 0$.

(e) You are going to use the graph of $y = x^2 - 2x - 4$ to solve the equation $x^2 - x - 8 = 0$.
What is the equation of the line you need to draw? AQA

9 (a) Copy and complete this table of values and use it to draw the graph of $y = 2x^2 - 4x + 3$ for values of x from -3 to 4.

x	-3	-2	-1	0	1	2	3	4
y	…	19	9	3	…	3	9	19

(b) Use your graph to find the value of x when $y = 6$ and $x < 0$.

(c) Explain how the graph shows that there are no values of x for which $2x^2 - 4x + 3 = 0$. AQA

10 The circle $x^2 + y^2 = 25$ and the line $y = x + 7$ intersect at the points P and Q.
Use a graphical method to find the coordinates of P and Q.

Direct and Inverse Proportion

Direct proportion

If x and y are quantities such that $y : x$ is always constant, then y is said to vary **directly** with x.

In this table, $y : x$ in its simplest form is always 2 : 5.

x	10	20	30	40
y	4	8	12	16

So, y **varies directly** with x.
This can be expressed in words as: y is **proportional** to x and
expressed in symbols as: $y \propto x$ where $\propto$ means "is proportional to".
The relationship between x and y can also be expressed as an equation.
$y = kx$ where k is the **constant of proportionality**.
This equation can be rearranged to make k the subject of the formula.

$$k = \frac{y}{x}$$

Remember:
A ratio in its **simplest form** has only whole numbers that have no common factor other than 1.
To **simplify a ratio** divide each part of the ratio by the highest common factor of the numbers.
Equivalent ratios can be found by multiplying, or dividing, each part of the ratio by the **same** number.

Finding k

Corresponding values of the proportional quantities can be used to find the value of k.
From the table: when $x = 10$, $y = 4$.
Substitute these values into the equation $y = kx$.
$4 = k \times 10$
Divide both sides by 10.
$k = 0.4$
The **equation connecting x and y** is $y = 0.4x$.

Look at the table of values of x and y:
What happens to y when x is doubled? What happens to y when x is trebled?

When two quantities are in direct proportion, as one quantity increases the other quantity **increases** at the **same rate**.

Examples of quantities in **direct proportion**:

When a vehicle travels at constant speed:
distance travelled (d) is proportional to time taken (t). $d \propto t$ $d = kt$

For any circle:
circumference (C) is proportional to radius (r). $C \propto r$ $C = kr$

What is the constant of proportionality, k, in each of these examples?

EXAMPLE

1 The table shows values of l and m. Show that m is proportional to l.

l	3	6	8
m	7.5	15	20

If $m \propto l$, the ratio $m : l$ must be equivalent for **all** pairs of values of m and l.
$7.5 \div 3 = 2.5$ $15 \div 6 = 2.5$ $20 \div 8 = 2.5$
So, $m : l$ is always equivalent to 1 : 2.5.
So, $m = 2.5 \times l$ which also means $m \propto l$.

EXAMPLE

2 The age of a tree, A years, is proportional to the radius of its trunk, r mm.
When the tree is 5 years old the radius of its trunk is 22 mm.
(a) Find a formula connecting A and r.
(b) Find A when $r = 242$.
(c) Find r when $A = 80$.

(a) You are given that $A \propto r$.
So, $A = kr$
When $A = 5$ years, $r = 22$ mm.
$5 = k \times 22$
Divide both sides by 22.
$k = \frac{5}{22}$
The formula connecting A and r is $A = \frac{5}{22}r$.

(b) $A = \frac{5}{22}r$
When $r = 242$.
$A = \frac{5}{22} \times 242$
$A = 55$

(c) $A = \frac{5}{22}r$
When $A = 80$.
$80 = \frac{5}{22} \times r$
$r = 352$

Exercise 18.1

Do not use a calculator.

1

a	2	5	10	30	50
b	3	7.5	15	45	75

c	4	8	20	60	200
d	0.2	0.4	1	3	10

e	10	25	40	55	70
f	2	3.5	5	6.5	8

g	10	20	40	60	80
h	0.35	0.7	1.4	2.1	2.8

Which of the following statements are true?
(a) $b \propto a$ (b) $d \propto c$ (c) $f \propto e$ (d) $h \propto g$
Give a reason for each answer.

2 In this table p varies directly with q.

p	1	a	12	c	70
q	5	10	b	40	d

(a) Find, in its simplest form, the ratio
(i) $1 : a$ (ii) $c : 40$
(b) Calculate the values of a, b, c and d.

3 $y = kx$ and $y = 18$ when $x = 3$.
(a) Find the value of the constant k.
Hence, find the equation connecting y and x.
(b) Find y when $x = 8$.

4 y varies directly as x, so, $y = kx$. When $y = 2$, $x = 10$.
(a) Find the equation connecting y and x.
(b) Find y when $x = 45$.

5 m varies directly as n, and $m = 6$ when $n = 4$.
(a) Find the equation connecting m and n.
(b) Find (i) m when $n = 6$, (ii) n when $m = 15$.

6 In this table p is proportional to q.
(a) Find an equation connecting p and q.
(b) Find the values of a and b.

p	4	12	b
q	2.5	a	20

7 In this table $y \propto x$.
(a) Find an equation expressing y in terms of x.
(b) Find the values of a and b.

x	25	100	b
y	1.25	a	20

8 The cost, £C, of building a wall is proportional to its area, $A\ \text{m}^2$.
A wall of area $20\ \text{m}^2$ costs £210.
(a) Find a formula giving C in terms of A.
(b) Find the cost of a wall of area $55\ \text{m}^2$.
(c) A wall of height 1.6 m costs £1155. How long is the wall?

9 The distance travelled by a fixed wheel bicycle, d, is proportional to n, the number of times the pedal rotates. In a journey of 400 m the pedal rotates 80 times.
(a) Find an equation expressing d in terms of n.
(b) How far does the bicycle travel if the pedal rotates 200 times?
(c) How many times does the pedal rotate if the bicycle travels 3 km?

10 (a) The extension, e, of a spring is proportional to w, the weight hung on the spring.
When $w = 2\ \text{kg}$, $e = 60\ \text{mm}$.
(i) Find an equation expressing e in terms of w.
(ii) Find e when $w = 5.3\ \text{kg}$.
(iii) Find w when $e = 96.6\ \text{mm}$.
(b) The extension of another spring is also proportional to the weight hung on it.
Weights of 2.5 kg and 8 kg are hung on this spring and the extensions measured.
What is the ratio of these two extensions in its simplest form?

Inverse proportion

If x and y are quantities such that $y : \frac{1}{x}$ is always constant, then y varies directly with $\frac{1}{x}$.
y is also said to vary **inversely** with x.

In this table, $y : \frac{1}{x}$ in its simplest form is always 1 : 60.

x	1	2	5	10	12	15
y	60	30	12	6	5	4

So, y **varies directly** with $\frac{1}{x}$.

This can be expressed in words as: y is **inversely proportional** to x and expressed in symbols as: $y \propto \frac{1}{x}$

The relationship between x and y can also be expressed as an equation.

$y = \frac{k}{x}$ or $xy = k$ where k is the **constant of proportionality**.

Finding k

Corresponding values of the proportional quantities can be used to find the value of k.
From the table: When $x = 1$, $y = 60$.
Substitute these values into the equation $y = \frac{k}{x}$.

$60 = \frac{k}{1}$, giving $k = 60$.

The equation connecting x and y is $y = \frac{60}{x}$. (This equation can also be written as $xy = 60$.)

Look at the table of values of x and y:
What happens to y when x is doubled? What happens to y when x is trebled?

When two quantities are in inverse proportion, as one quantity increases the other quantity **decreases** at the **same rate**.

Examples of quantities in **inverse proportion**:

For cars travelling the same distance:
average speed (s) is inversely proportional to time taken (t). $s \propto \frac{1}{t}$ $st = k$

For rectangles of constant area:
length (l) is inversely proportional to width (w). $l \propto \frac{1}{w}$ $lw = k$

EXAMPLE

In this table y is inversely proportional to x. Express y in terms of x and find the value of a.

x	5	a
y	20	4

$y \propto \frac{1}{x}$

So, $y = \frac{k}{x}$ and $xy = k$.

When $x = 5$, $y = 20$.

$k = 5 \times 20 = 100$

The equation connecting x and y is $y = \frac{100}{x}$

To find the value of a, $y = \frac{100}{x}$ can be rearranged as $x = \frac{100}{y}$.

$x = \frac{100}{y}$.

When $x = a$, $y = 4$.

$a = \frac{100}{4}$

$a = 25$

Exercise 18.2

Do not use a calculator for this exercise.

1

a	2	2.5	4	5	10
b	5	4	2.5	2	1

c	6	12	30	90	300
d	0.2	0.4	1	3	10

Which of the following statements are true? (a) $b \propto \frac{1}{a}$ (b) $d \propto \frac{1}{c}$

Give a reason for each answer.

2 The table shows values of p and q.

(a) Show that q is inversely proportional to p.

(b) Express q in terms of p.

p	2	10	50
q	0.25	0.05	0.01

3 Copy and complete the table for the relationship $b \propto \frac{1}{a}$ and find the constant of proportionality.

a	2	5	10	25	50
b	10				

4 (a) Given that $p \propto \frac{1}{q}$ and $p = 3$ when $q = 10$, find an equation connecting p and q.

(b) Hence, find p when $q = 6$.

5 In this table y is inversely proportional to x.

(a) Find an equation connecting x and y.

(b) Find the values of a and b.

x	5	12	b
y	9.6	a	0.5

6 y is inversely proportional to x. $y = 12$ when $x = 3$. Find y when $x = 4$.

7 s is inversely proportional to r. $s = 20$ when $r = 5$. Find r when $s = 4$.

8 In this table M varies inversely with L.

(a) Find an equation connecting L and M.

(b) Find the values of a and b.

L	8	12	b
M	1.8	a	0.9

9 The length of a rectangle, l, of constant area is inversely proportional to its width, w.
When $l = 20$ cm, $w = 10$ cm.
Find an equation expressing l in terms of w.
Find the value of the constant of proportionality. What does it represent?

10 A building contractor uses teams of workers to build sheds.
The time, t, for a team to build one shed is inversely proportional to the number of workers, n, in the team. A team of 2 take 6 days to build one shed.

(a) Find an equation expressing t in terms of n.

(b) Find the time it takes a team of 6 to build one shed.

(c) A team builds a shed in 4 days. How many workers are in the team?

Other forms of proportion

EXAMPLES

1 When $x = 4$, $y = 8$. Find an equation for y in terms of x if:

(a) y is proportional to x^2,

(b) y is proportional to the square root of x.

(a) $y \propto x^2$ so, $y = kx^2$.
When $x = 4$, $y = 8$.
$8 = k \times 4^2$ which gives $k = \frac{1}{2}$.
$y = \frac{1}{2}x^2$

(b) $y \propto \sqrt{x}$ so, $y = k\sqrt{x}$.
When $x = 4$, $y = 8$.
$8 = k\sqrt{4}$
$8 = 2k$ which gives $k = 4$.
$y = 4\sqrt{x}$

2 A magnet is at a distance, x cm, from a metal object.
The force, F Newtons, exerted by the magnet on the metal object is inversely proportional to the square of x. When $x = 10$, $F = 2$.

(a) Find an equation expressing F in terms of x.

(b) Find F when $x = 20$.

(c) Find x when $F = 8$.

(a) $F \propto \frac{1}{x^2}$ so, $F = \frac{k}{x^2}$.
When $x = 10$, $F = 2$.
$2 = \frac{k}{10^2}$
$k = 10^2 \times 2$ which gives $k = 200$.
$F = \frac{200}{x^2}$

(b) When $x = 20$. $F = \frac{200}{20^2} = 0.5$

(c) When $F = 8$. $8 = \frac{200}{x^2}$
$x^2 = \frac{200}{8} = 25$
$x = 5$

Exercise 18.3

You may use a calculator for this exercise.

1 Write each of the following statements in words.

(a) $V \propto t^2$ (b) $L \propto \frac{1}{a}$ (c) $y \propto x^3$ (d) $p \propto \frac{1}{q^2}$ (e) $L \propto \sqrt{m}$ (f) $b \propto \frac{1}{c^3}$

2 $y = kx^2$ and $y = 18$ when $x = 3$.
Find the constant, k, and hence, write an equation connecting y and x.

3 y is proportional to the square of x.

(a) What happens to y when x is doubled?

(b) y is multiplied by 25. What happens to x?

4 y is inversely proportional to x^3. When $x = 2$, $y = 8$.
Find an equation for y in terms of x.

5 Use the information given to find an equation connecting each of the following pairs of variables.

(a) V is proportional to the square of t. $V = 45$ when $t = 3$.

(b) R is inversely proportional to the cube of s. $R = 5$ when $s = 4$.

(c) y is proportional to the square root of x. $y = 40$ when $x = 16$.

(d) A is inversely proportional to b. $A = 20$ when $b = 5$.

(e) L is proportional to the cube of p. $L = 25$ when $p = 2$.

(f) Q is proportional to the square root of P. $Q = 100$ when $P = 64$.

6 y is inversely proportional to the square of x. When $x = 2$, $y = 20$.

(a) Find the constant of proportionality.

(b) Find y when $x = 4$.

(c) Find x when $y = 5$.

7 Copy and complete the table for the relationships $b \propto a^2$

a	1	2	10
b	3		

8 This table shows values of p and q where q is proportional to p^3.
Calculate A and B.

p	2	5	B
q	16	A	54

9 $y = \frac{k}{x^2}$. When $x = 2$, $y = 0.4$.
(a) Find the value of k.
(b) Find y when $x = 10$.

10 $y = k\sqrt{x}$. When $x = 0.16$, $y = 4$.
(a) Find the value of k.
(b) Find x when $y = 40$.

11 y is proportional to the cube of x. When $x = 10$, $y = 1$.
(a) Find y when $x = 20$.
(b) Find x when $y = 5$.

12 For each part of this question find an equation connecting the variables p and q.
Also find the values of the letters a to f.
(a) q varies inversely with the square of p.

p	4	10	b
q	2.5	a	0.1

(b) q varies inversely with the cube of p.

p	4	10	d
q	2.5	c	20

(c) q varies with the square of p.

p	4	10	f
q	2.5	e	10

13 y is inversely proportional to the square of x.
(a) What happens to y when x is doubled?
(b) y is multiplied by 25. What happens to x?

14 The time taken, t seconds, for a skier to slide from rest down a ski slope is proportional to the square root of the distance travelled, d metres. When $d = 100$, $t = 16$.
(a) Find an equation for t in terms of d.
(b) Find the value of t when $d = 25$.

15 The height, h m, that a stone reaches when it is thrown upwards varies directly with the square of its initial speed, s metres per second. When $s = 10$, $h = 5$.
(a) Find a formula expressing h in terms of s.
(b) Calculate s when $h = 20$.
(c) Calculate h when $s = 50$.

16 In cylinders of equal volume, the radius, r cm, is inversely proportional to the square root of the height, h cm. When $h = 4$, $r = 8$.
(a) Find h when $r = 4$.
(b) What does the constant of proportionality represent?

17 The radius, r, of a cone of fixed volume is inversely proportional to the square root of its height, h.
(a) r is doubled. What happens to h?
(b) h is multiplied by 16. What happens to r?

Proportion and graphs

The general form of the relationships dealt within this chapter is $y \propto x^n$.
Values of $n > 0$ give **direct** proportionality.
Values of $n < 0$ give **inverse** proportionality.

When n is unknown, the graph of y plotted against x can be used to help find its value.

When $n = 1$.

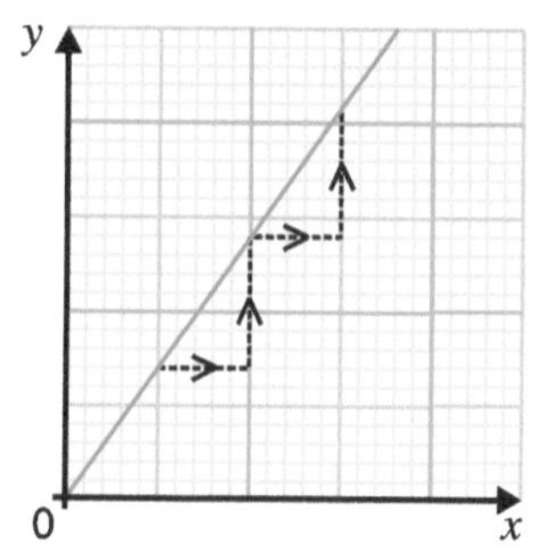

- This relationship gives a straight-line graph which passes through the origin.
- As x increases by a constant amount, y increases by a constant amount.
- k, the constant of proportionality, is given by the gradient of the graph.

When $n > 1$.

- This relationship gives a curved graph which passes through the origin.
- As x increases by a constant amount, y increases by amounts that increase.
- As k increases the graph gets steeper.

When $0 < n < 1$.

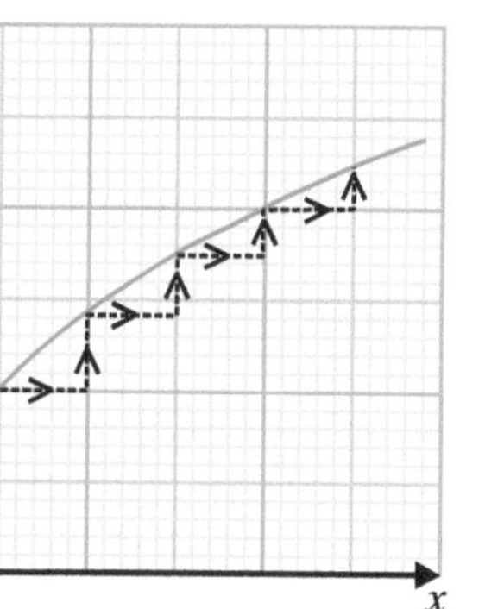

- This relationship gives a curved graph which passes through the origin.
- As x increases by a constant amount, y increases by amounts that decrease.
- As k increases the graph gets steeper.

When $n < 0$.

- This relationship gives a curved graph. The graph does not pass through the origin.
- As x increases by a constant amount, y decreases by amounts that decrease.
- As k increases the graph gets steeper.

EXAMPLE

Use this table of values of x and y to find a formula connecting x and y where $y \propto x^n$.

x	1	2	4
y	12	3	0.75

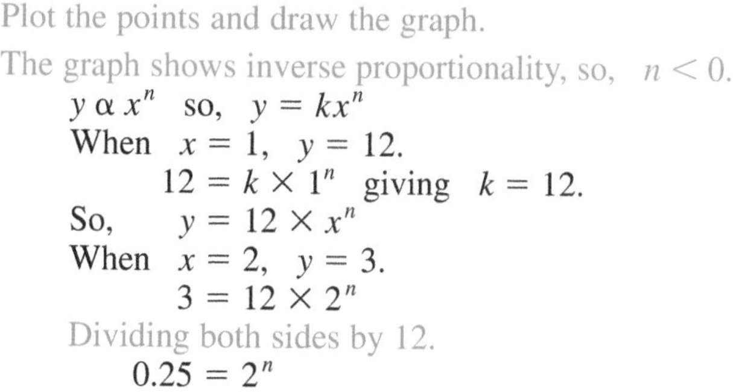

Plot the points and draw the graph.

The graph shows inverse proportionality, so, $n < 0$.

$y \propto x^n$ so, $y = kx^n$

When $x = 1$, $y = 12$.

$12 = k \times 1^n$ giving $k = 12$.

So, $y = 12 \times x^n$

When $x = 2$, $y = 3$.

$3 = 12 \times 2^n$

Dividing both sides by 12.

$0.25 = 2^n$

This gives $n = -2$.

So, $k = 12$, $n = -2$ and $y = 12x^{-2}$.

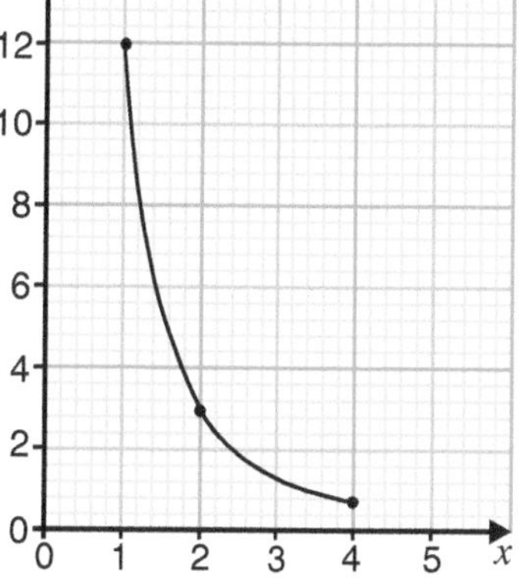

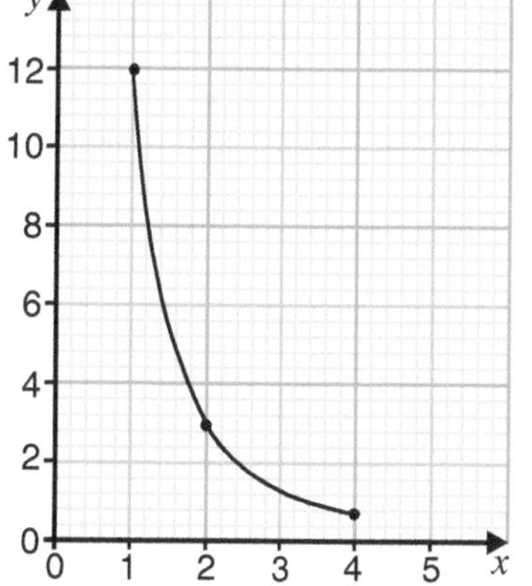

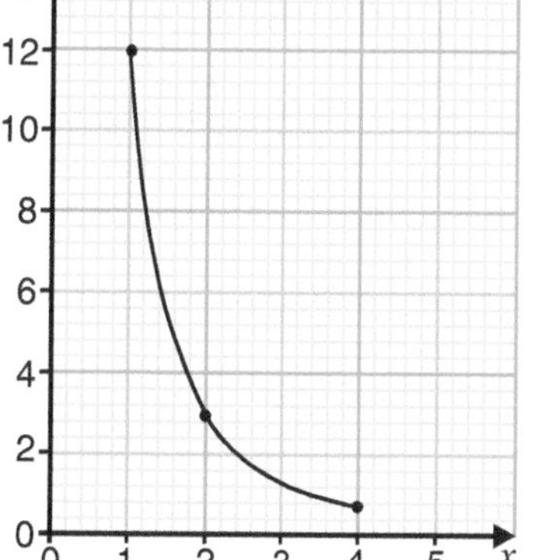

Exercise 18.4

1 Match each graph to a relationship.

Graphs

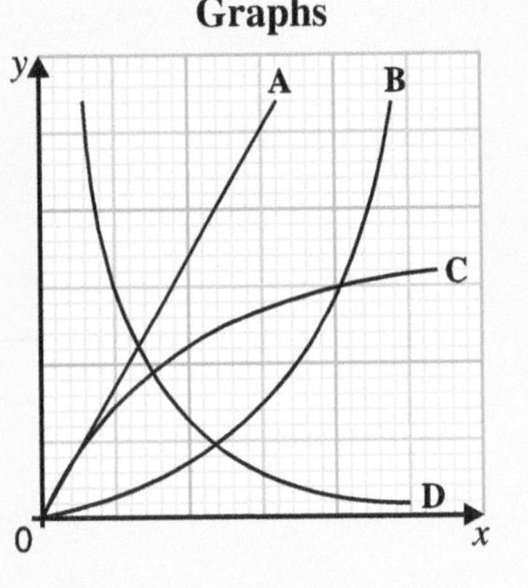

Relationships

(a) y varies directly with the square of x.

(b) y varies directly with x.

(c) y varies inversely with the cube of x.

(d) y varies directly with the square root of x.

2 (a) What is the value of n when $y \propto x^n$ and:
(i) y varies directly with the cube of x,
(ii) y varies inversely with x,
(iii) y varies inversely with the square root of x?

In each of the relationships in (a) the constant of proportionality is 10.
(b) Draw the graph of each relationship for positive values of x.
(c) What happens to your graphs in (b):
(i) when $k < 10$, (ii) when $k > 10$?

3 The relationship 'y is inversely proportional to the square of x' is of the form $y \propto x^n$.
Find the value of n.

4 The graph of the relationship $y \propto x^n$ passes through the points (0, 0), (1, 0.25) and (4, 4).
(a) Find the value of n.
(b) Find the constant of proportionality.
Repeat for a graph passing through the points (1, 10), (2, 2.5) and (5, 0.4).

5 Use this table of values of p and q to find a formula connecting p and q where $q \propto p^n$.

p	1	2	5
q	0.2	1.6	25

6 Use this table of values of s and t to find a formula connecting s and t where $s \propto t^n$.

t	1	4	25
s	20	10	4

7 The sketch shows a graph with equation $y = kx^n$.

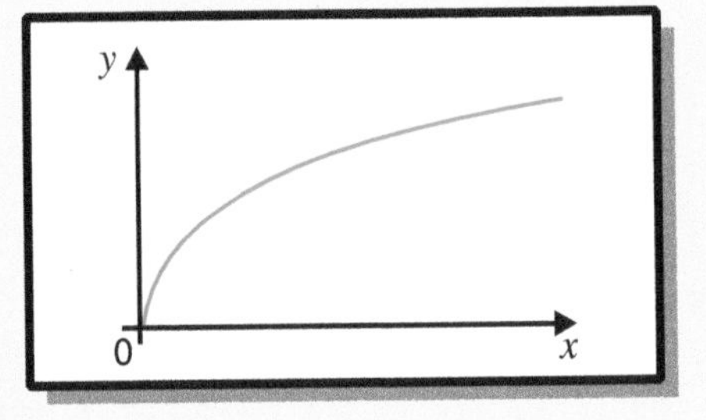

The graph passes through the points (1, 5) and (4, 10).
(a) Find the values of k and n.
(b) Find the value of x when $y = 15$.

What you need to know

- If x and y are quantities such that $y : x^n$ is always constant, then y varies **directly** with x^n.
- If x and y are quantities such that $y : \frac{1}{x^n}$ is always constant, then y varies **inversely** with x^n.
- Direct and inverse proportion can be expressed:

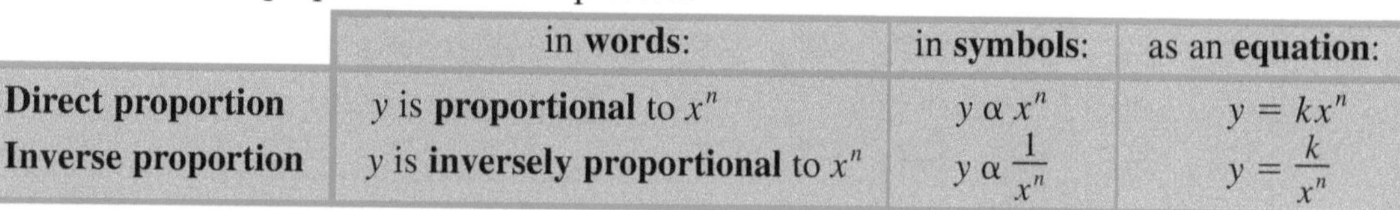

	in **words**:	in **symbols**:	as an **equation**:
Direct proportion	y is **proportional** to x^n	$y \propto x^n$	$y = kx^n$
Inverse proportion	y is **inversely proportional** to x^n	$y \propto \frac{1}{x^n}$	$y = \frac{k}{x^n}$

The symbol $\propto$ means 'is proportional to'. k is the constant of proportionality.

- The general form of a proportional relationship is $y \propto x^n$ or $y = kx^n$.

Direct proportion
$y = kx^n$, $n > 0$

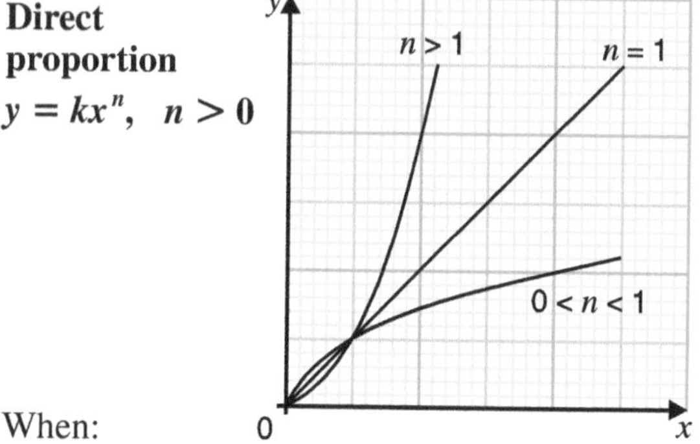

When:
$n = 1$: y increases at a constant rate.
$0 < n < 1$: y increases at a rate that decreases.
$n > 1$: y increases at a rate that increases.

Inverse proportion
$y = kx^n$, $n < 0$

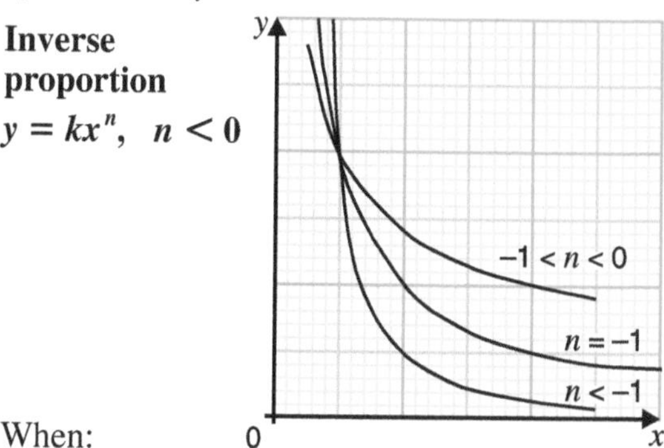

When:
$n = -1$: the graph is symmetrical about the line $y = x$.

Review Exercise 18

1. m is directly proportional to the square of n. When $m = 18$, $n = 3$. Find the value of m when $n = 5$.

2. C is inversely proportional to t^2. When $C = 16$, $t = 3$. Find t when $C = 9$. AQA

3. y is directly proportional to the square root of x. Copy and complete the table.

x	100	25	
y	3		6

AQA

4. y is inversely proportional to the square root of x. When $y = 6$ then $x = 4$.
   (a) What is the value of y when $x = 9$?
   (b) What is the value of x when $y = 10$? AQA

5. M and G are positive quantities. M is inversely proportional to G. When $M = 90$, $G = 40$. Find the value of M when $G = M$. AQA

6. In an electrical appliance, the power, P watts, is proportional to the square of the current, I amps, flowing through it. Sketch a graph of P against I. AQA

7. This diagram shows a graph with equation of the form $y = kx^n$. The graph passes through the points (1, 10) and (8, 20). Find the values of k and n.

y
0 x

AQA

CHAPTER 19

Quadratic Equations

Brackets

You can multiply out brackets, either by using a diagram or by expanding.

EXAMPLES

1 Multiply out $2(x + 3)$.

Diagram method

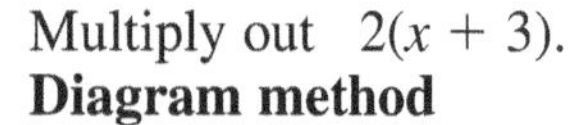
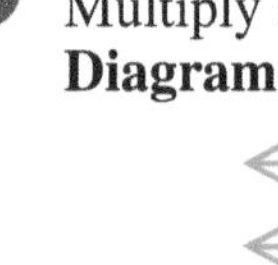
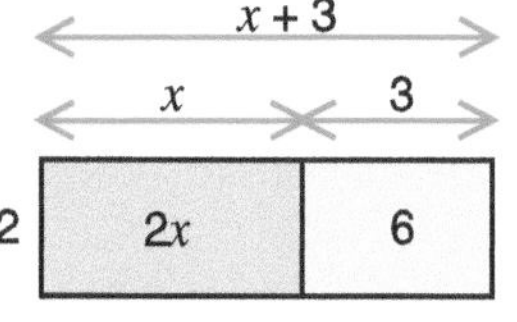

The areas of the two parts are $2x$ and 6.
The total area is $2x + 6$.
$2(x + 3) = 2x + 6$

Expanding

$2(x + 3) = 2 \times x + 2 \times 3$
$= 2x + 6$

2 Multiply out $x(x + 3)$.

The areas of the two parts are x^2 and $3x$.
$x(x + 3) = x^2 + 3x$

More brackets

This method can be extended to multiply out $(x + 2)(x + 3)$.

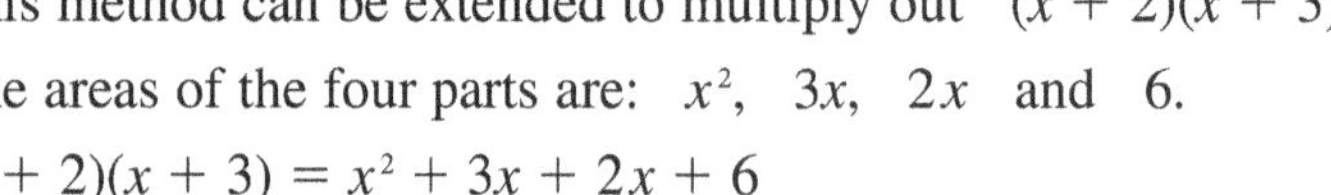
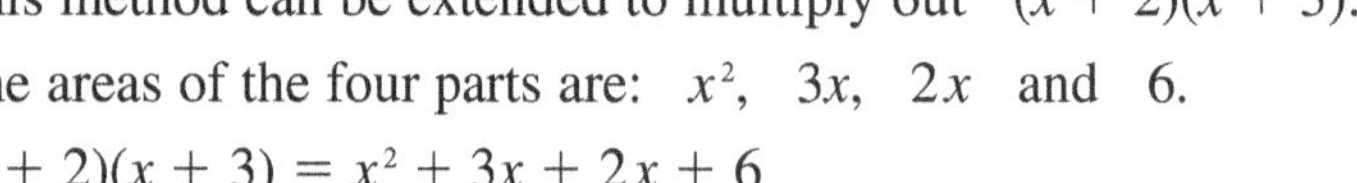

The areas of the four parts are: x^2, $3x$, $2x$ and 6.

$(x + 2)(x + 3) = x^2 + 3x + 2x + 6$

Collect like terms and simplify (i.e. $3x + 2x = 5x$)

$= x^2 + 5x + 6$

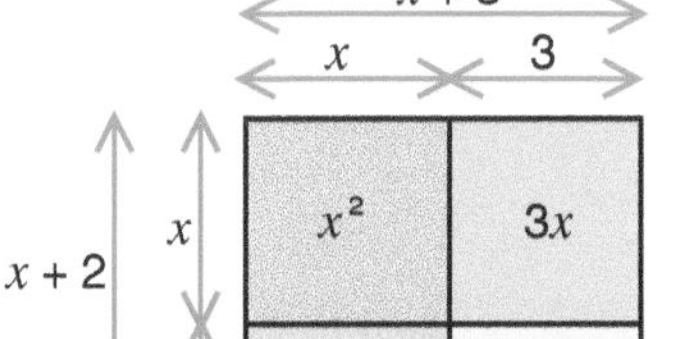
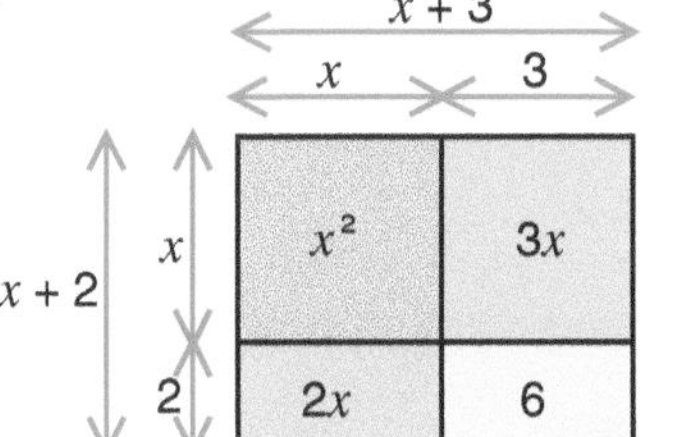
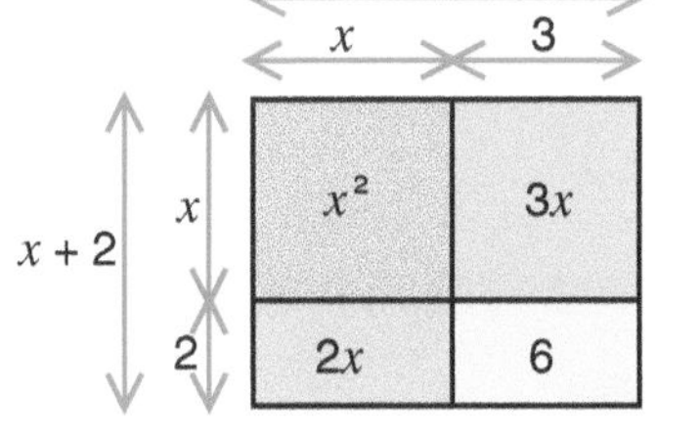
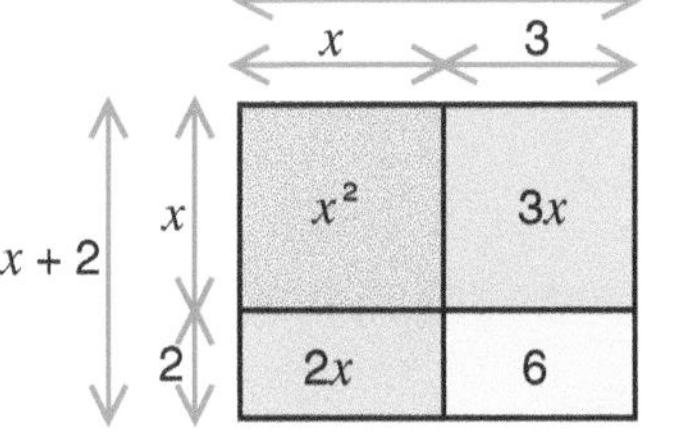
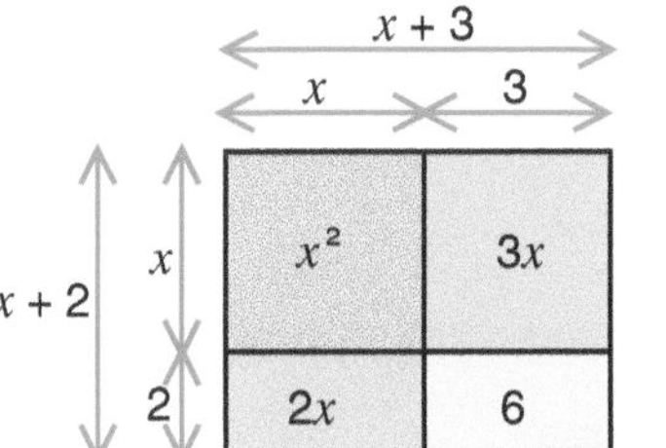
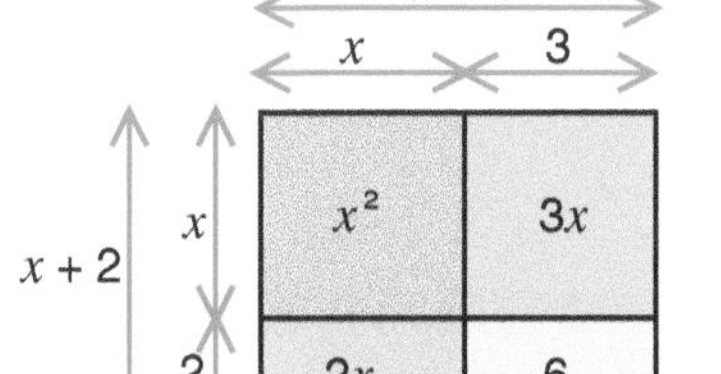

EXAMPLES

1 Expand $(2x + 3)(x + 4)$.

	x	4
$2x$	$2x^2$	$8x$
3	$3x$	12

$(2x + 3)(x + 4) = 2x^2 + 8x + 3x + 12$
$= 2x^2 + 11x + 12$

2 Expand $(x + 3)(x - 5)$.

	x	-5
x	x^2	$-5x$
3	$3x$	-15

The diagram method works with negative numbers.

$(x + 3)(x - 5) = x^2 - 5x + 3x - 15$
$= x^2 - 2x - 15$

EXAMPLES

3 Expand $(2x - 5)^2$.

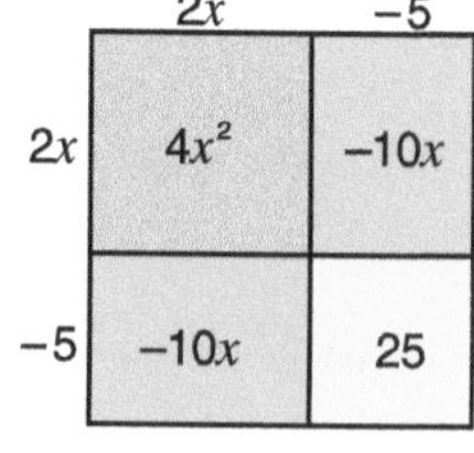

$$(2x - 5)^2 = (2x - 5)(2x - 5)$$
$$= 4x^2 - 10x - 10x + 25$$
$$= 4x^2 - 20x + 25$$

4 Expand $(x - 1)(2x + 3)$.
As you become more confident you may not need a diagram to expand the brackets.

$(x - 1)(2x + 3)$

1 $x \times 2x = 2x^2$
2 $x \times 3 = 3x$
3 $-1 \times 2x = -2x$
4 $-1 \times 3 = -3$

$$(x - 1)(2x + 3) = 2x^2 + 3x - 2x - 3$$
$$= 2x^2 + x - 3$$

Exercise 19.1

Questions 1 to 6. Use diagrams to multiply out the brackets.

1. $(x + 3)(x + 4)$
2. $(x + 1)(x + 5)$
3. $(x - 5)(x + 2)$
4. $(2x + 1)(x - 2)$
5. $(3x - 2)(x - 6)$
6. $(2x + 1)(3x + 2)$

Questions 7 to 24. Expand the following brackets. Only draw a diagram if necessary.

7. $(x + 8)(x - 2)$
8. $(x + 5)(x - 2)$
9. $(x - 1)(x + 3)$
10. $(x - 3)(x - 2)$
11. $(x - 4)(x - 1)$
12. $(x - 7)(x + 2)$
13. $(2x + 1)(x + 3)$
14. $(x + 5)(3x - 2)$
15. $(5x - 3)(x - 1)$
16. $(2x + 3)(2x - 1)$
17. $(3x - 1)(2x + 5)$
18. $(4x - 2)(3x + 5)$
19. $(x + 3)(x - 3)$
20. $(x + 5)(x - 5)$
21. $(x + 7)(x - 7)$
22. $(x + 5)^2$
23. $(x - 7)^2$
24. $(2x - 3)^2$

Factorising

Factorising is the opposite operation to removing brackets.
For example, $x^2 + 4x = x(x + 4)$
and $3x^2 + 6 = 3(x^2 + 2)$.
You can check that you have factorised an expression correctly by multiplying out the brackets.

Factorising quadratic expressions

$x^2 + 8x + 15$, $x^2 - 4$ and $x^2 + 7x$ are examples of **quadratic expressions**.

You will need to be able to factorise quadratic expressions in order to solve quadratic equations.

The general form of a quadratic expression is $ax^2 + bx + c$, where a cannot be equal to 0.

Common factors

A **common factor** is a factor which will divide into each term of an expression.
For example, $x^2 + 7x$ has a common factor of x.
$$x^2 + 7x = x(x + 7)$$

Difference of two squares

In the expression $x^2 - 4$,
$x^2 = x \times x$ and $4 = 2^2 = 2 \times 2$.

$x^2 - 4 = (x + 2)(x - 2)$
This result is called the **difference of two squares**.
In general: $a^2 - b^2 = (a + b)(a - b)$

Checking your work:
To check your work, expand the brackets and simplify, where necessary.
The result should be the same as the original expression.

EXAMPLES

Factorise the following.

1. $x^2 - 100$
   $= x^2 - 10^2$
   $= (x + 10)(x - 10)$

2. $25 - x^2$
   $= 5^2 - x^2$
   $= (5 + x)(5 - x)$

3. $s^2 - t^2$
   $= (s + t)(s - t)$

Quadratics of the form $x^2 + bx + c$

The expression $x^2 + 8x + 15$ can be factorised.

From experience, we know that the answer is likely to be of the form:
$x^2 + 8x + 15 = (x + ?)(x + ?)$
where the question marks represent numbers.
Replacing the question marks with letters, p and q, we get:
$x^2 + 8x + 15 = (x + p)(x + q)$

Multiply the brackets out, using either the diagram method or by expanding, and compare the results with the original expression.

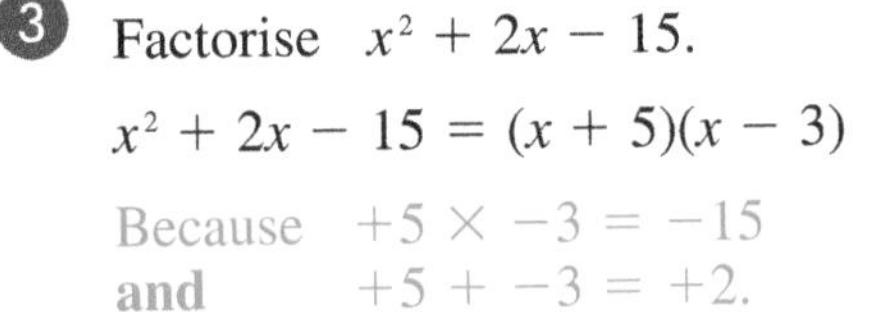

$px + qx = x(p + q) = 8x$
and $p \times q = 15$

1 $x \times x = x^2$
2 $x \times q = qx$
3 $p \times x = px$
4 $p \times q = pq$

$(x + p)(x + q) = x^2 + qx + px + pq$
$= x^2 + (q + p)x + pq$
$= x^2 + 8x + 15$

Two numbers are required which: when multiplied give $+15$, **and** when added give $+8$.
$+5$ and $+3$ satisfy **both** conditions.
$x^2 + 8x + 15 = (x + 5)(x + 3)$

EXAMPLES

1. Factorise $x^2 + 6x + 9$.
   $x^2 + 6x + 9 = (x + 3)(x + 3)$
   $= (x + 3)^2$
   Because $3 \times 3 = 9$
   and $3 + 3 = 6$.

2. Factorise $x^2 - 8x + 12$.
   $x^2 - 8x + 12 = (x - 6)(x - 2)$
   Because $-6 \times -2 = 12$
   and $-6 + -2 = -8$.

3. Factorise $x^2 + 2x - 15$.
   $x^2 + 2x - 15 = (x + 5)(x - 3)$
   Because $+5 \times -3 = -15$
   and $+5 + -3 = +2$.

4. Factorise $x^2 - 2x - 8$.
   $x^2 - 2x - 8 = (x + 2)(x - 4)$
   Because $+2 \times -4 = -8$
   and $+2 + -4 = -2$.

Exercise 19.2

1 Factorise these expressions.

(a) $x^2 + 5x$ (b) $x^2 - 7x$ (c) $y^2 - 6y$ (d) $2y^2 - 12y$
(e) $5t - t^2$ (f) $8y + y^2$ (g) $x^2 - 20x$ (h) $3x^2 - 60x$

2 Factorise.

(a) $x^2 - 9$ (b) $x^2 - 81$ (c) $y^2 - 25$ (d) $y^2 - 1$
(e) $x^2 - 64$ (f) $100 - x^2$ (g) $36 - x^2$ (h) $x^2 - a^2$

3 Copy and complete the following.

(a) $x^2 + 6x + 5 = (x + 5)(x + \dots)$ (b) $x^2 + 9x + 14 = (x + 7)(x + \dots)$
(c) $x^2 + 6x + 8 = (x + \dots)(x + 4)$ (d) $x^2 + 9x + 18 = (x + \dots)(x + 6)$
(e) $x^2 - 6x + 5 = (x - 5)(x - \dots)$ (f) $x^2 - 7x + 10 = (x - 5)(x - \dots)$
(g) $x^2 - 7x + 12 = (x - \dots)(x - 3)$ (h) $x^2 + 3x - 4 = (x + 4)(x - \dots)$
(i) $x^2 + 5x - 14 = (x + 7)(x - \dots)$ (j) $x^2 - 4x - 5 = (x - 5)(x + \dots)$

Multiply the brackets out mentally to check your answers.

4 Factorise.

(a) $x^2 + 3x + 2$ (b) $x^2 + 8x + 7$ (c) $x^2 + 8x + 15$ (d) $x^2 + 8x + 12$
(e) $x^2 + 12x + 11$ (f) $x^2 + 9x + 20$ (g) $x^2 + 10x + 24$ (h) $x^2 + 13x + 36$
(i) $x^2 + 15x + 14$ (j) $x^2 + 10x + 16$

5 Factorise.

(a) $x^2 - 6x + 9$ (b) $x^2 - 6x + 8$ (c) $x^2 - 11x + 10$ (d) $x^2 - 16x + 15$
(e) $x^2 - 8x + 15$ (f) $x^2 - 10x + 16$ (g) $x^2 - 12x + 20$ (h) $x^2 - 11x + 24$
(i) $x^2 - 13x + 12$ (j) $x^2 - 8x + 12$

6 Factorise.

(a) $x^2 - x - 6$ (b) $x^2 - 5x - 6$ (c) $x^2 + 2x - 24$ (d) $x^2 + 5x - 24$
(e) $x^2 - 2x - 15$ (f) $x^2 + 3x - 18$ (g) $x^2 - 3x - 40$ (h) $x^2 - 4x - 12$
(i) $x^2 + 3x - 10$ (j) $x^2 - x - 20$

7 Factorise.

(a) $x^2 - 4x + 4$ (b) $x^2 + 11x + 30$ (c) $x^2 + 2x - 8$ (d) $x^2 - 4x - 21$
(e) $x^2 + x - 20$ (f) $x^2 + 7x + 12$ (g) $x^2 + 8x + 16$ (h) $x^2 - 2x + 1$
(i) $x^2 - 49$ (j) $t^2 + 12t$ (k) $x^2 - 9x + 14$ (l) $x^2 - 7x + 6$
(m) $x^2 + 11x + 18$ (n) $x^2 + 11x + 24$ (o) $x^2 + 19x + 18$ (p) $x^2 - y^2$
(q) $x^2 + x - 6$ (r) $y^2 + 4y$ (s) $y^2 - 10y + 25$ (t) $x^2 - 12x + 36$

Quadratics of the form $ax^2 + bx + c$

The expression $2x^2 + 11x + 15$ can be factorised.

First look at the term in x^2.
The coefficient of x^2 is 2, so, the brackets start: $(2x \dots \dots)(x \dots \dots)$

Next, look for two numbers which when multiplied give $+15$.
The possibilities are: 1 and 15 or 3 and 5.

Use these pairs of numbers to complete the brackets.
Multiply out pairs of brackets until the correct factorisation of $2x^2 + 11x + 15$ is found.

$(2x + 1)(x + 15) = 2x^2 + 30x + x + 15 = 2x^2 + 31x + 15$
$(2x + 15)(x + 1) = 2x^2 + 2x + 15x + 15 = 2x^2 + 17x + 15$
$(2x + 3)(x + 5) = 2x^2 + 10x + 3x + 15 = 2x^2 + 13x + 15$
$(2x + 5)(x + 3) = 2x^2 + 6x + 5x + 15 = 2x^2 + 11x + 15$
$2x^2 + 11x + 15 = (2x + 5)(x + 3)$

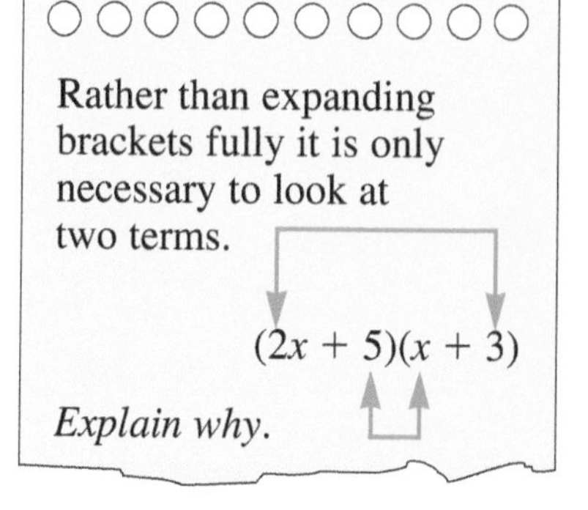

EXAMPLES

1. Factorise $3x^2 + 5x + 2$.

   $3x^2 + 5x + 2 = (3x + 2)(x + 1)$

2. Factorise $3a^2 + 7a + 2$.

   $3a^2 + 7a + 2 = (3a + 1)(a + 2)$

Multiply the brackets out mentally to check the answers.

3. Factorise $4x^2 + 19x - 5$.

   $4x^2 + 19x - 5 = (4x - 1)(x + 5)$
   The brackets contain different signs.
   Explain why.

4. Factorise $4x^2 - 16x + 15$.

   $4x^2 - 16x + 15 = (2x - 5)(2x - 3)$
   Each bracket contains a minus sign.
   Explain why.

Exercise 19.3

1. Copy and complete the following.
   (a) $2x^2 + 12x + 18 = (2x + 6)(\ldots\ \ldots\ \ldots)$
   (b) $2y^2 - 9y - 18 = (\ldots\ \ldots\ \ldots)(y - 6)$
   (c) $2a^2 + 5a - 18 = (2a + \ldots)(a - \ldots)$
   (d) $2m^2 - 20m + 18 = (2m\ \ldots\ \ldots)(m\ \ldots\ \ldots)$
   (e) $4x^2 - 13x + 3 = (4x\ \ldots\ \ldots)(x\ \ldots\ \ldots)$
   (f) $4d^2 + 4d - 3 = (2d - 1)(\ldots\ \ldots\ \ldots)$
   Multiply the brackets out mentally to check your answers.

2. Factorise.
   (a) $2x^2 + 7x + 3$ (b) $2x^2 + 11x + 12$ (c) $2a^2 + 25a + 12$
   (d) $4x^2 + 4x + 1$ (e) $6y^2 + 7y + 2$ (f) $6x^2 + 17x + 12$
   (g) $11m^2 + 12m + 1$ (h) $3x^2 + 17x + 10$ (i) $5k^2 + 12k + 4$

3. Factorise.
   (a) $2a^2 - 5a + 3$ (b) $2y^2 - 9y + 7$ (c) $4x^2 - 4x + 1$
   (d) $3x^2 - 5x + 2$ (e) $6d^2 - 11d + 3$ (f) $2x^2 - 3x + 1$
   (g) $2x^2 - 11x + 15$ (h) $9y^2 - 12y + 4$ (i) $4t^2 - 12t + 9$

4. Factorise.
   (a) $2x^2 + x - 6$ (b) $2a^2 - a - 3$ (c) $3x^2 - 14x - 5$
   (d) $3t^2 + 14t - 5$ (e) $2y^2 - 5y - 7$ (f) $4y^2 + 7y - 2$
   (g) $9x^2 + 3x - 2$ (h) $10x^2 + 27x - 9$ (i) $5m^2 - 2m - 3$

Further factorising

When factorising, work logically.
Does the expression have a common factor?
Is the expression a difference of two squares?
Will the expression factorise into two brackets?

EXAMPLES

Factorise the following.

1. $2x^2 - 14x = 2x(x - 7)$

2. $25 - 4y^2 = 5^2 - (2y)^2$
   $= (5 + 2y)(5 - 2y)$

3. $2x^2 - 18 = 2(x^2 - 9)$
   $= 2(x + 3)(x - 3)$

4. $2x^2 - 6x - 8 = 2(x^2 - 3x - 4)$
   $= 2(x + 1)(x - 4)$

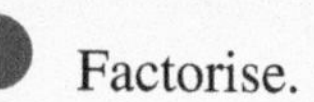

Exercise 19.4

1 Factorise.

(a) $3x + 12y$ (b) $t^2 - 16$ (c) $x^2 + 4x + 3$
(d) $y^2 - y$ (e) $2d^2 - 6$ (f) $p^2 - q^2$
(g) $a^2 - 2a$ (h) $x^2 - 2x + 1$ (i) $2y^2 - 8y$
(j) $a^2 - 6a + 9$ (k) $3m^2 - 12$ (l) $v^2 - v - 6$
(m) $ax^2 - ay^2$ (n) $-8 - 2x$ (o) $4 - 4k + k^2$
(p) $18 - 2x^2$ (q) $50 - 2x^2$ (r) $12x^2 - 27y^2$
(s) $6a^2 - 3a$ (t) $x^2 - 15x + 56$ (u) $x^2 + 2x - 15$

2 Factorise.

(a) $3a^2 - 2a - 8$ (b) $3x^2 + 31x + 10$ (c) $64y^2 - 49$
(d) $3x^2 + 11x + 10$ (e) $7x^2 - 19x + 10$ (f) $6x^2 - 25x - 25$
(g) $5m^2 - 18m + 9$ (h) $8x^2 + 2x - 15$ (i) $13x - 20 - 2x^2$
(j) $4y^2 + 22y + 24$ (k) $6x^2 + 3x - 18$ (l) $28n - 10 + 6n^2$

Solving quadratic equations

Activity

I am thinking of two numbers.
I multiply them together.
The answer is zero.

Write down 2 numbers which could be Jim's numbers.
Now write down **four** more pairs.
What can you say about Jim's numbers?

You should have discovered that at least one of Jim's numbers must be **zero**.
We can use this fact to solve **quadratic equations**.

EXAMPLES

1 Find x if:

(a) $x(x - 2) = 0$
Either $x = 0$ or $x - 2 = 0$
Because one of them must be zero.
$x = 0$ or $x = 2$

(b) $(x - 3)(x + 2) = 0$
$x - 3 = 0$ or $x + 2 = 0$
Because one of them must be zero.
$x = 3$ or $x = -2$

2 Solve $x^2 - 5x + 6 = 0$

$x^2 - 5x + 6 = 0$
Factorise the quadratic expression first.
$(x - 2)(x - 3) = 0$
$x = 2$ or $x = 3$

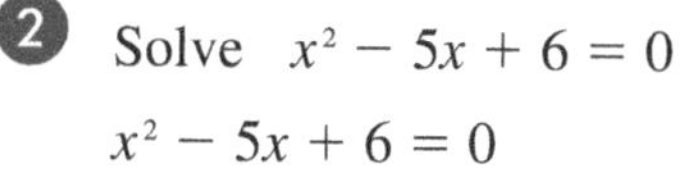

To check solutions:
Does $(2)^2 - 5 \times (2) + 6 = 0$
and $(3)^2 - 5 \times (3) + 6 = 0$?

3 Solve these equations.

(a) $x^2 - 9 = 0$
$(x - 3)(x + 3) = 0$
$x = 3$ or $x = -3$

(b) $y^2 + 7y = 0$
$y(y + 7) = 0$
$y = 0$ or $y = -7$

(c) $a^2 + 4a - 5 = 0$
$(a - 1)(a + 5) = 0$
$a = 1$ or $a = -5$

Exercise 19.5

1 Solve these equations.

(a) $(x - 2)(x - 3) = 0$ (b) $(x + 4)(x + 6) = 0$ (c) $(x - 3)(x + 1) = 0$
(d) $(x - 5)(x + 2) = 0$ (e) $x(x - 4) = 0$ (f) $3x(x + 2) = 0$

2 Solve.

(a) $x^2 - 3x + 2 = 0$ (b) $y^2 + 7y + 12 = 0$ (c) $m^2 - 2m - 8 = 0$
(d) $a^2 + a - 12 = 0$ (e) $n^2 - 5n - 36 = 0$ (f) $z^2 - 9z + 18 = 0$
(g) $k^2 + 8k + 15 = 0$ (h) $c^2 + 15c + 56 = 0$ (i) $b^2 + b - 20 = 0$
(j) $v^2 - 7v - 60 = 0$ (k) $w^2 + 8w - 48 = 0$ (l) $p^2 - p - 72 = 0$

3 Solve.

(a) $x^2 - 5x = 0$ (b) $y^2 + y = 0$ (c) $p^2 + 3p = 0$
(d) $4a - a^2 = 0$ (e) $t^2 - 6t = 0$ (f) $g^2 - 4g = 0$

4 Solve.

(a) $x^2 - 4 = 0$ (b) $y^2 - 144 = 0$ (c) $9 - a^2 = 0$
(d) $d^2 - 16 = 0$ (e) $x^2 - 100 = 0$ (f) $36 - x^2 = 0$
(g) $x^2 - 49 = 0$ (h) $2.25 - x^2 = 0$ (i) $2x^2 - 18 = 0$

5 Solve these equations.

(a) $x^2 + 6x = 0$ (b) $x^2 + 5x + 4 = 0$ (c) $x^2 - 64 = 0$
(d) $x^2 - 4x + 3 = 0$ (e) $5x^2 - 10x = 0$ (f) $x^2 - x - 6 = 0$
(g) $x^2 + 2x - 15 = 0$ (h) $2x^2 - 8 = 0$ (i) $x^2 - 15x + 56 = 0$

6 Solve these equations.

(a) $2x^2 + 7x + 5 = 0$ (b) $2x^2 - 11x + 5 = 0$ (c) $3x^2 - 17x - 28 = 0$
(d) $6y^2 + 7y + 2 = 0$ (e) $3x^2 + 7x - 6 = 0$ (f) $3z^2 - 5z - 2 = 0$
(g) $5m^2 - 8m + 3 = 0$ (h) $6a^2 + 3a - 63 = 0$ (i) $4y^2 - 3y - 10 = 0$

7 Rearrange these equations and then solve them.

(a) $y^2 = 4y + 5$ (b) $x^2 = x$ (c) $x^2 = 8x - 16$
(d) $x^2 = 2x + 15$ (e) $n^2 - 10n = 24$ (f) $7 = 8m - m^2$
(g) $a(a - 5) = 24$ (h) $x^2 - 5x + 6 = 3 - x$ (i) $x = 6x^2 - 1$
(j) $15 - 7m = 2m^2$ (k) $6a^2 = 2 - a$ (l) $8x^2 = 3 - 2x$

Using the quadratic formula

Some quadratic expressions do not factorise.

Consider the equation $x^2 - 5x + 2 = 0$.

It is not possible to find two numbers which:
when multiplied give $+2$ **and** when added give -5.

We could draw the graph of $y = x^2 - 5x + 2$ and use it to solve the equation $x^2 - 5x + 2 = 0$.
One solution lies between 0 and 1.
Another solution lies between 4 and 5.

Graphical solutions have limited accuracy and are time consuming. Explain why.

The solutions to a quadratic equation can be found using the **quadratic formula**.

$$\text{If } ax^2 + bx + c = 0 \text{ and } a \neq 0 \text{ then } x = \frac{-b \pm \sqrt{b^2 - 4ac}}{2a}$$

The quadratic formula can be used at any time, even if the quadratic expression factorises.

Solve $x^2 - 5x + 2 = 0$, giving the answers correct to three significant figures.

Using $x = \frac{-b \pm \sqrt{b^2 - 4ac}}{2a}$

Substitute: $a = 1$, $b = -5$ and $c = 2$.

$$x = \frac{-(-5) \pm \sqrt{(-5)^2 - 4(1)(2)}}{2(1)}$$

$$x = \frac{5 \pm \sqrt{25 - 8}}{2}$$

$$x = \frac{5 \pm \sqrt{17}}{2}$$

Either $x = \frac{5 + \sqrt{17}}{2}$ or $x = \frac{5 - \sqrt{17}}{2}$

$x = 4.5615\ldots$ or $x = 0.43844\ldots$

$x = 0.438$ or 4.56, correct to 3 sig. figs.

Check the answers by substituting the values of x into the original equation.

EXAMPLE

Solve $6x = 3 - 5x^2$, giving the answers correct to two decimal places.

Rearrange $6x = 3 - 5x^2$ to get $5x^2 + 6x - 3 = 0$.

Using $x = \frac{-b \pm \sqrt{b^2 - 4ac}}{2a}$

Substitute: $a = 5$, $b = 6$ and $c = -3$.

$$x = \frac{-6 \pm \sqrt{6^2 - 4(5)(-3)}}{2(5)}$$

$$x = \frac{-6 \pm \sqrt{36 + 60}}{10}$$

$$x = \frac{-6 \pm \sqrt{96}}{10}$$

Either $x = \frac{-6 + \sqrt{96}}{10}$ or $x = \frac{-6 - \sqrt{96}}{10}$

$x = 0.3797\ldots$ or $x = -1.5797\ldots$

$x = -1.58$ or 0.38, correct to 2 d.p.

Exercise 19.6

1 Use the quadratic formula to solve these equations.

(a) $x^2 + 4x + 3 = 0$ (b) $x^2 - 3x + 2 = 0$ (c) $x^2 + 2x - 3 = 0$

2 Solve these equations, giving the answers correct to 2 decimal places.

(a) $x^2 + 4x + 2 = 0$ (b) $x^2 + 7x + 5 = 0$ (c) $x^2 + x - 1 = 0$
(d) $x^2 - 3x - 2 = 0$ (e) $x^2 - 5x - 3 = 0$ (f) $x^2 + 5x + 3 = 0$
(g) $x^2 + 4x - 10 = 0$ (h) $x^2 - 3x + 1 = 0$ (i) $x^2 + 3x - 2 = 0$

3 Use the quadratic formula to solve these equations.

(a) $2x^2 + x - 3 = 0$ (b) $2x^2 - 3x + 1 = 0$ (c) $5x^2 + 2x - 3 = 0$

4 Solve these equations, giving the answers correct to 2 decimal places.

(a) $3x^2 - 5x + 1 = 0$ (b) $2x^2 - 7x + 4 = 0$ (c) $3x^2 + 23x + 8 = 0$
(d) $3x^2 - 4x - 5 = 0$ (e) $2z^2 + 6z + 3 = 0$ (f) $3x^2 - 2x - 20 = 0$

5 Rearrange the following to form quadratic equations.
Then solve the quadratic equations, giving answers correct to 2 decimal places.

(a) $x^2 = 6x - 3$ (b) $2x^2 - 6 = 3x$ (c) $3x = 2x^2 - 4$
(d) $2 = x + 5x^2$ (e) $3(x + 1) = x^2 - x$ (f) $x(2x + 5) = 10$
(g) $3x(x - 4) = 2(x - 1)$ (h) $(3x - 2)^2 = (x + 2)^2 + 12$ (i) $5 = 2x(x - 1)$

6 (a) Draw axes marked from -3 to 2 for x and from -2 to 10 for y.
Using the same axes draw the following graphs:

$y = 2x^2 + 4x + 1$ $y = 2x^2 + 4x + 2$ $y = 2x^2 + 4x + 3$

Use your graphs to solve the equations:

$2x^2 + 4x + 1 = 0$ $2x^2 + 4x + 2 = 0$ $2x^2 + 4x + 3 = 0$

(b) Use the quadratic formula to solve the following equations:

$2x^2 + 4x + 1 = 0$ $2x^2 + 4x + 2 = 0$ $2x^2 + 4x + 3 = 0$

(c) Compare your answers to parts (a) and (b).

Completing the square

16, x^2, $9y^2$, $(m-3)^2$ are all examples of **perfect squares**.

What must be added to $x^2 - 2x$ to make it a perfect square?
$(x-1)^2 = x^2 - 2x + 1$
So, if $+1$ is added to $x^2 - 2x$ the result is $(x-1)^2$.

The process of adding a constant term to a quadratic expression to make it a perfect square is called **completing the square**.

Consider the following.

$(x+1)^2 = x^2 + 2x + 1$ $\quad$ $(x-1)^2 = x^2 - 2x + 1$
$(x+2)^2 = x^2 + 4x + 4$ $\quad$ $(x-2)^2 = x^2 - 4x + 4$
$(x+3)^2 = x^2 + 6x + 9$ $\quad$ $(x-3)^2 = x^2 - 6x + 9$

In the expanded form the coefficient of x is always **twice** the constant in the brackets or, to put it another way, the constant in the brackets is **half the coefficient of x**.

$(x+a)^2 = x^2 + 2ax + a^2$ $\quad$ $(x-a)^2 = x^2 - 2ax + a^2$

Rearranging quadratics

Quadratics can be rearranged into a **square term** and a **constant value** (number).
This is sometimes called **completed square form** and can be used to solve quadratic equations.

> To write the quadratic $x^2 + bx + c$ in completed square form:
> write the **square term** as $(x + \frac{1}{2}$ the coefficient of $x)^2$ and then adjust for the **constant value**.

$x^2 + 6x + 10$ can be written in the form $(x+a)^2 + b$. Find a and b.

The coefficient of x is 6. $\frac{1}{2}$ of 6 is 3.
The **square term** is $(x+3)^2$.
$(x+3)^2 = x^2 + 6x + 9$
$x^2 + 6x + 10 = x^2 + 6x + 9 + 1 = (x+3)^2 + 1$.
The **constant value** is $+1$.
So, $a = 3$ and $b = 1$.

> In general, the **vertex** of the graph $y = (x+a)^2 + b$ is at the point $(-a, b)$.

This change of form can still be done when the coefficient of x^2 is not 1.
Consider the following.

$(2x+1)^2 = 4x^2 + 4x + 1$ $\quad$ $(2x-1)^2 = 4x^2 - 4x + 1$
$(2x+2)^2 = 4x^2 + 8x + 4$ $\quad$ $(2x-2)^2 = 4x^2 - 8x + 4$
$(2x+3)^2 = 4x^2 + 12x + 9$ $\quad$ $(2x-3)^2 = 4x^2 - 12x + 9$

In general: $(px+q)^2 = p^2x^2 + 2pqx + q^2$
This can be used to change the form of quadratics.

$4x^2 + 20x + 31$ can be written in the form $(px+q)^2 + r$. Find p, q and r.
Using $(px+q)^2 = p^2x^2 + 2pqx + q^2$.
$4x^2 + 20x + 31 = p^2x^2 + 2pqx + q^2$.

Comparing the coefficients of x^2.
$p^2 = 4$
$p = \sqrt{4} = 2$

Comparing the coefficients of x.
$2pq = 20$
But $p = 2$.
So, $4q = 20$
$q = 5$

$(2x+5)^2 = 4x^2 + 20x + 25$
$4x^2 + 20x + 31 = 4x^2 + 20x + 25 + 6 = (2x+5)^2 + 6$
The constant value, r, is $+6$.
So, $p = 2$, $q = 5$ and $r = 6$.

EXAMPLES

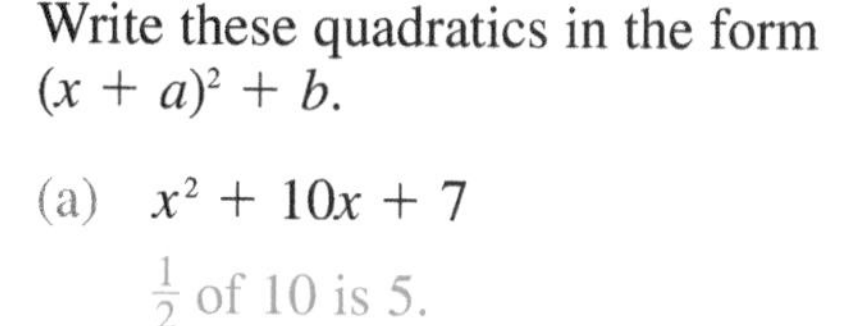

1 Write these quadratics in the form $(x + a)^2 + b$.

(a) $x^2 + 10x + 7$

$\frac{1}{2}$ of 10 is 5.

$$(x + 5)^2 = x^2 + 10x + 25$$
$$x^2 + 10x + 7 = x^2 + 10x + 25 - 18$$
$$= (x + 5)^2 - 18$$

(b) $x^2 - 8x - 5$

$\frac{1}{2}$ of -8 is -4.

$$(x - 4)^2 = x^2 - 8x + 16$$
$$x^2 - 8x - 5 = x^2 - 8x + 16 - 21$$
$$= (x - 4)^2 - 21$$

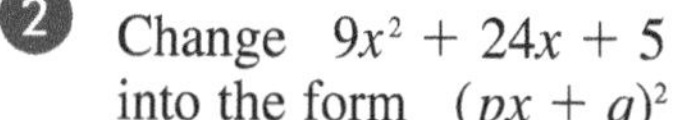

2 Change $9x^2 + 24x + 5$ into the form $(px + q)^2 + r$.

Using $(px + q)^2 = p^2x^2 + 2pqx + q^2$.

$$9x^2 + 24x + 5 = p^2x^2 + 2pqx + q^2$$

$p^2 = 9$

$p = \sqrt{9} = 3$

$2pq = 24$

But $p = 3$.

So, $6q = 24$

$q = 4$

$$(3x + 4)^2 = 9x^2 + 24x + 16$$
$$9x^2 + 24x + 5 = 9x^2 + 24x + 16 - 11$$
$$= (3x + 4)^2 - 11$$

Exercise 19.7

1 What must be added to each expression to make a perfect square?

(a) $x^2 + 2x$ (b) $x^2 + 4x$ (c) $x^2 - 6x$
(d) $y^2 + 8y$ (e) $p^2 - 2p$ (f) $a^2 - 3a$
(g) $m^2 + 10m$ (h) $x^2 - 5x$

2 Complete these squares.

(a) $x^2 + 6x + \ldots = (x + \ldots)^2$ (b) $a^2 - 4a + \ldots = (a - \ldots)^2$
(c) $b^2 + 2b + \ldots = (b + \ldots)^2$ (d) $m^2 - 8m + \ldots = (m - \ldots)^2$
(e) $n^2 - n + \ldots = (n - \ldots)^2$ (f) $x^2 + 5x + \ldots = (x + \ldots)^2$

3 Change these quadratics into the form $(x + a)^2 + b$.

(a) $x^2 + 6x + 20$ (b) $x^2 + 6x + 5$ (c) $x^2 + 10x - 4$
(d) $x^2 - 4x + 5$ (e) $x^2 - 4x + 2$ (f) $x^2 - 4x - 4$
(g) $x^2 - 6x + 4$ (h) $x^2 - 8x$ (i) $x^2 + 12x$

4 These quadratics can be written in the form $(x + a)^2 + b$.
State the values of a and b.

(a) $x^2 + 6x + 15$ (b) $x^2 + 10x + 5$ (c) $x^2 - 6x - 5$
(d) $x^2 - 8x + 4$ (e) $x^2 + 12x + 4$ (f) $x^2 + 6x + 9$

5 Change these quadratics into the form $(px + q)^2 + r$.
State the values of p, q and r.

(a) $4x^2 + 16x + 5$ (b) $9x^2 + 12x + 3$ (c) $25x^2 - 40x - 3$
(d) $16x^2 + 32x - 5$ (e) $100x^2 + 60x + 3$ (f) $16x^2 - 40x + 9$

6 The expression $x^2 - 4x - 1$ can be written in the form $(x + a)^2 + b$.

(a) Find the values of a and b.
(b) What is the minimum value of the expression $x^2 - 4x - 1$?

7 $x^2 - 6x + a = (x + b)^2 - 1$.
Find the values of a and b.

8 What is the minimum value of the expression $x^2 - 10x + 26$?

Solving quadratic equations by completing the square

When a quadratic equation is expressed in the form $(x + m)^2 = n$, with $n > 0$, it is possible to solve the equation by taking the square roots of both sides.

EXAMPLES

1 Solve $x^2 + 2x = 8$.

$x^2 + 2x = 8$

Write the left-hand side (LHS) of the equation in the form $(x + a)^2 + b$ by completing the square.

$x^2 + 2x = (x + 1)^2 - 1$

$(x + 1)^2 - 1 = 8$

Add 1 to both sides.

$(x + 1)^2 = 9$

Take the square root of both sides of the equation.

$x + 1 = \pm 3$

Either $x + 1 = 3$ or $x + 1 = -3$

$x = 2$ or $x = -4$

2 Solve $2x^2 - 12x = 9$, giving the answer correct to one decimal place.

$2x^2 - 12x = 9$

To solve a quadratic equation by completing the square we must first make the coefficient of x^2 unity (one).

Divide both sides of the equation by 2.

$x^2 - 6x = 4.5$

Complete the square on the LHS.

$(x - 3)^2 - 9 = 4.5$

$(x - 3)^2 = 4.5 + 9$

$(x - 3)^2 = 13.5$

$x - 3 = \pm\sqrt{13.5}$

Either $x = 3 + 3.7$ or $x = 3 - 3.7$

$x = 6.7$ or -0.7, correct to 1 d.p.

Exercise 19.8

1 Solve the following equations by completing the square.

(a) $x^2 - 2x = 3$ (b) $t^2 - 4t = 21$ (c) $x^2 + 6x = 27$

(d) $y^2 - 6y - 16 = 0$ (e) $m^2 - 10m - 24 = 0$ (f) $e^2 - 14e + 48 = 0$

2 Solve the following equations by completing the square.
Give your answers correct to one decimal place.

(a) $x^2 - 10x = 15$ (b) $m^2 + 4m = 7$ (c) $x^2 - 6x = -3$

(d) $y^2 - 3y - 2 = 0$ (e) $x^2 - 12x = 5$ (f) $a^2 + 6a + 4 = 0$

(g) $2n^2 + 2n = 1$ (h) $2x^2 + 2x = 3$ (i) $4z^2 - 12z + 3 = 0$

(j) $3x^2 - 12x = -11$ (k) $x^2 + 6x + 7 = 0$ (l) $4x^2 - 4x - 15 = 0$

3 (a) Write $x^2 - 4x + 1$ in the form $(x + a)^2 + b$.

(b) Hence, solve the equation $x^2 - 4x + 1 = 0$, leaving your answer in the form $c \pm \sqrt{d}$.

4 The solutions of the equation $x^2 - 6x + 4 = 0$ can be given as $a \pm \sqrt{b}$ where a and b are prime numbers.
Find the values of a and b.

5 Solve the equation $2x^2 - 4x - 5 = 0$.
Give your answers in the form $p \pm \sqrt{q}$ where p and q are rational.

6 The solutions of the equation $x^2 + mx + n = 0$ are $x = 3 \pm \sqrt{2}$.
Find the values of m and n.

7 The solutions of the equation $2x^2 + ax + b = 0$ are $x = \frac{4 \pm \sqrt{22}}{2}$.
Find the values of a and b.

8 Starting with $ax^2 + bx + c = 0$ prove the quadratic formula $x = \frac{-b \pm \sqrt{b^2 - 4ac}}{2a}$.

Problems solved using quadratic equations

Some mathematical problems involve the forming of equations which are quadratic.
The equation is solved and the solutions are analysed to answer the original problem.

EXAMPLES

1 The sum of two numbers is 15 and their product is 56. Find the numbers.

If x is one of the numbers then the other number is $(15 - x)$.

Their product is 56, so, $\quad x(15 - x) = 56$

Expand the brackets. $\quad 15x - x^2 = 56$

Rearrange to the form $ax^2 + bx + c = 0$. $\quad x^2 - 15x + 56 = 0$

Factorise. $\quad (x - 7)(x - 8) = 0$

Solve the equation. $\quad x = 7 \text{ or } x = 8$

The numbers are 7 and 8.

Check: Does $7 + 8 = 15$ **and** $7 \times 8 = 56$?

2 A rectangular lawn, 10 m by 8 m, is surrounded by a path which is x m wide.
The total area of the path and lawn is 143 m^2.
Calculate the width of the path.

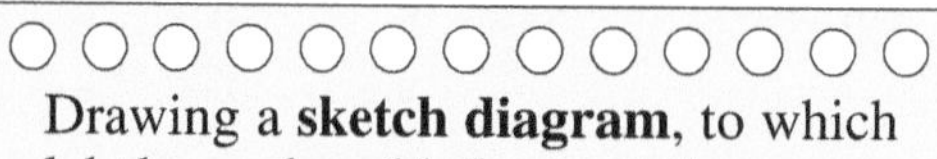

Drawing a **sketch diagram**, to which labels can be added, may help you to understand and solve problems.

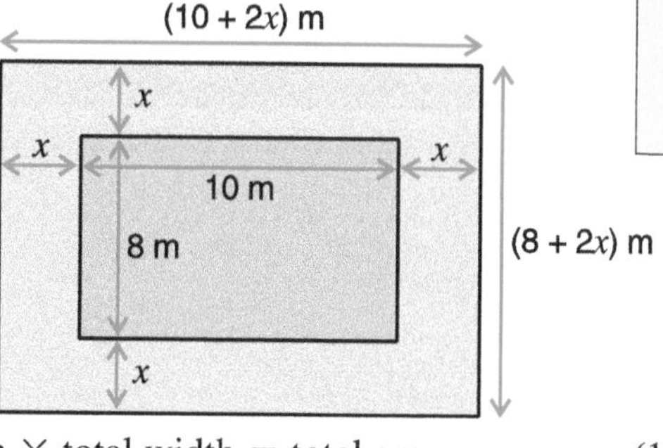

Total length $= (10 + 2x)$ m
Total width $= (8 + 2x)$ m
Total area $= 143$ m^2

Total length × total width = total area $\quad (10 + 2x)(8 + 2x) = 143$

Expand the brackets. $\quad 80 + 20x + 16x + 4x^2 = 143$

Rearrange to the form $ax^2 + bx + c = 0$. $\quad 4x^2 + 36x - 63 = 0$

$(2x - 3)(2x + 21) = 0$

Either $2x - 3 = 0$ or $2x + 21 = 0$

$2x = 3$ or $2x = -21$

$x = 1.5$ or $x = -10.5$

The quadratic equation has two solutions.
However, $x \neq -10.5$, as it is impossible to have a path of negative width in the context of the problem.
So, $x = 1.5$
The width of the path is 1.5 m.

Check: Does $(10 + 2 \times 1.5) \times (8 + 2 \times 1.5) = 143$?

Exercise 19.9

In this exercise form suitable equations and solve them using an appropriate method.

1 This rectangle has an area of 21 cm^2.
(a) Form an equation in x.
(b) By solving your equation, find the value of x.

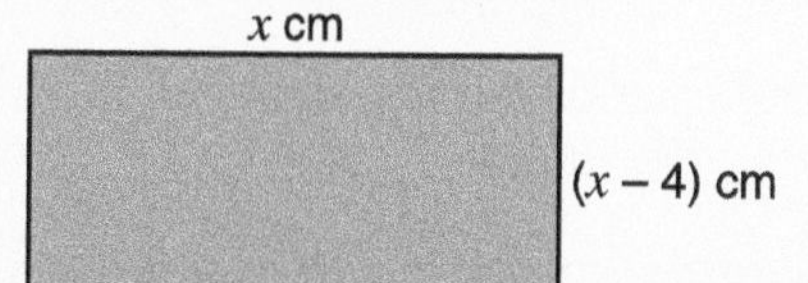

2 The product of x and $(x - 2)$ is 63. By forming an equation, find the value of x.

3 A right-angled triangle is cut from the corner of a rectangle as shown.
The shaded area is $7.5\,\text{cm}^2$.

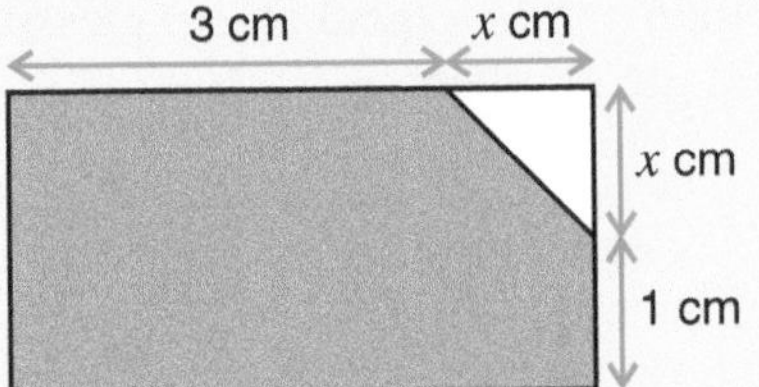

(a) Form an equation in x and show that it simplifies to $x^2 + 8x - 9 = 0$.
(b) By solving the equation $x^2 + 8x - 9 = 0$, find the value of x.

4 Garden A consists of a lawn 8 m by 5 m with a path x m wide on all sides.
Garden B consists of a lawn 15 m by 14.5 m with a path x m wide on all sides.
Garden B is three times the area of Garden A.

Garden A **Garden B**

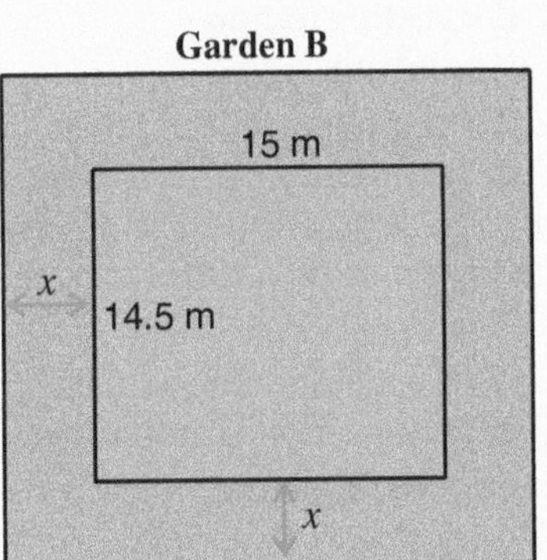

(a) Form an equation in x.
(b) Solve it to find x (the width of the path).
(c) Check that your value for x gives the area of Garden B as three times the area of Garden A.

5 The length of a rectangle is 5 cm greater than its width. The area of the rectangle is $40\,\text{cm}^2$.
Calculate the length and width of the rectangle.
Give your answer correct to two significant figures.

6 The perimeter of a rectangle is 62 cm.
The length of the diagonal is 25 cm.
Calculate the dimensions of the rectangle.

Hint: In a right-angled triangle
"The square on the hypotenuse is equal to the sum of the squares on the other two sides."

7 The perimeter of a rectangle is 20 m. The area is $20\,\text{m}^2$.
Calculate the dimensions of the rectangle. Give your answers correct to two decimal places.

8 (a) Two positive numbers differ by 5. Their product is 234. What are the two numbers?
(b) The sum of two numbers is 12. Their product is 30. What are the two numbers?

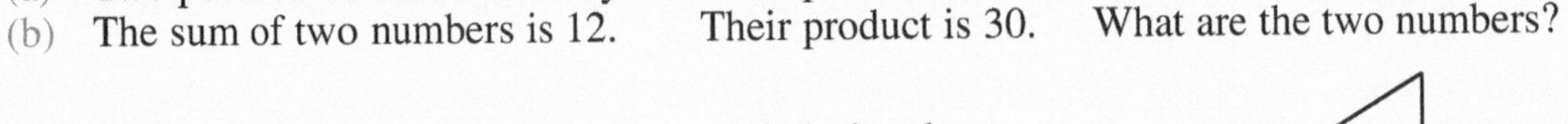

9 A canoe course is in the shape of a right-angled triangle.
The length of the course is 300 m.
The longest leg of the course is 125 m.
What are the lengths of the other two legs?

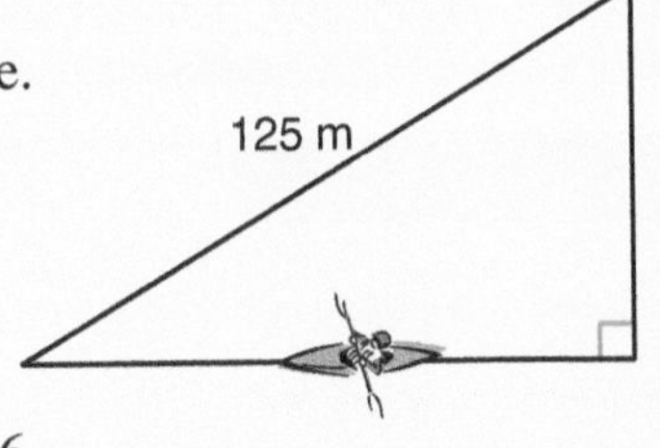

10 Three numbers are in this sequence: x, $x + 3$, $x + 6$.
The sum of the squares of the two smaller numbers is equal to the square of the largest number.
What are the three numbers? (There are two solutions.)

11 Consider these numbers: x, $x + 1$, $x + 3$ and $x + 7$.
The sum of the squares of the two smaller numbers is equal to the product of the two larger numbers.
(a) Form an equation and simplify it to a quadratic expression equal to 0.
(b) Solve the equation to find the numbers (two solutions).

12 The nth triangular number is $\frac{1}{2}n(n + 1)$.
(a) Which triangular number is equal to 66?
(b) Which triangular number is equal to 120?

13 Two numbers differ by 5. Their squares differ by 55. What are the two numbers?

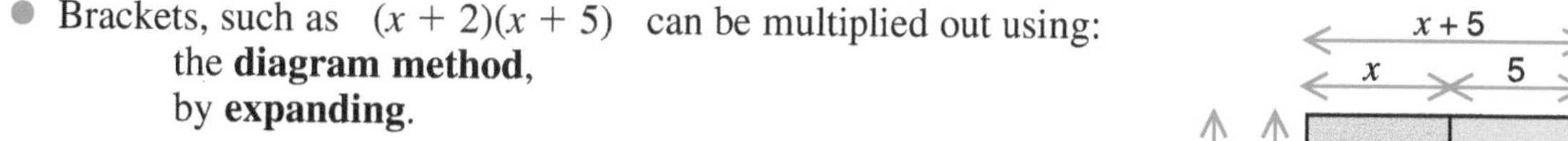

What you need to know

- Brackets, such as $(x + 2)(x + 5)$ can be multiplied out using:
  the **diagram method**,
  by **expanding**.

	x	5
x	x^2	$5x$
2	$2x$	10

$(x + 2)(x + 5) = x^2 + 5x + 2x + 10$
$= x^2 + 7x + 10$

- **Factorising** is the opposite operation to removing brackets.
  When factorising, work logically.
  Does the expression have a **common factor**?
  Is the expression a difference of **two squares**?
  Will the expression factorise into **two brackets**?
- A **common factor** is a factor which divides into two, or more, numbers (or terms). For example: $x^2 + 7x = x(x + 7)$
- **Difference of two squares**: $a^2 - b^2 = (a - b)(a + b)$
- **Quadratic equations** can be solved:
  by factorising, graphically, using the quadratic formula, by completing the square.
- The general form for a **quadratic equation** is $ax^2 + bx + c = 0$, where a cannot be zero.
- The solutions to a quadratic equation can be found using the **quadratic formula**.

  If $ax^2 + bx + c = 0$ and $a \neq 0$ then $x = \frac{-b \pm \sqrt{b^2 - 4ac}}{2a}$
- Quadratic expressions, such as $x^2 + 8x + 20$, can be written in the form $(x + a)^2 + b$, where a and b are integers.
  In **completed square form**, $(x + a)^2 + b$:
  the value of a is half the coefficient of x,
  the value of b is found by subtracting the value of a^2 from the constant term of the original expression.
  For example: $x^2 + 8x + 20 = (x + 4)^2 + 4$
- Quadratic expressions, such as $4x^2 + 16x + 5$,
  where the coefficient of x^2 is not 1 can be written in the form $(px + q)^2 + r$.
  In completed square form, $(px + q)^2 + r$:
  the value of p is the square root of the coefficient of x^2,
  the value of q is the coefficient of x divided by $2p$,
  the value of r is found by subtracting the value of q^2 from the constant term of the original expression.
  For example: $4x^2 + 16x + 5 = (2x + 4)^2 - 11$

Review Exercise 19

1 Expand and simplify $(x + 5)(x - 2)$. AQA

2 (a) Factorise $x^2 - 3x - 10$.
(b) Hence, or otherwise, solve the equation $x^2 - 3x - 10 = 0$. AQA

3 Multiply out the brackets and simplify your answer: $(2x - 3)(x + 4)$ AQA

4 Solve the equations:
(a) $2x + 3 = 15 - x$, (b) $(x + 3)(x - 4) = 0$. AQA

5 Solve $x^2 + 8x + 12 = 0$. AQA

6 (a) Factorise $x^2 - 2x - 8$. (b) Solve $a^2 + 4a = 0$.

7 These two rectangles have the same area.

(a) Form an equation in x and show that it can be simplified to $x^2 + x - 2 = 0$.
(b) Solve the equation $x^2 + x - 2 = 0$ to find the length of BC.

A (x + 3) cm B; x cm; D C
P 2 cm Q; (x + 1) cm; S R

AQA

8 The sum of two numbers is 13. Their product is 36.
By forming an equation, find the two numbers.

9 Solve the equation $2x^2 + 5x - 3 = 0$. You **must** show all your working. AQA

10 The expression $x^2 - 8x + 17$ can be written in the form $(x - p)^2 + q$.
Calculate the values of p and q. AQA

11 (a) Write $x^2 + 4x - 10$ in the form $(x + a)^2 + b$.
(b) Hence, or otherwise, solve $x^2 + 4x - 10 = 0$, leaving your answers in the form $c \pm \sqrt{d}$. AQA

12 Expand and simplify $(n + 2)^2 - 2(n + 2)$. AQA

13 (a) Factorise $6x^2 - x - 12$.
(b) Solve the equation $x^2 - 7x - 6 = 0$, giving your answers to 2 decimal places. AQA

14 The formula for the stopping distance of a car, d feet, in terms of its speed, v miles per hour, is $d = v + \frac{v^2}{20}$.
(a) Show that, when $d = 175$, the formula can be written as $v^2 + 20v - 3500 = 0$.
(b) Solve the equation $v^2 + 20v - 3500 = 0$. AQA

15 Solve the equation $3x^2 - 5x + 1 = 0$.
Give your answers correct to 2 decimal places. AQA

16 A rectangular lawn has a path of width x m on three sides, as shown.
The lawn is 5 m long and 3 m wide.
The total area of the lawn and path is $39\,\text{m}^2$.

(a) By forming an expression, in terms of x, for the total area of the lawn and path, show that $2x^2 + 13x - 24 = 0$.
(b) By solving the equation $2x^2 + 13x - 24 = 0$, find the value of x.

5 m; x m; 3 m; x m; x m

AQA

17 Rectangle A has length $(x + 5)$ cm and width $2x$ cm.
Rectangle B has length $(x - 5)$ cm and width $(x + 10)$ cm.
The combined areas of rectangle A and rectangle B total $130\,\text{cm}^2$.
(a) Show that $x^2 + 5x - 60 = 0$.
(b) By solving the equation $x^2 + 5x - 60 = 0$, find the value of x correct to 2 decimal places. AQA

Simultaneous Equations

$x + y = 10$ is an equation with two unknown quantities x and y.
Many pairs of values of x and y fit this equation.

For example.
$x = 1$ and $y = 9$, $x = 4$ and $y = 6$, $x = 2.9$ and $y = 7.1$, $x = 1.005$ and $y = 8.995$, ...
$x - y = 2$ is another equation with the **same** two unknown quantities x and y.
Again, many pairs of values of x and y fit this equation.

For example.
$x = 4$ and $y = 2$, $x = 7$ and $y = 5$, $x = 2.9$ and $y = 0.9$, $x = -1$ and $y = -3$, ...

There is only **one** pair of values of x and y which fit **both** of these equations ($x = 6$ and $y = 4$).
Pairs of equations like $x + y = 10$ and $x - y = 2$ are called **simultaneous equations**.

To solve simultaneous equations you need to find values which fit **both** equations simultaneously.
Simultaneous equations can be solved using different methods.

Using graphs to solve simultaneous equations

Consider the simultaneous equations $x + 2y = 5$ and $x - 2y = 1$.

Draw the graphs of $x + 2y = 5$ and $x - 2y = 1$.

For $x + 2y = 5$:
When $x = 1$, $y = 2$.
This gives the point (1, 2).

When $x = 5$, $y = 0$.
This gives the point (5, 0).

To draw the graph of $x + 2y = 5$ draw a line through the points (1, 2) and (5, 0).

For $x - 2y = 1$:
When $x = 1$, $y = 0$.
This gives the point (1, 0).

When $x = 5$, $y = 2$.
This gives the point (5, 2).

To draw the graph of $x - 2y = 1$ draw a line through the points (1, 0) and (5, 2).

The values of x and y at the point where the lines cross give the solution to the simultaneous equations.

The lines cross at the point (3, 1).

This gives the solution $x = 3$ and $y = 1$.

To solve a pair of simultaneous equations plot the graph of each of the equations on the same diagram.
The coordinates of the point where the two lines cross:

- fit **both equations** simultaneously,
- give the **graphical solution** of the equations.

20 Simultaneous Equations

EXAMPLE

Use a graphical method to solve this pair of simultaneous equations: $5x + 2y = 20$
$y = 2x + 1$

Find the points that fit the equations $5x + 2y = 20$ and $y = 2x + 1$.

For $5x + 2y = 20$:

When $x = 0$, $y = 10$.
This gives the point (0, 10).

When $y = 0$, $x = 4$.
This gives the point (4, 0).

Draw a line through the points (0, 10) and (4, 0).

For $y = 2x + 1$:

When $x = 0$, $y = 1$.
This gives the point (0, 1).

When $x = 4$, $y = 9$.
This gives the point (4, 9).

Draw a line through the points (0, 1) and (4, 9).

The lines cross at the point (2, 5).
This gives the solution $x = 2$ and $y = 5$.

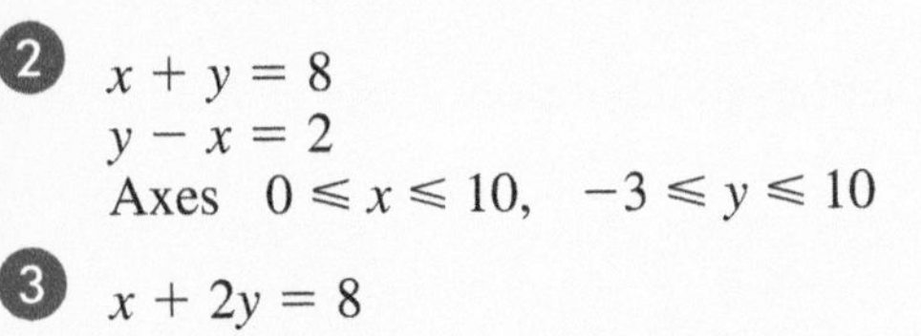

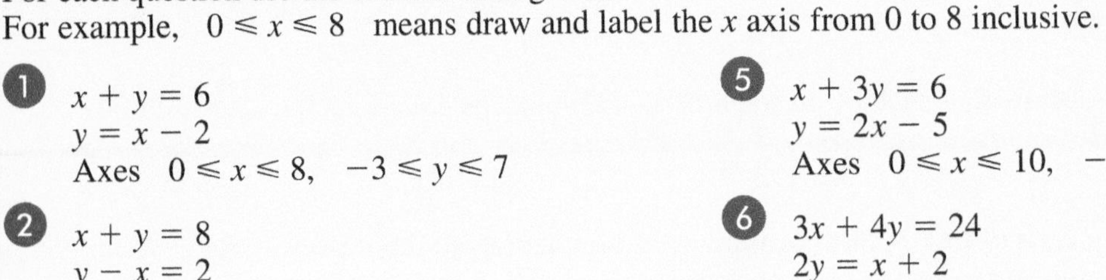

Check:
You can check a graphical solution by substituting the values of x and y into the original equations.
When $x = 2$ and $y = 5$

$5x + 2y = 5 \times 2 + 2 \times 5 = 20$

$y = 2x + 1 = 2 \times 2 + 1 = 5$

Exercise 20.1

Use a graphical method to solve each of these pairs of simultaneous equations.
For each question use the sizes of axes given.
For example, $0 \leqslant x \leqslant 8$ means draw and label the x axis from 0 to 8 inclusive.

1. $x + y = 6$
   $y = x - 2$
   Axes $0 \leqslant x \leqslant 8$, $-3 \leqslant y \leqslant 7$
2. $x + y = 8$
   $y - x = 2$
   Axes $0 \leqslant x \leqslant 10$, $-3 \leqslant y \leqslant 10$
3. $x + 2y = 8$
   $2x + y = 7$
   Axes $0 \leqslant x \leqslant 10$, $0 \leqslant y \leqslant 8$
4. $3x + 2y = 12$
   $y = x + 1$
   Axes $0 \leqslant x \leqslant 5$, $0 \leqslant y \leqslant 8$
5. $x + 3y = 6$
   $y = 2x - 5$
   Axes $0 \leqslant x \leqslant 10$, $-6 \leqslant y \leqslant 4$
6. $3x + 4y = 24$
   $2y = x + 2$
   Axes $-4 \leqslant x \leqslant 10$, $-2 \leqslant y \leqslant 8$
7. $2x + y = -1$
   $x + 2y = 4$
   Axes $-3 \leqslant x \leqslant 5$, $-2 \leqslant y \leqslant 4$
8. $x + 2y = 6$
   $x - 2y = 4$
   Axes $0 \leqslant x \leqslant 6$, $-2 \leqslant y \leqslant 3$

Simultaneous equations with no solution

Some pages of simultaneous equations do not have a solution.

EXAMPLE

Show that this pair of simultaneous equations do not have a solution.

$y - 2x = 4$
$2y = 4x - 1$

Method 1

Draw the graph of each equation.

$y - 2x = 4$
When $x = 0$, $y = 4$.
When $y = 0$, $x = -2$.
Plot and draw a line through the points $(0, 4)$ and $(-2, 0)$.

$2y = 4x - 1$
When $x = 0$, $y = -0.5$.
When $x = 2$, $y = 3.5$.
Plot and draw a line through the points $(0, -0.5)$ and $(2, 3.5)$.

The two lines are **parallel**.
This means they never cross and there are no values of x and y which fit both equations.
So, the simultaneous equations have no solution.

Method 2

Rearrange each equation to the form $y = mx + c$.

$y - 2x = 4$
Add $2x$ to both sides.
$y = 2x + 4$
The graph of this equation has a gradient (m) of 2 and a y-intercept (c) of 4.

$2y = 4x - 1$
Divide both sides by 2.
$y = 2x - 0.5$
The graph of this equation has a gradient (m) of 2 and a y-intercept (c) of -0.5.

Both lines have the same gradient (2) and different y-intercepts which shows that the lines are parallel.

Exercise 20.2

1. Draw graphs to show that each of these pairs of simultaneous equations have no solution.
   (a) $x + y = 6$, $y = 2 - x$
   (b) $y - 4x = 8$, $y = 4x + 2$
   (c) $3x + 4y = 12$, $8y = 24 - 6x$
   (d) $5y - 2x = 10$, $5y = 2x + 20$

2. By rearranging each of these pairs of simultaneous equations to the form $y = mx + c$ show that they do not have a solution.
   (a) $2x + y = 6$, $y = 3 - 2x$
   (b) $2y - 4x = 7$, $y - 2 = 2x$
   (c) $5x - 2y = 8$, $4y = 10x + 7$
   (d) $4y + 12x = 5$, $1 - 2y = 6x$

3. Two of these pairs of simultaneous equations have no solution.
   (a) $5x + y = 6$, $y - 5x = 2$
   (b) $5x = 8 + y$, $y - 5x = 2$
   (c) $2y - 10x = 5$, $4y + 20x = 5$
   (d) $y + 1 = -5x$, $2y + 10x = 5$

   Use an appropriate method to find which ones.
   Use a graphical method to solve the other two.

Using the elimination method to solve simultaneous equations

The graphical method of solving simultaneous equations can be quite time consuming.
Sometimes, due to the equations involved, the coordinates of the points where the lines intersect can be difficult to read accurately.
For these reasons, other methods of solving simultaneous equations are often used.

Consider again the simultaneous equations $x + 2y = 5$ and $x - 2y = 1$.
Both equations have the same number of x's and the same number of y's.
If the two equations are added together the y's will be **eliminated** as shown.

$$\begin{aligned} x + 2y &= 5 \\ x - 2y &= 1 \\ \hline \text{Adding gives} \quad 2x &= 6 \\ \text{So,} \quad x &= 3 \end{aligned}$$

Remember: $+2y + -2y = 2y - 2y = 0$

By **substituting** the value of this letter (x) into one of the original equations we can find the value of the other letter (y).

$$\begin{aligned} x + 2y &= 5 \\ 3 + 2y &= 5 \\ 2y &= 2 \\ y &= 1 \end{aligned}$$

This gives the solution $x = 3$ and $y = 1$.

Remember: If you do the same to both sides of an equation it is still true.

EXAMPLES

1 Use the elimination method to solve this pair of simultaneous equations: $2x - y = 1$, $3x + y = 9$

Each equation has the **same number** of y's but the **signs** are **different**.
To eliminate the y's the equations must be **added**.

$5x = 10$
$x = 2$

Substitute $x = 2$ into $3x + y = 9$.
$3 \times 2 + y = 9$
$6 + y = 9$
$y = 3$

The solution is $x = 2$ and $y = 3$.

Check:
Substitute $x = 2$ and $y = 3$ into $2x - y = 1$
$2 \times 2 - 3 = 1$
$4 - 3 = 1$
$1 = 1$
The equation is true, so, the solution $x = 2$ and $y = 3$ is correct.

2 Use the elimination method to solve this pair of simultaneous equations: $2x + 3y = 9$, $2x + y = 7$

Each equation has the **same number** of x's and the **signs** are the **same**.
To eliminate the x's one equation must be **subtracted** from the other.

Subtract $2x + y = 7$ from $2x + 3y = 9$.
$2y = 2$
$y = 1$

Substitute $y = 1$ into $2x + y = 7$.
$2x + 1 = 7$
$2x = 6$
$x = 3$

The solution is $x = 3$ and $y = 1$.

Check:
Substitute $x = 3$ and $y = 1$ into $2x + 3y = 9$.
Do this and make sure the solution is correct.

Exercise 20.3

Use the elimination method to solve each of these pairs of simultaneous equations.

1. $3x - y = 1$, $x + y = 3$
2. $2x - y = 2$, $x + y = 7$
3. $4x + y = 9$, $2x - y = 3$
4. $-x + 2y = 13$, $x + y = 8$
5. $2x + y = 7$, $x + y = 4$
6. $3x + y = 9$, $2x + y = 7$
7. $2x + y = 12$, $x + y = 7$
8. $x + 5y = 14$, $x + 2y = 8$
9. $x + 2y = 13$, $x + 4y = 21$
10. $x + 4y = 11$, $x + y = 5$
11. $2x + 5y = 13$, $2x + y = 9$
12. $5x + 3y = 26$, $2x + 3y = 14$
13. $5x + 4y = 22$, $5x + y = 13$
14. $2x - y = 10$, $3x + y = 10$
15. $5x - 2y = 13$, $3x + 2y = 3$
16. $x + 5y = 14$, $-x + 2y = 7$
17. $2x + 3y = 8$, $2x + y = -4$
18. $2x + y = 4$, $4x - y = 11$
19. $3x + 4y = -8$, $x + 4y = 4$
20. $3x + 2y = 6$, $x - 2y = 6$
21. $x - 3y = 8$, $x + 2y = -7$
22. $2x - 2y = 9$, $4x - 2y = 16$
23. $3x - y = 5$, $3x + y = 4$
24. $5x - 3y = 5$, $5x + y = -5$

Further use of the elimination method

Look at this pair of simultaneous equations:

$$5x + 2y = 11$$
$$3x - 4y = 4$$

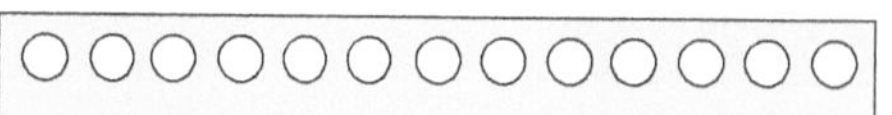

A useful technique is to use capital letters to label the equations.

$5x + 2y = 11$ Equation A

$3x - 4y = 4$ Equation B

These equations do not have the same number of x's or the same number of y's.

To make the number of y's the same we can multiply equation A by 2.

A × 2 gives $10x + 4y = 22$ Equation C

B × 1 gives $3x - 4y = 4$ Equation D

C + D gives $13x = 26$

$x = 2$

The number of y's in equations C and D is the **same** but the **signs** are **different**. To eliminate the y's the equations must be **added**.

Substitute $x = 2$ into $5x + 2y = 11$.

$$5 \times 2 + 2y = 11$$
$$10 + 2y = 11$$
$$2y = 1$$
$$y = 0.5$$

In this example eliminating the y's rather than the x's is less likely to produce an error.
Try to solve the equations by eliminating the x's.

The solution is $x = 2$ and $y = 0.5$.

Check the solution by substituting $x = 2$ and $y = 0.5$ *into* $3x - 4y = 4$.

EXAMPLE

Solve this pair of simultaneous equations:

$$3x + 7y = -2$$
$$4x + 9 = -3y$$

Both equations must be in the form $px + qy = r$ before the elimination method can be used.
You may have to **rearrange** the equations you are given.
$4x + 9 = -3y$ can be rearranged as $4x + 3y = -9$.

Rearrange and label the equations.

$3x + 7y = -2$ A
$4x + 3y = -9$ B

These equations do not have the same number of x's or the same number of y's.
So, the multiplying method can be used.

Method 1

Eliminating the x's.

A × 4 gives $12x + 28y = -8$ C
B × 3 gives $12x + 9y = -27$ D

C − D gives
$$19y = -8 - -27$$
$$19y = -8 + 27$$
$$19y = 19$$
$$y = 1$$

Substitute $y = 1$ into $3x + 7y = -2$.

$$3x + 7 \times 1 = -2$$
$$3x + 7 = -2$$
$$3x = -9$$
$$x = -3$$

The solution is $x = -3$ and $y = 1$.

Method 2

Eliminating the y's.

A × 3 gives $9x + 21y = -6$ C
B × 7 gives $28x + 21y = -63$ D

D − C gives
$$19x = -63 - -6$$
$$19x = -63 + 6$$
$$19x = -57$$
$$x = -3$$

Substitute $x = -3$ into $3x + 7y = -2$.

$$3 \times -3 + 7y = -2$$
$$-9 + 7y = -2$$
$$7y = 7$$
$$y = 1$$

Check the solution by substituting $x = -3$ and $y = 1$ into $4x + 9 = -3y$.

Exercise 20.4

Solve each of these pairs of simultaneous equations.

1. $3x + 2y = 8$, $2x - y = 3$
2. $x + y = 5$, $5x - 3y = 1$
3. $2x + 3y = 9$, $x + 4y = 7$
4. $x + 3y = 10$, $2x + 5y = 18$
5. $5x + 2y = 8$, $2x - y = 5$
6. $3x + y = 9$, $x - 2y = 10$
7. $3x - 4y = 10$, $x + 2y = 5$
8. $x + 6y = 0$, $3x - 2y = -10$
9. $2x + 3y = 11$, $3x + y = 13$
10. $2x + y = 10$, $-x + 2y = 9$
11. $2x + 3y = 9$, $4x - y = 4$
12. $2x + 3y = 8$, $3x + 2y = 7$
13. $3x + 4y = 23$, $2x + 5y = 20$
14. $2x - 3y = 8$, $x - 5y = 11$
15. $3x + 4y = 5$, $-2x + 5y = 12$
16. $3x - 2y = 4$, $x + 4y = 6$
17. $-3x + 2y = 5$, $4x + 3y = -1$
18. $3x + 4y = 6$, $3y = 7 - x$
19. $5x + 3y = 16$, $2y = 13 - x$
20. $5x - 4y = 24$, $2x = y + 9$
21. $2x + 3y = 14$, $8x - 5y = 5$
22. $4x - 7y = 15$, $5x - 12 = 2y$
23. $8x + 3y = 2$, $5x = 1 - 2y$
24. $9x = 4y - 20$, $5x = 6y - 13$

Using the substitution method to solve simultaneous equations

For some pairs of simultaneous equations a method using **substitution** is sometimes more convenient.

EXAMPLE

Solve this pair of simultaneous equations: $5x + y = 9$
$y = 4x$

$5x + y = 9$ Equation A
$y = 4x$ Equation B

Substitute $y = 4x$ into Equation A
$5x + 4x = 9$
$9x = 9$
$x = 1$

Substitute $x = 1$ into $y = 4x$.
$y = 4 \times 1$
$y = 4$

The solution is $x = 1$ and $y = 4$.

Check the solution by substituting $x = 1$ and $y = 4$ *into* $5x + y = 9$.

Exercise 20.5

Use the substitution method to solve these pairs of simultaneous equations.

1. $2x + y = 10$, $y = 3x$
2. $3x - y = 9$, $y = 2x$
3. $x + 5y = 18$, $x = 4y$
4. $x + 2y = 15$, $y = 2x$
5. $2x + y = 17$, $y = 6x + 1$
6. $3x + 2y = 4$, $x = y - 2$
7. $5x + 6y = 34$, $y = x + 2$
8. $5x - 2y = 23$, $x = y + 1$
9. $5x - y = 12$, $y = 32 - 6x$
10. $x + 5y = 13$, $x = 3y + 9$
11. $5x - 3y = 26$, $y = 2x + 14$
12. $x + 4y = 32$, $x = 2y - 4$

Solving problems using simultaneous equations

EXAMPLE

Billy buys 5 first class stamps and 3 second class stamps at a cost of £2.13.
Jane buys 3 first class stamps and 5 second class stamps at a cost of £1.95.
Calculate the cost of a first class stamp and the cost of a second class stamp.

Let x pence be the cost of a first class stamp, and let y pence be the cost of a second class stamp.

Billy's purchase of the stamps gives this equation. $5x + 3y = 213$ Equation A
Jane's purchase of the stamps gives this equation. $3x + 5y = 195$ Equation B

This gives a pair of simultaneous equations which can be solved using the elimination method.

$5x + 3y = 213$ A
$3x + 5y = 195$ B
A × 5 gives $25x + 15y = 1065$ C
B × 3 gives $9x + 15y = 585$ D
C − D gives $16x = 480$
$x = 30$

Substitute $x = 30$ into $5x + 3y = 213$.
$5 \times 30 + 3y = 213$
$3y = 63$
$y = 21$

So, the cost of a first class stamp is 30 pence and the cost of a second class stamp is 21 pence.

Check the solution by substituting the values for x and y into the original problem.

Exercise 20.6

1. Pencils cost x pence each and pens cost y pence each.
   Pam buys 6 pencils and 3 pens for 93 pence.
   Ray buys 2 pencils and 5 pens for 91 pence.
   (a) Write down two equations connecting x and y.
   (b) By solving these simultaneous equations find the cost of a pencil and the cost of a pen.

2. Apples are x pence per kg.
   Oranges are y pence each.
   5 kg of apples and 30 oranges cost £9.00.
   10 kg of apples and 15 oranges cost £12.60.
   (a) Write down two equations connecting x and y.
   (b) By solving these simultaneous equations find the cost of a kilogram of apples and the cost of an orange.

3. Standard eggs cost x pence per dozen.
   Small eggs cost y pence per dozen.
   10 dozen standard eggs and 5 dozen small eggs cost £13.60.
   5 dozen standard eggs and 8 dozen small eggs cost £11.31.
   By forming two simultaneous equations find the values of x and y.

4. A group of children and adults went on a coach trip to a theme park.
   Ticket prices for the theme park were £20 for adults and £15 for children.
   Ticket prices for the coach were £10 for adults and £6 for children.
   The total cost of the tickets for the theme park was £560.
   The total cost of the coach tickets was £232.
   How many children and adults went on the trip?

5. Jenny types at x words per minute.
   Stuart types at y words per minute.
   When Jenny and Stuart both type for 1 minute they type a total of 170 words.
   When Jenny types for 5 minutes and Stuart types for 3 minutes they type a total of 710 words.
   Calculate x and y.

6. At a café, John buys 3 coffees and 2 teas for £4.30 and Susan buys 2 coffees and 3 teas for £4.20.
   Calculate the price of a coffee and the price of a tea.

7. Standard coaches hold x passengers and first class coaches hold y passengers.
   A train with 5 standard coaches and 2 first class coaches carries a total of 1040 passengers.
   A train with 7 standard coaches and 3 first class coaches carries a total of 1480 passengers.
   By forming two simultaneous equations find the values of x and y.

8. Nick writes a number in each of these boxes: $x = \square$ $y = \square$
   When he adds 5 to x the result is 2 times y.
   When he adds 17 to y the result is 2 times x.
   Find the two numbers.

9. In a game you score p points if you win and q points if you lose.
   Matt plays. He wins 2 games and loses 3 games. He scores 4 points.
   Kath plays. She wins 3 games and loses 2 games. She scores 11 points.
   By forming two simultaneous equations find the values of p and q.

10. A local pop band hold two gigs.
    Tickets for the gigs cost £x and CDs cost £y.
    At the first gig they sell 92 tickets and 8 CDs for a total of £1186.
    At the second gig they sell 104 tickets and 16 CDs for a total of £1372.
    Find the cost of a ticket and the cost of a CD.

Simultaneous equations in which one equation is linear and one is quadratic

Using graphs

To solve a pair of simultaneous equations, where one is **linear** and one is **quadratic**, plot the graph of each equation on the same diagram.
The coordinates at the **points of intersection** give the solutions to the equations.

EXAMPLE

Solve the simultaneous equations $y = x + 2$ and $y = 4 - x^2$.

Draw the graph of $y = x + 2$.

Draw the graph of $y = 4 - x^2$.

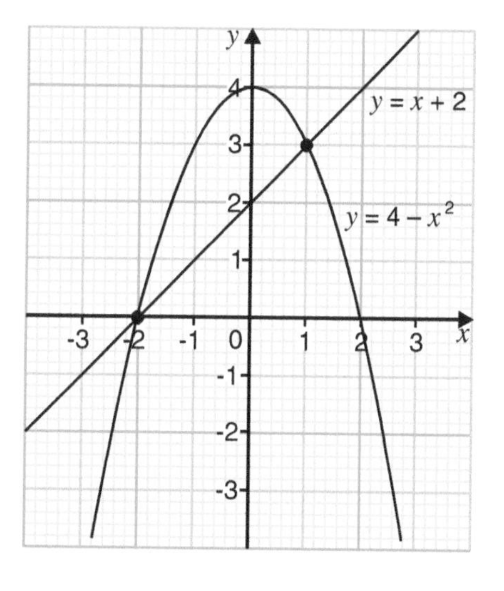

The graphs intersect (cross) at the points $(-2, 0)$ and $(1, 3)$.

This gives the solutions:

$x = -2, \ y = 0$

and $x = 1, \ y = 3$.

Exercise 20.7

Use a graphical method to solve each of these pairs of simultaneous equations.

1. $y = x + 2$
   $y = x^2$
2. $y = 1 - x$
   $y = x^2 - 1$
3. $y = x - 3$
   $y = 3x - x^2$
4. $y = 2x - 3$
   $y = x^2 - 3x + 3$
5. $y = x$
   $x^2 + y^2 = 8$
6. $y - 2x = 0$
   $x^2 + y^2 = 5$

Using the method of substitution

In Exercise 20.5 we used the method of substitution to find the solutions of simultaneous equations where both equations were linear.
We can now use the same method to solve simultaneous equations where one is **linear** and one is **quadratic**.

EXAMPLES

1 Solve the simultaneous equations $x - y = 2$ and $x^2 + 3y = 12$.

Rearrange the linear equation, $x - y = 2$, to make y the subject of the equation.

$x - y = 2$

Add y to both sides.

$x = 2 + y$

Subtract 2 from both sides.

$x - 2 = y$

Substitute $y = x - 2$ into the quadratic equation $x^2 + 3y = 12$.

$x^2 + 3(x - 2) = 12$

Simplify.

$x^2 + 3x - 6 = 12$

$x^2 + 3x - 18 = 0$

Solve the quadratic equation for x.

$(x + 6)(x - 3) = 0$

$x + 6 = 0$ or $x - 3 = 0$

$x = -6$ or $x = 3$.

Using $x - y = 2$.

When $x = -6$.

$-6 - y = 2$

$y = -8$

When $x = 3$.

$3 - y = 2$

$y = 1$

This gives the solutions: $x = -6$, $y = -8$ **and** $x = 3$, $y = 1$.

2 Solve the simultaneous equations $y = x - 2$ and $x^2 + y^2 = 10$.

Substitute $y = x - 2$ into $x^2 + y^2 = 10$.

$x^2 + (x - 2)^2 = 10$

$x^2 + x^2 - 4x + 4 = 10$

$2x^2 - 4x - 6 = 0$

Divide both sides by 2.

$x^2 - 2x - 3 = 0$

$(x + 1)(x - 3) = 0$

$x = -1$ or $x = 3$.

Using $y = x - 2$.

When $x = -1$.

$y = -1 - 2$

$y = -3$

When $x = 3$.

$y = 3 - 2$

$y = 1$

This gives the solutions: $x = -1$, $y = -3$ **and** $x = 3$, $y = 1$.

Exercise 20.8

Use the method of substitution to solve each of these pairs of simultaneous equations.

1. $x = 3$
   $x^2 + y^2 = 13$
2. $y = 6$
   $3x^2 = 2y$
3. $y = x + 1$
   $y = x^2 + x$
4. $y = x^2 + 2$
   $y + 3x = 0$
5. $y = 5x$
   $y = x^2 - x + 5$
6. $x^2 + y^2 = 18$
   $x + y = 0$
7. $x^2 + y^2 = 20$
   $x = 2y$
8. $y = \frac{4}{x}$
   $y = 5 - x$
9. $xy = 5$
   $y = 2x + 3$
10. $2x^2 - y = 5$
    $x + y = 1$
11. $5x^2 - xy = 6$
    $y - 2x = 3$
12. $3x^2 = y$
    $y - 2 = 5x$
13. $y = 2x$
    $y^2 + 7x = 2$
14. $y + 2 = 11x$
    $y = 5x^2$
15. $x^2 + y^2 = 13$
    $2x + y = 1$
16. $4y + x = 3$
    $x + \frac{2}{y} = 1$
17. $xy + 2x^2 = 5$
    $x + 2y = 1$
18. $2x^2 + 3xy - y^2 = 2$
    $x - 2 = y$

What you need to know

- A pair of **simultaneous equations** are linked equations with the same unknown letters in each equation.
- To solve a pair of simultaneous equations find values for the unknown letters that fit **both** equations.
- Simultaneous equations can be solved either **graphically** or **algebraically**.
- Solving simultaneous equations **graphically** involves:
  drawing the graphs of both equations, finding the point(s) where the graphs cross.
  When the graphs of the equations are parallel, the equations have no solution.
- Solving simultaneous equations **algebraically** involves using either:
  the **elimination** method or the **substitution** method.

Review Exercise 20

1. Use a graphical method to solve each of these simultaneous equations.
   For each question use the size of axes given.
   (a) $x + y = 10$
   $y = 2x + 1$
   Axes $0 \leqslant x \leqslant 11$, $0 \leqslant y \leqslant 11$
   (b) $5x + 6y = 30$
   $2y = x - 2$
   Axes $0 \leqslant x \leqslant 8$, $-3 \leqslant y \leqslant 8$

2. This graph shows the line $y - 2x = -1$.
   Copy the graph.

   By drawing another line, use the graph to solve the simultaneous equations:
   $y - 2x = -1$
   $x + 2y = 4$

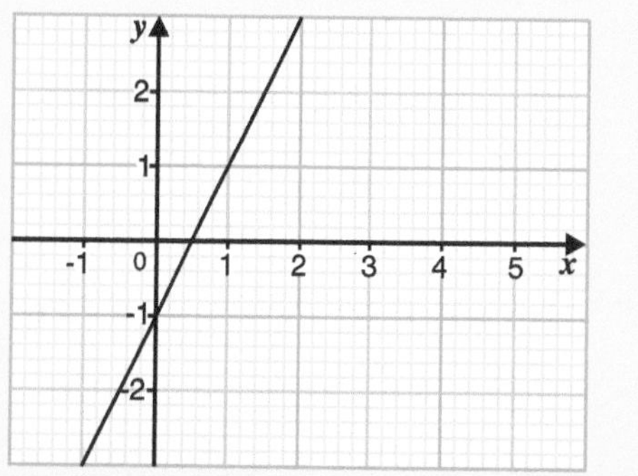

AQA

3 **Use a graphical method** to solve the simultaneous equations:
$y = x + 7$ and $y + 3x = 5$ AQA

4 (a) Show that the simultaneous equations $y = 2x - 2$ and $2y - 4x = 3$ have no solution.
(b) Show that the simultaneous equations $4y = x + 1$ and $8y - 2x = 3$ have no solution.
(c) The simultaneous equations $y = 3x + 2$ and $y = ax + b$ have no solution. What can you say about the values of a and b?
(d) The simultaneous equations $y = 3x + 2$ and $py + qx = r$ have no solution. Find some possible values for p, q and r.

5 Use an algebraic method to solve these simultaneous equations: $2x + y = 8$
$x + y = 5$

6 Solve each of these simultaneous equations.
(a) $2x + 3y = 11$, $2x + y = 5$ (b) $4x + y = 6$, $2x - y = 6$ (c) $x - y = -3$, $x + 4y = 7$ (d) $x + 2y = 9$, $5x - 2y = 3$

7 Solve the simultaneous equations: $x - 3y = 7$
$2x + y = 0$ AQA

8 Solve the simultaneous equations: $4x + 5y = 5$
$2x - y = 6$ AQA

9 Solve the simultaneous equations: $3x + y = -5$
$x - 4y = 7$ AQA

10 Peaches cost x pence each and oranges cost y pence each.
4 peaches and one orange cost 58p.
6 peaches and 2 oranges cost 92p.
(a) Write down two equations connecting x and y.
(b) Solve these simultaneous equations to find the values of x and y.

11 In a competition, a win scores x points and a draw scores y points.
Nayeem has won one game, drawn three games and has 16 points.
Freda has won two games, drawn one game and has 17 points.
By forming two simultaneous equations find the values of x and y. AQA

12 Use a graphical method to solve the simultaneous equations:
$2y + x = 4$ and $2y = x^2$.

13 Use an algebraic method to solve the simultaneous equations:
$3x + y = 17$ and $xy = 20$.

14 Solve the simultaneous equations:
$2x + y = 7$ and $x^2 + y^2 = 13$.

15 Solve the simultaneous equations:
$y = x + 2$ and $y = 3x^2$. AQA

16 Solve the simultaneous equations:
$4y = 5x + 3$ and $y = 6 - x^2$.

17 The graphs of $x^2 + y^2 = 25$ and $y = 7 - x$ intersect at points P and Q.
Find the coordinates of P and Q.

CHAPTER 21 Algebraic Methods

When solving problems or simplifying expressions it is necessary to be able to manipulate algebra.

Simplifying expressions

EXAMPLES

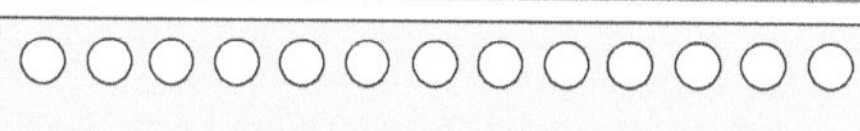

The effect of a minus sign in front of a bracket is to change **all** signs inside the bracket.
Explain why.

1 Expand and simplify $(2x + 1)^2 - (x - 2)^2$.

$(2x + 1)^2 - (x - 2)^2$
Expand the brackets.
$= 4x^2 + 4x + 1 - (x^2 - 4x + 4)$
The second bracket was expanded and kept within a bracket because of the minus sign.
Remove the bracket.
$= 4x^2 + 4x + 1 - x^2 + 4x - 4$
Simplify.
$= 3x^2 + 8x - 3$

2 Expand and simplify $(x + 2y)^2 + (2x - y)^2$.
$= x^2 + 4xy + 4y^2 + 4x^2 - 4xy + y^2$
$= 5x^2 + 5y^2$
$= 5(x^2 + y^2)$ (In factorised form.)

Identities

A house has a front and a back garden.
The front garden consists of a square flower bed, 3 m by 3 m, surrounded by a path, x m wide.
The back garden consists of a rectangular lawn, 9 m by 5 m, surrounded by a path, x m wide.
Show that the total area of both gardens is given by: $2(4x^2 + 20x + 27)$.

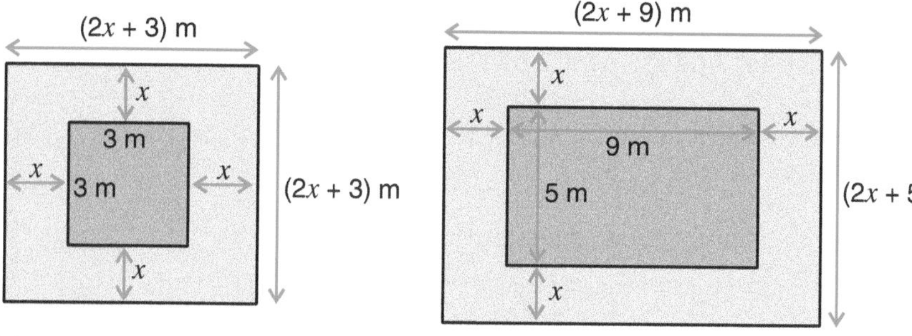

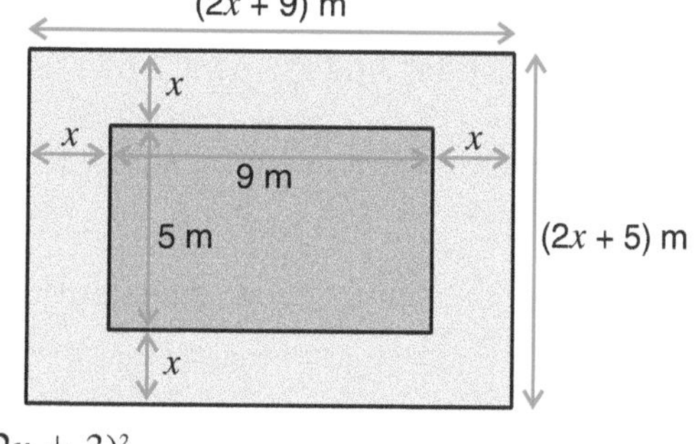

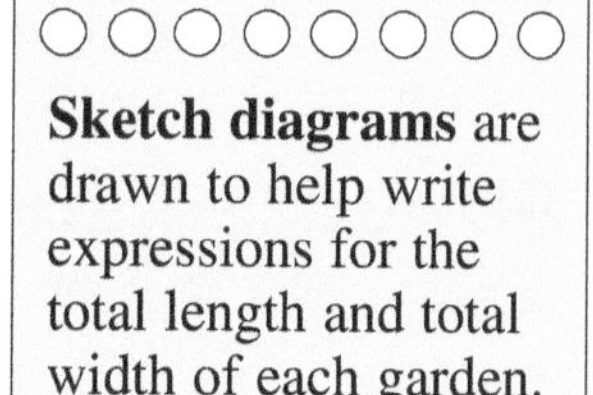

Sketch diagrams are drawn to help write expressions for the total length and total width of each garden.
Explain how each expression is obtained.

The area of the front garden is $(2x + 3)^2$.
The area of the back garden is $(2x + 9)(2x + 5)$.
The total area $= (2x + 3)^2 + (2x + 9)(2x + 5)$
$= (2x + 3)(2x + 3) + (2x + 9)(2x + 5)$
$= 4x^2 + 6x + 6x + 9 + 4x^2 + 10x + 18x + 45$
$= 8x^2 + 40x + 54$
$= 2(4x^2 + 20x + 27)$

This has shown that $(2x + 3)^2 + (2x + 9)(2x + 5) = 2(4x^2 + 20x + 27)$.
Although this looks like an equation it is called an **identity** because it is true for all values of x.
An identity is the same expression written in another form.

EXAMPLES

1 Show that $(x - 5)(x + 4) + (x - 3)(x - 2) = 2(x^2 - 3x - 7)$.

LHS $= (x - 5)(x + 4) + (x - 3)(x - 2)$
Expand the brackets.
$= x^2 + 4x - 5x - 20 + x^2 - 2x - 3x + 6$
Simplify.
$= 2x^2 - 6x - 14$
Factorise. Common factor 2.
$= 2(x^2 - 3x - 7)$
$=$ RHS

LHS = RHS, so, $(x - 5)(x + 4) + (x - 3)(x - 2) = 2(x^2 - 3x - 7)$.

Abbreviations:
LHS = left-hand side
RHS = right-hand side
To show that an identity is true, either: start with the LHS and show that it is equal to the RHS, or start with the RHS and show that it is equal to the LHS.

2 Show that $(x + y)^2 - (x - y)^2 = 4xy$.

LHS $= (x + y)^2 - (x - y)^2$
$= x^2 + 2xy + y^2 - (x^2 - 2xy + y^2)$
$= x^2 + 2xy + y^2 - x^2 + 2xy - y^2$
$= 4xy$
$=$ RHS

LHS = RHS, so, $(x + y)^2 - (x - y)^2 = 4xy$.

Exercise 21.1

1 Expand and simplify.
(a) $(x + 5)(x + 2) + (x - 4)(x - 6)$
(b) $(4x + 1)(x - 1) - (x + 2)(x - 3)$
(c) $(x - 5)(x - 5) - (x - 4)(x - 6)$
(d) $(2x + 3)(3x + 4) - (4x - 5)(5x - 6)$

2 Expand and simplify.
(a) $3(2x + 3y - 4z) - 4(3x - 5y + 3z)$
(b) $4(3x - 6y + 9z) - 3(4x - 8y + 12z)$
(c) $(x + 3y)^2 + (3x + y)^2$
(d) $(2x + 3y)^2 + (3x + 2y)^2$

3 Show that:
(a) $(x + 2)(x + 4) + (x - 3)(x - 4) = 2x^2 - x + 20$
(b) $(3x - 1)(x + 1) + (x + 4)(2x + 1) = 5x^2 + 11x + 3$
(c) $(x + 5)(x - 6) - (x - 4)(x - 3) = 6(x - 7)$
(d) $(2x + 1)(3x - 4) - (6x + 5)(x - 3) = 8x + 11$

4 Show that:
(a) $4(2x + 3y) - 2(3x - 4y) = 2(10y + x)$
(b) $3(s + 2t) + 2(3s + t) = 9s + 8t$
(c) $(x + y)^2 + (x - y)^2 = 2(x^2 + y^2)$
(d) $(x + 3y)^2 - (3x + y)^2 = 8(y + x)(y - x)$

5 Prove these identities.
(a) $(x + ay)^2 + (ax - y)^2 = (a^2 + 1)(x^2 + y^2)$
(b) $(x + ay)^2 - (ax + y)^2 = (a^2 - 1)(y^2 - x^2)$
(c) $(ax + by)^2 + (bx - ay)^2 = (a^2 + b^2)(x^2 + y^2)$

Algebraic fractions

Algebraic fractions have a numerator and a denominator (just as an ordinary fraction) but at least one of them is an expression involving an unknown.

e.g. $\frac{1}{x}$ $\quad \frac{x}{2}$ $\quad \frac{2}{x-5}$ $\quad \frac{x+1}{x-6}$ $\quad \frac{3}{x+7}$ $\quad \frac{x^2+2x+1}{x^2-1}$

Algebraic fractions can be simplified, added, subtracted, multiplied, divided and used in equations in the same way as numerical fractions.

Simplifying algebraic fractions

Fractions can be simplified if the numerator and the denominator have a common factor.
In its **simplest form**, the numerator and denominator of an algebraic fraction have no common factor other than 1.

The numerator and denominator of $\frac{15}{25}$ have a highest common factor of 5.

$$\frac{15}{25} = \frac{15 \div 5}{25 \div 5} = \frac{3}{5}$$

3 and 5 have no common factors, other than 1.

$\frac{15}{25} = \frac{3}{5}$ in its simplest form.

Algebraic fractions work in a similar way.

To write an algebraic fraction in its simplest form:

- factorise the numerator and denominator of the fraction,
- divide the numerator and denominator by their highest common factor.

This is sometimes called **cancelling** a fraction.

EXAMPLES

Write the following in their simplest form.

1. $\frac{3x-6}{3} = \frac{3(x-2)}{3} = x - 2$
2. $\frac{3x+9}{4x^2+12x} = \frac{3(x+3)}{4x(x+3)} = \frac{3}{4x}$
3. $\frac{x^2y+3xy^2}{xy-2x^2y^2} = \frac{xy(x+3y)}{xy(1-2xy)} = \frac{x+3y}{1-2xy}$
4. $\frac{x^2+5x+6}{x^2-2x-8} = \frac{(x+2)(x+3)}{(x+2)(x-4)} = \frac{x+3}{x-4}$
5. $\frac{x^2+2x+1}{x^2-1} = \frac{(x+1)^2}{(x+1)(x-1)} = \frac{x+1}{x-1}$
6. $\frac{9-3y}{y-3} = \frac{3(3-y)}{y-3} = \frac{-3(y-3)}{y-3} = -3$

Exercise 21.2

1. Simplify these algebraic fractions.

(a) $\frac{4d+6}{2}$ (b) $\frac{9x+6}{3}$ (c) $\frac{8a+10b}{2}$ (d) $\frac{15m-10n}{5}$

(e) $\frac{8x-4y}{2}$ (f) $\frac{ax+bx}{x}$ (g) $\frac{x^2-x}{x}$ (h) $\frac{2x^2-4}{2}$

(i) $\frac{12}{3x-9}$ (j) $\frac{5x+10}{15}$ (k) $\frac{2x+4y+6z}{4x-6y+2z}$ (l) $\frac{-2x-4}{-6x-4}$

(m) $\frac{2x}{6x-4}$ (n) $\frac{3x}{6x^2-3}$ (o) $\frac{3x}{6x^2-3x}$ (p) $\frac{2x-1}{6x^2-3x}$

(q) $\frac{3m-6}{2m-4}$ (r) $\frac{m^2+3m}{3m+9}$ (s) $\frac{x^2-3x}{x^2+2x}$ (t) $\frac{5x-10}{6-3x}$

2 Match the algebraic fractions with the values **A** to **E**.

(a) $\frac{3x+6}{4x+8}$ (b) $\frac{5x-15}{2x-6}$ (c) $\frac{8-4y}{6-3y}$ (d) $\frac{8x+10}{4x+5}$

(e) $\frac{-2x-4}{-4x-8}$ **A** 2 **B** $\frac{4}{3}$ **C** $2\frac{1}{2}$ **D** $\frac{1}{2}$ **E** $\frac{3}{4}$

3 Simplify these algebraic fractions.

(a) $\frac{x^2y+5xy^2}{3xy+4x^2y^2}$ (b) $\frac{x^2+xy+2xz}{2x^2-3xy+xz}$ (c) $\frac{a^2b^2-3ab}{ab^2+a^2b}$ (d) $\frac{ax+ab+ay}{a^2b-2ay}$

4 Simplify these algebraic fractions.

(a) $\frac{x^2+3x}{x^2+4x+3}$ (b) $\frac{x^2+3x+2}{x^2+4x+3}$ (c) $\frac{x^2-2x}{x^2+x-6}$

(d) $\frac{x^2-x-20}{x^2+7x+12}$ (e) $\frac{x^2-4x+4}{x^2-5x+6}$ (f) $\frac{x^2-5x}{x^2+4x}$

(g) $\frac{x^2-1}{x+1}$ (h) $\frac{x-2}{x^2-4}$ (i) $\frac{x^2-2x-3}{2x-6}$

(j) $\frac{2x^2-2x-12}{2x^2-18}$ (k) $\frac{2x^2-7x-15}{2x^2-5x-12}$ (l) $\frac{6x^2-13x-5}{9x^2-1}$

Arithmetic of algebraic fractions

The same methods used for adding, subtracting, multiplying and dividing numeric fractions can be applied to algebraic fractions.

Multiplication

Numeric

$$\frac{8}{9}\times\frac{3}{14}=\frac{\not{2}\times 4}{3\times\not{3}}\times\frac{\not{3}}{\not{2}\times 7}=\frac{4}{21}$$

Algebraic

$$\frac{2x+2}{5x-15}\times\frac{3x-9}{4x+4}=\frac{\not{2}(x+1)}{5(x-3)}\times\frac{3(x-3)}{_2\not{4}(x+1)}=\frac{3}{10}$$

Division

Numeric

$$\frac{3}{7}\div\frac{9}{14}=\frac{3}{7}\times\frac{14}{9}$$
$$=\frac{\not{3}}{\not{7}}\times\frac{2\times\not{7}}{3\times\not{3}}$$
$$=\frac{2}{3}$$

Algebraic

$$\frac{7x-7}{3x}\div\frac{4-4x}{x}=\frac{7x-7}{3x}\times\frac{x}{4-4x}$$
$$=\frac{7(x-1)}{3\not{x}}\times\frac{\not{x}}{-4(x-1)}$$
$$=-\frac{7}{12}$$

Addition

Numeric

$$\frac{4}{11}+\frac{2}{5}=\frac{4\times5+2\times11}{11\times5}$$
$$=\frac{42}{55}$$

Algebraic

$$\frac{4}{x-3}+\frac{2}{x+2}=\frac{4(x+2)+2(x-3)}{(x-3)(x+2)}$$
$$=\frac{6x+2}{(x-3)(x+2)}$$

Subtraction

Numeric

$$\frac{4}{7}-\frac{2}{5}=\frac{4\times5-2\times7}{7\times5}$$
$$=\frac{6}{35}$$

Algebraic

$$\frac{4}{x-3}-\frac{2}{2x+1}=\frac{4(2x+1)-2(x-3)}{(x-3)(2x+1)}$$
$$=\frac{8x+4-2x+6}{(x-3)(2x+1)}$$
$$=\frac{6x+10}{(x-3)(2x+1)}$$
$$=\frac{2(3x+5)}{(x-3)(2x+1)}$$

EXAMPLE

Simplify $\frac{x}{x-1} + \frac{2-x}{x+2}$

$$\frac{x}{x-1} + \frac{2-x}{x+2} = \frac{x(x+2) + (2-x)(x-1)}{(x-1)(x+2)}$$
$$= \frac{x^2 + 2x + 2x - 2 - x^2 + x}{(x-1)(x+2)}$$
$$= \frac{5x-2}{(x-1)(x+2)}$$

It is common practice to leave the denominators of algebraic fractions in their factorised forms.

Exercise 21.3

1. Simplify.
   (a) $\frac{4x-2}{15-3x} \times \frac{5-x}{2x-1}$ (b) $\frac{5+5y}{y-3} \times \frac{6-2y}{8y+8}$ (c) $\frac{7}{y} \times \frac{y^2}{14}$
2. Simplify.
   (a) $\frac{4x-2}{15-3x} \div \frac{2x-1}{5-x}$ (b) $\frac{4+y}{7-y} \div \frac{2y+8}{y-7}$ (c) $\frac{10-2x}{x^2} \div \frac{x-5}{x}$
3. Simplify.
   (a) $\frac{3}{x+5} + \frac{4}{x-4}$ (b) $\frac{5}{2x+3} + \frac{6}{5x-3}$ (c) $\frac{x+4}{x+2} + \frac{2x-1}{x-3}$
4. Simplify.
   (a) $\frac{4}{x+5} - \frac{3}{x-4}$ (b) $\frac{5}{2x+3} - \frac{6}{5x-3}$ (c) $\frac{2x-1}{x-3} - \frac{x+4}{x+2}$
5. If $\mathbf{p} = \frac{2}{x}$, $\mathbf{q} = \frac{3}{2x-6}$ and $\mathbf{r} = \frac{x-3}{2x}$, calculate
   (a) $\mathbf{p} + \mathbf{q}$ (b) $\mathbf{p} - \mathbf{r}$ (c) $\mathbf{q} + \mathbf{r}$ (d) $\mathbf{p} \div \mathbf{r}$ (e) $\mathbf{q} \times \mathbf{r}$ (f) $\mathbf{r} \div \mathbf{p}$

Solving equations involving algebraic fractions

EXAMPLE

1. Solve the equation $\frac{7}{x} + \frac{6}{x+5} = 2$.

$\frac{7}{x} + \frac{6}{x+5} = 2$

Multiply throughout by $x(x+5)$.

$$\frac{7}{x} \times x(x+5) + \frac{6}{(x+5)} \times x(x+5) = 2 \times x(x+5)$$
$$7(x+5) + 6x = 2x(x+5)$$
$$7x + 35 + 6x = 2x^2 + 10x$$
$$13x + 35 = 2x^2 + 10x$$
$$2x^2 - 3x - 35 = 0$$

Solve the quadratic equation.

$(2x+7)(x-5) = 0$

Either $2x + 7 = 0$ or $x - 5 = 0$

$x = -3\frac{1}{2}$ or $x = 5$

Solution: $x = -3\frac{1}{2}$ or 5

The final answers can always be checked by substituting the solutions, in turn, into the original equation.

$\frac{7}{x} + \frac{6}{x+5} = 2$

Substitute $x = -3\frac{1}{2}$

Does $\frac{7}{-3\frac{1}{2}} + \frac{6}{-3\frac{1}{2}+5} = 2$*?*

Check the solution $x = 5$.

EXAMPLE

2 Henry took part in a sponsored walk between Aylestone and Bedrock.
He walked from Aylestone to Bedrock, a distance of 12 km, at a steady speed of x km/h.
His average speed for the return part of the walk was 2 km/h slower.

(a) Write down, in terms of x, the time taken for the whole journey.

(b) Henry walked for a total of $3\frac{1}{2}$ hours.
Write an equation in x.
Show that the equation can be written as
$7x^2 - 62x + 48 = 0$.

(c) What was Henry's speed from Aylestone to Bedrock?

> Sometimes, questions are set in the form of a problem.
> Use the information given in the question to **form an equation.**
> **Solve** the equation.
> **Analyse** the solutions to answer the original problem.

(a) Using Time = Distance ÷ Speed

Total time $= \frac{12}{x} + \frac{12}{x-2}$

(b) $\frac{12}{x} + \frac{12}{x-2} = 3\frac{1}{2}$

Add the algebraic fractions.	$\frac{12(x-2) + 12x}{x(x-2)} = 3\frac{1}{2}$
Multiply both sides by $x(x-2)$.	$12x - 24 + 12x = 3\frac{1}{2}x(x-2)$
Simplify the left-hand side.	$24x - 24 = 3\frac{1}{2}x(x-2)$
Multiply both sides by 2.	$48x - 48 = 7x(x-2)$
Expand the right-hand side.	$48x - 48 = 7x^2 - 14x$
Rearrange into the form $ax^2 + bx + c = 0$.	$7x^2 - 62x + 48 = 0$

(c) $7x^2 - 62x + 48 = 0$
$(7x - 6)(x - 8) = 0$
Either $7x - 6 = 0$ or $x - 8 = 0$
$7x = 6$ or $x = 8$
$x = \frac{6}{7}$

The value of $\frac{6}{7}$ is a possible solution to the equation **but** it gives a negative speed for the return journey.
So, the solution which fits the problem is $x = 8$.
Henry's speed for the first part of the journey is 8 km/h.

Exercise 21.4

1 Solve the equation $\frac{4}{3x-1} - \frac{2}{x} = 1$.

2 (a) Show that the equation $\frac{12}{x-3} + \frac{7}{x+1} = 5$ can be rearranged to give the equation $5x^2 - 29x - 6 = 0$.

(b) Solve $5x^2 - 29x - 6 = 0$ to find solutions for x.

3 (a) Show that the equation $\frac{6}{x+1} - \frac{5}{x+2} = 2$ can be rearranged to give the equation $2x^2 + 5x - 3 = 0$.

(b) Hence, solve the equation $\frac{6}{x+1} - \frac{5}{x+2} = 2$.

4 Solve these equations, giving answers correct to 3 significant figures where necessary.

(a) $\frac{5}{x+1} + \frac{9}{x+2} = 8$

(b) $\frac{8}{x+4} - \frac{4}{x-2} = 5$

(c) $\frac{5}{2x+1} + \frac{4}{x+1} = 3$

(d) $\frac{10}{x-1} + \frac{12}{x+2} = 3\frac{1}{2}$

5 A motorist completes a 75 mile journey on country lanes at an average speed of v miles per hour.

(a) Write an expression, in terms of v, for the time taken in hours.

On his return journey the motorist uses the motorway and completes the 75 mile journey at an average speed of 5 miles per hour faster.

(b) Write an expression, in terms of v, for the time taken on his return journey in hours.

His return journey took 10 minutes less to complete.

(c) By forming an equation calculate his average speed on country lanes.

6 A car travels 60 km at a speed of x km/h.
It then travels a further 90 km at a speed 6 km/h faster.
The whole journey takes $5\frac{1}{2}$ hours.

(a) Form an equation involving x.

(b) Show that this equation can be written as $11x^2 - 234x - 720 = 0$.

(c) Solve this equation to find the speed, x.

Iteration

The solutions to a variety of equations can be found using a process called iteration.

The process of iteration has three stages.

1 Rearranging an equation to form an **iterative formula**. Sometimes the formula is given.

2 Choosing a **starting value**, x_1. Sometimes the starting value is given.

3 **Substituting** the value of x_1 into the iterative formula.
This produces a value called x_2 which is then substituted into the iterative formula.
Substituting a value for x_n produces the next value to be used, x_{n+1}.
The process is continued, producing values $x_3, x_4, \dots$ until the required degree of accuracy is obtained.

EXAMPLE

Find a solution to the equation $x^2 - 5x + 3 = 0$, correct to 3 significant figures, using **iteration**.
Use $x_1 = 4$ as a starting value.

$x^2 - 5x + 3 = 0$

$x^2 = 5x - 3$

Take the square root of both sides. $x = \sqrt{5x - 3}$

Write the iterative formula. $x_{n+1} = \sqrt{5x_n - 3}$

$x_1 = 4$ (this was given)

$x_2 = \sqrt{5 \times 4 - 3} = 4.1231\dots$

$x_3 = \sqrt{5 \times 4.1231\dots - 3} = 4.1970\dots$

$x_4 = \sqrt{5 \times 4.1970\dots - 3} = 4.2409\dots$

$x_5 = \sqrt{5 \times 4.2409\dots - 3} = 4.2666\dots$

$x_6 = \sqrt{5 \times 4.2666\dots - 3} = 4.2817\dots$

$x_7 = \sqrt{5 \times 4.2817\dots - 3} = 4.2905\dots$

$x_8 = \sqrt{5 \times 4.2905\dots - 3} = 4.2956\dots$

$x_9 = \sqrt{5 \times 4.2956\dots - 3} = 4.2986\dots$

$x_{10} = \sqrt{5 \times 4.2986\dots - 3} = 4.3003\dots$

At this stage the answer is not changing from 4.30 (correct to 3 sig. figs.).

To form an iterative formula, rearrange the equation so that one side has the variable on its own, e.g. '$x = \dots$'

The efficient use of the memory facilities of a calculator can help to reduce greatly the time taken to perform lengthy calculations.

Exercise 21.5

1. (a) Show how the equation $x^2 - 5x + 3 = 0$ can be rearranged to give:
   $x = \frac{5x - 3}{x}$ and hence, the iterative formula $x_{n+1} = \frac{5x_n - 3}{x_n}$.
   (b) Starting with the value $x = 4$ find a solution to the equation using iteration, correct to 3 significant figures. Show all calculations.
   (c) A solution to the equation $x^2 - 5x + 3 = 0$ was found in the example using a different iterative formula. Compare the two iterative processes.

2. (a) Show how the equation $6x^2 - 7x - 2 = 0$ can be rearranged to give:
   $x = \sqrt{\frac{7x + 2}{6}}$ and hence, the iterative formula $x_{n+1} = \sqrt{\frac{7x_n + 2}{6}}$.
   (b) Starting with the value $x_1 = 1$ find a solution to the equation using iteration, correct to 3 significant figures. Show all calculations.
   (c) Show how this equation can also be rearranged to give: $x = \frac{7x + 2}{6x}$
   and hence, the iterative formula $x_{n+1} = \frac{7x_n + 2}{6x_n}$.
   Use a starting value of $x_1 = 1$ and this formula.
   Which is the more efficient iterative formula?

3. When finding a solution to $x^2 + 7x - 10 = 0$ the following iterative formula is used:
   $x_{n+1} = \sqrt{10 - 7x_n}$. The starting value $x_1 = 1$ is used. What happens?

4. Use iteration to find a solution to the equation $x^2 - x - 5 = 0$.
   Show your iterative formula and use $x_1 = 3$ as a starting point.
   Give your answer correct to two decimal places.

5. Find a solution to the equation $x^2 + 7x = 100$ (correct to 3 sig. figs.) by using iteration.
   Show your rearrangement giving an iterative formula and a good starting value.

6. Find a solution to the equation $x^2 + 3x - 8 = 0$ by using iteration.
   Give your answer to an appropriate degree of accuracy.

Trial and improvement

Some equations cannot be solved directly.
Numerical solutions can be found by making a guess and improving the accuracy of the guess by **trial and improvement**.
This can be a time-consuming method of solving equations and is often used only as a last resort for solving equations which cannot be easily solved by algebraic or graphical methods.

A solution to the equation
$x^3 - 4x = 7$ lies between 2 and 3.
Find this solution to 1 decimal place.

Because the solution lies between 2 and 3, notice that:

when $x = 2$ $\quad 2^3 - 4(2) = 0$ $\quad$ Answer is less than 7.
when $x = 3$ $\quad 3^3 - 4(3) = 15$ $\quad$ Answer is greater than 7.

We are trying to find a value for x which produces the answer 7.

First guess: $x = 2.5$ $\quad 2.5^3 - 4(2.5) = 5.625$ $\quad$ Too small.
Second guess: $x = 2.6$ $\quad 2.6^3 - 4(2.6) = 7.176$ $\quad$ Too big.

The solution lies between 2.5 and 2.6, but 2.6 gives an answer closer to 7.
The answer is probably 2.6, correct to 1 d.p.
To be certain, try $x = 2.55$. $2.55^3 - 4(2.55) = 6.381375$ $\quad$ Too small.

The solution lies between 2.55 and 2.6.
2.6 is the solution, correct to 1 d.p.

Exercise 21.6

1 Use trial and improvement to solve $x^3 = 54$, correct to one decimal place.
The working can be shown in a table.

x	x^3	
3	27	Too small
4	64	Too big
3.5		

2 Use trial and improvement to solve these equations. Show all your trials in a table.
(a) $w^3 = 72$ (b) $4x^3 = 51$

3 A solution to the equation $x^3 + 2x = 40$ lies between 3 and 4.
Use trial and improvement to find this solution to 1 decimal place.

x	$x^3 + 2x$	
3	33	Too small
3.5	49.875	Too big

4 A solution to the equation $x^3 + 5x = 880$ lies between 9 and 10.
Use trial and improvement to find this solution to 1 decimal place. Show your trials.

5 Show that a solution to $x^3 - 5x^2 = 47$ lies between 6 and 7.
Use trial and improvement to find this solution, showing your trials,
(a) to 1 decimal place, (b) to 2 decimal places.

6 This cuboid has dimensions: x, $x + 2$ and $2x$.
(a) Write down an expression, in terms of x, for the volume of the cuboid.
(b) The volume of the cuboid is $400\,\text{cm}^3$.
Form an equation and use trial and improvement to find x, correct to 1 d.p.

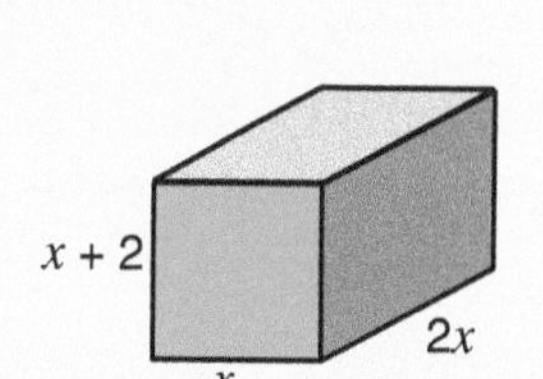

7 A solution to $x^4 + 5x - 20 = 0$ lies between 1 and 2.
Use trial and improvement to find this solution correct to 2 decimal places. Show your trials.

What you need to know

- An **identity** is true for all values of x. An identity is the same expression written in another form.
  For example: $(2x + 3)^2 + (2x + 9)(2x + 5) = 2(4x^2 + 20x + 27)$.
  To show that an identity is true, either:
  start with the LHS and show that it is equal to the RHS, or
  start with the RHS and show that it is equal to the LHS.
- **Algebraic fractions** have a numerator and a denominator (just as an ordinary fraction) but at least one of them is an expression involving an unknown.
- To write an algebraic fraction in its **simplest form**:
  factorise the numerator and denominator of the fraction,
  divide the numerator and denominator by their highest common factor.
- The same methods used for adding, subtracting, multiplying and dividing numeric fractions can be applied to algebraic fractions.
- The solutions to a variety of equations can be found using a process called **iteration**.
  The process of iteration has three stages.
  1 Rearranging an equation to form an **iterative formula**.
  2 Choosing a **starting value**, x_1.
  3 **Substituting** the starting value, and then values of x_n into the iterative formula.
  Continuing the process until the required degree of accuracy is obtained.
- **Trial and improvement** is a method used to solve equations. The accuracy of the value of the unknown letter is improved until the required degree of accuracy is obtained.

Review Exercise 21

1 (a) Factorise: (i) $x^2 - 4$ (ii) $3x^2 + 2x - 8$

(b) Hence, simplify as fully as possible $\dfrac{x^3 - 4x}{3x^2 + 2x - 8}$ AQA

2 Simplify fully the following expression. $\dfrac{6x - 18}{x^2 - 5x + 6}$ AQA

3 Show that the equation $\dfrac{2}{x + 3} - \dfrac{3}{3x - 1} = 2$ can be simplified to $6x^2 + 13x + 5 = 0$. AQA

4 Simplify the expression $\dfrac{6}{2x + 1} - \dfrac{3}{x + 1}$. AQA

5 Solve the equation $\dfrac{2}{x + 2} + \dfrac{3}{x} = 4$. AQA

6 I drive 20 miles to work each day.
My average speed for the journey is v miles per hour.

(a) Show that the time taken for the journey, t minutes, is given by $t = \dfrac{1200}{v}$.

If I leave home earlier, there is less traffic.
My average speed increases by 10 mph and my journey takes 4 minutes less.

(b) (i) Write down an expression in terms of v for the time taken for this faster journey.
(ii) Hence, write down an equation in terms of v only.
(iii) Show that the equation simplifies to $v^2 + 10v - 3000 = 0$.
(iv) Calculate the value of v. AQA

7 Kath discovers the following results when doing some fraction problems:

$\frac{2}{1} - \frac{1}{2} = \frac{3}{2}$ $\frac{3}{2} - \frac{2}{3} = \frac{5}{6}$ $\frac{4}{3} - \frac{3}{4} = \frac{7}{12}$ $\frac{5}{4} - \frac{4}{5} = \frac{9}{20}$

The nth line of this sequence is $\dfrac{(n + 1)}{n} - \dfrac{n}{(n + 1)} = \dfrac{2n + 1}{n(n + 1)}$

Prove that the left-hand side of this expression is equal to the right-hand side.
Show all your working clearly. AQA

8 Use a trial and improvement method to solve the equation $x^3 - x = 300$.
Give your answer to one decimal place.
Show all your trials in a table, as shown.

Trial x	$x^3 - x$	Too high/too low
8	504	Too high

AQA

9 A rectangular lawn has width x metres and length $2x$ metres.
The length is increased by 3 metres and the width by 1 metre.

(a) Multiply out the expression $(2x + 3)(x + 1)$ to give the area of the new larger lawn.

(b) The new lawn has an area of $66\,\text{m}^2$.
Form an equation and solve it to find the area of the original lawn.

(c) Another lawn is 3 metres longer than it is wide. Its area is $92\,\text{m}^2$.
To find the width, x, this iteration is used:

$$x_{n+1} = \frac{92}{x_n + 3}.$$

Starting with $x_1 = 8$, find the width, x, correct to **one** decimal place.

AQA

10 Solve the equation $\dfrac{5}{x + 2} + \dfrac{1}{2x - 5} = 1$.
Give your answer correct to 2 decimal places. AQA

Transforming Graphs

Transformations, such as **translations** and **stretches**, can be used to change the position and size of a graph.
The equation of the transformed (new) graph is related to the equation of the original graph.

Translating a graph

The graph of $y = x^2$ is transformed to the graph of $y = x^2 + 2$ by a **translation**.

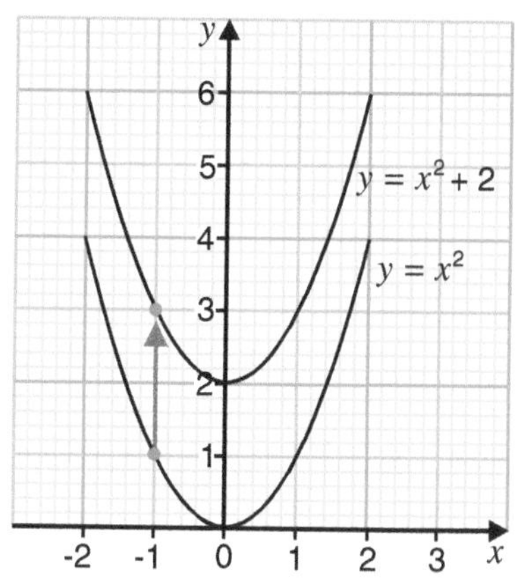

All points on the original graph are moved the **same distance** in the **same direction** without twisting or turning.

The translation has vector $\begin{pmatrix} 0 \\ 2 \end{pmatrix}$.

This means that the curve moves 2 units vertically up.

Original graph: $y = x^2$

Transformation: translation with vector $\begin{pmatrix} 0 \\ 2 \end{pmatrix}$.

New graph: $y = x^2 + 2$

> *Draw the following graphs:*
>
> $y = x^2$ *for* $-2 \leqslant x \leqslant 2$,
>
> $y = x^2 - 3$ *for* $-2 \leqslant x \leqslant 2$.
>
> *Describe how the graph of* $y = x^2$ *is transformed to the graph of* $y = x^2 - 3$.

The graph of $y = x^2$ is transformed to the graph of $y = (x + 3)^2$ by a **translation**.

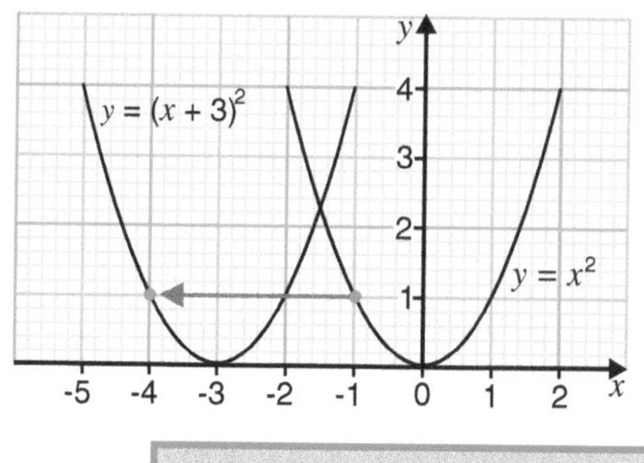

The translation has vector $\begin{pmatrix} -3 \\ 0 \end{pmatrix}$.

This means that the curve moves 3 units horizontally to the left.

Original graph: $y = x^2$

Transformation: translation with vector $\begin{pmatrix} -3 \\ 0 \end{pmatrix}$.

New graph: $y = (x + 3)^2$

> *Draw the following graphs:*
>
> $y = x^2$ *for* $-2 \leqslant x \leqslant 2$,
>
> $y = (x - 2)^2$ *for* $0 \leqslant x \leqslant 4$.
>
> *Describe how the graph of* $y = x^2$ *is transformed to the graph of* $y = (x - 2)^2$.

Stretching a graph

The graph of $y = x^2$ is transformed to the graph of $y = 2x^2$ by a **stretch**.

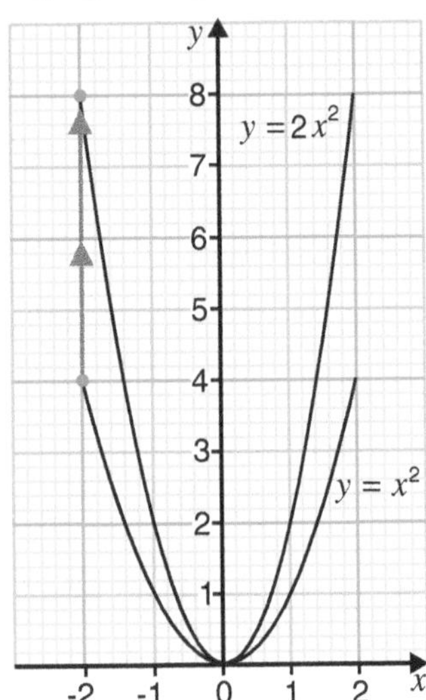

The stretch is **from** the x axis and is **parallel** to the y axis.
The stretch has scale factor 2.

This means that y coordinates on the graph of $y = x^2$ are doubled to obtain the corresponding y coordinates on the graph of $y = 2x^2$.

Original graph: $y = x^2$

Transformation: stretch from x axis, parallel to y axis, scale factor 2.

New graph: $y = 2x^2$

Draw the following graphs:

$y = x^2$ *for* $-2 \leqslant x \leqslant 2$,

$y = 3x^2$ *for* $-2 \leqslant x \leqslant 2$.

Describe how the graph of $y = x^2$ *is transformed to the graph of* $y = 3x^2$.

The graph of $y = x^2$ is transformed to the graph of $y = (2x)^2$ by a **stretch**.

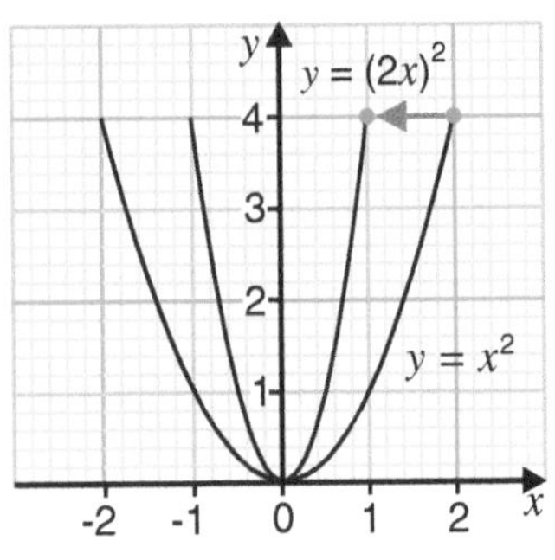

The stretch is **from** the y axis and is **parallel** to the x axis.
The stretch has scale factor $\frac{1}{2}$.

This means that x coordinates on the graph of $y = x^2$ are halved to obtain the corresponding x coordinates on the graph of $y = (2x)^2$.

Original graph: $y = x^2$

Transformation: stretch from y axis, parallel to x axis, scale factor $\frac{1}{2}$.

New graph: $y = (2x)^2$

Draw the following graphs:

$y = x^2$ *for* $-2 \leqslant x \leqslant 2$,

$y = (4x)^2$ *for* $-0.5 \leqslant x \leqslant 0.5$.

Describe how the graph of $y = x^2$ *is transformed to the graph of* $y = (4x)^2$.

In general

Original graph	New graph	Transformation
$y = x^2$	$y = x^2 + a$	**translation**, vector $\begin{pmatrix} 0 \\ a \end{pmatrix}$.
$y = x^2$	$y = (x + a)^2$	**translation**, vector $\begin{pmatrix} -a \\ 0 \end{pmatrix}$.
$y = x^2$	$y = ax^2$	**stretch**, from the x axis, parallel to the y axis, scale factor a.
$y = x^2$	$y = (ax)^2$	**stretch**, from the y axis, parallel to the x axis, scale factor $\frac{1}{a}$.

The general rules for translating and stretching the graph of $y = x^2$ apply to **all** graphs.

Exercise 22.1

You may find it helpful to use a graphical calculator.

1 (a) Complete tables of values for:

(i) $y = x^3$ for $-2 \leqslant x \leqslant 2$, (ii) $y = x^3 + 2$ for $-2 \leqslant x \leqslant 2$,

(iii) $y = (x - 3)^3$ for $1 \leqslant x \leqslant 5$, (iv) $y = 2x^3$ for $-2 \leqslant x \leqslant 2$.

(b) Use your tables of values to draw graphs of each of the equations in (a).

(c) Describe how the graph of $y = x^3$ is transformed to:

(i) the graph of $y = x^3 + 2$, (ii) the graph of $y = (x - 3)^3$,

(iii) the graph of $y = 2x^3$.

(d) Write down the equations of the graphs obtained when the graph of $y = x^3$ is transformed by:

(i) a translation with vector $\begin{pmatrix} 0 \\ -5 \end{pmatrix}$,

(ii) a translation with vector $\begin{pmatrix} 2 \\ 0 \end{pmatrix}$.

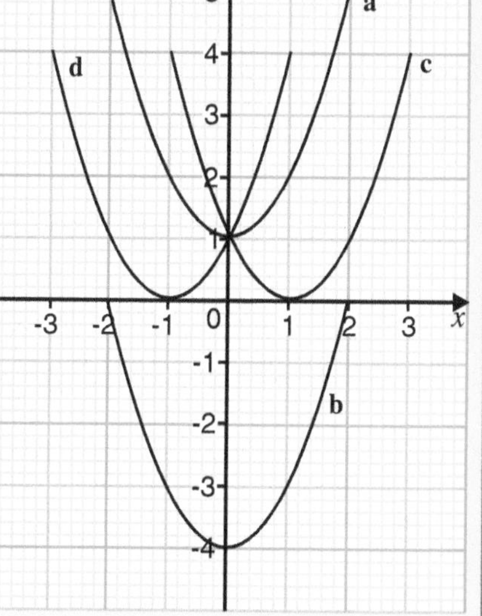
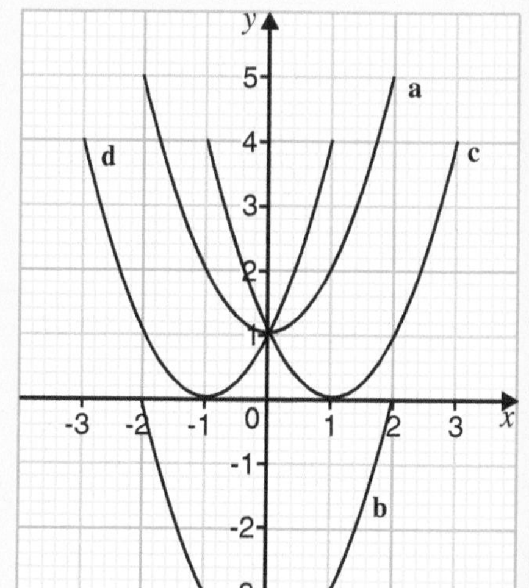

2 In this diagram, each of the graphs labelled **a**, **b**, **c** and **d** is a transformation of the graph $y = x^2$.

What are the equations of graphs **a**, **b**, **c** and **d**?

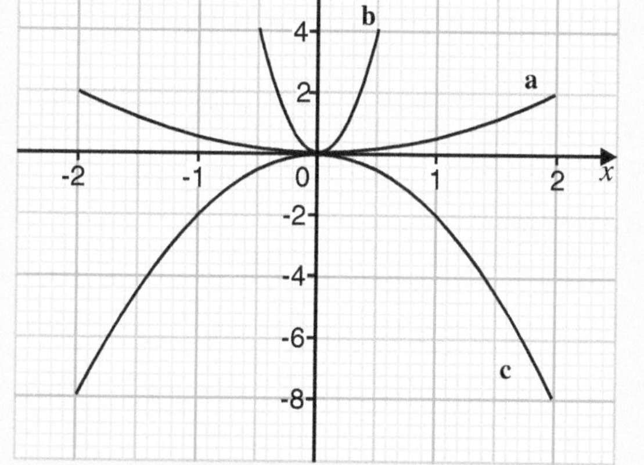

3 In this diagram, each of the graphs **a, b** and **c** is a stretch of the graph of $y = x^2$.

What are the equations of graphs **a, b** and **c**?

4 The graph of $y = 4 - x^2$ for $-2 \leqslant x \leqslant 2$ is transformed by:

(a) a translation of $\begin{pmatrix} 0 \\ 5 \end{pmatrix}$,

(b) a stretch from the x axis, parallel to the y axis, with scale factor 2,

(c) a stretch from the y axis, parallel to the x axis, with scale factor 2.

In each case draw the original graph and its transformation.
Write down the equation of each transformed graph.

5 The graph of $y = x^2 - x$ for $-2 \leqslant x \leqslant 2$ is transformed by:

(a) a translation of (i) $\begin{pmatrix} 0 \\ -5 \end{pmatrix}$, (ii) $\begin{pmatrix} 3 \\ 0 \end{pmatrix}$.

(b) a stretch from the x axis, parallel to the y axis, with scale factor $\frac{1}{2}$.

(c) a stretch from the y axis, parallel to the x axis, with scale factor 2.

In each case write down the equation of the transformed graph.
Check each equation by drawing its graph.

Function notation

Function notation is an alternative way of expressing the relationship between two variables.

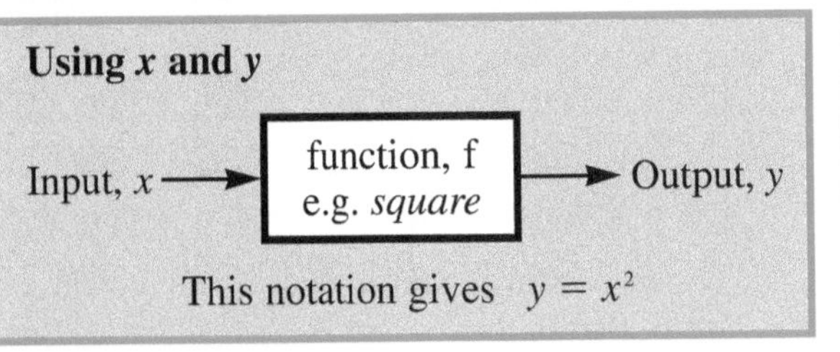

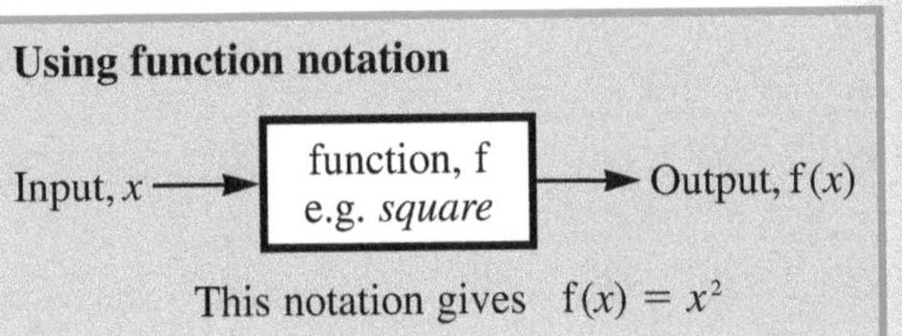

$f(x)$ means 'a function of x'.
In general, $y = f(x)$.
For example:
The equation $y = x^5$ can be expressed in function notation as $f(x) = x^5$, where $y = f(x)$.

Letters other than f can be used.
For example:
The equation $y = \frac{1}{x^2 + 1}$ can be expressed in function notation as $g(x) = \frac{1}{x^2 + 1}$, where $y = g(x)$.

EXAMPLES

1 The function f is defined as $f(x) = x^2$.
Describe how each of the following graphs are related to the graph of $y = f(x)$.

(a) $y = f(x) + 2$
(b) $y = f(2x)$

(a) Because, $f(x) = x^2$, $y = f(x) + 2$ is equivalent to $y = x^2 + 2$.
So, the graph of $y = f(x)$ is transformed to the graph of $y = f(x) + 2$ by:
a **translation** with vector $\begin{pmatrix} 0 \\ 2 \end{pmatrix}$.

(b) $y = f(2x)$ is equivalent to $y = (2x)^2$.
So, the graph of $y = f(x)$ is transformed to the graph of $y = f(2x)$ by:
a **stretch** from the y axis, parallel to the x axis, scale factor $\frac{1}{2}$.

> $f(x) = x^2$
> Substitute $(2x)$ for x.
> $f(2x) = (2x)^2$

2 The diagram shows the graph of $y = f(x)$.

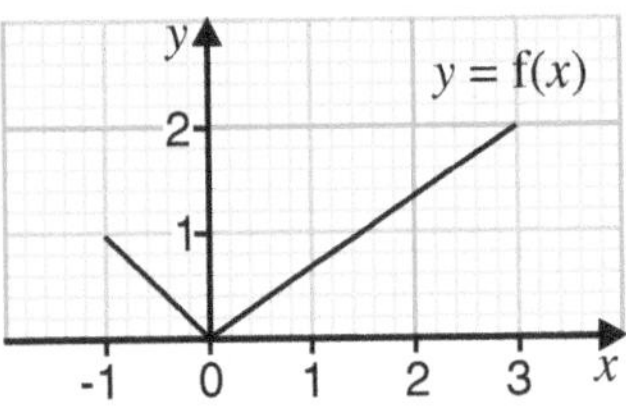

On separate diagrams show the graphs of:
(a) $y = f(x)$ and $y = f(x + 3)$,
(b) $y = f(x)$ and $y = 2f(x)$.

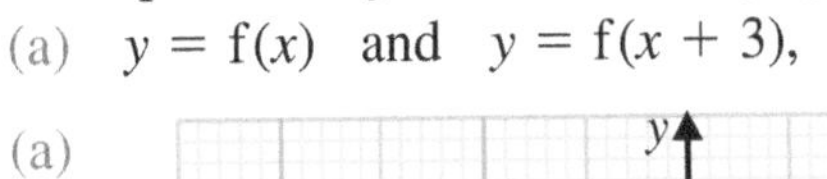

(a)

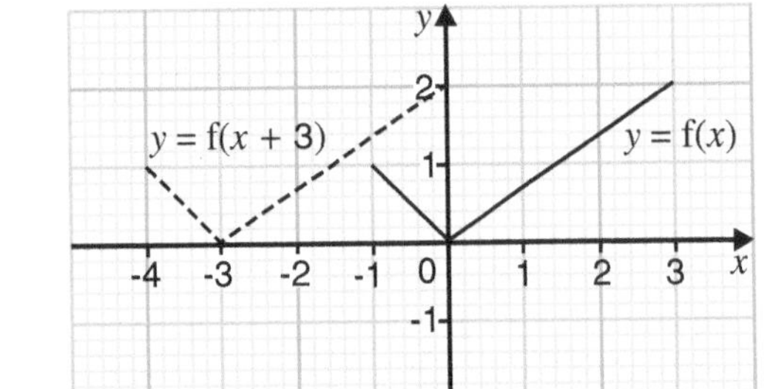

(b)

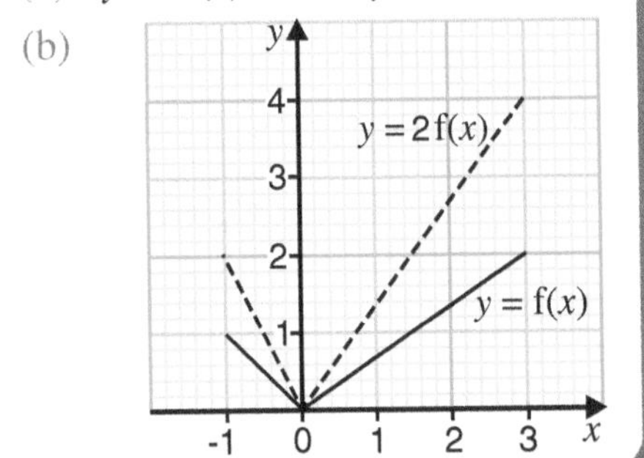

In general

Original	New graph	Transformation	Note
$y = f(x)$	$y = f(x) + a$	**translation**, vector $\begin{pmatrix} 0 \\ a \end{pmatrix}$.	If a is **positive**, curve moves a units **up**. If a is **negative**, curve moves a units **down**.
$y = f(x)$	$y = f(x + a)$	**translation**, vector $\begin{pmatrix} -a \\ 0 \end{pmatrix}$.	If a is **positive**, curve moves a units **left**. If a is **negative**, curve moves a units **right**.
$y = f(x)$	$y = af(x)$	**stretch**, from the x axis, parallel to the y axis, scale factor a.	The y coordinates on the graph of $y = f(x)$ are **multiplied** by a.
$y = f(x)$	$y = f(ax)$	**stretch**, from the y axis, parallel to the x axis, scale factor $\frac{1}{a}$.	The x coordinates on the graph of $y = f(x)$ are **divided** by a.

Reflecting a graph

The diagram shows the graphs of $y = f(x)$ and $y = -f(x)$.

The graph of $y = f(x)$ is transformed to the graph of $y = -f(x)$ by:
a **stretch** from the x axis, parallel to the y axis, scale factor -1.
This transformation is equivalent to a **reflection** in the x axis.

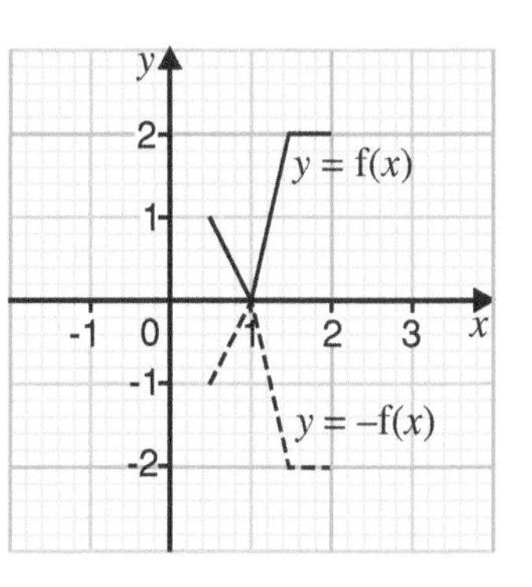

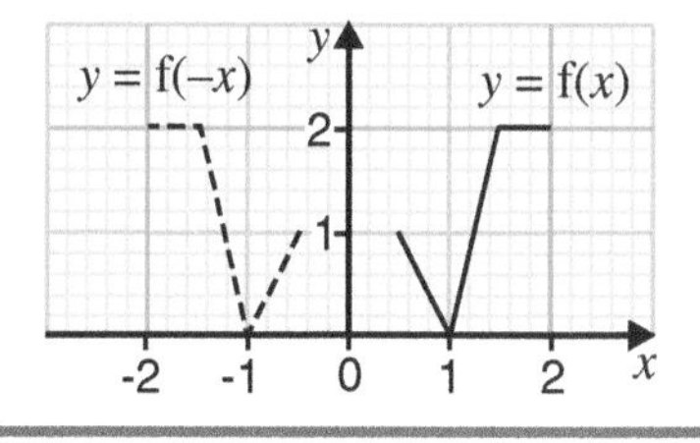

The diagram shows the graphs of $y = f(x)$ and $y = f(-x)$.

The graph of $y = f(x)$ is transformed to the graph of $y = f(-x)$ by:
a **stretch** from the y axis, parallel to the x axis, scale factor -1.
This transformation is equivalent to a **reflection** in the y axis.

EXAMPLE

1 The graph of the function f, where $f(x) = x^3 + 2$, is shown together with the graphs of two related functions g and h.
Find equations for $g(x)$ and $h(x)$.

The graph of $y = g(x)$ is a reflection of the graph of $y = f(x)$ in the x axis.

This means that $g(x) = -f(x)$

So, $g(x) = -(x^3 + 2)$

So, $g(x) = -x^3 - 2$

The graph of $y = h(x)$ is a reflection of the graph of $y = f(x)$ in the y axis.

This means that $h(x) = f(-x)$

So, $h(x) = (-x)^3 + 2$

So, $h(x) = -x^3 + 2$

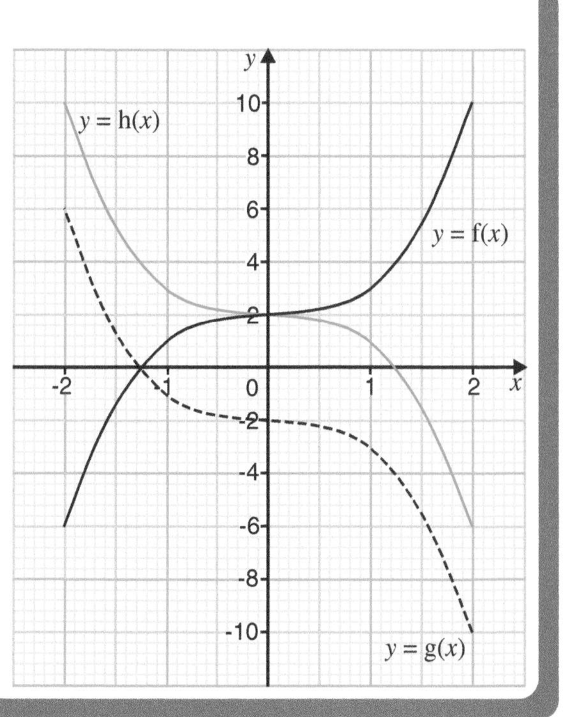

EXAMPLE

2 The diagram shows the graphs of $y = \sin x$ and two related functions **a** and **b**.
Find equations for the functions **a** and **b**.

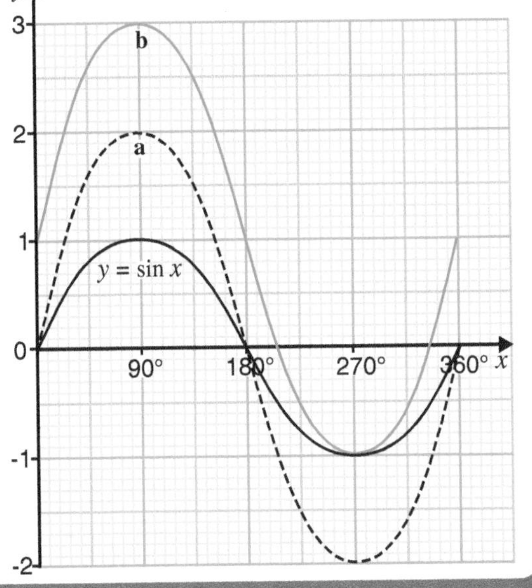
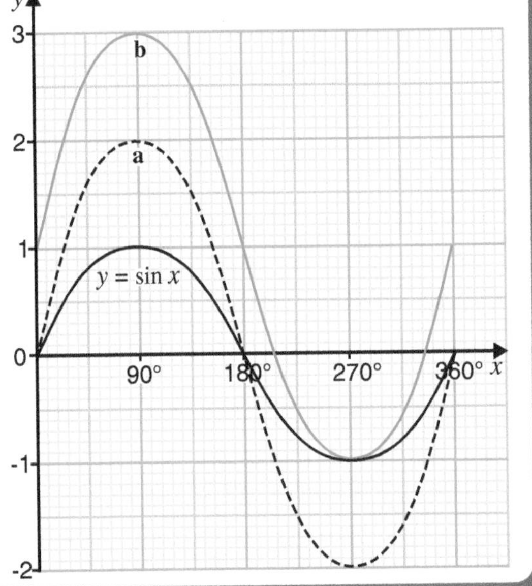

The graph of $y = \sin x$ is transformed to the graph of function **a** by:
a **stretch** from the x axis, parallel to the y axis, scale factor 2.
So, the equation of function **a** is $y = 2\sin x$.

The graph of function **a** is transformed to the graph of function **b** by:
a **translation** with vector $\begin{pmatrix} 0 \\ 1 \end{pmatrix}$.
So, the equation of function **b** is $y = 2\sin x + 1$.

Exercise 22.2

1 Given that $f(x) = x^2$, sketch on separate diagrams the graphs of
(a) $y = -f(x)$, (b) $y = f(2 - x)$, (c) $y = f(x) - 2$, (d) $y = f(2x)$, (e) $y = 2f(x)$.

2 The diagram shows the graph of the function $y = f(x)$.

(a) Copy the diagram, drawing axes marked from -4 to 4 for x and y.
Draw the graph of the function $y = f(x) + 2$.
(b) Describe fully how the graph of $y = f(x)$ is transformed to the graph of $y = f(x) + 2$.
(c) Repeat parts (a) and (b) for each of these functions:

(i) $y = f(x) - 1$ (ii) $y = f(x - 3)$ (iii) $y = f(x + 2)$
(iv) $y = 0.5f(x)$ (v) $y = 2f(x)$ (vi) $y = f(0.5x)$
(vii) $y = f(2x)$ (viii) $y = -f(x)$ (ix) $y = f(-x)$
(x) $y = f(2x) + 1$ (xi) $y = 2f(x + 1)$ (xii) $y = 1 - f(x)$

3 The graph of $y = f(x)$ is shown.
On separate diagrams sketch:

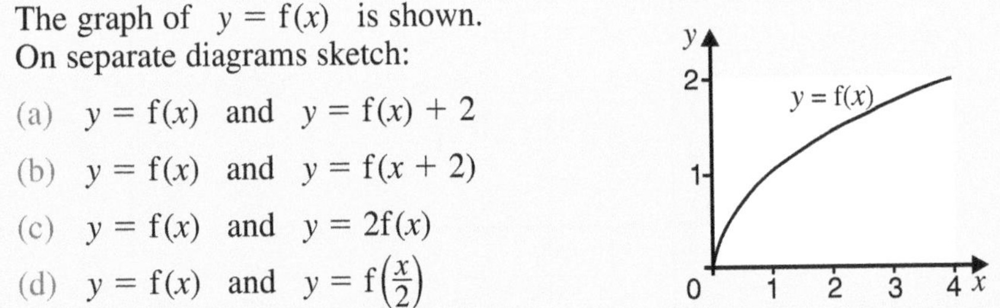

(a) $y = f(x)$ and $y = f(x) + 2$
(b) $y = f(x)$ and $y = f(x + 2)$
(c) $y = f(x)$ and $y = 2f(x)$
(d) $y = f(x)$ and $y = f\left(\frac{x}{2}\right)$
(e) $y = f(x)$ and $y = -f(x)$

4 The graph shows $y = \text{h}(x)$.
On separate diagrams sketch:
(a) $y = \text{h}(x) + 1$
(b) $y = \text{h}(x - 90)$
(c) $y = \text{h}(2x)$
(d) $y = 2\text{h}(x)$
(e) $y = -\text{h}(x)$

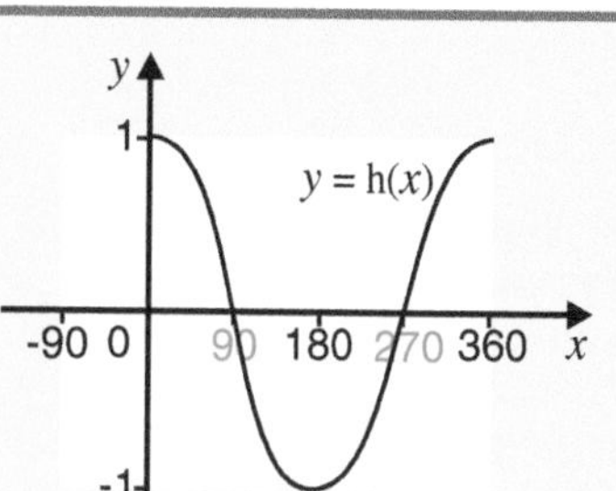

5 The graph of the function $y = \text{f}(x)$ is transformed to each of the graphs **a**, **b** and **c**.
Find, in terms of $\text{f}(x)$, equations for the graphs **a**, **b** and **c**.

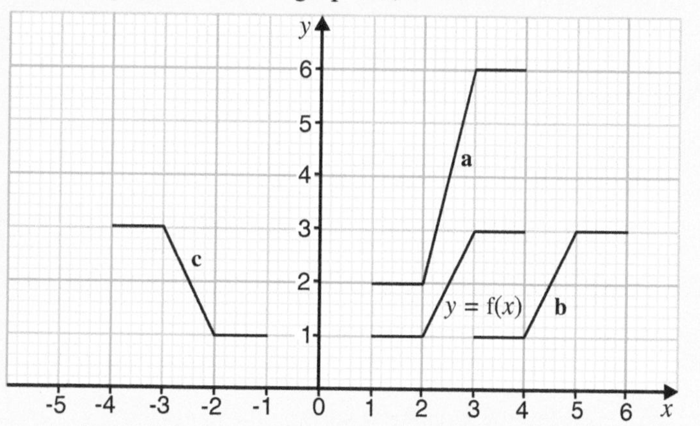

6 The diagram shows a sketch of $y = \sin x$ and three related functions.

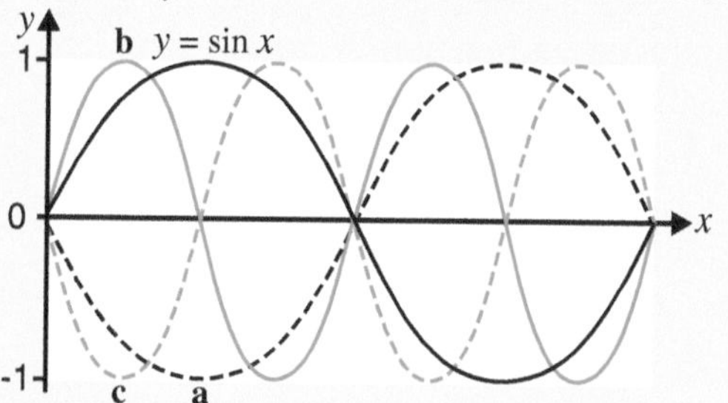

(a) Describe how the graph of $y = \sin x$ is transformed to
(i) the graph of function **a**,
(ii) the graph of function **b**.
(b) Describe how the graph of function **b** is transformed to the graph of function **c**.
(c) Describe how the graph of $y = \sin x$ is transformed to the graph of function **c**, by using a combination of **two** transformations.
(d) What are the equations of the graphs **a**, **b** and **c**?

7 The diagram shows the graphs of the functions $y = \text{f}(x)$, $y = \text{g}(x)$ and $y = \text{h}(x)$.
Express $\text{g}(x)$ and $\text{h}(x)$ in terms of $\text{f}(x)$.

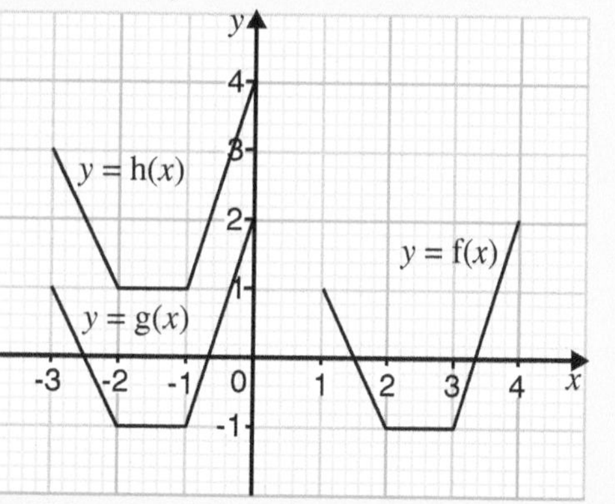

Using graphs to find relationships

Graphs can be used to help find the relationship between two variables.
This is particularly useful when looking for rules to fit the results of scientific experiments.

Linear functions

Linear functions have straight line graphs.
Straight line graphs have equations of the form $y = mx + c$, where m is the gradient and c is the y-intercept.

EXAMPLE

The table shows the speed, v (m/s), of a vehicle at various times, t (sec), after it starts to decelerate.

t (sec)	10	20	30	40	50
v (m/s)	35	27	19	11	3

(a) Find an equation connecting v and t.
(b) Use your equation to find:
(i) v when $t = 25$,
(ii) t when $v = 0$.

(a) **Step 1**
Plot the data and draw a line of best fit, by eye, to show the relationship between the variables speed and time.

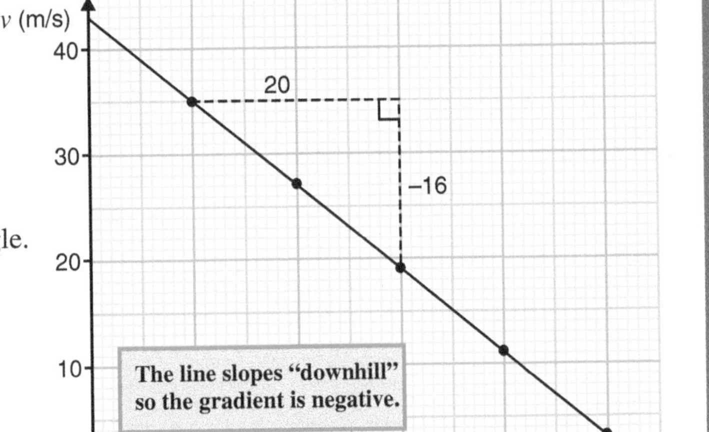

Step 2
Find the equation of the line.
The equation is of the form
$v = mt + c$.
Find the gradient, m.
Draw a suitable right-angled triangle.
$m = \frac{-16}{20} = -0.8$
Find the intercept, c.
The line passes through the point (0, 43).
$c = 43$.

Step 3
Substitute values for m and c into $v = mt + c$.
$m = -0.8$ and $c = 43$.
So, the equation connecting v and t is: $v = -0.8t + 43$.

(b) (i) Using $v = -0.8t + 43$.
When $t = 25$.
$v = -0.8 \times 25 + 43$
$v = 23$

(ii) Using $v = -0.8t + 43$.
When $v = 0$.
$0 = -0.8t + 43$
$0.8t = 43$
Divide both sides by 0.8.
$t = 53.75$

Non-linear functions

This section looks at **non-linear functions** of the form $y = ax^n + b$.

To find the values of a and b:

1. Write $y = ax^n + b$ as the **linear function** $y = az + b$, by substituting $z = x^n$.
2. Draw the graph of y against z.
3. Use the graph to find:
   the gradient, a,
   the y-intercept, b.

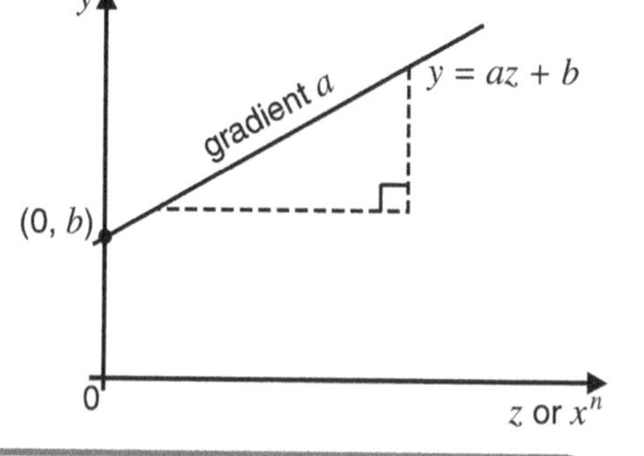

EXAMPLE

The table shows values of x and y which are connected by an equation of the form $y = ax^2 + b$.
Find the values of a and b.

x	0	0.5	1	1.5	2
y	-4	-3.25	-1	2.75	8

Substituting $z = x^2$ in $y = ax^2 + b$ gives the linear function $y = az + b$.
So, draw the graph of y against z.
Since $z = x^2$, values of z are found by squaring the given values of x in the table.

This table shows the values of z and y.

$z(=x^2)$	0	0.25	1	2.25	4
y	-4	-3.25	-1	2.75	8

The graph has gradient 3 and y-intercept -4.
So, $a = 3$ and $b = -4$.

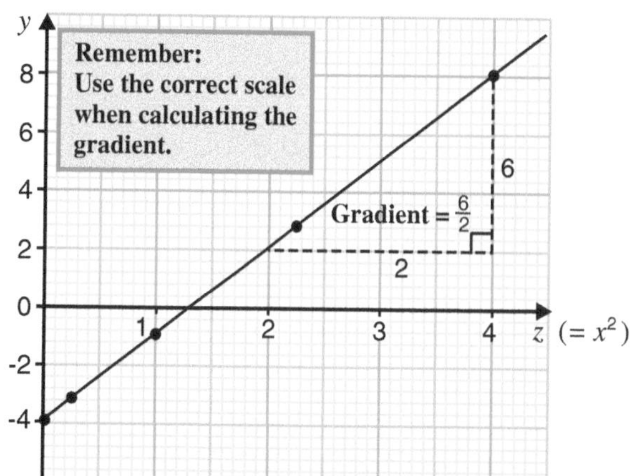

Exercise 22.3

1 These tables show the data collected in two experiments.

x	0.4	0.8	1.2	1.6	2
y	24.2	35.4	46.6	57.8	69.0

p	0.4	0.8	1.2	1.6	2
q	35.2	30.1	24.9	19.8	14.1

(a) In each case assume that the variables are connected by a linear function and use a graphical method to find its equation.

(b) Use your equations to find: (i) y when $x = 2.7$, (ii) p when $q = 7.6$.

2 In an experiment, Kym hangs weights, w grams, on the end of a spring and measures its length, l centimetres.
This graph shows Kym's results.

(a) Find an equation expressing l in terms of w.
(b) Use your equation to find:
(i) l when $w = 500$,
(ii) w when $l = 55$.

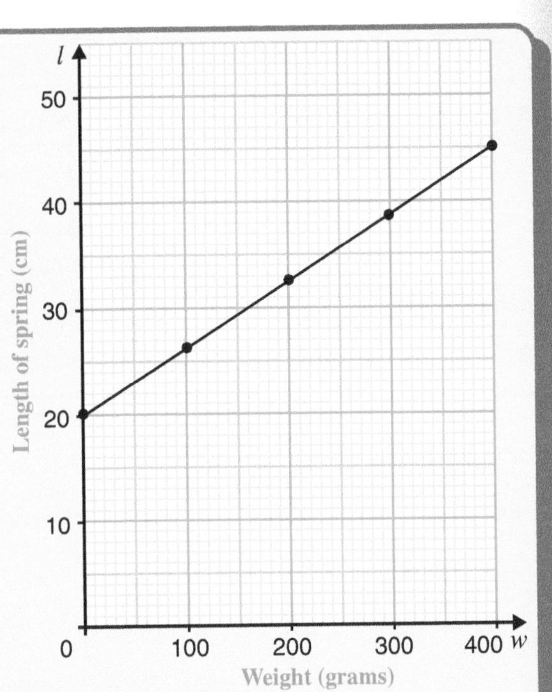

3 Two variables, x and y, are connected by the equation $y = ax^2 + b$.
Some pairs of values of x and y are shown in the table.

x	2	4	6	8	10
y	2.8	4	6	8.8	12.4

By drawing an appropriate graph find the values of a and b.

4 A vehicle starts from rest and accelerates with a constant acceleration of a m/s^2.
The formula $s = at^2$ gives the distance travelled, s metres, after time, t seconds.
The table shows pairs of values of s and t.

t	2	4	6	8	10
s	8.4	33.6	75.6	134.4	210

Use a graphical method to find the acceleration.

5 Wyn and Les perform an experiment. The results are shown in the table.

x	0.4	0.8	1.2	1.6	2
y	24.3	35.5	65.9	125.1	222.7

Wyn thinks that x and y are connected by a rule of the form $y = ax^2 + b$.
Les thinks the rule is of the form $y = ax^3 + b$.

Wyn uses the substitution $u = x^2$.

$u(=x^2)$	0.16	0.64	1.44	2.56	4
y	24.3	35.5	65.9	125.1	222.7

(a) Draw the graph of y against u and, if possible, use it to find the values of a and b.

Les uses the substitution $v = x^3$.

$v(=x^3)$	0.064	0.512	1.728	4.096	8
y	24.3	35.5	65.9	125.1	222.7

(b) Draw the graph of v against y and, if possible, use it to find the values of a and b.

6 Two variables, x and y, are connected by the equation $y = ax^3 + b$.
Some pairs of values of x and y are shown in the table.

x	2	4	6	8	10
y	2.8	6.0	13.2	28.0	52.4

By drawing an appropriate graph find the values of a and b.

What you need to know

- **Function notation** is a way of expressing a relationship between two variables.
  For example

  Input, x → function, f e.g. *cube* → Output, $f(x)$

  This notation gives $f(x) = x^3$

  $f(x)$ means 'a function of x'.
  In the example above, $f(x) = x^3$ is equivalent to the equation $y = x^3$ where $y = f(x)$.
- **Transformations,** such as **translations** and **stretches**, can be used to change the position and size of a graph.
  The equation of the transformed (new) graph is related to the equation of the original graph.

In general

Original	New graph	Transformation	Note
$y = f(x)$	$y = f(x) + a$	**translation**, vector $\begin{pmatrix} 0 \\ a \end{pmatrix}$.	If a is **positive**, curve moves a units **up**. If a is **negative**, curve moves a units **down**.
$y = f(x)$	$y = f(x + a)$	**translation**, vector $\begin{pmatrix} -a \\ 0 \end{pmatrix}$.	If a is **positive**, curve moves a units **left**. If a is **negative**, curve moves a units **right**.
$y = f(x)$	$y = af(x)$	**stretch**, from the x axis, parallel to the y axis, scale factor a.	The y coordinates on the graph of $y = f(x)$ are **multiplied** by a.
$y = f(x)$	$y = f(ax)$	**stretch**, from the y axis, parallel to the x axis, scale factor $\frac{1}{a}$.	The x coordinates on the graph of $y = f(x)$ are **divided** by a.
$y = f(x)$	$y = -f(x)$	**reflection** in the x axis.	The y coordinates on the graph of $y = f(x)$ **change signs**.
$y = f(x)$	$y = f(-x)$	**reflection** in the y axis.	The x coordinates on the graph of $y = f(x)$ **change signs**.

- **Finding relationships between variables**:
  Linear functions have straight line graphs, such as $y = ax + b$.
  From the graph of **y against x**, the gradient $= a$ and the y-intercept $= b$.
  Non-linear functions, such as $y = ax^n + b$, can be written as the linear function $y = az + b$ by substituting $z = x^n$.
  From the graph of **y against x^n**, the gradient $= a$ and the y-intercept $= b$.

Review Exercise 22

1 The graph of $y = \sin x$ is shown.
Sketch the graphs of
(a) $y = 2\sin x$,
(b) $y = \sin x + 2$,
(c) $y = \sin 2x$,
(d) $y = -\sin x$.

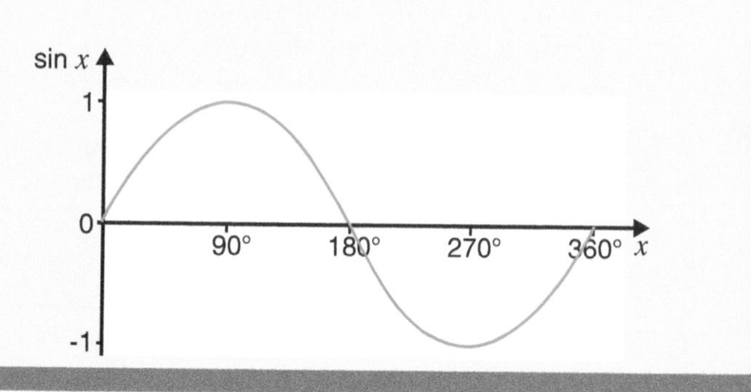

2 The function $y = f(x)$ is illustrated.

(a) On separate copies of the axes sketch

(i) $y = 2f(x)$,

(ii) $y = f(x - 1)$.

(b) Which one of the sketches below is of the form $y = f(x) + a$, where a is a constant?
What is the value of a?

A

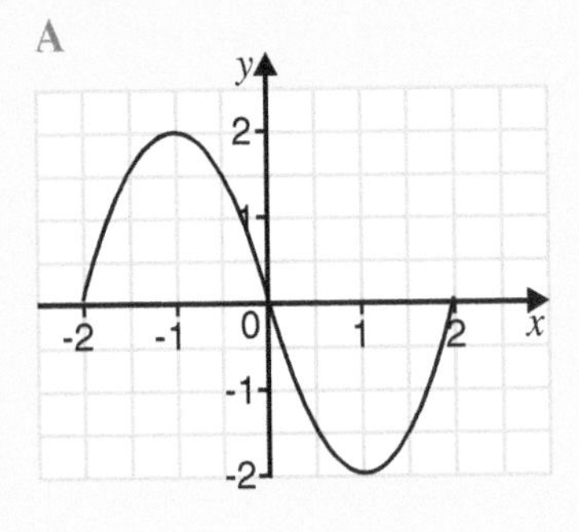

B

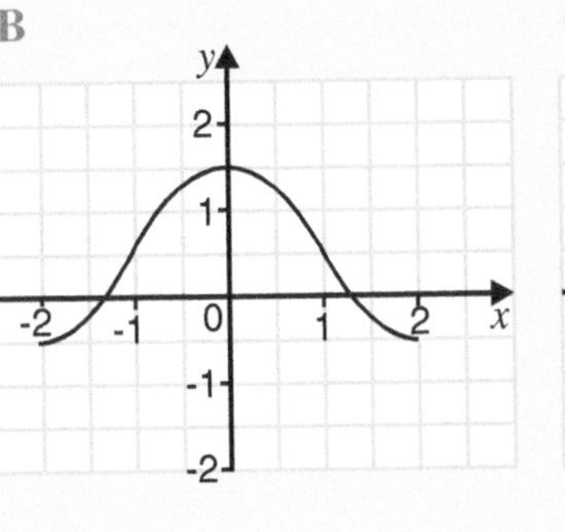

C

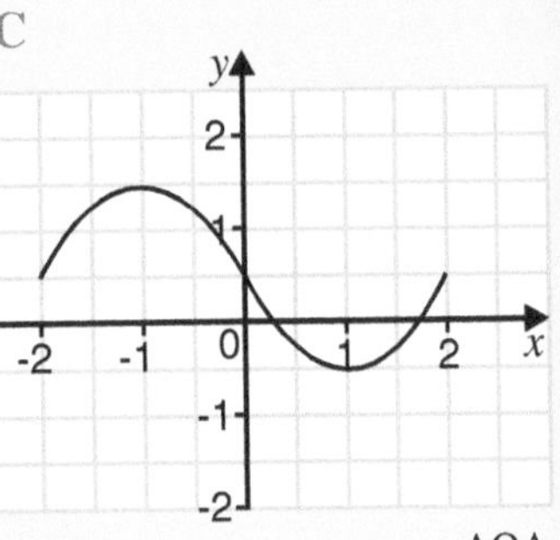

3 The diagram shows a sketch of the graph of $y = f(x)$ for $-3 \leqslant x \leqslant 3$.

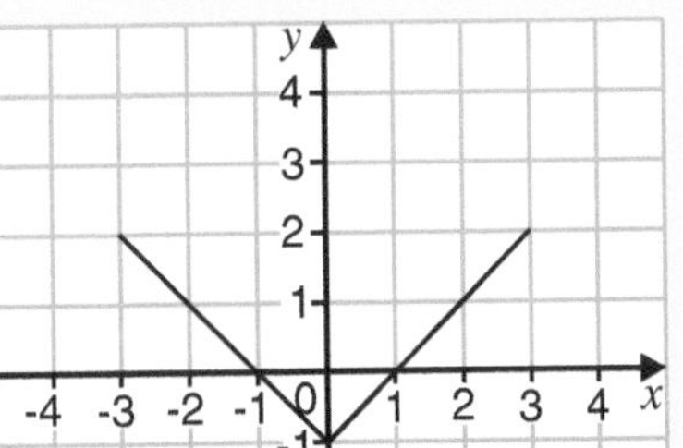

(a) Draw each of the graphs with these equations.

(i) $y = f(x) + 2$

(ii) $y = f(x - 2)$

(b)

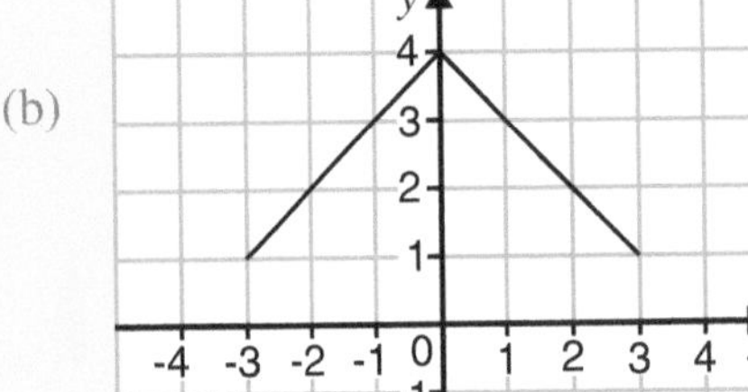

The graph opposite is a transformation of the graph $y = f(x)$.

Write down, in terms of $f(x)$, the equation of the graph.

AQA

4 The diagram shows a sketch of the function $y = f(x)$.
The curve cuts the x axis at (0, 0) and (3, 0).
$A(1, 2)$ is a point on the curve.
The function is transformed.
Sketch the transformed function.
State the coordinates of A on each new curve.

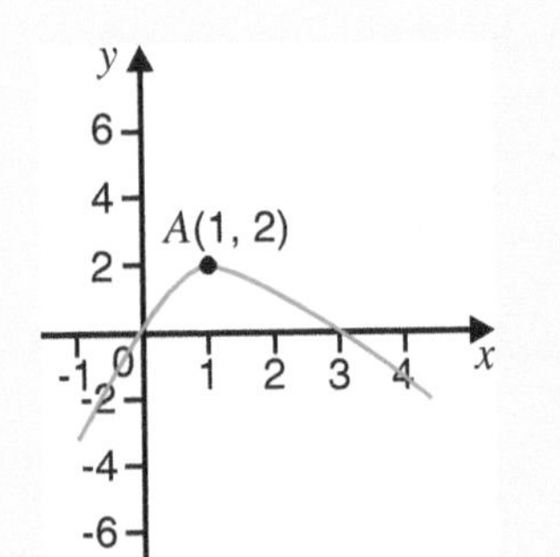

(a) $y = f(x) - 2$

(b) $y = 2f(x)$

(c) $y = f(2x)$

AQA

5 The following data are obtained from an experiment.

x	1	2	3	4	5	6
y	0.7	1.2	2.1	3.3	4.7	6.7

It is thought that they may obey the rule $y = ax^2 + b$.

(a) By plotting values of y against x^2, find approximate values of a and b.

(b) Find the value of y when $x = 4.5$.

AQA

Algebra
Non-calculator Paper

Do not use a calculator for this exercise.

1 (a) Work out the value of $2p^2$ when $p = 3$.
(b) Simplify $2(x - 3) + 5$.
(c) Solve $3y - 7 = 5$.

2 Mrs Crawley drove from her home to the supermarket, did her shopping and then returned home. The distance-time graph shows her journey.

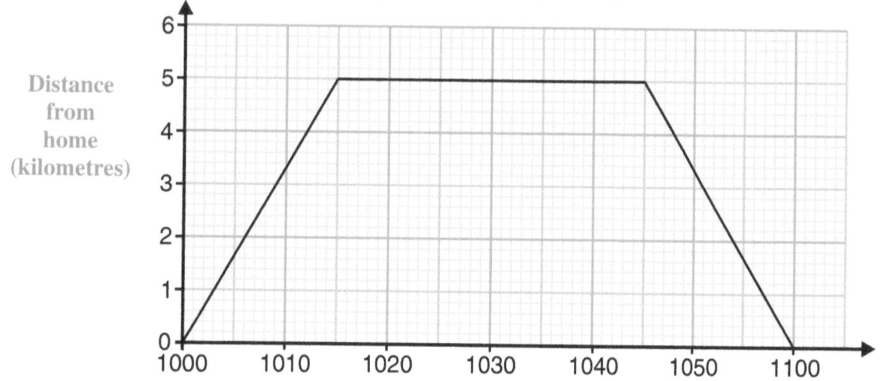

(a) At what time did Mrs Crawley leave home?
(b) How far is the supermarket from Mrs Crawley's home?
(c) How many minutes did Mrs Crawley spend at the supermarket?
(d) Work out Mrs Crawley's average speed on her journey home?
Give your answer in kilometres per hour.

3 (a) Using the values $u = 4$, $v = -3$ and $w = 5$, work out: (i) $u^2 + v^2$ (ii) $\frac{uv}{2w}$
(b) Simplify. $m \times n \times 3$
(c) Multiply out the brackets. (i) $7(3x + 2y)$ (ii) $a(a - 3)$

4 Solve these equations. (a) $4x - 7 = 5$ (b) $2(y + 5) = 28$ (c) $7z + 2 = 9 - 3z$

AQA

5 (a) Buns cost x pence each. How much will 2 buns cost?
(b) A doughnut costs 5 pence more than a bun. How much will 3 doughnuts cost?
(c) The cost of buying 2 buns and 3 doughnuts is 95 pence.
By forming an equation find the cost of a bun.

6 (a) Draw the graph of $y + 2x = 5$ for values of x from -1 to 4.
(b) Use your graph to find the value of x when $y = -2$.

7 P is the point (0, 4). Q is the point (4, −2).
Find the coordinates of the midpoint of the line segment PQ.

8 (a) Multiply out and simplify where possible. (i) $x(x - 5)$ (ii) $5(3p + 2) + 5p$
(b) Factorise. (i) $6n + 9$ (ii) $2m^2 + m$
(c) Solve. (i) $\frac{x}{4} = 25$ (ii) $3x + 5 = x - 2$

9 (a) Solve the inequality $3x - 5 < 7$.
(b) An equality is shown on the number line.

-3 -2 -1 0 1 2 3 4 5

Write down the inequality.

10 Simplify. (a) $c \times c \times c \times c$ (b) $d^3 \times d^2$ (c) $\frac{e}{e^8}$ AQA

11 (a) Use the formula $y = mx + c$ to find the value of y when $m = -4$, $x = -3$ and $c = -5$.

(b) Rearrange the formula $y = mx + c$ to make x the subject.

12 The area, y, of this rectangle is given by $y = x^2 + 2x$.

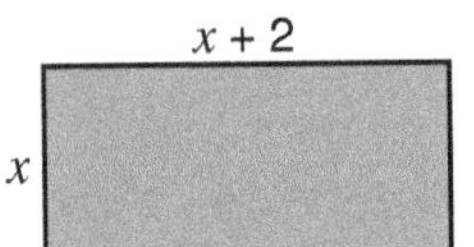

(a) Copy and complete this table of values for y.

x	0	1	2	3	4	5
y	0		8	15		35

(b) Draw the graph of $y = x^2 + 2x$.

(c) Use your curve to find the value of x if the area of the rectangle is $20\,\text{cm}^2$.

13 (a) Solve the equation $4(x + 3) = 22 + x$

(b) (i) Write down the integer values of n for which $2 < 5n \leqslant 12$.

(ii) Solve the inequality $5x + 3 \geqslant 4$.

(c) Solve the simultaneous equations: $3x + y = 13$
$2x - y = 7$

14 The graph shows the cost of printing wedding invitation cards.

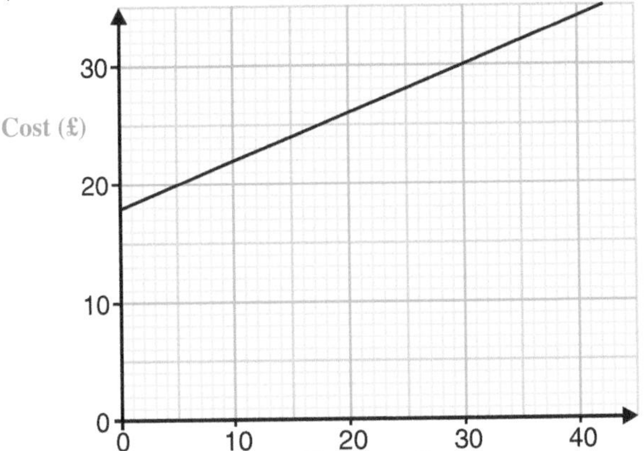

(a) Find the equation of the line in the form $y = mx + c$.

(b) For her wedding Charlotte needs 100 cards to be printed. How much will they cost? AQA

15 (a) Solve the inequality $x + 3 < 2 - x$.

(b) Multiply out and simplify $(x - 5)^2$.

(c) Factorise (i) $3m^2 - 6m$, (ii) $t^2 - t - 12$.

16 (a) Simplify the following expressions.

(i) $a^5 \times a^3$ (ii) $a^5 \div a^3$ (iii) $(a^5)^3$

(b) (i) Which expression is negative when $a = -1$?

$a^5 \times a^3$ $a^5 \div a^3$ $(a^5)^3$

(ii) Which expression has the greatest value when $a = 0.1$?

$a^5 \times a^3$ $a^5 \div a^3$ $(a^5)^3$ AQA

17 (a) Simplify. $\frac{a^2 + 2a}{5a + 10}$ (b) Factorise. $2x^2 - 18$

18 (a) Expand and simplify.

(i) $3(x + 2) + 5(2x - 1)$ (ii) $(x + 4)(x - 2)$

(b) (i) Factorise the expression $x^2 + 8x + 15$.

(ii) Hence, solve the equation $x^2 + 8x + 15 = 0$. AQA

19 Factorise. (a) $2x^2 - 50$ (b) $2x^2 + 3x - 5$

20 The area, A square metres, of a new logo is proportional to the square of its width, w metres.
The area of the logo is 12 square metres when its width is 4 metres.
(a) Find an equation connecting A and w.
(b) Find the area of a logo with a width of 5 metres.

AQA

21 (a) Factorise $x^2 - 10x + 25$.
(b) Hence, or otherwise, solve the equations $(y - 3)^2 - 10(y - 3) + 25 = 0$.

AQA

22 Solve the simultaneous equations $4x + 3y = 5$ and $2x - 5y = 9$.

AQA

23 A rectangle has length $(x + 2)$ cm and width $(x + 1)$ cm.

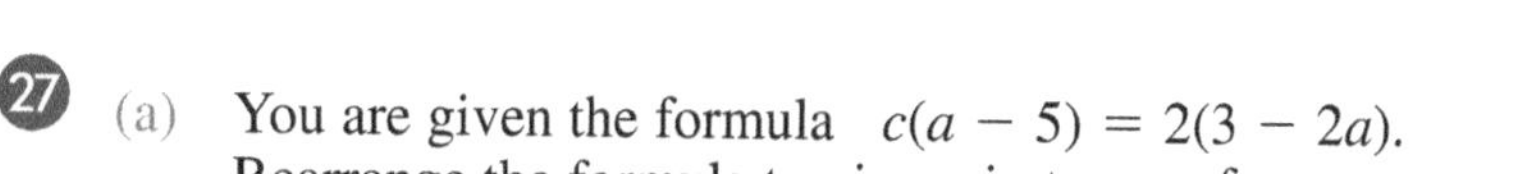

The rectangle has an area of $6\,\text{cm}^2$.
Form an equation and show that it can be simplified to $x^2 + 3x - 4 = 0$.

24 (a) Draw the graph of $y = x^2 + x - 2$ for values of x between -4 and 3.
(b) Use your graph to the find solutions of:
(i) $x^2 + x - 2 = 0$ (ii) $x^2 + x - 2 = 4$
(c) By drawing an appropriate linear graph on the same axes,
find the solutions of $x^2 - x - 6 = 0$, giving your answers to one decimal place.

25 R is inversely proportional to the square of P.
When $R = 36$, $P = 2$.
Find R when $P = 6$.

AQA

26 The diagram shows a square $ABCD$.
(a) Find the gradients of the line segments AC and BD.
(b) Find the equation of the line parallel to AC which passes through D.
(c) Find the equation of the line perpendicular to AC which passes through A.

27 (a) You are given the formula $c(a - 5) = 2(3 - 2a)$.
Rearrange the formula to give a in terms of c.
(b) Simplify the expression $\frac{x}{x - 1} - \frac{1}{x + 1}$.

AQA

28 Solve the simultaneous equations $y + x^2 = 2$ and $x = 2 - y$.

29 (a) Find the values of a and b such that $x^2 + 6x - 3 = (x + a)^2 + b$.
(b) Hence, or otherwise, solve the equation $x^2 + 6x - 3 = 0$,
giving your answers in surd form.

AQA

30 The graph of the function $y = f(x)$ is shown on the grid.
The point $P(-1, 5)$ lies on the curve.
Draw the graph of the transformed function.
In each case write down the coordinates of the transformed point P.
(a) $y = f(x + 3)$
(b) $y = 2f(x)$
(c) $y = -f(x)$

AQA

Algebra Calculator Paper

You may use a calculator for this exercise.

1 n represents any whole number.
(a) What type of whole number is $2n$?
(b) Which of the statements below describes the number $3n + 1$? Explain your answer.
always even **always odd** **could be even or odd** AQA

2 (a) Pencils cost 18p each. How much does Janet pay for x pencils?
(b) Graham goes to a different shop.
At this shop pencils cost n pence each. Rulers cost m pence each.
Use the letters n and m to write down the total cost of 3 pencils and 2 rulers. AQA

3 (a) Solve the equation $5x - 7 = 3x + 5$.
(b) Expand and simplify $2x + 3(x - 4)$. AQA

4 (a) Sarah thinks of a number. She doubles it and then subtracts 3.
(i) What was her number when the answer is 27?
(ii) What was her number when the answer is -1?
(b) Solve the equation $2(2x - 3) = 6$. AQA

5 (a) Write, as simply as possible, an expression for the total length of these rods.
(b) The total length of the rods is 23 cm.
By forming an equation find the value of a.

2a cm 3 cm 3a cm

6 The nth term of a sequence is $3n - 1$.
(a) Write down the first and second terms of the sequence.
(b) Which term in the sequence is equal to 32?
(c) Explain why 85 is not a term in this sequence. AQA

7 Multiply out and simplify. $5(x + 2) - 2(x + 3)$

8 (a) Write down the next term in the sequence: 2, 6, 10, 14, ...
(b) Write an expression, in terms of n, for the nth term of the sequence.

9 Using trial and improvement copy and continue the table to find a solution to the equation $x^3 + x = 20$.

x	$x^3 + x$	Comment
2	10	Too low

Give your answer correct to 1 decimal place. AQA

10 (a) Complete this table of values for the graph $y = x^2 - 7$.

x	-3	-2	-1	0	1	2	3
y	2			-7		-3	2

(b) Draw the graph of $y = x^2 - 7$.
(c) Use your graph to find the values of x for which $x^2 - 7 = 0$.

11 (a) Expand and simplify. $2(4x + 3) - 5x$
(b) Solve the following equations. (i) $\frac{x}{3} = 9$ (ii) $6x + 7 = x + 3$
(c) Simplify. (i) $m^2 \times m^3$ (ii) $\frac{n^6}{n^3}$

12 (a) Write down the integer values of n for which $-4 \leqslant 4n < 8$.
(b) Solve the inequality $3x + 7 \geqslant 22$.

13 (a) Factorise. (i) $8y + 4$ (ii) $x^3 - 5x$
(b) Simplify. (i) $x^2y^3 \times \frac{x^3}{y}$ (ii) $(p^2)^3$
(c) Rearrange the formula $y = 5x + 10$ to make x the subject.
(d) Solve. (i) $\frac{t}{3} = 7 - t$ (ii) $5(x - 3) = 3x + 1$

14 (a) Copy and complete the table for the equation $y = x^2 - 2x + 2$.

x	-1	0	1	2	3
y					

(b) Draw the graph of $y = x^2 - 2x + 2$ for $-1 \leqslant x \leqslant 3$.
(c) Use your graph to find the values of x when $y = 3$.
AQA

15 A sequence begins 7, 5, 3, 1, -1, ...
What is (a) the next term in the sequence, (b) the nth term of the sequence?

16 (a) x is a number which is greater than 1.
List the following four terms in order of size, smallest first.

x^{-2} x $x^{\frac{1}{2}}$ $\frac{1}{x}$

(b) If $0 < x < 1$, how should your list in part (a) be rearranged, if at all?
AQA

17 The nth term of a sequence is $(n + 1)(n + 2)$.
Explain why every term of the sequence is an even number.
AQA

18 (a) On the same diagram, draw and label the lines $y = x$ and $y = 3 - x$ for values of x from 0 to 3.
(b) Explain how you can use your graph to solve the equation $x = 3 - x$.
(c) Shade the region that is satisfied by all of the inequalities
$y \geqslant x$, $y \leqslant 3 - x$ and $x \geqslant 0$.

19 (a) Solve the equation $3(x + 2) = 4 - x$.
(b) Factorise $m^2 - 7m$.
(c) List the values of n, where n is an integer, such that $1 \leqslant n - 5 < 4$.
(d) Solve the simultaneous equations $3x + 4y = 0$
$4x - 2y = -11$.
AQA

20 The diagram shows a sketch of the line $2y + x = 10$.
(a) Find the coordinates of points G and H.

Copy the diagram.
(b) (i) On your diagram, sketch the graph of $y = 2x$.
(ii) Solve the simultaneous equations
$2y + x = 10$ and $y = 2x$.

AQA

21 Solve. (a) $9(x - 1) = 5(x - 2)$ (b) $\frac{x + 1}{2} + \frac{x - 3}{4} = 2$
AQA

22 Simplify fully. (a) $(2t^3u) \times (3tu^2)$ (b) $(2c^4)^3$
AQA

23 $V = \frac{4}{3}\pi r^3$.
(a) Rearrange the formula to give r in terms of V.
(b) Calculate the value of r when $V = 905$.

24 Solve the equation $y^2 + 5y = 0$.
AQA

25 (a) Which of the following equations are illustrated by the graphs shown?

$xy = 1$ $\quad y = 2 - x$ $\quad x^2 + y^2 = 3$ $\quad y = x^3$ $\quad y = 1 - x^2$ $\quad 2y = 2 + x$

(i) (ii) (iii) (iv)

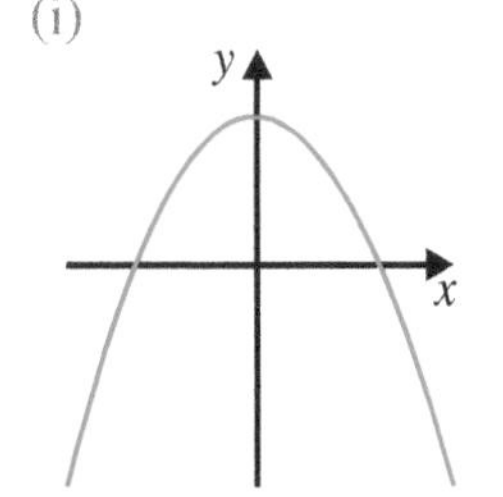

(b) Draw a graph of $y = 2^x$ for $-3 \leqslant x \leqslant 3$.

26 (a) The expression $x^2 - 6x + 7$ can be written in the form $(x + a)^2 + b$, where a and b are constants.
Determine the values of a and b and hence, find the minimum value of the expression.

(b) The solutions of the equation $x^2 - 6x + 7 = 0$ can be written in the form $x = p \pm q\sqrt{2}$, where p and q are integers.
Determine the values of p and q. AQA

27 (a) Express as a fraction in its simplest form $\frac{3}{x} - \frac{5}{2x}$

(b) Simplify this expression $\frac{y^2 - 2y - 15}{2y^2 - 11y + 5}$ AQA

28 The diagram shows the speed-time graph of an underground train between two stations.

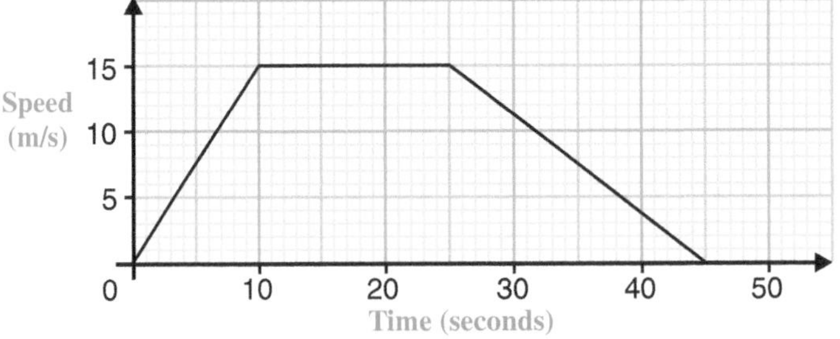

(a) Calculate the acceleration of the train between 0 and 10 seconds.

(b) Calculate the time it takes to travel half the distance between the stations.

29 A rectangular lawn has a path on 3 sides as shown.
The lawn has dimensions x metres by $3x$ metres.
The path is 1 metre wide.

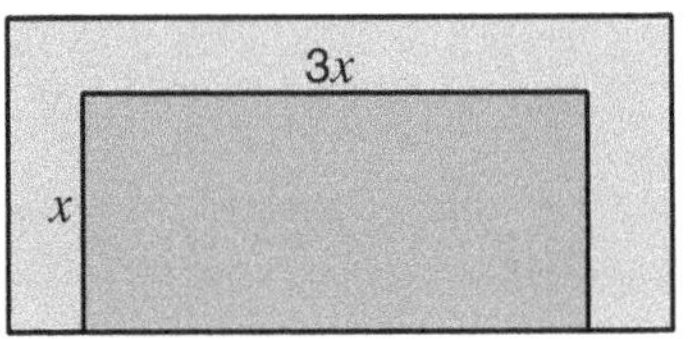

(a) The total area of the lawn and the path is $80\,\text{m}^2$.
Show that $3x^2 + 5x - 78 = 0$.

(b) By solving the equation $3x^2 + 5x - 78 = 0$ find the area of the lawn. AQA

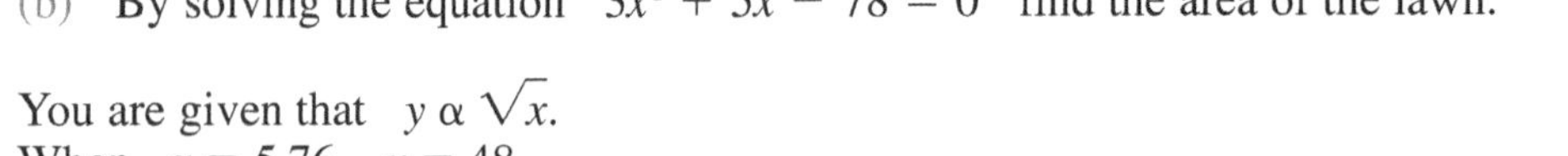

30 You are given that $y \propto \sqrt{x}$.
When $x = 5.76$, $y = 48$

(a) Find a formula for y in terms of x.

(b) Calculate x when $y = 85$. AQA

31 Find the solutions to the quadratic equation $x^2 + 4x - 6 = 0$.
Give your answers correct to 2 decimal places. AQA

32 Solve the equation $\frac{4}{2x - 1} - \frac{1}{x + 1} = 1$. AQA

33 a, b and x are connected by the expression $\frac{x}{a - x} = \frac{b - x}{x}$.

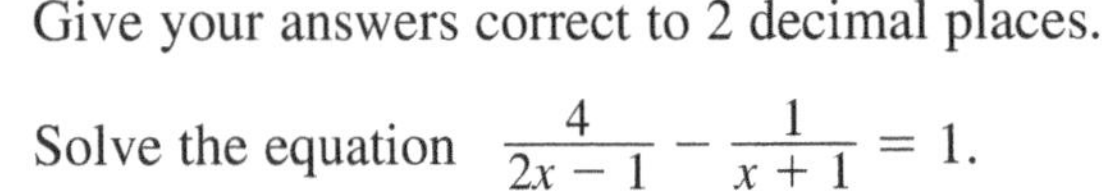

Rearrange this expression to give a formula for x in terms of a and b. AQA

Chapter 23 Angles and 2-D Shapes

An **angle** is a measure of turn.
Angles are measured in **degrees**.

Types and names of angles

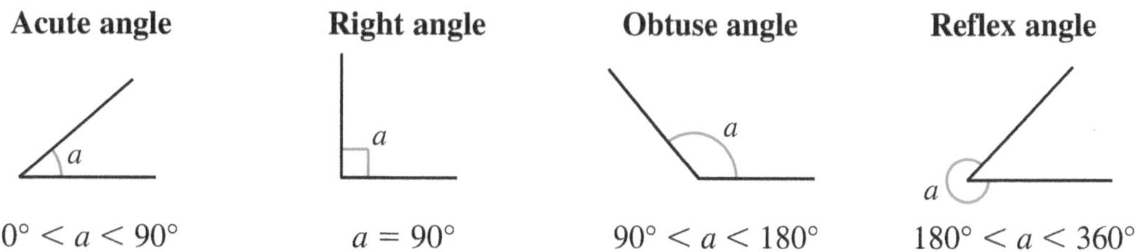

$0° < a < 90°$ | $a = 90°$ | $90° < a < 180°$ | $180° < a < 360°$

Angle properties

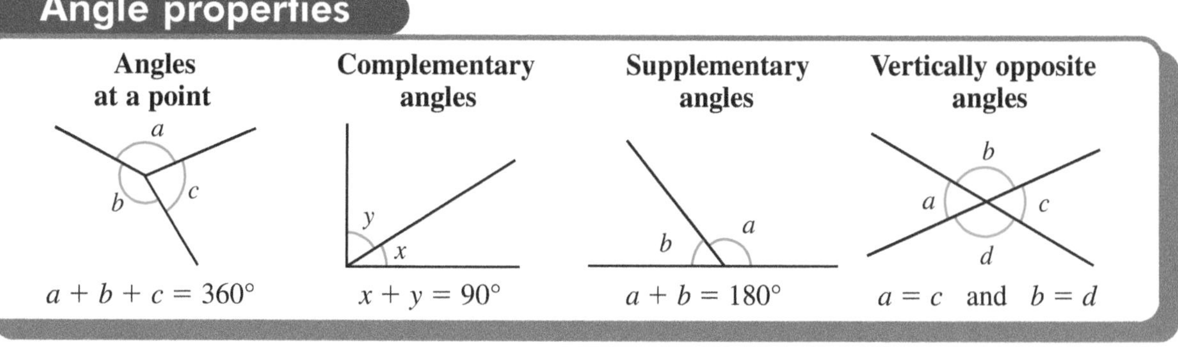

$a + b + c = 360°$ | $x + y = 90°$ | $a + b = 180°$ | $a = c$ and $b = d$

Lines

A straight line joining two points is called a **line segment**.

Perpendicular lines

Lines which meet at right angles are **perpendicular** to each other.

Parallel lines

Parallel lines are lines which never meet.

Arrowheads are used to show that lines are parallel.

Parallel lines and angles

Each diagram shows two parallel lines crossed by another straight line called a **transversal**.

Corresponding angles

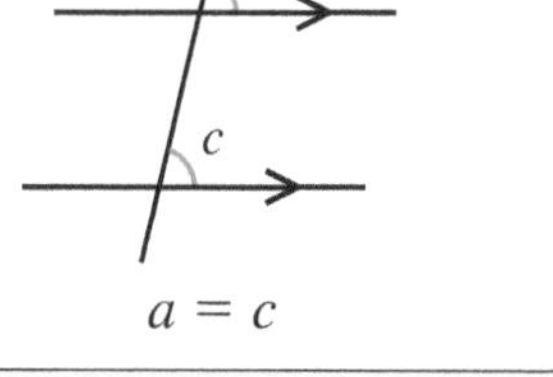

$a = c$

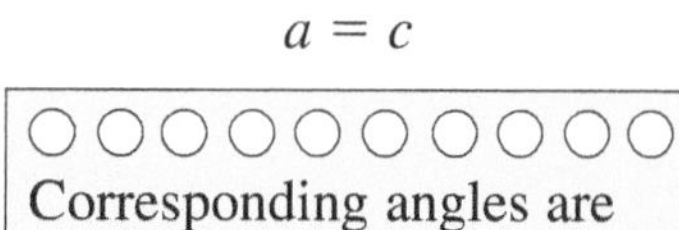

Corresponding angles are always on the same side of the transversal.

Alternate angles

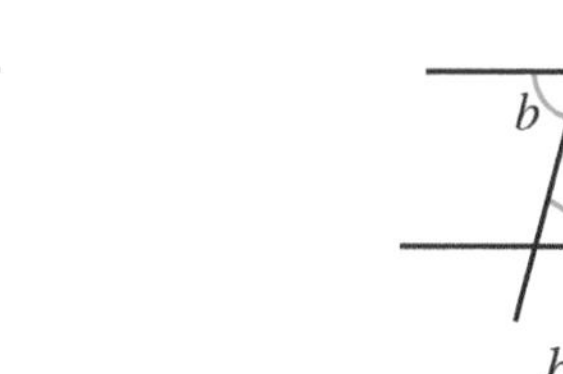
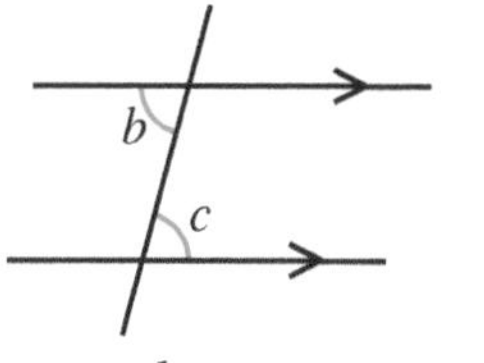
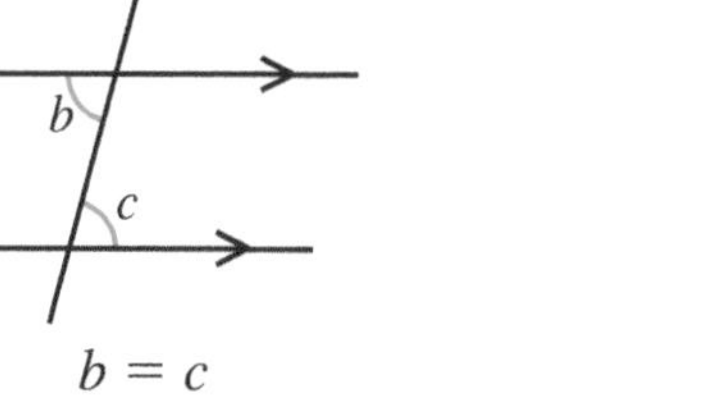

$b = c$

Alternate angles are always on opposite sides of the transversal.

Allied angles

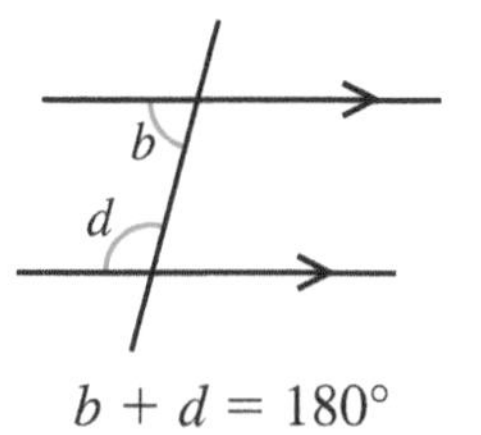

$b + d = 180°$

Allied angles are always between parallels on the same side of the transversal.

Naming angles

Angles can be identified by using small letters to mark the angle or by using three capital letters to show the sides that form the angle.

For example.

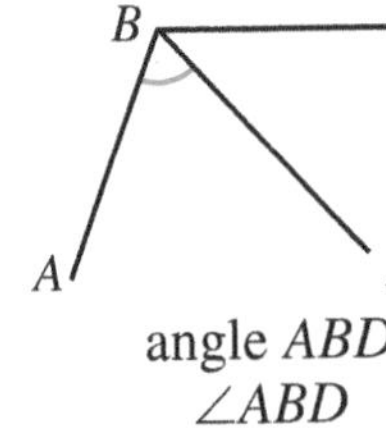
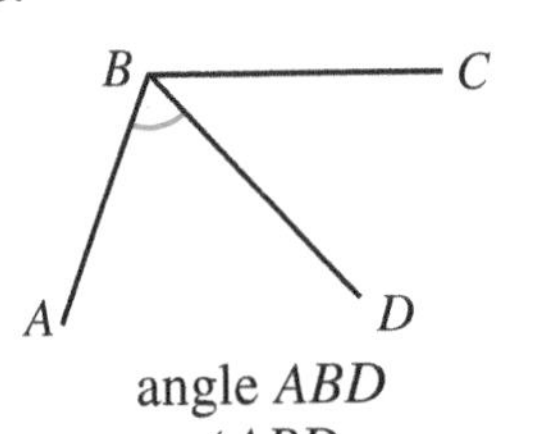
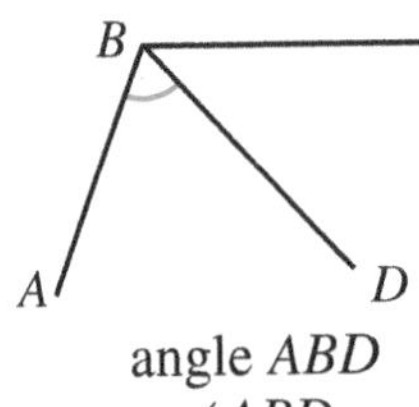

angle a

angle ABD
$\angle ABD$

> $\angle$ means 'angle'.
> $\angle CBA$ is the same as $\angle ABC$.
> Notice that the middle letter is where the angle is made.

EXAMPLES

Without measuring, work out the size of the angles marked with letters.

1

Angles at a point add up to 360°.

$a + 95° + 120° = 360°$

$a = 360° - 95° - 120°$

$a = 145°$

2

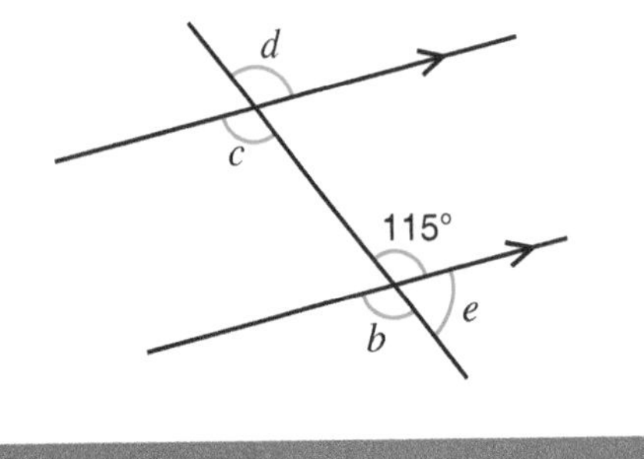

$b = 115°$ (vertically opposite angles)

$c = 115°$ (alternate angles)

$d = 115°$ (corresponding angles)

$e + 115° = 180°$ (supplementary angles)

$e = 180° - 115°$

$e = 65°$

$b = 115°, \quad c = 115°, \quad d = 115°, \quad e = 65°$

Exercise 23.1

The diagrams in this exercise are not drawn accurately.

1 Work out the size of the angles marked with letters. Give a reason for each answer.

(a)

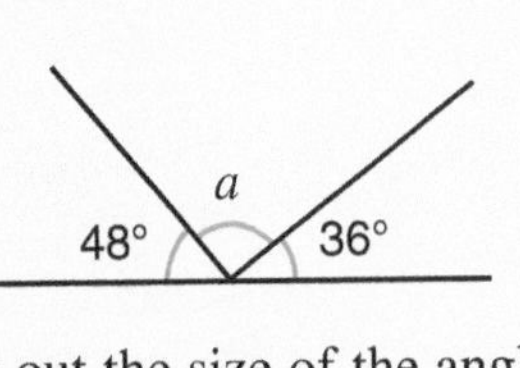

(b)

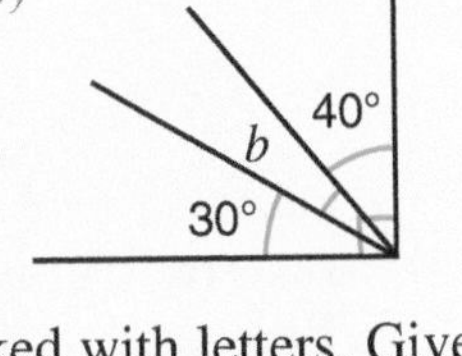

(c)

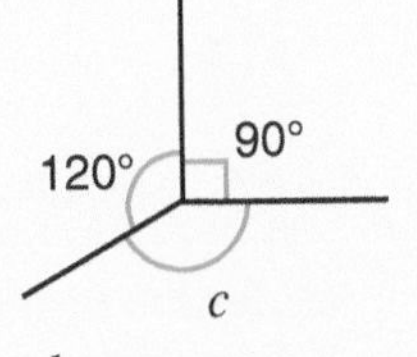

2 Work out the size of the angles marked with letters. Give a reason for each answer.

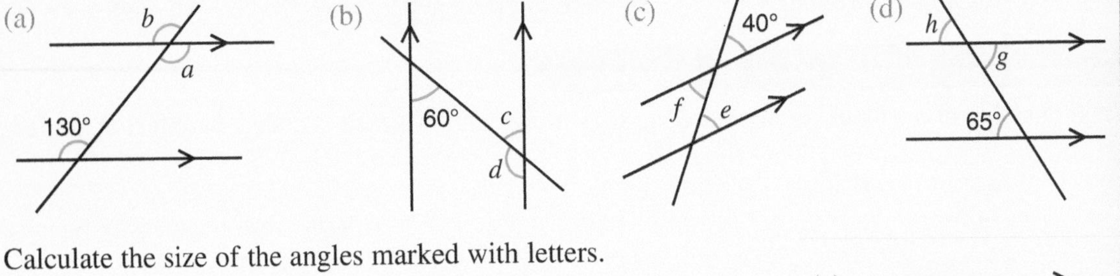

3 Calculate the size of the angles marked with letters.

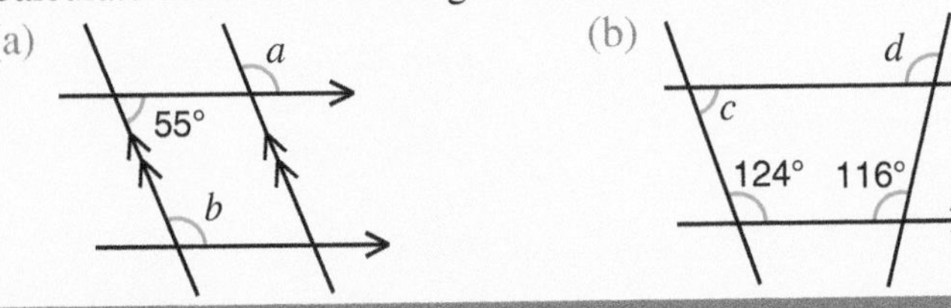

(c) 28°, e, 33°

4 Work out the size of the required angles.

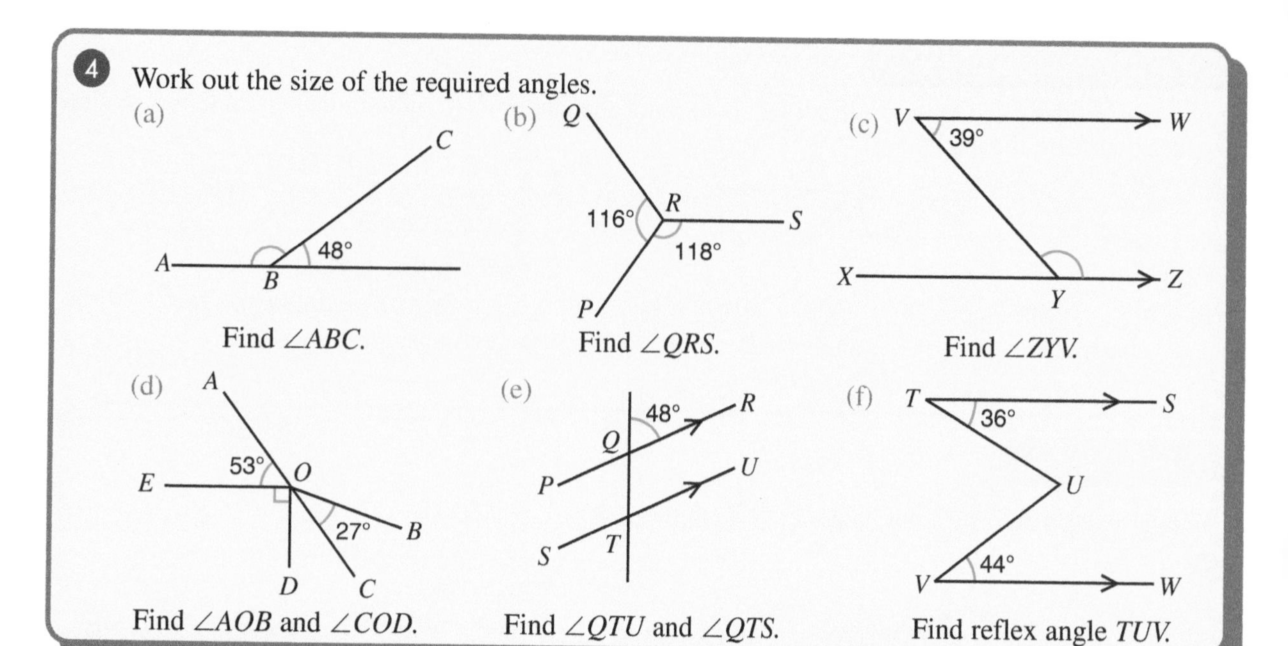

Triangles

A **triangle** is a shape made by three straight lines.
The sum of the three angles in a triangle is 180°.

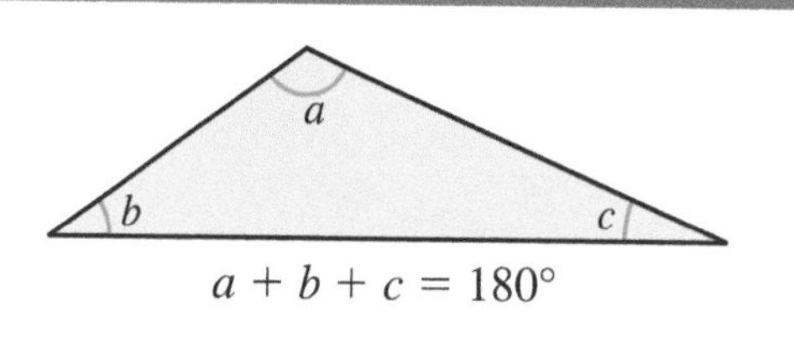

$a + b + c = 180°$

Types of triangle

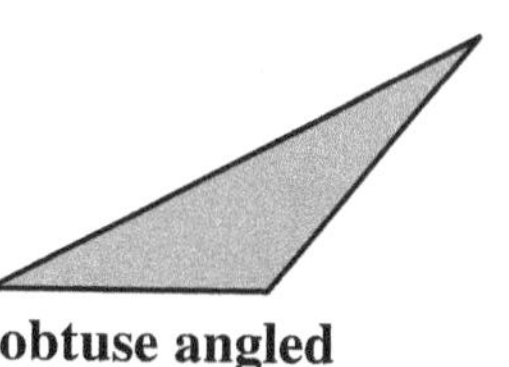

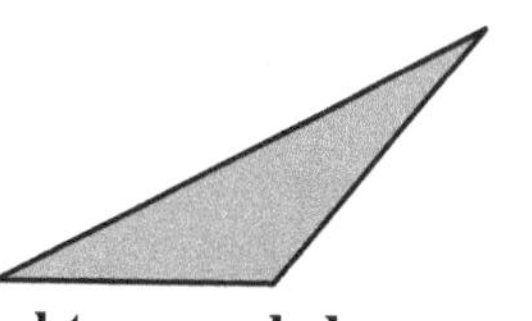

acute angled
Triangles with three acute angles.

obtuse angled
Triangles with an obtuse angle.

right angled
Triangles with a right angle.

Special triangles

scalene triangle
Sides have different lengths.
Angles are all different.

isosceles triangle
Two equal sides.
Two angles equal.

equilateral triangle
Three equal sides.
All angles are 60°.

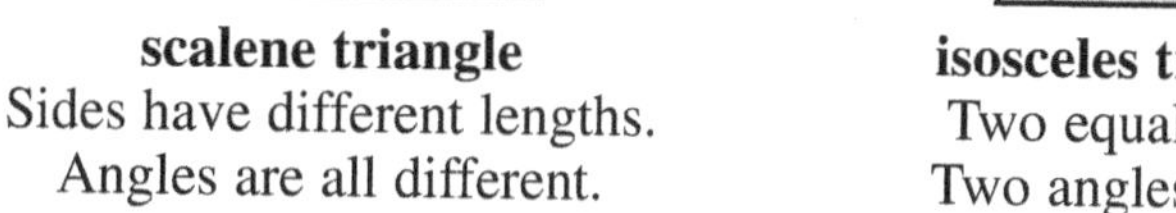

Exterior angle of a triangle

When one side of a triangle is extended, as shown, the angle formed is called an **exterior angle**.

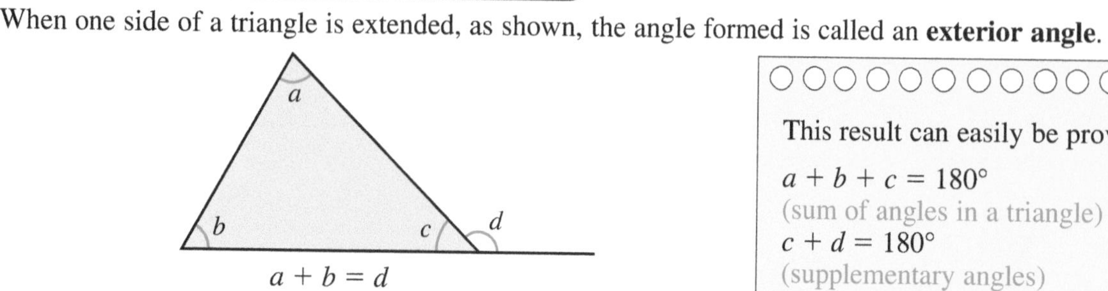

$a + b = d$

In any triangle the exterior angle is always equal to the sum of the two opposite interior angles.
Check this by measuring the angles a, b and d in the diagram.

This result can easily be proved.
$a + b + c = 180°$
(sum of angles in a triangle)
$c + d = 180°$
(supplementary angles)
$a + b + c = c + d$
$a + b = d$

EXAMPLE

Find the sizes of the angles marked a and b.

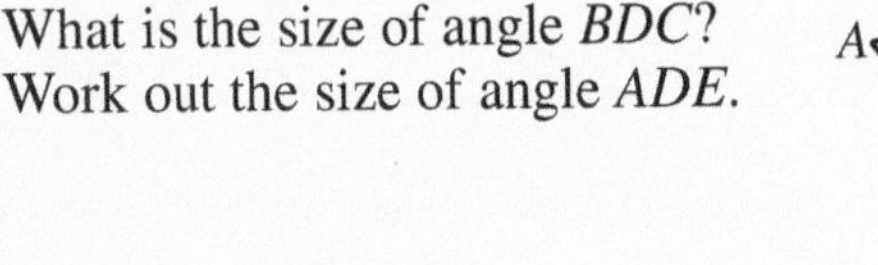

$a = 84° + 43°$ (ext. ∠ of a Δ)
$a = 127°$

$b + 127° = 180°$ (supp. ∠'s)
$b = 180° - 127°$
$b = 53°$

Short but accurate:
In Geometry we often abbreviate words and use symbols to provide the reader with full details using the minimum amount of writing.

Δ is short for triangle.
ext. ∠ of a Δ means exterior angle of a triangle.
supp. ∠'s means supplementary angles.

Exercise 23.2

The diagrams in this exercise have not been drawn accurately.

1 Is it possible to draw triangles with the following types of angles?
Give a reason for each of your answers.
(a) three acute angles,
(b) one obtuse angle and two acute angles,
(c) two obtuse angles and one acute angle,
(d) three obtuse angles,
(e) one right angle and two acute angles,
(f) two right angles and one acute angle.

2 Is it possible to draw a triangle with these angles?
If a triangle can be drawn, what type of triangle is it?
Give a reason for each of your answers.
(a) 95°, 78°, 7° (b) 48°, 62°, 90° (c) 48°, 62°, 70°
(d) 90°, 38°, 52° (e) 130°, 35°, 15°

3 Work out the size of the marked angles.

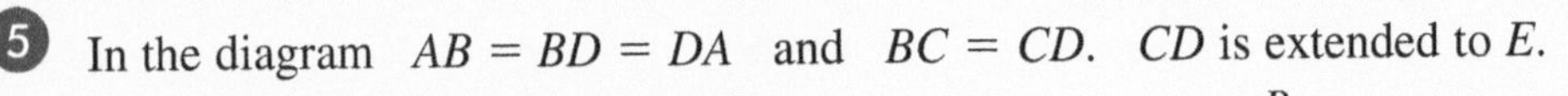

4 Work out the size of the angles marked with letters.

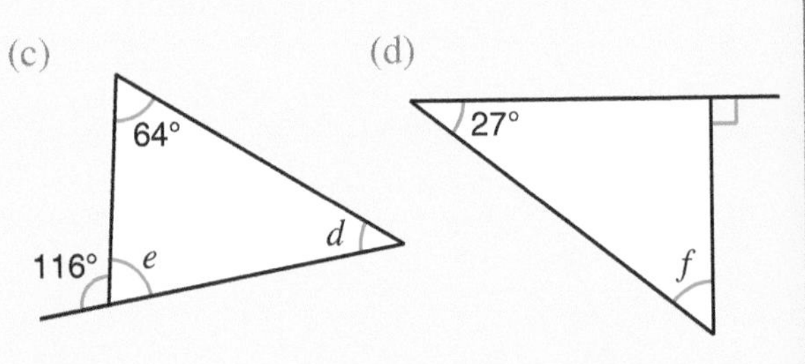

5 In the diagram $AB = BD = DA$ and $BC = CD$. CD is extended to E.
(a) What type of triangle is BCD?
(b) What is the size of angle BDC?
(c) Work out the size of angle ADE.

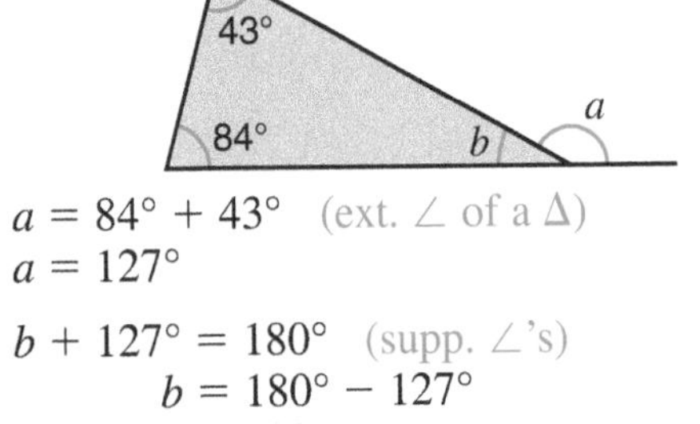

6 Work out the size of the required angles.

(a)

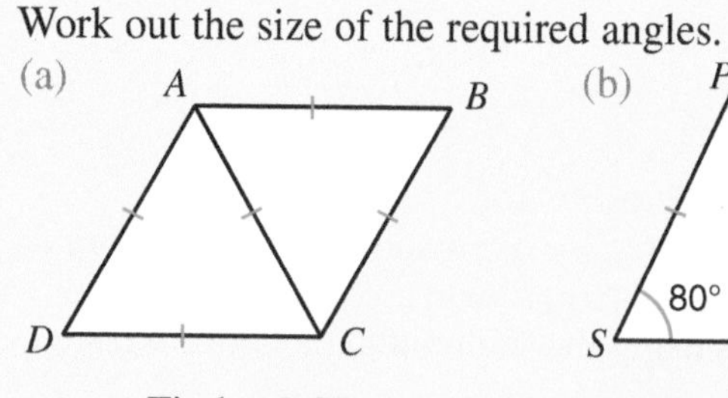

Find $\angle BCD$.

(b)

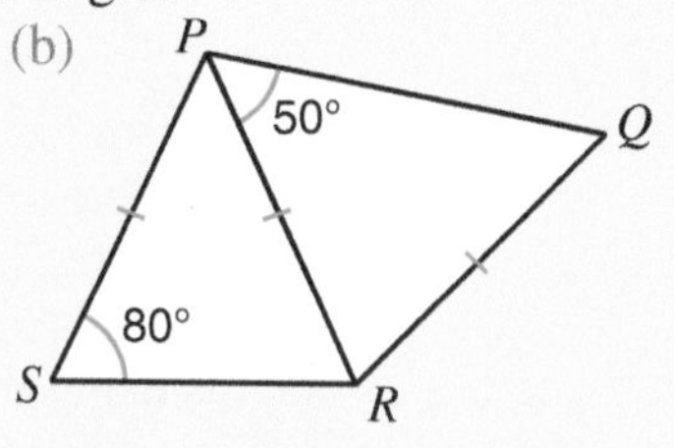

Find $\angle PRQ$ and $\angle QRS$.

(c)

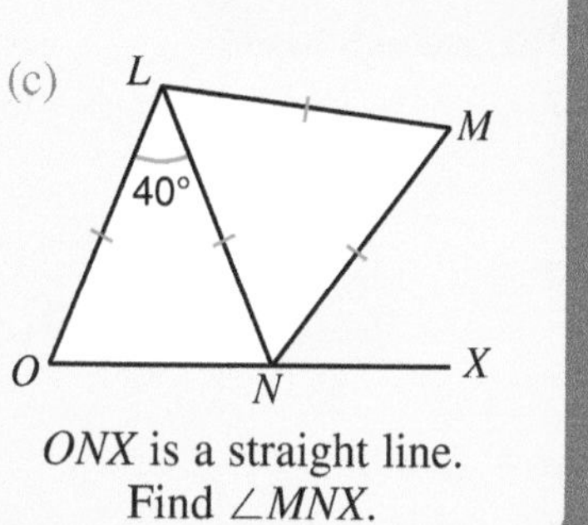

ONX is a straight line.
Find $\angle MNX$.

Quadrilaterals

A **quadrilateral** is a shape made by four straight lines.
The sum of the four angles of a quadrilateral is 360°.

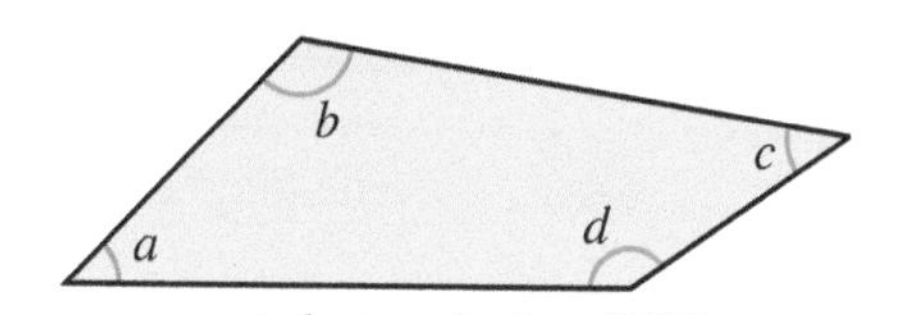

$a + b + c + d = 360°$

Special quadrilaterals

Square

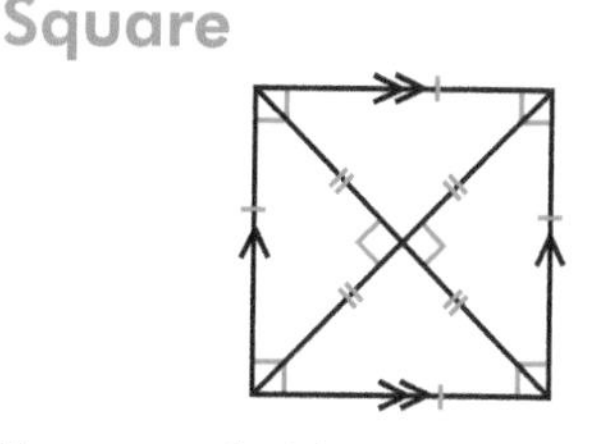

Four equal sides.
Opposite sides parallel.
Angles of 90°.
Diagonals bisect each other at 90°.

Rectangle

Opposite sides equal and parallel.
Angles of 90°.
Diagonals bisect each other.

Parallelogram

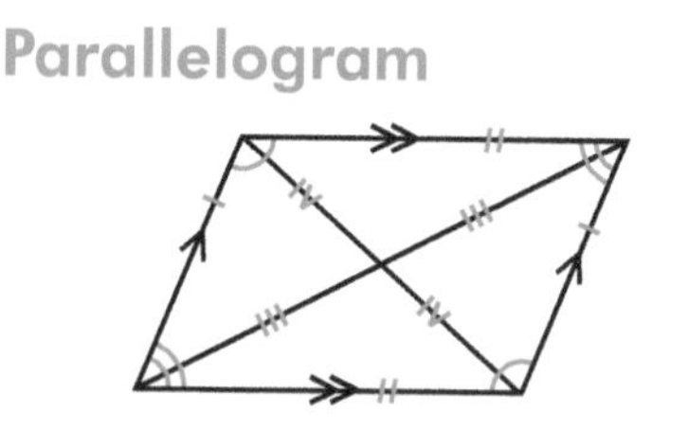

Opposite sides equal and parallel.
Opposite angles equal.
Diagonals bisect each other.

Rhombus

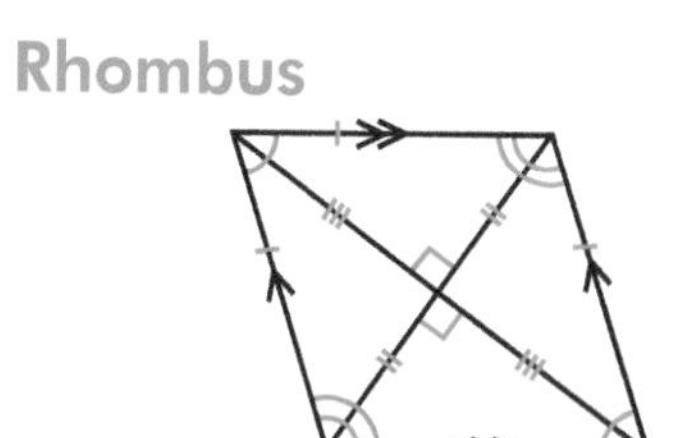

Four equal sides.
Opposite sides parallel.
Opposite angles equal.
Diagonals bisect each other at 90°.

Trapezium

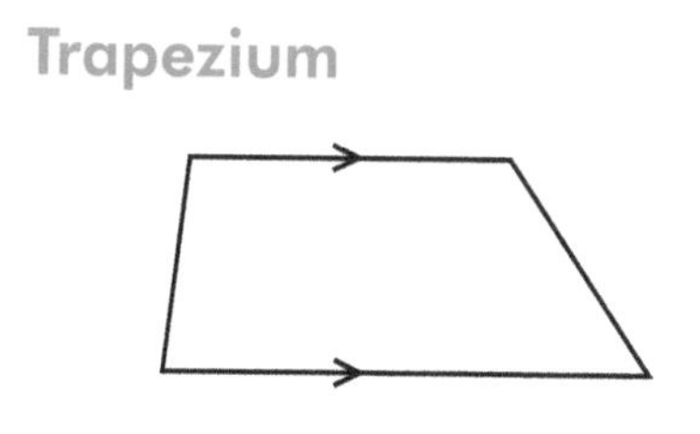

One pair of parallel sides.

Kite

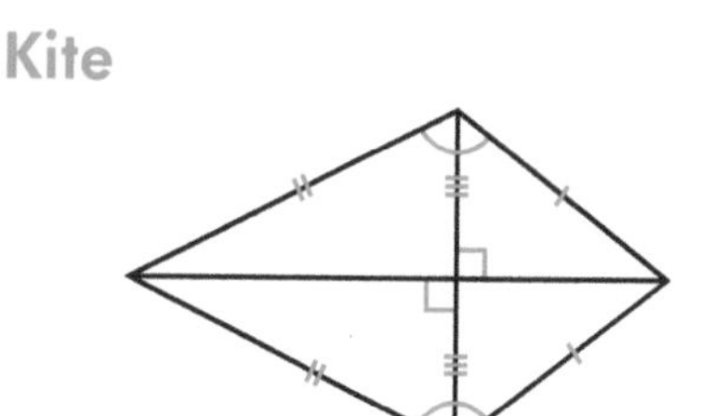

Two pairs of adjacent sides equal.
One pair of opposite angles equal.
One diagonal bisects the other at 90°.

EXAMPLE

PQRS is a parallelogram. Work out the size of the angle marked x.

The opposite angles of a parallelogram are equal.

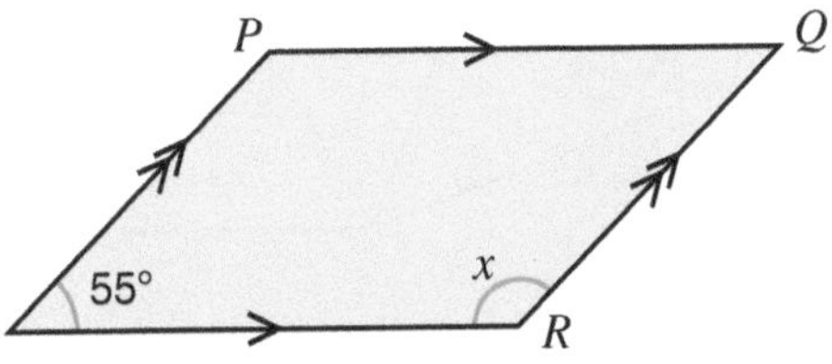

$$55° + 55° + x + x = 360°$$
$$110° + 2x = 360°$$
$$2x = 360° - 110°$$
$$2x = 250°$$
$$x = 125°$$

Exercise 23.3

Use squared paper to answer questions 1 to 4.

1. *PQRS* is a rectangle. *P* is the point (1, 3), *Q* (4, 6), *R* (6, 4).
   Find the coordinates of *S*.
2. *WXYZ* is a parallelogram. *W* is the point (1, 0), *X* (4, 1), *Z* (3, 3).
   Find the coordinates of *Y*.
3. *OABC* is a kite. *O* is the point (0, 0), *B* (5, 5), *C* (3, 1).
   Find the coordinates of *A*.
4. *STUV* is a square with *S* at (1, 3) and *U* at (5, 3).
   Find the coordinates of *T* and *V*.
5. The following diagrams have not been drawn accurately.
   Work out the size of the angles marked with letters.

(a) (b) 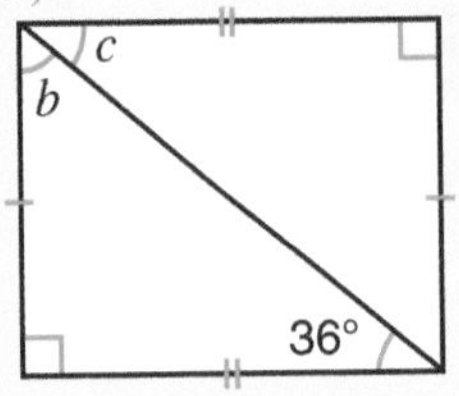 (c) 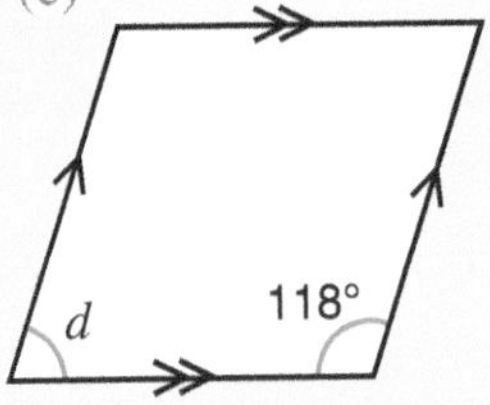 (d) 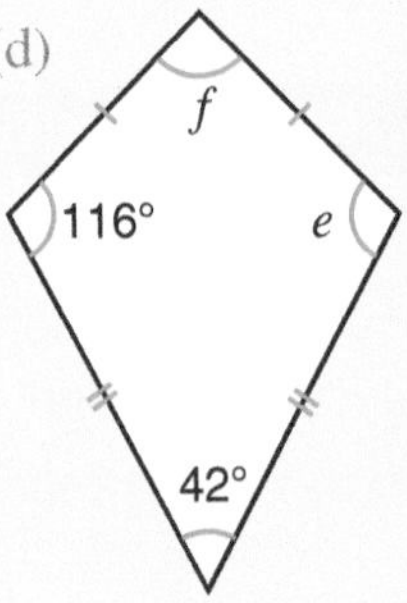 (e) 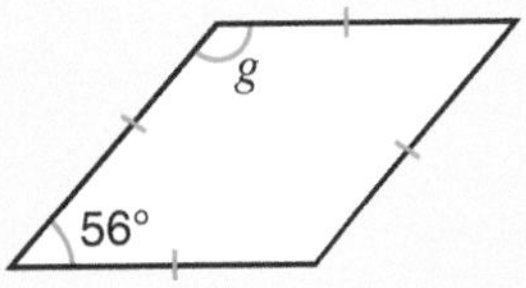 (f) 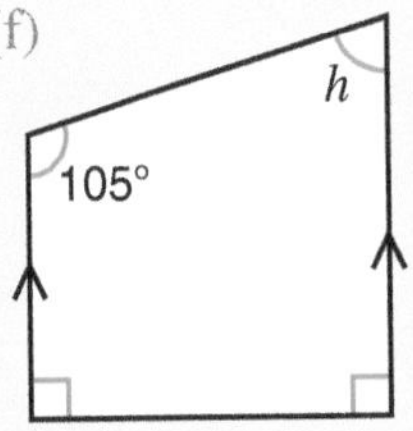 (g) 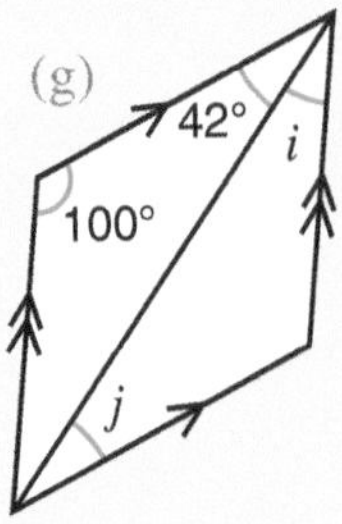 (h) 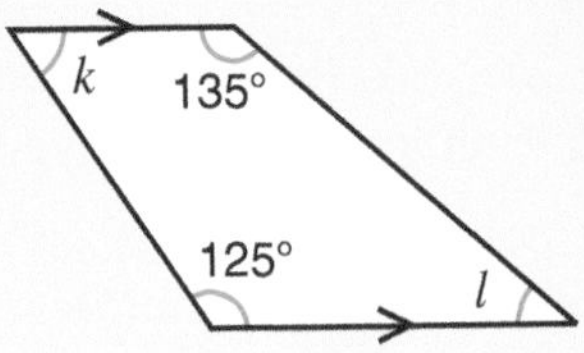 (i) 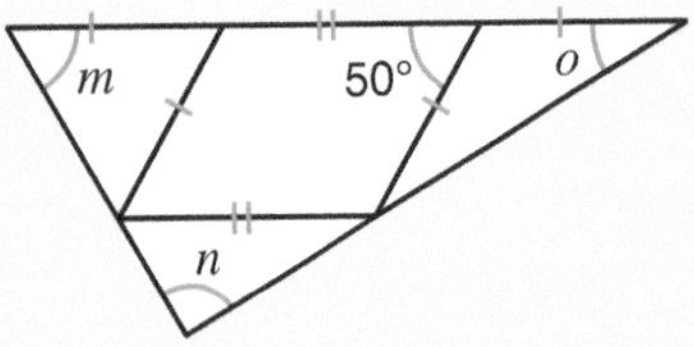 (j) 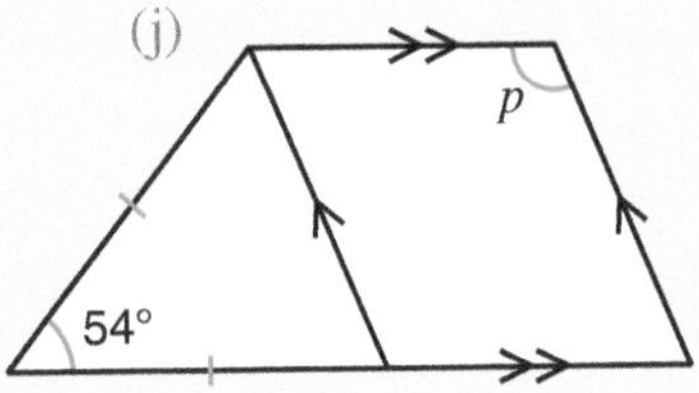 (k) 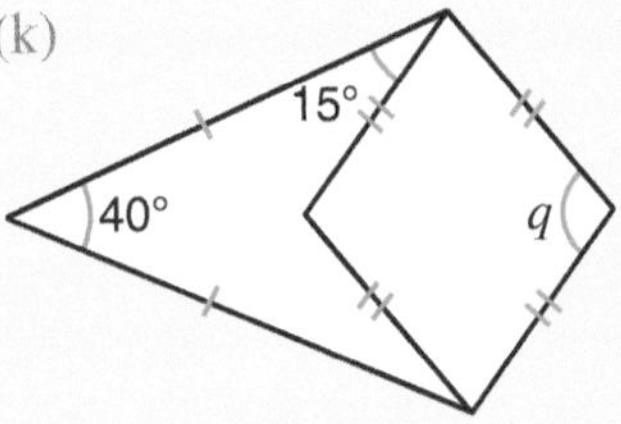 (l) 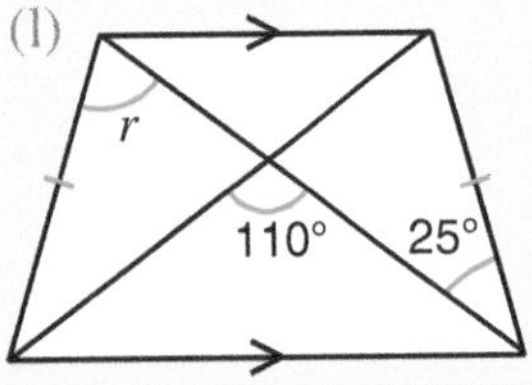 (m) 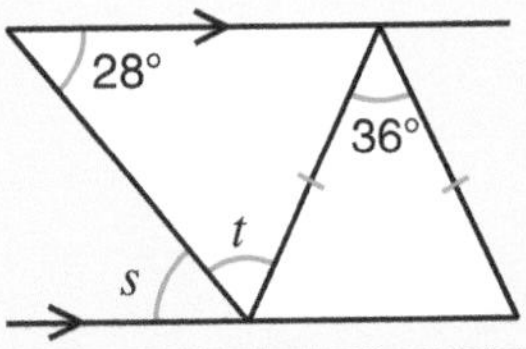 (n)

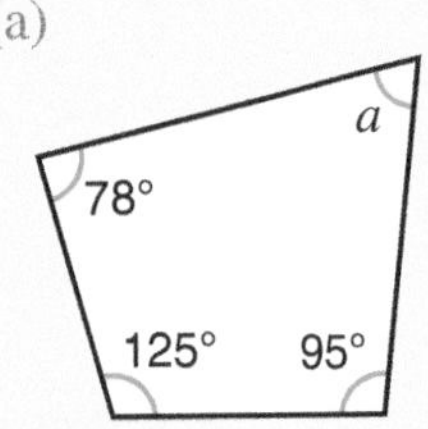

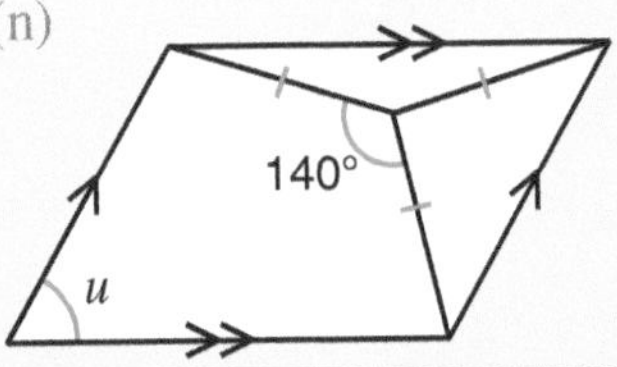

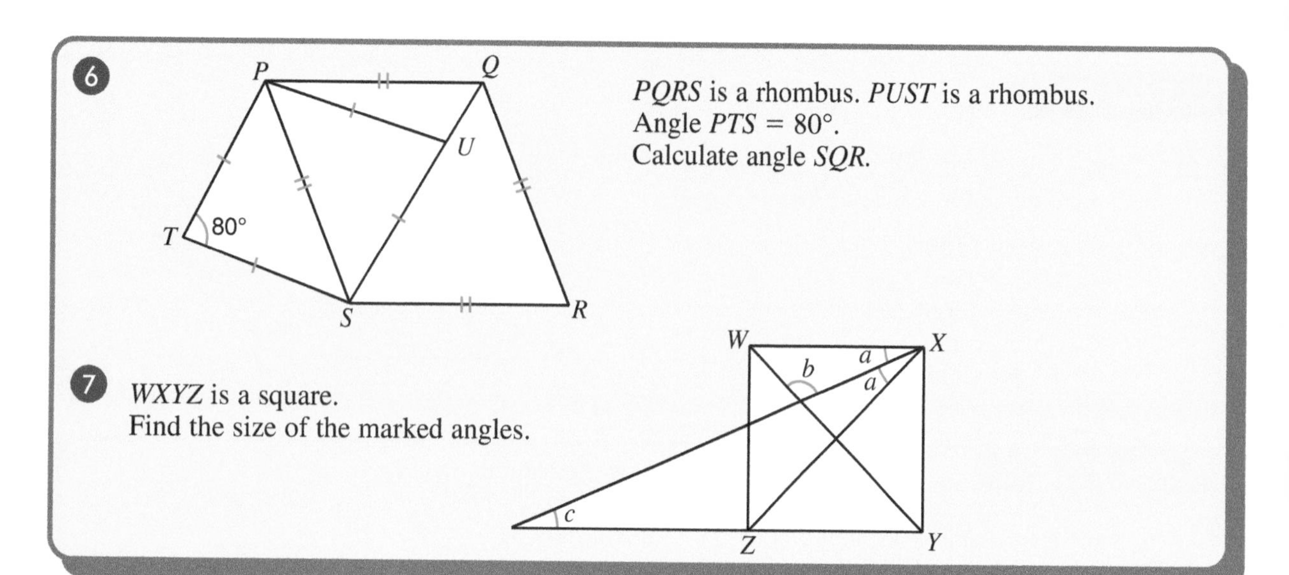

6 $PQRS$ is a rhombus. $PUST$ is a rhombus.
Angle $PTS = 80°$.
Calculate angle SQR.

7 $WXYZ$ is a square.
Find the size of the marked angles.

Polygons

A **polygon** is a shape made by straight lines.
A three-sided polygon is a **triangle**.
A four-sided polygon is called a **quadrilateral**.

A polygon is a many-sided shape.
Look at these polygons.

Pentagon 5 sides | **Hexagon** 6 sides | **Octagon** 8 sides

Interior and exterior angles of a polygon

Angles formed by sides inside a polygon are called **interior angles**.

When a side of a polygon is extended, as shown, the angle formed is called an **exterior angle**.

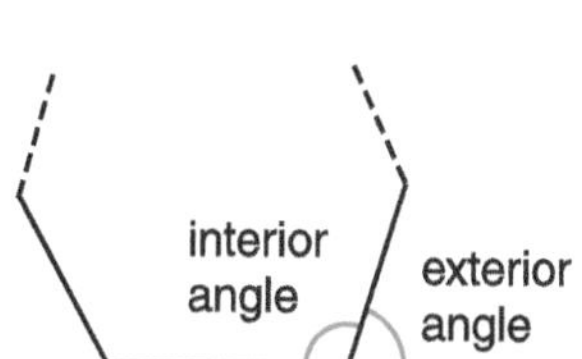

At each vertex of the polygon:
interior angle + exterior angle = 180°

Sum of the interior angles of a polygon

The diagram shows a polygon with the diagonals from one vertex drawn.

The polygon has 5 sides and is divided into 3 triangles.
So, the sum of the interior angles is $3 \times 180° = 540°$.

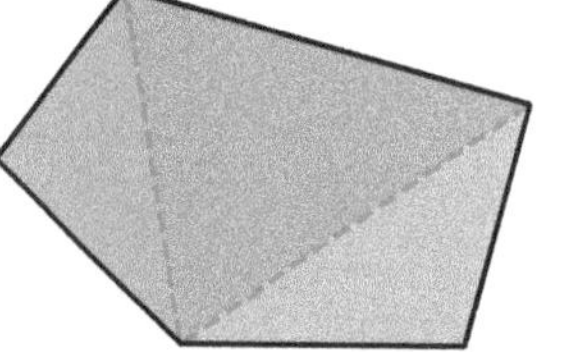

In general, for any n-sided polygon, the sum of the interior angles is $(n - 2) \times 180°$.

Sum of the exterior angles of a polygon

The sum of the exterior angles of **any** polygon is 360°.

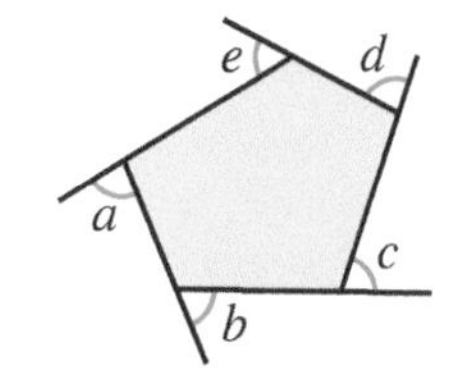

$$a + b + c + d + e = 360°$$

EXAMPLES

1 Find the size of angle x.

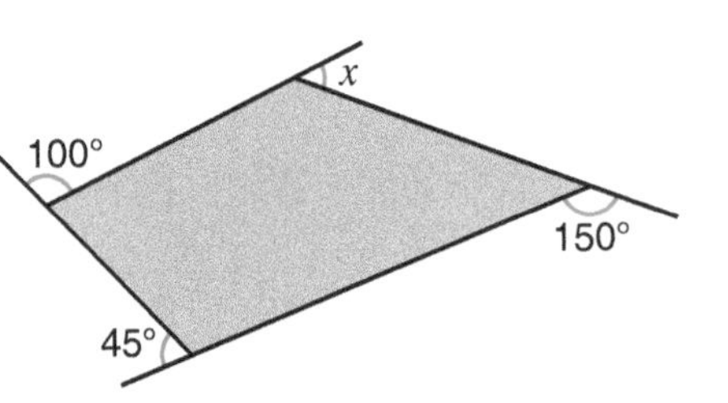

The sum of the exterior angles is 360°.
$x + 100° + 45° + 150° = 360°$
$x + 295° = 360°$
$x = 360° - 295°$
$x = 65°$

2 Find the sum of the interior angles of a pentagon.

To find the sum of the interior angles of a pentagon substitute $n = 5$ into $(n - 2) \times 180°$.
$(5 - 2) \times 180°$
$= 3 \times 180°$
$= 540°$

3 Find the size of the angles marked a and b.

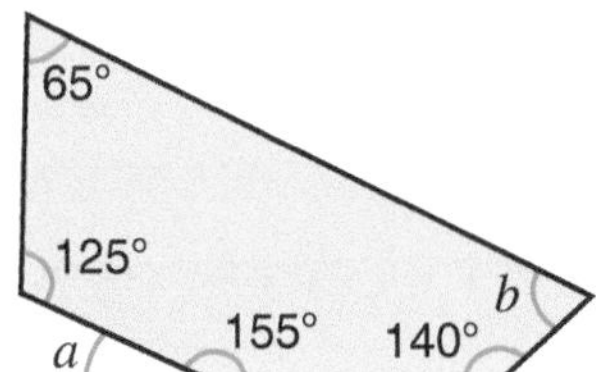

$155° + a = 180°$
(int. angle + ext. angle = 180°)
$a = 180° - 155°$
$a = 25°$

The sum of the interior angles of a pentagon is 540°.
$b + 140° + 155° + 125° + 65° = 540°$
$b + 485° = 540°$
$b = 540° - 485°$
$b = 55°$

Exercise 23.4

The diagrams in this exercise have not been drawn accurately.

1 In the diagram, ABC is a straight line.
(a) Explain why angle $x = 50°$.
(b) Show that angle $y = 60°$.

2 Work out the size of the angles marked with letters.
(a) (b) (c)

3 Work out the sum of the interior angles of these polygons.
(a) (b) (c)

4 Work out the size of the angles marked with letters.

(a) (b) (c)

5 The figure is made up of straight lines.
Find the sum of the angles a, b, c, d, e, f, g, h.

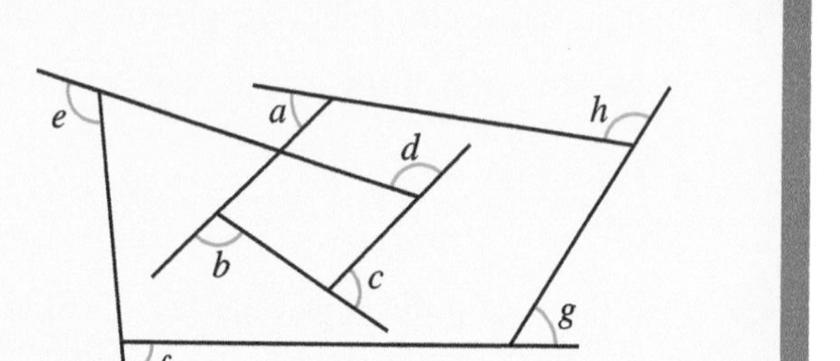

Regular polygons

A polygon with all sides equal and all angles equal is called a **regular polygon**.

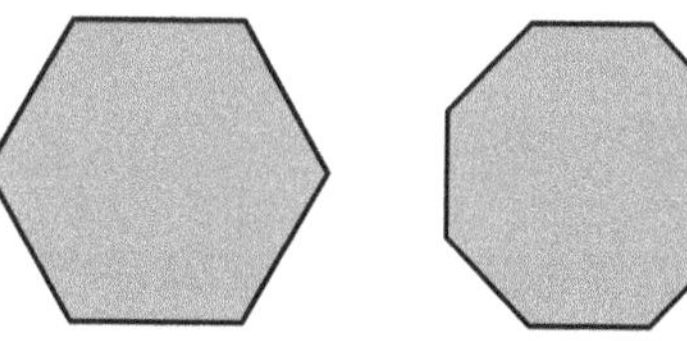

Regular hexagon **Regular octagon**

A regular triangle is usually called an **equilateral triangle**.
A regular quadrilateral is usually called a **square**.

Exterior angles of regular polygons

In general, for any regular n-sided polygon: exterior angle $= \frac{360°}{n}$

By rearranging the formula we can find the number of sides, n, of a regular polygon when we know the exterior angle.

$$n = \frac{360°}{\text{exterior angle}}$$

EXAMPLE

A regular polygon has an exterior angle of 30°.
(a) How many sides has the polygon?
(b) What is the size of an interior angle of the polygon?

(a) $n = \frac{360°}{\text{exterior angle}}$

$n = \frac{360°}{30°} = 12$

The polygon has 12 sides.

(b) interior angle + exterior angle = 180°
int. $\angle$ + 30° = 180°
int. $\angle$ = 180° − 30°
interior angle = 150°

Remember:
It is a good idea to write down the formula you are using.

Tessellations

Covering a surface with identical shapes produces a pattern called a **tessellation**.

To tessellate the shape must not overlap and there must be no gaps.

Regular tessellations

This pattern shows a tessellation of regular hexagons.

This pattern is called a **regular tessellation** because it is made by using a single regular polygon.

Exercise 23.5

1. Calculate (a) the exterior angle and (b) the interior angle of these regular polygons.

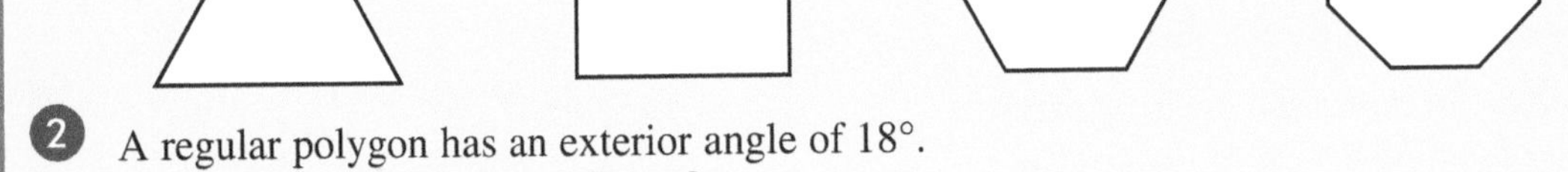

2. A regular polygon has an exterior angle of 18°.
How many sides has the polygon?

3. A regular polygon has an interior angle of 135°.
How many sides has the polygon?

4. (a) Calculate the size of an exterior angle of a regular pentagon.
(b) What is the size of an interior angle of a regular pentagon?
(c) What is the sum of the interior angles of a pentagon?

5. The following diagrams are drawn using regular polygons.
Work out the values of the marked angles.

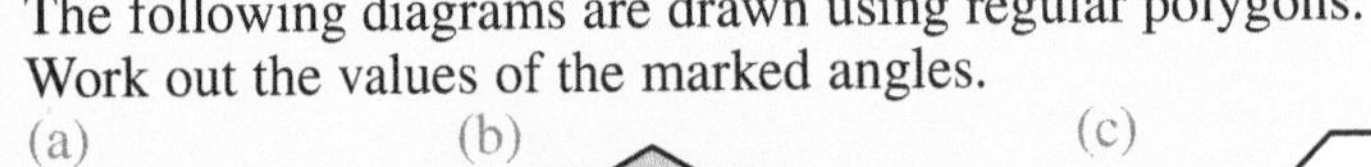

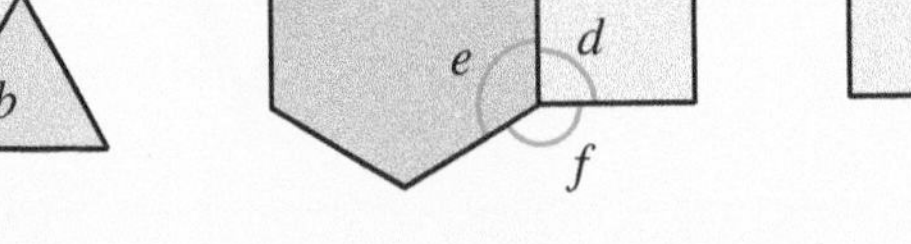

6. AB and BC are two sides of a regular octagon.
PAB is an isosceles triangle with $AP = PB$.
Angle $APB = 36°$.
Calculate angle PBC.

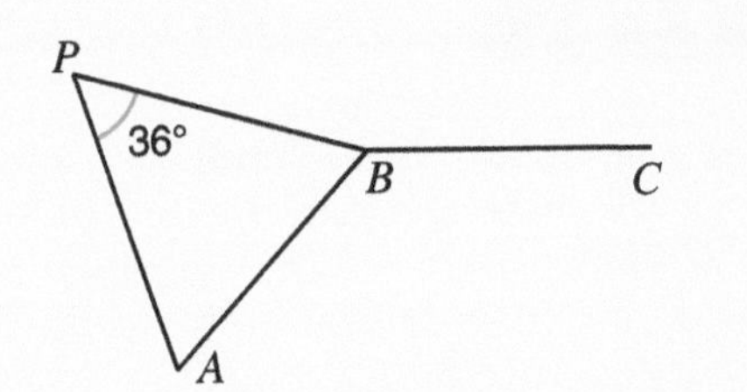

7. A regular polygon has 12 sides.
Calculate the size of an interior angle.

8. Explain why regular pentagons will not tessellate.

9. Any triangle can be used to make a tessellation.
Draw a triangle of your own, make copies, and show that it will tessellate.

Symmetry

Lines of symmetry

These shapes are **symmetrical**.

When each shape is folded along the dashed line one side will fit exactly over the other side.
The dashed line is called a **line of symmetry**.

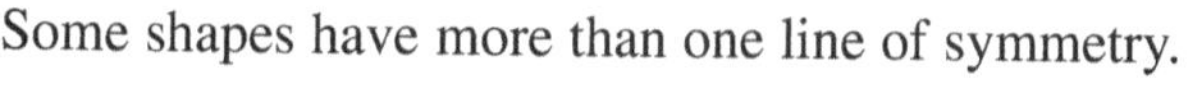

Some shapes have more than one line of symmetry.

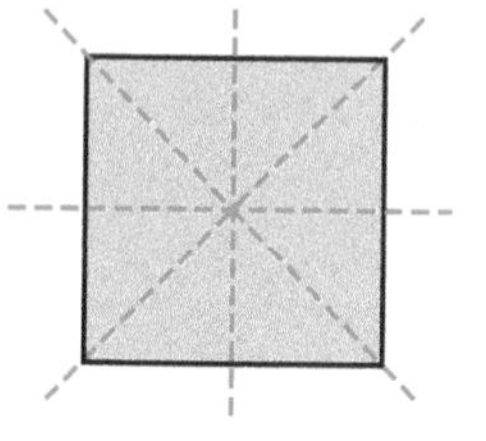

Square
4 lines of symmetry.

Circle
Infinite number of lines of symmetry.

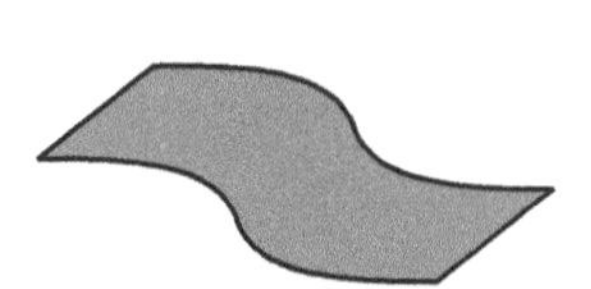

Shape with no lines of symmetry.

Rotational symmetry

Is this shape symmetrical?

The shape does not have line symmetry.

Try placing a copy of the shape over the original and rotating it about the centre of the circle.

After 180° (a half-turn) the shape fits into its own outline.
The shape has **rotational symmetry**.
The point about which the shape is rotated is called the **centre of rotation**.
The **order of rotational symmetry** is 2.
When rotating the shape through 360° it fits into its own outline twice (once after a half-turn and again after a full-turn).
A shape is only described as having rotational symmetry if the order of rotational symmetry is 2 or more.

A shape can have both line symmetry and rotational symmetry.

For example, the shape on the right has:
order of rotational symmetry 4, and 4 lines of symmetry.

Exercise 23.6

1. These shapes have **line symmetry**.
   Copy each shape and draw the line of symmetry.

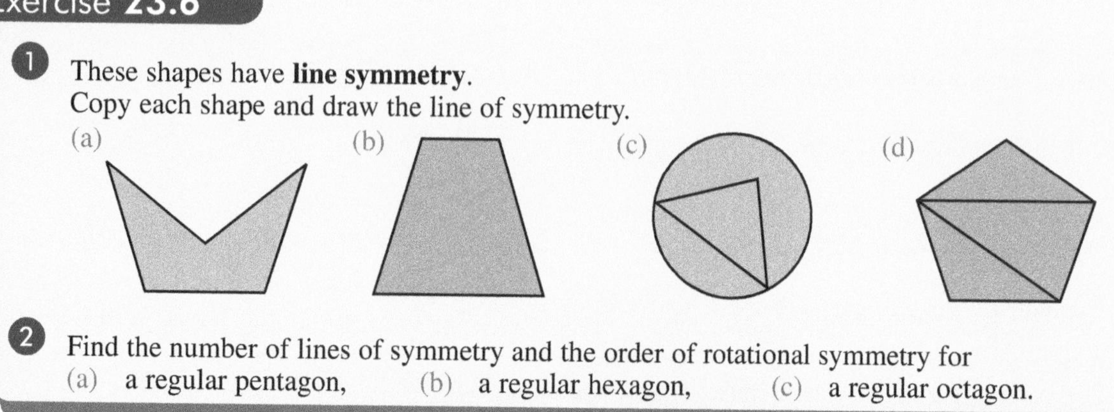

2. Find the number of lines of symmetry and the order of rotational symmetry for
   (a) a regular pentagon, (b) a regular hexagon, (c) a regular octagon.

3 Look at these triangles.

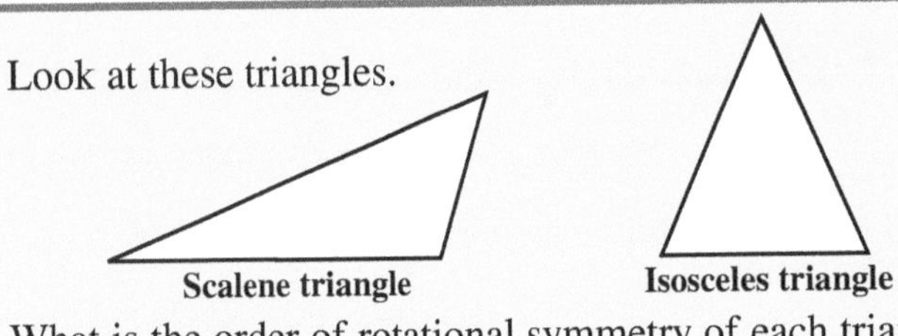
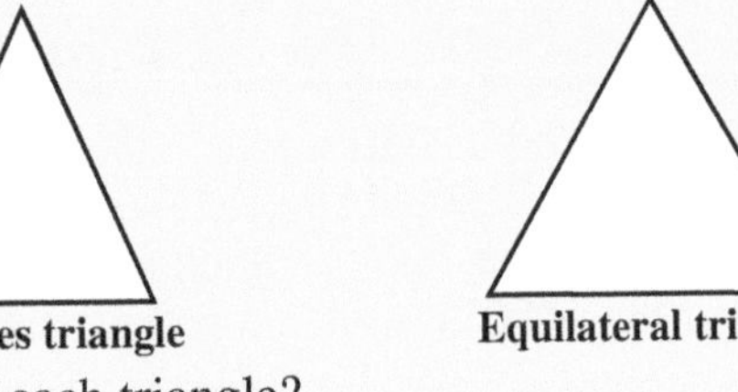
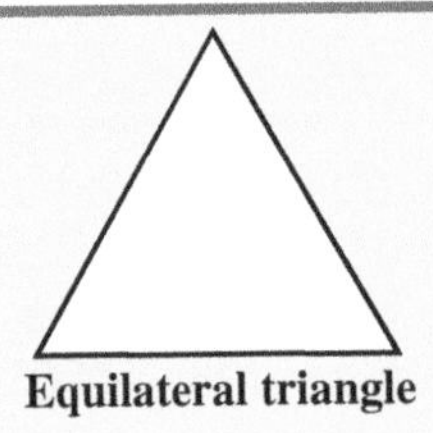
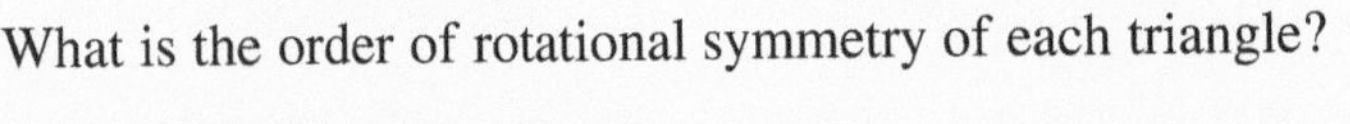

What is the order of rotational symmetry of each triangle?

4 Copy each of these diagrams.

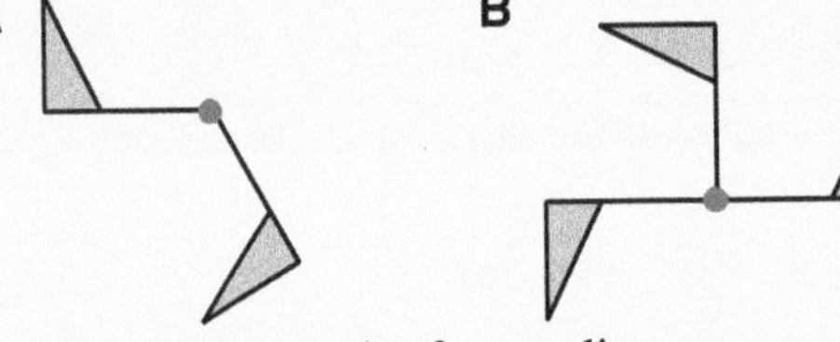

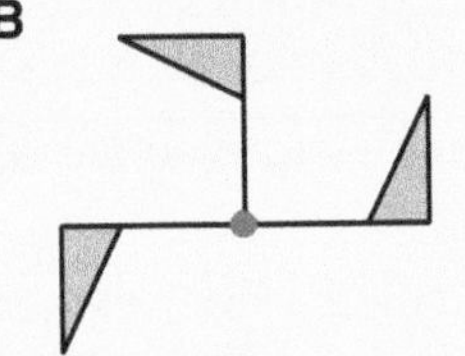

C

(a) Add **one** more flag to each of your diagrams so that the final diagrams have rotational symmetry.

(b) What is the order of rotational symmetry for each of your diagrams?

5 These quadrilaterals have been drawn on squared paper.

A B C D E F G H I

Copy and complete the table for each shape.

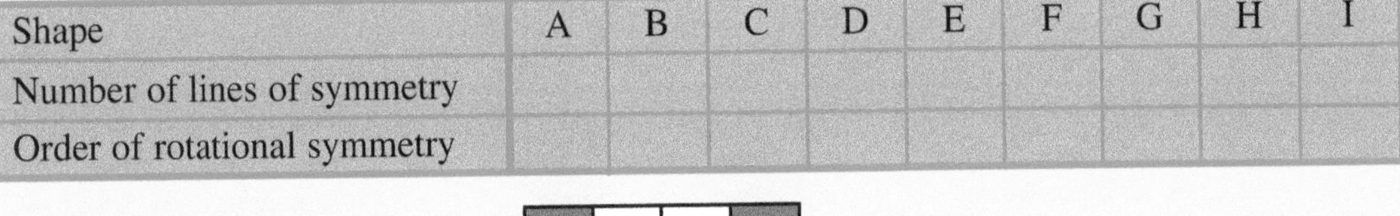

Shape	A	B	C	D	E	F	G	H	I
Number of lines of symmetry									
Order of rotational symmetry									

6 Make a copy of this shape.

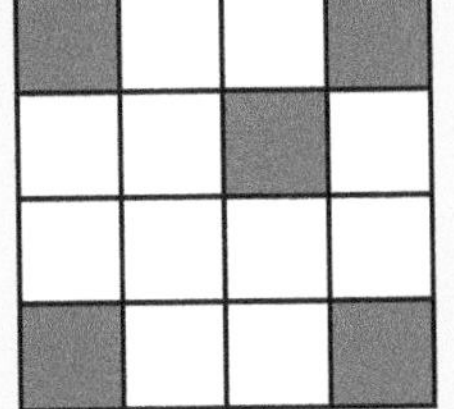

(a) How many lines of symmetry does the shape have?

(b) (i) Colour one more square so that your shape has rotational symmetry of order 2.

(ii) Mark the centre of rotational symmetry on your shape.

7 Make a copy of this shape.

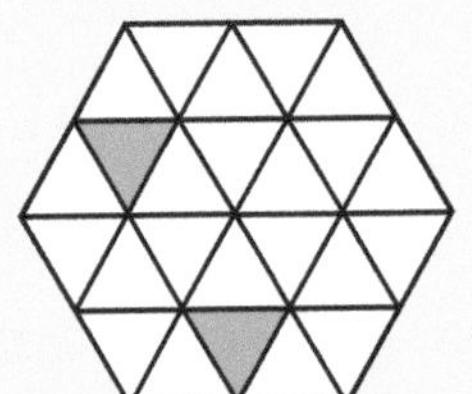

(a) How many lines of symmetry does the shape have?

(b) (i) Colour one more triangle so that your shape has rotational symmetry of order 3.

(ii) How many lines of symmetry does your shape have?

What you need to know

- An angle of 90° is called a **right angle**.
  An angle less than 90° is called an **acute angle**.
  An angle between 90° and 180° is called an **obtuse angle**.
  An angle greater than 180° is called a **reflex angle**.
- The sum of the angles at a point is 360°.
- Angles on a straight line add up to 180°.
  Angles which add up to 180° are called **supplementary angles**.
  Angles which add up to 90° are called **complementary angles**.
- When two lines cross, the opposite angles formed are equal and are called **vertically opposite angles**.
- A straight line joining two points is called a **line segment**.
- Lines which meet at right angles are **perpendicular** to each other.
- Lines which never meet and are always the same distance apart are **parallel**.
- When two parallel lines are crossed by a transversal the following pairs of angles are formed.

Corresponding angles

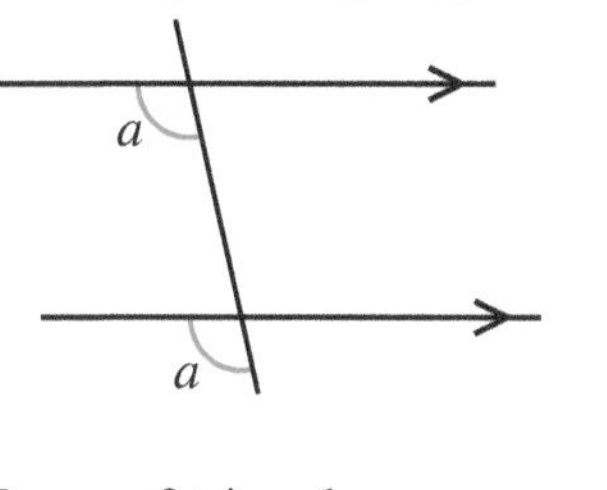

Alternate angles

Allied angles

$a + b = 180°$

- Types of triangle:

Scalene triangle

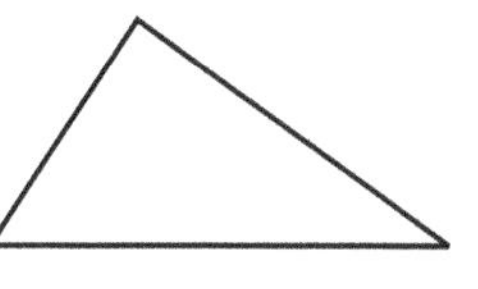

Sides have different lengths.
Angles are all different.

Isosceles triangle

Two equal sides.
Two equal angles.

Equilateral triangle

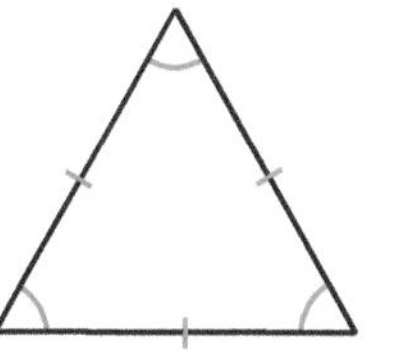

Three equal sides.
Three equal angles, 60°.

- The sum of the angles in a triangle is 180°.
  $a + b + c = 180°$
- The exterior angle is equal to the sum of the two opposite interior angles.
  $a + b = d$

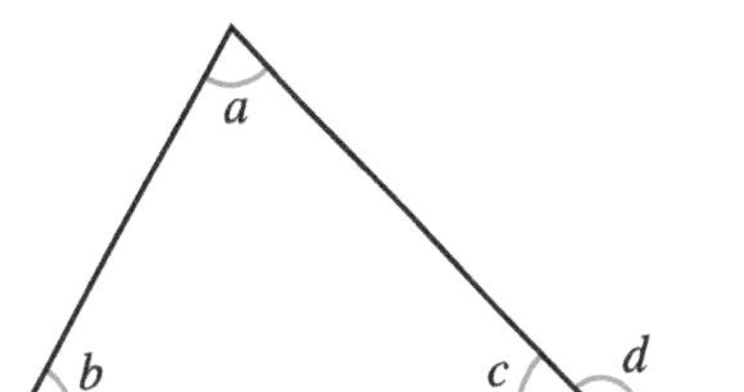

- A two-dimensional shape has **line symmetry** if the line divides the shape so that one side fits exactly over the other.
- A two-dimensional shape has **rotational symmetry** if it fits into a copy of its outline as it is rotated through 360°.

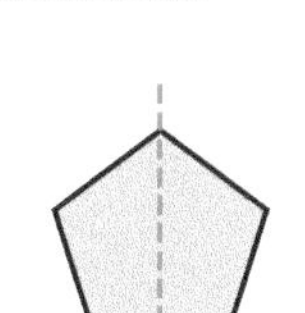

- A shape is only described as having rotational symmetry if the order of rotational symmetry is 2 or more.
- The number of times a shape fits into its outline in a single turn is the **order of rotational symmetry**.

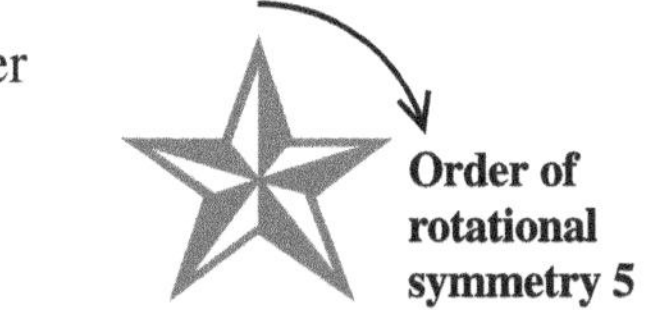

Order of rotational symmetry 5

- The sum of the angles in a quadrilateral is 360°.
- Facts about these special quadrilaterals:

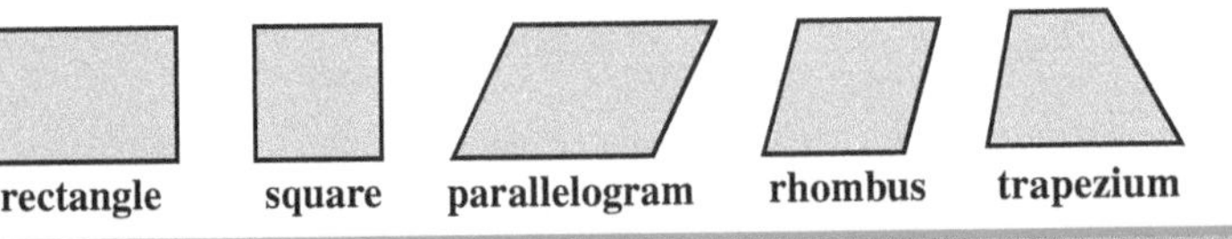

Quadrilateral	Sides	Angles	Diagonals	Lines of symmetry	Order of rotational symmetry
Rectangle	Opposite sides equal and parallel	All 90°	Bisect each other	2	2
Square	4 equal sides, opposite sides parallel	All 90°	Bisect each other at 90°	4	4
Parallelogram	Opposite sides equal and parallel	Opposite angles equal	Bisect each other	0	2
Rhombus	4 equal sides, opposite sides parallel	Opposite angles equal	Bisect each other at 90°	2	2
Trapezium	1 pair of parallel sides				
Isosceles trapezium	1 pair of parallel sides, non-parallel sides equal	2 pairs of equal angles	Equal in length	1	1*
Kite	2 pairs of adjacent sides equal	1 pair of opposite angles equal	One bisects the other at 90°	1	1*

*A shape is only described as having rotational symmetry if the order of rotational symmetry is 2 or more.

- A **polygon** is a many-sided shape made by straight lines.
- A polygon with all sides equal and all angles equal is called a **regular polygon**.
- Shapes you need to know: A 5-sided polygon is called a **pentagon**.
  A 6-sided polygon is called a **hexagon**.
  An 8-sided polygon is called an **octagon**.
- The sum of the exterior angles of any polygon is 360°.
- At each vertex of a polygon: interior angle + exterior angle = 180°
- The sum of the interior angles of an n-sided polygon is given by: $(n - 2) \times 180°$
- For a regular n-sided polygon: exterior angle $= \frac{360°}{n}$
- A shape will **tessellate** if it covers a surface without overlapping and leaves no gaps.
- All triangles tessellate.
- All quadrilaterals tessellate.

interior angle

exterior angle

Review Exercise 23

1. In the diagram AOB and POQ are straight lines.
   Angle $ROB = 90°$. Angle $AOP = 27°$.
   (a) (i) Work out the size of angle AOQ.
   (ii) Work out the size of angle POR.
   (b) What angle is the same size as angle AOP?

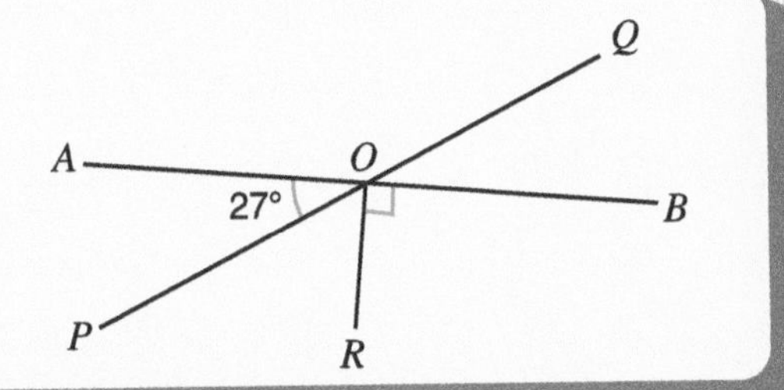

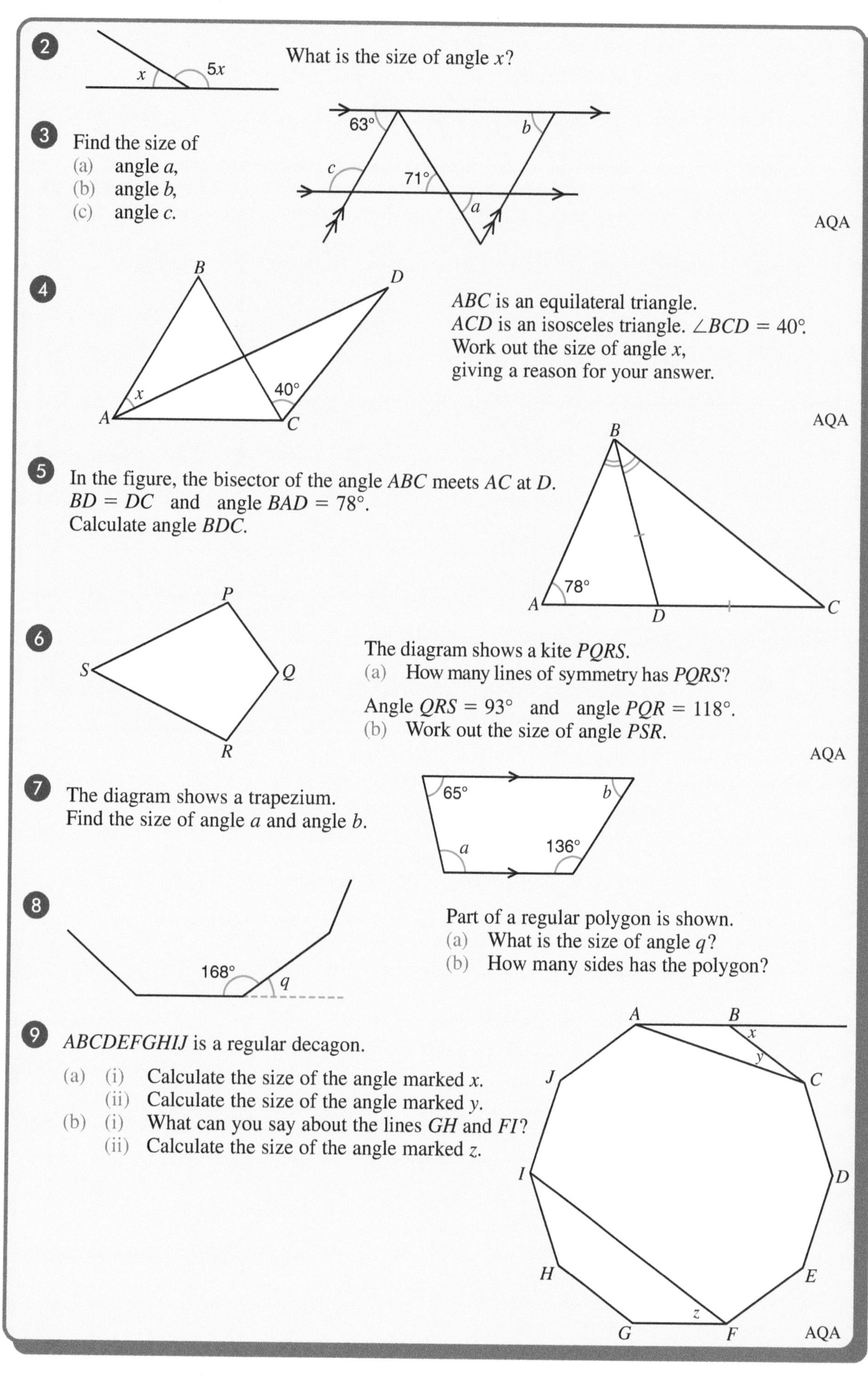

2 What is the size of angle x?

3 Find the size of
(a) angle a,
(b) angle b,
(c) angle c.

AQA

4 ABC is an equilateral triangle.
ACD is an isosceles triangle. $\angle BCD = 40°$.
Work out the size of angle x,
giving a reason for your answer.

AQA

5 In the figure, the bisector of the angle ABC meets AC at D.
$BD = DC$ and angle $BAD = 78°$.
Calculate angle BDC.

6 The diagram shows a kite $PQRS$.
(a) How many lines of symmetry has $PQRS$?

Angle $QRS = 93°$ and angle $PQR = 118°$.
(b) Work out the size of angle PSR.

AQA

7 The diagram shows a trapezium.
Find the size of angle a and angle b.

8 Part of a regular polygon is shown.
(a) What is the size of angle q?
(b) How many sides has the polygon?

9 $ABCDEFGHIJ$ is a regular decagon.
(a) (i) Calculate the size of the angle marked x.
(ii) Calculate the size of the angle marked y.
(b) (i) What can you say about the lines GH and FI?
(ii) Calculate the size of the angle marked z.

AQA

CHAPTER 24 Circle Properties

Circles

A **circle** is the shape drawn by keeping a pencil the same distance from a fixed point on a piece of paper. Compasses can be used to draw circles accurately.

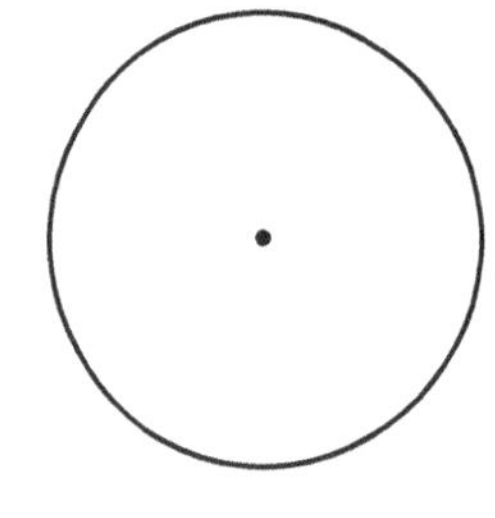

It is important that you understand the meaning of the following words:

Circumference – special name used for the perimeter of a circle.

Radius – distance from the centre of the circle to any point on the circumference.

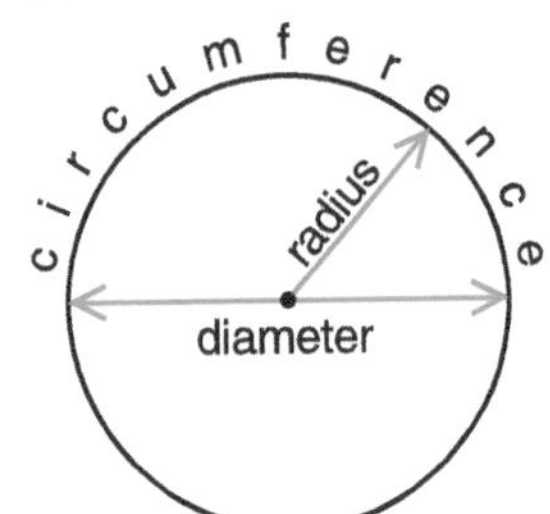

Diameter – distance right across the circle, passing through the centre point.
Notice that the diameter is twice as long as the radius.

Chord – a line joining two points on the circumference.
The longest chord of a circle is the diameter.

Tangent – a line which touches the circumference of a circle at one point only.

Arc – part of the circumference of a circle.

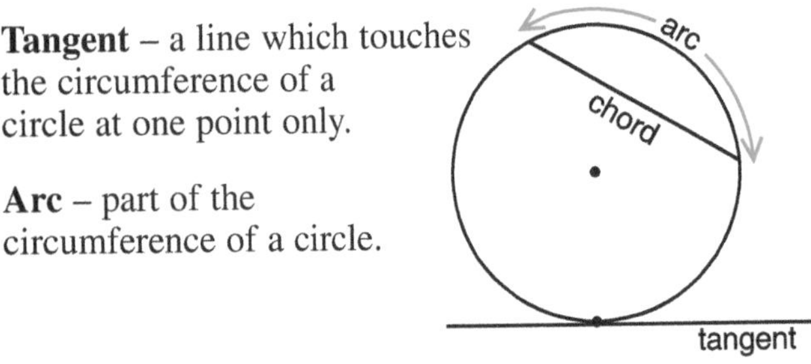

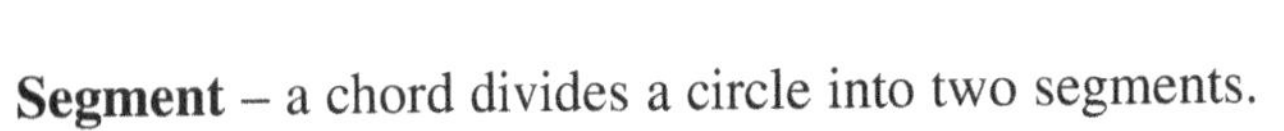

Segment – a chord divides a circle into two segments.

Sector – two radii divide a circle into two sectors.

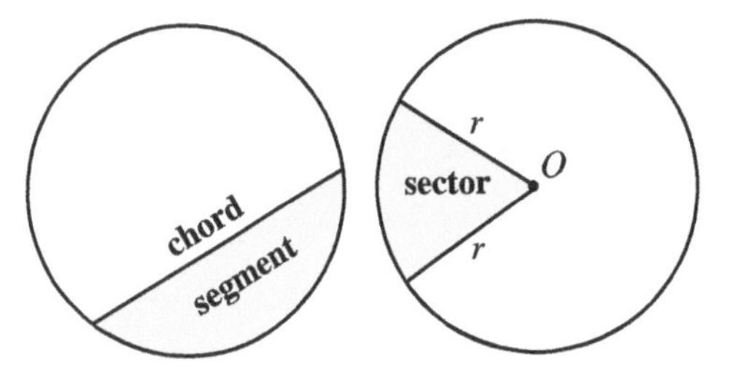

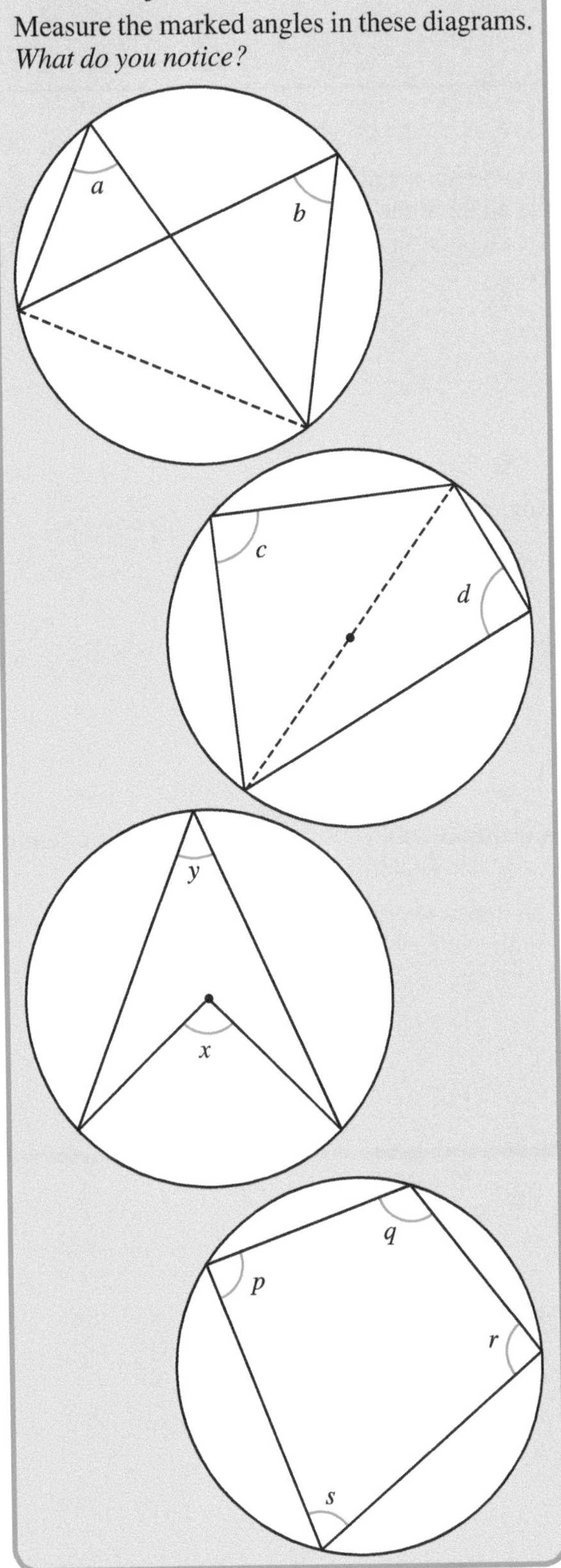

Circle properties

The angle in a semi-circle is a right angle

Angles which are:
at the circumference,
standing on a diameter,
are equal to 90°.

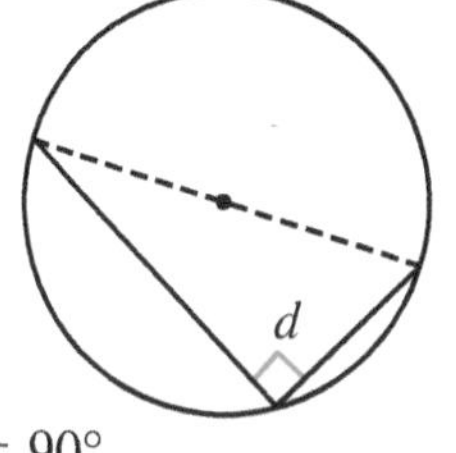

$c = d = 90°$

The angle at the centre is twice the angle at the circumference

If two angles are standing on the same chord, the angle at the centre of the circle is twice the angle at the circumference.

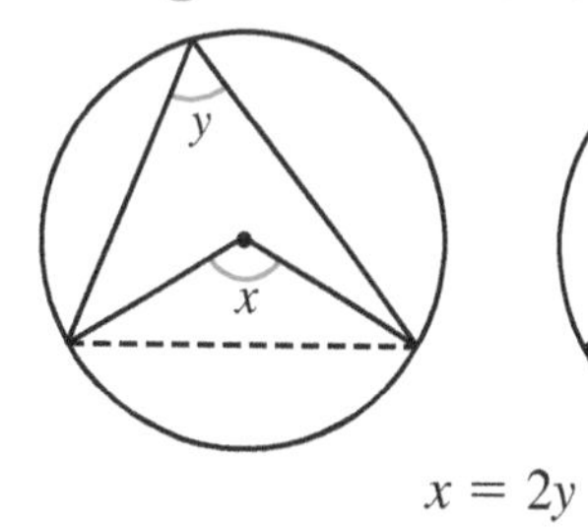

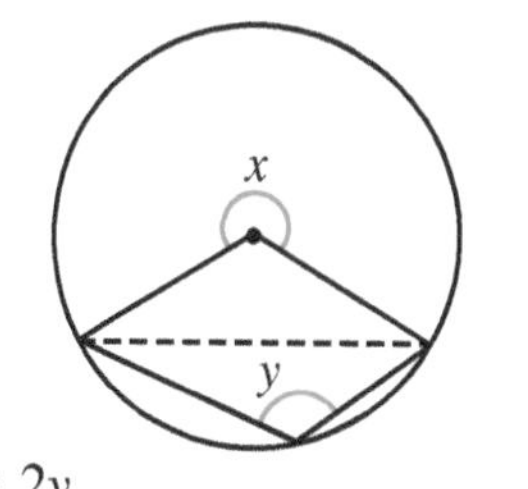

$x = 2y$

Angles in the same segment are equal

Angles which are:
at the circumference,
standing on the same chord,
in the same segment,
are equal.

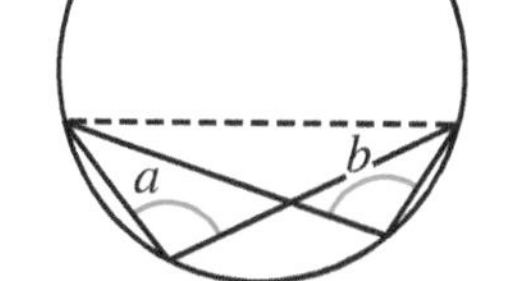

$a = b$

Opposite angles of a cyclic quadrilateral are supplementary

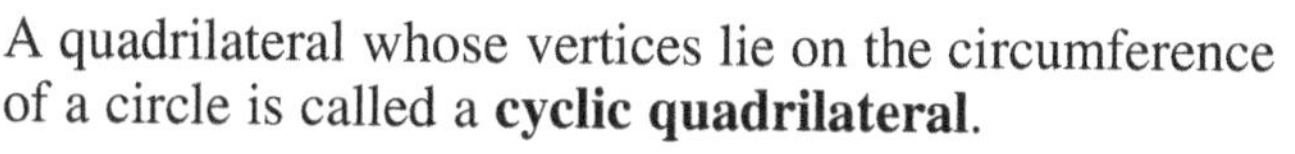

A quadrilateral whose vertices lie on the circumference of a circle is called a **cyclic quadrilateral**.

The opposite angles of a cyclic quadrilateral are supplementary (add up to 180°).

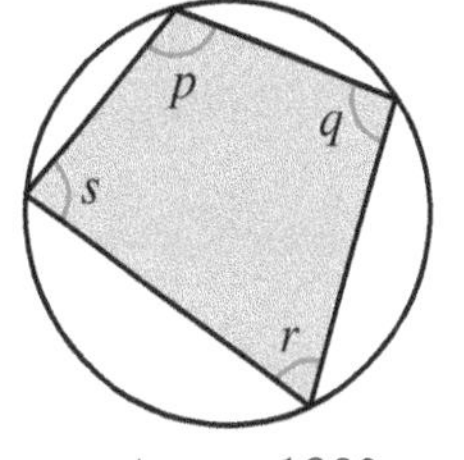

$p + r = 180°$
and
$q + s = 180°$

EXAMPLE

The diagram has not been drawn accurately.
O is the centre of the circle.
Find the marked angles.

$a = 56°$ (angles in the same segment)

$b = 2 \times 56°$ (∠ at centre = twice ∠ at circum.)
$b = 112°$

$c = 180° - 56°$ (opp. ∠'s of a cyclic quad.)
$c = 124°$

Exercise 24.1

The diagrams in this exercise have not been drawn accurately.

1. The following diagrams show triangles drawn in semi-circles.
   Work out the size of the marked angles.

(a) (b) (c) (d)

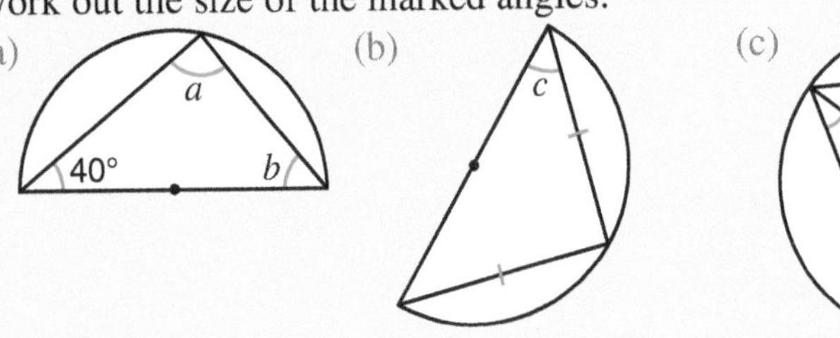

2. O is the centre of the circle.
   Work out the size of the marked angles.

(a) (b) (c) (d)

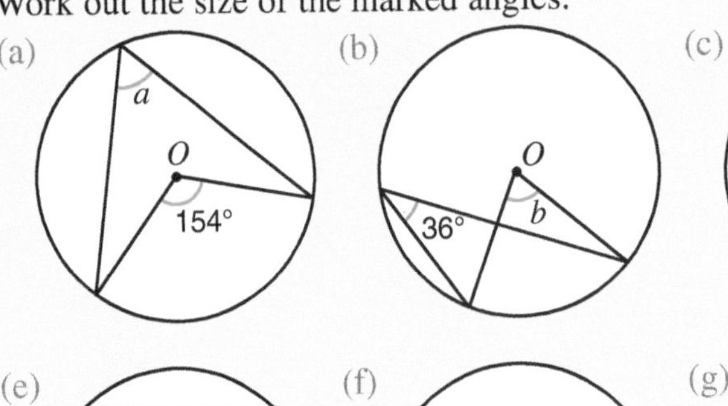

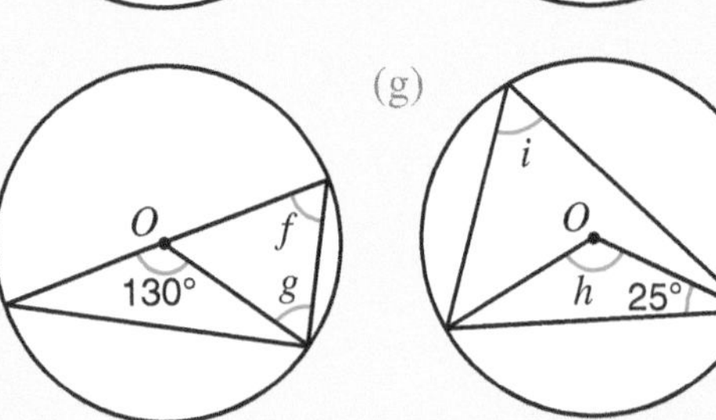

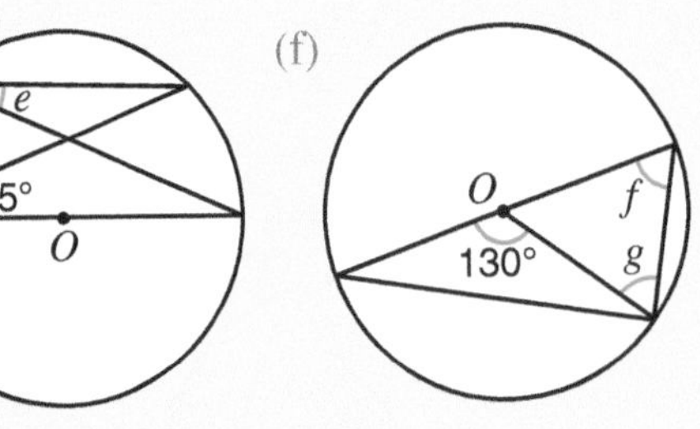

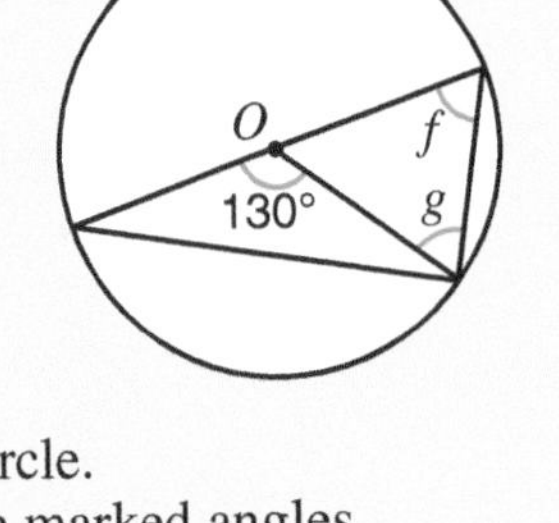

(e) (f) (g) (h)

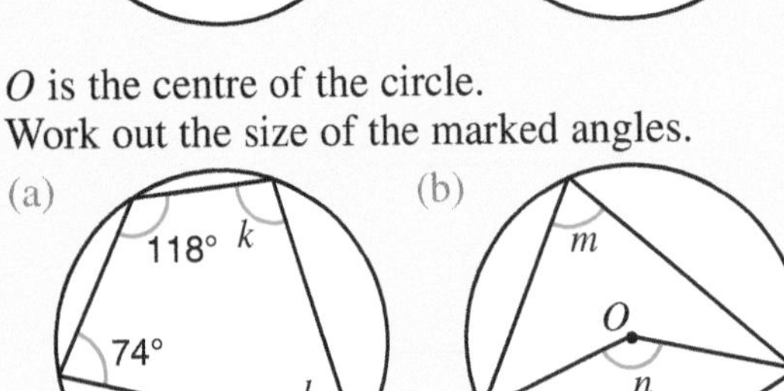

3. O is the centre of the circle.
   Work out the size of the marked angles.

(a) (b) (c) (d)

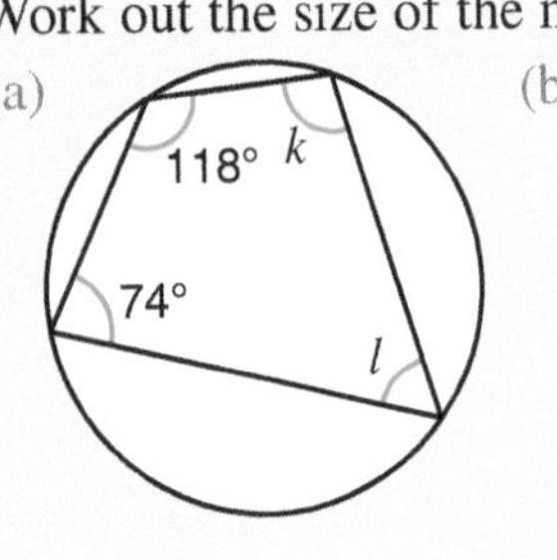

(e) (f) (g) (h)

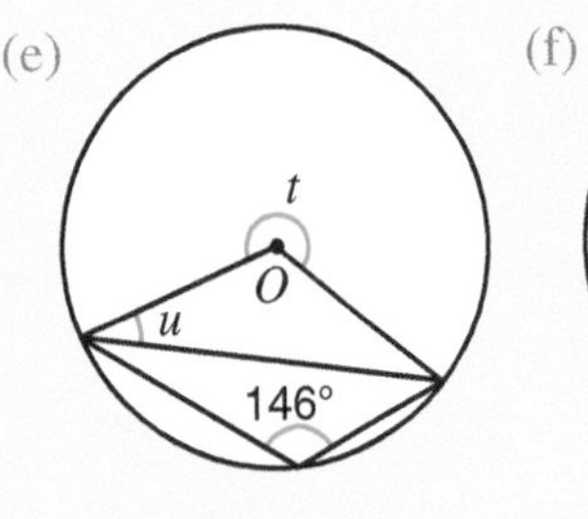

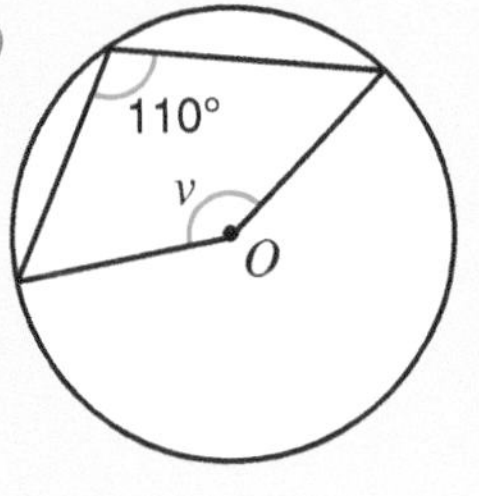

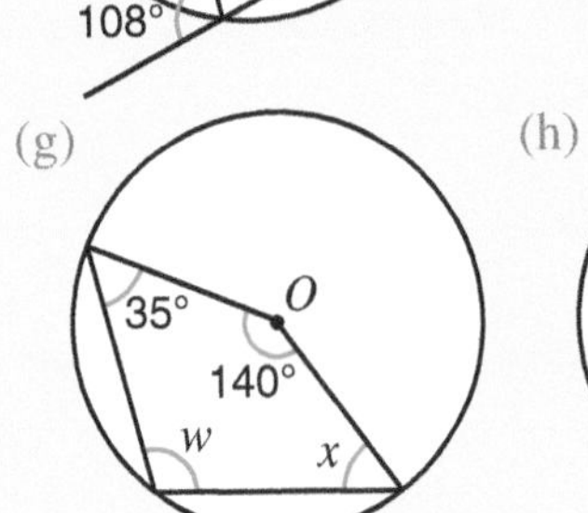

4. In the diagram BD is a diameter.
   Angle $ACD = 43°$.
   Calculate angle x.

5 $ABCD$ is a cyclic quadrilateral, with BA produced (extended) to E.
Angle $EAD = 70°$ and angle $CDB = 34°$.
Calculate angles BCD and CAD.

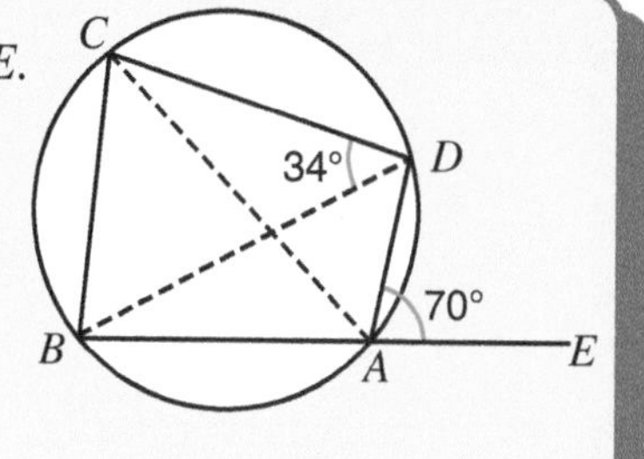

6 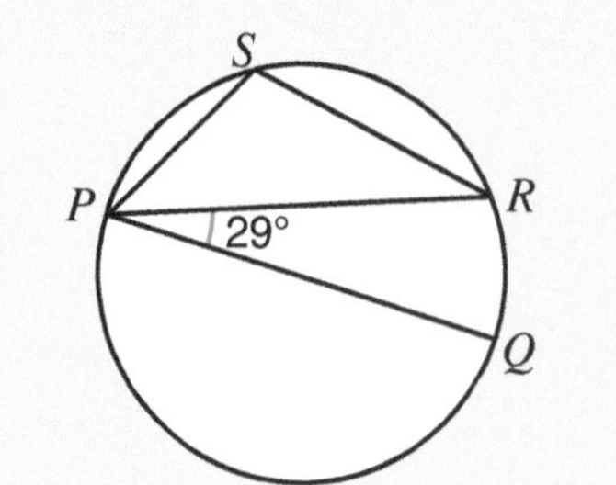

PQ is a diameter.
Angle $RPQ = 29°$.
Calculate angle PSR.

7 In the diagram, O is the centre of the circle.
Angle $PSQ = 28°$ and angle $QSR = 47°$.
Calculate angle PQR and angle QRS.

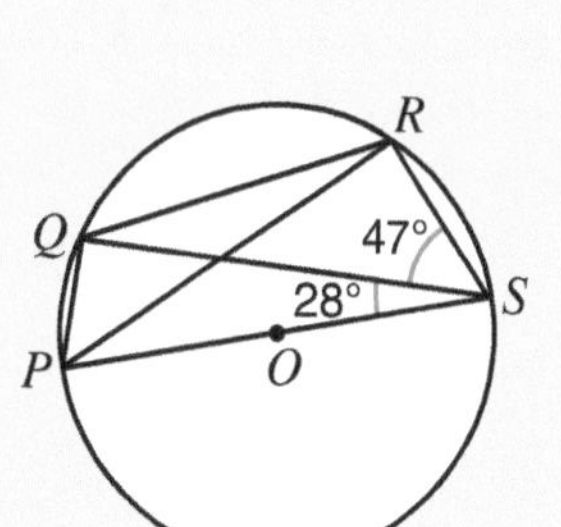

8 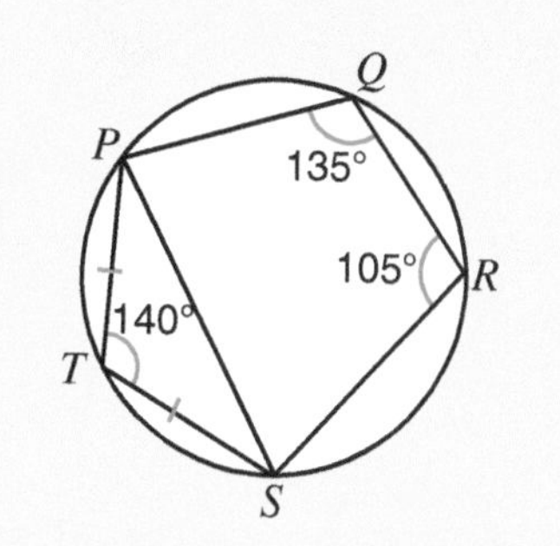

P, Q, R, S and T are points on a circle.
$\angle QRS = 105°$, $\angle PTS = 140°$, $\angle PQR = 135°$ and $PT = TS$.
Calculate $\angle PSR$ and $\angle TPQ$.

9 In the figure, O is the centre of the circle.
Angle $ORP = 20°$.
Calculate (a) angle POR, (b) angle PQR.

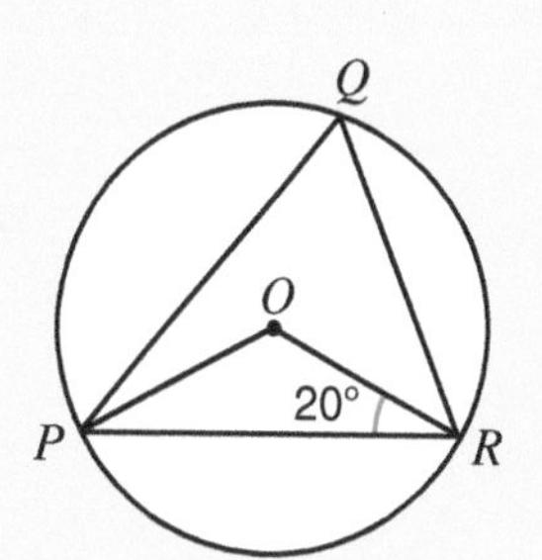

10 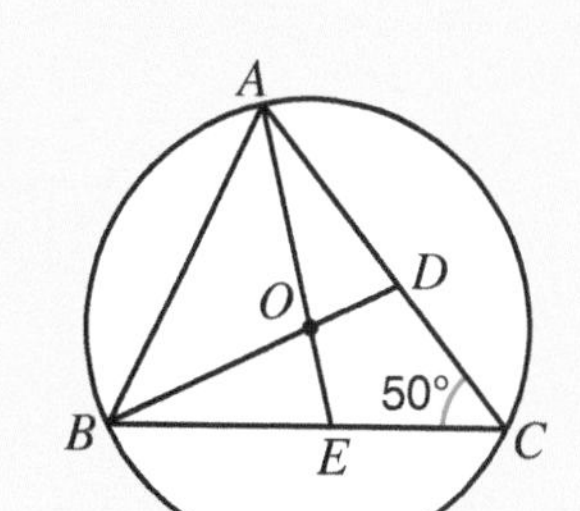

In the figure, O is the centre of the circle.
$AC = CB$.
Calculate the angles (a) DOE, (b) ABD, (c) DBE.

11 The centre of the circle is O.
Calculate the value of x.

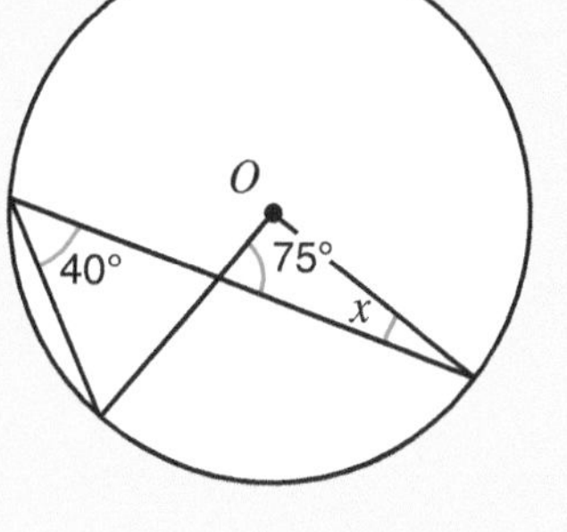
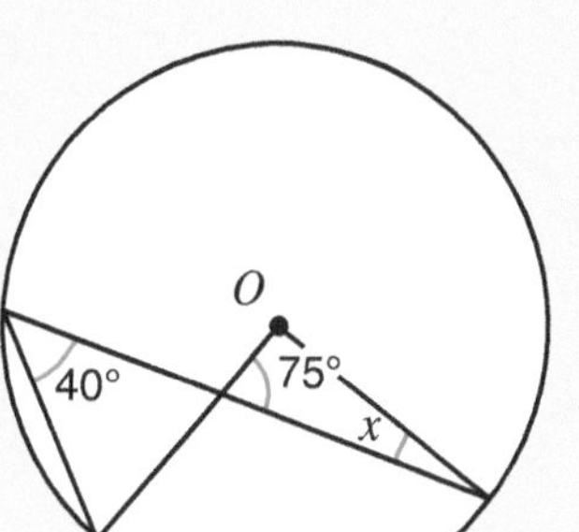

12 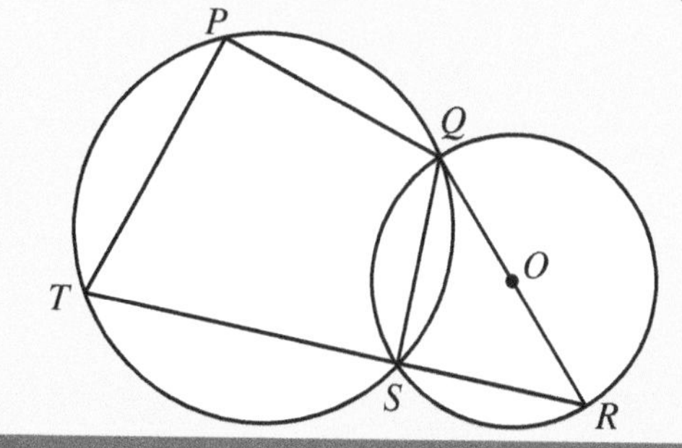

In the circle, centre O, QR is a diameter.
The line QS is a common chord of the two circles.
TR is a straight line.
Find angle QPT.

Tangents

The diagram shows a tangent drawn from a point P to touch the circle, centre O, at T.

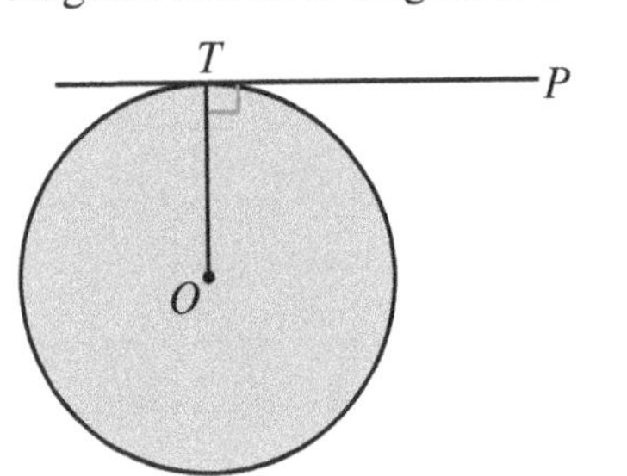

A tangent to a circle is a straight line which touches the circumference at one point only.

A tangent is perpendicular to the radius at the point of contact.

This diagram shows two tangents drawn from a point P to touch the circle, centre O, at points A and B.

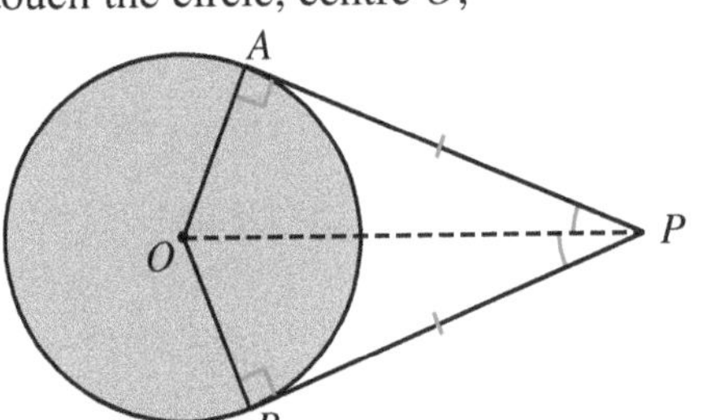

Tangents drawn to a circle from the same point are equal, $PA = PB$.
OP bisects angle APB.

What is the name given to the shape OAPB?

EXAMPLE

The diagram has not been drawn accurately.
O is the centre of the circle. Find the marked angles.

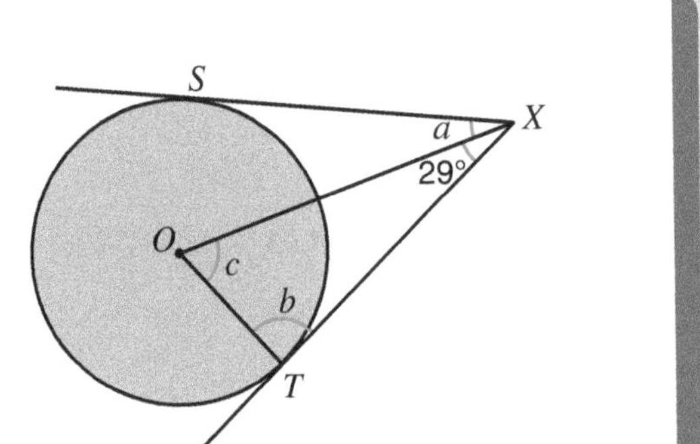

$a = 29°$ (SX and TX are tangents from X, so, $\angle TXO = \angle SXO$.)

$b = 90°$ (Tangent perpendicular to radius at T.)

$c = 180° - 90° - 29°$
$c = 61°$

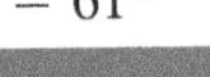

Activity

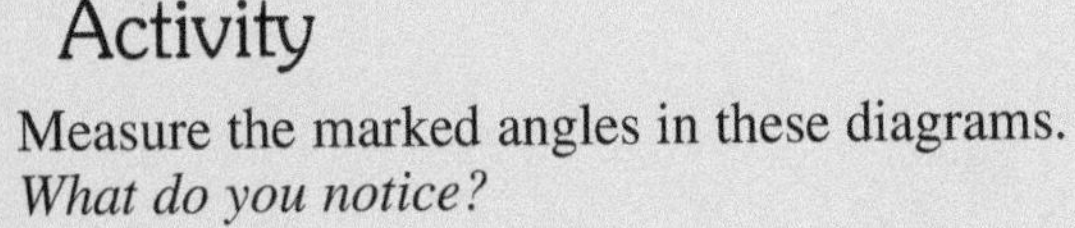

Measure the marked angles in these diagrams.
What do you notice?

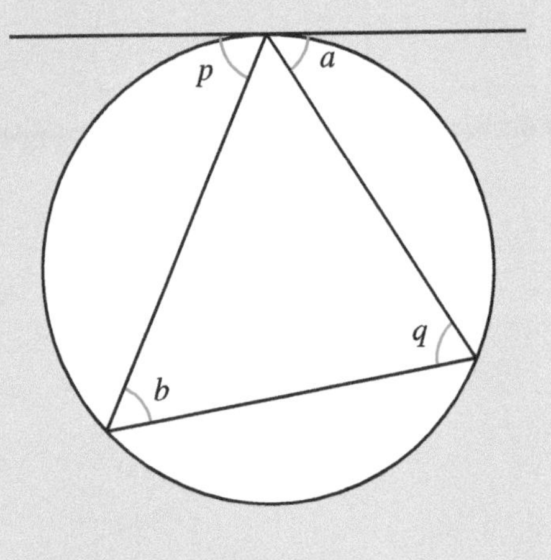

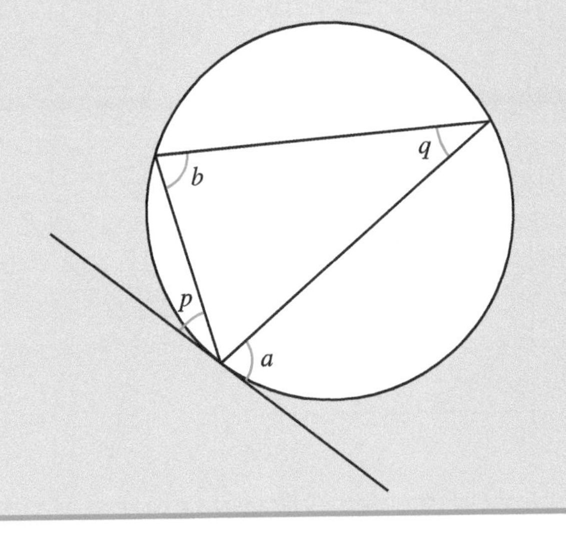

Alternate segment theorem

You should find that in each diagram, angle a = angle b and angle p = angle q.

The angle between a tangent and a chord is equal to any angle in the alternate segment, standing on the same chord.
This is known as the **alternate segment theorem**.

In each of these diagrams the shaded area shows the alternate (opposite) segment to the angle between the tangent and the chord.

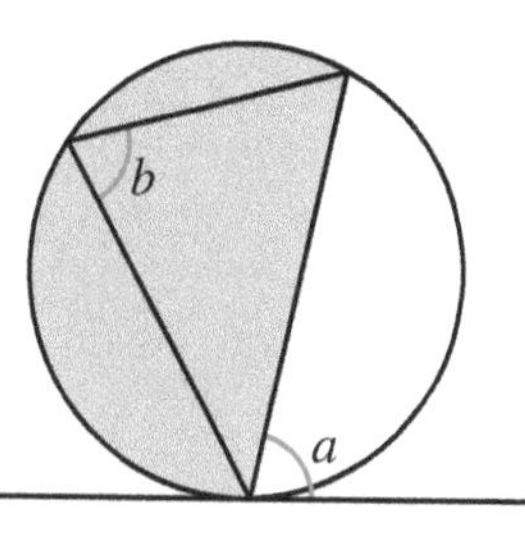

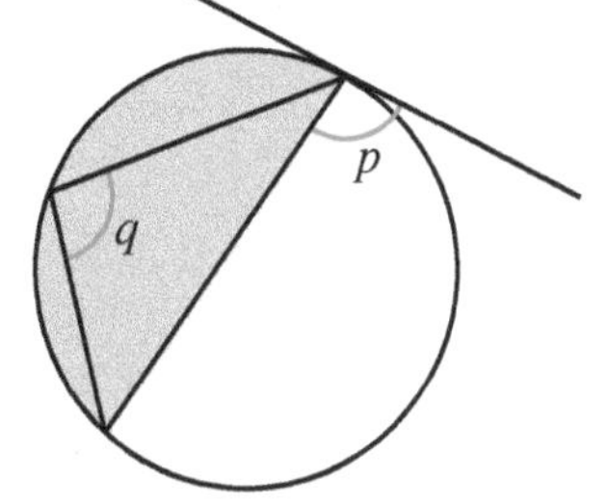

Exercise 24.2

The diagrams in this exercise have not been drawn accurately.

1 O is the centre of the circle.
Work out the size of the marked angles.

2 O is the centre of the circle.
Work out the size of the marked angles.

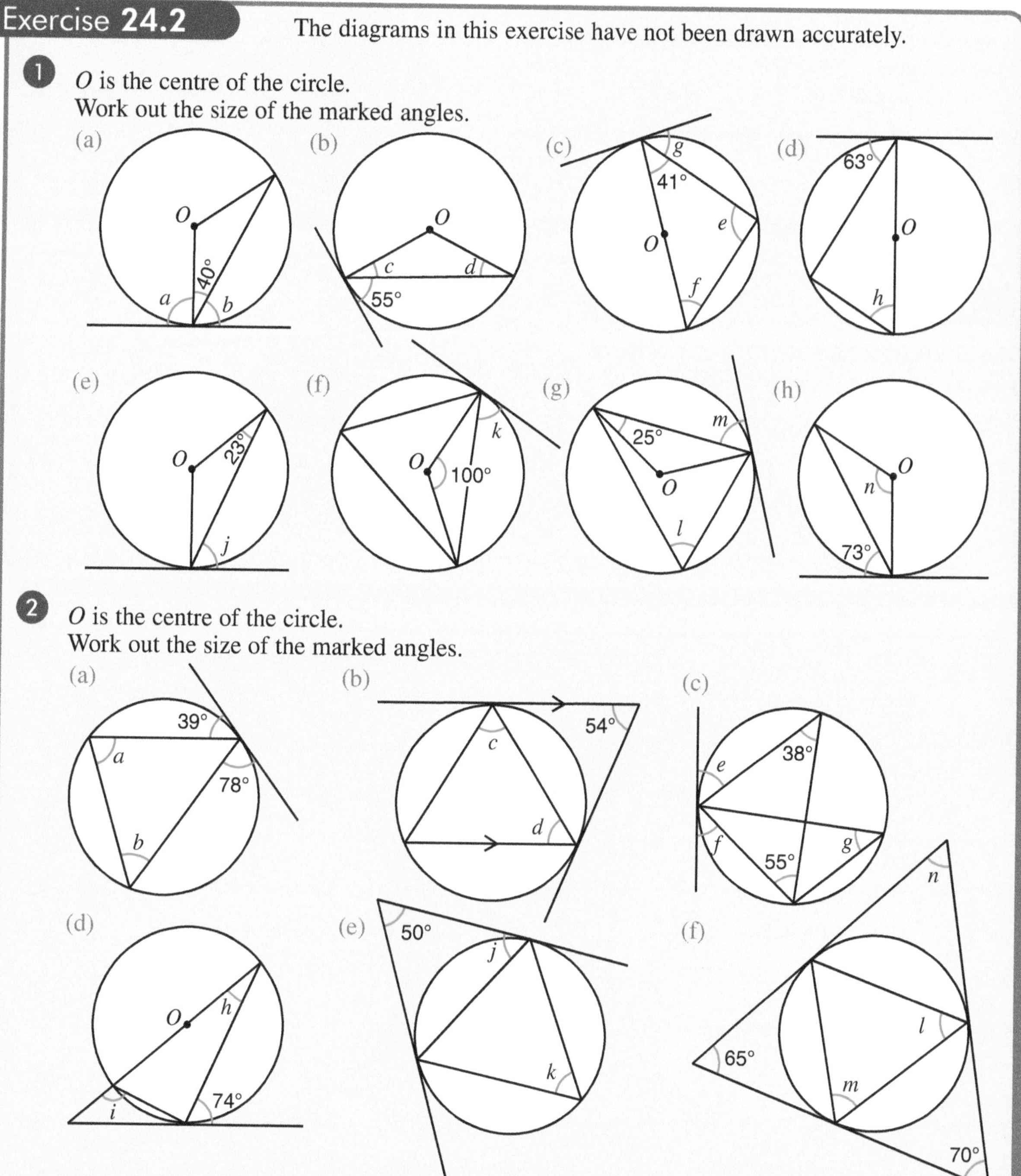

3

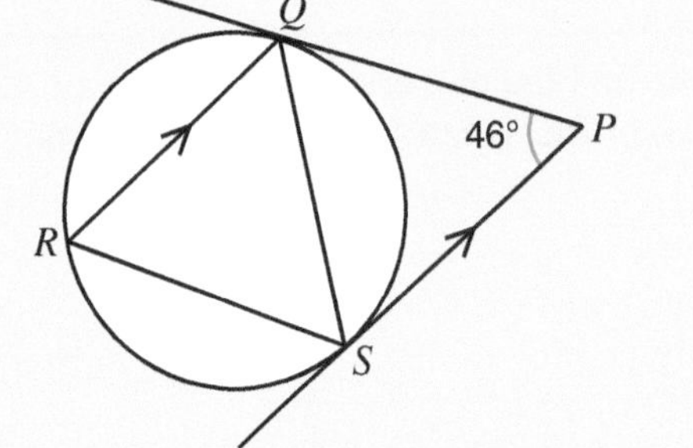

In the diagram, PQ and PS are tangents.
RQ is parallel to SP.
Angle $SPQ = 46°$.
Calculate (a) angle QRS,
(b) angle RSQ.

4 In the diagram, PQ and PS are tangents to the circle centre O.
POR is a straight line.
Angle $SRQ = 50°$.
Calculate (a) $\angle SOQ$,
(b) $\angle SPO$,
(c) $\angle RSO$.

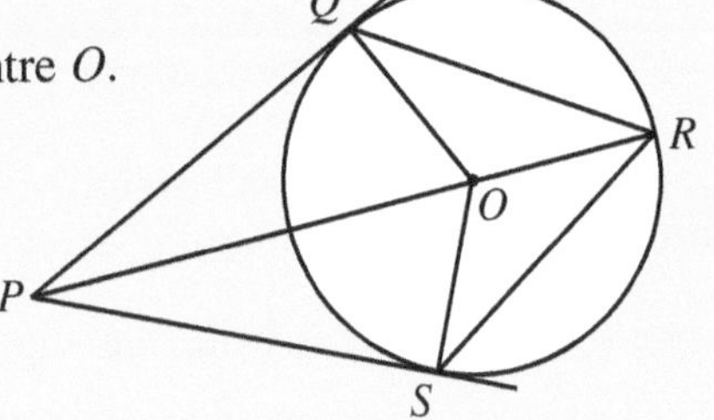

5

PQ and PS are tangents to the circle centre O.
POR is a straight line.
Angle $SPO = 18°$.
Calculate (a) $\angle POS$,
(b) $\angle OSR$,
(c) $\angle SQR$.

6 In the diagram, PT is the tangent at T to the circle.
Find the sizes of the angles marked x, y and z.

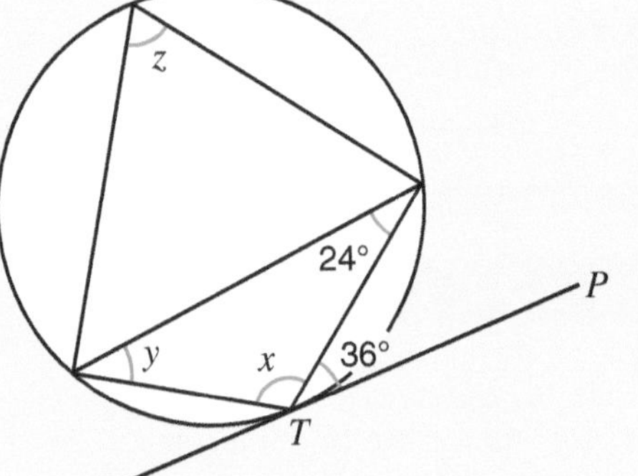
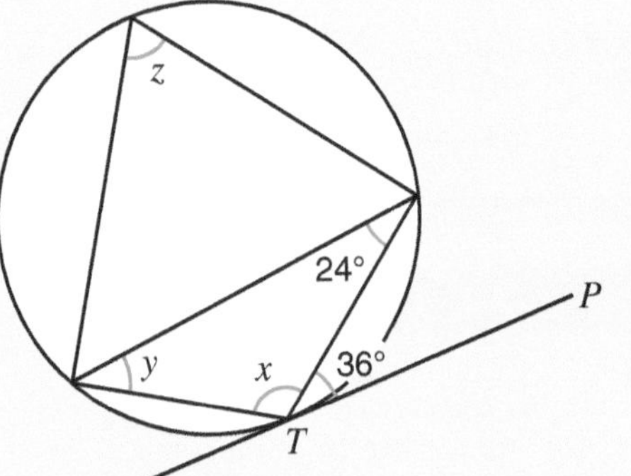
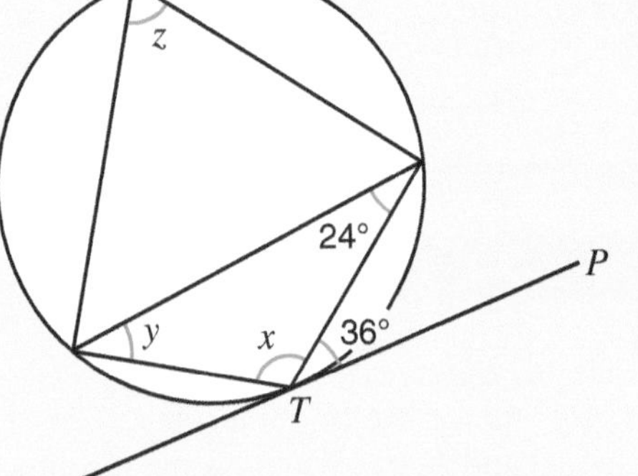

7

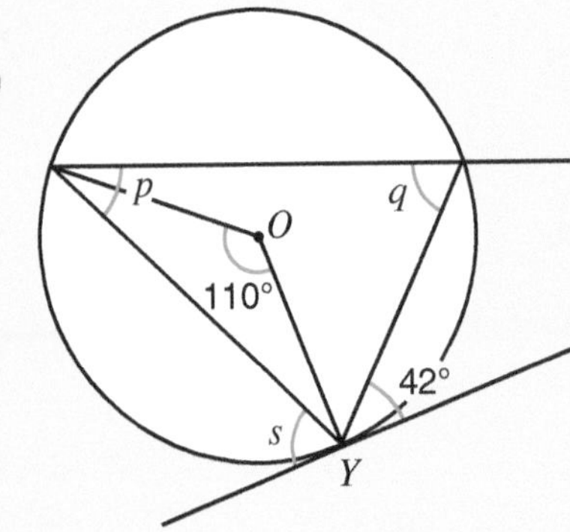

The line XY is a tangent to the circle and
O is the centre of the circle.
Find the sizes of the angles marked p, q, r and s.

8 In the figure, AB touches the circle at T and CD is a diameter.
Write down the value of $x + 2y$.

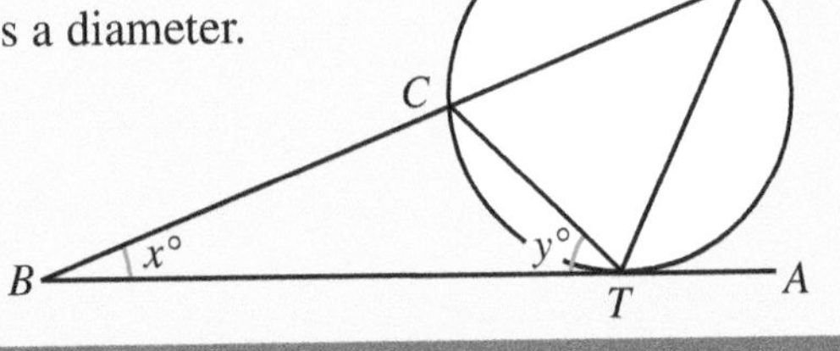

Geometric proofs

Earlier in the chapter you used the properties of circles to find the numerical values of required angles. In this section we will prove that these properties, or **facts**, can be applied to all situations and not just to particular situations.

To prove a geometrical fact involves:

- starting with basic facts that can be assumed to be true,
- using these facts to deduce other facts, eventually leading to the proof of the required fact.

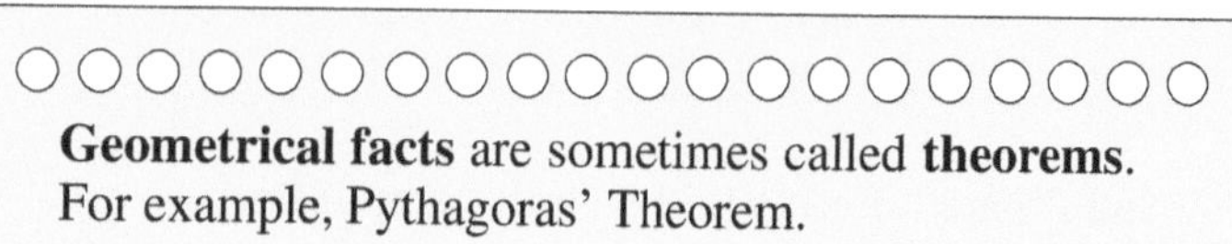

Geometrical facts are sometimes called **theorems**.
For example, Pythagoras' Theorem.

Once a geometrical fact has been proved it can be stated and used, without proof, in proving other required facts.

The angle at the centre is twice the angle at the circumference.

Prove that: $\angle AOB = 2 \times \angle APB$
Draw: Line from P, through O, to X.
Proof: Let $\angle APO = x$ and $\angle BPO = y$.

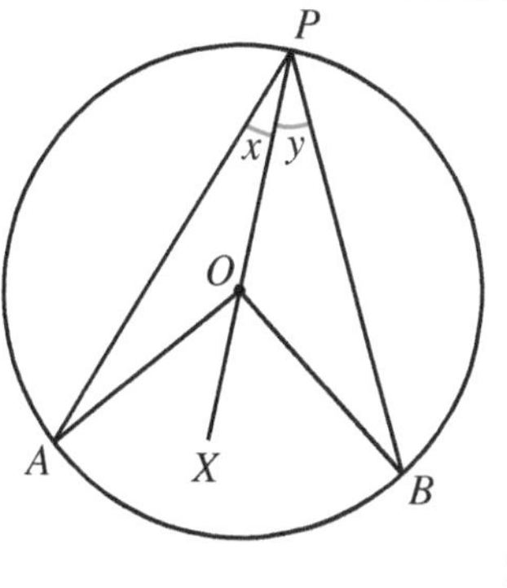

$\angle OAP = \angle APO = x$ (ΔAOP is isosceles)
$\angle AOX = \angle OAP + \angle APO = 2x$ (Ext. $\angle$ of a Δ – see page 230)
$\angle OBP = \angle BPO = y$ (ΔBOP is isosceles)
$\angle BOX = \angle OBP + \angle BPO = 2y$ (Ext. $\angle$ of a Δ)
$\angle AOB = \angle AOX + \angle BOX$ and $\angle APB = \angle APO + \angle BPO$
$= 2x + 2y$ $\quad = x + y$
$= 2(x + y)$
$= 2 \times \angle APB$

So, $\angle AOB = 2 \times \angle APB$

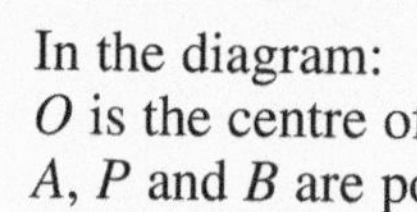

Exercise 24.3

1 In the diagram:
O is the centre of the circle.
A, P and B are points on the circumference.
The line PX has been drawn.

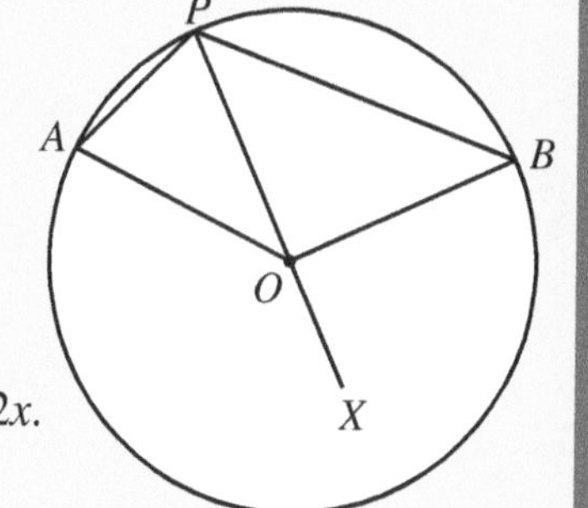

(a) If $\angle OPA = x$, show that ΔOAP is isosceles.
(b) If $\angle OPB = y$, show that ΔOBP is isosceles.
(c) Using the exterior angle of a triangle, show that $\angle AOX = 2x$.
(d) Find, in terms of x and y, the size of the reflex angle AOB.
(e) Show that reflex $\angle AOB = 2 \times \angle APB$.

2

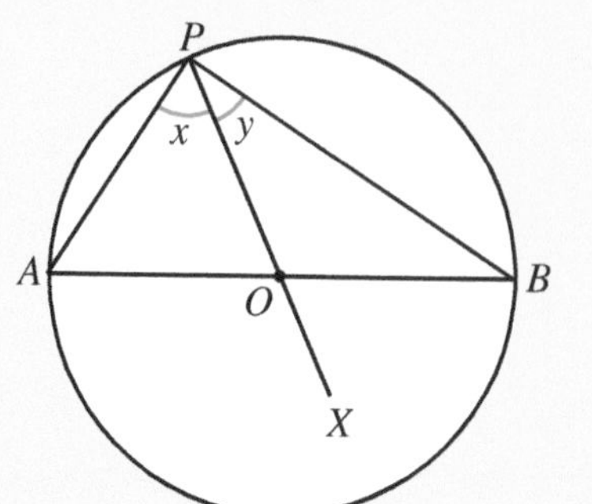

In the diagram:
AB is a diameter of the circle, centre O.
$\angle APO = x$ and $\angle BPO = y$.

(a) Show that $\angle AOB = 2 \times \angle APB$.
(b) Hence, prove that the angle in a semi-circle is a right angle.

3 (a) If $\angle APB = x$, write, in terms of x, the size of $\angle AOB$.
Give a reason for your answer.
(b) Write, in terms of x, the size of $\angle AQB$.
Give a reason for your answer.
(c) Hence, prove that angles in the same segment, standing on the same chord, are equal.

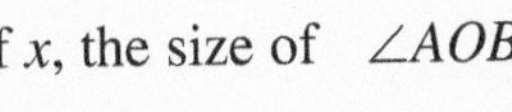
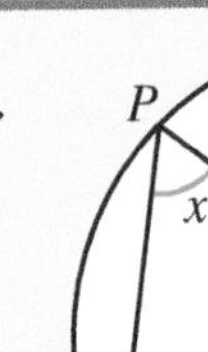
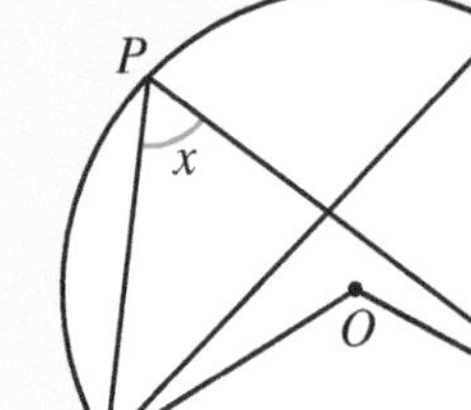
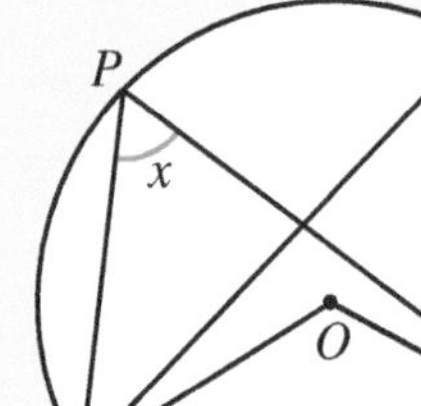
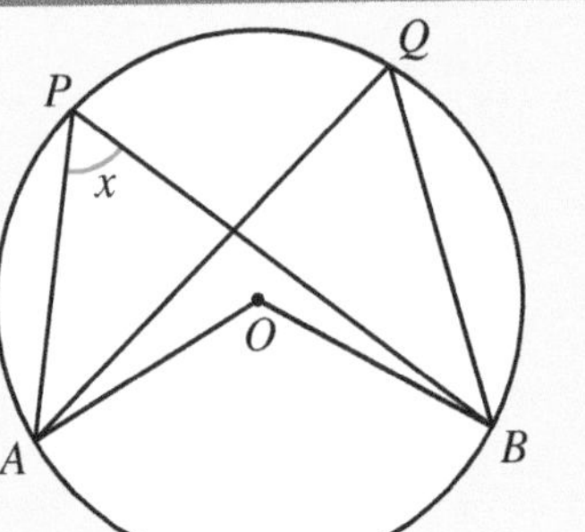

4 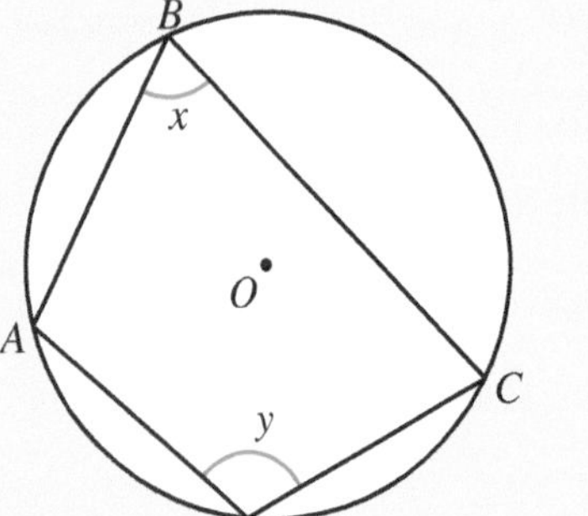

In the diagram:
$\angle ABC = x$ and $\angle ADC = y$.
(a) Copy the diagram.
Draw the lines AO and CO.
(b) Write, in terms of x and y, the size of
(i) $\angle AOC$, (ii) reflex $\angle AOC$.
Give reasons for your answers.
(c) $\angle AOC$ + reflex $\angle AOC = 360°$.
Show that $x + y = 180°$ and that opposite angles of a cyclic quadrilateral are supplementary.

5 In the diagram:
The line PQ is a tangent to the circle at T.
The line TX is a diameter of the circle, centre O.
$\angle BTQ = x$ and $\angle BTO = y$.
(a) Copy the diagram and draw the line BX.
(b) Explain why $x + y = 90°$.
(c) What is the size of $\angle BXT$?
(d) Explain why $\angle BXT = \angle BAT$.
(e) Show that $\angle BTQ = \angle BAT$ and hence, prove the alternate segment theorem.

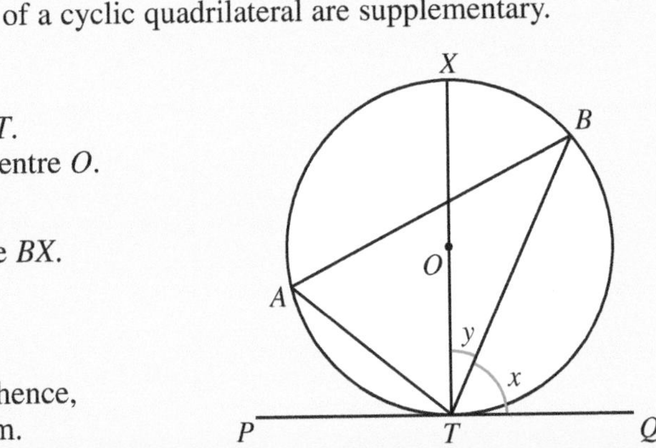

What you need to know

- A **circle** is the shape drawn by keeping a pencil the same distance from a fixed point on a piece of paper.
- The meaning of the following words:

Circumference – special name used for the perimeter of a circle.

Radius – distance from the centre of the circle to any point on the circumference.
The plural of radius is **radii**.

Diameter – distance right across the circle, passing through the centre point.
The diameter is twice as long as the radius.

Chord – a line joining two points on the circumference.
The longest chord is the diameter.

Tangent – a line which touches the circumference of a circle at one point only.

Arc – part of the circumference of a circle.

Segment – a chord divides a circle into two segments.

Sector – two radii divide a circle into two sectors.

- **Circle properties**

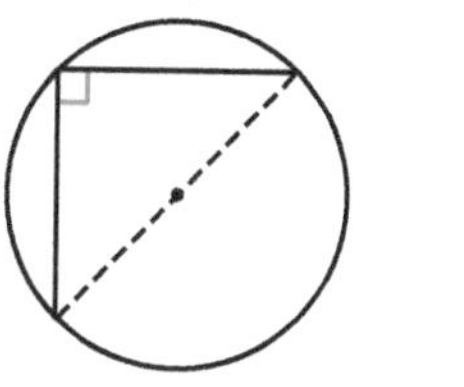

The angle in a semi-circle is a right angle.

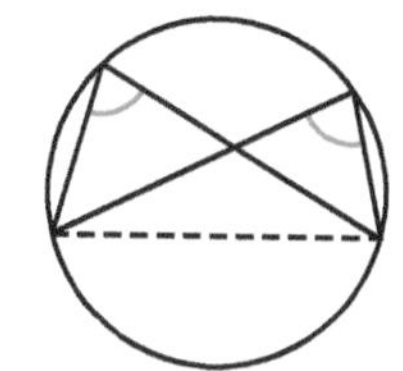

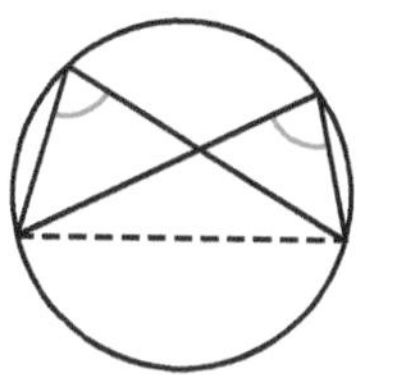

Angles in the same segment are equal.

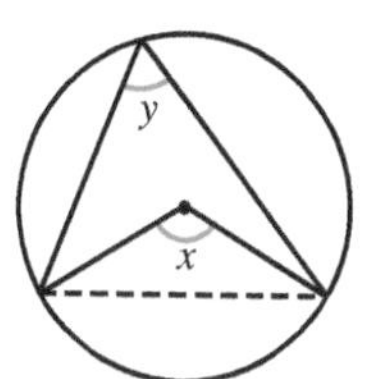

$x = 2y$
The angle at the centre is twice the angle at the circumference.

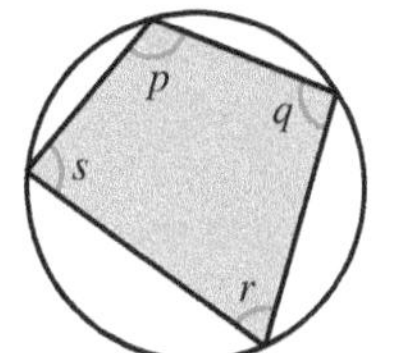

$p + r = 180°$ and $q + s = 180°$
Opposite angles of a cyclic quadrilateral are supplementary.

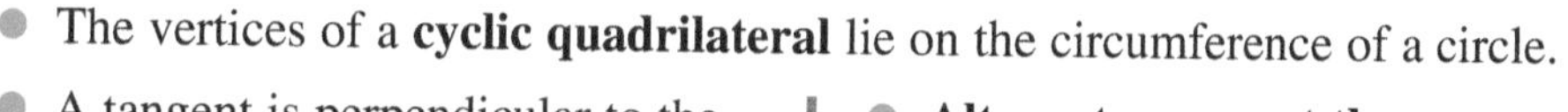

- The vertices of a **cyclic quadrilateral** lie on the circumference of a circle.
- A tangent is perpendicular to the radius at the point of contact.
- Tangents drawn to a circle from the same point are equal in length.

- **Alternate segment theorem.**
The angle between a tangent and a chord is equal to any angle in the alternate (opposite) segment.

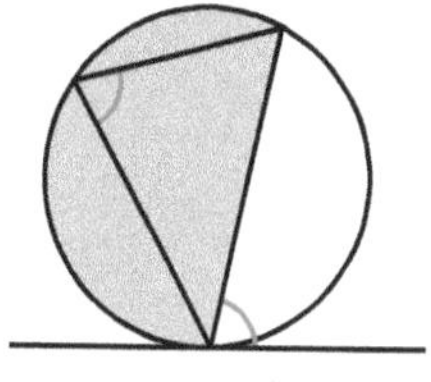

You should be able to prove that:

- the angle at the centre is twice the angle at the circumference,
- the angle in a semi-circle is a right angle,
- angles in the same segment are equal,
- opposite angles of a cyclic quadrilateral are supplementary,
- the angle between a tangent and a chord is equal to any angle in the alternate segment.

Review Exercise 24

The diagrams in this exercise have not been drawn accurately.

1 O is the centre of the circle.
Work out the size of the marked angles.

(a)

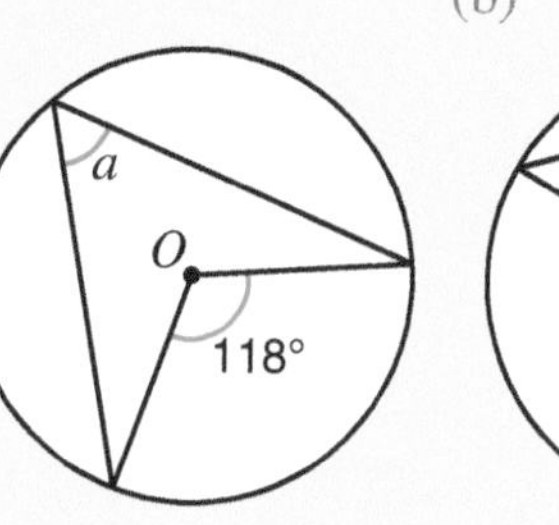

(b)

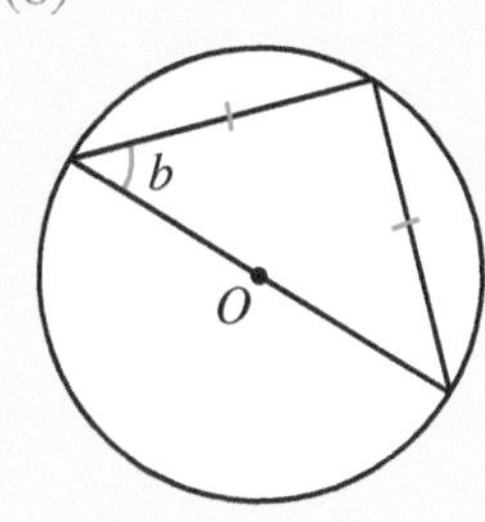

(c)

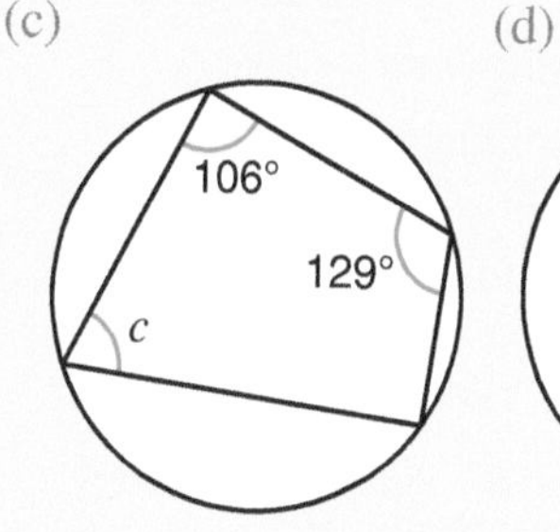

(d)

2 In the diagram, O is the centre of the circle.
Angle $OAC = 12°$ and angle $BOC = 80°$.

Calculate the size of the following angles, giving a geometrical reason for each of your answers.

(a) Angle OCA.
(b) Angle AOC.
(c) Angle ACB.
(d) Angle ABC.

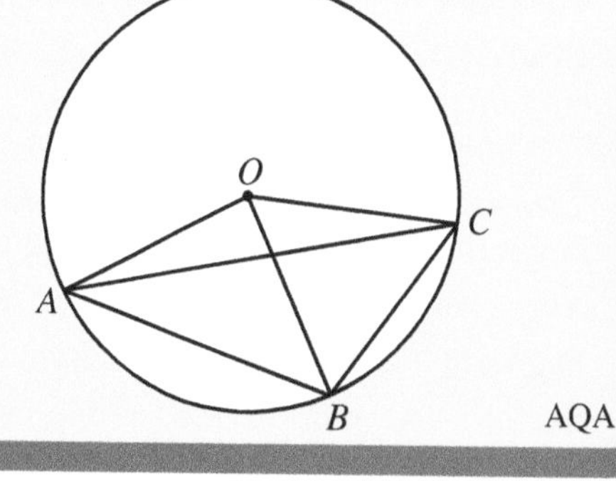

AQA

❸ In the diagram, O is the centre of the circle.
Find x.

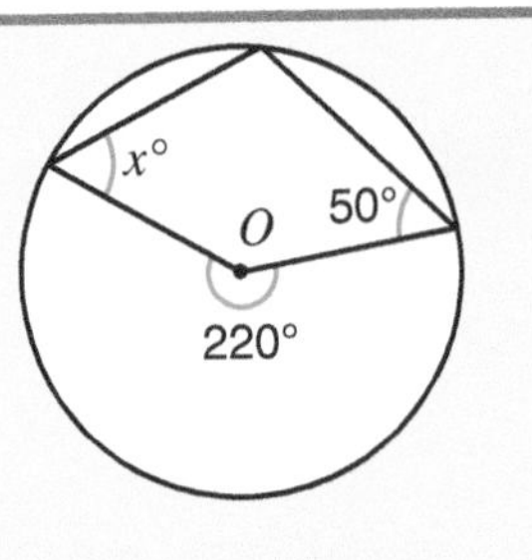

❹ 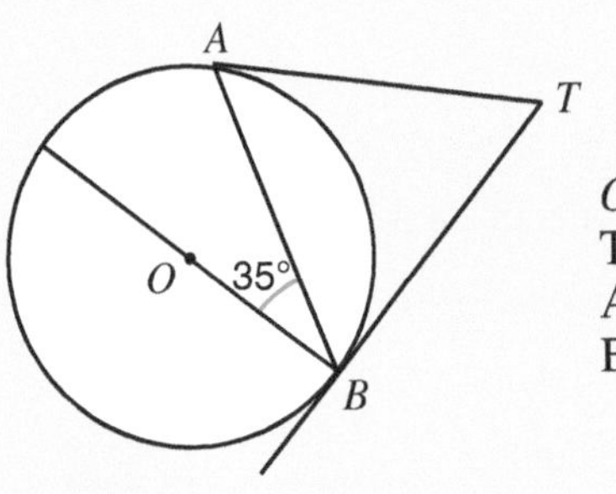

O is the centre of the circle.
Tangents are drawn from T to touch the circle at A and B.
Angle $OBA = 35°$.
Explain why angle $ATB = 70°$.

❺ In the diagram, O is the centre of the circle and TR is the tangent to the circle at T. Calculate x.

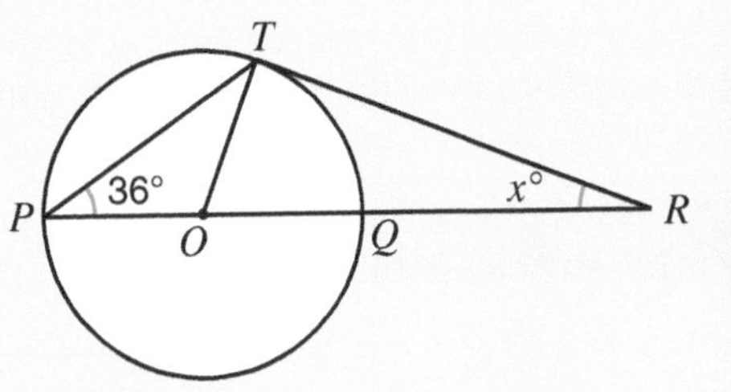

❻ 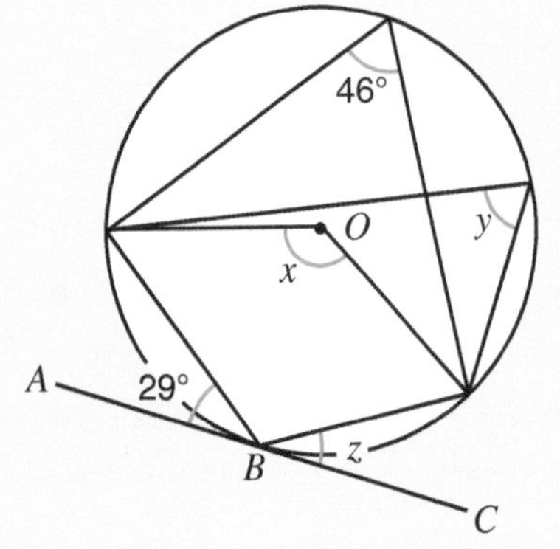

O is the centre of the circle.
ABC is a tangent to the circle at B.
Work out the size of angles x, y and z.

❼ In the diagram, O is the centre of the circle and PQR is a tangent to the circle.
Find the values of the angles marked x, y and z.

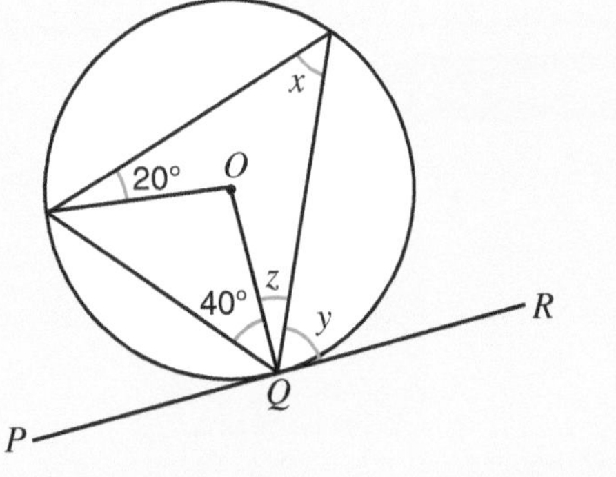

❽ 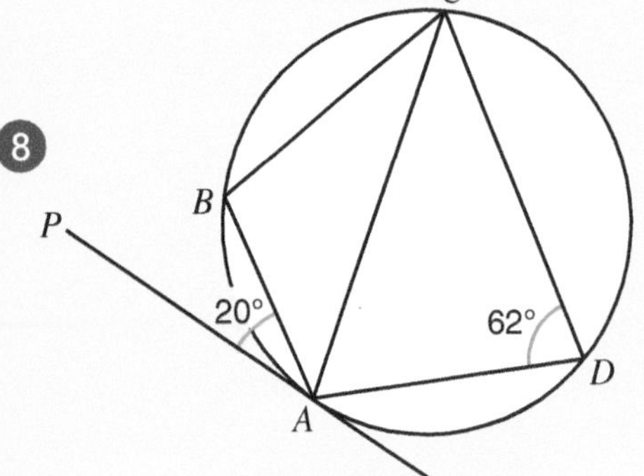

$ABCD$ is a cyclic quadrilateral and
PA is a tangent to the circle at A.
Angle $BAP = 20°$ and angle $ADC = 62°$.
Find angles ABC and BAC.

AQA

❾ The line PQR is a tangent to a circle with centre O.
QS is a diameter of the circle.
T is a point on the circumference of the circle such that POT is a straight line.
The angle OPQ is $34°$.
Calculate the size of angle TQR.

AQA

Perimeters and Areas

Perimeter

The **perimeter** is the distance round the outside of a shape.
The perimeter of a rectangle (or square) is the sum of the lengths of its four sides.

Area

Area is the amount of surface covered by a shape.
The standard unit for measuring area is the square centimetre, cm^2.
Small areas are measured using square millimetres, mm^2.
Large areas are measured using square metres, m^2, or square kilometres, km^2.

1 cm
1 cm^2 1 cm

Area formulae

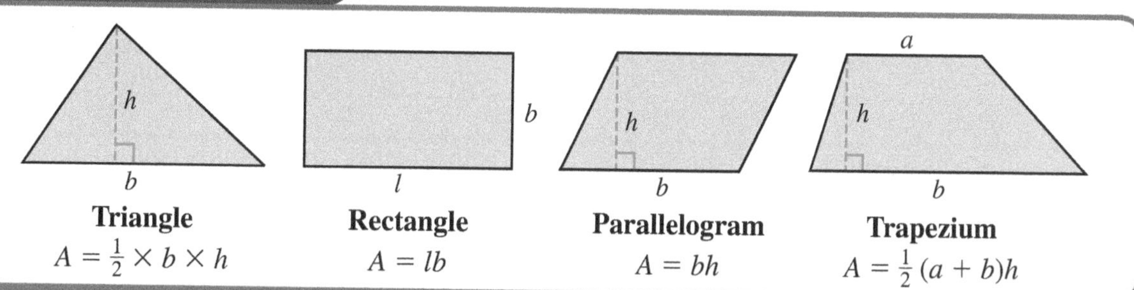

Triangle	Rectangle	Parallelogram	Trapezium
$A = \frac{1}{2} \times b \times h$	$A = lb$	$A = bh$	$A = \frac{1}{2}(a + b)h$

EXAMPLES

1 Find the perimeter and area of this rectangle.

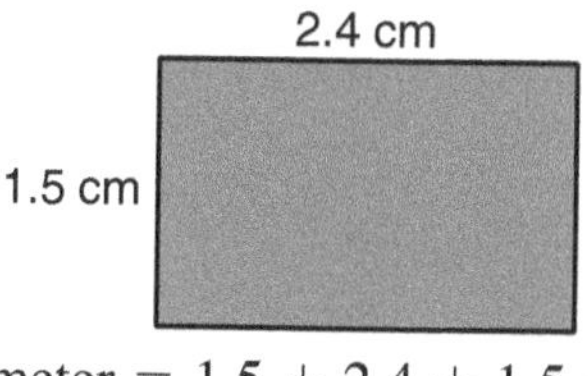

Perimeter $= 1.5 + 2.4 + 1.5 + 2.4$
$= 7.8\,\text{cm}$

Area $=$ length $\times$ breadth
$= 2.4 \times 1.5$
$= 3.6\,\text{cm}^2$

2 Calculate the area of this triangle.

7 cm
12 cm

$A = \frac{1}{2} \times b \times h$
$= \frac{1}{2} \times 12 \times 7$
$= 42\,\text{cm}^2$

3 Find the area of this trapezium.

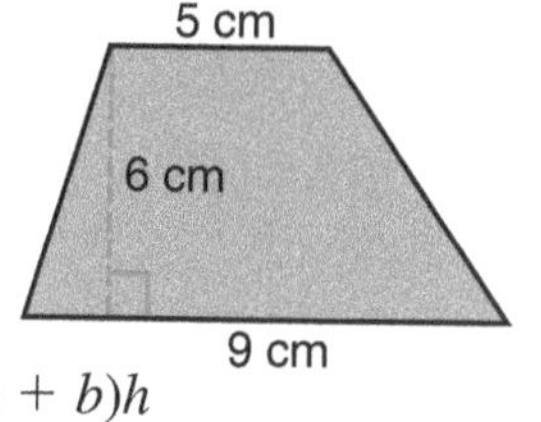

$A = \frac{1}{2}(a + b)h$
$= \frac{1}{2}(5 + 9)6$
$= \frac{1}{2} \times 14 \times 6$
$= 42\,\text{cm}^2$

4 The area of a rectangular room is $17.5\,\text{m}^2$.
The room is 5 m long.
Find the width of the room.

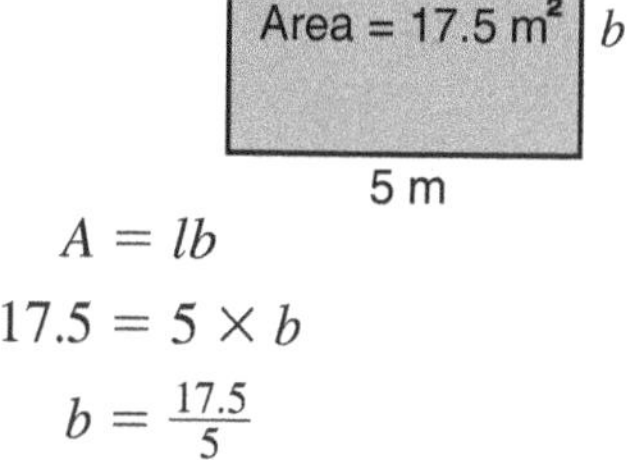

$A = lb$
$17.5 = 5 \times b$
$b = \frac{17.5}{5}$
$b = 3.5\,\text{m}$

Exercise 25.1

1. Calculate the perimeters and areas of these rectangles and squares.

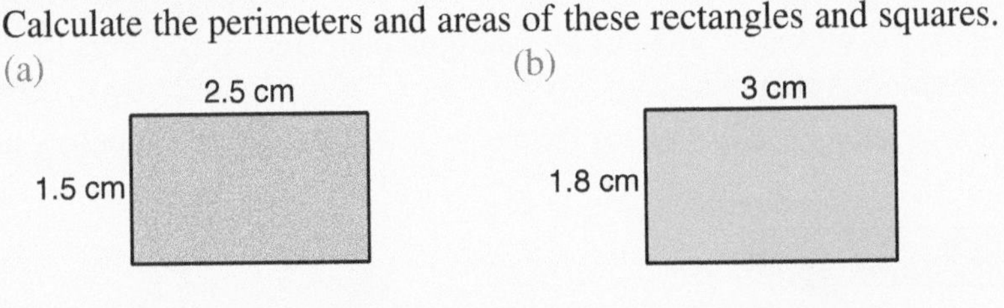

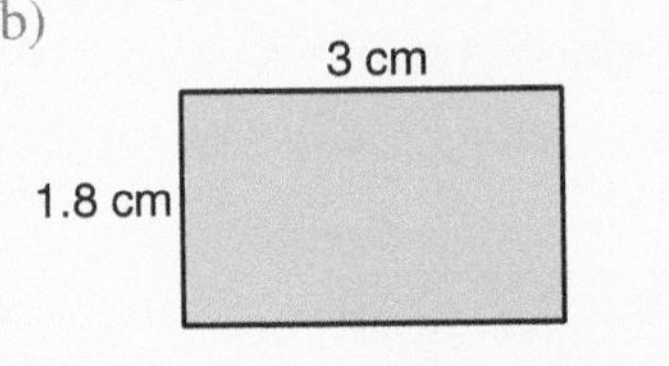

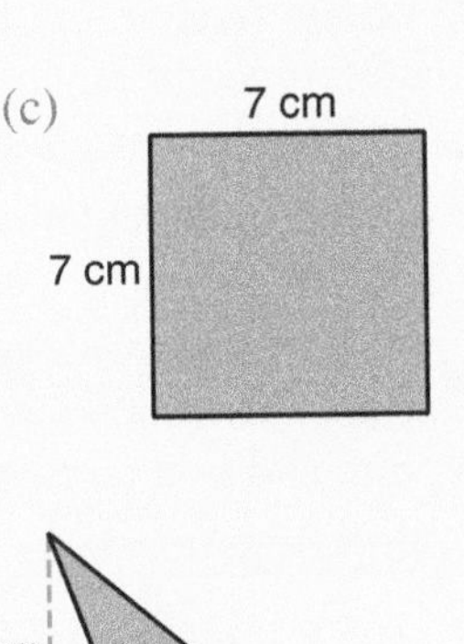

2. Calculate the areas of these triangles.

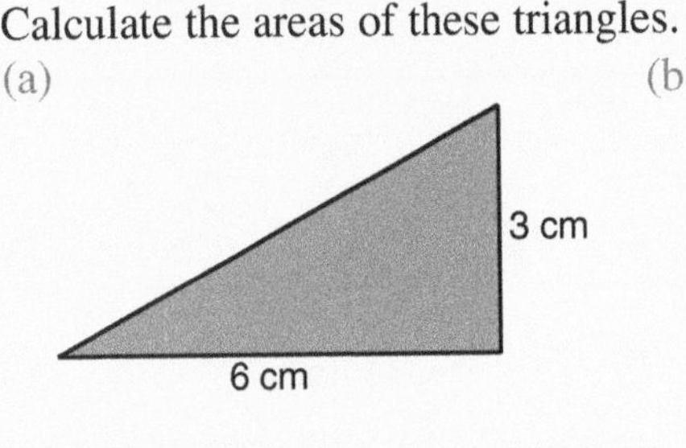

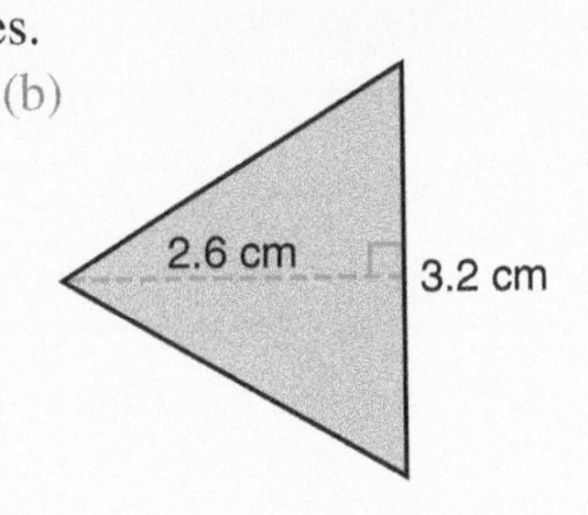

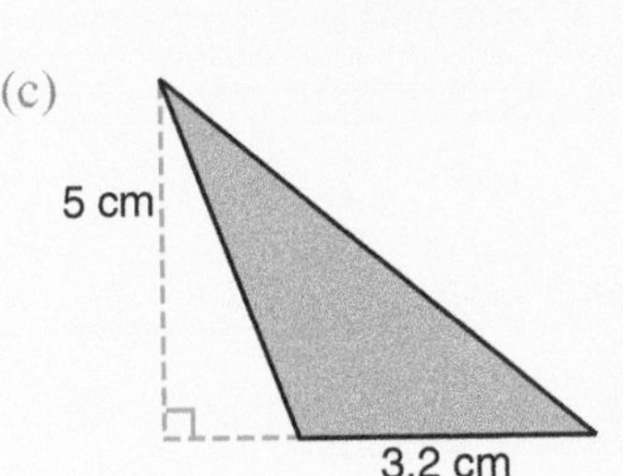

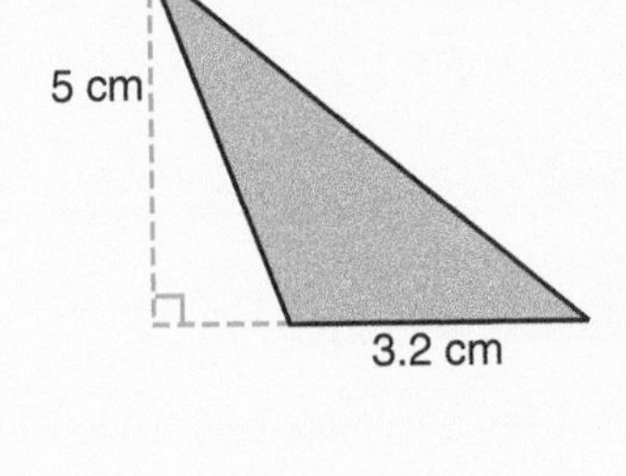

3. Calculate the areas of these shapes.

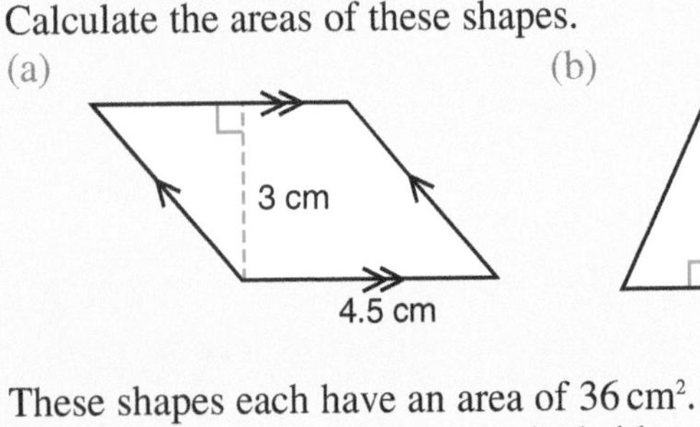

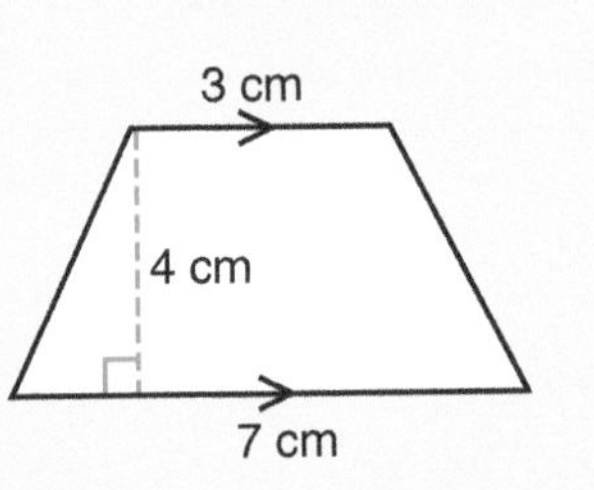

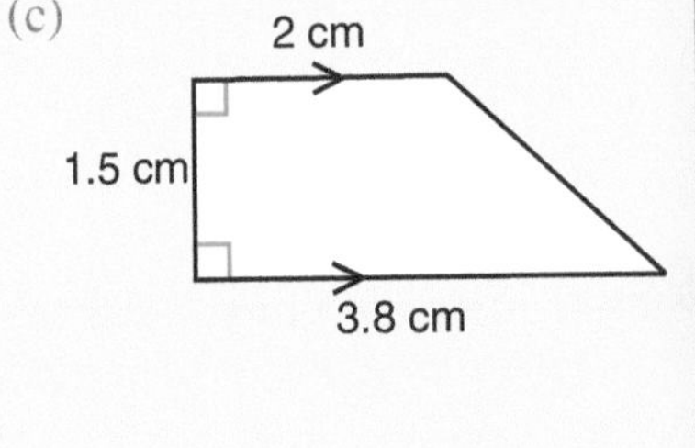

4. These shapes each have an area of $36\,\text{cm}^2$.
   Calculate the lengths of the marked sides.

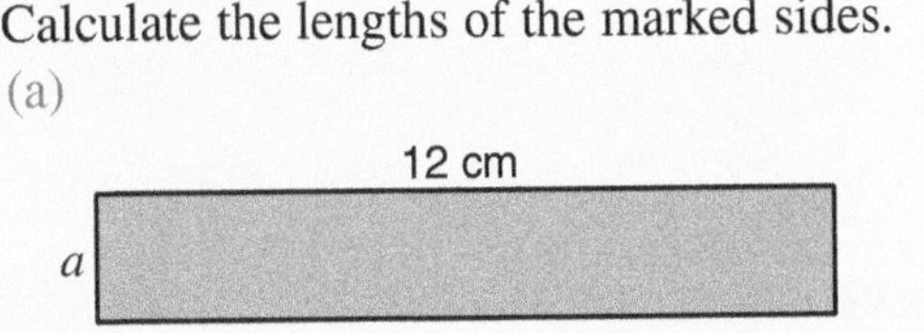

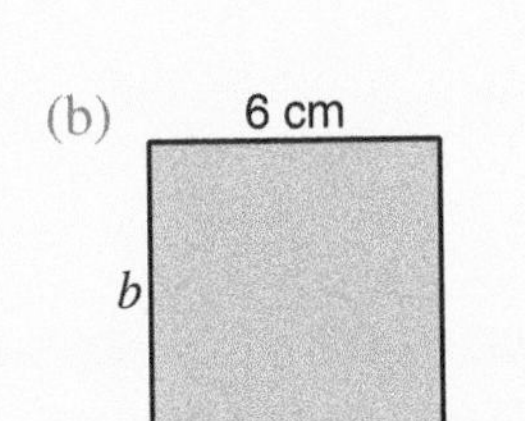

(c) 8 cm, c

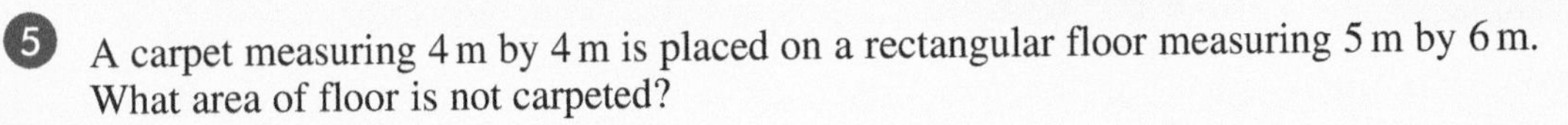

5. A carpet measuring 4 m by 4 m is placed on a rectangular floor measuring 5 m by 6 m.
   What area of floor is not carpeted?

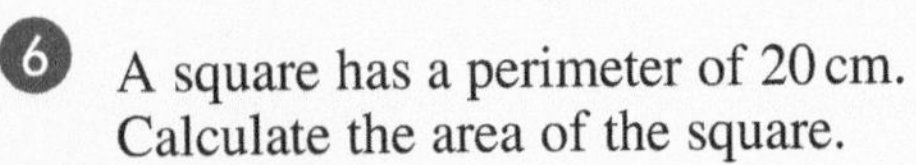

6. A square has a perimeter of 20 cm.
   Calculate the area of the square.

7. The diagram shows a picture in a rectangular frame.
   The outer dimensions of the frame are 18 cm by 10 cm.
   The frame is 2 cm wide.
   What is the area of the picture?

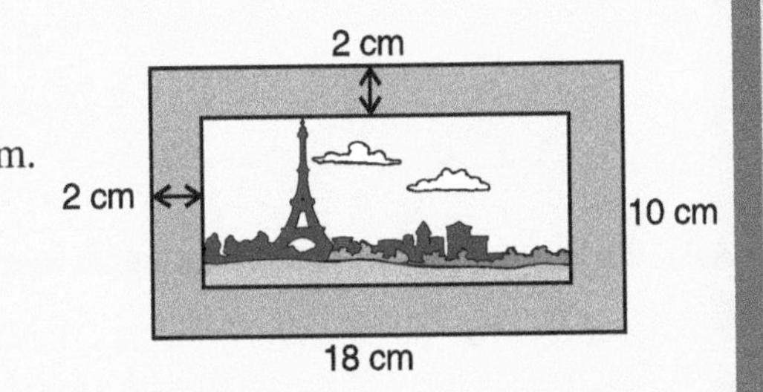

8. A trapezium has an area of $30\,\text{cm}^2$.
   The two parallel sides are 7 cm and 8 cm.
   What is the perpendicular distance between these sides?

9. This triangle has an area of $150\,\text{cm}^2$.
   Calculate the perimeter of the triangle.

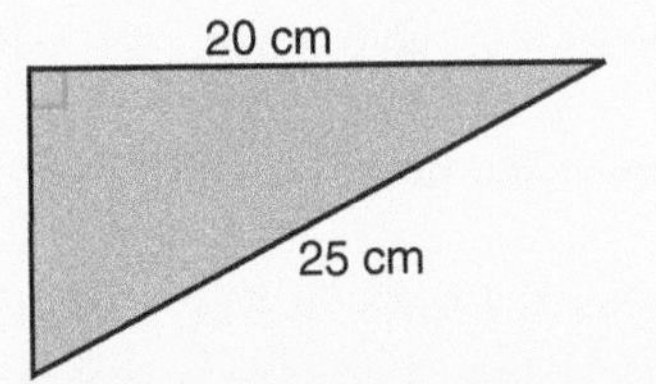

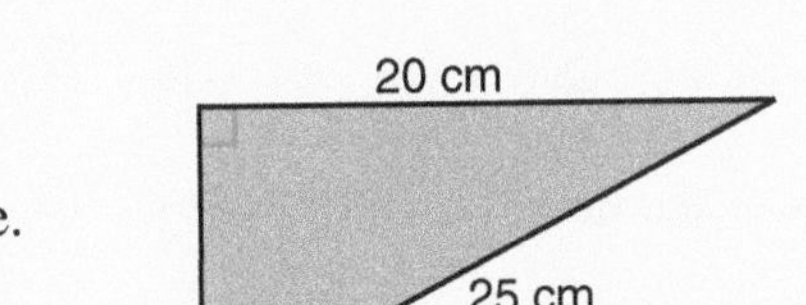

Circles

The Greek letter π

The circumference of any circle is just a bit bigger than three times the diameter of the circle.
The Greek letter π is used to represent this number.

We use an approximate value for π, such as 3, $3\frac{1}{7}$, 3.14, or the π key on a calculator, depending on the accuracy we require.

Circumference of a circle

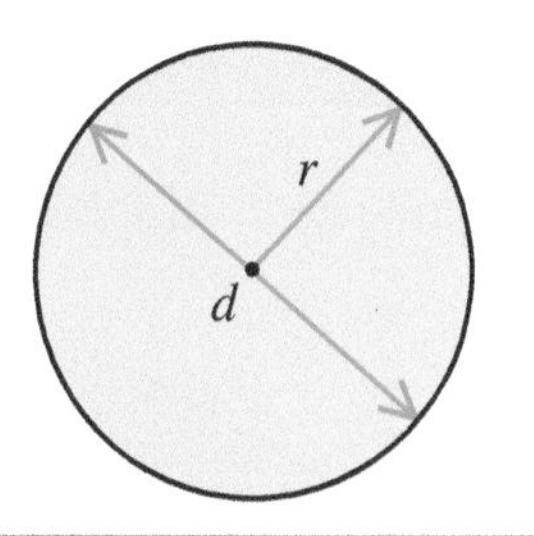

The diagram shows a circle with radius r and diameter d.

The **circumference**, C, of a circle can be found using the formulae:

$$C = \pi \times d \quad \text{or} \quad C = 2 \times \pi \times r$$

EXAMPLES

1 Find the circumference of a circle with diameter 80 cm. Take π to be 3.14.
Give your answer to the nearest centimetre.

$C = \pi \times d$
$= 3.14 \times 80\,\text{cm}$
$= 251.2\,\text{cm}$ Circumference is 251 cm to the nearest centimetre.

2 A circle has circumference 37.2 cm.
Find the radius of the circle, giving your answer to the nearest millimetre. Take $\pi = 3.14$.

$C = 2 \times \pi \times r$
$37.2 = 2 \times 3.14 \times r$
$37.2 = 6.28 \times r$
$r = \frac{37.2}{6.28}$
$r = 5.923\ldots$ $r = 5.9\,\text{cm}$, to the nearest millimetre.

Exercise 25.2

Take π to be 3.14 or use the π key on your calculator.

1 Calculate the circumference of these brackets.

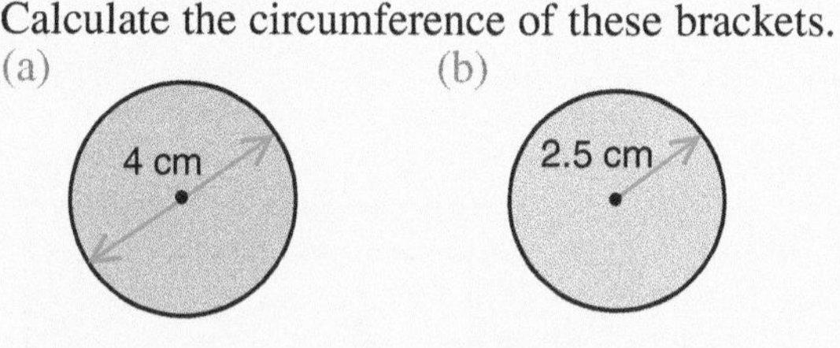
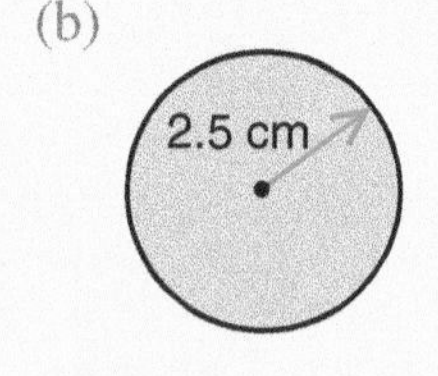
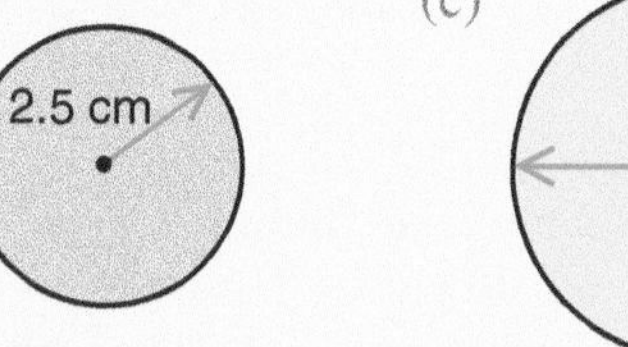
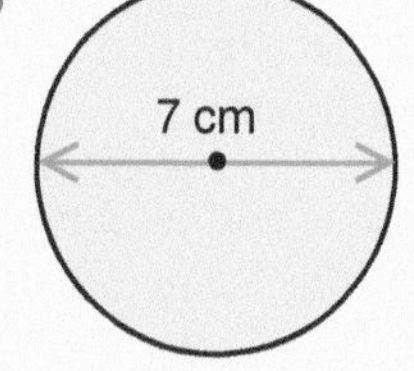

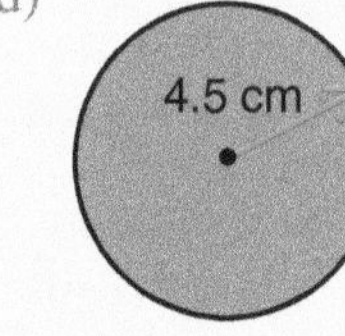
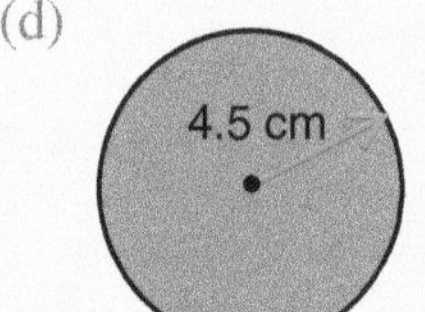

2 A circular biscuit tin has a diameter of 24 cm. What is the circumference of the tin?

3 A circle has a radius of 6.5 cm. Calculate the circumference of the circle.
Give your answer correct to one decimal place.

4 The circumference of a copper pipe is 94 mm.
Find, to the nearest millimetre, the diameter of the pipe.

5 The circumference of a bicycle wheel is 190 cm.
Find the diameter of the wheel,
giving your answer to the nearest centimetre.

6 The circumference of the London Eye is approximately 420 metres.
What is the radius? Give your answer correct to the nearest metre.

Area of a circle

The area of a circle can be found using the formula: $A = \pi \times r^2$

EXAMPLES

1 Estimate the area of a circle with radius of 6 cm.
Take π to be 3.

$A = \pi \times r \times r$
$= 3 \times 6 \times 6$
$= 108\,\text{cm}^2$

Area is approximately $108\,\text{cm}^2$.

2 Calculate the area of a circle with diameter 10 cm.
Give your answer in terms of π.

Remember: $r = \frac{d}{2}$

$A = \pi \times r \times r$
$= \pi \times 5 \times 5$
$= 25\pi\,\text{cm}^2$

The area of the circle is $25\pi\,\text{cm}^2$.

3 The top of a tin of cat food has an area of $78.5\,\text{cm}^2$.
What is the radius of the tin?
Take $\pi = 3.14$.

$A = \pi \times r^2$

Substitute values for A and π.

$78.5 = 3.14 \times r^2$

Solve this equation to find r.
Divide both sides of the equation by 3.14.

$\frac{78.5}{3.14} = r^2$

$r^2 = 25$

Take the square root of both sides.

$r = 5$

The radius of the tin is 5 cm.

Exercise 25.3

Take π to be 3.14 or use the π key on your calculator.

1 Calculate the areas of these circles. Give your answers correct to one decimal place.

(a) (b) (c) (d)

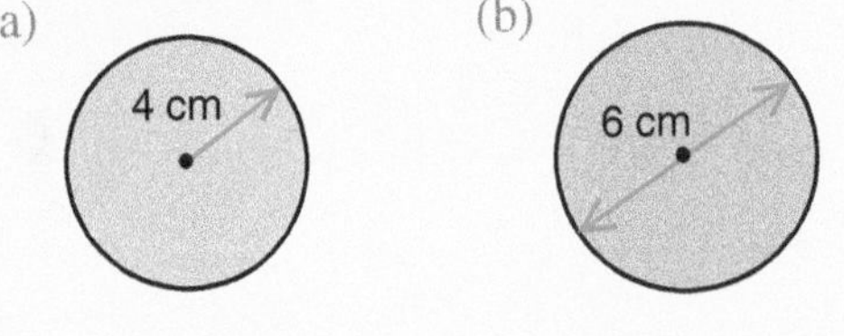

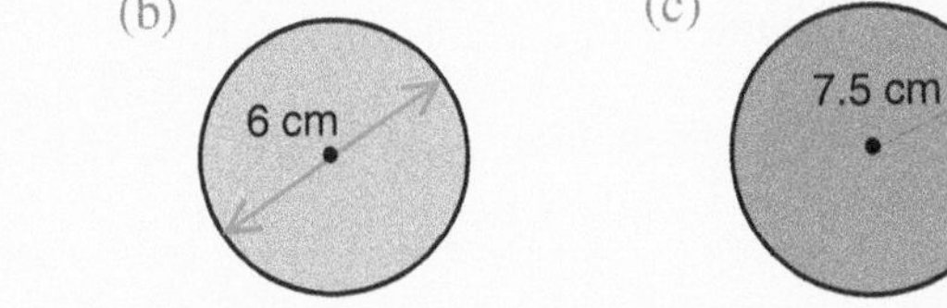

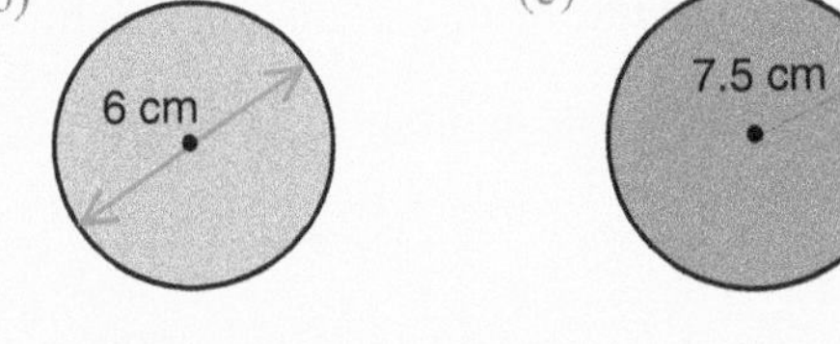

2 Calculate the shaded area in each of these diagrams.

(a) (b) (c)

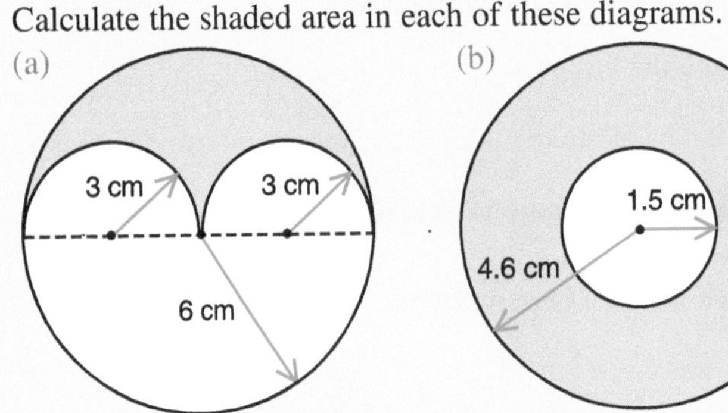

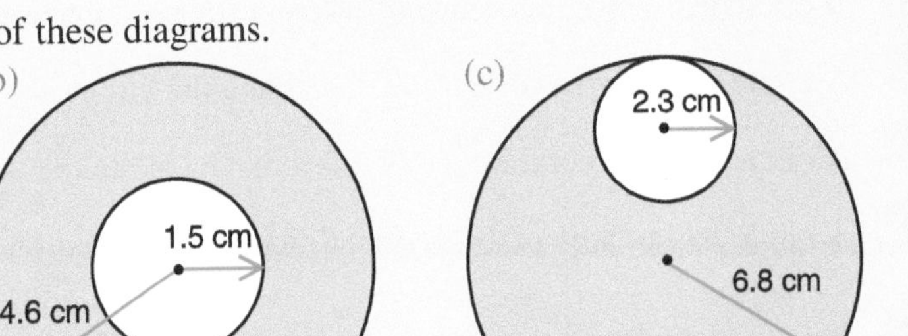

3 A circle has an area of $50\,\text{cm}^2$.
Calculate the radius of the circle.

4 A circular flower bed has an area of $40\,\text{m}^2$.
Calculate the diameter of the flower bed.
Give your answer to a suitable degree of accuracy.

Mixed questions involving circumferences and areas of circles

Some questions will involve finding the area and some the circumference of a circle.
Remember: Choose the correct formula for area or circumference.
You need to think about whether to use the radius or the diameter.

Exercise 25.4

In this exercise take π to be 3.14 or use the π key on your calculator.

1. Thirty students join hands to form a circle.
   The diameter of the circle is 8.5 m.
   (a) Find the circumference of the circle.
   Give your answer to the nearest metre.
   (b) What area is enclosed by the circle?

2. A semi-circle has a radius of 6 cm.
   Find the perimeter of a semi-circle with radius 6 cm.
   Give your answer to the nearest whole number.

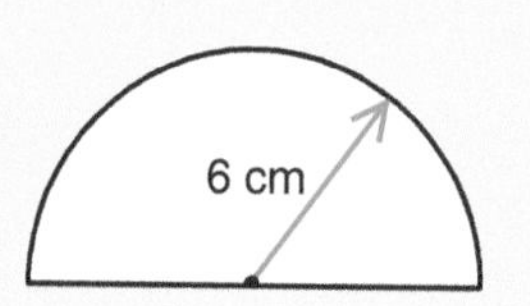

3. Which has the greater area:
   a circle with radius 4 cm, or a semi-circle with diameter 11 cm?
   You must show all your working.

4. The radius of a circular plate is 15 cm.
   (a) What is its area?
   (b) What is its circumference?
   Give your answers in terms of π.

5. The wheel of a wheelbarrow rotates 60 times when it is pushed a distance of 50 m.
   Calculate the radius of the wheel.

6. Calculate the area of a semi-circle with a diameter of 6.8 cm.

7. Calculate the perimeter of a semi-circle with a radius of 7.5 cm.

8. A pastry cutter is in the shape of a semi-circle.
   The straight side of the semi-circle is 12 cm long.
   How long is the curved side?

9. A circle has a circumference of 64 cm. Calculate the area of the circle.

10. A circle has an area of 128 cm^2. Calculate the circumference of the circle.

11. A circle has a circumference of 14π cm. Calculate the area of the circle in terms of π.

12. A circle has an area of 144π cm^2. Calculate the circumference of the circle in terms of π.

13. The diagram shows a small circle drawn inside a larger circle.
    The shaded area is 55π cm^2.
    The small circle has a radius of 3 cm.
    Calculate the radius of the larger circle.

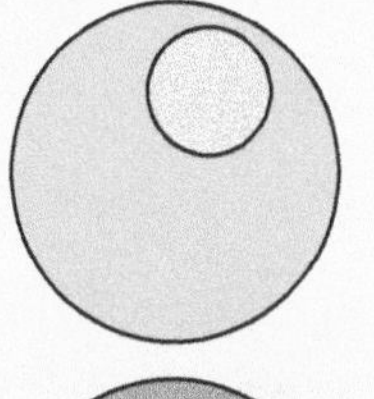

14. The diagram shows a small circle drawn inside a larger circle.
    The small circle has an area of 16π cm^2.
    The larger circle has a circumference of 18π cm.
    Calculate the shaded area.
    Give your answer in terms of π.

Compound shapes

Shapes formed by joining different shapes together are called **compound shapes**.
To find the area of a compound shape we must first divide the shape up into rectangles, triangles, circles, etc, and then find the area of each part.
Shapes can be divided in different ways, but they should all give the same answer.

EXAMPLE

Calculate the area of this shape.
Give the answer correct to three significant figures.

The shape can be split into: a rectangle $ABCD$,
a trapezium $BXYC$ and
a semi-circle XYZ.

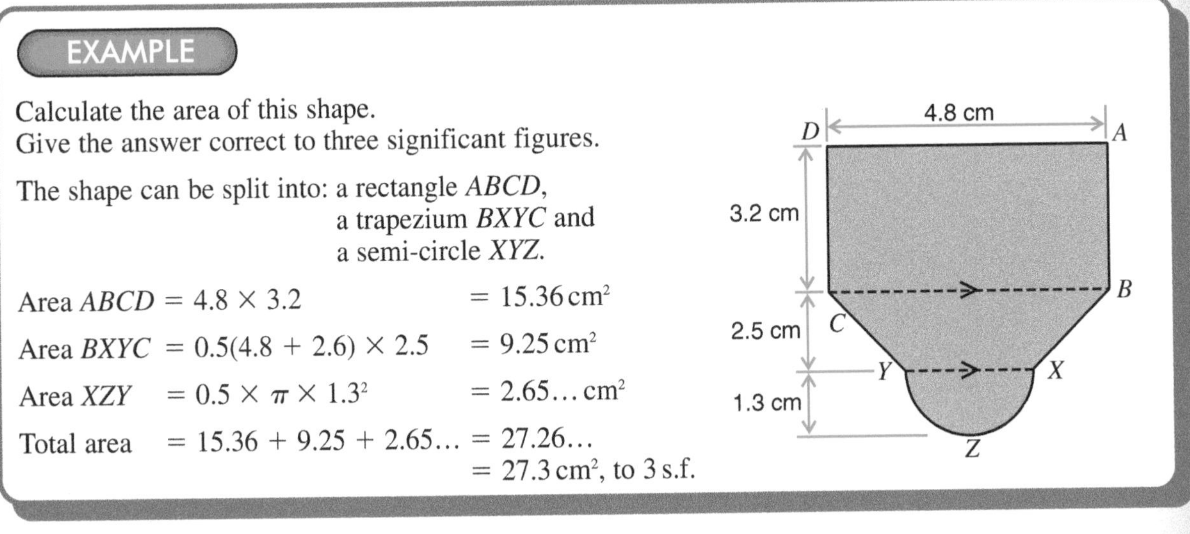

Area $ABCD = 4.8 \times 3.2 = 15.36\,\text{cm}^2$

Area $BXYC = 0.5(4.8 + 2.6) \times 2.5 = 9.25\,\text{cm}^2$

Area $XZY = 0.5 \times \pi \times 1.3^2 = 2.65\ldots\,\text{cm}^2$

Total area $= 15.36 + 9.25 + 2.65\ldots = 27.26\ldots$
$= 27.3\,\text{cm}^2$, to 3 s.f.

Exercise 25.5

1. Find the areas of these shapes which are made up of rectangles.
   (a) (b)

2. Find the areas of these shapes.
   (a) (b) (c)

3. Calculate the shaded areas in each of these shapes.
   (a) (b) (c)

Segments and sectors

A circle can be divided up in different ways.

Segment – a chord divides a circle into two segments.

Sector – two radii divide a circle into two sectors.

Lengths of arcs and areas of sectors

The lengths of arcs and the areas of sectors of circles are in proportion to the angle at the centre of the circle.

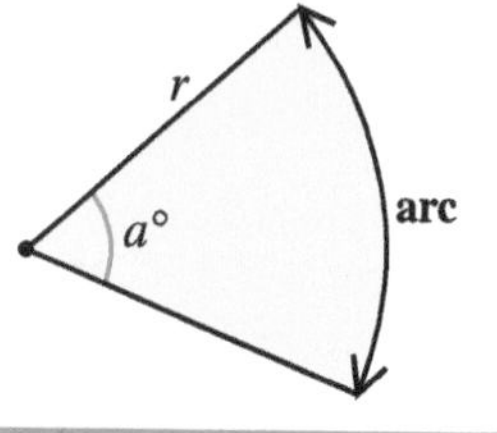

For a sector with angle $a°$ Length of arc $= \frac{a}{360} \times \pi d$ Area of sector $= \frac{a}{360} \times \pi r^2$

EXAMPLE

This shape is a sector of a circle with radius 9 cm and angle 80°.
Calculate (a) the length of arc AB,
(b) the area of sector OAB.

(a) Length $AB = \frac{a}{360} \times \pi d$
$= \frac{80}{360} \times \pi \times 18$
$= 12.56...$
$= 12.6$ cm, to 3 s.f.

(b) Area $OAB = \frac{a}{360} \times \pi r^2$
$= \frac{80}{360} \times \pi \times 9^2$
$= 56.54...$
$= 56.5$ cm², to 3 s.f.

Exercise 25.6

Take π to be 3.14 or use the π key on your calculator.
Give answers correct to three significant figures.

1 Calculate (a) the length of the arc of each sector and (b) the area of each sector.

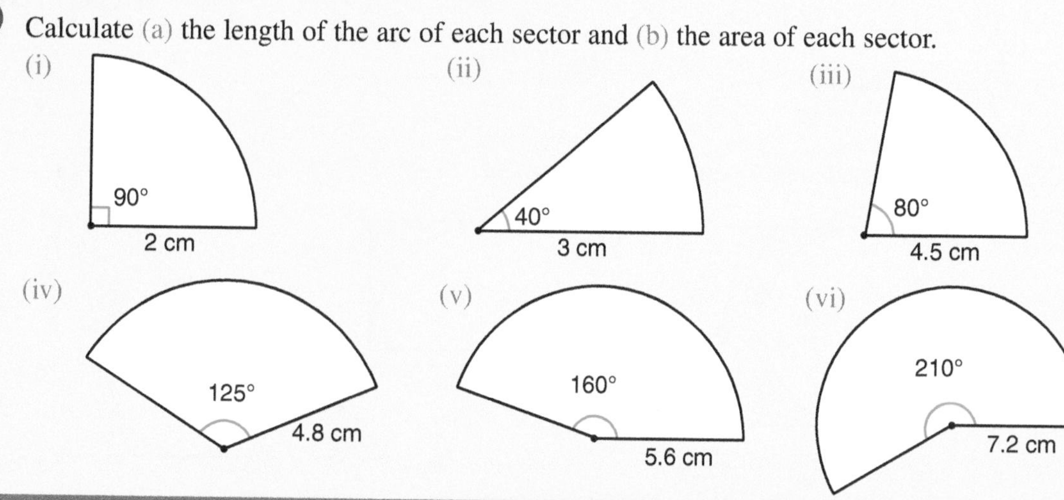

2 OPQ is a sector of a circle with radius 20 cm.
X is the midpoint of OP and Y is the midpoint of OQ.
Angle $POQ = 65°$.

Calculate
(a) the area of $XPQY$,
(b) the perimeter of $XPQY$.

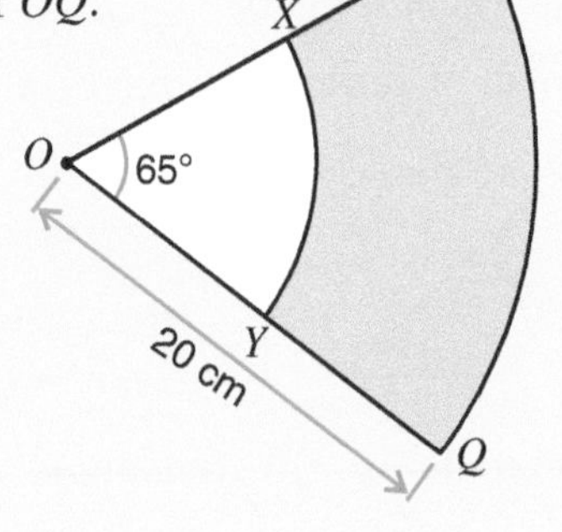

3 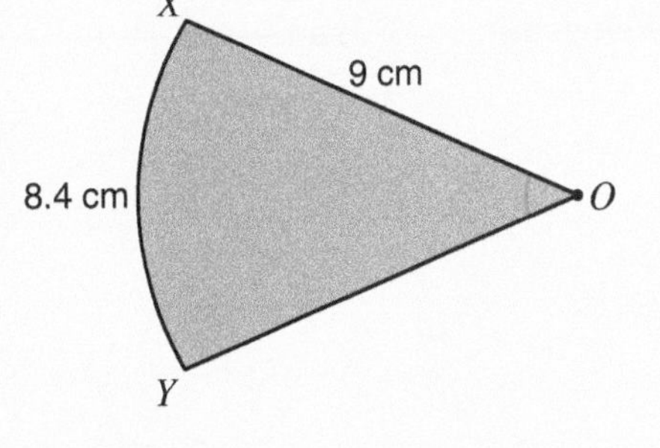

OXY is a sector of a circle, centre O, with radius 9 cm.
The arc length of the sector is 8.4 cm.

Calculate the size of angle XOY.

4 OPQ is a sector of a circle, centre O.
The arc length of the sector is 15.6 cm.
Angle $POQ = 135°$.

Calculate the radius of the circle.

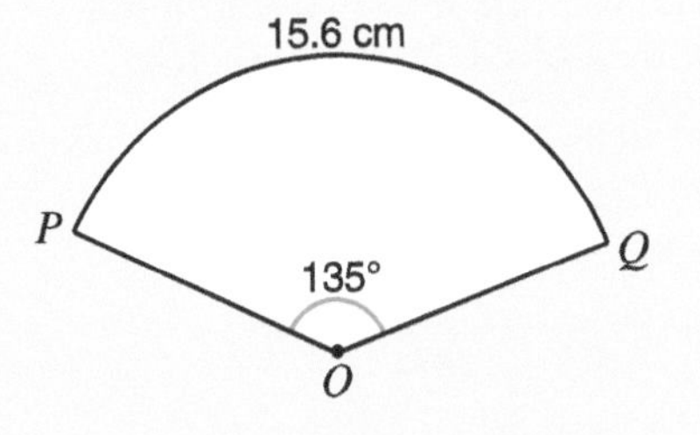

5 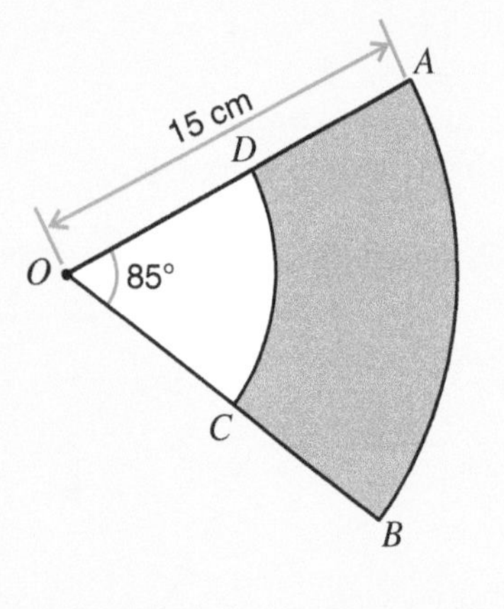

AB and DC are arcs of a circle, centre O.
$OA = 15$ cm and angle $AOB = 85°$.
The area of $ABCD$ is 100 cm^2.

Calculate
(a) the length of OD,
(b) the length of the arc DC.

6 The diagram shows a quadrant of a circle, centre O.
Calculate the area of the shaded segment.

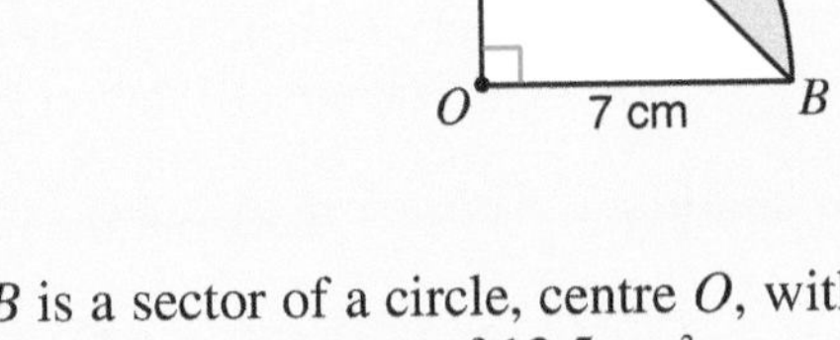

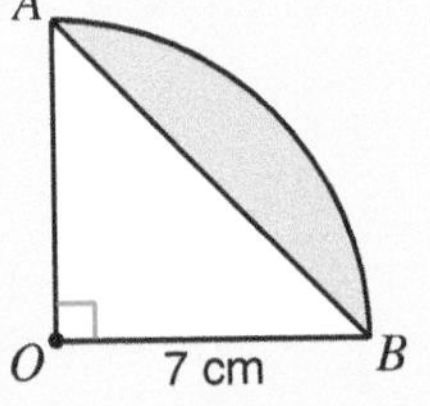

7 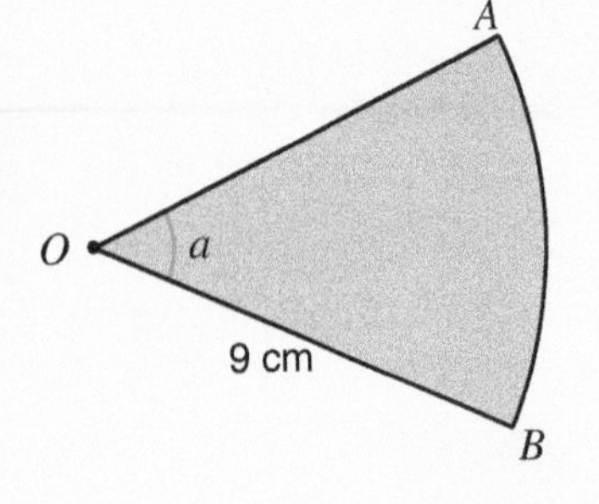

OAB is a sector of a circle, centre O, with radius 9 cm.
The sector has an area of 13.5 cm^2.

Calculate the size of angle a.

8 A sector of a circle has an area of 30 cm^2.
Angle $a = 72°$.

Calculate the radius of the circle.

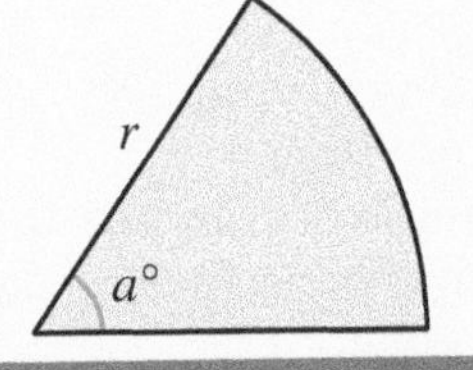

What you need to know

- **Perimeter** is the distance round the outside of a shape.
- **Area** is the amount of surface covered by a shape.
- **Area formulae**

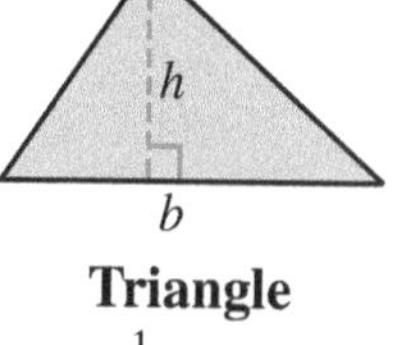
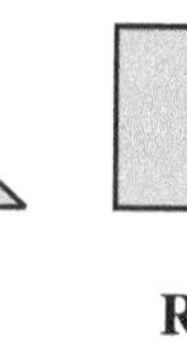
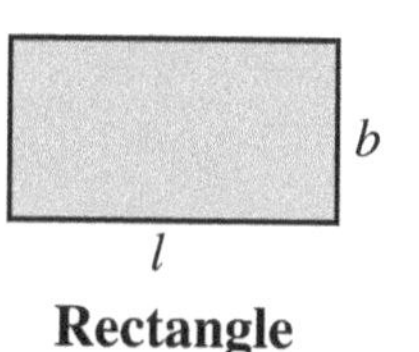
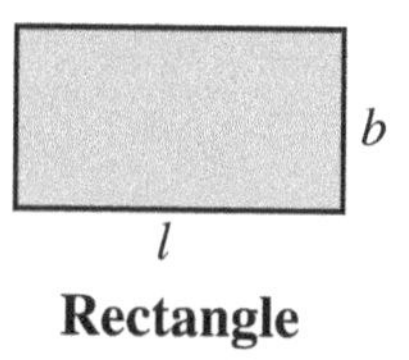
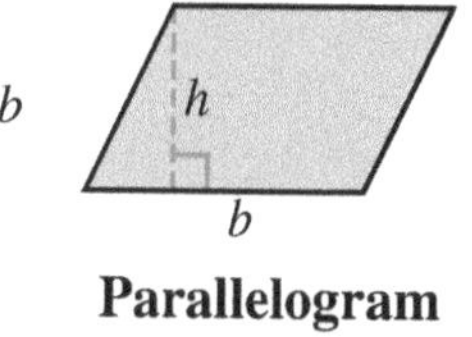
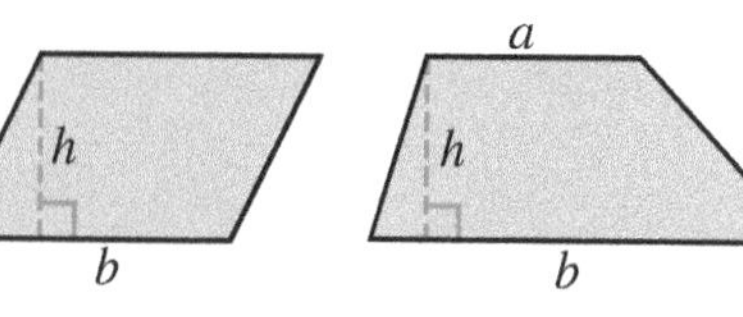

Triangle	Rectangle	Parallelogram	Trapezium
$A = \frac{1}{2} \times b \times h$	$A = lb$	$A = bh$	$A = \frac{1}{2}(a + b)h$

- The **circumference** of a circle is given by:
  $C = \pi \times d$ or $C = 2 \times \pi \times r$
- The **area** of a circle is given by: $A = \pi r^2$
- A **chord** divides a circle into two **segments**.
  Two radii divide a circle into two **sectors**.
- The **lengths of arcs** and the **areas of sectors** are proportional to the angle at the centre of the circle.
  For a sector with angle $a°$

  Length of arc $= \frac{a}{360} \times \pi d$

  Area of sector $= \frac{a}{360} \times \pi r^2$

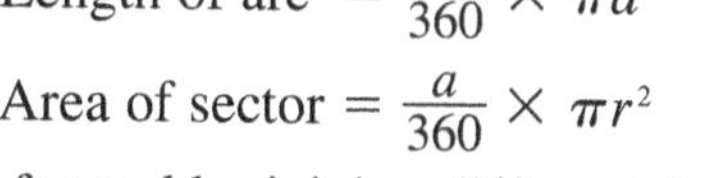

- Shapes formed by joining different shapes together are called **compound shapes**.
  To find the area of a compound shape we must first divide the shape up into rectangles, triangles, circles, etc, and then find the area of each part.

Review Exercise 25

Take π to be 3.14 or use the π key on your calculator.

1 $ABCD$ is a trapezium.
(a) Find the area of $ABCD$.
(b) Find the area of triangle BCD.

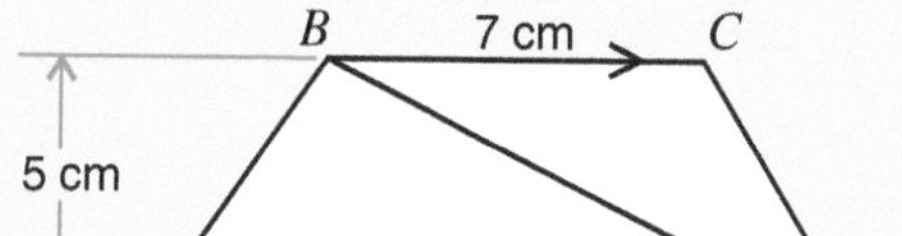
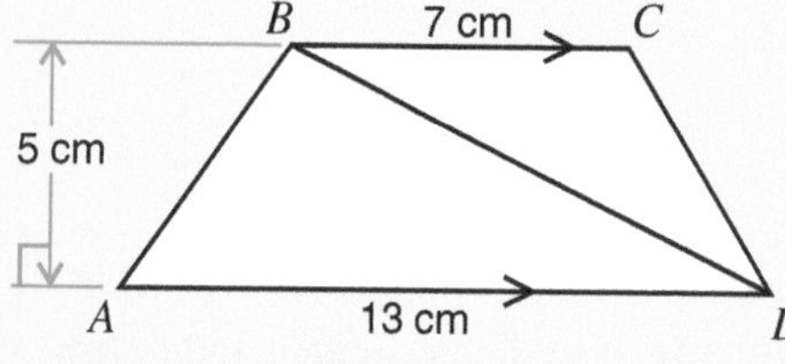

AQA

2

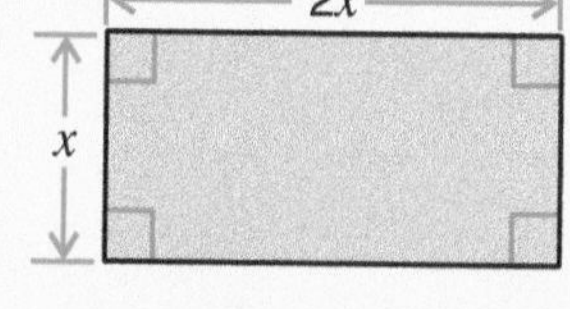

The length of a rectangle is twice its width.
The area of the rectangle is $50\,cm^2$.
(a) Show that the length of the rectangle is 10 cm.
(b) Work out the perimeter of the rectangle.

AQA

3 (a) Calculate the area of a circle of radius 1.2 cm.
(b) Calculate the circumference of a circle of diameter 3.5 cm.

AQA

4 Calculate the area of a circle with a diameter of 15 cm.
Give your answer to an appropriate degree of accuracy.

AQA

5 Andre is rolling a hoop along the ground.
The hoop has a diameter of 90 cm.
What is the minimum number of complete turns the hoop must make to cover a distance of 5 m?

AQA

6 A circular flower bed has an area of $84\,\text{m}^2$.
Explain why the radius of the flower bed must be bigger than 5 m. AQA

7 The diagram shows a window.
The arc AB is a semi-circle.
$BC = AD = 75\,\text{cm}$, $DC = 80\,\text{cm}$.
Calculate the area of the window.

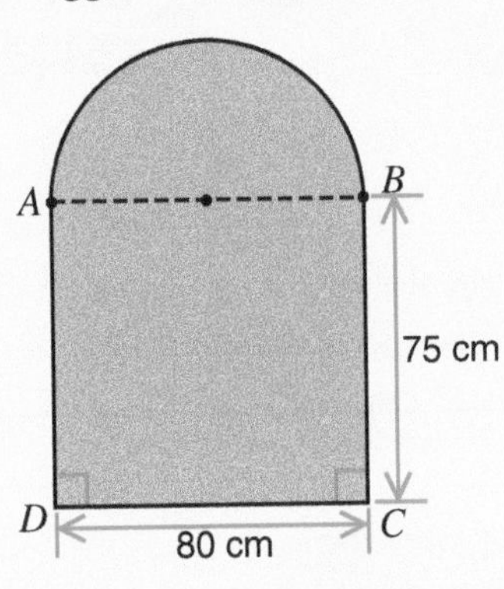
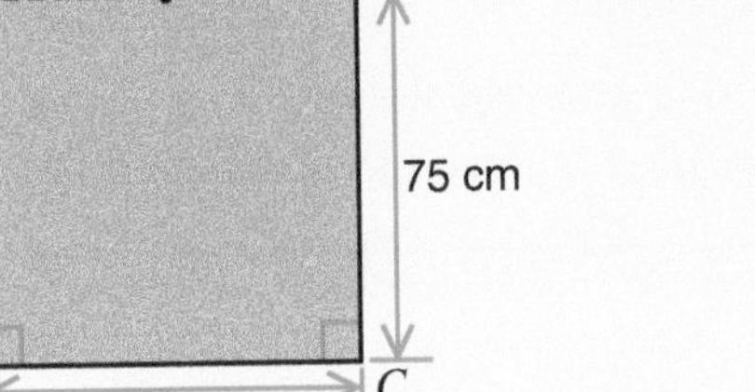

AQA

8 A circle has a circumference of 42 cm.
Calculate the area of the circle.
Give your answer correct to two significant figures.

9 The diagram shows a trapezium $PQRS$.
$PQ = 9.7\,\text{cm}$ and $SR = 4.1\,\text{cm}$.
The area of the trapezium is $41.4\,\text{cm}^2$.
Calculate the height, h cm, of the trapezium. AQA

10 A wiper blade on a windscreen cleans the clear area as shown.

Calculate the area of the windscreen cleaned by the wiper.

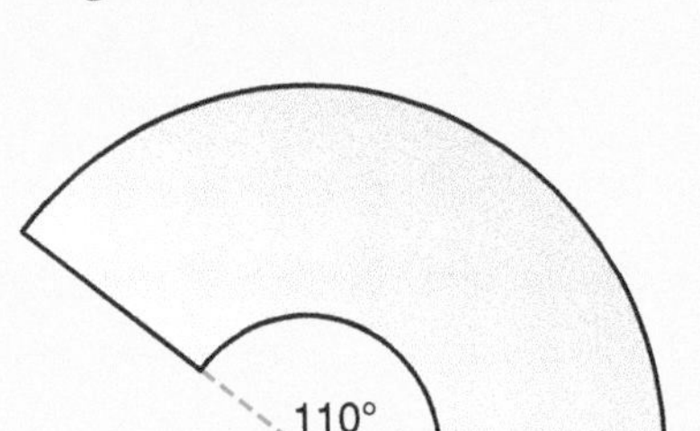
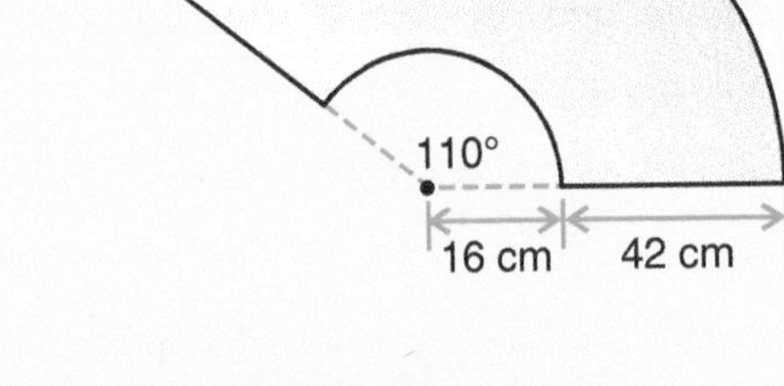
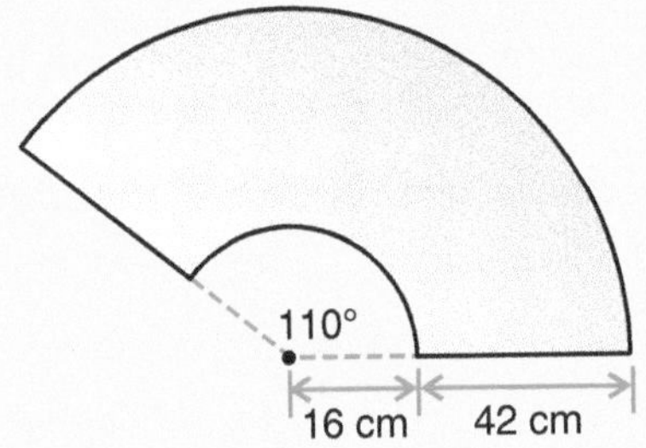

AQA

11 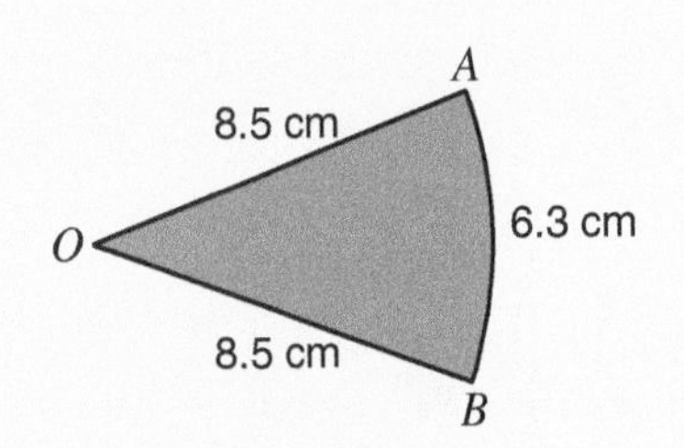

OAB is a sector of radius 8.5 cm.
The length of the arc, AB, is 6.3 cm.

Calculate angle AOB.

AQA

12 The diagram shows a Go Kart track.
The bend, AB, is the arc of a sector with centre P, radius 30 m and angle 60°.
The bend, BC, is the arc of a sector with centre Q, radius 22.5 m and angle 240°.

Calculate, in terms of π, the total distance along the track from A to B to C.

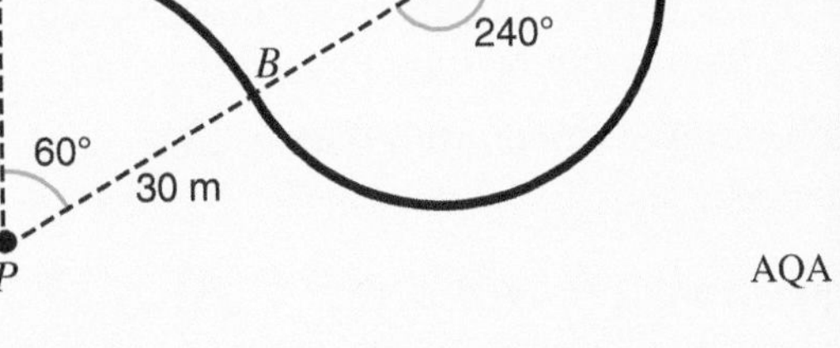
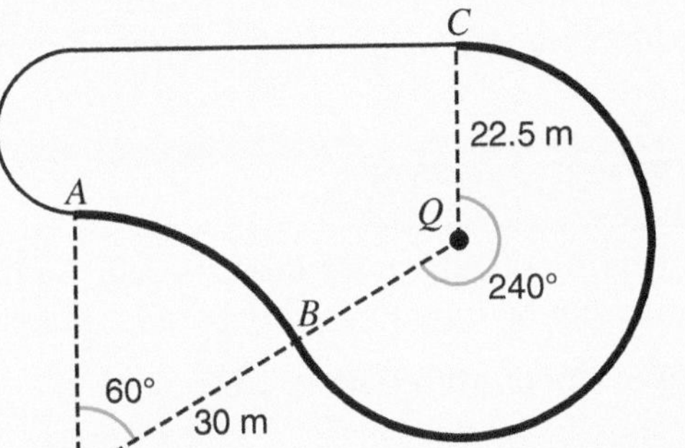

AQA

13 The area of a sector of a circle is $25\,\text{cm}^2$.
The radius of the sector is 6 cm.
Calculate the perimeter of the sector.

Maps, Loci and Constructions

Maps

When planning journeys we often use **maps**.
To interpret maps we need to understand:
angles in order to describe **direction** and **scales** in order to find **distances**.
Compass points and **three-figure bearings** are used to describe direction.

Compass points

The diagram shows the points of the compass.

The angle between North and East is 90°.

The angle between North and North-East is 45°.

Three-figure bearings

Bearings are used to describe the direction in which you must travel to get from one place to another.

A bearing is an angle measured from the North line in a clockwise direction.

The angle, which can be from 0° to 360°, is written as a three-figure number.

Bearings which are less than 100° include noughts to make up the three figures, e.g. 005°, 087°.

This diagram shows the positions of Bath and Poole.
The bearing of Poole from Bath is 162°.

If you are at Bath, facing North, and turn through 162° in a clockwise direction you will be facing in the direction of Poole.

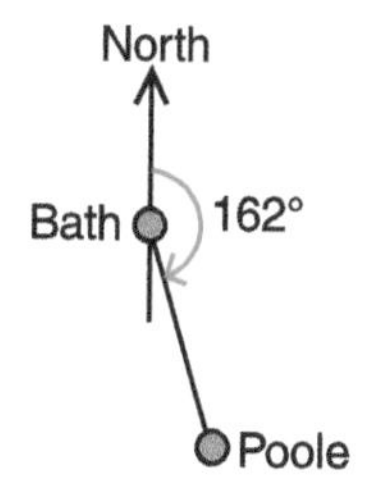

Back bearings

The return bearing of Bath from Poole is called a **back bearing**.
Back bearings can be found by using parallel lines and alternate angles.

The bearing of Poole from Bath is 162°.

$a = 162°$ (alternate angles)

Required angle = 180° + 162° = 342°.
The bearing of Bath from Poole (the back bearing) is 342°.

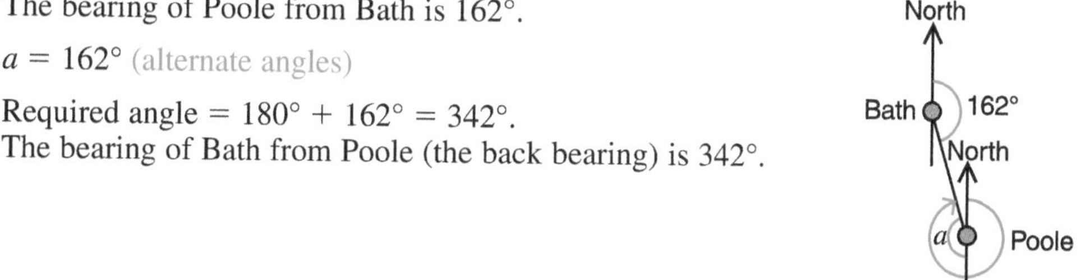

Scale drawing

Maps and plans are scaled down representations of real-life situations.
The **scale** used in drawing a map or plan determines the amount of detail that can be shown.

The distances between different points on a map are all drawn to the same scale.
There are two ways to describe a scale.

1. A scale of 1 cm to 10 km means that a distance of 1 cm on the map represents an actual distance of 10 km.
2. A scale of 1 : 10 000 means that all distances measured on the map have to be multiplied by 10 000 to find the real distance.

EXAMPLES

1. A road is 3.7 cm long on a map.
The scale given on the map is '1 cm represents 10 km'.
What is the actual length of the road?

1 cm represents 10 km.
Scale up, so multiply.
3.7 cm represents $3.7 \times 10\text{ km} = 37\text{ km}$
The road is 37 km long.

2. A plan of a field is to be drawn using a scale of 1 : 500.
Two trees in the field are 350 metres apart.
How far apart will they be on the plan?

Scale down, so divide.
Distance on plan $= 350\text{ m} \div 500$
Change 350 m to centimetres.
$= 35\,000\text{ cm} \div 500 = 70\text{ cm}$
The trees will be 70 cm apart on the plan.

Exercise 26.1

1. (a) Claire is facing South.
In which direction will she face after turning clockwise through an angle of 135°?
(b) Kevin turned anticlockwise through an angle of 270°.
He is now facing South-East.
In which direction was he facing?

2. Here is a map of an island.
(a) Use your protractor to find:
(i) the bearing of Q from P,
(ii) the bearing of P from Q.
(b) (i) Measure the distance between P and Q on the map.
(ii) What is the actual distance between P and Q?

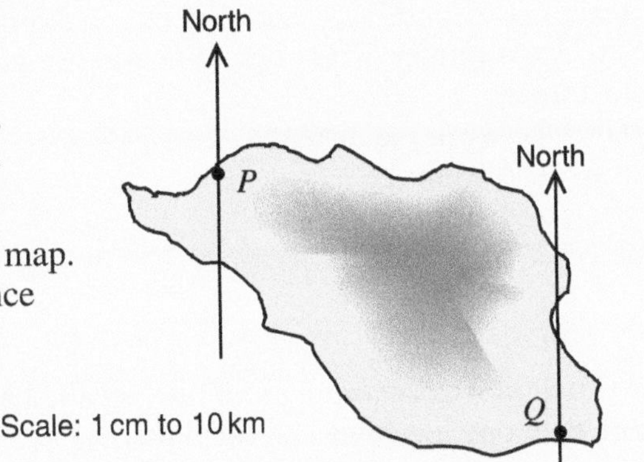

3. The scale of a map is 1 : 200.
(a) On the map a house is 3.5 cm long. How long is the actual house?
(b) A field is 60 m wide. How wide is the field on the map?

4 Helen drew a plan of her classroom using a scale of 5 cm to represent 1 m.
(a) Write the scale Helen used in the form 1 : n.
(b) On the plan, the length of the classroom is 29 cm.
What is the actual length of the classroom?
(c) The actual width of the classroom is 4.5 m.
What is the width of the classroom on the plan?

5 The diagram shows the plan of a cross-country course.
Runners have to go round markers at A, B and C.

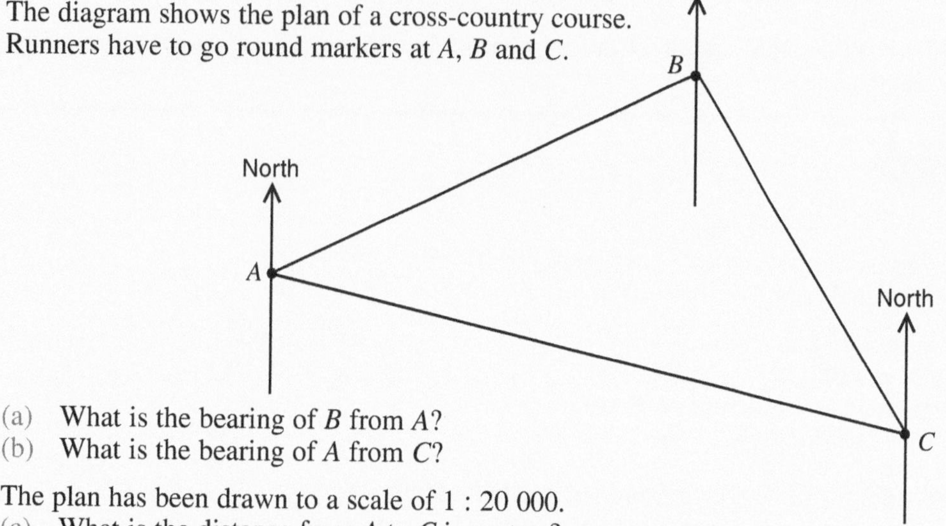

(a) What is the bearing of B from A?
(b) What is the bearing of A from C?

The plan has been drawn to a scale of 1 : 20 000.
(c) What is the distance from A to C in metres?

6 A building plot is a triangle with sides of length 60 m, 90 m and 100 m.
Construct an accurate scale drawing of the building plot.
Use a scale of 1 cm to 20 m.

7 A boat leaves port and sails on a bearing of 144° for 4 km.
It then changes course and sails due East for 5 km to reach an island.
Find by scale drawing:
(a) the distance of the island from the port,
(b) the bearing of the island from the port,
(c) the bearing on which the boat must sail to return directly to the port.

8 A yacht sails on a bearing of 040° for 5000 m and then a further 3000 m on a bearing of 120°.
Find by scale drawing:
(a) the distance of the yacht from its starting position,
(b) the bearing on which it must sail to return directly to its starting position.

9 An aircraft leaves an airport, at A, and flies on a bearing of 035° for 50 km and then on a bearing of 280° for a further 40 km before landing at an airport, at B.
Find by scale drawing:
(a) the distance between the airports,
(b) the bearing of B from A,
(c) the bearing of A from B.

Locus

The path of a point which moves according to a rule is called a **locus**.
If we talk about more than one locus we call them **loci**.

For example:
John walks so that he is always 2 metres from a lamp post.

The locus of his path is a circle, radius 2 metres.

2 m
lamp post

Exercise 26.2

1. Adam goes down this slide. Make a sketch of the slide as viewed from the side and show the locus of Adam's head as he goes down the slide.

2. The diagram shows part of a rectangular lawn.
Starting from the wall, Sally walks across the lawn so that she is always the same distance from both hedges. Draw a sketch to show the locus of Sally's path.

3. $PQRS$ is a square of side 8 cm.
A point X is inside the square.
X is less than 8 cm from P.
X is nearer to PQ than to SR.
Make a sketch showing where X could be.

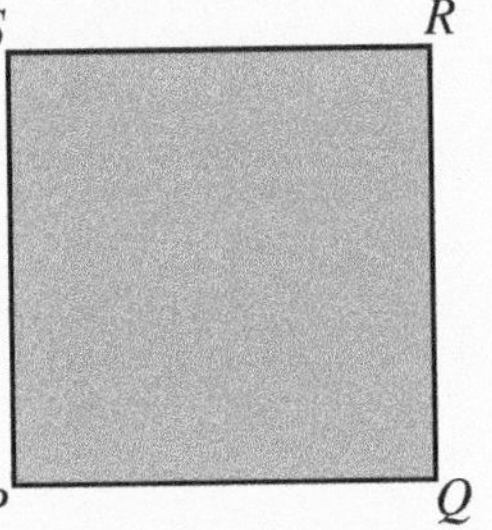

4. A ball is rolled down a step.
Copy the diagram, and sketch the locus of P, the centre of the ball, as it rolls from X to Y.

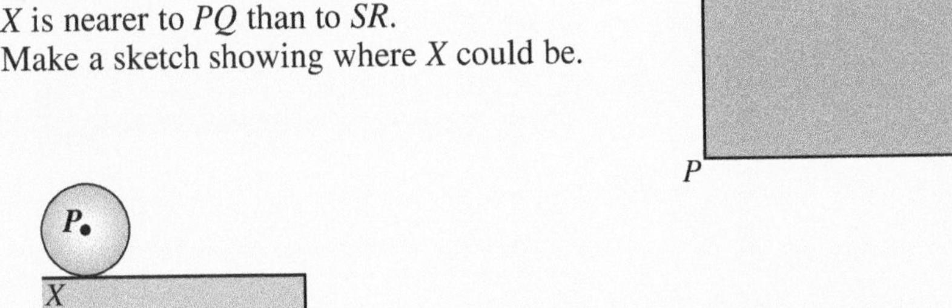

5. A wire is stretched between two posts.
A ring slides along the wire and a dog is attached to the ring by a rope.
Make a sketch to show where the dog can go.

6. A point P is 1 cm from this shape.
Copy the diagram, and draw an accurate locus of **all** the positions of P.

Accurate constructions

Sometimes it is necessary to construct loci accurately.
You are expected to use only a ruler and compasses.
Here are the methods for two constructions.

To draw the perpendicular bisector of a line

This means to draw a line at right angles to a given line dividing it into two equal parts.

1. Draw line AB.

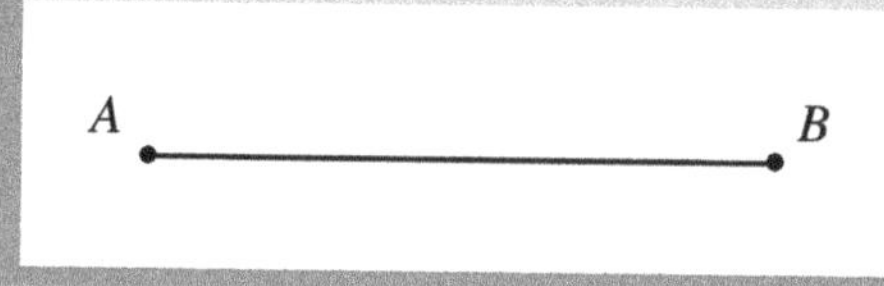

2. Open your compasses to just over half the distance AB. Mark two arcs which cross at C and D.

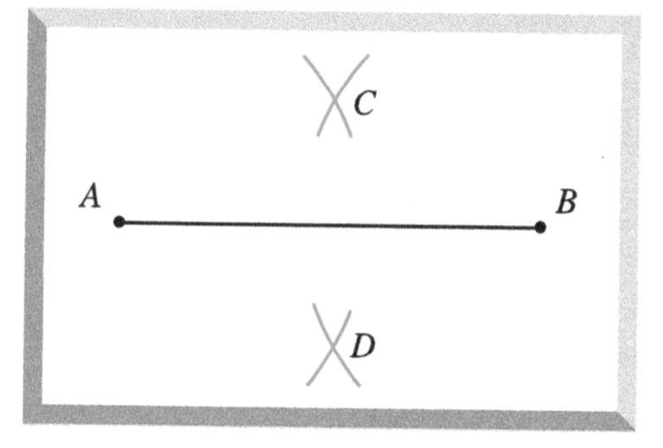

3. Draw a line which passes through the points C and D.

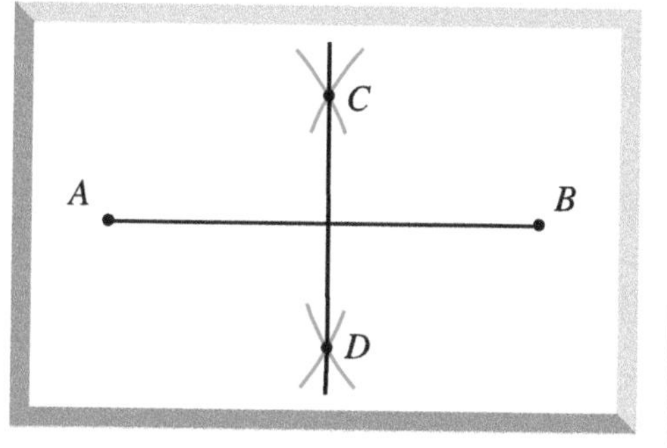

This line is the locus of a point which is the same distance from A and B.
Points on the line CD are **equidistant** (the same distance) from points A and B.
The line CD is at right angles to AB.
CD is sometimes called the **perpendicular bisector** of AB.

To draw the bisector of an angle

This means to draw a line which divides an angle into two equal parts.

1. Draw the angle A.
Use your compasses, centre A, to mark points B and C which are the same distance from A.

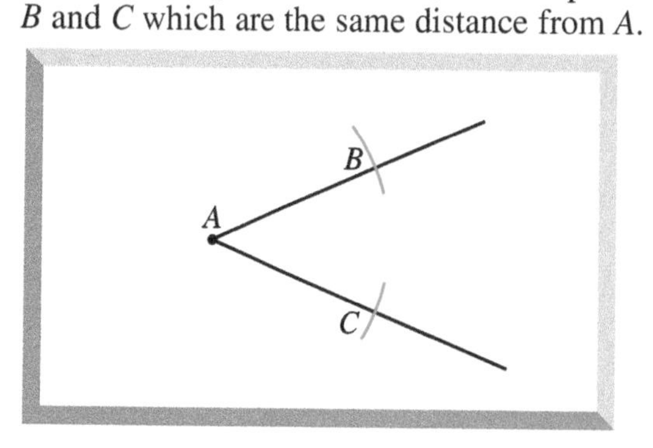

2. Use points B and C to draw equal arcs which cross at D.

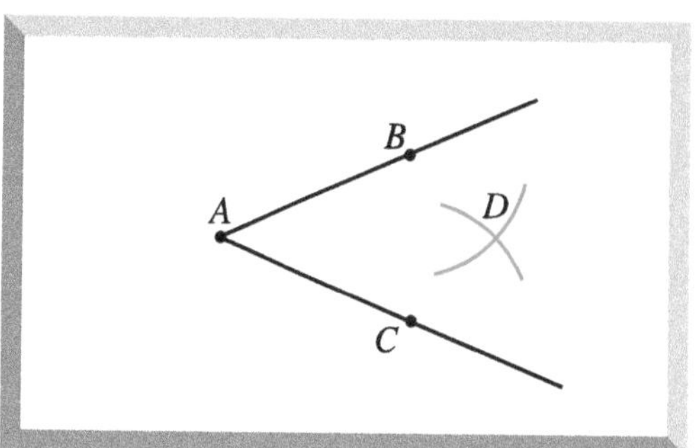

3. Draw a line which passes through the points A and D.

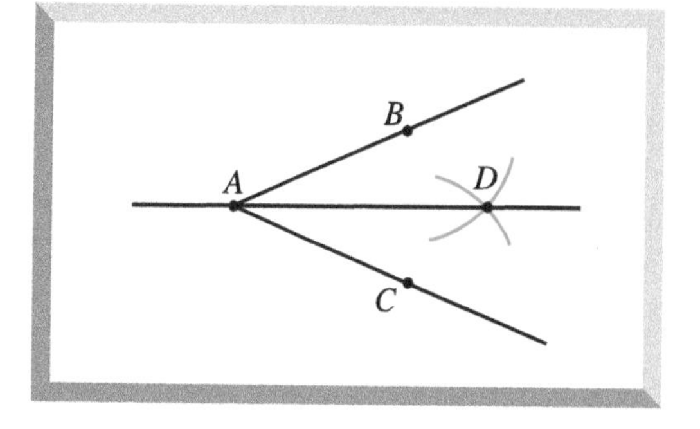

This line is the locus of a point which is the same distance from AB and AC.
Points on the line AD are **equidistant** from the lines through AB and AC.
The line AD cuts angle BAC in half.
AD is sometimes called the **bisector** of angle BAC.

Exercise 26.3

1 Mark two points, A and B, 10 cm apart. Construct the perpendicular bisector of AB.

2 Use a protractor to draw an angle of $60°$. Construct the bisector of the angle.
Check that both angles are $30°$.

3 Draw a triangle in the middle of a new page.
Construct the perpendicular bisectors of all three sides. They should meet at a point, Y.

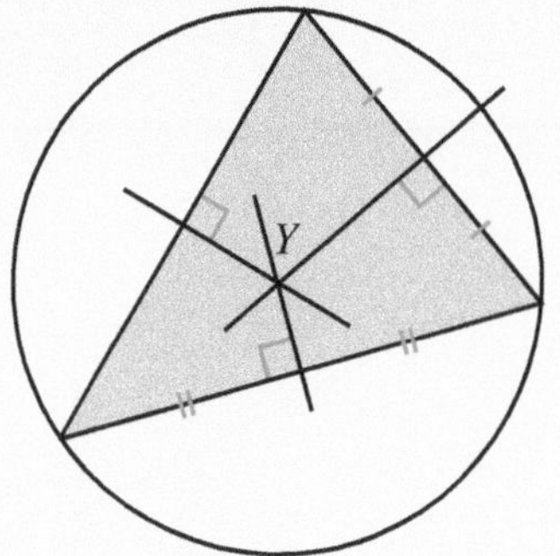

Put the point of your compasses on Y and draw the circle which goes through all three vertices of the triangle. This construction is sometimes called the **circumscribed circle of a triangle**.

4 Draw another triangle on a new page. Bisect each angle of the triangle.
The bisectors should meet at a point, X.

Put the point of your compasses on point X and draw the circle which just touches each side of the triangle.
This construction is sometimes called the **inscribed circle of a triangle**.

5 Using a circle of radius 4 cm, copy the diagram.
Draw the perpendicular bisectors of the chords WX and YZ.
What do you notice about the perpendicular bisectors?

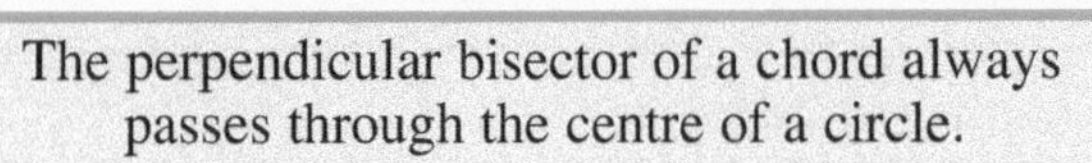
The perpendicular bisector of a chord always passes through the centre of a circle.

6 Two trees are 6 metres apart.
Alan walks so that he is always an equal distance from each tree.
Draw a scale diagram to show his path.

7 ABC is an equilateral triangle with sides 4 cm.

A point X is inside the triangle.
It is nearer to AB than to BC.
It is less than 3 cm from A.
It is less than 2 cm from BC.
Shade the region in which X could lie.

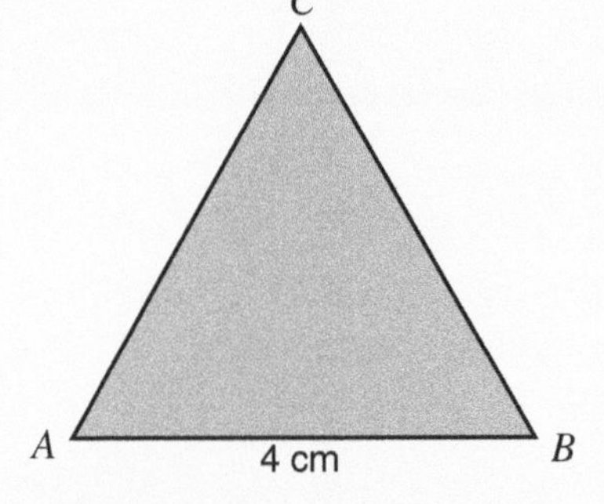

8 Draw a rectangle $ABCD$ with $AB = 6$ cm and $AD = 4$ cm.
(a) Mark, with a thin line, the locus of a point which is 1 cm from AB.
(b) Mark, with a dotted line, the locus of a point which is the same distance from A and B.
(c) Mark, with a dashed line, the locus of a point which is 3 cm from A.

9 Draw a right-angled triangle with sides of 6 cm, 8 cm and 10 cm.

A point X is in the triangle.
It is 4 cm from B.
It is the same distance from A and B.

Mark accurately, the position of X.

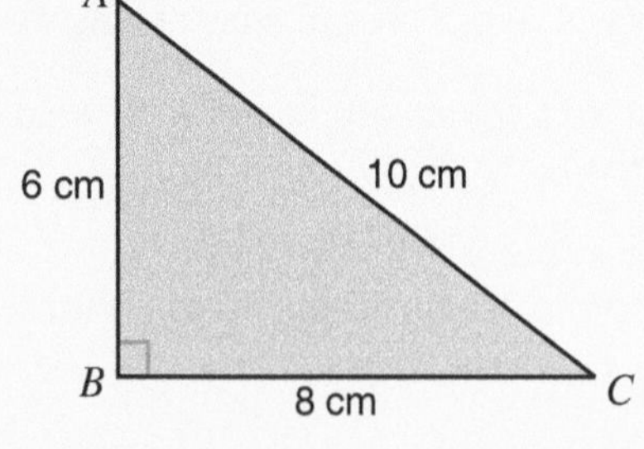

10 Copy the diagram and draw the locus of a point which is the same distance from PQ and RS.

11 Triangle ABC is isosceles with $AB = BC = 7$ cm and $AC = 6$ cm.

(a) Construct triangle ABC.

(b) Point X is equidistant from A, B and C. Mark accurately the position of X.

12 (a) The diagram shows the sketch of a field.
Make a scale drawing of the field using 1 cm to represent 100 m.

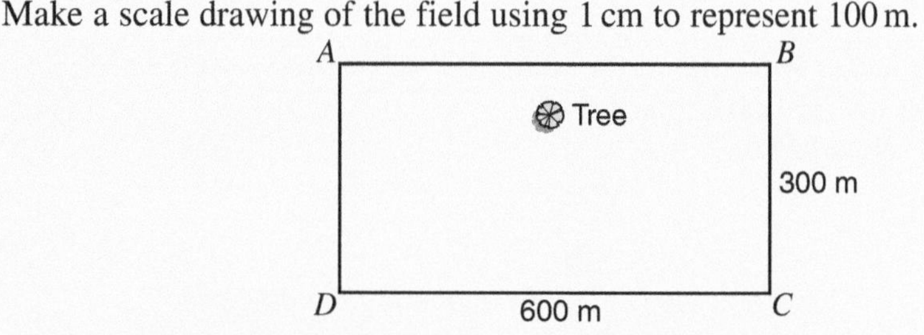

(b) A tree is 400 m from corner D and 350 m from corner C.
Mark the position of the tree on your drawing.

(c) John walks across the field from corner D, keeping the same distance from AD and CD.
Show his path on your diagram.

(d) Does John walk within 100 m of the tree in crossing the field?

13 Part of a coastline is shown.

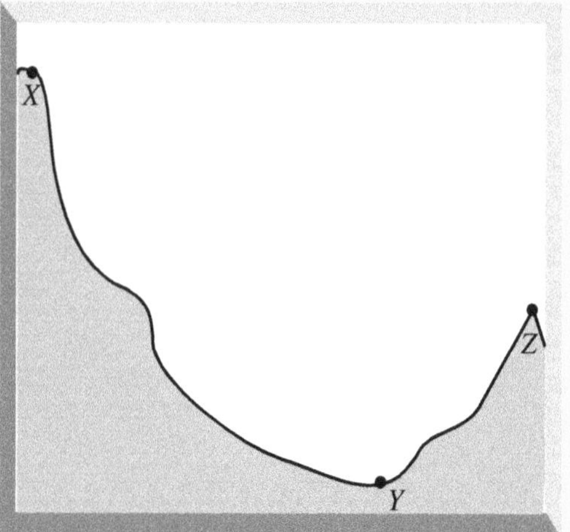

A boat is: (i) equidistant from X and Z, and (ii) equidistant from XY and YZ.
Copy the diagram and mark the position of the boat.

More constructions

To draw the perpendicular from a point to a line

This means to draw a line at right angles to a given line, from a point that **is not on the line**.

1. Open your compasses so that from point A you can mark two arcs on the line PQ.

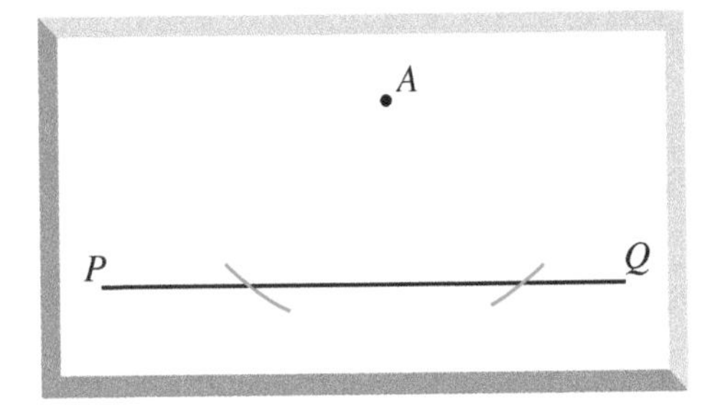

2. Use points B and C to draw equal arcs which cross at D.

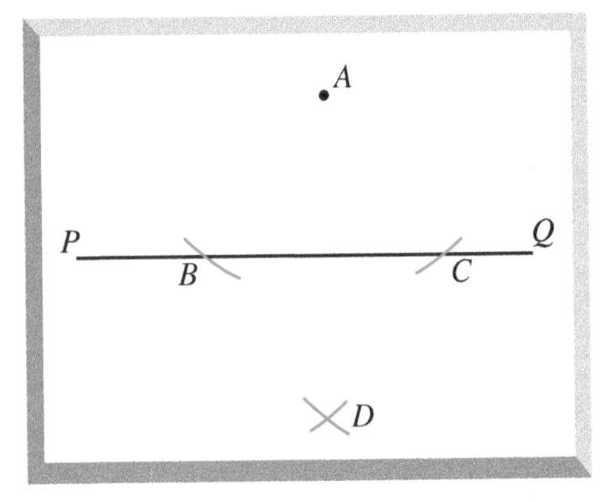

3. Draw a line from A to D.

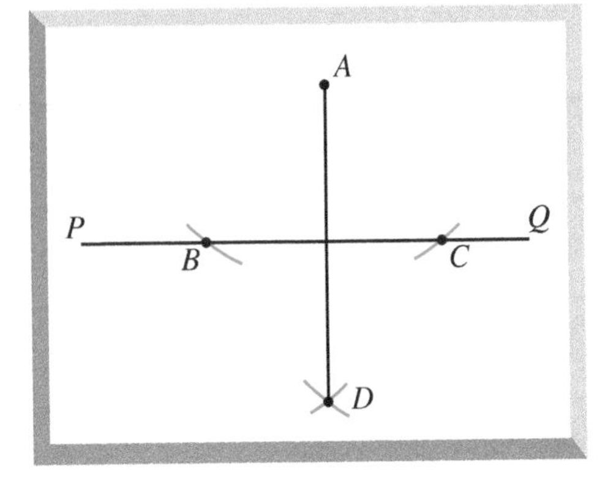

The line AD is perpendicular (at right angles) to the line PQ.

To draw the perpendicular from a point on a line

This means to draw a line at right angles to a given line, from a point that **is on the line**.

Keep your compasses at the same setting whilst doing this construction.

1. From A, draw an arc which cuts the line PQ.

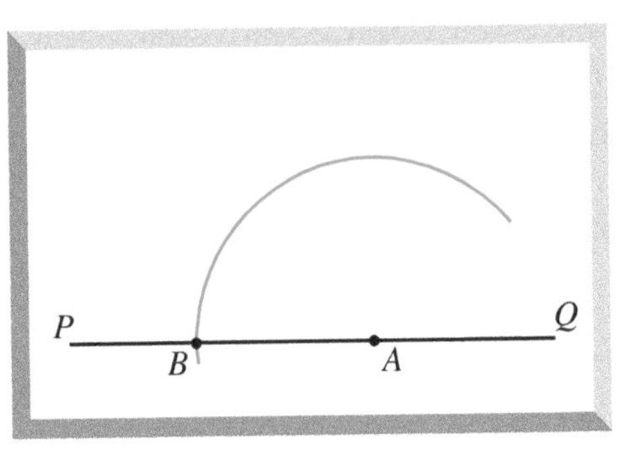

2. From B, draw an arc to cut the first arc at C. Then from C, draw an arc to cut the first arc at D.

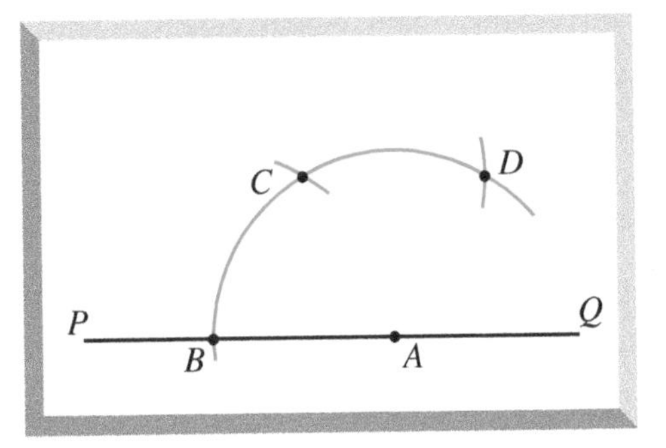

3. From C and D, draw arcs to meet at E. Draw the line AE.

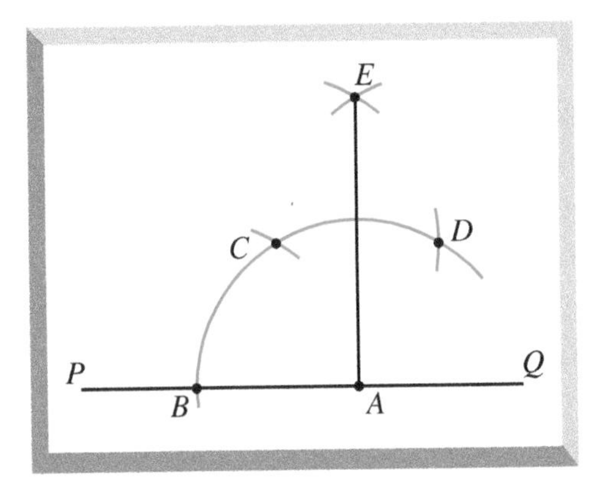

The line AE is perpendicular (at right angles) to the line PQ.

Exercise 26.4

1 Draw a line PQ, 8 cm long.
Mark a point A, about 5 cm above the line.
Draw the line which passes through A and is perpendicular to line PQ.

2 Draw a line PQ, 8 cm long. Mark a point A, somewhere on the line.

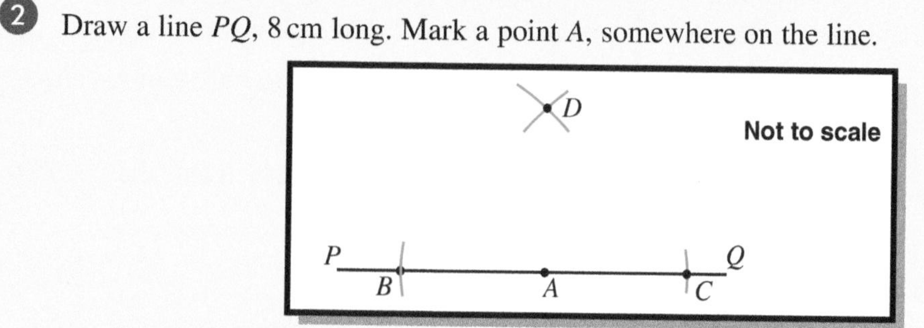

(a) Using your compasses, mark points B and C, which are 3 cm from A on the line PQ.
(b) Set your compasses to 5 cm.
Draw arcs from B and C which intersect at D.
(c) Draw the line AD.

3 Using ruler and compasses only, make an accurate drawing of the triangle shown in this sketch.

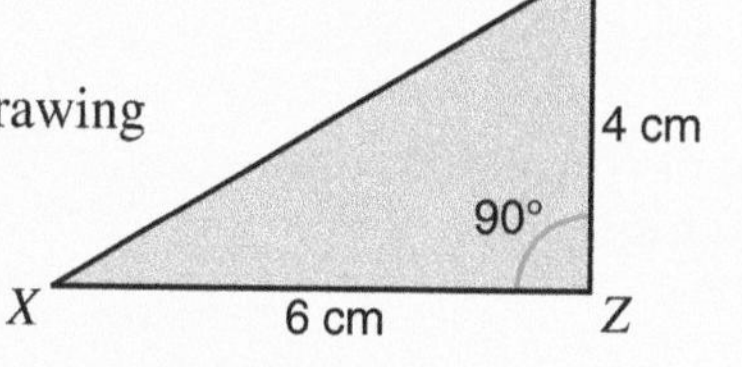

4

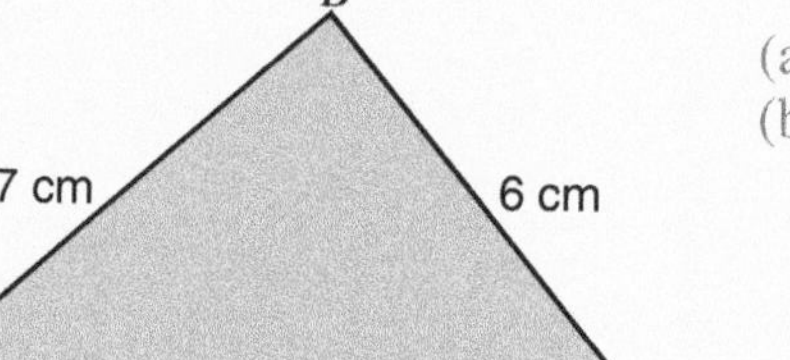
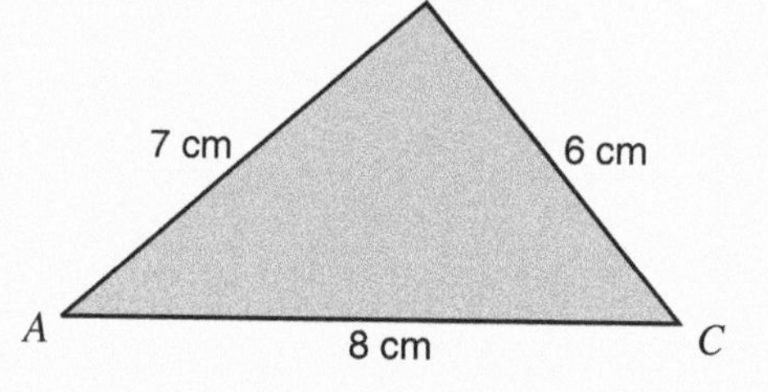

(a) Make an accurate drawing of this triangle.
(b) The **altitude** of the triangle is a line perpendicular to a side which passes through the opposite corner of the triangle.
Draw the altitude of the triangle ABC which passes through point B.

5 (a) Copy the diagram.
(b) Mark all points inside the rectangle that are less than 2 cm from the line AB.

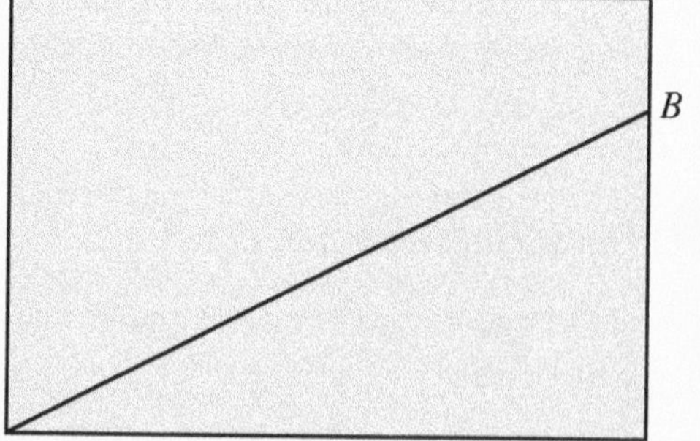

6 Draw a line PQ, 5 cm long.
Using compasses, draw an arc centre P to cut PQ at X.
With your compasses at the same setting draw another arc, centre X, to cut the first arc at Y.
Draw a line through PY.
Measure angle XPY.
What do you find?
Show how you can use this construction to draw angles of 30° and 120°.

Maps, Loci and Constructions

What you need to know

- **Compass points** and **three-figure bearings** are used to describe direction.
- A **bearing** is an angle measured from the North line in a clockwise direction.
- A bearing can be any angle from 0° to 360° and is written as a three-figure number.
- To find a bearing:
  measure angle a to find the bearing of Y from X,
  measure angle b to find the bearing of X from Y.

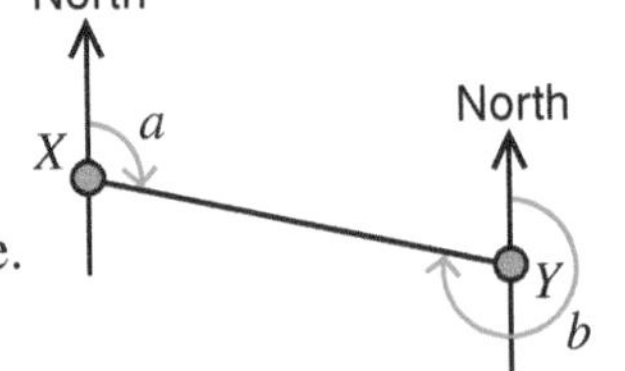

- **Scales**
  The distances between points on a map are all drawn to the same scale.
  There are two ways to describe a scale.
  1. A scale of 1 cm to 10 km means that a distance of 1 cm on the map represents an actual distance of 10 km.
  2. A scale of 1 : 10 000 means that all distances measured on the map have to be multiplied by 10 000 to find the real distance.
- The path of a point which moves according to a rule is called a **locus**.
- The word **loci** is used when we talk about more than one locus.

Using a ruler and compasses you should be able to:

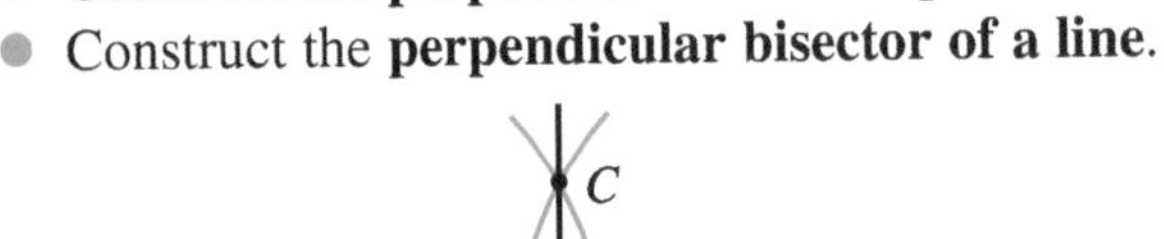

- Construct the **perpendicular from a point to a line**.
- Construct the **perpendicular from a point on a line**.
- Construct the **perpendicular bisector of a line**.

Points on the line CD are **equidistant** from the points A and B.

- Construct the **bisector of an angle**.

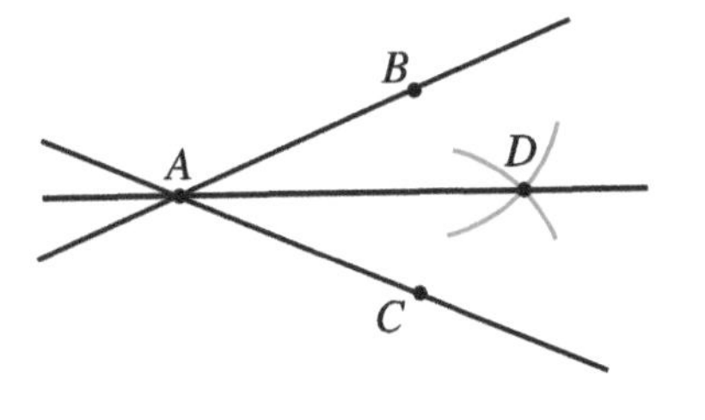

Points on the line AD are **equidistant** from the lines AB and AC.

1 A coin is rolled along a line.
Sketch the locus of a point which starts off at the bottom.

What is the locus of the point if the coin rolls around another coin, or if the coins are not round, or … ?

Investigate.

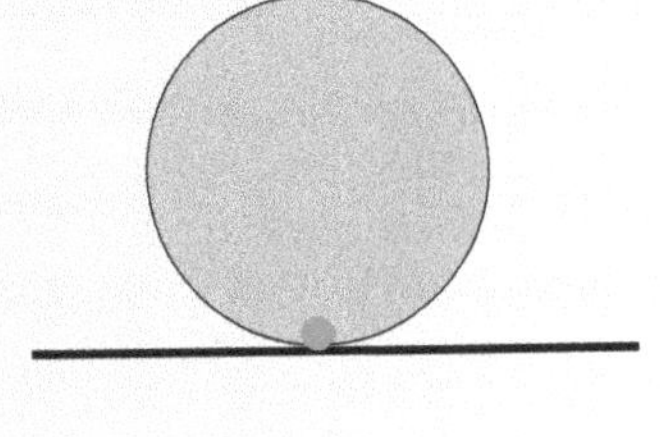

2 **3-dimensional coordinates**

One coordinate identifies a point on a line.
Two coordinates identify a point on a plane.
Three coordinates identify a point in space.

The diagram shows a cuboid drawn in 3-dimensions.

Using the axes x, y and z shown:
Point A is given as (0, 2, 0).
Point B is given as (2, 2, 0).
Point C is given as (2, 2, 3).

Give the 3-dimensional coordinates of points D, E, F, and G.

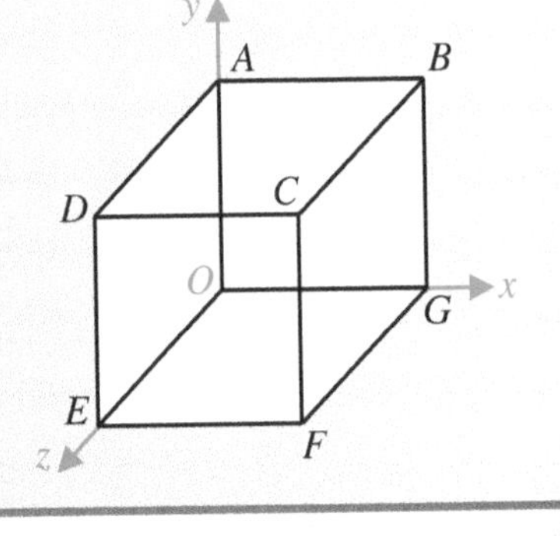

Review Exercise 26

1 The diagram shows the plan of a sailboard race.
The sailboards have to go round buoys at A, B and C.
Buoy B is on a bearing of 050° from buoy A. Angle ABC is 90°.

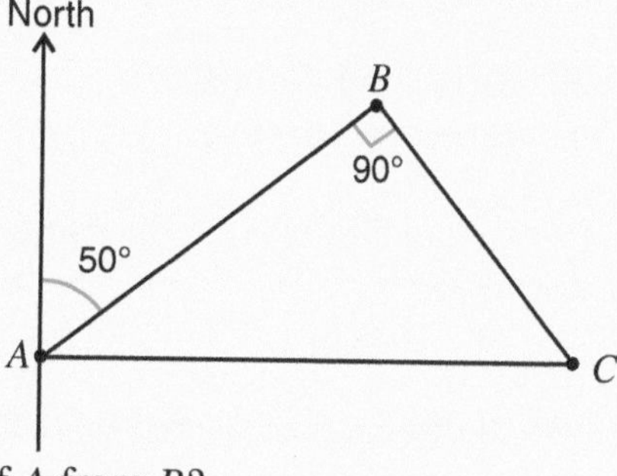

(a) What is the bearing of A from B?
(b) What is the bearing of C from B?

The plan has been drawn to a scale of 1 : 20 000.
(c) (i) Measure AB.
(ii) What is the distance from A to B in metres?

2 (a) Use ruler and compasses only to construct an equilateral triangle of side 4 cm.
(b) A point P is 1 cm from the edge of the triangle.
Draw an accurate locus of **all** the possible positions of P.

3 $PQRS$ shows a sketch of a park.
(a) Use ruler and compasses only to construct a plan of the park using a scale of 1 cm to represent 100 m.

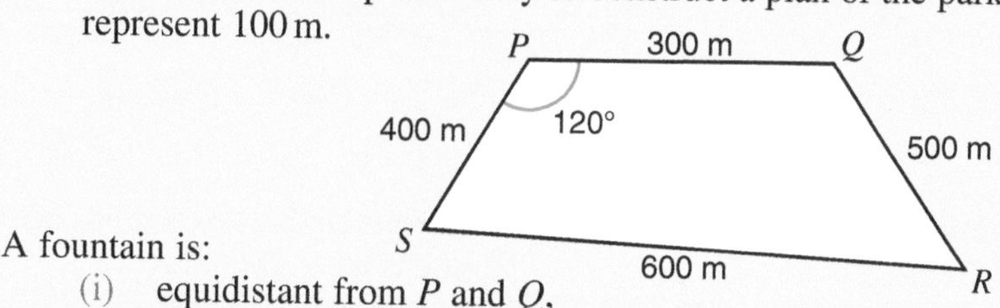

A fountain is:
(i) equidistant from P and Q,
(ii) equidistant from PS and SR.
(b) Draw the locus for (i) and (ii) on your plan, and hence, find the position of the fountain. Label it with the letter F.
(c) Find the distance, in metres, of the fountain from R.

4 Ceri and Diane want to find how far away a tower, T, is on the other side of a river.
To do this they mark a base line, AB, 100 metres long as shown on the diagram.
Next they measure the angles at the ends A and B between the base line and the lines of sight of the tower. These angles are 30° and 60°.

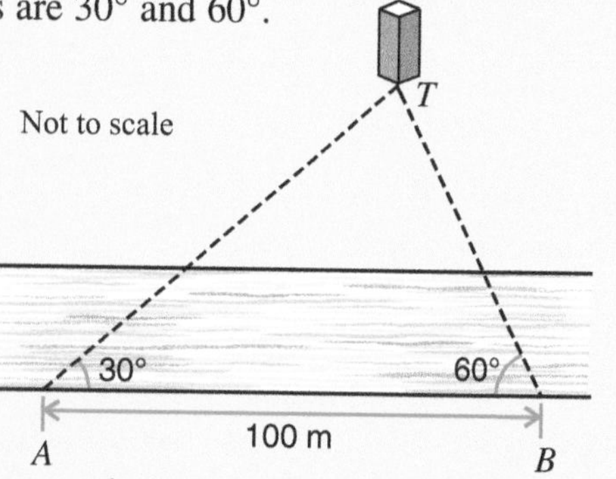

(a) Use ruler and compasses only to make a scale drawing of the situation.
Use a scale of 1 cm to represent 10 m.
Show clearly all your construction lines.
(b) Find the shortest distance of the tower, T, from the base line AB. AQA

Transformations

The movement of a shape from one position to another is called a **transformation**.
The change in position of the shape can be described in terms of a **reflection**, a **rotation** or a **translation**.
Later in the chapter you will meet another transformation, called an **enlargement**.

Reflection

Look at this diagram.
It shows the reflection of shape P in the line $x = 4$.

Shape P_1 is the **image** of P.
We also say that P is **mapped** onto P_1.

When a shape is reflected it stays the same shape and size.

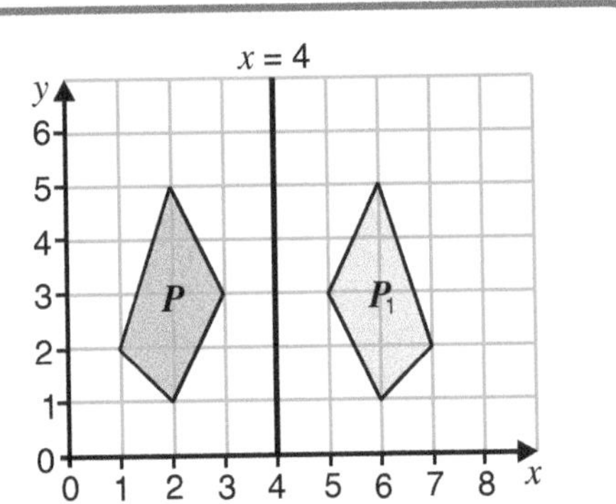

EXAMPLE

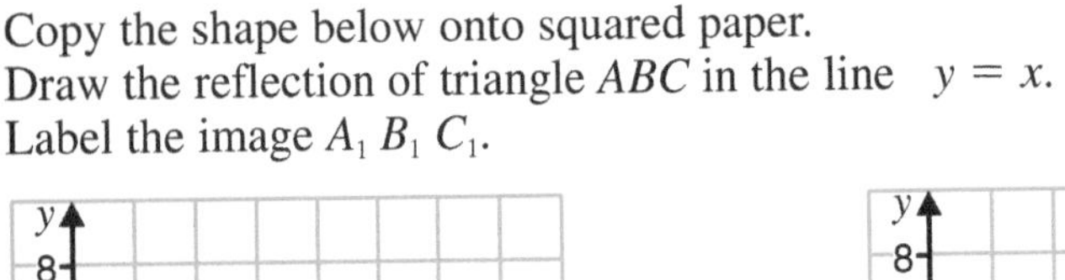

Copy the shape below onto squared paper.
Draw the reflection of triangle ABC in the line $y = x$.
Label the image A_1 B_1 C_1.

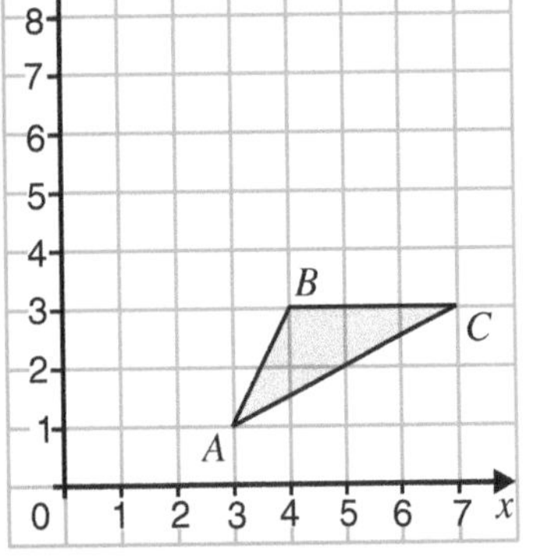

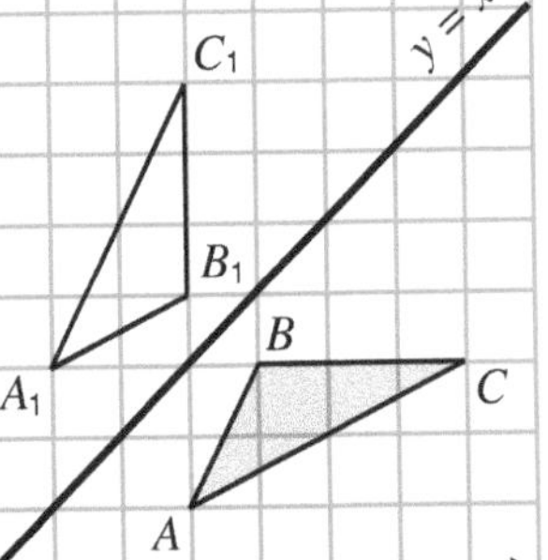

Notice that:
$A(3, 1) \rightarrow A_1(1, 3)$
$B(4, 3) \rightarrow B_1(3, 4)$
$C(7, 3) \rightarrow C_1(3, 7)$
Can you see a pattern?

Exercise 27.1

1 In the diagram P is the point (2, 1).

Find the coordinates of the image of P under a reflection in
(a) the x axis,
(b) the y axis,
(c) the line $x = 1$,
(d) the line $y = -1$,
(e) the line $y = x$.

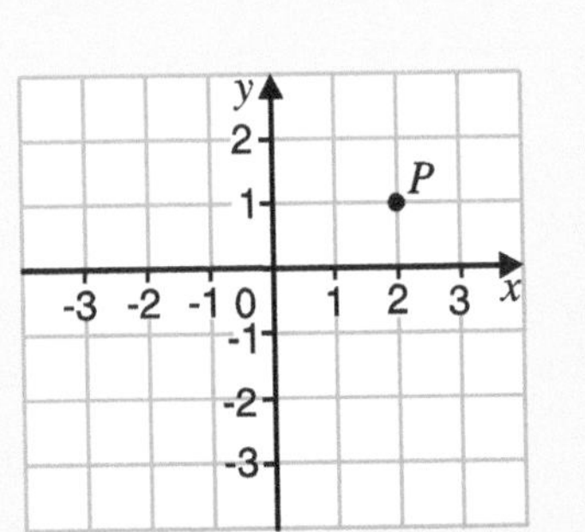

2 Copy each of the following diagrams onto squared paper and draw the reflection of each shape in the line given.

(a) Reflect in the y axis.

(b) Reflect in the x axis.

(c) Reflect in $y = 2$.

(d) Reflect in $x = 3$.

(e) Reflect in $y = -1$.

(f) Reflect in $y = x$.

3 The diagram shows a quadrilateral $ABCD$.
Give the coordinates of B after:

(a) a reflection in the x axis,
(b) a reflection in the y axis,
(c) a reflection in the line $x = 3$,
(d) a reflection in the line $x = -1$,
(e) a reflection in the line $y = x$.

Rotation

Look at this diagram.
It shows the **rotation** of a shape P through $\frac{1}{4}$ turn anticlockwise about centre X.

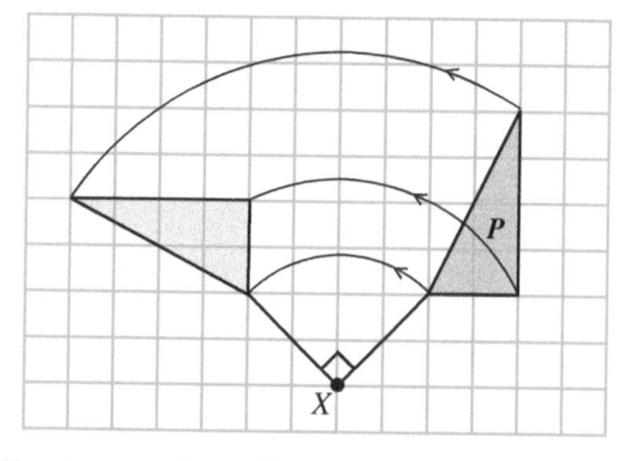

When describing rotations the direction of turn can be **clockwise** or **anticlockwise**.

Remember:

clockwise anticlockwise

All points on shape P are turned through the same angle about the same point.
This point is called the **centre of rotation**.

When a shape is rotated it stays the same shape and size but its **position** on the page changes.

For a rotation we need: a centre of rotation, an amount of turn, a direction of turn.

EXAMPLE

Copy triangle ABC onto squared paper.
Draw the image of triangle ABC after it has been rotated through 90° clockwise about the point $P(1, 1)$.
Label the image $A_1B_1C_1$.

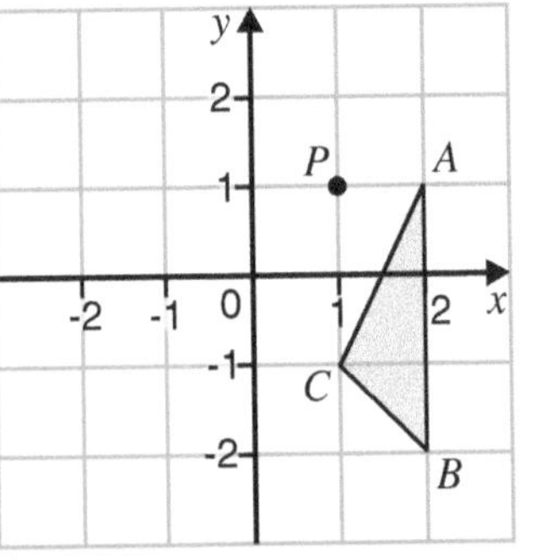

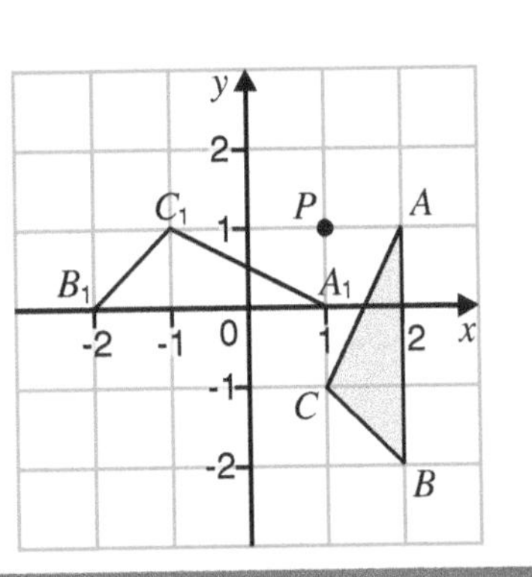

Exercise 27.2

1. Copy each of these shapes onto squared paper.
   Draw the new position of each shape after the rotation given.

   (a) $\frac{1}{4}$ **turn clockwise about centre X.**

   (b) $\frac{1}{2}$ **turn about centre X.**

   (c)

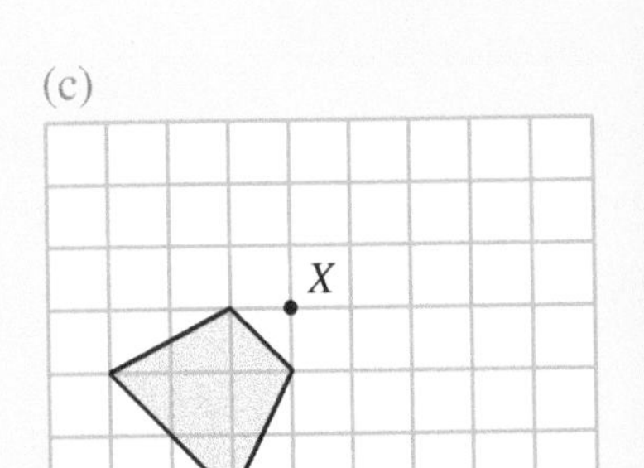

   $\frac{3}{4}$ **turn anticlockwise about centre X.**

2. Copy each of these shapes onto squared paper.
   Draw the new position of each shape after the rotation given.

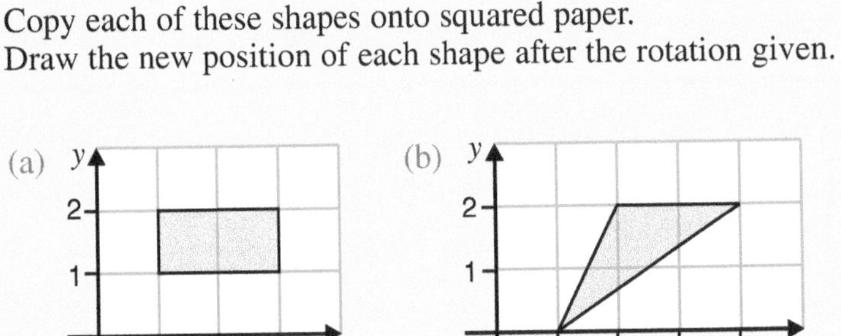

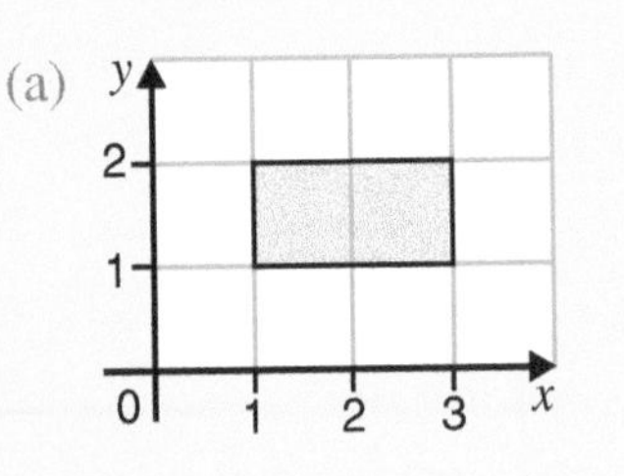

   (a) **90° clockwork about (0, 0).**

   (b) **90° anticlockwise about (0, 0).**

   (c) **180° about (0, 0).**

3. The diagram shows a quadrilateral $ABCD$.

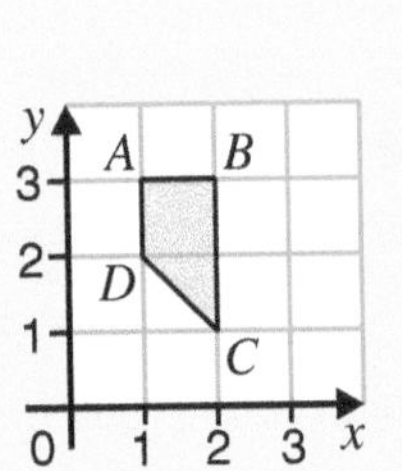

   Give the coordinates of B after:
   (a) a rotation through 90°, clockwise about (0, 0),
   (b) a rotation through 90°, anticlockwise about (0, 0),
   (c) a rotation through 180°, about (0, 0),
   (d) a rotation through 90°, clockwise about (2, 1),
   (e) a rotation through 90°, anticlockwise about (2, 1),
   (f) a rotation through 180°, about (2, 1).

Translation

Look at this diagram. It shows a **translation** of a shape P.

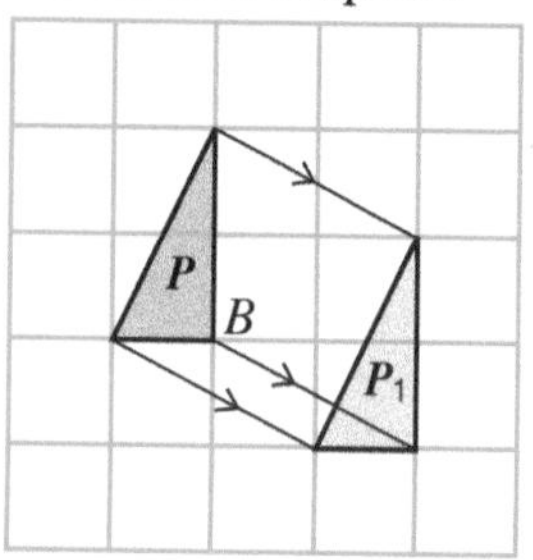

P is mapped onto P_1.
All points on the shape P are moved the same distance in the same direction without turning.
A translation can be given:

- in terms of a **distance** and a **direction**, e.g. 2 units to the right and 1 unit down.
- with a vector, e.g. $\begin{pmatrix} 2 \\ -1 \end{pmatrix}$.

When a **vector** is used to describe a translation:
the top number describes the **horizontal** part of the movement:
$+$ = to the right, $-$ = to the left
the bottom number describes the **vertical** part of the movement:
$+$ = upwards, $-$ = downwards

When a shape is translated it stays the same shape and size and has the same orientation.

EXAMPLE

Copy triangle P onto squared paper.

The translation $\begin{pmatrix} -3 \\ 2 \end{pmatrix}$ maps P onto P_1.

Draw and label P_1.

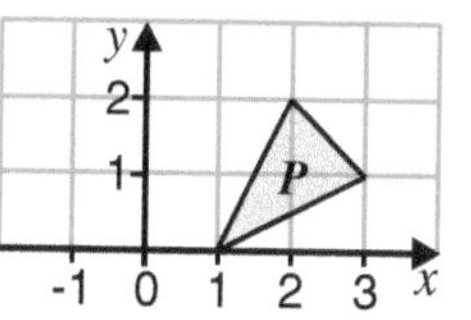

y 4 2 1 P1 P -2 -1 0 1 2 3 x

Vector notation:

$\begin{pmatrix} -3 \\ 2 \end{pmatrix}$ means move triangle P 3 units to the left and 2 units up.

Exercise 27.3

1. Copy the shape onto squared paper.
   Draw the new position of the shape after each of the following translations:
   (a) 2 units to the right and 3 units up,
   (b) 1 unit to the right and 2 units down,
   (c) 3 units to the left and 2 units up,
   (d) 1 unit to the left and 3 units down.

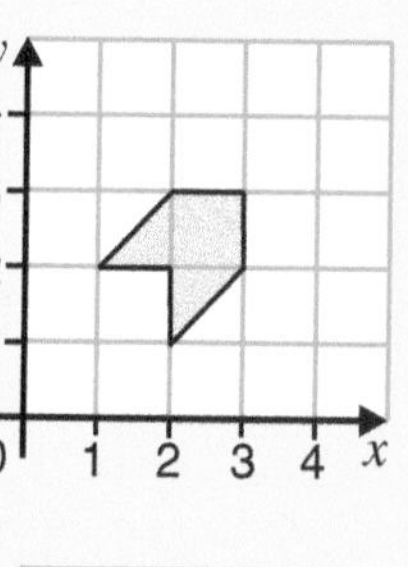

2. Copy the shape onto squared paper.
   Draw the new position of the shape after each of the following translations.
   (a) $\begin{pmatrix} 3 \\ 2 \end{pmatrix}$ (b) $\begin{pmatrix} 2 \\ -3 \end{pmatrix}$ (c) $\begin{pmatrix} -2 \\ 3 \end{pmatrix}$ (d) $\begin{pmatrix} -2 \\ -3 \end{pmatrix}$

y 3 2 1 0 1 2 3 x

3. The translation $\begin{pmatrix} 2 \\ -1 \end{pmatrix}$ maps $S(5, 3)$ onto T.
   What are the coordinates of T?

4 Write down the translation which maps:
(a) $X(1, 1)$ onto $P(3, 2)$, (b) $X(1, 1)$ onto $Q(2, -1)$,
(c) $X(1, 1)$ onto $R(-2, 2)$, (d) $X(1, 1)$ onto $S(-2, -1)$.

5 The diagram shows quadrilateral S.
Copy S onto squared paper.
(a) The translation $\begin{pmatrix} 3 \\ 2 \end{pmatrix}$ maps S onto T.
Draw and label T.
(b) Write down the translation which maps T onto S.

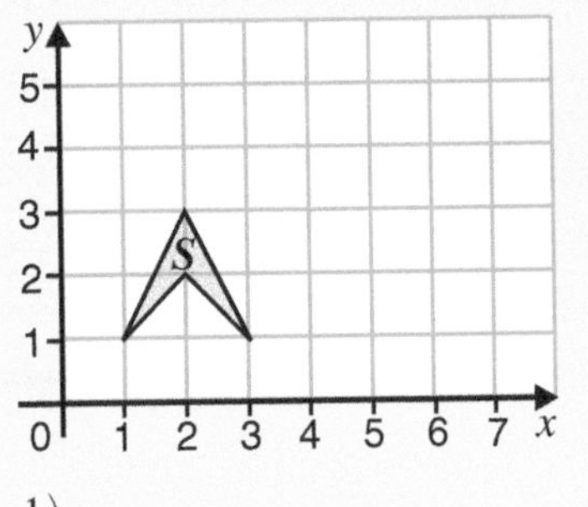

6 The translation $\begin{pmatrix} -2 \\ -3 \end{pmatrix}$ maps A onto B and the translation $\begin{pmatrix} -1 \\ 2 \end{pmatrix}$ maps C onto B.
(a) B has coordinates $(-1, 2)$.
What are the coordinates of A?
(b) Write down the translation which maps A onto C.

Enlargement

This diagram shows another transformation, called an **enlargement**.
It shows an enlargement of a shape P with scale factor 2 and centre O.

P is mapped onto P_1.

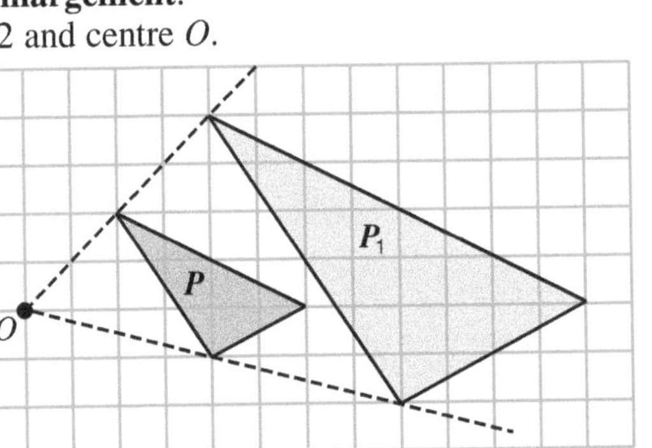

When a shape is enlarged:
angles remain unchanged,
all **lengths** are multiplied by a **scale factor**.

Scale factor $= \frac{\text{new length}}{\text{original length}}$

For an enlargement we need:
a centre of enlargement,
a scale factor.

Using a centre of enlargement

Draw a line from the centre of enlargement, P, to one corner, A.
Extend this line to A' so that the length of PA' = the scale factor $\times$ the length of PA.
Do the same for other corners of the shape.
Join up the corners to make the enlarged shape. Label the diagram.

EXAMPLES

1 Draw an enlargement of triangle ABC, centre $P(0, 1)$ and scale factor 3.

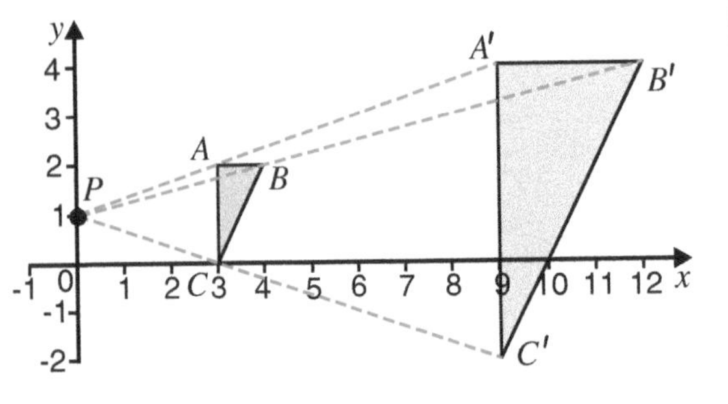

2 Use centre $P(3, 2)$ and a scale factor of 2 to enlarge triangle ABC.

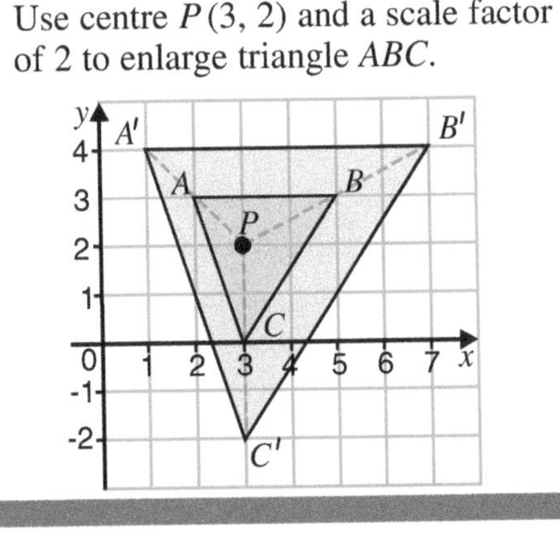

Using a scale factor which is a fraction

When the scale factor is a value between 0 and 1, such as 0.5 or $\frac{1}{3}$, the new shape is smaller than the original shape.
Even though the shape gets smaller it is still called an enlargement.

EXAMPLE

Draw an enlargement of this shape with centre (0, 1) and scale factor $\frac{1}{3}$.

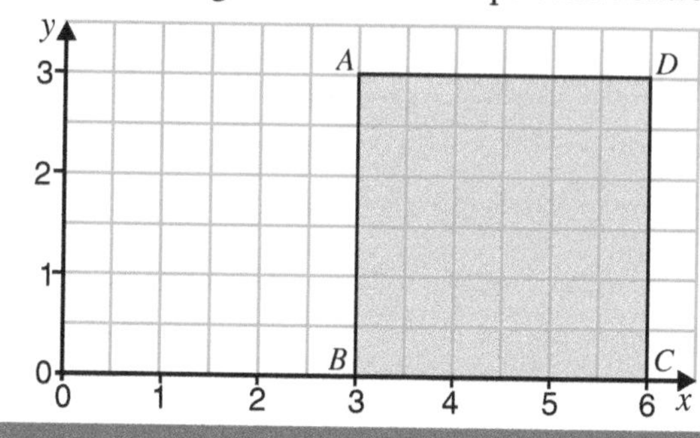

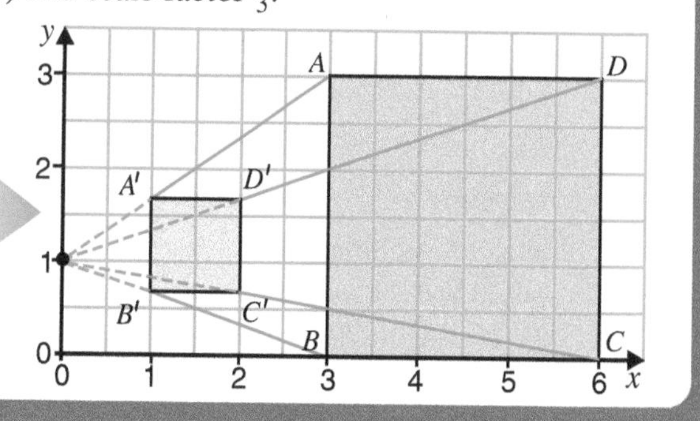

Exercise 27.4

1. Copy this shape onto squared paper.
   Enlarge the shape with scale factor 2, centre X.

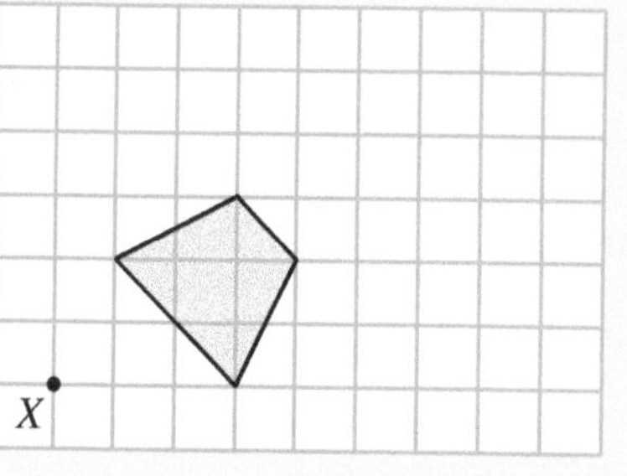

2. Copy each diagram onto squared paper and enlarge it using the centre and scale factor given.
   You will need longer axes than those shown below.

(a)

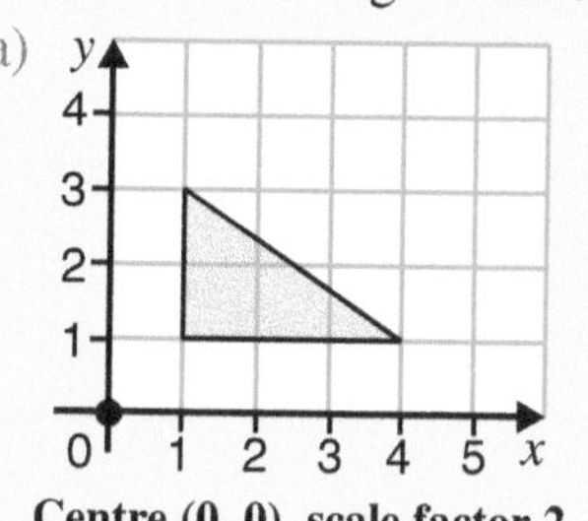

Centre (0, 0), scale factor 2.

(b)

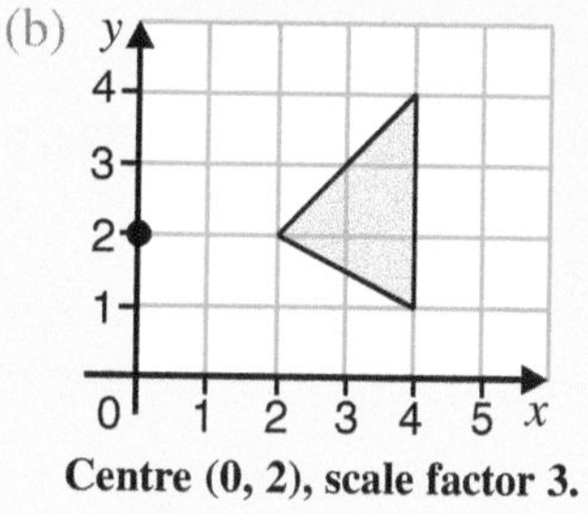

Centre (0, 2), scale factor 3.

(c)

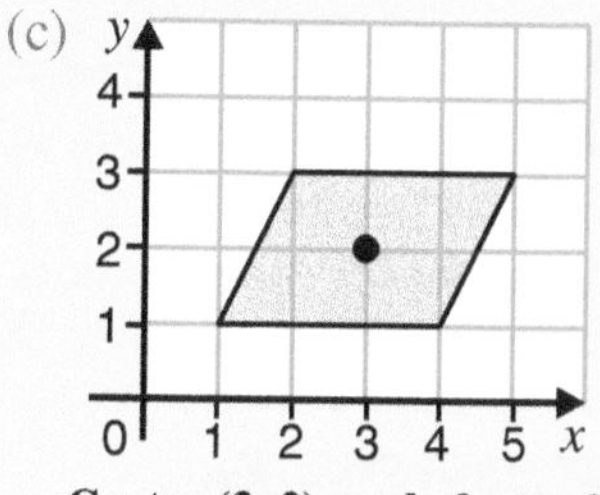

Centre (3, 2), scale factor 2.

3. Copy the following shapes onto squared paper and draw the enlargement given.

(a) Scale factor $\frac{1}{2}$, centre (1, 2).

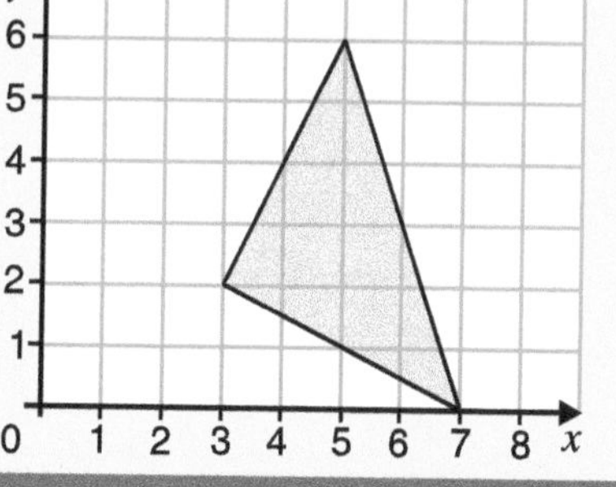

(b) Scale factor $\frac{1}{3}$, centre (0, 0).

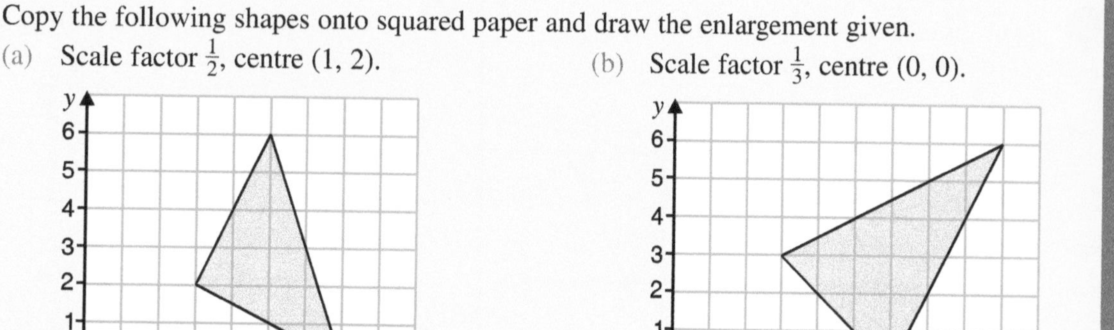

4. The diagram shows a quadrilateral $ABCD$.
Give the coordinates of B after an enlargement:
(a) scale factor 2, centre (0, 0),
(b) scale factor 3, centre (0, 0),
(c) scale factor 2, centre (0, 2),
(d) scale factor 3, centre C(2, 1),
(e) scale factor 2, centre D(1, 2),
(f) scale factor 2, centre A(1, 3).

5. For each of the following, give the coordinates of B after the enlargement given.
(a) Scale factor $\frac{1}{3}$, centre (0, 1).
(b) Scale factor $\frac{1}{2}$, centre (4, 3).

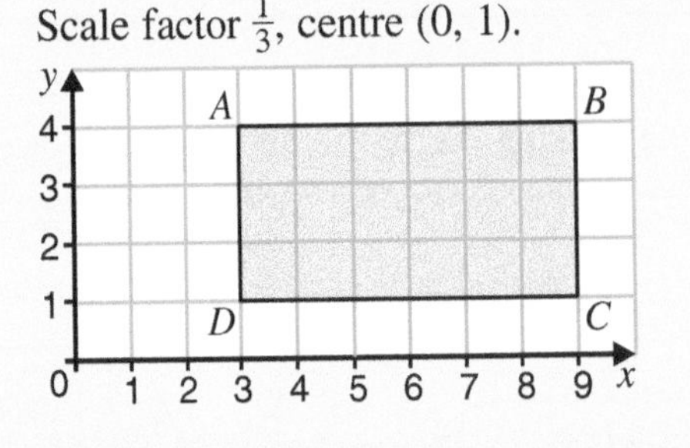

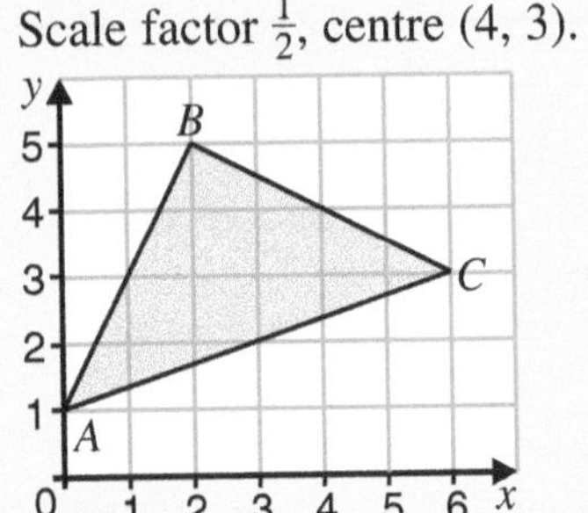

Enlargement with a negative scale factor

Draw a line from one corner, A, to the centre of enlargement, P.
Extend this line to A' so that:
the length of PA' = the scale factor × the length of PA.
Do the same for other corners of the shape.
Join up the corners to make the enlarged shape.
Label the diagram.

EXAMPLE

Draw an enlargement of triangle ABC, centre $P(-2, -1)$ and scale factor -2.

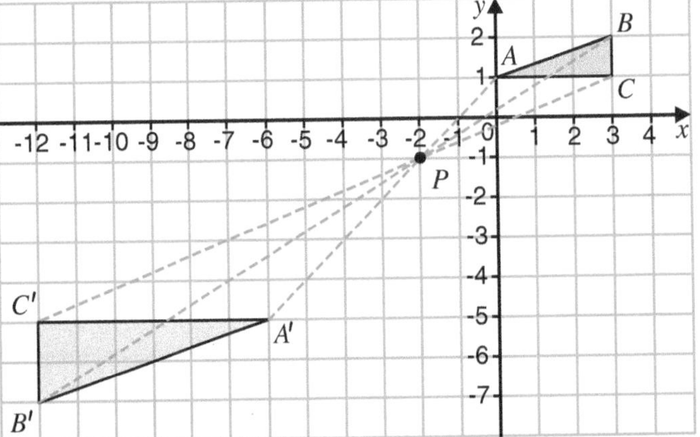

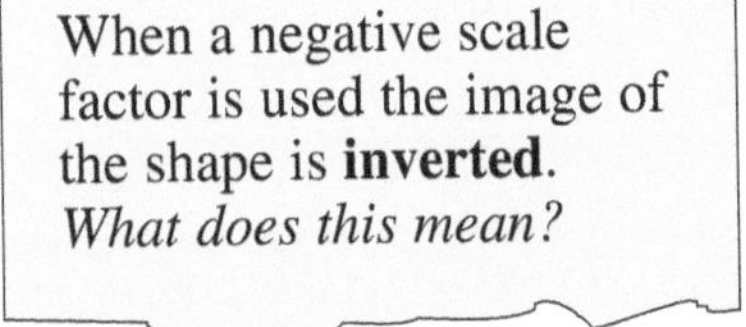

When a negative scale factor is used the image of the shape is **inverted**.
What does this mean?

Exercise 27.5

1. Copy this shape onto squared paper.
Enlarge the shape with scale factor -2, centre X.

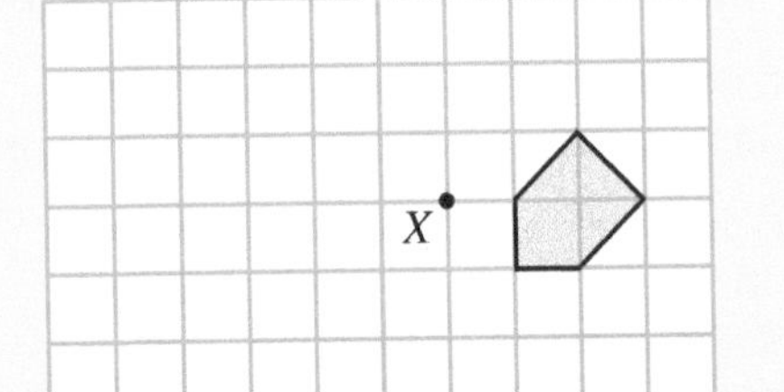

2 Copy the following shapes onto squared paper and draw the enlargement given.

(a) Scale factor -1, centre $(0, 0)$.

(b) Scale factor -2, centre $(-1, 0)$.

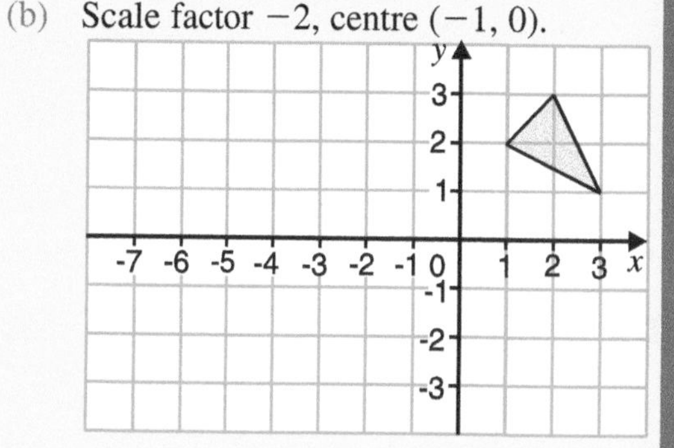

3 Copy each diagram onto squared paper and enlarge it using the centre and scale factor given. You will need longer axes than those shown below.

(a)

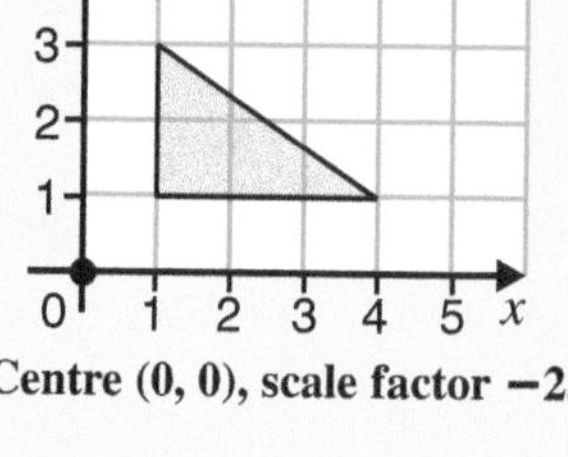

Centre (0, 0), scale factor −2.

(b)

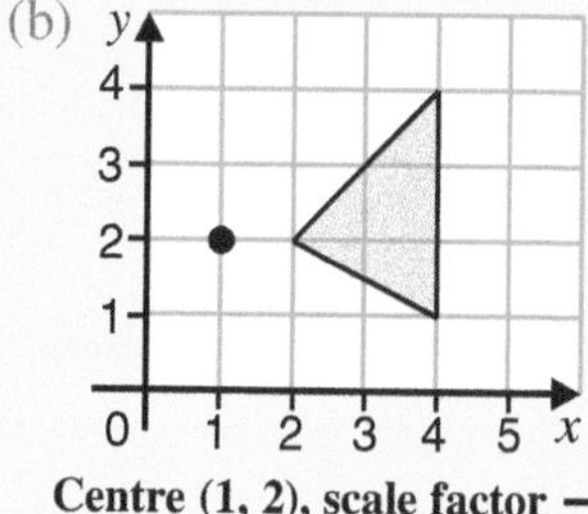

Centre (1, 2), scale factor −3.

(c)

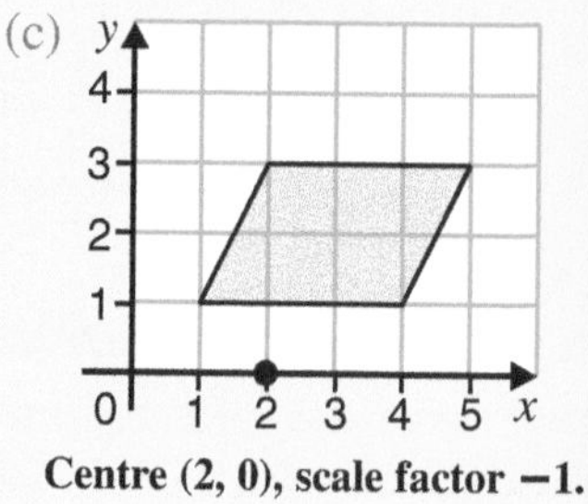

Centre (2, 0), scale factor −1.

4 Copy this shape onto squared paper.
Enlarge the shape with scale factor $-\frac{1}{2}$, centre X.

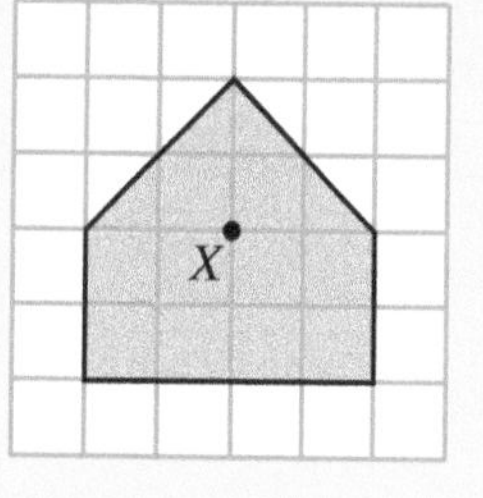

5 Describe the images of shapes which have been enlarged using:

(a) $-1 <$ scale factor < 0,

(b) scale factor $= -1$,

(c) scale factor < -1.

6 The diagram shows triangle ABC. Give the coordinates of B after an enlargement:

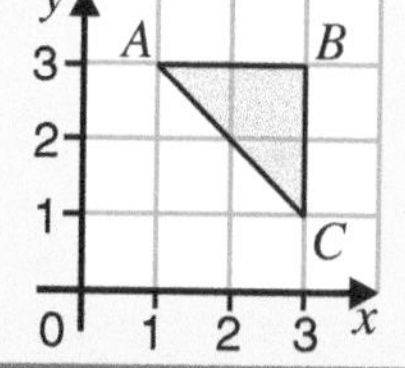
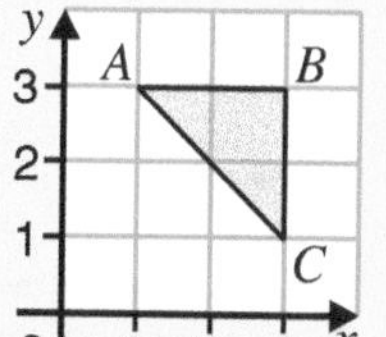

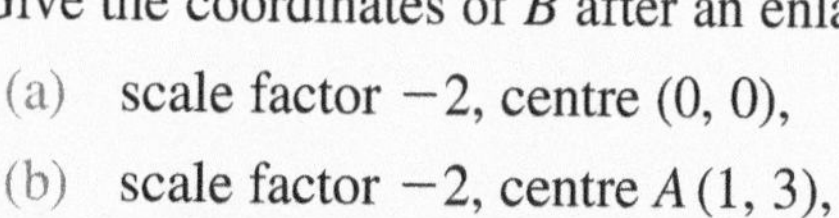

(a) scale factor -2, centre $(0, 0)$,

(b) scale factor -2, centre $A\,(1, 3)$,

(c) scale factor $-\frac{1}{2}$, centre $A\,(1, 3)$,

(d) scale factor $-\frac{1}{3}$, centre $(0, 0)$.

Describing transformations

Look at the shapes in this diagram.

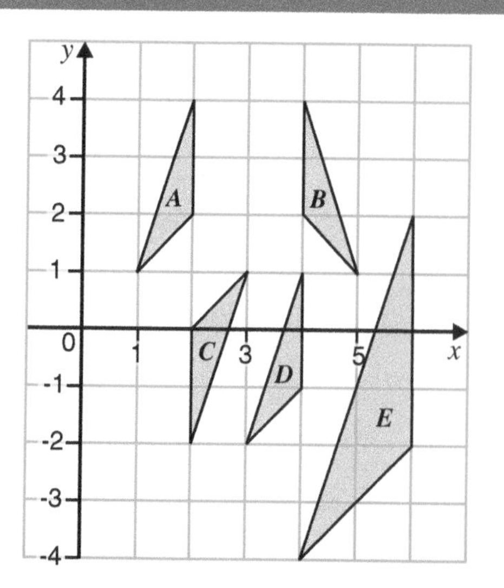

We can describe the single transformation which maps A onto B as a **reflection** in the line $x = 3$.

We can describe the single transformation which maps A onto C as a **rotation** of 180° about $(2, 1)$.

We can describe the single transformation which maps A onto D as a **translation** with vector $\begin{pmatrix} 2 \\ -3 \end{pmatrix}$.

We can describe the single transformation which maps D onto E as an **enlargement**, scale factor 2, centre $(2, 0)$.

The flow chart below can be used to decide what type of transformation has taken place. The details required to fully describe each type of transformation are also given.

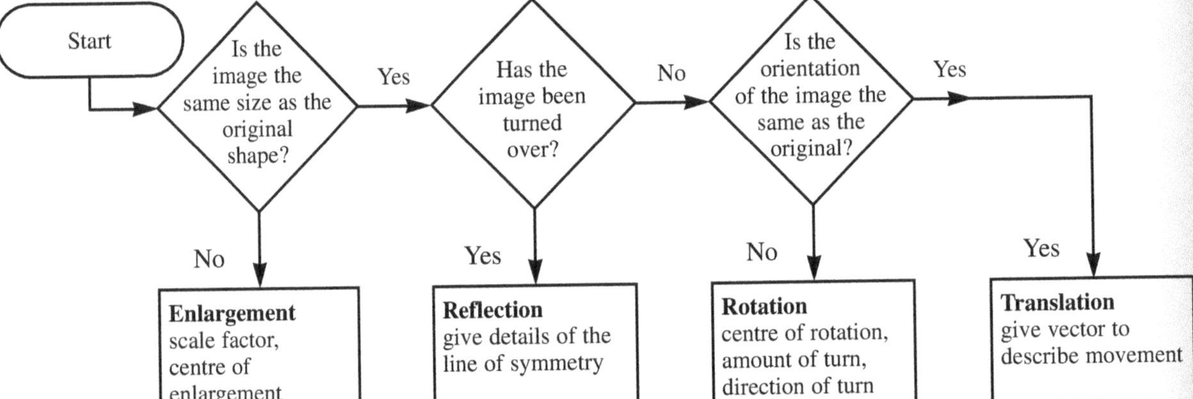

Use the flow chart to check the transformations described at the bottom of page 282.

To find a line of reflection

1 Join each point to its image point.
2 Put a mark halfway along each line.
3 Use a ruler to join the marks.

To find the centre and angle of rotation

1 Join each point to its image point.
2 Put a mark halfway along each line.
3 Use a set-square to draw a line at right angles to each line. The point where the lines cross is the centre of rotation, R.
4 Join one point and its image to the centre of rotation.
5 The angle of rotation is given by the size of the angle ARA_1.

To find the centre and scale factor of an enlargement

1 Join each point to its image point.
2 Extend these lines until they meet. This point is the centre of enlargement.
3 Measure a pair of corresponding lengths.

$$\text{Scale factor} = \frac{\text{new length}}{\text{original length}}$$

EXAMPLE

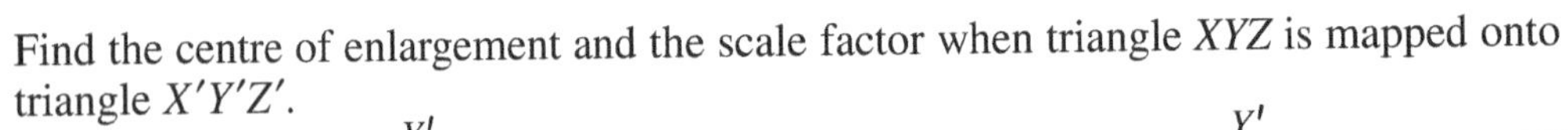

1 Find the centre of enlargement and the scale factor when triangle XYZ is mapped onto triangle $X'Y'Z'$.

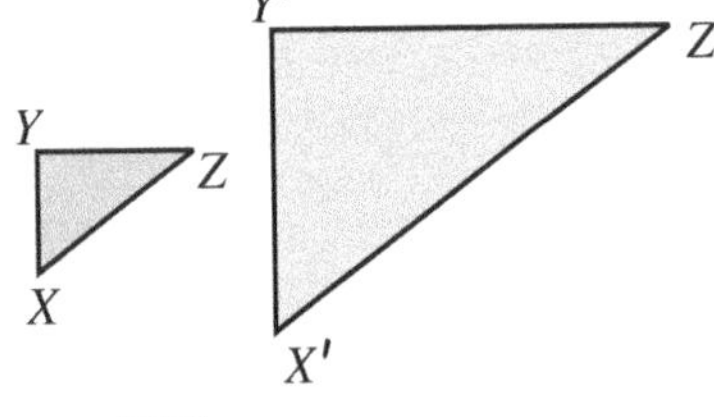

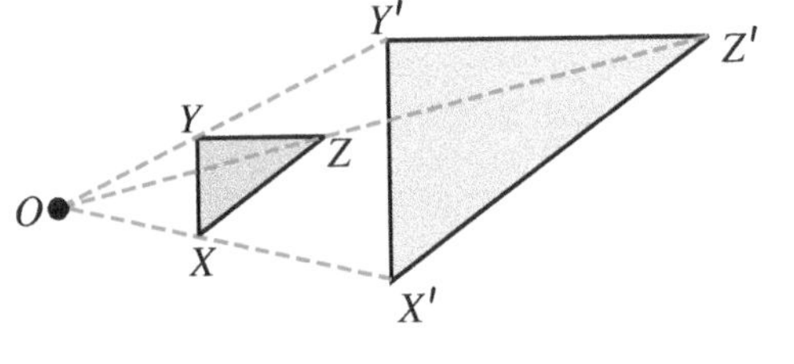

$$\text{Scale factor} = \frac{X'Y'}{XY} = \frac{2.0}{0.8} = 2.5$$

The centre of enlargement is the point O.

EXAMPLE

2 Find the centre of enlargement and the scale factor when triangle XYZ is mapped onto triangle $X'Y'Z'$.

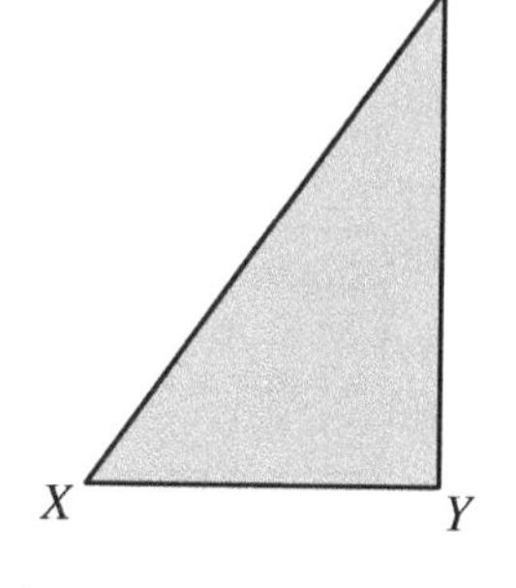

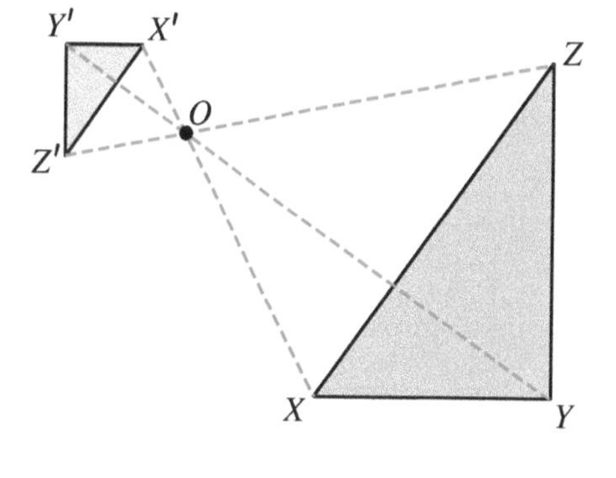

Scale factor $= \frac{X'Y'}{XY} = -\frac{0.7}{2.1} = -\frac{1}{3}$

The scale factor is negative because the image is inverted.

The centre of enlargement is the point O.

Exercise 27.6

1 Describe fully the single transformation which takes $\mathbf{L}_1$ onto $\mathbf{L}_2$, $\mathbf{L}_3$, $\mathbf{L}_4$, $\mathbf{L}_5$ and $\mathbf{L}_6$.

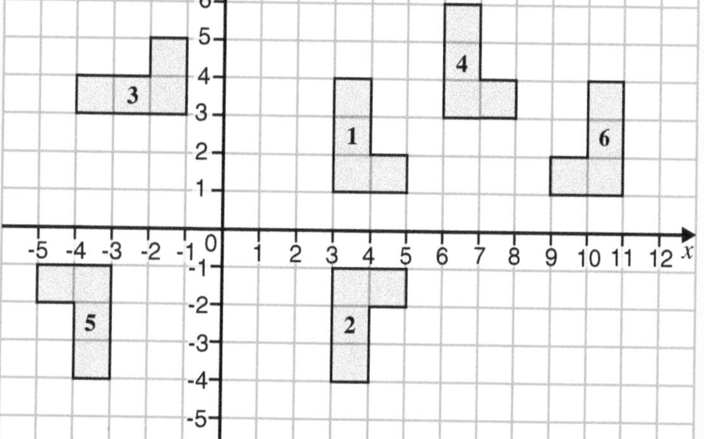

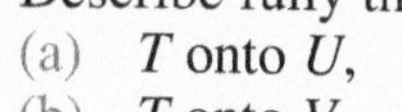

2 Describe fully the single transformation which maps

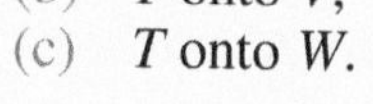

(a) T onto U,
(b) T onto V,
(c) T onto W.

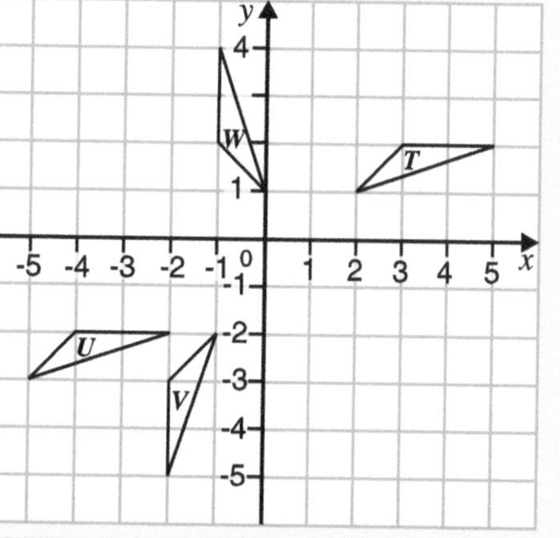

3 Describe fully the single transformation which maps
(a) A onto B,
(b) A onto C,
(c) A onto D.

4 For each of the following diagrams, describe fully the single transformation which maps A onto B.

(a) (b) (c) (d) (e) (f)

5 For each of the following diagrams, describe fully the single transformation which maps A onto B.

(a) (b)

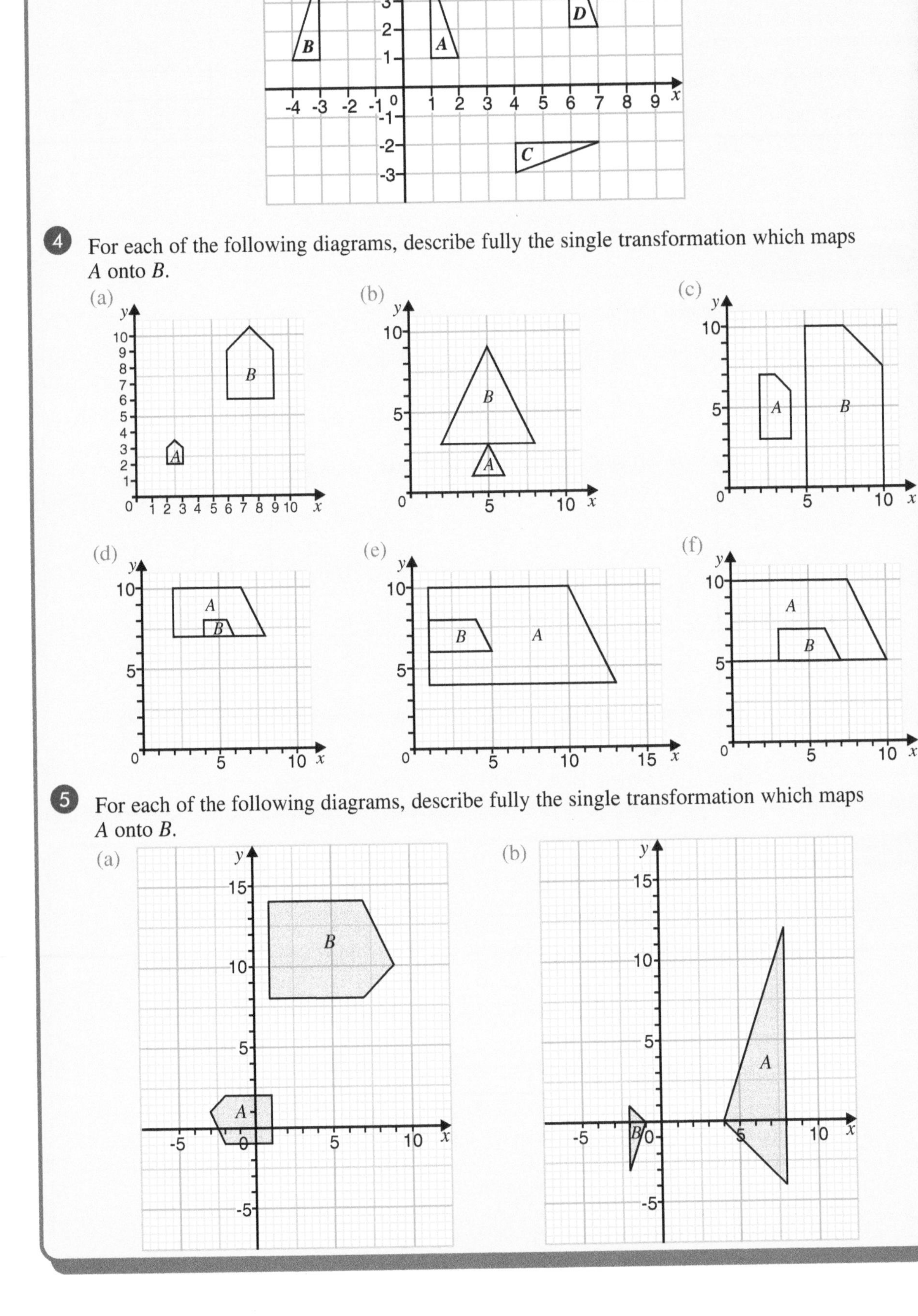

Combinations of transformations

So far we have looked at **single** transformations only.
There is no reason why the image of a transformation cannot be transformed.
The result of applying more than one transformation is called a **combined transformation**.

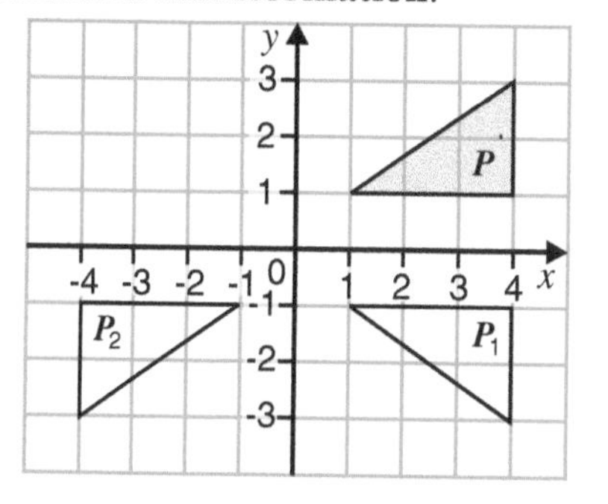

For example, in the diagram P has been mapped onto P_1 by a reflection in the x axis.

Then P_1 has been mapped onto P_2 by a reflection in the y axis.

The result of mapping P onto P_1 and then P_1 onto P_2 is a combined transformation.

EXAMPLE

Copy triangle Q onto squared paper.

(a) Q is mapped onto Q_1 by a rotation through 90°, clockwise about (0, 0).
Draw and label Q_1.

(b) Q_1 is mapped onto Q_2 by a reflection in the line $y = 0$.
Draw and label Q_2.

(c) Describe the single transformation which maps Q onto Q_2.

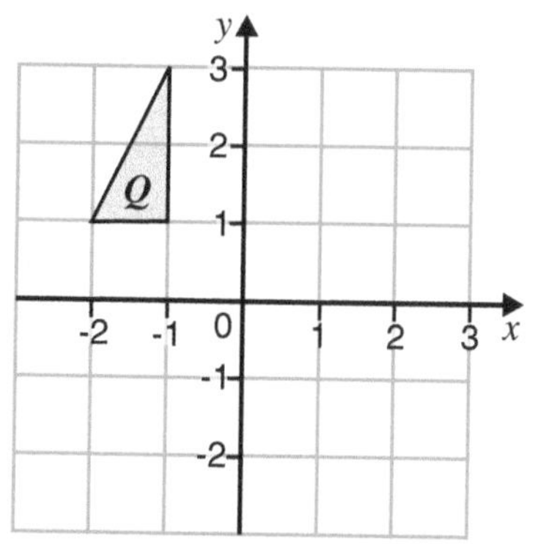

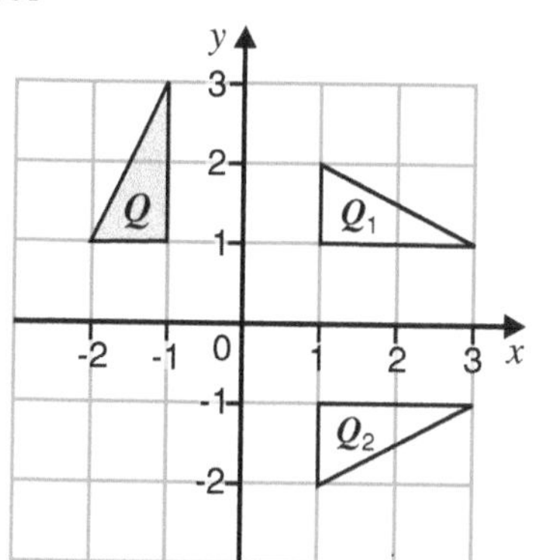

(c) The single transformation which maps Q onto Q_2 is a reflection in the line $y = x$.

Exercise 27.7

1 The diagram shows a quadrilateral labelled A. Copy the diagram onto squared paper.

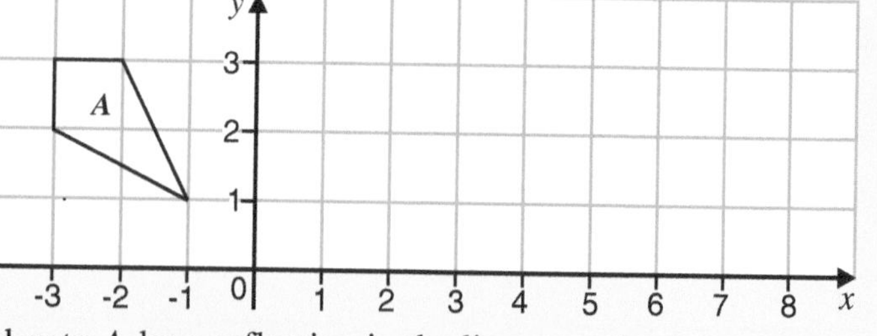

(a) A is mapped onto A_1 by a reflection in the line $x = 0$.
Draw and label A_1.

(b) A_1 is mapped onto A_2 by a reflection in the line $x = 4$.
Draw and label A_2.

(c) Describe fully the single transformation which maps A onto A_2.

2 The diagram shows a triangle labelled P. Copy the diagram onto squared paper.

(a) P is mapped onto P_1 by a reflection in the line $y = x$.
Draw and label P_1.

(b) P_1 is mapped onto P_2 by a reflection in the line $x = 5$.
Draw and label P_2.

(c) Describe fully the single transformation which maps P onto P_2.

3 The diagram shows a triangle labelled T. Copy the diagram.

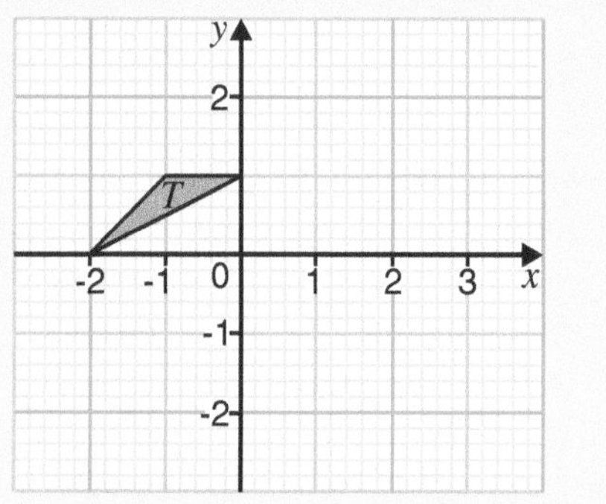

(a) Rotate T through 90° clockwise about (0, 0) to T_1.
Draw and label T_1.

(b) Reflect T_1 in the line $y = 0$ to T_2.
Draw and label T_2.

(c) Reflect T_2 in the line $x = 0$ to T_3.
Draw and label T_3.

(d) Describe fully the single transformation which maps T onto T_3.

4 The diagram shows a quadrilateral labelled Q. Copy the diagram.

(a) Q is mapped onto Q_1 by a rotation through 90°, anticlockwise about (0, 0).
Draw and label Q_1.

(b) Q_1 is mapped onto Q_2 by a rotation through 90°, anticlockwise about (2, 0).
Draw and label Q_2.

(c) Describe fully the single transformation which maps Q onto Q_2.

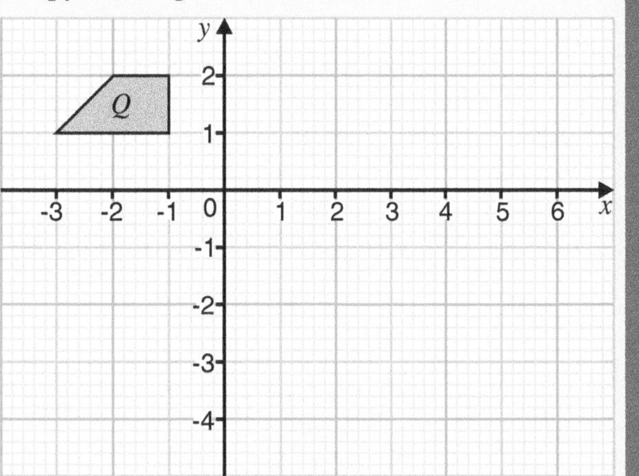

5 The diagram shows a shape labelled S. Copy the diagram.

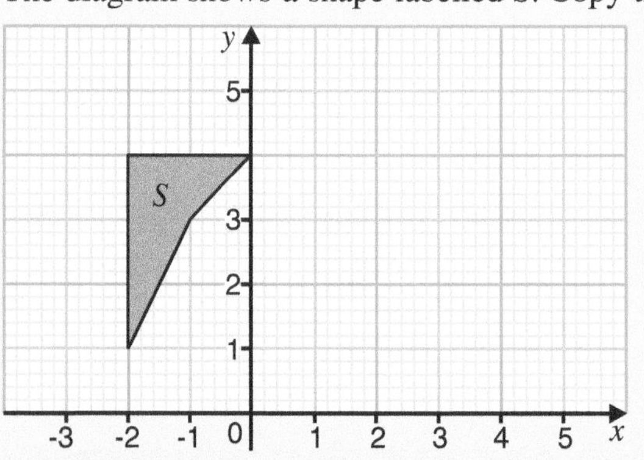

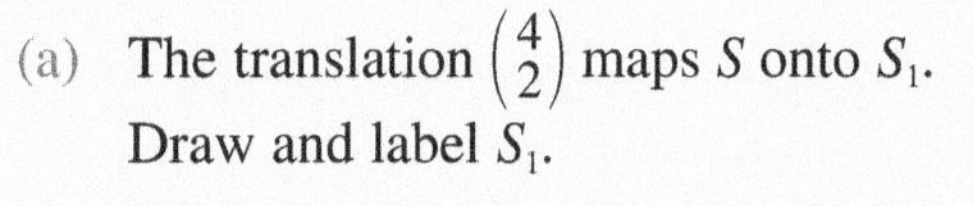

(a) The translation $\begin{pmatrix} 4 \\ 2 \end{pmatrix}$ maps S onto S_1.
Draw and label S_1.

(b) The translation $\begin{pmatrix} -8 \\ 1 \end{pmatrix}$ maps S_1 onto S_2.
Draw and label S_2.

(c) Describe fully the single transformation which maps S onto S_2.

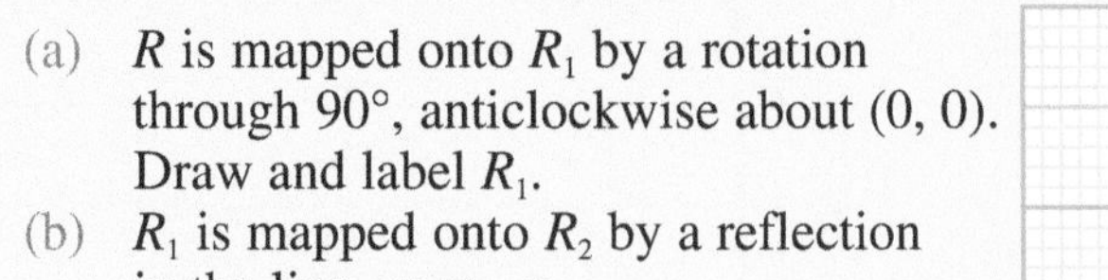

6 The diagram shows a triangle labelled R. Copy the diagram.

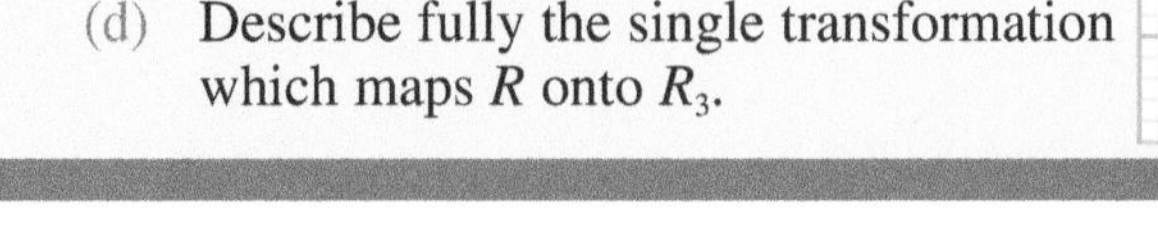

(a) R is mapped onto R_1 by a rotation through 90°, anticlockwise about (0, 0). Draw and label R_1.

(b) R_1 is mapped onto R_2 by a reflection in the line $y = -x$. Draw and label R_2.

(c) R_2 is mapped onto R_3 by a reflection in the line $x = 3$. Draw and label R_3.

(d) Describe fully the single transformation which maps R onto R_3.

What you need to know

- The movement of a shape from one position to another is called a **transformation**.
- **Single transformations** can be described in terms of a reflection, a rotation, a translation or an enlargement.
- **Reflection**: The image of the shape is the same distance from the mirror line as the original.
- **Rotation**: All points are turned through the same angle about the same point, called a centre of rotation.
- **Translation**: All points are moved the same distance in the same direction without turning.
- **Enlargement**: All lengths are multiplied by a scale factor.

  $\text{Scale factor} = \frac{\text{new length}}{\text{original length}}$ New length = scale factor × original length

  The size of the original shape is:

  increased by using a scale factor greater than 1,

  reduced by using a scale factor which is a fraction, i.e. between 0 and 1.
- When a shape is enlarged using a **negative scale factor** the image is **inverted**.
- How to fully describe a transformation.

Transformation	Image same shape and size?	Details needed to describe the transformation
Reflection	Yes	Mirror line, sometimes given as an equation.
Rotation	Yes	Centre of rotation, amount of turn, direction of turn.
Translation	Yes	Horizontal movement and vertical movement. Vector: top number = horizontal movement, bottom number = vertical movement.
Enlargement	No	Centre of enlargement, scale factor.

Review Exercise 27

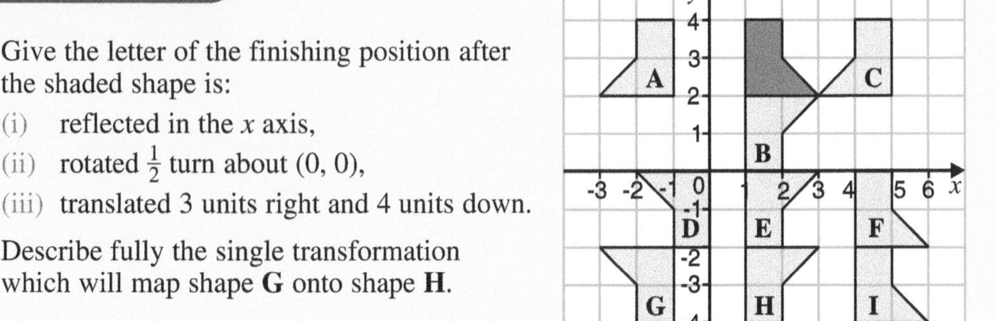

1 (a) Give the letter of the finishing position after the shaded shape is:

(i) reflected in the x axis,

(ii) rotated $\frac{1}{2}$ turn about (0, 0),

(iii) translated 3 units right and 4 units down.

(b) Describe fully the single transformation which will map shape **G** onto shape **H**.

AQA

2 The diagram shows triangles P, Q, R, S and T.
Describe fully the single transformation which maps
(a) P onto Q,
(b) T onto Q,
(c) R onto S,
(d) S onto R,
(e) R onto T,
(f) T onto R.

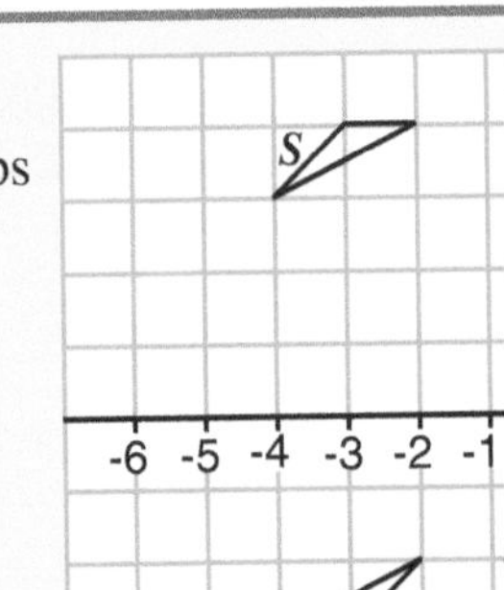

3 Triangle T has vertices (5, 0), (1, 0) and (4, 3).
(a) Draw triangle T on squared paper.
(b) Triangle T is enlarged by scale factor -2, from centre of enlargement (0, 0).
Draw the new triangle. AQA

4

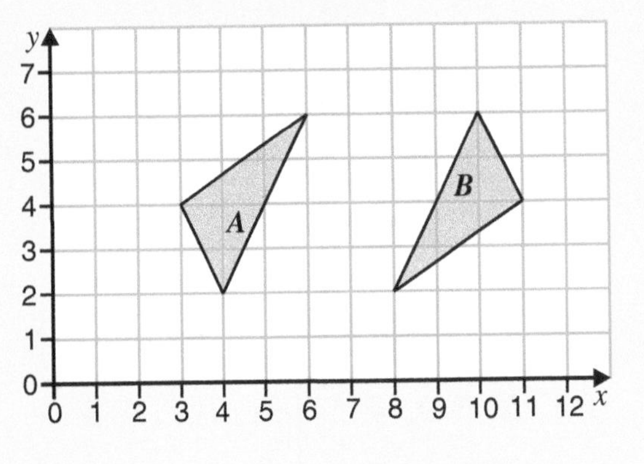

Triangles $\boldsymbol{A}$ and $\boldsymbol{B}$ are shown.
(a) Describe fully the **single** transformation which maps $\boldsymbol{A}$ onto $\boldsymbol{B}$.
(b) Copy the diagram.
$\boldsymbol{A}$ maps onto $\boldsymbol{C}$ by a reflection in the line $y = x$.
Show the position of $\boldsymbol{C}$ on your diagram. AQA

5 Copy the diagram onto squared paper and draw an enlargement of the shape using centre (2, 2) and scale factor -1.

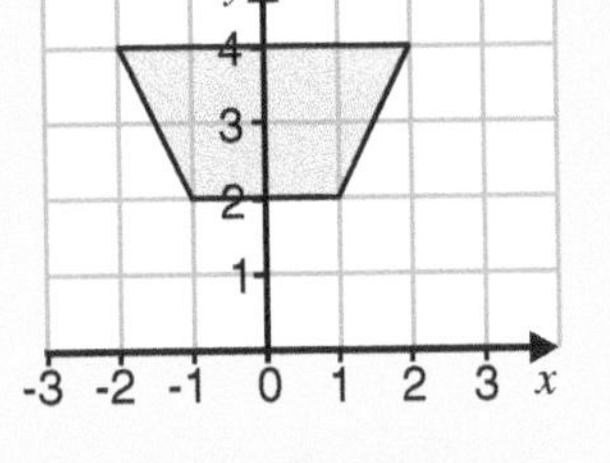

6

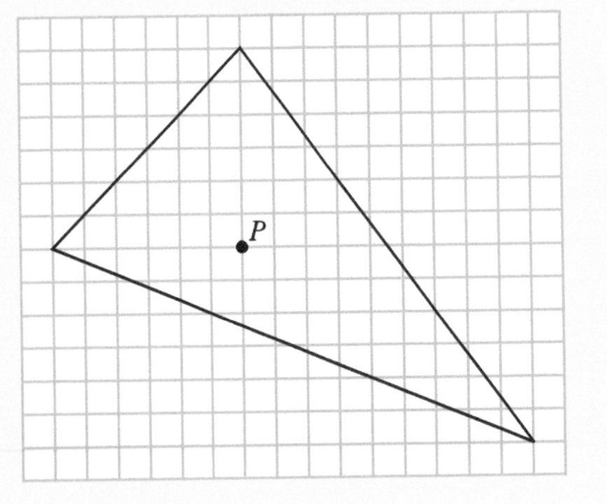

Copy the diagram onto squared paper and enlarge the triangle with scale factor $\frac{1}{3}$, centre P. AQA

7 The diagram shows shape A.
Copy the diagram onto squared paper, drawing the x axis from -6 to 8 and the y axis from -3 to 6.
(a) Reflect shape A in the line $x = 3$.
Label its new position B.
(b) Rotate shape A through 90° anticlockwise about centre (1, 0).
Label its new position C.
(c) Another triangle, D, has vertices $(-3, 3)$, $(-5, 5)$ and $(-2, 6)$.
Describe fully the **single** transformation which will take A onto D. AQA

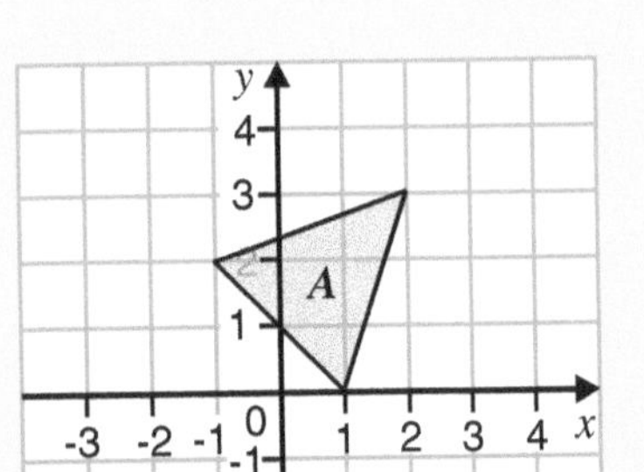

8 The diagram shows a trapezium labelled Q.
Copy the diagram onto squared paper, drawing both x and y axes from -2 to 9.

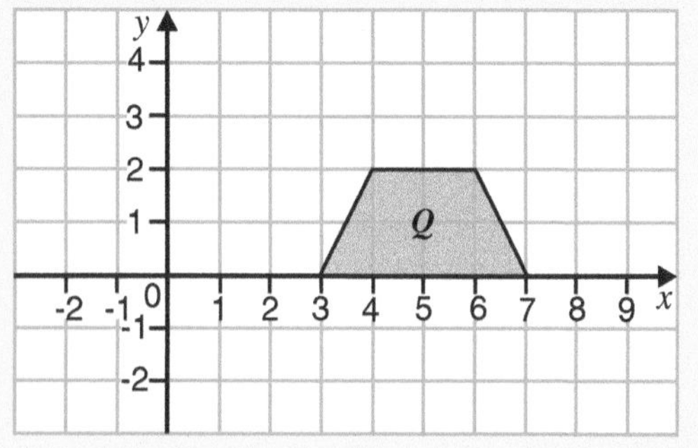

(a) Q is mapped onto Q_1 by a reflection in the x axis. Draw and label Q_1.

(b) Q_1 is mapped onto Q_2 by a translation with vector $\begin{pmatrix} 2 \\ 4 \end{pmatrix}$. Draw and label Q_2.

(c) Q_2 is mapped onto Q_3 by a reflection in the line $y = x$. Draw and label Q_3.

(d) Describe fully the single transformation which maps Q onto Q_3.

9 The grid shows a triangle ABC and a triangle $A'B'C'$.

(a) Draw the triangle $A''B''C''$ which is an enlargement of ABC with a scale factor $-\frac{1}{2}$ with centre (2, 1).

(b) Describe fully the transformation that takes triangle $A''B''C''$ to triangle $A'B'C'$.

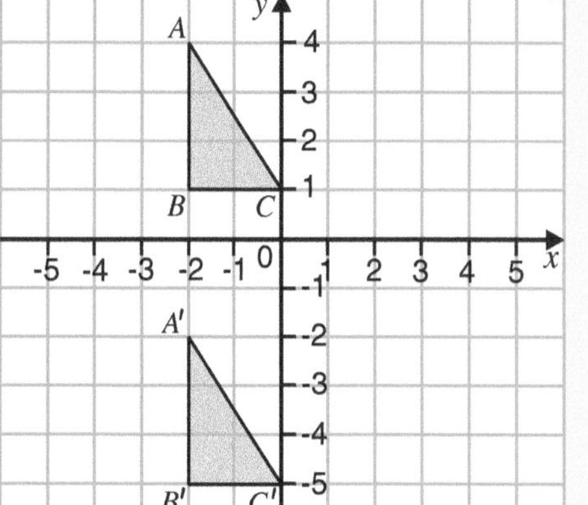

AQA

10 The diagram shows three triangles A, B and C.

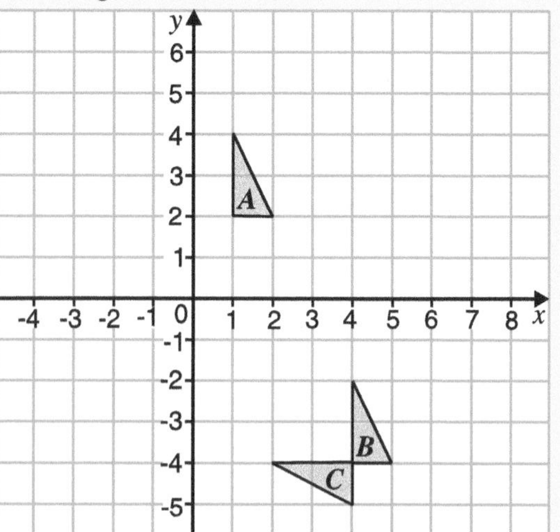

(a) Describe fully the single transformation which maps triangle A onto triangle B.

(b) Describe fully the single transformation which maps triangle B onto triangle C.

(c) P is a clockwise rotation of 90° with centre $(3, -2)$.
Q is a reflection in the line $y = x - 1$.
Triangle B maps onto triangle D by P followed by Q.
Copy triangle B and draw triangle D on the same diagram.

AQA

Pythagoras' Theorem

The longest side in a right-angled triangle is called the **hypotenuse**.

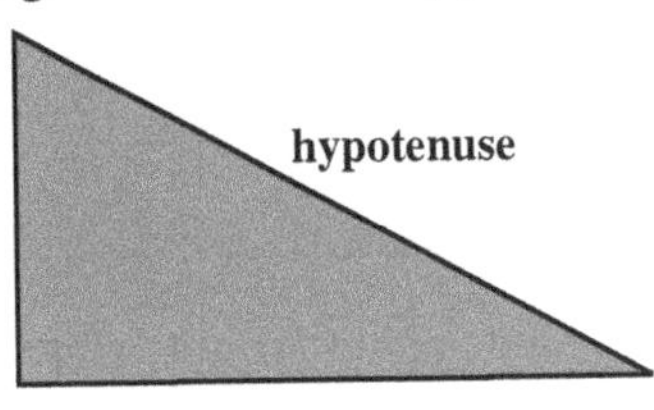

In any right-angled triangle it can be proved that:

"The square on the hypotenuse is equal to the sum of the squares on the other two sides."

This is known as the **Theorem of Pythagoras**, or **Pythagoras' Theorem**.

Checking the Theorem of Pythagoras

Look at this triangle.

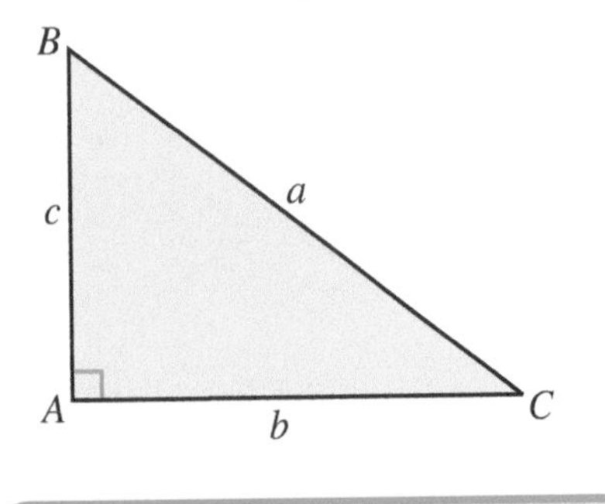

Notice that: the side opposite angle A is labelled a,
the side opposite angle B is labelled b,
the side opposite angle C is labelled c.

ABC is a right-angled triangle because $\angle BAC = 90°$.

$a = 5\text{ cm}$, so, $a^2 = 25\text{ cm}^2$.
$b = 4\text{ cm}$, so, $b^2 = 16\text{ cm}^2$.
$c = 3\text{ cm}$, so, $c^2 = 9\text{ cm}^2$.

$a^2 = b^2 + c^2$

Activity

Use a ruler and a pair of compasses to draw the following triangles accurately.

(a)

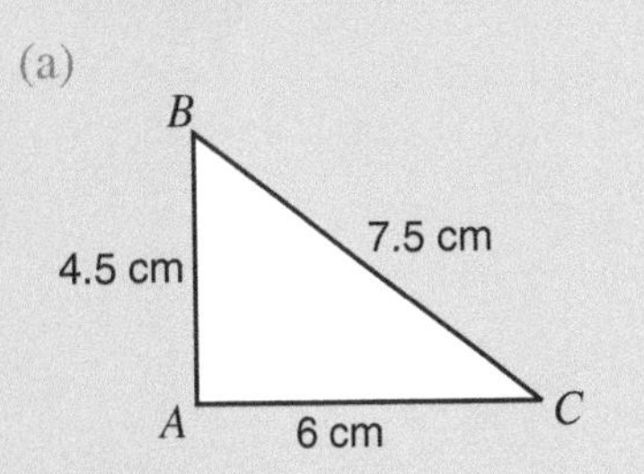

(b)

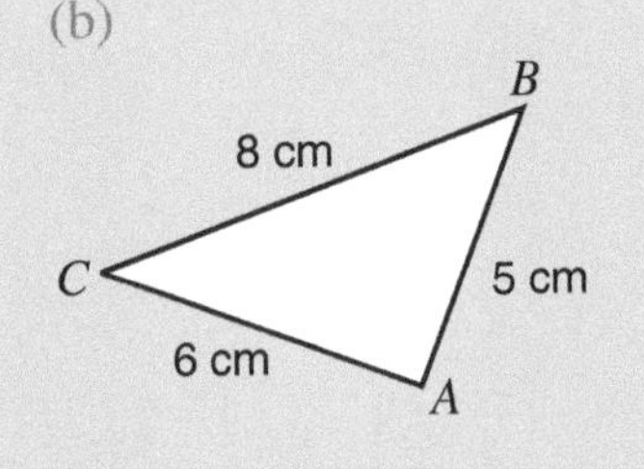

(c)

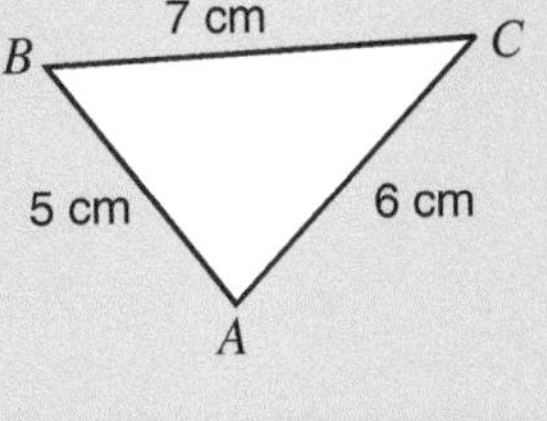
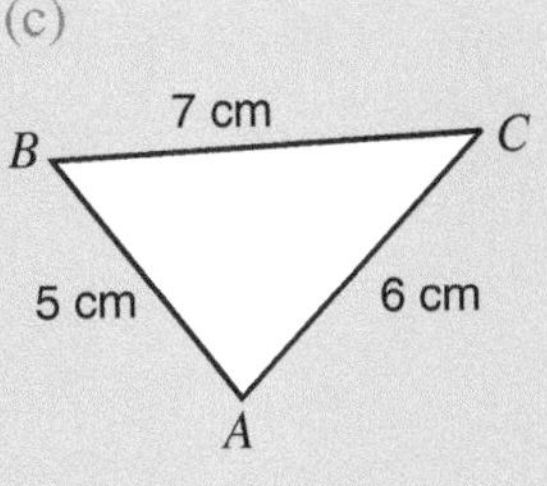

(d)

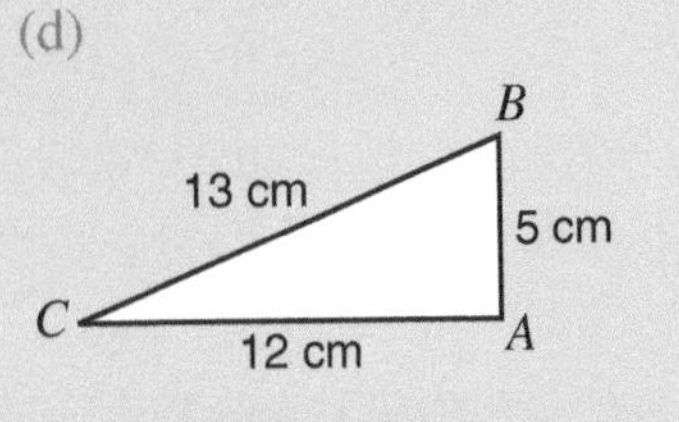

(e)

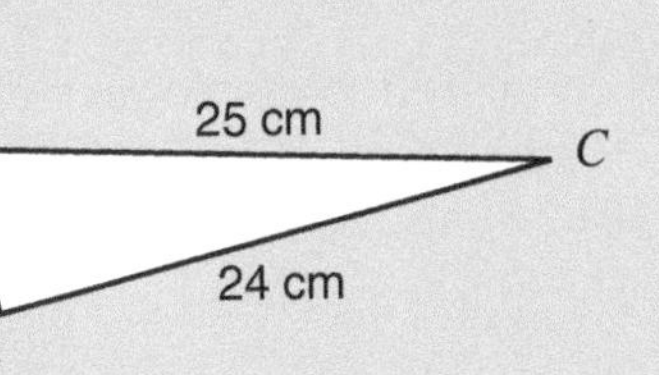

(f)

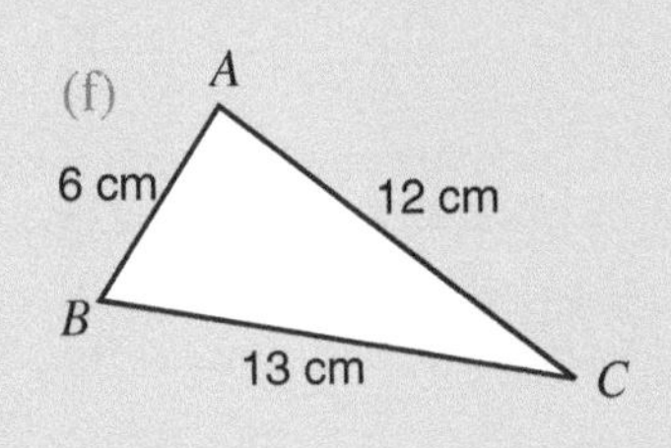

For each triangle: Measure angle BAC.
Is angle $BAC = 90°$?
Does $a^2 = b^2 + c^2$?

Explain your answers.

When we know the lengths of two sides of a right-angled triangle, we can use the Theorem of Pythagoras to find the length of the third side.

Finding the hypotenuse

EXAMPLE

The roof of a house is 12 m above the ground.
What length of ladder is needed to reach the roof, if the foot of the ladder has to be placed 5 m away from the wall of the house?

Using Pythagoras' Theorem.

$l^2 = 5^2 + 12^2$

$l^2 = 25 + 144$

$l^2 = 169$

Take the square root of both sides.

$l = \sqrt{169}$

$l = 13$ m

The ladder needs to be 13 m long.

Exercise 28.1

1 These triangles are right-angled. Calculate the length of the hypotenuse.

(a) 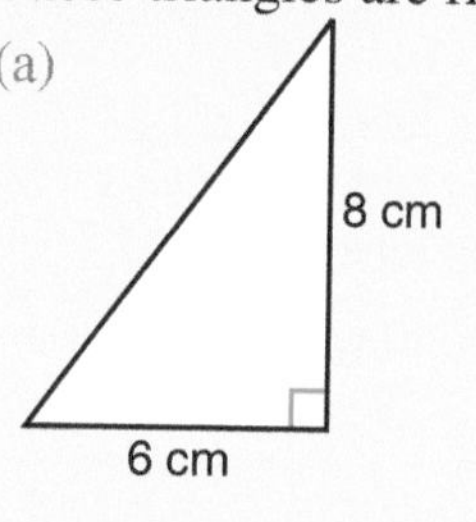

(b) 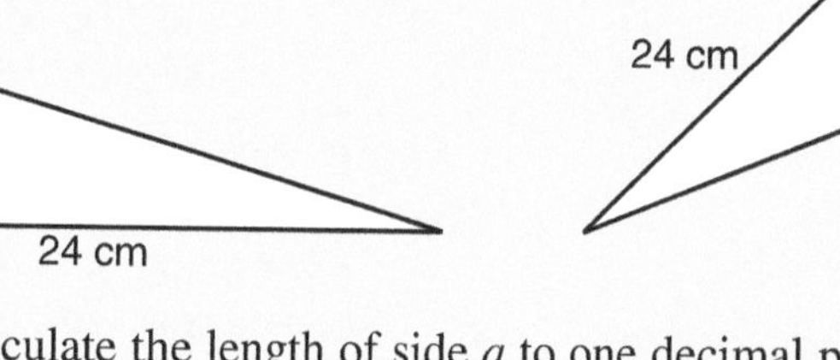

(c)

2 These triangles are right-angled. Calculate the length of side a to one decimal place.

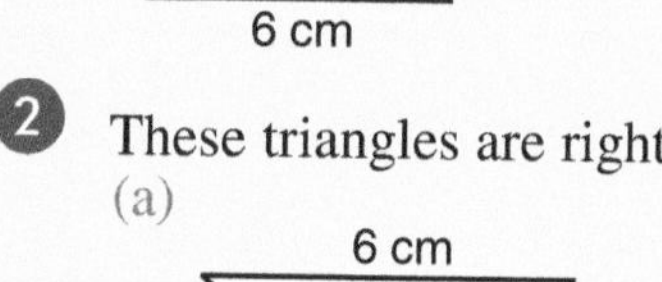

(a) 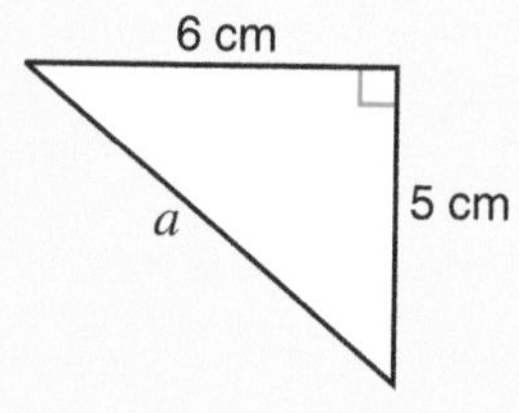

(b) 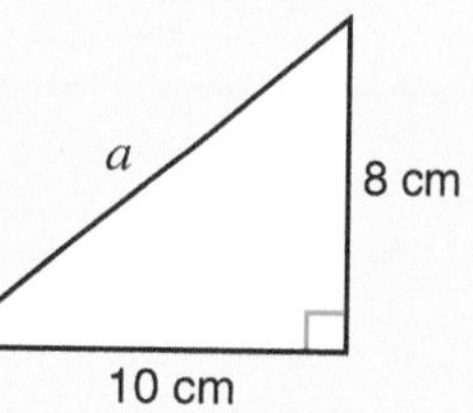

(c) 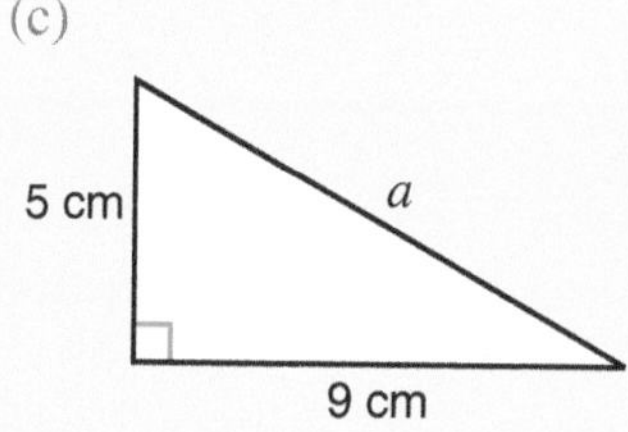

3 AB and CD are line segments, drawn on a centimetre-squared grid.
Calculate the exact length of

(a) AB, (b) CD.

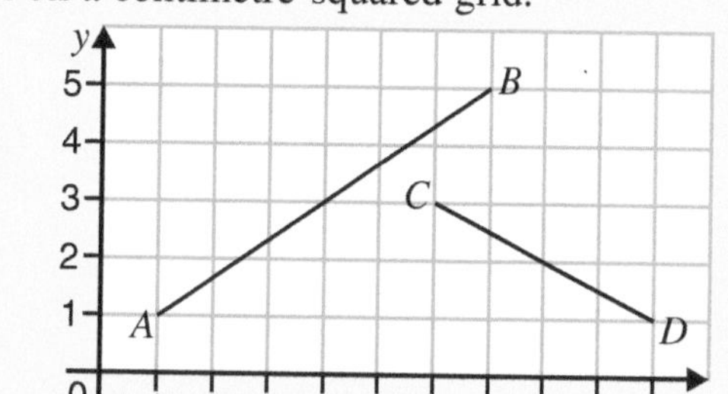
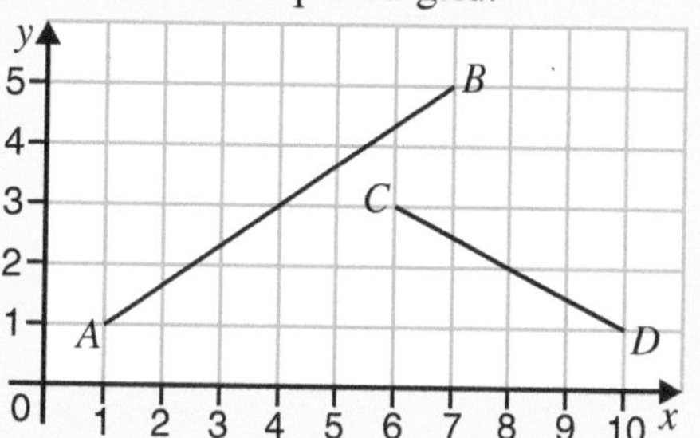

4 Calculate the distance between the following points.

(a) $A(2, 0)$ and $B(6, 3)$.
(b) $C(6, 3)$ and $D(0, 10)$.
(c) $E(2, 2)$ and $F(-3, -10)$.
(d) $G(-2, -2)$ and $H(-6, 5)$.
(e) $I(3, -1)$ and $J(-3, -5)$.

5 The coordinates of the vertices of a parallelogram are $P(1, 1)$, $Q(3, 5)$, $R(x, y)$ and $S(7, 3)$.

(a) Find the coordinates of R.
(b) X is the midpoint of PQ. Find the coordinates of X.
(c) Y is the midpoint of PS. Find the coordinates of Y.
(d) Calculate the distance XY.

Finding one of the shorter sides

To find one of the shorter sides we can rearrange the Theorem of Pythagoras.

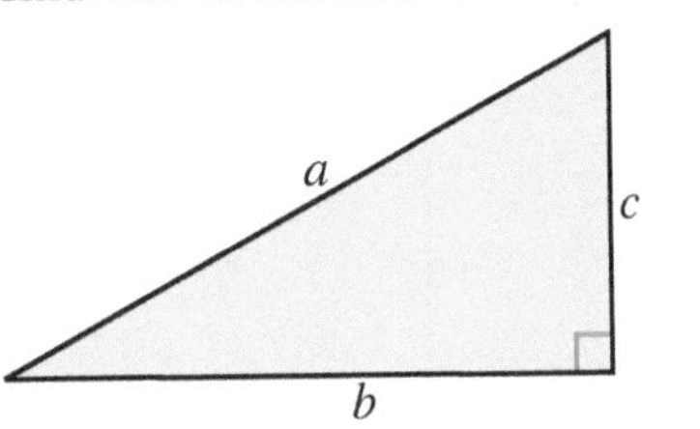

To find b we use: $b^2 = a^2 - c^2$

To find c we use: $c^2 = a^2 - b^2$

To find the length of a shorter side of a right-angled triangle:
Subtract the square of the known short side from the square on the hypotenuse.
Take the square root of the result.

EXAMPLE

A wire used to keep a radio aerial steady is 9 metres long.
The wire is fixed to the ground 4.6 metres from the base of the aerial.
Find the height of the aerial, giving your answer correct to one decimal place.

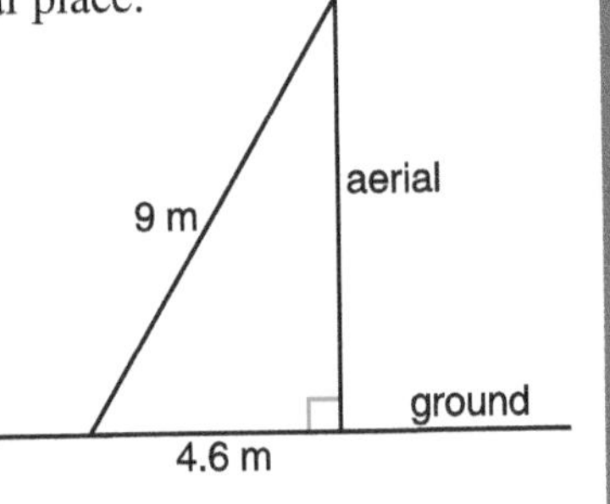

Using Pythagoras' Theorem. $9^2 = h^2 + 4.6^2$

Rearranging this we get: $h^2 = 9^2 - 4.6^2$

$h^2 = 81 - 21.16$

$h^2 = 59.84$

Take the square root of both sides. $h = \sqrt{59.84}$

$h = 7.735...$

$h = 7.7$ m, correct to 1 d.p.

The height of the aerial is 7.7 m, correct to 1 d.p.

Exercise 28.2

1 Work out the length of side b.

(a)

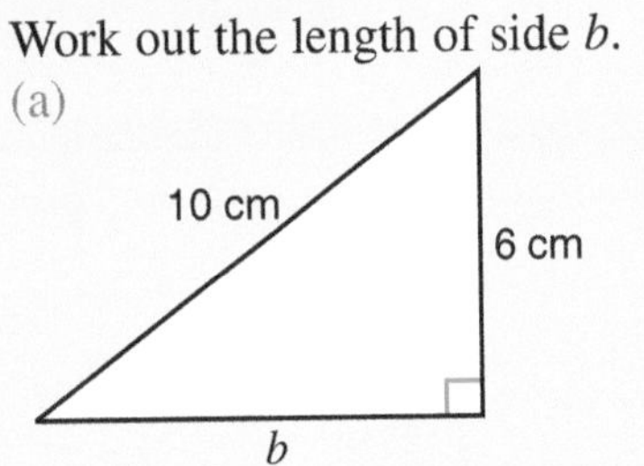

(b)

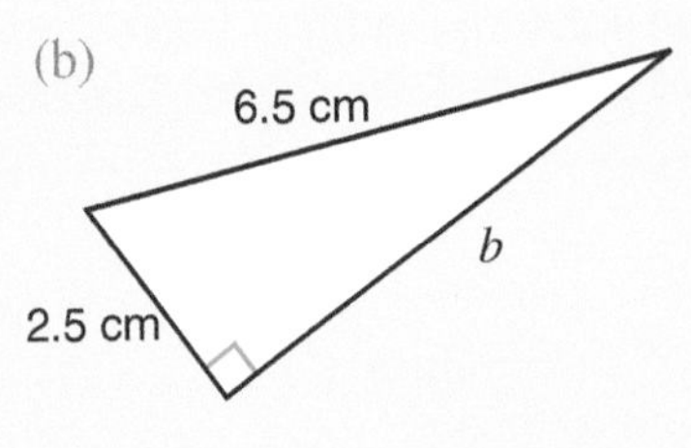

(c)

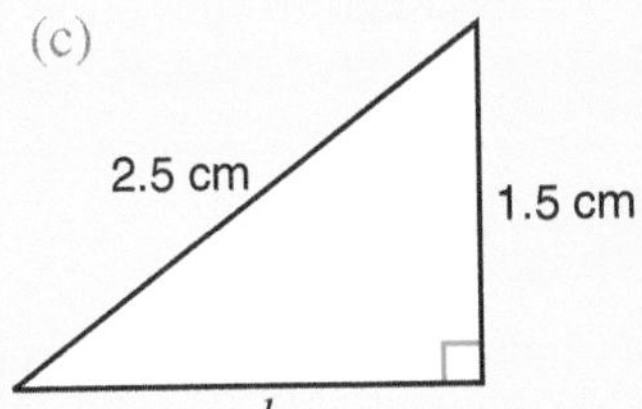

2 Work out the length of side c, correct to one decimal place.

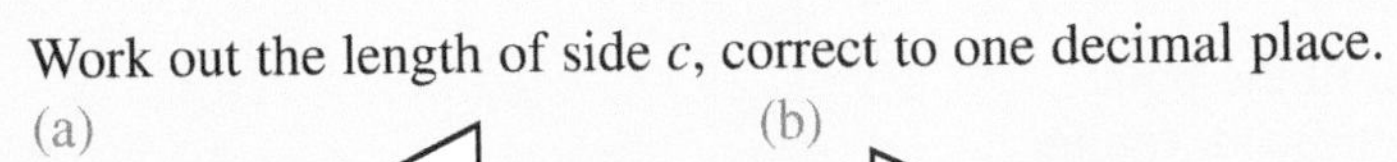

(a)

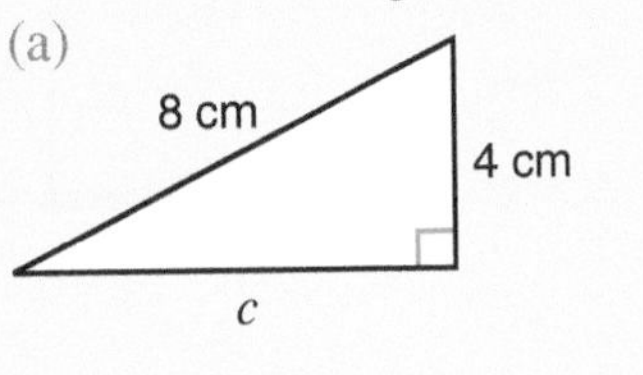

(b)

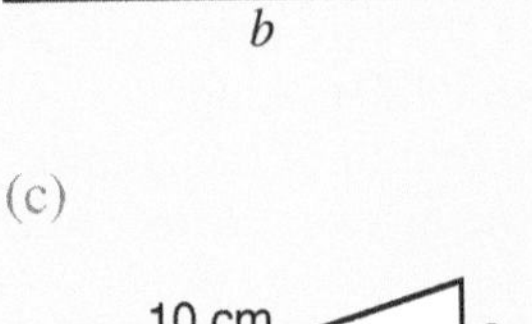

(c)

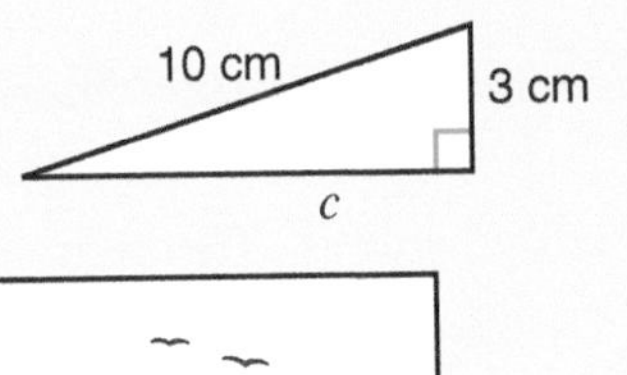

3 Two boats A and B are 360 m apart.
Boat A is 120 m due east of a buoy.
Boat B is due north of the buoy.
How far is boat B from the buoy?

4 The diagram shows a right-angled triangle, ABC, and a square, $ACDE$.

$AB = 2.5\,\text{cm}$ and $BC = 6.5\,\text{cm}$.

Calculate the area of the square $ACDE$.

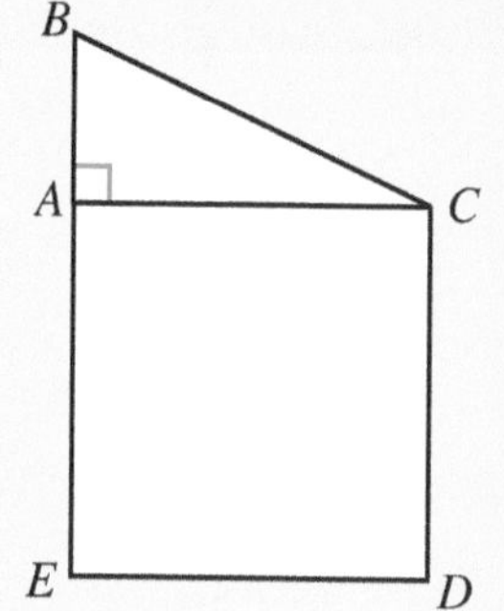
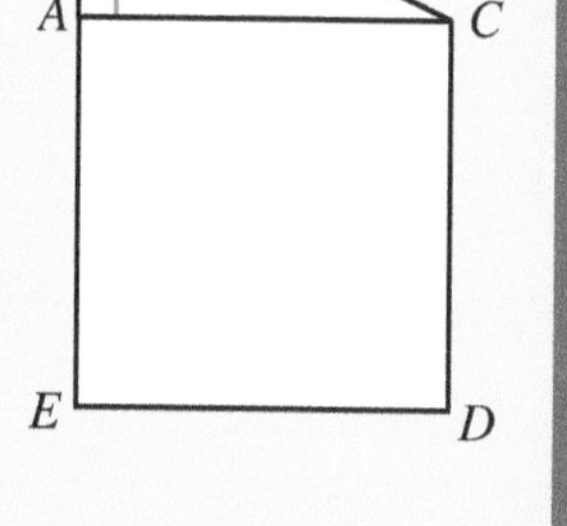

5 The diagram shows a right-angled triangle, ABC, and a square, $XYBA$.

$BC = 6\,\text{cm}$.

The square $XYBA$ has an area of $23.04\,\text{cm}^2$.

Calculate the length of AC.

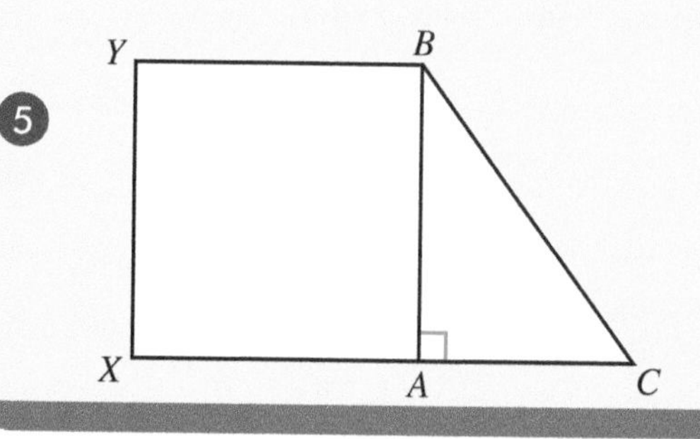

Problems involving the use of Pythagoras' Theorem

Questions leading to the use of Pythagoras' Theorem often involve:

Understanding the problem.

What information is given?
What are you required to find?

Drawing diagrams.

In some questions a diagram is not given.
Drawing a diagram may help you to understand the problem.

Selecting a suitable right-angled triangle.

In more complex problems you will have to select a right-angled triangle which can be used to answer the question. It is a good idea to draw this triangle on its own, especially if it has been taken from a three-dimensional drawing.

EXAMPLE

The diagram shows the side view of a swimming pool.
It slopes steadily from a depth of 1 m to 3.6 m.
The pool is 20 m long.
Find the length of the sloping bottom of the pool, giving the answer correct to three significant figures.

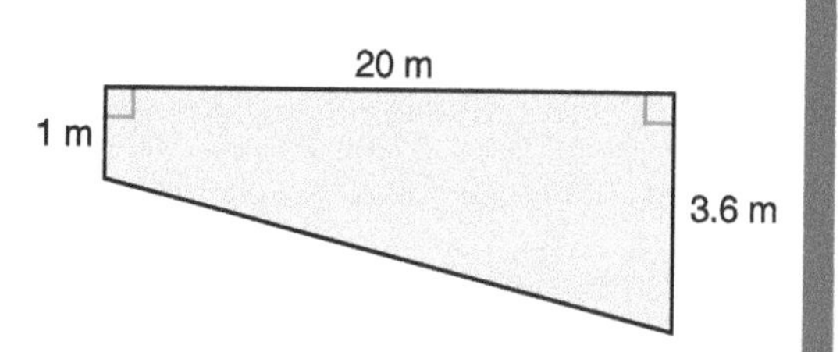

ΔCDE is a suitable right-angled triangle.

$CD = 3.6 - 1 = 2.6\,\text{m}$

Using Pythagoras' Theorem in ΔCDE.

$DE^2 = CD^2 + CE^2$

$DE^2 = 2.6^2 + 20^2$

$DE^2 = 6.76 + 400$

$DE^2 = 406.76$

$DE = \sqrt{406.76}\ \text{m}$

$DE = 20.1682...\ \text{m}$

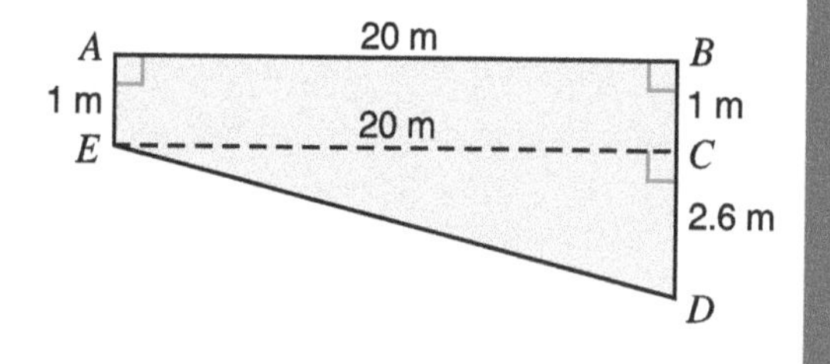

The length of the sloping bottom of the pool is 20.2 m, correct to 3 sig. figs.

Exercise 28.3

1. In each of the following, work out the length of the side marked x.

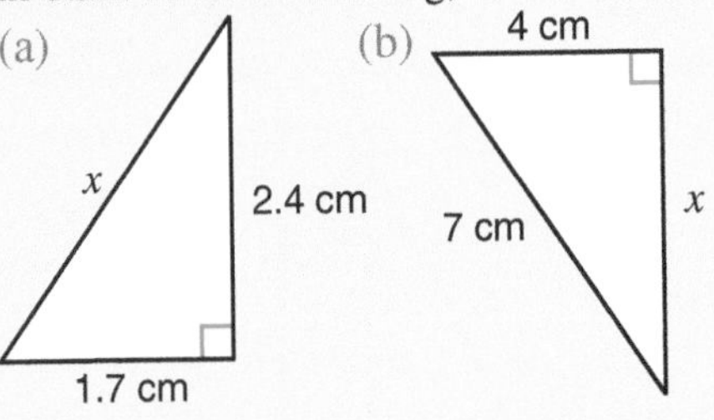

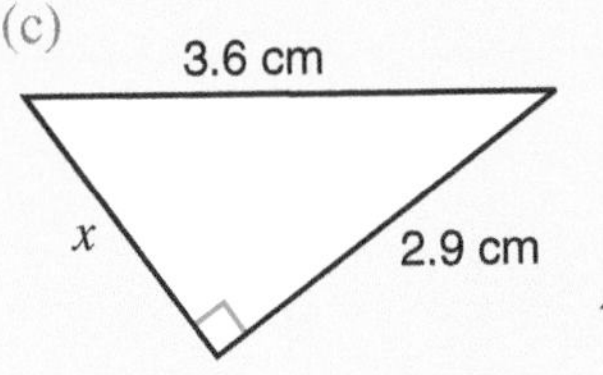

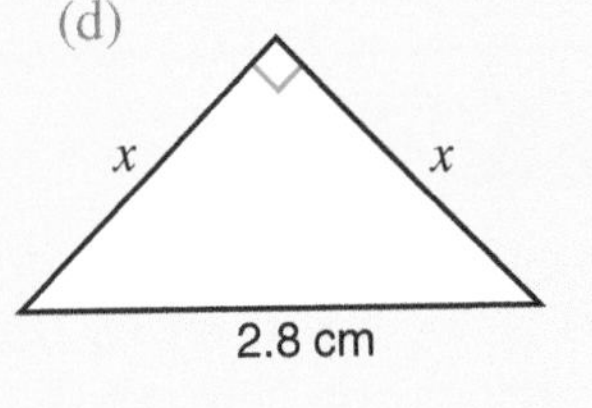

2. A rectangle is 8 cm wide and 15 cm long.
Work out the length of its diagonals.

3. The length of a rectangle is 24 cm. The diagonals of the rectangle are 26 cm.
Work out the width of the rectangle.

4. A square has sides of length 6 cm.
Work out the length of its diagonals.

5. The diagonals of a square are 15 cm.
Work out the length of its sides.

6. The height of an isosceles triangle is 12 cm. The base of the triangle is 18 cm.
Work out the length of the equal sides.

7. An equilateral triangle has sides of length 8 cm.
Work out the height of the triangle.

8. The diagram shows the side view of a car ramp.
The ramp is 110 cm long and 25 cm high.
The top part of the ramp is 40 cm long.
Calculate the length of the sloping part of the ramp.

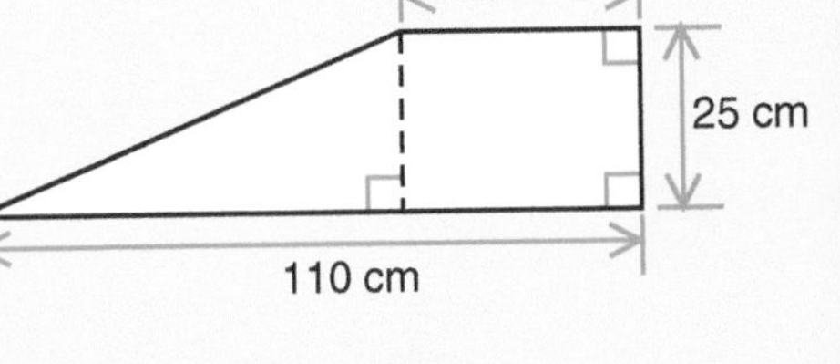

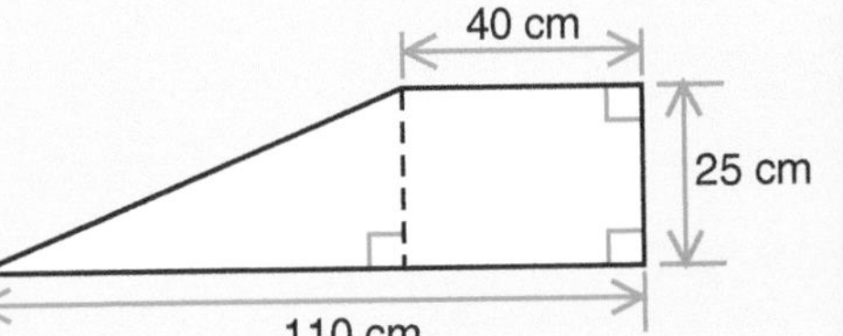

9. The top of a lampshade has a diameter of 10 cm.
The bottom of the lampshade has a diameter of 20 cm.
The height of the lampshade is 12 cm.
Calculate the length, l, of the sloping sides.

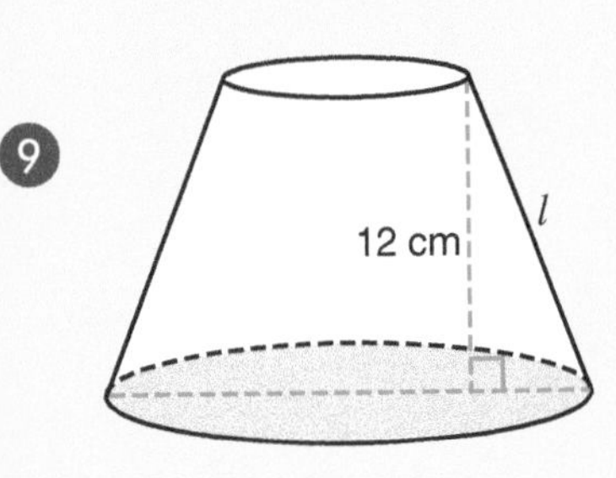

10. The top of a bucket has a diameter of 30 cm.
The bottom of the bucket has a diameter of 16 cm.
The sloping sides are 25 cm long.
How deep is the bucket?

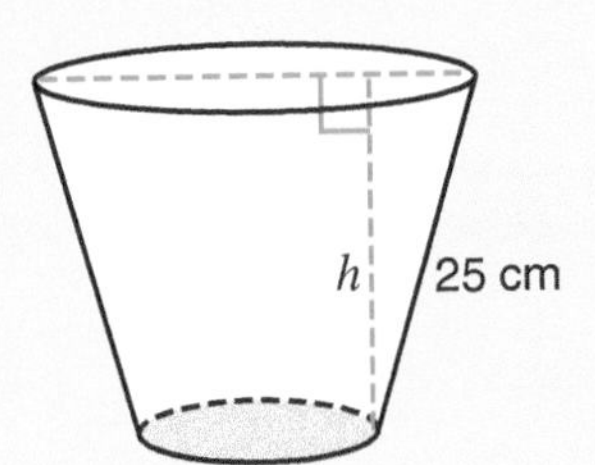

11. $ABCD$ is a kite.
$AB = 8.5$ cm, $BC = 5.4$ cm and $BD = 7.6$ cm.
(a) Calculate the length of AC.
(b) Calculate the area of the kite.

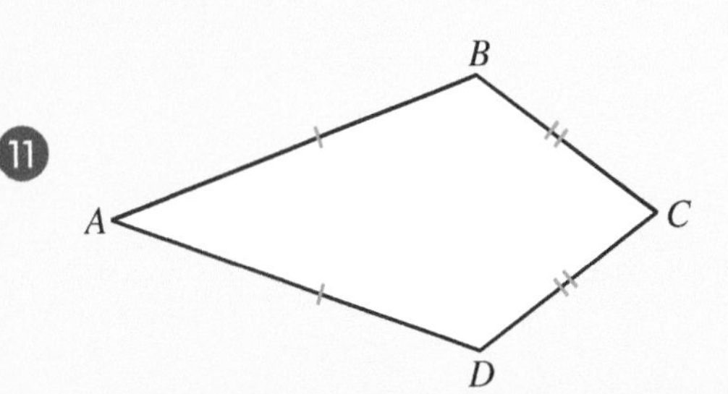

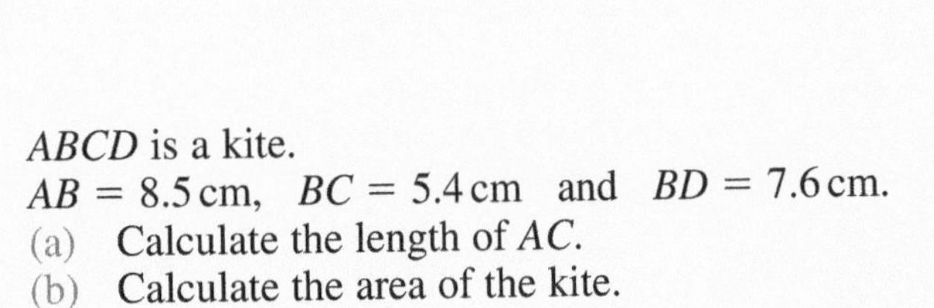

Solving problems in three dimensions

When we solve problems in three dimensions we often need to use more than one triangle to solve the problem.

EXAMPLE

This cube has sides of length 5 cm.
What is the distance from D to F?
Give the answer correct to 1 decimal place.

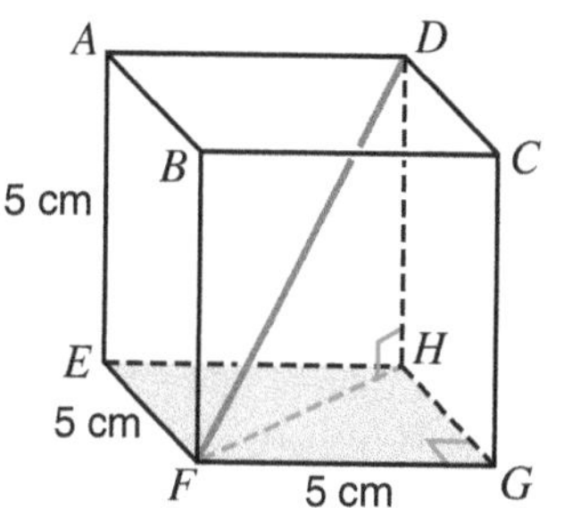
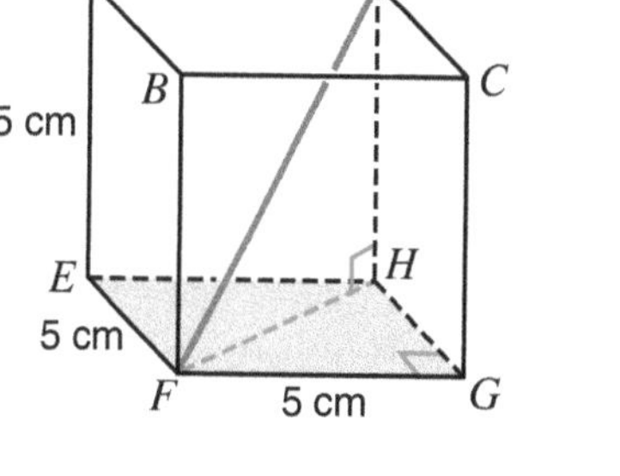

ΔDFH is a suitable right-angled triangle,
but we only know the length of the side DH.
The length of FH can be found by using ΔFGH.

Using Pythagoras' Theorem in ΔFGH.

$FH^2 = FG^2 + GH^2$
$FH^2 = 5^2 + 5^2$
$FH^2 = 25 + 25$
$FH^2 = 50$
$FH = \sqrt{50}\,\text{cm}$

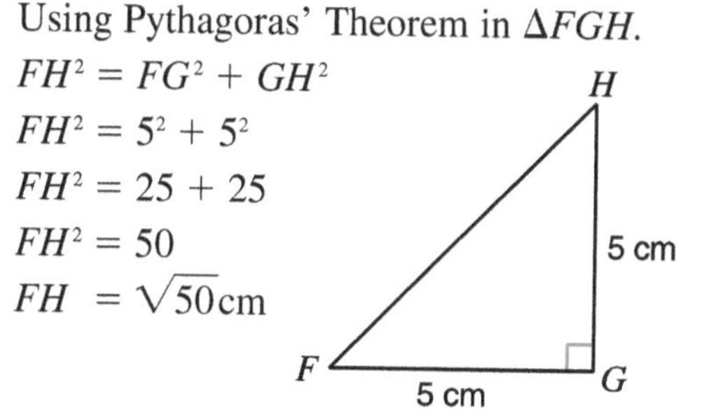

Using Pythagoras' Theorem in ΔDFH.

$DF^2 = DH^2 + FH^2$
$DF^2 = 5^2 + (\sqrt{50})^2$
$DF^2 = 25 + 50$
$DF^2 = 75$
$DF = \sqrt{75}\,\text{cm}$
$DF = 8.66\ldots\,\text{cm}$
$DF = 8.7\,\text{cm}$, correct to 1 d.p.

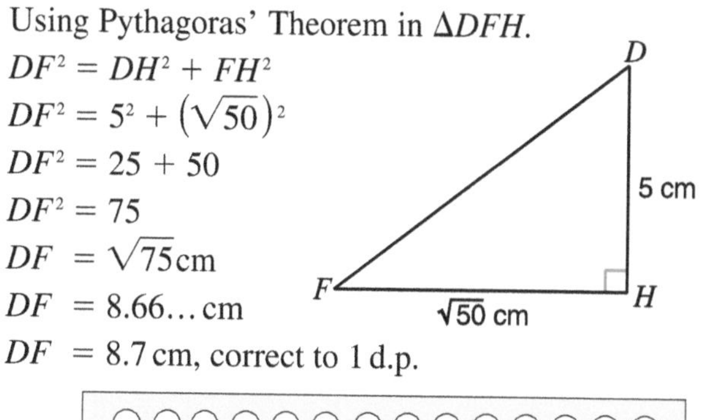

For accuracy, the exact value of FH, $(\sqrt{50})$, should be used.

Exercise 28.4

1. The cube has sides of length 6 cm.
   Calculate the length of the diagonal PQ.

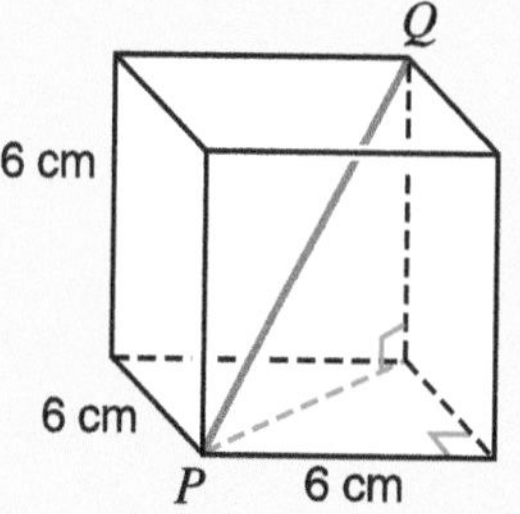

2. The diagram shows a wedge $ABCDEF$.
   $AD = 13\,\text{cm}$, $DE = 15\,\text{cm}$ and $EC = 6\,\text{cm}$.
   Calculate the length of the line AC.

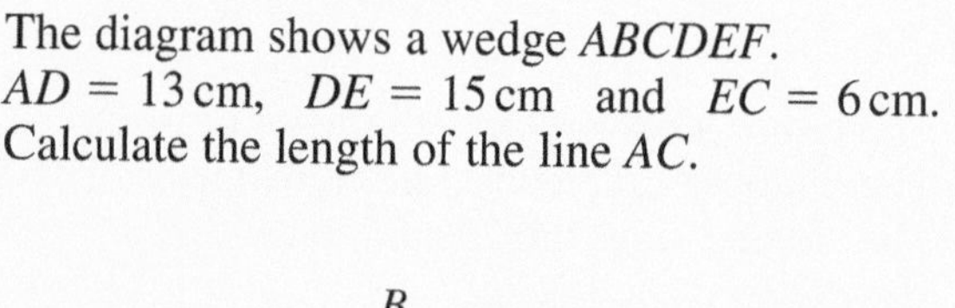
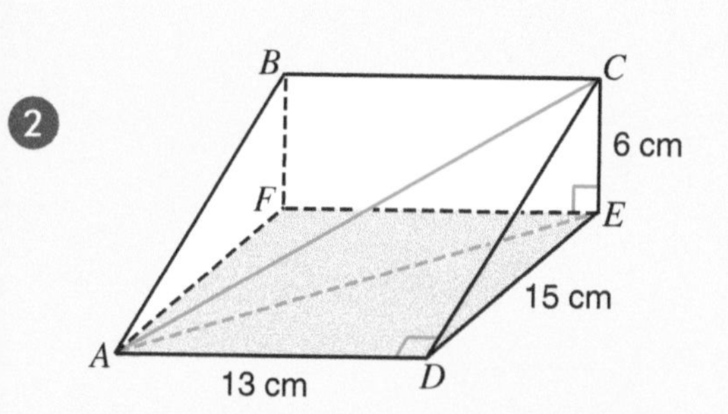

3. $ABCDEFGH$ is a cuboid.
   $AE = 5\,\text{cm}$, $EH = 9\,\text{cm}$ and $HG = 6\,\text{cm}$.
   Calculate the length of the line AG.

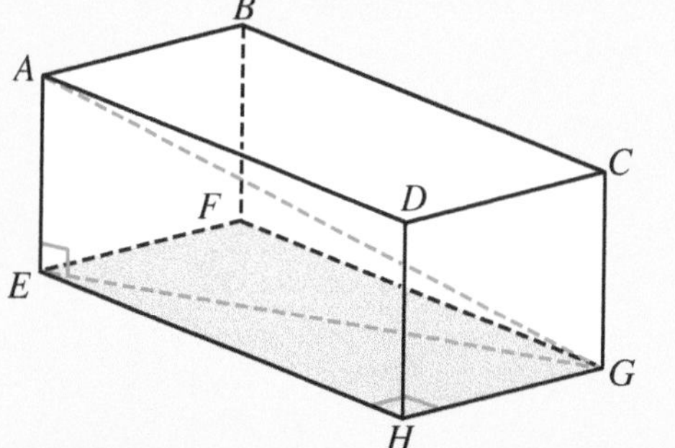

4 The diagram shows a pyramid $PABCD$.
X is at the centre of the base.
The base $ABCD$ is a rectangle with
$AB = CD = 12\text{ cm}$ and $BC = DA = 9\text{ cm}$.
The height of the pyramid, PX, is 18 cm.
Calculate the length of the edge PA.

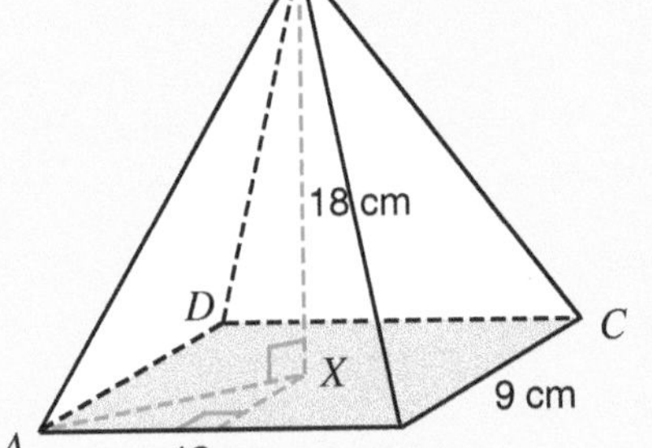

5 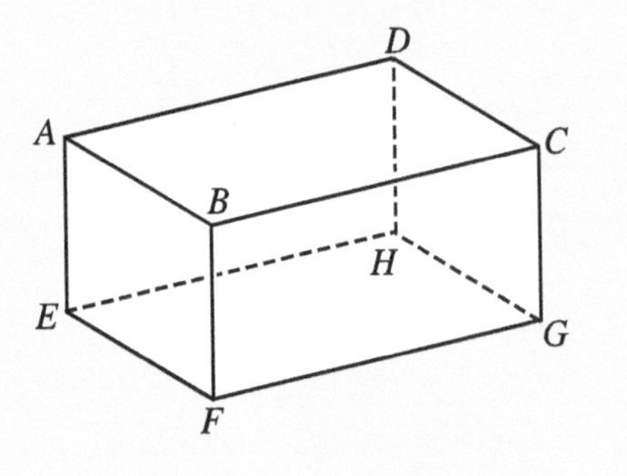

The diagram of a cuboid is shown.

(a) Calculate the length of EC, when
$EF = 6\text{ cm}$, $FG = 8\text{ cm}$ and $CG = 5\text{ cm}$.

(b) Calculate the length of AG, when
$AD = 10\text{ cm}$, $DC = 8\text{ cm}$ and $CG = 6\text{ cm}$.

(c) Calculate the length of FD, when
$FG = 7\text{ cm}$, $GH = 5\text{ cm}$ and $HD = 4\text{ cm}$.

6 The diagram shows a square based pyramid.
All edges are 7 cm in length.
Calculate the height of the pyramid, PX.

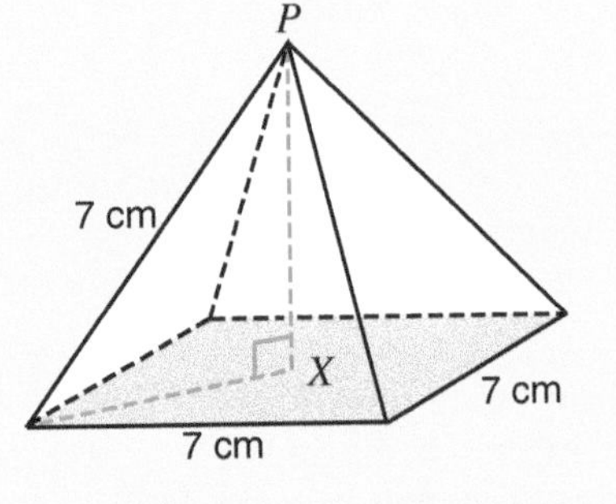

7 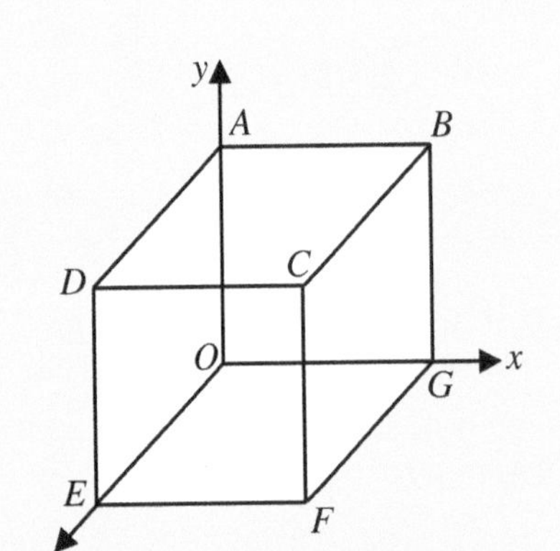

The diagram shows a cuboid drawn in 3-dimensions.
Using the x, y and z axes shown:

Point A has coordinates (0, 3, 0).
Point B has coordinates (2, 3, 0).
Point C has coordinates (2, 3, 4).

Find the lengths of the following.

(a) AB (b) BC (c) AC
(d) DF (e) AE (f) OF

Give your answers correct to one decimal place.

What you need to know

- The longest side in a right-angled triangle is called the **hypotenuse**.
- The **Theorem of Pythagoras** states:
"In any right-angled triangle the square on the hypotenuse is equal to the sum of the squares on the other two sides."

$$a^2 = b^2 + c^2$$

Rearranging gives:

$$b^2 = a^2 - c^2$$
$$c^2 = a^2 - b^2$$

- When we know the lengths of two sides of a right-angled triangle, we can use the Theorem of Pythagoras to find the length of the third side.

Investigate the relationship between the areas of the semi-circles A, B and C.

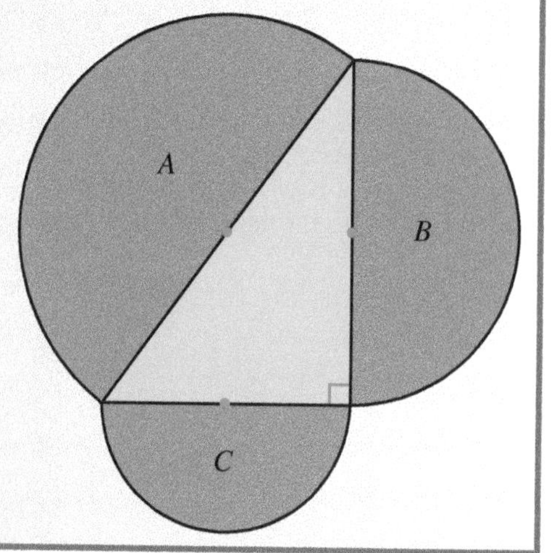

Review Exercise 28

1. The diagram shows a right-angled triangle PQR.
   $PR = 12\,\text{cm}$ and $QR = 9\,\text{cm}$
   Calculate the length of PQ.

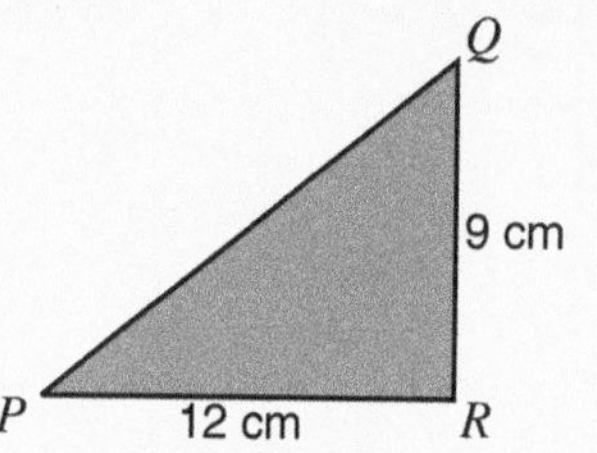

2. The coordinates of the points A and B are (6, 8) and (1, 1).
   Work out the length of AB.

   AQA

3. The sketch shows triangle ABC.
   $AB = 40\,\text{cm}$, $AC = 41\,\text{cm}$ and $CB = 9\,\text{cm}$.
   By calculation, show that triangle ABC is a right-angled triangle.

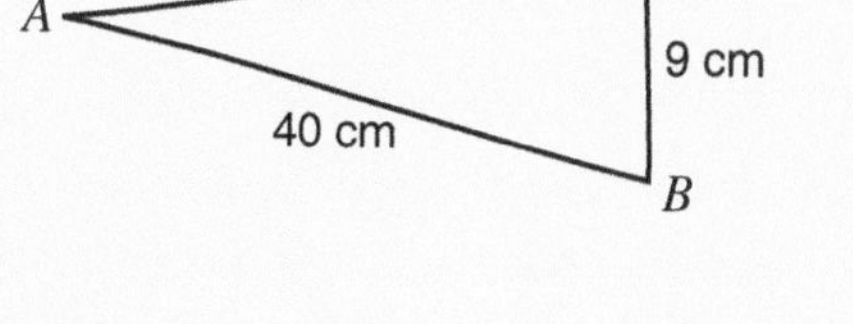
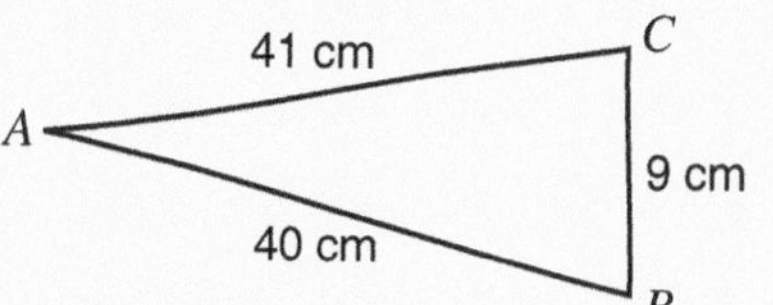
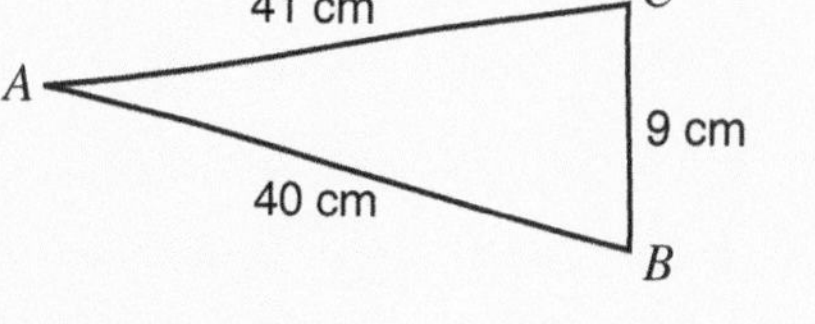
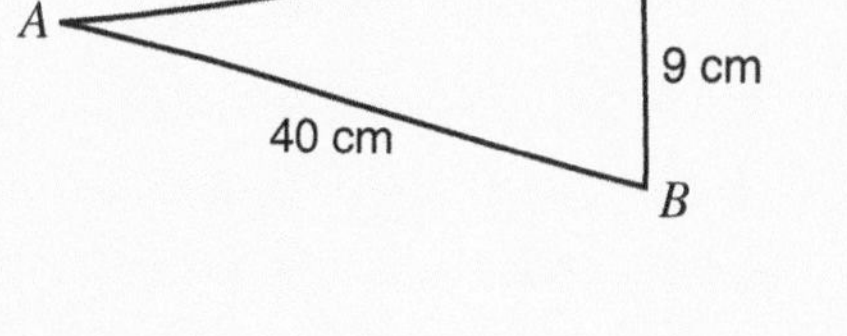
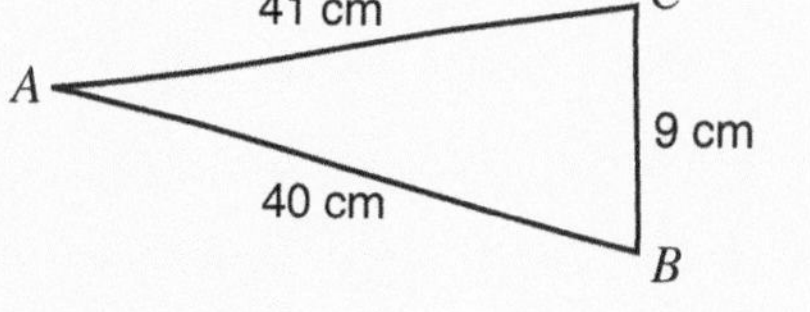

   AQA

4. The diagram consists of three right-angled triangles.
   $AB = 7\,\text{cm}$, $BC = DE = 5\,\text{cm}$ and $AE = 12\,\text{cm}$.
   (a) Calculate the length AC.
   (b) Calculate the length AD.

   AQA

5. The diagram shows a trapezium $ABCD$.
   Calculate the **exact** length of the line AB.

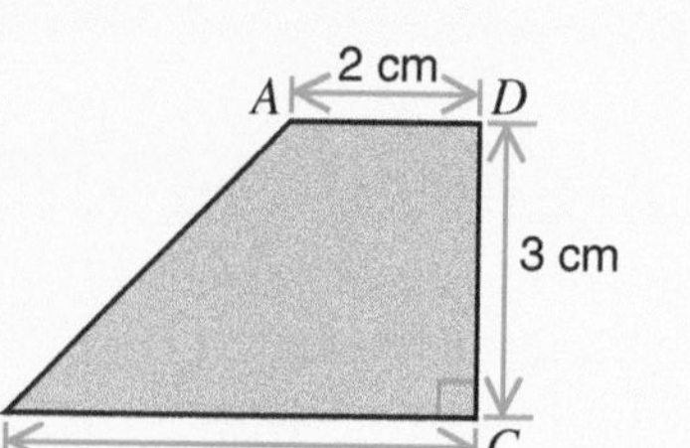
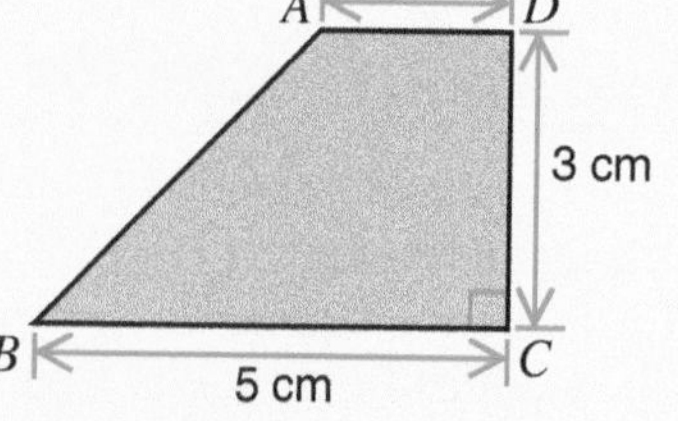

6. Mike is standing 200 m due west of a power station and 300 m due north of a pylon.
   Calculate the distance of the power station from the pylon.

7. A helicopter flies from its base on a bearing of 045° for 20 km before landing.
   How far east of its base is the helicopter when it lands?

8. 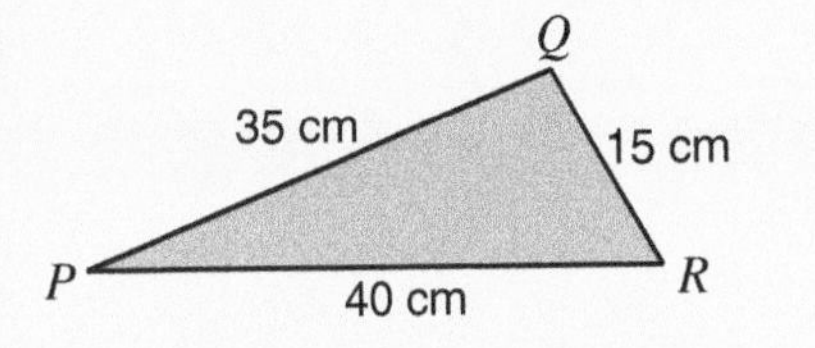

The diagram shows a triangle PQR.
$PQ = 35$ cm, $QR = 15$ cm and $PR = 40$ cm.
Show that angle $PQR > 90°$.

9. The diagram shows a cuboid drawn in 3 dimensions.
Using the axes x, y and z shown:
Point A is given as $(-2, 4, 1)$.
Point B is given as $(4, 4, 1)$.
Point C is given as $(4, 4, 9)$.
The length of AB is 6 units.
Find the exact lengths of the lines AC, AE and AF.

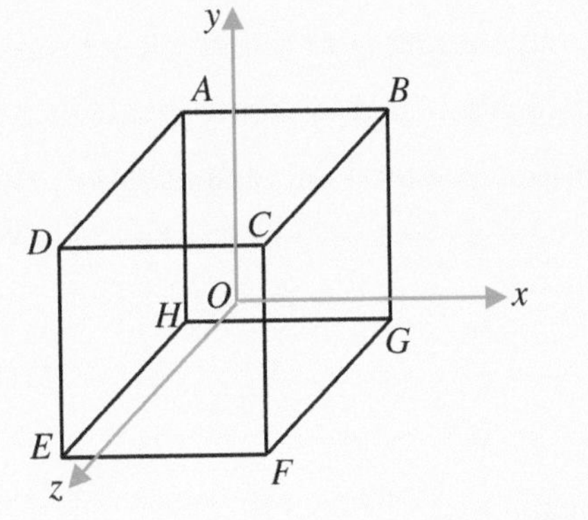

10. 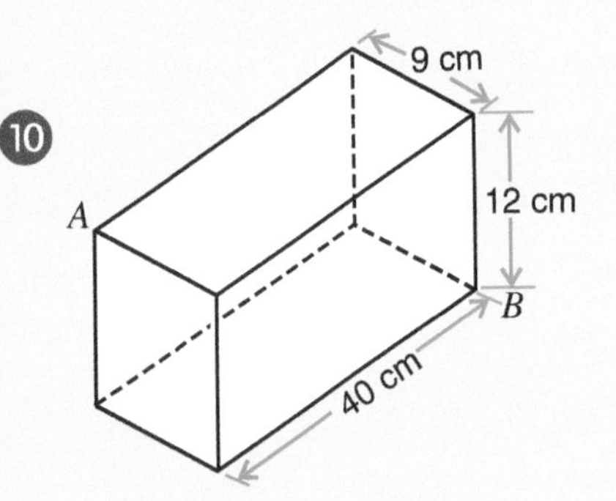

An open box has internal dimensions 9 cm by 12 cm by 40 cm.
What is the length of the diagonal AB?

AQA

11. The cube has sides of length 10 cm.
Calculate the length of the diagonal PQ.

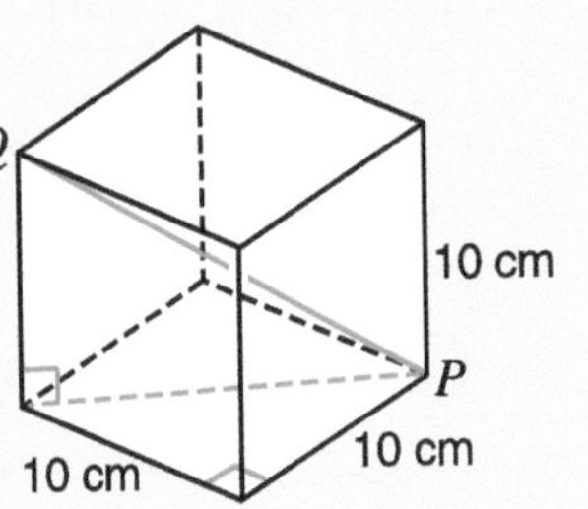
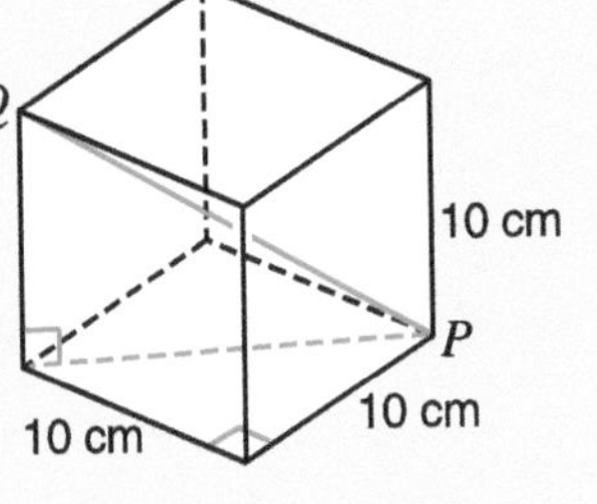

12. The diagram shows a wedge $ABCDEF$.
$AC = 13$ cm, $DE = 7$ cm and $EC = 5$ cm.
Calculate the length of the line AD.

13. The diagram shows a square based pyramid.
All edges are 8 cm in length.
Calculate the height of the pyramid, PX.

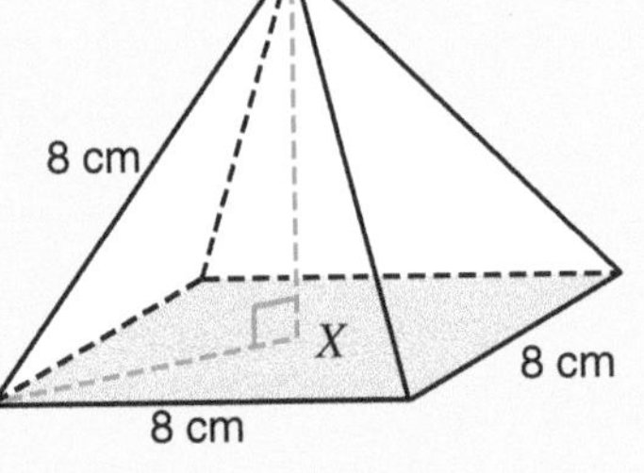
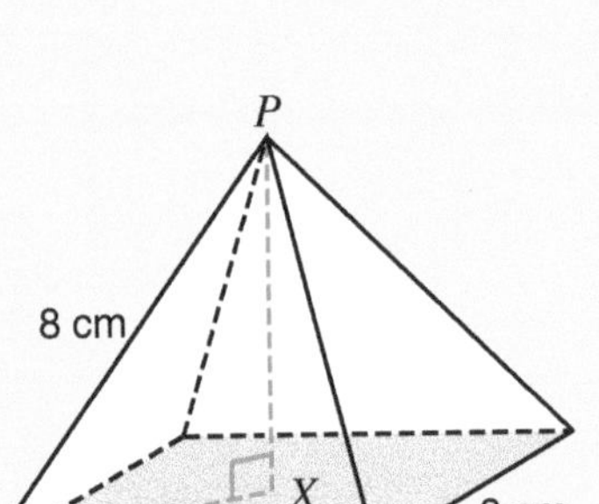

14. The 3-dimensional coordinates of M and N are $(2, 0, -1)$ and $(-1, 2, 3)$.
Calculate the length of MN, correct to one decimal place.

Trigonometry

We use **trigonometry** to find the lengths of sides and the sizes of angles in right-angled triangles.

We already know that the longest side of a right-angled triangle is called the **hypotenuse**.

In order to understand the relationships between sides and angles the other sides of the triangle also need to be named.

The **opposite** side is the side directly opposite the angle being used and the **adjacent** side is the side next to the angle.

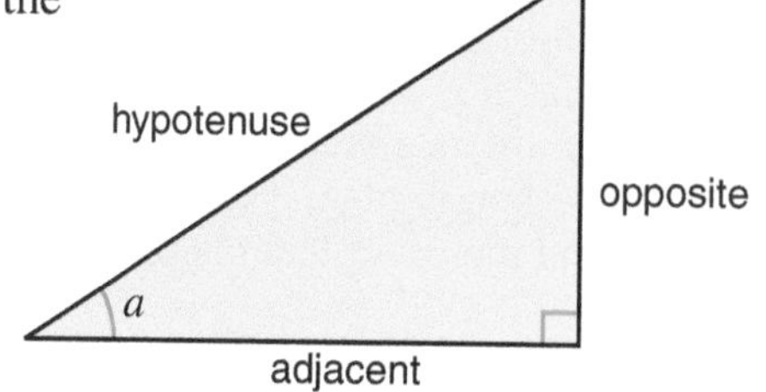

Look at this diagram.
It shows how the height of a kite changes as more and more string is let out.

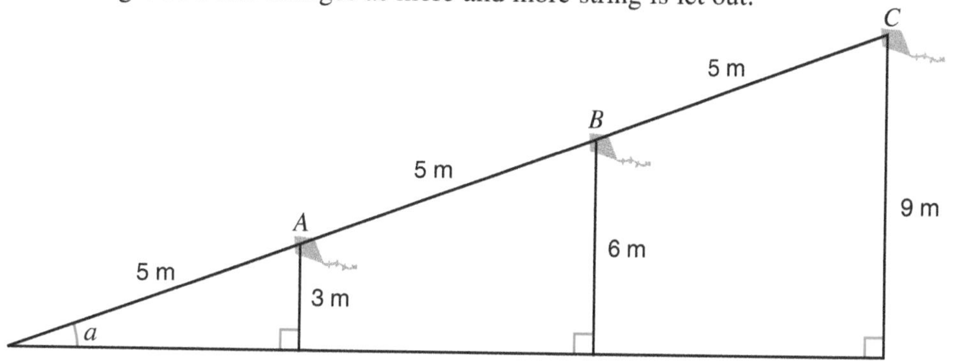

When the kite is at A, the string is 5 m long and the kite is 3 m high.

At A, the ratio $\frac{\text{height of kite}}{\text{length of string}}$, is therefore $\frac{3}{5} = 0.6$.

Calculate the value of the same ratio at B and C.

What do you notice?

When the kite is flying at angle a, the ratio $\frac{\text{height of kite}}{\text{length of string}}$ will always be the same whatever the length of the kite string and is called the **sine** of angle a.

The sine ratio

For any right-angled triangle: $\sin a = \frac{\text{opposite}}{\text{hypotenuse}}$

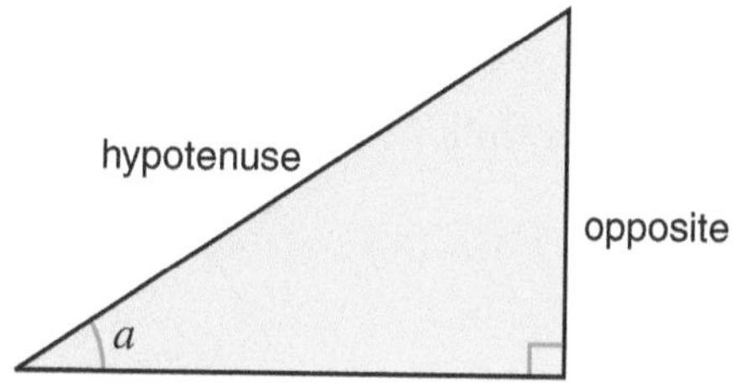

The sine ratio links three pieces of information:

the size of an **angle**,
the length of the side **opposite** the angle,
the length of the **hypotenuse**.

If we are given the values for two of these we can find the value of the third.

EXAMPLE

Find the height of a kite when it is flying at an angle of 40° and the kite string is 12 m long.
Give the answer correct to 3 significant figures.

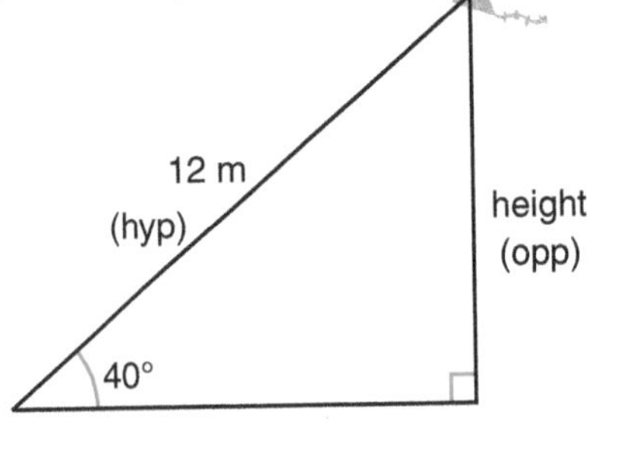

Word	Abbreviation
sine	sin
opposite	opp
hypotenuse	hyp

Mathematical shorthand:

$\sin a = \frac{\text{opp}}{\text{hyp}}$

Substitute known values.

$\sin 40° = \frac{h}{12}$

Multiply both sides by 12.

$h = 12 \times \sin 40°$

Using your calculator, press: 1 2 × sin 4 0 =

$h = 7.713...$

$h = 7.71$ m, correct to 3 s.f.

The height of the kite is 7.71 m, correct to 3 s.f.

Exercise 29.1

1. Find the height, h, of these kites.
Give your answers correct to 3 significant figures.

(a)

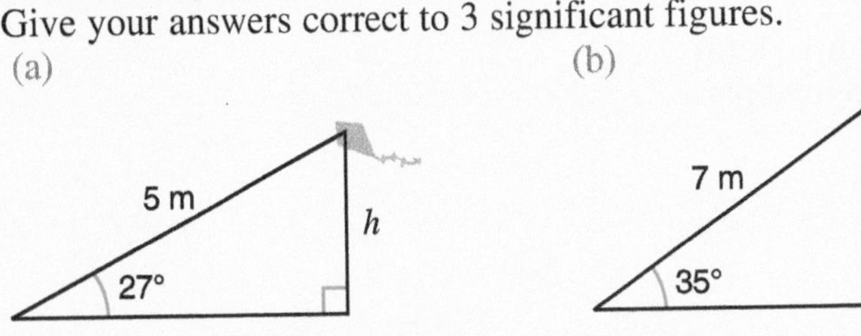

(b)

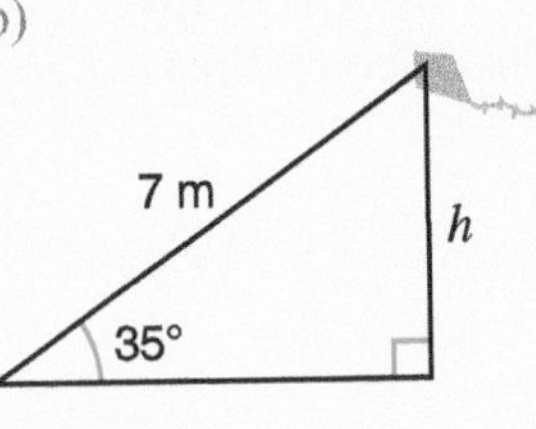

(c)

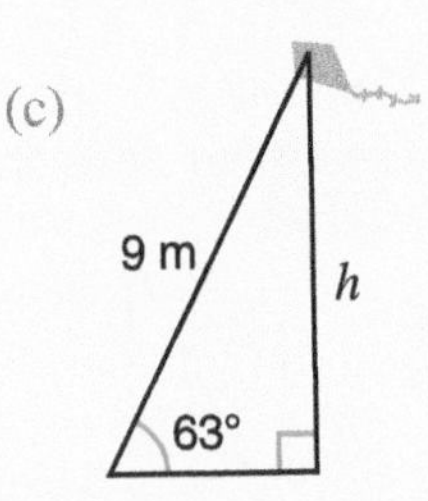

2. Calculate the lengths marked x.
Give your answers correct to 3 significant figures.

(a)

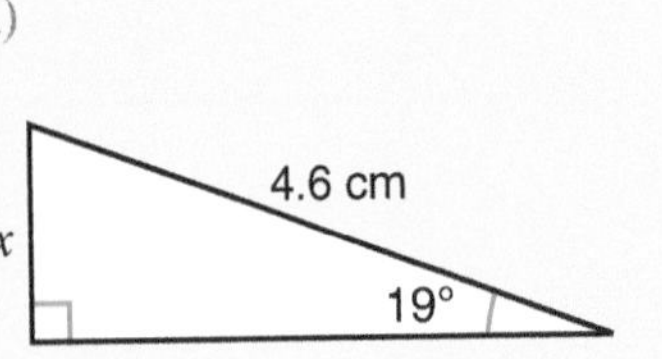

(b)

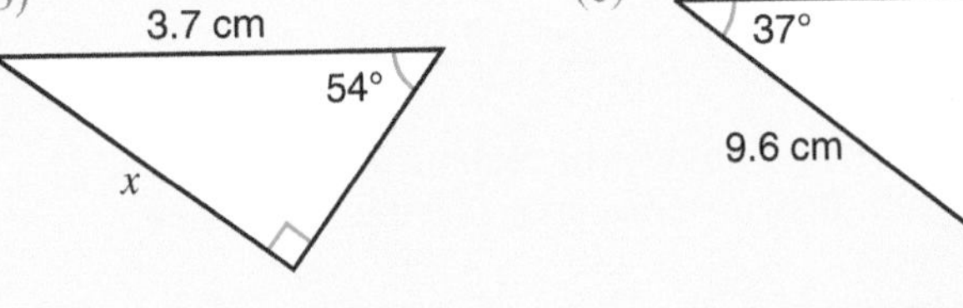

(c)

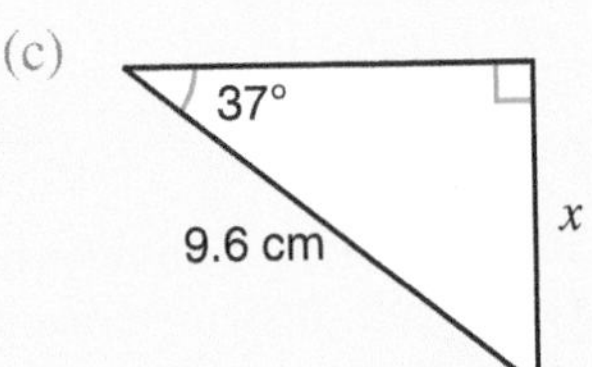

3. In ΔABC, angle $ACB = 90°$.

(a) If $\angle BAC = 47.5°$ and $AB = 4.6$ m find BC.
(b) If $\angle ABC = 67.4°$ and $AB = 12.4$ m find AC.
(c) If $\angle BAC = 15.8°$ and $AB = 17.4$ cm find BC.
(d) If $\angle BAC = 35°$ and $AB = 8.5$ cm find the size of $\angle ABC$ and then find AC.
Give your answers correct to 3 significant figures.

Finding an angle

If you are given the sine of an angle and asked to find the angle, use the inverse sine function, $\sin^{-1}$, on your calculator.

Using your calculator, press: sin 3 0

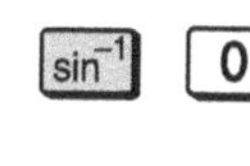

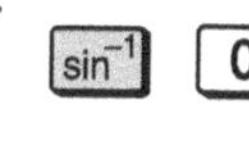

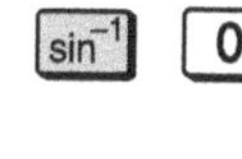

The display should read 0.5.

Clear the display and press: $\sin^{-1}$ 0 . 5 =

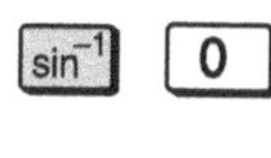

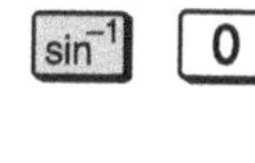

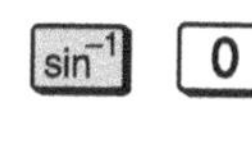

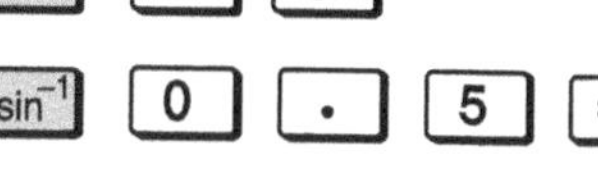

What do you notice?

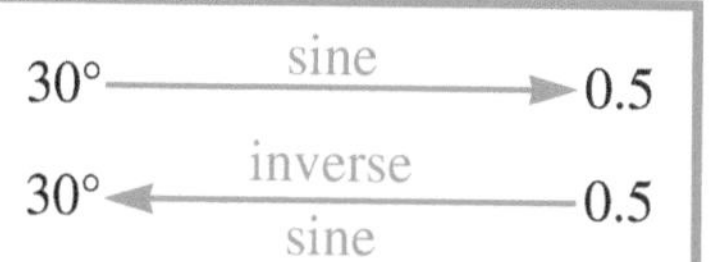

EXAMPLE

Find the size of angle a when the kite string is 12 m long and the kite is flying 7 m above the ground.
Give the answer correct to one decimal place.

$$\sin a = \frac{\text{opp}}{\text{hyp}}$$

Substitute known values.

$$\sin a^\circ = \frac{7}{12}$$

$$a = \sin^{-1}\frac{7}{12}$$

Using your calculator, press: $\sin^{-1}$ (7 ÷ 1 2) =

$a = 35.685\ldots$

$a = 35.7°$, correct to 1 d.p.

Exercise 29.2

1 Find the size of angle a for each of these kites.
Give your answers correct to one decimal place.

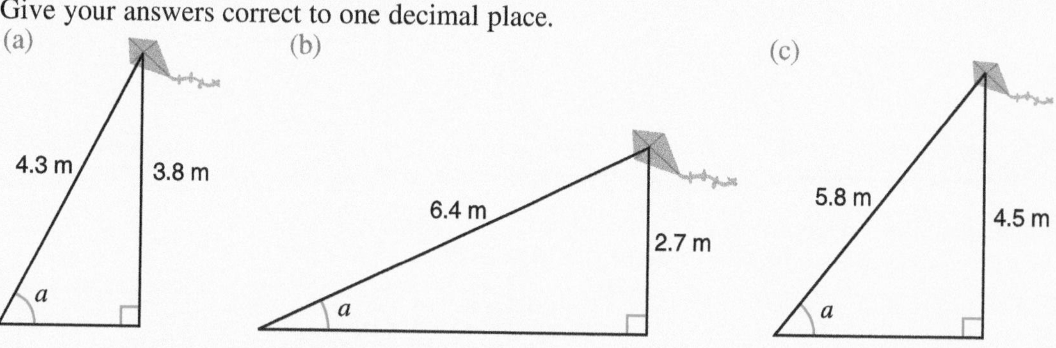

2 Find the size of angle x in each of these triangles.
Give your answers correct to one decimal place.

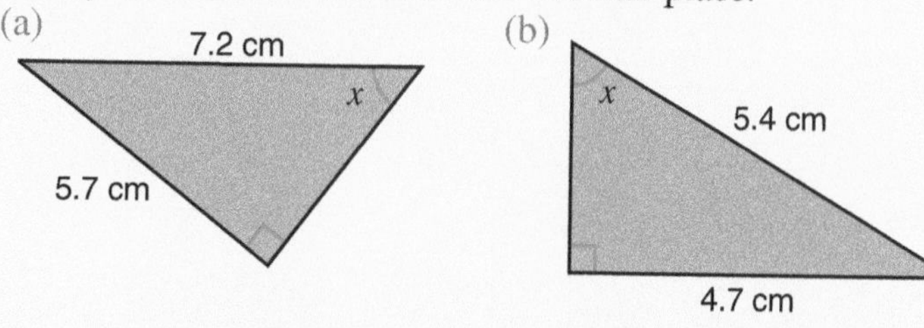

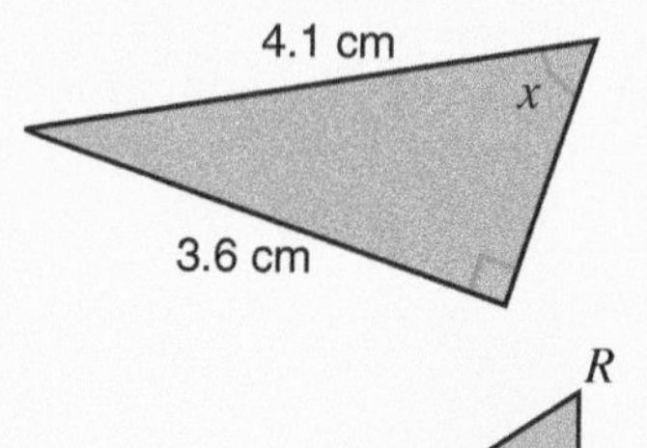

3 In ΔPQR angle $PQR = 90°$.

(a) If $QR = 4$ m and $PR = 10$ m find the size of $\angle QPR$.
(b) If $PQ = 4.7$ cm and $PR = 5.2$ cm find the size of $\angle PRQ$.
(c) If $QR = 7.2$ m and $PR = 19.4$ m find the size of $\angle QPR$.
(d) If $PQ = 3.7$ cm and $PR = 9.1$ cm find the size of $\angle QRP$ and then find the size of $\angle QPR$.

Give your answers correct to one decimal place.

EXAMPLE

Find the length of the string, l, when a kite is 6 m high and the string makes an angle of 50° with the ground.
Give the answer correct to 3 significant figures.

$$\sin a = \frac{\text{opp}}{\text{hyp}}$$

Substitute known values.

$$\sin 50° = \frac{6}{l}$$

Multiply both sides by l.

$$l \times \sin 50° = 6$$

Divide both sides by sin 50°.

$$l = \frac{6}{\sin 50°}$$

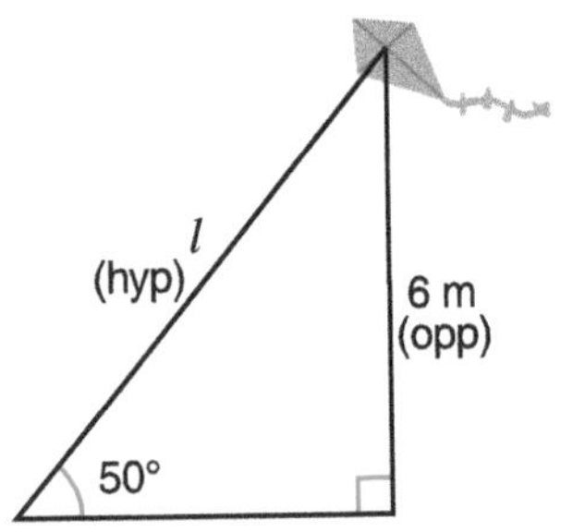

Using your calculator, press: 6 ÷ sin 5 0 =

$l = 7.832\ldots$

$l = 7.83$ m, correct to 3 s.f.

The length of the string is 7.83 m, correct to 3 s.f.

Exercise 29.3

1 Find the lengths, l, of these kite strings.
Give your answers correct to 3 significant figures.

(a) (b) (c)

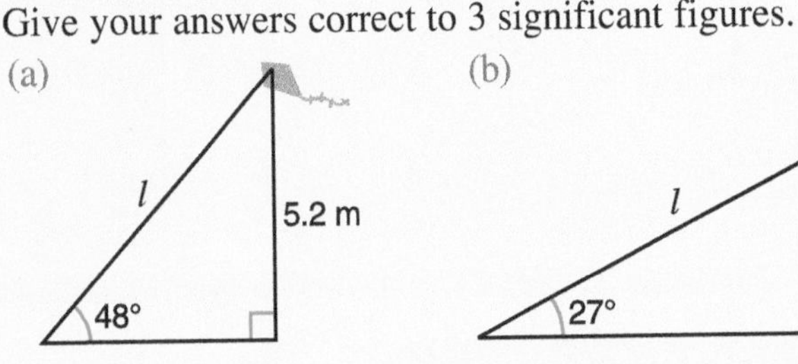

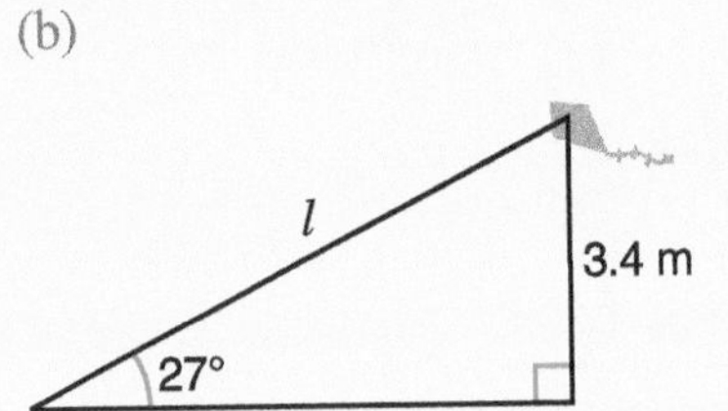

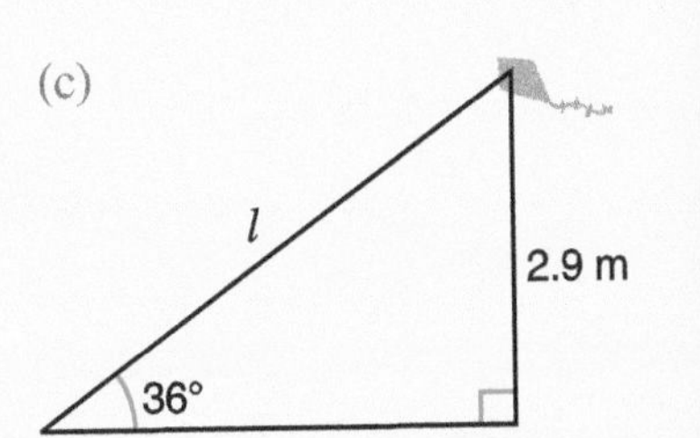

2 Calculate the length of side x in each of these triangles.
Give your answers correct to two decimal places.

(a) (b) (c)

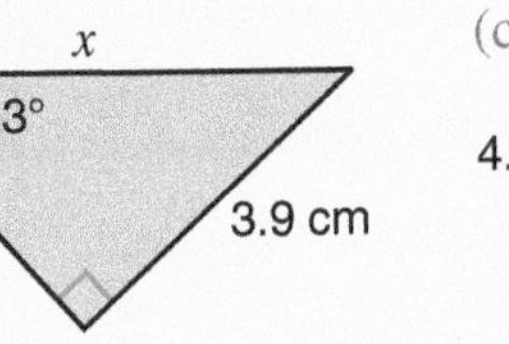

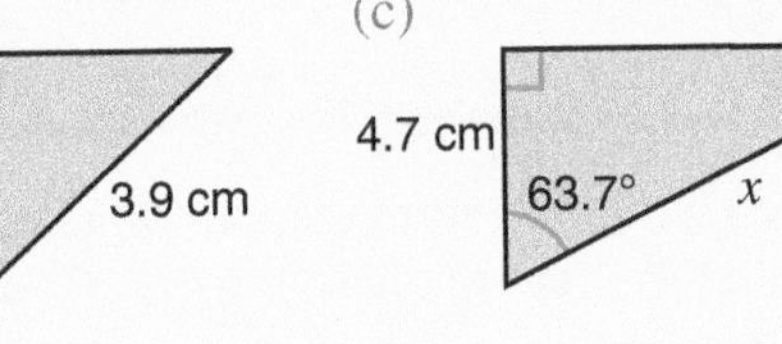
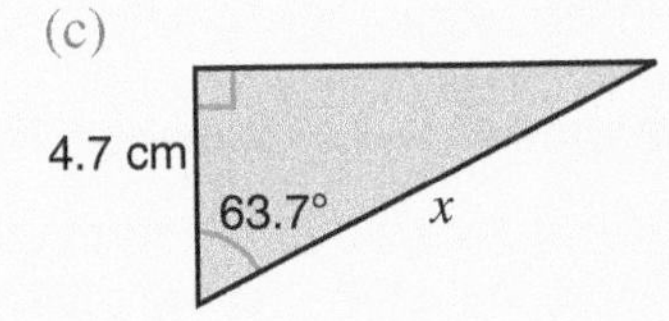
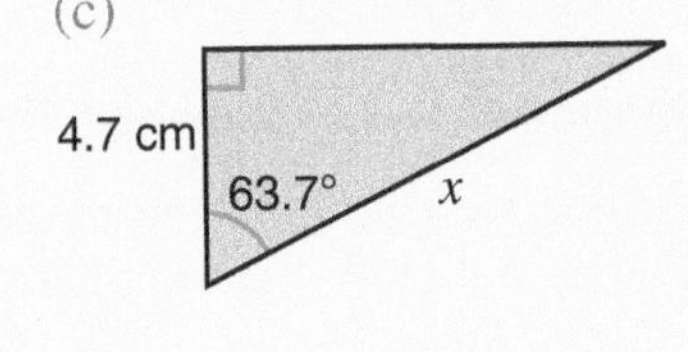

3 In ΔABC angle $ACB = 90°$.

(a) If $\angle BAC = 36.2°$ and $BC = 4.5$ m find AB.
(b) If $\angle ABC = 64.7°$ and $AC = 15.8$ cm find AB.
(c) If $\angle BAC = 12.7°$ and $BC = 14.7$ cm find AB.
(d) If $\angle BAC = 72.8°$ and $AC = 7.6$ m
find the size of $\angle ABC$ and then find AB.

Give your answers correct to 3 significant figures.

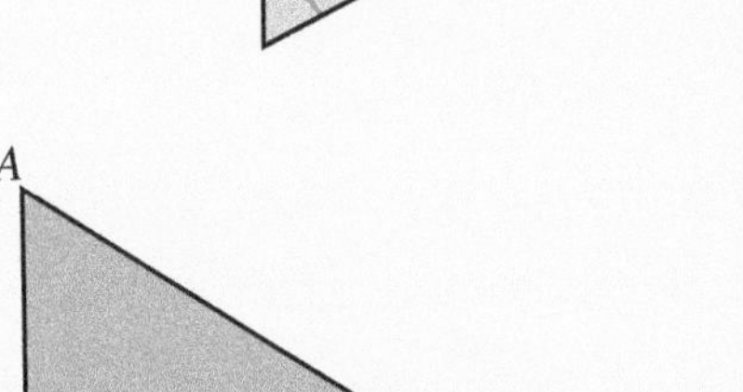

The cosine and tangent ratios

We have found that **sine** is the ratio $\frac{\text{opposite}}{\text{hypotenuse}}$.

In a similar way we can find two other ratios, the **cosine** of angle a and the **tangent** of angle a.

The cosine ratio

For any right-angled triangle:

$$\cos a = \frac{\text{adjacent}}{\text{hypotenuse}}$$

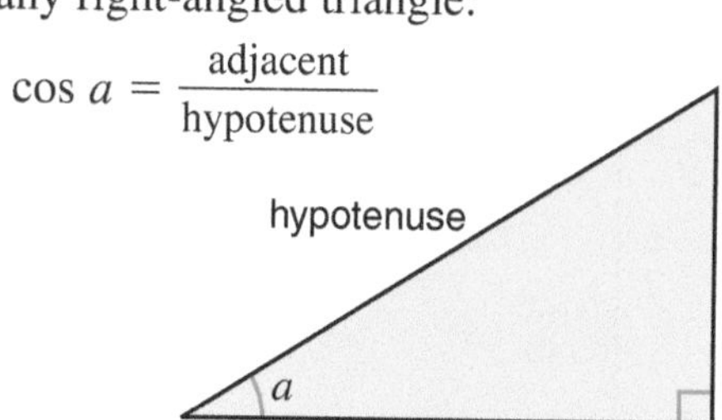

The cosine ratio links three pieces of information:

the size of an **angle**,
the length of the side **adjacent** to the angle,
the length of the **hypotenuse**.

If we are given the values of two of these we can find the value of the third.

The tangent ratio

For any right-angled triangle:

$$\tan a = \frac{\text{opposite}}{\text{adjacent}}$$

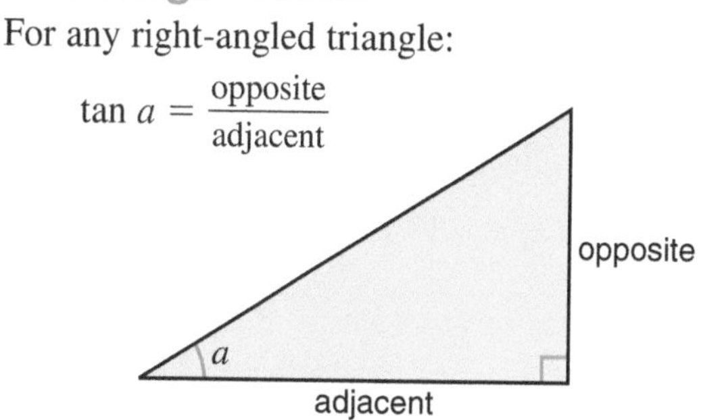

The tangent ratio links three pieces of information:

the size of an **angle**,
the length of the side **opposite** to the angle,
the length of the side **adjacent** to the angle.

If we are given the values of two of these we can find the value of the third.

EXAMPLE

1. Write down the sin, cos and tan ratios for angle a in the triangle.

$$\sin a = \frac{\text{opp}}{\text{hyp}} = \frac{8}{17}$$

$$\cos a = \frac{\text{adj}}{\text{hyp}} = \frac{15}{17}$$

$$\tan a = \frac{\text{opp}}{\text{adj}} = \frac{8}{15}$$

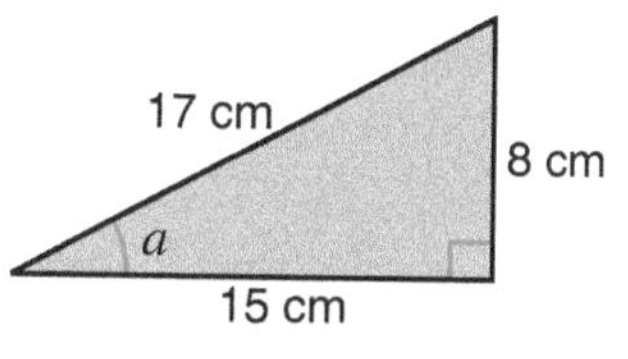

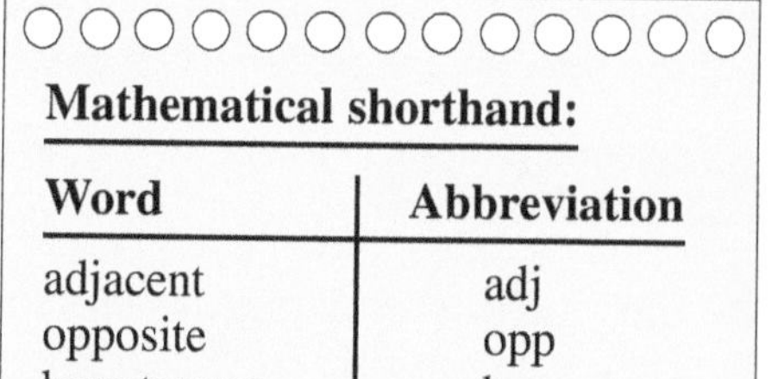

Mathematical shorthand:

Word	Abbreviation
adjacent	adj
opposite	opp
hypotenuse	hyp
sine	sin
cosine	cos
tangent	tan

How to select and use the correct ratio

There are only 3 different types of question for each of the ratios.
Selecting the correct ratio is most important.
To do this:

1. Go to the angle you know (or want to find).
2. Name sides (opp, adj, hyp).
   If you are trying to find the length of a side, name that side first together with one other side of known length.
   If you are trying to find the size of an angle, name two sides of known length.
3. Select the correct ratio and write it down.
   $\sin a = \frac{\text{opp}}{\text{hyp}}$ $\cos a = \frac{\text{adj}}{\text{hyp}}$ $\tan a = \frac{\text{opp}}{\text{adj}}$
   One way to remember the ratios is to use the initial letters, SOHCAHTOA.
   You may know another method.
4. Substitute known values from the question.
5. Rearrange to isolate the angle, or side, you are trying to find.
6. Use your calculator to find the size of the angle, or side, writing down more figures than you need for the final answer.
7. Correct to the required degree of accuracy.
8. Give the answer, stating the degree of approximation and giving the correct units. When giving the answer to a problem you should use a short sentence.

EXAMPLES

2 Find the length, h, giving the answer to 3 significant figures

$\tan a = \frac{\text{opp}}{\text{adj}}$

Substitute known values.

$\tan 28° = \frac{h}{12.4}$

Multiply both sides by 12.4.

$h = 12.4 \times \tan 28°$

Using your calculator, press: 1 2 . 4 × tan 2 8 =

$h = 6.593\ldots$

$h = 6.59$ m, correct to 3 s.f.

3 Find the size of angle a, correct to one decimal place.

$\cos a = \frac{\text{adj}}{\text{hyp}}$

Substitute known values.

$\cos a° = \frac{11}{16}$

$a = \cos^{-1} \frac{11}{16}$

Using your calculator, press: cos⁻¹ (1 1 ÷ 1 6) =

$a = 46.56\ldots$

$a = 46.6°$, correct to 1 d.p.

Exercise 29.4

1 Write down the sin, cos and tan ratios for angle p in each of the following triangles.

(a)

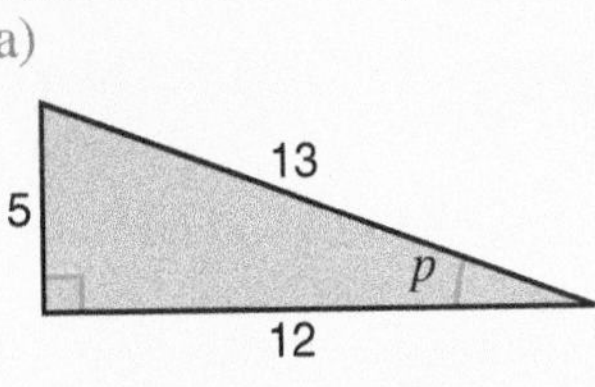

(b)

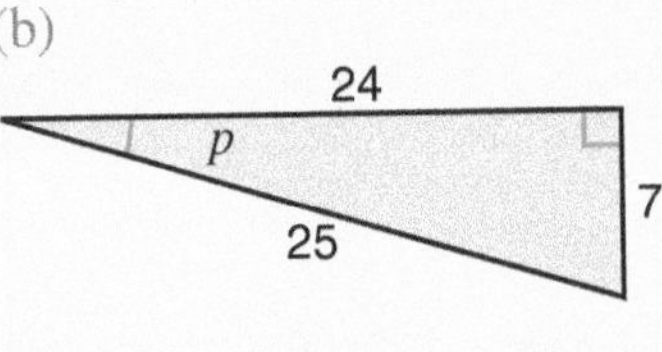

(c)

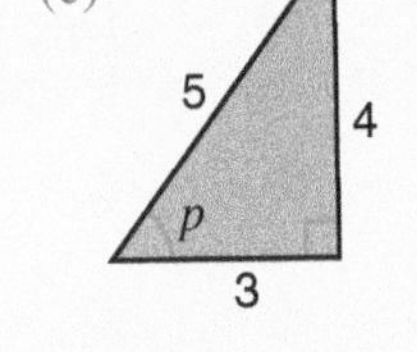

2 By choosing the correct ratio, calculate angle p in each of the following triangles. Give your answers correct to one decimal place.

(a)

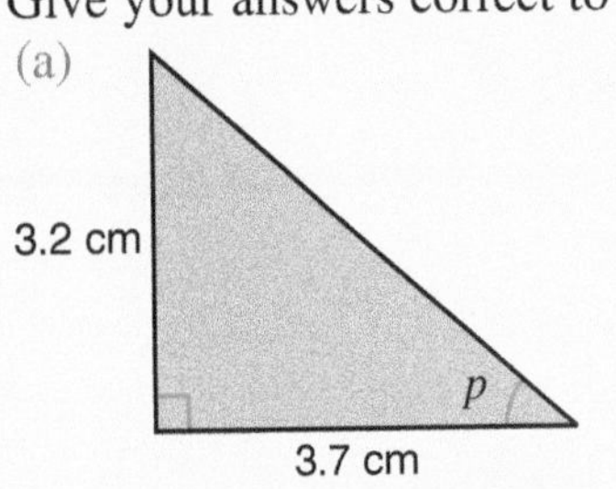

(b)

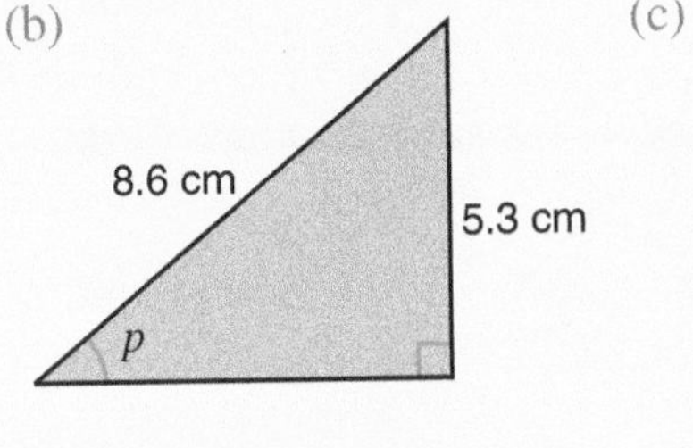

(c)

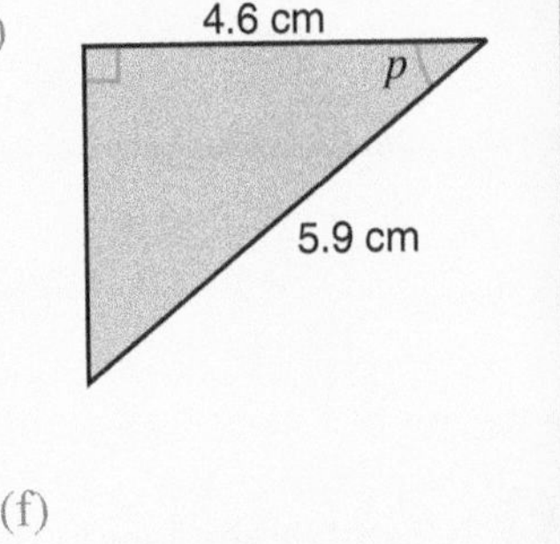

(d)

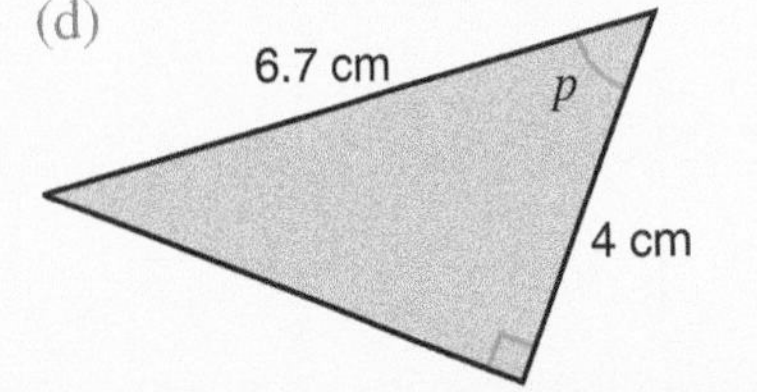

(e)

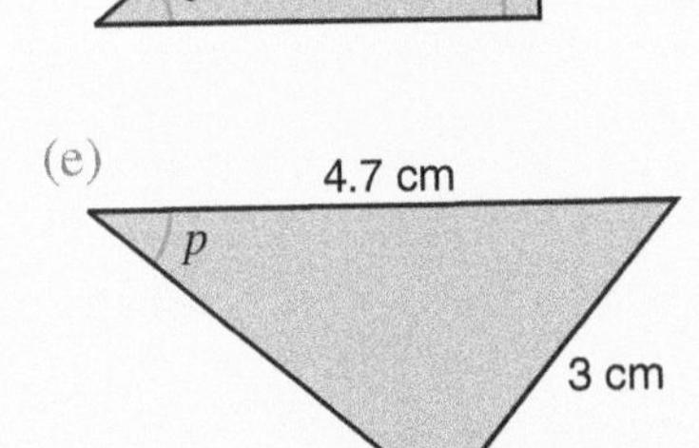

(f)

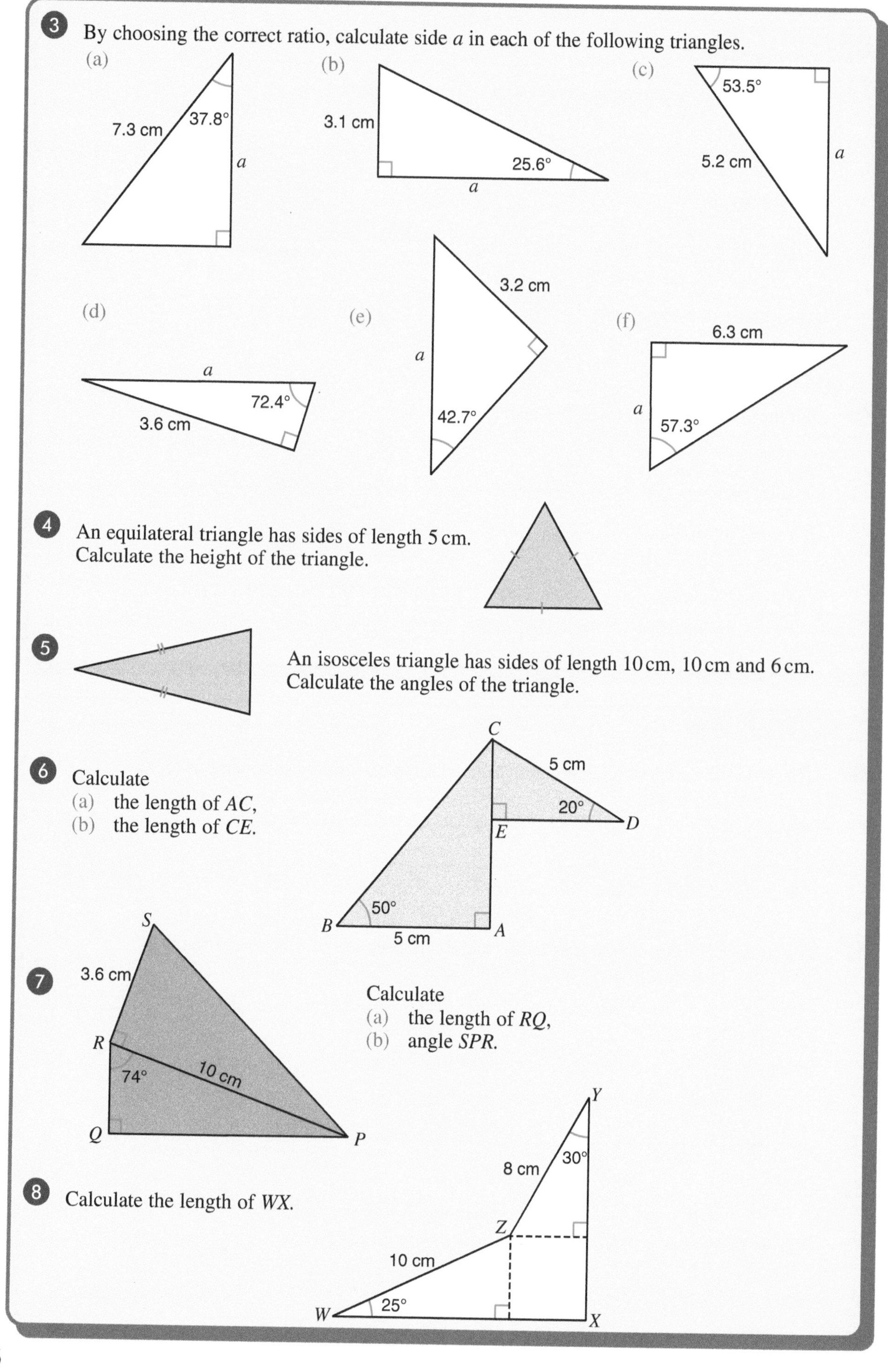

3
By choosing the correct ratio, calculate side a in each of the following triangles.
(a)
7.3 cm
37.8°
a
(b)
3.1 cm
25.6°
a
(c)
53.5°
5.2 cm
a
(d)
a
72.4°
3.6 cm
(e)
3.2 cm
a
42.7°
(f)
6.3 cm
a
57.3°
4
An equilateral triangle has sides of length 5 cm.
Calculate the height of the triangle.
5
An isosceles triangle has sides of length 10 cm, 10 cm and 6 cm.
Calculate the angles of the triangle.
6
Calculate
(a) the length of AC,
(b) the length of CE.
C
5 cm
20°
D
E
50°
B
5 cm
A
7
S
3.6 cm
R
74°
10 cm
Q
P
Calculate
(a) the length of RQ,
(b) angle SPR.
8
Calculate the length of WX.
Y
8 cm
30°
Z
10 cm
W
25°
X

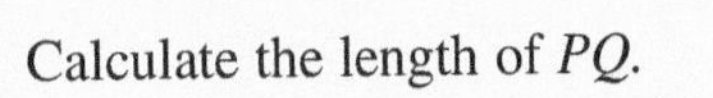

9. Calculate the length of PQ.

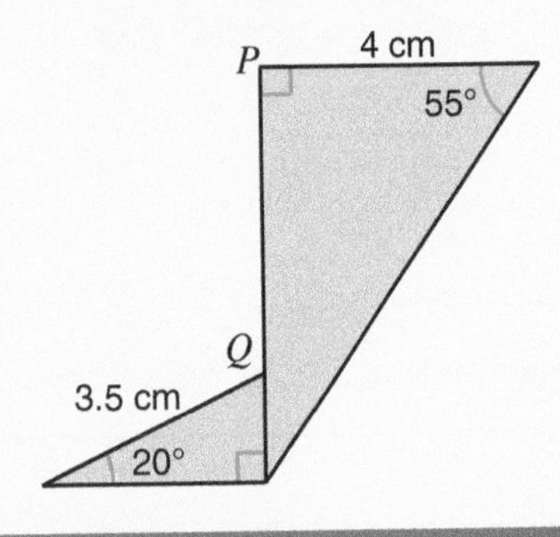

10. Calculate the length of CD.

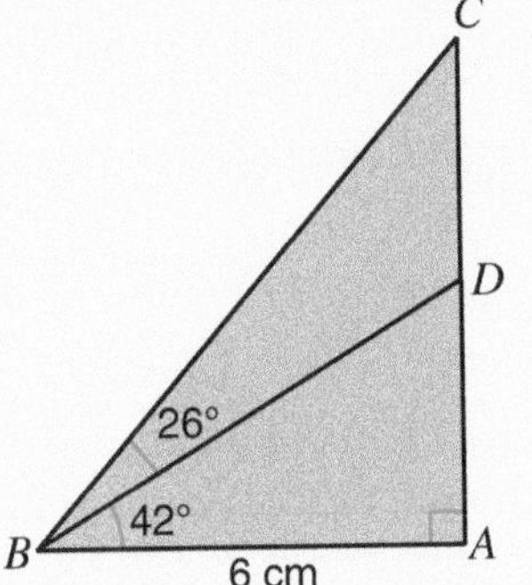

Angles of elevation and depression

When we look **up** from the horizontal the angle we turn through is called an **angle of elevation**.

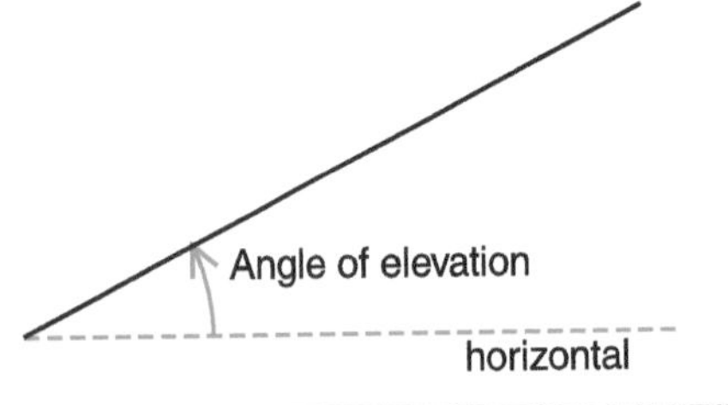

When we look **down** from the horizontal the angle we turn through is called an **angle of depression**.

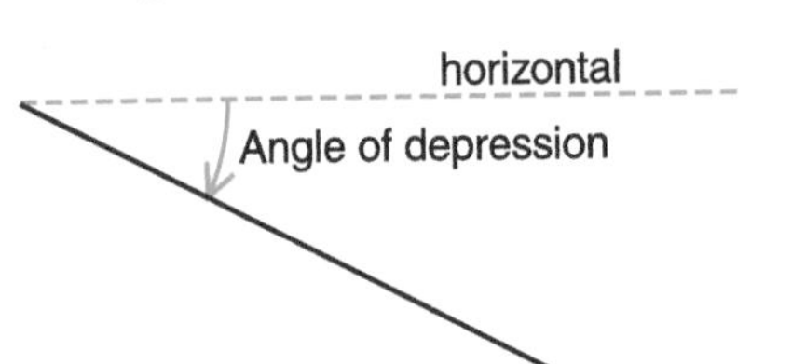

EXAMPLES

1. From a point on the ground, 30 m from the base of a pylon, the angle of elevation to the top of the pylon is 50°.

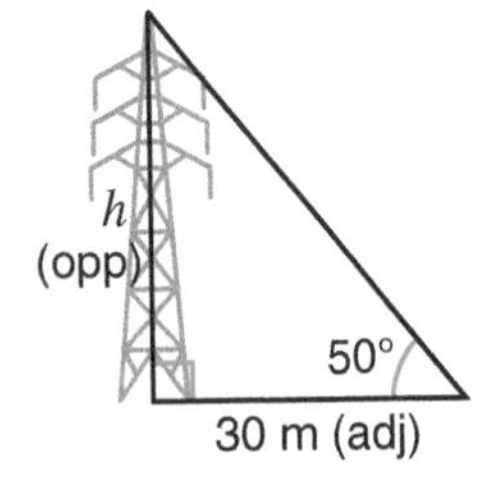

Find the height of the pylon.

$$\tan a = \frac{\text{opp}}{\text{adj}}$$

$$\tan 50° = \frac{h}{30}$$

$h = 30 \times \tan 50°$
$h = 35.75\ldots$
$h = 35.8$ m, correct to 3 s.f.

The height of the pylon is 35.8 m, correct to 3 s.f.

2. Staten Island Ferry is 270 m away from the base of the Statue of Liberty.
The ferry can be seen from a viewing point in the lantern, 85 m above the ground.

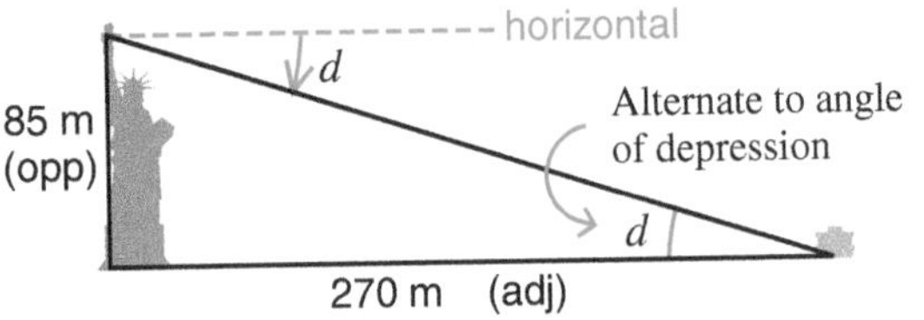

What is the angle of depression to the ferry from the viewing point?

$$\tan a = \frac{\text{opp}}{\text{adj}}$$

$$\tan d° = \frac{85}{270}$$

$d = \tan^{-1} \frac{85}{270}$
$d = 17.47\ldots$
$d = 17.5°$, correct to 1 d.p.

The angle of depression to the ferry from the viewing point is 17.5°, correct to 1 d.p.

Exercise 29.5

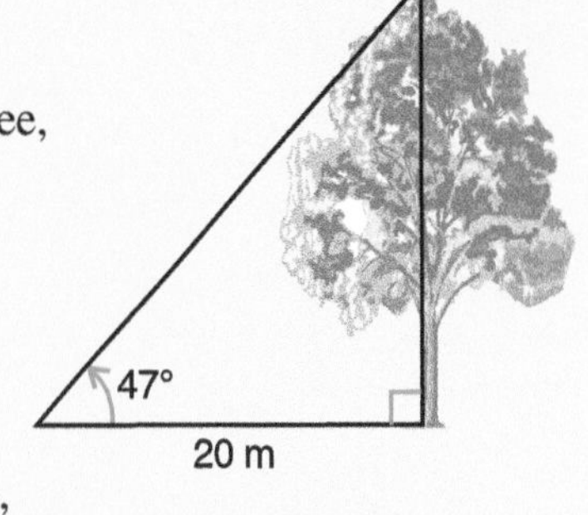

1. From a point on the ground 20 m from the base of a tree, the angle of elevation to the top of the tree is 47°. Calculate the height of the tree.

2. From a point on the ground 10 m from a block of flats, the angle of elevation to the top of the block is 76°. Calculate the height of the block of flats.

3. A fishing boat is 200 m from the bottom of a vertical cliff. From the top of the cliff the angle of depression to the fishing boat is 34°.
   (a) Calculate the height of the cliff.
   (b) A buoy is 100 m from the bottom of the cliff. Calculate the angle of depression to the buoy from the top of the cliff.

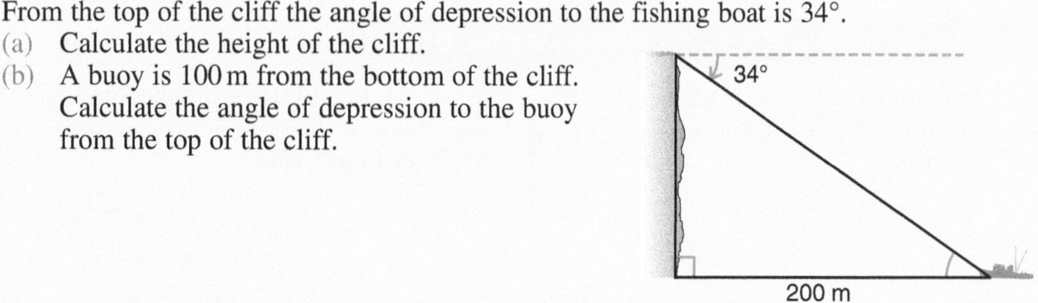

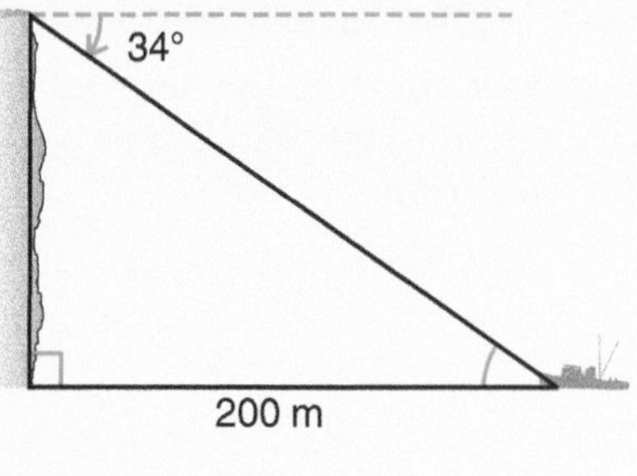

4. A yacht is 20 m from the bottom of a lighthouse. From the top of the lighthouse the angle of depression to the yacht is 48°. Calculate the height of the lighthouse.

5. A cat is on the ground 25 m from the foot of a house. A bird is perched on the gutter of the house 15 m from the ground. Calculate the angle of elevation from the cat to the bird.

6. A tree, 6 m high, casts a shadow of 4.8 m on horizontal ground. Calculate the angle of elevation to the sun.

7. From a point, *A*, on the ground, the angle of elevation to a hot air balloon is 9°. The balloon is 150 m above the ground. Calculate the distance from *A* to the balloon.

8. From the top of a cliff, 36 m high, the angles of depression of two boats at sea are 17° and 25°. The boats are in a straight line from the foot of the cliff. Calculate the distance between the two boats.

Bearings

Bearings are used to describe the direction in which you must travel to get from one place to another.
They are measured from the North line in a clockwise direction.
A bearing can be any angle from 0° to 360° and is written as a three-figure number.

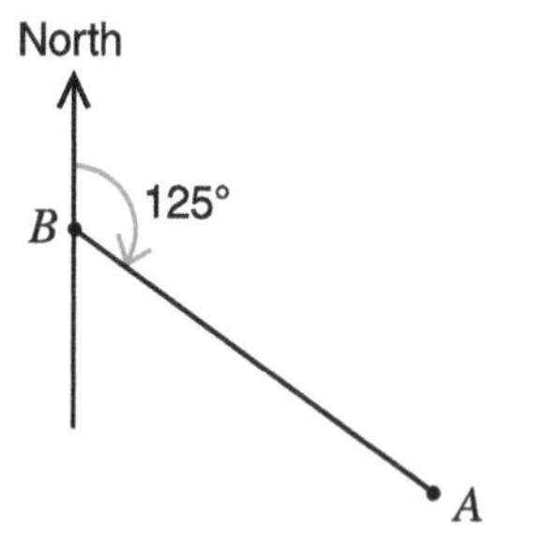

A is on a bearing of 125° from *B*.

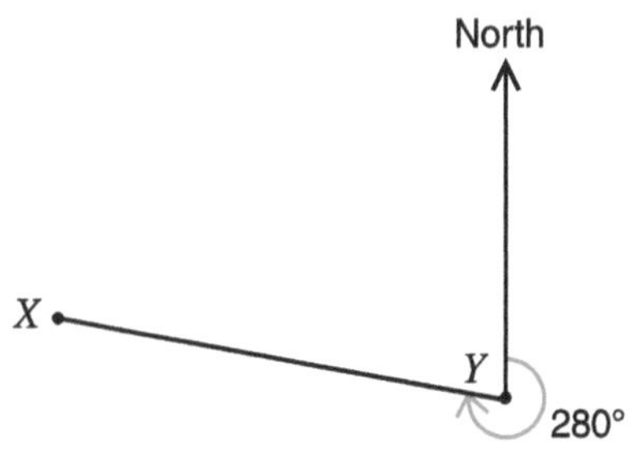

X is on a bearing of 280° from *Y*.

EXAMPLE

A plane flies 300 km on a bearing of 132° from an airport.
How far South and East is it from the airport?
Give the answers correct to 3 significant figures.

x is the distance South.
y is the distance East.
Using supplementary angles:
$\angle PAB = 180° - 132° = 48°$.

N, A, 132°, 48°, 300 km, x, B, y, P

To find x

A, 48°, 300 km (hyp), x (adj), B, P

$\cos a = \frac{\text{adj}}{\text{hyp}}$

$\cos 48° = \frac{x}{300}$

$x = 300 \times \cos 48°$

$x = 200.73\ldots$ km

$x = 201$ km, correct to 3 s.f.

To find y

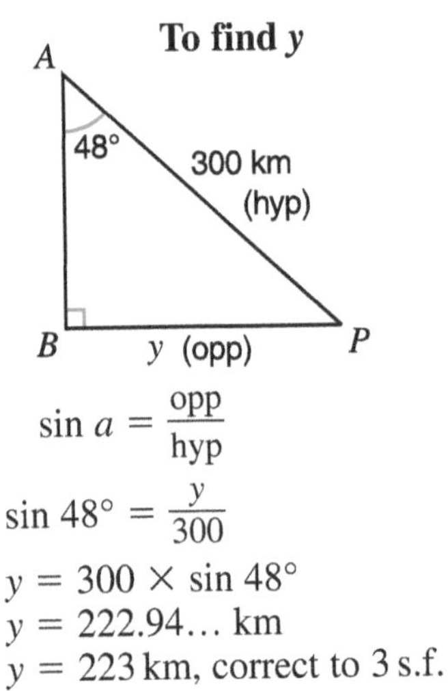

$\sin a = \frac{\text{opp}}{\text{hyp}}$

$\sin 48° = \frac{y}{300}$

$y = 300 \times \sin 48°$

$y = 222.94\ldots$ km

$y = 223$ km, correct to 3 s.f.

Sketch diagrams:
Drawing a sketch diagram may help you to understand the question.
More information can be added to the diagram as you answer the question.

The plane is 201 km South and 223 km East of the airport, correct to 3 s.f.

How can you use Pythagoras' Theorem to check the answer?

Exercise 29.6

1. A plane flies 250 km on a bearing of 050°.
   (a) How far north is it from its original position?
   (b) How far east is it from its original position?
2. A helicopter leaves its base and flies 23 km on a bearing of 285°.
   How far west is it from its base?
3. A ship at A is 3.8 km due north of a lighthouse.
   A ship at B is 2.7 km due east of the same lighthouse.
   What is the bearing of the ship at B from the ship at A?
4. A helicopter has flown from its base on a bearing of 153°. Its distance east of base is 19 km.
   How far has the helicopter flown?
5. A fishing boat leaves port and sails on a straight course.
   After 2 hours its distance south of port is 24 km and its distance east of port is 7 km.
   On what bearing did it sail?
6. A yacht sails 15 km on a bearing of 053°, then 7 km on a bearing of 112°.
   How far north is the yacht from its starting position?
7. A plane flies 307 km on a bearing of 234°, then 23 km on a bearing of 286°.
   How far south is the plane from its starting position?
8. Jayne sails 1.5 km on a bearing of 050°.
   She then changes course and sails 2 km on a bearing of 140°.
   On what bearing must she sail to return to her starting position?

Solving problems in three-dimensions

When we solve problems in three-dimensions the first task is to identify the length, or angle, we are asked to calculate.
This length, or angle, will always form part of a triangle.
Having identified a suitable triangle it is a good idea to draw it in two dimensions, so that lengths of sides and sizes of angles can clearly be seen.

Most problems in three-dimensions involve:
finding lengths,
finding the size of an angle between a line and a plane.

A suitable triangle includes the side, or angle, you are trying to calculate **and** either:
two other sides of known length, or one side of known length and an angle of known size.
Sometimes we need to use more than one triangle to solve a problem.

The angle between a line and a plane

A straight line meets a plane at a **point**, *P*.

To find the angle between a line and a plane:
Take a point, *X*, on the line.
From *X*, draw a line, *XT*, which is at right angles to the plane.
The angle *XPT* is the angle between the line and the plane.

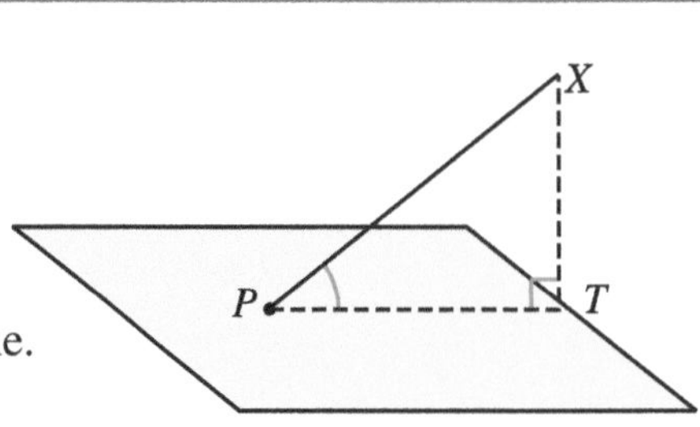

EXAMPLE

PQRST is a pyramid. The perpendicular height, *PX*, is 10 cm.
The base is a square of side 8 cm and *X* is at the centre of the square.

Calculate the angle between the line *PQ* and the base *QRST*.

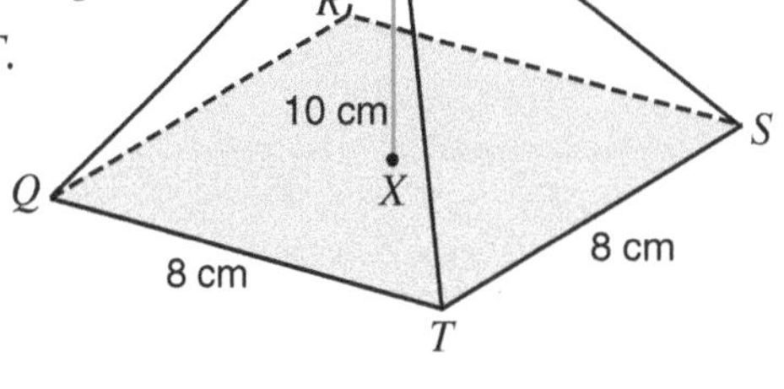

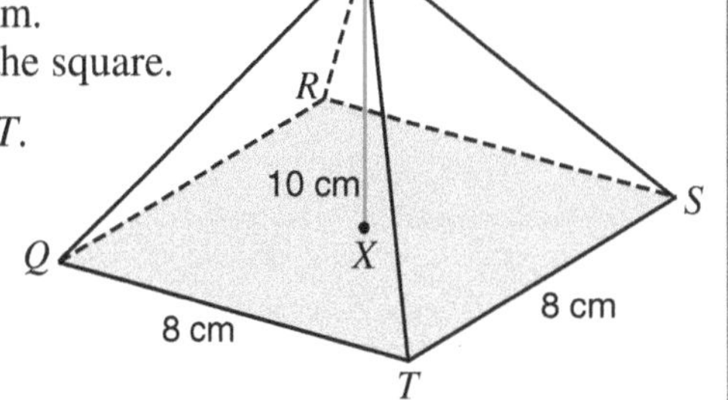

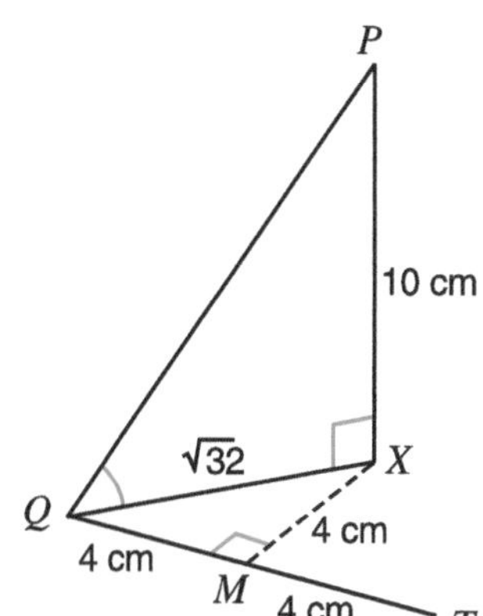

The angle between the line *PQ* and the base *QRST* is $\angle PQX$.

ΔPQX is a suitable triangle, but we only know the length of the side *PX*.

To find *QX* we use ΔQMX, where *M* is the midpoint of *QT*.
Using Pythagoras' Theorem in ΔQMX.

$QX^2 = MQ^2 + MX^2$
$= 4^2 + 4^2$
$= 16 + 16$
$= 32$
$QX = \sqrt{32}$

Mark this length on your diagram.

Using the tangent ratio in ΔPQX.

$\tan a = \frac{\text{opp}}{\text{adj}}$

$\tan \angle PQX = \frac{PX}{QX} = \frac{10}{\sqrt{32}}$

$\angle PQX = \tan^{-1} \frac{10}{\sqrt{32}}$
$= 60.503...$
$\angle PQX = 60.5°$

The angle between the line *PQ* and the base *QRST* is 60.5°, correct to 1 decimal place.

Exercise 29.7

1 The diagram shows a wedge $ABCDXY$.
The base $ABCD$ and the face $CBXY$ are rectangles.
Angle $XAB = 90°$.
$BC = 9\,\text{cm}$, $BA = 18\,\text{cm}$ and $AX = 15\,\text{cm}$.

Calculate
(a) angle XBA,
(b) the angle between the line XC and the base $ABCD$.

2 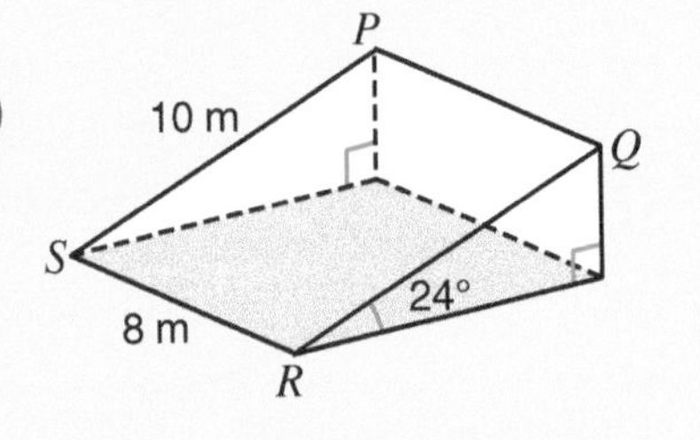

The diagram shows a ramp.
The ramp, $PQRS$, is 8 m wide and 10 m long.
The ramp slopes at 24° to the horizontal.
Find the slope of the diagonal QS to the horizontal.

3 The diagram shows a cube $ABCDEFGH$.
$AB = 6\,\text{cm}$.

Calculate
(a) angle DCE,
(b) the angle between the line AH and the base $EFGH$.

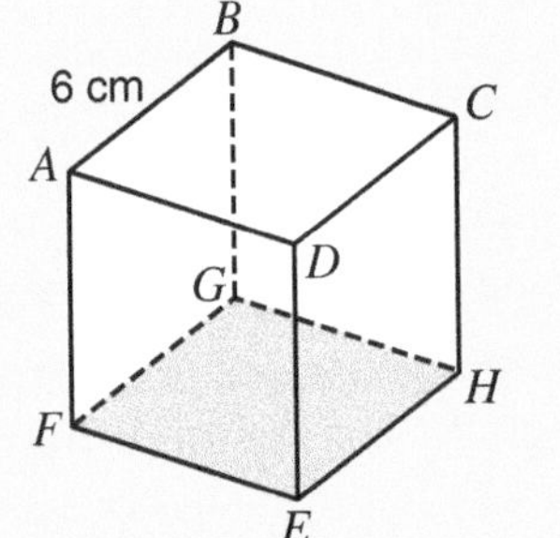

4 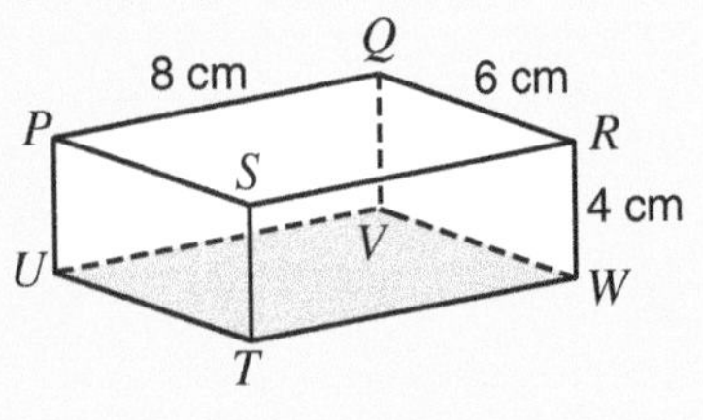

$PQRSTUVW$ is a cuboid.
$PQ = 8\,\text{cm}$, $QR = 6\,\text{cm}$ and $RW = 4\,\text{cm}$.

Calculate
(a) angle PVU,
(b) the angle between the line PW and the base $UVWT$.

5 $PLMNO$ is a pyramid of height 8 cm.
The base of the pyramid is a square.
$OL = LM = MN = NO = 10\,\text{cm}$.
P is directly above the centre of the base.

Calculate the angle between the line PN and the base $LMNO$.

6 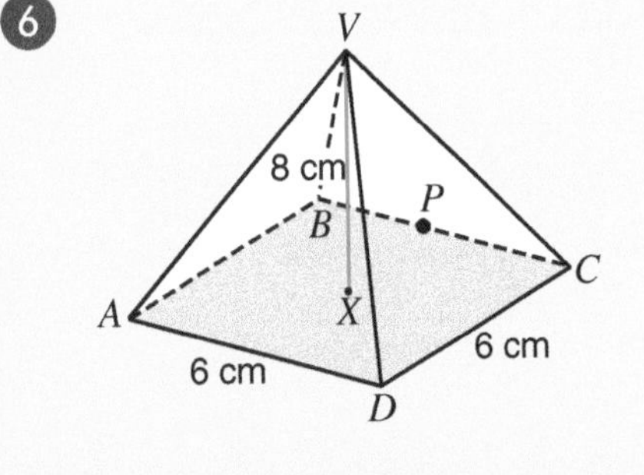

$VABCD$ is a square based pyramid.
The perpendicular height, VX, is 8 cm.
$AB = BC = CD = DA = 6\,\text{cm}$.
V is directly above the centre of the base.
P is the midpoint of the line BC.

Calculate
(a) the length VP,
(b) the angle XPV,
(c) the angle between the line VA and the base $ABCD$.

What you need to know

- **Trigonometry** is used to find the lengths of sides and the sizes of angles in right-angled triangles.
- You must learn the **sine**, **cosine** and **tangent** ratios.

  $\sin a = \dfrac{\text{opposite}}{\text{hypotenuse}}$ $\cos a = \dfrac{\text{adjacent}}{\text{hypotenuse}}$ $\tan a = \dfrac{\text{opposite}}{\text{adjacent}}$

  hypotenuse, opposite, adjacent, a

- Each ratio links the size of an angle with the lengths of two sides. If we are given the values for two of these we can find the value of the third.
- When we look **up** from the horizontal the angle we turn through is called the **angle of elevation**.
- When we look **down** from the horizontal the angle we turn through is called the **angle of depression**.

  Angle of elevation, horizontal, Angle of depression

- **Bearings** are used to describe the direction in which you must travel to get from one place to another.
  They are measured from the North line in a clockwise direction.
  A bearing can be any angle from 0° to 360° and is written as a three-figure number.
- When working in three-dimensions the first task is to identify the length, or angle, that you are trying to find. The length, or angle, will always form part of a triangle together with either:
  two other sides of known length, or
  one side of known length and an angle of known size.
  Sometimes, more than one triangle is needed to solve a problem.

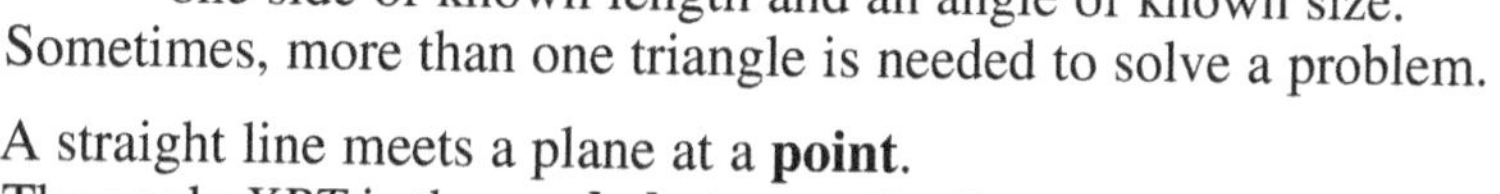

- A straight line meets a plane at a **point**.
  The angle XPT is the **angle between the line and the plane**.
  The line XT is perpendicular to the plane.

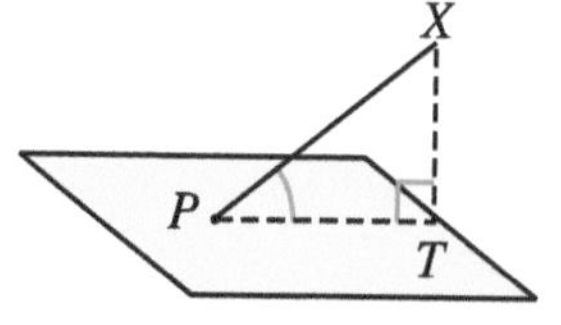

Review Exercise 29

1 When Colin is 10 m from a haystack, the angle of elevation to the top of the haystack is 18°.
Calculate h, the height of the haystack.

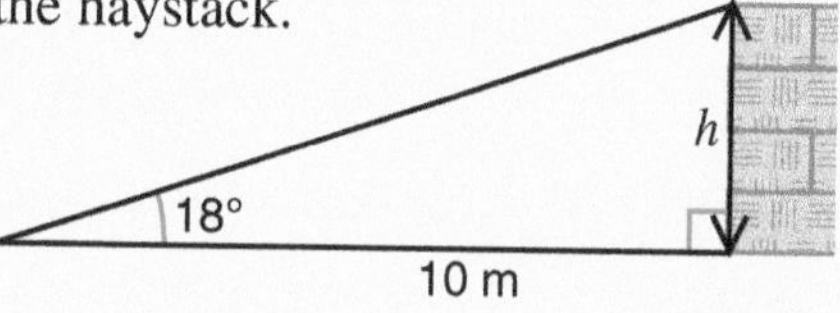

2 The design for a children's slide, $ABCD$, for a playground is shown.
The height of the slide, BD, is 136 cm.
The distance DC is 195 cm.
(a) Calculate the angle BCD.
(b) Angle ABD is 28°.
Calculate the distance AB.

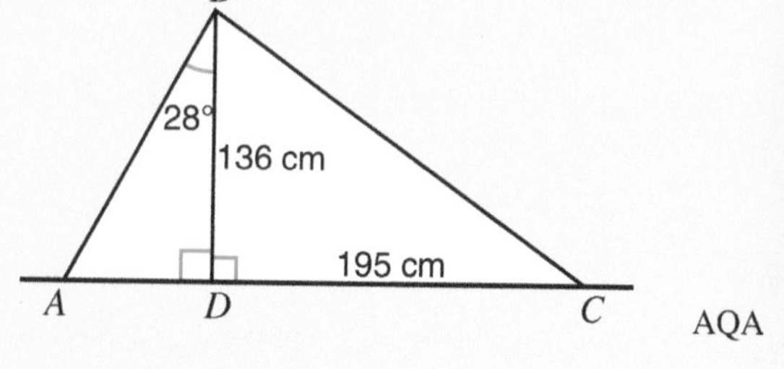

AQA

3 The diagram shows the positions of Leeds and Newcastle.
Calculate how far south Leeds is from Newcastle, the distance marked x on the diagram.
Give your answer to an appropriate degree of accuracy.

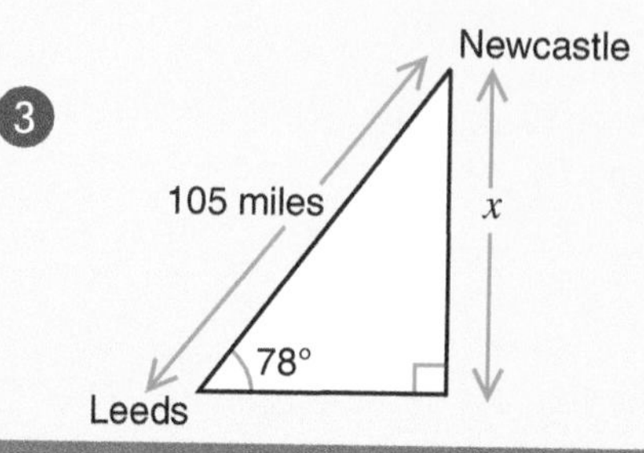

AQA

4 This quadrilateral is made from two right-angled triangles.

(a) Calculate the length x.

(b) Calculate the angle A.

AQA

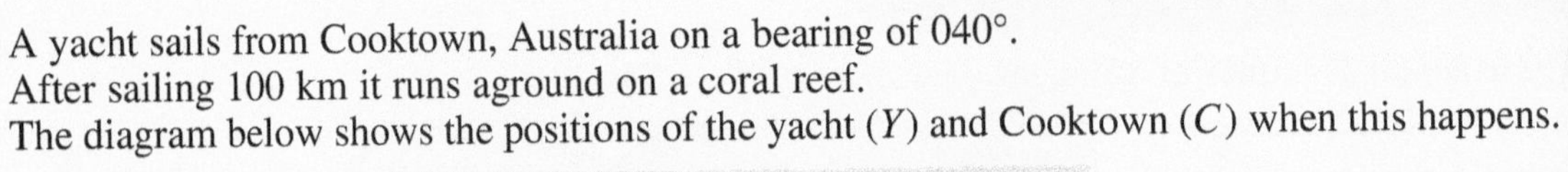

5 A yacht sails from Cooktown, Australia on a bearing of 040°.
After sailing 100 km it runs aground on a coral reef.
The diagram below shows the positions of the yacht (Y) and Cooktown (C) when this happens.

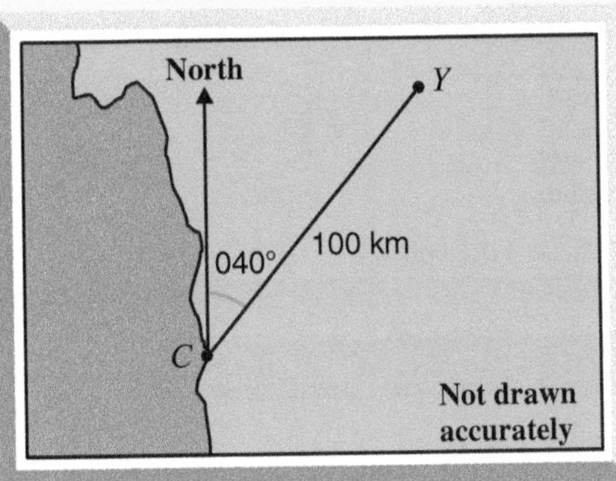

(a) Calculate how far the yacht (Y) is north of Cooktown (C) when it runs aground.

(b) At the time the yacht runs aground the nearest ship in the area is on a bearing of 130° from the yacht and on a bearing of 067° from Cooktown.

(i) Draw a sketch to show the positions of the ship (S), Cooktown (C) and the yacht (Y).

(ii) The angle CYS is 90°.
Calculate the distance of the ship from the yacht.

(iii) The ship sails towards the yacht. On what bearing does it sail?

AQA

6 $\text{Cos } PQR = \frac{12}{13}$

(a) Find (i) $\tan PQR$,
(ii) $\sin PQR$.

(b) $PQ = 6\text{ cm}$.
What are the lengths of QR and PR?

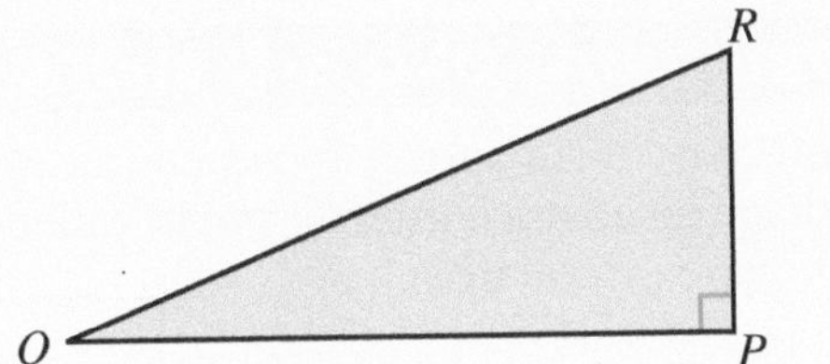

7

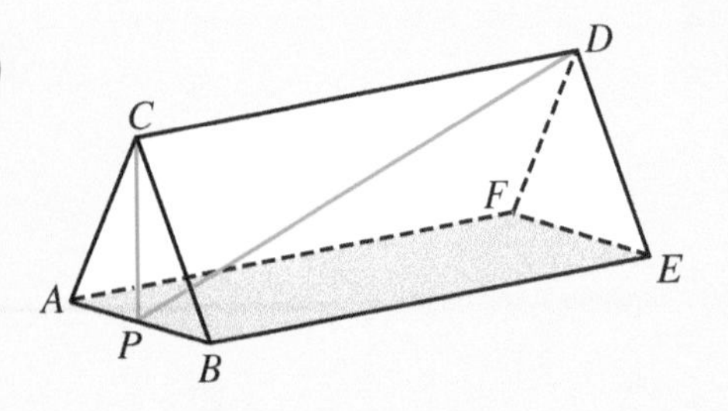

$ABCDEF$ is a triangular prism, 10 cm long.
ABC is an equilateral triangle of side 3 cm.
P is the foot of the perpendicular from C to AB.

(a) Calculate the length of PD.

(b) Calculate the size of the angle between CE and PE.

AQA

8 The diagram shows a square based pyramid.
P is directly above the centre of the base.
$PA = 6\text{ cm}$ and $AD = 4\text{ cm}$.

(a) Calculate

(i) the height of the pyramid,

(ii) the angle between the line PA and the base $ABCD$.

X is the midpoint of the line DC.

(b) Calculate the angle between PX and the base $ABCD$.

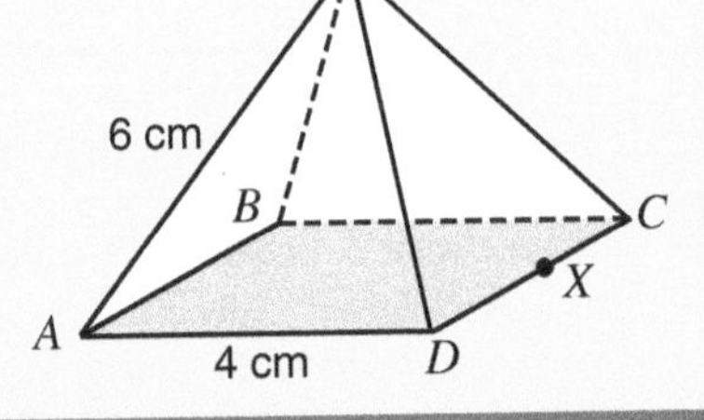

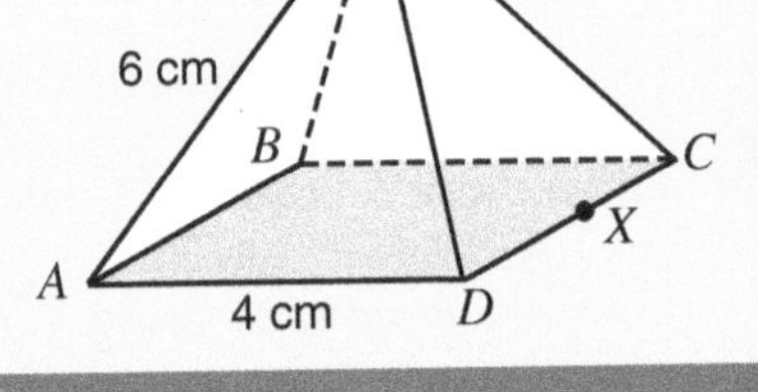

3-D Shapes

3-D shapes (or solids)

These are all examples of 3-dimensional shapes.

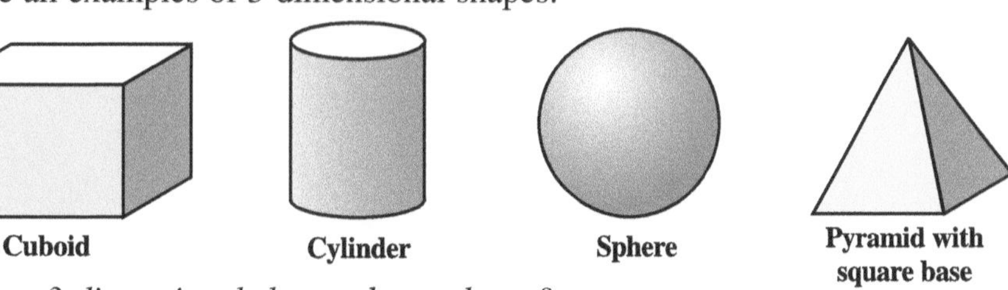

What other 3-dimensional shapes do you know?

Making and drawing 3-D shapes

Nets

3-dimensional shapes can be made using **nets**. This is the net of a cube.

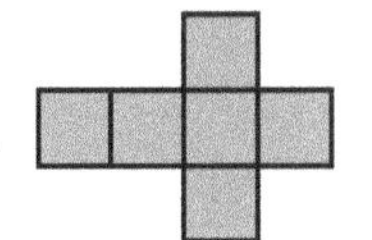

2-D drawings of 3-D shapes

Isometric drawings are used to draw 3-dimensional shapes.
Here is an isometric drawing of a cube.

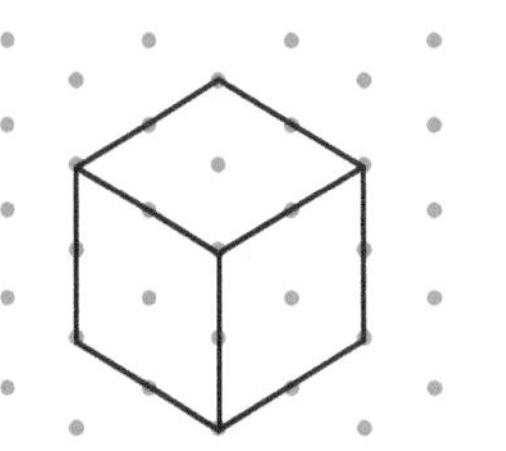

Naming parts of a 3-D shape

Each flat surface is called a **face**. Two faces meet at an **edge**.
Edges of a shape meet at a corner, or point, called a **vertex**.
The plural of vertex is **vertices**.

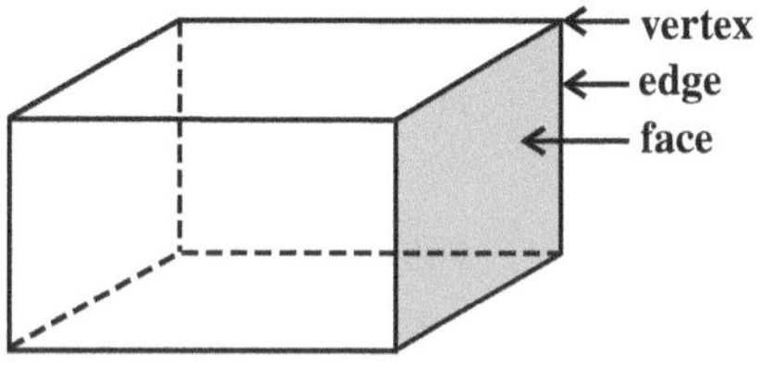

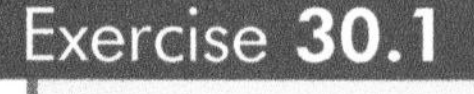

Exercise 30.1

1 (a) Draw an accurate net for each of these 3-D shapes.

(i) (ii)

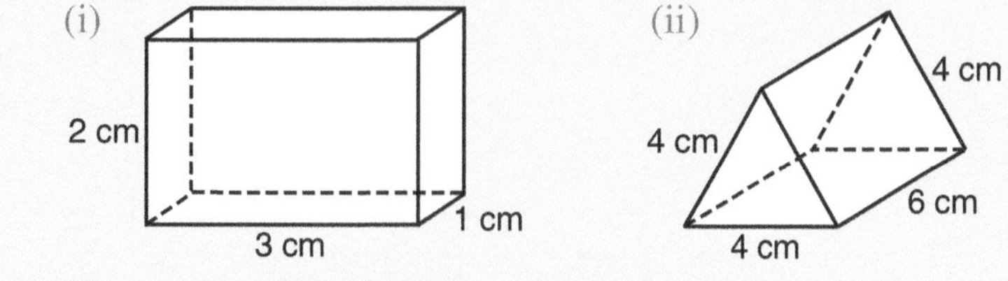

(iii)

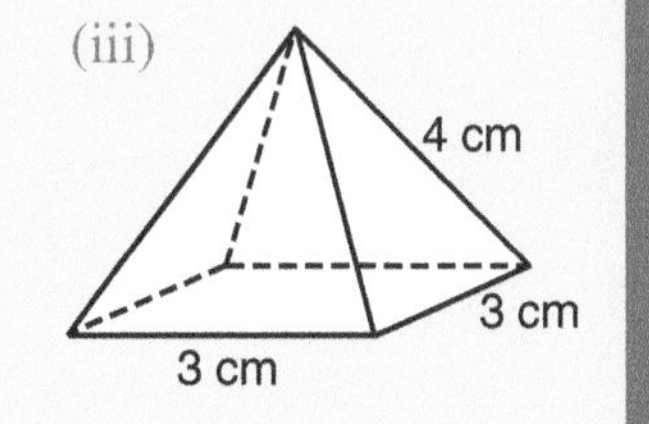

(b) How many edges does the pyramid have?

2 Draw these 3-D shapes on isometric paper.

(a) A cube of side 3 cm.

(b) A cuboid measuring 3 cm by 2 cm by 1 cm.

Symmetry in three-dimensions

Planes of symmetry

Two-dimensional shapes can have line symmetry.
Three-dimensional objects can have **plane symmetry**.
A **plane of symmetry** slices through an object so that one half is the mirror image of the other half.

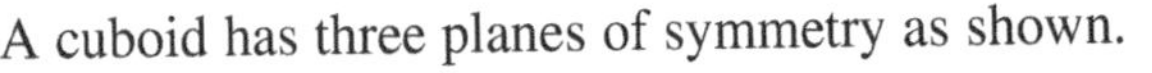

A cuboid has three planes of symmetry as shown.

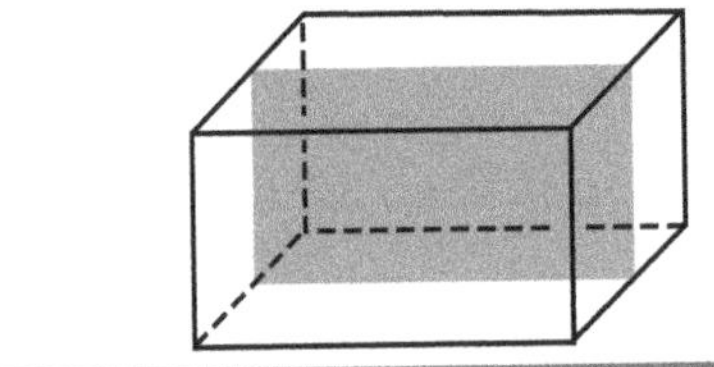

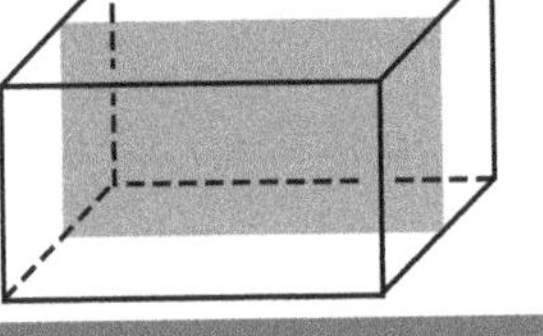

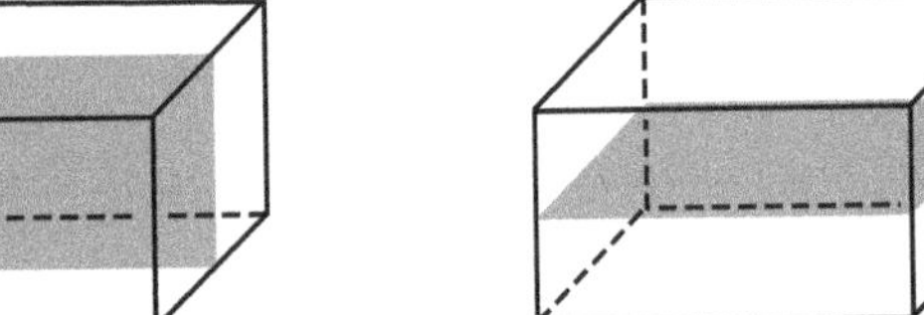

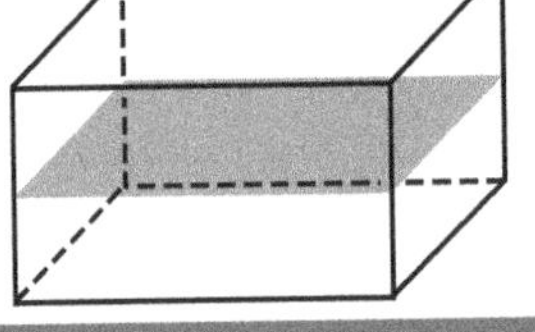

Axes of symmetry

A wall is built using cuboids.

In how many different ways can the next cuboid be placed in position?

If the cuboid can be placed in more than one way, it must have rotational symmetry about one or more **axes**.

A cuboid has three axes of symmetry.
The diagram shows one **axis of symmetry**.
The order of rotational symmetry about this axis is two.

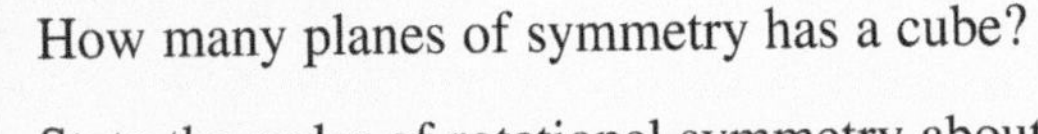

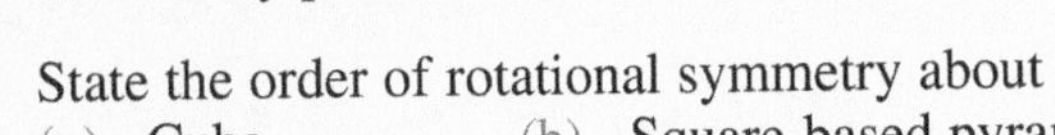

Exercise 30.2

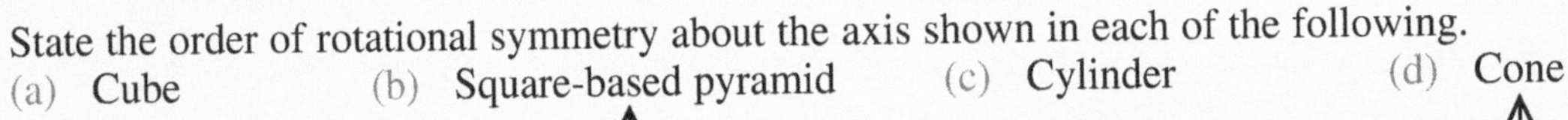

1. How many planes of symmetry has a cube?

2. State the order of rotational symmetry about the axis shown in each of the following.
   (a) Cube (b) Square-based pyramid (c) Cylinder (d) Cone

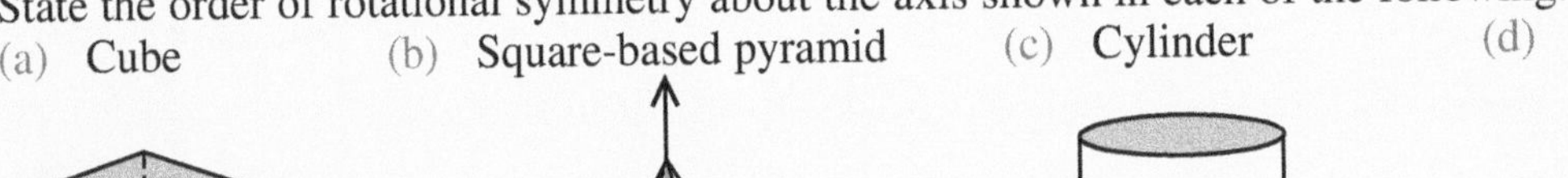

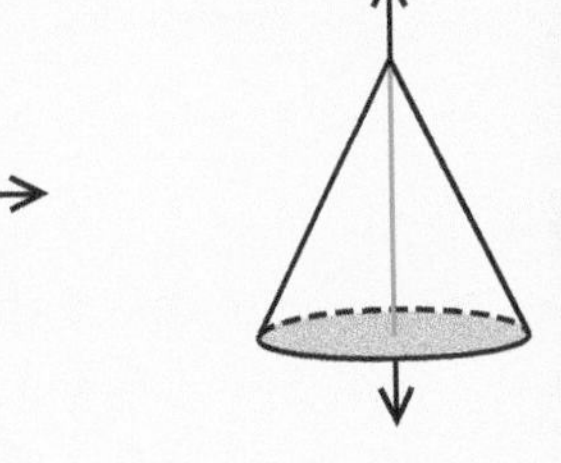

3. Each end of this cuboid is a square.
   The axes of symmetry are labelled a, b and c.
   What is the order of rotational symmetry about:
   (a) axis a,
   (b) axis b,
   (c) axis c?

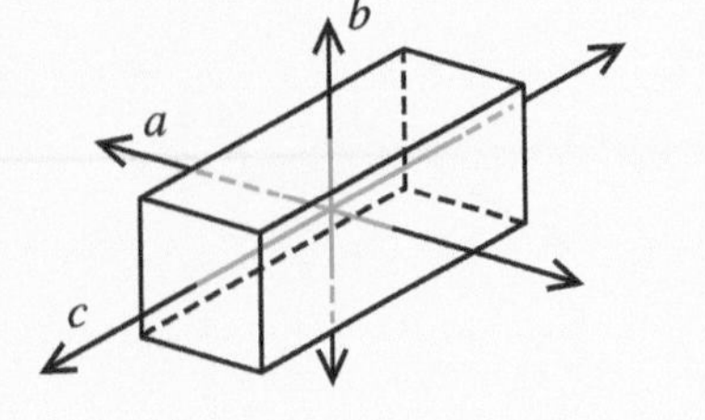

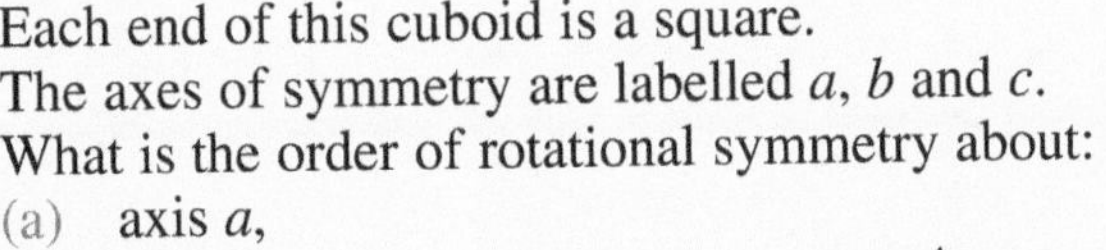

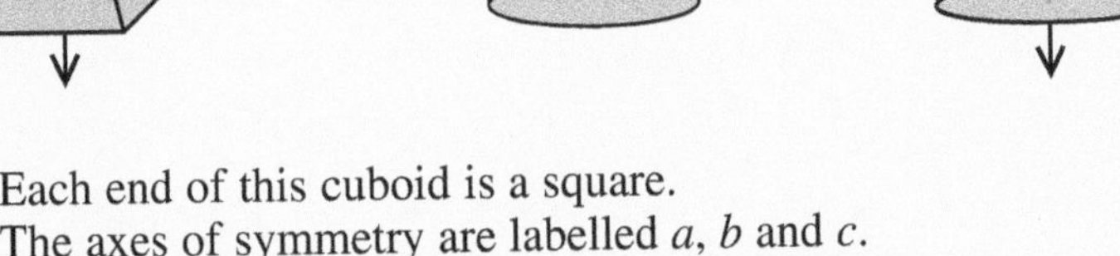

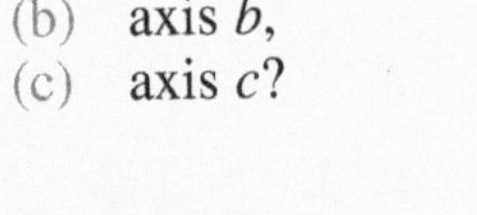

4. The diagram shows a cuboid, with a square base.
   On top of the cuboid is a square-based pyramid with vertex A above the centre of the top of the cuboid.
   (a) How many planes of symmetry has the figure?
   (b) How many axes of symmetry has the figure?
   Give the order of rotational symmetry about each axis.

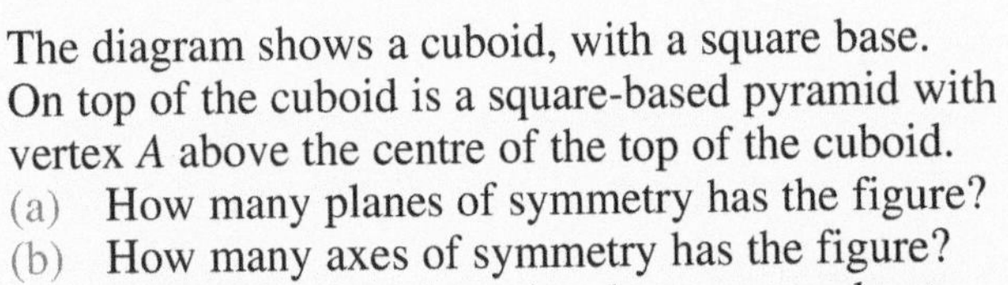

Plans and Elevations

When an architect designs a building he has to draw diagrams to show what the building will look like from different directions.
These diagrams are called **plans and elevations.**

The view of a building looking from above is called the **plan.**
The views of a building from the front or sides are called **elevations.**

To show all the information about a 3-dimensional shape we often need to draw several diagrams.

EXAMPLE

The diagram shows a 3-dimensional shape.
Draw the plan and the elevations **A**, **B** and **C**.

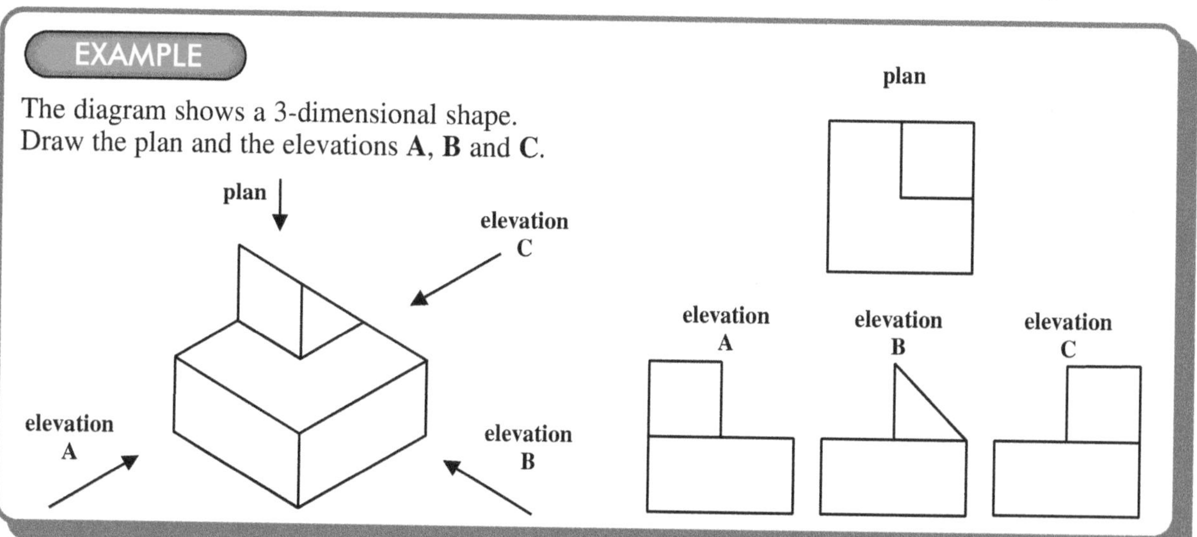

Exercise 30.3

1 Draw a sketch to show the plan view of each of these 3-dimensional shapes.

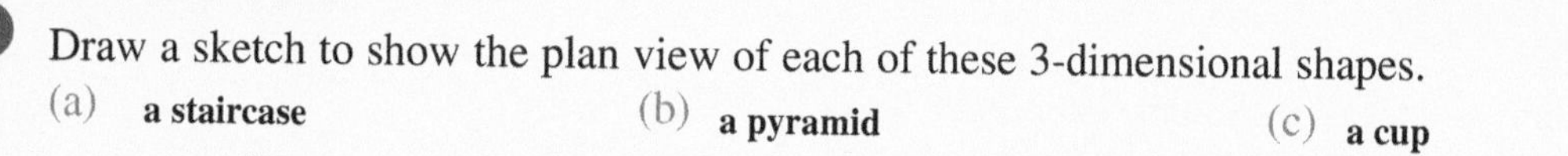

(a) **a staircase** (b) **a pyramid** (c) **a cup**

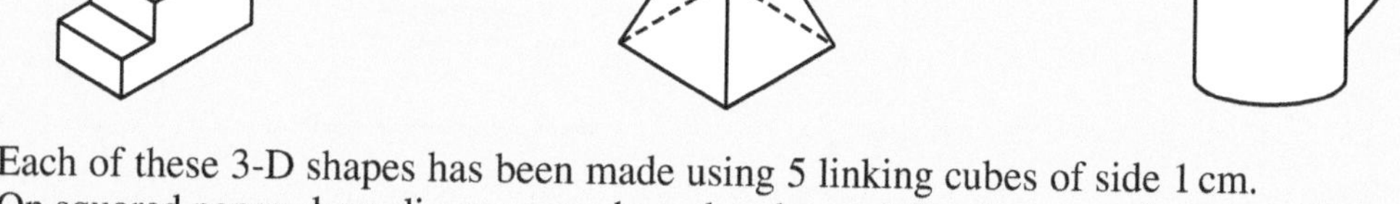

2 Each of these 3-D shapes has been made using 5 linking cubes of side 1 cm.
On squared paper, draw diagrams to show the plan and the elevations **A**, **B** and **C** of each shape.

(a) (b) (c)

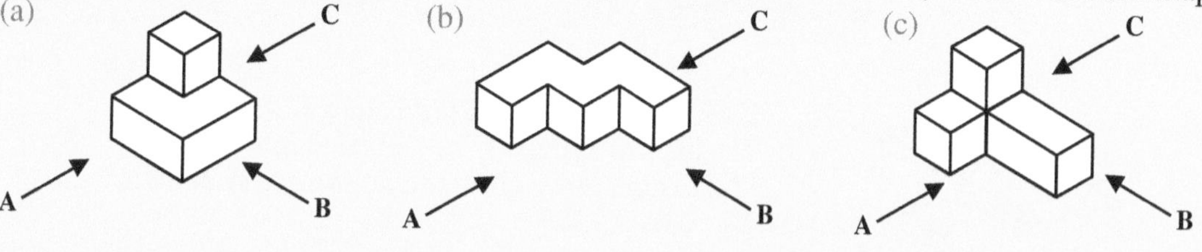

3 The diagram shows a plastic cylinder of height 3 cm and radius 2 cm with a hole of radius 1 cm drilled through the centre.

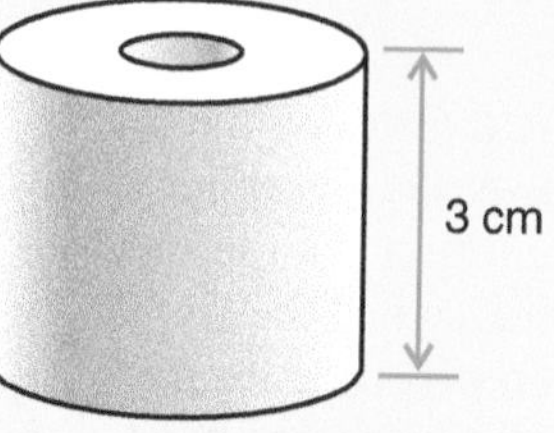

Draw the plan and a side elevation of the cylinder.

4 The plans and elevations of two 3-D shapes made from linking cubes of side 1 cm are shown.
Draw both of these 3-D shapes on isometric paper.

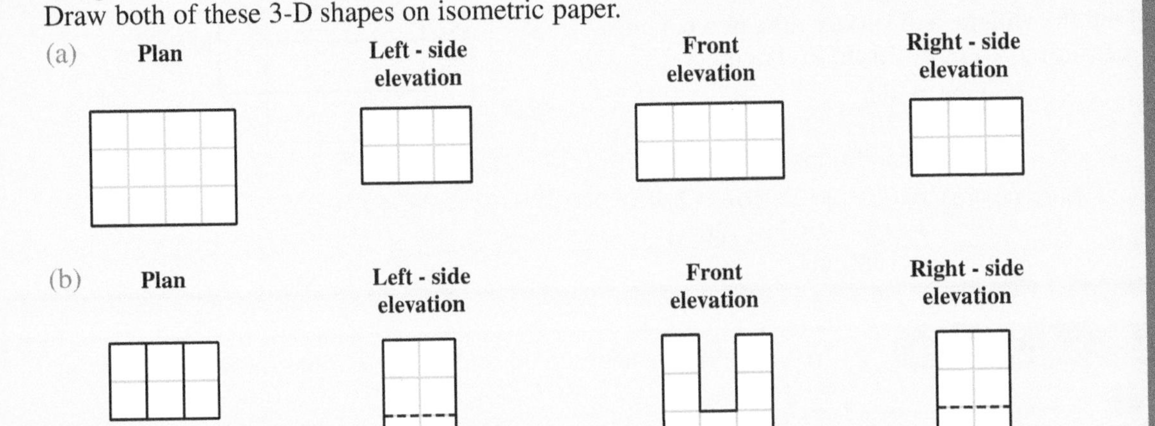

Surface area of a cuboid

Opposite faces of a cuboid are the same shape and size.

To find the surface area of a cuboid find the areas of the six rectangular faces and add the answers.

The surface area of a cuboid can also be found by finding the area of its net.

Volume

Volume is the amount of space occupied by a three-dimensional shape.

This **cube** is 1 cm long, 1 cm wide and 1 cm high.
It has a volume of **1 cubic centimetre**.
The volume of this cube can be written as 1 cm^3.

Small volumes can be measured using cubic millimetres (mm^3).
Large volumes can be measured using cubic metres (m^3).

Volume = 1 cm^3

Volume of a cuboid

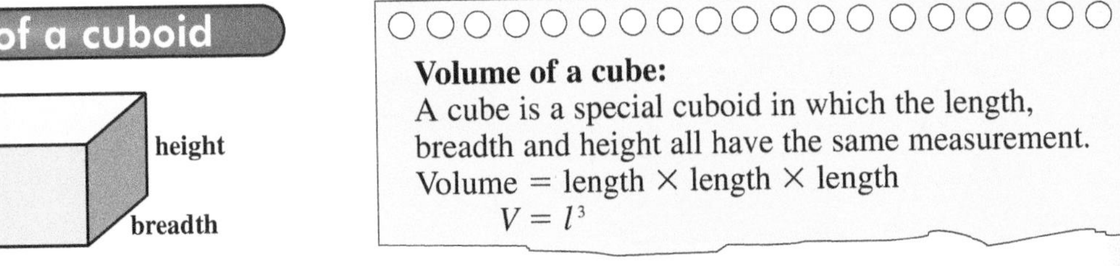

Volume of a cube:
A cube is a special cuboid in which the length, breadth and height all have the same measurement.
Volume = length × length × length
$V = l^3$

The formula for the volume of a cuboid is: Volume = length × breadth × height
This formula can be written using letters as: $V = lbh$

EXAMPLE

Find the volume and surface area of a cuboid measuring 30 cm by 15 cm by 12 cm.

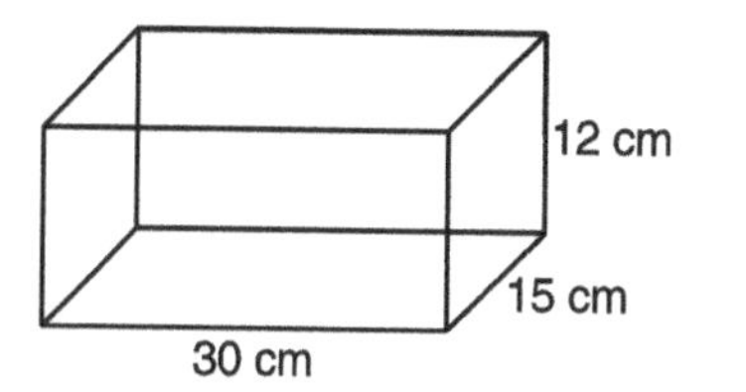

Volume $= lbh$

$= 30\,\text{cm} \times 15\,\text{cm} \times 12\,\text{cm}$

$= 5400\,\text{cm}^3$

Surface area $= (2 \times 30 \times 15) + (2 \times 15 \times 12) + (2 \times 30 \times 12)$

$= 900 + 360 + 720$

$= 1980\,\text{cm}^2$

Exercise 30.4

You should be able to do questions 1 to 3 without using a calculator.

1. A cuboid measures 2 cm by 4 cm by 3 cm.
   (a) Draw a net of the cuboid on one-centimetre squared paper.
   (b) Find the surface area of the cuboid.

2. Calculate the volumes and surface areas of these cubes and cuboids.

(a)

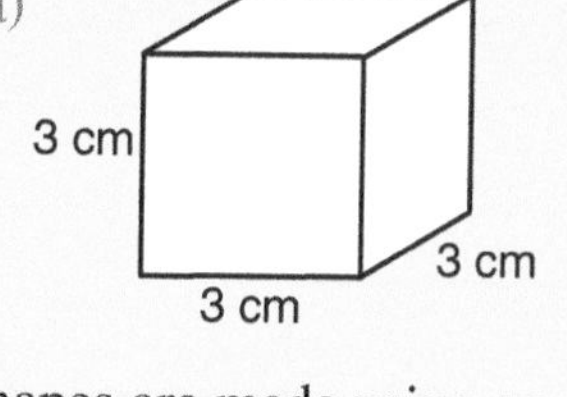

(b)

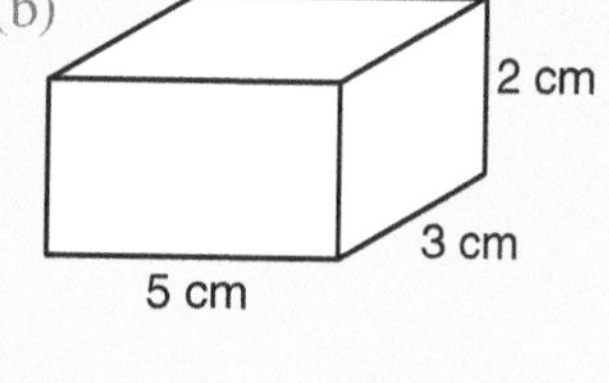

(c)

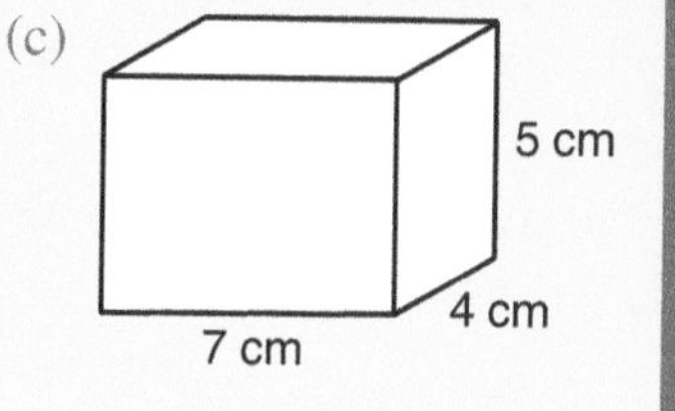

3. Shapes are made using one-centimetre cubes.
   Find the volume and surface area of each shape.

(a) (b) (c) (d) (e) (f)

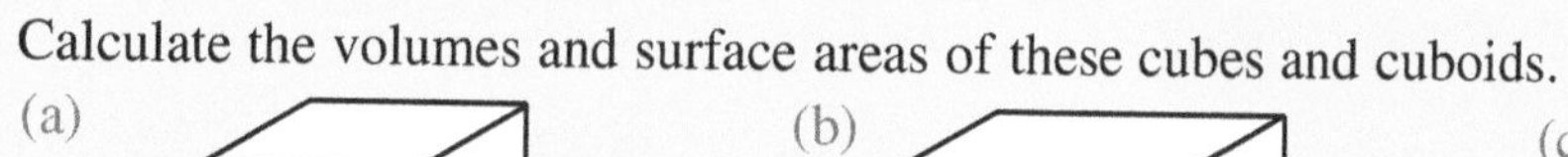

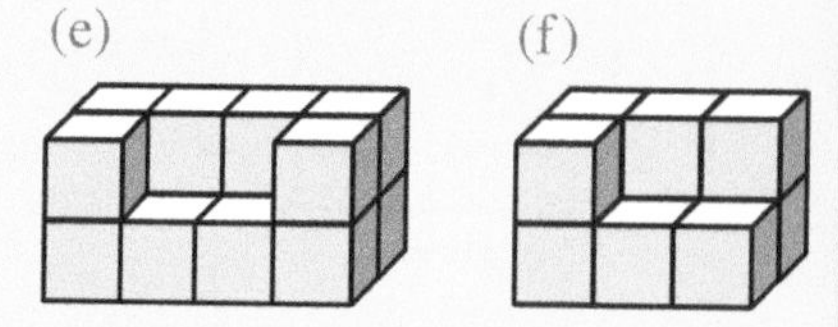

4. A cuboid measures 3.2 cm by 4.8 cm by 6.3 cm.
   Calculate the volume and surface area of the cuboid.

5. A cuboid has a volume of $76.8\,\text{cm}^3$.
   The length of the cuboid is 3.2 cm. The breadth of the cuboid is 2.4 cm.
   What is the height of the cuboid?

6. A cuboid has a volume of $2250\,\text{cm}^3$.
   The length of the cuboid is 25 cm. The height of the cuboid is 12 cm.
   Calculate the surface area of the cuboid.

7. The surface area of a cuboid is $294\,\text{cm}^2$.
   Calculate the volume of a cube with the same surface area.

Prisms

These shapes are all **prisms**.

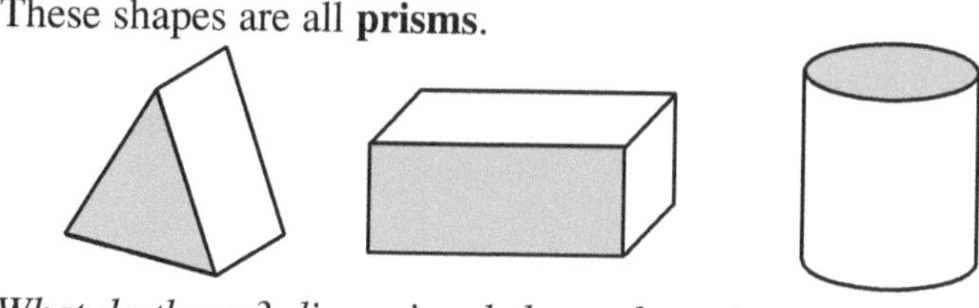

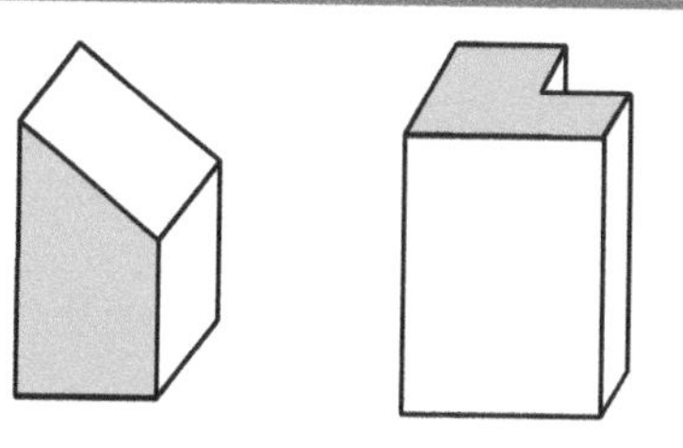

What do these 3-dimensional shapes have in common?
Draw a different 3-dimensional shape which is a prism.

Volume of a prism

The formula for the volume of a prism is:

Volume = area of cross-section × length.

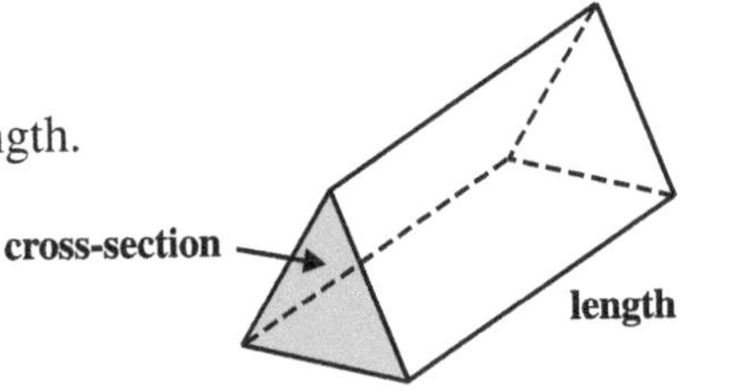

Volume of a cylinder

A **cylinder** is a prism.

The **volume of a cylinder** can be written as:

Volume = area of cross-section × height

$V = \pi r^2 h$

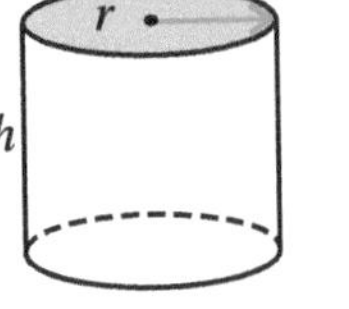

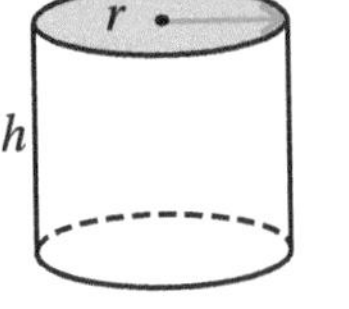

Notice that length has been replaced by height.

EXAMPLES

Find the volumes of these prisms.

(a) Area 18 cm², 10 cm

Volume = area of cross-section × length
= 18 × 10
= 180 cm³

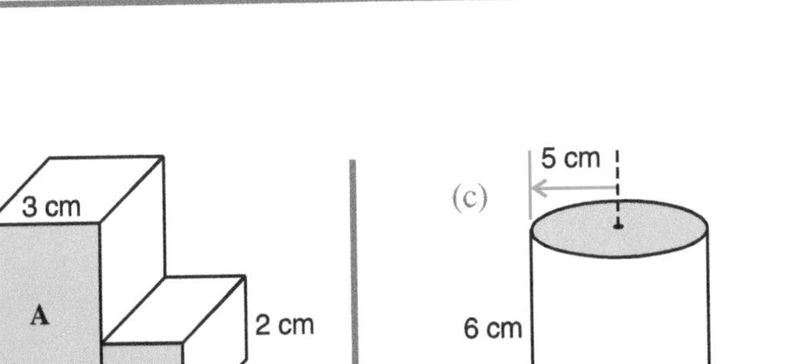

(b)
Area A = 8 × 3 = 24 cm²
Area B = 3 × 2 = 6 cm²
Total area = 30 cm²
Volume = 30 × 5 = 150 cm³

(c)
$V = \pi r^2 h$
$= \pi \times 5^2 \times 6$
= 471.238…
= 471 cm³, correct to 3 s.f.

Exercise 30.5

1 Find the volumes of these prisms.

(a)

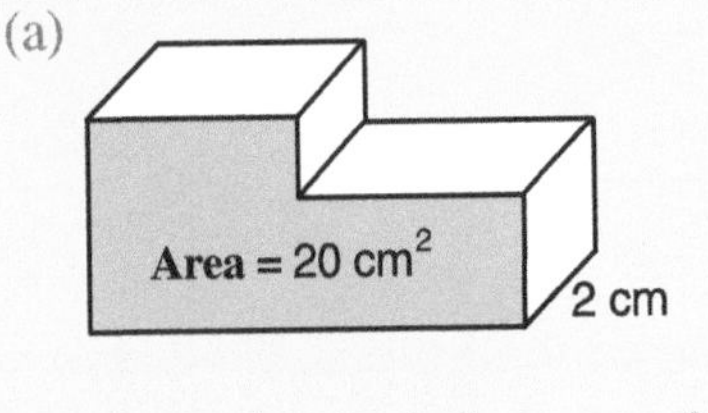

(b)

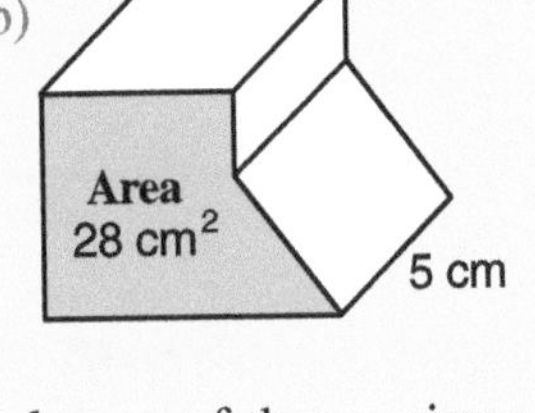

(c) Area 9.6 cm², 10 cm

2 Calculate the shaded areas and the volumes of these prisms.

(a) 1.5 cm, 2 cm, 4 cm, 1.4 cm, 4 cm
(b) 4 cm, 3 cm, 2.5 cm
(c) 10 cm, 20 cm. Take π to be 3.14
(d) 5 cm, 2 cm. Take π to be 3.14

3 Find the volumes of these prisms.
Where necessary take π to be 3.14 or use the π key on your calculator.

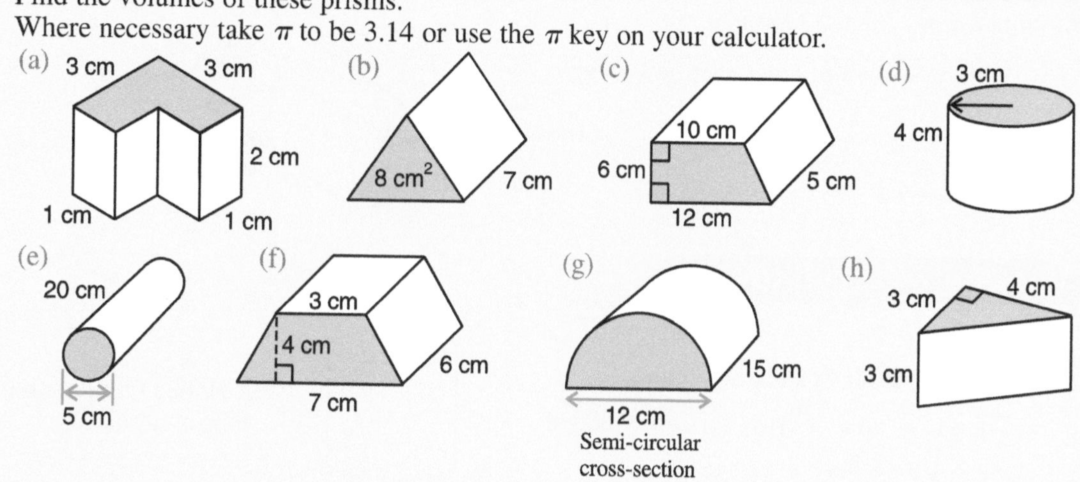

4 Sylvia says, "A cylinder with a radius of 5 cm and a height of 10 cm has the same volume as a cylinder with a radius of 10 cm and a height of 5 cm."
Is she right? Explain your answer.

5 The radius of a cylinder is 5 cm. It has a volume of $900\,\text{cm}^3$.
Calculate the height of the cylinder, giving your answer correct to 1 decimal place.

6 A cylinder is 8 cm high. It has a volume of $183\,\text{cm}^3$.
Calculate the radius of the cylinder correct to 1 decimal place.

7 A cylinder has a diameter of 10.6 cm. The volume of the cylinder is $1060\,\text{cm}^3$.
Calculate the height of the cylinder. Give your answer to an appropriate degree of accuracy.

8 A cylinder with a radius of 3 cm and a height of 8 cm is full of water.
The water is poured into another cylinder with a diameter of 8 cm.
Calculate the height of the water.

Surface area of a cylinder

The top and bottom of a cylinder are circles. The curved surface of a cylinder is a rectangle.

The rectangle has the same height, h, as the cylinder.
The length of the rectangle must be just long enough to "wrap around" the circle.
The lid of the cylinder has radius r and circumference $2\pi r$.
So, the length of the rectangle is also $2\pi r$.

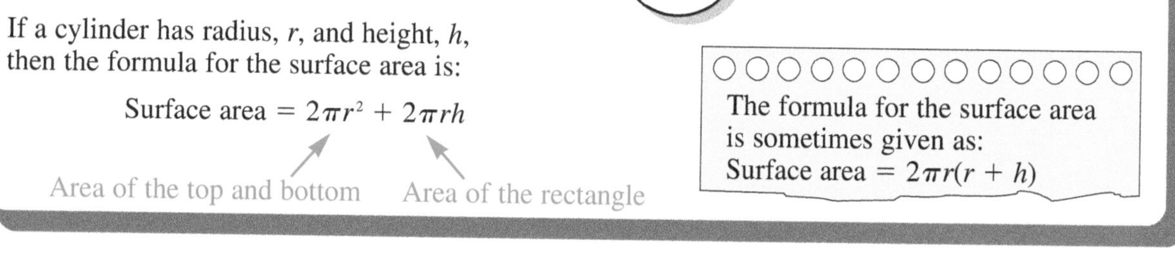

$$\begin{aligned}\text{Area of lid} &= \pi r^2\\ \text{Area of base} &= \pi r^2\\ \text{Area of lid and base} &= 2\pi r^2\\ \text{Area of rectangle} &= \text{length} \times \text{breadth}\\ &= 2\pi r \times h\\ &= 2\pi rh\end{aligned}$$

If a cylinder has radius, r, and height, h, then the formula for the surface area is:

$$\text{Surface area} = 2\pi r^2 + 2\pi rh$$

Area of the top and bottom — Area of the rectangle

The formula for the surface area is sometimes given as:
Surface area $= 2\pi r(r + h)$

EXAMPLE

Find the surface area of a cylinder with radius 4 cm and height 6 cm.
Use the π key on your calculator.

$$\begin{aligned} \text{Area} &= 2\pi rh + 2\pi r^2 \\ &= 2 \times \pi \times 4 \times 6 + 2 \times \pi \times 4^2 \\ &= 150.796 \ldots + 100.530 \ldots \\ &= 251.327 \ldots \\ &= 251.3\,\text{cm}^2\text{, correct to 1 d.p.} \end{aligned}$$

Exercise 30.6

Take π to be 3.14 or use the π key on your calculator.

1. Find the surface areas of these cylinders.

(a) 3 cm, 15 cm
(b) 3 cm, 5 cm
(c) 6.5 cm, 1.2 cm

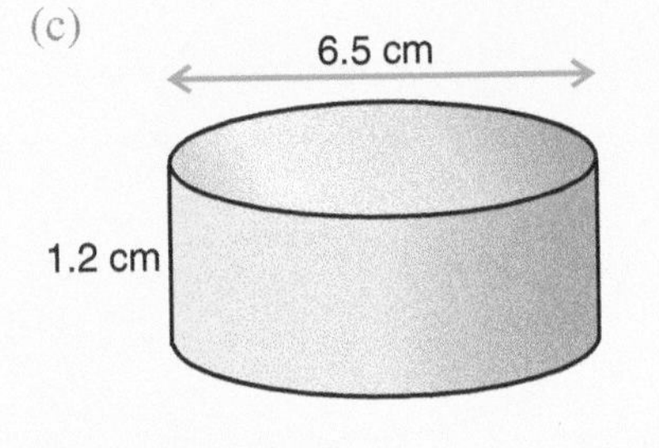

2. Show that the curved surface area of this can is approximately 75 cm².

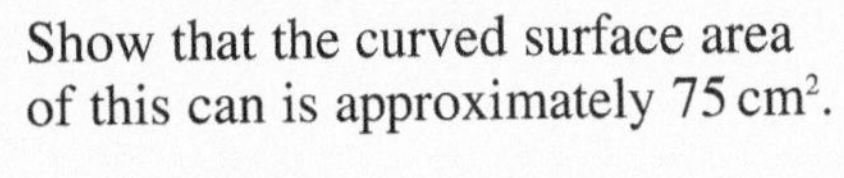

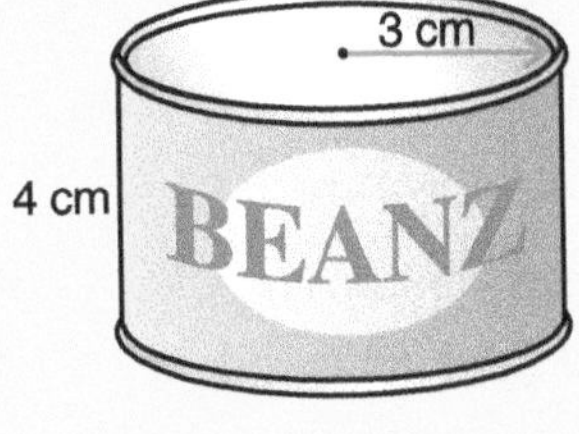

3. A bucket is in the shape of a cylinder.
   (a) Calculate the area of the bottom of the bucket.
   (b) Calculate the curved surface area of the bucket.

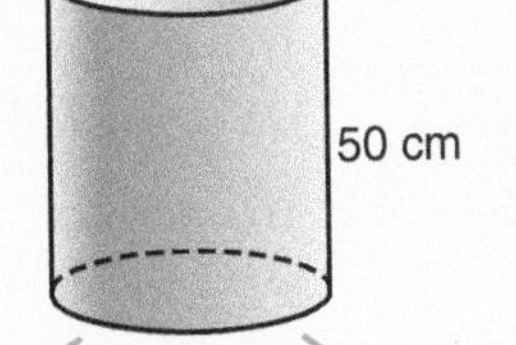

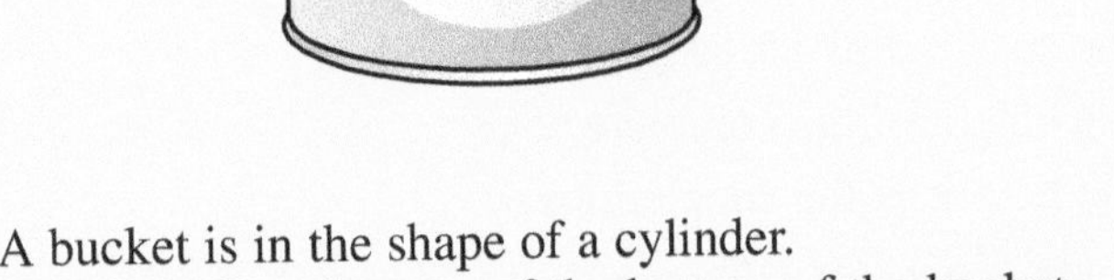

4. A cylinder has a radius of 3.6 cm. The length of the cylinder is 8.5 cm.
   Calculate the total surface area of the cylinder.
   Give your answer to an appropriate degree of accuracy.

5. A concrete pipe is 150 cm long.
   It has an internal radius of 15 cm and an external radius of 20 cm.
   Calculate, giving your answers to 3 significant figures,
   (a) the area of the curved surface inside of the pipe,
   (b) the curved surface area of the outside of the pipe.

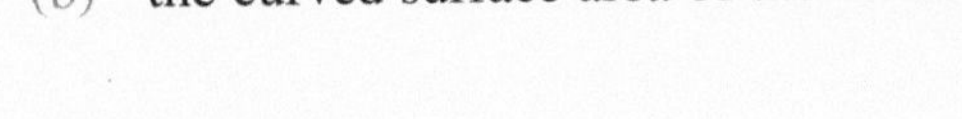

6. A cylinder is 15 cm high.
   The curved surface area of the cylinder is 377 cm².
   Calculate the radius of the cylinder.

Cones

The diagram shows a cone with: circular base, radius r, slant height l, perpendicular height h.

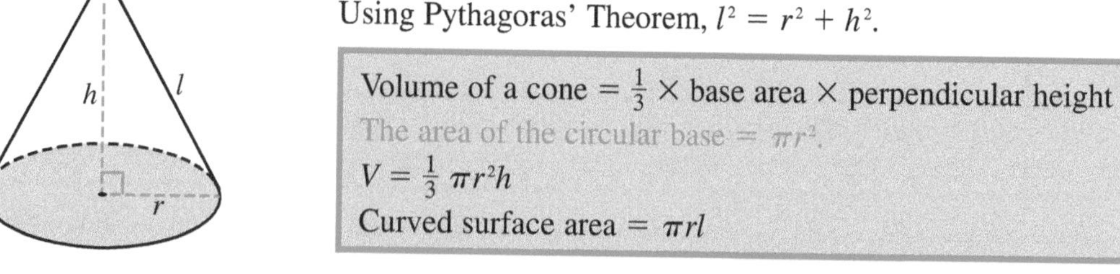

Using Pythagoras' Theorem, $l^2 = r^2 + h^2$.

Volume of a cone $= \frac{1}{3} \times$ base area $\times$ perpendicular height
The area of the circular base $= \pi r^2$.
$V = \frac{1}{3}\pi r^2 h$
Curved surface area $= \pi r l$

A **frustum of a cone** is formed by removing the top of a cone with a cut parallel to its circular base.

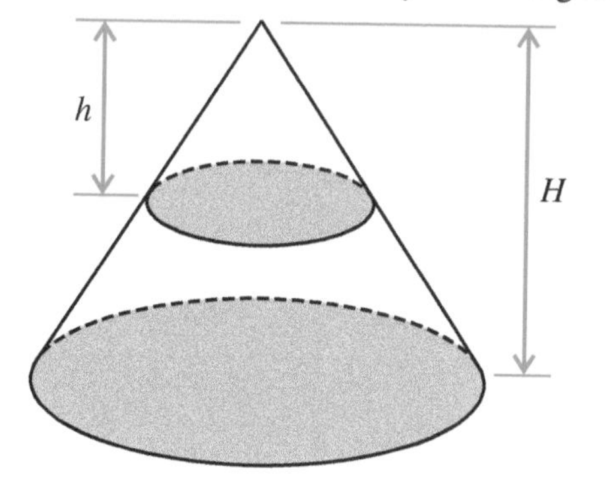

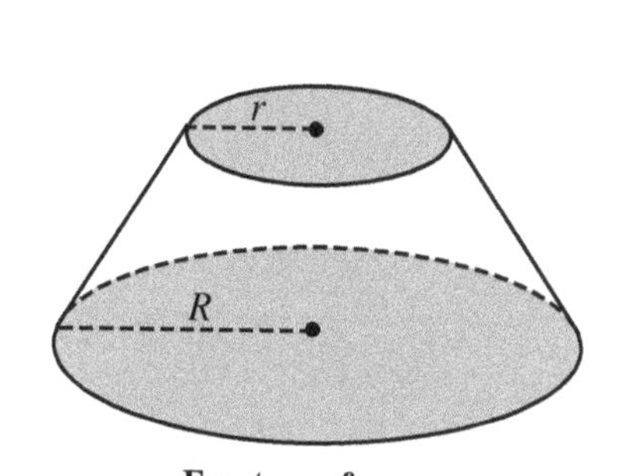

Frustum of a cone

The **volume of a frustum** = Volume of complete cone − Volume of cone removed

$$= \frac{1}{3}\pi R^2 H - \frac{1}{3}\pi r^2 h$$

where R is the base radius of the complete cone,
r is the base radius of the cone removed,
H is the height of the complete cone,
h is the height of the cone removed.

The **curved surface area** of a frustum is given by:

Curved surface area of complete cone − Curved surface area of cone removed

How could you find the total surface area of a solid frustum of a cone?

Pyramids

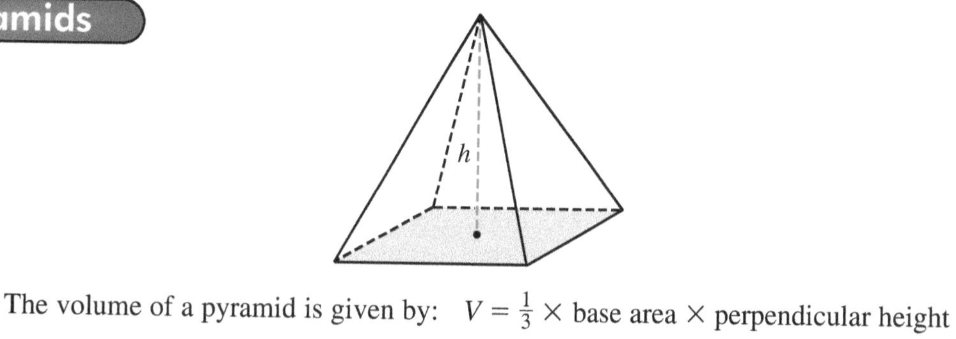

The volume of a pyramid is given by: $V = \frac{1}{3} \times$ base area $\times$ perpendicular height

Spheres

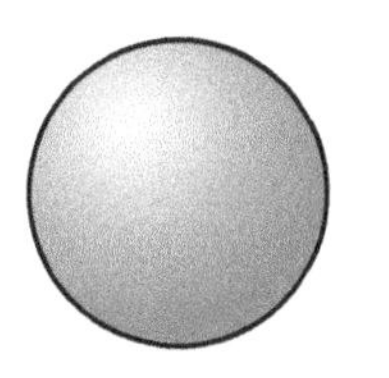

Volume $= \frac{4}{3}\pi r^3$ Surface area $= 4\pi r^2$

EXAMPLE

A child's toy is made from a cone with base 3 cm and height 4 cm joined to a hemisphere with radius 3 cm.

(a) Calculate the volume of the toy.
(b) Calculate the total surface area.

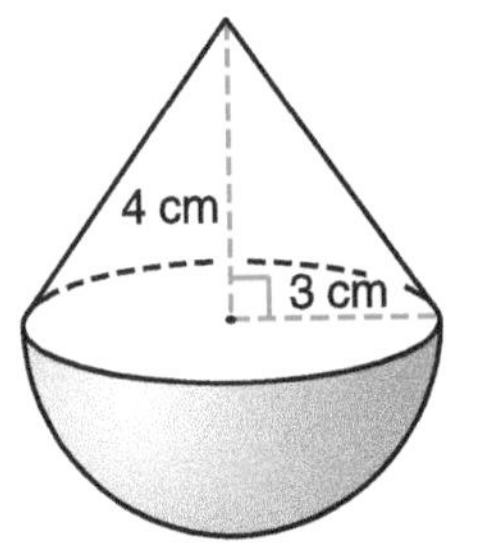

(a) Volume of hemisphere

$= \frac{1}{2} \times \frac{4}{3} \pi r^3$

$= \frac{1}{2} \times \frac{4}{3} \times \pi \times 3^3$

$= 56.548\ldots \text{ cm}^3$

Volume of cone

$= \frac{1}{3} \pi r^2 h$

$= \frac{1}{3} \times \pi \times 3^2 \times 4$

$= 37.699\ldots \text{ cm}^3$

Total volume

$= 56.548\ldots + 37.699\ldots$

$= 94.247\ldots \text{ cm}^3$

$= 94.2 \text{ cm}^3$, correct to 3 s.f.

(b) Surface area of hemisphere

$= \frac{1}{2} \times 4 \pi r^2$

$= \frac{1}{2} \times 4 \times \pi \times 3^2$

$= 56.548\ldots \text{ cm}^2$

Curved surface area of cone $= \pi r l$

By Pythagoras: $l^2 = r^2 + h^2$

$l^2 = 3^2 + 4^2$

$l^2 = 25$

$l = \sqrt{25} = 5 \text{ cm}$

Curved surface area of cone

$= \pi \times 3 \times 5$

$= 47.123\ldots \text{ cm}^2$

Total surface area

$= 56.548\ldots + 47.123\ldots$

$= 103.672\ldots$

$= 104 \text{ cm}^2$, correct to 3 s.f.

Exercise 30.7

Use the π key on your calculator and give answers correct to 3 significant figures where appropriate.

1 Find the volumes of these solids.

(a)

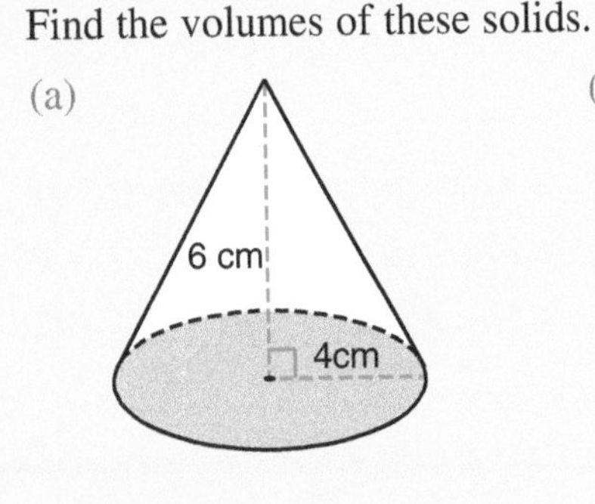

(b)

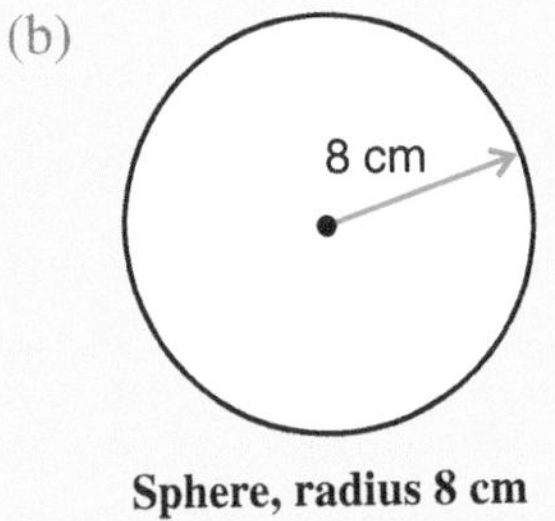

Sphere, radius 8 cm

(c)

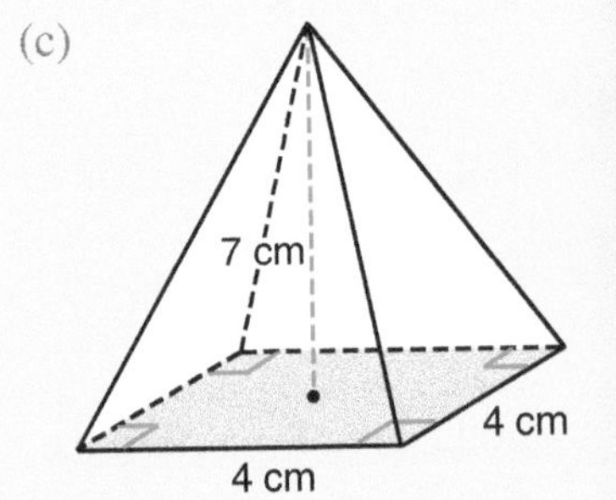

2 Find the areas of the curved surfaces for these cones.

(a)

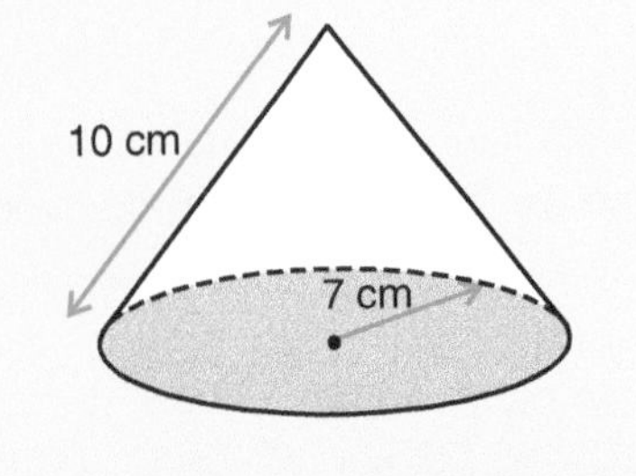

(b) 4 cm, 5 cm, 3 cm

(c)

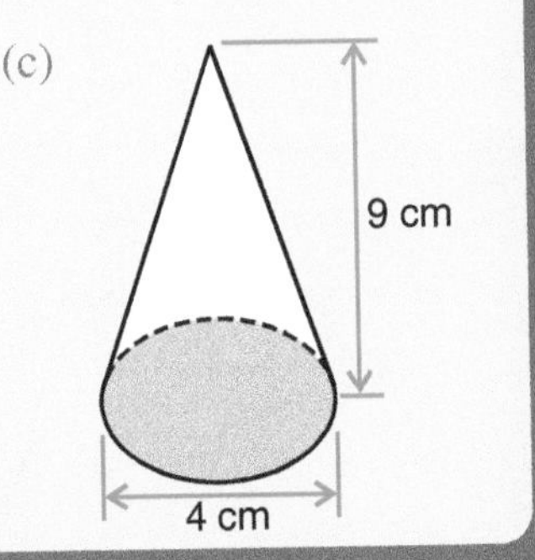

3 Which of these containers has the greater volume?
Show all your working.

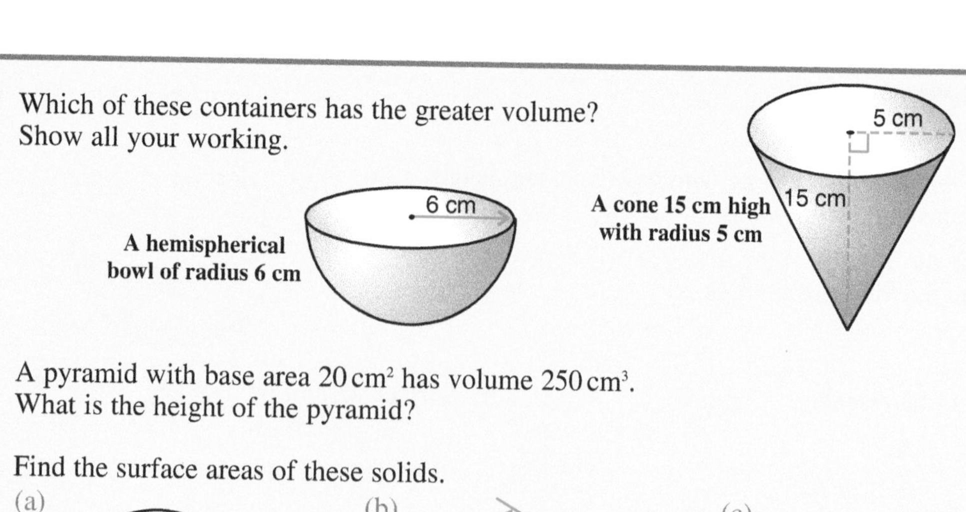

4 A pyramid with base area $20\,cm^2$ has volume $250\,cm^3$.
What is the height of the pyramid?

5 Find the surface areas of these solids.

(a) Sphere, radius 5 cm (b) (c) Hemisphere, radius 15 cm

6 For each of these solids, find,
(i) the total volume, (ii) the total surface area.

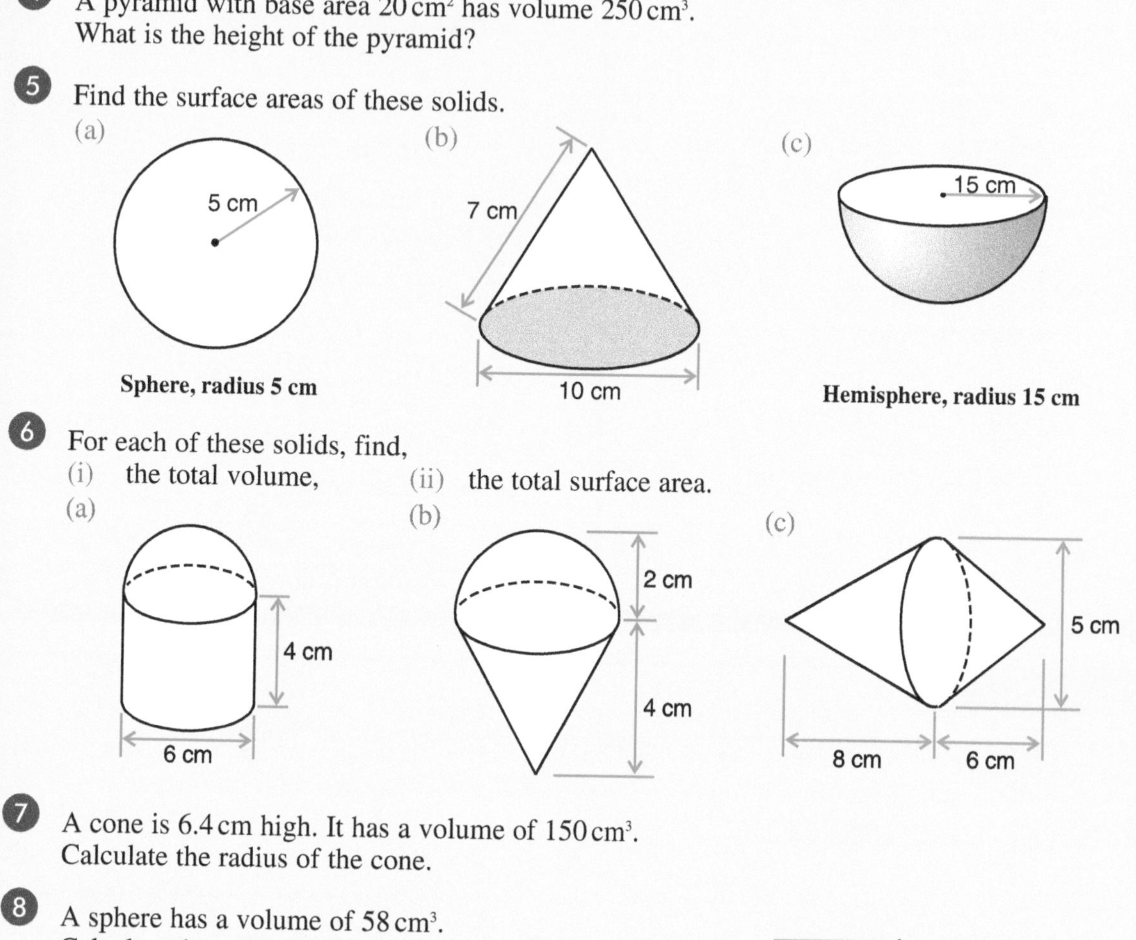

7 A cone is 6.4 cm high. It has a volume of $150\,cm^3$.
Calculate the radius of the cone.

8 A sphere has a volume of $58\,cm^3$.
Calculate the radius of the sphere.

9 Rubber bungs are made by removing the tops of cones.
Starting with a cone of radius 10 cm and height 16 cm, a rubber bung is made by cutting a cone of radius 5 cm and height 8 cm from the top.
Find the volume and total surface area of the rubber bung.

10 *ABCDX* is a pyramid.
The base *ABCD* is a rectangle measuring 18 cm by 10 cm and *X* is 12 cm vertically above the midpoint of the base.

Calculate
(a) the volume of the pyramid,
(b) the total surface area of the pyramid.

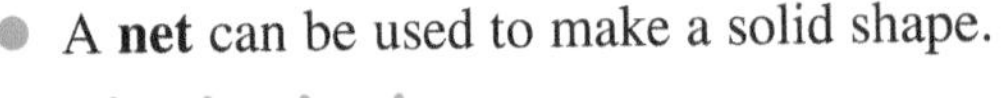

What you need to know

- **Faces**, **vertices** (corners) and **edges**.
  For example, a cube has 6 faces, 8 vertices and 12 edges.

vertex
edge
face

- A **net** can be used to make a solid shape.

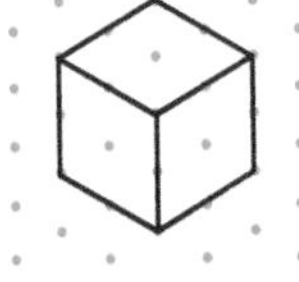

- **Isometric paper** is used to make 2-D drawings of 3-D shapes.

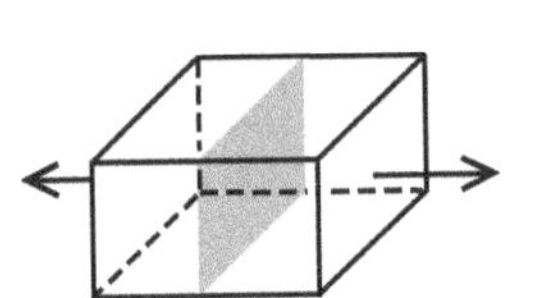

- A **plane of symmetry** slices through a 3-D shape so that one half is the mirror image of the other half.
- Three-dimensional shapes can have **axes of symmetry**.

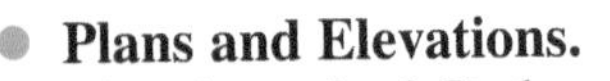

- **Plans and Elevations.**
  The view of a 3-D shape looking from above is called a **plan**.
  The view of a 3-D shape from the front or sides is called an **elevation**.
- **Volume** is the amount of space occupied by a 3-D shape.

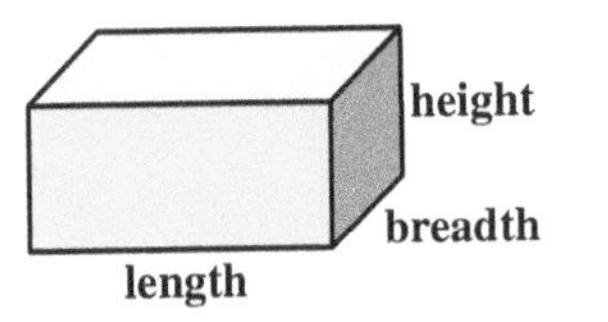

- The formula for the volume of a **cuboid** is:
  Volume = length × breadth × height
  $V = l \times b \times h$

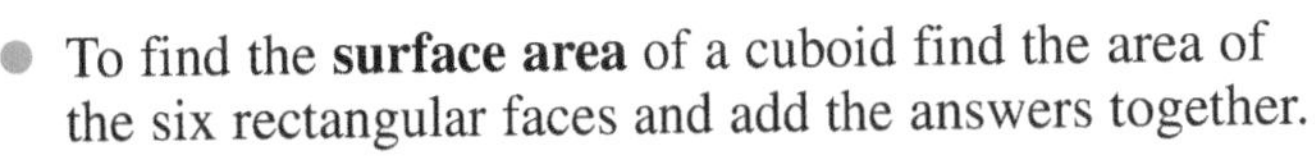

- To find the **surface area** of a cuboid find the area of the six rectangular faces and add the answers together.
- Volume of a **cube** is: Volume = (length)^3
  $V = l^3$

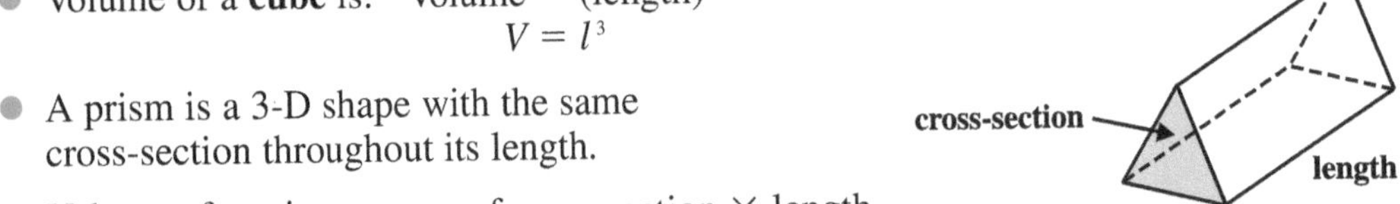

- A prism is a 3-D shape with the same cross-section throughout its length.

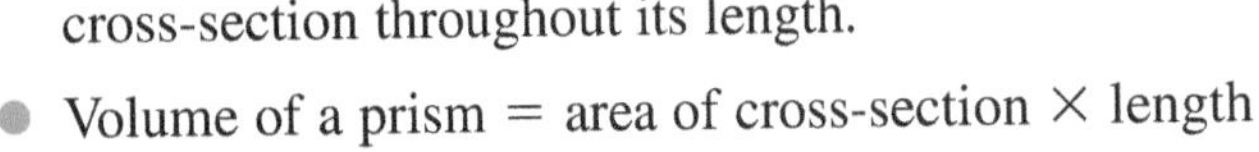

- Volume of a prism = area of cross-section × length

cross-section
length
Triangular prism

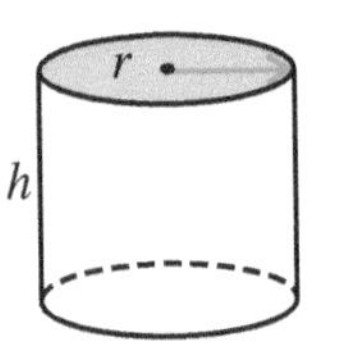

- A **cylinder** is a prism.
  Volume of a cylinder is: $V = \pi \times r^2 \times h$
  Surface area of a cylinder is: Surface area = $2\pi r^2 + 2\pi rh$
- These formulae are used in calculations involving **cones, pyramids,** and **spheres.**

Cone

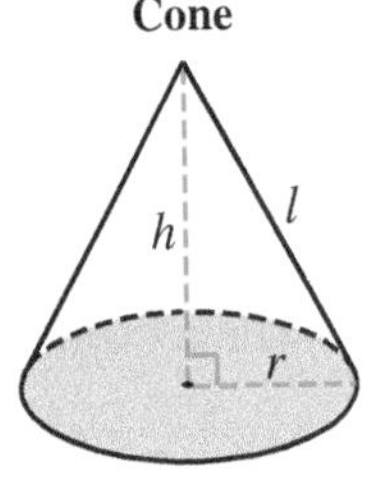

$V = \frac{1}{3} \times$ base area $\times$ height
$V = \frac{1}{3}\pi r^2 h$
Curved surface area = πrl

Pyramid

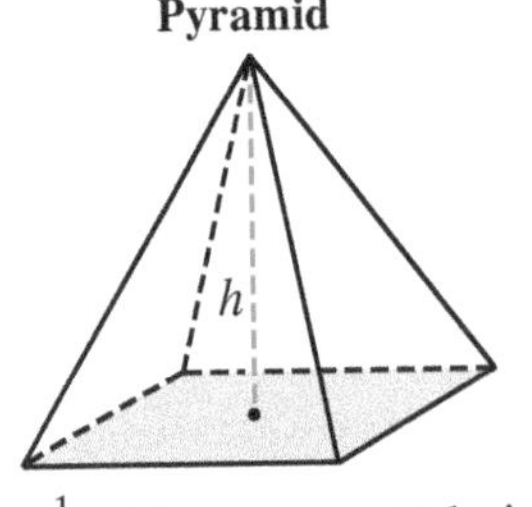

$V = \frac{1}{3} \times$ base area $\times$ height

Sphere

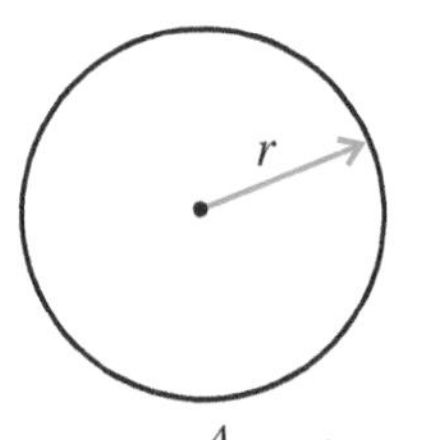

Volume = $\frac{4}{3}\pi r^3$
Surface area = $4\pi r^2$

- A **frustum of a cone** is formed by removing the top of a cone with a cut parallel to its circular base.

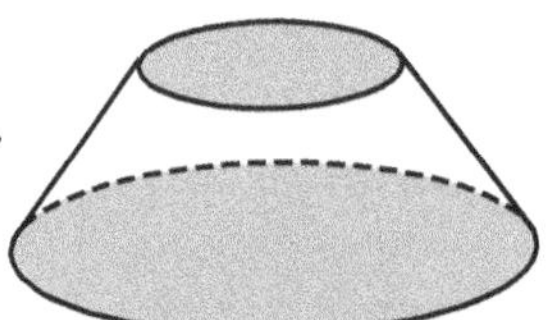

Review Exercise 30

1 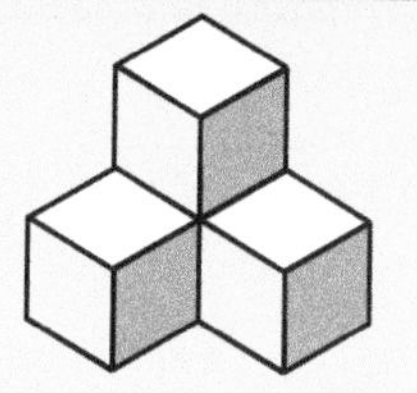
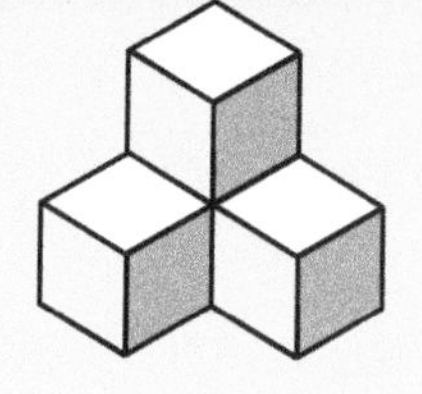

This 3-D shape has been made using 4 linking cubes of side 1 cm.
Draw this shape on isometric paper.

2 A box contains 2 balls of radius 2 cm, as shown.

(a) Draw a plan of the box.
(b) Draw an elevation of the box from the direction marked **X**.

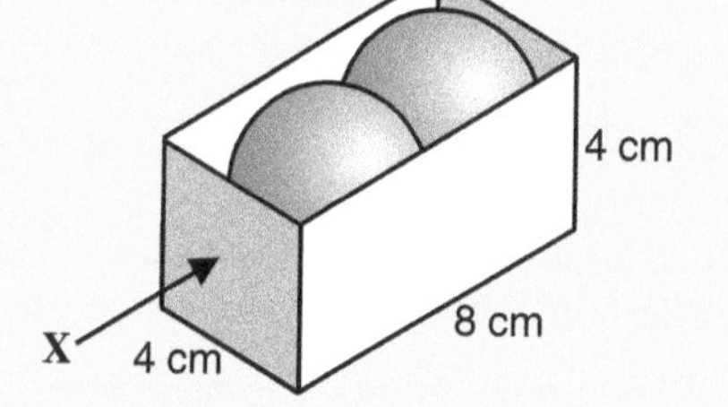

3 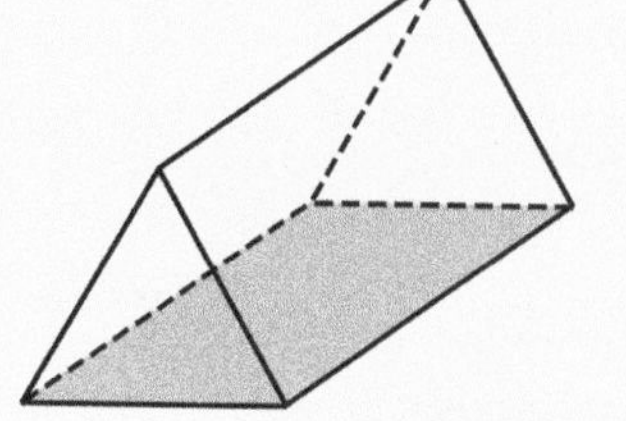
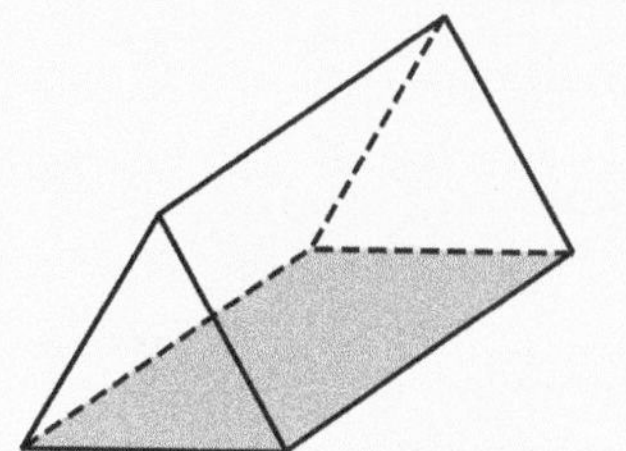

The diagram shows a triangular prism.
The ends of the prism are equilateral triangles.

(a) How many axes of symmetry has the prism?
(b) How many planes of symmetry has the prism?

4 This block of jelly measures 2 cm by 6 cm by 4.5 cm.

(a) What is the volume of this block of jelly?
(b) A shopkeeper buys a box of these jellies.
The box measures 6 cm by 12 cm by 9 cm.
How many blocks of jelly fill the box?

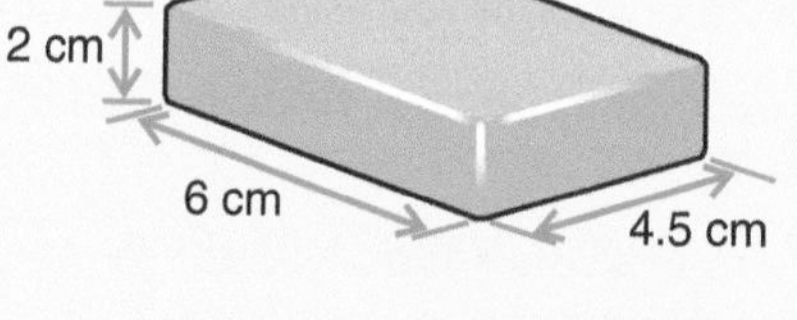
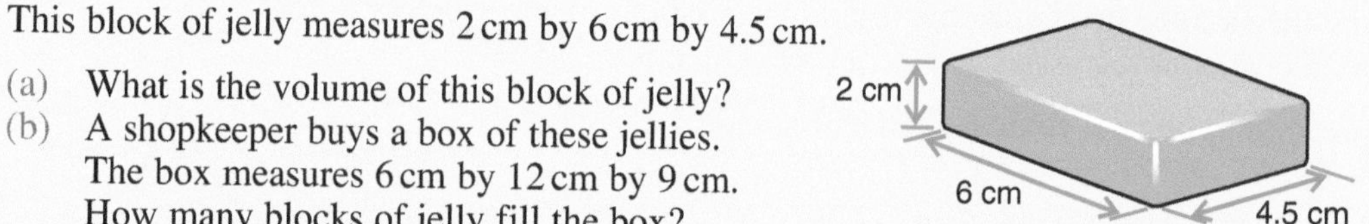

AQA

5 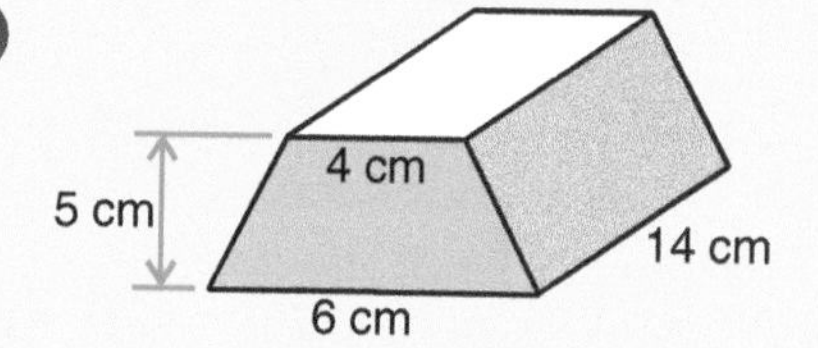

The diagram shows a prism.
The cross section of the prism is a trapezium.
Calculate the volume of the prism.

AQA

6 The diagram shows a cuboid which is just big enough to hold six tennis balls.
Each tennis ball has a diameter of 6.8 cm.
Calculate the volume of the cuboid.

AQA

7 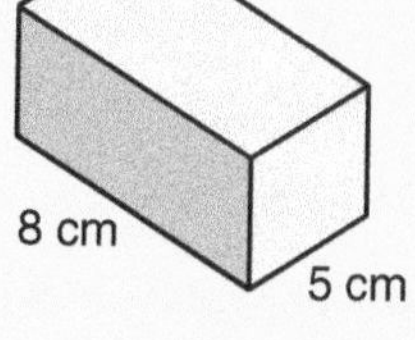
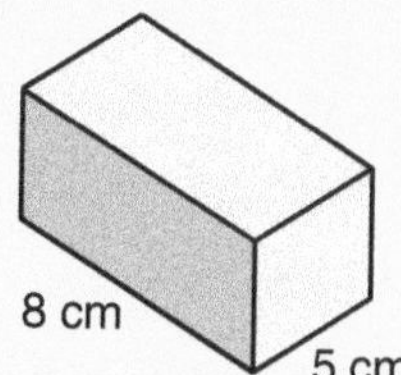

The surface area of this cuboid is 197 cm^2.
Calculate the volume of the cuboid.

8 The diagram shows a block of wood.
The block is a cuboid measuring 8 cm by 13 cm by 16 cm.
A cylindrical hole of radius 5 cm is drilled through the block of wood.
Find the volume of wood remaining.

9 The volume of a cube is given by the formula $V = L^3$.

(a) A cube has volume $5832\,\text{cm}^3$.
What is the length of each edge?

(b) The area of each face of a different cube is $625\,\text{cm}^2$.
What is the volume of this cube?

AQA

10

Lentil soup is sold in cylindrical tins.
Each tin has a base radius of 3.8 cm and a height of 12.6 cm.

(a) Calculate the total surface area of a tin.

(b) (i) Calculate the volume of soup in a full tin.

(ii) Luke has a full tin of lentil soup for dinner.
He pours the soup into a cylindrical bowl of radius 7 cm.
What is the depth of the soup in the bowl?

Give your answers to a suitable degree of accuracy.

11 This diagram shows a sphere of radius 4.5 cm fitting tightly inside a box.
The box is a cube.

Calculate the volume of the space around the sphere, inside the cube.

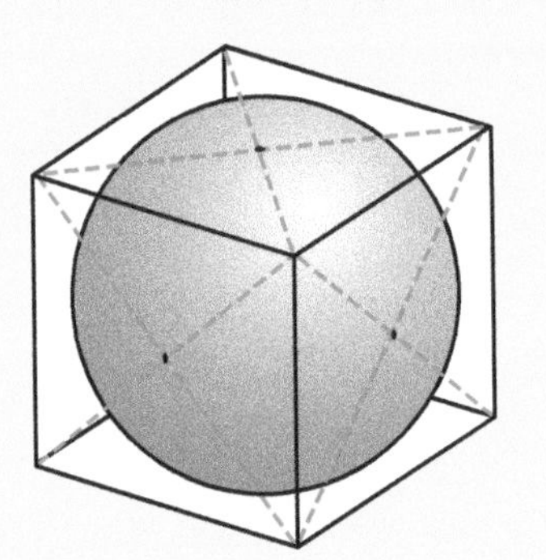

12 A sector of a circle of radius 15 cm is cut out of card.

(a) Calculate the arc length AB.

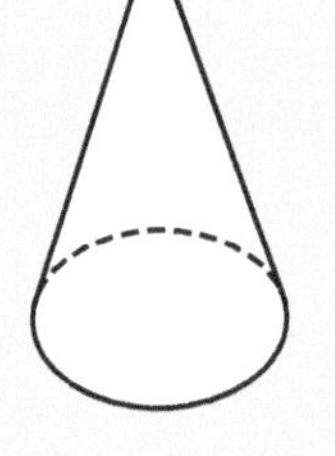

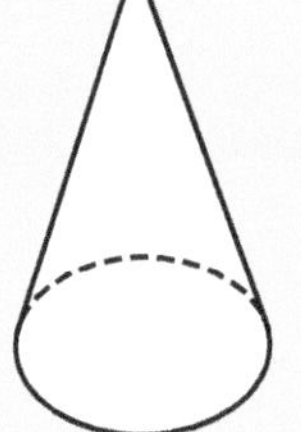

The piece of card is used to create a cone by joining OA to OB with no overlap.

(b) Calculate the volume of the cone.

AQA

13 A spinning top consists of a cone of base radius 5 cm, height 9 cm and a hemisphere of radius 5cm, as illustrated.

(a) Calculate the volume of the spinning top.

(b) Calculate the total surface area of the spinning top.

AQA

14 The diagram shows a pepper pot.
The pot consists of a cylinder and a hemisphere.
The cylinder has a diameter of 5 cm and a height of 7 cm.
The pepper takes up half the **total** volume of the pot.
Find the depth of pepper in the pot, marked x in the diagram.

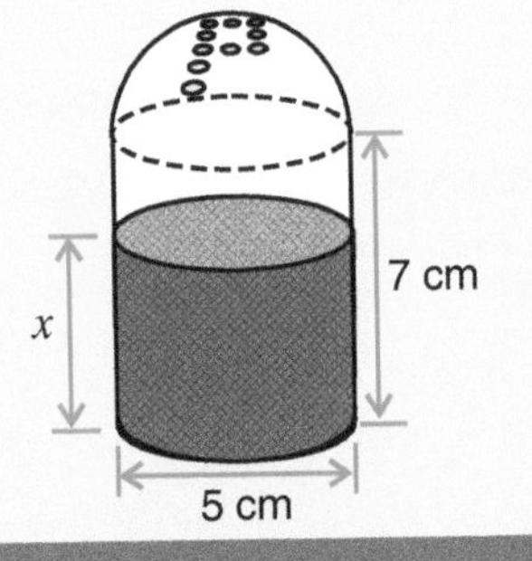

AQA

Understanding and Using Measures

Units of measurement

Different units can be used to measure the same quantity.
For example:

The same **length** can be measured using centimetres, kilometres, inches, miles, …
The same **mass** can be measured using grams, kilograms, pounds, ounces, …
The same **capacity** can be measured using litres, millilitres, gallons, pints, …

There are two sorts of units in common use – **metric** units and **imperial** units.

Metric units

The common metric units used to measure length, mass (weight) and capacity (volume) are shown below.

Length	**Mass**	**Capacity and volume**
1 kilometre (km) = 1000 metres (m)	1 tonne (t) = 1000 kilograms (kg)	1 litre = 1000 millilitres (ml)
1 m = 100 centimetres (cm)	1 kg = 1000 grams (g)	1 cm^3 = 1 ml
1 cm = 10 millimetres (mm)		

Kilo means thousand, 1000. So, a kilogram is one thousand grams.

For example: 3 kilograms = 3000 grams.

Centi means hundredth, $\frac{1}{100}$. So, a centimetre is one hundredth of a metre.

For example: 2 centimetres = $\frac{2}{100}$ metre = 0.02 metres.

Milli means thousandth, $\frac{1}{1000}$. So, a millilitre is one thousandth of a litre.

For example: 5 millilitres = $\frac{5}{1000}$ litre = 0.005 litres.

Changing from one metric unit to another

Changing from one metric unit to another involves multiplying, or dividing, by a power of 10 (10, 100 or 1000).

EXAMPLES

1 Change 6.3 cm into millimetres.

1 cm = 10 mm

To change centimetres into millimetres, multiply by 10.

$6.3 \times 10 = 63$
6.3 cm = 63 mm

2 Change 245 g into kilograms.

1 kg = 1000 g

To change grams into kilograms, divide by 1000.

$245 \div 1000 = 0.245$
245 g = 0.245 kg

Changing units - areas and volumes

1 m = 100 cm

1 m^2 = 100×100 cm^2 = 10 000 cm^2

1 m^3 = $100 \times 100 \times 100$ cm^3 = 1 000 000 cm^3

How many mm equal 1 cm?

How many mm^2 equal 1 cm^2?

How many mm^3 equal 1 cm^3?

Exercise 31.1

Do not use a calculator.

1. Copy and complete each of the following.
   (a) 320 000 ml = ... l
   (b) 0.32 t = ... kg = ... g
   (c) 320 mm = ... cm = ... m
   (d) 32 000 cm^2 = ... m^2
2. Find the number of kilograms in 9300 g.
3. Find the number of metres in 8000 mm.
4. Find the number of millilitres in 0.03 litres.
5. Which two lengths are the same? 2000 m 20 km 200 m 2 km 0.02 km
6. Which two weights are the same? 8 g 8 kg 8000 g 0.8 kg 80 kg
7. Which length is longest? 0.5 km 50 m 5000 mm 500 cm
8. Which weight is heaviest? 0.3 t 3000 g 3 kg 30 kg
9. Which length is the shortest? 0.02 km 200 mm 2 cm 20 m
10. How many:
    (a) metres are there in 3123 mm,
    (b) centimetres are there in 4.5 m,
    (c) metres are there in 3.24 km,
    (d) grams are there in 1 tonne,
    (e) litres are there in 400 ml?
11. Which area is larger? 0.5 m^2 or 500 cm^2 Give a reason for your answer.
12. Which volume is larger? 0.08 m^3 or 800 000 cm^3 Give a reason for your answer.
13. A can of coke contains 330 ml. How many litres of coke are there in 6 cans?
14. One lap of a running track is 400 m. How many laps are run in an 8 km race?
15. Twenty children at a party share equally 1 kg of fruit pastilles.
    How many grams of pastilles does each child receive?
16. A recipe for a dozen biscuits uses 240 g of flour. James has 1.2 kg of flour.
    How many biscuits can he make?
17. Ben takes two 5 ml doses of medicine four times a day for 5 days.
    Originally, there was $\frac{1}{4}$ of a litre of medicine.
    How much medicine is left?
18. A table mat has an area of 360 cm^2. What area does it cover in square metres?
19. The floor of a corridor is a rectangle, 5.6 m long and 120 cm wide.
    Calculate the area of the floor in: (a) square centimetres, (b) square metres.
20. A box has a volume of 0.6 m^3. What is the volume of the box in cubic centimetres?

Estimating with sensible units using suitable degrees of accuracy

It is a useful skill to be able to estimate length, mass and capacity. These facts might help you.

Length
Most adults are between 1.5 m and 1.8 m tall. The door to your classroom is about 2 m high.
Find some more facts which will help you to estimate length and distance.

Mass
A biro weighs about 5 g. A standard bag of sugar weighs 1 kg.
Find some more facts which will help you to estimate weight or mass.

Capacity
A teaspoon holds about 5 ml. A can of pop holds about 330 ml.
Find some more facts which will help you to estimate volume and capacity.

EXAMPLES

1. The Great Wall of China, the longest man-made structure in the world, is about 2350 km long.
   Degree of accuracy: nearest 50 km.
2. Earthquake shock waves travel through rock at a speed of approximately 25 000 km/hour.
   Degree of accuracy: nearest 1000 km/hour.
3. The smallest mammal is the Kitti's hog-nosed bat. It weighs about 1.5 g.
   Degree of accuracy: nearest 0.1 g.
4. The current Olympic record for the men's 100 m is 9.92 seconds.
   Degree of accuracy: nearest 0.01 seconds.

Exercise 31.2

1. Give the most appropriate metric unit that you would use to measure the following.
   (a) The distance from London to York. (b) The distance across a road.
   (c) The length of your foot. (d) The length of your little finger nail.
   (e) The weight of a bag of potatoes. (f) The weight of an egg.
   (g) The capacity of a bucket. (h) The capacity of a medicine bottle.
2. Which of the following is the best estimate for the mass of a banana?
   1 kg 5 g 250 g 30 g 3 kg 750 g
3. Which of the following is the best estimate for the diameter of a football?
   2 m 50 mm 30 cm 1.5 m 0.6 m 800 mm
4. Which of the following would be the best estimate for the capacity of a mug?
   15 m 1200 ml 2 *l* 0.5 *l* 200 ml 800 ml
5. Give a sensible estimate using an appropriate unit for the following measures:
   (a) the length of a matchstick, (b) the length of a football pitch,
   (c) the weight of a 30 cm ruler, (d) the weight of a double decker bus,
   (e) the volume of drink in a glass.
   In each case state the degree of accuracy you have chosen for your estimate.
6. "My teacher's height is about 1.7 mm."
   This statement is incorrect.

   It can be corrected by changing the unit: "My teacher's height is about 1.7 m."
   It can also be corrected by changing the quantity: "My teacher's height is about 1700 mm."

   Each of these statements is also incorrect.
   "Tyrannosaurus, a large meat-eating dinosaur, is estimated to have been about 12 cm long."
   "The tallest mammal is the giraffe which grows up to about 5.9 mm tall."
   "My car used 5 ml of petrol on a journey of 35 miles."
   "The area of the school hall is about 500 mm^2."

   Correct each statement:
   (a) by changing the unit, (b) by changing the quantity.

Imperial units

The following imperial units of measurement are in everyday use.

Length	Mass	Capacity and volume
1 foot = 12 inches	1 pound = 16 ounces	1 gallon = 8 pints
1 yard = 3 feet	14 pounds = 1 stone	

Metric and imperial conversions

In order to convert to and from metric and imperial units you need to know these facts.

Length	Mass	Capacity and volume
5 miles is about 8 km	1 kg is about 2.2 pounds	1 litre is about 1.75 pints
1 inch is about 2.5 cm		1 gallon is about 4.5 litres
1 foot is about 30 cm		

EXAMPLES

1 Convert 40 cm to inches.

1 inch is about 2.5 cm.
40 cm is about $40 \div 2.5$ inches.
40 cm is about 16 inches.

2 Tim is 6 feet 2 inches tall.
Estimate Tim's height in centimetres.

6 feet 2 inches $= 6 \times 12 + 2 = 74$ inches.
74 inches is about 74×2.5 cm.
6 feet 2 inches is about 185 cm.

Exercise 31.3

1. Convert each quantity to the units given.
   (a) 15 kg to pounds. (b) 20 litres to pints. (c) 5 metres to inches.
   (d) 6 inches to millimetres. (e) 50 cm to inches. (f) 6 feet to centimetres.
2. How far is 32 km in miles?
3. How many pints are there in a 4-litre carton of milk?
4. Kelvin is 5 feet 8 inches tall. Estimate Kelvin's height in centimetres.
5. Paddy weighs 9 stones 12 pounds. Estimate Paddy's weight in kilograms.
6. Estimate the number of:
   (a) metres in 2000 feet, (b) kilometres in 3 miles, (c) feet in 150 centimetres,
   (d) pounds in 1250 grams, (e) litres in 10 gallons.
7. The capacity of a car's petrol tank is 54 litres. How much does the petrol tank hold in gallons?
8. A box contains 200 balls. Each ball weighs 50 g.
   Estimate the total weight of the balls in pounds.
9. Jemma has a 4-litre carton of milk and 20 glasses. Each glass holds one third of a pint.
   Can Jemma fill all the glasses with milk? Show your working.
10. Lauren says 10 kg of potatoes weighs the same as 20 lb of sugar.
    Is she correct? Show all your working.
11. Alfie cycles 6 miles. Jacob cycles 10 kilometres.
    Alfie claims that he has cycled further than Jacob. Is he correct? Show all your working.
12. One litre of water weighs 1 kg. Estimate the weight of one gallon of water in pounds.
13. A sheet of card measures 12 inches by 20 inches.
    What is the area of the card in square centimetres?
14. Convert the following speeds to kilometres per hour.
    (a) 30 miles per hour. (b) 50 miles per hour. (c) 30 metres per second.
15. Convert the following speeds to miles per hour.
    (a) 60 km per hour. (b) 40 metres per second.

16. Convert a speed of 60 miles per hour to metres per second.
Give your answer to a suitable degree of accuracy.

17. (a) A car does 40 miles to the gallon. How many kilometres does it do per litre?
(b) A car does 9.6 kilometres to the litre. How many miles does it do per gallon?

18. 30 g of grass seed is needed to sow 1 m^2 of lawn.
What weight of grass seed is needed to sow a rectangular lawn measuring 40 foot by 30 foot?

19. Concrete is sold by the cubic metre.
A path, 31 feet long, 5 feet wide and 1 foot 6 inches deep, is to be made of concrete.
How many cubic metres of concrete are needed?

Discrete and continuous measures

Discrete measures

Discrete measures can only take particular values.

For example: The number of people on a bus is 42.
Two ice skating judges give scores of 5.1 and 5.2.

> Accuracy in data and measurement, lower and upper bounds (limits) were first covered in Chapter 3.

The number of people on a bus must be a whole number.
The ice skating scores show that a discrete measure does not need to be a whole number.
They are discrete because scores are not given between numbers like 5.1 and 5.2.

Continuous measures

James was 14 on the day of his 14th birthday.
He will still be called 14 years old right up to the day before his 15th birthday.
So, although James is 14, his actual age is any age within a range of 1 year.

I am 14 years and 3 months.

Jenny is **not** exactly 14 years and 3 months old.
However, Jenny's age is given to a greater degree of accuracy than James' age because the range of possible ages in her case is smaller.
What is the range of possible ages in Jenny's measurement of her age?

Measures which can lie within a range of possible values are called **continuous measures**.
The value of a continuous measure depends on the accuracy of whatever is making the measurement.

> If a **continuous measure**, c, is recorded to the nearest x, then:
> the **limits** of the possible values of c can be written as: $c \pm \frac{1}{2}x$.

EXAMPLES

1. The number of words on a page of a book is 400 to the nearest 100.
What is the smallest and largest possible number of words on the page?

The smallest whole number that rounds to 400 is 350.
The largest whole number that rounds to 400 is 449.
The smallest number of words on the page is 350 and the largest number of words is 449.

2. The length of this page is 26 cm to the nearest centimetre.
What are the limits between which the true length of the page must lie?

Length of page = 2 cm ± 0.5 cm $\quad$ 25.5 cm $\leqslant$ length of page $<$ 26.5 cm.

Exercise 31.4

1 State whether each of the following are discrete or continuous measures.
(a) The volume of wine in a wine glass.
(b) The votes cast for the Independent Party candidate in a local election.
(c) The number of pages in a newspaper.
(d) The time it takes to walk to school.

2 For each of the following measures state whether the value given is an exact value or give the limits within which it could lie.
(a) A book has 224 pages. (b) A bus was 5 minutes late. (c) I weigh 63 kg.

3 What are the slowest and fastest possible times:
(a) of Derek who runs exactly 100 metres in 13 seconds measured to the nearest second,
(b) of Jan who swims exactly 100 metres in 82.6 seconds measured to the nearest tenth of a second?

4 Fifty flowers, to the nearest 10, each grow to a height of 50 cm, to the nearest 10 cm.

45 50 55

(a) Show on one copy of the number line the possible numbers of flowers.
(b) Show on another copy of the number line the possible heights of the flowers.

5 Tina is 1.53 m tall.
(a) To what degree of accuracy is Tina's height given?
(b) What are the limits between which her true height lies?

6 (a) What is the minimum weight of a 62 kg parcel measured to the nearest kilogram?
(b) What is the minimum length of a 2.3 m shelf measured to the nearest 0.1 m?
(c) What is the minimum time for a race timed at 12.63 seconds measured to the nearest one hundredth of a second?

7 What degree of accuracy has been used to make these estimates of the length of a corridor:
(a) About 120 m. The smallest possible length is 115 m.
(b) About 100 m. The smallest possible length is 75 m.
(c) About 200 m. The smallest possible length is 150 m.

8 Write down the largest and smallest values of the following discrete measures.
(a) 10 000 measured to the nearest 1000.
(b) 10 000 measured to the nearest 100.
(c) 10 000 measured to the nearest 50.
(d) 500 measured to 3 significant figures.
(e) 500 measured to 2 significant figures.

Calculations involving continuous measures

EXAMPLE

Gail's best time for running 100 metres is 12.7 seconds.
The length is accurate to the nearest metre. The time is accurate to the nearest tenth of a second.
Calculate her maximum possible average speed.

Time
Upper bound = 12.7 s + 0.05 s = 12.75 s
Lower bound = 12.7 s − 0.05 s = 12.65 s

Distance
Upper bound = 100 m + 0.5 m = 100.5 m
Lower bound = 100 m − 0.5 m = 99.5 m

$$\text{Gail's maximum possible average speed} = \frac{\text{greatest distance}}{\text{shortest time}} = \frac{100.5 \text{ metres}}{12.65 \text{ seconds}} = 7.944\ldots \text{ m/s}$$

Gail's maximum possible average speed = 7.94 m/s, correct to 3 significant figures.
What bounds should be used to calculate Gail's minimum possible average speed?

Exercise 31.5

1 In the diagram, all angles are measured to the nearest degree.
Find the maximum possible size of angle a.

2 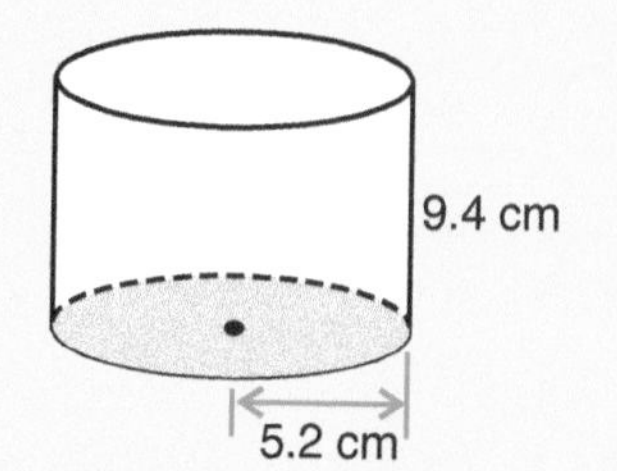

A cylinder has radius 5.2 cm and height 9.4 cm.
Both measurements are given to the nearest 0.1 cm.
Find the difference between the minimum and maximum possible volumes of the cylinder.

3 A water tank, in the shape of a cuboid, is full of water.
Water is drained from the tank at a rate of 8 litres per minute.
The dimensions of the tank are given to the nearest 10 cm.
The rate at which the water is drained from the tank is given to the nearest 0.5 litres per minute.

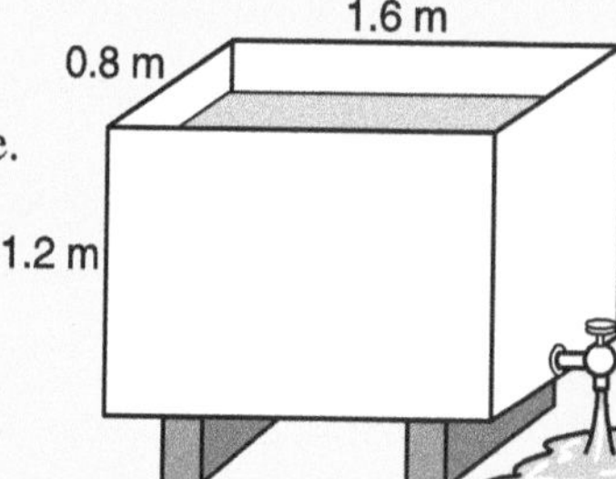

(a) Calculate the smallest possible time to drain the tank.
(b) Calculate the greatest possible time to drain the tank.

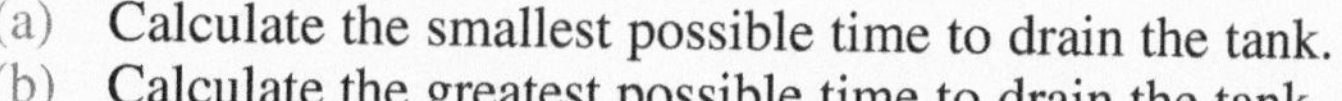

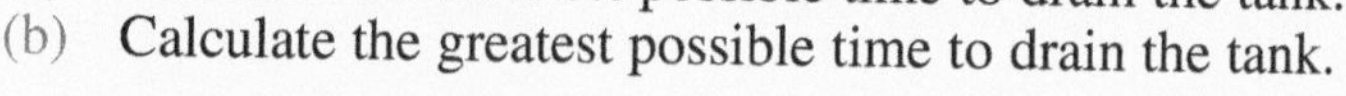

4 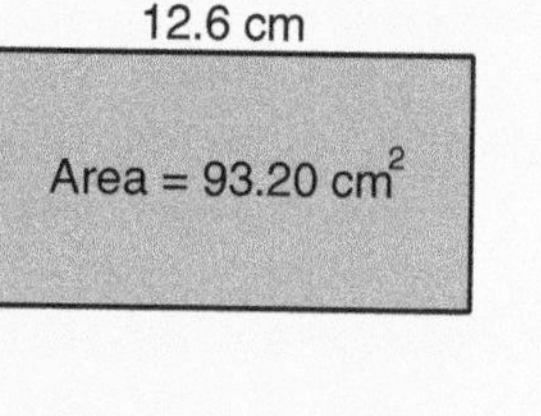

A rectangle has length 12.6 cm and area 93.20 cm^2.
The length is given to the nearest millimetre.
The area is given to the nearest 0.05 cm^2.
Calculate the upper bound of the width of the rectangle.

5 The diameter of a concrete sphere is given as 20 cm, to the nearest centimetre.
The density of concrete is 2.2 g/cm^3.
Calculate the minimum mass of the sphere.

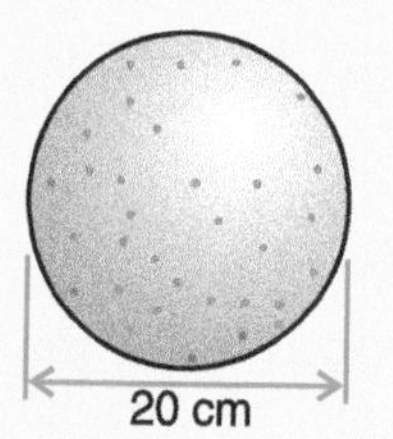

6 The distance between A and B is 12.2 km to the nearest 100 m.
The distance between B and C is 14.34 km to the nearest 10 m.
Trevor walks from A to B and then from B to C.
He takes 8.5 hours to the nearest half an hour.

(a) Calculate the upper and lower bounds of the total distance that Trevor walks.
(b) Calculate the lower bound of Trevor's average speed on the walk.

7 The diagram shows a rectangular flag, $ABCD$.
The flag is blue with a central white circle.
$AB = 3.2$ m and $BC = 2.1$ m,
both accurate to the nearest 10 cm.
The diameter of the circle is 0.95 m to the nearest centimetre.
Calculate the lower bound of the area coloured blue on the flag.

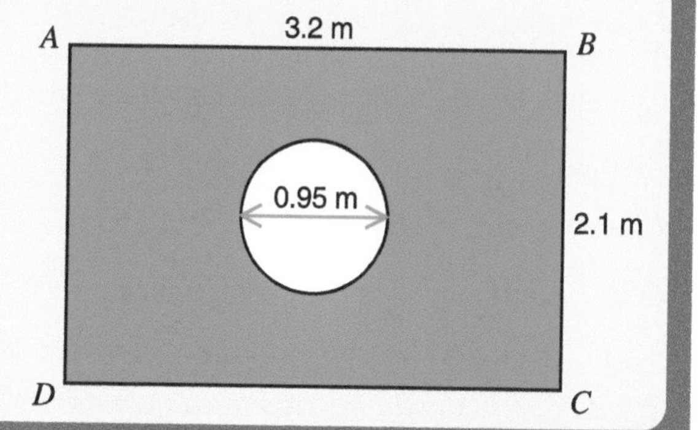

Dimensions and formulae

Formulae can be used to calculate perimeters, areas and volumes of various shapes.
By analysing the **dimensions** involved it is possible to decide whether a given formula represents a perimeter, an area or a volume.

Length (L) has **dimension 1.**
Length (L) × Length (L) = **Area** (L^2) has **dimension 2.**
Length (L) × Length (L) × Length (L) = **Volume** (L^3) has **dimension 3.**

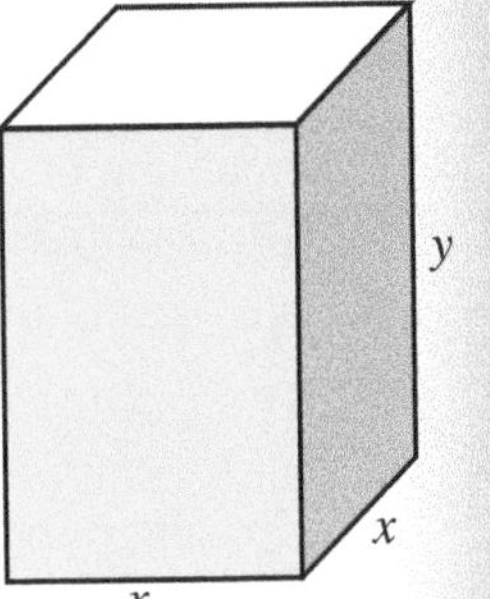

The size of this square based cuboid depends on:
x, the length of the side of the square base, y, the height of the cuboid.

The total **length** of the edges of the cuboid is given by the formula: $E = 8x + 4y$
This formula involves: Numbers: 8 and 4
Lengths (L): x and y
The formula has **dimension 1**.

The total **surface area** of the cuboid is given by the formula: $S = 2x^2 + 4xy$
This formula involves: Numbers: 2 and 4
Areas (L^2): $x \times x$ and $x \times y$
The formula has **dimension 2**.

The **volume** of the cuboid is given by the formula: $V = x^2y$
This formula involves: Volume (L^3): $x \times x \times y$
This formula has **dimension 3**.

EXAMPLE

In each of these expressions the letters a, b and c represent lengths.
Use dimensions to check whether the expressions could represent a perimeter, an area or a volume.

(a) $2a + 3b + 4c$ (b) $3a^2 + 2b(a + c)$
(c) $2a^2b + abc$ (d) $3a + 2ab + c^3$

Note:
When checking formulae and expressions, numbers can be ignored because they have no dimension.
≡ means 'is equivalent to'.

(a) $2a + 3b + 4c$
Write this using dimensions.
$L + L + L \equiv 3L \equiv L$
$2a + 3b + 4c$ has dimension 1 and could represent a perimeter.

(b) $3a^2 + 2b(a + c)$
Write this using dimensions.
$L^2 + L(L + L)$
$\equiv L^2 + L(2L)$
$\equiv L^2 + 2L^2$
$\equiv 3L^2$
$\equiv L^2$
$3a^2 + 2b(a + c)$ has dimension 2 and could represent an area.

(c) $2a^2b + abc$
Write this using dimensions.
$L^2 \times L + L \times L \times L$
$\equiv L^3 + L^3$
$\equiv 2L^3$
$\equiv L^3$
$2a^2b + abc$ has dimension 3 and could represent a volume.

(d) $3a + 2ab + c^3$
Write this using dimensions.
$L + L \times L + L^3$
$\equiv L + L^2 + L^3$
The dimensions are **inconsistent**.
$3a + 2ab + c^3$ does not represent a perimeter, an area or a volume.

Exercise 31.6

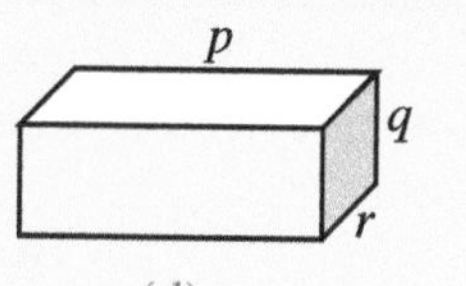

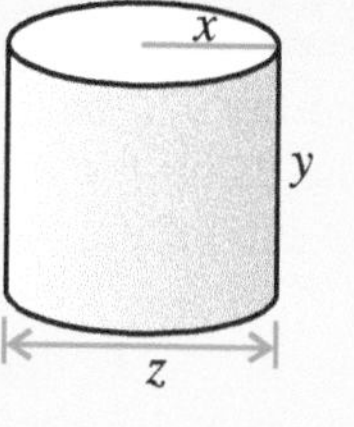

1 p, q, r and x, y, z represent lengths.
For each formula state whether it represents a length, an area or a volume.

(a) pq (b) $2\pi x$ (c) $p + q + r$ (d) πz
(e) pqr (f) $2(pq + qr + pr)$ (g) πx^2y (h) $2\pi x(x + y)$

2 In each of the expressions below, x, y and z represent lengths.
By using dimensions decide whether each expression could represent a perimeter, an area, a volume or none of these. Explain your answer in each case.

(a) $x + y + z$ (b) $xy + xz$ (c) xyz (d) $x^2(y^2 + z^2)$
(e) $x(y + z)$ (f) $\frac{x^2}{y}$ (g) $\frac{xz}{y}$ (h) $x + y^2 + z^3$
(i) $xy(y + z)$ (j) $x^3 + x^2(y + z)$ (k) $xy(y^2 + z)$ (l) $x(y + z) + z^2$

3 The diagram shows a discus. x and y are the lengths shown on the diagram.
These expressions could represent certain quantities relating to the discus.

$\pi(x^2 + y^2)$ πx^2y^2 πxy $2\pi(x + y)$

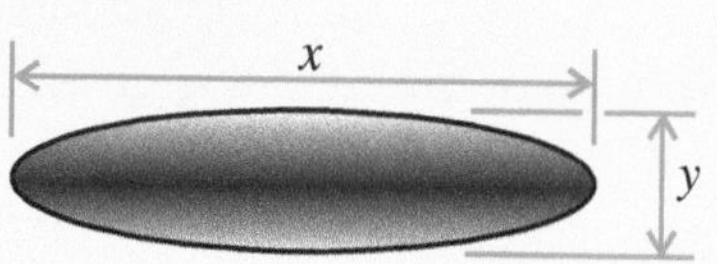

(a) Which of them could be an expression for:
  (i) the longest possible distance around the discus,
  (ii) the surface area of the discus?
(b) Use dimensions to explain your answers to part (a).

4 x, y and z represent lengths.

(a) $A = xyz + z(x - y) + 2y$
This is not a formula for perimeter, area or volume. Use dimensions to explain why.
(b) $P = 3z(x + y)$
This could be a formula for area. Use dimensions to explain why.
(c) $V = x^2y + z^2(2x - y) + 2y^3$
This could be a formula for volume. Use dimensions to explain why.

5 p, q, r and s represent the lengths of the edges of this triangular prism.
Match the formulas to the measurements.

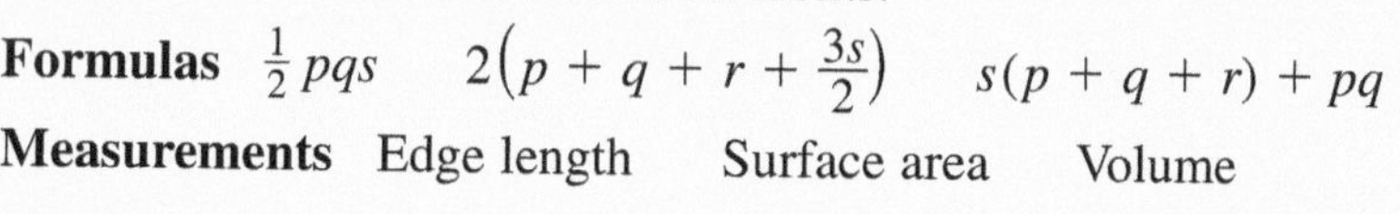

Formulas $\frac{1}{2}pqs$ $2\left(p + q + r + \frac{3s}{2}\right)$ $s(p + q + r) + pq$

Measurements Edge length Surface area Volume

6 These arrows are similar.

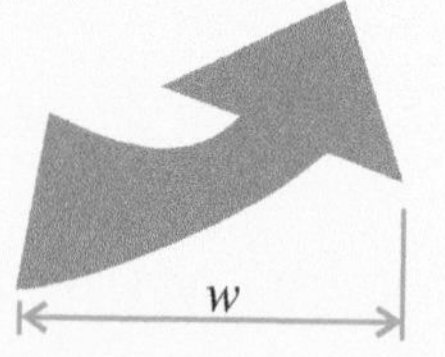

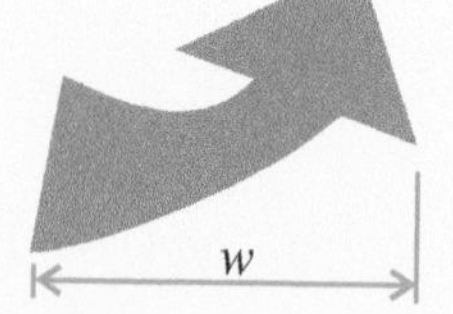

w represents the width of any arrow.
k and c are numbers.
H represents the height of the arrow and A its area.
Which of the following statements could be correct and which **must** be wrong?

(a) $H = \text{k}w$ (b) $H = \text{ck}w$ (c) $H = \text{k}w + \text{c}$
(d) $A = \text{c}w$ (e) $A = \text{k}w^2$ (f) $A = \text{k}w^3$

Give a reason for each of your answers and where you think the formula **must** be wrong suggest what it might be for.

7 In these formulae a, b and c represent lengths and A represents an area.

(a) $a = b + c$ (b) $a^2 = bc$ (c) $A = a^2 + bc$
(d) $c = A + ab$ (e) $Ab = a^3$ (f) $A = \frac{a^2}{c} + b^2$

Which of the formulae have consistent dimensions?

31

What you need to know

- The common units – both **metric** and **imperial** – used to measure **length, mass** and **capacity**.
- How to convert from one unit to another. This includes knowing the connection between one metric unit and another and the approximate equivalents between metric and imperial units.

Metric Units

Length
1 kilometre (km) = 1000 metres (m)
1 m = 100 centimetres (cm)
1 cm = 10 millimetres (mm)

Mass
1 tonne (t) = 1000 kilograms (kg)
1 kg = 1000 grams (g)

Capacity and volume
1 litre = 1000 millilitres (ml)
1 cm³ = 1 ml

Imperial Units

Length
1 foot = 12 inches
1 yard = 3 feet

Mass
1 pound = 16 ounces
14 pounds = 1 stone

Capacity and volume
1 gallon = 8 pints

Conversions

Length
5 miles is about 8 km
1 inch is about 2.5 cm
1 foot is about 30 cm

Mass
1 kg is about 2.2 pounds

Capacity and volume
1 litre is about 1.75 pints
1 gallon is about 4.5 litres

- How to change between units of area. For example $1\,m^2 = 10\,000\,cm^2$.
- How to change between units of volume. For example $1\,m^3 = 1\,000\,000\,cm^3$.
- Be able to recognise limitations on the accuracy of measurements.
  A **discrete measure** can only take a particular value and a **continuous measure** lies within a range of possible values which depends upon the degree of accuracy of the measurement.

> If a **continuous measure**, c, is recorded to the nearest x, then:
> the **limits** of the possible values of c can be written as: $c \pm \frac{1}{2}x$.

- By analysing the **dimensions** of a formula it is possible to decide whether a given formula represents a **length** (dimension 1), an **area** (dimension 2) or a **volume** (dimension 3).

Review Exercise 31

1 (a) How many centimetres are there in 39 millimetres?
(b) The distance from London to Edinburgh is 400 miles.
How many kilometres is this? AQA

2 Ben is 5 feet 10 inches tall and weighs 72 kg.
Sam is 165 cm tall and weighs 11 stone 7 pounds.
Who is taller? Who is heavier?

3 The diagram shows the petrol gauge on a car.
The car's petrol tank holds 60 litres when full.
Estimate how many gallons are in the petrol tank.

0 ¼ ½ ¾ Full

4 Change $1.5\,m^2$ to cm^2.

5 Dilip has a space in his living room which is $2\frac{1}{2}$ feet wide.
He has a bookcase which is 80 cm wide.
Will the bookcase fit into the space?
You **must** show your working. AQA

6 A rectangular strip of card measures 0.8 metres by 5 cm.
Calculate the area of the card in:
(a) square metres,
(b) square centimetres.

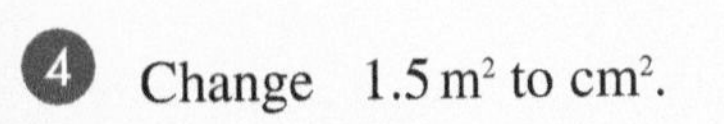

7 Convert a speed of 30 miles per hour into feet per second. 1 mile = 5280 feet. AQA

8 James runs at an average speed of 12 miles/hour.
Tim runs at an average speed of 320 m/minute.
Who runs the fastest?

9 According to the instructions Vinyl Matt paint covers about 13 m²/litre.
(a) Estimate the area covered with 2 gallons of paint?
(b) Estimate how many litres of paint is needed to cover a wall which measures 18 feet by 10 feet.

10 A blue whale weighs 140 tonnes to the nearest 10 tonnes.
What is the smallest possible weight of a blue whale?

11 1 pint = 568 ml 8 pints = 1 gallon
Sally uses the approximate rule:

1 gallon = 4.5 litres

Show that this rule is correct to within 1%. AQA

12 One of these formulae gives the volume of an egg of height H cm and width w cm.

Formula **A**: $\frac{1}{6}\pi H^2 w^2$ Formula **B**: $\frac{1}{6}\pi H w^2$ Formula **C**: $\frac{1}{6}\pi H w$

w cm
H cm

(a) Which is the correct formula?
(b) Explain how you can tell that this is the correct formula. AQA

13 In the following expressions, r, a and b represent lengths.
For each expression, state whether it represents a **length**, an **area**, a **volume**, or **none** of these.
(a) πab (b) $\pi r^2 a + 2\pi r$ (c) $\frac{\pi r a^3}{b}$ AQA

14 Which of these measurements is discrete and which is continuous?
(a) Two pints of milk. (b) Two 1 pint bottles of milk.
Explain your answer.

15 Jean gives her height as 166 cm, to the nearest centimetre.
What are the limits between which her true height lies?

16 Bob weighs 76.9 kg, to the nearest 100 g. What is Bob's minimum weight?

17 What is the greatest possible difference between the times of two runners who both run a race in 12.2 seconds timed to the nearest tenth of a second?

18 A shot putt weighs 3.25 kg, to the nearest 10 g.
What are the upper and lower bounds for the actual weight of the shot putt?
Give your answers in grams. AQA

19 A runner completes a 100 m sprint in 11.61 seconds.
The time is recorded to the nearest one hundredth of a second.
What are the upper and lower bounds of the possible times? AQA

20 A company makes rectangular sheets of tinplate for use in cans.

DIMENSIONS OF RECTANGULAR SHEETS

Thickness	*Length*	*Width*
0.15 mm	**830 mm**	**635 mm**

The length and width are given to the nearest mm, and the thickness to the nearest 0.01 mm.

Calculate the percentage saving in volume to the company if it produces the sheets to minimum dimensions rather than maximum dimensions. AQA

Congruent Triangles and Similar Figures

Congruent shapes

When two shapes are the same shape and size they are said to be **congruent**.
A copy of one shape would fit exactly over the second shape.
Sometimes it is necessary to turn the copy over to get an exact fit.
These shapes are all congruent.

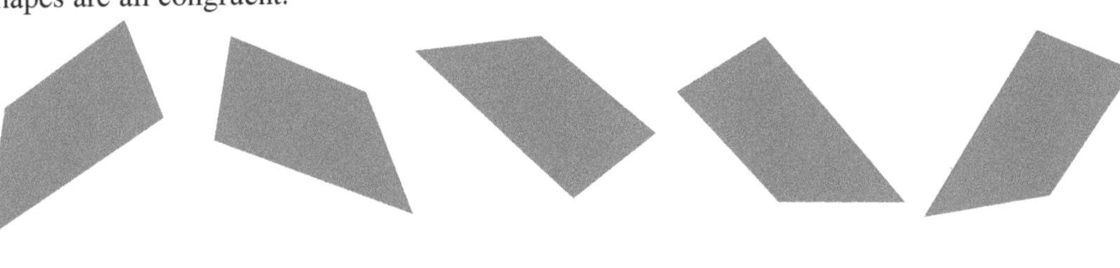

Congruent triangles

There are four ways to show that a pair of triangles are congruent.

1. Three sides.
SSS

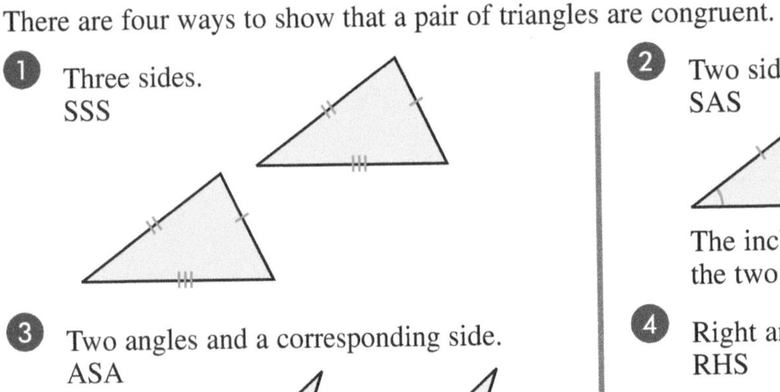

2. Two sides and the included angle.
SAS

The included angle is the angle between the two sides.

3. Two angles and a corresponding side.
ASA

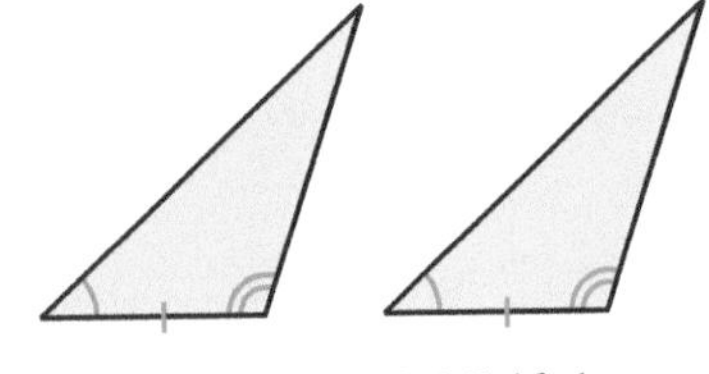

This can be written as AAS if the corresponding side is not between the angles.

4. Right angle, hypotenuse and one side.
RHS

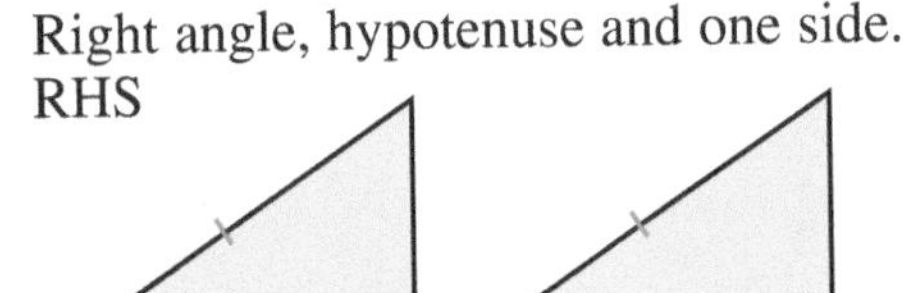

The hypotenuse is the side opposite the right angle and is the longest side in a right-angled triangle.

To show that two triangles are congruent you will need to state which pairs of sides and/or angles are equal, to match one of the four conditions for congruency given above.

EXAMPLE

Show that triangles ABC and PQR are congruent.

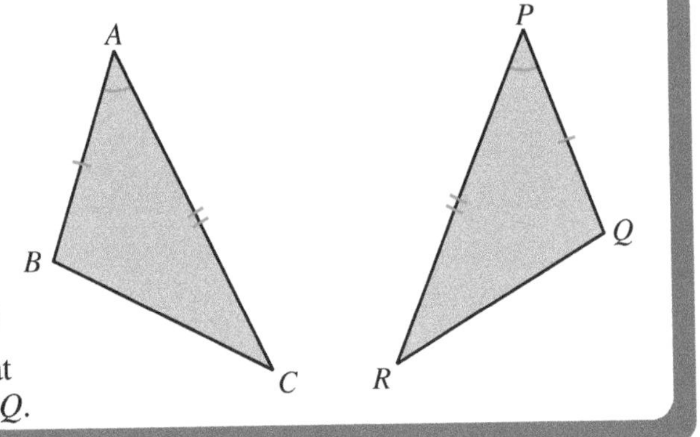

$AB = PQ$ (equal lengths, given)
$AC = PR$ (equal lengths, given)
$\angle BAC = \angle QPR$ (equal angles, given)

So, triangles ABC and PQR are congruent.
Reason: SAS (Two sides and the included angle.)

Since the triangles are congruent we also know that
$BC = QR$, $\angle ABC = \angle PQR$ and $\angle ACB = \angle PRQ$.

Exercise 32.1

1. Which of these shapes are congruent to each other?

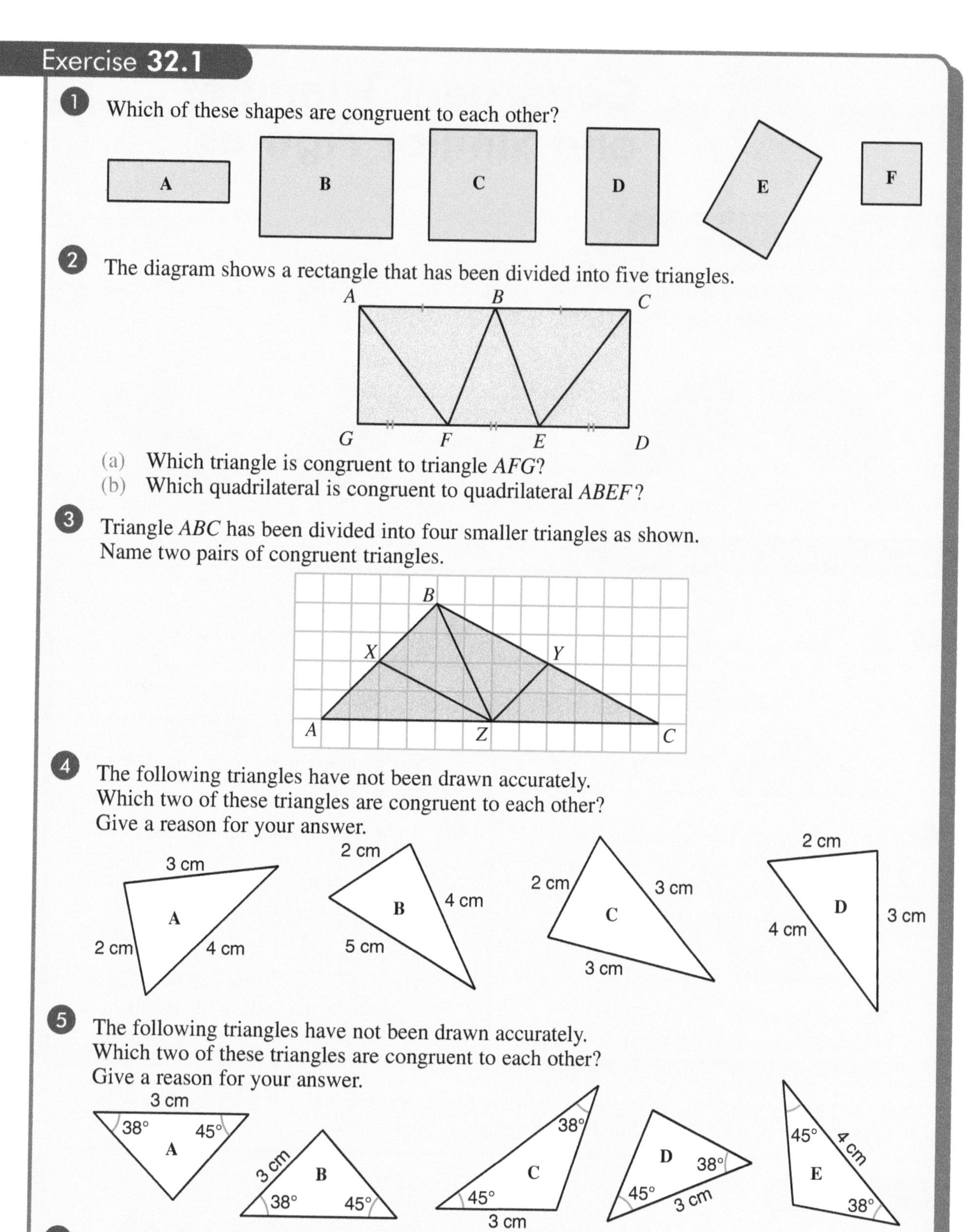

2. The diagram shows a rectangle that has been divided into five triangles.

   (a) Which triangle is congruent to triangle AFG?

   (b) Which quadrilateral is congruent to quadrilateral $ABEF$?

3. Triangle ABC has been divided into four smaller triangles as shown.
   Name two pairs of congruent triangles.

4. The following triangles have not been drawn accurately.
   Which two of these triangles are congruent to each other?
   Give a reason for your answer.

5. The following triangles have not been drawn accurately.
   Which two of these triangles are congruent to each other?
   Give a reason for your answer.

6. The following triangles have not been drawn accurately.
   State whether each pair of triangles is congruent or not.
   Where triangles are congruent give the reason.

   (a) (b) (c)

7 The following triangles have not been drawn accurately.
State whether each pair of triangles is congruent or not.
Where triangles are congruent give the reason.

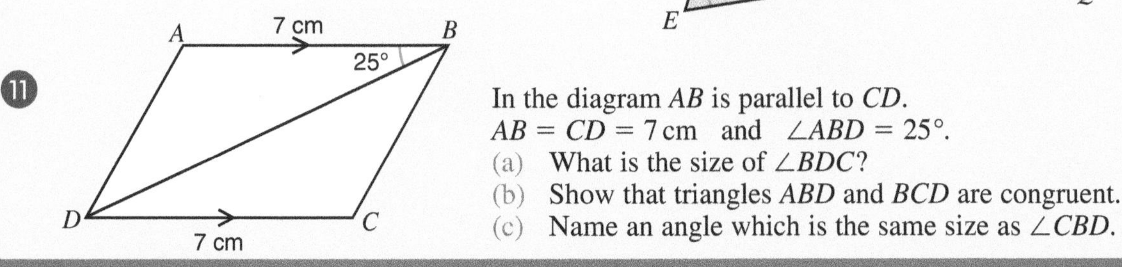

(d) 7, 52°, 5 — 5, 52°, 7

(e) 70°, 60°, 50° — 70°, 60°, 50°

(f) 3, 62°, 54° — 62°, 54°, 3

8 For each of the following, is it possible to draw a congruent triangle without taking any other measurements from the original triangle?
If a triangle can be drawn give the reason for congruence which applies.

(a) 30°, 115°, 35°

(b) 5 cm, 9 cm

(c) 6 cm, 3 cm

(d) 105°, 37°

(e) 85°, 4 cm, 55°

(f) 35°, 45°, 3 cm

(g) 4 cm, 3 cm, 6 cm

(h) 70°, 5 cm, 4 cm

(i) 27°, 3 cm

9 Triangle ABC has angles 90°, 50° and 40°.
Triangle XYZ also has angles 90°, 50° and 40°.
The triangles are not congruent.
Can you explain why?

10 Show that triangles DEF and PQR are congruent.

D, E, F, 65°, 45° — P, Q, R, 70°, 65°

11

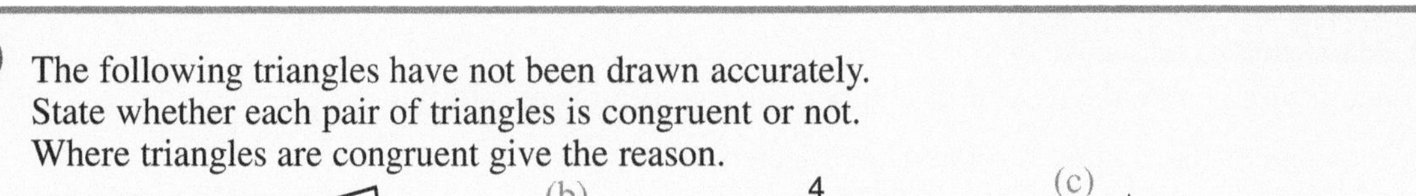

In the diagram AB is parallel to CD.
$AB = CD = 7\text{ cm}$ and $\angle ABD = 25°$.
(a) What is the size of $\angle BDC$?
(b) Show that triangles ABD and BCD are congruent.
(c) Name an angle which is the same size as $\angle CBD$.

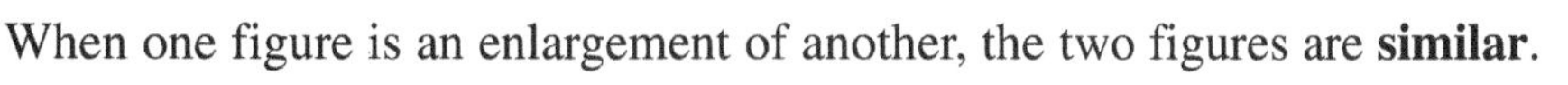

Similar figures

When one figure is an enlargement of another, the two figures are **similar**.

Sometimes one of the figures is rotated or reflected.

For example:
Figures **C** and **E** are enlargements of figure **A**.
Figures **A**, **C** and **E** are similar.

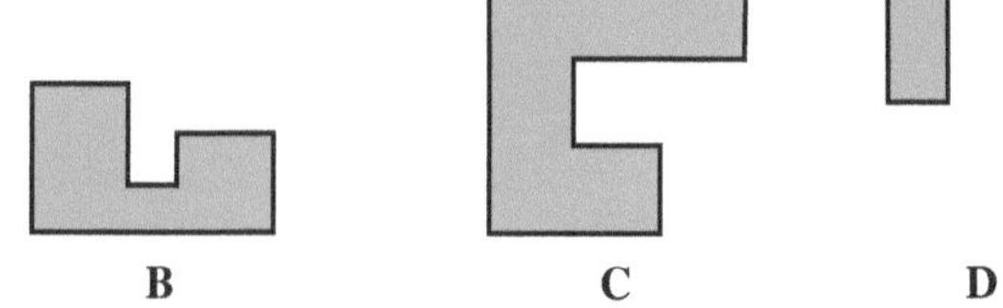

E

When two figures are **similar**:
their **shapes** are the same,
their **angles** are the same,
corresponding **lengths** are in the same ratio,
this ratio is the **scale factor** of the enlargement.

Scale factor $= \dfrac{\text{new length}}{\text{original length}}$
This can be rearranged to give
new length = original length × scale factor.

EXAMPLES

1 A photo has width 6 cm and height 9 cm.
An enlargement is made, which has width 8 cm.
Calculate the height of the enlargement.

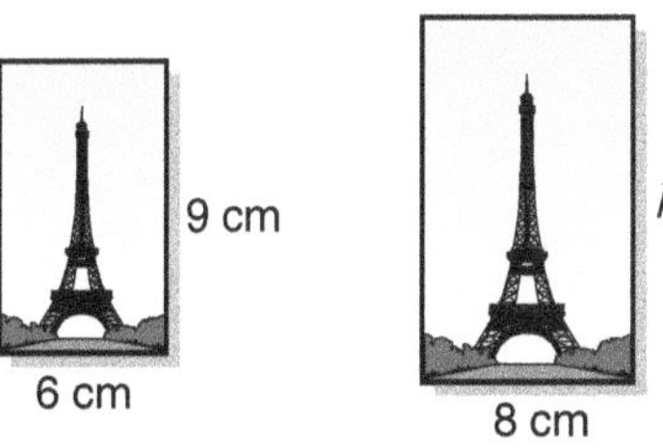

Scale factor $= \frac{8}{6}$
$h = 9 \times \frac{8}{6}$
$h = 12\,\text{cm}$

2 These two figures are similar.
Calculate the lengths of x and y.
Write down the size of the angle marked a.

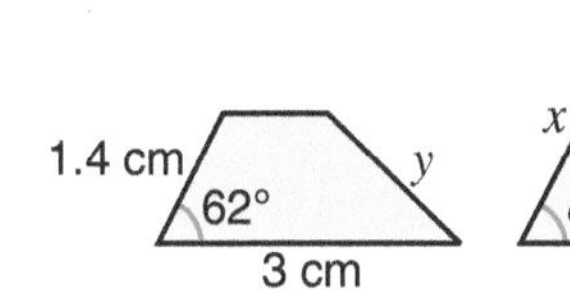

The scale factor $= \frac{4.5}{3} = 1.5$

Lengths in the large figure are given by: length in small figure × scale factor
$x = 1.4 \times 1.5$
$x = 2.1\,\text{cm}$

Lengths in the small figure are given by: length in large figure ÷ scale factor
$y = 2.7 \div 1.5$
$y = 1.8\,\text{cm}$

The angles in similar figures are the same, so, $a = 62°$.

Exercise 32.2

1 The shapes in this question have been drawn accurately.
Explain why these two shapes are not similar to each other.

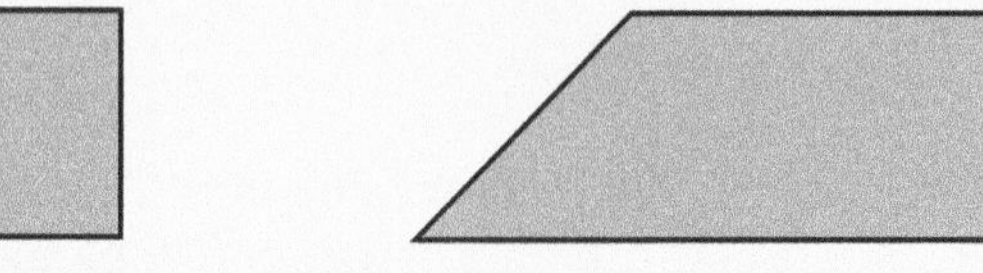

2 Which of the following must be similar to each other?
(a) Two circles. (b) Two kites. (c) Two parallelograms.
(d) Two squares. (e) Two rectangles.

3 These rectangles are all similar. The diagrams have not been drawn accurately.
Work out the lengths of the sides marked a and b.

8 cm, a — 16 cm, 8 cm — b, 12 cm

4 These two kites are similar.
(a) What is the scale factor of their lengths?
(b) Find the length of the side marked x.
(c) What is the size of angle a?

2 cm, 1.2 cm, a — 3 cm, x, 120°

5 A shape has width 8 cm and length 24 cm.
It is enlarged to give a new shape with width 10 cm.
Calculate the length of the new shape.

6 In each part, the two figures are similar. Lengths are in centimetres.
Calculate the lengths and angles marked with letters.

(a) y, 70°, 1, 2 — 3.6, a, x, 3

(b) y, a, 4.5, 3 — 2.5, 53°, 7.5, x

(c) 12, 21, 18, x — z, y, 15, 25

7 These two tubes are similar.
The width of the small size is 2.4 cm and the height of the small size is 10 cm.
The width of the large size is 3.6 cm.
Calculate the height of the large size.

10 cm, SMALL, 2.4 cm — ?, LARGE, 3.6 cm

8 A motor car is 4.2 m long and 1.4 m high.
A scale model of the car is 8.4 cm long.
What is the height of the model?

9 The smallest angle in triangle T is 18°.
Triangle T is enlarged by a scale factor of 2.
How big is the smallest angle in the enlarged triangle?

10 A castle has height 30 m.
The height of the castle wall is 6 m.
A scale model of the castle has height 25 cm.
Calculate the height of the castle wall in the scale model.

11 The dimensions of three sizes of paper are given.

Length (cm)	24	30	y
Width (cm)	x	20	32

All the sizes are similar.
Calculate the values of x and y.

Similar triangles

For any pair of similar triangles:
corresponding lengths are opposite equal angles,
the scale factor is the ratio of corresponding sides.

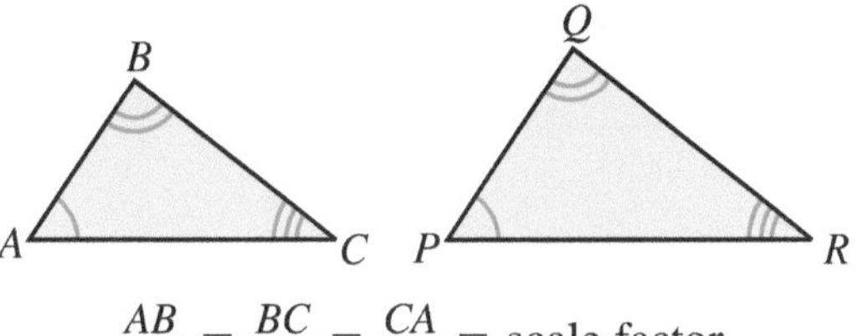

$$\frac{AB}{PQ} = \frac{BC}{QR} = \frac{CA}{RP} = \text{scale factor}$$

EXAMPLES

1 These two triangles are similar, with the equal angles marked.
Calculate the lengths x and y.

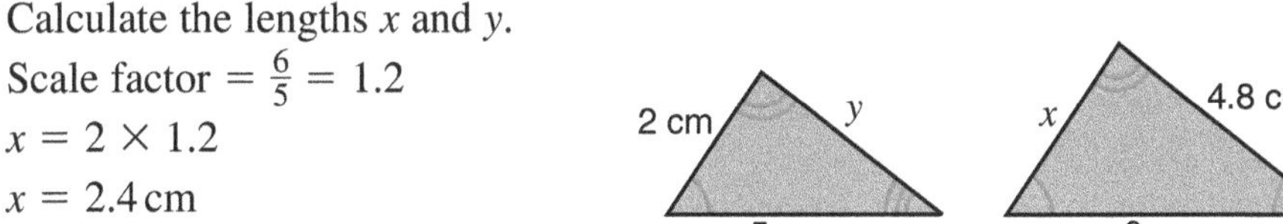
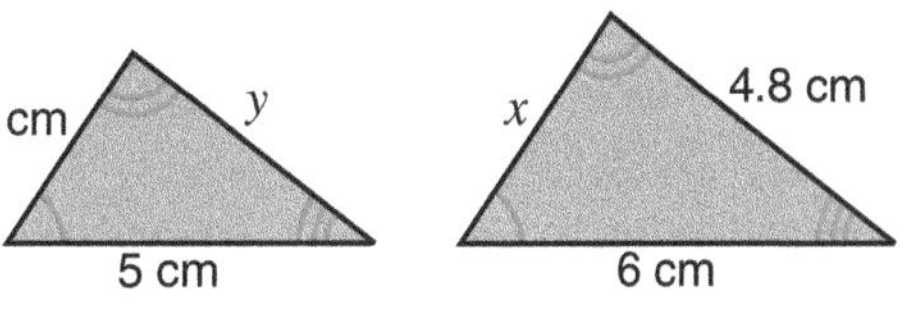

Scale factor $= \frac{6}{5} = 1.2$
$x = 2 \times 1.2$
$x = 2.4\,\text{cm}$
$y = 4.8 \div 1.2$
$y = 4\,\text{cm}$

2 Triangles ABC and PQR are similar.
Calculate the lengths of AC and PQ.

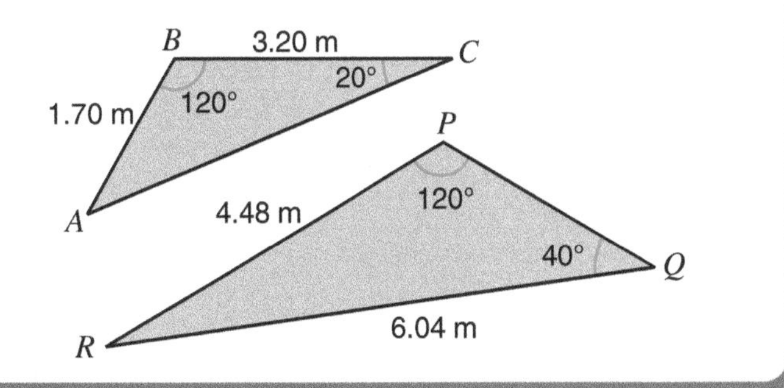

BC and PR are corresponding sides.
Scale factor $= \frac{4.48}{3.20} = 1.4$
$AC = 6.04 \div 1.4$
$AC = 4.31\,\text{m}$, correct to 2 d.p.
$PQ = 1.70 \times 1.4$
$PQ = 2.38\,\text{m}$

Exercise 32.3

Question 1 should be done without a calculator.

1 In each part, the triangles are similar, with equal angles marked.
Lengths are in centimetres.
Calculate lengths x and y.

(a)

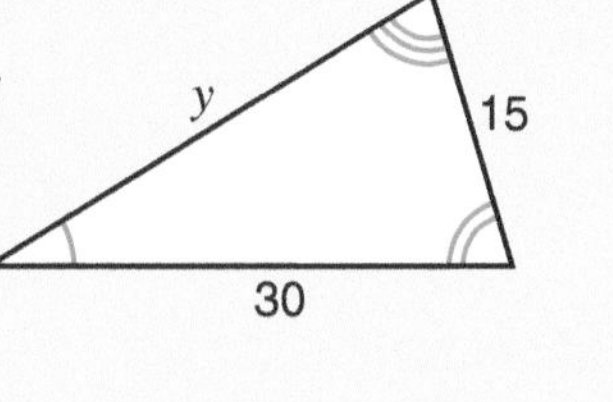

(b)

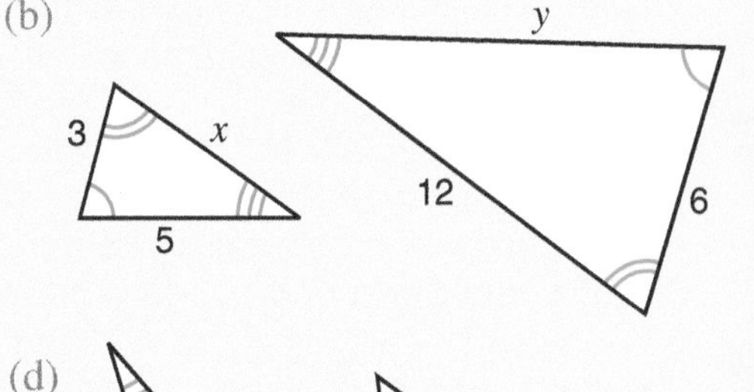

(c)

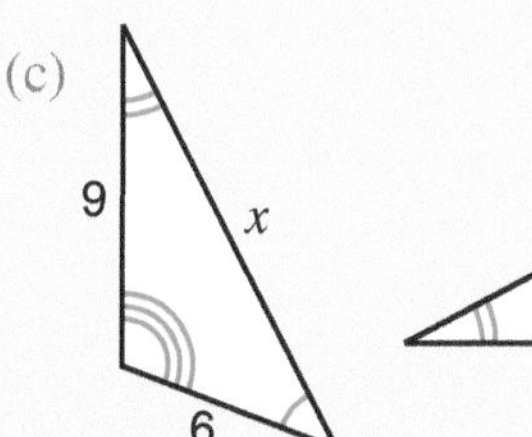

(d)

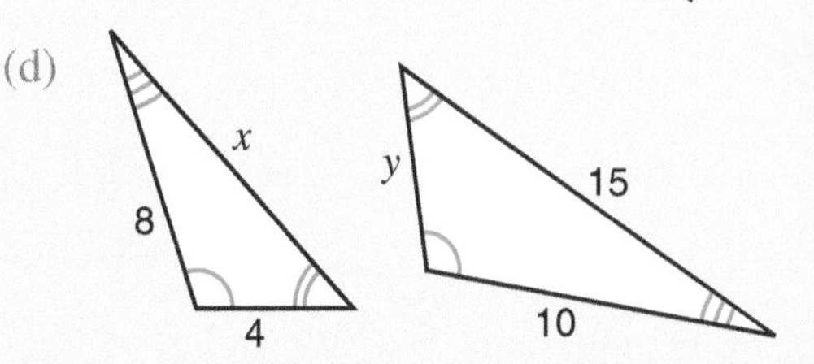

2 Triangles ABC and ADE are similar.
Lengths are in centimetres.
$\angle AED = \angle ACB$.

Calculate the lengths of AB and AE.

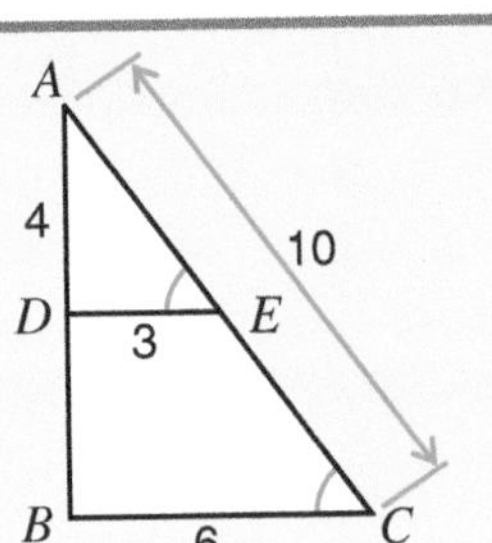

3 In each part the triangles are similar.
Calculate the unknown lengths in both triangles.

(a) (b)

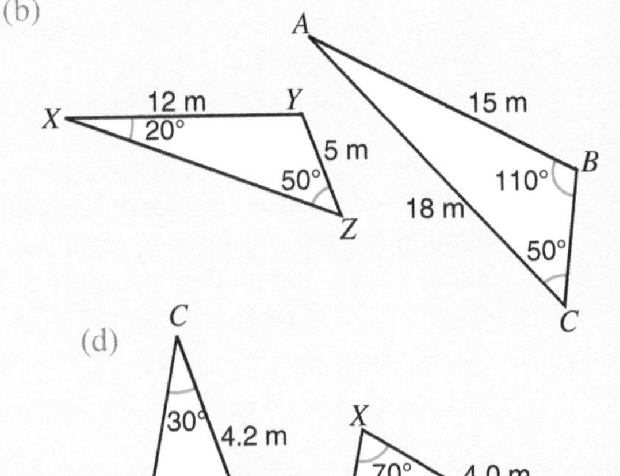

(c)

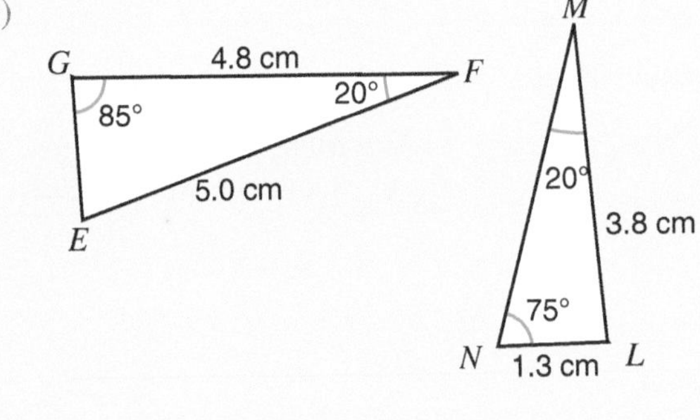

(d)

4 Triangles PST and PQR are similar.
Lengths are in centimetres.
$\angle PTS = \angle PRQ$.

(a) Write down the length of PR.
(b) Calculate QR, PQ and QS.

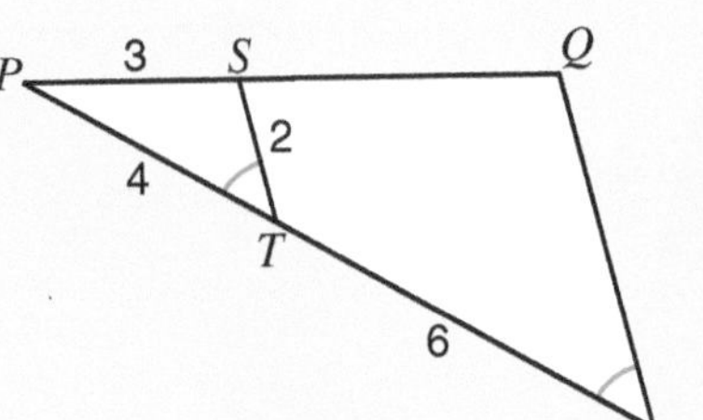
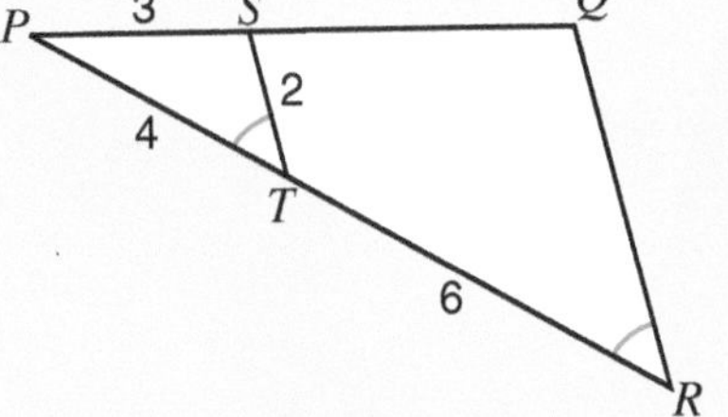

5 Triangles ABC and ADE are similar.
$\angle ADE = \angle ABC$.

(a) Write down the length of AB.
(b) Calculate the lengths of BC and AC.
(c) What is the length of EC?

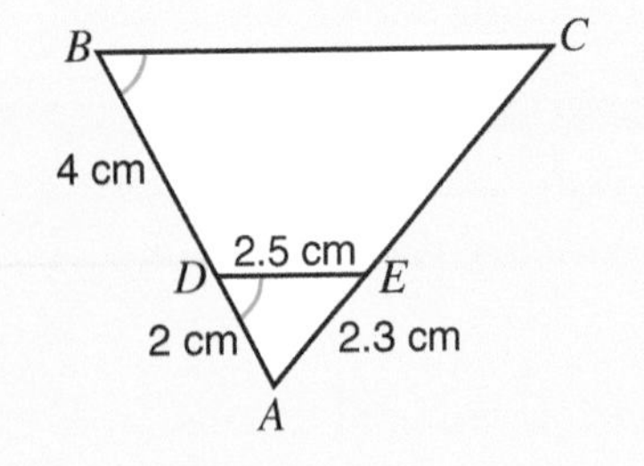

6 Triangles DEF and DGH are similar.
$\angle DGH = \angle DEF$.

Calculate GH and FH.

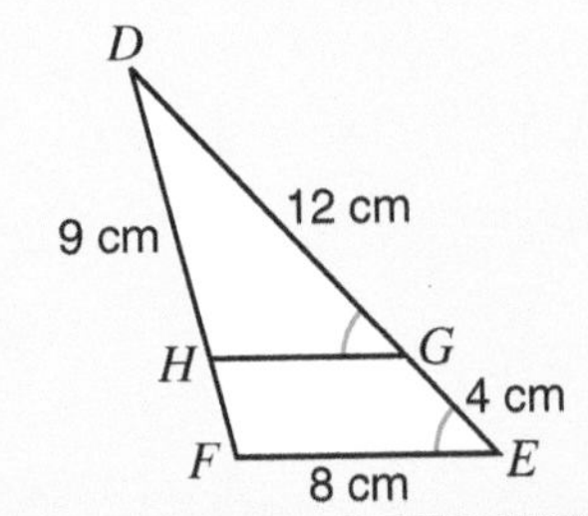

Showing that two triangles are similar

To show that two triangles are similar you have to show that:

either they have equal angles **or** corresponding lengths are all in the same ratio.

If you can show that one of these conditions is true then the other one is also true.

EXAMPLES

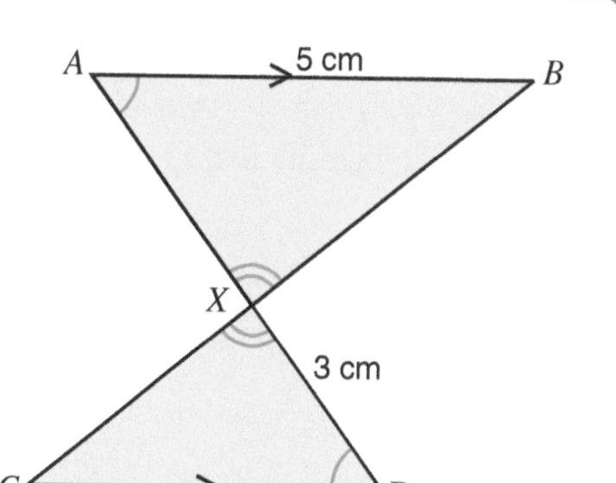

1 AB and CD are parallel lines. AD and BC meet at X.

(a) Prove that triangles ABX and DCX are similar.

(b) Which side in triangle DCX corresponds to AX in triangle ABX?

(c) Calculate the length of AX.

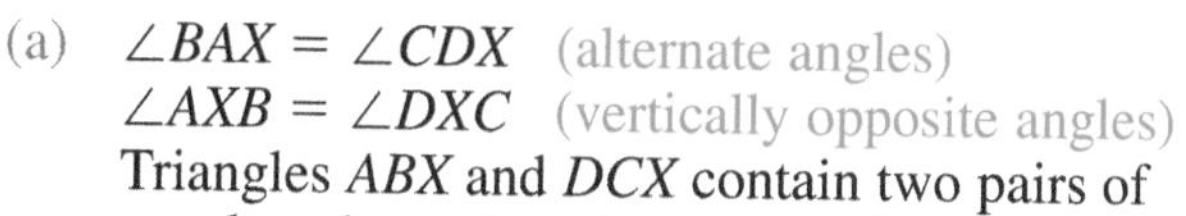

(a) $\angle BAX = \angle CDX$ (alternate angles)

$\angle AXB = \angle DXC$ (vertically opposite angles)

Triangles ABX and DCX contain two pairs of equal angles and so they are similar.

If two pairs of angles are equal then the third pair must be equal. Why?

(b) $\angle ABX = \angle DCX$.

Sides AX and DX are opposite these equal angles.

So, DX corresponds to AX.

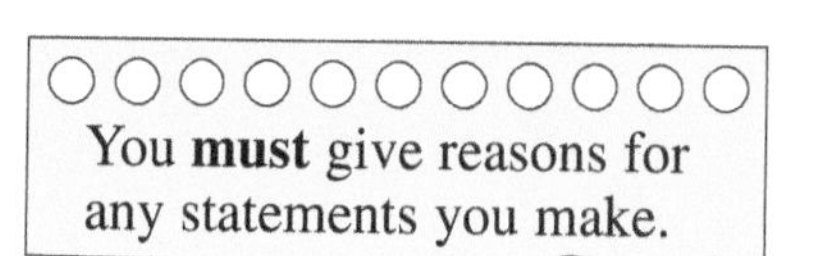

(c) $\frac{AX}{3} = \frac{5}{4}$ $\left(\text{or scale factor} = \frac{5}{4}\right)$

$AX = \frac{5}{4} \times 3$

$AX = 3.75\,\text{cm}$

2 Show that these two triangles are similar.

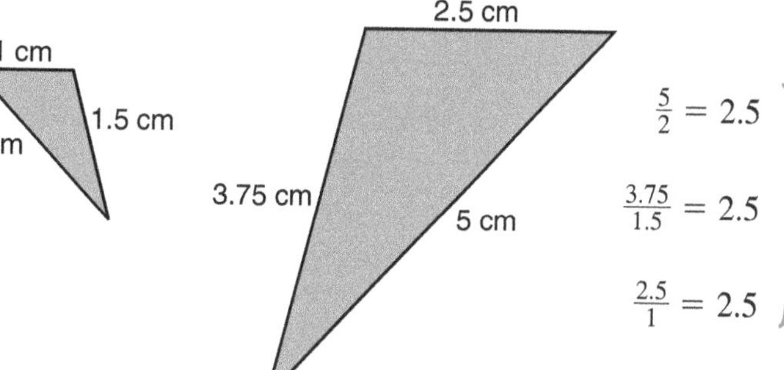

$\frac{5}{2} = 2.5$

$\frac{3.75}{1.5} = 2.5$

$\frac{2.5}{1} = 2.5$

All three pairs of corresponding sides are in the same ratio, so, the triangles are similar.

Exercise 32.4

1 BC is parallel to PQ.

Show that triangles ABC and APQ are similar and calculate the required lengths.

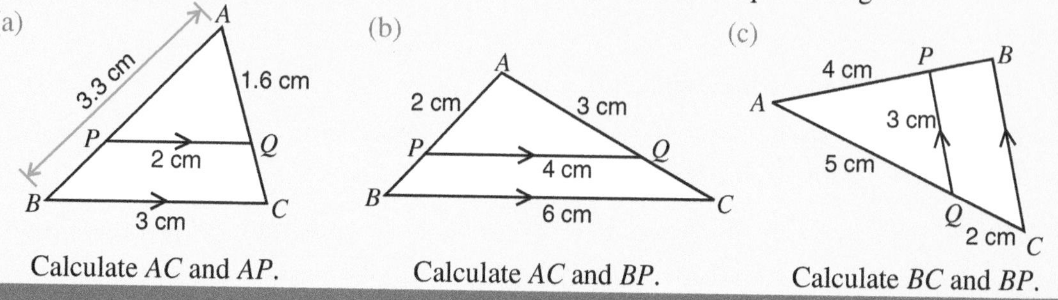

2 BC is parallel to PQ.
Show that triangles ABC and APQ are similar and calculate the required lengths.

(a)

Calculate AQ and BC.

(b)

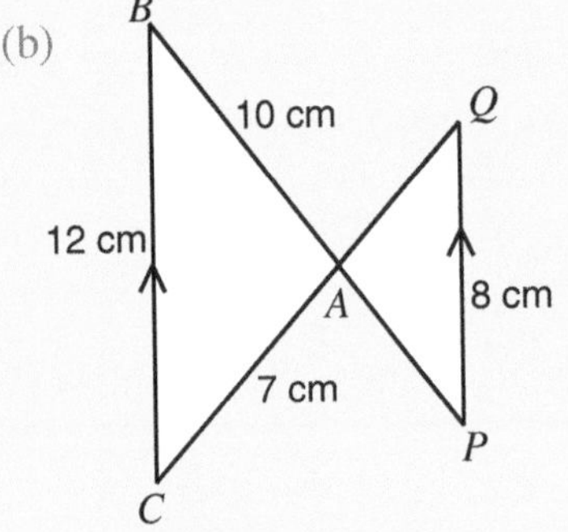

Calculate AQ and BP.

3 Show that these pairs of triangles are similar and find angle x.

(a)

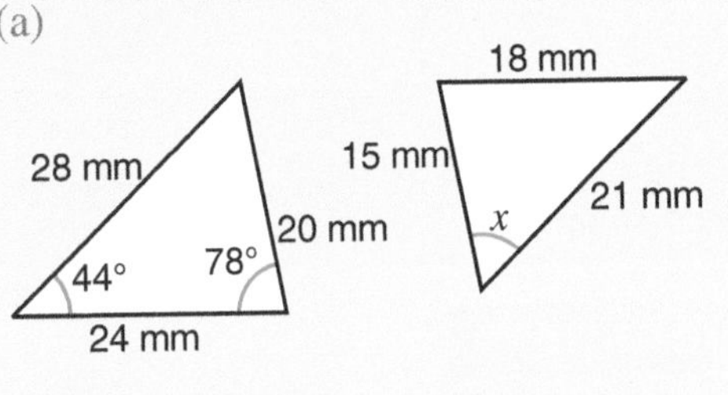

(b)

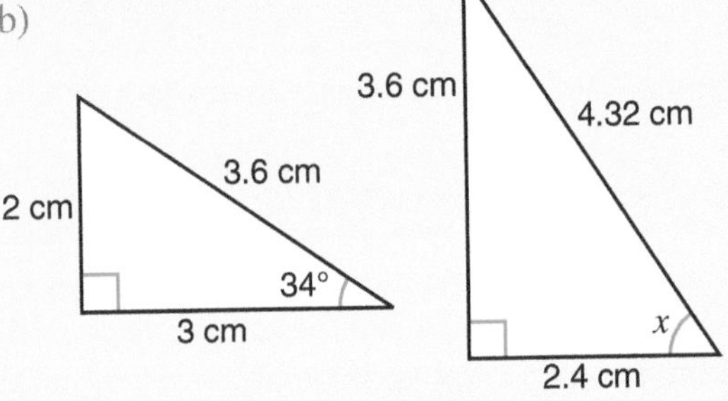

4 All marked lengths are in centimetres.

(a) In each part show that triangles ABC and APQ are similar and find angle x.

(i)

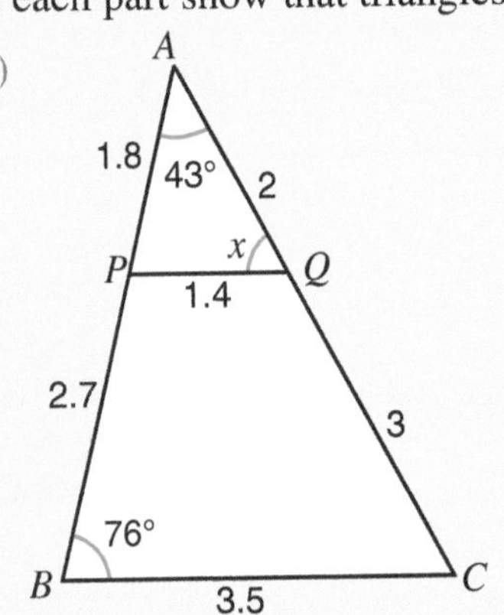

(ii)

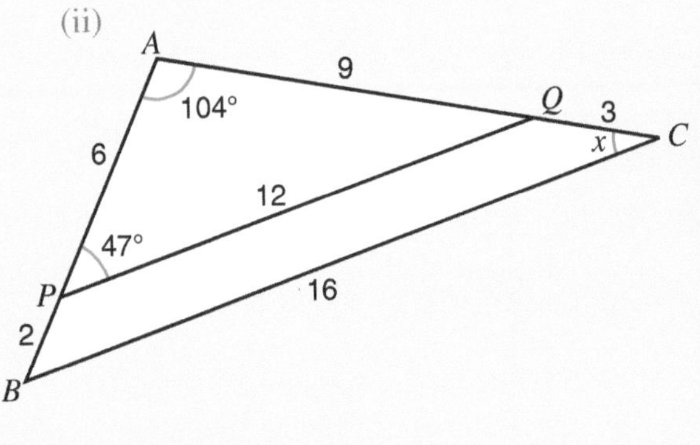

(b) In each part find the perimeters of triangles ABC and APQ.
What do you notice about the ratio of the perimeters of the triangles and the ratio of the lengths of corresponding sides?

5 (a) Explain why triangles ABC and PQR are similar.
(b) Calculate the length of AB.
(c) The perimeter of triangle ABC is 7.5 cm.
Find the perimeter of triangle PQR.

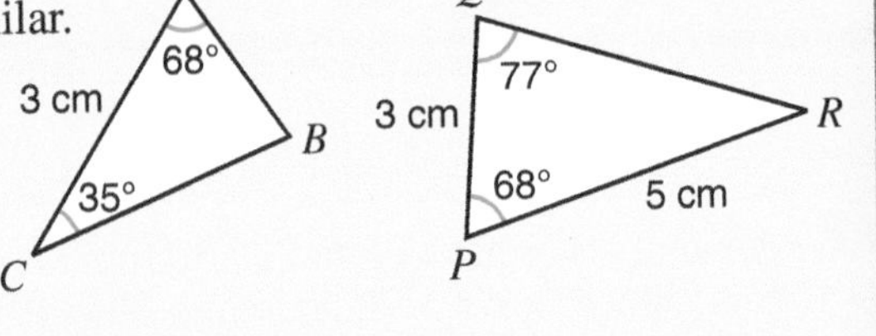

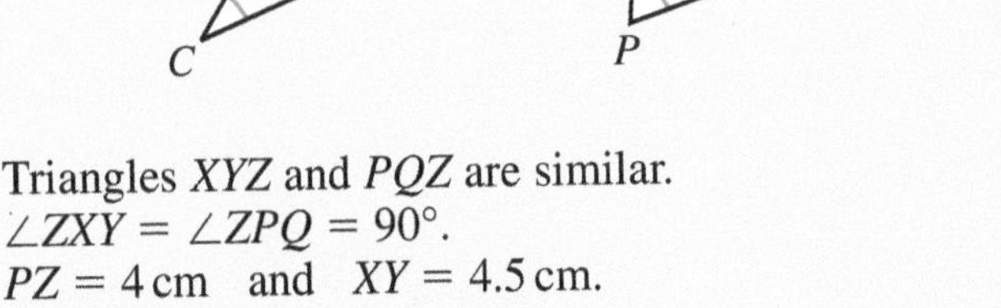

6 Triangles XYZ and PQZ are similar.
$\angle ZXY = \angle ZPQ = 90°$.
$PZ = 4\text{ cm}$ and $XY = 4.5\text{ cm}$.
The ratio $QZ : YZ$ is $2 : 3$.
Calculate the area of triangle XYZ.

Lengths, areas and volumes of similar figures

Activity

Some cubes have side 2 cm.
They are built together to make a larger cube with side 6 cm.
This represents an enlargement with scale factor 3.

Copy and complete the table.

	Length of side (cm)	Area of face (cm^2)	Volume of cube (cm^3)
Small cube	2		8
Large cube	6	36	
Scale factor	$\frac{6}{2} = 3$	$\frac{36}{} =$	$\frac{}{8} =$

What do you notice about the three scale factors?

Repeat this activity with a scale factor of 4.
Now repeat it with a scale factor of k.

Spheres

Three spheres have radii r cm, $2r$ cm and $3r$ cm.
Compare the surface areas and volumes of the three spheres.

Scale factors for length, area and volume

When the length scale factor $= k$
the area scale factor $= k^2$
the volume scale factor $= k^3$

For solids made of the same material, **mass is proportional to volume.**
So, the mass scale factor is k^3.

EXAMPLES

1 A prototype for a new plane is made.
The real plane will be an enlargement of the prototype with scale factor 5.
(a) The area of the windows on the prototype is 0.18 m^2.
Find the area of the real windows.
(b) The volume of the real fuel tank is 4000 litres.
Find the volume of the fuel tank on the prototype.

It can be assumed that "scale factor" refers to the length scale factor, unless specified differently in a question.

(a) Area scale factor $= 5^2 = 25$.
Real area $= 0.18 \times 25 = 4.5$ m^2

(b) Volume scale factor $= 5^3 = 125$.
Prototype volume $= \frac{4000}{125} = 32$ litres.

Corresponding lengths, areas and volumes:

Prototype plane — multiply by scale factor → Real plane

Prototype plane ← divide by scale factor — Real plane

2 A metal ingot has volume 20 000 cm^3.
It is melted down and made into identical smaller ingots.
Each small ingot is similar to the original ingot and has volume 25 cm^3.
The length of the large ingot is 50 cm.
Calculate the length of the small ingots.

Volume scale factor $= \frac{20\,000}{25} = 800$ Length scale factor $= \sqrt[3]{800} = 9.28...$

Length of small ingot $= \frac{25}{\sqrt[3]{800}} = 2.693... = 2.69$ cm, correct to 3 s.f.

Exercise 32.5

1. A model of a train is 60 cm long. It is made on a scale of 1 to 50.
What is the length of the actual train in metres?

2. A rectangle has length 8 cm and width 6 cm. A similar rectangle has length 12 cm.
   (a) What is the scale factor of their lengths?
   (b) What is the width of the larger rectangle?

3. A motor car is 4.2 m long and 1.4 m high. A scale model of the car is 8.4 cm long.
   (a) What is the scale of the model?
   (b) What is the height of the model?

4. The lengths of the sides of a kite are doubled. What happens to its area?

5. A rectangular vegetable plot needs 10 kg of fertiliser.
How much fertiliser is needed for a plot with double the dimensions?

6. Circle P has a radius of 3.5 cm. Circle Q has a radius of 35 cm.
How many times larger than the area of circle P is the area of circle Q?

7. A company logo is printed on all its stationery.
On small sheets of paper the logo is 1.2 cm high and covers an area of 3.5 cm^2.
On large sheets of paper the logo covers an area of 14 cm^2.
What is the height of the logo on large sheets of paper?

8. The lengths of the sides of a square are halved. What happens to its area?

9. A picture is 30 cm high and has an area of 360 cm^2.
Another print of the same picture is 15 cm high.
What is its area?

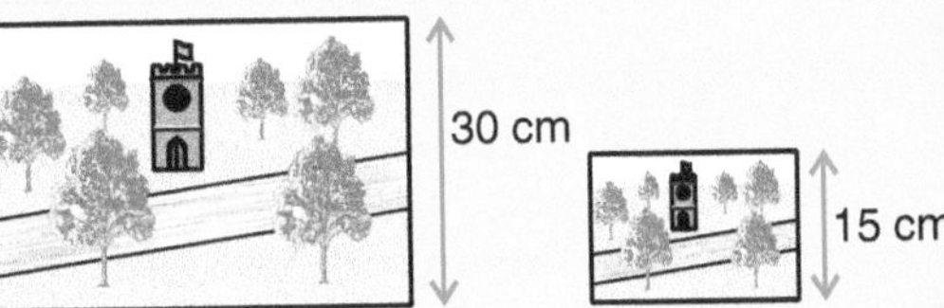

10. A king-size photograph is 18 cm long and 12 cm wide.
A standard size photograph is 12 cm long.
    (a) What is the width of a standard size photograph?
    (b) What is the area of a standard size photograph?

11. Two rugs are similar. The larger rug is 3.6 m in length and has an area of 9 m^2.
The smaller rug has an area of 4 m^2. What is the length of the smaller rug?

12. The scale of a map is 1 to 50 000.
    (a) The distance between two junctions on the map is 3 cm.
    What is the actual distance between the junctions, in kilometres?
    (b) A lake covers 20 cm^2 on the map.
    How many square kilometres does the lake actually cover?

13. A map has a scale of 1 : 25 000.
    (a) The length of a road is 3.5 km. Calculate its length, in centimetres, on the map.
    (b) The area of a field on the map is 12 cm^2. Calculate the true area in square metres.
    (c) A park has an area of 120 000 m^2. Calculate the area of the park on the map.

14. The measurements of a rabbit hutch are all doubled.
How many times bigger is its volume?

15. A teapot has a volume of 500 ml.
A similar teapot is double the height.
What is the volume of this teapot?

16. A box of height 4 cm has a surface area of 220 cm^2 and a volume of 200 cm^3.
    (a) What is the surface area and volume of a similar box of height 8 cm?
    (b) What is the surface area and volume of a similar box of height 2 cm?

17 Two garden ponds are similar.
The dimensions of the larger pond are three times as big as the smaller pond.
The smaller pond holds 20 litres of water.
How many litres of water does the larger pond hold?

18 Two solid spheres are made of the same material.
The smaller sphere has a radius of 4 cm and weighs 1.5 kg.
The larger sphere has a radius of 8 cm. How much does it weigh?

19 The measurements of a box are each halved. What happens to its volume?

20 A cylinder with a height of 6 cm has a volume of 200 cm^3.
What is the volume of a similar cylinder with a height of 3 cm?

21 A box with a height of 4 cm has a volume of 120 cm^3.
What is the volume of a similar box with a height of 6 cm?

22 The volumes of two similar statues are 64 cm^3 and 125 cm^3.
The height of the smaller statue is 8 cm.
(a) What is the height of the larger statue?
(b) What is the ratio of their surface areas?

23 Two fish tanks are similar. The smaller tank is 12 cm high and has a volume of 3.6 litres.
The larger tank has a volume of 97.2 litres. What is the height of the larger tank?

24 Pop and Fizzo come in similar cans. Cans of Pop are 8 cm tall and cans of Fizzo are 10 cm tall.
A can of Pop holds 200 ml. How much does a can of Fizzo hold?

25 (a) A scale model of a house is made using a scale of 1 : 200.
The roof area on the model is 82 cm^2. Find the real roof area, in square metres.
(b) The volume of the real roof space is 360 m^3.
Find the volume of the roof space in the model, in cubic centimetres.

26 Jane makes a scale model of her village.
A fence of length 12 m is represented by a length of 4 cm on her model.
(a) Calculate the scale which Jane is using.
(b) Calculate, in cubic centimetres, the volume on the model of a pool which has a volume of 20 m^3.
(c) Calculate the actual area of a playground which has an area of 320 cm^2 on the model.

27 Joe wants to enlarge a picture so that its area is doubled.
What length scale factor should he use?

28 Two balls have radii of 2 cm and 5 cm.
(a) Calculate the volumes of the two balls and show that the ratio of the volumes is 8 : 125.
(b) Show that the surface area of the smaller ball is 16% of the surface area of the larger ball.

29 Coffee filters are paper cones.
The cones are made in these similar sizes; small, medium and large.
The slant height of a small cone is 5 cm and the surface area is 15π cm^2.
(a) A large cone has a surface area of 135π cm^2.
Calculate the length of its slant height.
(b) A medium cone has a volume of $\frac{81\pi}{2}$ cm^3.
Calculate the surface area of a medium cone.

30 A bronze statue is made in two sizes.
The taller statue is 15 cm high and the shorter one is 9 cm high.
The taller statue weighs 3.75 kg.
What is the weight of the shorter statue?

What you need to know

- When two shapes are the same shape and size they are said to be **congruent**.
- There are four ways to show that a pair of triangles are congruent:
  - SSS Three equal sides.
  - SAS Two sides and the included angle.
  - ASA Two angles and a corresponding side.
  - RHS Right angle, hypotenuse and one other side.
- When two figures are **similar**:
  their **shapes** are the same,
  their **angles** are the same,
  corresponding **lengths** are in the same ratio, this ratio is the **scale factor** of the enlargement.

$$\text{Scale factor} = \frac{\text{new length}}{\text{original length}}$$

- All circles are similar to each other.
- All squares are similar to each other.
- For **similar triangles**:
  corresponding lengths are opposite equal angles, the scale factor is the ratio of the corresponding sides.
- For **similar areas and volumes**:
  when the **length** scale factor $= k$
  the **area** scale factor $= k^2$
  the **volume** scale factor $= k^3$

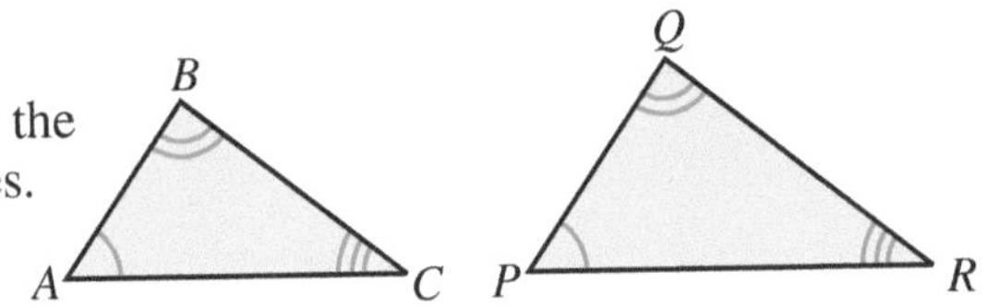

$$\frac{AB}{PQ} = \frac{BC}{QR} = \frac{CA}{RP} = \text{scale factor}$$

Review Exercise 32

1. These triangles have not been drawn accurately.
   Which two triangles are congruent to each other?
   Give a reason for your answer.

   A: 5 cm, 6 cm, 3 cm
   B: 5 cm, 30°, 6 cm
   C: 3 cm, 5 cm, 4 cm
   D: 6 cm, 5 cm, 3 cm
   E: 12 cm, 10 cm, 6 cm

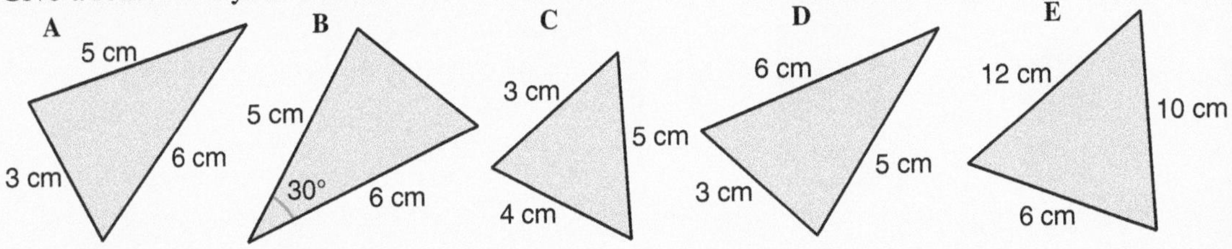

2. A teacher sketches five triangles.

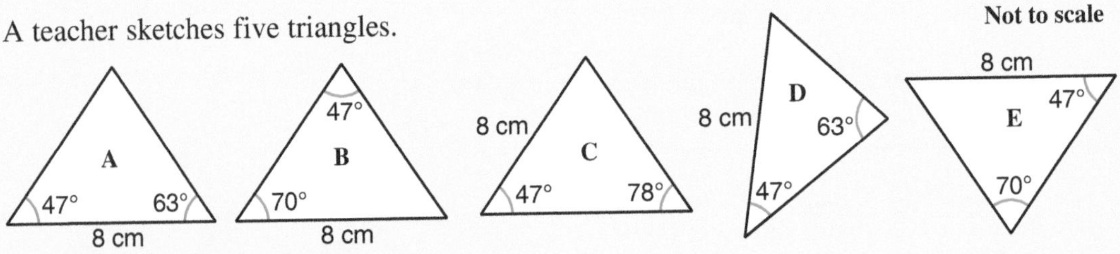

   Not to scale

   A: 47°, 63°, 8 cm
   B: 47°, 70°, 8 cm
   C: 8 cm, 47°, 78°
   D: 8 cm, 63°, 47°
   E: 8 cm, 47°, 70°

   Which two of these triangles are congruent?
   Give a reason for your answer. AQA

3. These triangles are congruent.
   What is the size of (a) x, (b) y?

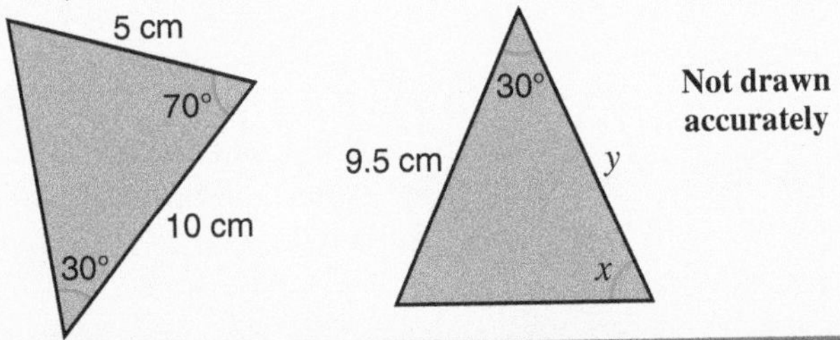

   5 cm, 70°, 10 cm, 30°
   30°, 9.5 cm, y, x

   Not drawn accurately

   AQA

4 Explain why triangles ABX and PQX are similar.

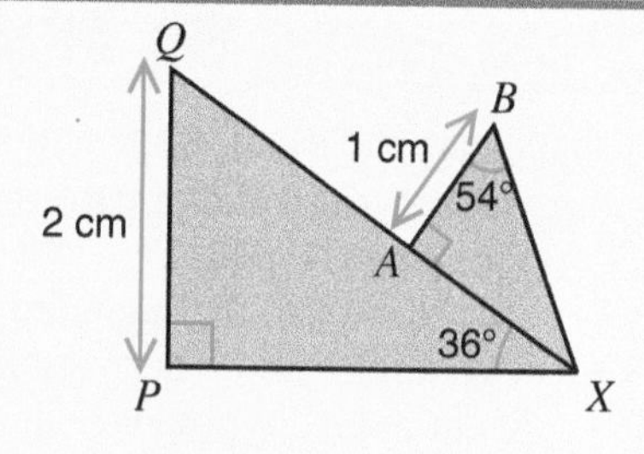

5 In the diagram AB is parallel to CD.
$AB = OC = 12\,\text{cm}$. $OB = 10\,\text{cm}$.

Use similar triangles to calculate the length of CD.

AQA

6 The diagram shows a triangle ABC.
DE is parallel to BC.
Calculate the lengths
(a) DE,
(b) AC.

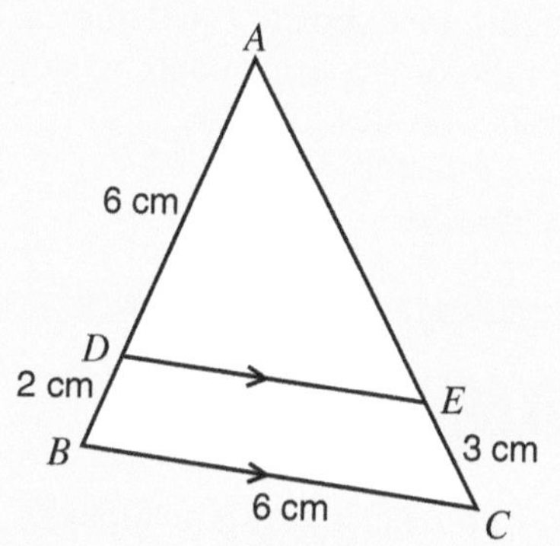

AQA

7 These candles are similar.
P has a surface area of $24\,\text{cm}^2$.
Q has a surface area of $54\,\text{cm}^2$.
P is 4 cm high.
(a) How high is **Q**?
(b) What is the ratio of their volumes?

8 A scale model of a ship is made, using a scale of 1 : 40.
(a) The area of the real deck is $500\,\text{m}^2$.
Find, in square centimetres, the area of the deck on the model.
(b) The volume of the hold on the model is $187\,500\,\text{cm}^3$.
Find, in cubic metres, the volume of the real hold.

9 Two similar solid shapes are made.
The height of the smaller shape is 7 cm.
The width of the smaller shape is 6 cm.
The width of the larger shape is 9.6 cm.
(a) Calculate the height of the larger shape.
(b) The volume of the larger shape is $695\,\text{cm}^3$.
Find the volume of the smaller shape.

AQA

10

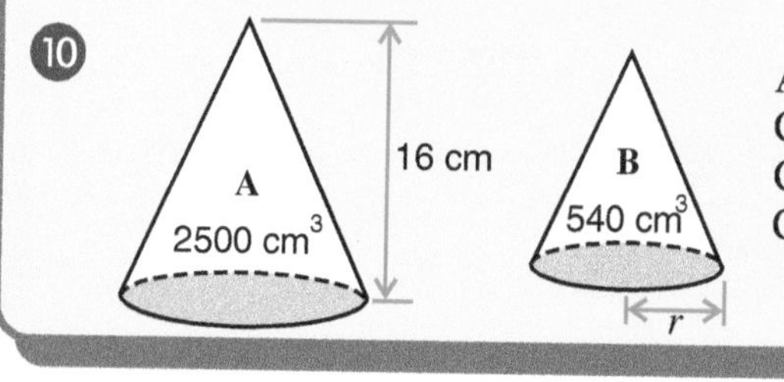

A and **B** are two similar cones.
Cone **A** has a height of 16 cm and a volume of $2500\,\text{cm}^3$.
Cone **B** has a volume of $540\,\text{cm}^3$.
Calculate the radius, r, of the base of Cone **B**.

AQA

Vectors

Vectors and scalars

Quantities which have both **size** and **direction** are called **vectors**.
Examples of vector quantities are:

Displacement	**Velocity**
A combination of distance and direction	A combination of speed and direction

Quantities which have size only are called **scalars**.
Examples of scalar quantities are: distance, area, speed, mass, volume, temperature, etc.

Vector notation

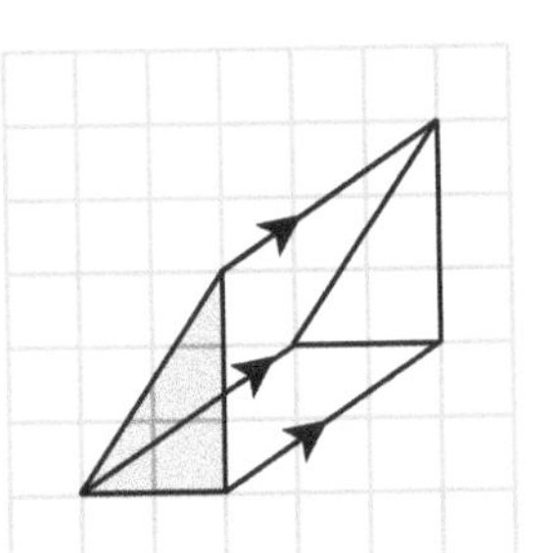

Column vectors

The diagram shows the translation of a triangle by the vector $\begin{pmatrix}3\\2\end{pmatrix}$.

The triangle has been displaced 3 units to the right and 2 units up.

$\begin{pmatrix}3\\2\end{pmatrix}$ gives information about the size and direction of the displacement.

The vector $\begin{pmatrix}3\\2\end{pmatrix}$ is sometimes called a **column vector**, because it consists of one column of numbers.

Directed line segments
Vectors can be represented in diagrams using **line segments**.
The **length** of a line represents the **size of the vector**.
The **direction of the vector** is shown by an **arrow**.

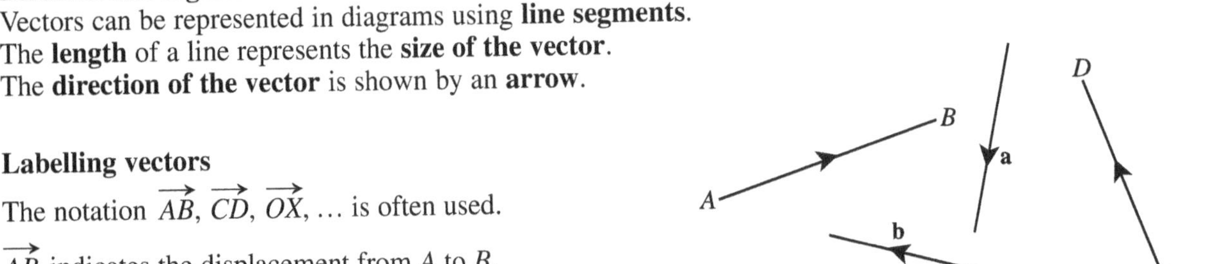

Labelling vectors

The notation $\overrightarrow{AB}$, $\overrightarrow{CD}$, $\overrightarrow{OX}$, ... is often used.

$\overrightarrow{AB}$ indicates the displacement from A to B.

$\overrightarrow{AB}$ should be read as "vector AB".

Bold lower case letters such as **a**, **b**, **c**, ... are also used.
In hand-written work **a** should be written as $\underline{a}$.
a and $\underline{a}$ should both be read as "vector a".

EXAMPLE

Draw and label the following vectors.

(a) $\mathbf{a} = \begin{pmatrix}-2\\3\end{pmatrix}$ (b) $\mathbf{b} = \begin{pmatrix}3\\-2\end{pmatrix}$ (c) $\overrightarrow{AB} = \begin{pmatrix}-4\\-2\end{pmatrix}$

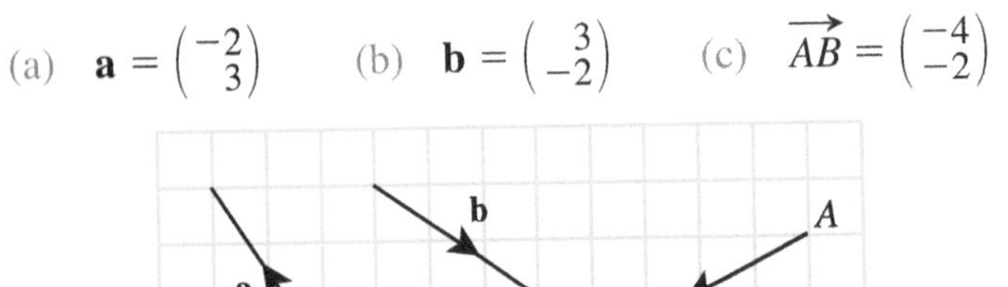

Remember:
The top number describes the **horizontal** part of the movement:
$+$ = to the right $-$ = to the left
The bottom number describes the **vertical** part of the movement:
$+$ = upwards $-$ = downwards

Exercise 33.1

1. On squared paper, draw and label the following vectors.

   (a) $\mathbf{a} = \begin{pmatrix} -5 \\ 2 \end{pmatrix}$ (b) $\mathbf{b} = \begin{pmatrix} 5 \\ -2 \end{pmatrix}$ (c) $\mathbf{c} = \begin{pmatrix} -5 \\ -2 \end{pmatrix}$ (d) $\mathbf{d} = \begin{pmatrix} 5 \\ 2 \end{pmatrix}$

2. Mark the point O on squared paper.
   Draw and label the following vectors.

   (a) $\overrightarrow{OA} = \begin{pmatrix} 2 \\ 4 \end{pmatrix}$ (b) $\overrightarrow{OB} = \begin{pmatrix} 4 \\ 2 \end{pmatrix}$ (c) $\overrightarrow{OC} = \begin{pmatrix} 4 \\ -2 \end{pmatrix}$ (d) $\overrightarrow{OD} = \begin{pmatrix} 2 \\ -4 \end{pmatrix}$

   (e) $\overrightarrow{OE} = \begin{pmatrix} -2 \\ -4 \end{pmatrix}$ (f) $\overrightarrow{OF} = \begin{pmatrix} -4 \\ -2 \end{pmatrix}$ (g) $\overrightarrow{OG} = \begin{pmatrix} -4 \\ 2 \end{pmatrix}$ (h) $\overrightarrow{OH} = \begin{pmatrix} -2 \\ 4 \end{pmatrix}$

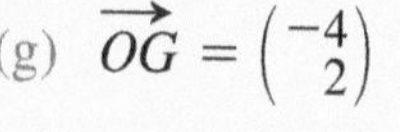
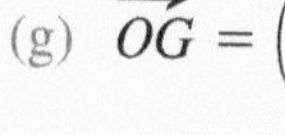

3. Write each of these vectors as column vectors.

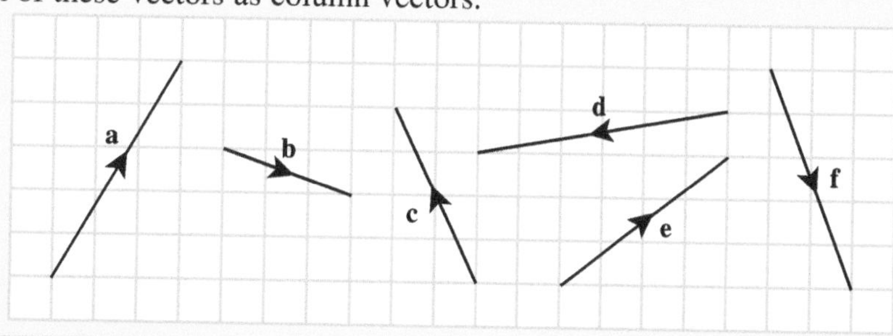

Equal vectors

The diagram shows three vectors, $\overrightarrow{AB}$, $\overrightarrow{PQ}$ and $\overrightarrow{XY}$.

$\overrightarrow{AB} = \begin{pmatrix} 5 \\ 2 \end{pmatrix}$ $\overrightarrow{PQ} = \begin{pmatrix} 5 \\ 2 \end{pmatrix}$ $\overrightarrow{XY} = \begin{pmatrix} 5 \\ 2 \end{pmatrix}$

The column vectors are equal, so, $\overrightarrow{AB} = \overrightarrow{PQ} = \overrightarrow{XY}$.

Vectors are **equal** if: they have the same length, **and** they are in the same direction.

Multiplying a vector by a scalar

The diagram shows three vectors, **a**, **b** and **c**.

$\mathbf{a} = \begin{pmatrix} 4 \\ -2 \end{pmatrix}$ $\mathbf{b} = \begin{pmatrix} 8 \\ -4 \end{pmatrix}$ $\mathbf{c} = \begin{pmatrix} 2 \\ -1 \end{pmatrix}$

a and **b** are in the same direction.

$\mathbf{b} = \begin{pmatrix} 8 \\ -4 \end{pmatrix} = \begin{pmatrix} 2 \times 4 \\ 2 \times -2 \end{pmatrix}$

This can be written as $\mathbf{b} = 2 \times \begin{pmatrix} 4 \\ -2 \end{pmatrix}$

So, $\mathbf{b} = 2\mathbf{a}$.

This means that **b** is twice the length of **a**.

b and **c** are in the same direction.

$\mathbf{b} = \begin{pmatrix} 8 \\ -4 \end{pmatrix} = 4 \times \begin{pmatrix} 2 \\ -1 \end{pmatrix} = 4\mathbf{c}$

This means that **b** is 4 times as long as **c**.

a and **c** are in the same direction.

$\mathbf{c} = \frac{1}{2}\mathbf{a}$.

This means that **c** is half the length of **a**.

a and n**a** are vectors in the same direction.
The length of vector $n\mathbf{a} = n \times$ the length of vector **a**.

Vectors in opposite directions

The diagram shows two vectors, **a** and **b**.
The vectors are equal in length, **but** are in opposite directions.

$\mathbf{a} = \begin{pmatrix} 5 \\ 2 \end{pmatrix}$ $\quad \mathbf{b} = \begin{pmatrix} -5 \\ -2 \end{pmatrix}$ $\quad$ So, $\mathbf{b} = -\mathbf{a}$

> Vectors **a** and −**a** have the same length, **but** are in opposite directions.

EXAMPLE

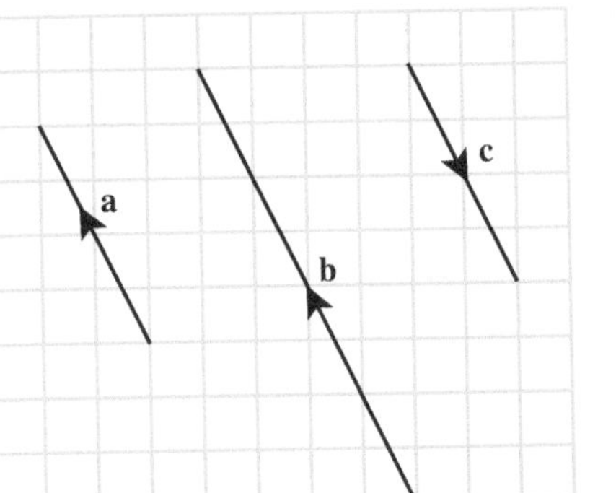

The diagram shows three vectors: **a**, **b** and **c**.
(a) Write each of **a**, **b** and **c** as column vectors.
(b) Express **b** and **c** in terms of **a**.

(a) $\mathbf{a} = \begin{pmatrix} -2 \\ 4 \end{pmatrix}$ $\quad \mathbf{b} = \begin{pmatrix} -4 \\ 8 \end{pmatrix}$ $\quad \mathbf{c} = \begin{pmatrix} 2 \\ -4 \end{pmatrix}$

(b) **b** is twice the length of **a** and in the same direction.
$\mathbf{b} = 2\mathbf{a}$

c is the same length as **a** but in the opposite direction.
$\mathbf{c} = -\mathbf{a}$

Exercise 33.2

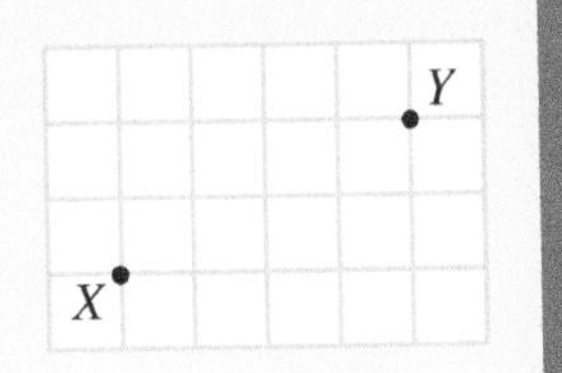

1. The points X and Y are marked on the grid.
On squared paper, draw each of the vectors **a** to **d**, where
$\mathbf{a} = 2\overrightarrow{XY}$, $\quad \mathbf{b} = -\overrightarrow{XY}$, $\quad \mathbf{c} = \frac{1}{2}\overrightarrow{XY}$, $\quad \mathbf{d} = \overrightarrow{YX}$.

2. This diagram shows three vectors, labelled **p**, **q** and **r**.

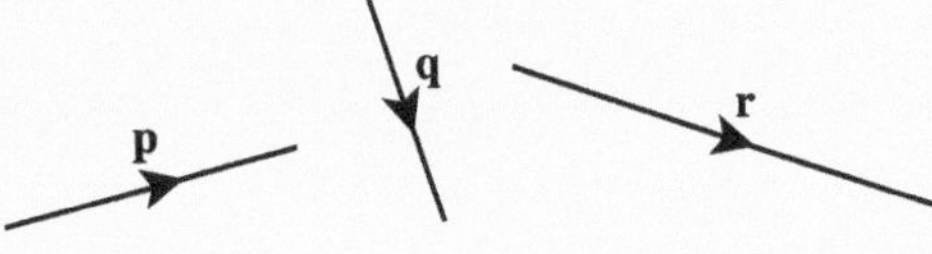

In the diagram below, which vectors can be expressed in the form:
(i) n**p**, (ii) n**q**, (iii) n**r**?
For each vector give the value of n.

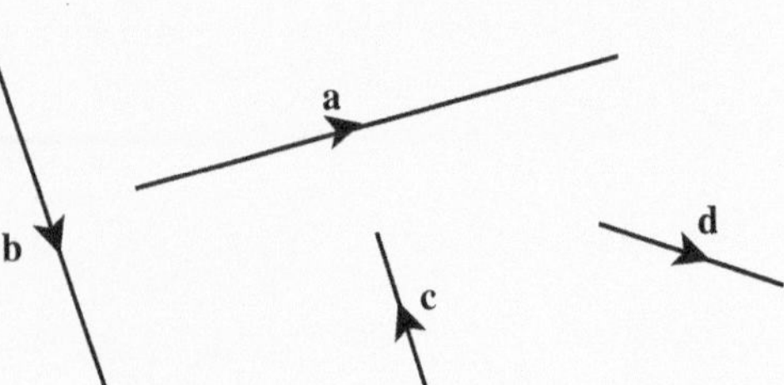

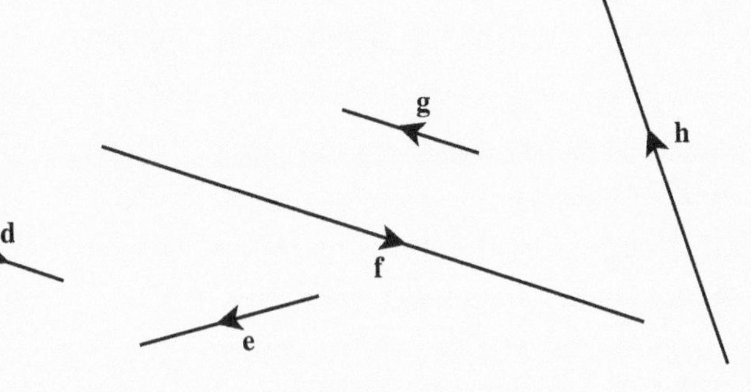

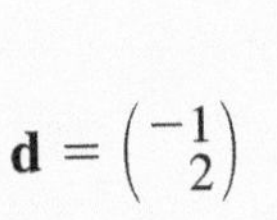

3. $\mathbf{a} = \begin{pmatrix} 2 \\ -1 \end{pmatrix}$ $\quad \mathbf{b} = \begin{pmatrix} -1 \\ 4 \end{pmatrix}$ $\quad \mathbf{c} = \begin{pmatrix} -3 \\ -1 \end{pmatrix}$ $\quad \mathbf{d} = \begin{pmatrix} -1 \\ 2 \end{pmatrix}$

On squared paper draw vectors to represent the following.
(a) 2**a** (b) 2**b** (c) 3**a** (d) −**a**
(e) 2**c** (f) −2**d** (g) −4**b** (h) 5**d**

Vector addition

The diagram shows the points A, B and C.
Start at A and move to B.
Then move from B to C.

The combination of these two displacements is equivalent to a total displacement from A to C.
This can be written using vectors as:

$$\overrightarrow{AB} + \overrightarrow{BC} = \overrightarrow{AC}$$

$\overrightarrow{AC}$ is called the **resultant vector**.

Using column vectors: $\overrightarrow{AB} = \begin{pmatrix} 5 \\ 3 \end{pmatrix}$, $\overrightarrow{BC} = \begin{pmatrix} 3 \\ -4 \end{pmatrix}$ and $\overrightarrow{AC} = \begin{pmatrix} 8 \\ -1 \end{pmatrix}$

The resultant vector, $\overrightarrow{AC}$, is a combination of: $\begin{pmatrix} \text{5 units right followed by 3 units right} \\ \text{3 units up followed by 4 units down} \end{pmatrix} = \begin{pmatrix} \text{8 units right} \\ \text{1 unit down} \end{pmatrix}$

$$\overrightarrow{AB} + \overrightarrow{BC} = \overrightarrow{AC}$$

$$\begin{pmatrix} 5 \\ 3 \end{pmatrix} + \begin{pmatrix} 3 \\ -4 \end{pmatrix} = \begin{pmatrix} 5 + 3 \\ 3 + -4 \end{pmatrix} = \begin{pmatrix} 8 \\ -1 \end{pmatrix}$$

Use vector addition to show that: $\overrightarrow{AB} + \overrightarrow{BC} = \overrightarrow{BC} + \overrightarrow{AB} = \overrightarrow{AC}$.

Vector diagrams

Diagrams can be drawn to show the combination of any number of vectors.
To combine vectors when the vectors are not joined, draw equal vectors so that the second vector joins the end of the first vector.

Note:
With the arrow = +
Against the arrow = −

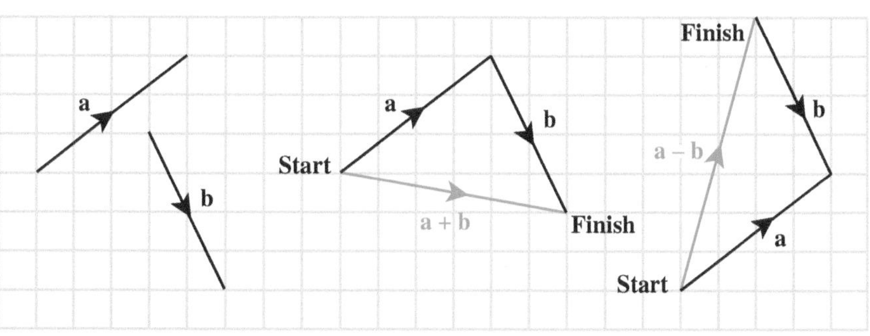

With the arrow, **+a**
With the arrow, **+b**
Resultant vector, **a + b**

With the arrow, **+a**
Against the arrow, **–b**
Resultant vector, **a – b**

EXAMPLE

a, **b** and **c** are three vectors as shown in the diagram.

Draw diagrams to show:

(a) **a + 2b**

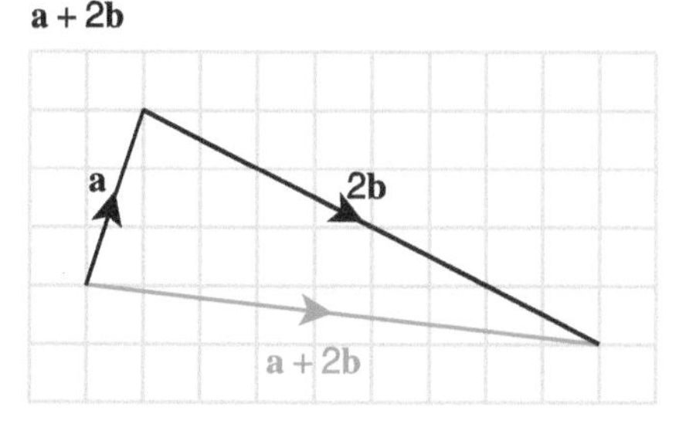

(b) **a – c**

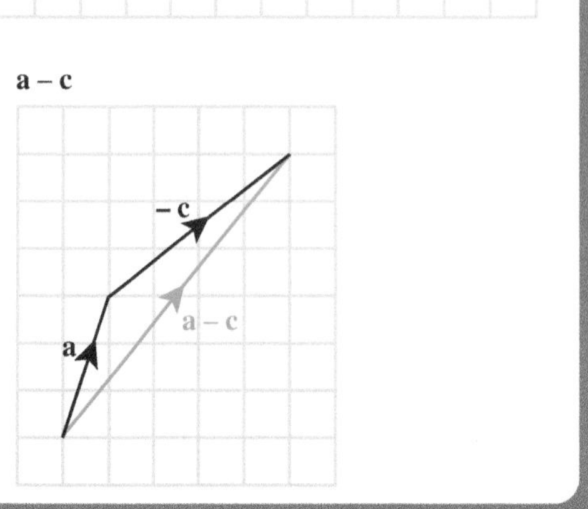

Draw a vector diagram to show that: **a** + **b** = **b** + ***a***

Exercise 33.3

1 $\overrightarrow{OA} = \begin{pmatrix} -1 \\ 4 \end{pmatrix}$, $\overrightarrow{OB} = \begin{pmatrix} 4 \\ 3 \end{pmatrix}$, $\overrightarrow{OC} = \begin{pmatrix} 5 \\ 1 \end{pmatrix}$.

(a) (i) Draw $\overrightarrow{OD}$ where $\overrightarrow{OD} = \overrightarrow{OA} + \overrightarrow{OB}$.

(ii) Draw $\overrightarrow{OE}$ where $\overrightarrow{OE} = 2\overrightarrow{OA} + \overrightarrow{OC}$.

(iii) Draw $\overrightarrow{OF}$ where $\overrightarrow{OF} = 2\overrightarrow{OB} - \overrightarrow{OC}$.

(iv) Draw $\overrightarrow{OG}$ where $\overrightarrow{OG} = 2\overrightarrow{BO} + 3\overrightarrow{OA}$.

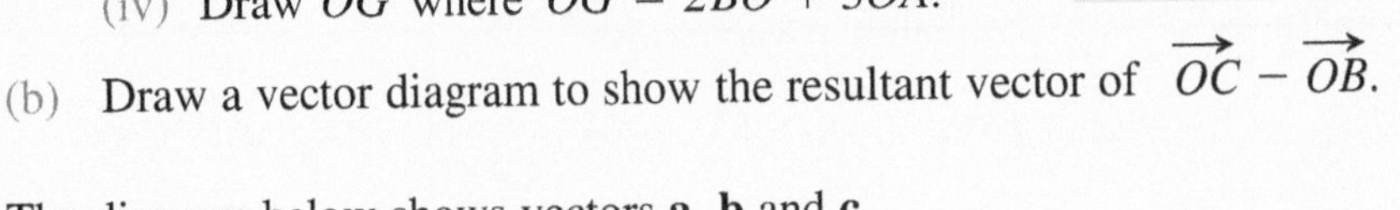

(b) Draw a vector diagram to show the resultant vector of $\overrightarrow{OC} - \overrightarrow{OB}$.

2 The diagram below shows vectors **a**, **b** and **c**.

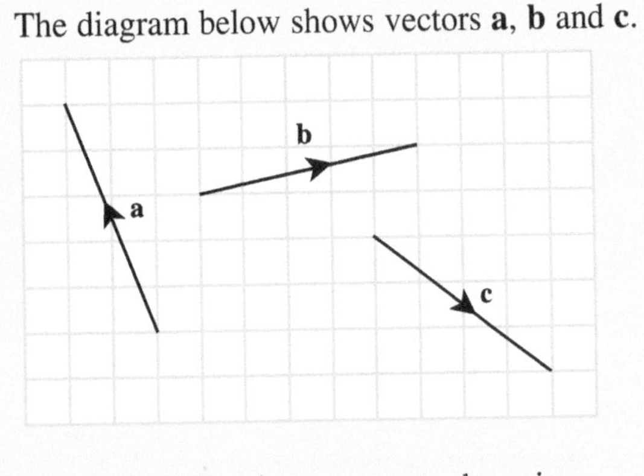

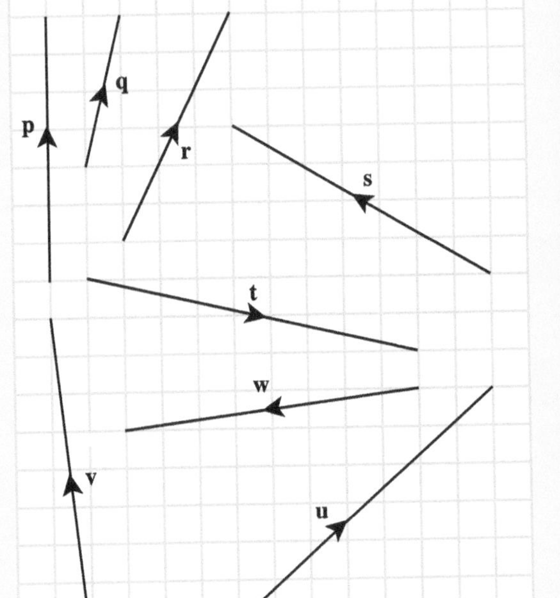

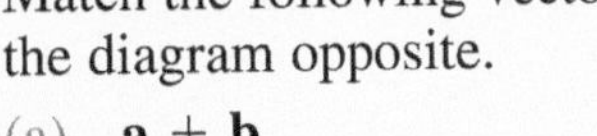

Match the following vectors to those in the diagram opposite.

(a) $\mathbf{a} + \mathbf{b}$

(b) $\mathbf{a} - \mathbf{b}$

(c) $\mathbf{b} + \mathbf{c}$

(d) $\mathbf{b} - \mathbf{c}$

(e) $\mathbf{a} + \mathbf{b} - \mathbf{c}$

(f) $\mathbf{a} + 2\mathbf{b}$

(g) $-2\mathbf{a} - 3\mathbf{c}$

(h) $2\mathbf{a} + \mathbf{c}$

3 This is a grid of congruent parallelograms.

$\overrightarrow{OA} = \mathbf{x}$ and $\overrightarrow{OB} = \mathbf{y}$.

(a) On a copy of the grid mark each of the points C to H where:

(i) $\overrightarrow{OC} = 3\mathbf{y}$

(ii) $\overrightarrow{OD} = \mathbf{x} + 2\mathbf{y}$

(iii) $\overrightarrow{OE} = 4\mathbf{x} + 3\mathbf{y}$

(iv) $\overrightarrow{EF} = -2\mathbf{x} + \mathbf{y}$

(v) $\overrightarrow{FG} = -2\mathbf{x} - 2\mathbf{y}$

(vi) $\overrightarrow{GH} = 4\mathbf{x} + 2\mathbf{y}$

(b) Using vectors, explain why $OBHE$ is a parallelogram.

4 In the diagram $ABCD$ is a parallelogram and C is the midpoint of BE.
P, Q, R and S are the midpoints of AB, BC, CD and DA respectively.

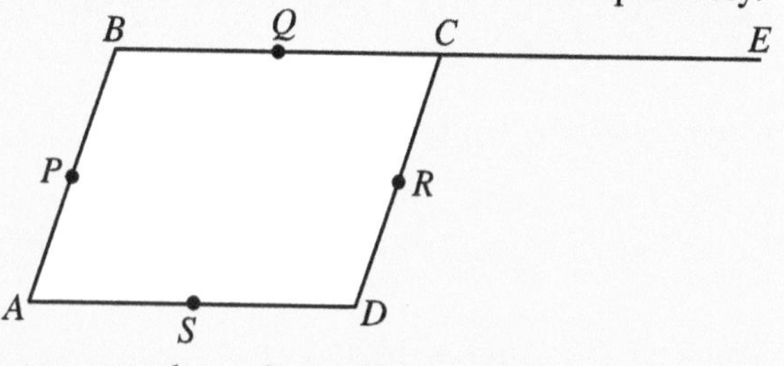

(a) Draw vector diagrams to show the resultant vectors for:

(i) $\overrightarrow{AQ} + \overrightarrow{QR}$ (ii) $\overrightarrow{AQ} - \overrightarrow{QR}$ (iii) $\overrightarrow{PR} - \overrightarrow{RA}$ (iv) $\overrightarrow{AP} + \overrightarrow{PQ} + \overrightarrow{QR}$

$\overrightarrow{AB} = \mathbf{a}$ and $\overrightarrow{AD} = \mathbf{b}$.

(b) Write each of the following vectors in terms of $\mathbf{a}$ and $\mathbf{b}$.

(i) $\overrightarrow{PC}$ (ii) $\overrightarrow{AR}$ (iii) $\overrightarrow{AE}$

(c) What do your answers to (b) tell you about the lines:

(i) PC and AR, (ii) PC and AE?

Vectors and geometry

The relationships between vectors can be used to solve geometrical problems.

EXAMPLE

1 $ABCD$ is a parallelogram.
E is the midpoint of the diagonal AC.
Show that E is also the midpoint of the diagonal BD.

Let $\overrightarrow{AB} = \mathbf{x}$ and $\overrightarrow{AD} = \mathbf{y}$.

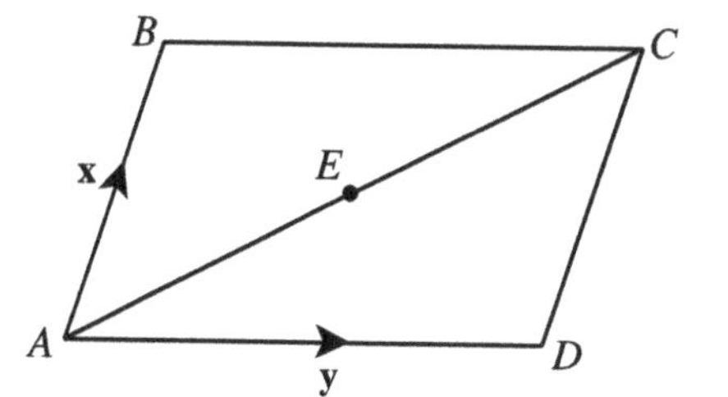

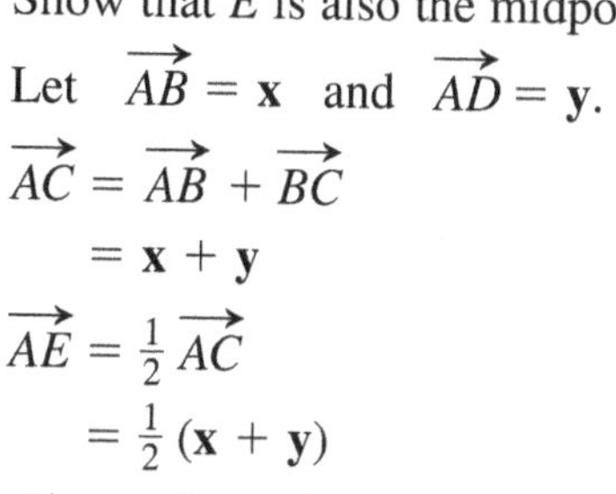

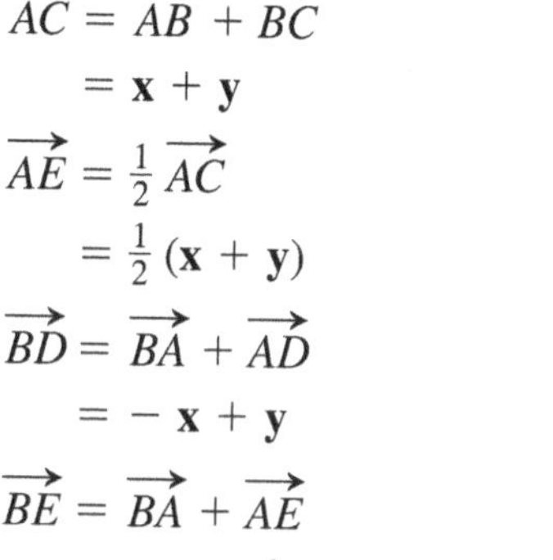

$\overrightarrow{AC} = \overrightarrow{AB} + \overrightarrow{BC}$

$= \mathbf{x} + \mathbf{y}$

$\overrightarrow{AE} = \frac{1}{2}\overrightarrow{AC}$

$= \frac{1}{2}(\mathbf{x} + \mathbf{y})$

$\overrightarrow{BD} = \overrightarrow{BA} + \overrightarrow{AD}$

$= -\mathbf{x} + \mathbf{y}$

$\overrightarrow{BE} = \overrightarrow{BA} + \overrightarrow{AE}$

$= -\mathbf{x} + \frac{1}{2}(\mathbf{x} + \mathbf{y})$

Multiply out brackets and simplify.

$= -\frac{1}{2}\mathbf{x} + \frac{1}{2}\mathbf{y}$

$= \frac{1}{2}(-\mathbf{x} + \mathbf{y})$

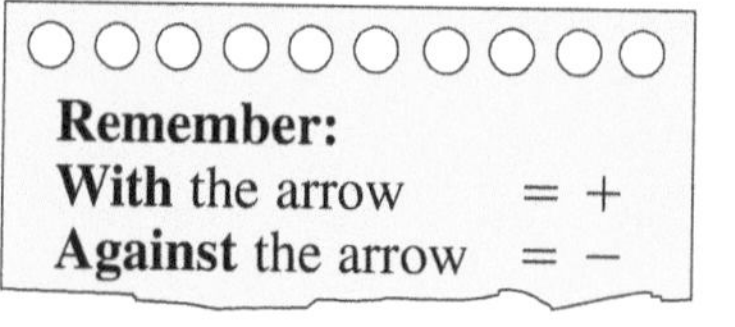

So, $\overrightarrow{BE} = \frac{1}{2}\overrightarrow{BD}$.

BE is in the same direction as BD and is half its length.
Because B is common to both BE and BD, BED must be a straight line.
So, E is the midpoint of BD.

EXAMPLE

2 $\overrightarrow{AB} = 2\mathbf{a} + n\mathbf{b}$ $\quad \overrightarrow{PQ} = 5\mathbf{a} - 10\mathbf{b}$

Calculate n if AB and PQ are parallel.

The number of **a**'s and the number of **b**'s in $\overrightarrow{AB}$ and $\overrightarrow{PQ}$ are in the same ratio.

$\frac{5}{2} = -\frac{10}{n}$

$5n = -20$

$n = -4$

$\overrightarrow{AB}$ and $\overrightarrow{PQ}$ are parallel.
So, $\overrightarrow{PQ}$ must be a multiple of $\overrightarrow{AB}$.
Why?

Exercise 33.4

The diagrams in this exercise have not been drawn accurately.

1 In the diagram, $\overrightarrow{OA} = \mathbf{a}$ and $\overrightarrow{OB} = \mathbf{b}$.

M is the midpoint of AB.

Express in terms of **a** and **b** in its simplest form,

(a) $\overrightarrow{AO}$, (b) $\overrightarrow{AB}$, (c) $\overrightarrow{MB}$, (d) $\overrightarrow{MO}$.

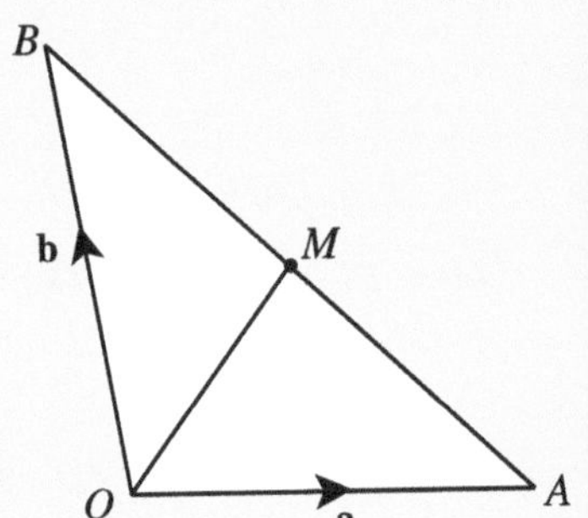

2 $OXYZ$ is a parallelogram.
P is the midpoint of OX and Q is the midpoint of OZ.

$\overrightarrow{OP} = \mathbf{p}$ and $\overrightarrow{OQ} = \mathbf{q}$

(a) Express in terms of **p** and **q** in its simplest form,

(i) $\overrightarrow{PQ}$, (ii) $\overrightarrow{OX}$,

(iii) $\overrightarrow{OY}$, (iv) $\overrightarrow{XZ}$.

(b) What can you say about the lines PQ and XZ?

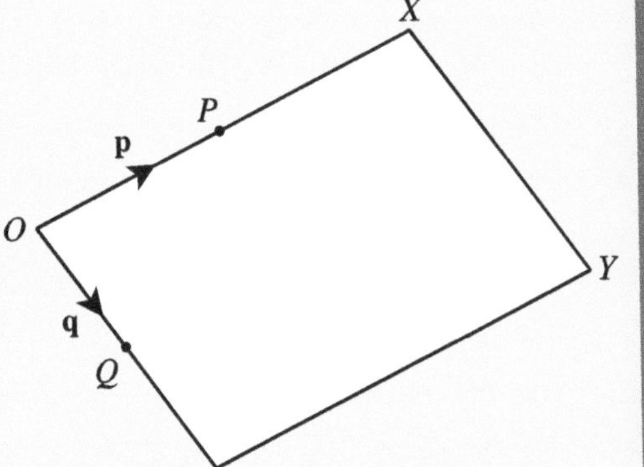

3 $OABC$ is a square.
X is the midpoint of BC.

Y is a point on XA such that $\overrightarrow{XA} = 3\overrightarrow{XY}$.

$\overrightarrow{OA} = \mathbf{a}$ and $\overrightarrow{OC} = \mathbf{b}$.

(a) Express in terms of **a** and **b**:

(i) $\overrightarrow{CX}$ (ii) $\overrightarrow{OX}$

(iii) $\overrightarrow{XA}$ (iv) $\overrightarrow{XY}$

(v) $\overrightarrow{OY}$ (vi) $\overrightarrow{OB}$

(b) What do $\overrightarrow{OY}$ and $\overrightarrow{OB}$ tell you about the points O, Y and B?

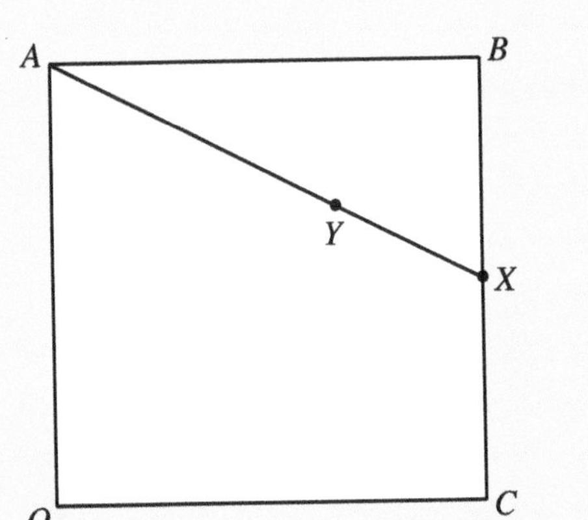

4 OXY is a triangle.
P, Q and R are the midpoints of the sides YO, OX and XY respectively.

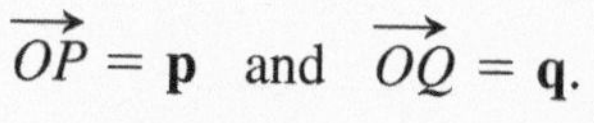

$\overrightarrow{OP} = \mathbf{p}$ and $\overrightarrow{OQ} = \mathbf{q}$.

(a) Express in terms of **p** and **q** in its simplest form:

(i) $\overrightarrow{OX}$ (ii) $\overrightarrow{OY}$

(iii) $\overrightarrow{PQ}$ (iv) $\overrightarrow{XY}$

(v) $\overrightarrow{PR}$ (vi) $\overrightarrow{RQ}$

(b) What can you say about the triangles OXY and PQR?

5 $OABC$ is an irregular quadrilateral.
W is the midpoint of OA.
X is the midpoint of AB.
Y is the midpoint of BC.
Z is the midpoint of CO.

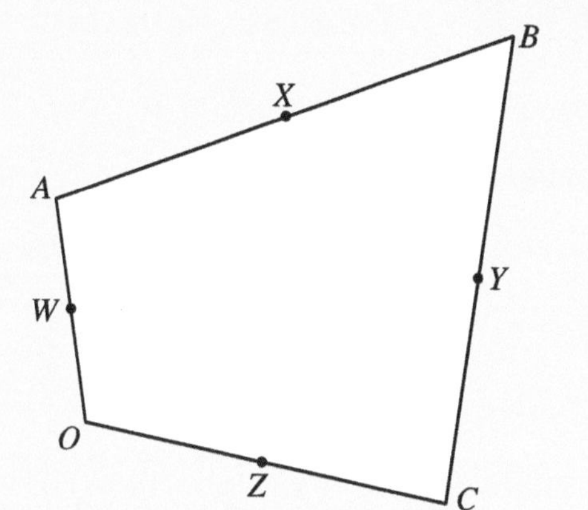

$\overrightarrow{OA} = \mathbf{a}$, $\overrightarrow{OB} = \mathbf{b}$, $\overrightarrow{OC} = \mathbf{c}$.

(a) Find in terms of **a**, **b** and **c**:

(i) $\overrightarrow{CB}$ (ii) $\overrightarrow{YZ}$

(iii) $\overrightarrow{CX}$ (iv) $\overrightarrow{YX}$

(v) $\overrightarrow{ZW}$

(b) Use your answers to (a) to show that $WXYZ$ is a parallelogram.

6 In the trapezium $OABC$, AB is parallel to OC.

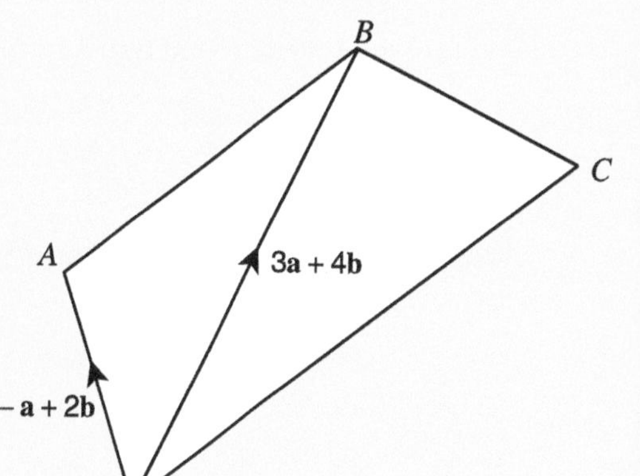

$\overrightarrow{OA} = -\mathbf{a} + 2\mathbf{b}$ and $\overrightarrow{OB} = 3\mathbf{a} + 4\mathbf{b}$.

(a) Calculate $\overrightarrow{AB}$ in terms of **a** and **b**.

$\overrightarrow{BC} = 3\mathbf{a} + n\mathbf{b}$.

(b) Calculate $\overrightarrow{OC}$ in terms of **a**, **b** and n, and hence, calculate the value of n.

7 $\overrightarrow{OA} = -2\mathbf{a} + 4\mathbf{b}$ $\overrightarrow{OB} = -\mathbf{a} + 6\mathbf{b}$

$\overrightarrow{OD} = \mathbf{a} + 2\mathbf{b}$ $\overrightarrow{OC} = 5\mathbf{a} + 2\mathbf{b}$

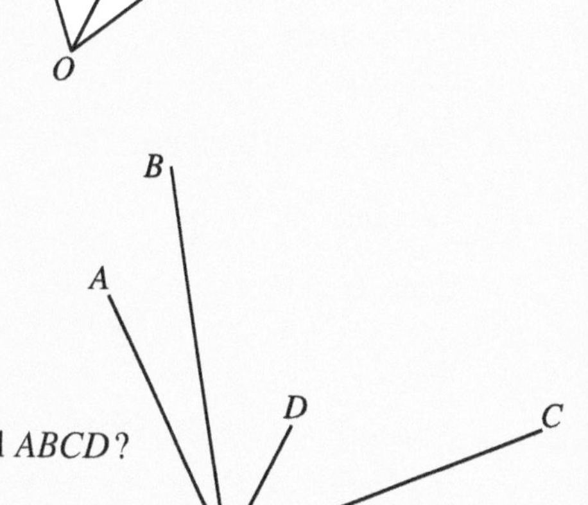

(a) Find, in terms of **a** and **b**,

$\overrightarrow{AD}$ and $\overrightarrow{BC}$.

(b) What do $\overrightarrow{AD}$ and $\overrightarrow{BC}$ tell you about quadrilateral $ABCD$?

What you need to know

- **Vector quantities**
  Quantities which have both **size** and **direction** are called **vectors**.
  Examples of vector quantities are:

Displacement
A combination of distance and direction

Velocity
A combination of speed and direction

- **Vector notation**
  Vectors can be represented by **column vectors** or by **directed line segments**.
  Vectors can be labelled using:
  capital letters to indicate the start and finish of a vector,
  bold lower case letters.

  In a **column vector:**
  The top number describes the **horizontal** part of the movement:
  $+$ = to the right $-$ = to the left
  The bottom number describes the **vertical** part of the movement:
  $+$ = upwards $-$ = downwards

  $\overrightarrow{AB} = \mathbf{a} = \begin{pmatrix} 5 \\ 3 \end{pmatrix}$

- Vectors are **equal** if they have the same length **and** they are in the same direction.
  Vectors **a** and $-$**a** have the same length **but** are in **opposite directions**.
  The vector $n\mathbf{a}$ is parallel to the vector **a**.
  The length of vector $n\mathbf{a} = n \times$ the length of vector **a**.
- **Vector addition**
  The combination of the displacement from A to B followed by the displacement from B to C is equivalent to a total displacement from A to C.

  This can be written using vectors as $\overrightarrow{AB} + \overrightarrow{BC} = \overrightarrow{AC}$

  $\overrightarrow{AC}$ is called the **resultant vector**.
- Combinations of vectors can be shown on **vector diagrams**.

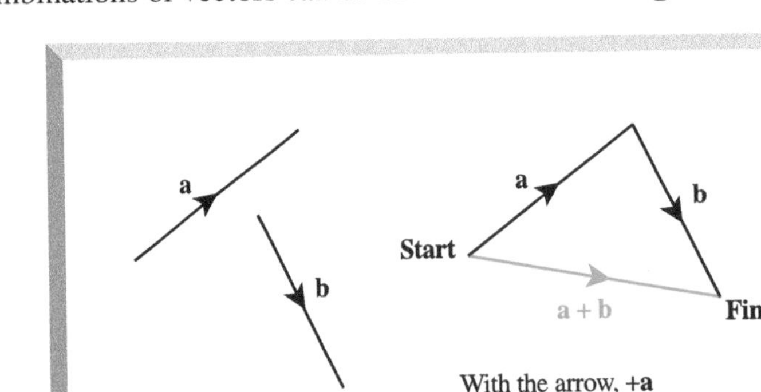

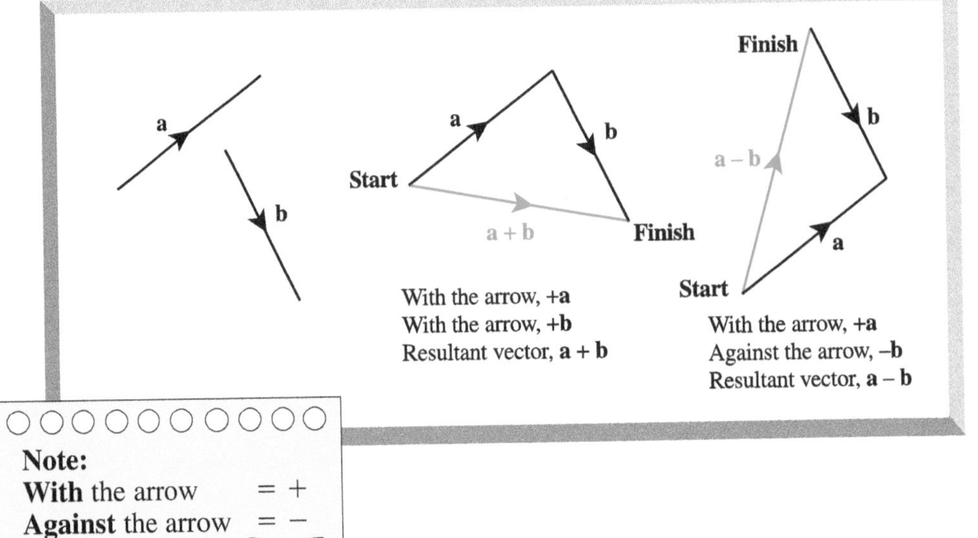

Note:
With the arrow $= +$
Against the arrow $= -$

- **Vector geometry** uses vectors to solve simple geometrical problems, which often involve parallel lines.

Review Exercise 33

The diagrams in this exercise have not been drawn accurately.

1 OAB is a triangle. E is the midpoint of AB.

$\overrightarrow{OA} = \mathbf{a}$ and $\overrightarrow{OB} = \mathbf{b}$.

Find in terms of **a** and **b**:

(a) $\overrightarrow{AB}$ (b) $\overrightarrow{AE}$ (c) $\overrightarrow{OE}$

2 $\overrightarrow{OA} = \begin{pmatrix} 3 \\ 2 \end{pmatrix}$ $\overrightarrow{OC} = \begin{pmatrix} 3 \\ 6 \end{pmatrix}$ $\overrightarrow{BC} = \begin{pmatrix} -4 \\ 1 \end{pmatrix}$

(a) (i) Write down a different column vector which is parallel to $\overrightarrow{BC}$.

(ii) Write down a column vector which is perpendicular to $\overrightarrow{OA}$.

(b) Find $\overrightarrow{AB}$ as a column vector.

AQA

3 $OABC$ is a parallelogram. M is the midpoint of BC. $\overrightarrow{OA} = \mathbf{a}$ and $\overrightarrow{OC} = \mathbf{c}$.

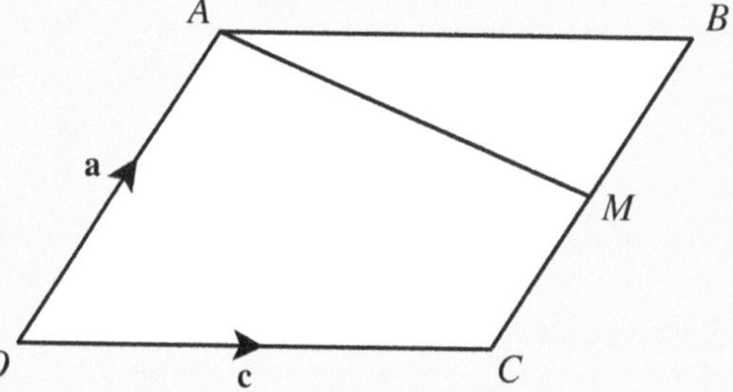

(a) Find, in terms of **a** and **c**, expressions for the following vectors.

(i) $\overrightarrow{OB}$ (ii) $\overrightarrow{CM}$ (iii) $\overrightarrow{AM}$

P is a point on AM such that AP is $\frac{2}{3}AM$.

(b) Find, in terms of **a** and **c**, expressions for

(i) $\overrightarrow{AP}$ (ii) $\overrightarrow{OP}$

(c) Describe as fully as possible what your answer to (b)(ii) tells you about the position of P.

AQA

4 $OABC$ is a quadrilateral. M, N, P and Q are the midpoints of OA, OB, AC and BC respectively.

$\overrightarrow{OA} = \mathbf{a}$, $\overrightarrow{OB} = \mathbf{b}$, $\overrightarrow{OC} = \mathbf{c}$.

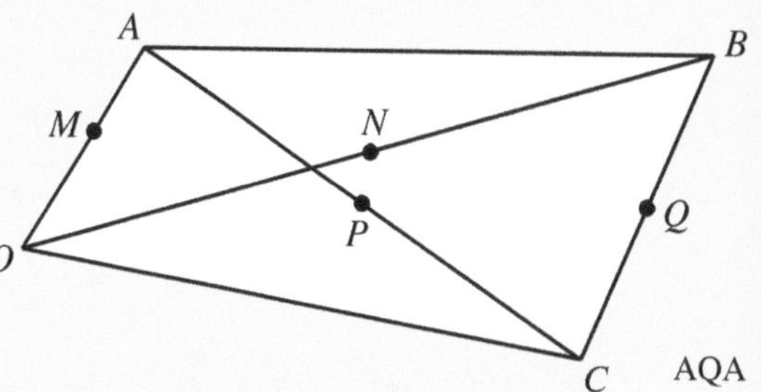

(a) Find, in terms of **a**, **b** and **c**, expressions for

(i) $\overrightarrow{BC}$ (ii) $\overrightarrow{NQ}$ (iii) $\overrightarrow{MP}$

(b) What can you deduce about the quadrilateral $MNQP$?
Give a reason for your answer.

AQA

5 The diagram shows the quadrilateral $OPQR$.

$\overrightarrow{OR} = \mathbf{a}$ and $\overrightarrow{OP} = \mathbf{b}$.

S is a point on OQ such that $\overrightarrow{OS} = \frac{1}{3}\overrightarrow{OQ}$.

$\overrightarrow{PQ} = 2\overrightarrow{OR}$.

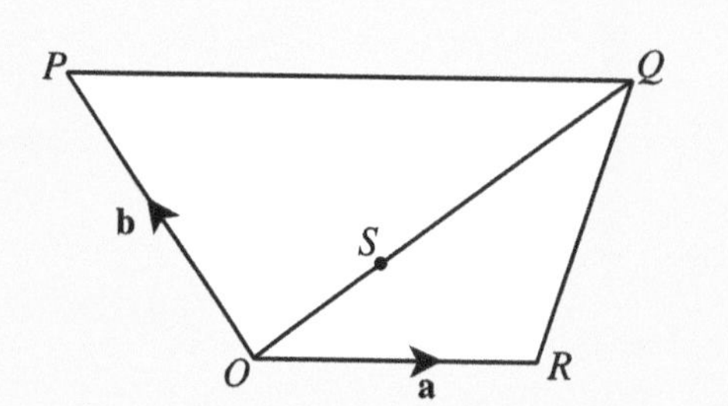

(a) Find in terms of **a** and **b**:

(i) $\overrightarrow{OQ}$ (ii) $\overrightarrow{PS}$ (iii) $\overrightarrow{RS}$

(b) (i) What is the ratio $PS : SR$?

(ii) Explain why vectors $\overrightarrow{PS}$ and $\overrightarrow{RS}$ indicate that the points P, S and R lie on a straight line.

AQA

Further Trigonometry

So far we have found sines, cosines and tangents of angles between 0° and 90°, but it is possible to find the sine, cosine and tangent of any angle.

Trigonometric functions and their graphs

Activity

The sine function

Copy and complete this table.
Use your calculator to find the value of $\sin x°$, correct to two decimal places, where necessary.

$x°$	0	30	60	90	120	150	180	210	240	270	300	330	360
$\sin x°$	0	0.5	0.87										

Plot these values on graph paper.
Draw the graph of the sine function by joining the points with a smooth curve.

What is the maximum value of $\sin x°$?
What is the minimum value of $\sin x°$?

Describe what happens to the sine function when $x > 360°$.
Describe what happens to the sine function when $x < 0°$.

You might also try this activity using a graphical calculator, or a computer and suitable software.

In a similar way, draw the graphs of the **cosine function** and the **tangent function**.
What do you notice about the graphs of each of these functions?

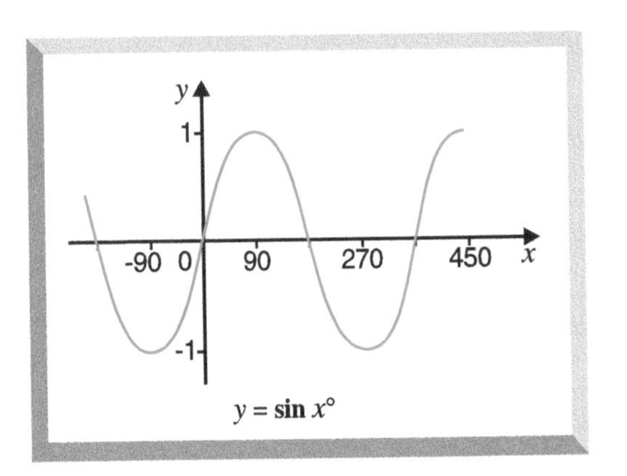

$y = \sin x°$

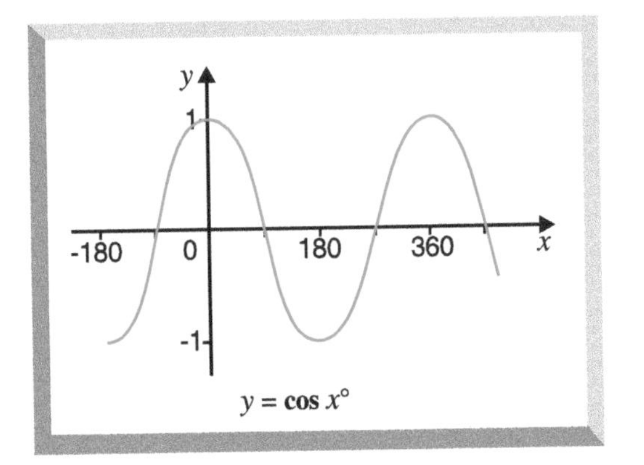

$y = \cos x°$

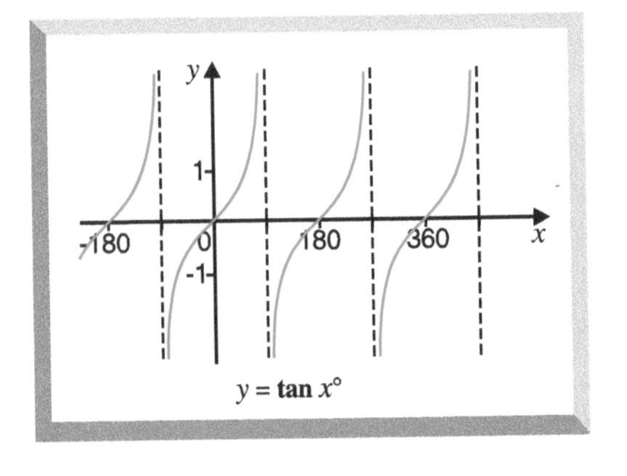

$y = \tan x°$

For every angle $x°$, the signs of $\sin x°$, $\cos x°$ and $\tan x°$ can be summarised as follows:

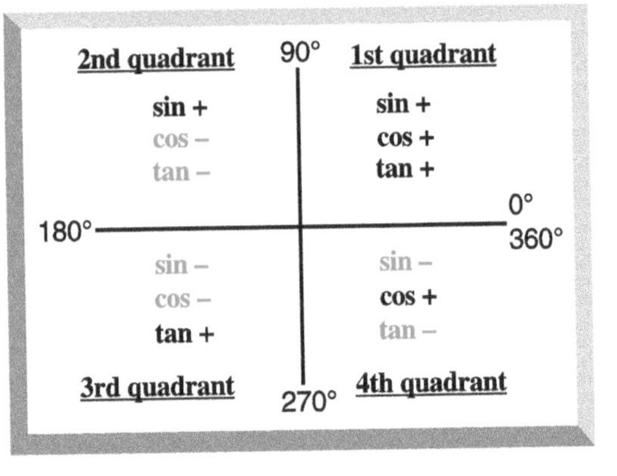

Sines, cosines and tangents of angles between 0° and 360°

Consider a point P. Line OP makes an angle $x°$ with line OX.
If $OP = 1$ unit, then in triangle OMP:

$\sin x° = \frac{\text{opp}}{\text{hyp}}$	$\cos x° = \frac{\text{adj}}{\text{hyp}}$	$\tan x° = \frac{\text{opp}}{\text{adj}}$
$\sin x° = \frac{PM}{OP}$	$\cos x° = \frac{OM}{OP}$	$\tan x° = \frac{PM}{OM}$
$\sin x° = \frac{PM}{1}$	$\cos x° = \frac{OM}{1}$	$\tan x° = \frac{\sin x°}{\cos x°}$
$PM = \sin x°$	$OM = \cos x°$	*Explain why.*

Positive angles are measured **anticlockwise** from the axis OX.

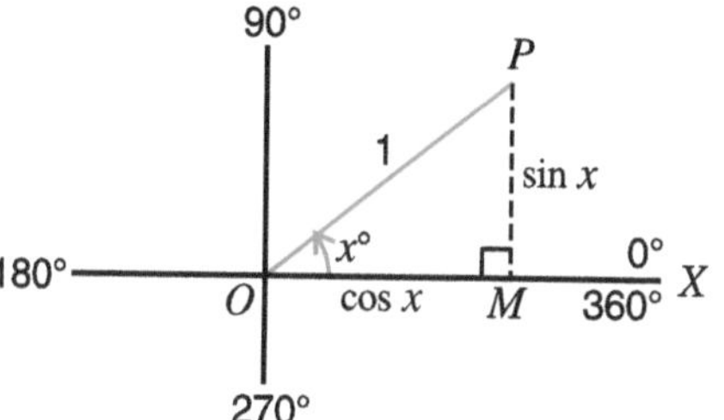

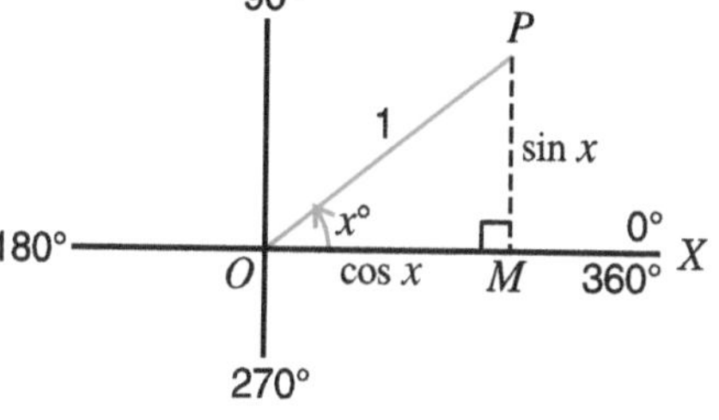

2nd quadrant

$\angle POM = (180 - x)°$

When $90° < x < 180°$,
$\sin x$ is positive.

We can write:
$\sin x° = \sin (180 - x)°$

Example
$\sin 150° = \sin (180 - 150)°$
$= \sin 30°$
$= 0.5$

3rd quadrant

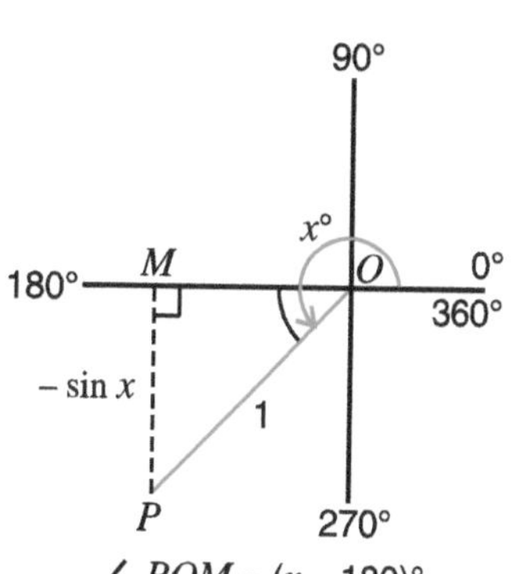

$\angle POM = (x - 180)°$

When $180° < x < 270°$,
$\sin x$ is negative.

We can write:
$\sin x° = -\sin (x - 180)°$

Example
$\sin 210° = -\sin (210 - 180)°$
$= -\sin 30°$
$= -0.5$

4th quadrant

$\angle POM = (360 - x)°$

When $270° < x < 360°$,
$\sin x$ is negative.

We can write:
$\sin x° = -\sin (360 - x)°$

Example
$\sin 330° = -\sin (360 - 330)°$
$= -\sin 30°$
$= -0.5$

Activity

In a similar way investigate the values of $\cos x$ and $\tan x$ for values of x between 0° and 360°.

EXAMPLE

Express the following in terms of the sin, cos or tan of an acute angle.
(a) $\tan 100°$ (b) $\sin 200°$ (c) $\cos 300°$ (d) $\sin (-10)°$
(e) $\cos (-20)°$ (f) $\tan 400°$ (g) $\sin 500°$

(a) $\tan 100° = -\tan (180 - 100)° = -\tan 80°$
(b) $\sin 200° = -\sin (200 - 180)° = -\sin 20°$
(c) $\cos 300° = \cos (360 - 300)° = \cos 60°$

(d) $\sin (-10)° = -\sin 10°$
(e) $\cos (-20)° = \cos 20°$

Negative angles are measured **clockwise** from the axis OX.

(f) $\tan 400° = \tan (400 - 360)° = \tan 40°$
(g) $\sin 500° = \sin (500 - 360)° = \sin 140°$
$= \sin (180 - 140)° = \sin 40°$

Angles greater than 360°
Subtract 360°, or multiples of 360°, to get the equivalent angle between 0° and 360°.

Exercise 34.1

Do not use a calculator for this exercise.

1 State whether the following are positive or negative.

(a) sin 50° (b) cos 50° (c) tan 50° (d) sin 100°
(e) cos 100° (f) tan 100° (g) cos 150° (h) sin 200°
(i) tan 250° (j) cos 300° (k) sin 350° (l) tan 87°
(m) cos 143° (n) sin 117° (o) tan 162° (p) cos 296°
(q) tan 321° (r) cos 196° (s) sin 218° (t) cos 400°
(u) tan 500° (v) sin 500° (w) cos (−30)° (x) sin (−100)°
(y) cos (−100)° (z) tan (−100)°

2 Express each of the following in terms of the sin, cos or tan of an acute angle.

(a) tan 100° (b) cos 150° (c) sin 200° (d) tan 250°
(e) cos 300° (f) sin 120° (g) tan 170° (h) cos 210°
(i) sin 290° (j) tan 330° (k) cos 370° (l) sin 260°
(m) tan 370° (n) sin 480° (o) cos 600° (p) sin (−50)°
(q) cos (−100)° (r) tan (−150)°

3 The exact values of sin 45°, cos 45° and tan 45° can be found by using this triangle.

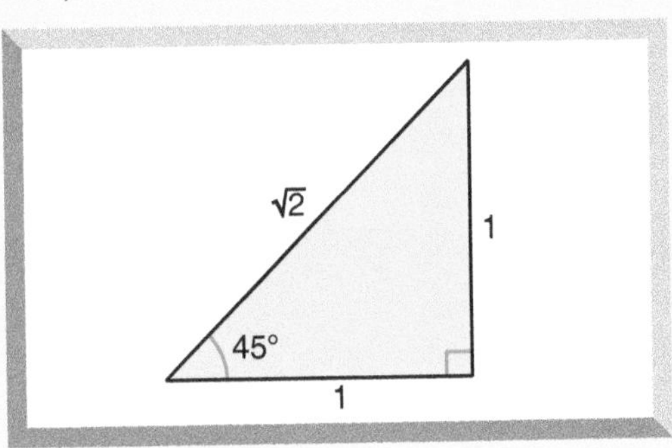

E.g. $\sin 45° = \frac{1}{\sqrt{2}}$

(a) Write down the exact value of
(i) cos 45°, (ii) tan 45°.

(b) Hence, write down the exact values of
(i) cos 135°, (ii) tan 225°, (iii) sin 315°,
(iv) sin 135°, (v) cos 225°, (vi) tan 315°.

4 The exact values of sin 30°, cos 30°, tan 30° and sin 60°, cos 60°, tan 60° can be found by using this triangle.

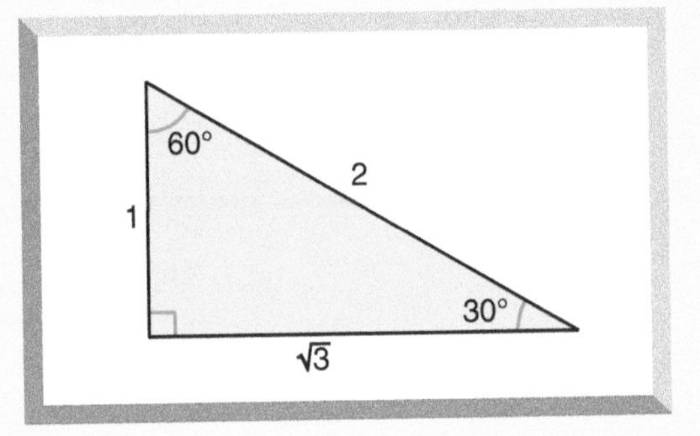

(a) Write down the exact values of sin 30°, cos 30°, tan 30° and hence, give the exact values of
(i) sin 150°, (ii) cos 210°, (iii) tan 330°, (iv) cos 330°.

(b) Write down the exact values of sin 60°, cos 60°, tan 60° and hence, give the exact values of
(i) cos 120°, (ii) tan 240°, (iii) cos 240°, (iv) sin 300°.

Finding angles

For which angles, between 0° and 360°, is $\cos p = 0.412$?

$\cos p = 0.412$

$\cos p$ is positive, so, possible angles are in the 1st and 4th quadrants.
This can be shown on a diagram.
Because of the symmetries of the graphs of trigonometric functions, angles formed between the lines and the horizontal axis are always **equal**.

$p = \cos^{-1}(0.412)$

Using a calculator.

$p = 65.669...$

$p = 65.7°$ to 1 d.p.

Positive angles are measured anticlockwise.

$p = 65.7°$ and $p = 360° - 65.7° = 294.3°$

So, $p = 65.7°$ and $294.3°$, correct to 1 d.p.

sin + 90° all +
65.7°
180° 0° 360°
65.7°
tan + 270° cos +

EXAMPLE

For which angles, between 0° and 360°, is $\tan p = -1.5$?

$\tan p = -1.5$

$\tan p$ is negative, so, possible angles are in the 2nd and 4th quadrants.

$p = \tan^{-1}(-1.5)$

$p = -56.309...$

$p = -56.3°$, to 1 d.p.

$p = 180° - 56.3° = 123.7°$, and $p = 360° - 56.3° = 303.7°$

So, $p = 123.7°$ and $303.7°$, correct to 1 d.p.

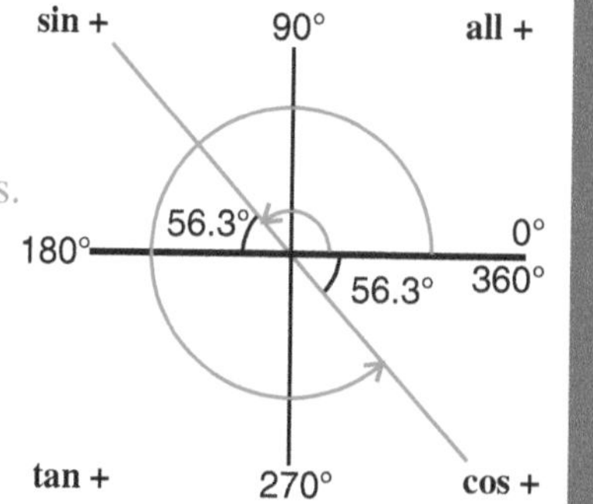

Exercise 34.2

Use a calculator for this exercise.

1 Find the values of the following. Give your answers to three decimal places.

(a) sin 100° (b) cos 150° (c) tan 200° (d) sin 250°
(e) cos 300° (f) tan 350° (g) tan 187° (h) cos 187°
(i) cos 143° (j) sin 117° (k) tan 162° (l) cos 296°
(m) tan 321° (n) cos 196° (o) sin 218° (p) cos 400°
(q) tan 500° (r) sin 500° (s) cos (−30)° (t) sin (−100)°

2 For each of the following, find all the values of p between 0° and 360°.

(a) $\cos p = 0.5$ (b) $\sin p = -0.5$ (c) $\tan p = 1$ (d) $\sin p = 0.5$
(e) $\tan p = -1$ (f) $\cos p = -0.5$ (g) $\sin p = -0.766$ (h) $\cos p = 0.766$
(i) $\sin p = 0.866$ (j) $\tan p = -2.050$ (k) $\tan p = 0.193$ (l) $\cos p = 0.565$
(m) $\sin p = 0.342$ (n) $\cos p = -0.866$ (o) $\tan p = 0.700$

3 Find the values of x, between 0° and 720°, where $\sin x = -0.75$.
Give your answers to one decimal place.

4 One solution of the equation $\cos x = -0.42$ is $x = 115°$, to the nearest degree.
Find **all** the other solutions to the equation $\cos x = -0.42$ for values of the x between −360° and 360°.

5 Solve the equation $\sin x = 0.8$ for $-360° \leqslant x \leqslant 360°$.
Give your answers to the nearest degree.

Another look at the graphs of trigonometric functions

At the beginning of the chapter we looked at the graphs of the sine, cosine and tangent functions.
Each of the functions is called a **periodic function**.
After a **period** (distance along the x axis) each graph repeats itself.

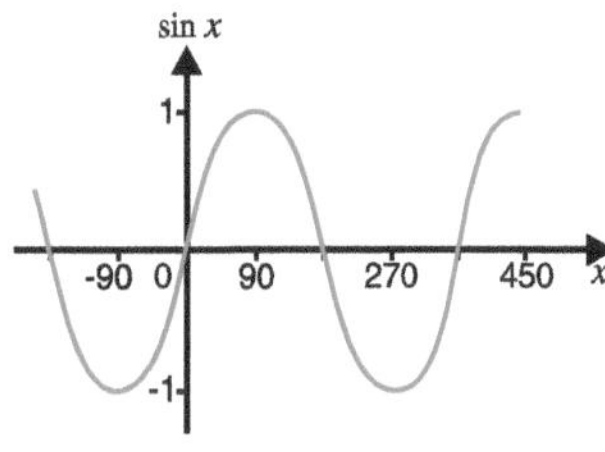

The **sine function** is a periodic function with period 360°.
$-1 \leqslant \sin x \leqslant 1$

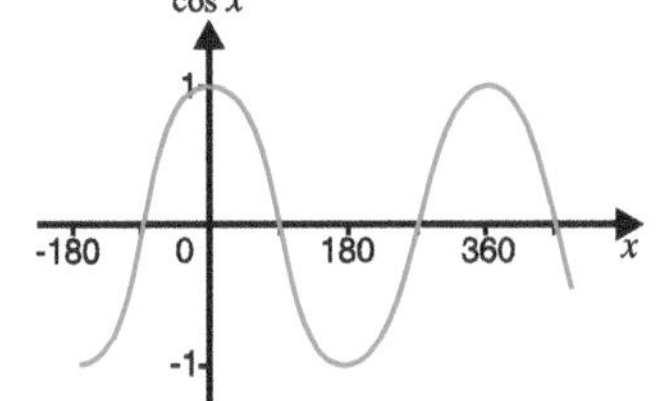

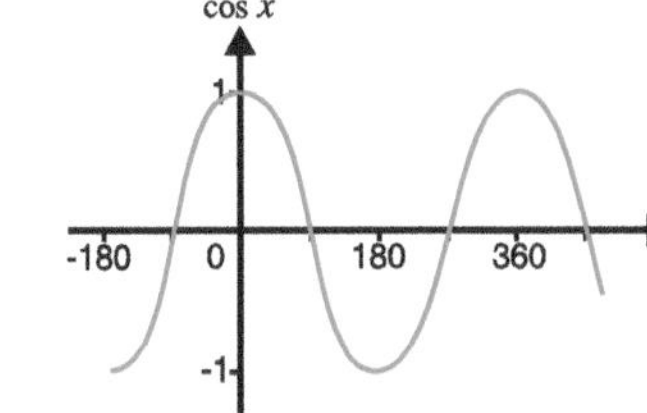

The **cosine function** is a periodic function with period 360°.
$-1 \leqslant \cos x \leqslant 1$

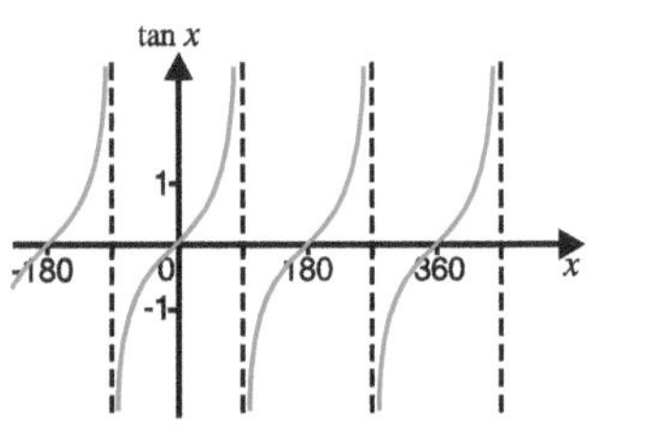

The **tangent function** is a periodic function with period 180°.
Tan x is undefined at 90°, 270°, ...

EXAMPLE

Compare the following graphs: $y = \sin x$, $y = 2\sin x$, $y = \sin 2x$.

x	0	45	90	135	180	225	270	315	360
$y = \sin x$	0	0.71	1	0.71	0	−0.71	−1	−0.71	0
$y = 2\sin x$	0	1.41	2	1.41	0	−1.41	−2	−1.41	0
$y = \sin 2x$	0	1	0	−1	0	1	0	−1	0

$y = \sin x$ Period: 360° $-1 \leqslant \sin x \leqslant 1$
$y = 2\sin x$ Period: 360° $-2 \leqslant 2\sin x \leqslant 2$
$y = \sin 2x$ Period: 180° $-1 \leqslant \sin 2x \leqslant 1$

Exercise 34.3

1. Use the graphs of the sine function and the cosine function, to find the values of x, between 0° and 360°, when:
   (a) (i) $\sin x = 1$ (ii) $\sin x = 0$ (iii) $\sin x = -1$
   (b) (i) $\cos x = 1$ (ii) $\cos x = 0$ (iii) $\cos x = -1$

2 Given that $\tan 45° = 1$, use the graph of the tangent function to find all the values of x, between 0° and 360°, when

(a) $\tan x = 1$ (b) $\tan x = 0$ (c) $\tan x = -1$

3 (a) Draw the graph of $y = 3\sin x$ for values of x between 0° and 360°.
(b) Use your graph to find:
(i) x when $y = 1.50$ (ii) y when $x = 240°$

4 (a) Draw the graph of $y = 3\cos x$ for values of x between 0° and 360°.
(b) Use your graph to solve the equation $3\cos x = 0.6$.

In questions 5 to 8, draw graphs for values of x between 0° and 360°.

5 Compare the graph of $y = 3\sin x$ with $y = \sin x$
and the graph of $y = 3\cos x$ with $y = \cos x$.
What do you notice?
Hence, sketch the graphs of:
(a) (i) $y = 2\sin x$ (ii) $y = 5\sin x$ (iii) $y = \frac{1}{2}\sin x$
(b) (i) $y = 2\cos x$ (ii) $y = 5\cos x$ (iii) $y = \frac{1}{2}\cos x$

6 (a) Draw the graph of $y = \sin 2x$.
(b) Compare this graph with the graph of $y = \sin x$.
What do you notice?
(c) Hence, sketch the graph of $y = \sin 3x$.

7 (a) Draw the graph of $y = \cos 2x$.
(b) Compare this graph with the graph of $y = \cos x$.
What do you notice?
(c) Hence, sketch the graph of $y = \cos 3x$.

8 (a) Draw graphs of the following.
(i) $y = \sin x + 1$ (ii) $y = \sin x - 2$
(b) Compare these graphs with the graph of $y = \sin x$.
What do you notice?
(c) Hence, sketch the graphs of:
(i) $y = \cos x + 2$ (ii) $y = \cos x - 1$

Triangles which are not right-angled

In Chapter 29 we used trigonometry to find sides and angles in right-angled triangles. However, not all triangles are right-angled and to solve problems involving acute-angled and obtuse-angled triangles we need to know further rules.

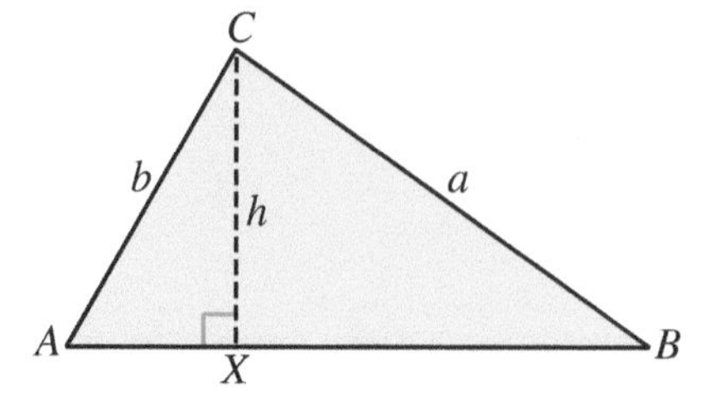

Remember:
Angles are labelled with **capital letters** and the **sides** opposite the angles with **lower case letters**.

The **altitude**, or height of the triangle, is shown by the line CX.
The line CX divides triangle ABC into two right-angled triangles.

In triangle CAX.

$\sin A = \frac{h}{b}$

This can be rearranged to give:
$h = b \sin A$

In triangle CBX.

$\sin B = \frac{h}{a}$

This can be rearranged to give:
$h = a \sin B$

$h = b \sin A$ and $h = a \sin B$,
so, $b \sin A = a \sin B$,
which can be rearranged as:

$$\frac{a}{\sin A} = \frac{b}{\sin B}$$

Activity

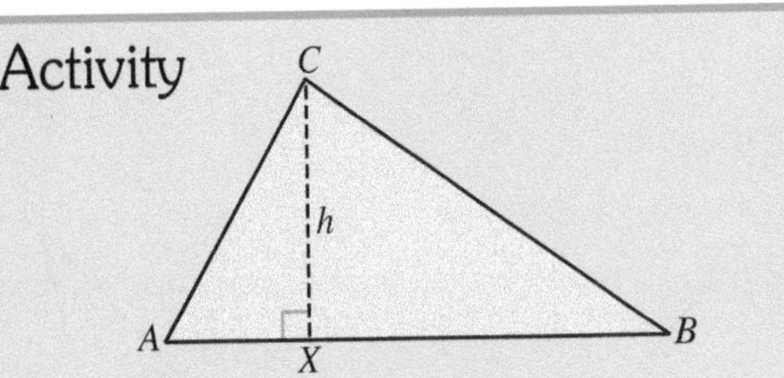

Find rules to connect:
$\sin A$ and $\sin C$,
$\sin B$ and $\sin C$.

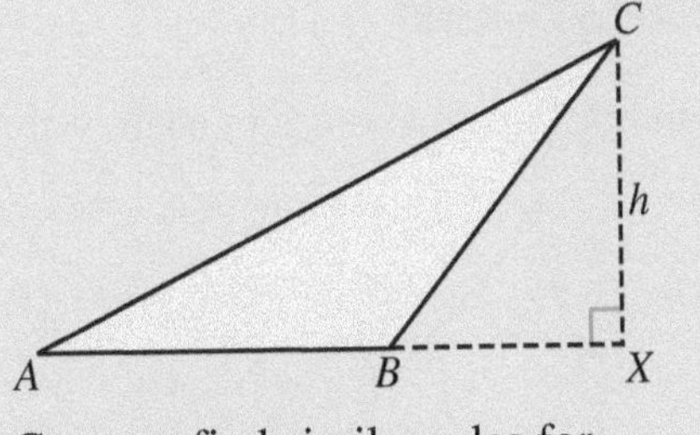

Can you find similar rules for obtuse-angled triangles?

The Sine Rule

In any triangle labelled ABC it can be proved that:

$$\frac{a}{\sin A} = \frac{b}{\sin B} = \frac{c}{\sin C}$$

The Sine Rule can also be written as:

$$\frac{\sin A}{a} = \frac{\sin B}{b} = \frac{\sin C}{c}$$

Finding sides

To find a side using the Sine Rule you need:
two angles of known size, **and**
the length of a side which is opposite one of the known angles.

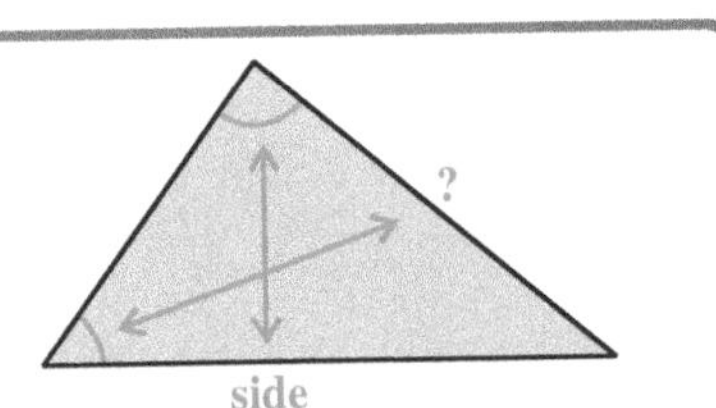

EXAMPLE

In triangle ABC, angle $A = 37°$, angle $C = 72°$ and $b = 12$ cm.
Calculate a, correct to 3 significant figures.

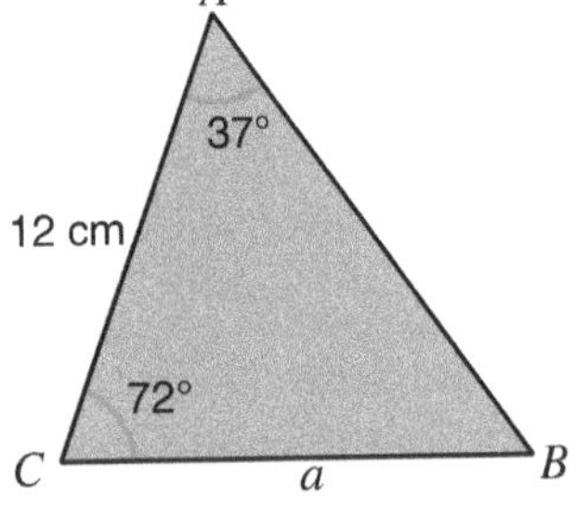

Using the Sine Rule:

$$\frac{a}{\sin A} = \frac{b}{\sin B} = \frac{c}{\sin C}$$

Substitute known values.

$$\frac{a}{\sin 37°} = \frac{12}{\sin 71°} = \frac{c}{\sin 72°}$$

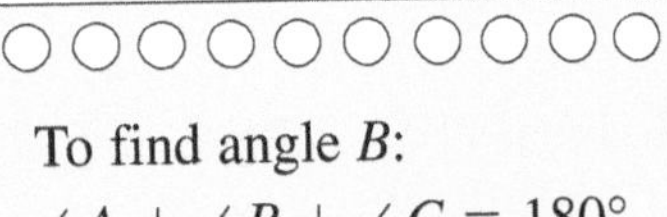

To find angle B:
$\angle A + \angle B + \angle C = 180°$
$37° + \angle B + 72° = 180°$
$\angle B = 71°$

Using:

$$\frac{a}{\sin 37°} = \frac{12}{\sin 71°}$$

Multiply both sides by $\sin 37°$.

$$a = \frac{12 \times \sin 37°}{\sin 71°}$$

$= 7.637\ldots$

$a = 7.64$ cm, correct to 3 s.f.

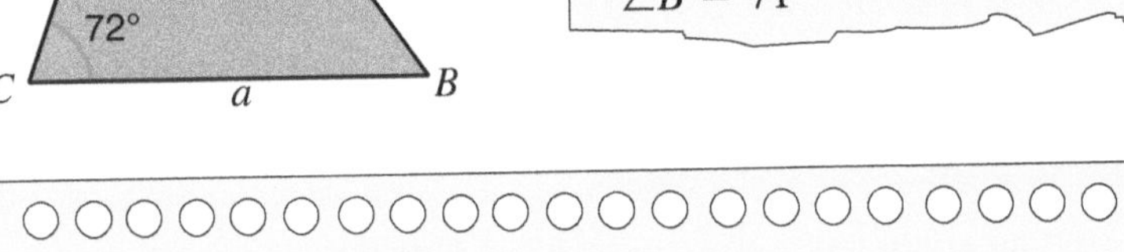

Alternative method:
First rearrange the equation, then substitute known values.

Use $\frac{a}{\sin A} = \frac{b}{\sin B}$

Rearrange to give $a = \frac{b \sin A}{\sin B}$

Substitute known values $a = \frac{12 \times \sin 37°}{\sin 71°}$

Exercise 34.4

1 Find the side marked a in each of the following triangles.

(a)

(b)

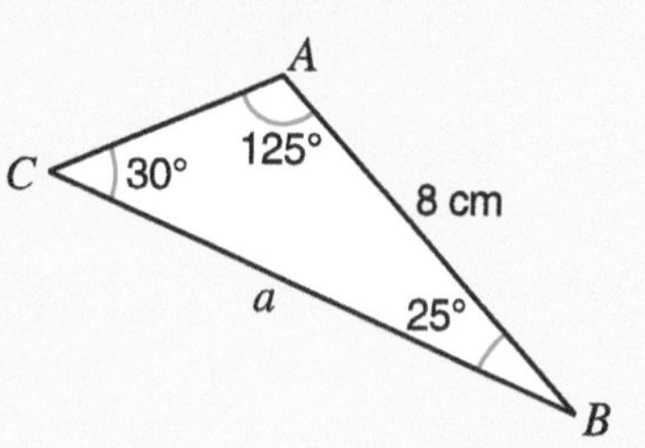

(c)

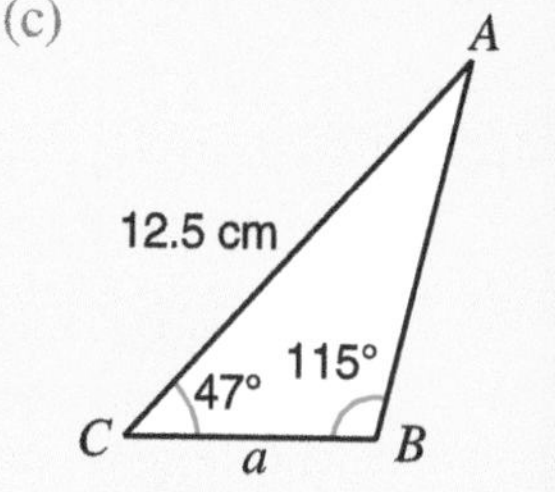

2 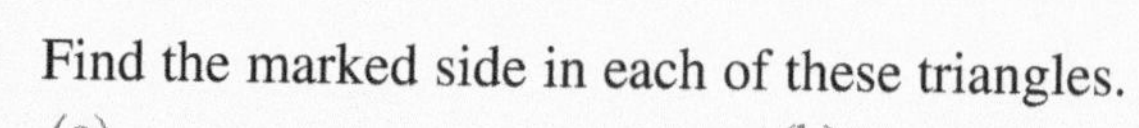
Find the marked side in each of these triangles.

(a)

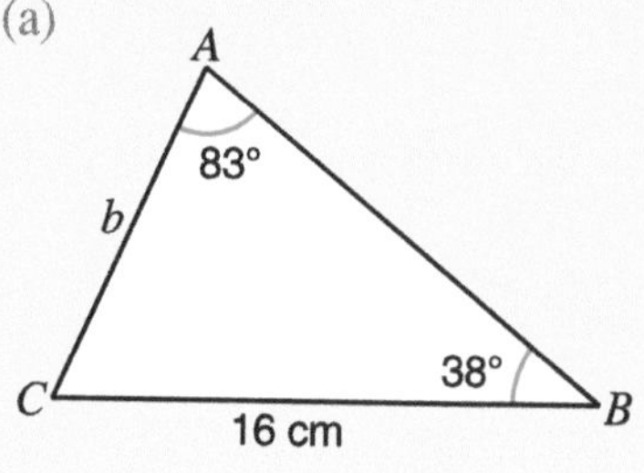

(b)

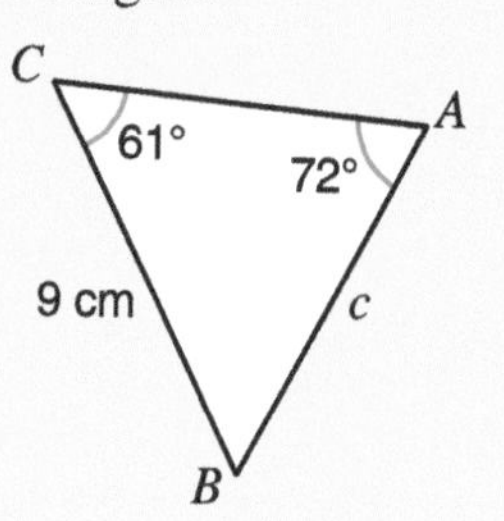

(c)

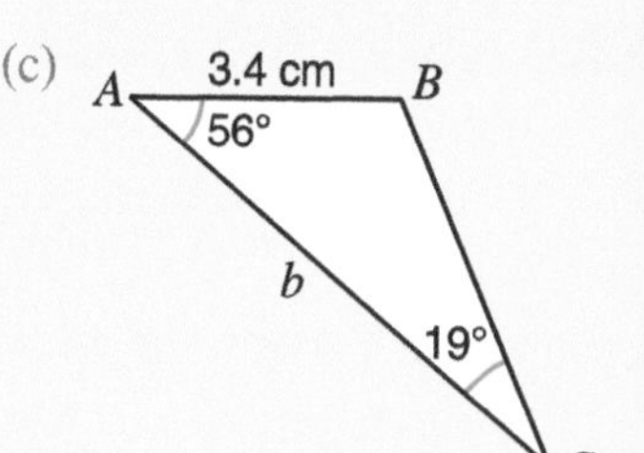

(d)

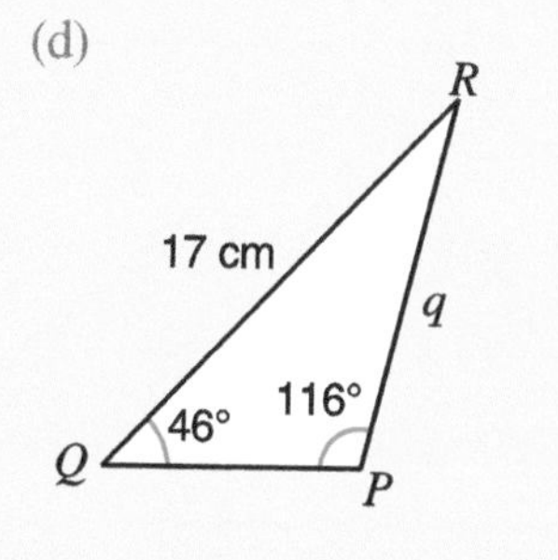

(e)

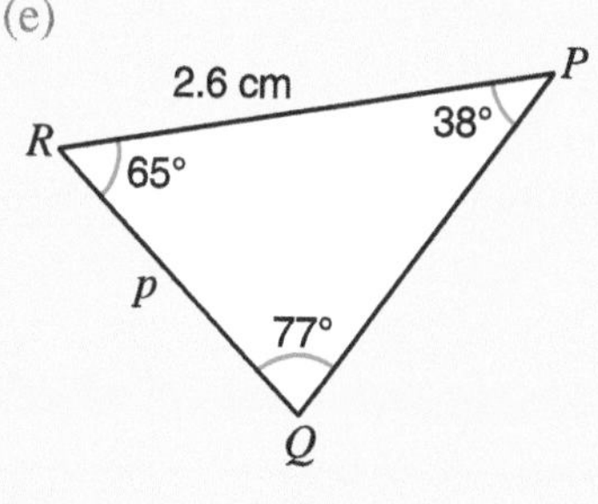

(f)

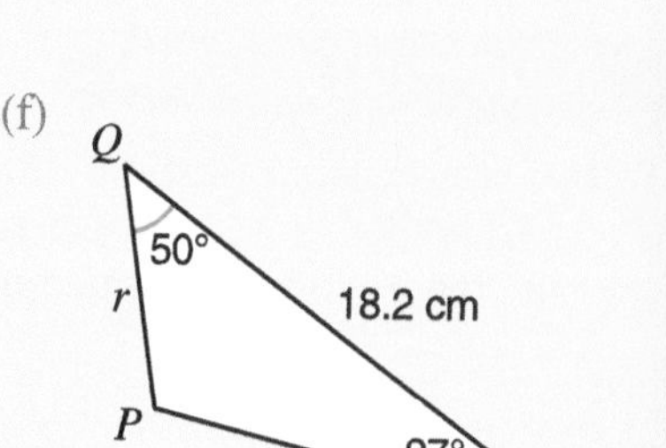

3 Find the remaining angles and sides of these triangles.

(a) ΔABC, when $\angle BAC = 57°$, $\angle ABC = 68°$ and $BC = 6.7$ cm.
(b) ΔLMN, when $\angle LMN = 33.2°$, $\angle LNM = 75.6°$ and $LN = 3.3$ cm.
(c) ΔPQR, when $\angle PQR = 62.8°$, $\angle PRQ = 47.4°$ and $PQ = 12.3$ cm.
(d) ΔSTU, when $\angle TSU = 94.9°$, $\angle STU = 53.3°$ and $SU = 19.4$ cm.
(e) ΔXYZ, when $\angle XYZ = 108.6°$, $\angle YXZ = 40.2°$ and $XZ = 13.8$ cm.

4 In ΔPQR, $\angle PQR = 110°$, $\angle QRP = 29°$ and $PQ = 5$ cm.
Calculate
(a) the length of PR,
(b) the length of QR.
Give your answers correct to one decimal place.

Finding angles

To find an angle using the Sine Rule you need:
the length of the side opposite the angle you are trying to find, **and**
the length of a side opposite an angle of known size.

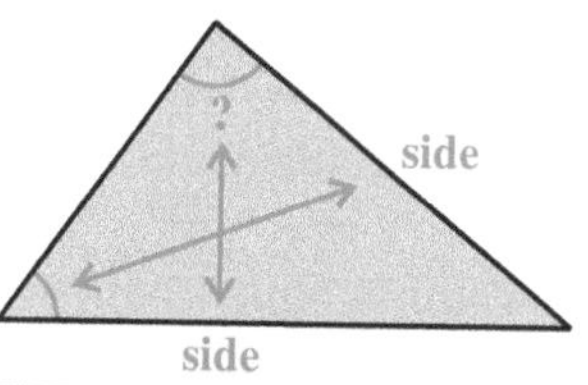

EXAMPLE

Angle $A = 42°$, $a = 6$ cm and $c = 4$ cm.
Calculate angle C, correct to one decimal place.

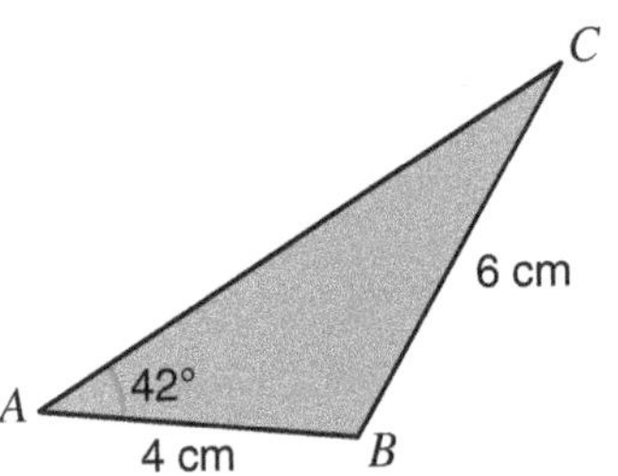

Using the Sine Rule:

$$\frac{\sin A}{a} = \frac{\sin B}{b} = \frac{\sin C}{c}$$

Substitute known values.

$$\frac{\sin 42°}{6} = \frac{\sin B}{b} = \frac{\sin C}{4}$$

Using:

$$\frac{\sin 42°}{6} = \frac{\sin C}{4}$$

Multiply both sides by 4.

$$\sin C = \frac{4 \times \sin 42°}{6}$$

$\sin C = 0.446...$
$C = \sin^{-1}(0.446...)$
$C = 26.49...$
$C = 26.5°$, correct to 1 d.p.

$C = \sin^{-1}(0.446...)$
gives possible angles of 26.5° and 153.5°.
$C \neq 153.5°$, since in an obtuse-angled triangle, the obtuse angle is always opposite the longest side of the triangle.

The ambiguous case

There are two angles between 0° and 180° which have the same sine.
When we use the sine rule to find an angle we must therefore look at the information to see if there are two possible values for the angle.

EXAMPLE

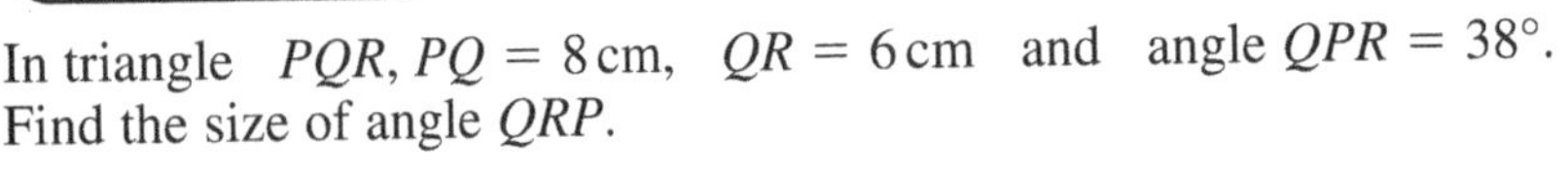

In triangle PQR, $PQ = 8$ cm, $QR = 6$ cm and angle $QPR = 38°$.
Find the size of angle QRP.

Using this information **two** triangles can be drawn.
Use a ruler and compasses to construct the triangles.

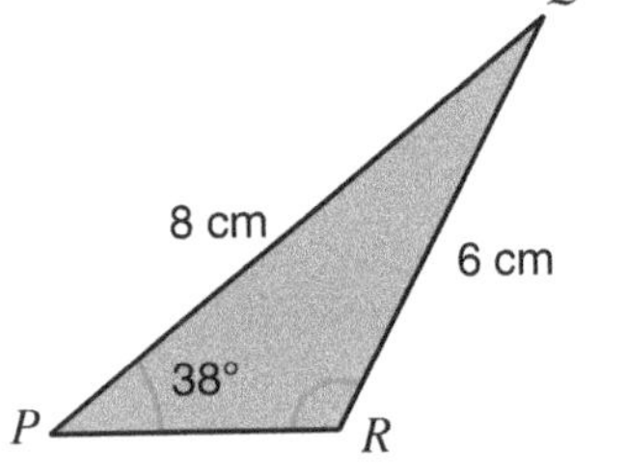

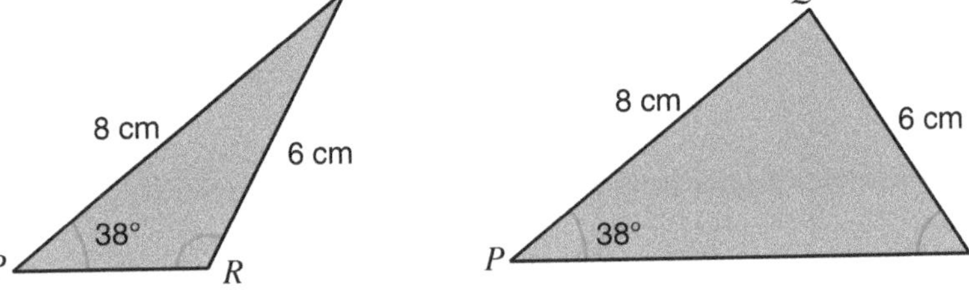

Using the Sine Rule:

$$\frac{\sin P}{p} = \frac{\sin Q}{q} = \frac{\sin R}{r}$$

Substitute known values.

$$\frac{\sin 38°}{6} = \frac{\sin Q}{q} = \frac{\sin R}{8}$$

Using:

$$\frac{\sin 38°}{6} = \frac{\sin R}{8}$$

Multiply both sides by 8.

$$\sin R = \frac{8 \times \sin 38°}{6}$$

$\sin R = 0.820...$
$R = \sin^{-1}(0.820...)$
$R = 55.17...$
$R = 55.2°$
Also, $R = 180° - 55.2° = 124.8°$
$\angle QRP = 55.2°$ or $124.8°$, correct to 1 d.p.

Exercise 34.5

1 Find angle A in each of these acute-angled triangles.

(a)

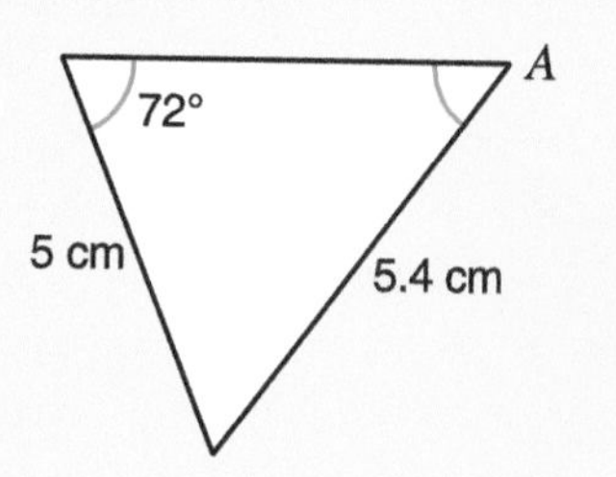

(b)

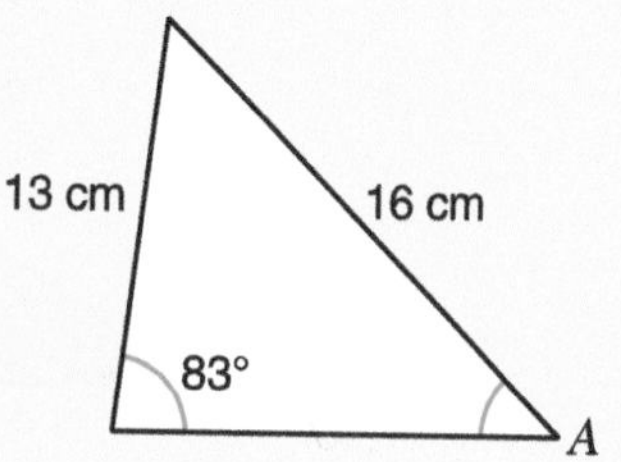

(c)

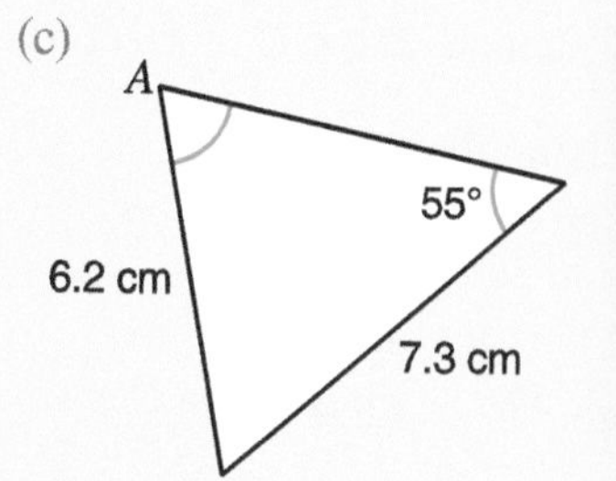

2 Find angle A in each of these obtuse-angled triangles.

(a)

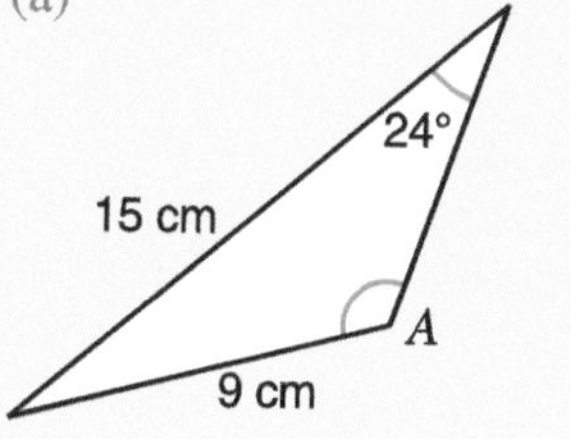

(b)

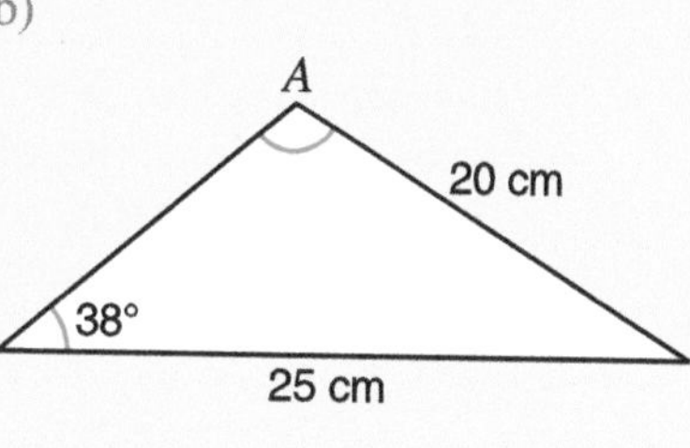

(c)

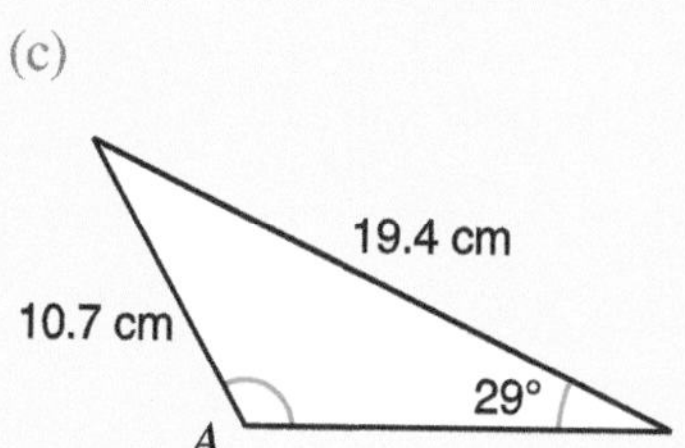

3 In triangle ABC, $AB = 6\,\text{cm}$, $BC = 4\,\text{cm}$ and angle $BAC = 30°$.
There are two possible triangles that can be constructed.
Calculate the two possible values of angle BCA.

4 In triangle PQR, $QR = 7.5\,\text{cm}$, $RP = 7\,\text{cm}$ and angle $PQR = 60°$.
There are two possible triangles that can be constructed.
Find the missing angles and sides of both triangles.

5 Find the missing angles and sides in each of the following triangles.

(a) ΔABC, when $AB = 8.3\,\text{cm}$, $AC = 8.9\,\text{cm}$ and angle $ABC = 69.3°$.

(b) ΔDEF, when $DE = 8.5\,\text{cm}$, $DF = 16.34\,\text{cm}$ and angle $DEF = 125°$.

Deriving the Cosine Rule

$AB = c$

Let $AX = x$

So, $BX = c - x$

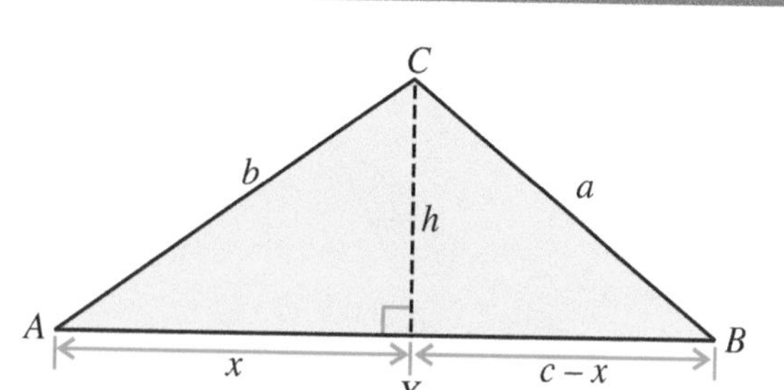

In triangle ACX.
Using Pythagoras' Theorem.

$b^2 = x^2 + h^2$

Using the cosine ratio.

$\cos A = \dfrac{\text{adj}}{\text{hyp}}$

$\cos A = \dfrac{x}{b}$

$x = b \cos A$

In triangle BCX.
Using Pythagoras' Theorem.

$a^2 = (c - x)^2 + h^2$

$a^2 = c^2 - 2cx + x^2 + h^2$

Replace: $x^2 + h^2$ by b^2 and x by $b \cos A$.

$a^2 = c^2 - 2c(b \cos A) + b^2$

This can be rearranged as:
$a^2 = b^2 + c^2 - 2bc \cos A$, and is called the **Cosine Rule**.

The Cosine Rule

In any triangle labelled ABC it can be proved that:

$$a^2 = b^2 + c^2 - 2bc \cos A$$

In a similar way, we can find the Cosine Rule for b^2 and c^2.

$$b^2 = a^2 + c^2 - 2ac \cos B$$
$$c^2 = a^2 + b^2 - 2ab \cos C$$

When using the Cosine Rule to find the size of an angle it is sometimes easier to rearrange the above formulae as:

$$\cos A = \frac{b^2 + c^2 - a^2}{2bc} \qquad \cos B = \frac{a^2 + c^2 - b^2}{2ac} \qquad \cos C = \frac{a^2 + b^2 - c^2}{2ab}$$

Finding sides

To find a side using the Cosine Rule you need:
two sides of known length, **and**
the size of the angle between the known sides.

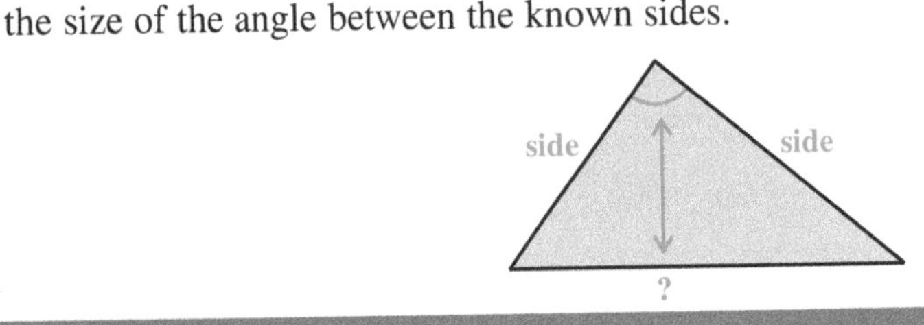

EXAMPLE

Calculate the length of side a.

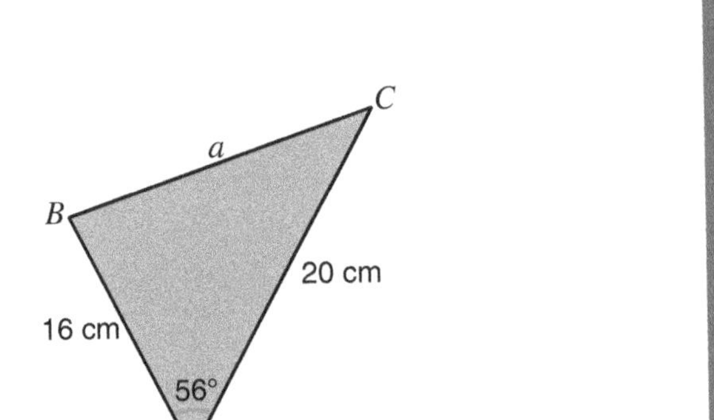

Using the Cosine Rule:
$a^2 = b^2 + c^2 - 2bc \cos A$
Substitute known values.
$a^2 = 20^2 + 16^2 - 2 \times 20 \times 16 \times \cos 56°$
$= 400 + 256 - 357.88...$
$= 298.116...$
Take the square root.
$a = 17.266...$
$a = 17.3$ cm, correct to 3 s.f.

Exercise 34.6

1. Calculate the lengths of the sides marked a.
Give your answers correct to three significant figures.

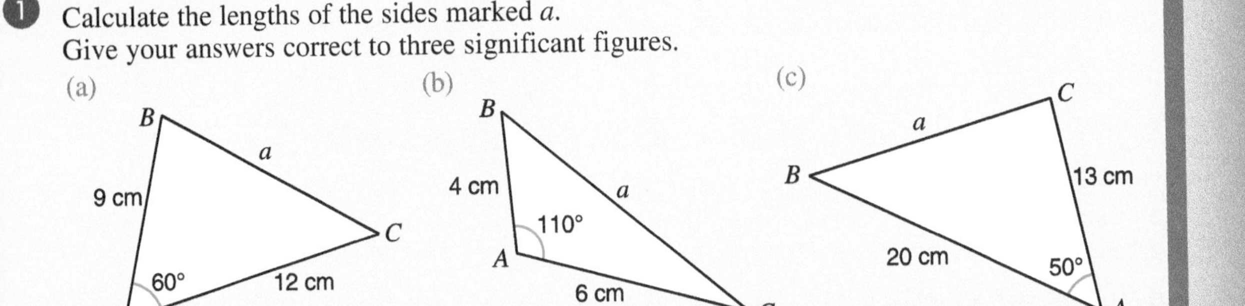

2 Calculate the lengths of the sides marked with letters.
Give your answers correct to three significant figures.

(a)

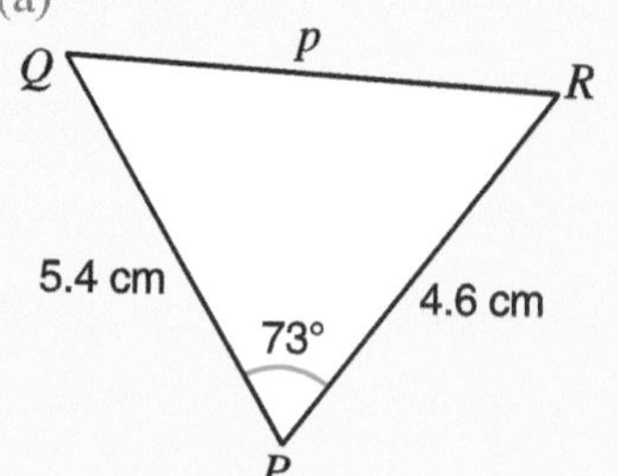

(b)

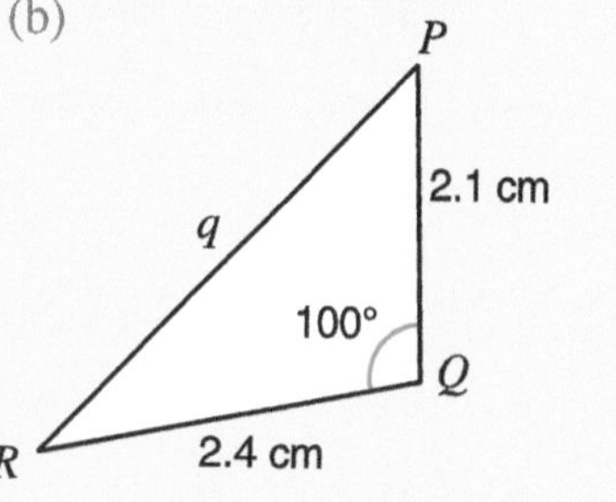

(c)

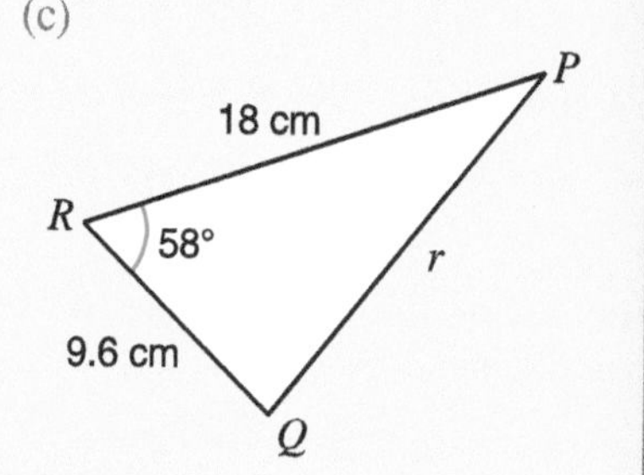

3 In triangle ABC, $AB = 8$ cm, $AC = 13$ cm and $\angle BAC = 70°$.
Calculate BC.

4 In triangle LMN, $LM = 16$ cm, $MN = 9$ cm and $\angle LMN = 46.8°$.
Calculate LN.

5 In triangle PQR, $PQ = 9$ cm, $PR = 5.4$ cm and $\angle RPQ = 135°$.
Calculate QR.

6 In triangle XYZ, $XY = 7.5$ cm, $YZ = 13$ cm and $\angle XYZ = 120°$.
Calculate XZ.

7 L is 19 km from K on a bearing of 068°.
K is 13 km due north of M.
Calculate the distance of L from M.

Finding angles

To find an angle using the Cosine Rule you need:
three sides of known length.

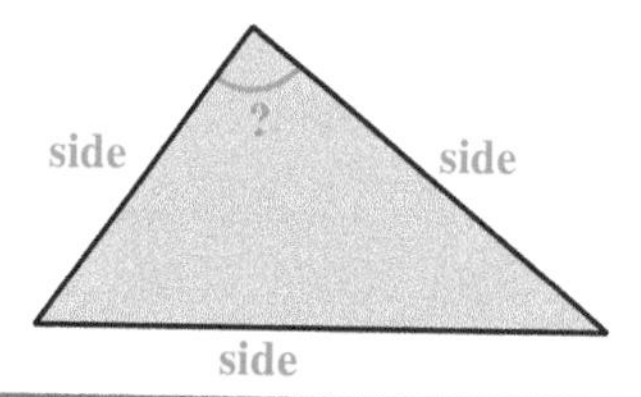

EXAMPLE

Calculate the size of angle A.

Using the Cosine Rule:

$\cos A = \frac{b^2 + c^2 - a^2}{2bc}$

Substitute known values.

$\cos A = \frac{5^2 + 4^2 - 6^2}{2 \times 5 \times 4}$

$\cos A = 0.125$

$A = \cos^{-1}(0.125)$

$A = 82.819...$

$A = 82.8°$, correct to 1 d.p.

Exercise 34.7

1. Calculate angle A in each of the following.
Give your answers correct to one decimal place.

(a)

(b)

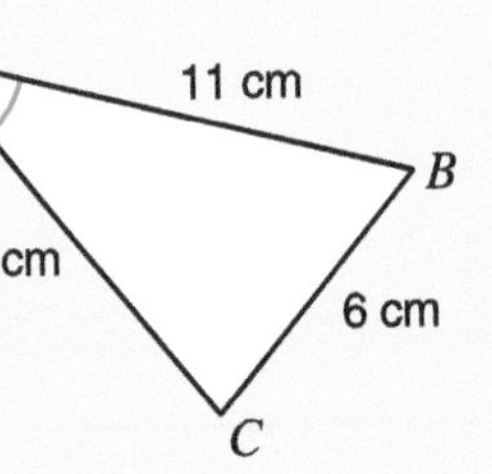

(c)

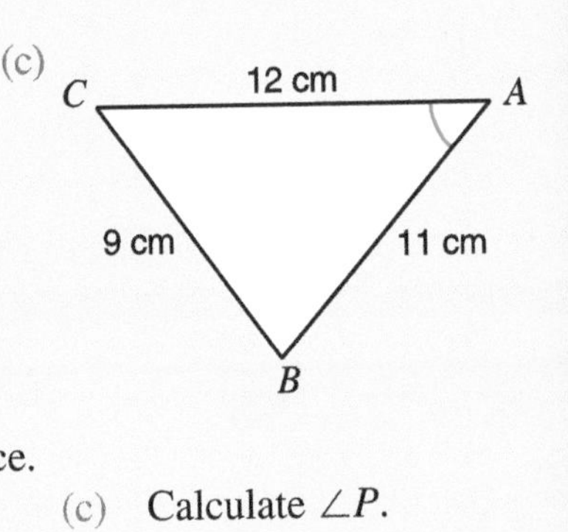

2. In this question give your answers correct to one decimal place.

(a) Calculate $\angle B$.

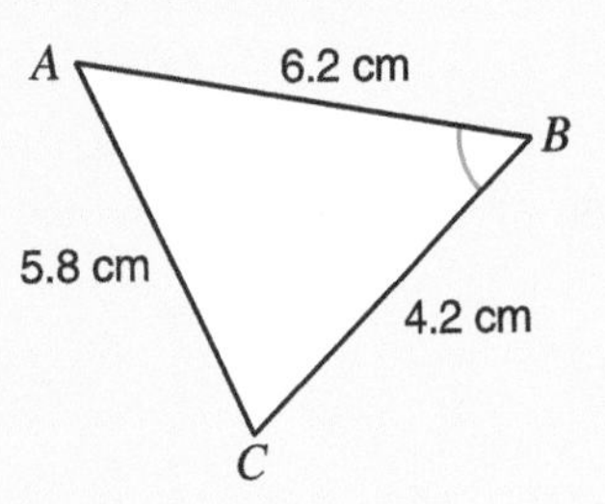

(b) Calculate $\angle C$.

(c) Calculate $\angle P$.

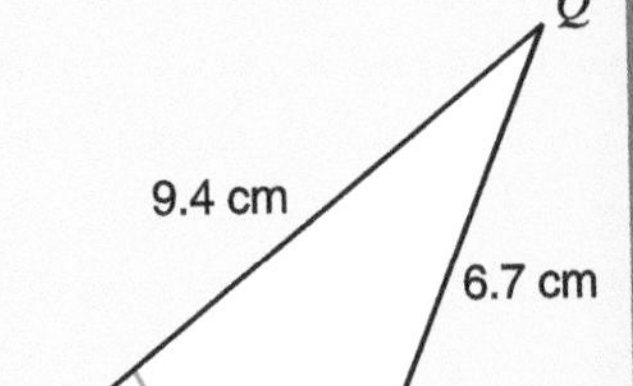

3. In triangle ABC, $AB = 18$ cm, $BC = 23$ cm and $CA = 24$ cm.
Calculate $\angle ACB$.

4. In triangle PQR, $PQ = 10$ cm, $QR = 12$ cm and $RP = 20$ cm.
Calculate $\angle PQR$.

5. In triangle LMN, $LM = 8$ cm, $MN = 6$ cm and $NL = 11$ cm.
Calculate $\angle LMN$.

6. In triangle XYZ, $XY = 9.6$ cm, $YZ = 13.4$ cm and $XZ = 20$ cm.
Calculate $\angle XYZ$.

Area of a triangle

Look at these triangles.
Both are labelled ABC.

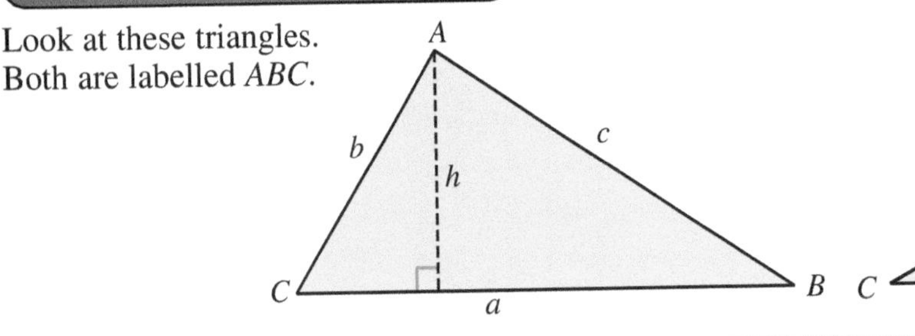

The height of each triangle is labelled h.
In both triangles $h = b \sin C$.

The area of triangle $ABC = \frac{1}{2} \times \text{base} \times \text{height}$

$= \frac{1}{2} \times a \times b \sin C$

$= \frac{1}{2} ab \sin C$

For any triangle ABC: Area $= \frac{1}{2} ab \sin C$

To use this formula to find the area of a triangle we need:
two sides of known length, **and** the size of the angle between the known sides.

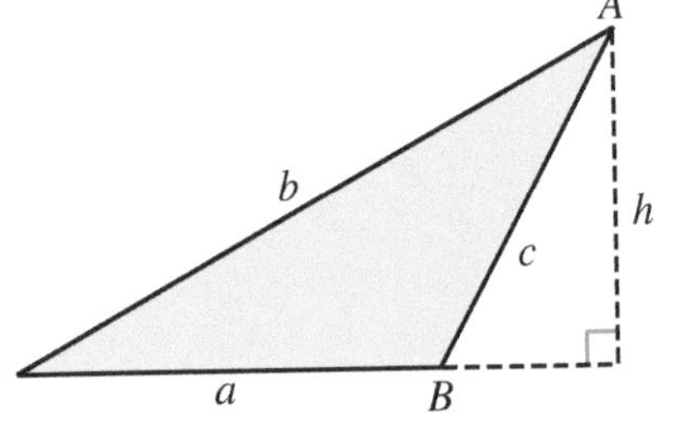

Can you find a rule which gives the area of a parallelogram?

EXAMPLE

Calculate the area of triangle PQR.

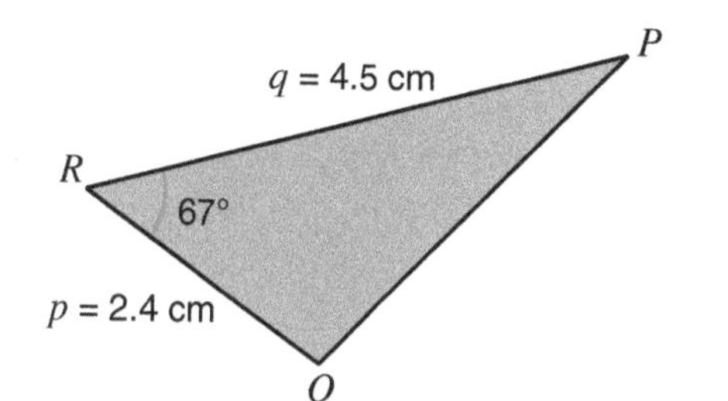

Area $PQR = \frac{1}{2} \times p \times q \times \sin R$

$= 0.5 \times 2.4 \times 4.5 \times \sin 67°$

$= 4.9707\ldots$

$= 4.97\,\text{cm}^2$, correct to 3 s.f.

Exercise 34.8

1. Calculate the areas of these triangles.
   Give your answers correct to three significant figures.

(a)

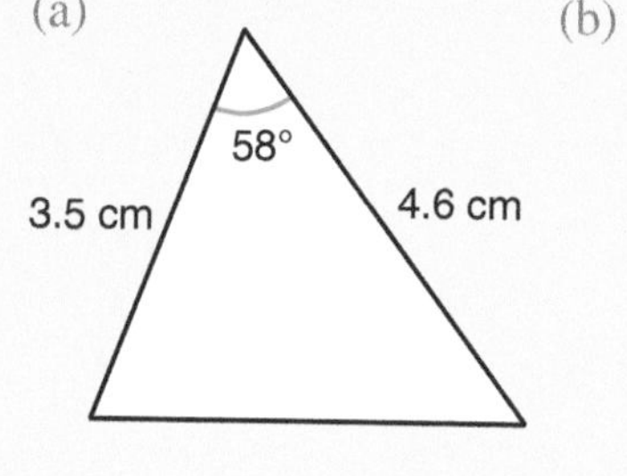

(b)

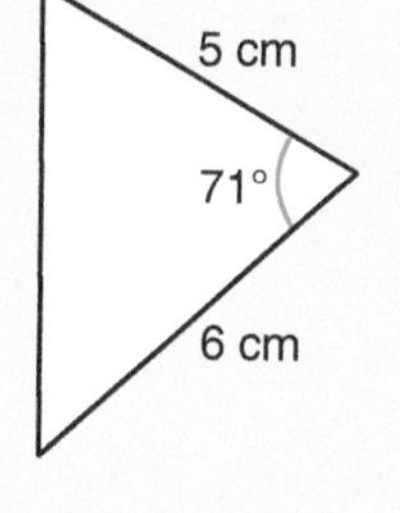

(c) (d)

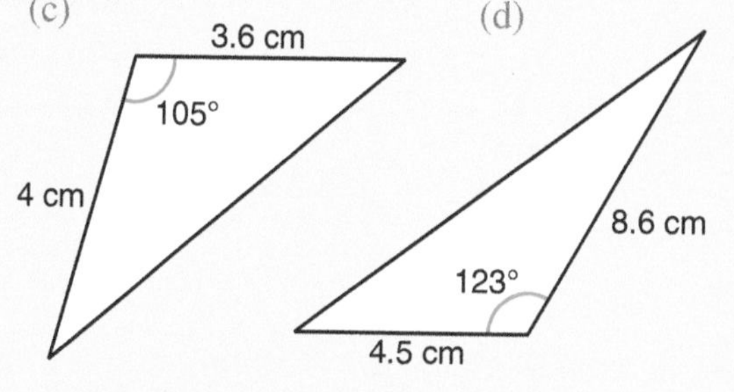

2. $PQRS$ is a parallelogram. $PQ = 9.2\,\text{cm}$, $PS = 11.4\,\text{cm}$ and angle $QPS = 54°$.
   Calculate the area of the parallelogram.

3. $ABCD$ is a quadrilateral.
   (a) Calculate the length of BD.
   (b) Calculate the area of $ABCD$.

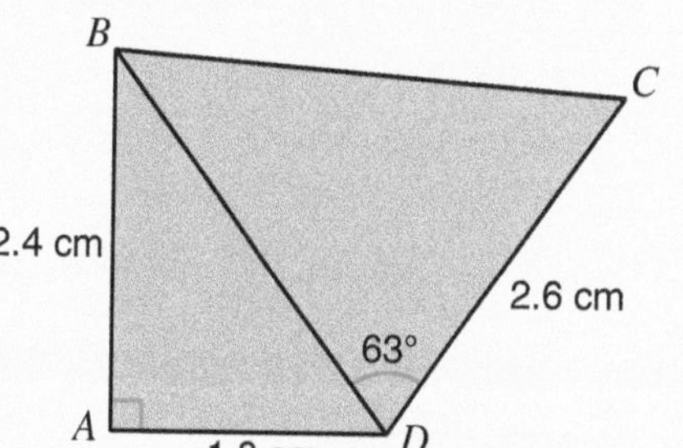

4. A regular hexagon has sides of length 5 cm.
   Calculate the area of the hexagon.

5. The area of triangle PQR is $31.7\,\text{cm}^2$. $PQ = 9.6\,\text{cm}$ and angle $PQR = 48°$.
   Calculate the length of QR.
   Give your answer correct to one decimal place.

6. $PQRS$ is a kite.
   $PQ = SP = 35\,\text{cm}$.
   $QR = RS = 24\,\text{cm}$.
   Angle $QPS = 48°$, angle $QRS = 72°$.
   Calculate the area of the kite.

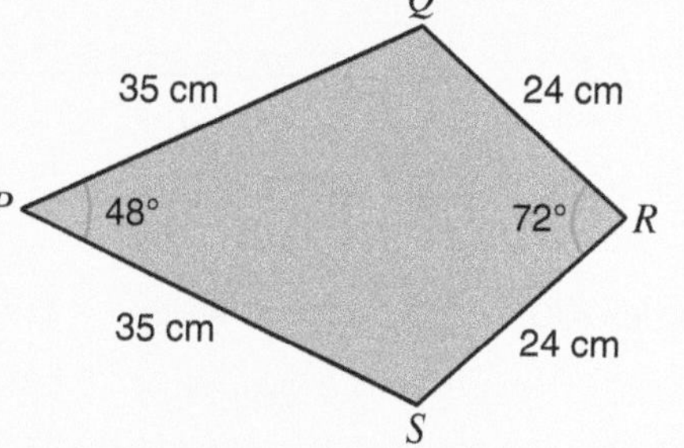

7. Triangle XYZ is isosceles, with $XY = YZ$. Angle $XYZ = 54°$.
   The area of triangle XYZ is $22\,\text{cm}^2$. Calculate the lengths of XY and YZ.

8. The area of triangle ABC is $25.6\,\text{cm}^2$. $AB = 6.7\,\text{cm}$, $AC = 9.2\,\text{cm}$.
   Calculate the two possible sizes of angle A.

Solving problems involving triangles

Right-angled triangles
Use the **trigonometric ratios** (sin, cos and tan) or **Pythagoras' Theorem**.

Triangles which are not right-angled

The Sine Rule
To find a **side** you need:
two angles of known size, **and**
the length of a side which is opposite one of the known angles.

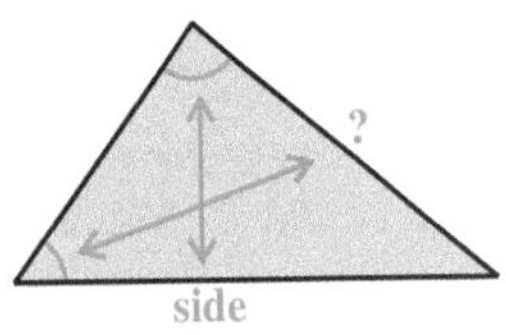

The Sine Rule
To find an **angle** you need:
the length of the side opposite the angle you are trying to find, **and** the length of a side opposite an angle of known size.

side
?
side

The Cosine Rule
To find a **side** you need:
two sides of known length, **and**
the size of the angle between the known sides.

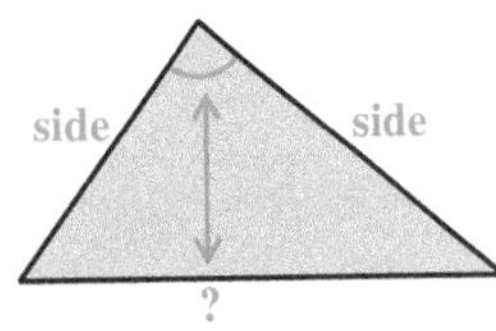

The Cosine Rule
To find an **angle** you need:
three sides of known length.

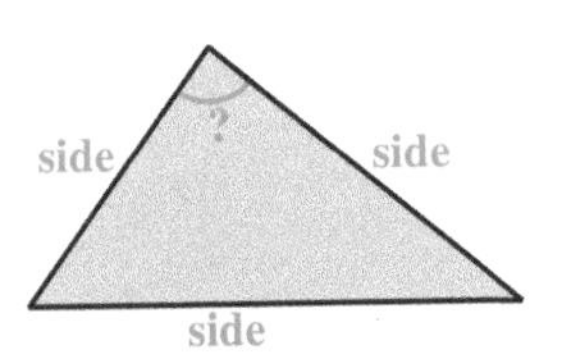

To find the **area of a triangle**, when you know two sides and the angle between the sides, use:

$$\text{Area of triangle} = \frac{1}{2}ab \sin C$$

EXAMPLE

Find the unknown sides and angles in this triangle.

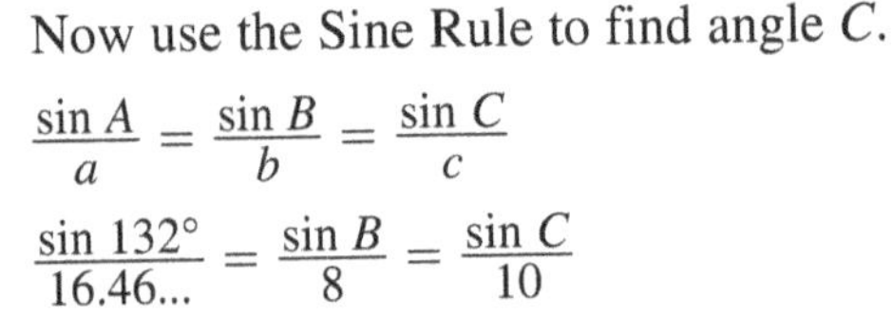

Using the Cosine Rule to find a.

$a^2 = b^2 + c^2 - 2bc \cos A$

$a^2 = 8^2 + 10^2 - 2 \times 8 \times 10 \times \cos 132°$

$a^2 = 271.06...$

$a = 16.46...$

Store this accurate value of a in the memory of your calculator.

$a = 16.5$ cm, correct to 3 s.f.

Now use the Sine Rule to find angle C.

$$\frac{\sin A}{a} = \frac{\sin B}{b} = \frac{\sin C}{c}$$

$$\frac{\sin 132°}{16.46...} = \frac{\sin B}{8} = \frac{\sin C}{10}$$

Using:

$$\frac{\sin 132°}{16.46...} = \frac{\sin C}{10}$$

$$\sin C = \frac{10 \times \sin 132°}{16.46...}$$

$\sin C = 0.451...$

$C = \sin^{-1}(0.451...)$

$C = 26.83...$

$C = 26.8°$, correct to 1 d.p.

$B = 180° - 26.8° - 132°$

$B = 21.2°$

Exercise 34.9

1 (a) Find the unknown sides and angles in these triangles.
(b) Calculate the area of each triangle.

(i) 9 cm, 60°, 12 cm

(ii) 3.6 cm, 110°, 6.5 cm

2 The diagram shows the positions of a tree, at T, and a pylon, at P, on opposite sides of a river. Two bridges cross the river at X and Y.

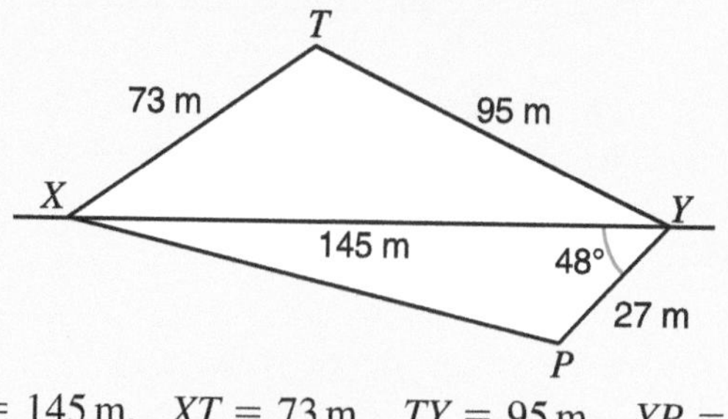

Remember:
The **longest side** is opposite the **largest angle**.
The **shortest side** is opposite the **smallest angle**.

$XY = 145\,\text{m}$, $XT = 73\,\text{m}$, $TY = 95\,\text{m}$, $YP = 27\,\text{m}$ and $\angle XYP = 48°$.

(a) Calculate the angles of triangle XTY.
(b) Calculate XP and $\angle XPY$.

3 Two ships, P and Q, leave port at 2 pm.
P travels at 24 km/h on a bearing of 085°.
Q travels at 32 km/h on a bearing of 120°.
Calculate the distance between the ships at 2.30 pm.

4

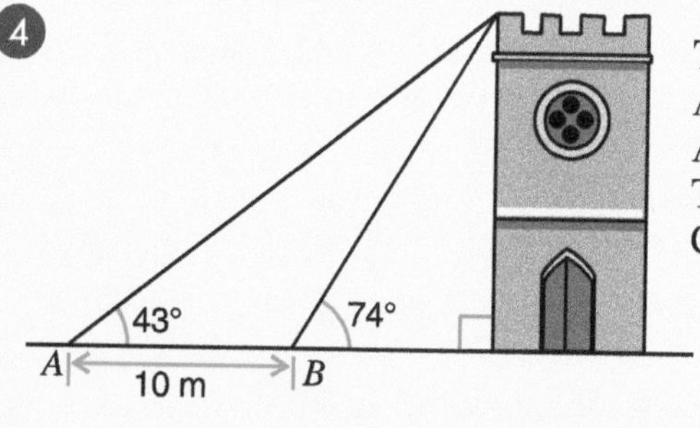

The diagram shows a tower.
At A the angle of elevation to the top of the tower is 43°.
At B the angle of elevation to the top of the tower is 74°.
The distance AB is 10m.
Calculate the height of the tower.

5 The diagram shows part of a steel framework, ABC.
B is on horizontal ground. A is 10 m vertically above B.
$BC = 16\,\text{m}$. Angle $ABC = 23°$.

(a) Calculate the length of AC.
(b) Calculate angle BCA.
(c) Calculate the vertical height of C above B.

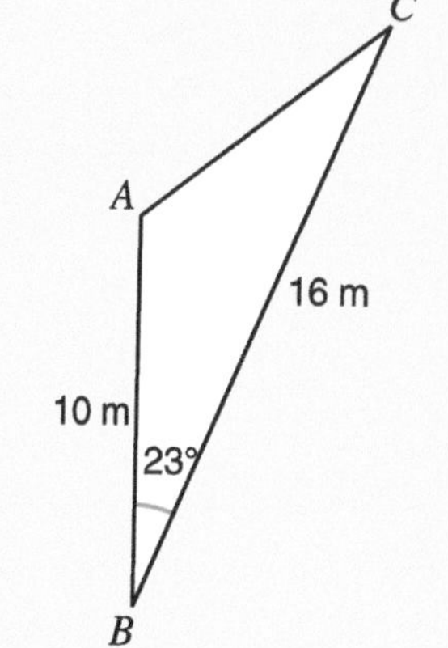

6 In triangle LMN, $MN = 7.5\,\text{cm}$,
angle $MLN = 104°$ and angle $LMN = 34°$.

(a) Calculate the length of LM.
(b) Calculate the area of the triangle.

7 The diagram shows a quadrilateral $PQRS$.
$PQ = 4\,\text{cm}$, $QR = 5\,\text{cm}$, $RS = 7\,\text{cm}$,
$QS = 8\,\text{cm}$ and angle $SPQ = 110°$.
(a) Calculate angle QSP.
(b) Calculate angle QSR.
(c) Calculate the area of $PQRS$.

8 An isosceles triangle has an area of $25.6\,\text{cm}^2$.
The two equal sides are 8.4 cm long.
Calculate the two possible lengths of the third side.

What you need to know

- The graphs of the trigonometric functions.

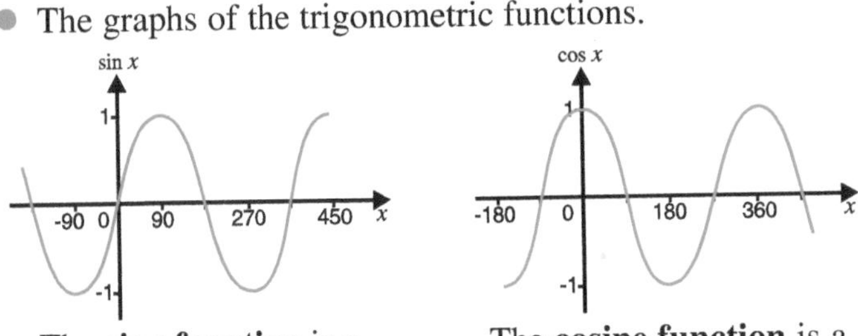

The **sine function** is a periodic function with period 360°.
$-1 \leqslant \sin x \leqslant 1$

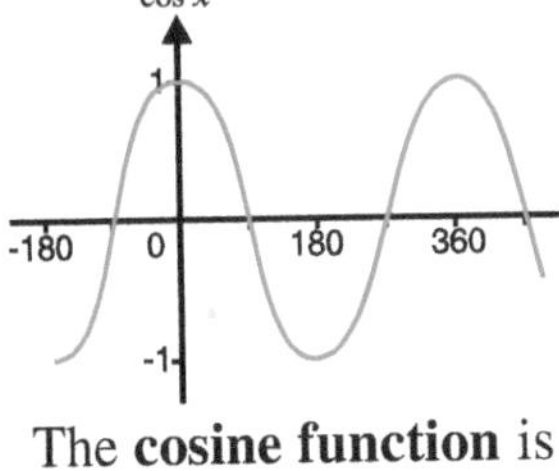

The **cosine function** is a periodic function with period 360°.
$-1 \leqslant \cos x \leqslant 1$

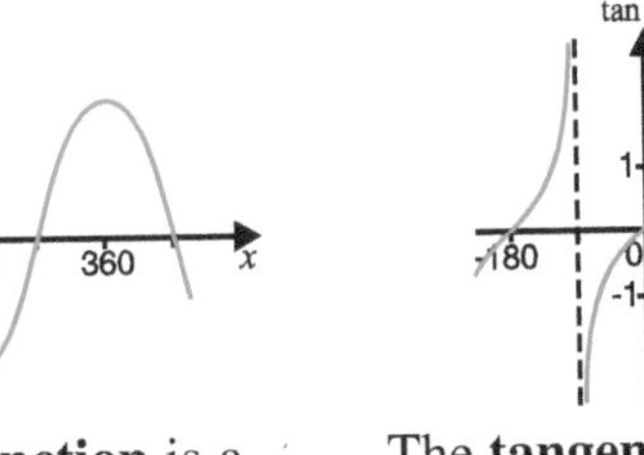

The **tangent function** is a periodic function with period 180°.
Tan x is undefined at 90°, 270°, ...

- For every angle $x°$, the signs of $\sin x°$, $\cos x°$ and $\tan x°$ can be shown on a diagram.
  Positive angles are measured **anticlockwise**.
  Negative angles are measured **clockwise**.
  For angles greater than 360°: subtract 360°, or multiples of 360°, to get the equivalent angle between 0° and 360°.

90°
Sin positive | All positive
180° | 0° 360°
Tan positive | Cos positive
270°

- The **exact values** of the trigonometric ratios for the angles 30°, 45° and 60° can be found from the triangles below.

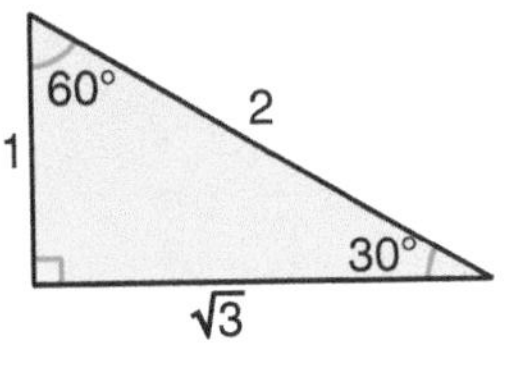

	30°	45°	60°
sin	$\frac{1}{2}$	$\frac{1}{\sqrt{2}}$	$\frac{\sqrt{3}}{2}$
cos	$\frac{\sqrt{3}}{2}$	$\frac{1}{\sqrt{2}}$	$\frac{1}{2}$
tan	$\frac{1}{\sqrt{3}}$	1	$\sqrt{3}$

- **The Sine Rule**

$$\frac{a}{\sin A} = \frac{b}{\sin B} = \frac{c}{\sin C}$$

This can be also written as: $\frac{\sin A}{a} = \frac{\sin B}{b} = \frac{\sin C}{c}$

- **The Cosine Rule**

$a^2 = b^2 + c^2 - 2bc \cos A$
$b^2 = a^2 + c^2 - 2ac \cos B$
$c^2 = a^2 + b^2 - 2ab \cos C$

When using the Cosine Rule to find the size of an angle it is sometimes easier to rearrange the above formulae as:

$$\cos A = \frac{b^2 + c^2 - a^2}{2bc} \qquad \cos B = \frac{a^2 + c^2 - b^2}{2ac} \qquad \cos C = \frac{a^2 + b^2 - c^2}{2ab}$$

- To find the **area of a triangle** when you know two sides and the angle between the two sides use:

  Area of triangle $= \frac{1}{2}ab \sin C$

- Solving problems involving triangles.

  Right-angled triangles

  Use the **trigonometric ratios** (sin, cos and tan) or **Pythagoras' Theorem.**

  Triangles which are not right-angled

  The Sine Rule

  To find a **side** you need:
  two angles of known size, **and** the length of a side which is opposite one of the known angles.

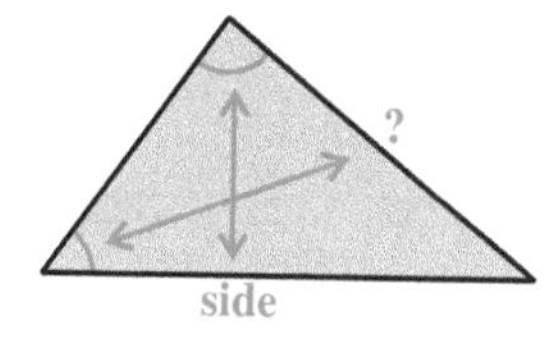

  To find an **angle** you need:
  the length of the side opposite the angle you are trying to find, **and** the length of a side opposite an angle of known size.

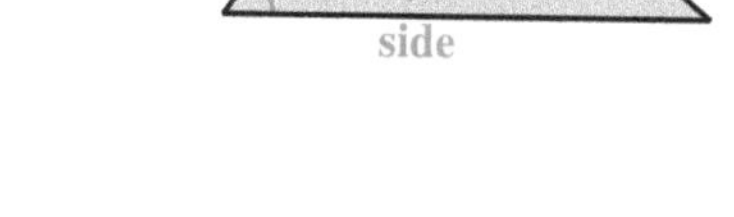

  The Cosine Rule

  To find a **side** you need:
  two sides of known length, **and** the size of the angle between the known sides.

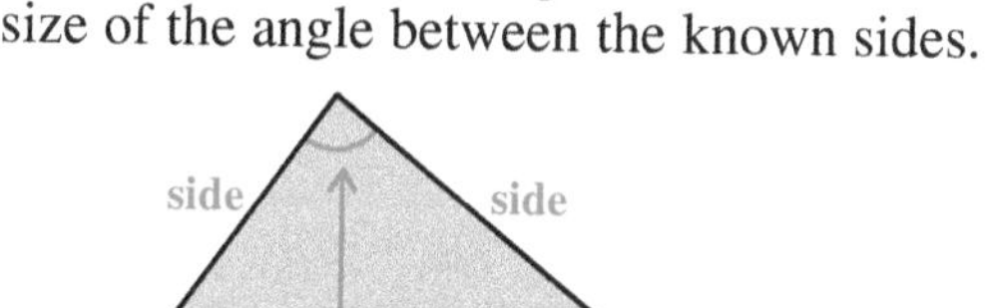

  To find an **angle** you need:
  three sides of known length.

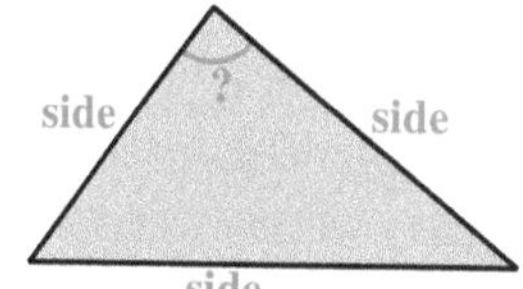

Investigate graphs of the form $y = a \sin bx$ and $y = a \cos bx$.

What can you say about the values of a and b and the graphs of the functions?

Review Exercise 34

1. Angles x and y lie between 0° and 360°.
   (a) Find the **two** values of x which satisfy the equation $\cos x = 0.5$.
   (b) Solve the equation $\sin y = \cos 240°$. AQA

2. (a) Sketch the graph of $y = \sin x$ for $-360° \leqslant x \leqslant 360°$.
   (b) One solution of the equation $\sin x = 0.4$ is $x = 24°$ to the nearest degree.
   Find all the other solutions to the equation $\sin x = 0.4$ for $-360° \leqslant x \leqslant 360°$. AQA

3. Find two values of x, between 0° and 360°, when $\cos x = -0.75$. AQA

4. In the triangle ABC, $AB = 6$ cm, $BC = 5$ cm and angle $BAC = 45°$.
   There are two possible triangles that can be constructed.
   Calculate the **two** possible values of the angle BCA. AQA

5 On a particular day the height, h metres, of the tide at Weymouth, relative to a certain point, can be modelled by the equation $h = 5\sin(30t)°$, where t is the time in hours after midnight.

(a) Copy the axes and sketch the graph of h against t for $0 \leqslant t \leqslant 12$.
(b) At what time is low tide?

AQA

6 A clock has a minute hand that is 10 cm long.
The hour hand is 7 cm long.
Calculate the distance between the ends of the hands when the clock shows 4 o'clock.

AQA

7 A radio mast, CM, stands at a corner of a horizontal field, ABC.
The angle of elevation of M from B is 5°.
$CM = 35$ m, $AB = 180$ m and angle $ABC = 107°$.
(a) Calculate the length of AC.
(b) Calculate the area of ABC.

AQA

8 In triangle ABC, angle ABC is **obtuse**.
Angle $BAC = 32°$, $AC = 10$ cm and $BC = 6$ cm.
Calculate the area of triangle ABC.

AQA

9

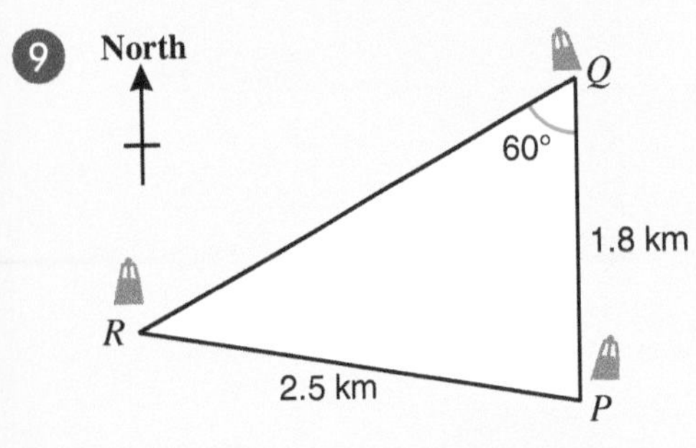

In the Olympic sailing regatta one of the courses is in the shape of a triangle.
Buoys are fixed at the three corners of the triangle PQR.
$PQ = 1.8$ km. $PR = 2.5$ km.
Q is due north of P. Angle $PQR = 60°$.
Calculate the bearing of P from R.

AQA

10 The diagram shows the plan of a garden $PQRS$.
QPS is an obtuse angle.
Calculate angle QPS.

AQA

Shape, Space and Measures Non-calculator Paper

Do not use a calculator for this exercise.

1 (a)

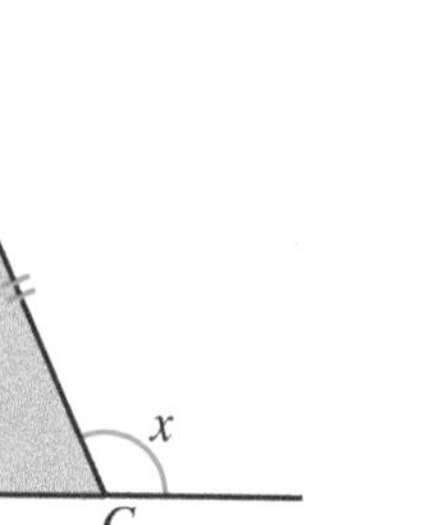

Triangle ABC is isosceles. $AB = AC$.
Work out the size of angle x.

(b)

BE is parallel to CD.
(i) Write down the size of angle p.
(ii) Work out the size of angle q. AQA

2 This 3-dimensional shape has been made using linking cubes of side 1 cm.
On squared paper, draw diagrams to show:
(a) the plan of the shape,
(b) the elevation of the shape from **X**.

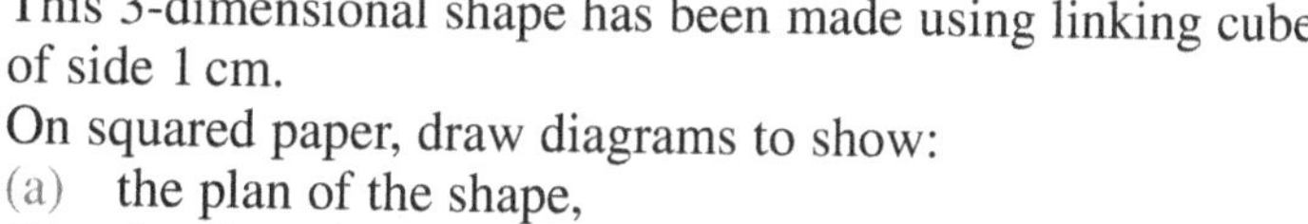

3 A cuboid has a volume of $540\,\text{cm}^3$.
The base of the cuboid measures 5 cm by 9 cm.
Find the height of the cuboid.

4

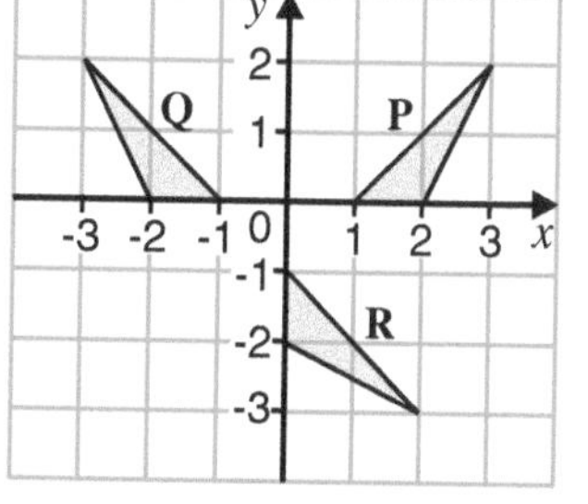

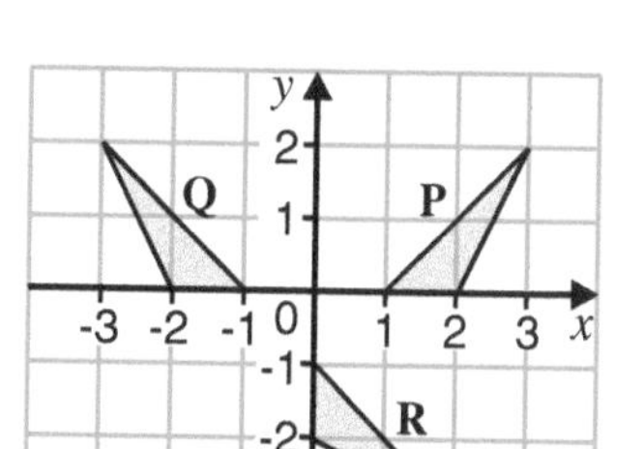

The diagram shows triangles **P**, **Q** and **R**.
(a) Describe the single transformation which takes **P** onto **Q**.
(b) Describe the single transformation which takes **P** onto **R**.

Copy triangle **P** onto squared paper.
(c) Draw an enlargement of triangle **P** with scale factor 2, centre (0, 0).

5 The diagram shows the shape of a car park.
(a) Calculate the perimeter of the car park.
(b) Calculate the area of the car park.

6 A table top is a circle of radius 50 cm.
(a) Calculate the circumference of the table top.
(b) Calculate the area of the table top.
Give your answers in terms of π.

7 Each shape is a regular polygon.
Work out the size of each lettered angle.

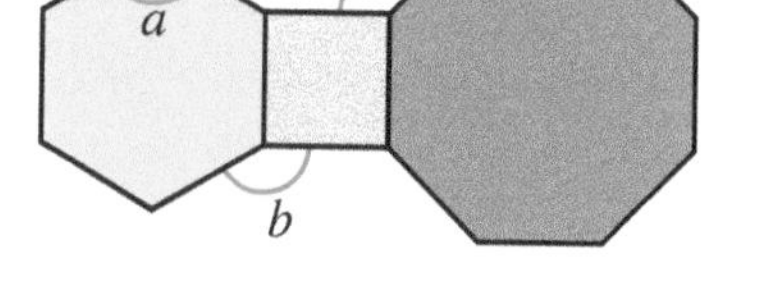

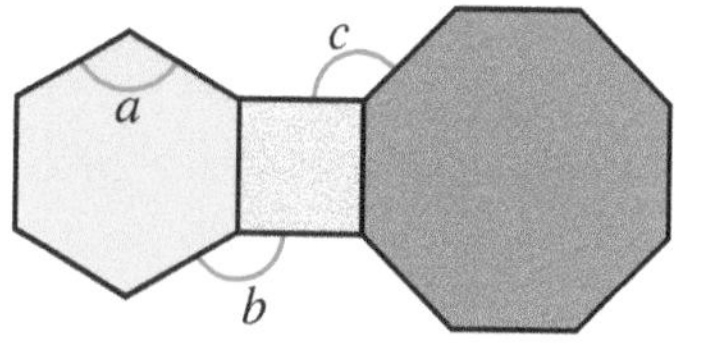

8 Y is 50 m from X on a bearing of 080°. Z is 70 m from Y on a bearing of 110°.
(a) Make a scale drawing to show the positions of X, Y and Z. Use a scale of 1 cm to 10 m.
(b) Find by measurement the distance and bearing of X from Z.

9 Two straight roads are shown on the diagram.
A new gas pipe is to be laid from Bere equidistant from the two roads.
The diagram is drawn to a scale of 1 cm to 1 km.

(a) Copy the diagram and construct the path of the gas pipe.
(b) The gas board needs a construction site depot.
The depot must be equidistant from Bere and Cole.
The depot must be less than 2 km from Alton.
Draw loci on your diagram to represent this information.
(c) The depot must be nearer the road through Cole than the road through Alton.
Mark on your diagram, with a cross, a possible position for the site depot.

10 Colin and Dexter are standing next to each other in the sunshine.
Colin is 150 cm tall and his shadow is 240 cm long.
Dexter is 165 cm tall. How long is his shadow?

11 The diagram shows the plan of a children's playground.

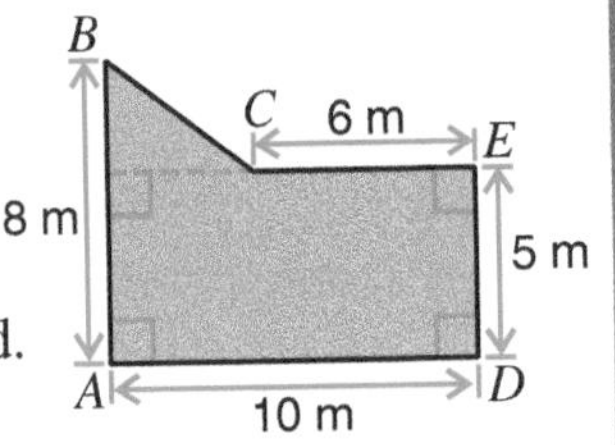

(a) What is the area of the playground?
(b) (i) Find the length of BC.
(ii) A fence is to be put around the perimeter of the playground.
Find the length of the fence.

12

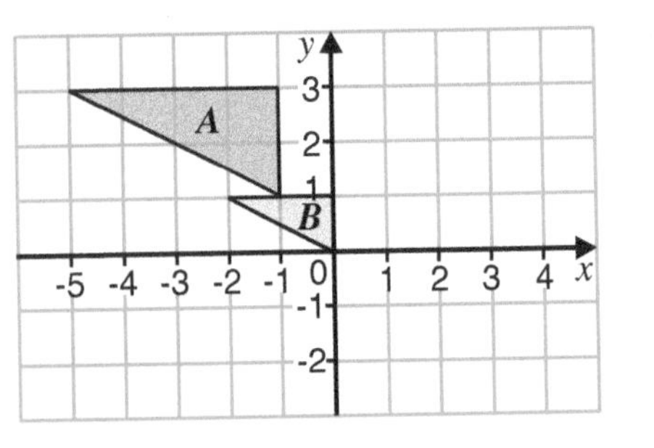

(a) Describe fully the single transformation which maps A onto B.
(b) B is mapped onto C by a translation with vector $\begin{pmatrix} 3 \\ -2 \end{pmatrix}$.
Draw a diagram to show the positions of B and C.
(c) B is mapped onto D by a rotation, 90° clockwise about (2, 1).
Show the position of D on your diagram.

13 The following formulae represent certain quantities connected with containers, where a, b and c are dimensions.

$\pi a^2 b \qquad 2\pi a(a + b) \qquad 2a + 2b + 2c \qquad \frac{1}{2}(a + b)c \qquad \sqrt{a^2 + b^2}$

(a) Which of these formulae represent area?
(b) Which of these formulae represent volume?

AQA

14 The diagram shows a right-angled triangle, ABC.
Angle $ABC = 90°$. Tan $x = \frac{3}{4}$.
(a) Calculate sin x.

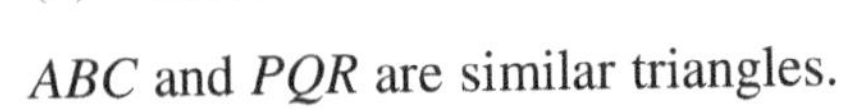

ABC and PQR are similar triangles.

PQ is the shortest side of triangle PQR.
Angle $PQR = 90°$ and $PR = 15$ cm.
(b) (i) What is the value of cos y?
(ii) What is the length of PQ?

AQA

15 These boxes are similar.

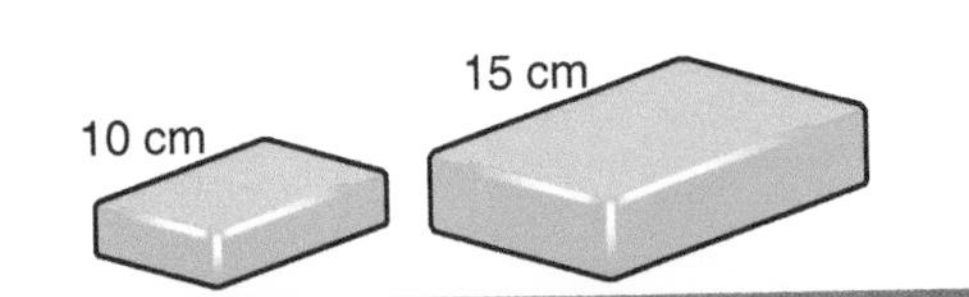

(a) The volume of the smaller box is 240 cm³.
Work out the volume of the larger box.
(b) The surface area of the larger box is 558 cm².
Work out the surface area of the smaller box.

16 In the diagram, O is the centre of the circle.
SAT is a tangent to the circle at A.
Angle $BAC = 80°$ and angle SAB = angle TAC.

(a) Calculate (i) angle BOC,
(ii) angle OBC,
(iii) angle ABO,
(iv) angle ACO.

(b) Explain why triangles OAB and OAC are congruent.

AQA

17

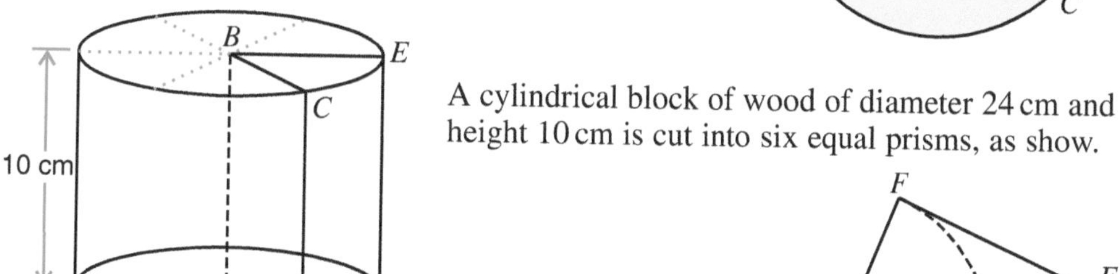

A cylindrical block of wood of diameter 24 cm and height 10 cm is cut into six equal prisms, as show.

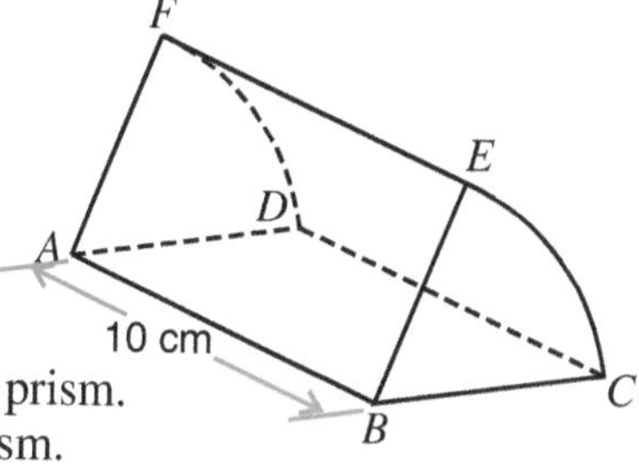

(a) Calculate the area of sector BEC, the cross-section of the prism.
(b) Calculate the area of $CDFE$, the curved surface of the prism.
(c) Calculate the volume of the prism.

Give your answers in terms of π.

AQA

18 Tan $45° = 1$.

(a) Write down the value of tan 135°.
(b) Tan x = tan 45°, $x \neq 45°$, $0 \leqslant x \leqslant 360°$. Write down the value of x.
(c) Tan $y = -1$. Write down the possible values of y between 0° and 360°.

19 The diagram shows the points O, A and B where $\overrightarrow{OA} = 6\mathbf{a}$ and $\overrightarrow{OB} = 9\mathbf{b}$.
OAD is a straight line and $ABCD$ is a parallelogram. A is the midpoint of OD.

(a) Find in terms of **a** and **b**. (i) $\overrightarrow{AD}$ (ii) $\overrightarrow{AB}$

(b) E and F are two points such that

$\overrightarrow{AE} = \frac{1}{3}\overrightarrow{AB}$ and $\overrightarrow{DF} = \frac{2}{3}\overrightarrow{DC}$.

Find in terms of **a** and **b**. (i) $\overrightarrow{OE}$ (ii) $\overrightarrow{OF}$

(c) What do $\overrightarrow{OE}$ and $\overrightarrow{OF}$ indicate about the points O, E and F?

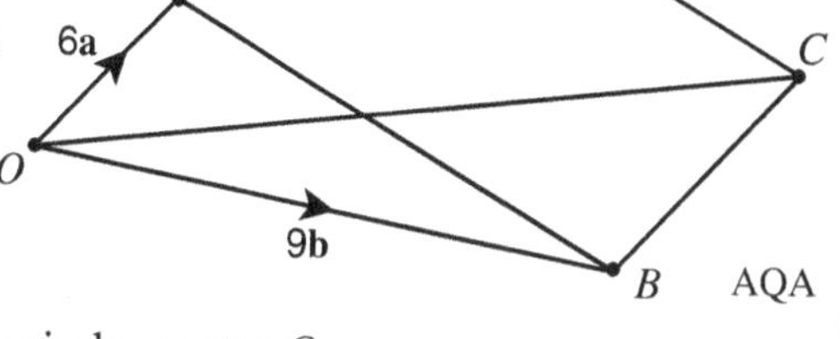

AQA

20 A, B and C are three points on the circumference of a circle, centre O.
CT is the tangent to the circle at C. ABT is a straight line.
Angle $AOB = x°$ and angle $TBC = y°$.
Give angle BCT in terms of x and y.

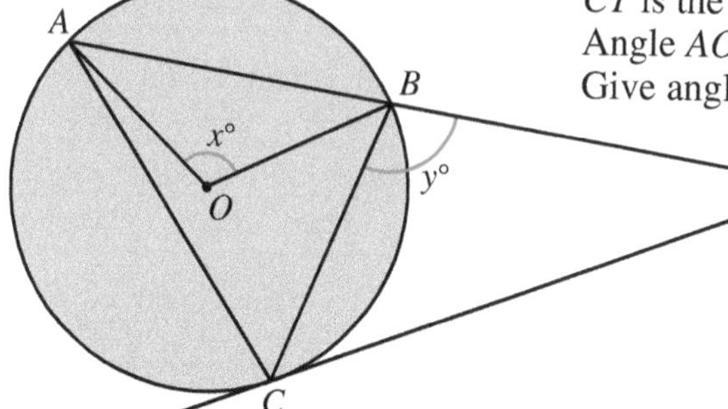

AQA

Shape, Space and Measures Calculator Paper

You may use a calculator for this exercise.

1 (a) Work out the size of the angle marked x.
Give a reason for your answer.

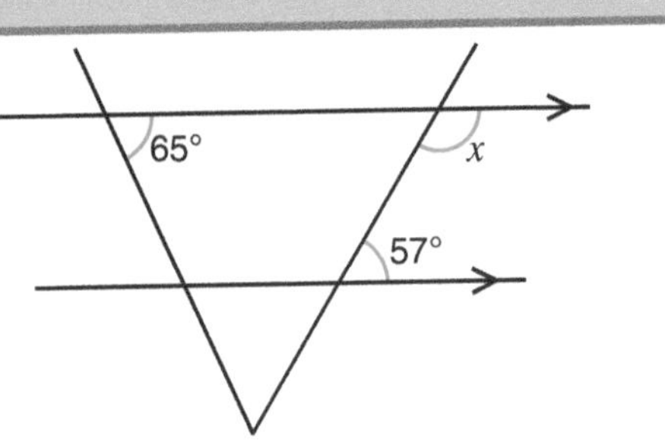

(b)

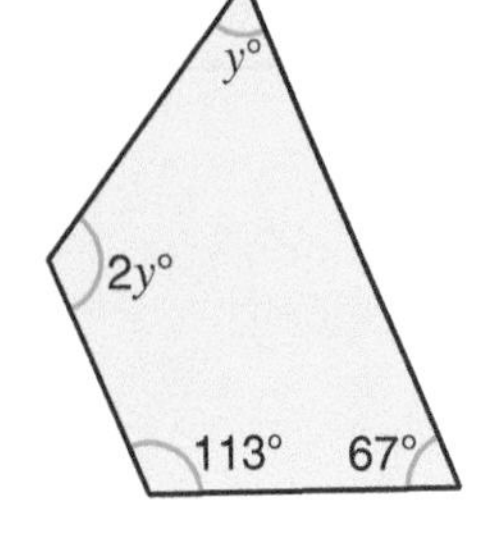
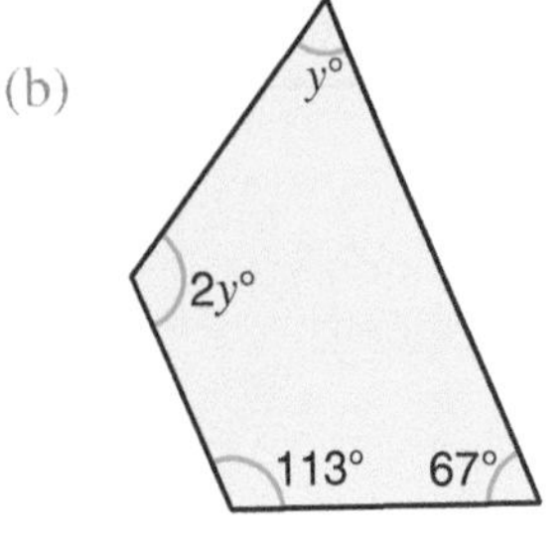

(i) What special name is given to this shape?
Give a reason for your answer.
(ii) Find the value of y.

2 OW, OX and OY meet at O, as shown.
Reflection in OW followed by reflection in OX followed by reflection in OY gives the same transformation as a reflection in a line OZ.
Find the angle WOZ.

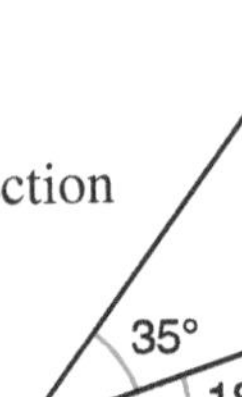
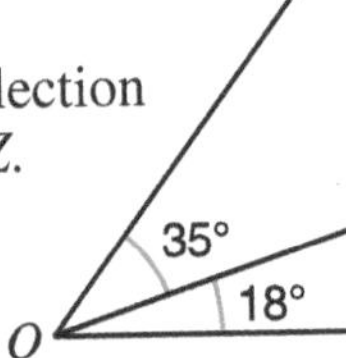

3 Y is 50 m from X on a bearing of 070°.
Z is 50 m from Y on a bearing of 130°.
What is the bearing of X from Z?

4 The diagram shows a rectangular doormat.
What is the area of the doormat
(a) in cm^2,
(b) in m^2?

5 A triangle has sides of length 7 cm, 5 cm and 4 cm.
(a) Make an accurate drawing of the triangle.
(b) Work out the area of the triangle.

6 (a) Adrian is 6 feet 3 inches tall. Work out Adrian's height in centimetres.
(b) Adrian weighs 78 kg. Work out Adrian's weight in pounds.

7 What is the size of one exterior angle of a regular pentagon?

8 The front wheels of a tractor each have diameter 100 cm. The tractor is driven 100 metres.
How many complete turns do each of the front wheels make?

9 (a) Calculate the area of this trapezium.
(b) Calculate the perimeter of this trapezium.

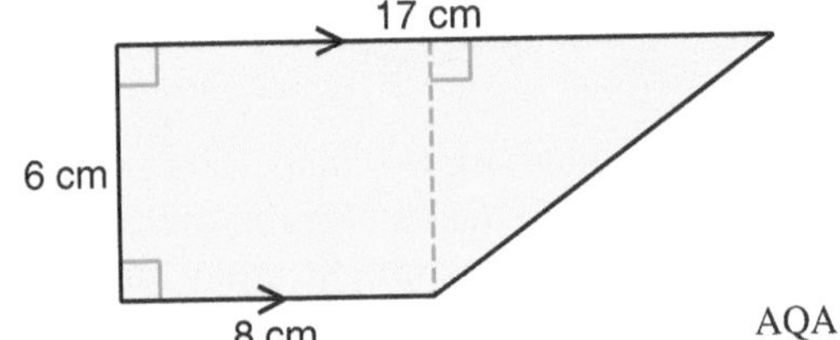
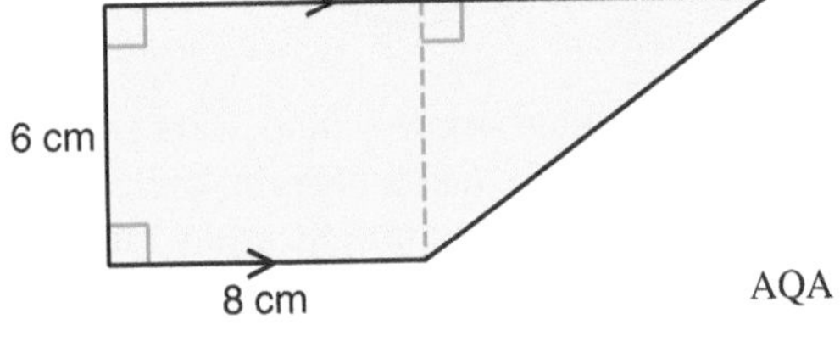

AQA

10

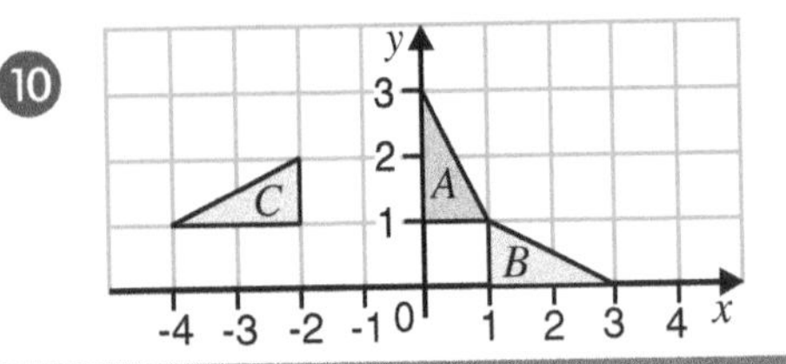

Describe the single transformation which maps
(a) A onto B,
(b) A onto C.

11 The diagram shows a circular paddling pool with a vertical side.
The radius of the pool is 35 cm.
The water in the pool is 8 cm deep.
Calculate the volume of water in the pool.

AQA

12 This shape is made up of a right-angled triangle and a semi-circle.
Calculate the total area of the shape.
Give your answer to a suitable degree of accuracy.

13 (a) Construct parallelogram $PQRS$, with $PQ = 4\text{ cm}$, $SP = 3\text{ cm}$ and angle $SPQ = 60°$.
(b) A point X is 2 cm from the parallelogram and outside the parallelogram.
Draw the locus of all the possible positions of X.

14 Triangles ABC, PQR and HIJ are all similar.
(a) Calculate the length of AB.
(b) What is the size of angle B?
(c) Calculate the length of HJ.

AQA

15 Wendy sketches a toy boat.
(a) **Using ruler and compasses only,**
draw an accurate diagram of sail A.
You must show your construction lines.
(b) Calculate the height h of sail B.
(c) Calculate the angle x.

AQA

16

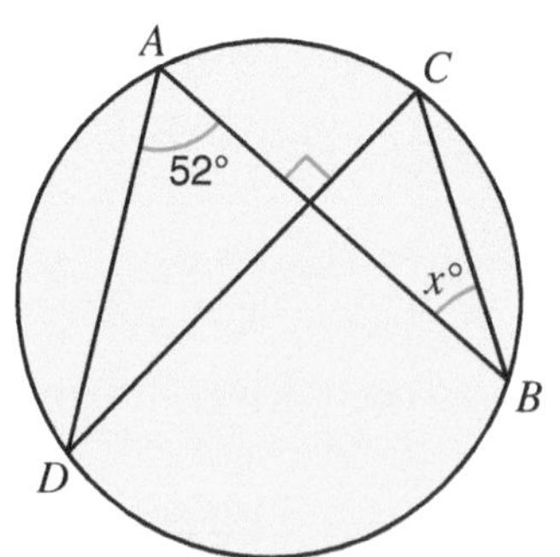

In the diagram, the chords AB and CD are perpendicular.
Angle $DAB = 52°$.
Calculate the value of x.
You must give a reason for your answer.

AQA

17 On the diagram, the shaded rectangle $ABCD$ represents a ski-slope.
Rectangle $ABEF$ is at right angles to rectangle $ECDF$.
Alison skis in a straight line from C to A.
Her average speed is 12 metres per second.
How long, in seconds, does it take her to ski from C to A?

AQA

18 $ABCD$ and $PQRS$ are squares.
Angle $DAP = y$.

Prove that triangles ABQ and DAP are congruent.

AQA

19 A minor sector AOB is cut from a circle of radius 20 cm.
Angle $AOB = 45°$.

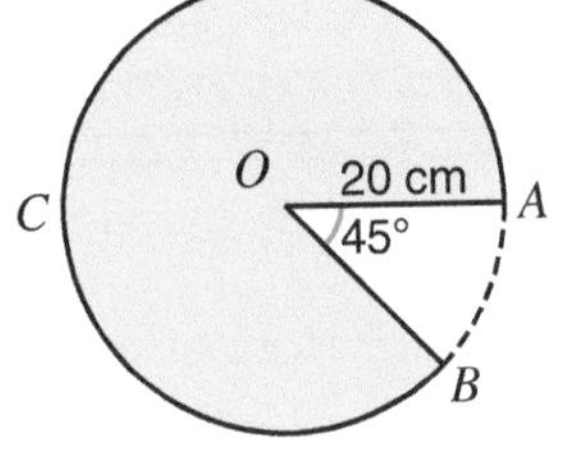

(a) Calculate the length of the arc ACB.

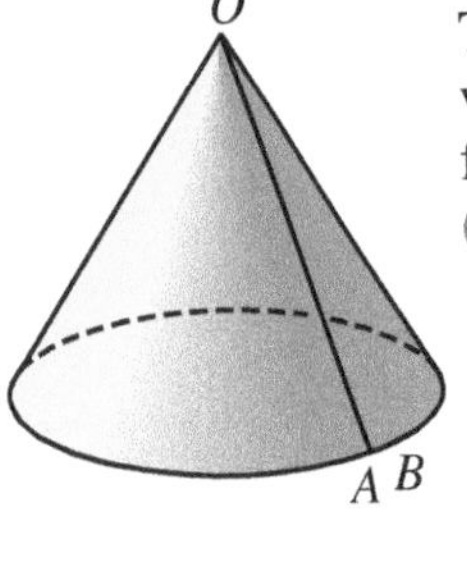

The radii OA and OB are joined without overlap so that a cone is formed, as shown in the diagram.

(b) Calculate
 (i) the base radius of the cone,
 (ii) the volume of the cone.

AQA

20 $ABCD$ is a quadrilateral with diagonal AC.
$AB = 4.3$ m and $CD = 5.2$ m.
Angle $BAC = 48°$, angle $BCA = 19°$
and angle $ACD = 79°$.
Calculate the length of AD.

AQA

21 The angles in triangle ABC are 45°, 53° and 82°, as shown in the diagram. $BC = 11$ cm.

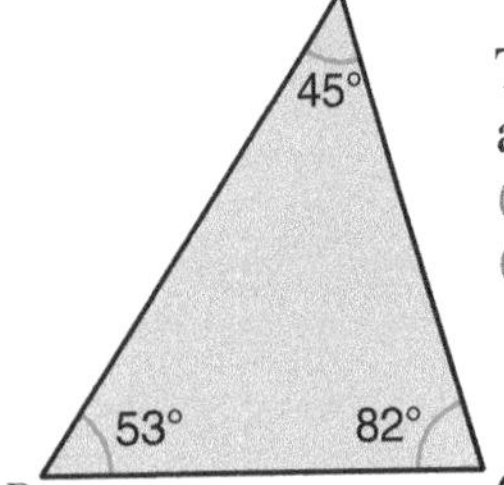

(a) Calculate the length of AB, correct to 1 decimal place.
(b) Hence, or otherwise, find the area of triangle ABC.

A circle, centre O, is drawn through the points A, B and C of the **same** triangle.

(c) Calculate the radius, r, of the circle.

AQA

22 (a) Sketch the graph of $y = \sin x$ for $-360° \leqslant x \leqslant 360°$.
(b) How many solutions are there to the equation $\sin x = -0.8$ between $-360°$ and $360°$?
(c) Solve the equation $\sin x = 0.6$ for $-360° \leqslant x \leqslant 360°$.
Give your answers to the nearest degree.

23 A hoist uses a cable with a breaking strain of 1400 kg, measured to 2 significant figures.
It is used to lift boxes with a weight of 40 kg, measured to 2 significant figures.
What is the greatest number of boxes that can be lifted at one time to be sure that the cable does not break?

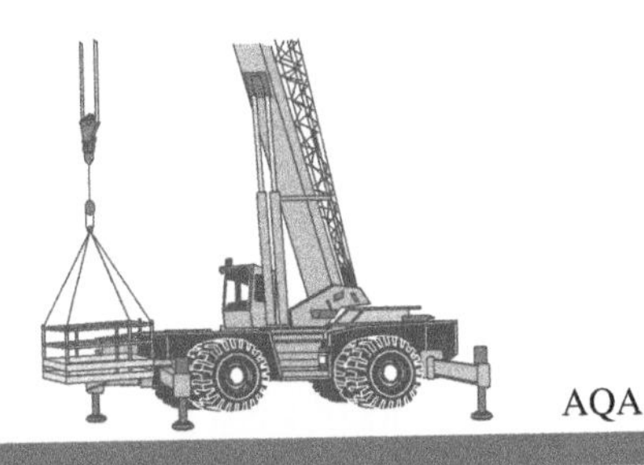

AQA

Collection and Organisation of Data

To answer questions such as: Which is the most popular colour of car?
Is it going to rain tomorrow?
Which team won the World Cup in 2002?

we need to collect data.

Primary and secondary data

When data is collected by an individual or organisation to use for a particular purpose it is called **primary data**.
Primary data is obtained from experiments, investigations, surveys and by using questionnaires.

Data which is already available or has been collected by someone else for a different purpose is called **secondary data**.
Sources of secondary data include the Annual Abstract of Statistics, Social Trends and the Internet.

Data

Data is made up of a collection of **variables**. Each variable can be described, numbered or measured.

Data which can only be **described** in words is **qualitative**.
Such data is often organised into categories, such as make of car, colour of hair, etc.

Data which is given **numerical** values, such as shoe size or height, is **quantitative**.
Quantitative data is either **discrete** or **continuous**.

Discrete data can only take certain values, usually whole numbers, but may include fractions (e.g. shoe sizes).
Continuous data can take any value within a range and is measurable (e.g. height, weight, temperature, etc.).

EXAMPLES

The taste of an orange is a qualitative variable.
The number of pips in an orange is a discrete quantitative variable.
The surface area of an orange is a continuous quantitative variable.

Exercise 35.1

State whether the following data is qualitative or quantitative.
If the data is quantitative state whether it is discrete or continuous.

1. The colours of cars in a car park.
2. The weights of eggs in a carton.
3. The numbers of desks in classrooms.
4. The names of students in a class.
5. The sizes of spanners in a toolbox.
6. The depths that fish swim in the sea.
7. The numbers of goals scored by football teams on a Saturday.
8. The brands of toothpaste on sale in supermarkets.
9. The sizes of ladies dresses in a store.
10. The heights of trees in a wood.

Collection of data

Data can be collected in a variety of ways:

by observation, by interviewing people and by using questionnaires.

The method of collection will often depend on the type of data to be collected.

Data collection sheets

Data collection sheets are used to record data.
To answer the question, "Which is the most popular colour of car?" we could draw up a simple data collection sheet and record the colours of passing cars by observation.

EXAMPLE

A **data collection sheet** for colour of car is shown, with some cars recorded.

Colour of car	Tally	Frequency
Black	||	2
Blue	~~||||~~ ~~||||~~ |||	13
Green	||||	4
Red	~~||||~~ ~~||||~~ |	
Silver	~~||||~~ ||	
White	~~||||~~ ~~||||~~ ||||	
	Total	

The colour of each car is recorded in the **tally** column by a single stroke.

To make counting easier, groups of 5 are recorded as ~~||||~~.

How many red cars are recorded?
How many cars are recorded altogether?

The total number of times each colour appears is called its **frequency**.
A table for data with the totals included is called a **frequency distribution**.

For large amounts of discrete data, or for continuous data, we organise the data into **groups** or **classes**. When data is collected in groups it is called a **grouped frequency distribution** and the groups you put the data into are called **class intervals**.

EXAMPLE

The weights of 20 boys are recorded in the grouped frequency table shown below.

Weight w kg	Tally	Frequency
$50 \leqslant w < 55$	|	1
$55 \leqslant w < 60$	|||	3
$60 \leqslant w < 65$	~~||||~~ ||||	9
$65 \leqslant w < 70$	~~||||~~ |	6
$70 \leqslant w < 75$	|	1
	Total	20

Weights are grouped into class intervals of equal width.

$55 \leqslant w < 60$

means 55 kg, or more, but less than 60 kg.

John weighs 54.9 kg. *In which class interval is he recorded?*
David weighs 55.0 kg. *In which class interval is he recorded?*

What is the width of each class interval?

Databases

If we need to collect data for more than one type of information, for example: the make, colour, number of doors and mileage of cars, we will need to collect data in a different way.

We could create a **data collection card** for each car.

Car	1
Make	Vauxhall
Colour	Grey
Number of doors	3
Mileage	18 604

Alternatively, we could use a data collection sheet and record all the information about each car on a separate line.

This is an example of a simple **database**.

Car	Make	Colour	Number of doors	Mileage
1	Vauxhall	Grey	3	18 604
2	Ford	Blue	2	33 216
3	Ford	White	5	27 435
4	Nissan	Red	4	32 006

When all the data has been collected, separate frequency or grouped frequency tables can be drawn up.

Exercise 35.2

1 The tally chart shows the number of glass bottles put into a bottle bank one day.

Colour of glass	Tally
Clear	~~IIII~~ ~~IIII~~ ~~IIII~~ II
Brown	~~IIII~~ ~~IIII~~ ~~IIII~~ ~~IIII~~ I
Green	~~IIII~~ ~~IIII~~ ~~IIII~~

How many **more** brown bottles than clear bottles were put into the bottle bank?

2 The heights, in centimetres, of 36 girls are recorded as follows.

148	161	175	156	155	160	178	159	170
163	147	150	173	169	170	174	166	163
162	158	155	165	168	154	156	163	167
172	170	165	160	164	172	157	173	161

(a) Copy and complete the grouped frequency table for the data.

Height h cm	Tally	Frequency
$145 \leqslant h < 150$		
$150 \leqslant h < 155$		

(b) What is the width of each class interval?

(c) How many girls are in the class interval $155 \leqslant h < 160$?

(d) How many girls are less than 160 cm?

(e) How many girls are 155 cm or taller?

3 The database gives information about the babies born at a maternity hospital one day.

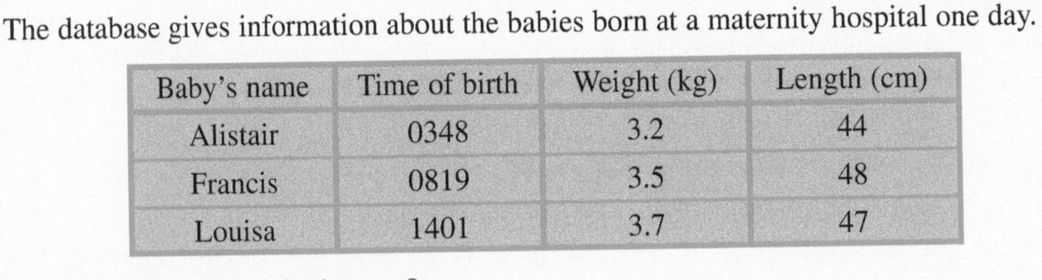

Baby's name	Time of birth	Weight (kg)	Length (cm)
Alistair	0348	3.2	44
Francis	0819	3.5	48
Louisa	1401	3.7	47

(a) Which baby was the longest?
(b) Which baby was the heaviest?
(c) Which baby was born first?

4 A database of cars is shown.

Car	Make	Colour	Number of doors	Mileage
1	Vauxhall	Grey	3	18 604
2	Ford	Blue	2	33 216
3	Ford	White	5	27 435
4	Nissan	Red	4	32 006
5	Vauxhall	Blue	4	31 598
6	Ford	Green	3	37 685
7	Vauxhall	Red	3	21 640
8	Nissan	White	2	28 763
9	Ford	White	3	30 498
10	Vauxhall	White	5	9 865
11	Nissan	Red	3	7 520
12	Vauxhall	Grey	5	16 482

(a) (i) Draw up separate frequency tables for make, colour and number of doors.
(ii) Draw up a grouped frequency table for mileage.
Use class intervals of 5000 miles,
starting at $0 \leqslant m < 5000$, $5000 \leqslant m < 10\,000$, ...
(b) (i) Which make of car is the most popular?
(ii) How many Ford cars are white?
(iii) How many cars have a mileage of 30 000 or more?
(iv) How many cars have exactly 3 doors?

5 Use data collection cards to collect information about students in your class.
Include gender, height, shoe size and pulse rate.
(a) What is the smallest shoe size for students in your class?
(b) What is the gender of the student with the highest pulse rate?
(c) What is the difference between the highest pulse rate and the lowest pulse rate?
(d) What is the height of the tallest student?
(e) What differences are there in the data collected for male and female students?

6 (a) Design a data collection card to collect information on the leisure time activities of students.
(b) Draw up frequency or grouped frequency tables for the data.
(c) Which leisure time activity is the most popular?
(d) What differences are there in the leisure time activities of male and female students?

Questionnaires

Questionnaires are frequently used to collect data.
In business they are used to get information about products or services and in politics they are frequently used to test opinion on a range of issues and personalities.

When constructing questions for a questionnaire you should:

(1) use simple language, so that everyone can understand the question;
(2) ask short questions which can be answered precisely, with a "yes" or "no" answer, a number, or a response from a choice of answers;
(3) provide tick boxes, so that questions can be answered easily;
(4) avoid open-ended questions, like: "What do you think of education?" which might produce long rambling answers which would be difficult to collate or process;
(5) avoid leading questions, like: "Don't you agree that there is too much bad language on television?" and ask instead:
"Do you think that there is too much bad language on television?" Yes ☐ No ☐
(6) ask questions in a logical order.

Multiple-response questions

In many instances a choice of responses should be provided.

Instead of asking, "How old are you?" which does not indicate the degree of accuracy required and many people might consider personal, we could ask instead:

Which is your age group?

under 18 ☐ 18 to 40 ☐ 41 to 65 ☐ over 65 ☐

Notice there are no gaps and only **one** response applies to each person.

Sometimes we invite **multiple responses** by asking questions, such as:

Which of these soaps do you watch?

Coronation Street ☐ EastEnders ☐ Emmerdale ☐ Hollyoaks ☐

Tick as many as you wish.

Hypothesis

A **hypothesis** is a statement that may or may not be true.
To test a hypothesis we can construct a questionnaire, carry out a survey and analyse the results.

EXAMPLE

A questionnaire to test the hypothesis,

"People think it is better to give than to receive,"

could include questions like these.

1. **Gender:** male ☐ female ☐
2. **Age (years):** 11 - 16 ☐ 17 - 21 ☐ 22 - 59 ☐ 60 & over ☐
3. **Do you think it is better to give than to receive?**
   Yes ☐ No ☐
4. **To which of the following have you given in the last year?**
   School ☐ Charities ☐ Church ☐
   Hospital ☐ Special appeals ☐ Homeless ☐

Other (please list) ______________________

Suggest another question which could be included.

Exercise 35.3

1. Susan wants to find out what people think about the Health Service.
   Part of the questionnaire she has written is shown.

   Q4. What is your date of birth?
   Q5. Don't you agree that waiting lists for operations are too long?
   Q6. How many times did you visit your doctor last year?
   ☐ less than 5 ☐ 5 - 10 ☐ 10 or more

   (a) Why should Q4 not be asked?
   (b) Give a reason why Q5 is unsuitable.
   (c) (i) Explain why Q6 is unsuitable in its present form.
       (ii) Rewrite the question so that it could be included in the questionnaire.

2. In preparing questions for a survey on the use of a library the following questions were considered. Explain why each question in its present form is unsuitable and rewrite the question.
   (a) How old are you?
   (b) How many times have you used the library?
   (c) Which books do you read?
   (d) How could the library be improved?

3. A mobile phone company wants to carry out a survey.
   It wants to find out the distribution of the age and sex of customers and the frequency with which they use the phone. The company intends to use a questionnaire.
   Write three questions and responses that will enable the company to carry out the survey.

4. A survey of reading habits is to be conducted. Suggest five questions which could be included.

5. Phil included this question in a questionnaire:
   Don't you agree that the radio gives the best news reports?
   Explain why this question is unsuitable.

6. Design a questionnaire to test the hypothesis:
   "People think that everyone should take part in sport."

7. Design a questionnaire to test the hypothesis:
   "People think that animals should not be used to test drugs."

8. Design a questionnaire to test the hypothesis: ***"Children have too much homework."***

Two-way tables

We have already seen that the results of a survey can be recorded on data collection sheets and then collated using frequency or grouped frequency tables. We can also illustrate data using **two-way tables**.
A two-way table is used to illustrate the data for two different features (variables) in a survey.

EXAMPLE

The two-way table shows the results of a survey.

(a) How many boys wear glasses?
(b) How many children wear glasses?
(c) Do the results prove or disprove the hypothesis:
    "More boys wear glasses than girls"?

	Wear Glasses	
	Yes	No
Boys	4	14
Girls	3	9

(a) 4 (b) 7 (c) Disprove. Boys: $\frac{4}{18} \times 100 = 22\%$ Girls: $\frac{3}{12} \times 100 = 25\%$

Exercise 35.4

1 The two-way table shows information about the ages of people in a retirement home.

	Age (years) 60 - 64	65 - 69	70 - 74	75 - 79	80 and over
Men	0	2	5	8	1
Women	2	5	6	5	6

(a) How many men are aged 75 - 79?
(b) How many men are included?
(c) How many people are aged 75 or more?
(d) How many people are included?
(e) What percentage of these people are aged 75 or more?

2 The two-way table shows information about a class of pupils.

	Can swim	Cannot swim
Boys	14	6
Girls	8	2

(a) How many boys can swim?
(b) How many boys are in the class?
(c) What percentage of the boys can swim?
(d) What percentage of the girls can swim?
(e) Do the results prove or disprove the hypothesis:
"More boys can swim than girls"?
Explain your answer.

3 The two-way table shows the number of boys and girls in families taking part in a survey.

Number of girls					
3	1		2		
2	1	2	3		
1	5	9		1	1
0		3		2	
Number of boys	0	1	2	3	4

(a) (i) How many families have two children?
(ii) Does the data support the hypothesis:
"More families have less than 2 children than more than 2 children"?
Explain your answer.
(b) (i) How many girls are included in the survey?
(ii) Does the data support the hypothesis:
"More boys are born than girls"?
Explain your answer.

4 In a survey, 100 people were asked: "Would you like to be taller?"
58 of the people asked were men.
65 of the people asked said, "Yes."
24 of the women asked said, "No."
(a) Construct a two-way table to show the results of the survey.
(b) A newspaper headline stated:

Over 80% of men would like to be taller.

Do the results of the survey support this headline?
Give a reason for your answer.

Sampling

When information is required about a small group of people it is possible to survey everyone.
When information is required about a large group of people it is not always possible to survey everyone and only a **sample** may be asked.
The sample chosen should be large enough to make the results meaningful and representative of the whole group or the results may be **biased**.
For example, to test the hypothesis,

"Girls are more intelligent than boys",

you would need to ask equal numbers of boys and girls from various age groups.
The results of a survey may also be biased if there is any form of deliberate selection or if the sample is incomplete.

Sampling methods

In a **simple random sample** everyone has an equal chance of being selected.
For example, the names of the whole group could be written on identical pieces of paper and placed in a hat.
Names could then be taken from the hat, without looking, until the sample is complete.

A **census** is when the whole of a population is surveyed.
The only true sample is when 100% of the population is surveyed.
A large sample allows more reliable inferences to be made about the whole population.

In a **systematic random sample** people are selected according to some rule.
For example, names could be listed in alphabetical order and every tenth person selected.

However, samples chosen using these methods may not be representative of the whole group.
For example, a sample taken from a school population may consist of girls only, or pupils from Year 10 only.

Quota samples are often used in market research.
The population is divided into groups (gender, age, etc).
A given number (quota) is surveyed from each group.
This type of sample is not random, but is cheap to carry out and can be done quickly.
How reliable are such samples?

A **stratified random sample** is used to overcome the possible bias of random samples by taking into account the composition of all the people in the original group.
The original group is divided up into separate categories or strata, such as male/female, age group, etc, before a random sample is taken.
A simple random sample is then taken from each category in proportion to the size of the category.

EXAMPLE

Jayne is investigating the spending habits of girls in her school.
The table shows the numbers of girls in each part of the school.

Year Group	Lower School	Upper School	Sixth Form
Number of girls	140	100	60

Jayne wants a stratified random sample of 30 girls.
How many girls should be chosen from each part of the school?

Sample size   Lower School: $\frac{140}{300} \times 30 = 14$

Upper School: $\frac{100}{300} \times 30 = 10$

Sixth Form: $\frac{60}{300} \times 30 = 6$

Exercise 35.5

1. George is investigating the cost of return journeys by train.
He plans to ask ten passengers who are waiting at a station at midday the cost of their return journeys. Give two reasons why he might get biased results.

2. Judy is investigating shopping habits.
She plans to interview 50 women at her local supermarket on a Tuesday morning.
Give three reasons why she might get biased results.

3. State one advantage and one disadvantage of a postal survey.

4. The two-way table shows the age and gender of people taking part in a survey.

	Age (years)				
	Under 18	18 - 25	26 - 40	41 - 64	65 and over
Female	0	2	7	9	7
Male	0	4	17	19	10

Give two reasons why the data collected may not be representative of the whole population.

5. In a school there are 320 pupils in the lower school and 240 pupils in the upper school.
How many pupils from each part of the school should be included in a stratified random sample of size 40?

6. The number of pupils in each year group at a boys school is shown.

Year Group	7	8	9	10	11
Number of boys	90	110	110	100	90

The school carries out a homework survey.
(a) Explain why a simple random sample of pupils should not be used.
(b) Calculate how many pupils should be included from each year group in a stratified random sample of size 50.

7. The table shows the numbers of employees in each section of a company.

Department	Managerial	Clerical	Technical	Manual
Number of employees	26	65	637	572

A survey on job satisfaction is to be carried out.
(a) Explain why a simple random sample of employees is unsuitable.
(b) A stratified random sample of size 100 is used.
How many employees from the technical department will be included?

8. The table shows the age distribution of a club's membership.

Age (years)	Under 21	21 to 65	Over 65
Number	49	76	13

How many members should be included from each of these age groups in a stratified random sample of 25?

9. An industrial concern employs personnel as shown in the table.

Management	Sales	Shop floor
48	261	2691

How many employees from each section should be included in a stratified random sample of 60 employees?

What you need to know

- **Primary data** is data collected by an individual or organisation to use for a particular purpose. Primary data is obtained from experiments, investigations, surveys and by using questionnaires.
- **Secondary data** is data which is already available or has been collected by someone else for a different purpose. Sources of secondary data include the Annual Abstract of Statistics, Social Trends and the Internet.
- **Qualitative** data – Data which can only be described in words.
- **Quantitative** data – Data that has a numerical value.
  Quantitative data is either **discrete** or **continuous**.
  Discrete data can only take certain values.
  Continuous data has no exact value and is measurable.
- **Data Collection Sheets** – Used to record data during a survey.
- **Tally** – A way of recording each item of data on a data collection sheet.
  A group of five is recorded as 𝍸.
- **Frequency Table** – A way of collating the information recorded on a data collection sheet.
- **Grouped Frequency Table** – Used for continuous data or for discrete data when a lot of data has to be recorded.
- **Database** – A collection of data.
- **Class Interval** – The width of the groups used in a grouped frequency distribution.
- **Questionnaire** – A set of questions used to collect data for a survey.
  Questionnaires should:
  (1) use simple language,
  (2) ask short questions which can be answered precisely,
  (3) provide tick boxes,
  (4) avoid open-ended questions,
  (5) avoid leading questions,
  (6) ask questions in a logical order.
- **Hypothesis** – A hypothesis is a statement which may or may not be true.
- When information is required about a large group of people it is not always possible to survey everyone and only a **sample** may be asked.
  The sample chosen should be large enough to make the results meaningful and representative of the whole group (population) or the results may be **biased**.
- **Two-way Tables** – A way of illustrating two features of a survey.
- In a **simple random sample** everyone has an equal chance of being selected.
- In a **systematic random sample** people are selected according to some rule.
- In a **stratified random sample** the original group is divided up into separate categories or strata, such as male/female, age group, etc, before a random sample is taken.
  A simple random sample is then taken from each category in proportion to the size of the category.

Review Exercise 35

1. Claire carried out a survey about smoking.
   She included the following questions.

   > 1. How many cigarettes do you smoke?
   > 2. Do you agree that smoking is bad for your health?

   (a) Make **one** criticism of each question.
   (b) Rewrite question 2 so that it is suitable for use in Claire's survey. AQA

2 The table shows the age and gender of people taking part in a survey to test the hypothesis:

"Children have too much homework."

	Age (years)				
	Under 11	11 - 16	17 - 25	26 - 50	Over 50
Male	0	4	6	5	5
Female	0	0	0	0	0

Give three reasons why the sample is biased.

3 A travel agent says, "More women prefer holidays abroad than men."
The table shows the results of a survey to test this statement.

	Men	Women
Prefer holidays abroad	18	21
Prefer holidays in the UK	6	7

Do these results support the statement made by the travel agent? Explain your answer.

4 Two students, Lyn and Pat, decide to do a survey on school dinners.
Their school has 200 pupils in each of Years 7 to 11.
There are the same number of boys as girls in each year group.
Lyn and Pat each decide to survey 50 pupils.

(a) (i) Lyn visits a different Year 7 form each morning for a week and surveys 5 boys and 5 girls each day. Comment on this method of sampling.

(ii) Pat gets an alphabetical list of all 1000 pupils in the school and selects every 20th name on the list. She then surveys these pupils.
Comment on this method of sampling.

(b) Harry also decides to do a survey of 50 pupils.
Explain how he can choose a stratified sample of the school population.

AQA

5 A school is carrying out a survey about the sporting activities of pupils.
The numbers of pupils in each year are shown on the right.

Year	Number
8	212
9	176
10	183
11	143
Total	714

(a) Give **one** reason why a simple random sample of pupils may not be representative of the whole school.

The school decides to interview a stratified sample of 40 pupils.

(b) How many pupils from Year 11 should be selected?

AQA

6 The table shows the number of people working in different sections of a paint manufacturing company.

Work Section	Number of men	Number of women
Manufacturing workshop	500	100
Storage and Distribution	200	100
Purchasing	30	50
Marketing and management	10	10

The owner wants to question 200 workers on how to improve production.
He proposes to allocate each person a number and then select 200 numbers at random.

(a) State **one** disadvantage of such a selection process.

It was suggested that a Stratified Sample would be more representative of the workers in the company.

(b) (i) Calculate the number of people in a Stratified Sample of 200 who would represent the purchasing section.

(ii) How many women should be included within the people chosen from the purchasing section?

AQA

CHAPTER 36

Averages and Range

Types of average

There are three types of **average**: the **mode**, the **median** and the **mean**.
The **mode** is the most common value.
The **median** is the middle value (or the mean of the two middle values) when the values are arranged in order of size.

Mean $= \dfrac{\text{Total of all values}}{\text{Number of values}}$

Range

The **range** is a measure of **spread**, and is the difference between the highest and lowest values.
Range = highest value − lowest value.

EXAMPLE

The price, in pence, of a can of cola in eight different shops is shown.

35, 39, 39, 32, 37, 35, 35, 40.

Find (a) the mode,
(b) the median,
(c) the mean,
(d) the range of these prices.

(a) The **mode** is the most common value.
The most common price is 35.
The mode is 35 pence.
We sometimes say the modal price is 35p.

(b) The **median** is found by arranging the values in order of size and taking the middle value.
Arrange the prices in order of size.
32, 35, 35, 35, 37, 39, 39, 40.
The middle price is $\frac{35 + 37}{2} = 36$
The median is 36 pence.

Where there are an even number of values the median is the average of the middle two.

(c) The **mean** is found by finding the total of all the values and dividing the total by the number of values.
Add the prices.
$35 + 39 + 39 + 32 + 37 + 35 + 35 + 40 = 292$
The mean $= \frac{292}{8} = 36.5$
The mean is 36.5 pence.

(d) The **range** = highest value − lowest value.
The highest price = 40p, the lowest price = 32p
The range = 40 − 32 = 8 pence.

Exercise 36.1

Do not use a calculator for questions 1 and 2.

1. Claire recorded the number of e-mail messages she received each day last week.
5 2 1 7 4 1 1
Find the mode, median, mean and range of the number of messages received each day.

2 Gail noted the number of stamps on parcels delivered to her office.

3 2 4 5 7 3 4 3 5 3

(a) Write down the mode.
(b) Find the range of the number of stamps on a parcel.
(c) Find the median number of stamps on a parcel.
(d) Calculate the mean number of stamps on a parcel.

3 The boot sizes of players in a rugby team are:

9 8 10 11 9 8 10 9 10 9 10 10 8 9 10

(a) Which boot size is the mode?
(b) Find the median boot size.
(c) Calculate the mean boot size. Give your answer correct to one decimal place.

4 The weights of 8 men in a rowing team are:

86 kg 95 kg 89 kg 93 kg 84 kg 78 kg 91 kg 79 kg

Calculate the mean of their weights correct to one decimal place.

5 The mean of six numbers is 5. Five of the numbers are: 2, 3, 7, 8 and 6.
What is the other number?

6 The mean length of 8 rods is 75 cm. An extra rod is added.
The total length of the 9 rods is 729 cm.
What is the length of the extra rod?

7 A group of 2 boys and 3 girls take a test.
The mean mark for the boys is 14.5
The mean mark for the girls is 16.
Calculate the mean mark for the whole group.

Frequency distributions

After data is collected it is often presented using a **frequency distribution table**.
To find measures of average for a **frequency distribution** we have to take into account the frequency of each amount.

EXAMPLE

The table shows the boot sizes of players in a rugby team.

Boot size	8	9	10	11
Frequency	3	5	6	1

Find (a) the mode, (b) the median, (c) the mean boot size.

(a) The mode is the boot size with the largest frequency.
The modal boot size is 10.

(b) The median is the boot size of the middle player.
There are 15 players altogether.
The middle one is the 8th player.
The first 3 players wear boot size 8,
the next 5 wear boot size 9.
So, the 8th player wears boot size 9.
The median boot size is 9.

Mathematical shorthand:
Σ is the Greek letter 'sigma'.
Σf means the sum of frequencies.
Σfx means the sum of the values of fx.
Mean $= \frac{\Sigma fx}{\Sigma f}$

(c) The mean $= \frac{\text{Total of all boot sizes}}{\text{Total number of players}} = \frac{8 \times 3 + 9 \times 5 + 10 \times 6 + 11 \times 1}{3 + 5 + 6 + 1} = \frac{140}{15} = 9.333\ldots$

The mean boot size is 9.3, correct to 1 d.p.

Calculating averages from diagrams

EXAMPLE

Calculate the range, mode, median and mean of the ages for the data shown in the bar chart.

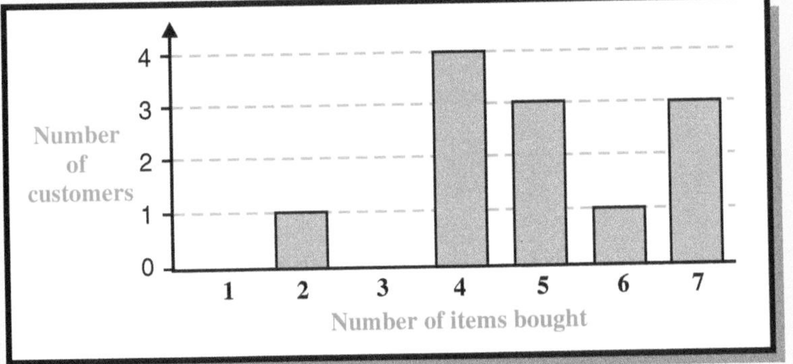

The range is the difference between the highest and lowest ages.
Range = 17 - 14
= 3 years

The most common age is shown by the tallest bar.
So, the modal age is 17 years.

Use a table to find the median and the mean.

Age (years) x	Frequency f	Frequency × Age $f \times x$
14	30	420
15	40	600
16	36	576
17	41	697
Totals	$\Sigma f = 147$	$\Sigma fx = 2293$

The middle student is given by: $\frac{147 + 1}{2} = 74$
The 74th student in the list has the median age.
The first 70 students are aged 14 or 15 years.
The 74th student has age 16 years.
Median age is 16 years.

$$\text{Mean} = \frac{\text{Total of all ages}}{\text{Number of students}} = \frac{\Sigma fx}{\Sigma f}$$

$$\text{Mean} = \frac{2293}{147} = 15.598\ldots$$

Mean age is 15.6 years, correct to 1 d.p.

Exercise 36.2

1. Find the mode, median and mean for the following data.

Number of letters delivered	1	2	3	4	5	6
Number of days taken to deliver	6	9	6	6	2	1

2. Hilary observed customers using the express checkout at a supermarket.

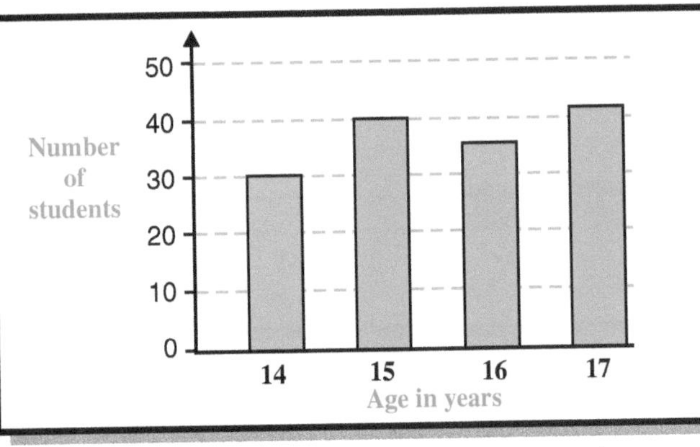

(a) Find the range of the number of items bought.
(b) What is the mode of the number of items bought?
(c) Work out the median number of items bought.
(d) Calculate the mean number of items bought.

3 A group of students took part in a quiz on the Highway Code. The bar chart shows their scores.

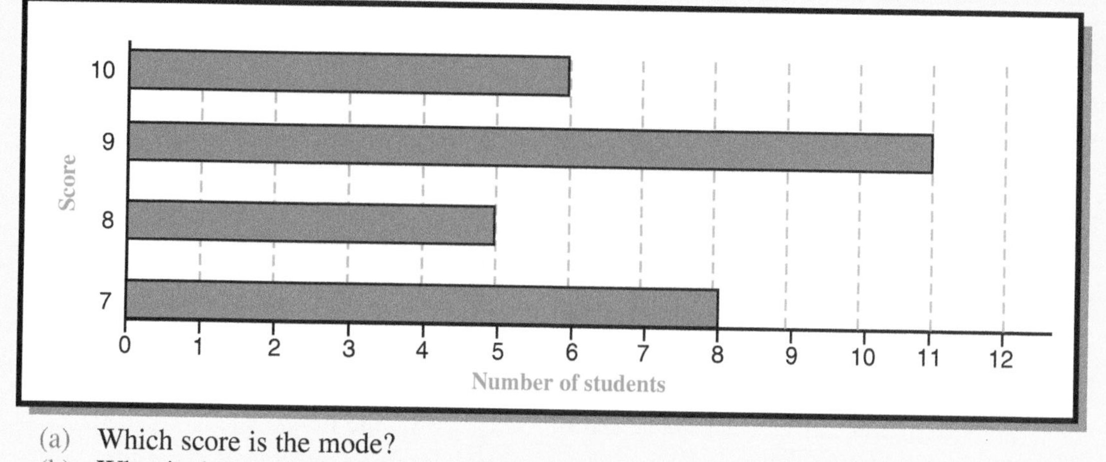

(a) Which score is the mode?
(b) What is the median score?
(c) How many students took part in the quiz?
(d) Calculate the mean score.

Grouped frequency distributions

When there is a lot of data, or the data is continuous, **grouped frequency distributions** are used.

Calculating the mean

For a grouped frequency distribution the true value of the mean cannot be found as the actual values of the data are not known.

To **estimate the mean**, we assume that all the values in each class are equal to the midpoint of the class.

$$\text{Estimated mean} = \frac{\Sigma\,(\text{frequency} \times \text{midpoint})}{\text{Total frequency}} = \frac{\Sigma fx}{\Sigma f}$$

Modal class

For a grouped frequency distribution with equal class width intervals, the **modal class** is the class (or group) with the highest frequency.

EXAMPLE

The table shows the masses of a group of children.

(a) Calculate an estimate of the mean mass.
(b) Find the modal class.

Mass (m kg)	Frequency
$40 \leqslant m < 50$	3
$50 \leqslant m < 60$	10
$60 \leqslant m < 70$	6
$70 \leqslant m < 80$	12

(a)

Mass (m kg)	Midpoint x	Frequency f	Frequency × Midpoint $f \times x$
$40 \leqslant m < 50$	45	3	135
$50 \leqslant m < 60$	55	10	550
$60 \leqslant m < 70$	65	6	390
$70 \leqslant m < 80$	75	12	900
	Totals	$\Sigma f = 31$	$\Sigma fx = 1975$

Midpoint of the class $40 \leqslant m < 50$ is given by: $\frac{40 + 50}{2} = \frac{90}{2} = 45$

Estimate of mean $= \frac{\Sigma fx}{\Sigma f} = \frac{1975}{31} = 63.709\ldots$

Estimate of mean mass $= 63.7$ kg, correct to 3 sig. figs.

(b) The modal class is $70\,\text{kg} \leqslant m < 80\,\text{kg}$

Exercise 36.3

0 - means 0 or more but less than 10.

1. Give the modal class and calculate an estimate of the mean for each of the following.

(a)

Salary (s) (£000's)	Number of employees
$10 \leqslant s < 20$	79
$20 \leqslant s < 30$	32
$30 \leqslant s < 40$	14
$40 \leqslant s < 50$	0
$50 \leqslant s < 60$	2

(b)

Time spent watching TV per week (hours)	Number of students
0 -	2
10 -	8
20 -	5
30 -	14
40 - 50	7

2. The table shows the distribution of the weights of some turkeys.

Weight (w kg)	$2 \leqslant w < 4$	$4 \leqslant w < 6$	$6 \leqslant w < 8$	$8 \leqslant w < 10$
Frequency	7	9	5	3

Calculate an estimate of the mean weight of these turkeys.
Give your answer correct to one decimal place.

3. The table shows the distribution of the prices of houses for sale in a particular neighbourhood.

Price (p £000's)	$60 \leqslant p < 80$	$80 \leqslant p < 100$	$100 \leqslant p < 120$	$120 \leqslant p < 140$
Number of houses	3	7	4	1

Calculate an estimate of the mean price of these houses.
Give your answer to an appropriate degree of accuracy.

4. The table shows the distribution of marks in a test.

Mark	0 - 19	20 - 29	30 - 34	35 - 39	40 - 50
Number of students	12	23	25	14	3

Calculate an estimate of the mean mark. *Notice that the class intervals are not all equal.*

Comparing distributions

The table shows the marks gained in a test. Compare the marks obtained by the boys and the girls.

Mark (out of 10)	7	8	9	10
Number of boys	2	5	3	0
Number of girls	4	0	2	1

To compare the marks we can use the range and the mean.

Boys: Range $= 9 - 7 = 2$
Girls: Range $= 10 - 7 = 3$

The girls had the higher range of marks.

Boys: Mean $= \frac{2 \times 7 + 5 \times 8 + 3 \times 9 + 0 \times 10}{10} = \frac{81}{10} = 8.1$

Girls: Mean $= \frac{4 \times 7 + 0 \times 8 + 2 \times 9 + 1 \times 10}{7} = \frac{56}{7} = 8$

The boys had the higher mean mark.

Overall the boys did better as:
the girls' marks were more spread out with a lower average mark,
the boys' marks were closer together with a higher average mark.

Note:
To compare the overall standard, the median could be used instead of the mean.

Exercise 36.4

1 The times, in minutes, taken by 8 boys to swim 50 metres are shown.

1.8	2.0	1.7	2.2	2.1	1.9	1.8	2.1

(a) (i) What is the range of these times?
(ii) Calculate the mean time.

The times, in minutes, taken by 8 girls to swim 50 metres are shown.

2.1	1.9	1.8	2.3	1.6	2.0	2.6	1.9

(b) Comment on the times taken by these boys and girls to swim 50 m.

2 Use the mean and the range to compare the number of visits to the cinema by these women and men.

Number of visits to the cinema last month	0	1	2	3	4	5	6	More than 6
Number of women	8	9	7	3	2	1	1	0
Number of men	0	12	7	1	0	0	0	0

3 Deepak thought that the girls in his class wore smaller shoes than the boys on average, but that the boys' shoe sizes were less varied than the girls'.
He did a survey to test his ideas.
The table shows his results. Was he correct?

Shoe size	$4\frac{1}{2}$	5	$5\frac{1}{2}$	6	$6\frac{1}{2}$	7	$7\frac{1}{2}$	8	$8\frac{1}{2}$	9	$9\frac{1}{2}$
Number of boys	1	0	5	4	4	2	1	0	1	0	0
Number of girls	0	2	0	2	3	0	2	0	3	1	1

4 (a) Find the modal class for the ages of customers in each of these two restaurants.
(b) Which restaurant attracts more younger people?
(c) Why is it only possible to find an approximate value for the age range of customers?

Age (years)	0 - 9	10 - 19	20 - 29	30 - 39	40 - 49	50 - 59	60 - 69	70 - 79	80 - 89
MacQuick	8	9	10	7	1	1	3	1	0
Pizza Pit	2	4	12	15	5	3	3	2	1

5 The graphs show the monthly sales of bicycles before and after a marketing campaign.
Calculate the medians and the ranges.
Use your results to compare 'Before' with 'After'.

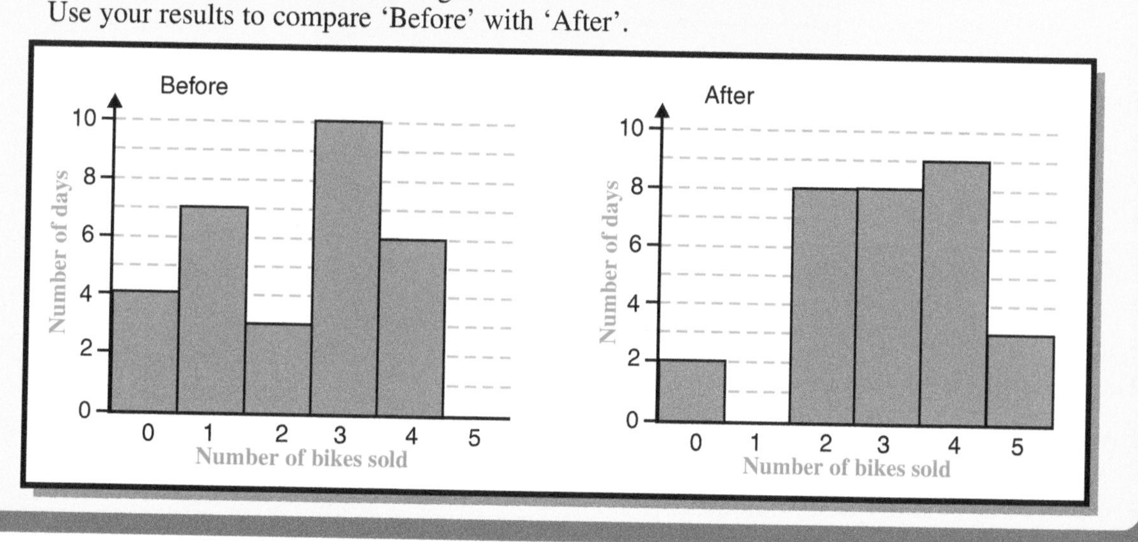

Which is the best average to use?

Many questions in mathematics have definite answers. This one does not.
Sometimes the mean is best, sometimes the median and sometimes the mode.
It all depends on the situation and what you want to use the average for.

EXAMPLE

A youth club leader gets a discount on cans of drinks if she buys all one size.
She took a vote on which size people wanted.
The results were as follows:

Size of can (ml)	100	200	330	500
Number of votes	9	12	19	1

Mode = 330 ml
Median = 200 ml
Mean = 245.6 ml, correct to one decimal place.

Which size should she buy?

The mean is no use at all because she can't buy cans of size 245.6 ml.
Even if the answer is rounded to the nearest whole number (246 ml), it's still no use.
The median is possible because there is an actual 200 ml can.
However, only 12 out of 41 people want this size.
In this case the **mode** is the best average to use, as it is the most popular size.

Exercise 36.5

In questions 1 and 2 find all the averages possible.
State which is the most sensible and why.

1. On a bus: 23 people are wearing trainers,
   10 people are wearing boots,
   8 people are wearing lace-up shoes.

2. 20 people complete a simple jigsaw. Their times, in seconds, are recorded.
   5, 6, 8, 8, 9, 10, 11, 11, 12, 12,
   12, 15, 15, 15, 15, 18, 19, 20, 22, 200.

3. The times for two swimmers to complete each of ten 25 m lengths are shown below.

Swimmer A	30.1	30.1	30.1	30.6	30.7	31.1	31.1	31.5	31.7	31.8
Swimmer B	29.6	29.7	29.7	29.9	30.0	30.0	30.1	30.1	30.1	44.6

   Which is the better swimmer?
   Explain why.

4. A teacher sets a test.
   He wants to choose a minimum mark for a distinction so that 50% of his students get this result.
   Should he use the modal mark, the median mark or the mean mark?
   Give a reason for your answer.

5. The cost of Bed and Breakfast at 10 different hotels is given.
   £39.50 £55 £60 £50 £49 £42 £95 £59 £39.50 £45
   (a) Wyn says, "The average cost of Bed and Breakfast is £39.50."
   Which average is he using?
   Give a reason why this is not a sensible average to use for this data.
   (b) Which of the mode, median and mean best describes the average cost of Bed and Breakfast?
   Give a reason for your answer.

What you need to know

- There are three types of **average**: the **mode**, the **median** and the **mean**.
  The **mode** is the most common value.
  The **median** is the middle value (or the mean of the two middle values) when the values are arranged in order of size.
  Mean $= \dfrac{\text{Total of all values}}{\text{Number of values}}$
- The **range** is a measure of **spread**.
  Range = highest value − lowest value
- To find the mean of a **frequency distribution** use:
  $$\text{Mean} = \frac{\text{Total of all values}}{\text{Number of values}} = \frac{\Sigma fx}{\Sigma f}$$
- To find the mean of a **grouped frequency distribution**, first find the value of the midpoint of each class.
  Then use: $$\text{Estimated mean} = \frac{\text{Total of all values}}{\text{Number of values}} = \frac{\Sigma fx}{\Sigma f}$$
- Choosing the best average to use:
  When the most **popular** value is wanted use the **mode**.
  When **half** of the values have to be above the average use the **median**.
  When a **typical** value is wanted use either the **mode** or the **median**.
  When all the **actual** values have to be taken into account use the **mean**.
  When the average should not be distorted by a few very small or very large values do **not** use the mean.

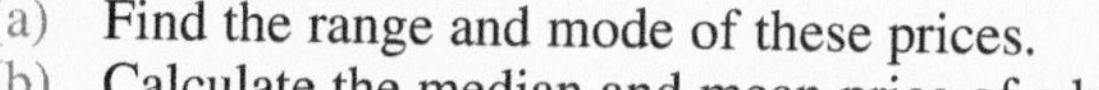

Review Exercise 36

1 (a) Find the range and mode of these prices.
(b) Calculate the median and mean price of a bottle of milk.

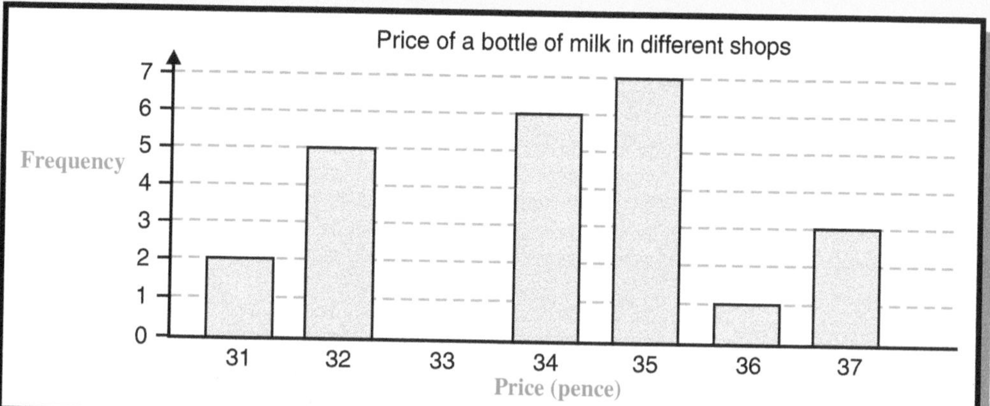

2 David is playing cricket.
The table shows the number of runs he has scored off each ball so far.

Number of runs	0	1	2	3	4	5	6
Number of balls	3	8	4	3	5	0	2

(a) (i) What is the median number of runs per ball?
(ii) Calculate the mean number of runs per ball.

Off the next five balls, David scores the following runs: 4, 4, 5, 3 and 6.

(b) (i) Calculate the new median.
(ii) Calculate the new mean.

(c) Give a reason why the mean is used, rather than the median, to give the average number of runs scored per ball.

AQA

3 In an experiment 50 people were asked to estimate the length of a rod to the nearest centimetre. The results were recorded.

Length (cm)	20	21	22	23	24	25	26	27	28	29
Frequency	0	4	6	7	9	10	7	5	2	0

(a) Find the value of the median.
(b) Calculate the mean length.
(c) In a second experiment another 50 people were asked to estimate the length of the same rod.
The most common estimate was 23 cm.
The range of the estimates was 13 cm.
Make **two** comparisons between the results of the two experiments. AQA

4 The weekly wages of employees are recorded.

Wage (£)	100 -	200 -	300 -	400 -	500 -	600 - 1000
Frequency	2	13	4	0	2	12

(a) Which is the modal group?
(b) In which group is the median value?
(c) Without calculating, state which of the mean, mode or median is the largest.
Explain your answer. AQA

5 On holiday Val records the length of time people stay in the pool.
The results are shown in the table.

Time (t mins)	Number of people
$0 < t \leqslant 10$	4
$10 < t \leqslant 20$	7
$20 < t \leqslant 30$	3
$30 < t \leqslant 40$	2
	16

Calculate an estimate of the mean time spent in the pool.
Give your answer to an appropriate degree of accuracy. AQA

6 Mrs Wilson wants to sell her herd of dairy cows.
A buyer will need to know the herd's average daily yield of milk.
The daily milk yield, p litres, is monitored over 5 weeks.
The table shows the results of this survey.

Milk yield (p litres)	Frequency (f)
$140 \leqslant p < 145$	3
$145 \leqslant p < 150$	5
$150 \leqslant p < 155$	9
$155 \leqslant p < 160$	6
$160 \leqslant p < 165$	8
$165 \leqslant p < 170$	4
Total	35

(a) Mrs Wilson finds the modal class for the daily yield. What is this value?
(b) Calculate the estimated mean daily milk yield.
(c) Which is the more suitable average for Mrs Wilson to use?
Give a reason for your answer.

Presentation of Data 1

Most people find numerical information easier to understand if it is presented in pictorial form. For this reason, television reports, newspapers and advertisements frequently use graphs and diagrams to present and compare information.

Bar charts

Bar charts are a simple but effective way of displaying data.
Bars can be drawn either horizontally or vertically.
Bar charts can also be used to compare data.

The height of each bar gives the **frequency**.
The most frequently occurring variable is called the **mode** or **modal category**.

EXAMPLE

The table shows how a class of children travelled to school one day.

Method of travel	Bus	Cycle	Car	Walk
Boys	2	7	1	5
Girls	3	1	5	6

To make it easier to compare information for boys and girls we can draw both bars on the same diagram, as shown.

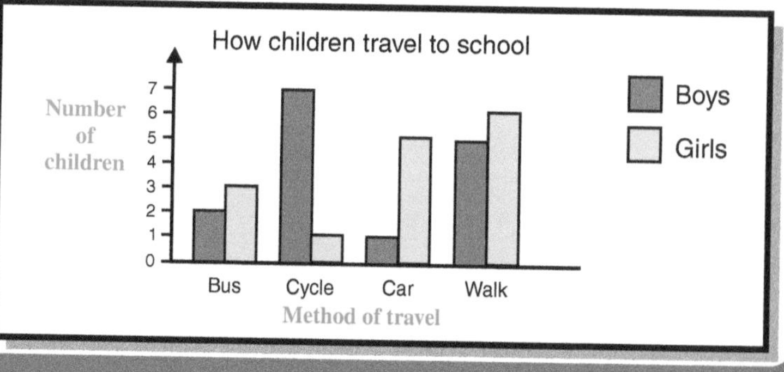

Cycling it the most popular method of travelling to school for boys.
Which method of travelling to school is the most popular for girls?

Compare and comment on the method of travel of these boys and girls.

Instead of drawing bars to show frequency we could draw horizontal or vertical lines.
Such graphs are called **bar-line graphs**.

Pie charts

Pie charts are useful for showing and comparing proportions of data.
However, they do not show frequencies. Such information can be found by interpreting the pie chart.
The whole circle represents the total frequency and each sector represents the frequency of one part (category) of the data.

EXAMPLE

The pie chart shows the makes of 120 cars.

(a) Which make of car is the mode?

(b) How many of the cars are Ford?

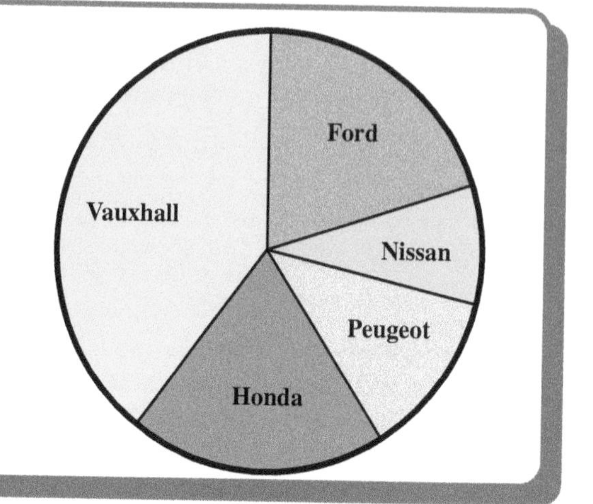

(a) The sector representing Vauxhall is the largest.
Therefore, Vauxhall is the mode.

(b) The angle of the sector representing Ford is 72°.
The number of Ford cars $= \frac{72}{360} \times 120 = 24$.

Exercise 37.1

1 The table shows the amount of pocket money given each week to a number of girls.

Amount (£)	1	2	3	4	5	6	7	8	9	10
Number of girls	0	0	1	5	10	4	0	3	0	7

(a) Draw a bar-line graph of the data.
(b) What is the range of the amount of pocket money given each week?
(c) Calculate the mean amount of pocket money given each week.
(d) What percentage of the girls got less than £5?

2 The breakfast cereal preferred by some adults is shown.

Breakfast cereal	Corn flakes	Muesli	Porridge	Bran flakes
Number of adults	25	20	12	15

(a) Show the information in a pie chart.
(b) Which breakfast cereal is the mode?

3 A group of children were asked how many hours they had spent watching television on a particular Sunday.
The bar chart shows the results.

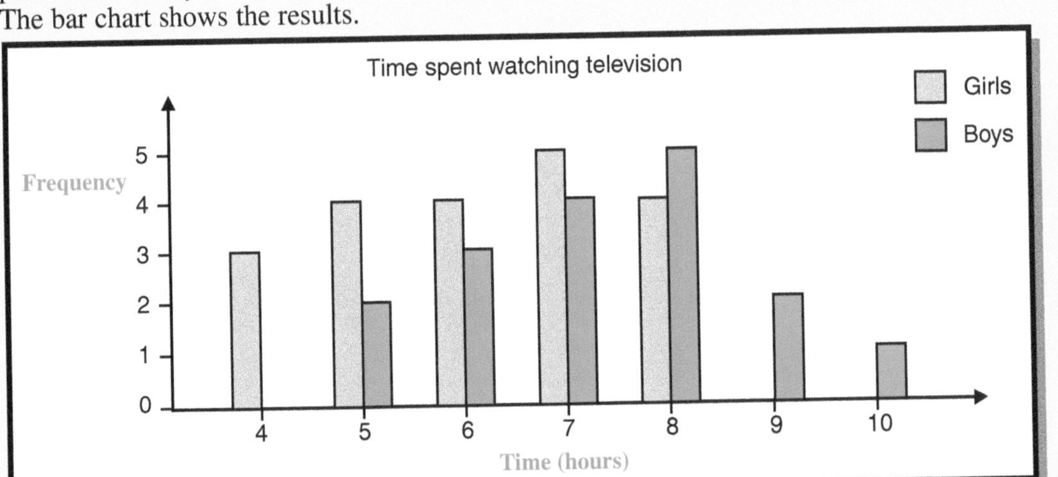

(a) What was the modal time for the girls?
(b) What was the range in time for the boys?
(c) How many boys watched television for more than 8 hours?
(d) (i) How many girls were included in the survey?
(ii) What percentage of the girls watched television for 6 hours?
(e) Compare and comment on the time spent watching television for these boys and girls.

4 The pie chart shows the different types of tree in a forest.
There are 54 oak trees and these are represented by a sector with an angle of 27°.

(a) The pine trees are represented by an angle of 144°.
How many pine trees are there?

(b) There are 348 silver birch trees.
Calculate the angle of the sector representing silver birch trees.

(c) Write, in its simplest form, the ratio of ash trees to oak trees in the forest.

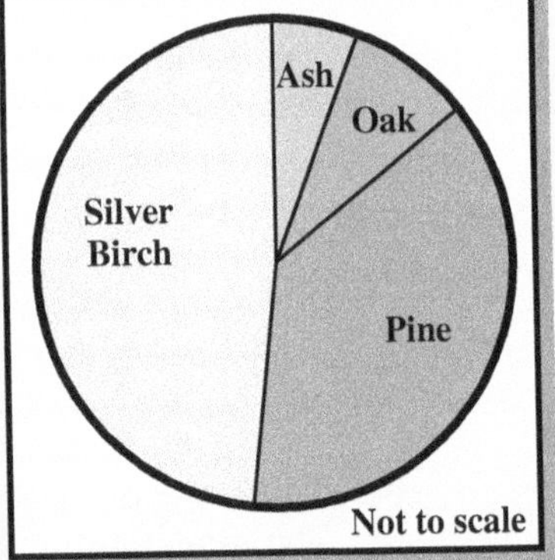

Stem and leaf diagrams

Data can also be represented using a **stem and leaf diagram**.
Back to back stem and leaf diagrams can be used to compare two sets of data.

EXAMPLES

1 The times, in seconds, taken by 14 students to complete a puzzle are shown.

15 9 23 32 17 12 27 19 26 15 11 24 31 10

Construct a stem and leaf diagram to represent this information.

A stem and leaf diagram is made by splitting each number into two parts.
As the data uses 'tens' and 'units', the stem will represent the 'tens' and the leaf will represent the 'units'.
To draw the stem and leaf diagram begin by drawing a vertical line.
The digits to the left of the line make the **stem**.
The digits to the right of the line are the **leaves**.

9 is recorded as 0 9

The first number is 15.
Next to the stem of 1 record 5.

Stem	Leaves
0	9
1	5 7 2 9 5 1 0
2	3 7 6 4
3	2 1

Once the data has been recorded, it is usual to redraw the diagram so that the leaves are in numerical order.

1 | 5 means 15 seconds

Stem	Leaves
0	9
1	0 1 2 5 5 7 9
2	3 4 6 7
3	1 2

Stem and leaf diagrams are often drawn without column headings, in which case a key is necessary.
e.g. 1 | 5 means 15 seconds

2 The results for examinations in Mathematics and English for a group of students are shown. The marks are given as percentages.

Mathematics:	91	27	55	69	83	25	45	53	67	71
	30	52	45	59	86	73	65	47	54	38
English:	45	40	48	65	75	55	36	85	76	69
	64	58	47	64	67	72	83	74	62	51

(a) Construct a back to back stem and leaf diagram for this data.
(b) Compare and comment on the results in Mathematics and English.

(a)

3 | 6 means 36%

Mathematics		English
7 5	2	
8 0	3	6
7 5 5	4	0 5 7 8
9 5 4 3 2	5	1 5 8
9 7 5	6	2 4 4 5 7 9
3 1	7	2 4 5 6
6 3	8	3 5
1	9	

For Mathematics:
5 | 2 means 25%

(b) The range of marks in Mathematics is larger than in English.
The modal group in English is 60 to 69, in Mathematics it is 50 to 59.

Exercise 37.2

1. The amount of petrol, in litres, bought by 20 motorists is shown.

16	23	27	10	35	42	26	25	24	17
23	41	33	35	25	19	16	31	12	29

(a) Construct a stem and leaf diagram to represent this information.
(b) What is the range in the amount of petrol bought?

2. The times, in seconds, taken to answer 24 telephone calls are shown.

3.2	5.6	2.4	3.5	4.3	3.6	2.8	5.8	3.3	2.6	3.2	2.8
5.6	3.5	4.2	1.5	2.7	2.5	3.7	3.1	2.9	4.2	2.4	3.0

(a) Copy and complete the stem and leaf diagram to represent this information.

3 | 2 means 3.2 seconds

Stem	Leaf
1	
2	
3	2
4	
5	

For this data:
the stem represents 'units',
the leaf represents 'tenths'.

(b) What is the median time taken to answer a telephone call?

3. The heights, in centimetres, of the heels on 20 different pairs of shoes are shown.

2.7	3.4	2.0	6.0	4.5	3.6	3.1	2.4	4.2	1.8
3.5	2.5	2.6	2.1	4.0	3.5	4.2	2.6	3.9	5.4

Construct a stem and leaf diagram to represent this information.

4. David did a survey to find the cost, in pence, of a loaf of bread.
The stem and leaf diagram shows the results of his survey.

2 | 7 means 27 pence

Stem	Leaf
2	7 9
3	1 1 2 9 9 9
4	2 5 5 9
5	0 4 5
6	0 5

(a) How many loaves of bread are included in the survey?
(b) What is the range of the prices?
(c) Which price is the mode?

5. The time taken to complete a computer game is recorded to the nearest tenth of a minute.
The times for a group of 20 adults and 20 children are shown.

Adults

7.9	8.2	7.3	9.2	6.4	6.5	6.1	8.2	7.8	7.0
9.4	8.0	7.3	5.4	7.7	10.1	5.9	6.7	7.3	6.0

Children

6.4	5.4	4.9	6.6	7.1	5.1	6.5	6.3	7.4	6.5
8.2	7.7	5.9	6.8	7.6	5.3	6.2	8.0	4.7	7.9

(a) Construct a back to back stem and leaf diagram for this data.
(b) Compare and comment on the times for adults and children.

Scatter graphs

When we investigate statistical information we often find there are connections between sets of data, for example, height and weight.
In general taller people weigh more than shorter people.

To see if there is a connection between two sets of data we can plot a **scatter graph**.
The scatter graph below shows information about the heights and weights of ten boys.

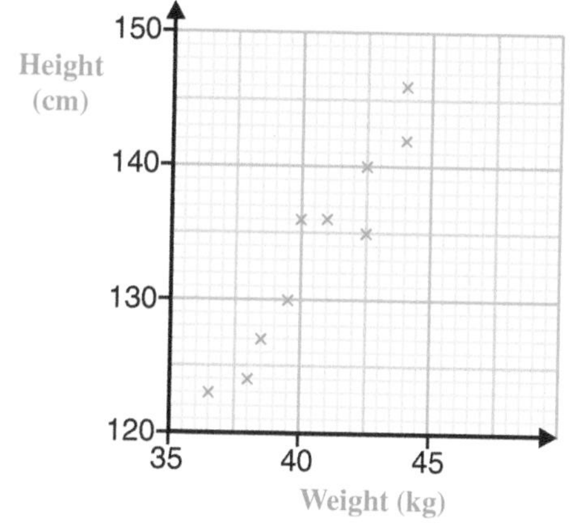

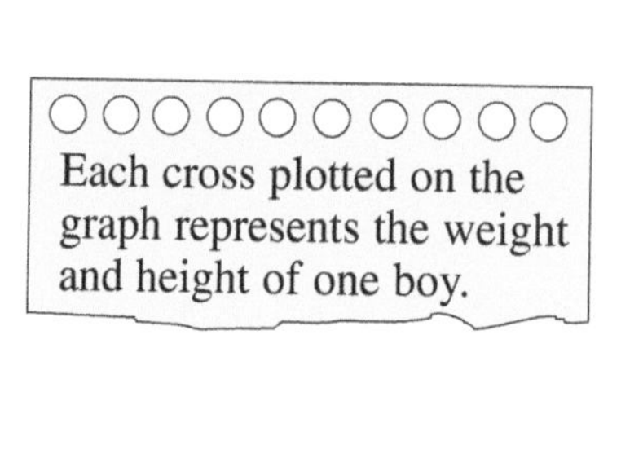

The diagram shows that taller boys generally weigh more than shorter boys.

Correlation

The relationship between two sets of data is called **correlation**.

In general the scatter graph of the heights and weights shows that as height increases, weight increases. This type of relationship shows there is a **positive correlation** between height and weight.
But if as the value of one variable increases the value of the other variable decreases, then there is a **negative correlation** between the variables.

When no linear relationship exists between two variables there is **zero correlation**. This does not necessarily imply "no relationship", but merely "no linear relationship".

The following graphs show types of correlation.

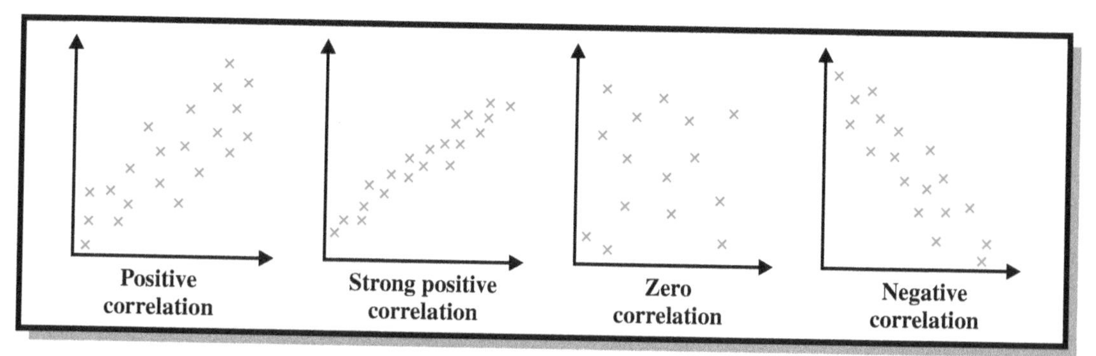

As points get closer to a straight line the stronger the correlation.
Perfect correlation is when all the points lie on a straight line.

Exercise 37.3

1. Describe the type of correlation you would expect between:
   (a) the age of a car and its secondhand selling price,
   (b) the heights of children and their ages,
   (c) the shoe sizes of children and the distance they travel to school,
   (d) the number of cars on the road and the number of road accidents,
   (e) the engine size of a car and the number of kilometres it can travel on one litre of fuel.

2 (a) Which of these graphs shows the strongest positive correlation?
(b) Which of these graphs shows perfect negative correlation?
(c) Which of these graphs shows the weakest correlation?

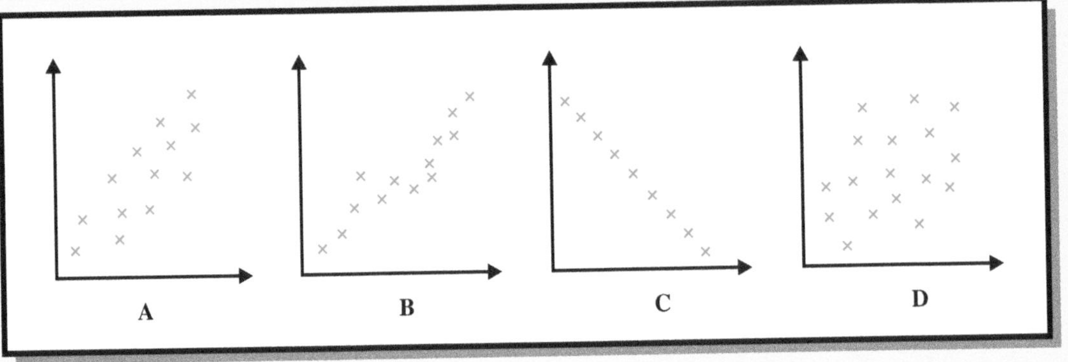

3 The scatter graph shows the shoe sizes and heights of a group of girls.

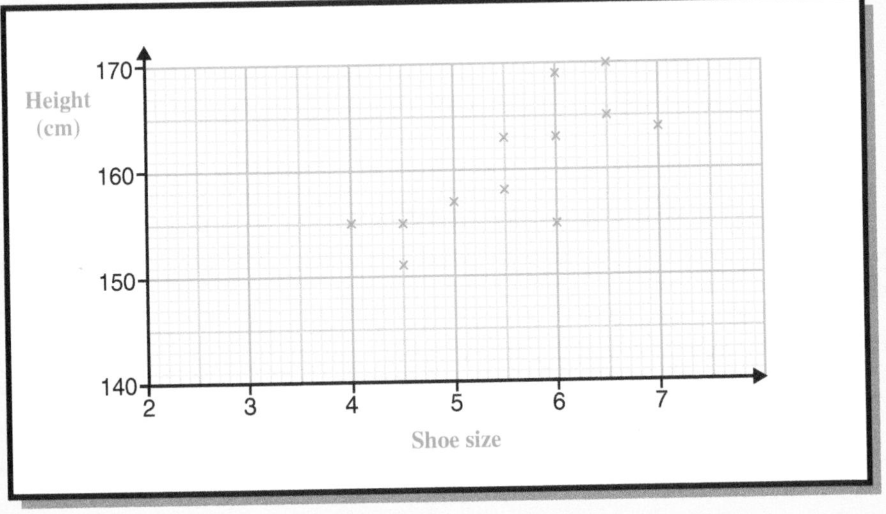

(a) How many girls wear size $6\frac{1}{2}$ shoes?
(b) How tall is the girl with the largest shoe size?
(c) Does the shortest girl wear the smallest shoes?
(d) What do you notice about the shoe sizes of taller girls compared to shorter girls?

4 The table shows the distance travelled and time taken by motorists on different journeys.

Distance travelled (km)	30	45	48	80	90	100	125
Time taken (hours)	0.6	0.9	1.2	1.2	1.3	2.0	1.5

(a) Draw a scatter graph for the data.
(b) What do you notice about distance travelled and time taken?
(c) Give one reason why the distance travelled and the time taken are not perfectly correlated.

5 Tyres were collected from a number of different cars.
The table shows the distance travelled and depth of tread for each tyre.

Distance travelled (1000 km)	4	5	9	10	12	15	18	25	30
Depth of tread (mm)	9.2	8.4	7.6	8	6.5	7.4	7	6.2	5

(a) Draw a scatter graph for the data.
(b) What do you notice about the distance travelled and the depth of tread?
(c) Explain how you can tell that the relationship is quite strong.

Line of best fit

We have seen that **scatter graphs** can be used to illustrate two sets of data and, from the distribution of points plotted, an indication of the relationship which exists between the data can be seen.

The scatter graph of heights and weights has been redrawn below and a **line of best fit** has been drawn, by eye, to show the relationship between height and weight.

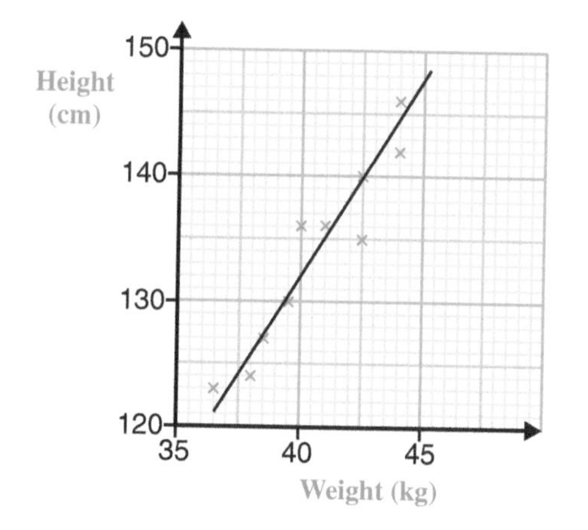

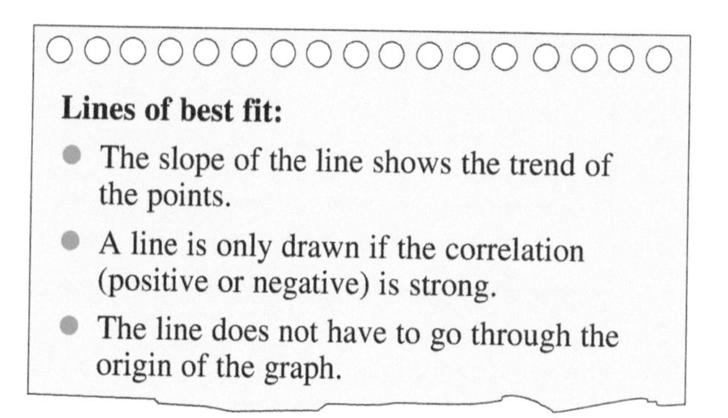

Where there is a relationship between the two sets of data the line of best fit can be used to estimate other values.

A boy is 132 cm tall.
Using the line of best fit an estimate of his weight is 40 kg.

Exercise 37.4

1 The table shows the ages and weights of ten babies.

Age (weeks)	2	4	9	7	13	5	6	1	10	12
Weight (kg)	3.5	3.3	4.2	4.7	5	3.8	4	3	5	5.5

(a) Use this information to draw a scatter graph.
(b) What type of correlation is shown on the scatter graph?
(c) Draw a line of best fit.
(d) Mrs Wilson's baby is 11 weeks old.
Use the graph to estimate the weight of her baby.

2 The table shows the temperature of water as it cools in a freezer.

Time (minutes)	5	10	15	20	25	30
Temperature (°C)	36	29	25	20	15	8

(a) Use this information to draw a scatter graph.
(b) What type of correlation is shown?
(c) Draw a line of best fit.
(d) Use the graph to estimate the time when the temperature of the water reaches 0°C.

3 The table shows the times taken by some boys to run 200 metres and their inside-leg measurements.

Time (seconds)	31	33	34	38	38	38	42	43	45	47
Inside-leg (cm)	69	65	72	63	69	75	70	65	74	69

(a) Plot a scatter graph of these data.
(b) Explain why a line of best fit for these data would not be useful in estimating the time for a different boy to run 200 metres by taking his inside-leg measurement.

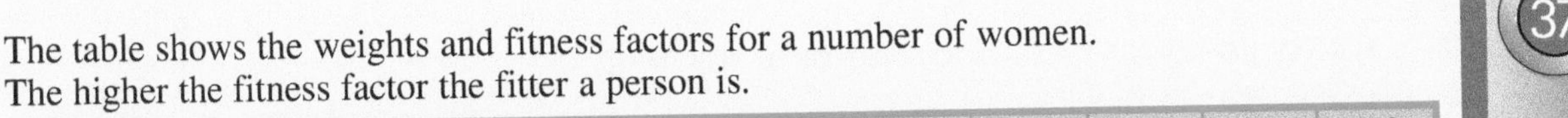

4 The table shows the weights and fitness factors for a number of women.
The higher the fitness factor the fitter a person is.

Weight (kg)	45	48	50	54	56	60	64	72	99	112
Fitness Factor	41	48	40	40	35	40	34	30	17	15

(a) Use this information to draw a scatter graph.
(b) What type of correlation is shown on the scatter graph?
(c) Draw a line of best fit.
(d) Use the graph to estimate:
  (i) the fitness factor for a woman whose weight is 80 kg,
  (ii) the weight of a woman whose fitness factor is 22.

5 The following table gives the marks obtained by some candidates taking examinations in French and German.

Mark in French	53	35	39	53	50	59	36	43
Mark in German	64	32	44	70	56	68	40	48

(a) (i) Use this information to draw a scatter graph.
  (ii) Draw the line of best fit by eye.
(b) Use the graph to estimate:
  (i) the mark in German for a candidate who got 70 in French,
  (ii) the mark in French for a candidate who got 58 in German.
(c) Which of the two estimates in (b) is likely to be more reliable?
  Give a reason for your answer.

What you need to know

- **Bar chart**. Used for data which can be counted.
  Often used to compare quantities of data in a distribution.
  Bars can be drawn horizontally or vertically.
  Bars are the same width and there are gaps between bars.
  The length of each bar represents frequency.
  The longest bar represents the **mode**.
  The difference between the largest and smallest variable is called the **range**.
- **Bar-line graph**. Instead of drawing bars, horizontal or vertical lines are drawn to show frequency.
- **Pie chart**. Used for data which can be counted.
  Often used to compare proportions of data, usually with the total.
  The whole circle represents all the data.
  The size of each sector represents the frequency of data in that sector.
  The largest sector represents the **mode**.
- **Stem and leaf diagrams**. Used to represent data in its original form.
  Data is split into two parts.
  The part with the higher place value is the stem,
  e.g. 15 stem 1 leaf 5.

  The data is shown in numerical order on the diagram.

  e.g. 2 | 3 5 9 represents 23, 25, 29.

  A key is given to show the value of the data.

  e.g. 3 | 4 means 34 cm or 3 | 4 means 3.4 cm, etc.

  Back to back stem and leaf diagrams can be used to compare two sets of data.

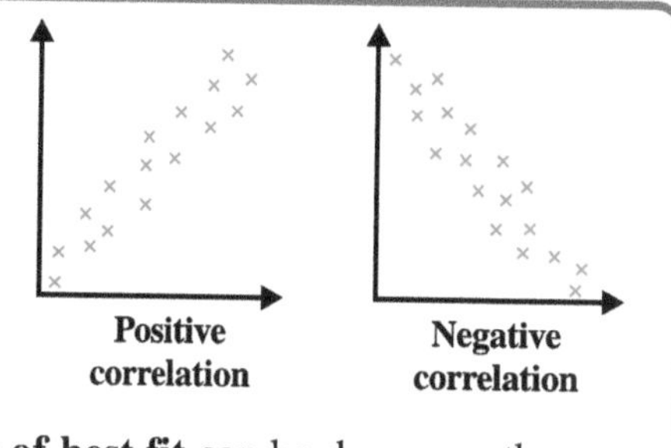

- A **scatter graph** can be used to show the relationship between two sets of data.
- The relationship between two sets of data is referred to as **correlation.**
- You should be able to recognise **positive** and **negative** correlation.
- When there is a relationship between two sets of data a **line of best fit** can be drawn on the scatter graph.
  The correlation is stronger as points get closer to a straight line.
  Perfect correlation is when all the points lie on a straight line.
- The line of best fit can be used to **estimate** the value from one set of the data when the corresponding value of the other set is known.

Review Exercise 37

1 The weights in grams of 20 cherry tomatoes are shown.

5.4 4.6 6.7 3.9 4.2 5.0 6.3 5.4 4.8 3.5
4.6 5.6 5.8 6.0 2.8 4.4 4.7 5.6 5.1 4.8

Draw a stem and leaf diagram to represent this information.

2 In one week Ronnie rents out 90 items from his shop as shown in the table below.

Item	Televisions	Videos	Computers	Other equipment
Frequency	35	30	17	8

Draw a pie chart to show all the week's rentals.

AQA

3 The data from a survey of cars was used to plot several scatter graphs.

1 **2** **3**

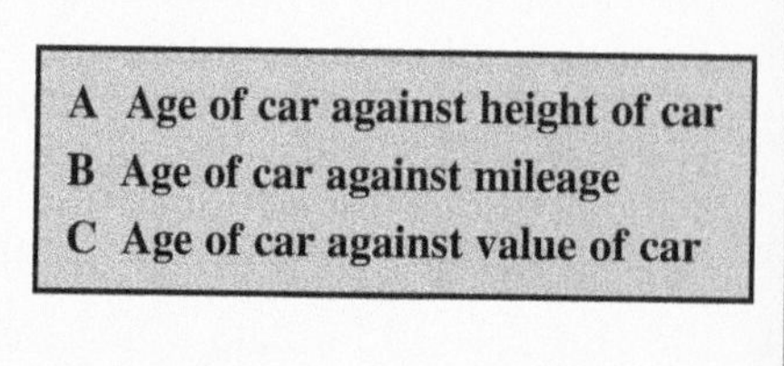
A Age of car against height of car
B Age of car against mileage
C Age of car against value of car

Match each scatter graph to the correct description.

4 The stem and leaf diagram shows the distribution of marks for a test marked out of 50.

1 | 7 means 17 marks

Boys	Stem	Girls
	0	9
6 2	1	0 1 2 7
7 6 4 3	2	1 3 5 5 6 7 8
9 5 3 2 0	3	2 5 9
5 1	4	1
0	5	

(a) What is the lowest mark for the girls?
(b) What is the highest mark for the boys?
(c) How many of these boys and girls scored more than 25 marks?
(d) Compare and comment on the marks for boys and girls.

5 The frequency diagram shows the distribution of shoe sizes for a class of 40 pupils.

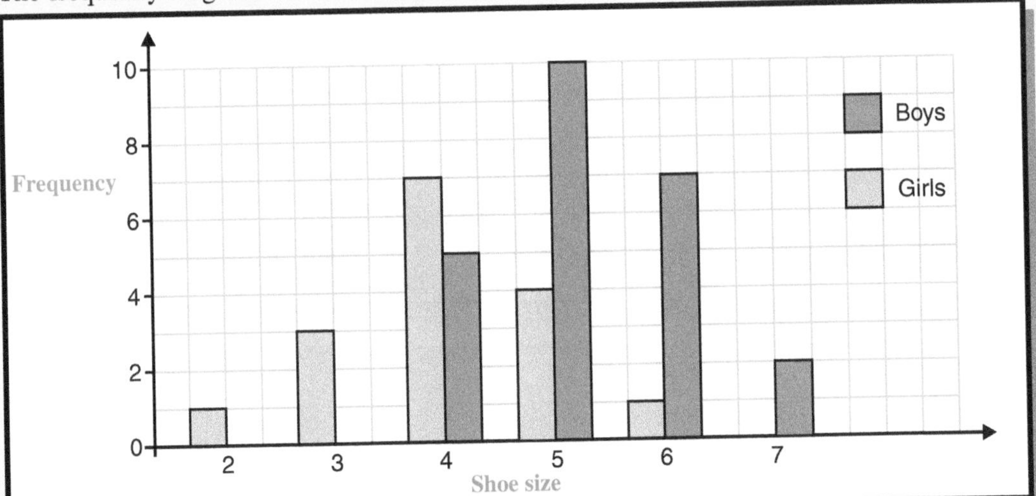

(a) What is the ratio of girls to boys in the class?
Give your answer in its simplest form.

(b) What percentage of the class have shoe size 6?

(c) By working out the range and mode for the boys and for the girls, compare and comment on the shoe sizes for boys and girls.

AQA

6 (a) The table shows the fuels used for heating in all the houses in a large town.

Fuel	Number of houses (in 1000's)
Solid fuel	8
Electricity	42
Gas	70
TOTAL	120

Draw a clearly labelled pie chart to represent this information.

(b) This pie chart shows the fuels used for heating in all the houses in a small village.

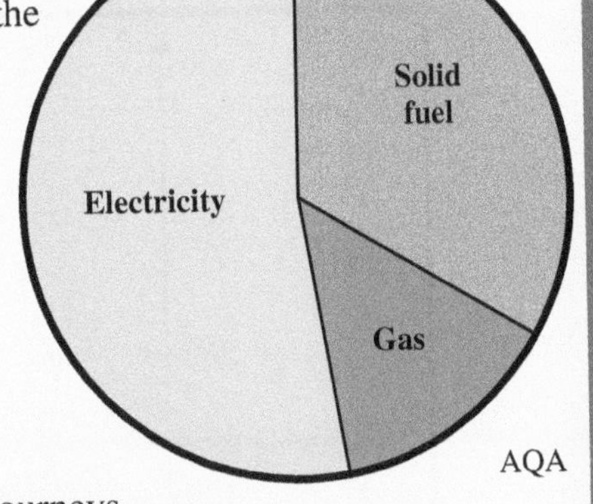

(i) What fraction of these houses use gas?

(ii) Solid fuel is used in 24 houses.
How many houses are in the village?

(c) Use the information from these two pie charts to say which fuel is most likely to be the mode.
Give a reason for your answer.

AQA

7 The table gives information about the petrol used for car journeys.

Petrol (litres)	3	5	6	8	4	11	10	9	2
Distance (km)	28	50	70	110	50	110	120	130	24

(a) Draw a scatter graph for this information.

(b) Draw a line of best fit.

(c) Use your graph to estimate:

(i) the distance a car travels on 7 litres of petrol,

(ii) the number of litres of petrol used by a car travelling a distance of 150 km.

(d) Which of the estimates in (c) is likely to be more reliable?
Give a reason for your answer.

AQA

Presentation of Data 2

Time series

The money spent on shopping **each day**, the gas used **each quarter** and the rainfall **each month** are all examples of **time series**. A time series is a set of readings taken at time intervals.
A time series is often used to monitor progress and to show the **trend** (increases and decreases) so that future performance can be predicted. The type of graph used in this situation is called a **line graph**.

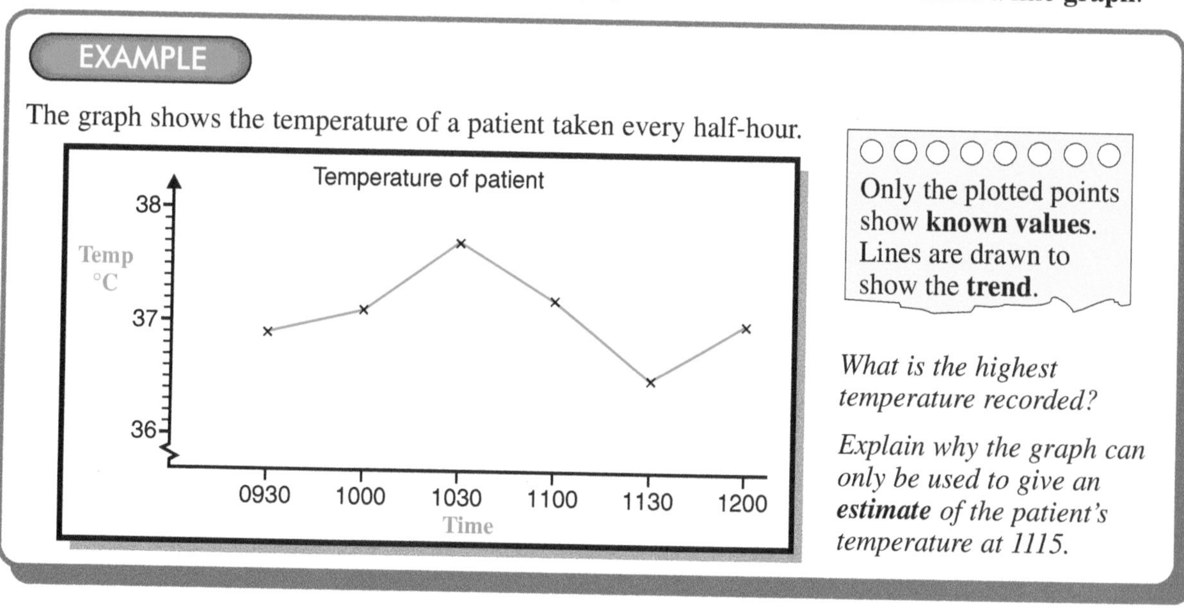

EXAMPLE

The graph shows the temperature of a patient taken every half-hour.

Only the plotted points show **known values**. Lines are drawn to show the **trend**.

What is the highest temperature recorded?

*Explain why the graph can only be used to give an **estimate** of the patient's temperature at 1115.*

Seasonal variation

The graph below shows the amount of gas used by a householder each quarter over a period of 3 years.

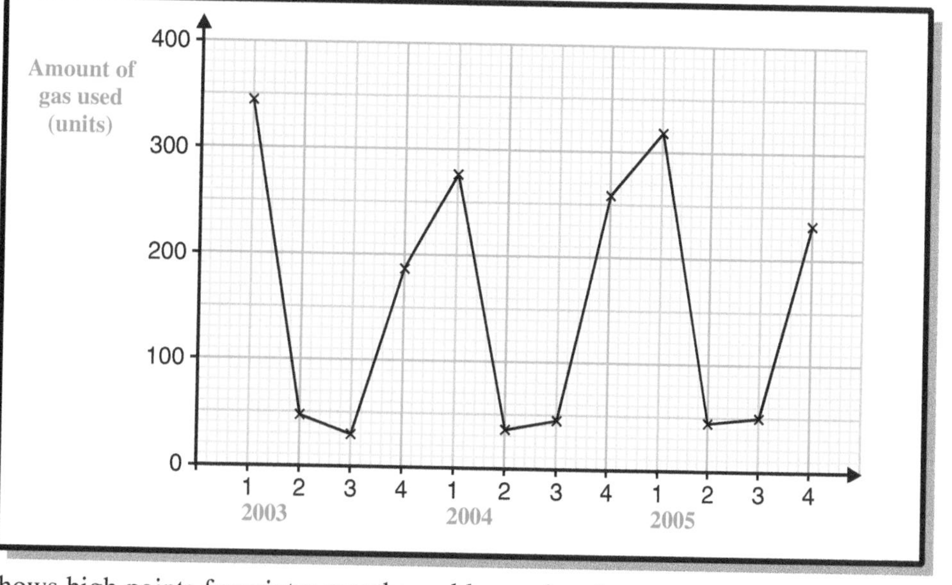

The graph shows high points for winter months and low points for summer months.
These ups and downs occur because more (or less) gas is used at certain times of the year.
As these ups and downs vary with the seasons of the year they are called **seasonal variations**.
To get a better idea of the average number of units of gas used each quarter we need to smooth out these seasonal variations.

Moving averages

To smooth out seasonal variations we need to find the average for every four quarters moving forward one quarter at a time. These are called **moving averages**.
The first moving average is found by calculating the mean for the four quarters of 2003.

$$\frac{345 + 49 + 30 + 188}{4} = \frac{612}{4} = 153$$

The second moving average is found by calculating the mean of the last three quarters in 2003 and the first quarter in 2004.

$$\frac{49 + 30 + 188 + 277}{4} = \frac{544}{4} = 136$$

This process is repeated moving forward one quarter at a time.

Because these averages are calculated for four consecutive quarters at a time they are called **4-point moving averages**.

The table opposite shows a useful way of setting out the calculations.

Copy and complete the table.

The number of points chosen for the moving average depends on the pattern in the data.
In the example given a 4-point moving average was chosen because the pattern was repeated every 4 quarters.

Gas used			Moving average calculations			
Year	Quarter	Units				
2003	1	345	345			
	2	49	49	49		
	3	30	30	30	30	
	4	188	188	188	188	
2004	1	277		277	277	
	2	37			37	
	3	46				
	4	260				
2005	1	319				
	2	45				
	3	50				
	4	232				
4-point moving sum			612	544	532	
4-point moving average			153	136		

Exercise 38.1

1. Each year, on his birthday, a teenager records his height. The table shows the results.

Age (yrs)	13	14	15	16	17	18	19
Height (cm)	145	151	157	165	174	179	180

(a) Draw a line graph to represent this information.
(b) Use your graph to estimate:
(i) the height of the teenager when he was $14\frac{1}{2}$ years of age,
(ii) the age of the teenager when he reached 160 cm in height.

2. These values have been taken at equal time intervals.
(a) Calculate the 3-point moving averages for: 17 58 30 14 61 33 23 55 42
(b) Calculate the 5-point moving averages for: 27 26 15 34 42 26 24 17 44 54

3. The number of new cars sold each quarter by a garage for the last three years is shown.

Year	2003				2004				2005			
Quarter	1	2	3	4	1	2	3	4	1	2	3	4
Number of cars sold	20	15	25	12	18	13	21	8	14	9	17	5

Calculate the 4-quarterly moving averages.

4. Calculate the value of the third average in a 5-point moving average for these values.

5 7 10 4 1 7 11 13 6 7

5. These values have been taken at equal time intervals.

160 640 310 145 670 355 175 640 385

Calculate an appropriate moving average.

Moving average graphs

Moving averages are calculated in order to investigate the general **trend** in a time series.
To investigate the trend the moving averages are plotted on the same diagram as the original graph.
Each moving average is plotted at the midpoint of the period to which it refers, so, in the case of the gas used, the first moving average is plotted midway between the 2nd and 3rd quarters of 2003, the second is plotted midway between the 3rd and 4th quarters of 2003 and so on.
When all the points have been plotted a line of best fit is drawn for the moving averages to show the general **trend**.
The graph below shows the moving average values and trend line for the amount of gas used over a three-year period.

The trend shows a slight increase in the amount of gas used.

Exercise 38.2

1. The table shows the amount of water used every 6 months by a householder over a period of 5 years.

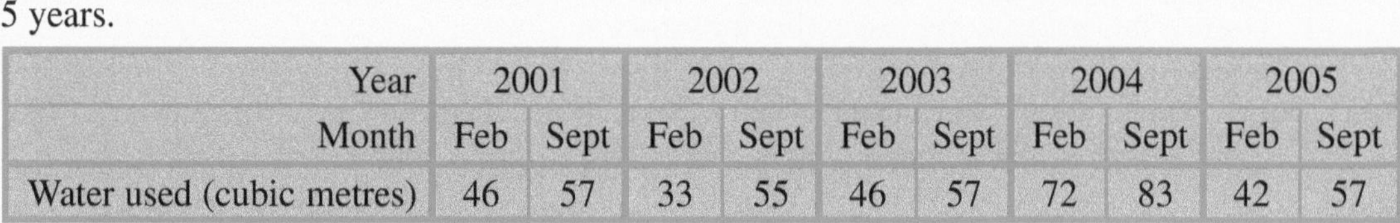

Year	2001		2002		2003		2004		2005	
Month	Feb	Sept	Feb	Sept	Feb	Sept	Feb	Sept	Feb	Sept
Water used (cubic metres)	46	57	33	55	46	57	72	83	42	57

(a) Plot these values on graph paper.
(b) Calculate a 2-point moving average.
(c) Plot the moving averages on your graph.
(d) Comment on the trend in the amount of water used.

2. The table shows the quarterly electricity bills for a householder.

Quarterly electricity bills

Year	Feb	May	Aug	Nov
2003	£64.75	£72.16	£57.71	£67.80
2004	£60.51	£74.43	£46.33	£67.39
2005	£81.39	£75.69	£48.24	£61.65

(a) Use the data to draw a graph.
(b) Calculate a 4-point moving average.
(c) Plot the moving averages on your graph.
(d) Comment on the trend in the cost of electricity.

3 The table shows the termly absences from school for a class over a four-year period.

Termly absences

Year		Autumn	Spring	Summer
	Year 8	19	54	32
	Year 9	22	132	35
	Year 10	28	57	47
	Year 11	25	51	62

(a) Use the data to draw a graph.
(b) Calculate the values of a suitable moving average.
(c) Plot the moving averages on your graph.
(d) Comment on the trend in the number of absences for this class.

4 The table gives the daily sales (in £100's) for a new shop during the first three weeks of trading.

Daily sales

	Mon	Tue	Wed	Thu	Fri	Sat
Week 1	57	39	27	34	56	74
Week 2	38	34	28	32	61	83
Week 3	43	37	31	43	67	92

(a) Plot these values on graph paper.
(b) Calculate a 6-point moving average.
(c) Plot the moving averages on the same graph.
(d) Draw a trend line.
(e) Comment on the trend in sales.
(f) Estimate the daily sales for Monday in Week 4.

5 The number of new homes completed each quarter by a builder for the last three years is shown.

Year	2003				2004				2005			
Quarter	1	2	3	4	1	2	3	4	1	2	3	4
Number of new homes	21	35	36	18	19	23	28	14	15	19	22	9

(a) Plot these values on graph paper.
(b) Calculate the values of a suitable moving average.
(c) Plot the moving averages on the same graph.
(d) Draw a trend line by eye.
(e) Comment on the trend.
(f) Estimate the number of new homes completed by the builder in the first quarter of 2006.

Frequency diagrams

We use **bar charts** when data can be counted and there are only a few different items of data.
If there is a lot of data, or the data is continuous, we draw a **histogram** or **frequency polygon**.

Histograms with equal class width intervals

Histograms are used to present information contained in **grouped frequency distributions**.
Histograms can have equal or unequal class width intervals.
In a histogram with equal class intervals, **frequency** is proportional to the **heights** of the bars.

A histogram with equal class width intervals looks like a bar chart with no gaps.

Frequency polygons

Frequency polygons are often used instead of histograms when we need to compare two, or more, groups of data.
To draw a frequency polygon:
- plot the frequencies at the midpoint of each class interval,
- join successive points with straight lines.

To compare data, frequency polygons for different groups of data can be drawn on the same diagram.

EXAMPLE

The frequency distribution of the heights of some boys is shown.

Height (cm)	130 -	140 -	150 -	160 -	170 -	180 -
Frequency	1	7	12	9	3	0

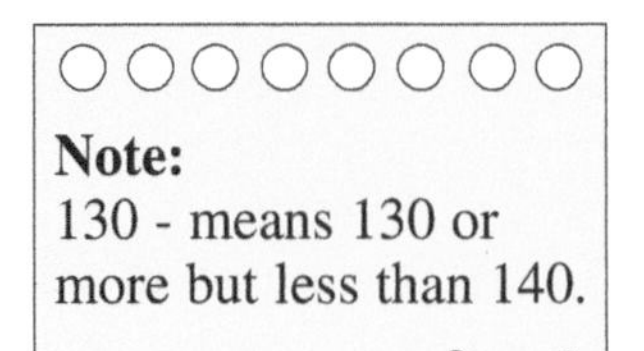
Note:
130 - means 130 or more but less than 140.

Draw a histogram and a frequency polygon to illustrate the data.

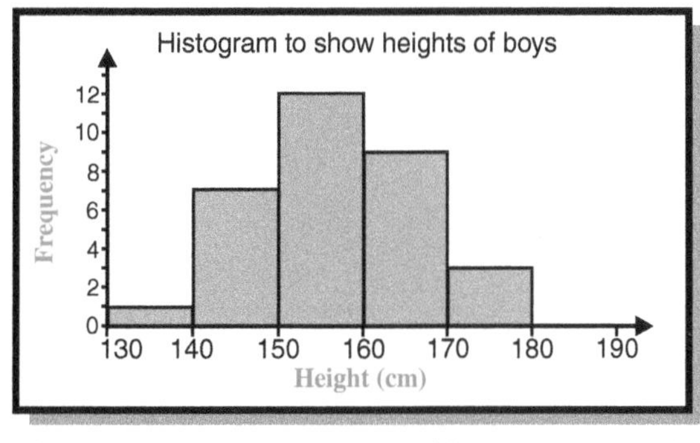

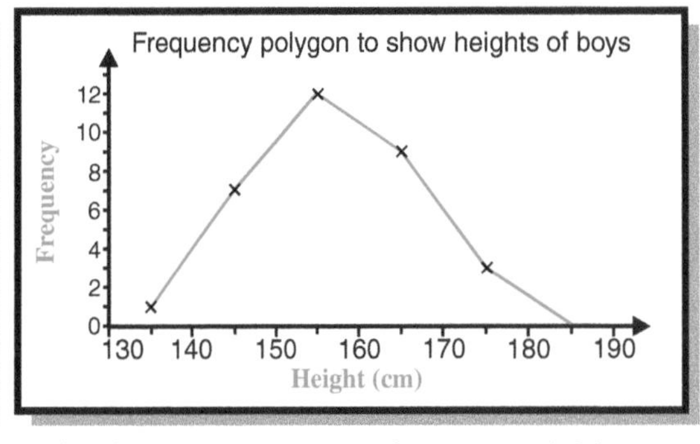

Individual bars are not labelled because the horizontal axis represents a **continuous** variable.

Exercise 38.3

1. A supermarket opens at 0800.
The frequency diagram shows the distribution of the times employees arrive for work.

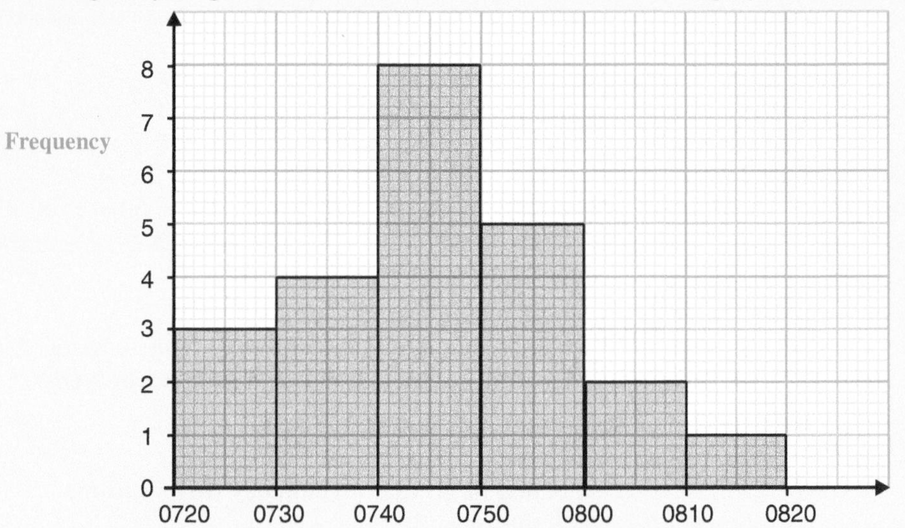

(a) How many employees arrive before 0730?
(b) How many employees arrive between 0730 and 0800?
(c) How many employees arrive after 0800?
(d) What is the modal class?

2 The frequency polygon shows the distribution of the distances travelled to work by the employees at a supermarket.

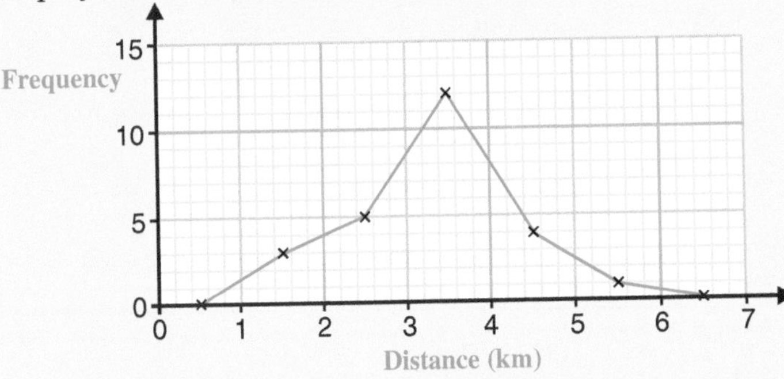

When drawing frequency polygons, frequencies are plotted at the midpoints of the class intervals.

(a) How many employees travel between 2 km and 3 km to work?
(b) How many employees travel more than 4 km to work?

3 The distances, in metres, recorded in a long jump competition are shown.

5.46 5.80 5.97 5.43 6.72 5.93 6.26 6.64 5.13 6.05 6.36 6.88
6.11 5.50 6.38 5.71 6.55 6.10 5.84 5.49 6.20 5.67 6.34 6.00

(a) Copy and complete the following frequency distribution table.

Distance (m metres)	Frequency
$5.00 \leqslant m < 5.50$	
$5.50 \leqslant m < 6.00$	
$6.00 \leqslant m < 6.50$	
$6.50 \leqslant m < 7.00$	

(b) Draw a histogram to illustrate the data.
(c) Which is the modal class?

4 The table shows the grouped frequency distribution of the marks of 200 students.

Mark (%)	1 - 10	11 - 20	21 - 30	31 - 40	41 - 50	51 - 60	61 - 70	71 - 80	81 - 90	91 - 100
Number of Students	0	2	16	24	44	50	35	20	8	1

Draw a frequency diagram to show these results.

5 The frequency polygon shows the distribution of the ages of pupils who attend a village school.

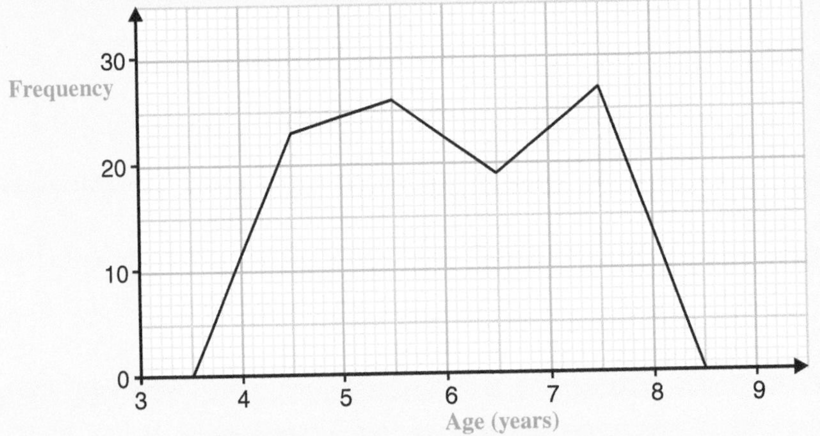

(a) How many pupils are under 4 years of age?
(b) How many pupils are between 5 and 6 years of age?
(c) How many pupils are over 7 years of age?
(d) How many pupils attend the school?

6 A frequency distribution of the heights of some girls is shown.
Draw a histogram to illustrate the data.

Height (h cm)	Frequency
$130 \leqslant h < 140$	3
$140 \leqslant h < 150$	5
$150 \leqslant h < 160$	12
$160 \leqslant h < 170$	4
$170 \leqslant h < 180$	1

7 The table shows the distances travelled to school by 100 children.

Distance (k km)	$0 \leqslant k < 2$	$2 \leqslant k < 4$	$4 \leqslant k < 6$	$6 \leqslant k < 8$	$8 \leqslant k < 10$
Frequency	27	35	22	10	6

Draw a frequency polygon to illustrate this information.

8 The frequency distribution of the weights of some students is shown.

Weight (kg)	40 -	50 -	60 -	70 -	80 -	90 - 100
Number of males	0	6	11	5	2	1
Number of females	3	14	8	0	0	0

(a) On the same diagram draw a frequency polygon for the males and a frequency polygon for the females.
(b) Compare and comment on the weights of male and female students.

9 The table shows the results of students in tests in English and Mathematics.

Marks	0 and less than 10	10 and less than 20	20 and less than 30	30 and less than 40	40 and less than 50
English	0	4	9	12	0
Mathematics	1	4	9	7	4

(a) Draw a frequency polygon for the English marks.
(b) On the same diagram draw a frequency polygon for the Mathematics marks.
(c) Compare and comment on the marks of the students in these two tests.

10 The table shows the results for competitors in the 2004 and 2005 Schools' Javelin Championship. Only the best distance thrown by each competitor is shown.

Distance thrown (m metres)	Number of competitors 2004	Number of competitors 2005
$10 \leqslant m < 20$	0	1
$20 \leqslant m < 30$	3	4
$30 \leqslant m < 40$	14	19
$40 \leqslant m < 50$	21	13
$50 \leqslant m < 60$	7	11
$60 \leqslant m < 70$	0	2

(a) On the same diagram draw a frequency polygon for the 2004 results and then a frequency polygon for the 2005 results.
(b) Compare and comment on the results.

Histograms with unequal class width intervals

Data is sometimes grouped using **unequal** class width intervals.
In a histogram with unequal class intervals the vertical axis is labelled **frequency density**.
Frequency is proportional to the **areas** of the bars.

$$\text{frequency} = \text{frequency density} \times \text{class width interval}$$

EXAMPLES

1 The table shows the frequency distribution of the heights of some women.

Height	150 -	165 -	175 -	180 -	185 - 195
Frequency	15	20	20	10	4

Draw a histogram for the data.

The data has unequal class intervals.
Before a histogram can be drawn, the height of each bar must be calculated.
The height of each bar is given by the frequency density for each group, where:

$$\text{Frequency density} = \frac{\text{frequency}}{\text{class width interval}}$$

Frequency	15	20	20	10	4
Class width interval	15	10	5	5	10
Frequency density	1	2	4	2	0.4

The histogram is shown below.

Frequency density
0
1
2
3
4
150
160
170
180
190
200
Height (cm)

When drawing histograms mark a **continuous scale** on the horizontal axis.
Draw bars, between the lower and upper class boundaries, for each class interval.

Remember:
The **area** of each bar is proportional to the **frequency**.

2 The histogram represents the heights of some plants.

Frequency density
0
10
20
30
40
50
Height (cm)

Five plants are less than 20 cm high.

(a) How many plants are at least 30 cm high?

(b) How many plants are there altogether?

Find the frequency density for the plants $10 \leq h < 20$.

Frequency density is given by:

$$\frac{\text{frequency}}{\text{class width interval}} = \frac{5}{10} = 0.5$$

The frequency density axis can now be labelled.

(a) Plants at least 30 cm high are in the interval $30 \leq h < 45$.
Frequency = frequency density × class width interval
Frequency density = 0.2 (from vertical axis)
Class width interval = $45 - 30 = 15$ (from horizontal axis)
Frequency = $0.2 \times 15 = 3$
3 plants are at least 30 cm high.

(b) Total frequency $= 5 + 1.6 \times 5 + 2 \times 5 + 3$
$= 5 + 8 + 10 + 3 = 26$
There are 26 plants altogether.

Exercise 38.4

1. The ages of people in a coach party is shown.

Age (y years)	$10 \leqslant y < 20$	$20 \leqslant y < 40$	$40 \leqslant y < 45$	$45 \leqslant y < 50$	$50 \leqslant y < 60$
Frequency	4	16	13	9	6

Draw a histogram for the data.

2. The frequency distribution of the weights of some children is shown in the table.

Weight (w kg)	Frequency
$40 \leqslant w < 50$	9
$50 \leqslant w < 55$	10
$55 \leqslant w < 60$	17
$60 \leqslant w < 65$	32
$65 \leqslant w < 75$	28
$75 \leqslant w < 85$	4

Draw a histogram for the data.

3. The frequency distribution of the annual wages of all the employees of a small firm is shown in the table.

Annual Wage (£1000's)	6 -	10 -	15 -	20 -	30 - 45
Frequency	6	8	13	17	3

Represent this information as a histogram.

4. The histogram shows the age distribution of students delivering newspapers in a town in 2005.

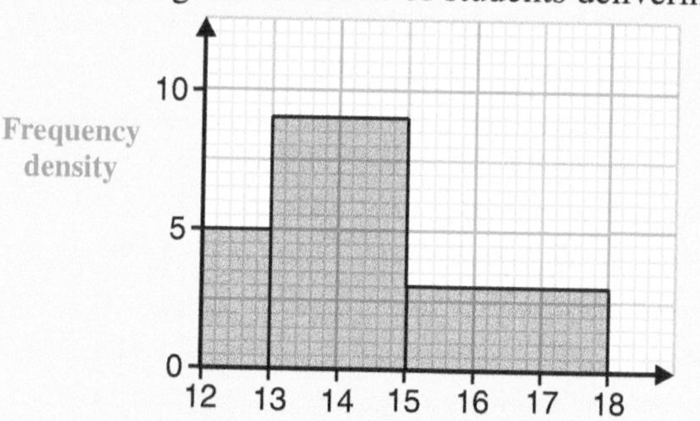

How many students delivered newspapers in the town in 2005?

5. The histogram shows the distribution of distances thrown by competitors in a throwing competition.

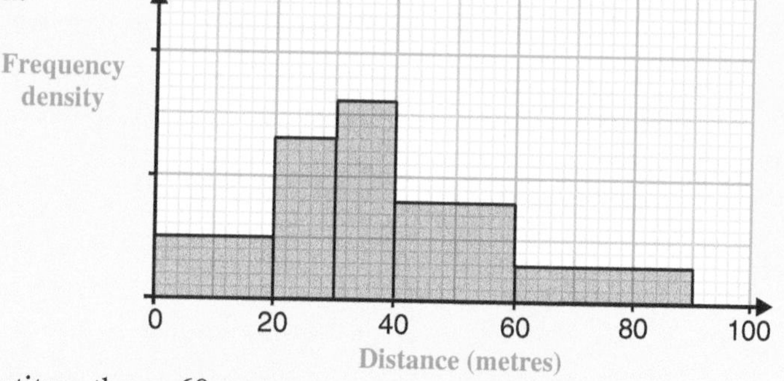

Nine competitors threw 60 metres or more.
(a) How many competitors threw less than 20 metres?
(b) How many competitors took part in the competition?

6 The histogram represents the age distribution of the employees in a small company.

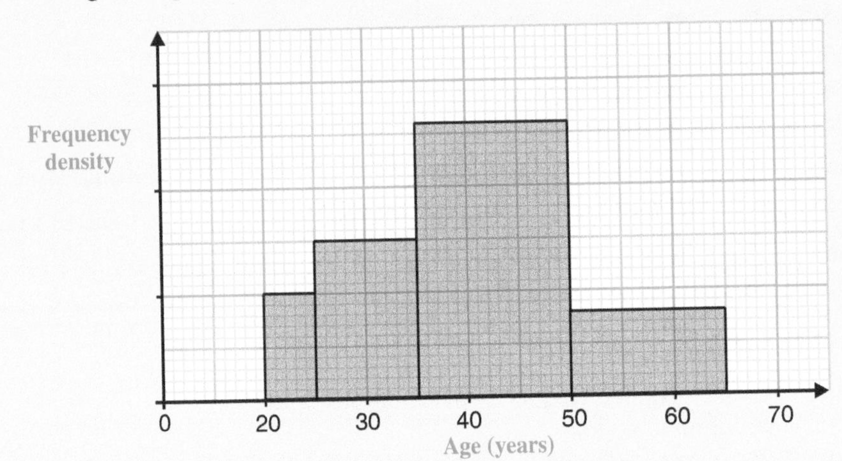

There are 5 employees in the 20 to 25 age range.

(a) How many employees are in the 50 to 65 age range?

(b) How many employees has the company?

7 The table shows the distribution of the heights of all the sixth formers in a school.

Height (h cm)	Number of students
$150 \leqslant h < 165$	75
$165 \leqslant h < 175$	100
$175 \leqslant h < 180$	100
$180 \leqslant h < 185$	50
$185 \leqslant h < 195$	20

(a) Represent this information as a histogram.

(b) Francis is 178 cm tall.
Estimate how many of these sixth formers are taller than him.

8 The table shows the distribution of the heights of some plants.
The heights were measured to the nearest centimetre.

Height (cm)	1 - 5	6 - 10	11 - 15
Frequency	0	15	10

(a) (i) What is the height of the smallest plant that could be placed in the 6 - 10 class?

(ii) What is the width of the 6 - 10 class?

(b) Draw a histogram of the distribution.

Note:
When the classes have gaps between them the upper class boundary is usually halfway between the end of one class and the beginning of the next.

9 The table below shows the amount of milk, measured to the nearest litre, produced by some cows on a given day.

Milk (litres)	5 - 14	15 - 19	20 - 24	25 - 34
Number of cows	5	10	7	3

(a) Draw a histogram to represent these data.

(b) Estimate the percentage of these cows which produced less than $17\frac{1}{2}$ litres of milk.

What you need to know

- A **time series** is a set of readings taken at time intervals.
- A **line graph** is used to show a time series.
  Only the plotted points represent actual values.
  Points are joined by lines to show the **trend**.
- Variations in a time series which recur with the seasons of the year are called **seasonal variations**.
- **Moving averages** are used to smooth out variations in a time series so that the trend can be seen.
- **Frequency polygon**. Used to illustrate grouped frequency distributions.
  Often used to compare two or more distributions on the same diagram.
  Frequencies are plotted at the midpoints of the class intervals and joined with straight lines.
  The horizontal axis is a continuous scale.
- **Histograms**. Used to illustrate grouped frequency distributions.
  The horizontal axis is a continuous scale.
  Bars are drawn between the lower and upper class boundaries for each class interval.
  When the classes have gaps between them the upper class boundary is usually halfway between the end of one class and the beginning of the next.
- Histograms can have equal or unequal class width intervals.
  With **equal** class width intervals:
  frequency is proportional to the **heights** of the bars.
  With **unequal** class width intervals:
  frequency is proportional to the **areas** of the bars.

$$\text{frequency} = \text{frequency density} \times \text{class width interval}$$

- Before a histogram can be drawn, the height of each bar must be calculated.
  The height of each bar is given by the **frequency density** for each group, where:

$$\text{frequency density} = \frac{\text{frequency}}{\text{class width interval}}$$

Review Exercise 38

1 Jim bought his house in 1990.
The table shows the value of Jim's house on January 1st at 5-yearly intervals.

Year	1990	1995	2000	2005
Value of house (£)	95 000	68 000	84 000	136 000

(a) Draw a line graph to show this information.
(b) Estimate the value of Jim's house on July 1st, 2003.
(c) Jim uses the graph to estimate the value of his house on January 1st, 2010.
Give a reason why his estimate may not be very accurate.

2 A factory operates 5 days a week.
The table shows the number of workers absent each day over the last 3 weeks.

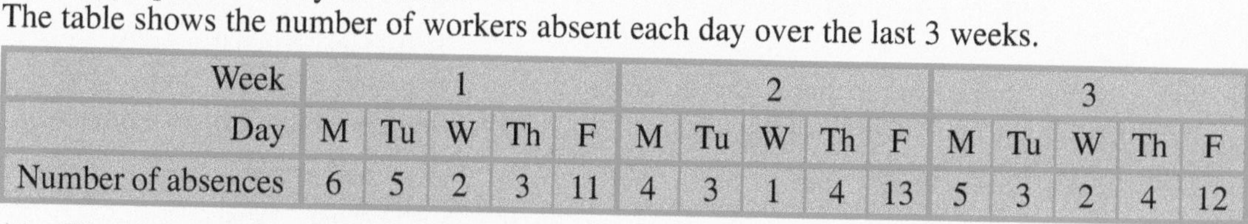

Week	1					2					3				
Day	M	Tu	W	Th	F	M	Tu	W	Th	F	M	Tu	W	Th	F
Number of absences	6	5	2	3	11	4	3	1	4	13	5	3	2	4	12

(a) Plot these values on graph paper.
(b) Calculate a 5-point moving average.
(c) Plot the moving averages on the same graph.
(d) Draw a trend line by eye.
(e) Comment on the trend.

3 The following graph appeared in a newspaper advert.

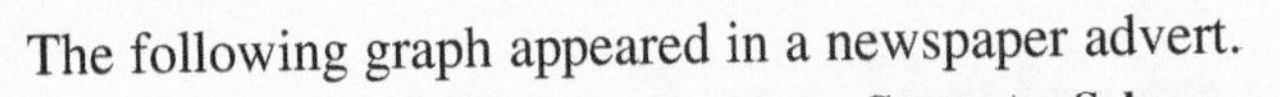
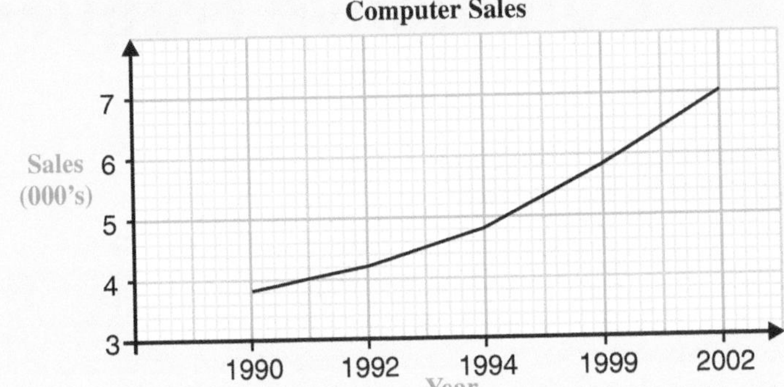

Write down **two** ways in which the graph is misleading. AQA

4 The table gives information about the weight of the potato crop produced by 100 potato plants of two different types.

Weight of potatoes per plant w kg	Number of plants Type X	Number of plants Type Y
$0 \leqslant w < 0.5$	0	0
$0.5 \leqslant w < 1.0$	3	0
$1.0 \leqslant w < 1.5$	12	6
$1.5 \leqslant w < 2.0$	55	39
$2.0 \leqslant w < 2.5$	23	32
$2.5 \leqslant w < 3.0$	7	23
$3.0 \leqslant w < 3.5$	0	0

(a) On the same diagram draw a frequency polygon for each type of potato.
(b) Which type of potato produces the heavier crop?
(c) (i) Which type of potato has more variation in the weight of the crop?
(ii) Give a reason for your answer. AQA

5 The amount of money, £x, given to charity each month by 80 people is shown below.

Amount, £x	$0 < x \leqslant 1$	$1 < x \leqslant 5$	$5 < x \leqslant 10$	$10 < x \leqslant 20$
Number of people	16	32	20	12

(a) Draw a histogram to represent this information.
(b) Use your histogram to estimate the median amount given by these people each month.
(c) Calculate an estimate of the percentage of these people who give between £3 and £8 each month to charity. AQA

6 The speeds of 100 cars travelling along a motorway are shown in the table.

Speed (s km/h)	Frequency
$50 \leqslant s < 65$	6
$65 \leqslant s < 80$	9
$80 \leqslant s < 90$	15
$90 \leqslant s < 100$	38
$100 \leqslant s < 120$	32

(a) Draw a histogram to show this information.
(b) The speed limit on motorways is 112 km/h.
Estimate the number of cars that exceed the speed limit. AQA

Cumulative Frequency

Cumulative frequency tables

This **frequency table** shows the masses of some stones recorded in an experiment.

Mass (kg)	20 -	30 -	40 -	50 -	60 - 70
Frequency	3	5	10	8	4

The first class is 20 - 30 kg. The **upper class boundary** of this class is 30 kg.
What are the upper class boundaries of the other classes?

Note: If the question does not give the upper class boundaries, then the upper class boundary of each class is equal to the lower class boundary of the next class.

The information given in a frequency table can be used to make a **cumulative frequency table.**

To find the cumulative frequencies we must add together the frequencies which are '**less than**' each of the upper class boundaries.

For example, using the data in the experiment above.

Mass of stone (less than)	Number of stones
20 kg	0
30 kg	0 + 3 = 3
40 kg	0 + 3 + 5 = 8

This gives the **cumulative frequency table**:

Mass (kg), less than	20	30	40	50	60	70
Cumulative frequency	0	3	8	18	26	30

The cumulative frequency table can be used to draw a cumulative frequency graph.

Cumulative frequency graph

Here is a cumulative frequency graph for the masses of stones in the experiment.

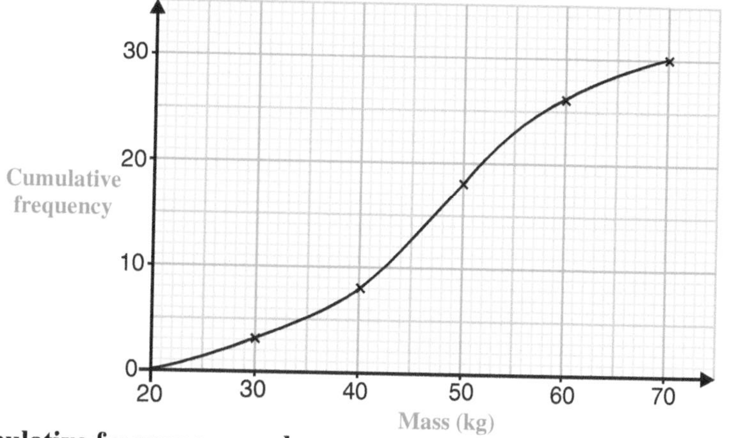

To draw a **cumulative frequency graph**:

1. Draw and label: the **variable** on the horizontal axis, **cumulative frequency** on the vertical axis.
2. Plot the cumulative frequency against the upper class boundary of each class.
3. Join the points with a smooth curve.

Using a cumulative frequency graph

Median

The median of 13 numbers is the 7th number.
The median of 14 numbers is between the 7th and 8th numbers.

We could call this the "$7\frac{1}{2}$th" number.

For n numbers, the rule for finding the median is:

Median $= \frac{1}{2}(n + 1)$th number

> The median of a frequency distribution is the value of the middle number.
> On a cumulative frequency graph this value is read from the horizontal axis.

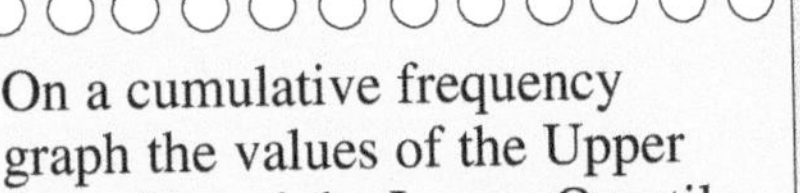

Interquartile range

The **range** of a set of data was covered in Chapter 36.
It measures how spread out the data is.
Range = highest value − lowest value.
The range is influenced by extreme high or low values of data and can be misleading.

A better way to measure spread is to find the range of the **middle 50%** of the data.
This is called the **interquartile range**.
Interquartile range IQR = Upper Quartile − Lower Quartile

For n numbers the rules for finding the quartiles are:

Lower Quartile $= \frac{1}{4}(n + 1)$th number

Upper Quartile $= \frac{3}{4}(n + 1)$th number

> On a cumulative frequency graph the values of the Upper Quartile and the Lower Quartile are read from the horizontal axis.

EXAMPLE

Using the cumulative frequency graph from the previous page, estimate:

(a) the median mass of the stones,
(b) the interquartile range of the masses of the stones.

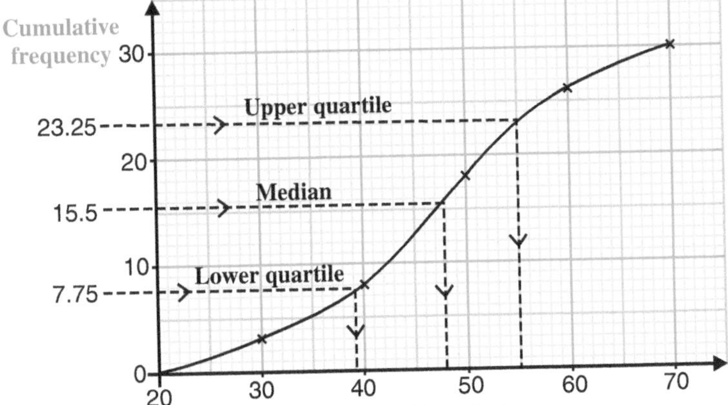

(a) There are 30 stones.
$\frac{1}{2}(30 + 1) = 15.5$,
so, the median is the mass of the 15.5th stone.
Read along from 15.5 on the vertical axis and down to the horizontal axis.
This is shown on the graph.
Median = 48 kg.

(b) $\frac{1}{4}(30 + 1) = 7.75$, so, the lower quartile is the mass of the 7.75th stone.
$\frac{3}{4}(30 + 1) = 23.25$, so, the upper quartile is the mass of the 23.25th stone.
Read these values from the graph.
IQR = Upper Quartile − Lower Quartile
= 55 − 39
= 16 kg
The interquartile range is 16 kg.

> **When n is large:**
> When the total frequency, n, is 'large' you need not bother with '+1' in the rules for the median, lower quartile and upper quartile.
> Instead use: median $= \frac{1}{2}n$th number
> lower quartile $= \frac{1}{4}n$th number
> upper quartile $= \frac{3}{4}n$th number
> 'Large' means greater than 50.

Exercise 39.1

1 Silvia made a record of the times students took to walk to school in the morning.

Time (minutes)	0 -	5 -	10 -	15 -	20 -	25 - 30
Number of students	4	12	18	15	8	3

(a) Copy and complete this cumulative frequency table.

Time (minutes), less than	0	5	10	15	20	25	30
Cumulative frequency							

(b) Draw a cumulative frequency graph for the data.
(c) Use your graph to find:
  (i) the median time,
  (ii) the upper quartile time,
  (iii) the lower quartile time.
(d) Calculate the interquartile range of the times.

2 A secretary weighed a sample of letters to be posted.

Weight (g)	20 -	30 -	40 -	50 -	60 -	70 -	80 - 90
Number of letters	2	4	12	7	8	16	3

(a) How many letters were in the sample?
(b) Draw a cumulative frequency graph for the data.
(c) Use your graph to find:
  (i) the median weight of a letter, (ii) the interquartile range of the weights.

3 The times spent by students on mobile phones one day is shown.

Time (t minutes)	Number of students
$0 \leqslant t < 5$	34
$5 \leqslant t < 10$	22
$10 \leqslant t < 15$	12
$15 \leqslant t < 20$	8
$20 \leqslant t < 25$	4

(a) Draw a cumulative frequency graph for the data.
(b) Use your graph to find:
  (i) the median time, (ii) the interquartile range of the times.

Another look at cumulative frequency graphs

Some variables are discrete and can only take certain values.
In many cases these are whole numbers. For these variables there are gaps between the classes.

For example, this frequency distribution table shows the number of spelling mistakes found in some essays.

Number of mistakes	0 - 5	6 - 10	11 - 15	16 - 20
Number of essays	9	17	8	3

The first class ends at 5. The highest number of spelling mistakes in the class 0 - 5 is 5.
The second class starts at 6.
We take the upper class boundary for the class 0 - 5 to be halfway between 5 and 6, at 5.5.
What are the upper class boundaries for the other classes?

In the following example there are gaps between the classes, because the lengths are measured to the nearest centimetre.

EXAMPLE

In an experiment Sophie measured and recorded the longest roots of plants.

Length (to nearest cm)	0 - 2	3 - 5	6 - 8	9 - 11	12 - 14
Number of plants	0	14	33	23	10

(a) Make a cumulative frequency table for the data Sophie collected.

(b) Draw a cumulative frequency graph for the data.

(c) Use your graph to find:
(i) the median length,
(ii) the interquartile range of lengths.

(d) How many plants had roots at least 11.2 cm long?

The class 3 - 5 includes measurements from 2.5 cm to 5.5 cm.
The upper class boundary is 5.5 cm.

(a)

Length (cm) less than	2.5	5.5	8.5	11.5	14.5
Cumulative frequency	0	14	47	70	80

(b)

Remember:
Cumulative frequency is plotted against the upper class boundary for each class.

(c) (i) There are 80 plants.
$\frac{1}{2}$ of 80 = 40, so, the median is the length of the roots of the 40th plant.
Median = 8.0 cm

(ii) For the lower quartile use $\frac{1}{4}$ of 80 = 20.
Lower quartile = 6.2 cm
For the upper quartile use $\frac{3}{4}$ of 80 = 60.
Upper quartile = 10.0 cm
IQR = Upper Quartile − Lower Quartile = 10.0 − 6.2 = 3.8 cm

(d) From 11.2 cm on the horizontal axis read upwards to the graph and across to the vertical axis.
There are 68 plants of length less than 11.2 cm.
So, the number of plants with roots at least 11.2 cm long is 80 − 68 = 12 plants.

Exercise 39.2

1 The times taken by competitors to complete the crossword in an annual competition were recorded to the nearest minute.

Time (minutes)	10 - 14	15 - 19	20 - 24	25 - 29	30 - 34
Frequency	7	21	37	12	3

(a) Copy and complete the cumulative frequency table.

Time (minutes)	< 9.5	< 14.5	< 19.5	< 24.5	< 29.5	< 34.5
Cumulative frequency						

(b) Draw a cumulative frequency graph for the data.
(c) Use your graph to find:
(i) the interquartile range of times,
(ii) the number of competitors who took less than last year's winning time of 16 minutes to complete the crossword.

2 Draw a cumulative frequency graph for the results shown in this table.

Number of marks	0 - 20	21 - 25	26 - 30	31 - 40
Number of students	10	17	23	14

(a) Find the median mark.
(b) Find the interquartile range of marks.
(c) The minimum mark for a Grade A is 33.
What percentage of students gained Grade A?
(d) 10% of students failed.
What was the minimum mark for a pass?

3 A survey was made of the heights of plants produced by a batch of seed.
(a) How many plants were measured in the survey?
(b) Draw a cumulative frequency graph for the data.
Label the horizontal axis from 70 cm to 110 cm.
(c) Use your graph to estimate:
(i) the median height,
(ii) the interquartile range of heights,
(iii) the number of plants taller than 97.5 cm.
(d) The shortest 10 plants are used for testing.
Estimate the height of the tallest of these 10 plants.

Height (h cm)	Frequency
$80 \leqslant h < 85$	20
$85 \leqslant h < 90$	35
$90 \leqslant h < 95$	15
$95 \leqslant h < 100$	11
$100 \leqslant h < 105$	14

4 The times spent listening to the radio last week by some students is shown.

Time (t hours)	Number of students
$0 \leqslant t < 5$	7
$5 \leqslant t < 10$	15
$10 \leqslant t < 15$	18
$15 \leqslant t < 20$	24
$20 \leqslant t < 30$	12
$30 \leqslant t < 40$	4

(a) Draw a cumulative frequency graph for the data.
(b) Use your graph to find the interquartile range of times.
(c) What percentage of the students spent more than 25 hours per week listening to the radio?

5 The heights of a number of men are shown in the table.

Height (h cm)	Frequency
$140 \leqslant h < 150$	1
$150 \leqslant h < 160$	6
$160 \leqslant h < 170$	8
$170 \leqslant h < 180$	21
$180 \leqslant h < 190$	14

Draw a cumulative frequency graph for the data. Use your graph to find:

(a) the median and interquartile range of heights,
(b) the number of men less than 155 cm,
(c) the number of men at least 163 cm,
(d) the maximum height of the shortest 20 men,
(e) the minimum height of the tallest 10% of men.

Comparing distributions

EXAMPLE

A firm tested a sample of electric motors produced by an assembly line.
They kept a running total of the numbers of motors which had failed at any time.
This cumulative frequency graph was plotted using the data collected.

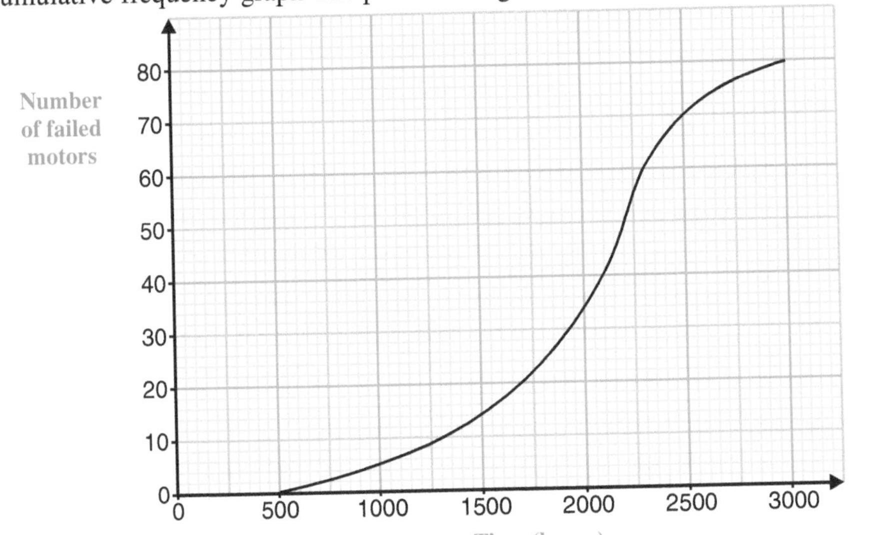

(a) Find the median lifetime of a motor.
(b) Find the interquartile range of lifetimes.
(c) A sample of motors, produced by another assembly line, was tested and found to have:
Median lifetime = 1900 hours, Upper quartile = 2000 hours, Lower quartile = 1700 hours.
Compare the two samples of motors.

(a) Median lifetime = 2100 hours
(b) IQR = Upper Quartile − Lower Quartile = 2300 − 1700 = 600 hours
(c) For the second sample: Median = 1900 hours
IQR = 2000 − 1700 = 300 hours

The first sample has a greater median time.
On average, they lasted 200 hours longer than motors in the second sample.
The spread of the second sample (given by the IQR) is much smaller.
It is 300 hours, compared with 600 hours for the first sample.

Exercise 39.3

1 The heights of a group of boys and a group of girls were recorded separately. The results are shown by the cumulative frequency graphs.

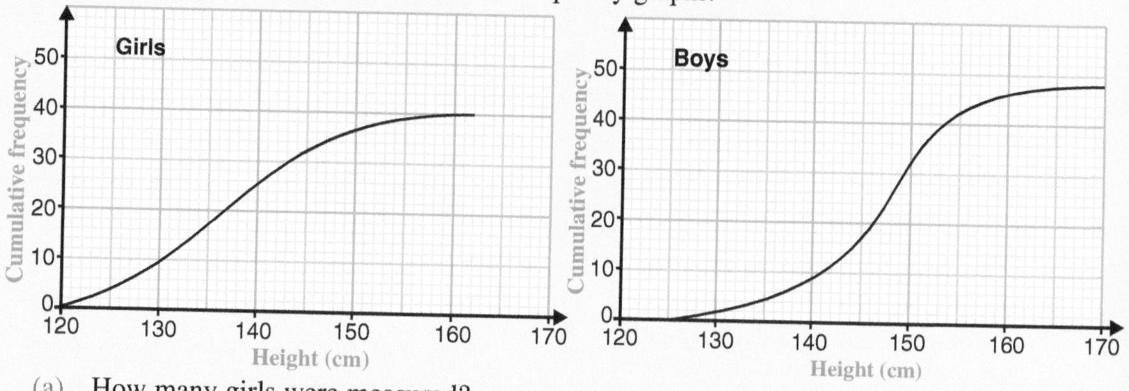

(a) How many girls were measured?
(b) Find the interquartile range of heights of the girls.
(c) Find the interquartile range of heights of the boys.
(d) Use your answers to parts (b) and (c) to comment on the heights of the girls compared with the boys.
(e) How many boys are taller than the tallest girl?

2 The milk yields of a herd of cows is shown in the table below.

Milk yield (x litres)	$5 \leqslant x < 10$	$10 \leqslant x < 15$	$15 \leqslant x < 20$	$20 \leqslant x < 25$	$25 \leqslant x < 30$
Number of cows	15	28	37	26	25

(a) Use the data to draw a cumulative frequency graph.
(b) Use your graph to estimate:
  (i) the median milk yield,
  (ii) the interquartile range of milk yields.
(c) A neighbouring farmer calculated the following results for his herd of cows.
Median yield = 22 litres, Lower quartile = 9 litres, Upper quartile = 28 litres.
Compare and comment on the data for the two herds.

3 A sample of potatoes of Variety X and a sample of potatoes of Variety Y are weighed. The cumulative frequency graphs show information about the weight distribution of each sample.

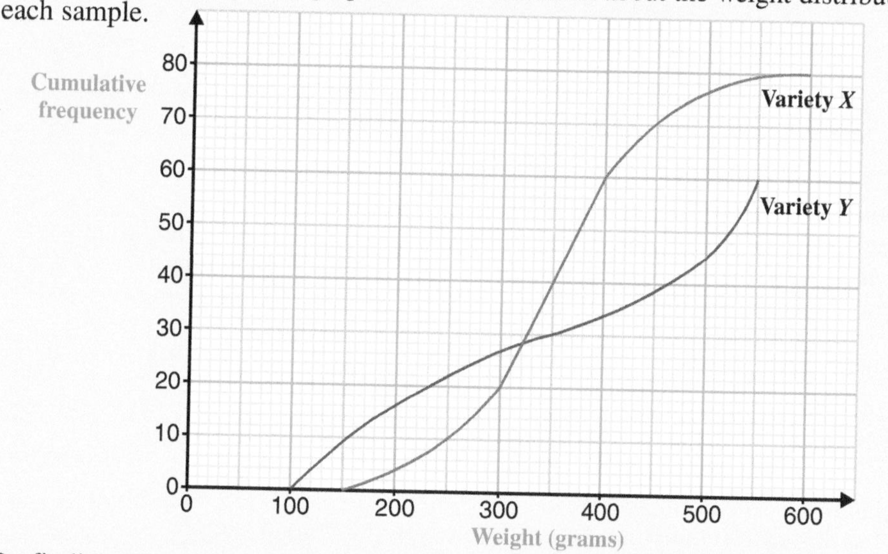

By finding the median and interquartile range of each sample, compare and comment on the weights of these varieties of potato.

Box plots

Box plots (or **box and whisker diagrams**) provide a useful way of representing the range, the median and the quartiles of a set of data.
They are also useful for comparing two (or more) distributions.

The graph shows the cumulative frequency distribution of the masses of 80 fish.

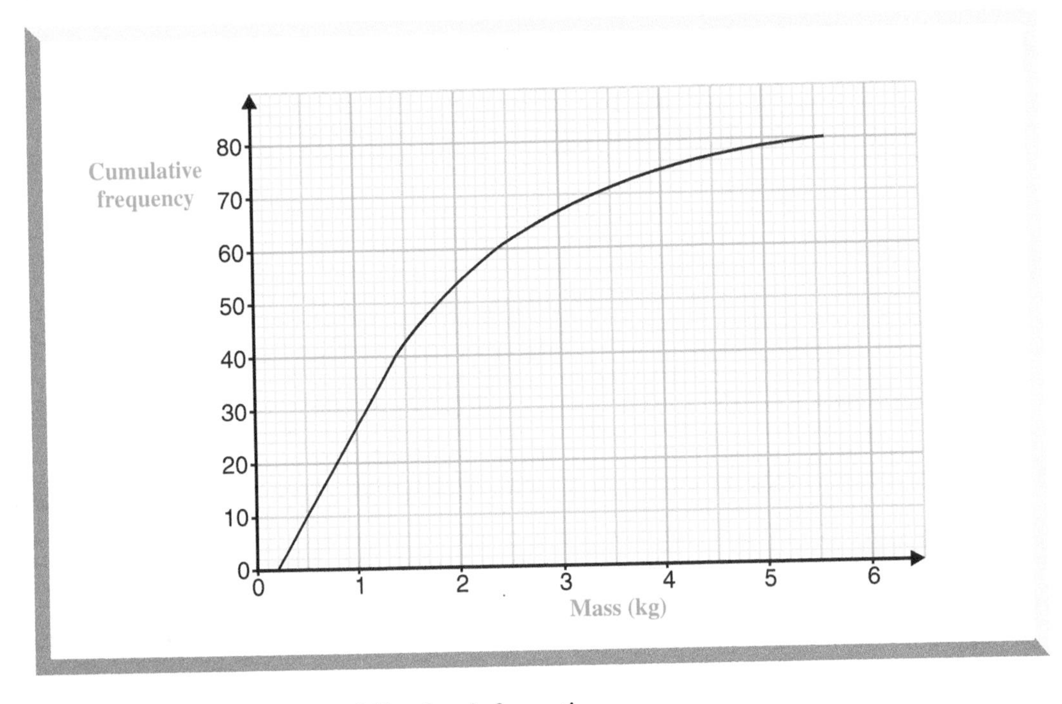

From the graph we can read off the following information:

- the minimum mass is 0.2 kg,
- the maximum mass is 5.6 kg,
- the median mass is 1.4 kg,
- the lower quartile is 0.8 kg,
- the upper quartile is 2.4 kg.

This information can now be represented as a box plot.

Begin by drawing a horizontal line and marking a scale from 0 to 6 kg.
Above your line, draw a box from the lower quartile to the upper quartile and mark in the median with a line across the box.
Draw lines (sometimes called whiskers) from the lower end of the box to 0.2 kg and from the upper end of the box to 5.6 kg to represent the range.

A box plot for the masses of these fish is shown below.

Minimum value
LQ
Median
UQ
Maximum value
0 1 2 3 4 5 6
Mass (kg)

The box plot shows how the masses of these fish are spread out and how the middle 50% are clustered.

EXAMPLE

The number of points scored by the 11 players in a basketball team during a competition are shown.

13 12 5 34 8 10 11 46 25 23 4

Draw a box plot to illustrate the data.

Begin by putting the data in order and then locate the median, lower quartile and upper quartile.

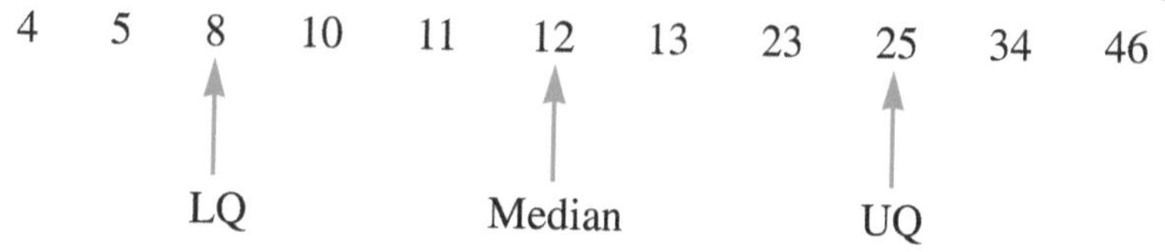

Then use these values to draw the box plot.

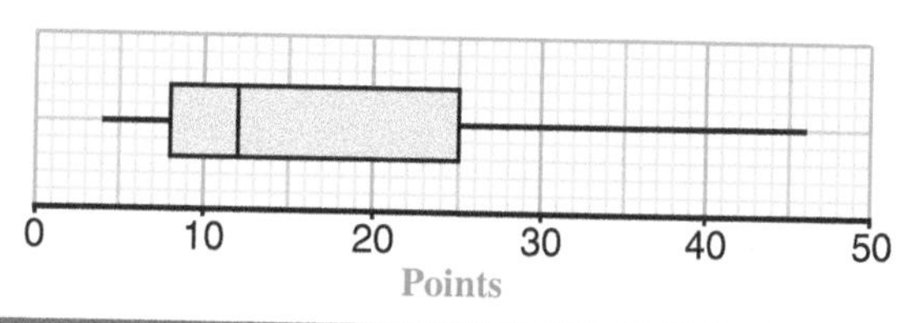

Exercise 39.4

1. The midday temperatures, in °C, for 11 cities around the world are:

   9 12 20 24 25 28 28 30 31 32 35

   Draw a box plot to represent these temperatures.

2. A group of 15 people were asked to estimate the weight of a large tortoise. Their estimates, in kilograms, are shown.

2.8	3.0	3.2	3.3	3.5	3.6	3.8	3.8
4.0	4.0	4.2	4.2	4.4	4.5	4.7	

   Draw a box plot to represent these estimates.

3. The cumulative frequency graph shows information about the distances travelled by children to get to school.

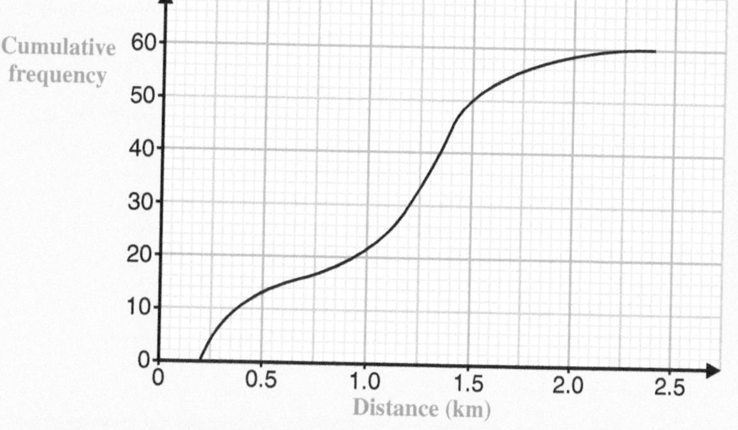

   Draw a box plot to illustrate the data.

4 The table gives information about the heights of players in a rugby team.

Median	188 cm	Tallest	198 cm
Lower quartile	185 cm	Shortest	175 cm
Upper quartile	193 cm		

Draw a box plot to represent this information.

5 The box plot illustrates the reaction times of a group of people.

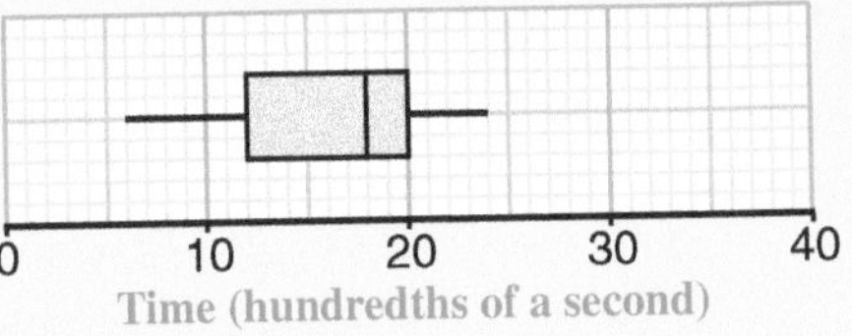

(a) What was the minimum reaction time?
(b) What is the value of the interquartile range?

6 A group of students took examinations in Mathematics and English.
The box plots illustrate the results.

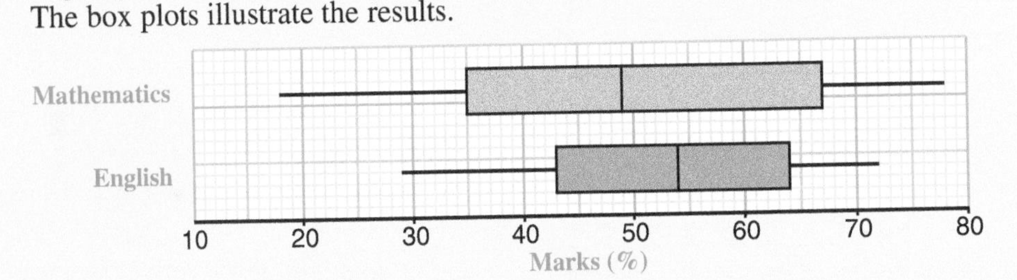

(a) What was the highest mark scored in English?
(b) What was the lowest mark scored in Mathematics?
(c) Which subject has the higher median mark?
(d) What is the value of the interquartile range for English?
(e) Comment on the results of these examinations.

7 The cumulative frequency graphs show the times taken by students to run 100 m.

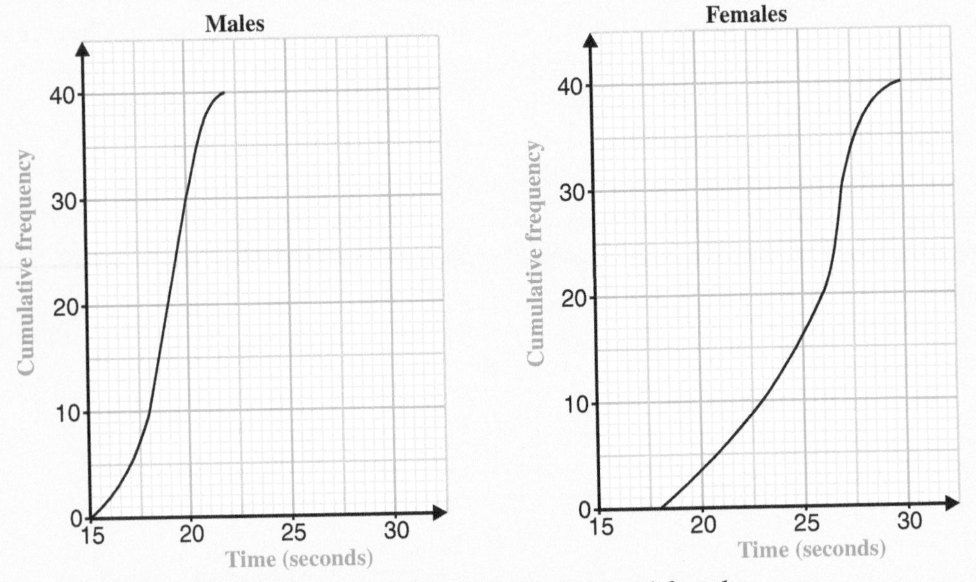

(a) Draw box plots to compare the times for males and females.
(b) Comment on the times for males and females.

8 A sample of 23 people were asked to record the amount they spent on food last week.
The amounts, in £s, are shown.

57	66	75	83	39	83	36	42
62	67	43	38	71	58	47	53
62	84	76	68	77	57	42	

(a) Find the median and quartiles of this distribution.
(b) Draw a box plot to represent the data.

What you need to know

- The information given in a frequency table can be used to make a **cumulative frequency table**.
- To draw a **cumulative frequency graph**:

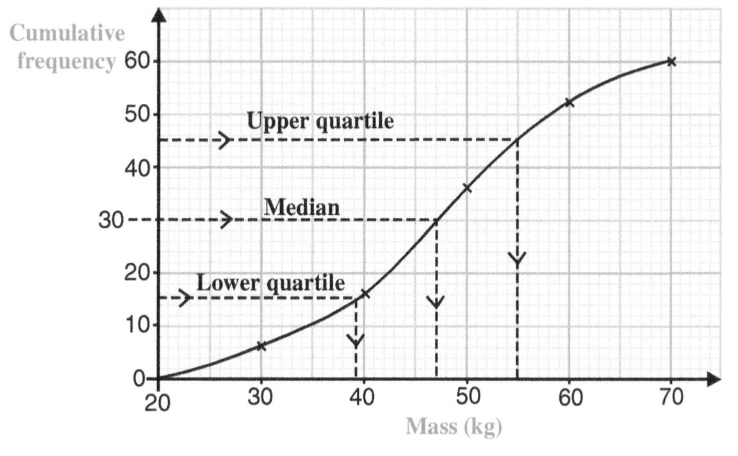

1. Draw and label: the variable on the horizontal axis,
cumulative frequency on the vertical axis.
2. Plot the cumulative frequency against the upper class boundary of each class.
3. Join the points with a smooth curve.

- If the question does not give the upper class boundaries, then the upper class boundary of each class is equal to the lower class boundary of the next class.
- When the classes have gaps between them then the upper class boundary is halfway between the end of one class and the beginning of the next.
- The **median** is the value of the middle number.
  The **lower quartile** is the value located at $\frac{1}{4}$ of the total frequency.
  The **upper quartile** is the value located at $\frac{3}{4}$ of the total frequency.
  The **interquartile range** measures the spread of the middle 50% of the data.
  Interquartile range = Upper Quartile − Lower Quartile
- A **box plot** is used to represent the range, the median and the quartiles of a distribution.

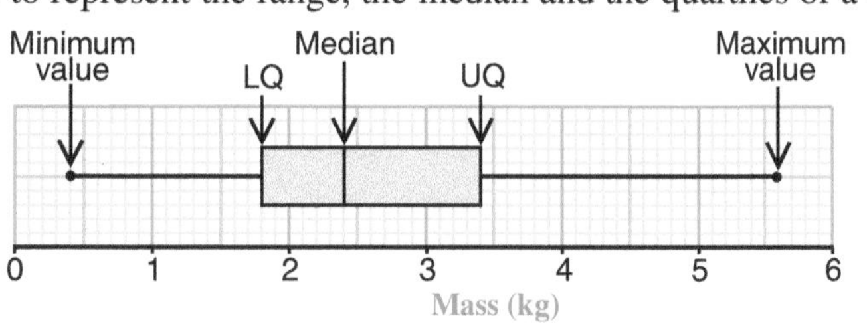

- The box plot shows how the data is spread out and how the middle 50% of data is clustered.
- Box plots can be used to compare two (or more) distributions.

Review Exercise 39

1. A sample of 40 trout is taken at a fish farm.

Mass (g)	40 -	50 -	60 -	70 -	80 -	90 - 100
Frequency	2	3	8	9	13	5

(a) Copy and complete the cumulative frequency table.

Mass (g)	< 50	< 60	< 70	< 80	< 90	< 100
Cumulative frequency						

(b) Draw the cumulative frequency graph.
(c) Find the median mass of the trout.
(d) Find the interquartile range of the mass of the trout.
(e) A second sample of trout has a median mass of 75 g, an upper quartile of 93 g and a lower quartile of 55 g.
Compare and comment on the spread of the data in these two samples. AQA

2. In a school examination, 96 students each took two maths papers.
Each paper was marked out of 100. The results for Paper 1 are given in the table.

Exam mark (M)	Number of students
$0 \leqslant M \leqslant 20$	0
$20 < M \leqslant 30$	2
$30 < M \leqslant 40$	9
$40 < M \leqslant 50$	19
$50 < M \leqslant 60$	25
$60 < M \leqslant 70$	20
$70 < M \leqslant 80$	10
$80 < M \leqslant 90$	8
$90 < M \leqslant 100$	3

(a) Draw a cumulative frequency diagram.
(b) Showing your method clearly, use your diagram to estimate:
(i) the median mark,
(ii) the interquartile range.
(c) The results for Paper 2 have a median of 61 marks and an interquartile range of 17 marks.
Which exam paper did the students find easier? Give a reason for your answer. AQA

3. The box plot shows information about the price of a pint of milk in a number of shops.

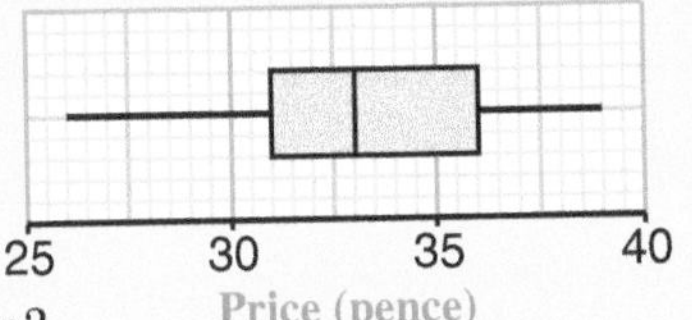

(a) What is the range in price?
(b) What is the median price?
(c) What is the value of the interquartile range?

4. The times taken by students to access a website gave the following information.

Maximum time	5.1 minutes	Lower quartile	2.8 minutes
Minimum time	1.5 minutes	Upper quartile	4.2 minutes
Median time	3.8 minutes		

Draw a box plot to represent this information.

Probability

What is probability?

Probability, or **chance**, involves describing how likely something is to happen.

For example: How likely is it to rain tomorrow?

We often make forecasts, or judgements, about how likely things are to happen.
When trying to forecast tomorrow's weather the following **outcomes** are possible:

sun, cloud, wind, rain, snow, …

We are interested in the particular **event**, rain tomorrow.

> In any situation, the possible things that can happen are called **outcomes**.
> An outcome of particular interest is called an **event**.

The chance of an event happening can be described using these words:
Impossible **Unlikely** **Evens** **Likely** **Certain**

Use one of the words in the box to describe the chance of rain tomorrow.

Probability and the probability scale

Estimates of probabilities can be shown on a **probability scale**.
The scale goes from 0 to 1.
A **probability of 0** means that an event is **impossible**.
A **probability of 1** means that an event is **certain**.
Probabilities are written as a fraction, a decimal or a percentage.

Less likely ← → More likely

Impossible — Certain

0 — $\frac{1}{2}$ — 1

Describe the likelihood that an event will occur if it has a probability of $\frac{1}{2}$.

EXAMPLES

Estimate the probability of each of the following events happening.
Show your estimate on a probability scale.

1. It will snow in London next July.
   This is possible but **very unlikely**.
   So, the probability is very close to 0.

   0 — $\frac{1}{2}$ — 1

2. You will eat some vegetables today.
   This is **very likely**.
   So, the probability is close to 1.

   0 — $\frac{1}{2}$ — 1

Exercise 40.1

1. Describe each of the following events as:

   Impossible **Unlikely** **Evens** **Likely** **Certain**

   (a) You roll a normal dice and get an odd number.
   (b) You roll a normal dice and get a 7.
   (c) A 6 is scored at least 80 times when a normal dice is rolled 600 times.
   (d) It will rain on three days running in August.
   (e) Somewhere in the world it is raining today.
   (f) A coin is tossed five times and lands heads up on each occasion.

2. The probability scale shows the probabilities of events **P**, **Q**, **R**, **S** and **T**.

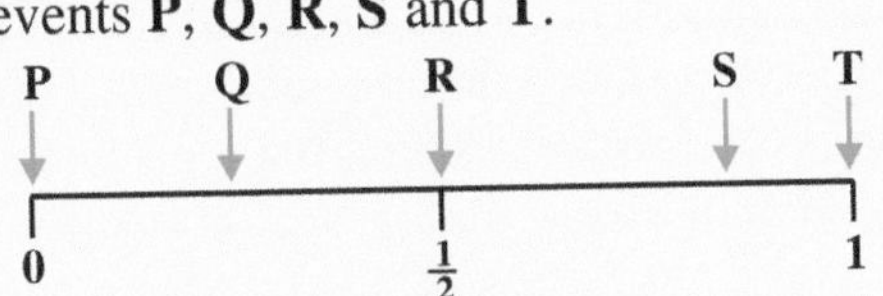

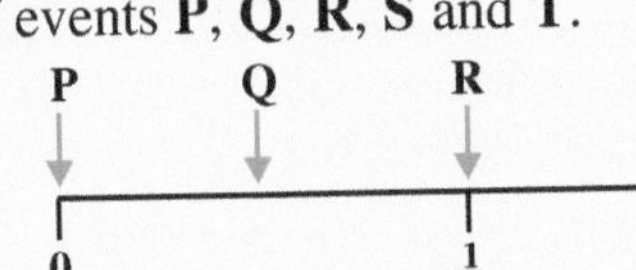

   Which of the five events
   (a) is certain to happen,
   (b) is impossible,
   (c) has an evens chance of happening,
   (d) is more likely to happen than to not happen, but is not certain to happen?

3. The probabilities of four events have been marked on a probability scale.
   Copy the probability scale.

   Event V A coin lands 'heads' up.
   Event W A person is over 3 metres tall.
   Event X Rolling an ordinary dice and getting a score less than 7.
   Event Y There is a 35% chance that it will rain tomorrow.

   Label the arrows on your diagram to show which event they represent.

Calculating probabilities using equally likely outcomes

Probabilities can be **calculated** where all outcomes are **equally likely**.

The probability of an event X happening is given by:

$$\text{Probability (X)} = \frac{\text{Number of outcomes in the event}}{\text{Total number of possible outcomes}}$$

Random and Fair

In many probability questions words such as '**random**' and '**fair**' are used.
These are ways of saying that all outcomes are equally likely.
For example:
A card is taken at **random** from a pack of cards.
This means that each card has an equal chance of being taken.
A **fair** dice is rolled.
This means that the outcomes 1, 2, 3, 4, 5 and 6 are equally likely.

EXAMPLE

A fair dice is rolled. What is the probability of getting:
(a) a 6, (b) an odd number, (c) a 2 or a 3?

Total number of possible outcomes is 6. (1, 2, 3, 4, 5 and 6).
The dice is fair, so each of these outcomes is equally likely.

(a) 1 of the possible outcomes is a 6. $P(6) = \frac{1}{6}$

(b) 3 of the possible outcomes are odd numbers. $P(\text{an odd number}) = \frac{3}{6} = \frac{1}{2}$

(c) 2 of the possible outcomes are 2 or 3. $P(2 \text{ or } 3) = \frac{2}{6} = \frac{1}{3}$

Exercise 40.2

1. A fair dice is rolled. What is the probability of getting a number less than five?

2. A bag contains a red counter, a blue counter and two green counters.
A counter is taken from the bag at random. What is the probability of taking:
(a) a red counter, (b) a green counter, (c) a counter that is not blue?

3. A bag contains 3 red sweets and 7 black sweets.
A sweet is taken from the bag at random. What is the probability of taking:
(a) a red sweet, (b) a black sweet?

4. The eleven letters of the word M I S S I S S I P P I are written on separate tiles.
The tiles are placed in a bag and mixed up. One tile is selected at random.
What is the probability that the tile selected shows:
(a) the letter M, (b) the letter I, (c) the letter P?

5. A card is taken at random from a full pack of 52 playing cards with no jokers.
What is the probability that the card:
(a) is red, (b) is a heart, (c) is the ace of hearts?

6. A bag contains 4 red counters, 3 white counters and 3 blue counters.
A counter is taken from the bag at random. What is the probability that the counter is:
(a) red, (b) white or blue, (c) red, white or blue, (d) green?

7. In a hat there are twelve numbered discs.
Nina takes a disc from the hat at random.
What is the probability that Nina takes a disc:
(a) with at least one 4 on it,
(b) that has not got a 4 on it,
(c) that has a 3 or a 4 on it?

43 44 45 46
47 48 49 50
51 52 53 54

8. This table shows how fifty counters are numbered either 1 or 2 and coloured red or blue.
One of the counters is chosen at random.
What is the probability that the counter is:
(a) a 1, (b) blue, (c) blue and a 1?

	Red	Blue
1	12	8
2	8	22

A blue counter is chosen at random.
(d) What is the probability that it is a 1?
A counter numbered 1 is chosen at random.
(e) What is the probability that it is blue?

9. Tim plays a friend at Noughts and Crosses.
He says: "I can win, draw or lose, so the probability that I will win must be $\frac{1}{3}$."
Explain why Tim is wrong.

10. The table shows the number of boys and girls in a class of 30 pupils who wear glasses.

	Boy	Girl
Wears glasses	3	1
Does not wear glasses	11	15

A pupil from the class is picked at random.
(a) What is the probability that it is a boy?
(b) What is the probability that it is a girl who does not wear glasses?
A girl from the class is picked at random.
(c) What is the probability that she wears glasses?
A pupil who wears glasses is picked at random.
(d) What is the probability that it is a boy?

11 The table shows the way that 120 pupils from Year 7 travel to Linfield School.

	Walk	Bus	Car	Bike
Boys	23	15	12	20
Girls	17	20	8	5

A pupil from Year 7 is chosen at random.
What is the probability that the pupil:
(a) walks to school,
(b) is a girl who travels by car,
(c) is a boy who does not travel by bus?

A girl from Year 7 is chosen at random.
What is the probability that:
(d) she walks to school,
(e) she does not travel by car?

A Year 7 pupil who travels by bike is chosen at random.
(f) What is the probability that the pupil is a boy?

Estimating probabilities using relative frequency

In question 9 in Exercise 40.2, probabilities **cannot** be calculated using equally likely outcomes.
In such situations probabilities can be estimated using the idea of **relative frequency**.

It is not always necessary to perform an experiment or make observations.
Sometimes the information required can be found in past records.

The relative frequency of an event is given by:

$$\text{Relative frequency} = \frac{\text{Number of times the event happens in an experiment (or in a survey)}}{\text{Total number of trials in the experiment (or observations in the survey)}}$$

EXAMPLES

1 Jamie does the following experiment with a bag containing 2 red and 8 blue counters.

Take a counter from the bag at random.
Record the colour then put the counter back in the bag. Repeat this for 100 trials.

Jamie calculates the relative frequency of getting a red counter every 10 trials and shows his results on a graph. Draw a graph showing the results that Jamie might get.
This is the sort of graph that Jamie might get.

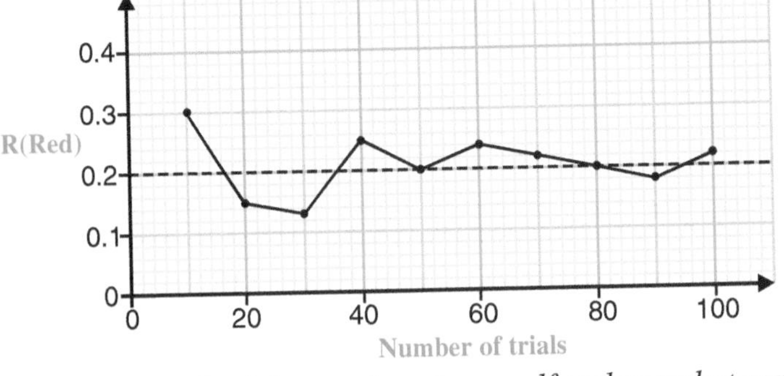

The dotted line shows the **calculated probability**.

P(Red) $= \frac{2}{10} = 0.2$

As the number of trials increases, relative frequency gives a better estimate of calculated probability.

Try Jamie's experiment yourself and see what sort of results you get.

2 In an experiment a drawing pin is dropped for 100 trials.
The drawing pin lands "point up" 37 times.
What is the relative frequency of the drawing pin landing "point up"?

Relative frequency $= \frac{37}{100} = 0.37$

Relative frequency gives a better estimate of probability the larger the number of trials.

Exercise 40.3

1. 50 cars are observed passing the school gate. 14 red cars are observed.
What is the relative frequency of a red car passing the school gate?

2. In an experiment a gardener planted 40 daffodil bulbs of which 36 grew to produce flowers.
Use these results to find the relative frequency that a daffodil bulb will produce a flower.

3. The results from 40 spins of a numbered spinner are:

2	1	4	3	2	1	3	4	5	2
1	2	2	3	2	1	2	4	5	2
1	5	3	4	2	3	3	3	2	4
2	3	4	2	1	5	3	3	5	3

Use these results to estimate the probability of getting a 2 with the next spin.

4. A counter is taken from a bag at random.
Its colour is recorded and the counter is then put back in the bag.
This is repeated 300 times.
The number of red counters taken from the bag after every 100 trials is shown in the table.

Number of trials	Number of red counters
100	52
200	102
300	141

(a) Calculate the relative frequency after each 100 trials.
(b) Which is the best estimate of taking a red counter from the bag?

5. Gemma keeps a record of her chess games with Helen.
Out of the first 10 games, Gemma wins 6. Out of the first 30 games Gemma wins 21.
Based on these results, estimate the probability that Gemma will win her next game of chess with Helen.

6. Rachel selects 40 holiday brochures at random.
The probability of a brochure being for a holiday in Italy is found to be 0.2.
How many brochures did Rachel select for holidays in Italy?

7. A counter was taken from a bag of counters and replaced.
The relative frequency of getting a red counter was found to be 0.3.
There are 60 counters in the bag.
Estimate the number of red counters.

8. 500 tickets are sold for a prize draw.
Greg buys some tickets.
The probability that Greg wins first prize is $\frac{1}{20}$.
How many tickets did he buy?

9. A bypass is to be built to avoid a town.
There are three possible routes that the road can take.
A survey was carried out in the town.

Route	A	B	C
Relative frequency	0.4	0.5	0.1

30 people opted for Route C.
(a) How many people were surveyed altogether?
(b) How many people opted for Route A?
(c) How many people opted for Route B?

Mutually exclusive events

Events which **cannot happen at the same time** are called **mutually exclusive events**.
For example, the event 'Heads' cannot occur at the same time as the event 'Tails'.

> When A and B are events which cannot happen at the same time:
> P(A or B) = P(A) + P(B)

The probability of an event not happening

> The events A and not A cannot happen at the same time.
> Because the events A and not A are certain to happen:
> P(not A) = 1 − P(A)

EXAMPLES

1 A bag contains 3 red (R) counters, 2 blue (B) counters and 5 green (G) counters.
A counter is taken from the bag at random.
What is the probability that the counter is:
(a) red, (b) green, (c) red or green?

Find the total number of counters in the bag.
5 + 2 + 3 = 10
Total number of possible outcomes = 10.

(a) There are 3 red counters. $P(R) = \frac{3}{10}$

(b) There are 5 green counters. $P(G) = \frac{5}{10} = \frac{1}{2}$

(c) Events R and G cannot happen at the same time.
$P(R \text{ or } G) = P(R) + P(G) = \frac{3}{10} + \frac{5}{10} = \frac{8}{10} = \frac{4}{5}$

2 A bag contains 10 counters. 3 of the counters are red (R).
A counter is taken from the bag at random.
What is the probability that the counter is:
(a) red, (b) not red?

Total number of possible outcomes = 10.

(a) There are 3 red counters. $P(R) = \frac{3}{10}$

(b) $P(\text{not } R) = 1 - P(R) = 1 - \frac{3}{10} = \frac{7}{10}$

Exercise 40.4

1 A fish is taken at random from a tank.
The probability that the fish is black is $\frac{2}{5}$.
What is the probability that the fish is not black?

2 Tina has a bag of beads.
She takes a bead from the bag at random.
The probability that the bead is white is 0.6.
What is the probability that the bead is not white?

3 The probability of a switch working is 0.96.
What is the probability of a switch not working?

4 Six out of every 100 men are taller than 1.85 m.
A man is picked at random.
What is the probability that he is not taller than 1.85 m?

5 A bag contains red, white and blue balls.
A ball is taken from the bag at random.
The probability of taking a red ball is 0.4.
The probability of taking a white ball is 0.35.
What is the probability of taking a white ball or a blue ball?

6 Tom and Sam buy some tickets in a raffle.
The probability that Tom wins 1st prize is 0.03.
The probability that Sam wins 1st prize is 0.01.
(a) What is the probability that Tom or Sam win 1st prize?
(b) What is the probability that Tom does not win 1st prize?

7 A spinner can land on red, white or blue.
The probability of the spinner landing on red is 0.2.
The probability of the spinner landing on red or on blue is 0.7.
The spinner is spun once.
What is the probability that the spinner lands:
(a) on blue,
(b) on white?

8 A bag contains red, green, blue, yellow and white counters.
The table shows the probabilities of obtaining each colour when a counter is taken from the bag at random.

Red	Green	Blue	Yellow	White
30%	25%	20%	20%	10%

(a) (i) How can you tell that there is a mistake in the table?
(ii) The probability of getting a white counter is wrong.
What should it be?

A counter is taken from the bag at random.
(b) (i) What is the probability that it is either green or blue?
(ii) What is the probability that it is red, green or blue?
(iii) What is the probability that it is not yellow?

9 Some red, white and blue cubes are numbered 1 or 2.
The table shows the probabilities of obtaining each colour and number when a cube is taken at random.

	Red	White	Blue
1	0.1	0.3	0
2	0.3	0.1	0.2

A cube is taken at random.
(a) What is the probability of taking a red cube?
(b) What is the probability of taking a cube numbered 2?
(c) State whether or not the following pairs of events are mutually exclusive.
Give a reason for each answer.
(i) Taking a cube numbered 1 and taking a blue cube.
(ii) Taking a cube numbered 2 and taking a blue cube.
(d) (i) What is the probability of taking a cube which is blue or numbered 1?
(ii) What is the probability of taking a cube which is blue or numbered 2?
(iii) What is the probability of taking a cube which is numbered 2 or red?

EXAMPLES

1 A fair coin is thrown twice.
Identify all of the possible outcomes and write down their probabilities.

Method 1
List the outcomes systematically.

1st throw	2nd throw
Head (H)	Head (H)
Head (H)	Tail (T)
Tail (T)	Head (H)
Tail (T)	Tail (T)

Method 2
Use a **possibility space diagram.**

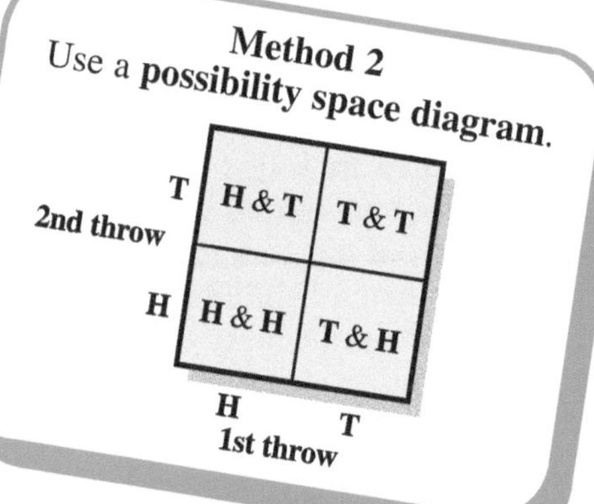

Method 3
Use a **tree diagram.**

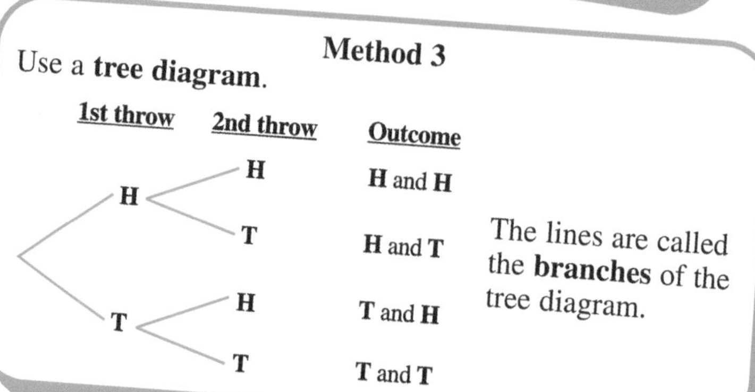

When a fair coin is tossed twice, there are four possible outcomes.
Because the coin is fair all the possible outcomes are **equally likely**.
Because all the outcomes are equally likely their probabilities can be worked out.

$$P(\text{H and H}) = P(\text{H and T}) = P(\text{T and H}) = P(\text{T and T}) = \frac{1}{4}.$$

2 A fair dice is rolled twice.
Use a possibility space diagram to show all the possible outcomes.
What is the probability of getting a 'double six'?
What is the probability of getting any 'double'?
What is the probability that exactly one 'six' is obtained?

2nd roll						
6	1 and 6	2 and 6	3 and 6	4 and 6	5 and 6	6 and 6
5	1 and 5	2 and 5	3 and 5	4 and 5	5 and 5	6 and 5
4	1 and 4	2 and 4	3 and 4	4 and 4	5 and 4	6 and 4
3	1 and 3	2 and 3	3 and 3	4 and 3	5 and 3	6 and 3
2	1 and 2	2 and 2	3 and 2	4 and 2	5 and 2	6 and 2
1	1 and 1	2 and 1	3 and 1	4 and 1	5 and 1	6 and 1
	1	2	3	4	5	6
	1st roll					

The dice is fair so there are 36 equally likely outcomes.

P(double 6)
There is one outcome in the event (6 and 6).
P(double 6) $= \frac{1}{36}$

P(any double)
The 6 outcomes in the event are shaded blue.
P(any double) $= \frac{6}{36} = \frac{1}{6}$

P(exactly one six)
The 10 outcomes in the event are shaded grey.
P(exactly one six) $= \frac{10}{36} = \frac{5}{18}$

Exercise 40.5

1 A red car (R), a blue car (B) and a green car (G) are parked on a narrow drive, one behind the other.

(a) List all the possible orders in which the three cars could be parked.

The cars are parked on the drive at random.

(b) What is the probability that the blue car is the first on the drive?

2 Two fair dice are rolled and the numbers obtained are added.

(a) Draw a possibility space diagram to show all of the possible outcomes.

(b) Use your diagram to work out:

(i) the probability of obtaining a total of 10,

(ii) the probability of obtaining a total greater than 10,

(iii) the probability of obtaining a total less than 10.

(c) Explain why the probabilities you worked out in (b) should add up to 1.

3 A fair coin is tossed and a fair dice is rolled.
Copy and complete the table to show all the possible outcomes.

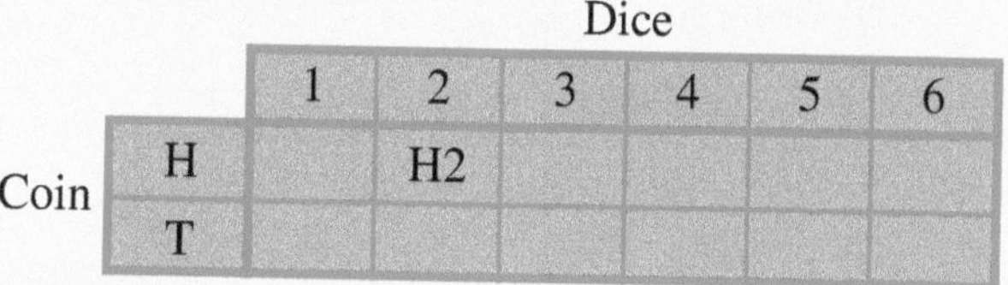

		Dice					
		1	2	3	4	5	6
Coin	H		H2				
	T						

What is the probability of obtaining:

(a) a head and a 5,

(b) a tail and an even number,

(c) a tail and a 6,

(d) a tail and an odd number,

(e) a head and a number more than 4,

(f) an odd number?

4 Sanjay has to travel to school in two stages.

Stage 1: he can go by bus or train or he can get a lift.

Stage 2: he can go by bus or he can walk.

(a) List all the different ways that Sanjay can travel to school.

Sanjay decides the way that he travels on each stage at random.

(b) What is the probability that he goes by bus in both stages?

5 The diagram shows a fair spinner.
It is divided into four equal sections numbered as shown.
The spinner is spun twice and the numbers the arrow lands on each time are added to obtain a score.

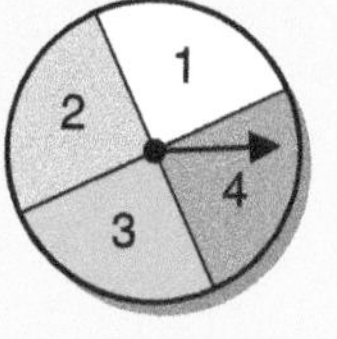

(a) Copy and complete this table to show all the possible scores.

		2nd spin			
		1	**2**	**3**	**4**
1st spin	**1**	2	3		
	2	3			
	3				
	4				

(b) Calculate the probability of getting a score of:

(i) 2, (ii) 3, (iii) 6.

6 Students at a college must choose to study two subjects from the list:

Maths **English** **Science** **Art**

(a) Write down all the possible pairs of subjects that the students can choose.

David chooses both subjects at random.

(b) What is the probability that one of the subjects he chooses is Maths?

James chooses Maths and one other subject at random.

(c) What is the probability that he chooses Maths and Science?

7 Bag A contains 2 red balls and 1 white ball.
Bag B contains 2 white balls and 1 red ball.
A ball is drawn at random from each bag.

(a) Copy and complete the table to show all possible pairs of colours.

(b) Explain why the probability of each outcome is $\frac{1}{9}$.

(c) Calculate the probability that the two balls are the same colour.

		Bag A		
		R	R	W
Bag B	W	RW		
	W			
	R			

8 The diagram shows two sets of cards A and B.

One card is taken at random from set A.
One card is taken at random from set B.

(a) List all the possible outcomes.

The two numbers are added together.

(b) (i) What is the probability of getting a total of 5?
(ii) What is the probability of getting a total that is not 5?

All the cards are put together and one of them is taken at random.

(c) What is the probability that it is labelled A or 2?

9 A spinner has an equal probability of landing on red, green, blue, yellow or white.
The spinner is spun twice.

(a) List all the possible outcomes.

(b) (i) What is the probability that, on both spins, the spinner lands on white?
(ii) What is the probability that, on both spins, the spinner lands on white at least once?
(iii) What is the probability that, on both spins, the spinner lands on the same colour?

10 The diagram shows two fair spinners.
Each spinner is divided into equal sections and numbered as shown.
Each spinner is spun and the numbers that each arrow lands on are added together.

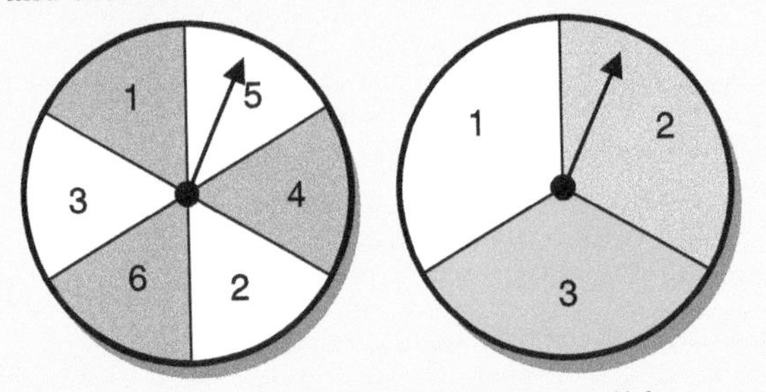

(a) Draw a possibility space diagram to show all the possible outcomes.

(b) Calculate the probability of getting a total of 2.

(c) Calculate the probability of getting a total of 6.

Independent events

When two coins are tossed, the outcome of the first toss has no effect on the outcome of the second toss.
One person being left-handed does not influence another person being left-handed.
These are examples of events which can happen together but which do not affect each other.
Events like this are called **independent** events.

When A and B are **independent** events then the probability of A and B occurring is given by:
P(A and B) = P(A) × P(B)

This rule can be extended to any number of independent events.
For example:
P(A and B and C) = P(A) × P(B) × P(C)

Using tree diagrams to work out probabilities

A **tree diagram** can be used to find all the possible outcomes when two or more events are combined.
It can also be used to help calculate probabilities when outcomes are not equally likely.

Box A contains 1 red ball (R) and 1 blue ball (B).
Box B contains 3 red balls (R) and 2 blue balls (B).

Box A Box B

A ball is taken at random from Box A.
A ball is then taken at random from Box B.

(a) Draw a tree diagram to show all the possible outcomes.
(b) Calculate the probability that two red balls are taken.

(a)

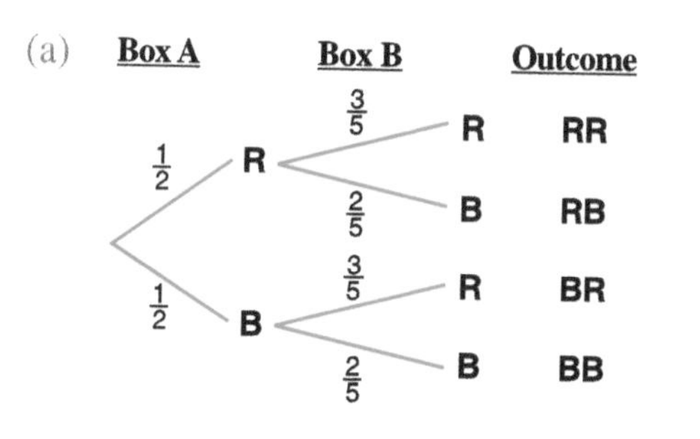

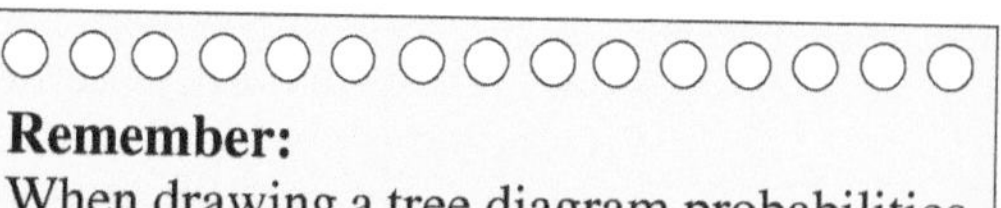

Remember:
When drawing a tree diagram probabilities must be put on the branches.
Using equally likely outcomes:
In Box A
$P(R) = \frac{1}{2}$ $P(B) = \frac{1}{2}$
In Box B
$P(R) = \frac{3}{5}$ $P(B) = \frac{2}{5}$

(b) The numbers of red and blue balls are unequal in Box B.
This means that the outcomes RR, RB, BR and BB are not equally likely.
Multiply the probabilities along the branches of the tree diagram.

$P(RR) = \frac{1}{2} \times \frac{3}{5} = \frac{1 \times 3}{2 \times 5} = \frac{3}{10}$

The probability that two red balls are taken is $\frac{3}{10}$.

EXAMPLE

2 The probability that Amanda is late for school is 0.4.
Use a tree diagram to find the probability that on two days running:
(a) she is late twice,
(b) she is late exactly once.

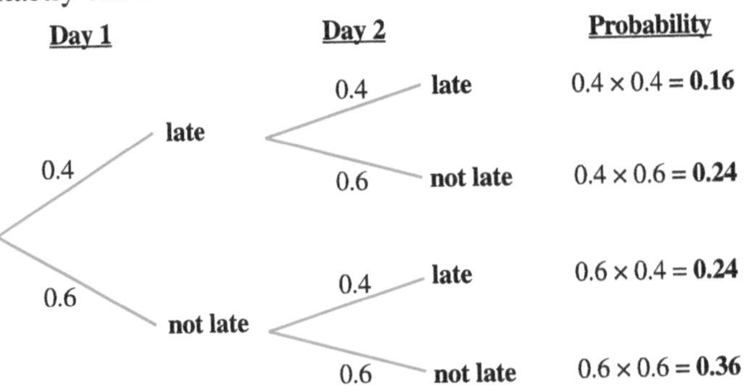

(a) The probability that Amanda is late twice.
The outcome included in this event is: (late **and** late) P(late twice) = 0.16

(b) The probability that Amanda is late exactly once.
The outcomes included in this event are: (late **and** not late) **or** (not late **and** late)
These outcomes are mutually exclusive.
P(late exactly once) = 0.24 + 0.24 = 0.48

Exercise 40.6

1 A bag contains 3 red counters and 2 blue counters.
A counter is taken at random from the bag and then replaced.
Another counter is taken at random from the bag.

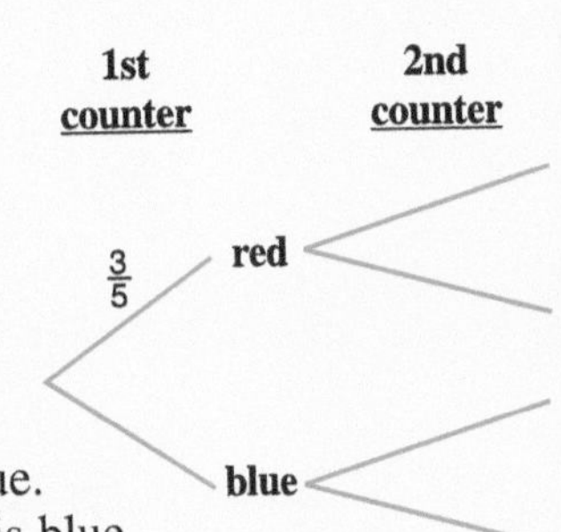

(a) Copy and complete the tree diagram to show all the possible outcomes. Write the probability of each of the events on the branches of the tree diagram.
(b) (i) Calculate the probability that both counters taken are blue.
(ii) Calculate the probability that at least one counter taken is blue.

2 A manufacturer makes an electrical circuit which contains two switches.
The probability that a switch is faulty is 0.1.
(a) Copy and complete the tree diagram.

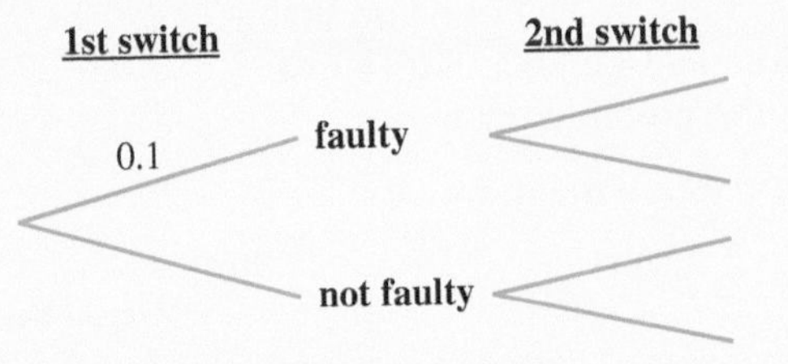

(b) (i) Calculate the probability that both switches are faulty.
(ii) Calculate the probability that exactly one switch is faulty.

The circuit works if both switches are not faulty.
(c) The manufacturer makes 1000 circuits.
Estimate the number that work.

3 Five people in a group of 50 people are left handed.
There are 20 females in the group.
A person is picked at random from the group.

(a) (i) What is the probability that the person is left handed?
(ii) What is the probability that the person is right handed?
(iii) What is the probability that the person is female?
(iv) What is the probability that the person is male?

(b) Copy and complete the tree diagram.

left handed

right handed

(c) (i) Calculate the probability that the person picked is a left handed female.
(ii) Calculate the probability that the person picked is a left handed female or is a right handed male.

4 Bag A contains 3 blue counters, 5 red counters and 2 white counters.
Bag B contains 2 blue counters and 3 red counters.
A ball is taken at random from Bag A.
A ball is taken at random from Bag B.

(a) Copy and complete the tree diagram to show the possible pairs of colours.
Write the probability of each of the events on the branches.

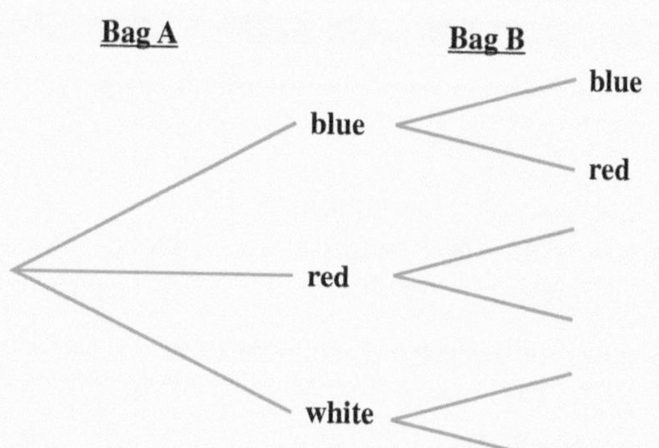

(b) Calculate the probability that a blue counter is taken from Bag A and a red counter is taken from Bag B.

(c) Calculate the probability that both counters are the same colour.

5 The probability that a car fails an MOT test on lights is $\frac{1}{10}$.

The probability that a car fails an MOT test on brakes is $\frac{1}{5}$.
A car is stopped at random and given an MOT test.
The tree diagram shows the possible outcomes.

(a) Copy and complete the tree diagram.

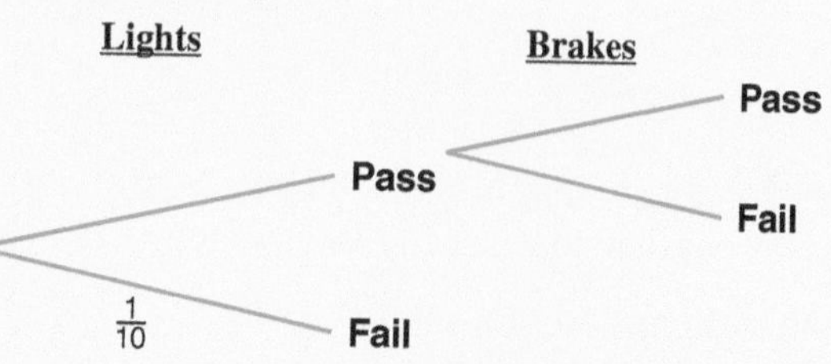

(b) Calculate the probability that the car will fail the MOT test on lights or brakes.

6 A fair spinner can land on either black or white.
The probability that it lands on white when it is spun is 0.3.
(a) What is the probability that it lands on black when it is spun?

The spinner is spun twice.
(b) Find the probability of getting two blacks.
(c) Find the probability of getting the same colour on both spinners.

7 In box A there are 3 red and 5 blue counters.
In box B there are 2 red and 3 blue counters.
A counter is taken at random from each box.
Calculate the probability that:
(a) both counters are red,
(b) at least one of the counters is red.

8 Colin takes examinations in Maths and in English.
The probability that he passes Maths is 0.7.
The probability that he passes English is 0.8.
The results in each subject are independent of each other.
Calculate the probability that:
(a) he passes Maths and fails English,
(b) he fails both subjects.

9 Dice A and Dice B are two normal dice. Dice A is a fair dice.
Dice B is biased, so that the probability of getting an even number is $\frac{2}{3}$.
Both of the dice are tossed.
Find the probability that:
(a) an odd number is scored on dice A and an even number on dice B,
(b) an odd number is scored on one dice and an even number on the other.

10 A box contains cubes which are coloured red (R), white (W) or blue (B) and numbered 1, 2 or 3.
The table shows the probabilities of obtaining each colour and each number when a cube is taken from the box at random.

		Colour of cube		
		R	W	B
Number on cube	1	0.2	0	0.1
	2	0.1	0.3	0
	3	0	0.1	0.2

A single cube is taken from the box at random.
(a) What is the probability that the cube is:
(i) red and numbered 2,
(ii) white or numbered 1,
(iii) white or numbered 3?

A cube is taken from the box at random and then replaced.
Another cube is then taken from the box at random.
(b) Calculate the probability that:
(i) both cubes are blue and numbered 1,
(ii) both cubes are blue.

11 A bag contains red, white and blue cubes.
The cubes are numbered 1, 2, 3 and 4.
The probabilities of taking cubes from the box at random are shown in the table.

		Colour of cube		
		Red	White	Blue
Number on cube	1	0.1	0	0
	2	0.1	0.1	0.1
	3	0	0.2	0.2
	4	0.1	0	0.1

A single cube is taken from the box at random.
(a) What is the probability that:
(i) the cube is white,
(ii) the cube is red and numbered 4,
(iii) the cube is white or numbered 3?

A cube is taken from the box at random and then replaced.
Another cube is then taken from the box at random.
(b) Calculate the probability that both cubes are the same colour.

Conditional probability

In some situations the probability of a particular event occurring can be affected by other events.
For example:
If it rains, the probability of a particular driver winning a Formula 1 Grand Prix might change.
When probabilities depend on other events they are called **conditional probabilities**.

EXAMPLE

A bag contains six red counters and four blue counters.
Three counters are taken from the bag at random and put in a box.

(a) What is the probability that the **second** counter put in the box is blue:
  (i) if the **first** counter put in the box is red,
  (ii) if the **first** counter put in the box is blue?

(b) What is the probability that the **third** counter put in the box is red:
  (i) if the **first two** counters put in the box are red,
  (ii) if the **first** counter put in the box is blue and the **second** counter put in the box is red?

(c) Show all the possible ways that the counters can be put in the box on a tree diagram. Label the branches of the tree diagram with their probabilities.

(d) What is the probability that there are at least two blue counters in the box?

(a) (i) If the first counter put in the box is red, then 4 of the 9 counters left in the bag are blue. So, P(Blue) $= \frac{4}{9}$
  (ii) If the first counter put in the box is blue, then 3 of the 9 counters left in the bag are blue. So, P(Blue) $= \frac{3}{9} = \frac{1}{3}$

(b) (i) If the first two counters put in the box are red, then 4 of the 8 counters left in the bag are red. So, P(Red) $= \frac{4}{8} = \frac{1}{2}$
  (ii) If the first counter put in the box is blue and the second counter put in the box is red, then 5 of the 8 counters left in the bag are red. So, P(Red) $= \frac{5}{8}$

(c)

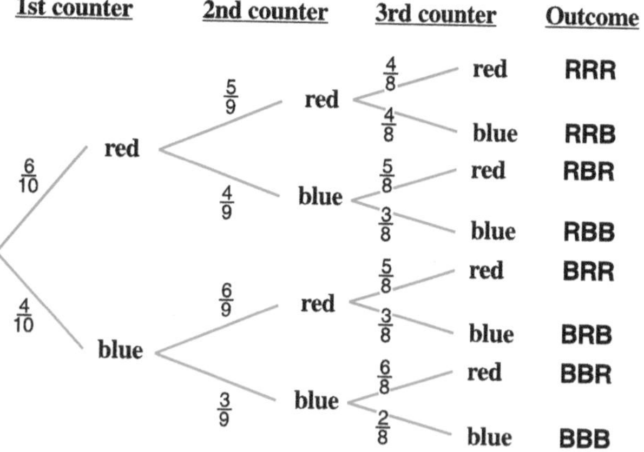

(d) The outcomes included in the event at least two blue counters in the box are:

Red (R) and Blue (B) and Blue (B) P(R and B and B) $= \frac{6}{10} \times \frac{4}{9} \times \frac{3}{8} = \frac{1}{10}$

Blue (B) and Red (R) and Blue (B) P(B and R and B) $= \frac{4}{10} \times \frac{6}{9} \times \frac{3}{8} = \frac{1}{10}$

Blue (B) and Blue (B) and Red (R) P(B and B and R) $= \frac{4}{10} \times \frac{3}{9} \times \frac{6}{8} = \frac{1}{10}$

Blue (B) and Blue (B) and Blue (B) P(B and B and B) $= \frac{4}{10} \times \frac{3}{9} \times \frac{2}{8} = \frac{1}{30}$

So, P(at least two blue counters) $= \frac{1}{10} + \frac{1}{10} + \frac{1}{10} + \frac{1}{30}$
$= \frac{1}{3}$

Exercise 40.7

1. Three female students and two male students are nominated for a school committee.
The committee has two vacancies which are to be filled at random from the nominees.

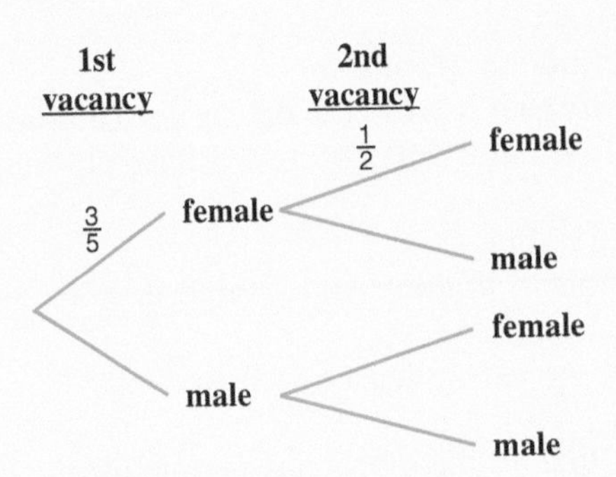

(a) Copy and complete the tree diagram showing all the probabilities.
(b) Calculate the probability that both vacancies are filled by females.
(c) Calculate the probability that the vacancies are filled by one female and one male.
(d) Calculate the probability that at least one vacancy is filled by a male.

2. Mila has a box of chocolates.
The box contains seven milk chocolates and four plain chocolates.
She takes two chocolates at random from the box.

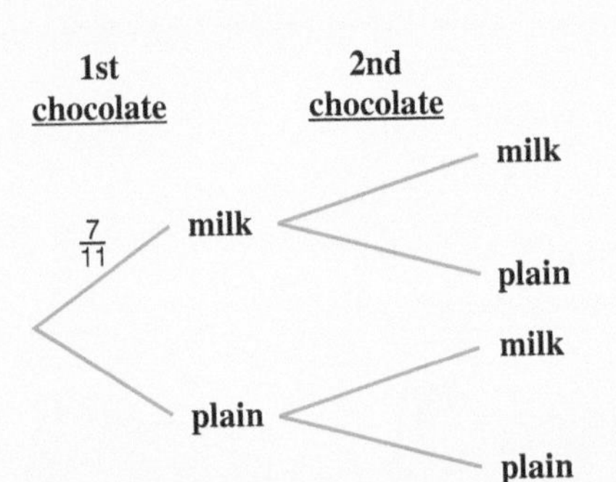

(a) Copy and complete the tree diagram showing all the probabilities.
(b) Calculate the probability that Mila takes one plain chocolate and one milk chocolate.

3. A pencil case contains 4 red pens and 6 blue pens.
Leo takes two pens at random from the pencil case without looking.
(a) Draw a tree diagram to show all the possible outcomes and probabilities.
(b) Calculate the probability that both pens are the same colour.

4. Vinnie has a box of flags.
The box contains 3 red flags, 4 blue flags and 5 green flags.
Vinnie takes two flags from the box at random.
(a) Draw a tree diagram to show all the possible outcomes and probabilities.
(b) What is the probability that Vinnie takes two flags of the same colour from the box?

5. Lydia and Chloe play two games of squash.
The probability of Lydia winning the first game is 0.7.
If Lydia wins the first game the probability of her winning the second game is 0.8.
If Chloe wins the first game the probability of her winning the second game is 0.5.
(a) Draw a tree diagram to show all the possible outcomes and probabilities.
(b) Calculate the probability that Chloe wins both games.
(c) Calculate the probability that they win one game each.

6. Ian has a photocopier in his office.
If the photocopier is working on a certain day then the probability that it will be working the next day is 0.9.
If the photocopier is not working on a certain day then the probability that it will not be working the next day is 0.3.
The photocopier is not working on Monday.
   (a) Calculate the probability that it will be working on Tuesday **and** Wednesday.
   (b) Calculate the probability that it will **not** be working on Wednesday.

7. Anne-Marie has a bag of mixed nuts.
The bag contains 8 Brazil nuts, 5 Walnuts and 7 Filberts.
Anne-Marie eats two nuts at random.
What is the probability that she eats at least one Brazil nut?

8. After a flood, Debbie finds that all the labels have come off the tins in her cupboard.
Debbie knows that she had 5 tins of tomatoes and 7 tins of baked beans.
She opens three tins at random.
What is the probability that she has opened 2 tins of baked beans and one tin of tomatoes?

9. Ashnil has a security code on his mobile phone.
Each time he switches his phone on he has to enter the code.
The probability that he enters the code correctly on his first attempt is 0.9.
If he enters the code **incorrectly**, the probability that he enters the code correctly on his next attempt is 0.7.
Each time he turns his phone on, Ashnil is allowed 3 attempts to enter the correct code.
Ashnil turns his phone on.
Calculate the probability that Ashnil enters the correct code on either his first attempt, his second attempt or his third attempt.

10. Bag A contains 2 blue balls and 3 white balls.
Bag B contains 5 blue balls and 3 white balls.

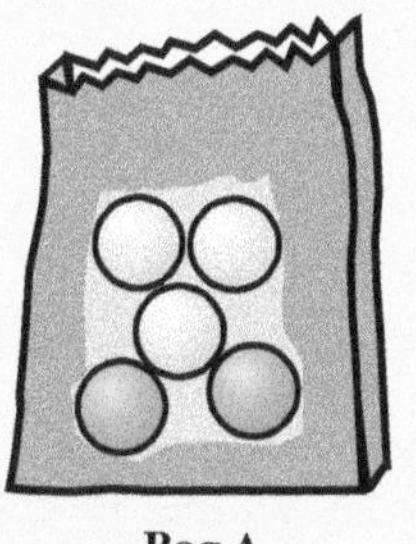
Bag A

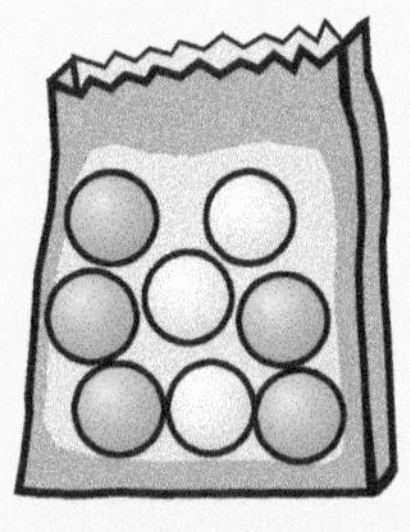
Bag B

Ros takes two balls at random from bag A and places them in bag B.
She then takes two balls at random from bag B and places them in bag A.
Calculate the probability that bag A now contains 5 white balls.

11. A restaurant offers a choice of a free bottle of red or white wine with lunchtime-special meals of chicken, fish or beef.
Analysis of orders over a month shows that:
   if a customer orders chicken then the probability of choosing white wine is 0.7,
   if a customer orders fish then the probability of choosing white wine is 0.8,
   if a customer orders beef then the probability of choosing red wine is 0.9.

In any one week approximately 800 customers order a lunchtime-special meal.
Of these, the probability that a customer orders chicken is 0.5 and the probability that a customer orders fish is 0.3.
Estimate the number of bottles of white wine that are required.

What you need to know

- You need to know the meaning of these terms: impossible, unlikely, evens, likely, certain.
- **Probability** describes how likely or unlikely it is that an event will occur.
  Probabilities can be shown on a probability scale.
  Probability **must** be written as a fraction, a decimal or a percentage.

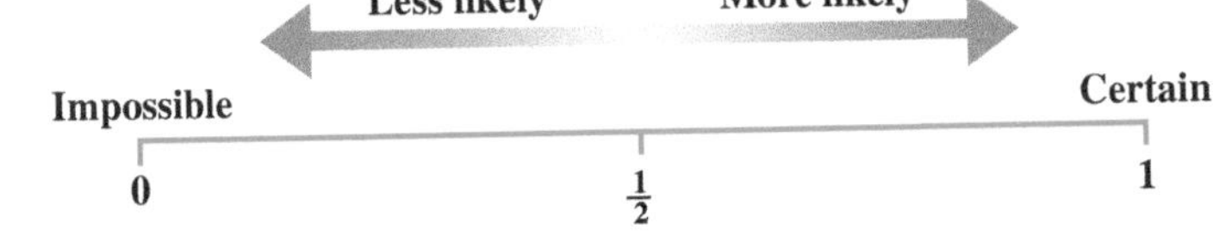

- How to work out probabilities using **equally likely outcomes**.
  The probability of an event X happening is given by:

$$\text{Probability (X)} = \frac{\text{Number of outcomes in the event}}{\text{Total number of possible outcomes}}$$

- How to estimate probabilities using **relative frequency**.
  The relative frequency of an event is given by:

$$\text{Relative frequency} = \frac{\text{Number of times the event happens in an experiment (or in a survey)}}{\text{Total number of trials in the experiment (or observations in the survey)}}$$

- How to use probabilities to **estimate** the number of times an event occurs in an **experiment** or **observation**.
  Estimate = total number of trials (or observations) × probability of event
- **Mutually exclusive events** cannot occur at the same time.
  When A and B are mutually exclusive events:

$$P(A \text{ or } B) = P(A) + P(B)$$

- A general rule for working out the probability of an event, A, **not happening** is:

$$P(\text{not } A) = 1 - P(A)$$

- How to find all the possible outcomes when two events are combined.
  By **listing** the outcomes systematically.
  By using a **possibility space diagram.**
  By using a **tree diagram.**
- The outcomes of **independent events** do not influence each other.
  When A and B are independent events:

$$P(A \text{ and } B) = P(A) \times P(B)$$

- **Conditional probabilities** arise when the probabilities of particular events occurring are affected by other events.

Review Exercise 40

1 Dennis has five coins in his pocket.
He has two 10p coins and three 2p coins.

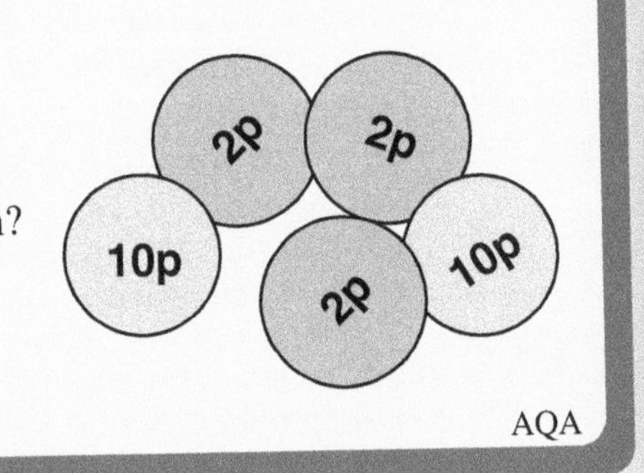

He chooses a coin at random.
(a) What is the probability that Dennis chooses a 10p coin?

Dennis puts the coin back in his pocket.
He chooses another coin at random.
(b) Is it more likely to be a 10p coin or a 2p coin?
Explain your answer.

AQA

2 The probability that someone gets flu next winter is 0.3.
What is the probability that someone **does not** get flu next winter?

AQA

3 Mandy has a bag containing packets of crisps.
The bag contains: 5 packets of plain crisps;
3 packets of salt and vinegar crisps;
4 packets of bacon flavour crisps.

Mandy takes one packet of crisps out of the bag at random.
(a) What is the probability that it is a packet of salt and vinegar crisps?
(b) What is the probability that the packet will **not** be bacon flavour crisps?

AQA

4 In a raffle 100 tickets are sold. Only one prize can be won.
(a) Nicola buys one ticket.
What is the probability that she wins the prize?
(b) Dee buys five tickets.
What is the probability that she wins the prize?
(c) Keith buys some tickets.
The probability that he wins the prize is $\frac{3}{20}$.
(i) What is the probability that he does **not** win the prize?
(ii) How many tickets did he buy?

AQA

5 A fair spinner has eight sides. The sides are numbered 1, 2, 2, 3, 3, 4, 5 and 6.
The spinner is spun once.
(a) What is the probability that the spinner lands on a 3?

Michael and Sheila play a game using the spinner.
The spinner is spun once.
Michael wins if the spinner lands on 1 or 2 or 3.
Sheila wins if it lands on 4 or 5 or 6.
(b) What is the probability that Sheila will win the game?
(c) Explain why this game is **not** fair.

AQA

6 A card is chosen at random from Set 1 and another card is chosen at random from Set 2.

Set 1 A B C **Set 2** A B C

(a) List all the possible outcomes.
(b) What is the probability that a C is not one of the two cards chosen.

AQA

7 The table shows information about a group of adults.

	Can drive	Cannot drive
Male	32	8
Female	38	12

(a) One of these adults is chosen at random.
What is the probability that the adult can drive?
(b) A man in the group is chosen at random.
What is the probability that he can drive?
(c) A woman in the group is chosen at random.
The probability that she can drive is 0.76.
What is the probability that she cannot drive?
(d) Does the information given support the statement: "More women can drive than men"?
Explain your answer.

AQA

8 Two ordinary six-sided dice are thrown.
What is the probability that two fives are thrown.

AQA

9 There are two parts to a driving test. The first part is a theory test.
You must pass the theory test before you take the practical test.
Rob takes his driving test.
The probability that he passes the theory is 0.8.
The probability that he passes the practical is 0.6.

(a) The tree diagram shows the possible outcomes.
Copy and complete the tree diagram.

Theory Test		Practical Test	
0.8	Pass	.....	Pass
		.....	Fail
.....	Fail		

(b) (i) Calculate the probability that Rob passes both parts of the driving test.
(ii) Calculate the probability that he fails the driving test. AQA

10 An office has two photocopiers, A and B.
On any one day: the probability that A is working is 0.8,
the probability that B is working is 0.9.

(a) Calculate the probability that, on any one day, both photocopiers will be working.
(b) Calculate the probability that, on any one day, only one of the photocopiers will be working. AQA

11 Boxes P and Q each contain five numbered balls.
The balls in each box are numbered as shown.
A ball is taken from each box at random.

Box P: 1, 3, 2, 5, 4 Box Q: 3, 1, 4, 2, 5

(a) What is the probability that both balls are numbered 2?
(b) What is the probability that both balls have the same number?
(c) What is the probability that the number on the ball from box P is greater than the number on the ball from box Q? AQA

12 A pack of 10 pens contains 5 blue, 3 red and 2 green. Two pens are selected at random.
What is the probability that **at least one** pen is red? AQA

13 A football team has to play two games.
The first game is played away. The second game is played at home.
The probability that the team will win the away game is 0.3.
If the team wins the away game, then the probability that it will win the home game is 0.6.
If the team does **not** win the away game, then the probability that it will win the home game is 0.45.
Calculate the probability that the team:

(a) wins both games,
(b) wins only one game. AQA

14 Simone wants to estimate the number of deer in a forest.
She catches 60 deer and puts an orange mark on each of them.
She then releases them back into the forest.
A week later she catches 30 deer and finds that 9 of them have an orange mark.
Estimate the number of deer in the forest. AQA

15 A bag contains ten counters. Four of the counters are red.
In an experiment, three counters are taken from the bag at random and put in a box.

(a) Calculate the probability that there are **exactly two** red counters in the box.

The same experiment is carried out 600 times.

(b) How many times would you expect there to be **at least two** red counters in the box? AQA

16 A bag contains 4 black buttons and 2 white buttons.
Two buttons are taken from the bag at random.
What is the probability that they are both the same colour?

AQA

Handling Data
Non-calculator Paper

Do not use a calculator for this exercise.

1 A group of students were each asked how many books they had read last month.
The frequency diagram shows the results.

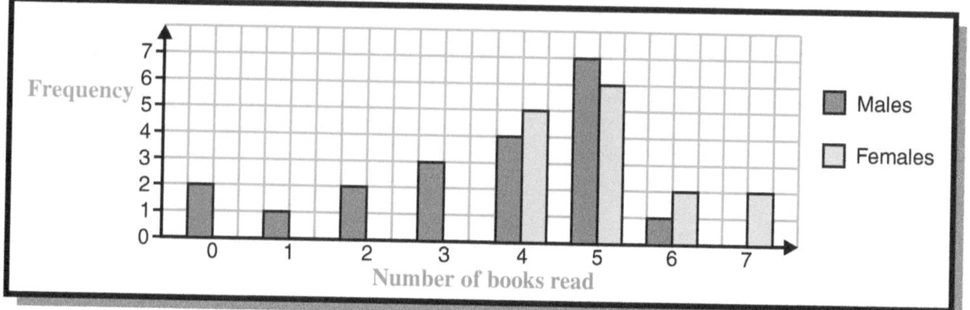

(a) What is the range in the number of books read by females?
(b) Calculate the mean number of books read by females.
(c) Compare and comment on the number of books read by males and the number of books read by females.

2 A fair six-sided dice and a fair coin are thrown together.
(a) List all the possible outcomes.
(b) What is the probability of getting a tail and a six?

3 In an opinion poll, 100 women in Manchester are asked how they intend to vote in a General Election.
(a) Give three reasons why this is an unreliable way of predicting the outcome of a General Election.
(b) Give three ways in which the opinion poll could be improved.

4 A box contains 20 plastic ducks. 3 of the ducks are green, 10 are blue and the rest are yellow.
A duck is taken from the box at random.
(a) What is the probability that it is green?
(b) What is the probability that it is yellow?

5 A spinner is labelled as shown.
The results of the first 30 spins are given below.

1 2 3 3 5 1 3 2 2 4 5 3 2 1 2
5 2 4 1 5 1 5 2 2 4 2 5 4 2 3

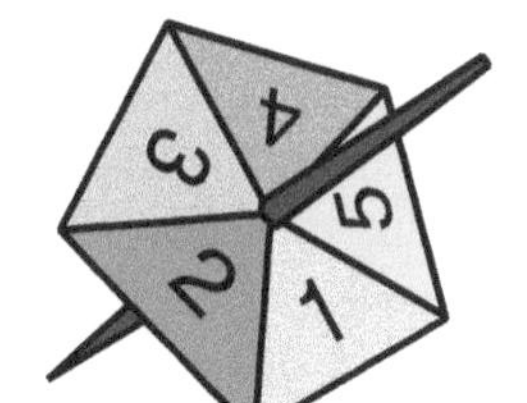

(a) What is the relative frequency of getting the number 1?
(b) Is the spinner fair?
Give a reason for your answer.

6 The table shows information about a group of children.

	Can swim	Cannot swim
Boys	16	4
Girls	19	6

(a) One of these children is chosen at random. What is the probability that the child can swim?
(b) A girl in the group is chosen at random. What is the probability that she cannot swim?
(c) Tony says, "These results show a higher proportion of girls can swim."
Is he correct? Give reasons for your answer.

7 The table shows the height and trunk diameter of each of 8 trees.

Height (m)	1.5	2.5	4	4.5	5.5	6	6.5	7
Trunk diameter (cm)	10	10	15	20	20	25	25	30

(a) Plot a scatter graph of these data.
(b) Draw a line of best fit through the points on the scatter graph.
(c) Describe the relationship shown in the scatter graph.
(d) A tree is 3 metres tall. The trunk diameter is given as 31 centimetres.
Explain why 13 centimetres is more likely.

AQA

8 Sue does a survey about the time that people spend watching TV and reading.
Two questions on her questionnaire are:

Question 1: How much do you read?
Question 2: There are more and more TV channels.
More and more TV programmes are being made.
We are watching more and more TV, so now we don't read enough.
Don't you agree?

(a) Explain why each question is **not** a good one for this questionnaire.
(b) Rewrite Question 1 so that it is more suitable for the questionnaire.

AQA

9 The maximum load for a lift is 1200 kg.
The table shows the distribution of the weights of 22 people waiting for the lift.
Will the lift be overloaded if all of these people get in?
You **must** show working to support your answer.

Weight (w kg)	Frequency
$30 \leqslant w < 50$	8
$50 \leqslant w < 70$	10
$70 \leqslant w < 90$	4

AQA

10 The weights of 80 bags of rice are measured.
The table summarises the results.

Minimum	480 g
Lower quartile	500 g
Median	540 g
Upper quartile	620 g
Maximum	720 g

(a) Draw a box plot to show this information.
(b) Write down the interquartile range for these data.
(c) How many bags weigh:
(i) less than 480 g,
(ii) less than 500 g?
(d) Draw a cumulative frequency diagram to show the information.

AQA

11 The diagram shows a fair five-sided spinner.
The spinner has two black sections and three white sections.
The sections are numbered as shown.

The spinner is spun twice.
(a) What is the probability of getting two black sections?
(b) What is the probability of getting a black section on the first spin and a 3 on the second spin?
(c) What is the probability of getting a black 3 and a white 3?

AQA

12 Simon kept a record of the length of time he spent on the Internet each day for 360 days.
The table summarises the results.

Time, t (minutes)	$0 \leqslant t < 10$	$10 \leqslant t < 30$	$30 \leqslant t < 60$	$60 \leqslant t < 120$
Frequency	50	100	90	120

(a) Draw a histogram to represent these times.
(b) Estimate the number of days that Simon was on the Internet for more than 50 minutes.

AQA

13 (a) Describe how to obtain a stratified random sample of 50 people from 380 men and 320 women.

(b) State **one** advantage of stratified random sampling over simple random sampling.

14 One hundred females were asked how many hours of television they watched in one week.

(a) This histogram shows the results.

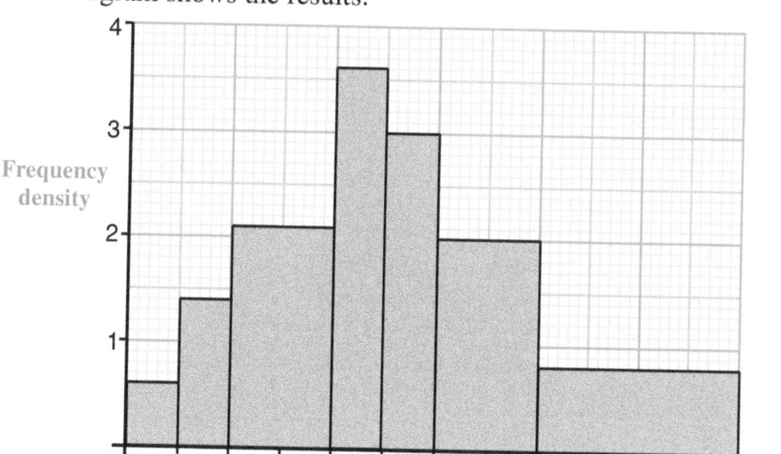

(i) How many of the females watched television for between 10 and 25 hours in the week?

(ii) Estimate the number of females who watched television for more than 35 hours in the week.

The one hundred females were chosen, at random, from customers leaving a supermarket on a Saturday morning.

(b) Give one reason why this method of choosing a sample is unsatisfactory. AQA

15 Jamil is playing a game with this set of cards.

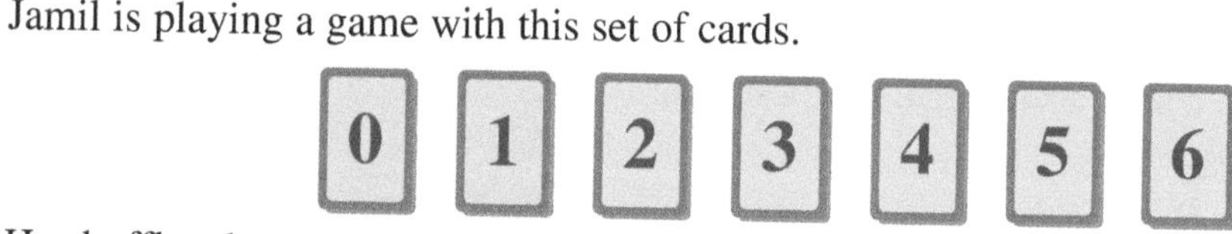

He shuffles the cards and turns one over.
He can then decide between:

Option A: Score the number on the card.
Option B: Turn over a second card from the remaining six.

If Option B is used:

When the number on the second card is less than the number on the first card the score is 0.
Otherwise the score is the number on the second card.

Jamil turns over the first card and it shows 4.
He chooses Option B.

(a) Copy and complete the table to show the scores that Jamil could now get, and their probabilities.

Score			
Probability			

(b) Jamil wants a high score.
His first card is a 4.
Would he be better to choose Option A or Option B?
You must explain your answer.

AQA

Handling Data Calculator Paper

You may use a calculator for this exercise.

1 The lengths, in centimetres, of a sample of leaves are shown.

4.7	5.0	6.4	5.6	6.1	4.9	5.8	6.5	7.2	6.7	6.5	5.3

(a) Calculate the mean length of these leaves.
(b) Draw a stem and leaf diagram to represent this information.

2 A class survey found the probability of having brown eyes is 0.6.
(a) What is the probability of not having brown eyes?
(b) There are 30 children in the class.
Estimate the number with brown eyes.

3 The two-way table shows the age and sex of a sample of 20 pupils at a school.

	Age (years)			
	12	13	14	15
Number of boys	4	3	3	2
Number of girls	2	2	1	3

There are 1000 pupils in the school altogether.
(a) Use the values in the table to estimate the number of boys in the school.
(b) How could a better estimate be obtained?

AQA

4 Jamal, Des, William and Sue play a board game.
There is only one winner.
The probability that Jamal wins is 0.3.
The probability that Des wins is 0.1.
(a) Calculate the probability that either William or Sue wins.
(b) William is twice as likely to win as Sue.
What is the probability that Sue wins?

AQA

5 The two-way table shows the results of a survey among adults at an out of town supermarket.

	Can drive	Cannot drive
Male	22	4
Female	15	4

A town has 27 953 adults.
(a) Give a reason why the sample may not be representative of the town.
(b) Use the results of the survey to estimate the number of adults in the town who can drive.
Give your answer to a suitable degree of accuracy.

AQA

6 A fair coin is thrown 20 times. It lands heads 12 times.
(a) What is the relative frequency of throwing a head?

The coin continues to be thrown.
The table shows the number of heads recorded for 20, 40, 60, 80 and 100 throws.

Number of throws	20	40	60	80	100
Number of heads	12	18	30	42	49

(b) Draw a graph to show the relative frequency of throwing a head for these data.
(c) Estimate the relative frequency of throwing a head for 1000 throws.

AQA

7 The table shows the ages and weights of chickens.

Age (days)	10	20	40	50	70	80	100
Weight (g)	100	300	1000	1300	2000	2000	2400

(a) Use this information to draw a scatter graph.
(b) Describe the correlation between the age and weight of these chickens.
(c) Draw a line of best fit.
(d) Explain how you know the relationship is quite strong.

8 A survey was made of the time taken by 100 people to do their shopping at a local supermarket on a Monday.

Time taken (t minutes)	Number of people
$0 \leqslant t < 10$	43
$10 \leqslant t < 20$	35
$20 \leqslant t < 30$	17
$30 \leqslant t < 40$	5

(a) Calculate an estimate of the mean time taken by these people to do their shopping.
(b) Draw a frequency polygon to illustrate this information.

9 Nick travels to work by train on two days.
The probability that the train is late on any day is 0.3.
(a) Draw a tree diagram to show all the possible outcomes for the two days.
(b) What is the probability that the train is late on at least one of the two days?

10 A college records the number of people who sign up for adult educational classes each term.
The table shows the numbers from Autumn 2004 to Summer 2006.

Term	Autumn 2004	Spring 2005	Summer 2005	Autumn 2005	Spring 2006	Summer 2006
Number of people	520	300	380	640	540	500

(a) Calculate the first value of the three-point moving average for these data.
(b) Explain why a **three-point** moving average is appropriate.

AQA

11 Pupils in Year 7 are all given the same test.
The test is marked out of 100.
The results for the boys are:

Lowest score	7 marks
Highest score	98 marks
Lower quartile	36 marks
Median	53 marks
Upper quartile	66 marks

0 10 20 30 40 50 60 70 80 90 100
Girls

(a) The diagram shows the box plot for the girls' results.
Draw a box plot to show the information for the boys.
(b) Comment on the differences between the boys' and girls' results.

AQA

12 Sue has two bags of sweets.
Bag A contains 6 toffees and 3 chocolates.
Bag B contains 5 toffees and 2 chocolates.
Sue takes one sweet from each bag.
What is the probability that she gets two sweets of the same sort?

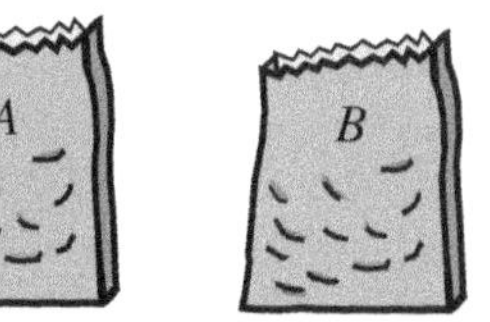

AQA

13 The number of pupils in each of three schools is given in the table.

School	Number of pupils
Ashworth Comprehensive School	1500
Brigholm Grammar School	640
Corchester School	360

A stratified random sample of 200 pupils is to be taken.
Calculate the number of pupils that should be chosen from each of the schools. AQA

14 The histogram summarises the marks out of 100 obtained on a test by a group of students.

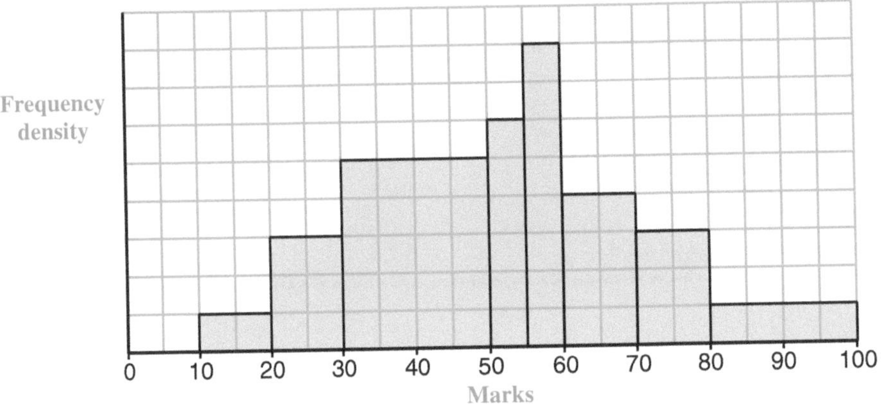

Twelve students scored between 70 and 80 marks.

(a) How many students scored between 80 and 100 marks?
(b) How many students took the test?
(c) Estimate the mean score of the students on this test. AQA

15 The speeds of 100 cars travelling along a road are shown in this table.

Speed (s km/h)	$20 \leqslant s < 35$	$35 \leqslant s < 45$	$45 \leqslant s < 55$	$55 \leqslant s < 65$	$65 \leqslant s < 85$
Frequency	6	19	34	26	15

(a) Draw a histogram to show this information.
(b) The speed limit along this road is 48 km/h.
Estimate the number of cars exceeding the speed limit. AQA

16 Sarah has a part-time job and works on Mondays and Tuesdays.
She often oversleeps and arrives late for work.
She has estimated the probability that she oversleeps on Monday to be 0.25.
If Sarah oversleeps on Monday, the probability that she will oversleep on Tuesday is 0.16, but if she does not oversleep on Monday, the probability that she oversleeps on Tuesday is 0.35.

(a) Calculate the probability that Sarah does not oversleep on either day.
(b) Calculate the probability that Sarah oversleeps on Tuesday. AQA

17 Steve has 6 socks in a drawer.
4 of the socks are black and 2 of the socks are white.

Apart from the colour, the socks are identical.
He takes 2 socks out of the drawer at random.
What is the probability that he gets a pair of socks of the same colour? AQA

Non-calculator Paper

Do not use a calculator for this exercise.

1 In the diagram, PQR is an isosceles triangle.
The lines PQ and RS are parallel.
(a) Work out the size of angle x.
(b) (i) What is the size of angle y?
(ii) Give a reason for your answer.

2 (a) What is the value of $2x^2$ when $x = -5$?
(b) Simplify $3t - t - 3$.
(c) Solve (i) $2(x - 3) = 8$, (ii) $3t + 1 = 7 - t$.

3 The graph shows Zak's journey to work.
He starts on the motorway but is held up by an accident.
The police move traffic onto side roads where Zak continues his journey.
He arrives at work at 8.45 am.

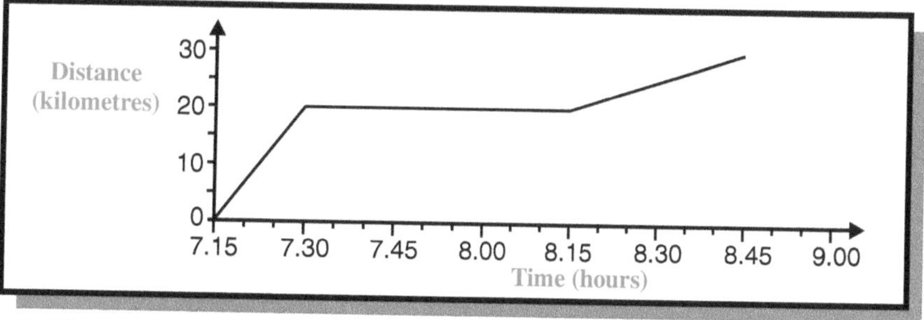

(a) What is his average speed before he is held up by the accident?
(b) How long was he held up by the accident?
(c) How far was his journey in kilometres?
(d) What was Zak's average speed for the whole journey?

AQA

4 A circle has a radius of 4.87 cm.
Use approximation to estimate the area of the circle in terms of π.

5 Two girls, Anne and Margaret, and two boys, Brian and Colin, play a computer game.
Each game has only one winner. The probability that Anne wins is 0.3.
The probability that Brian wins is 0.15.
The probability that Colin wins is 0.45.
(a) What is the probability that Colin does not win?
(b) What is the probability that Margaret wins?
(c) What is the probability that one of the boys wins?

AQA

6 Simplify. (a) $a^2 \times a^3$ (b) $b^6 \div b^2$ (c) $(c^3)^2$

7 Phil has 80 birds; some are blue, and the rest are yellow.
Phil sells 30% of his birds. The new ratio of blue birds to yellow birds is 4 : 3.
How many blue birds has he got left?

AQA

8 Use approximations to estimate the value of $\dfrac{316 \times 4.03}{0.198}$

AQA

9 Find the nth term of the following sequences. (a) 5, 8, 11, 14, … (b) 2, 5, 8, 11, …

10 (a) Factorise the expression $2n^2 - n$.
(b) Multiply out and simplify $3(1 - x) - x(x + 1)$.
(c) Solve the equation $2(x + 5) = 3(2 - x)$.

Exam Practice: Non-calculator Paper

EP

11 The diagram shows the positions of shapes $\boldsymbol{T}$, $\boldsymbol{M}$ and $\boldsymbol{N}$.

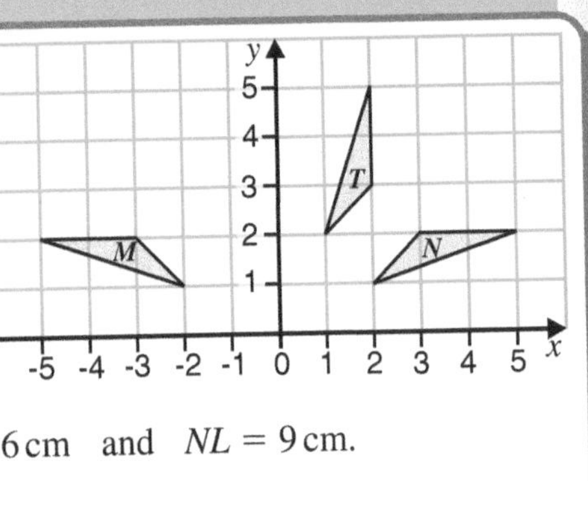

(a) Describe fully the single transformation which maps $\boldsymbol{T}$ onto $\boldsymbol{M}$.

(b) Describe fully the single transformation which maps $\boldsymbol{T}$ onto $\boldsymbol{N}$.

(c) $\boldsymbol{M}$ is mapped onto $\boldsymbol{P}$ by an enlargement, scale factor 2, centre $(-5, 1)$.
Draw a diagram to show the positions of $\boldsymbol{M}$ and $\boldsymbol{P}$.

12 LMN is a right-angled triangle. $\angle LMN = 90°$, $MN = 6\,\text{cm}$ and $NL = 9\,\text{cm}$.
Calculate the **exact** length of LM.

13 $ABCD$ is a rhombus with sides of length 6 cm. $\angle ABC = 120°$.

(a) Draw accurately the rhombus $ABCD$.

(b) A point X is (i) equidistant from BC and CD, and (ii) equidistant from C and D.
By drawing the loci of (i) and (ii) mark the position of X on your diagram.

(c) Write down the ratio of $CX : XA$ in its simplest form.

14 (a) Express 96 as a product of its prime factors.

(b) Find the Highest Common Factor (HFC) of 36 and 96. AQA

15 (a) Solve the equation $\frac{m}{2} - \frac{m}{3} = 5$

(b) Multiply out (i) $3y(x^2 - 2y)$ (ii) $(x - 2)(3x + 5)$

16 In triangle LMN, $\cos L = \frac{3}{5}$.
Calculate the length of NL.

N L M 9 cm

17 You are given the formula $v = u + at$.

(a) Work out the value of v when $u = 20$, $a = -6$ and $t = \frac{9}{5}$.

(b) Rearrange the formula to give t in terms of v, u and a. AQA

18 (a) List the values of n, where n is a whole number, such that $3 < 2n + 1 \leqslant 7$.

(b) Solve the simultaneous equations $x + 2y = -1$ and $4x - 2y = -9$.

(c) Solve the equation $x^2 - 5x = 0$.

19 The cumulative frequency table gives the age distribution of people living in a village.

Age (years)	< 10	< 20	< 30	< 40	< 50	< 60	< 70	< 80	< 100
Cumulative frequency	5	30	70	120	180	260	350	380	400

(a) Draw a cumulative frequency graph to illustrate the data.

(b) Estimate the number of people less than 25 years old.

(c) Use your graph to estimate: (i) the median, (ii) the interquartile range.

(d) The age distribution of people living in a town has a median of 45 years and an interquartile range of 20 years.
Compare and comment on the ages of people living in the village and the town. AQA

20 The table shows values for the equation $y = x^2 - 2x - 4$.

x	−3	−2	−1	0	1	2	3	4	5
y	11	4	−1	−4	−5	−4	−1	4	11

(a) Draw the graph of $y = x^2 - 2x - 4$ for values of x from -3 to $+5$.

(b) Use the graph to write down the solutions of the equation $x^2 - 2x - 4 = 0$.

(c) Write down the value of k for which the equation $x^2 - 2x - 4 = k$ has just one solution.

(d) By drawing an appropriate straight line graph on the grid, solve the equation $x^2 - x - 5 = 0$. AQA

21 Work out. (a) $(5 \times 10^4) \times (8 \times 10^6)$ (b) $\frac{2 \times 10^4}{8 \times 10^6}$
Give your answers in standard form.
AQA

22 (a) Simplify (i) $5p^3 \times 3p^2$, (ii) $8p^6 \div 4p^3$, (iii) $\frac{x^2 - 4}{x^2 + 2x}$.
(b) Factorise (i) $3x^2 - 6xy$, (ii) $2n^2 - 5n + 3$.

23 (a) These two triangles are congruent.
Write down the values of x and y.
(b) The length of the side PQ was given as 9 centimetres.
This was measured to the nearest millimetre.
What is the maximum length of PQ?
Give your answer in millimetres.
AQA

24 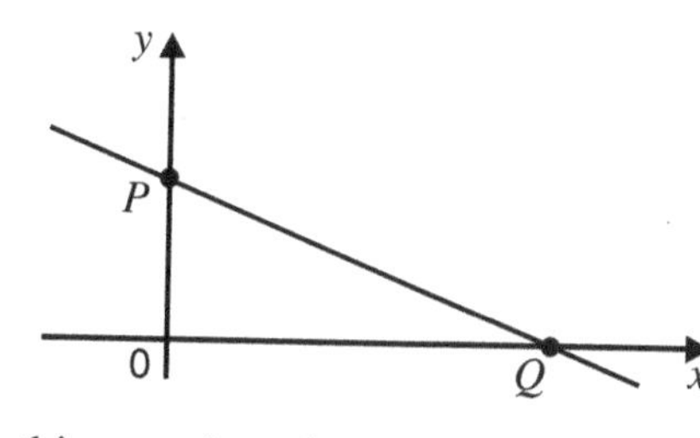

In this sketch graph, the line $3y + 2x = 12$ crosses the axes at the points P and Q.
(a) Write down the coordinates of P.
(b) Another line, l, also goes through the point P.
The gradient of line l is 2.
Write down the equation of the line l.
AQA

25 In this question, the letters a, b and c represent lengths.
(a) Which of the following expressions **cannot** represent area?
πab $\frac{1}{2}c^2$ $a^2 + b^2$ $\frac{4}{3}abc$ $b(a + c)$
(b) Ilesh has worked out this formula Volume $= 2\pi(a^3 + bc)$
Explain how you know that he has made a mistake.
AQA

26 Solve these equations (a) $4y - 9 = 2(5 - 3y)$ (b) $x^2 - 6x + 8 = 0$
AQA

27 You are give the formula $P = \left(\frac{m}{n}\right)^2$
(a) Calculate the value of P when $m = -0.12$ and $n = \frac{2}{5}$.
(b) Rearrange the formula to give n in terms of P and m.

28 Triangle PQR has vertices at $P(-5, -3)$, $Q(-1, -3)$ and $R(-1, -1)$.
The triangle is enlarged with centre (3, 2) and scale factor $-\frac{1}{2}$.
On a single diagram draw triangle PQR and its enlargement.

29 State **two** conditions that must be satisfied when collecting data for a stratified sample.
AQA

30 (a) Rationalise the denominator and simplify fully $\frac{1}{\sqrt{12}}$
(b) By simplifying $\sqrt{32} - \sqrt{18}$, write $\sqrt{3}(\sqrt{32} - \sqrt{18})$ in its simplest form.
AQA

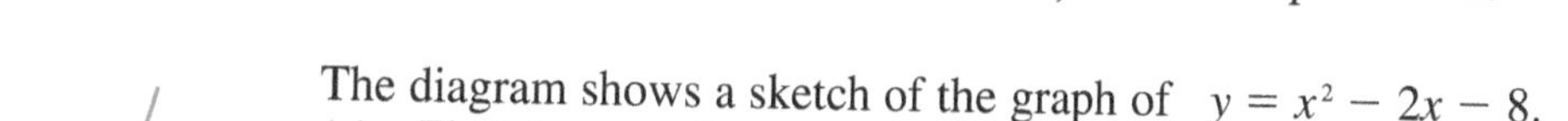

31 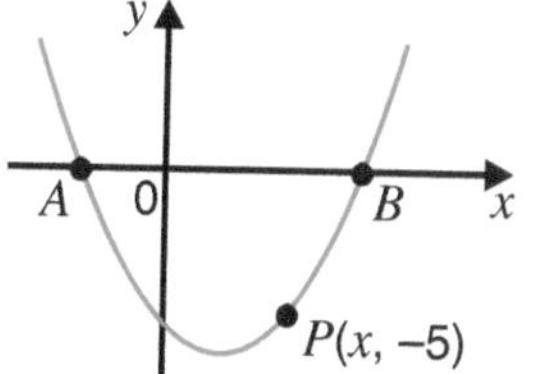

The diagram shows a sketch of the graph of $y = x^2 - 2x - 8$.
(a) Find the coordinates of points A and B.
(b) The point P has coordinates $P(x, -5)$.
Find the value of x.

32 The diagram shows the circle $x^2 + y^2 = 25$ and the line $y = x + 7$.
The line and the circle intersect at the points A and B.
(a) By substituting the equation $y = x + 7$ into the equation $x^2 + y^2 = 25$, show that $x^2 + 7x + 12 = 0$.
(b) Hence, find the coordinates of A and B.
AQA

Exam Practice: Non-calculator Paper

33 In triangle PQR, $PQ = 4\,\text{cm}$, $QR = 5\,\text{cm}$ and $PR = 7\,\text{cm}$.
Use the cosine rule to show that the angle PQR is greater than $90°$.

34 (a) Work out $49^{\frac{1}{2}} \times 5^{-3}$. Give your answer as a fraction.
(b) Calculate $\frac{4^7}{4^{-2}}$, giving your answer in the form 2^n.
(c) Work out the value of $81^{-\frac{3}{4}}$. Give your answer as a fraction. AQA

35 In the diagram the two circles intersect at points A and B.
PBS, PQX and SRX are straight lines.
Prove that $XQAR$ is a cyclic quadrilateral.

36 A solid cone has a height of 12 cm and a slant height of 13 cm.
Calculate the **total** surface area of the cone.
Give your answer in terms of π.

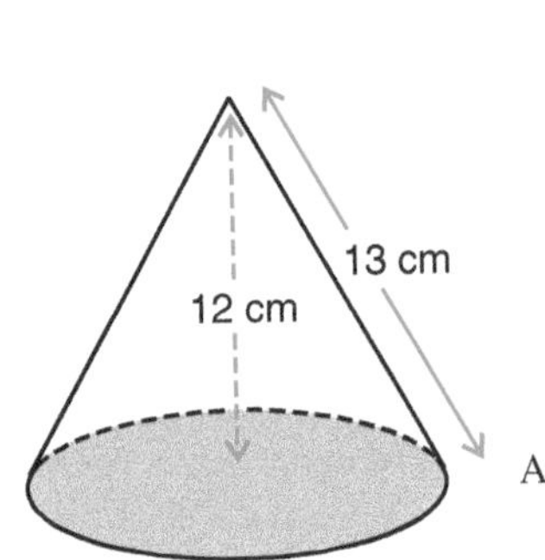

AQA

37 Shereen has two bags of marbles.
Bag A contains 3 red marbles and 4 green marbles.
Bag B contains 2 red marbles and 3 green marbles.
Shereen throws a fair six-sided dice.
If the dice lands on a six, she takes a marble at random from bag A.
If the dice lands on any other number, she takes a marble at random from bag B.
(a) Draw a fully labelled tree diagram showing the above information.
(b) Calculate the probability that a red marble is selected. AQA

38 Solve the equation $x^2 + 4x + 2 = 0$.
Express your answers in the form $p \pm q\sqrt{2}$, where p and q are integers. AQA

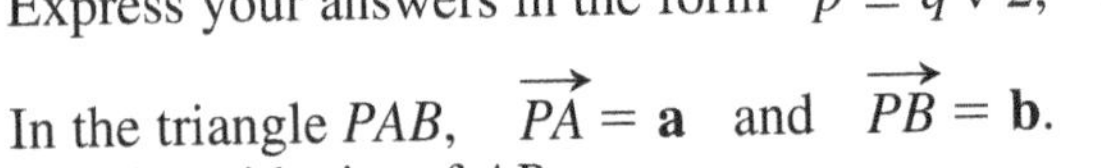

39 In the triangle PAB, $\overrightarrow{PA} = \mathbf{a}$ and $\overrightarrow{PB} = \mathbf{b}$.
X is the midpoint of AB.
(a) Express $\overrightarrow{AB}$ in terms of $\mathbf{a}$ and $\mathbf{b}$.
(b) Express $\overrightarrow{PX}$ in terms of $\mathbf{a}$ and $\mathbf{b}$.
X is also the midpoint of the line PY.
(c) Show that $\overrightarrow{PY} = \mathbf{a} + \mathbf{b}$.
(d) Find an expression for $\overrightarrow{BY}$.
(e) Explain why your answer to (d) proves that $AYBP$ is a parallelogram. AQA

40 The diagrams, which are not drawn to scale, show two transformations of the graph $y = x^3 - 5x$.
In each case the points A and B are transformed into the points A' and B'.
Write down the equation of each transformed graph.

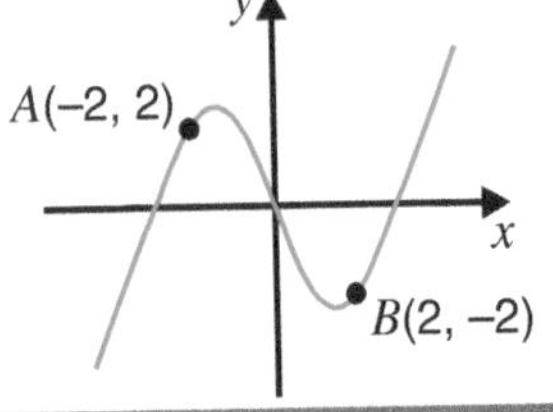
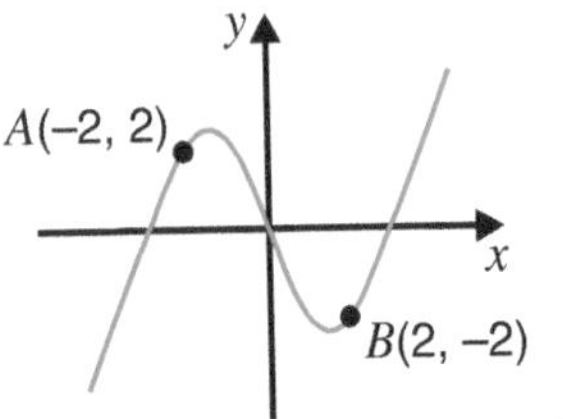

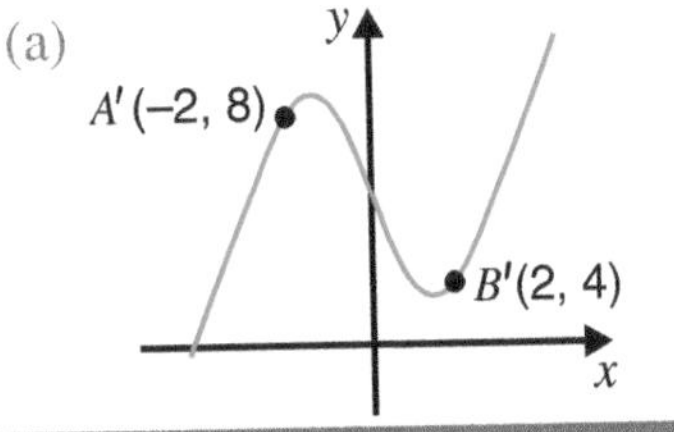

(b)

AQA

Exam Practice

Calculator Paper

You may use a calculator for this exercise.

1 A garden is a rectangle measuring 26 m by 11.5 m. Grass covers 67% of the area of the garden. Calculate the area of grass. Give your answer to a suitable degree of accuracy.

2 A shop sells flour in two sizes.
Which size gives better value for money?
You **must** show all your working.

Size 1: weight 500 g, cost 39 pence.
Size 2: weight 800 g, cost 59 pence.

3 A rucksack costs £36 plus VAT at $17\frac{1}{2}\%$.
What is the total price of the rucksack?

4 Calculate $3.4^2 + \sqrt{3.4}$. Give your answer correct to 1 d.p.

5 A penny farthing bicycle is shown.
(a) The large wheel has a diameter of 160 cm.
Find the circumference of the wheel.
(b) The small wheel has a circumference of 137 cm.
The large wheel rotates once.
How many complete rotations does the small wheel make?

AQA

6 Change $0.6\,m^2$ to cm^2.

7 A bowl of strawberries and cream weighs 210 grams.
The ratio, by weight, of strawberries to cream is 5 : 1.
What is the weight of the cream?

8 The diagram shows the shape of a playground in a park.
Calculate the area of the playground.

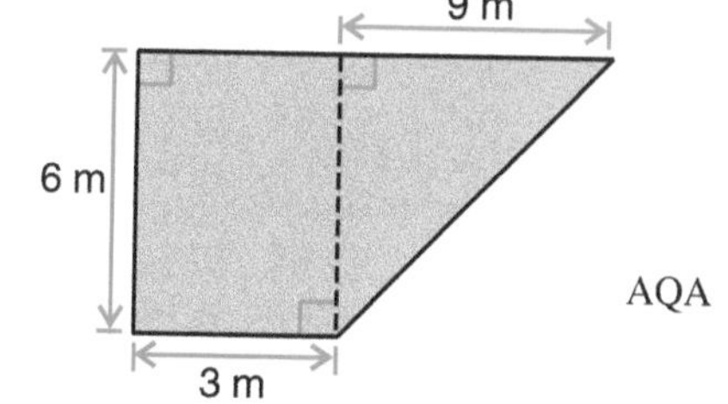

AQA

9 (a) Solve (i) $5x + 7 = 2$,
(ii) $3x - 9 = 7 - 2x$.
(b) Factorise $3a - 6b$.
(c) Simplify $5x - 2(x + 3)$.

10 A candle weighs x grams.
(a) Write an expression, in terms of x, for the weight of 20 candles.
(b) A box of 20 candles weighs 3800 g. The box weighs 200 g.
By forming an equation, find the value of x.

11 A police officer records the speeds of 60 cars on a dual carriageway.

Speed (mph)	40 to less than 50	50 to less than 60	60 to less than 70	70 to less than 80
Frequency	9	27	21	3

(a) Write down the modal class.
(b) Use the class midpoints to calculate an estimate of the mean speed of these cars. AQA

12 Mike took 400 books to sell at a Saturday market.
By 3 pm, he had sold 310 books at 80 pence each.
Mike then reduced the selling price of the remaining books to 50 pence each.
He was left with 24 unsold books which he gave away.
(a) Find the total amount Mike received from selling the books.
(b) Mike had spent £150 buying the books. What was Mike's percentage profit? AQA

EP

13 (a) Calculate $\dfrac{89.6 \times 10.3}{19.7 + 9.8}$

(b) By using approximations show that your answer to (a) is about right.
You must show all your working. AQA

14 Harry drives 182 miles. His average speed is 35 miles per hour.
How long does the journey take? Give your answer in hours and minutes. AQA

15 (a) Write 36 as a product of its prime factors.

(b) Find the least common multiple of 36 and 45.

16 A sequence begins: 5, 3, 1, −1, …
Write an expression, in terms of n, for the nth term of the sequence.

17 (a) This diagram shows a kite $ABCD$.
$AB = 15$ cm, $BC = 36$ cm and $AC = 39$ cm.
Explain why angle $B = 90°$.

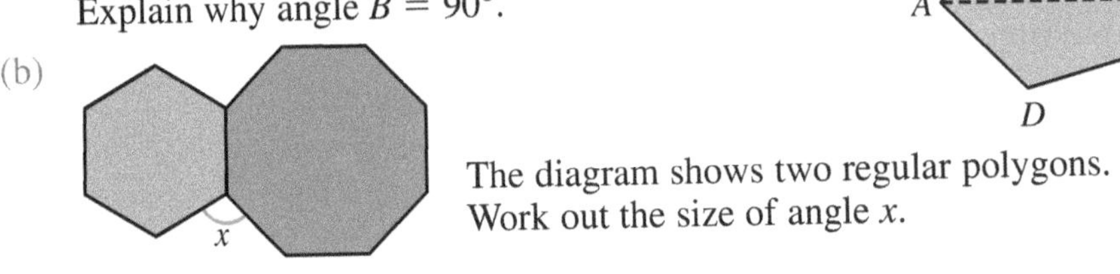

(b) The diagram shows two regular polygons.
Work out the size of angle x. AQA

18 (a) Solve (i) $\dfrac{x-7}{3} = -2$, (ii) $8 - 3x = 2(x + 5)$.

(b) Use trial and improvement to solve the equation $x^3 + 2x = 400$.
Show all your trials. Give your answer correct to one decimal place.

19 £4500 is invested at 3.2% compound interest per annum.
How many years will it take for the investment to exceed £5000? AQA

20 A water trough has a semi-circular cross-section, as shown.
The diameter of the end of the trough is 38 cm.
The trough is 3 m long.
Calculate the volume of water in the trough when it is full.
Give your answer in litres. AQA

21 (a) Solve the inequality $3m - 5 > 7$.

(b) Write down all the solutions of the inequality $-6 < 3n \leqslant 5$ where n is an integer.

(c) Solve the simultaneous equations $3x - 2y = 15$ and $2x - 4y = 2$.

22

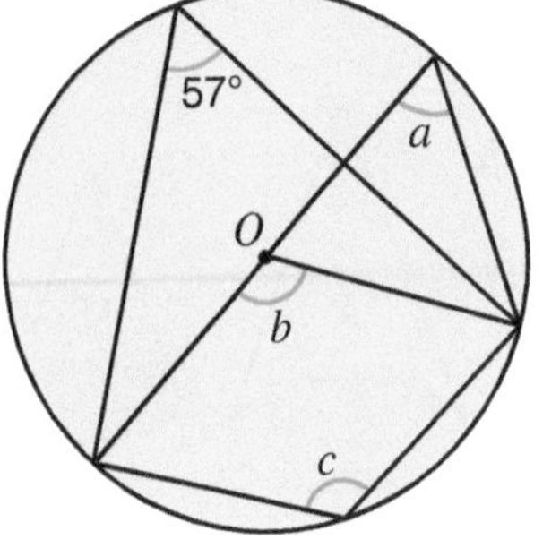

O is the centre of the circle.
Work out the size of angles a, b and c.

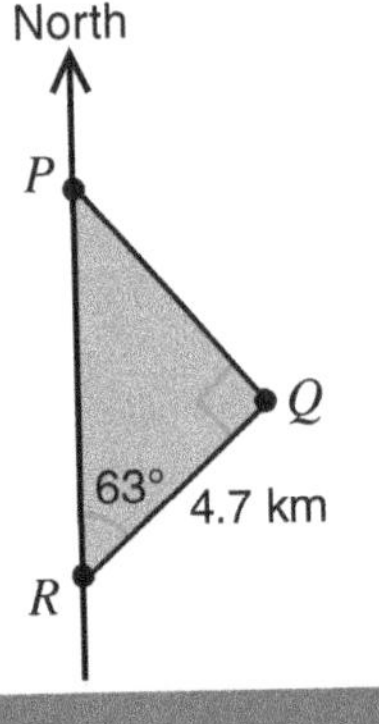

23 A piece of land is bounded by three straight roads PQ, QR and RP.
P is due north of R. Q is on a bearing of 063° from R.
PQR is a right angle.

(a) What is the bearing of P from Q?

The length of QR is 4.7 km.

(b) Calculate the area of the piece of land. AQA

24 Philip and Abdul run in different races.
The probability that Philip wins his race is 0.7.
The probability that Abdul wins his race is 0.6.

(a) Copy and complete the tree diagram.

(b) Calculate the probability that only one of the boys wins his race.

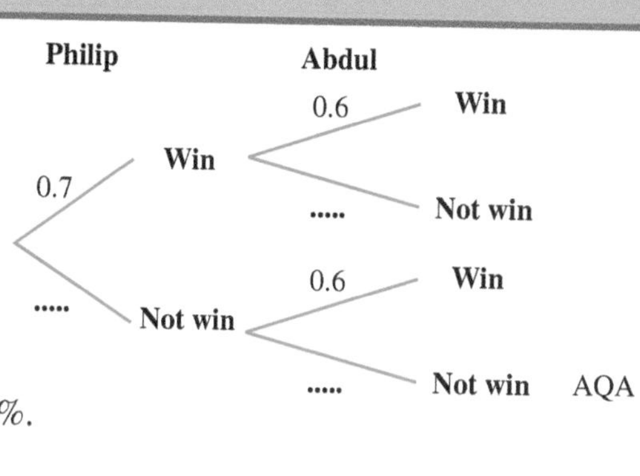

AQA

25 A ring costs £329. The price includes VAT at $17\frac{1}{2}\%$.
How much is the VAT?

26 (a) Find the equation of the straight line drawn through the points $(0, -4)$ and $(4, 8)$.
(b) Draw a graph to show the region where: $3x + y \geqslant 6$, $x \leqslant 2$ and $y \leqslant 4$.

27 AOB is a sector of a circle, centre O.
The radius of the circle is 12 cm.
Angle $AOB = 72°$.
Calculate the area of the shaded segment.

28 (a) The expression $x^2 - 10x + a$ can be written in the form $(x + b)^2$.
Find the values of a and b.
(b) Solve the equation $x^2 - 10x + 20 = 0$. Give your answers to two decimal places.
(c) State the minimum value of y if $y = x^2 - 10x + 20$.

AQA

29 Glass marbles are made in different sizes. The weight of each marble, W grams, is directly proportional to the cube of its radius, r millimetres.
A marble of weight 9 g has a radius of 9 mm.
(a) Find an equation connecting W and r.
(b) Calculate the weight of a marble with a radius of 6 mm.
(c) Another marble weighs 20 g. Calculate its radius.

30 (a) Express $0.\dot{4}\dot{2}$ as a fraction in its simplest form.
(b) Hence, or otherwise, express $0.7\dot{4}\dot{2}$ as a fraction in its simplest form.

AQA

31

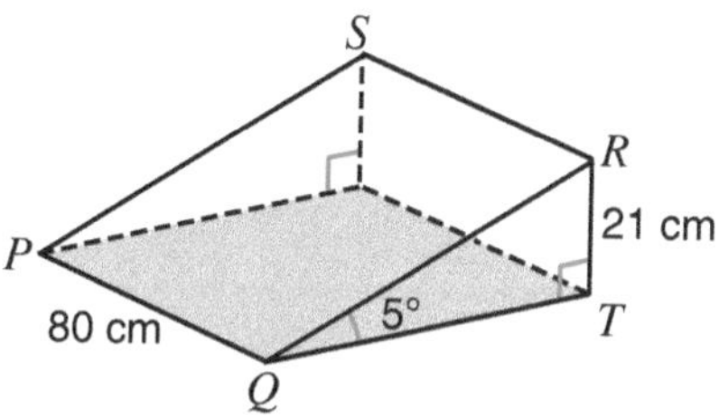

The diagram shows a rectangular ramp, $PQRS$, inclined at 5° to the horizontal.
The height, RT, of the ramp is 21 cm.
The ramp is 80 cm wide.
Calculate the length of the diagonal PR.

AQA

32 (a) The volume of a cone is given by the formula $V = \frac{1}{3}\pi r^2 h$.
Rearrange the formula to give r, in terms of V and h.
(b) These cones are similar.
The ratio of the height of P to the height of Q is 3 : 2.
The volume of P is $5.4\,\text{cm}^3$.
Calculate the volume of Q.

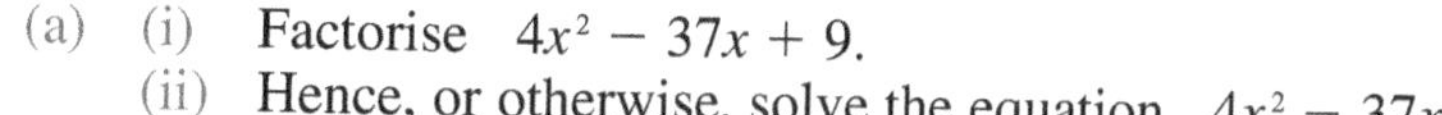

AQA

33 (a) (i) Factorise $4x^2 - 37x + 9$.
(ii) Hence, or otherwise, solve the equation $4x^2 - 37x + 9 = 0$.
(b) By considering your answers to part (a), find all the solutions to the equation $4y^4 - 37y^2 + 9 = 0$.

AQA

34 Angles x and y lie between 0° and 360°.
(a) Find the **two** values of x which satisfy the equation $\cos x = 0.5$.
(b) Solve the equation $\sin y = \cos 240°$.

AQA

35 The table shows the age, in years, of workers in a factory.

Age, x (years)	$15 \leqslant x < 20$	$20 \leqslant x < 25$	$25 \leqslant x < 30$	$30 \leqslant x < 40$	$40 \leqslant x < 60$
Number of workers	4	10	6	22	8

(a) Draw a histogram to represent these ages.
(b) Calculate an estimate of the number of workers who are aged under 21. AQA

36 Solve the equation $2x^2 - 3x - 1 = 0$, giving your answers correct to 2 d.p.

37 A water tank is 50 cm long, 34 cm wide and 24 cm high.
It contains water to a depth of 18 cm.

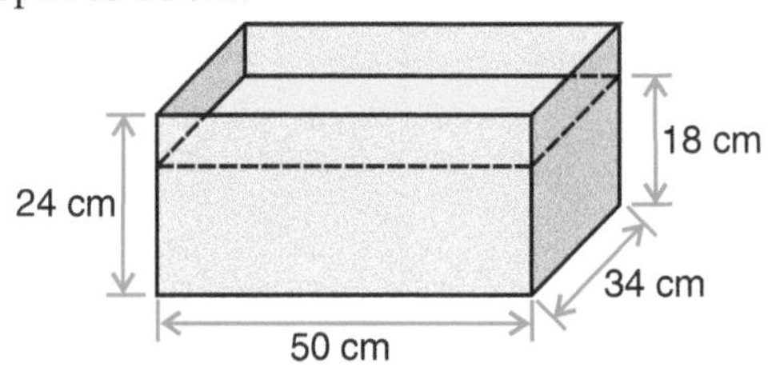

Four identical spheres are placed in the tank and are fully submerged.
The water level rises by 4.5 cm.
Calculate the radius of the spheres. AQA

38

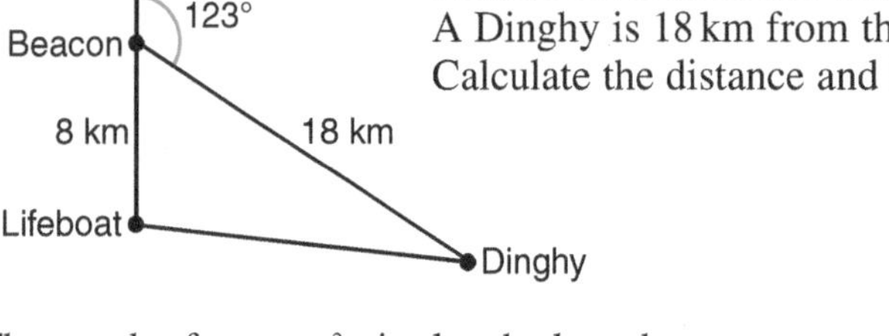

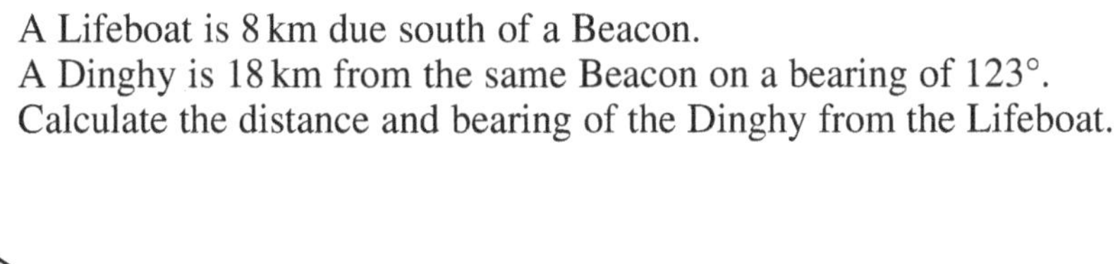
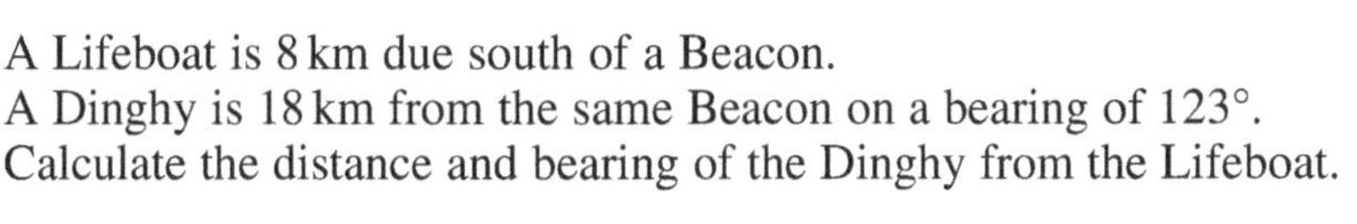

A Lifeboat is 8 km due south of a Beacon.
A Dinghy is 18 km from the same Beacon on a bearing of 123°.
Calculate the distance and bearing of the Dinghy from the Lifeboat.

39 The graph of $y = x^2$ is sketched on the axes.
Copy the diagram.
On the same axes sketch and label the graphs of
(a) $y = (x - 2)^2$, (b) $y = x^2 - 2$. AQA

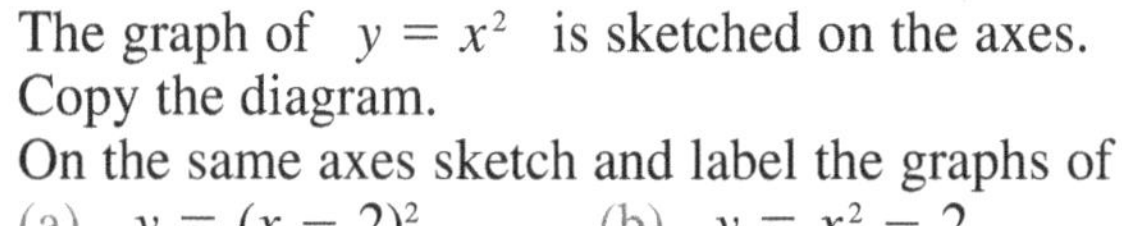

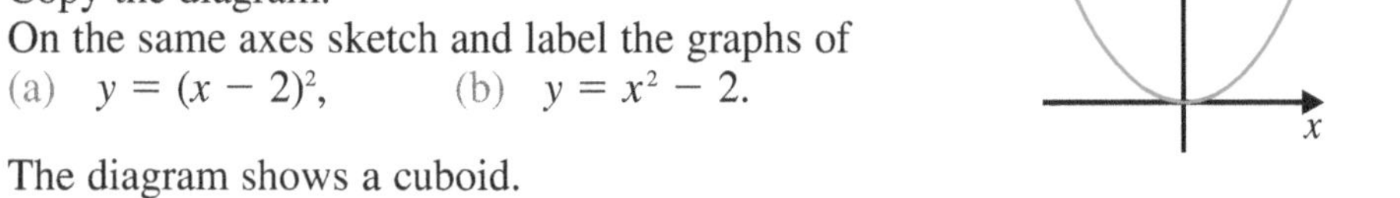

40 The diagram shows a cuboid.

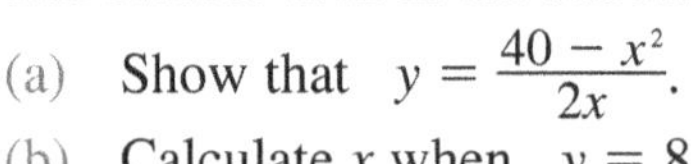

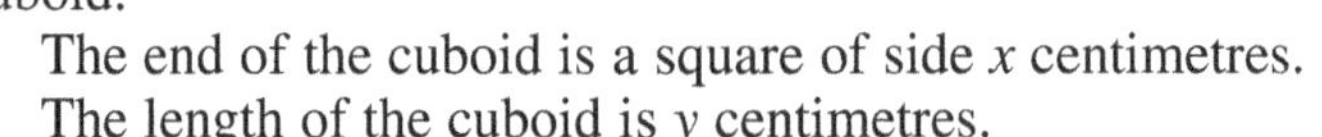

The end of the cuboid is a square of side x centimetres.
The length of the cuboid is y centimetres.
The surface area of the cuboid is 80 cm².

(a) Show that $y = \frac{40 - x^2}{2x}$.
(b) Calculate x when $y = 8$.
Hence, calculate the volume of the cuboid.
Give your answer correct to one decimal place. AQA

41 $ABCD$ is a rectangular region of length 10.2 m and area 55.25 m².

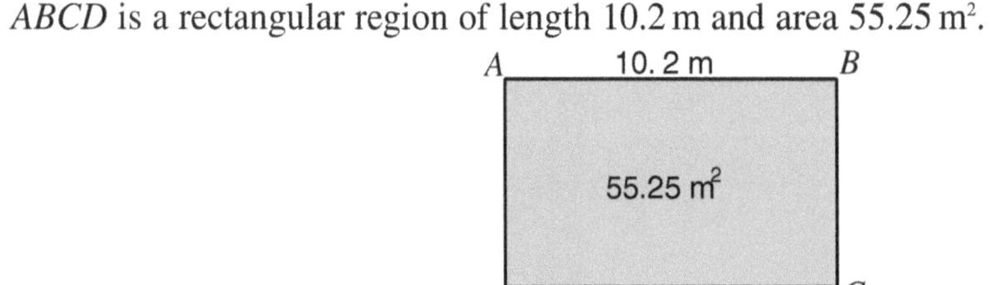

The length is given to the nearest 10 cm.
The area is given to the nearest 0.05 m².
Calculate the lower bound of the width of the region. AQA

Answers

CHAPTER 1

Exercise 1.1 — Page 1

1. (a) 546 (b) 70 209 (c) 1 200 052
2. (a) twenty-three thousand five hundred and ninety
   (b) six thousand and forty-nine
   (c) ninety-three million one hundred and forty-five thousand six hundred and seventy
   (d) nine million eighty thousand and four
3. 7 in 745
4. 3801, 3842, 3874, 4765, 5814
5. 9951, 9653, 9648, 9646, 9434
6. (a) 6512 Digits largest to smallest with the smallest even digit last.
   (b) 1265 Digits smallest to largest with the largest odd digit last.
7. (a) 27 (b) 35 (c) 74 (d) 70
   (e) 112 (f) 17 (g) 49 (h) 105
   (i) 575 (j) 4
8. (a) 952 (b) 2002 (c) 12 203 (d) 1541
   (e) 158 (f) 469 (g) 6268 (h) 3277
9. 40 710
10. 1030 grams
11. 81 030
12. 31 runs
13. 89
14. (a) 17p (b) 6p
15. Car C.
    Car A 8479, Car B 11 643, Car C 13 859

Exercise 1.2 — Page 3

1. (a) 549
   (b) 252
   (c) 2112
   (d) 15 895
2. (a) 17
   (b) 136
   (c) 206
   (d) 1098
   (e) 20 140
3. 72
4. 29
5. 34 hours
6. (a) 21 480
   (b) 13 000
   (c) 32 960
   (d) 21 510
7. £1450
8. 10 000 m
9. 60 000 staples
10. (a) 253 (b) 79
    (c) 537 (d) 126
11. £35
12. 20 minutes
13. (a) 2432
    (b) 13 338
    (c) 55 311
    (d) 606 375
14. £5191
15. £43 848
16. (a) 43 (b) 504
    (c) 410 (d) 654
17. 23
18. (a) 16 rem 10
    (b) 25 rem 7
    (c) 17 rem 35
    (d) 14 rem 7
19. 14 pints, 10p change
20. (a) 41 (b) 16

Exercise 1.3 — Page 4

1. 2
2. (a) 37 (b) 3 (c) 9 (d) 58 (e) 6
   (f) 30 (g) 5 (h) 19 (i) 14 (j) 20
   (k) 24 (l) 4 (m) 0 (n) 5 (o) 6
3. (a) $(7 - 2) \times 3 = 15$
   (b) $(3 + 5) \div 2 = 4$
   (c) $(4 + 1) \times (7 - 2) = 25$
4. Many answers, for example:
   $6 - 3 \times 2 + 1 = 1$ $6 - 3 - 2 + 1 = 2$
   $6 \div 3 + 2 - 1 = 3$ $6 \div 3 + 2 \times 1 = 4$
   $6 - 3 + 2 \times 1 = 5$ $6 + 3 - 2 - 1 = 6$
   $6 + 3 - 2 \times 1 = 7$ $6 \times 3 \div 2 - 1 = 8$
   $6 \times 3 \div 2 \times 1 = 9$ $6 + 3 + 2 - 1 = 10$
5. 148 cm
6. 15 g
7. (a) 36 (b) 43

Exercise 1.4 — Page 5

1. −28°C, −13°C, −3°C, 19°C, 23°C
2. −78, −39, −16, −9, 11, 31, 51
3. 5°C
4. 8 kg
5. (a) 2 (b) 1 (c) −9 (d) 8
   (e) 4 (f) −5 (g) −7 (h) 7
   (i) 1 (j) −15 (k) −3 (l) −6
6. (a) 13 (b) 6 (c) 7 (d) 7
   (e) 5 (f) −12 (g) −1 (h) 13
   (i) −11 (j) 11 (k) 9 (l) 0
7. (a) 6 (b) −8 (c) −28 (d) 0
   (e) −35 (f) 19
8. (a) 7 (b) −1 (c) 36 (d) −6
   (e) 38 (f) 15 (g) −15 (h) 25
9. (a) 10°C (b) 10°C (c) 5°C (d) 37°C
10. −22°C
11. (a) −35 (b) 35 (c) −24 (d) −45
    (e) 64 (f) −42 (g) −80 (h) 32
    (i) −20 (j) 60 (k) −30 (l) 60
    (m) −60 (n) −100 (o) −189
12. (a) −4 (b) 4 (c) 5 (d) −5
    (e) −5 (f) 5 (g) 6 (h) −6
    (i) 4 (j) −8 (k) 6 (l) −5
    (m) 4 (n) −4 (o) −1

Review Exercise 1 — Page 7

1. 4587
2. E.g. 3, 4, 5: 3 + 4 + 5 = 12, which is even.
3. 298 miles
4. 184 cm
5. (a) 25 (b) (i) $(1 + 3) \times (5 + 5) = 40$
   (ii) $1 + 3 \times (5 + 5) = 31$
6. 10 120
7. (a) −1 (b) 40
   (c) −10
8. (a) 166 (b) 4
9. £3427
10. 32
11. 50
12. 9700
13. (a) 21°C (b) −0.4°F
14. 4

CHAPTER 2

Exercise 2.1 — Page 8

1. (a) A: 5.2, B: 5.6 (b) C: 0.54, D: 0.59
2. 3.567, 3.576, 3.652, 3.657, 3.675
3. 1.55 m
4. (a) 10.9 (b) 1.65 (c) 0.9 (d) 11 (e) 1.2 (f) 0.6 (g) 0.24 (h) 8.9
5. (a) 6.84 (b) 5.22 (c) 5.003 (d) 3 (e) 1.24 (f) 15.781
6. £3.32, £1.68 change
7. 4.88 m

Exercise 2.2 — Page 10

1. (a) £2.50 (b) £25
2. (a) £7.95 (b) £0.12 (c) 86.9p
3. 12.3×1000 and $12.3 \div 0.001$
   $12.3 \div 100$ and 12.3×0.01
   12.3×0.1 and $12.3 \div 10$
   $12.3 \div 0.01$ and 12.3×100
   12.3×10 and $12.3 \div 0.1$
   12.3×0.001 and $12.3 \div 1000$
4. £7.60
5. (a) £14.95 (b) £5.05
6. (a) 8.5 (b) 1.3 (c) 0.48 (d) 0.06 (e) 15 (f) 4 (g) 50 (h) 24
7. £10.35
8. £1.35
9. (a) 0.075 (b) 8.75 (c) 16.53 (d) 1.025 (e) 3.888 (f) 9.38 (g) 0.0432 (h) 0.028
10. (a) 21p (b) £1.84 (c) 78p
11. (a) 1.6 (b) 1.75 (c) 1.125 (d) 12.3 (e) 2.92 (f) 1430 (g) 12.5 (h) 37.5
12. (a) (i) 3 (ii) 1.5 (iii) 0.24
    (b) Each answer is less than the original number.
13. (a) (i) 10 (ii) 6 (iii) 0.3
    (b) Each answer is greater than the original number.
14. £60.75
15. (a) (i) £5.32 (ii) £7.68 (iii) £3.60 (iv) £2.79
    (b) £1.38
16. £1.75
17. 67
18. 45

Exercise 2.3 — Page 11

1. (a) 4.768
2. (a) $2.\dot{1}$
3. (a) 12.2
4. (a) **C** (b) **B** (c) **D** (d) **A**
5. (a) 0.077 836 … (b) 0.355 960 … (c) 11.045 … (d) 3.421 98 … (e) 322.528 571 … (f) −386 (g) 0.155

Exercise 2.4 — Page 12

1. E.g. $\frac{15}{24}$, $\frac{30}{48}$, $\frac{45}{72}$, …, $\frac{10}{16}$, $\frac{20}{32}$, …
   simplest form $\frac{5}{8}$
2. $\frac{8}{15}$, $\frac{3}{5}$, $\frac{2}{3}$, $\frac{7}{10}$
3. $\frac{4}{5}$
4. (a) $\frac{4}{5}$ (b) $\frac{2}{3}$ (c) $\frac{2}{3}$ (d) $\frac{2}{5}$
5. (a) $1\frac{1}{2}$ (b) $2\frac{1}{8}$ (c) $3\frac{3}{4}$ (d) $4\frac{3}{5}$
6. (a) $\frac{27}{10}$ (b) $\frac{8}{5}$ (c) $\frac{35}{6}$
7. $\frac{3}{5}$
8. $\frac{30}{50} = \frac{3}{5}$
9. (a) 3 (b) 4 (c) 3 (d) 8 (e) 8 (f) 9 (g) 12 (h) 20 (i) 40 (j) 12
10. (a) 6 (b) 24
11. £5
12. £148.40
13. £3187.50
14. (a) 9 squares (b) 10 squares (c) $\frac{5}{24}$

Exercise 2.5 — Page 14

1. (a) $\frac{3}{8}$ (b) $\frac{7}{12}$ (c) $\frac{7}{10}$ (d) $\frac{8}{15}$ (e) $\frac{9}{14}$
2. (a) $\frac{1}{8}$ (b) $\frac{1}{12}$ (c) $\frac{3}{10}$ (d) $\frac{2}{15}$ (e) $\frac{5}{14}$
3. (a) $1\frac{1}{4}$ (b) $1\frac{1}{2}$ (c) $1\frac{11}{20}$ (d) $1\frac{8}{21}$ (e) $1\frac{5}{24}$
4. (a) $\frac{1}{8}$ (b) $\frac{8}{15}$ (c) $\frac{5}{8}$ (d) $\frac{1}{15}$ (e) $\frac{1}{3}$
5. (a) $4\frac{1}{4}$ (b) $3\frac{5}{6}$ (c) $4\frac{3}{8}$ (d) $5\frac{17}{20}$ (e) $6\frac{13}{30}$
6. (a) $1\frac{1}{10}$ (b) $\frac{5}{12}$ (c) $1\frac{3}{8}$ (d) $3\frac{3}{10}$ (e) $2\frac{1}{4}$
7. $\frac{4}{15}$
8. $\frac{13}{60}$
9. (a) $\frac{5}{12}$ (b) $\frac{11}{12}$
10. (a) $\frac{1}{4}$ (b) Billy

Exercise 2.6 — Page 15

1. (a) $3\frac{1}{2}$ (b) $1\frac{4}{5}$ (c) $6\frac{1}{4}$ (d) 9 (e) $10\frac{1}{2}$
2. (a) $\frac{1}{20}$ (b) $\frac{1}{10}$ (c) $\frac{3}{8}$ (d) $\frac{1}{16}$ (e) $\frac{1}{20}$
3. (a) $7\frac{1}{2}$ (b) $7\frac{1}{2}$ (c) $4\frac{4}{5}$ (d) $7\frac{1}{3}$ (e) $7\frac{4}{5}$ (f) $2\frac{1}{4}$ (g) $\frac{3}{4}$ (h) $\frac{4}{5}$ (i) $\frac{2}{3}$ (j) $\frac{5}{8}$
4. (a) $\frac{1}{2}$ (b) $\frac{3}{10}$ (c) $\frac{1}{3}$ (d) $\frac{2}{7}$ (e) $\frac{1}{4}$
5. (a) $1\frac{1}{8}$ (b) 2 (c) $3\frac{3}{4}$ (d) $3\frac{17}{20}$ (e) $4\frac{7}{8}$ (f) $1\frac{13}{15}$ (g) 14 (h) $13\frac{1}{2}$ (i) $2\frac{7}{10}$ (j) $8\frac{3}{4}$
6. (a) $\frac{4}{15}$ (b) $\frac{2}{5}$
7. (a) $\frac{4}{15}$ (b) 15

Exercise 2.7 — Page 16

1. (a) $\frac{1}{10}$ (b) $\frac{3}{8}$ (c) $\frac{1}{3}$ (d) $\frac{1}{10}$ (e) $\frac{2}{7}$
2. (a) $\frac{3}{4}$ (b) $\frac{1}{3}$ (c) $\frac{4}{7}$ (d) $\frac{4}{9}$ (e) $\frac{2}{3}$
3. (a) 2 (b) $\frac{2}{5}$ (c) $2\frac{5}{8}$ (d) $3\frac{1}{3}$ (e) 6
4. (a) $\frac{5}{6}$ (b) $\frac{9}{16}$ (c) $\frac{4}{5}$ (d) $1\frac{1}{3}$ (e) $\frac{2}{3}$
   (f) $1\frac{1}{2}$ (g) $\frac{2}{3}$ (h) $1\frac{1}{6}$ (i) $1\frac{1}{2}$ (j) $1\frac{4}{5}$
5. (a) 14 (b) 5 (c) 2 (d) 6 (e) 6
   (f) $\frac{4}{5}$ (g) $1\frac{1}{2}$ (h) $1\frac{1}{4}$ (i) $1\frac{1}{7}$ (j) $\frac{21}{25}$
6. $\frac{1}{10}$ 7. $\frac{1}{9}$ 8. 23

Exercise 2.8 — Page 17

1. £3.20 2. 70 km 3. £3.20 4. £6 5. 96 cm

Exercise 2.9 — Page 18

1. (a) $\frac{3}{25}$ (b) $\frac{3}{5}$ (c) $\frac{8}{25}$ (d) $\frac{7}{40}$ (e) $\frac{9}{20}$
   (f) $\frac{13}{20}$ (g) $\frac{11}{50}$ (h) $\frac{101}{500}$ (i) $\frac{7}{25}$ (j) $\frac{111}{200}$
   (k) $\frac{5}{8}$ (l) $\frac{21}{25}$
2. (a) 0.25 (b) 0.1 (c) 0.4 (d) 0.75
   (e) 0.7 (f) 0.8
3. (a) 0.15 (b) 0.16 (c) 0.07 (d) 0.95
   (e) 0.36 (f) 0.53
4. (a) 0.125 (b) 0.625
   (c) 0.225 (d) 0.725
5. (a) $0.\dot{7}$ (b) $0.\dot{1}$ (c) $0.\dot{3}\dot{6}$
   (d) $0.\dot{8}\dot{2}$ (e) $0.\dot{1}3\dot{5}$ (f) $0.\dot{2}1\dot{6}$
   (g) $0.1\dot{6}$ (h) $0.\dot{2}8571\dot{4}$
6. (a) $0.\dot{8}$ (b) $0.\dot{4}$ (c) $0.\dot{5}\dot{1}$
   (d) $0.\dot{7}\dot{2}$ (e) $0.\dot{1}4285\dot{7}$ (f) $0.\dot{8}5714\dot{2}$
   (g) $0.0\dot{3}$ (h) $0.4\dot{6}$ (i) $0.8\dot{3}$
   (j) $0.7\dot{7}\dot{2}$

Review Exercise 2 — Page 19

1. (a) $\frac{9}{20}$ (b) (i) 0.28 (ii) $0.\dot{5}$
2. (a) 12.41 (b) 4.33 (c) 20 (d) 0.08
3. (a) 64 (b) $\frac{7}{16}$
4. $\frac{7}{10}$
5. (a) 0.143
   (b) 1.014, 1.14, $1\frac{1}{7}$, 1.41, 11.14
6. (a) $\frac{11}{12}$ (b) $\frac{11}{40}$ (c) $\frac{3}{7}$ (d) $\frac{5}{6}$
7. $\frac{4}{15}$
8. (a) 1.1275 (b) 27.5
9. 208 km
10. $\frac{1}{10}$
11. (a) $4\frac{10}{63}$ (b) $2\frac{19}{30}$ (c) 10 (d) $1\frac{5}{7}$

12. £80
13. (a) Press [=] after 1.76.
    (b) 3.12 (3 sig. figs.)
14. 240.159574…
15. 24.86052632…
16. 68.72 euros
17. £200
18. Yes

CHAPTER 3

Exercise 3.1 — Page 20

1. (a) 50 (b) 50 (c) 70
2. (a) 7480 (b) 7500 (c) 7000
3. (a) 3456 (b) 3500
4. (a) 745, 746, 747, 748, 749
   (b) 750, 751, 752, 753, 754
   (c) Any number from 8450 to 8499
   (d) Any number from 8500 to 8549
5. (a) £50 (nearest pound)
   (b) 24 100 (nearest hundred)
   (c) 309 000 km^2 (nearest thousand km^2)
   (d) 190 km (nearest ten kilometres)
6. Nearest 1000.
7. (a) 35 (b) 44
8. (a) 1950 (b) 2049
9. 42 500
10. Smallest: 135, greatest: 144
11. 2749

Exercise 3.2 — Page 21

1. 4 2. 5 3. 9 4. 24 5. 16 6. 7 7. 29

Exercise 3.3 — Page 22

1. (a) 3.962 (b) 3.96 (c) 4.0
2. (a) 567.65 (b) 567.7 (c) 568
3. 4.86
4. 68.8 kg
5. Missing entries are:
   0.96, 0.97, 15.281, 0.06, 4.99, 5.00
6. (a) (i) 46.1 (ii) 59.7 (iii) 569.4 (iv) 17.1 (v) 0.7
   (b) (i) 46.14 (ii) 59.70 (iii) 569.43 (iv) 17.06 (v) 0.66
   (c) (i) 46.145 (ii) 59.697 (iii) 569.434 (iv) 17.059 (v) 0.662
7. (a) 40.9 litres, nearest tenth of a litre
   (b) £2.37, nearest penny
   (c) 35.7 cm, nearest millimetre
   (d) £1.33, nearest penny
   (e) £14.26, nearest penny

Exercise 3.4 — Page 24

1. (a) 20 (b) 500 (c) 400
   (d) 2000 (e) 20 (f) 0.08
   (g) 0.09 (h) 0.009 (i) 0.01

2.

Number	sig. fig.	Answer
456 000	2	460 000
454 000	2	450 000
7 981 234	3	7 980 000
7 981 234	2	8 000 000
1290	2	1300
19 602	1	20 000

3.

Number	sig. fig.	Answer
0.000567	2	0.00057
0.093748	2	0.094
0.093748	3	0.0937
0.093748	4	0.09375
0.010245	2	0.010
0.02994	2	0.030

4. 490

5. (a) (i) 80 000 (ii) 80 (iii) 1000
(iv) 0.007 (v) 0.002
(b) (i) 83 000 (ii) 83 (iii) 1000
(iv) 0.0073 (v) 0.0019
(c) (i) 82 700 (ii) 82.7 (iii) 1000
(iv) 0.00728 (v) 0.00190

6. 472 m² (3 s.f.)

7. (a) 157 cm² (3 s.f.) (b) 6100 m² (2 s.f.)
(c) 1.23 m (nearest cm)
(d) 15.9 m² (round up 1 d.p.)

Exercise 3.5 — Page 25

1. $40 \times 50 = 2000$

2. (a) (i) $40 \times 20 = 800$
(ii) $100 \times 20 = 2000$
(iii) $800 \times 50 = 40\,000$
(iv) $900 \times 60 = 54\,000$
(b) (i) $80 \div 20 = 4$
(ii) $600 \div 30 = 20$
(iii) $900 \div 60 = 15$
(iv) $4000 \div 80 = 50$

3. $40 \times 50\text{p} = £20$

4. (a) 30 is bigger than 29 and 50 is bigger than 48. So, 30×50 is bigger than 29×48.
(b) $200 \div 10 = 20$, 14, estimate is bigger.

5. (a) £8000 + £1000 = £9000 (b) £9312

6. $400 \times 0.5 = 200$
Kath's answer is 10 times too small.

7. (a) 2.047 009 …
(b) $(50 + 10) \div (10 \times 3) = 2$

8. (a) 6.4, 6.41875 (b) 20, 18.709677…
(c) 20, 20.45631… (d) 2000, 2132.9583…

9. (a) 410.08316…
(b) $80 \div (0.04 \times 5) = 400$

10. $6\,000\,000 \div 0.03 = 200\,000\,000$
Niamh's answer is 10 times too big.

Exercise 3.6 — Page 27

1. 167.5 cm

2. 8.5 m $\leqslant$ height of building < 9.5 m

3. Minimum weight: 835 g
Maximum weight: 845 g

4. 11.55 seconds

5. 9.35 kg

6. Least length: 2.65 m, greatest length: 2.75 m

7. 94 ml

8. Minimum weight: 7.5 kg
Maximum weight: 8.5 kg

9. (a) 38.5 cm (b) 3.45 kg

10. 189 g

11. 85.5 km

12. (a) 0.525 kg $\leqslant$ weight of book < 0.535 kg
(b) 5.35 kg

Exercise 3.7 — Page 28

1. 1260 cm

2. Lower bound: 26.485 km
Upper bound: 26.595 km

3. (a) 505 kg (b) 3.4 kg

4. 10

5. £1.13

6. (a) (i) 7.45 (ii) 3.675
(b) Lower bound: 3.665, upper bound: 3.775
(c) 27.01125
(d) 2.02721…

Review Exercise 3 — Page 29

1. 1 500 000

2. (a) $90 \times 2 = 180$ (b) $2000 \div 50 = 40$

3. 18.7 is less than 19, but rounds to 19.
0.96 is less than 1, but rounds to 1.
$19 \times 1 = 19$,
so, 18.7×0.96 must be less than 19.

4. $40 \times 0.03 = 1.2$

5. $100 \times £30 = £3000$

6. 12 packs

7. Smallest: 85, largest: 94

8. (a) 25.57 (b) 25.6 (c) 30

9. (a) 9.2 (b) 9.18

10. (a) 1 (b) 1.5

11. (a) 2550 (b) 2649

12. (a) 17.5 m
(b) 17.5 m $\leqslant$ length of bus < 18.5 m

13. 390

14. (a) 31.284 …
(b) $(90 \times 10) \div (20 + 10) = 30$

15. 5.18 (2 d.p.)

16. (a) $30 \times 40 \times 10 = £12\,000$
(b) Too large. All values rounded up.

17. $0.4 \times 90 \div 6 = 6$, not correct

18. (a) 24.8605263… (b) $\frac{5 \times 20}{6 - 2} = \frac{100}{4} = 25$

19. (a) 19 m²
(b) Round up to the nearest square metre for enough carpet to cover whole floor.
20. Lower bound: 1750 kg, upper bound: 1850 kg
21. 18.9 kg
22. (a) 2.5 hours (b) 137.5 miles
23. (a) (i) 550 (ii) 55
(b) (i) 705 (ii) 24 750
(c) 10
24. 6037.5 m²
25. 54.5 m²

CHAPTER 4

Exercise 4.1 — Page 32

1. (a) $\frac{1}{10}$ (b) $\frac{1}{4}$ (c) $\frac{1}{20}$
(d) $\frac{7}{20}$ (e) $\frac{12}{25}$ (f) $\frac{1}{8}$
2. (a) 0.2 (b) 0.15 (c) 0.01
(d) 0.72 (e) 0.875 (f) 1.5
3. (a) 34% (b) 48% (c) 15% (d) 80%
(e) 27% (f) 65% (g) $66\frac{2}{3}$% (h) $22\frac{2}{9}$%
4. (a) 15% (b) 32% (c) 12.5%
(d) 7% (e) 112% (f) 1.5%
5. (a) $\frac{2}{5}$, $\frac{1}{2}$, 0.55, 60%
(b) 0.42, 43%, $\frac{11}{25}$, $\frac{9}{20}$
(c) 28%, 0.2805, $\frac{57}{200}$, $\frac{23}{80}$
6. (a) 60% (b) 21%
7. 30%
8. (a) 40% (b) 60%
9. 12.5%
10. 37.5%
11. 3%
12. Maths. Geog 80%, English 84%, Maths 85%
13. Team A
14. (a) 10% (b) 60% (c) 15%
15. 9.5%
16. 25%

Exercise 4.2 — Page 33

1. (a) £16 (b) £15 (c) £66
(d) £52.50 (e) £25 (f) 27 kg
(g) £11.25 (h) 12 m
2. 270
3. £20
4. £2.70
5. £42
6. (a) £80 (b) £13
(c) £16.50 (d) £57.50
7. (a) £510 (b) £5.50
(c) £33.60 (d) £40.95
8. 40p per minute

9. £215
10. 759 g
11. £244.64
12. £4706
13. 40p per pint
14. 87.7p per litre
15. £10 530
16. £109 520

Exercise 4.3 — Page 35

1. 20%
2. 20%
3. 28%
4. Becky.
Hinn's increase = 32%
Becky's increase = 40%
5. (a) $12\frac{1}{2}$% (b) 12%
Rent went up by a greater percentage.
6. Car A 13.8%, Car B 18.2%
7. 19.5%
8. 18.6%
9. 5.9% decrease
10. 8% increase
11. 19%
12. 3.5%

Exercise 4.4 — Page 36

1. 600 ml
2. £600
3. £220
4. 1.89 m
5. £312 500
6. 67.5 mg
7. £1420
8. £8000
9. School A 425
School B 450
10. £400
11. £150

Exercise 4.5 — Page 37

1. 36 hours
2. £694.40
3. £347.48
4. £10.20
5. (a) £1185
(b) £118.50
6. £3947.90
7. £432.03
8. £3837.46
9. £11 911.20
10. £704.60

Exercise 4.6 — Page 39

1. £5.60
2. £40.50
3. 700 g
4. £67.28
5. £537.40
6. £198
7. 678 units
8. £12 400

Exercise 4.7 — Page 40

1. £6
2. (a) £3.50
(b) £73.50
3. (a) £15.75
(b) £105.75
4. £291.40
5. £77.55
6. £27 025
7. £428.87
8. £92.82
9. £183.75
10. £3.20
11. £293.75
12. £2928.10

Exercise 4.8 — Page 42

1. £10
2. (a) £30 (b) £15
3. £225
4. £48
5. £225
6. £242
7. (a) earns 78p more interest
8. £11 910.16
9. (a) £2420 (b) £1938.66
10. 81 m
11. (a) (i) £8268.73 (ii) 63.6%
(b) (i) £12 721.12 (ii) 63.6%
(c) Same percentage reduction.
12. 9 years

Review Exercise 4 — Page 43

1. (a) 18 (b) 20%
2. £780
3. 90%
4. £5080
5. 98p per kg
6. £3.66
7. £586.32
8. £157.50
9. Large: 9.34 g per penny.
   Small: 9.66 g per penny.
   Small pot is better value.
10. £36
11. 4 hours
12. (a) 166.4 cm (b) 13.7%
13. 44.4%
14. (a) £2434.50 (b) 73.9%
15. (a) 6900
    (b) Yes. $6000 \times 1.15^4 = 10\,500$
16. £85
17. £5.76
18. £2450.09
19. £1645
20. £8470.96
21. £704.90
22. (a) 29 000 (b) 61 000
    (c) 1989

CHAPTER 5

Exercise 5.1 — Page 45

1. (a) 35 (b) 48
2. E.g. 2, 3, 4, 5: $2 + 3 + 4 + 5 = 14$
   14 is not a multiple of 4.
3. (a) 14, 35 (b) 4, 5, 20 (c) 3, 5
4. (a) Second (b) Second
   (c) Fifth (d) Third
5. (a) 1, 2, 4, 8, 16
   (b) 1, 2, 4, 7, 14, 28
   (c) 1, 2, 3, 4, 6, 9, 12, 18, 36
   (d) 1, 3, 5, 9, 15, 45
   (e) 1, 2, 3, 4, 6, 8, 12, 16, 24, 48
   (f) 1, 2, 5, 10, 25, 50
6. (a) 1, 5 (b) 1, 2, 4 (c) 1, 2
   (d) 1, 2, 3, 4, 6, 12 (e) 1, 2, 3, 6
7. (a) 31, 37 (b) No. 7 is a factor of 49.

Exercise 5.2 — Page 47

1. (a) $2^2 \times 3^2$ (b) $2 \times 3^3 \times 5$
   (c) $2 \times 3 \times 5^2$ (d) $2^3 \times 3 \times 5^2$
   (e) $3^3 \times 5^3$ (f) 7^4
2. (a) 2, 3 (b) 2, 5 (c) 2, 7
   (d) 3, 5 (e) 2, 3, 11 (f) 2, 3
3. (a) $2^2 \times 3$ (b) $2^2 \times 5$ (c) $2^2 \times 7$
   (d) $3^2 \times 5$ (e) $2 \times 3 \times 11$ (f) $2^2 \times 3^3$
4. 2^7
5. $2^3 \times 5^3$
6. (a) 25 000 (b) $2^3 \times 5^6$
7. $2^3 \times$ (consecutive prime numbers)
   104, 136, 152
8. (a) 24 (b) 160 (c) 20 (d) 90
   (e) 90 (f) 24 (g) 40 (h) 630

9. (a) 6 (b) 8 (c) 2 (d) 4
   (e) 11 (f) 4 (g) 3 (h) 15
10. (a) $2^3 \times 3$ (b) 2×3^3 (c) 6 (d) 216
11. (a) 3 (b) 2×3^4 (c) 54 (d) 324
12. 9.18 am

Exercise 5.3 — Page 49

1. (a) 49 (a) 125 (c) 0.5
2. Yes. $7^2 = 49$ and $8^2 = 64$
   55 is between 49 and 64.
   So, $\sqrt{55}$ lies between 7 and 8.
3. 64
4. No. $2^2 = 4$, $3^2 = 9$ and $5^2 = 25$
   $2^2 + 3^2 = 4 + 9 = 13$
5. (a) (i) 9 (ii) −8 (iii) 16 (iv) −125
   (b) The result of squaring a negative number is a positive number. The result of cubing a negative number is a negative number.
6. (a) 8 (b) 7 (c) 5 (d) 5
7. (a) 2 (b) 4 (c) 5
8. (a) 6.25 (b) 0.64 (c) 3375
   (d) 13.824 (e) 16 807 (f) 0.0625
   (g) 75 418.89…
9. (a) 0.25 (b) 0.05 (c) 0.04 (d) 2
   (e) 4 (f) 2.5 (g) 6.25
10. E.g. $5 \times \frac{1}{5} = 1$
11. 9.8
12. (a) 2.3 (b) 17.3 (c) 4.0
13. (a) 27 000 (b) 800 (c) 2.98
    (d) 10.648 (e) 70.56 (f) 31.25
    (g) 25.215 (h) 1.28 (i) 126 216
    (j) 134.4
14. (a) E.g. $16 = 4^2$, so, $16^3 = (4^2)^3 = 4^6$
    (b) $x = 12$
15. $x = 1.43$

Exercise 5.4 — Page 50

1. (a) 2^5 (b) 3^7 (c) 5^8 (d) 7^4 (e) 9^5
2. (a) 2 (b) 3^3 (c) 5^4 (d) 7^2 (e) 9^5
3. (a) 3^6 (b) 10^2 (c) 4^5 (d) 5^3
   (e) 2^3 (f) $5^0 = 1$ (g) 7^{-2} (h) 3^3
4. 2^{15}
5. (a) 2^6 (b) 3^{10} (c) 5^6 (d) 7^9 (e) $9^0 = 1$
6. (a) 8 (b) 7 (c) 11 (d) 3
   (e) 1 (f) 4 (g) 6 (h) 20
   (i) 2 (j) 8 (k) 6 (l) 3

Exercise 5.5 — Page 51

1. (a) 9^0 (b) 2^{-2} (c) 5^{-2} (d) 8^{-5}
   (e) 2^{-4} (f) 5^{12} (g) 11^{-5} (h) 7^{-1}
2. (a) 8^2 (b) 7^{-5} (c) 2.5^{-1}
   (d) 4^0 (e) 10^{-1} (f) 6^{-4}
   (g) 0.1^{-12} (h) 5^{-15} (i) 4^5
   (j) $4^2 \times 8^7$ (k) 4×5^{-2} (l) $2^{-2} \times 5^5$
   (m) 3 (n) 5^3 (o) 2^{-6}

3. (a) $\frac{1}{3^2}$ (b) $\frac{1}{2^3}$ (c) $\frac{1}{3}$ (d) $\frac{1}{5^3}$ (e) 3^2
(f) 5^2 (g) 2^4 (h) $\frac{1}{3^2}$ (i) $\frac{1}{5^6}$ (j) 3^6

4. (a) 1 (b) $\frac{1}{3}$ (c) $\frac{1}{8}$ (d) $\frac{1}{9}$ (e) 8 (f) 27

5. (a) $11\frac{1}{10}$ (b) $2\frac{7}{8}$ (c) $1\frac{6}{25}$ (d) $\frac{5}{6}$
(e) $\frac{29}{100}$ (f) $\frac{1}{100}$ (g) $\frac{4}{25}$ (h) $1\frac{7}{30}$

6. (a) -4 (b) -2 (c) -2 (d) -3
(e) -1 (f) -2 (g) -3 (h) -1

7. 2^{14}

Exercise 5.6

Page 53

1. (a) 20 (b) 3 (c) 10 (d) 2 (e) 2 (f) 2.5
2. (a) 8 (b) 2 (c) 3 (d) 2 (e) 5 (f) 6
3. (a) $\frac{1}{10}$ (b) $\frac{1}{7}$ (c) $\frac{1}{2}$ (d) $\frac{1}{5}$ (e) $\frac{1}{4}$ (f) $\frac{1}{3}$
4. (a) 100 (b) 27 (c) 8 (d) 4 (e) 32
(f) 243 (g) 25 (h) 32 (i) 81 (j) 216
5. (a) $\frac{1}{100}$ (b) $\frac{1}{64}$ (c) $\frac{1}{4}$ (d) $\frac{1}{8}$ (e) $\frac{1}{8}$
(f) $\frac{1}{100\,000}$ (g) $\frac{1}{125}$ (h) $\frac{1}{8}$ (i) $\frac{1}{32}$ (j) $\frac{1}{25}$
6. (a) (i) $\frac{1}{2}$ (ii) $\frac{3}{7}$ (b) (i) 25 (ii) 9
7. (a) 3 (b) 4 (c) 5 (d) $\frac{1}{2}$ (e) $\frac{2}{3}$
(f) $\frac{1}{4}$ (g) $\frac{3}{4}$ (h) 256 (i) 125 (j) 1.5
8. (a) $\frac{3}{4}$ (b) $\frac{1}{2}$ (c) 3 (d) $\frac{15}{56}$ (e) $6\frac{3}{4}$
10. (a) 19.300 259 07… (b) 291.866 535 1…
(c) 3.169 674 155… (d) 3.024 773 331…
(e) 8.556 307 844… (f) 6.695 526 189…

Review Exercise 5

Page 55

1. (a) 9, 25, 100 (b) 20, 25, 100 (c) 3, 29
2. (a) 8 (b) $13^3 = 13 \times 13 \times 13 = 2197$
3. (a) 3 (b) 10 000
4. (a) 108 (b) 1, 3, 37, 111
(c) 103 (d) 2
5. (a) 125 (b) 64
6. (a) 4 (b) $2^3 \times 3^2$ (c) 24 (d) 144
7. 90 seconds
8. (a) 5 (b) 5 (c) 1
9. (a) $x = 5$ (b) $x = 1$
(c) $x = 6$ (d) $x = 4$
10. (a) $8^4 = (2^3)^4 = 2^{12}$ (b) $x = 6$
11. (a) 2^0 (b) 2^6
12. (a) 2 (b) $\frac{1}{16}$ (c) $\frac{1}{3}$
13. (a) $w = 0$ (b) $x = -2$
(c) $y = \frac{1}{5}$ (d) $z = -4$
14. (a) 9 (b) $\frac{1}{125}$ (c) $\frac{1}{27}$
15. 46.8
16. 0.167
17. 2 significant figures. $2^{10} = 1024$, $10^3 = 1000$
18. (a) 209 953 (b) 3830.32…
19. (a) 343 (b) 0.232 (c) 2.82 (d) 0.0346

CHAPTER 6

Exercise 6.1

Page 57

1. (a) 6×10^5 (b) 9 000 000
2. Missing entries are:
(a) 7.5×10^4 (b) 800 000 000, 8×10^8
(c) 35 000 000 000 000,
$3.5 \times 10\,000\,000\,000\,000$
(d) $6.23 \times 10\,000\,000\,000\,000$, 6.23×10^{13}
3. (a) 3×10^{11} (b) 8×10^7
(c) 7×10^8 (d) 2×10^9
(e) 4.2×10^7 (f) 2.1×10^{10}
(g) 3.7×10^9 (h) 6.3×10^2
4. (a) 600 000 (b) 2000
(c) 50 000 000 (d) 900 000 000
(e) 3 700 000 000 (f) 28
(g) 99 000 000 000 (h) 71 000
5. (a) (i) 4.5×10^3 (ii) 7.8×10^7
(iii) 5.3×10^5 (iv) 3.25×10^4
(b) (i) 4500 (ii) 78 000 000
(iii) 530 000 (iv) 32 500
6. (a) (i) Brazil (ii) 250 000 000
(b) 1.288×10^9

Exercise 6.2

Page 58

1. (a) 3.5×10^{-4} (b) 0.0025
2. Missing entries are:
(a) 7.5×10^{-3}
(b) $8.75 \times 0.000\,001$, 8.75×10^{-6}
(c) 0.000 000 003 5, $3.5 \times 0.000\,000\,001$
(d) $6.23 \times 0.000\,000\,000\,001$, 6.23×10^{-12}
3. (a) 7×10^{-3} (b) 4×10^{-2}
(c) 5×10^{-9} (d) 8×10^{-4}
(e) 2.3×10^{-9} (f) 4.5×10^{-8}
(g) 2.34×10^{-2} (h) 2.34×10^{-9}
(i) 6.7×10^{-3}
4. (a) 0.35 (b) 0.0005 (c) 0.000 072
(d) 0.0061 (e) 0.000 000 000 117
(f) 0.000 000 813 5 (g) 0.064 62
(h) 0.000 000 004 001
5. (a) (i) 3.4×10^{-3} (ii) 5.65×10^{-5}
(iii) 7.2×10^{-4} (iv) 9.13×10^{-1}
(b) (i) 0.0034 (ii) 0.000 056 5
(iii) 0.000 72 (iv) 0.913

Exercise 6.3

Page 59

1. (a) 6 E 13, 6×10^{13}, 60 000 000 000 000
(b) 9.6 E 12, 9.6×10^{12}, 9 600 000 000 000
(c) 1.05 E 13, 1.05×10^{13},
10 500 000 000 000
(d) 1.3 E 14, 1.3×10^{14},
130 000 000 000 000
(e) 2.4 E 14, 2.4×10^{14},
240 000 000 000 000
(f) 2.5 E 12, 2.5×10^{12}, 2 500 000 000 000

2. (a) 6 E −12, 6×10^{-12}, 0.000 000 000 006
(b) 1.35 E −10, 1.35×10^{-10}, 0.000 000 000 135
(c) 3 E −11, 3×10^{-11}, 0.000 000 000 03
(d) 1.15 E −10, 1.15×10^{-10}, 0.000 000 000 115
(e) 4.24 E −09, 4.24×10^{-9}, 0.000 000 004 24
(f) 9.728 E −11, 9.728×10^{-11},

Exercise 6.4 — Page 60

1. (a) 262 500
(b) 105 000 000 000 000
(c) 72 250 000 000 000
(d) 0.000 000 125
(e) 30 000 000 000
(f) 263 160 000 000 000 000 000
2. (a) 9.38×10^{19} (b) 1.6×10^{1}
(c) 2.25×10^{26} (d) 1.25×10^{-37}
(e) 2.4×10^{20} (f) 5×10^{-2}
3. (a) 2.088×10^{9} (b) $1.525\,965 \times 10^{16}$ km
4. (a) Pluto, Saturn, Jupiter
(b) 1.397×10^{5} km, 139 700 km
5. (a) 9.273×10^{6} km² (b) 4.93×10^{5} km²
6. 2.52×10^{12}
7. 2.2×10^{-5}
8. 1.49×10^{9} square miles
9. 30%
10. 9×10^{-28} grams
11. (a) 4.55×10^{9} years
(b) About 6.6×10^{3} times

Exercise 6.5 — Page 61

1. (a) 5.2×10^{3} (b) 8.4×10^{6}
(c) 4×10^{4} (d) 5×10^{4}
2. (a) 2.8×10^{3} (b) 9×10^{5}
(c) 3×10^{7} (d) 9.5×10^{8}
3. (a) 8×10^{7} (b) 6×10^{7}
(c) 2.4×10^{8} (d) 2.7×10^{15}
4. (a) 3×10^{3} (b) 3×10^{3}
(c) 5×10^{1} (d) 2×10^{11}
(e) 4×10^{5} (f) 3×10^{-3}
5. (a) 1.5×10^{0} (b) 2.7×10^{13}
(c) 4.5×10^{4} (d) 1.25×10^{-13}
(e) 6×10^{8}
6. (a) 1.5×10^{11} (b) 2.4×10^{3}
(c) 4.2×10^{-8} (d) 6.4×10^{7}
(e) 5×10^{1} (f) 3×10^{9}
(g) 6.25×10^{-6} (h) 2.5×10^{4}

Review Exercise 6 — Page 62

1. (a) (i) 3.9×10^{5} (ii) 390 000
(b) (i) 6.7×10^{-3} (ii) 0.0067
2. (a) 4×10^{4} (b) 1.6×10^{7}
3. (a) 3.45×10^{10} (b) 5.43×10^{-7}
(c) 1.125×10^{2}
4. (a) 4.746×10^{7}
(b) 2.67×10^{8}
5. £1.81×10^{4}
6. 1.35×10^{9} km³
7. 3.51%
8. (a) £4.01×10^{10}
(b) £1070 (3 s.f.)

CHAPTER 7

Exercise 7.1 — Page 63

1. (a) 1 : 2 (b) 1 : 3 (c) 3 : 4 (d) 2 : 5
(e) 3 : 4 (f) 2 : 5 (g) 3 : 7 (h) 9 : 4
(i) 4 : 9 (j) 7 : 3
2. (a) 12 (b) 28 (c) 100 (d) 20
3. 198 cm
4. 400 g
5. 64
6. 18 years old
7. 2 : 3
8. 5 : 2
9. 1 : 250
10. 1 : 1500
11. (a) 4 : 1 (b) 2 : 25
(c) 11 : 2 (d) 5 : 2
(e) 4 : 1 (f) 40 : 17
(g) 9 : 20 (h) 25 : 1
12. 40 : 9
13. 2 : 3
14. 1 : 2
15. 3 : 2

Exercise 7.2 — Page 65

1. (a) 6, 3
(b) 15, 5
(c) 7, 28
(d) 90, 10
2. 8
3. 18
4. Sunny £36, Chandni £12
5. (a) £14, £21
(b) £32, £24
(c) £3.50, £2
(d) £1.80, £3
6. 45
7. £192
8. $\frac{1}{4}$
9. 80%
10. $\frac{5}{8}$
11. 60%
12. 48%
13. 170 000 km²
14. (a) Jenny 50, Tim 30
(b) 10
15. (a) 5 : 2
(b) 6
(c) 55
(d) 32 is not a multiple of $5 + 2 = 7$
16. 8 cm, 12 cm, 18 cm
17. 200 kg
18. 40°, 60°, 80°
19. 45%

Exercise 7.3 — Page 66

1. (a) 16p
(b) £1.28
2. £2.85
3. 4
4. (a) 120 cm
(b) 2 kg
5. £201.60
6. 18 minutes
7. (a) 250 g
(b) 150 g
8. £2.85
9. 15
10. £89.28
11. £9.10
12. (a) 12 minutes
(b) 16 miles
13. (a) 12 m²
(b) 12 litres
14. 24
15. (a) £3.08
(b) 12 minutes
16. (a) £260
(b) 32
17. 50 minutes
18. $17\frac{1}{2}$ minutes
19. £8.35
20. £4 213 100

Review Exercise 7 — Page 68

1. 1 : 3
2. 9
3. £262.50, £157.50
4. £4900
5. £16.92
6. 25 minutes
7. £4
8. 1 : 200 000
9. (a) $\frac{3}{4}$ (b) $12\frac{1}{2}\%$ (c) $200\,cm^3$
10. (a) 35 (b) 40
11. 35%
12. (a) 37.5% (b) 24

CHAPTER 8

Exercise 8.1 — Page 70

1. 8 miles per hour
2. 7 km/h
3. 25 metres per minute
4. (a) 20 km/h (b) 50 km/h (c) 4 km/h
5. 10 km/h
6. 8 km
7. 30 miles
8. 3 km
9. (a) 150 km (b) 86 km (c) 40 km
10. $\frac{1}{2}$ hour
11. 20 seconds
12. $1\frac{1}{2}$ hours
13. (a) 3 hours (b) 2 hours (c) $3\frac{1}{2}$ hours
14. $2\frac{1}{4}$ hours
15. 6 km
16. $1\frac{1}{4}$ hours
17. (a) $2\frac{1}{2}$ hours (b) $1\frac{1}{4}$ hours
18. (a) 60 km/h (b) 1 hour
19. 11.20 am

Exercise 8.2 — Page 71

1. (a) 300 km (b) 5 hours (c) 60 km/h
2. 5 m/s
3. 10 km/h
4. 10.30 am
5. 11.09 am
6. (a) 13.8 km/h (b) 1.10 pm
7. (a) 8 hours (b) 111 km/h
8. (a) 4.81 m/s (b) 17.3 km/h
9. 150 m
10. 300 000 000 m/s

Exercise 8.3 — Page 72

1. $8\,g/cm^3$
2. $9\,g/cm^3$
3. $2.5\,g/cm^3$
4. $28.6\,g/cm^3$
5. 7200 g
6. $800\,cm^3$
7. $2000\,cm^3$
8. $0.8\,g/cm^3$
9. (a) $350\,cm^3$ (b) $2.57\,g/cm^3$
10. 118.3 people/km^2
11. (a) 30 530 km^2 (b) 104.2 people/km^2 (c) 57 400 000
12. 1150 g

Review Exercise 8 — Page 74

1. 94 km/h
2. 14 km
3. 60 miles per hour
4. 40 miles per hour
5. 45 km/h
6. 2 hours 40 minutes
7. 1315
8. 8 km/h
9. 36 minutes
10. 5 hours 48 minutes
11. 25 km/h
12. 24 km/h
13. 125 m
14. $2.92\,g/cm^3$
15. $4.35\,cm^3$
16. 2 900 000 people
17. 40 minutes

CHAPTER 9

Exercise 9.1 — Page 76

1. (a) $\frac{7}{9}$ (b) $\frac{1}{9}$ (c) $\frac{8}{9}$ (d) $\frac{4}{11}$ (e) $\frac{82}{99}$ (f) $\frac{5}{37}$ (g) $\frac{8}{37}$ (h) $\frac{425}{999}$ (i) $\frac{6}{37}$ (j) $\frac{2}{7}$
2. (a) $\frac{1}{6}$ (b) $\frac{1}{30}$ (c) $\frac{11}{18}$ (d) $\frac{7}{15}$ (e) $\frac{7}{30}$ (f) $\frac{5}{6}$ (g) $\frac{181}{990}$ (h) $\frac{3}{22}$ (i) $\frac{31}{36}$ (j) $\frac{17}{22}$
3. $\frac{2}{17}$
4. 2, 4, 5, 8, 10, 16, 20

Exercise 9.2 — Page 77

1. (a) $\frac{1}{2}$ (b) $\frac{9}{20}$ (c) $\frac{3}{10}$ (d) $\frac{5}{8}$ (e) $\frac{9}{4}$ (f) $-\frac{1}{4}$ (g) $\frac{3}{11}$ (h) 7 (i) $\frac{7}{20}$ (j) $\frac{25}{111}$ (k) 2 (l) $-\frac{1}{5}$ (m) 4 (n) $\frac{3}{4}$ (o) $\frac{11}{90}$
2. (a) Rational, $\frac{1}{4}$ (b) Rational, $\frac{5}{9}$ (c) Irrational (d) Rational, 5 (e) Rational, $\frac{3}{5}$ (f) Rational, $-\frac{1}{2}$ (g) Irrational (h) Rational, $\frac{19}{9}$ (i) Rational, 2 (j) Rational, $\frac{4}{3}$

Exercise 9.3 — Page 78

1. (a), (d)
2. (a) $2\sqrt{3}$ (b) $3\sqrt{3}$ (c) $3\sqrt{5}$ (d) $4\sqrt{3}$ (e) $4\sqrt{2}$ (f) $5\sqrt{2}$ (g) $3\sqrt{6}$ (h) $2\sqrt{6}$ (i) $7\sqrt{2}$ (j) $4\sqrt{5}$ (k) $3\sqrt{7}$ (l) $10\sqrt{2}$ (m) $8\sqrt{2}$ (n) $4\sqrt{7}$ (o) $5\sqrt{7}$
3. (a) $2\sqrt{2}$ (b) $\sqrt{5}$ (c) $7\sqrt{3}$ (d) $5\sqrt{2}$ (e) $\sqrt{2}$ (f) $7\sqrt{5}$ (g) $3\sqrt{3}$ (h) $6\sqrt{3}$ (i) $6\sqrt{2}$ (j) $12\sqrt{5}$ (k) $14\sqrt{3}$ (l) $5\sqrt{5}$
4. (a) $10\sqrt{2}$ (b) $5\sqrt{3}$ (c) $6\sqrt{2}$
5. (a) $\frac{3}{2}$ (b) $\frac{5}{4}$ (c) $\frac{3}{2}$ (d) $\frac{2\sqrt{6}}{3}$ (e) $\frac{\sqrt{6}}{2}$ (f) $\frac{2\sqrt{5}}{5}$
6. (a) 3 (b) 6 (c) 30 (d) 4 (e) 6 (f) $5\sqrt{2}$ (g) $6\sqrt{2}$ (h) $10\sqrt{2}$ (i) 12 (j) $6\sqrt{6}$ (k) $8\sqrt{6}$ (l) $12\sqrt{6}$
7. (a) $2 + \sqrt{2}$ (b) $3\sqrt{2} - 3$ (c) $5\sqrt{2} - 5$

Exercise 9.4 — Page 79

1. (a) $\frac{\sqrt{3}}{3}$ (b) $\frac{\sqrt{5}}{5}$ (c) $\frac{\sqrt{7}}{7}$
   (d) $\sqrt{2}$ (e) $\sqrt{5}$ (f) $2\sqrt{2}$
   (g) $2\sqrt{3}$ (h) $2\sqrt{7}$ (i) $\frac{\sqrt{6}}{2}$
   (j) $3\sqrt{5}$ (k) $3\sqrt{3}$ (l) $\frac{\sqrt{15}}{3}$
   (m) $3\sqrt{6}$ (n) $7\sqrt{5}$
2. (a) $\frac{3\sqrt{2}}{2}$ (b) $\frac{\sqrt{6}}{2}$ (c) $\sqrt{2}$
   (d) $\frac{3\sqrt{2}}{2}$ (e) $\frac{\sqrt{2}}{2}$ (f) 2
   (g) 3 (h) $\frac{1}{2}$ (i) $\sqrt{3}$
   (j) $2\sqrt{2}$ (k) 1 (l) $\sqrt{5}$
   (m) $\frac{\sqrt{2}}{2}$ (n) $\frac{\sqrt{3}}{3}$ (o) $\frac{5\sqrt{3}}{6}$
   (p) $\frac{\sqrt{10}}{5}$ (q) $\frac{3\sqrt{10}}{10}$ (r) $\frac{2\sqrt{3}}{3}$

Review Exercise 9 — Page 80

1. (a) $0.\dot{5}$ (b) $\frac{23}{90}$
2. $\frac{16}{33}$
3. (a) $\frac{5}{11}$ (b) $\frac{71}{110}$ (c) $\frac{325}{999}$
4. (a) $3\sqrt{2}$ (b) $7\sqrt{2}$
5. (a) $2\sqrt{5}$ (b) $\frac{6}{5}$ (c) 6
6. (a) $2\sqrt{3}$ (b) $5\sqrt{3}$ (c) $\frac{\sqrt{5}}{2}$
7. (a) $2\sqrt{5}$ (b) $\frac{5\sqrt{5} - 3\sqrt{5}}{5\sqrt{5} + 3\sqrt{5}} = \frac{2\sqrt{5}}{8\sqrt{5}} = \frac{1}{4}$
8. $10 \times 0.4\dot{7} = 4.7\dot{7}$
   $-\ 1 \times 0.4\dot{7} = 0.4\dot{7}$
   $9 \times 0.4\dot{7} = 4.3$
   $0.4\dot{7} = \frac{4.3}{9} = \frac{43}{90}$
9. $9\sqrt{2}$
10. $\sqrt{12}(\sqrt{75} - \sqrt{48}) = 2\sqrt{3}(5\sqrt{3} - 4\sqrt{3})$
    $= 2\sqrt{3} \times \sqrt{3}$
    $= 2 \times 3 = 6$
11. (a) $7\sqrt{2}$ (b) $70\sqrt{3}$

Number — Section Review

Non-calculator Paper — Page 81

1. (a) 100 000 (b) 7 (c) 8
   (d) 0.09 (e) 0.45
2. (a) 25 (b) 75%
3. (a) 36
   (b) (i) $1\frac{5}{12}$ (ii) $\frac{1}{10}$ (iii) $\frac{3}{10}$
   (c) Any decimal between 0.25 and 0.33…
4. 6
5. Large (500 g) = 2 × Standard (250 g)
   2 × £1.39 = £2.78 and £2.78 > £2.69
   Large size is better value for money.
6. (a) $\frac{70}{10 - 3}$ (b) 10
7. 40 miles per hour
8. (a) 220 (b) 60%
9. £90
10. 76p
11. 3^4 is bigger. $2^6 = 64$, $3^4 = 81$.
12. Minimum: 395, maximum 404
13. (a) 17.1911 (b) 171.911
14. (a) 4.2 (b) 0.08 (c) $1\frac{17}{20}$
15. 15
16. No. Year 1: £500 × 0.04 = £20
    Year 2: £520 × 0.04 = £20.80
    Total interest = £20 + £20.80 = £40.80
17. Least: 74.5 m, greatest: 75.5 m
18. 7 g/cm^3
19. (a) $2^3 \times 3^2$ (b) 24
20. Andi is correct.
    Increase of 10% = 110% of original speed
    = 1.1 × original speed
    Decrease of 10% = 90% × 1.1 = 0.99
    = 99% of original speed
21. (a) $10^2 \div 0.4 = 250$
    (b) (i) $p = 2$, $q = 3$
    (ii) $2 \times 3 \times 3 = 2 \times 3^2$
    (iii) 72
22. A: False. The prime numbers will be factors of the product.
    B: True. Even number + Odd number
    = Odd number + Even number
    = Odd number.
23. 3 hours 30 minutes
24. 42 miles per hour
25. (a) $\frac{600 \times 5}{0.2} = 15\,000$ (b) $4\frac{4}{15}$
    (c) 1 (d) 8×10^{13}
26. £40
27. 36 km/h
28. $3\frac{1}{4}$
29. (a) £434.70 (b) 2.5 kg
30. Sam has 5 less counters
31. (a) 2.4×10^3 (b) 8×10^3
    (c) 1.6×10^{11}
32. 2.5 m
33. (a) $\frac{34}{99}$ (b) $\frac{256}{2475}$
34. (a) $5\sqrt{2}$ (b) $12\sqrt{6}$
35. (a) 5 (b) $\frac{1}{7}$ (c) 3^{-2} (d) 4
36. 68%
37. $\frac{2}{7}$

38. (a) If n is even, then $n + 2$ is even, so, 2 even numbers. If n is odd, then $n + 1$ is even, so, 1 even number.
(b) One value is a multiple of 3, e.g. 1, 2, 3 or 5, 6, 7, etc.
One value is even.
Multiple of 2 × multiple of 3 = multiple of 6.

39. $\frac{90 \times 10^6}{2 \times 10^5} = 450$ seconds ≈ 8 minutes

40. (a) $\frac{\sqrt{2}}{4}$
(b) $\sqrt{12} + \sqrt{108} = 8\sqrt{3}$ $\quad \frac{8\sqrt{3}}{\sqrt{8}} = 2\sqrt{6}$

41. (a) $\frac{1}{9}$ (b) 2^5 (c) $\frac{\sqrt{6}}{2}$

42. $\frac{7}{110}$

Number — Section Review

Calculator Paper — Page 84

1. 45 miles per hour
2. 35 pence
3. (a) 5.29 (b) 30 (c) 6.2
4. (a) Yum. Yum 8.4 g/p, Core 8.3 g/p (b) 39p
5. (a) 62.4 kg (b) 10%
6. £65.80
7. (a) 75 g (b) 40%
8. £1.92
9. If 1st number is odd, last number is odd.
Odd + Odd = Even
If 1st number is even, last number is even.
Even + Even = Even
10. (a) 39.520… (b) $\frac{20^2}{7 + 3} = \frac{400}{10} = 40$
11. 95.8%
12. (a) 1 953 125 (b) 0.143 (c) 7.8
13. (a) Always even.
(b) Could be either odd or even.
14. Yes, she does have enough money.
400 000 dollars = £291 971
450 000 euros = £288 462
15. 2 hours 40 minutes
16. Angela £20, Fran £35, Dan £45
17. 16.4%
18. (a) 220 (b) 2.401×10^7
19. (a) 18 (b) 7560
20. 4.78×10^{-1}
21. (a) (i) 40 (ii) 1 : 1.25
(b) 7×10^6 (c) £11
22. £27
23. (a) (i) 1 (ii) 1.67 (iii) 26.62
(b) $(1 + 2 + 3)^2 - (1^2 + 2^2 + 3^2)$
$= 36 - 14 = 22$
24. 50.9 lb/cu.ft. (3 s.f.)
25. 0.30 euros
26. 58 quarts
27. 19 cm
28. 6 weeks
29. (a) 3 906 250
(b) (i) 0.110 (3 s.f.)
(ii) Accuracy of numbers in the question only 3 sig. figs.
(c) 0.153 (3 s.f.)
30. (a) 2 hours 51 minutes (b) 530 mph
31. (5.92×10^{-4})%
32. (a) $\frac{1}{64}$ (b) $\frac{3}{5}$ (c) $\frac{1}{16}$
33. 44%
34. (a) 1.441
(b) No. Maximum safe load is 1450 kg.
Maximum weight of boxes:
$3 \times 141.5 + 7 \times 150.5 = 1478$ kg
$1478 > 1450$
35. $a^3 + a = a(a^2 + 1)$
If a is odd, a^2 is odd, so, $(a^2 + 1)$ is even.
Odd × Even = Even. If a is even, a^2 is even, so, $(a^2 + 1)$ is odd. Even × Odd = Even
36. (a) (i) $10\sqrt{3}$ (ii) $11\sqrt{3}$ (iii) $6\sqrt{6}$
(b) $\frac{\sqrt{10}}{15}$
37. (a) 0.4 seconds (b) 7.384 m/s (4 s.f.)
38. (a) E.g. If $q = 4\sqrt{a}$
then $pq = 4\sqrt{a} \times \sqrt{a} = 4a$,
which is an integer
(b) $110 - 12\sqrt{6}$

CHAPTER 10

Exercise 10.1 — Page 87

1. $m + 6$
2. $m - 12$
3. $8m$
4. $p - 1$
5. $p + 5$
6. $25p$
7. $6k$
8. $5b$ pence
9. $\frac{c}{3}$ pence
10. $\frac{a}{5}$ pence
11. $\frac{36}{g}$

Exercise 10.2 — Page 88

1. (a) $3n$ (b) $5y$ (c) $10g$ (d) $4y$
(e) $8m$ (f) $-4x$ (g) $-14a$
2. (a) Can be simplified, $2v$.
(b) Cannot be simplified, different terms.
(c) Can be simplified to $3v + 4$.
(d) Cannot be simplified, different terms.
3. (a) $8x + y$ (b) $w + 2v$ (c) $2a - 2b$
(d) $a + b$ (e) $5 - 5k$ (f) $a + 3$
(g) $3b - 2a$ (h) $2x + 3y$ (i) $a + 2b$
(j) $-2f$ (k) $v - 4w$ (l) $-2 - 5t$
4. (a) $2xy$ (b) $2pq$
(c) $3ab - a + b$ (d) $2x^2$
(e) $9y^2 - y$ (f) $4a^2 + a$
(g) $2d^2 - 3g^2$ (h) $5t^2 - t$
(i) $6m - m^2$
5. (a) $4x + 2$ (b) $4a + 6b$
(c) $3x$ (d) $6y + 9$

Exercise 10.3 — Page 89

1. (a) $3a$ (b) $8c$ (c) $9d$ (d) $8f$ (e) $6p$ (f) $15q$ (g) r^2 (h) $2g^2$ (i) $6g^2$ (j) $12t^2$ (k) $15u^2$ (l) $9d^2$

2. (a) $-3y$ (b) $-5y$ (c) $2y$ (d) $-6y$ (e) $-t^2$ (f) $-2t^2$ (g) $-10t^2$ (h) $10t^2$

3. (a) $5a$ (b) 12 (c) 20 (d) 2 (e) 18 (f) $3p$ (g) 3 (h) 9 (i) 4 (j) 5 (k) 4 (l) 9

4. (a) $-2y$ (b) $-3y$ (c) -5 (d) m (e) -3 (f) -2 (g) -3 (h) 1

5. (a) ab (b) y^2 (c) $2pq$ (d) $2a^2$ (e) $12gh$ (f) abc (g) m^3 (h) $2d^3$ (i) $6x^3$ (j) $10m^2n$ (k) $3abc$ (l) $18pqr$

6. (a) y^3 (b) t^5 (c) g^{10} (d) m^6 (e) $6d^5$ (f) $8x^5$ (g) $6t^5$ (h) $24r^6$ (i) m^3n^2 (j) a^4b^2 (k) $6r^3s^4$ (l) $10x^5y^3$

7. (a) t^6 (b) y^6 (c) g^9 (d) x^8 (e) $9a^2$ (f) $8h^3$ (g) $2m^6$ (h) $4m^6$ (i) $3d^6$ (j) $9d^6$ (k) $27a^6$ (l) $8k^9$ (m) x^2y^2 (n) m^3n^6 (o) $8s^3t^3$ (p) $9p^4q^2$

8. (a) y^2 (b) a (c) 1 (d) g^{-1} (e) $6b^2$ (f) $5m$ (g) $4t$ (h) $3h^{-1}$ (i) y (j) m^2n^{-1} (k) $\frac{2}{3}p^2q^{-1}$ (l) $3rs^2$

9. (a) t^2 (b) x^2 (c) $\frac{12}{b^3}$ (d) $15pq^2$ (e) x^{10} (f) $\frac{1}{a}$ (g) $3x^4$ (h) $\frac{2w^4}{t^5}$ (i) $\frac{1}{a^6}$ (j) y^4 (k) $\frac{5}{m^9}$ (l) $\frac{n^4}{9}$

10. (a) t (b) g^{-1} (c) m^2 (d) y (e) y^2 (f) m^{-1} (g) $2t^2$ (h) 3

11. (a) $2v$ (b) $4p$ (c) $\frac{ac^3}{2}$ (d) $\frac{4}{m}$ (e) $\frac{16}{w^2}$ (f) 1 (g) 4 (h) $4ac^{-3}$

Exercise 10.4 — Page 91

1. (a) $2x + 10$ (b) $3a + 18$ (c) $4y + 12$

2. (a) $3x + 6$ (b) $2y + 10$ (c) $4x + 2$ (d) $3p + 3q$

3. (a) $a^2 + a$ (b) $2d + d^2$ (c) $2x^2 + x$

4. (a) $2x + 8$ (b) $3t - 6$ (c) $6a + 2b$ (d) $15 - 10d$ (e) $x^2 + 3x$ (f) $t^2 - 3t$ (g) $2g^2 + 3g$ (h) $2m - 3m^2$

5. (a) $2p^2 + 6p$ (b) $6d - 9d^2$ (c) $m^2 - mn$ (d) $5x^2 + 10xy$

6. (a) $2x + 5$ (b) $3a + 11$ (c) $6w - 17$ (d) $10 + 2p$ (e) $3q$ (f) $7 - 3t$ (g) $5z + 8$ (h) $8t + 15$ (i) $2c - 6$ (j) $5a - 9$ (k) $3y - 10$ (l) $2x + 6$ (m) $8a + 23$ (n) $10x - 12$ (o) $2p - 11$ (p) $5a + 2b$ (q) $3x + y$ (r) $2p - 5q$ (s) $5x - x^2$ (t) $a^2 - 2a$ (u) $2y$

7. (a) $5x + 8$ (b) $5a + 13$ (c) $9y + 23$ (d) $9a + 5$ (e) $26t + 30$ (f) $5z + 13$ (g) $12q + 16$ (h) $11x - 3$ (i) $20e - 16$ (j) $6d^2 + 3d$ (k) $3m^2 - 3m$ (l) $5a^2 - 4a$

8. (a) $-3x + 6$ (b) $-3x + 6$ (c) $-2y + 10$ (d) $-6 + 2x$ (e) $-15 + 3y$ (f) $-4 - 4a$ (g) $3 - 2a$ (h) $2d + 6$ (i) $2b - 6$ (j) $-6p - 9$ (k) $m - 6$ (l) $5d + 1$ (m) $-a^2 + 2a$ (n) $d - d^2$ (o) $2x^2$ (p) $-6g^2 - 9g$ (q) $7t^2 - 6t$ (r) $8m - 2m^2$

9. (a) $3a - 1$ (b) $y - 5$ (c) $5m - 1$ (d) $3x - 4$ (e) $1 - d$ (f) $t - 10$ (g) $4m + 9$ (h) $-x - 18$ (i) $23 - 6a$

Exercise 10.5 — Page 93

1. (a) $2(x + y)$ (b) $3(a - 2b)$ (c) $2(3m + 4n)$ (d) $x(x - 2)$ (e) $a(b + 1)$ (f) $x(2 - y)$ (g) $2a(b - 2)$ (h) $2x(2x + 3)$ (i) $dg(1 - g)$

2. (a) $2(a + b)$ (b) $5(x - y)$ (c) $3(d + 2e)$ (d) $2(2m - n)$ (e) $3(2a + 3b)$ (f) $2(3a - 4b)$ (g) $4(2t + 3)$ (h) $5(a - 2)$ (i) $2(2d - 1)$ (j) $3(1 - 3g)$ (k) $5(1 - 4m)$ (l) $4(k + 1)$

3. (a) $x(y - z)$ (b) $g(f + h)$ (c) $b(a - 2)$ (d) $q(3 + p)$ (e) $a(1 + b)$ (f) $g(h - 1)$ (g) $a(a + 3)$ (h) $t(5 - t)$ (i) $d(1 - d)$ (j) $m(m + 1)$ (k) $r(5r - 3)$ (l) $x(3x + 2)$

4. (a) $3(y + 2 - 3x)$ (b) $t(t^2 - t + 1)$
(c) $2d(d + 2)$ (d) $3m(1 - 2n)$
(e) $2g(f + 2g)$ (f) $4q(p - 2)$
(g) $3y(2 - 5y)$ (h) $2x(3x + 2y)$
(i) $2n(3n - 1)$ (j) $2b(2a + 3)$
(k) $\frac{1}{2}a(1 - a)$ (l) $4x(5 + y)$
(m) $a^2(a + a^3 + 1)$ (n) $\pi r(2 + r)$
(o) $4ab(5a + 3b)$ (p) $3pq(1 - 3p)$

Review Exercise 10 — Page 94

1. $6t$ pence
2. $(x + 3)$ years old
3. (a) $(n - 2)$ years old (b) $3n$ years old
(c) $(3n + 4)$ years old (d) $(8n + 2)$ years
4. (a) £1.50 (b) $25n$ pence (c) $x = 25y$
5. (a) $3ab$ (b) $a^2 + 2a$
(c) $2x - 6$ (d) $5x + 2$
6. (a) (i) $2x$ units (ii) $(2x - 3)$ units
(b) $(5x - 3)$ units
7. (a) $y^2 - 4y$ (b) $7y + 4$
8. (a) $3x - 6y$ (b) $2x^2 + 6x$ (c) $x^3 - 3x^2$
9. (a) p^3 (b) $24abc$ (c) 1
10. (a) $5x + 10y$ (b) $6x^2 - 3x + 4$
11. (a) $2x^2 - x$ (b) $1 - 3x$
12. (a) (i) $3(2x - 5)$ (ii) $y(y + 7)$
(b) $y + 12$
13. (a) t^8 (b) p^4 (c) a^4
14. (a) $6y^5$ (b) $2t^3$ (c) $8a^3$ (d) a^5b^2
15. (a) (i) $3y(2x - y)$ (ii) $2m(2m + 3)$
(b) (i) $10y^5$ (ii) $3x^4$
16. $2a^2b^2c$

CHAPTER 11

Exercise 11.1 — Page 95

1. (a) 3 (b) 4 (c) 9 (d) 16
2. (a) $x = 4$ (b) $a = 3$ (c) $y = 8$
(d) $t = 6$ (e) $h = 22$ (f) $d = 1$
(g) $z = 30$ (h) $p = 0$ (i) $c = 99$
3. (a) 5 (b) 5 (c) 18 (d) 21
4. (a) $a = 4$ (b) $e = 6$ (c) $p = 4$
(d) $y = \frac{1}{2}$ (e) $d = 10$ (f) $t = 9$
(g) $m = 28$ (h) $x = 100$
5. (a) 1 (b) 4 (c) 4 (d) 2 (e) 3 (f) 5

Exercise 11.2 — Page 96

1. 5 **3.** 6 **5.** 4 **7.** 4
2. 14 **4.** 11 **6.** 3 **8.** 7
9. (a) 2 (b) $2(x + 3) = 2x + 6$
10. (a) 9 (b) $3(x - 2) = 3x - 6$

Exercise 11.3 — Page 98

1. (a) $y = 3$ (b) $n = 16$ (c) $x = 2$
(d) $y = 19$ (e) $b = 7$ (f) $x = 29$
(g) $m = 4$ (h) $k = 5$ (i) $y = 7$
2. (a) $c = 4$ (b) $a = 4$ (c) $f = 3$
(d) $d = 30$ (e) $e = 14$ (f) $m = 20$
3. (a) $p = 4$ (b) $t = 3$ (c) $h = 7$
(d) $b = 2$ (e) $d = 10$ (f) $x = 6$
(g) $c = 5$ (h) $n = 3$ (i) $x = 2$

Exercise 11.4 — Page 98

1. (a) $k = \frac{1}{2}$ (b) $a = -3$ (c) $d = -4$
(d) $n = -\frac{1}{2}$ (e) $t = -5$ (f) $n = 1$
(g) $m = 1\frac{1}{2}$ (h) $x = 2\frac{1}{3}$ (i) $y = -\frac{1}{2}$
2. (a) $x = -2$ (b) $y = -3$ (c) $t = -2$
(d) $a = -2$ (e) $d = -3$ (f) $g = -3$
(g) $t = \frac{1}{2}$ (h) $x = 7\frac{1}{2}$ (i) $d = 1\frac{2}{5}$
(j) $a = 1\frac{1}{2}$ (k) $g = \frac{1}{5}$ (l) $b = 4\frac{1}{2}$
3. (a) $x = -2$ (b) $n = -\frac{1}{2}$ (c) $a = -1$
(d) $y = -3$ (e) $x = -1$ (f) $d = -3$
(g) $x = -1\frac{1}{2}$ (h) $x = -4$ (i) $x = -1\frac{1}{2}$
(j) $n = 1\frac{1}{2}$ (k) $z = 10$ (l) $m = -3$
(m) $n = 4$ (n) $p = -1\frac{1}{2}$ (o) $v = 9$

Exercise 11.5 — Page 99

1. (a) $x = 3$ (b) $a = 2$ (c) $t = 2$
(d) $p = 5$ (e) $c = 6$ (f) $x = 3$
(g) $d = 9$ (h) $e = 5$ (i) $f = 4$
2. (a) $w = 2$ (b) $s = 3$ (c) $t = 3$
(d) $y = 1$ (e) $x = 3$ (f) $y = 5$
3. (a) $p = -1$ (b) $d = -2$ (c) $g = -2$
(d) $x = 8\frac{1}{2}$ (e) $y = \frac{2}{5}$ (f) $t = \frac{1}{2}$
(g) $t = 1\frac{3}{4}$ (h) $a = 2\frac{1}{2}$ (i) $m = 2\frac{3}{5}$
4. (a) $x = 4$ (b) $a = 2$ (c) $m = -\frac{1}{2}$
(d) $y = 2\frac{1}{2}$ (e) $w = -3$ (f) $e = 2$
(g) $a = 1\frac{1}{2}$ (h) $t = -11$ (i) $x = 1$

Exercise 11.6 — Page 100

1. (a) $x = 5$ (b) $q = 2$ (c) $t = 3$
(d) $e = 3$ (e) $g = 4$ (f) $y = 1$
(g) $x = 2$ (h) $k = 1$ (i) $a = 4$
(j) $p = 6$ (k) $m = 2$ (l) $d = 5$
(m) $y = 5$ (n) $u = 3$ (o) $q = 0$
2. (a) $d = 8$ (b) $q = 3$ (c) $c = 2$
(d) $t = 3$ (e) $w = 2$ (f) $e = 3$
(g) $g = 5$ (h) $z = 4$ (i) $m = 6$
(j) $a = 5$ (k) $x = 4$ (l) $y = 3$

3. (a) $m = -4$ (b) $t = -2$ (c) $p = -2$
(d) $x = 3\frac{1}{2}$ (e) $a = \frac{1}{2}$ (f) $b = \frac{4}{5}$
(g) $y = \frac{4}{5}$ (h) $d = \frac{3}{4}$ (i) $f = -3\frac{1}{2}$

4. (a) $x = 4$ (b) $a = 1\frac{2}{3}$ (c) $m = \frac{1}{2}$
(d) $a = -5$ (e) $y = 5\frac{1}{2}$ (f) $n = -1\frac{1}{2}$
(g) $d = -4\frac{1}{2}$ (h) $k = -11$ (i) $t = -4$
(j) $q = 2$ (k) $x = \frac{2}{3}$ (l) $a = 2\frac{1}{2}$

5. (a) $h = 2$ (b) $x = -3$ (c) $w = 3$
(d) $y = 1\frac{1}{2}$ (e) $v = -8$ (f) $c = -1\frac{1}{2}$
(g) $x = -18$ (h) $x = -7$

Exercise 11.7 Page 101

1. (a) $x = 6$ (b) $d = -10$ (c) $a = 6$
(d) $m = -2$ (e) $t = \frac{4}{9}$

2. (a) $h = 11$ (b) $x = 8$ (c) $a = -3$
(d) $d = -3$ (e) $a = \frac{1}{10}$ (f) $h = 1\frac{1}{2}$
(g) $x = \frac{7}{9}$ (h) $a = 5$ (i) $x = 1\frac{1}{4}$

3. (a) $x = 1\frac{1}{3}$ (b) $x = 12$ (c) $x = -1\frac{3}{5}$
(d) $x = -4$ (e) $x = 1$ (f) $x = 19$
(g) $x = -1$ (h) $x = -1$ (i) $x = 3\frac{1}{2}$

Exercise 11.8 Page 102

1. (a) $6k$ kg (b) $2\frac{1}{2}$ kg
2. (a) $6x + 3$ (b) 7
3. (a) $(n + 7)$ years old
(b) Dominic is 18 years old
Marcie is 25 years old
4. (a) $(4y - 2)$ cm (b) 19 cm
5. (a) (i) £$(p + 4)$ (ii) £$(p - 3)$
(iii) £$(3p + 1)$
(b) Aimee £12, Grace £8, Lydia £5
6. (a) $(x + 10)$ pence (b) $(3x + 20)$ pence
(c) 15 pence
7. $3x + 2(x - 4) = 77$, $x = 17$
Cream biscuit 13 pence.
8. (a) $(7x - 4)$ cm (b) $7x - 4 = 59$, $x = 9$
(c) 13 cm, 17 cm, 29 cm
9. (a) $(x - 4)$ cm (b) $(4x - 8)$ cm
(c) $x = 7$
10. (a) $18x = 540°$ (b) largest angle = 150°
11. Cake costs 37 pence
12. 105 cm^2
13. (a) $(6x + 16)$ cm (b) $x = 2$ (c) 28 cm
14. (a) $5x + 12$
(b) $5x + 12 = 2(x + 12)$ so, $x = 4$.
Sarah has 32 CDs now.

Review Exercise 11 Page 104

1. (a) 7 (b) 4
2. (a) $x = 20$ (b) $x = 4$ (c) $n = 2$
3. (a) $y = 2$ (b) $t = -3$
(c) $g = \frac{1}{2}$ (d) $x = \frac{3}{5}$
4. (a) $x = 7$ (b) $x = -2$
5. (a) $x = 8$ (b) $x = 1\frac{3}{5}$ (c) $x = 5$
6. (a) $x = 1\frac{1}{4}$ (b) $x = 1\frac{1}{2}$
7. (a) $x = 6$ (b) $x = \frac{1}{2}$ (c) $x = 23$
8. (a) $x = -1$ (b) $x = 2\frac{4}{5}$
9. (a) $q = -2$ (b) $t = 1\frac{1}{2}$ (c) $x = 10\frac{1}{2}$
10. (a) 13 years old (b) 26 years old
11. $2\frac{1}{2}$ litres
12. (a) $(x - 15)$ pence (b) 42 pence
13. (a) $(3x + 2)$ cm (b) $x = 7$
14. $x = -1\frac{1}{2}$
15. (a) $x = -1$ (b) $x = 2\frac{4}{5}$
16. $x = -1$
17. $x = 2\frac{3}{5}$

CHAPTER 12

Exercise 12.1 Page 105

1. (a) −2 (b) 10 (c) 11 (d) 25
2. (a) −1 (b) −4 (c) −12 (d) 3
3. (a) 24 (b) −3 (c) 2 (d) 18 (e) 18
4. (a) 27 (b) −3 (c) $2\frac{1}{2}$
(d) 90 (e) 54
5. (a) −5 (b) −20 (c) −2
(d) −100 (e) 3
6. (a) 3 (b) −21 (c) $-2\frac{1}{2}$
(d) −9 (e) 54
7. (a) −24 (b) 3 (c) 2
(d) 33 (e) −2
8. −17

Exercise 12.2 Page 106

1. (a) $(a + 1)$ years old (b) $(a - 4)$ years old
(c) $(a + n)$ years old
2. $12e$
3. $b - 3$
4. $(h + 12)$ cm
5. (a) $P = 4g$ (b) $P = 4y + 4$
(c) $P = 3x - 1$ (d) $P = 2a + 2b$
6. $C = 25d$
7. (a) $P = y - 2$ (b) $P = 2y$
8. (a) £80 (b) $C = 12x + 8$
9. (a) 115 (b) 175 (c) 45
(d)

n	$n + 1$	$n + 2$
	$n + 11$	
	$n + 21$	

(e) $S_n = 5n + 35$

Exercise 12.3 — Page 108

1. 17 points
2. 26 cm
3. $T = 97$
4. (a) $M = -7$ (b) $n = 4\frac{1}{2}$
5. (a) $H = 2.5$ (b) $g = 6$
6. $F = 75$
7. $V = 2$
8. $P = 12$
9. $C = -40$
10. $S = 6$
11. $T = -13\frac{1}{2}$
12. $K = 13$
13. $L = 5$
14. 33
15. 96 m
16. 86°F
17. 138 minutes
18. 25°C
19. 240 volts
20. $F = 4100$
21. £45.14
22. 0.3

Exercise 12.4 — Page 109

1. $S = 4$
2. (a) $R = 15$ (b) $R = 3$
3. $K = 36$
4. (a) $S = 18$ (b) $S = 18$
5. (a) $S = 36$ (b) $S = 36$
6. $T = 39$
7. (a) $S = 32$ (b) $S = 32$
8. (a) $S = 16$ (b) $S = 16$
9. (a) $A = 8$ (b) $A = -27$
10. $\frac{3}{4}$
11. (a) 4 (b) 1.5 (c) -3
12. (a) $L = 10$ (b) $L = 0.5$
13. $F = 144$
14. $R = 2.25$
15. (a) $v = 6.3$ (b) $v = 7.4$

Exercise 12.5 — Page 110

1. (a) £13 (b) $C = 2k + 3$ (c) 2 km
2. (a) 140 minutes (b) $C = 40k + 20$ (c) 2 kg
3. (a) $F = 42$ (b) $F = 2C + 30$ (c) $C = 14$
4. (a) 410 (b) $b = 3n + 50$ (c) 140

Exercise 12.6 — Page 111

1. (a) $m = a - 5$ (b) $m = a - x$ (c) $m = a + 2$ (d) $m = a + b$
2. (a) $x = \frac{y}{4}$ (b) $x = \frac{y}{a}$ (c) $x = 2y$ (d) $x = ay$ (e) $x = \frac{5y}{3}$
3. (a) $p = \frac{1}{2}y - 3$ (b) $p = \frac{t - q}{5}$ (c) $p = \frac{m + 2}{3}$ (d) $p = \frac{q + r}{4}$
4. $n = \frac{C - 35}{24}$
5. $R = \frac{V}{I}$
6. (a) $c = \pm\sqrt{y}$ (b) $c = y^2$ (c) $c = \pm\sqrt{\frac{y}{d}}$ (d) $c = 9y^2$ (e) $c = \pm\sqrt{y - x}$ (f) $c = (y - x)^2$ (g) $c = \pm\sqrt{dy - dx}$ (h) $c = (ax + ay)^2$
7. (a) $a = b - c^2$ (b) $a = \pm\sqrt{b}$ (c) $a = \frac{p - d}{m}$ (d) $a = \pm\sqrt{\frac{F}{m}}$
8. (a) $x = \frac{3a}{2}$ (b) $x = \frac{3b - 2a}{5}$
9. (a) $a = \frac{3x}{2}$ (b) $a = \frac{b}{1 - x}$ (c) $a = \frac{b + 2}{1 - x}$
10. $t = \pm s\sqrt{R}$
11. $q = \frac{p}{2}$
12. (a) $a = \frac{3x - 2}{1 - x}$ (b) $a = \frac{5y + 3}{1 + y}$ (c) $a = \frac{2b + x}{x - b}$ (d) $a = \frac{6 + y}{y + 3}$ (e) $a = \frac{5x}{3}$ (f) $a = \frac{4x^3}{1 + 4x^2}$

Exercise 12.7 — Page 112

1. (a) $d = \frac{P}{4}$ (b) $d = 0.7$ cm
2. (a) $l = \frac{A}{b}$ (b) $l = 6$ cm
3. (a) (i) $D = ST$ (ii) $D = 96$ km (b) (i) $T = \frac{D}{S}$ (ii) $T = 2.5$ hours
4. (a) $b = \frac{1}{2}P - l$ (b) $b = 4.2$ cm
5. (a) $x = \frac{y - c}{m}$ (b) $x = 3$
6. (a) $b = \frac{2A}{h}$ (b) $b = 6.4$
7. (a) $q = \pm\sqrt{p^2 - r^2}$ (b) $q = \pm 6.8$
8. (a) $r = \sqrt[3]{\frac{3V}{4\pi}}$ (b) $r = 33.7$
9. (a) $x = \pm\sqrt{ab}$ (b) $x = \pm 4.0 \times 10^4$
10. (a) $n = \frac{1 - m}{1 + m}$ (b) $n = 8.4 \times 10^{-1}$

Review Exercise 12 — Page 113

1. $V = 7$
2. $S = -14$
3. $P = 7.5$
4. 6
5. $T = 60$
6. (a) 340 (b) $N = 2T + 20$ (c) 140
7. (a) £276 (b) £$(3M + 60)$ (c) (i) $3M + 60 = 186$ (ii) 42 miles
8. 7 rolls
9. $p = \frac{t + 50}{7}$
10. (a) $v = 9\frac{1}{5}$ (b) $t = \frac{v - u}{a}$
11. (a) $y = 27$ (b) $x = \pm 7$ (c) $x = \pm\sqrt{\frac{y}{3}}$
12. $x = \sqrt{\frac{V}{h}}$

13. $r = st^2$
14. (a) $D = 101.6\,\text{m}$ (b) $v = 68.7\,\text{km/h}$
15. $a = \dfrac{4b + 4c}{c + 2b}$
16. $b = \dfrac{am}{ca + m}$

CHAPTER 13

Exercise 13.1 — Page 114

1. (a) 17, 21, 25 (b) 14, 16, 18
(c) 16, 13, 10 (d) 28, 33, 38
(e) 48, 96, 192 (f) $1\frac{1}{2}$, $1\frac{3}{4}$, 2
(g) 2, 1, $\frac{1}{2}$ (h) 0.9, 1.0, 1.1
(i) 2, 0, −2 (j) 5, 2.5, 1.25
(k) 21, 28, 36 (l) 29, 47, 76
2. (a) 8, 14 (b) 10, 22 (c) 8, 32
(d) 16, −2 (e) 16, 36
(f) 8, 21 (g) 2, 20, 26
3. (a) Add 7; 37, 44
(b) Add 2; 13, 15
(c) Add 4; 21, 25
(d) Subtract 5; 11, 6
(e) Divide by 2; 2, 1
(f) Multiply by 3; 81, 243
(g) Subtract 2; −10, −12
(h) Subtract 3; −5, −8
4. (a) 28
(b) Keep on adding 3 to the last term until you get to the 10th term, 28.
5. (a) David multiplies the last term by 2, $4 \times 2 = 8$.
Tony adds the next counting number, $4 + 3 = 7$.
(b) 512 (c) 46
6. No. To find the next number, add 6 to the last term. All numbers in the sequence will be odd.

Exercise 13.2 — Page 116

1. (a) (i) 21 (ii) 3 (b) 37
2. (a) (i) 36 (ii) 123 (b) 10
3. (a) 37, 60 (b) 1, 4
4. −6, −27
5. (a) 49, 97 (b) 1537

Exercise 13.3 — Page 118

1. (a) 3 (b) 3 (c) 6
(d) 8 (e) −2 (f) −4
2. (a) $3n$ (b) $8n$
3. (a) 3 (c) $3n + 1$ (d) 25
4. (a) 2 (c) $2n + 7$ (d) 47
5. (a) −4 (c) $24 - 4n$
6. (a) $3n - 2$ (b) $22 - 3n$ (c) $4n + 1$
(d) $4n$ (e) $2n - 1$ (f) $4n + 3$
(g) $8 - 2n$ (h) $3n + 2$ (i) $5n - 2$
(j) $45 - 5n$ (k) $n - 1$ (l) $2n - 3$
7. 4, 7, 12.
Substitute $n = 1$, $n = 2$, etc., into $n^2 + 3$.

Exercise 13.4 — Page 120

1. (a) (i) 15 (ii) 30 (iii) 300 (b) $3n$
2. (a) $22\,\text{cm}^2$ (b) $42\,\text{cm}^2$ (c) $4n + 2\,\text{cm}^2$
3. (a) 27 (b) $T = 5n + 2$
(c) 52 (d) 15
4. (a) 11 (b) Pattern 18
(c) $m = 2p + 1$
5. (a) 25 (b) $4^2 = 16$, $5^2 = 25$
(c) n^2 (d) 225
6. (a) (i) 37, add 2 more than last time.
(ii) 35, add 2 more than last time.
(iii) 39, add 2 more than last time.
(iv) 72, add 4 more than last time.
(v) 42, add 2 more than last time.
(b) (i) $n^2 + 1$ (ii) $n^2 - 1$ (iii) $n^2 + 3$
(iv) $2n^2$ (v) $n^2 + n$
7. (a) $T = n^2 + n = n(n + 1)$ (b) 110
8. (a) 15 (b) $\frac{4 \times 5}{2} = 10$, $\frac{5 \times 6}{2} = 15$
(c) $\frac{n(n + 1)}{2}$ (d) 55
9. (a) 32 (b) $2^4 = 16$, $2^5 = 32$
(c) 2^n (d) 1024 (e) 8th
10. (a) 10, 100, 1000, 10 000, 100 000
(b) 10^n (c) 10^6
11. (a) 3^n (b) 59 049

Review Exercise 13 — Page 122

1. (a) 38 (b) $\frac{1}{4}$
2. (a) 81 (b) Yes. $196 = 14^2$
3. (a) 6, 7 (b) No. $\frac{8 + 4}{2} = 6$, $\frac{4 + 6}{2} = 5$
4. (a) 39 (b) $4n - 1$
(c) $4n - 1 = 1997$, so, $4n = 1998$.
1998 is not divisible by 4.
(d) 100
5. (a) (i) 17 (ii) $x + 4$ (b) 23
6. (a) 14 (b) 59 (c) $3n - 1$
7. (a) 33, 45
(b) (i) 65
(ii) Add twice the difference between the two previous terms.
(c) $2n + 1$
8. (a) $3n$ (b) $3n - 2$
9. (a) 30 (b) 40 is not a multiple of 3.
(c) $3p$
10. $2n + 3$

CHAPTER 14

Exercise 14.1 — Page 124

1. $P(-2, -3)$, $Q(3, -2)$, $R(-3, 1)$
2. (b) $(-1, 1)$
3. (c) $D(-1, 2)$

4. (a) (1) $y = 4$ (2) $x = -3$ (3) $x = 1$
(4) $y = -1$ (5) $y = x$
(b) $(1, 4)$ (c) $(-3, -1)$

5.

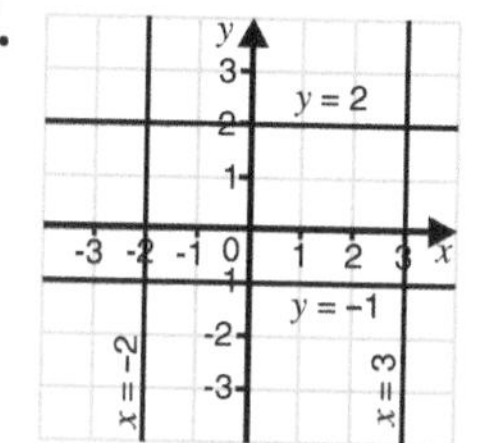

6. (a)

	x	0	1	2	3
(i)	y	2	3	4	5
(ii)	y	0	2	4	6
(iii)	y	0	−1	−2	−3
(iv)	y	2	1	0	−1

7.

	x	−1	0	1	2	3
(a)	y	−2	−1	0	1	2
(b)	y	−1	1	3	5	7
(c)	y	4	3	2	1	0
(d)	y	8	6	4	2	0

8. (a) Missing entries are: 6, 4, 3
(b) (i) $y = 2.5$ (ii) $y = 4.5$

9. (b) (i) $y = -5$ (ii) $x = 0.5$

10. (b) $x = 0.5$

11. (b) $x = -1.5$

12. (b) $(2, 1)$

Exercise 14.2

Page 127

1. (b) Same slope, parallel.
y-intercept is different.

2. (a) gradient 3, y-intercept -1

3. $y = 3x$, $y = 3x + 2$

4. gradients: 3, 2, −2, $\frac{1}{2}$, 2, 0
y-intercepts: 5, −3, 4, 3, 0, 3

5. (a) $y = 5x - 4$ (b) $y = -\frac{1}{2}x + 6$

6. **(1) C** **(2) D** **(3) B** **(4) A**

7. (a) $y = x - 2$ (b) $y = 2x - 2$
(c) $y = -2x - 2$

8. (c) 1 (d) $(0, -4)$ (e) $y = x - 4$

9. (c) $y = -2x - 1$

10. (a) Line slopes up from left to right.
(b) Line slopes down from left to right.
(c) Line is horizontal.

11. $y = 3x$

12. (a) £25 (b) £15 per hour
(c) $y = 15x + 25$ (d) £145

13. (a) £3 (b) £2
(c) $f = 2d + 3$ (d) £13

Exercise 14.3

Page 129

1. (a) $(0, 7)$ (c)
(b) $(7, 0)$

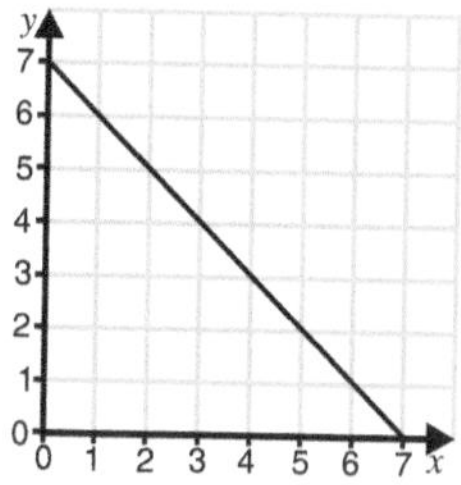

2. (a) 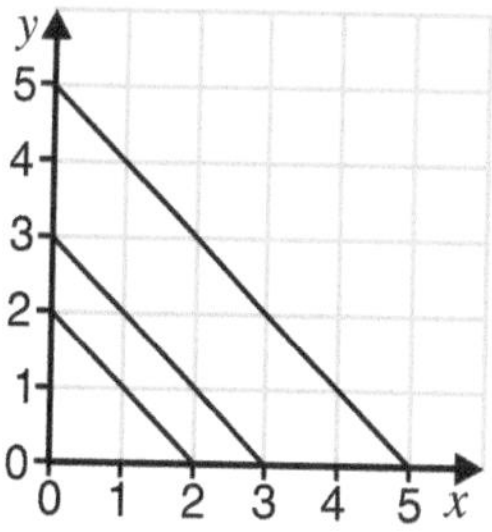 (b) Parallel lines, same gradient.

3. (a) $(0, 6)$ (c)
(b) $(2, 0)$

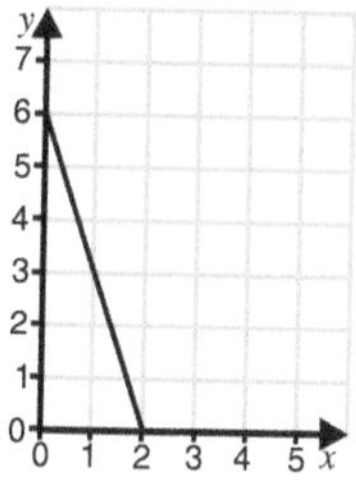

4. (a) 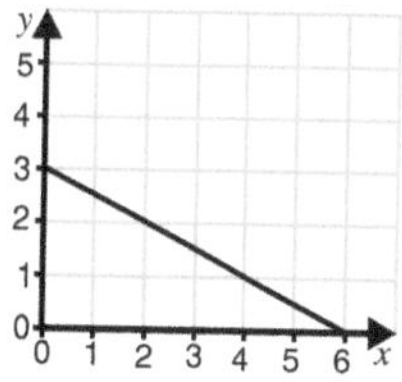 (b) 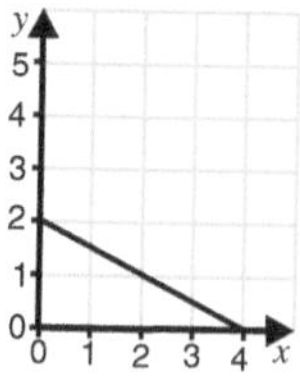
(c)

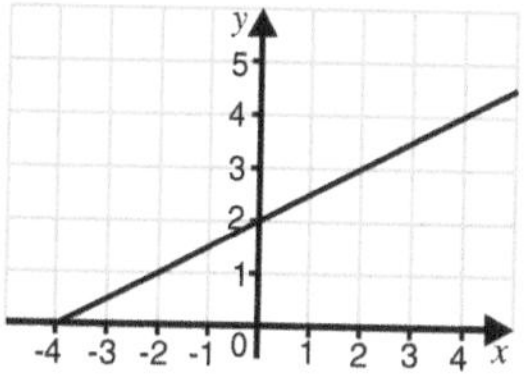

5. (a) $(0, 5)$ (c)
(b) $(3, 0)$

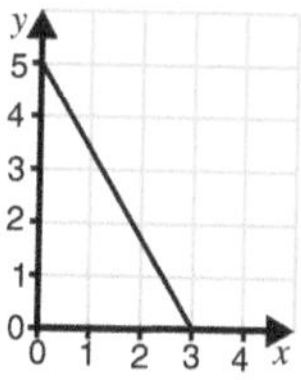

6. (a) 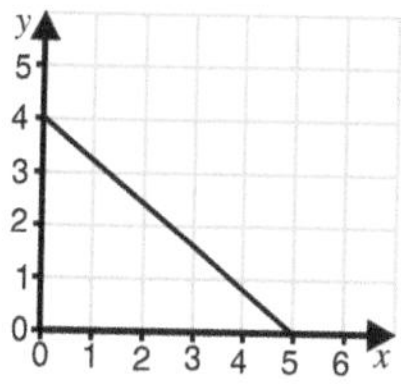 (b) 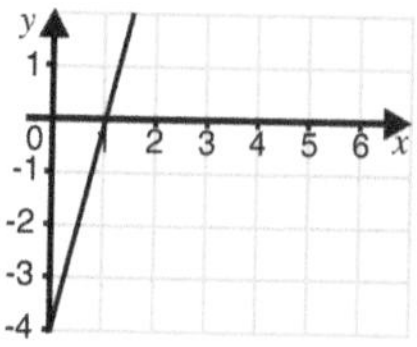
(c)

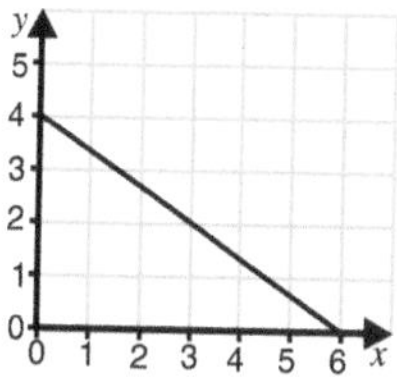

Exercise 14.4 — Page 130

1. (a)

x	1	2	3
$y = x + 2$	3	4	5
$y = 5 - x$	4	3	2

(c) (1.5, 3.5) (d) $x = 1.5$

2. (b) (2.5, 8.5) (c) $x = 2.5$
3. (b) $x = 0.5$
4. $x = 1$
5. (b) $y = 9$ (c) $x = 3$

Exercise 14.5 — Page 131

1. **B, C**
2. (a) $-\frac{1}{3}$ (b) -4 (c) $\frac{1}{2}$
3. (a) (i) $y = -x$ (ii) $y = 2 - x$
   (b) (i) $y = 1 - 2x$ (ii) $y = 2 - 2x$
   (c) $y = 3 - \frac{1}{2}x$
4. Gradient $PR = 2$, gradient $QS = -\frac{1}{2}$
   $2 \times \left(-\frac{1}{2}\right) = -1$,
   so, PR is perpendicular to QS.
5. (a) $y = 3x - 14$ (b) $y = -\frac{1}{3}x - \frac{2}{3}$
6. $AB : y = -\frac{1}{2}x - \frac{1}{2}$ $BC : y = 2x - 3$

Exercise 14.6 — Page 133

1. $y = \frac{3x}{2} + 3$
2. (a) $y = -\frac{1}{2}x + 2$ (b) $y = -\frac{4}{5}x + 4$
   (c) $y = -\frac{2}{3}x + \frac{4}{3}$ (d) $y = \frac{2}{7}x - 2$
3. (a) $\frac{1}{2}$ (b) 1 (c) 2
   (d) $-\frac{1}{3}$ (e) $\frac{3}{4}$ (f) $-\frac{2}{5}$
4. (a) $a = 0$ (b) $a = \frac{1}{2}$ (c) $a = -2$
   (d) $a = 2\frac{1}{2}$ (e) $a = 2$ (f) $a = \frac{2}{5}$
5. (a) $y = x +$ (any number)
   (b) $y = 2x +$ (any number)
   (c) $y = \frac{1}{2}x +$ (any number)
6. $y = -\frac{3}{2}x +$ (any number)
7. (b) $3y = x - 6$ and $y = 1 - 3x$
8. (a) $A(0, 2)$, $B(-4, 0)$ (b) $\frac{1}{2}$
   (c) (ii) $y = \frac{1}{2}x +$ (any number)
9. (a) $P(0, 3)$, $Q(6, 0)$ (b) $-\frac{1}{2}$
   (c) (ii) $y = -\frac{1}{2}x +$ (any number)
10. (a) (i) $y = x + 2$ (ii) $y = 2x + 1$
    (iii) $3y = x + 16$ (iv) $4y + x = -8$
    (b) (i) $y = 2 - x$ (ii) $2y = 7 - x$
    (iii) $y = 2 - 3x$ (iv) $y = 4x + 15$
11. $P(0, 6)$, $Q(-7, 0)$

Review Exercise 14 — Page 134

1. (a) (1) $y = 2$ (2) $x = -4$ (3) $y = x$
   (b) $(-4, 2)$
2. (a) Missing entries are: 3, 2, 0
   (b)
   (c) $P(2.5, 1.5)$
3. (b) (2, 4)
4.

x	-2	0	2	4
y	3	2	1	0

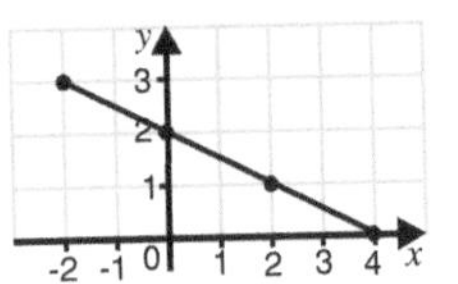

5. (a) 8 mm (b) 0.06
   (c) $l = 0.06w + 8$ (d) 26 mm
6. $y = -1\frac{1}{2}x + 6$
7. (a) $y = 3x + 5$ (b) (i) $\frac{2}{3}$ (ii) $(0, -2)$
8. $y = 4x - 5$
9. $2y = x + 2$ has gradient $\frac{1}{2}$.
   $y = 6 - 2x$ has gradient -2.
   $\frac{1}{2} \times (-2) = -1$, so, lines are perpendicular.
10. $y = -\frac{3}{2}x + 5$
11. (a) $y = -\frac{2}{3}x + 4$ (b) $y = -\frac{2}{3}x + 2$
    (c) $y = 1\frac{1}{2}x - 2\frac{1}{2}$

CHAPTER 15

Exercise 15.1 — Page 136

1. **B**
2. **1: C, 2: B, 3: E, 4: A**
3.
4. (d) B (e) A (f) 6 days
5. (a) (b) (c) (d)

Exercise 15.2 — Page 138

1. (a) 1042 (b) 28 km
   (c) (i) 1115 (ii) 8 km
   (d) 16 km/h
2. (a) 50 km/h (b) 20 m/s (c) 9 miles/hour
3. (a) 10 km/h (b) 6.7 km/h (c) 8 km/h
4. (a) 48 miles/hour (b) 30 miles/hour
5. (a) (b) (i)
   (b) (ii) 1110
6. (a) 1118 (b) 1230
   (c) Twice (d) 42 minutes
   (e) Between 1200 and 1230 (f) 4 km/h
7. (a) 9 miles/hour
   (b) (i)
   (ii) 1415
8. (a)
   (b) 15 km/h
9. (a)
   (b) (i) 1220 (ii) 1102 (iii) 14 km/hour

Exercise 15.3 — Page 142

1. $0.8\,\text{m/s}^2$
2. (a) (i) B and C, D and E (ii) A and B
   (iii) B and C, D and E (iv) F and G
   (b) (i) $0.53\,\text{m/s}^2$ (ii) $-0.5\,\text{m/s}^2$
3. (a) 8 m/s (b) $0.8\,\text{m/s}^2$ (c) $1.6\,\text{m/s}^2$ (d) 260 m
4. (a) 2 minutes (b) $1\frac{1}{2}$ minutes
   (c) $5\,\text{km/min} = 5000\,\text{m/min} = \frac{5000}{60}\,\text{m/s}$
   $1.5\,\text{min} = 1.5 \times 60$ seconds
   $\frac{5000}{60} \div (1.5 \times 60) \simeq 0.925\ldots \simeq 0.9\,\text{m/s}^2$
5. (a)
   (b) 960 m
6. (a) $0.8\,\text{m/s}^2$ (b) 27.25 seconds

Review Exercise 15 — Page 144

1. (a) Nina
   (b) Polly passed Nina and went into the lead.
   (c) Nina stopped running.
   (d) Nina
2. $A : W$, $B : Z$, $C : X$
3. (a) (i) (ii) (b)
4. (a) Twice
   (b) (i) 2 km (ii) 12 km/h
   (c) Between D and E
5. (a) 18 km (b) 20.6 km/h
6.
7. (a) $0.5\,\text{m/s}^2$ (b) 525 m

CHAPTER 16

Exercise 16.1 — Page 147

1. (a) True (b) True (c) True (d) False
   (e) True (f) False (g) False (h) True
2. (a) E.g. 5, 4, 3, … (b) E.g. −2, −1, 0 …
   (c) E.g. 5, 4, 3, … (d) E.g. 7, 6, 5, …
   (e) E.g. 11, 12, 13, …
   (f) −1
   (g) 5
   (h) One of: −6, −5, −4, −3, −2
3. (a) 2, 3, 4 (b) −1, 0, 1, 2, 3
   (c) −4, −3, −2, −1 (d) −1, 0, 1, 2
4. (a) $x \geqslant 2$ (b) $x < 1$
   (c) $-1 \leqslant x \leqslant 5$ (d) $-6 \leqslant x < -2$
   (e) $-2 < x < 1$ (f) $x < 5$ and $x \geqslant 8$
5. (a) (b) (c) (d) (e) (f) (g) (h)

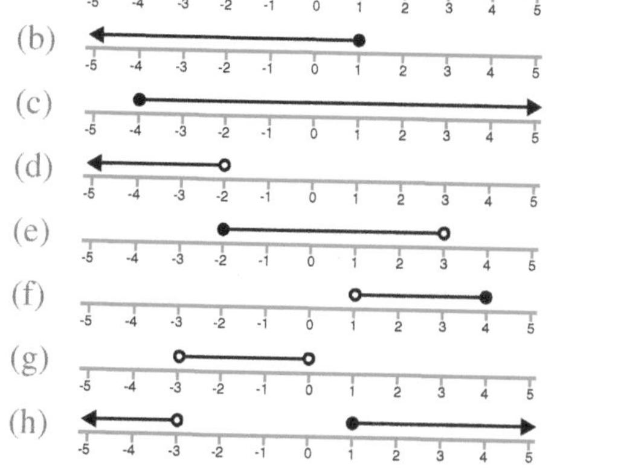
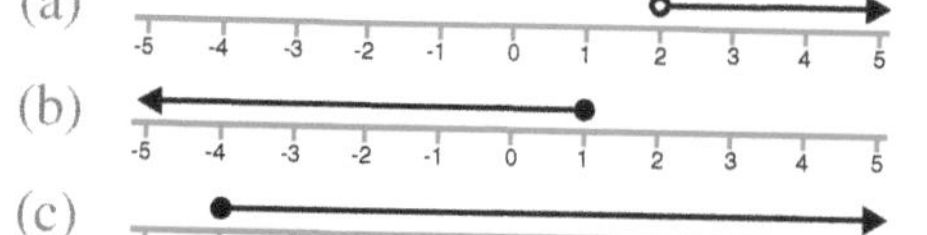
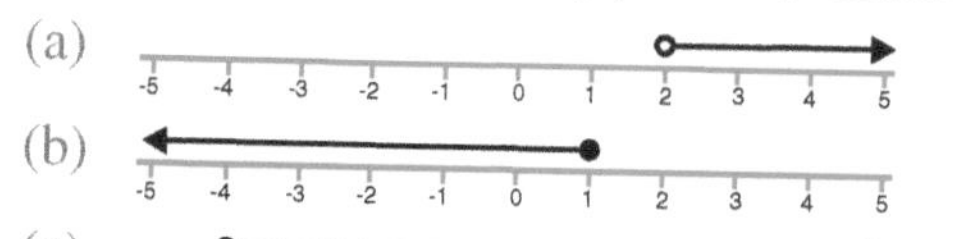
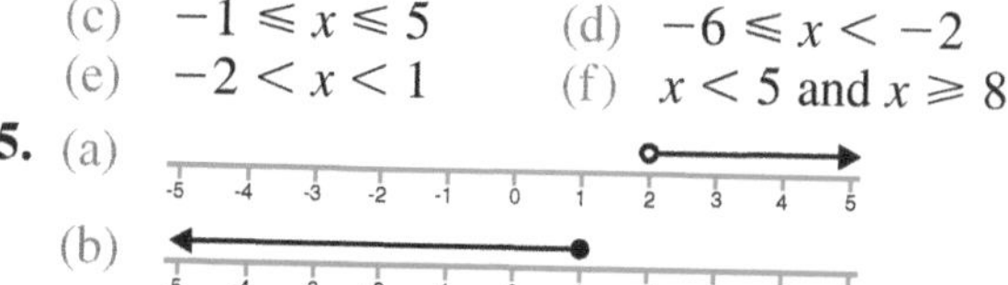
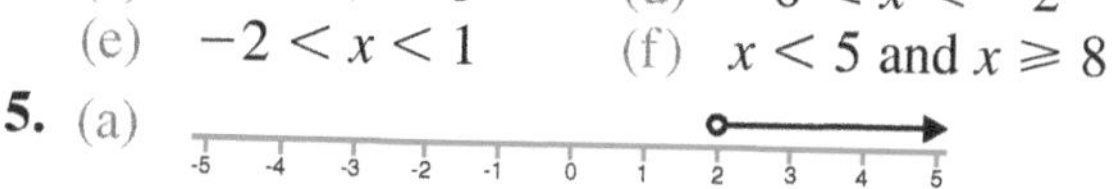

Exercise 16.2 Page 148

1. (a) $n > 2$ (b) $x < -2$ (c) $a < 4$
(d) $a < 4$ (e) $d \leqslant 3$ (f) $t < -3$
(g) $g > -2$ (h) $y \geqslant 0$

2. (a) $a < 4$ (b) $x \geqslant -2$ (c) $y < -3$
(d) $c > 5$ (e) $d < -3$ (f) $b \geqslant 1$
(g) $b \leqslant 1$ (h) $c \leqslant 3$ (i) $d > 4$
(j) $f < -2$ (k) $g \leqslant \frac{1}{2}$ (l) $h < 2$
(m) $x < -3$ (n) $j \geqslant 2\frac{1}{2}$ (o) $k > -4$
(p) $m \leqslant 1\frac{2}{5}$

3. (a) (i) $x \geqslant -1$ (ii) $x < 3$
(b) $-1, 0, 1, 2$

4. (a) $a \geqslant 6$ (b) $b < 15$ (c) $m > -6\frac{1}{2}$
(d) $n \leqslant 2$ (e) $p > 8$ (f) $q > 1$
(g) $r \geqslant -1\frac{1}{2}$ (h) $t > 13$ (i) $u \leqslant 9$
(j) $v < -1\frac{1}{2}$ (k) $w > 8$ (l) $x < 1\frac{2}{5}$

Exercise 16.3 Page 149

1. $a < -2$ **6.** $f < 1$ **11.** $m < -3$
2. $b \geqslant 3$ **7.** $g > -4$ **12.** $n > -4$
3. $c \leqslant -4$ **8.** $h \geqslant 2$ **13.** $p \geqslant \frac{5}{9}$
4. $d > -1$ **9.** $j \leqslant 1$ **14.** $q > -2\frac{1}{5}$
5. $e \geqslant 2$ **10.** $k > -\frac{1}{3}$ **15.** $n < 9$

Exercise 16.4 Page 150

1. (a) $1 < x \leqslant 5$ (b) $-1 \leqslant x < 9$
(c) $-7 < x \leqslant 4$

2. (a) $1 < x \leqslant 3$ (b) $-2 \leqslant x < 4$
(c) $3 < x < 4\frac{1}{2}$ (d) $-\frac{1}{3} \leqslant x \leqslant 4$
(e) $2 < x \leqslant 5$ (f) $-3 \leqslant x < 6$
(g) $-2 < x < 2$ (h) $3 \leqslant x < 6$
(i) $-4\frac{1}{2} \leqslant x \leqslant -1$

3. (a) 6, 7, 8 (b) −2, −1, 0, 1, 2, 3, 4
(c) 0, 1, 2 (d) 4, 5, 6, 7
(e) 5 (f) 2, 3
(g) −2, −1, 0 (h) −1, 0, 1, 2, 3
(i) −4, −3, −2, −1, 0, 1, 2, 3, 4, 5, 6
(j) 8, 9, 10, 11, 12, 13, 14, 15, 16
(k) −4, −3, −2, −1, 0, 1, 2, 3
(l) −6

Exercise 16.6 Page 152

3. **A** : ③, **B** : ①, **C** : ④, **D** : ②.

4. $y \geqslant 1$, $y \leqslant x$

5. (a)

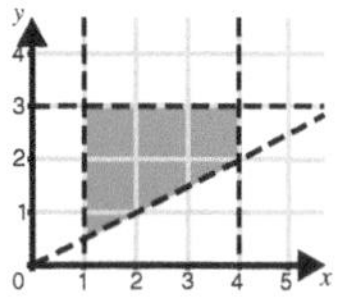

(b) (2, 2), (3, 2)

6. (a) $x \geqslant 1$, $y < 5$ and $y > x + 2$
(b) $y < 4$, $x \leqslant 3$, $y \leqslant 2x$ and $y > x$
(c) $x \leqslant 2$, $x + y > 3$, $y \leqslant x + 3$
(d) $y > 1$, $1 \leqslant x < 3$, $x + y \leqslant 5$

7. (b) (2, 3)

Review Exercise 16 Page 154

1. (a)

(b)

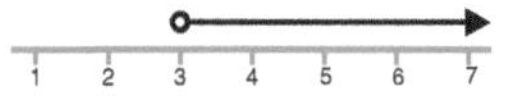

(c)

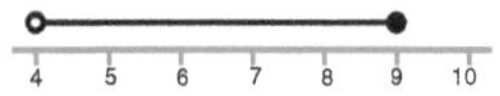

(d)

2. (a) $x < 3$ (b) $x \geqslant -5$
(c) $x \geqslant 4$ (d) $x \leqslant 3$

3. (a) $x \geqslant 2$ (b) $x < -1$

4. −3, −2, −1, 0, 1, 2, 3, 4

5. −1, 0, 1

6. (a) $x < 4$ (b) $-1 < x \leqslant 2$
(c) $-1 \leqslant x < 1$ (d) $-3 < x < -1$

7. (a) −1, 0, 1, 2, 3, 4 (b) −3, −2, −1, 0, 1
(c) −2, −1

8. (a) −4, −3, −2, −1, 0, 1
(b) $x < \frac{1}{2}$

9. $x > 2$

10. $a > 5$

11. (a) $x < 6$ (b) $x > 0$

12. (a)

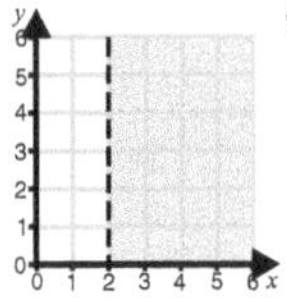

(b)

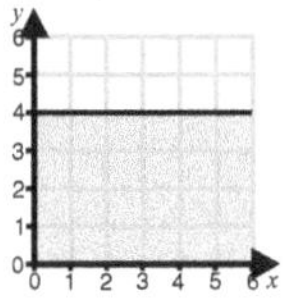

(c)

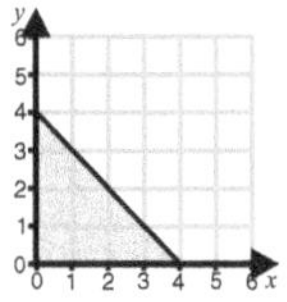

13. (a) $x \leqslant 1$
(b) (i) $x \geqslant -1\frac{2}{5}$ (ii) −1

14. **A** : ②, **B** : ④, **C** : ①, **D** : ③.

15.

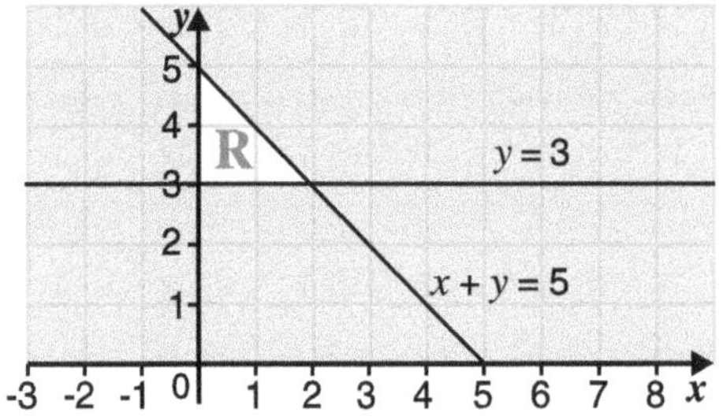

16. (a)

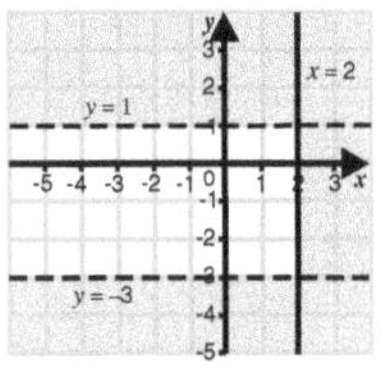

(b)

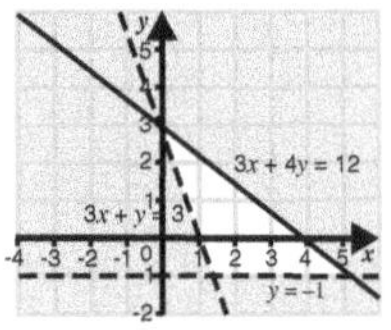

CHAPTER 17

Exercise 17.1 — Page 156

1. (a)

x	-3	-2	-1	0	1	2	3
y	7	2	-1	-2	-1	2	7

(c) $y = 0.25$

2. (b) $y = 7.25$
(c) $x = -1.7$ and $x = 1.7$
(d) (0, 1)

3. (a)

x	-3	-2	-1	0	1	2	3
y	-3	2	5	6	5	2	-3

(c) $(-2.4, 0), (2.4, 0)$
(d) (0, 6)

4.

x	-2	-1	0	1	2
y	8	2	0	2	8

5.

x	-2	-1	0	1	2	3
y	8	4	2	2	4	8

Exercise 17.2 — Page 157

1. $x = -3.6$ or $x = 0.6$

2. (b) $x = -1$ or $x = 0$
(c) $(-0.5, -0.25)$

3. (a) Entries are: 9, 4, 1, 0, 1, 4
(c) $x = 1$

4. (b) $x = \pm 3.2$
(c) (0, 10)

5. (a) $x = \pm 3.2$ (b) $x = \pm 2.2$
(c) $x = 1$ or $x = 2$ (d) $x = \pm 2.4$

6. (a) Missing entries are: 6, 0, 3
(c) $x = 0.5$ or $x = 1$
(d) (i) $y = 3$ (ii) $x = -0.5$ or $x = 2$

7. (b) $x = 0$ or 1
(c) (i) $x = -2.6$ or -1.6
(ii) $x = 0$ or 3

8. (b) $x = -0.6$ or 1.6

9. (b) (i) $x = -2.7$ or 0.7
(ii) $x = -4.3$ or 2.3

10. (b) $x = 0.3$ or 3.7
(c) (i) $y = 3 - x$ (ii) $x = 0$ or 3

11. (b) Draw $y = 3x + 2$, $x = -1$ or 3

12. (b) $x = 0.3$ or 3.7
(c) Draw $y = x - 3$, $x = 1$ or 4

Exercise 17.3 — Page 158

1.

2. (a) Entries are: $-9, 4, 5, 0, -5, -4, 9$
(c) (i) $x = -2.8$
(ii) $x = -2.3$, -0.3 or 2.6

Exercise 17.4 — Page 159

1. (c) $x = 0.27$ or 3.73

2.

3.

4. (a) Entries are: 0.04, 0.11, 0.33, 1, 3, 9, 27
(c) (i) $y = 5.2$ (ii) $x = 2.7$

Exercise 17.5 — Page 160

1. (a) 3
(b) (3, 0), (0, 3), $(-3, 0)$, $(0, -3)$

2. (b) $x^2 + y^2 = 25$
(d) $(-3, 4)$, (3, 4)

3. (c) $(-2.2, -3.3)$, (1.4, 3.7)

4. (c) (0.6, 0.8), (1, 0)

5. (c) $x = -1.2$

Exercise 17.6 — Page 161

1. **(1)** **F** **(2)** **E**
(3) **C** **(4)** **B**
(5) **A** **(6)** **D**

2. (a) (b) (c) (d)

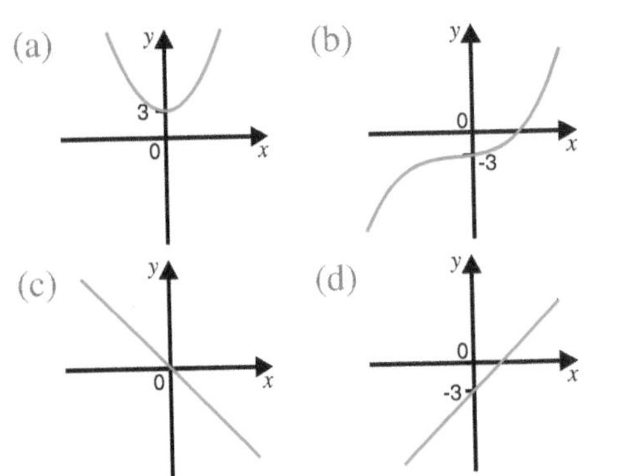

3. (a) $y = x + 2$ (b) $y = 2 - x$
(c) $y = 2x$ (d) $y = \frac{1}{2}x$

Review Exercise 17 — Page 162

1. (a) $x = 0$ or $x = 4$ (b) $(2, -4)$
2. (a) Entries are: 6, 1, −2, −3, −2, 1, 6
(c) $x = -0.7$ or $x = 2.7$
3. (a) Missing entries are: 5, −3
(c) (i) Where the graph crosses the x axis.
(ii) $x = -0.2$
4. **A : R, B : S, C : Q, D : P**
5. (a) Missing entry: 0.51 (c) $x = 1.2$
6. (c) $x = -2$ or 1
7. (a) Entries for y: 3, 0, −2, −3, −3, −2, 0, 3
(c) $x = -1.35$ or 1.85
8. (c) $x = -1$ or 5 (e) $y = 4 - x$
9. (a) Missing entries for y: 33, 1
(b) $x = -0.6$
(c) The graph does not cross $y = 0$ (the x axis)
10. $(-4, 3)$, $(-3, 4)$

CHAPTER 18

Exercise 18.1 — Page 165

1. (a) True, $a : b = 2 : 3$
(b) True, $c : d = 20 : 1$
(c) False, $e : f$ not the same
(d) True, $h : g = 200 : 7$
2. (a) (i) 1 : 2 (ii) 1 : 5
(b) $a = 2$, $b = 60$, $c = 8$, $d = 350$
3. (a) $k = 6$, $y = 6x$ (b) $y = 48$
4. (a) $y = \frac{1}{5}x$ (b) $y = 9$
5. (a) $m = 1\frac{1}{2}n$ (b) (i) $m = 9$ (ii) $n = 10$
6. (a) $q = 0.625p$ (or $p = 1.6q$)
(b) $a = 7.5$, $b = 32$
7. (a) $y = 0.05x$ (b) $a = 5$, $b = 400$
8. (a) $C = 10.5A$ (b) £577.50 (c) 68.75 m
9. (a) $d = 5n$ (b) 1 km (c) 600 times
10. (a) (i) $e = 30w$ (ii) 159 mm (iii) 3.22 kg
(b) 5 : 16

Exercise 18.2 — Page 167

1. (a) True, $ab = 10$ (b) False
2. (a) $pq = 0.5$ (b) $q = \frac{0.5}{p}$
3. $k = 20$

a	2	5	10	25	50
b	10	4	2	0.8	0.4

4. (a) $pq = 30$ or $p = \frac{30}{q}$ (b) $p = 5$
5. (a) $y = \frac{48}{x}$ or $xy = 48$
(b) $a = 4$, $b = 96$
6. $y = 9$
7. $r = 25$
8. (a) $LM = 14.4$ or $M = \frac{14.4}{L}$
(b) $a = 1.2$, $b = 16$
9. $lw = 200$, $k = 200$,
k represents the constant area
10. (a) $nt = 12$ or $t = \frac{12}{n}$ (b) 2 days
(c) 3 workers

Exercise 18.3 — Page 168

1. (a) V is proportional to the square of t.
(b) L is inversely proportional to a.
(c) y is proportional to the cube of x.
(d) p is inversely proportional to the square of q.
(e) L is proportional to the square root of m.
(f) b is inversely proportional to the cube of c.
2. $k = 2$, $y = 2x^2$
3. (a) y is multiplied by 4.
(b) x is multiplied by 5.
4. $y = \frac{64}{x^3}$
5. (a) $V = 5t^2$ (b) $R = \frac{320}{s^3}$
(c) $y = 10\sqrt{x}$ (d) $A = \frac{100}{b}$
(e) $L = 3.125p^3$ (f) $Q = 12.5\sqrt{P}$
6. (a) 80 (b) $y = 5$ (c) $x = 4$
7.

a	1	2	10
b	3	12	300

8. $A = 250$, $B = 3$
9. (a) $k = 1.6$ (b) $y = 0.016$
10. (a) $k = 10$ (b) $x = 16$
11. (a) $y = 8$ (b) $x = 17.1$
12. (a) $q = \frac{40}{p^2}$, $a = 0.4$, $b = 20$
(b) $q = \frac{160}{p^3}$, $c = 0.16$, $d = 2$
(c) $q = 0.15625p^2$, $e = 15.625$, $f = 8$
13. (a) y is divided by 4 (is 4 times smaller).
(b) x is divided by 5 (is 5 times smaller).

14. (a) $t = 1.6\sqrt{d}$ (b) $t = 8$ seconds

15. (a) $h = 0.05s^2$ (b) 20 (c) 125

16. (a) $h = 16$ (b) $\sqrt{\frac{\text{Volume}}{\pi}}$

17. (a) h is divided by 4 (is 4 times smaller).
(b) r is divided by 4 (is 4 times smaller).

Exercise 18.4 — Page 171

1. (a) **B** (b) **A** (c) **D** (d) **C**

2. (a) (i) 3 (ii) -1 (iii) $-\frac{1}{2}$
(b) (i) (ii) (iii)

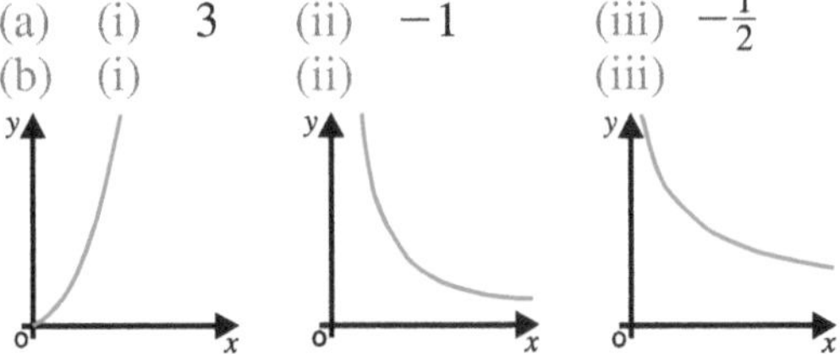
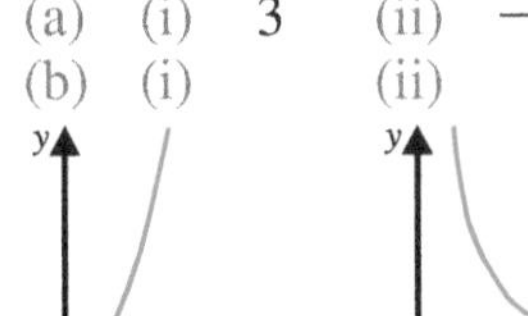
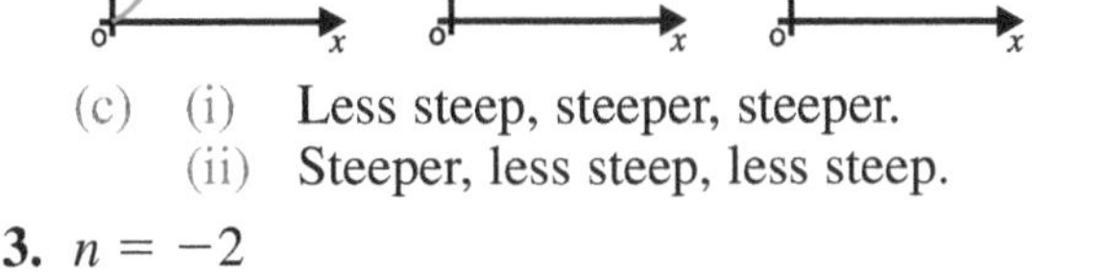

(c) (i) Less steep, steeper, steeper.
(ii) Steeper, less steep, less steep.

3. $n = -2$

4. (a) $n = 2$ (b) 0.25 $k = 10, \; n = -2$

5. $q = 0.2p^3$

6. $s = \frac{20}{\sqrt{t}}$

7. (a) $k = 5, \; n = \frac{1}{2}$ (b) $x = 9$

Review Exercise 18 — Page 172

1. $m = 50$

2. $t = 4$

3.

x	100	25	400
y	3	1.5	6

4. (a) $y = 4$
(b) $x = 1.44$

5. $M = 60$

6.

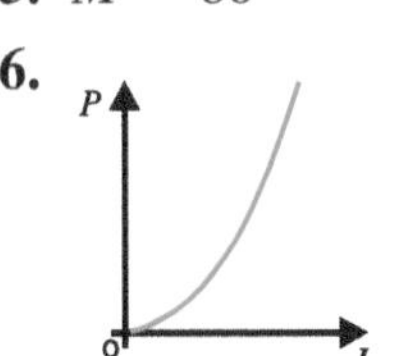

7. $k = 10, \; n = \frac{1}{3}$

CHAPTER 19

Exercise 19.1 — Page 174

1. $x^2 + 7x + 12$
2. $x^2 + 6x + 5$
3. $x^2 - 3x - 10$
4. $2x^2 - 3x - 2$
5. $3x^2 - 20x + 12$
6. $6x^2 + 7x + 2$
7. $x^2 + 6x - 16$
8. $x^2 + 3x - 10$
9. $x^2 + 2x - 3$
10. $x^2 - 5x + 6$
11. $x^2 - 5x + 4$
12. $x^2 - 5x - 14$
13. $2x^2 + 7x + 3$
14. $3x^2 + 13x - 10$
15. $5x^2 - 8x + 3$
16. $4x^2 + 4x - 3$
17. $6x^2 + 13x - 5$
18. $12x^2 + 14x - 10$
19. $x^2 - 9$
20. $x^2 - 25$
21. $x^2 - 49$
22. $x^2 + 10x + 25$
23. $x^2 - 14x + 49$
24. $4x^2 - 12x + 9$

Exercise 19.2 — Page 176

1. (a) $x(x + 5)$ (b) $x(x - 7)$ (c) $y(y - 6)$
(d) $2y(y - 6)$ (e) $t(5 - t)$ (f) $y(8 + y)$
(g) $x(x - 20)$ (h) $3x(x - 20)$

2. (a) $(x - 3)(x + 3)$ (b) $(x - 9)(x + 9)$
(c) $(y - 5)(y + 5)$ (d) $(y - 1)(y + 1)$
(e) $(x - 8)(x + 8)$ (f) $(10 - x)(10 + x)$
(g) $(6 - x)(6 + x)$ (h) $(x - a)(x + a)$

3. (a) $(x + 5)(x + 1)$ (b) $(x + 7)(x + 2)$
(c) $(x + 2)(x + 4)$ (d) $(x + 3)(x + 6)$
(e) $(x - 5)(x - 1)$ (f) $(x - 5)(x - 2)$
(g) $(x - 4)(x - 3)$ (h) $(x + 4)(x - 1)$
(i) $(x + 7)(x - 2)$ (j) $(x - 5)(x + 1)$

4. (a) $(x + 1)(x + 2)$ (b) $(x + 1)(x + 7)$
(c) $(x + 3)(x + 5)$ (d) $(x + 2)(x + 6)$
(e) $(x + 1)(x + 11)$ (f) $(x + 4)(x + 5)$
(g) $(x + 4)(x + 6)$ (h) $(x + 4)(x + 9)$
(i) $(x + 1)(x + 14)$ (j) $(x + 2)(x + 8)$

5. (a) $(x - 3)^2$ (b) $(x - 4)(x - 2)$
(c) $(x - 10)(x - 1)$ (d) $(x - 15)(x - 1)$
(e) $(x - 5)(x - 3)$ (f) $(x - 8)(x - 2)$
(g) $(x - 10)(x - 2)$ (h) $(x - 8)(x - 3)$
(i) $(x - 1)(x - 12)$ (j) $(x - 2)(x - 6)$

6. (a) $(x - 3)(x + 2)$ (b) $(x - 6)(x + 1)$
(c) $(x - 4)(x + 6)$ (d) $(x - 3)(x + 8)$
(e) $(x - 5)(x + 3)$ (f) $(x - 3)(x + 6)$
(g) $(x - 8)(x + 5)$ (h) $(x - 6)(x + 2)$
(i) $(x - 2)(x + 5)$ (j) $(x + 4)(x - 5)$

7. (a) $(x - 2)^2$ (b) $(x + 5)(x + 6)$
(c) $(x - 2)(x + 4)$ (d) $(x - 7)(x + 3)$
(e) $(x - 4)(x + 5)$ (f) $(x + 3)(x + 4)$
(g) $(x + 4)^2$ (h) $(x - 1)^2$
(i) $(x - 7)(x + 7)$ (j) $t(t + 12)$
(k) $(x - 2)(x - 7)$ (l) $(x - 6)(x - 1)$
(m) $(x + 2)(x + 9)$ (n) $(x + 3)(x + 8)$
(o) $(x + 1)(x + 18)$ (p) $(x - y)(x + y)$
(q) $(x + 3)(x - 2)$ (r) $y(y + 4)$
(s) $(y - 5)^2$ (t) $(x - 6)^2$

Exercise 19.3 — Page 177

1. (a) $(2x + 6)(x + 3)$ (b) $(2y + 3)(y - 6)$
(c) $(2a + 9)(a - 2)$ (d) $(2m - 2)(m - 9)$
(e) $(4x - 1)(x - 3)$ (f) $(2d - 1)(2d + 3)$

2. (a) $(2x + 1)(x + 3)$ (b) $(2x + 3)(x + 4)$
(c) $(2a + 1)(a + 12)$ (d) $(2x + 1)^2$
(e) $(3y + 2)(2y + 1)$ (f) $(3x + 4)(2x + 3)$
(g) $(11m + 1)(m + 1)$ (h) $(3x + 2)(x + 5)$
(i) $(5k + 2)(k + 2)$

3. (a) $(2a-3)(a+1)$ (b) $(2y-7)(y-1)$
(c) $(2x-1)^2$ (d) $(3x-2)(x-1)$
(e) $(3d-1)(2d-3)$ (f) $(2x-1)(x-1)$
(g) $(2x-5)(x-3)$ (h) $(3y-2)^2$
(i) $(2t-3)^2$

4. (a) $(2x-3)(x+2)$ (b) $(2a-3)(a+1)$
(c) $(3x+1)(x-5)$ (d) $(3t-1)(t+5)$
(e) $(2y-7)(y+1)$ (f) $(4y-1)(y+2)$
(g) $(3x-1)(3x+2)$ (h) $(10x-3)(x+3)$
(i) $(5m+3)(m-1)$

Exercise 19.4 — Page 178

1. (a) $3(x+4y)$
(b) $(t-4)(t+4)$
(c) $(x+1)(x+3)$
(d) $y(y-1)$
(e) $2(d^2-3)$
(f) $(p-q)(p+q)$
(g) $a(a-2)$
(h) $(x-1)^2$
(i) $2y(y-4)$
(j) $(a-3)^2$
(k) $3(m-2)(m+2)$
(l) $(v-3)(v+2)$
(m) $a(x-y)(x+y)$
(n) $-2(4+x)$
(o) $(2-k)^2$
(p) $2(3-x)(3+x)$
(q) $2(5-x)(5+x)$
(r) $3(2x-3y)(2x+3y)$
(s) $3a(2a-1)$
(t) $(x-7)(x-8)$
(u) $(x-3)(x+5)$

2. (a) $(3a+4)(a-2)$ (b) $(3x+1)(x+10)$
(c) $(8y+7)(8y-7)$ (d) $(3x+5)(x+2)$
(e) $(7x-5)(x-2)$ (f) $(6x+5)(x-5)$
(g) $(5m-3)(m-3)$ (h) $(2x+3)(4x-5)$
(i) $(5-2x)(x-4)$ (j) $2(y+4)(2y+3)$
(k) $3(x+2)(2x-3)$ (l) $2(3n-1)(n+2)$

Exercise 19.5 — Page 179

1. (a) $x=2$ or 3 (b) $x=-4$ or -6
(c) $x=3$ or -1 (d) $x=5$ or -2
(e) $x=0$ or 4 (f) $x=0$ or -2

2. (a) $x=1$ or 2 (b) $y=-3$ or -4
(c) $m=4$ or -2 (d) $a=3$ or -4
(e) $n=9$ or -4 (f) $z=6$ or 3
(g) $k=-3$ or -5 (h) $c=-7$ or -8
(i) $b=4$ or -5 (j) $v=12$ or -5
(k) $w=4$ or -12 (l) $p=9$ or -8

3. (a) $x=0$ or 5 (b) $y=0$ or -1
(c) $p=0$ or -3 (d) $a=0$ or 4
(e) $t=0$ or 6 (f) $g=0$ or 4

4. (a) $x=2$ or -2 (b) $y=12$ or -12
(c) $a=3$ or -3 (d) $d=4$ or -4
(e) $x=10$ or -10 (f) $x=6$ or -6
(g) $x=7$ or -7 (h) $x=1.5$ or -1.5
(i) $x=3$ or -3

5. (a) $x=-6$ or $x=0$
(b) $x=-1$ or $x=-4$
(c) $x=-8$ or $x=8$
(d) $x=1$ or $x=3$
(e) $x=0$ or $x=2$
(f) $x=-2$ or $x=3$
(g) $x=-5$ or $x=3$
(h) $x=-2$ or $x=2$
(i) $x=7$ or $x=8$

6. (a) $x=-2\frac{1}{2}$ or -1 (b) $x=\frac{1}{2}$ or 5
(c) $x=7$ or $-1\frac{1}{3}$ (d) $y=-\frac{1}{2}$ or $-\frac{2}{3}$
(e) $x=-3$ or $\frac{2}{3}$ (f) $z=-\frac{1}{3}$ or 2
(g) $m=\frac{3}{5}$ or 1 (h) $a=-3\frac{1}{2}$ or 3
(i) $y=2$ or $-1\frac{1}{4}$

7. (a) $y=5$ or -1 (b) $x=0$ or 1
(c) $x=4$ (d) $x=5$ or -3
(e) $n=12$ or -2 (f) $m=7$ or 1
(g) $a=8$ or -3 (h) $x=1$ or 3
(i) $x=\frac{1}{2}$ or $-\frac{1}{3}$ (j) $m=-5$ or $1\frac{1}{2}$
(k) $a=\frac{1}{2}$ or $-\frac{2}{3}$ (l) $x=\frac{1}{2}$ or $-\frac{3}{4}$

Exercise 19.6 — Page 180

1. (a) $x=-1$ or $x=-3$
(b) $x=1$ or $x=2$
(c) $x=-1$ or $x=3$

2. (a) $x=-0.59$ or -3.41
(b) $x=-6.19$ or -0.81
(c) $x=-1.62$ or 0.62
(d) $x=-0.56$ or 3.56
(e) $x=-0.54$ or 5.54
(f) $x=-4.30$ or -0.70
(g) $x=-5.74$ or 1.74
(h) $x=0.38$ or 2.62
(i) $x=-3.56$ or 0.56

3. (a) $x=-1.5$ or $x=1$
(b) $x=0.5$ or $x=1$
(c) $x=-1$ or $x=0.6$

4. (a) $x = 0.23$ or 1.43
(b) $x = 0.72$ or 2.78
(c) $x = -7.30$ or -0.37
(d) $x = -0.79$ or 2.12
(e) $z = -2.37$ or -0.63
(f) $x = -2.27$ or 2.94

5. (a) $x = 0.55$ or 5.45
(b) $x = -1.14$ or 2.64
(c) $x = -0.85$ or 2.35
(d) $x = -0.74$ or 0.54
(e) $x = -0.65$ or 4.65
(f) $x = -3.81$ or 1.31
(g) $x = 0.15$ or 4.52
(h) $x = -0.58$ or 2.58
(i) $x = -1.16$ or 2.16

6. (a) $x = -1.7$ or -0.3,
$x = -1$, no solutions
(b) $x = -1.71$ or -0.293,
$x = -1$, no solutions

Exercise 19.7 — Page 182

1. (a) 1 (b) 4 (c) 9 (d) 16
(e) 1 (f) 2.25 (g) 25 (h) 6.25

2. (a) $x^2 + 6x + 9 = (x + 3)^2$
(b) $a^2 - 4a + 4 = (a - 2)^2$
(c) $b^2 + 2b + 1 = (b + 1)^2$
(d) $m^2 - 8m + 16 = (m - 4)^2$
(e) $n^2 - n + \frac{1}{4} = \left(n - \frac{1}{2}\right)^2$
(f) $x^2 + 5x + 6.25 = (x + 2.5)^2$

3. (a) $(x + 3)^2 + 11$ (b) $(x + 3)^2 - 4$
(c) $(x + 5)^2 - 29$ (d) $(x - 2)^2 + 1$
(e) $(x - 2)^2 - 2$ (f) $(x - 2)^2 - 8$
(g) $(x - 3)^2 - 5$ (h) $(x - 4)^2 - 16$
(i) $(x + 6)^2 - 36$

4. (a) $a = 3,\ b = 6$
(b) $a = 5,\ b = -20$
(c) $a = -3,\ b = -14$
(d) $a = -4,\ b = -12$
(e) $a = 6,\ b = -32$
(f) $a = 3,\ b = 0$

5. (a) $(2x + 4)^2 - 11;\ p = 2, q = 4, r = -11$
(b) $(3x + 2)^2 - 1;\ p = 3, q = 2, r = -1$
(c) $(5x - 4)^2 - 19;\ p = 5, q = -4, r = -19$
(d) $(4x + 4)^2 - 21;\ p = 4, q = 4, r = -21$
(e) $(10x + 3)^2 - 6;\ p = 10, q = 3, r = -6$
(f) $(4x - 5)^2 - 16;\ p = 4, q = -5, r = -16$

6. (a) $a = -2,\ b = -5$ (b) -5

7. $a = 8,\ b = -3$

8. 1

Exercise 19.8 — Page 183

1. (a) $x = -1$ or 3 (b) $t = -3$ or 7
(c) $x = -9$ or 3 (d) $y = -2$ or 8
(e) $m = -2$ or 12 (f) $e = 6$ or 8

2. (a) $x = -1.3$ or 11.3
(b) $m = -5.3$ or 1.3
(c) $x = 0.6$ or 5.4
(d) $y = -0.6$ or 3.6
(e) $x = -0.4$ or 12.4
(f) $a = -5.2$ or -0.8
(g) $n = -1.4$ or 0.4
(h) $x = -1.8$ or 0.8
(i) $z = 0.3$ or 2.7
(j) $x = 1.4$ or 2.6
(k) $x = -4.4$ or -1.6
(l) $x = -1.5$ or -2.5

3. (a) $(x - 2)^2 - 3$ (b) $x = 2 \pm \sqrt{3}$

4. $a = 3,\ b = 5$

5. $x = 1 \pm \sqrt{\frac{7}{2}}$

6. $m = -6,\ n = 7$

7. $a = -8,\ b = -3$

Exercise 19.9 — Page 184

1. (a) $x(x - 4) = 21$
(b) $x = 7$ (x cannot equal -3)

2. $x = 9$ or -7

3. (a) $(x + 3)(x + 1) - \frac{1}{2}x^2 = 7.5$
$x^2 + 4x + 3 - \frac{1}{2}x^2 = 7.5$
$\frac{1}{2}x^2 + 4x - 4.5 = 0$
$x^2 + 8x - 9 = 0$
(b) $x = 1$ (x cannot equal -9)

4. (a) $3(8 + 2x)(5 + 2x) = (15 + 2x)(14.5 + 2x)$
(b) 2.5 m

5. 9.3 cm and 4.3 cm

6. 24 cm by 7 cm

7. 7.24 m by 2.76 m

8. (a) 13 and 18 (b) 3.55 and 8.45

9. 75 m and 100 m

10. 9, 12 and 15; −3, 0 and 3

11. (a) $x^2 - 8x - 20 = 0$
(b) 10, 11, 13 and 17; −2, −1, 1 and 5

12. (a) 11th (b) 15th

13. 3 and 8

Review Exercise 19 — Page 186

1. $x^2 + 3x - 10$

2. (a) $(x - 5)(x + 2)$ (b) $x = 5$ or -2

3. $2x^2 + 5x - 12$

4. (a) $x = 4$ (b) $x = -3$ or 4

5. $x = -2$ or -6

6. (a) $(x - 4)(x + 2)$ (b) $a = 0$ or -4

7. (a) $x(x + 3) = 2(x + 1)$ (b) 1 cm

8. $x(13 - x) = 36$, 4 and 9

9. $x = -3$ or $\frac{1}{2}$

10. $p = 4$, $q = 1$

11. (a) $(x + 2)^2 - 14$ (b) $x = -2 \pm \sqrt{14}$

12. $n^2 + 2n$

13. (a) $(3x + 4)(2x - 3)$

(b) $x = -0.77$ or 7.77

14. (b) $v = -70$ or 50

15. $x = 0.23$ or 1.43

16. (a) $(x + 5)(2x + 3) = 39$

$2x^2 + 13x + 15 = 39$

$2x^2 + 13x - 24 = 0$

(b) $x = 1.5$ (x cannot equal -8)

17. (a) $2x(x + 5) + (x - 5)(x + 10) = 130$

$2x^2 + 10x + x^2 + 5x - 50 = 130$

$3x^2 + 15x - 180 = 0$

$x^2 + 5x - 60 = 0$

(b) $x = 5.64$ (x cannot equal -10.64)

CHAPTER 20

Exercise 20.1 — Page 189

1. $x = 4$, $y = 2$

2. $x = 3$, $y = 5$

3. $x = 2$, $y = 3$

4. $x = 2$, $y = 3$

5. $x = 3$, $y = 1$

6. $x = 4$, $y = 3$

7. $x = -2$, $y = 3$

8. $x = 5$, $y = 0.5$

Exercise 20.2 — Page 190

1. (a) Both lines have gradient -1.

(b) Both lines have gradient 4.

(c) Both lines are the same.

(d) Both lines have gradient 0.4.

2. (a) $y = -2x + 6$, $y = -2x + 3$

(b) $y = 2x + 3.5$, $y = 2x + 2$

(c) $y = 2.5x - 4$, $y = 2.5x + 1.75$

(d) $y = -3x + 1.25$, $y = -3x + 0.5$

3. (b) and (d) have no solution.

(a) $x = 0.4$, $y = 4$

(c) $x = -0.125$, $y = 1.875$

Exercise 20.3 — Page 192

1. $x = 1$, $y = 2$

2. $x = 3$, $y = 4$

3. $x = 2$, $y = 1$

4. $x = 1$, $y = 7$

5. $x = 3$, $y = 1$

6. $x = 2$, $y = 3$

7. $x = 5$, $y = 2$

8. $x = 4$, $y = 2$

9. $x = 5$, $y = 4$

10. $x = 3$, $y = 2$

11. $x = 4$, $y = 1$

12. $x = 4$, $y = 2$

13. $x = 2$, $y = 3$

14. $x = 4$, $y = -2$

15. $x = 2$, $y = -1.5$

16. $x = -1$, $y = 3$

17. $x = -5$, $y = 6$

18. $x = 2.5$, $y = -1$

19. $x = -6$, $y = 2.5$

20. $x = 3$, $y = -1.5$

21. $x = -1$, $y = -3$

22. $x = 3.5$, $y = -1$

23. $x = 1.5$, $y = -0.5$

24. $x = -0.5$, $y = -2.5$

Exercise 20.4 — Page 193

1. $x = 2$, $y = 1$

2. $x = 2$, $y = 3$

3. $x = 3$, $y = 1$

4. $x = 4$, $y = 2$

5. $x = 2$, $y = -1$

6. $x = 4$, $y = -3$

7. $x = 4$, $y = 0.5$

8. $x = -3$, $y = 0.5$

9. $x = 4$, $y = 1$

10. $x = 2.2$, $y = 5.6$

11. $x = 1.5$, $y = 2$

12. $x = 1$, $y = 2$

13. $x = 5$, $y = 2$

14. $x = 1$, $y = -2$

15. $x = -1$, $y = 2$

16. $x = 2$, $y = 1$

17. $x = -1$, $y = 1$

18. $x = -2$, $y = 3$

19. $x = -1$, $y = 7$

20. $x = 4$, $y = -1$

21. $x = 2.5$, $y = 3$

22. $x = 2$, $y = -1$

23. $x = 1$, $y = -2$

24. $x = -2$, $y = 0.5$

Exercise 20.5 — Page 194

1. $x = 2$, $y = 6$

2. $x = 9$, $y = 18$

3. $x = 8$, $y = 2$

4. $x = 3$, $y = 6$

5. $x = 2$, $y = 13$

6. $x = 0$, $y = 2$

7. $x = 2$, $y = 4$

8. $x = 7$, $y = 6$

9. $x = 4$, $y = 8$

10. $x = 10.5$, $y = 0.5$

11. $x = -68$, $y = -122$

12. $x = 8$, $y = 6$

Exercise 20.6 — Page 195

1. (a) $6x + 3y = 93$, $2x + 5y = 91$

(b) Pencil 8p, pen 15p

2. (a) $5x + 30y = 900$, $10x + 15y = 1260$

(b) Apples £1.08 per kg, oranges 12p each

3. $x = 95$, $y = 82$

4. 32 children, 4 adults

5. $x = 100$, $y = 70$

6. Coffee 90p, Tea 80p

7. $x = 160$, $y = 120$

8. $x = 13$, $y = 9$

9. $p = 5$, $q = -2$

10. Ticket £12.50, CD £4.50

Exercise 20.7 — Page 196

1. $x = -1$, $y = 1$ or $x = 2$, $y = 4$

2. $x = -2$, $y = 3$ or $x = 1$, $y = 0$

3. $x = -1$, $y = -4$ or $x = 3$, $y = 0$

4. $x = 2$, $y = 1$ or $x = 3$, $y = 3$

5. $x = -2$, $y = -2$ or $x = 2$, $y = 2$

6. $x = -1$, $y = -2$ or $x = 1$, $y = 2$

Exercise 20.8 — Page 198

1. $x = 3,\ y = -2$ or $x = 3,\ y = 2$
2. $x = -2,\ y = 6$ or $x = 2,\ y = 6$
3. $x = -1,\ y = 0$ or $x = 1,\ y = 2$
4. $x = -2,\ y = 6$ or $x = -1,\ y = 3$
5. $x = 1,\ y = 5$ or $x = 5,\ y = 25$
6. $x = -3,\ y = 3$ or $x = 3,\ y = -3$
7. $x = -4,\ y = -2$ or $x = 4,\ y = 2$
8. $x = 1,\ y = 4$ or $x = 4,\ y = 1$
9. $x = -2.5,\ y = -2$ or $x = 1,\ y = 5$
10. $x = -2,\ y = 3$ or $x = 1.5,\ y = -0.5$
11. $x = -1,\ y = 1$ or $x = 2,\ y = 7$
12. $x = -\frac{1}{3},\ y = \frac{1}{3}$ or $x = 2,\ y = 12$
13. $x = -2,\ y = -4$ or $x = \frac{1}{4},\ y = \frac{1}{2}$
14. $x = \frac{1}{5},\ y = \frac{1}{5}$ or $x = 2,\ y = 20$
15. $x = -1.2,\ y = 3.4$ or $x = 2,\ y = -3$
16. $x = 5,\ y = -0.5$ or $x = -1,\ y = 1$
17. $x = -2,\ y = 1\frac{1}{2}$ or $x = 1\frac{2}{3},\ y = -\frac{1}{3}$
18. $x = -1,\ y = -3$ or $x = 1.5,\ y = -\frac{1}{2}$

Review Exercise 20 — Page 198

1. (a) $x = 3,\ y = 7$ (b) $x = 4.5,\ y = 1.25$
2. $x = 1.2,\ y = 1.4$
3. $x = -0.5,\ y = 6.5$
4. (a) Same gradient, 2
   $y = 2x - 2,\ y = 2x + \frac{3}{2}$
   (b) Same gradient, $\frac{1}{4}$
   $y = \frac{1}{4}x + \frac{1}{4},\ y = \frac{1}{4}x + \frac{3}{8}$
   (c) $a = 3$, $b =$ any number, except 2
   (d) $\frac{q}{p} = -3,\ \frac{r}{p} \neq 2$
5. $x = 3,\ y = 2$
6. (a) $x = 1,\ y = 3$ (b) $x = 2,\ y = -2$
   (c) $x = -1,\ y = 2$ (d) $x = 2,\ y = 3\frac{1}{2}$
7. $x = 1,\ y = -2$
8. $x = 2.5,\ y = -1$
9. $x = 1,\ y = -2$
10. (a) $4x + y = 58,\ 6x + 2y = 92$
    (b) $x = 12,\ y = 10$
11. $x + 3y = 16,\ 2x + y = 17$
    $x = 7,\ y = 3$
12. $x = -2.6,\ y = 3.3$ or $x = 1.6,\ y = 1.2$
13. $x = 1\frac{2}{3},\ y = 12$ or $x = 4,\ y = 5$
14. $x = 2,\ y = 3$ or $x = 3.6,\ y = -0.2$
15. $x = 1,\ y = 3$ or $x = -\frac{2}{3},\ y = \frac{4}{3}$
16. $x = -3,\ y = -3$ or $x = 1.75,\ y = 2.94$
17. $(3, 4), (4, 3)$

CHAPTER 21

Exercise 21.1 — Page 201

1. (a) $2x^2 - 3x + 34$ (b) $3x^2 - 2x + 5$
   (c) 1 (d) $-14x^2 + 66x - 18$
2. (a) $-6x + 29y - 24z$
   (b) 0
   (c) $10x^2 + 12xy + 10y^2$
   (d) $13x^2 + 24xy + 13y^2$

Exercise 21.2 — Page 202

1. (a) $2d + 3$ (b) $3x + 2$
   (c) $4a + 5b$ (d) $3m - 2n$
   (e) $4x - 2y$ (f) $a + b$
   (g) $x - 1$ (h) $x^2 - 2$
   (i) $\frac{4}{x - 3}$ (j) $\frac{x + 2}{3}$ (k) $\frac{x + 2y + 3z}{2x - 3y + z}$
   (l) $\frac{x + 2}{3x + 2}$ (m) $\frac{x}{3x - 2}$ (n) $\frac{x}{2x^2 - 1}$
   (o) $\frac{1}{2x - 1}$ (p) $\frac{1}{3x}$ (q) $\frac{3}{2}$
   (r) $\frac{m}{3}$ (s) $\frac{x - 3}{x + 2}$ (t) $-\frac{5}{3}$
2. (a) **E** (b) **C** (c) **B** (d) **A** (e) **D**
3. (a) $\frac{x + 5y}{3 + 4xy}$ (b) $\frac{x + y + 2z}{2x - 3y + z}$
   (c) $\frac{ab - 3}{b + a}$ (d) $\frac{x + b + y}{ab - 2y}$
4. (a) $\frac{x}{x + 1}$ (b) $\frac{x + 2}{x + 3}$ (c) $\frac{x}{x + 3}$
   (d) $\frac{x - 5}{x + 3}$ (e) $\frac{x - 2}{x - 3}$ (f) $\frac{x - 5}{x + 4}$
   (g) $x - 1$ (h) $\frac{1}{x + 2}$ (i) $\frac{x + 1}{2}$
   (j) $\frac{x + 2}{x + 3}$ (k) $\frac{x - 5}{x - 4}$ (l) $\frac{2x - 5}{3x - 1}$

Exercise 21.3 — Page 204

1. (a) $\frac{2}{3}$ (b) $-1\frac{1}{4}$ (c) $\frac{y}{2}$
2. (a) $\frac{2}{3}$ (b) $-\frac{1}{2}$ (c) $-\frac{2}{x}$
3. (a) $\frac{7x + 8}{(x + 5)(x - 4)}$ (b) $\frac{37x + 3}{(2x + 3)(5x - 3)}$
   (c) $\frac{3x^2 + 4x - 14}{(x + 2)(x - 3)}$
4. (a) $\frac{x - 31}{(x + 5)(x - 4)}$ (b) $\frac{13x - 33}{(2x + 3)(5x - 3)}$
   (c) $\frac{x^2 + 2x + 10}{(x - 3)(x + 2)}$
5. (a) $\frac{7x - 12}{x(2x - 6)}$ (b) $\frac{7 - x}{2x}$ (c) $\frac{x^2 - 3x + 9}{x(2x - 6)}$
   (d) $\frac{4}{x - 3}$ (e) $\frac{3}{4x}$ (f) $\frac{x - 3}{4}$

Exercise 21.4 Page 205

1. $x = -1$ or $\frac{2}{3}$

2. (b) $x = -0.2$ or 6

3. (b) $x = -0.5$ or 3

4. (a) $x = -1.5$ or 0.25
(b) $x = -2$ or 0.8
(c) $x = -0.721$ or 1.39
(d) $x = -\frac{5}{7}$ or 6

5. (a) $\frac{75}{v}$
(b) $\frac{75}{v + 5}$
(c) $\frac{75}{v} - \frac{75}{v + 5} = \frac{1}{6}$ 45 miles per hour

6. (a) $\frac{60}{x} + \frac{90}{x + 6} = 5\frac{1}{2}$ (c) 24 km/h

Exercise 21.5 Page 207

1. (b) $x = 4.30$
(c) $x_{n+1} = \frac{5x_n - 3}{x_n}$ is much faster.

2. (b) $x = 1.40$
(c) $x_{n+1} = \frac{7x_n + 2}{6x_n}$ is quicker.

3. $x_2 = 1.7320\ldots$, x_3 does not exist.

4. $x_{n+1} = \sqrt{x_n + 5}$ leads to $x = 2.79$

5. E.g. $x_{n+1} = \sqrt{100 - 7x_n}$ with $x_1 = 7$ leads to $x = 7.09$

6. E.g. $x_{n+1} = \frac{8 - 3x_n}{x_n}$ leads to $x = -4.70$
Other root is $x = 1.70$

Exercise 21.6 Page 208

1. $x = 3.8$

2. (a) $w = 4.2$ (b) $x = 2.3$

3. $x = 3.2$

4. $x = 9.4$

5. (a) $x = 6.2$ (b) $x = 6.22$

6. (a) $2x^2(x + 2)$ (b) $x = 5.3$

7. $x = 1.82$

Review Exercise 21 Page 209

1. (a) (i) $(x + 2)(x - 2)$
(ii) $(3x - 4)(x + 2)$
(b) $\frac{x(x - 2)}{3x - 4}$

2. $\frac{6}{x - 2}$

3. $2(3x - 1) - 3(x + 3) = 2(x + 3)(3x - 1)$
$6x - 2 - 3x - 9 = 6x^2 + 16x - 6$
$6x^2 + 13x + 5 = 0$

4. $\frac{3}{(2x + 1)(x + 1)}$

5. $x = -1.66$ or $x = 0.906$

6. (b) (i) $\frac{1200}{v + 10}$ minutes
(ii) $\frac{1200}{v} - \frac{1200}{v + 10} = 4$
(iii) $1200(v + 10) - 1200v = 4v(v + 10)$
(iv) $v = 50$ mph

8. $x = 6.7$

9. (a) $2x^2 + 5x + 3$
(b) 40.5 m^2
(c) 8.2 m

10. $x = 1.42$ or $x = 4.58$

CHAPTER 22

Exercise 22.1 Page 212

1. (a)

x	−2	−1	0	1	2
y	−8	−1	0	1	8

(b)

x	−2	−1	0	1	2
y	−6	1	2	3	10

(c)

x	1	2	3	4	5
y	−8	−1	0	1	8

(d)

x	−2	−1	0	1	2
y	−16	−2	0	2	16

(c) (i) Translation, vector $\begin{pmatrix} 0 \\ 2 \end{pmatrix}$
(ii) Translation, vector $\begin{pmatrix} 3 \\ 0 \end{pmatrix}$
(iii) Stretch, from x axis, parallel to y axis, scale factor 2
(d) (i) $y = x^3 - 5$ (ii) $y = (x - 2)^3$

2. **a:** $y = x^2 + 1$
b: $y = x^2 - 4$
c: $y = (x - 1)^2$
d: $y = (x + 1)^2$

3. **a:** $y = \frac{1}{2}x^2$
b: $y = (4x)^2$
c: $y = -2x^2$

4. (a) $y = 9 - x^2$
(b) $y = 2(4 - x^2)$
(c) $y = 4 - \left(\frac{1}{2}x\right)^2$

5. (a) (i) $y = x^2 - x - 5$
(ii) $y = (x - 3)^2 - (x - 3)$
(b) $y = \frac{1}{2}(x^2 - x)$
(c) $y = \left(\frac{1}{2}x\right)^2 - \left(\frac{1}{2}x\right)$

Exercise 22.2

1.

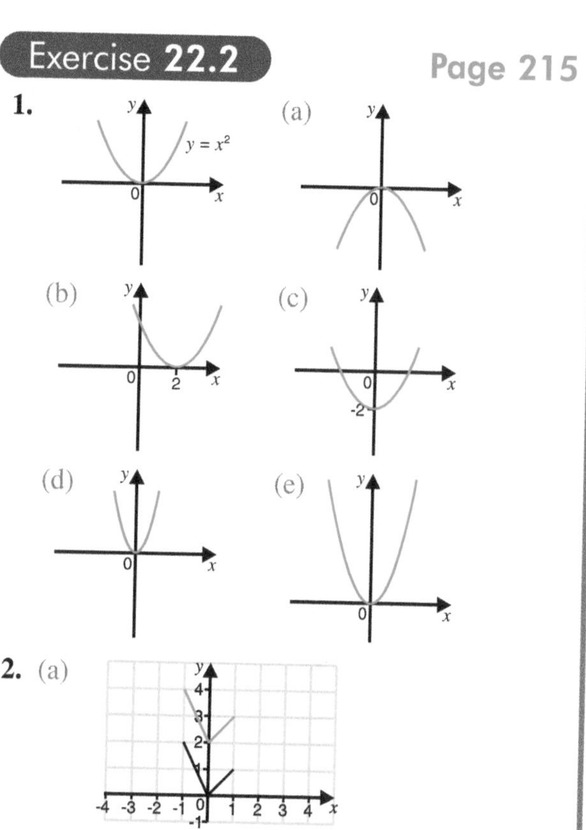

2. (a)

(b) Translation, vector $\begin{pmatrix} 0 \\ 2 \end{pmatrix}$

(c) (i)

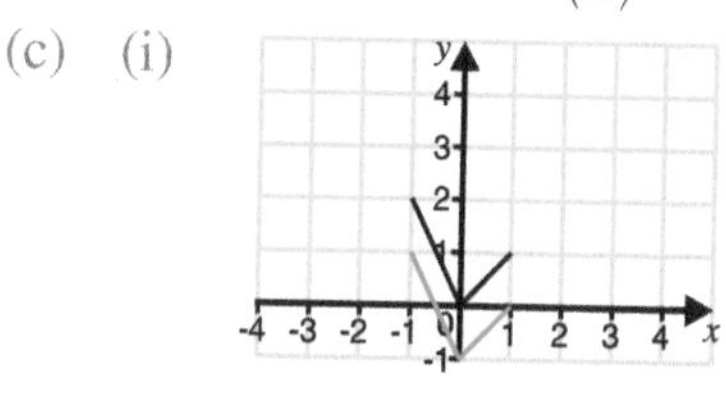

Translation, vector $\begin{pmatrix} 0 \\ -1 \end{pmatrix}$

(ii)

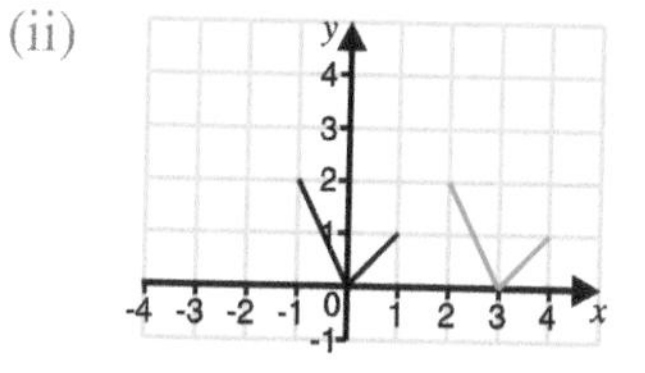

Translation, vector $\begin{pmatrix} 3 \\ 0 \end{pmatrix}$

(iii)

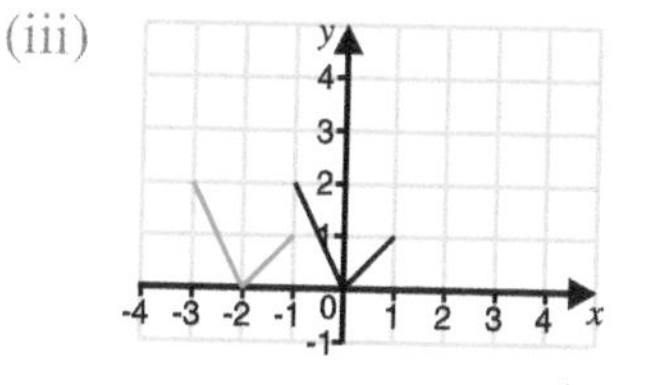

Translation, vector $\begin{pmatrix} -2 \\ 0 \end{pmatrix}$

(iv)

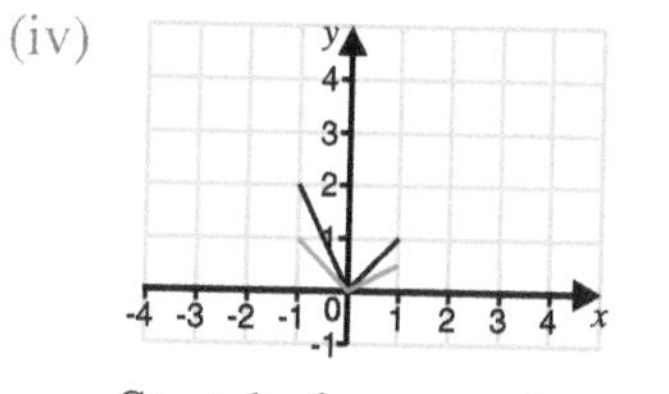

Stretch, from x axis, parallel to y axis, scale factor 0.5

(v)

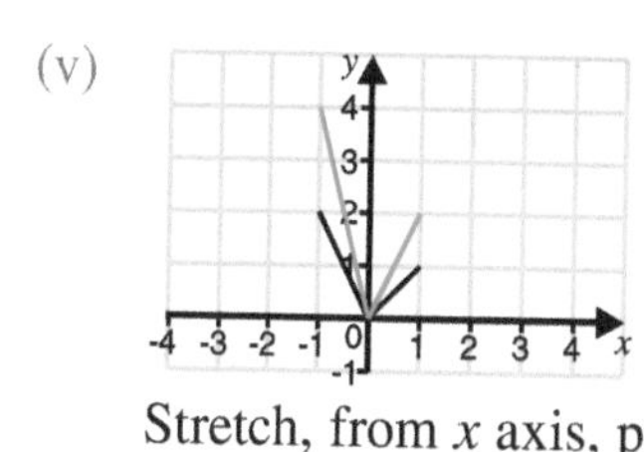

Stretch, from x axis, parallel to y axis, scale factor 2

(vi)

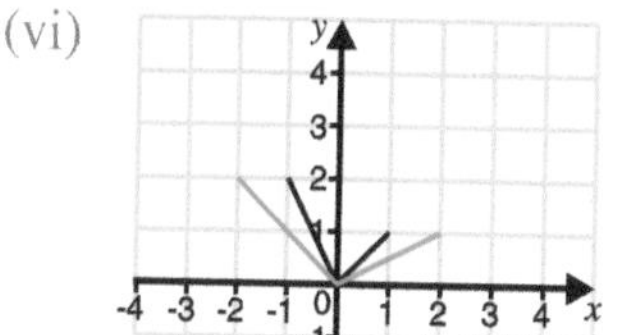

Stretch, from y axis, parallel to x axis, scale factor 2

(vii)

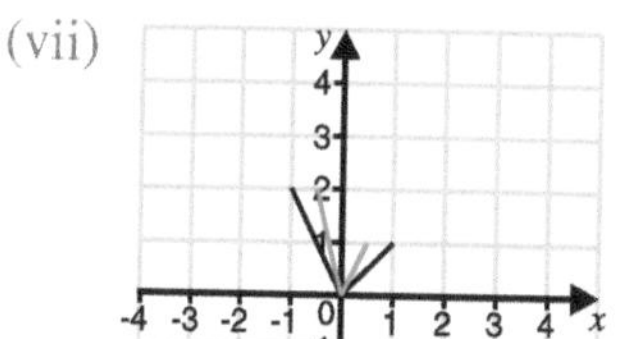

Stretch, from y axis, parallel to x axis, scale factor 0.5

(viii)

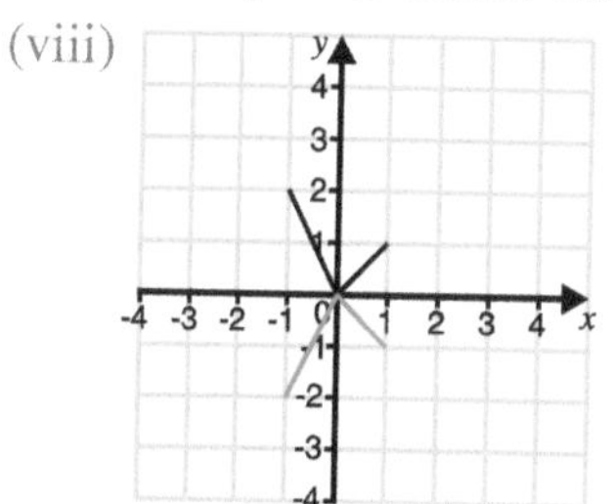

Reflection in x axis

(ix)

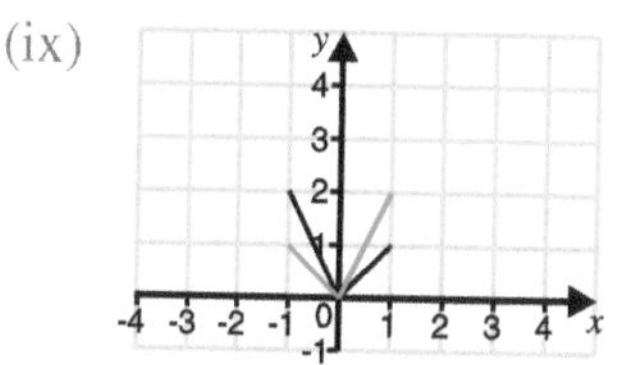

Reflection in y axis

(x)

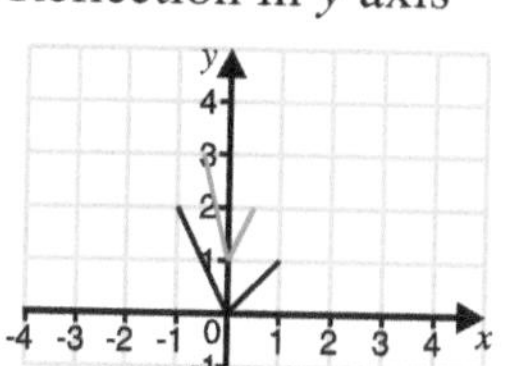

Stretch, from y axis, parallel to x axis, scale factor 0.5, followed by, translation, vector $\begin{pmatrix} 0 \\ 1 \end{pmatrix}$

(xi)

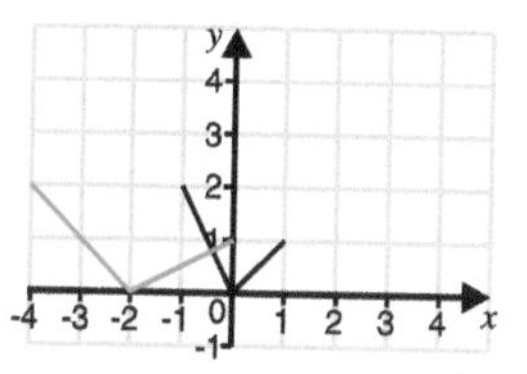

Translation, vector $\begin{pmatrix} -1 \\ 0 \end{pmatrix}$, followed by, stretch, from y axis, parallel to x axis, scale factor 2

(xii)

Reflection in x axis, followed by, translation, vector $\begin{pmatrix} 0 \\ 1 \end{pmatrix}$

3. (a) (b) (c) (d) (e)

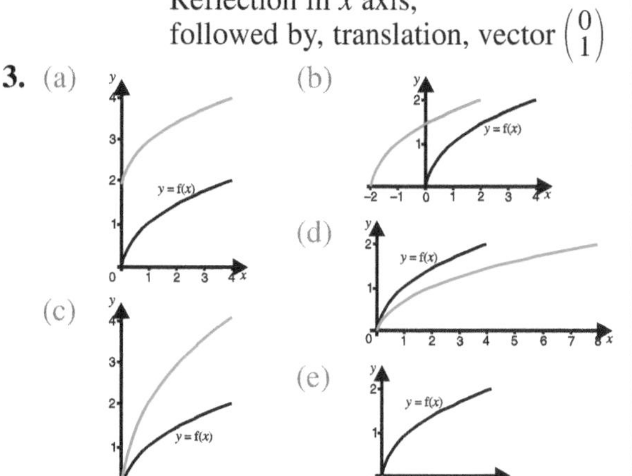

4. (a) (b) (c) (d) (e)

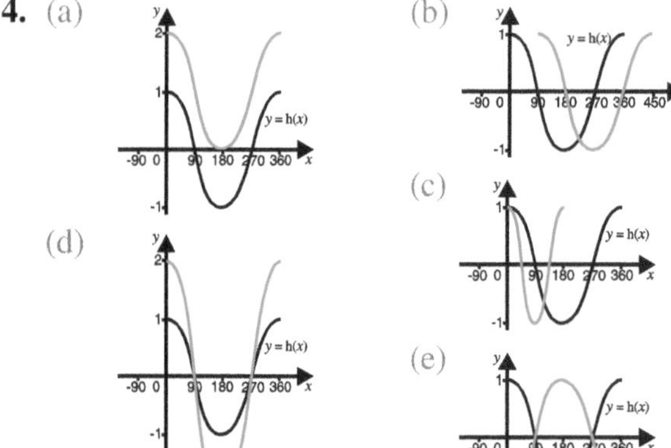

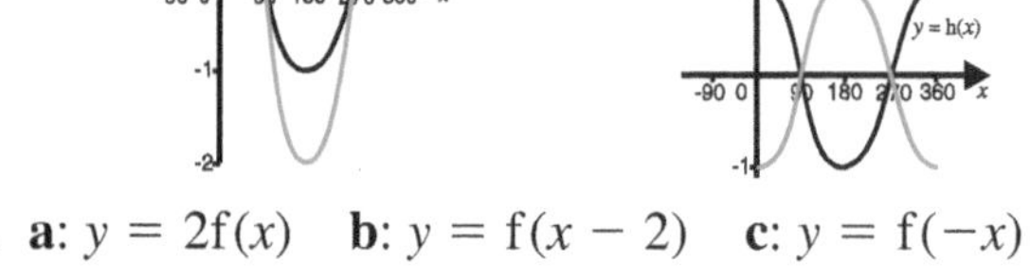

5. **a**: $y = 2f(x)$ **b**: $y = f(x - 2)$ **c**: $y = f(-x)$

6. (a) (i) Reflection in x axis.
(ii) Stretch, from y axis, parallel to x axis, scale factor $\frac{1}{2}$.
(b) Reflection in x axis.
(c) Reflection in x axis, followed by, stretch, from y axis, parallel to x axis, scale factor $\frac{1}{2}$.
(Transformations in either order.)
(d) **a**: $y = -\sin x$ **b**: $y = \sin 2x$
c: $y = -\sin 2x$

7. $g(x) = f(x + 4)$ $h(x) = f(x + 4) + 2$

Exercise 22.3 — Page 218

1. (a) $y = 28x + 13$, $q = 40.3 - 12.8p$
(b) (i) $y = 88.6$ (ii) $p = 2.55$

2. (a) $l = 0.0625w + 20$
(b) (i) $l = 51.25$ (ii) $w = 560$

3. $a = 0.1$, $b = 2.4$

4. 2.1 m/s^2

5. (a) Non-linear graph
(b) Linear graph, $y = 25x^3 + 22.7$, $a = 25$, $b = 22.7$

6. $a = 0.05$, $b = 2.4$

Review Exercise 22 — Page 220

1. (a) (b)

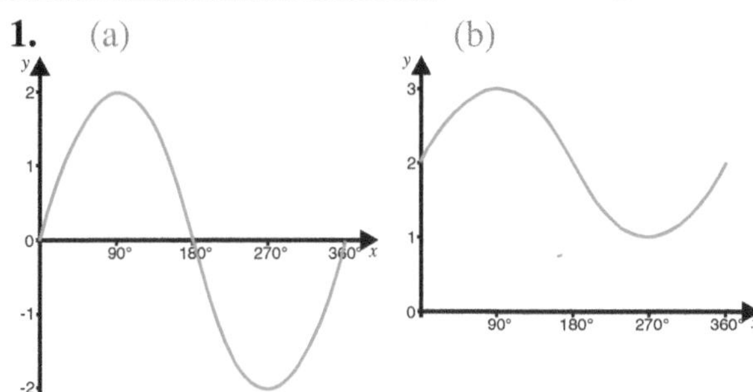

(c) (d)

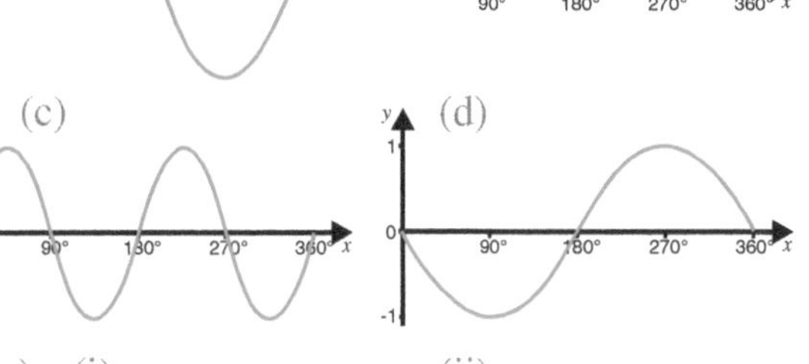

2. (a) (i) (ii)

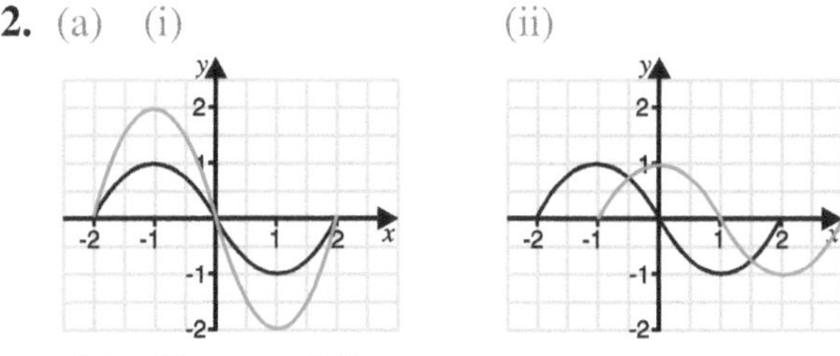

(b) **C**, $a = 0.5$

3. (a) (i) (ii)

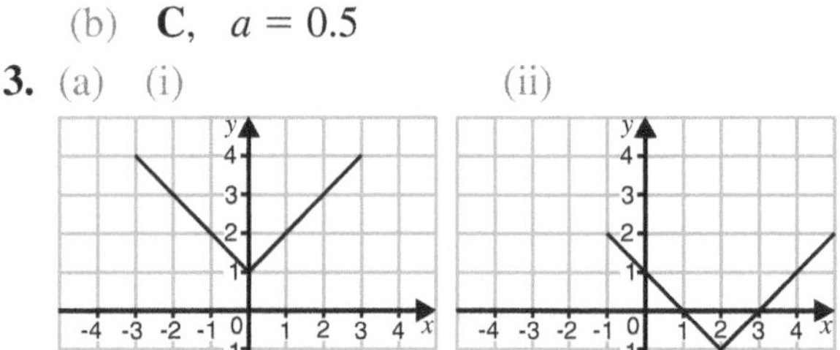

(b) $y = -f(x) + 3$

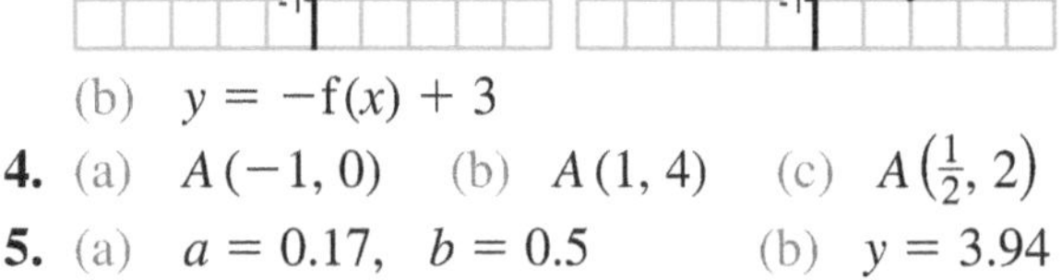

4. (a) $A(-1, 0)$ (b) $A(1, 4)$ (c) $A\left(\frac{1}{2}, 2\right)$

5. (a) $a = 0.17$, $b = 0.5$ (b) $y = 3.94$

Algebra

Non-calculator Paper — Page 222

1. (a) 18 (b) $2x - 1$ (c) $y = 4$

2. (a) 1000 (b) 5 km
(c) 30 minutes (d) 20 km/h

3. (a) (i) 25 (ii) -1.2
(b) $3mn$
(c) (i) $21x + 14y$ (ii) $a^2 - 3a$

4. (a) $x = 3$ (b) $y = 9$ (c) $z = 0.7$

5. (a) $2x$ pence (b) $3(x + 5)$ pence
(c) 16 pence

6. (b) $x = 3.5$

7. $(2, 1)$

8. (a) (i) $x^2 - 5x$ (ii) $20p + 10$
   (b) (i) $3(2n + 3)$ (ii) $m(2m + 1)$
   (c) (i) $x = 100$ (ii) $x = -3\frac{1}{2}$

9. (a) $x < 4$ (b) $x \leqslant 1$

10. (a) c^4 (b) d^5 (c) e^{-7}

11. (a) $y = 7$ (b) $x = \frac{y - c}{m}$

12. (a) Missing entries are: 3, 24
    (c) $x = 3.6$ cm

13. (a) $x = 3\frac{1}{3}$ (b) (i) 1, 2 (ii) $x \geqslant \frac{1}{5}$
    (c) $x = 4, \ y = 1$

14. (a) $y = 0.4x + 18$ (b) £58

15. (a) $x < -0.5$ (b) $x^2 - 10x + 25$
    (c) (i) $3m(m - 2)$ (ii) $(t - 4)(t + 3)$

16. (a) (i) a^8 (ii) a^2 (iii) a^{15}
    (b) (i) $(a^5)^3$ (ii) $a^5 \div a^3$

17. (a) $\frac{a}{5}$ (b) $2(x - 3)(x + 3)$

18. (a) (i) $13x + 1$ (ii) $x^2 + 2x - 8$
    (b) (i) $(x + 3)(x + 5)$ (ii) $x = -3$ or -5

19. (a) $2(x + 5)(x - 5)$ (b) $(2x + 5)(x - 1)$

20. (a) $A = 0.75w^2$ (b) $A = 18.75\,\text{m}^2$

21. (a) $(x - 5)(x - 5)$ (b) $y = 8$

22. $x = 2, \ y = -1$

23. $(x + 2)(x + 1) = 6$ $\quad x^2 + 3x + 2 = 6$
    $x^2 + 3x - 4 = 0$

24. (b) (i) $x = -2$ or 1 (ii) $x = -3$ or 2
    (c) Draw $y = 2x + 4$, $x = -2$ or 3

25. $R = 4$

26. (a) $m_{AC} = -1, \ m_{BD} = 1$
    (b) $y = x + 3$ (c) $y = x + 5$

27. (a) $a = \frac{5c + 6}{c + 4}$ (b) $\frac{x^2 + 1}{(x - 1)(x + 1)}$

28. $x = 1, \ y = 1$ or $x = 0, \ y = 2$

29. (a) $a = 3, \ b = -12$
    (b) $x = 2\sqrt{3} - 3$ or $x = -2\sqrt{3} - 3$

30. (a) $P(-4, 5)$
    (b) $P(-1, 10)$
    (c) $P(-1, -5)$

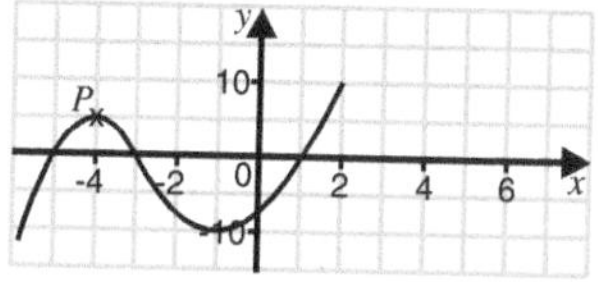

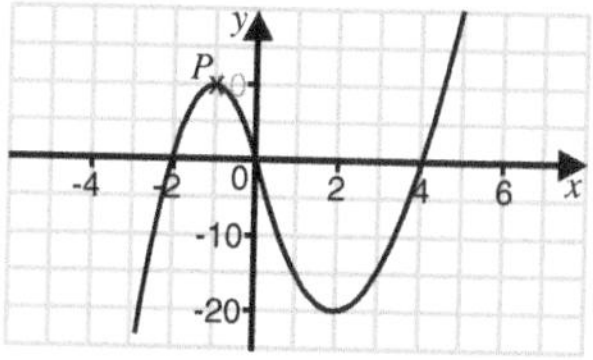

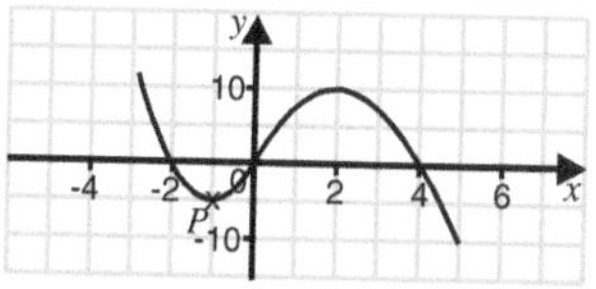

Algebra

Calculator Paper — Page 225

1. (a) Even number.
   (b) Could be even or odd.
   When n is even, $3n$ is even, $3n + 1$ is odd.
   When n is odd, $3n$ is odd, $3n + 1$ is even.

2. (a) $18x$ pence (b) $(3n + 2m)$ pence

3. (a) $x = 6$ (b) $5x - 12$

4. (a) (i) 15 (ii) 1
   (b) $x = 3$

5. (a) $(5a + 3)$ cm (b) $5a + 3 = 23, \ a = 4$

6. (a) 2, 5 (b) 11th term
   (c) $(85 + 1)$ is not divisible by 3.

7. $3x + 4$

8. (a) 18 (b) $4n - 2$

9. $x = 2.6$

10. (a) Missing entries are: $-3, -6, -6$
    (c) $x = \pm 2.6$

11. (a) $3x + 6$
    (b) (i) $x = 27$ (ii) $x = -\frac{4}{5}$
    (c) (i) m^5 (ii) n^3

12. (a) $-1, 0, 1$ (b) $x \geqslant 5$

13. (a) (i) $4(2y + 1)$ (ii) $x(x^2 - 5)$
    (b) (i) x^5y^2 (ii) p^6
    (c) $x = \frac{y - 10}{5}$
    (d) (i) $t = 5\frac{1}{4}$ (ii) $x = 8$

14. (a)

x	-1	0	1	2	3
y	5	2	1	2	5

    (c) $x = -0.4$ and 2.4

15. (a) -3 (b) $9 - 2n$

16. (a) $x^{-2}, \ \frac{1}{x}, \ x^{\frac{1}{2}}, \ x$ (b) $x, \ x^{\frac{1}{2}}, \ \frac{1}{x}, \ x^{-2}$

17. If n is odd, $(n + 1)$ is even and $(n + 2)$ is odd.
    If n is even, $(n + 1)$ is odd and $(n + 2)$ is even.
    Odd number × Even number = Even number.
    So, every term is an even number.

18. (a) (c) 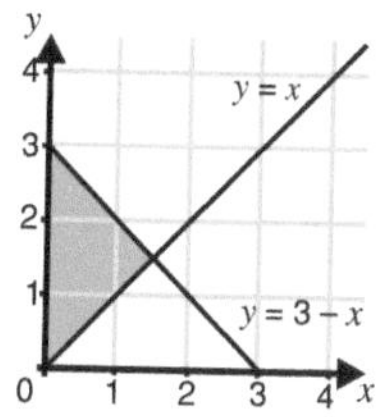
    (b) Where graphs cross.

19. (a) $x = -\frac{1}{2}$ (b) $m(m - 7)$
    (c) 6, 7, 8 (d) $x = -2, \ y = 1.5$

20. (a) $G(0, 5)$, $H(10, 0)$

(b) (i)

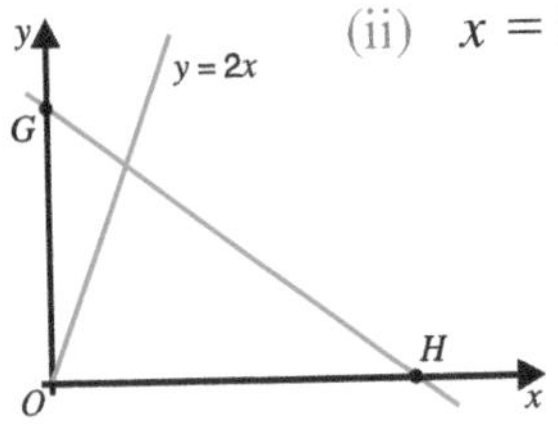

(ii) $x = 2, \quad y = 4$

21. (a) $x = -\frac{1}{4}$ (b) $x = 3$

22. (a) $6t^4u^3$ (b) $8c^{12}$

23. (a) $r = \sqrt[3]{\frac{3V}{4\pi}}$ (b) $r = 6$

24. $y = 0$ or $y = -5$

25. (a) (i) $y = 1 - x^2$ (ii) $2y = 2 + x$
(iii) $x^2 + y^2 = 3$ (iv) $xy = 1$

(b)

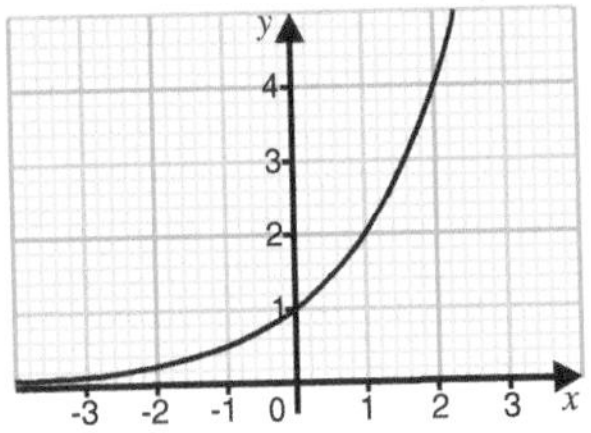

26. (a) $a = -3, \quad b = -2$, minimum value $= -2$
(b) $p = 3, \quad q = 1$

27. (a) $\frac{1}{2x}$ (b) $\frac{y + 3}{2y - 1}$

28. (a) $1.5\,\text{m/s}^2$ (b) 20 seconds

29. (a) $(3x + 2)(x + 1) = 80$
$3x^2 + 5x - 78 = 0$
(b) $x = 4\frac{1}{3}$, area of lawn $= 56.3\,\text{m}^2$

30. (a) $y = 20\sqrt{x}$
(b) $x = 18.0625$

31. $x = -5.16$ or 1.16

32. $x = -1\frac{1}{2}$ or 2

33. $x = \frac{ab}{a + b}$

CHAPTER 23

Exercise 23.1 Page 229

1. (a) $a = 96°$, supplementary angles.
(b) $b = 20°$, complementary angles.
(c) $c = 150°$, angles at a point.

2. (a) $a = 130°$, alt. ∠'s, $b = 130°$, corres. ∠'s
(b) $c = 60°$, alt. ∠'s, $d = 120°$, allied ∠'s
(c) $e = 40°$, corres. ∠'s,
$f = 40°$, vert. opp. ∠'s
(d) $g = 65°$, alt. ∠'s, $h = 65°$, corres. ∠'s

3. (a) $a = 125°, \quad b = 125°$
(b) $c = 56°, \quad d = 116°$
(c) $e = 61°$

4. (a) $\angle ABC = 132°$ (b) $\angle QRS = 126°$
(c) $\angle ZYV = 141°$
(d) $\angle AOB = 153°, \quad \angle COD = 37°$
(e) $\angle QTU = 48°, \quad \angle QTS = 132°$
(f) reflex $\angle TUV = 280°$

Exercise 23.2 Page 231

1. (a) Yes (b) Yes (c) No (d) No (e) Yes (f) No

2. (a) Yes, obtuse-angled (b) No
(c) Yes, acute-angled (d) Yes, right-angled
(e) Yes, obtuse-angled

3. (a) $a = 35°$
(b) $b = 148°, \quad c = 32°$
(c) $d = 52°, \quad e = 64°$
(d) $f = 63°$

4. (a) $a = 18°, \quad b = 144°$
(b) $c = 26.5°, \quad d = 153.5°$

5. (a) Isosceles (b) 74° (c) 46°

6. (a) $\angle BCD = 120°$
(b) $\angle PRQ = 80°, \quad \angle QRS = 160°$
(c) $\angle MNX = 50°$

Exercise 23.3 Page 233

1. $S(3, 1)$

2. $Y(6, 4)$

3. $A(1, 3)$

4. (3, 1), (3, 5)

5. (a) $a = 62°$ (b) $b = 54°, \quad c = 36°$
(c) $d = 62°$ (d) $e = 116°, \quad f = 86°$
(e) $g = 124°$ (f) $h = 75°$
(g) $i = 38°, \quad j = 42°$
(h) $k = 55°, \quad l = 45°$
(i) $m = 65°, \quad n = 90°, \quad o = 25°$
(j) $p = 117°$ (k) $q = 70°$
(l) $r = 85°$ (m) $s = 28°, \quad t = 80°$
(n) $u = 70°$

6. $\angle SQR = 50°$

7. $a = 22\frac{1}{2}°$ $b = 112\frac{1}{2}°$ $c = 22\frac{1}{2}°$

Exercise 23.4 Page 235

1. (a) $x + 130° = 180°$ (supp. ∠'s)
$x = 180° - 130° = 50°$
(b) $y + 130° + 70° + 100° = 300°$
(sum of ∠'s in a quad. = 360°)
$y = 360° - 300° = 60°$

2. (a) $a = 62°, \quad b = 55°$ (b) $c = 62°$
(c) $d = 76°$

3. (a) 900° (b) 1080° (c) 1260°

4. (a) $a = 130°$ (b) $b = 250°$ (c) $c = 85°$

5. 720°

Exercise 23.5 Page 237

1. (a) (i) 120° (ii) 90° (iii) 60° (iv) 45°
(b) (i) 60° (ii) 90° (iii) 120° (iv) 135°

2. 20

3. 8

4. (a) 72° (b) 108° (c) 540°

5. (a) $a = 90°, \quad b = 60°, \quad c = 210°$
(b) $d = 90°, \quad e = 120°, \quad f = 150°$
(c) $g = 90°, \quad h = 135°, \quad i = 135°$
(d) $j = 105°$

6. 153°
7. 150°
8. At any vertex, sum of angles cannot equal 360°.

Exercise 23.6 — Page 238

2. (a) 5, 5 (b) 6, 6 (c) 8, 8
3. 1, 1, 3
4. (a) A B C
   (b) **A**: 3, **B**: 4, **C**: 2
5.

A	B	C	D	E	F	G	H	I
1	0	1	4	2	2	0	0	1
1	1	1	4	2	2	1	2	1

6. (a) 1
   (b) (i) (ii)
7. (a) 1
   (b) (i)
   (ii) 3

Review Exercise 23 — Page 241

1. (a) (i) 153° (ii) 63° (b) $\angle BOQ$
2. $x = 30°$
3. (a) $a = 71°$ (b) $b = 63°$ (c) $c = 117°$
4. $x = 20°$, $\angle DAC = 40°$
5. $\angle BDC = 112°$
6. (a) 1 (b) 56°
7. $a = 115°$, $b = 44°$
8. (a) $q = 12°$ (b) 30 sides
9. (a) (i) $x = 36°$ (ii) $y = 18°$
   (b) (i) parallel (ii) $z = 36°$

CHAPTER 24

Exercise 24.1 — Page 245

1. (a) $a = 90°$, $b = 50°$ (b) $c = 45°$
   (c) $d = 40°$, $e = 54°$ (d) $f = 20°$
2. (a) $a = 77°$ (b) $b = 72°$
   (c) $c = 38°$ (d) $d = 40°$
   (e) $e = 25°$ (f) $f = 65°$, $g = 65°$
   (g) $h = 130°$, $i = 65°$
   (h) $j = 94°$, $k = 43°$
3. (a) $k = 106°$, $l = 62°$
   (b) $m = 70°$, $n = 140°$
   (c) $p = 108°$, $q = 105°$
   (d) $r = 117°$, $s = 84°$
   (e) $t = 292°$, $u = 56°$
   (f) $v = 140°$
   (g) $w = 110°$, $x = 75°$
   (h) $y = 130°$, $z = 82°$

4. 47°
5. $\angle BCD = 70°$ $\angle CAD = 76°$
6. 119°
7. $\angle PQR = 105°$ $\angle QRS = 118°$
8. $\angle PSR = 45°$ $\angle TPQ = 95°$
9. (a) 140° (b) 70°
10. (a) 100° (b) 40° (c) 25°
11. 25°
12. 90°

Exercise 24.2 — Page 248

1. (a) $a = 90°$, $b = 50°$
   (b) $c = 35°$, $d = 35°$
   (c) $e = 90°$, $f = 49°$, $g = 49°$
   (d) $h = 63°$ (e) $j = 67°$
   (f) $k = 50°$ (g) $l = 65°$, $m = 65°$
   (h) $n = 146°$
2. (a) $a = 78°$, $b = 39°$
   (b) $c = 54°$, $d = 63°$
   (c) $e = 55°$, $f = 38°$, $g = 38°$
   (d) $h = 16°$, $i = 106°$
   (e) $j = 65°$, $k = 65°$
   (f) $l = 57.5°$, $m = 67.5°$, $n = 45°$
3. (a) 67° (b) 46°
4. (a) 100° (b) 40° (c) 25°
5. (a) 72° (b) 36° (c) 54°
6. $x = 120°$, $y = 36°$, $z = 60°$
7. $p = 42°$, $q = 55°$, $r = 13°$, $s = 55°$
8. 90

Exercise 24.3 — Page 250

1. (d) $2x + 2y$
3. (a) $2x$ ($\angle$ at centre = twice $\angle$ at circum.)
   (b) x (angles in same segment are equal)
4. (b) (i) $2x$ (ii) $2y$
   ($\angle$ at centre = twice $\angle$ at circum.)
5. (b) Tangent PTQ perpendicular to radius OT.
   (c) x
   (d) Angles in the same segment are equal.

Review Exercise 24 — Page 252

1. (a) $a = 59°$ (b) $b = 45°$
   (c) $c = 51°$ (d) $d = 23°$
2. (a) $\angle OCA = 12°$ (b) $\angle AOC = 156°$
   (c) $\angle ACB = 38°$ (d) $\angle ABC = 102°$
3. $x = 60°$
4. $\angle TBO = 90°$ (tangent perpendicular to radius)
   $\angle TBA = 55°$ (complementary $\angle$'s)
   ΔABT is isosceles
   $\angle ATB = 180 - (55 + 55) = 70°$
5. $x = 18°$
6. $x = 92°$, $y = 46°$, $z = 17°$
7. $x = 50°$, $y = 60°$, $z = 30°$
8. $\angle ABC = 118°$, $\angle BAC = 42°$
9. $\angle TQR = 62°$

CHAPTER 25

Exercise 25.1 — Page 255

1. (a) Perimeter 8 cm, area 3.75 cm^2.
   (b) Perimeter 9.6 cm, area 5.4 cm^2.
   (c) Perimeter 28 cm, area 49 cm^2.
2. (a) 9 cm^2 (b) 4.16 cm^2 (c) 8 cm^2
3. (a) 13.5 cm^2 (b) 20 cm^2 (c) 4.35 cm^2
4. (a) 3 cm (b) 6 cm (c) 9 cm
5. 14 m^2
6. 25 cm^2
7. 84 cm^2
8. 4 cm
9. 60 cm

Exercise 25.2 — Page 256

1. (a) 12.6 cm (b) 15.7 cm (c) 22 cm (d) 28.3 cm
2. 75.4 cm
3. 40.8 cm
4. 30 mm
5. 60 cm
6. 67 m

Exercise 25.3 — Page 257

1. (a) 50.3 cm^2 (b) 28.3 cm^2 (c) 176.7 cm^2 (d) 530.9 cm^2
2. (a) 28.3 cm^2 (b) 59.4 cm^2 (c) 128.6 cm^2
3. 3.99 cm
4. 7.14 m

Exercise 25.4 — Page 258

1. (a) 27 m (b) 56.7 m^2
2. 31 cm
3. The circle is bigger.
   Circle: 50.3 cm^2. Semi-circle: 47.5 cm^2.
4. (a) $225\pi cm^2$ (b) $30\pi cm$
5. 13.3 cm
6. 18.2 cm^2
7. 38.6 cm
8. 18.8 cm
9. 326 cm^2
10. 40.1 cm
11. $49\pi cm^2$
12. $24\pi cm$
13. 8 cm
14. $65\pi cm^2$

Exercise 25.5 — Page 259

1. (a) 134 m^2 (b) 396 km^2
2. (a) 129 cm^2 (b) 41.1 cm^2 (c) 116 m^2
3. (a) 22.7 cm^2 (b) 17.6 cm^2 (c) 23.9 cm^2

Exercise 25.6 — Page 260

1. (a) (i) 3.14 cm (ii) 2.09 cm (iii) 6.28 cm (iv) 10.5 cm (v) 15.6 cm (vi) 26.4 cm
   (b) (i) 3.14 cm^2 (ii) 3.14 cm^2 (iii) 14.1 cm^2 (iv) 25.1 cm^2 (v) 43.8 cm^2 (vi) 95 cm^2
2. (a) 170 cm^2 (b) 54 cm
3. 53.5°
4. 6.62 cm
5. (a) 9.5 cm (b) 14.1 cm
6. 14.0 cm^2
7. $a = 19.1°$
8. 6.91 cm

Review Exercise 25 — Page 262

1. (a) 13.5 m^2 (b) 8.1 m^2
2. (a) 50 cm^2 (b) 17.5 cm^2
3. (a) 4.5 cm^2 (b) 11 cm
4. 177 cm^2
5. 2
6. $\pi \times 5^2 = 78.5$ (less than 84)
   So, radius must be bigger than 5 m.
7. 8510 cm^2
8. 140 cm^2
9. 6 cm
10. 2983 cm^2
11. 42.5°
12. 40π metres
13. 20.3 cm

CHAPTER 26

Exercise 26.1 — Page 265

1. (a) North-West (b) North-East
2. (a) (i) 128° (ii) 308°
   (b) (i) 4.7 cm (ii) 47 km
3. (a) 7 m (b) 30 cm
4. (a) 1 : 20 (b) 5.8 m (c) 22.5 cm
5. (a) 065° (b) 282° (c) 1800 m
7. (a) 8 km (b) 114° (c) 294°
8. (a) 6300 km (b) 248°
9. (a) 50 km (b) 347° (c) 167°

Exercise 26.2 — Page 267

4.

5.

6.

Exercise 26.3 — Page 269

5. Perpendicular bisectors pass through the centre of the circle.

7.

8.

9.

12.

(d) Yes

Review Exercise 26 — Page 274

1. (a) 230° (b) 140°
(c) (i) 3.7 cm (ii) 740 m

2.

3. (a) (b)

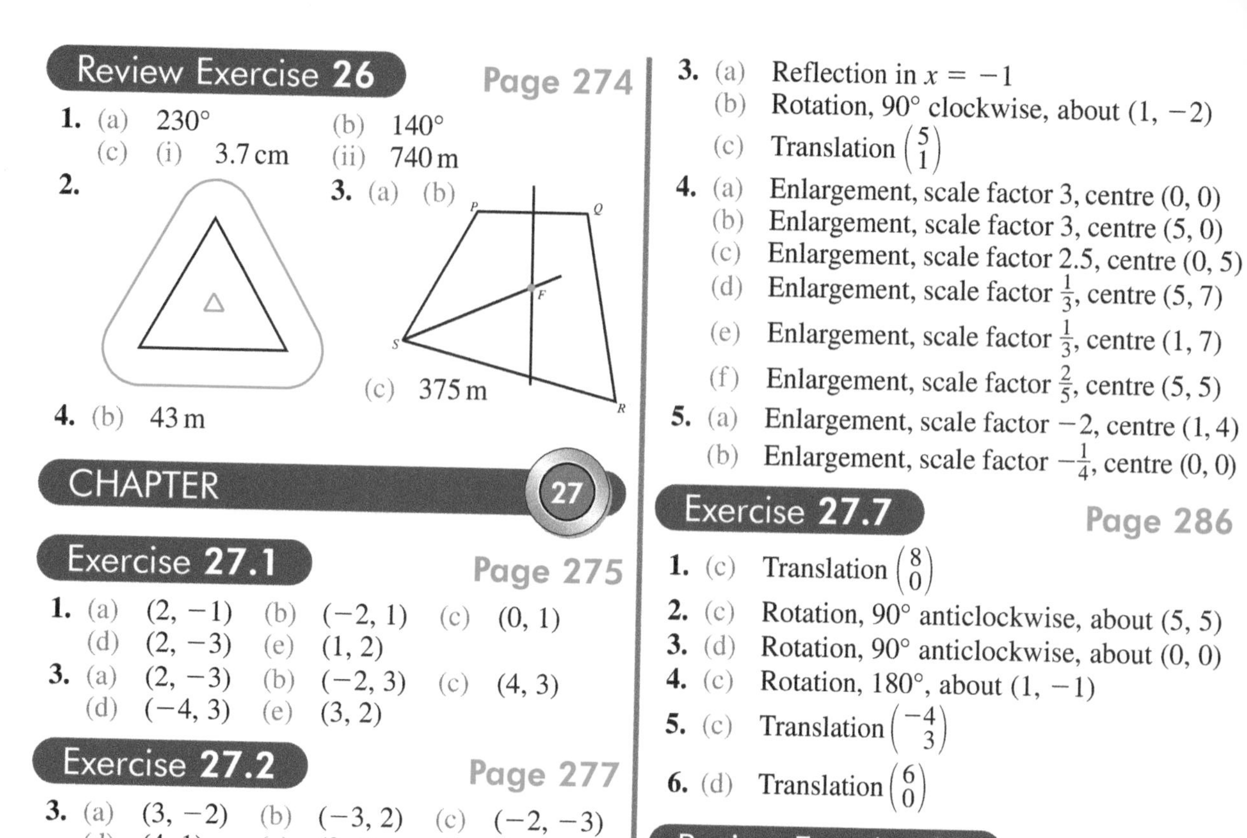

(c) 375 m

4. (b) 43 m

CHAPTER 27

Exercise 27.1 — Page 275

1. (a) (2, −1) (b) (−2, 1) (c) (0, 1)
(d) (2, −3) (e) (1, 2)

3. (a) (2, −3) (b) (−2, 3) (c) (4, 3)
(d) (−4, 3) (e) (3, 2)

Exercise 27.2 — Page 277

3. (a) (3, −2) (b) (−3, 2) (c) (−2, −3)
(d) (4, 1) (e) (0, 1) (f) (2, −1)

Exercise 27.3 — Page 278

3. $T(7, 2)$

4. (a) $\begin{pmatrix} 2 \\ 1 \end{pmatrix}$ (b) $\begin{pmatrix} 1 \\ -2 \end{pmatrix}$ (c) $\begin{pmatrix} -3 \\ 1 \end{pmatrix}$ (d) $\begin{pmatrix} -3 \\ -2 \end{pmatrix}$

5. (b) $\begin{pmatrix} -3 \\ -2 \end{pmatrix}$

6. (a) $A(1, 5)$ (b) $\begin{pmatrix} -1 \\ -5 \end{pmatrix}$

Exercise 27.4 — Page 280

4. (a) (4, 6) (b) (6, 9) (c) (4, 4)
(d) (2, 7) (e) (3, 4) (f) (3, 3)

5. (a) (3, 2) (b) (3, 4)

Exercise 27.5 — Page 281

5. (a) Inverted and smaller
(b) Inverted and same size
(c) Inverted and larger

6. (a) (−6, −6) (b) (−3, 3)
(c) (0, 3) (d) (−1, −1)

Exercise 27.6 — Page 284

1. $\mathbf{L_2}$: Reflection in x axis
$\mathbf{L_3}$: Rotation, 90° anticlockwise, about (0, 0)
$\mathbf{L_4}$: Translation, 3 units right and 2 units up
$\mathbf{L_5}$: Rotation, 180°, about (0, 0)
$\mathbf{L_6}$: Reflection in $x = 7$

2. (a) Translation $\begin{pmatrix} -7 \\ -4 \end{pmatrix}$
(b) Reflection in $y = -x$
(c) Rotation, 90° anticlockwise, about (1, 0)

3. (a) Reflection in $x = -1$
(b) Rotation, 90° clockwise, about (1, −2)
(c) Translation $\begin{pmatrix} 5 \\ 1 \end{pmatrix}$

4. (a) Enlargement, scale factor 3, centre (0, 0)
(b) Enlargement, scale factor 3, centre (5, 0)
(c) Enlargement, scale factor 2.5, centre (0, 5)
(d) Enlargement, scale factor $\frac{1}{3}$, centre (5, 7)
(e) Enlargement, scale factor $\frac{1}{3}$, centre (1, 7)
(f) Enlargement, scale factor $\frac{2}{5}$, centre (5, 5)

5. (a) Enlargement, scale factor −2, centre (1, 4)
(b) Enlargement, scale factor $-\frac{1}{4}$, centre (0, 0)

Exercise 27.7 — Page 286

1. (c) Translation $\begin{pmatrix} 8 \\ 0 \end{pmatrix}$

2. (c) Rotation, 90° anticlockwise, about (5, 5)

3. (d) Rotation, 90° anticlockwise, about (0, 0)

4. (c) Rotation, 180°, about (1, −1)

5. (c) Translation $\begin{pmatrix} -4 \\ 3 \end{pmatrix}$

6. (d) Translation $\begin{pmatrix} 6 \\ 0 \end{pmatrix}$

Review Exercise 27 — Page 288

1. (a) (i) **H** (ii) **G** (iii) **F**
(b) Reflection in the y axis

2. (a) Reflection in $x = -1$
(b) Reflection in $y = -1$
(c) Translation $\begin{pmatrix} -4 \\ 1 \end{pmatrix}$ (d) Translation $\begin{pmatrix} 4 \\ -1 \end{pmatrix}$
(e) Enlargement, scale factor 2, centre (0, 4)
(f) Enlargement, scale factor $\frac{1}{2}$, centre (0, 4)

3. New triangle has vertices at:
(−10, 0), (−2, 0), (−8, −6)

4. (a) Rotation, 180°, about (7, 4)
(b)

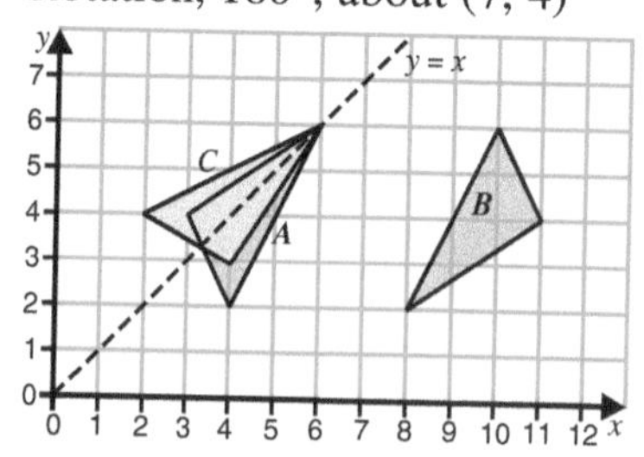

5. Enlarged shape has vertices at:
(3, 2), (5, 2), (6, 0), (2, 0)

7. (c) Translation $\begin{pmatrix} -4 \\ 3 \end{pmatrix}$

8. (a) (b) (c)

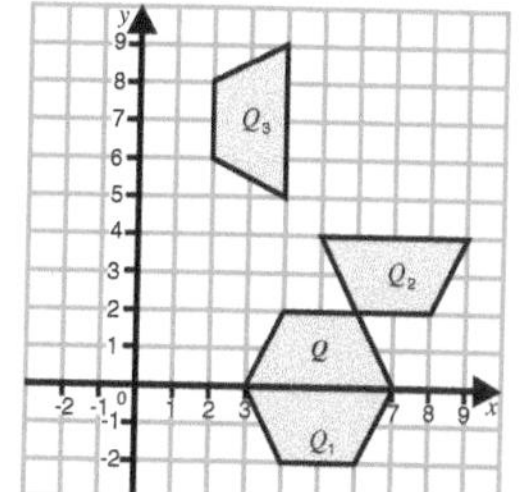

(d) Rotation, 90° anticlockwise, about (1, 3)

9. (a)

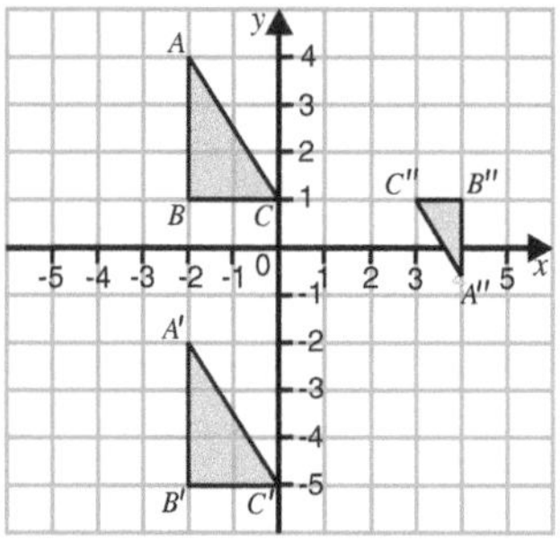

(b) Enlargement, scale factor -2, centre $(2, -1)$

10. (a) Translation $\begin{pmatrix} 3 \\ -6 \end{pmatrix}$ (b) Reflection in $y = -x$

CHAPTER 28

Exercise 28.1 — Page 292

1. (a) 10 cm (b) 25 cm (c) 26 cm
2. (a) 7.8 cm (b) 12.8 cm (c) 10.3 cm
3. (a) $\sqrt{52}$ cm (b) $\sqrt{20}$ cm
4. (a) 5 (b) 9.22 (c) 13 (d) 8.06 (e) 7.21
5. (a) $R(9, 7)$ (b) $X(2, 3)$ (c) $Y(4, 2)$ (d) 2.24

Exercise 28.2 — Page 293

1. (a) 8 cm (b) 6 cm (c) 2 cm
2. (a) 6.9 cm (b) 10.9 cm (c) 9.5 cm
3. 339 m
4. 36 cm^2
5. 3.6 cm

Exercise 28.3 — Page 295

1. (a) 2.9 cm (b) 5.7 cm (c) 2.1 cm (d) 2.0 cm
2. 17 cm
3. 10 cm
4. 8.5 cm
5. 10.6 cm
6. 15 cm
7. 6.9 cm
8. 74.3 cm
9. 13 cm
10. 24 cm
11. (a) 11.4 cm (b) 43.5 cm^2

Exercise 28.4 — Page 296

1. 10.4 cm
2. 20.7 cm
3. 11.9 cm
4. 19.5 cm
5. (a) 11.2 cm (b) 14.1 cm (c) 9.49 cm
6. 4.95 cm
7. (a) 2 units (b) 4 units (c) 4.5 units (d) 3.6 units (e) 5 units (f) 4.5 units

Review Exercise 28 — Page 298

1. $PQ = 15$ cm
2. 8.60 units
3. $41^2 = 40^2 + 9^2$ ΔABC is right-angled at C.
4. (a) 8.60 cm (b) 10.9 cm
5. $\sqrt{18}$ cm
6. 361 m
7. 14.1 km
8. $40^2 > 35^2 + 15^2$, so, $\angle PQR > 90°$

9. $AC = 10$ units, $AE = \sqrt{80}$ units
$AF = \sqrt{116}$ units
10. 42.7 cm
11. 17.3 cm
12. 9.75 cm
13. 5.66 cm
14. 5.4 units

CHAPTER 29

Exercise 29.1 — Page 301

1. (a) $h = 2.27$ m (b) $h = 4.02$ m (c) $h = 8.02$ m
2. (a) $x = 1.50$ cm (b) $x = 2.99$ cm (c) $x = 5.78$ cm
3. (a) $BC = 3.39$ m (b) $AC = 11.4$ m (c) $BC = 4.74$ cm (d) $\angle ABC = 55°$, $AC = 6.96$ cm

Exercise 29.2 — Page 302

1. (a) $a = 62.1°$ (b) $a = 25.0°$ (c) $a = 50.9°$
2. (a) $x = 52.3°$ (b) $x = 60.5°$ (c) $x = 61.4°$
3. (a) $\angle QPR = 23.6°$ (b) $\angle PRQ = 64.7°$ (c) $\angle QPR = 21.8°$ (d) $\angle QRP = 24.0°$, $\angle QPR = 66.0°$

Exercise 29.3 — Page 303

1. (a) $l = 7.00$ m (b) $l = 7.49$ m (c) $l = 4.93$ m
2. (a) $x = 6.49$ cm (b) $x = 5.31$ cm (c) $x = 10.61$ cm
3. (a) $AB = 7.62$ m (b) $AB = 17.5$ cm (c) $AB = 66.9$ cm (d) $\angle ABC = 17.2°$, $AB = 25.7$ m

Exercise 29.4 — Page 305

1.

	$\sin p$	$\cos p$	$\tan p$
(a)	$\frac{5}{13}$	$\frac{12}{13}$	$\frac{5}{12}$
(b)	$\frac{7}{25}$	$\frac{24}{25}$	$\frac{7}{24}$
(c)	$\frac{4}{5}$	$\frac{3}{5}$	$\frac{4}{3}$

2. (a) $p = 40.9°$ (b) $p = 38.0°$ (c) $p = 38.8°$ (d) $p = 53.3°$ (e) $p = 39.7°$ (f) $p = 35.5°$
3. (a) $a = 5.77$ cm (b) $a = 6.47$ cm (c) $a = 4.18$ cm (d) $a = 3.78$ cm (e) $a = 4.72$ cm (f) $a = 4.04$ cm
4. 4.33 cm
5. 72.5°, 72.5°, 34.9°
6. (a) 5.96 cm (b) 1.71 cm
7. (a) 2.76 cm (b) 19.8°
8. 13.06 cm
9. 4.52 cm
10. 9.45 cm

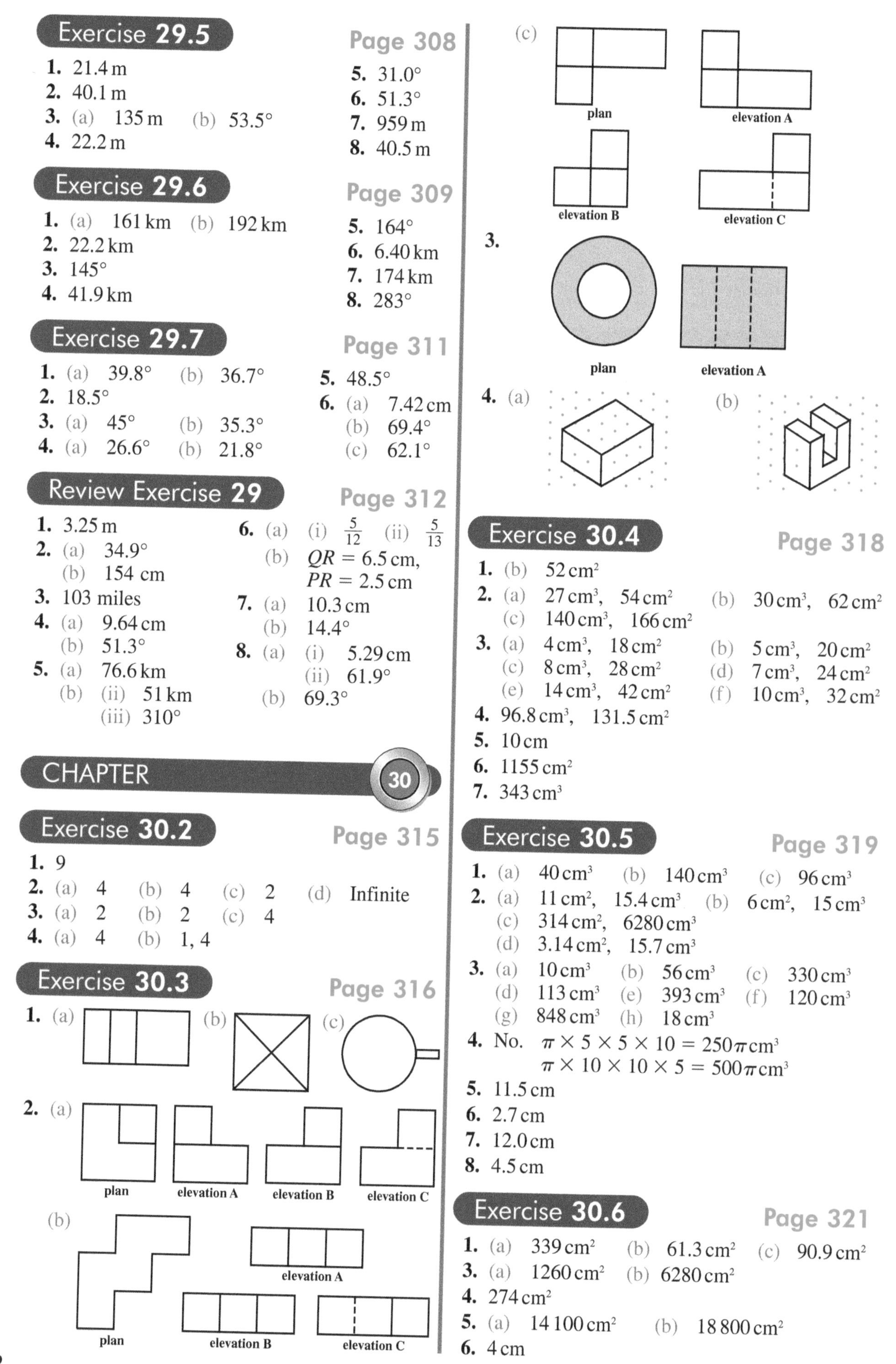

Exercise 29.5

Page 308

1. 21.4 m
2. 40.1 m
3. (a) 135 m (b) 53.5°
4. 22.2 m
5. 31.0°
6. 51.3°
7. 959 m
8. 40.5 m

Exercise 29.6

Page 309

1. (a) 161 km (b) 192 km
2. 22.2 km
3. 145°
4. 41.9 km
5. 164°
6. 6.40 km
7. 174 km
8. 283°

Exercise 29.7

Page 311

1. (a) 39.8° (b) 36.7°
2. 18.5°
3. (a) 45° (b) 35.3°
4. (a) 26.6° (b) 21.8°
5. 48.5°
6. (a) 7.42 cm (b) 69.4° (c) 62.1°

Review Exercise 29

Page 312

1. 3.25 m
2. (a) 34.9° (b) 154 cm
3. 103 miles
4. (a) 9.64 cm (b) 51.3°
5. (a) 76.6 km (b) (ii) 51 km (iii) 310°
6. (a) (i) $\frac{5}{12}$ (ii) $\frac{5}{13}$ (b) $QR = 6.5$ cm, $PR = 2.5$ cm
7. (a) 10.3 cm (b) 14.4°
8. (a) (i) 5.29 cm (ii) 61.9° (b) 69.3°

CHAPTER 30

Exercise 30.2

Page 315

1. 9
2. (a) 4 (b) 4 (c) 2 (d) Infinite
3. (a) 2 (b) 2 (c) 4
4. (a) 4 (b) 1, 4

Exercise 30.3

Page 316

1. (a) (b) (c)
2. (a)
   (b)
   (c)
3.
4. (a) (b)

Exercise 30.4

Page 318

1. (b) 52 cm²
2. (a) 27 cm³, 54 cm² (b) 30 cm³, 62 cm² (c) 140 cm³, 166 cm²
3. (a) 4 cm³, 18 cm² (b) 5 cm³, 20 cm² (c) 8 cm³, 28 cm² (d) 7 cm³, 24 cm² (e) 14 cm³, 42 cm² (f) 10 cm³, 32 cm²
4. 96.8 cm³, 131.5 cm²
5. 10 cm
6. 1155 cm²
7. 343 cm³

Exercise 30.5

Page 319

1. (a) 40 cm³ (b) 140 cm³ (c) 96 cm³
2. (a) 11 cm², 15.4 cm³ (b) 6 cm², 15 cm³ (c) 314 cm², 6280 cm³ (d) 3.14 cm², 15.7 cm³
3. (a) 10 cm³ (b) 56 cm³ (c) 330 cm³ (d) 113 cm³ (e) 393 cm³ (f) 120 cm³ (g) 848 cm³ (h) 18 cm³
4. No. $\pi \times 5 \times 5 \times 10 = 250\pi$ cm³
   $\pi \times 10 \times 10 \times 5 = 500\pi$ cm³
5. 11.5 cm
6. 2.7 cm
7. 12.0 cm
8. 4.5 cm

Exercise 30.6

Page 321

1. (a) 339 cm² (b) 61.3 cm² (c) 90.9 cm²
3. (a) 1260 cm² (b) 6280 cm²
4. 274 cm²
5. (a) 14 100 cm² (b) 18 800 cm²
6. 4 cm

Exercise 30.7 — Page 323

1. (a) 101 cm^3 (b) 2140 cm^3 (c) 37.3 cm^3
2. (a) 220 cm^2 (b) 47.1 cm^2 (c) 57.9 cm^2
3. Hemispherical bowl has greater volume. Cone 393 cm^3, hemisphere 452 cm^3.
4. 37.5 cm
5. (a) 314 cm^2 (b) 188 cm^2 (c) 2120 cm^2
6. (a) (i) 170 cm^3 (ii) 160 cm^2
   (b) (i) 33.5 cm^3 (ii) 53.2 cm^2
   (c) (i) 91.6 cm^3 (ii) 117 cm^2
7. 4.73 cm
8. 2.40 cm
9. 1470 cm^3, 837 cm^2
10. (a) 720 cm^3 (b) 564 cm^2

Review Exercise 30 — Page 326

2. (a) (b)
3. (a) 4 (b) 4
4. (a) 54 cm^3 (b) 12
5. 350 cm^3
6. 1890 cm^3
7. 180 cm^3
8. 1040 cm^3
9. (a) 18 cm (b) $15\,625\text{ cm}^3$
10. (a) 392 cm^2 (b) (i) 572 cm^3 (ii) 3.7 cm
11. 347 cm^3
12. (a) 32.7 cm (b) 400 cm^3
13. (a) 497 cm^3 (b) 319 cm^2
14. $x = 4.33$ cm

CHAPTER 31

Exercise 31.1 — Page 329

1. (a) 320 000 ml = 320*l*
   (b) 0.32t = 320 kg = 320 000 g
   (c) 320 mm = 32 cm = 0.32 m
   (d) $32\,000\text{ cm}^2 = 3.2\text{ m}^2$
2. 9.3 kg
3. 8 m
4. 30 ml
5. 2000 m and 2 km
6. 8 kg and 8000 g
7. 0.5 km
8. 0.3 t
9. 2 cm
10. (a) 3.123 m (b) 450 cm (c) 3240 km (d) 1 000 000 g (e) 0.4 *l*
11. 0.5 m^2 is larger than 500 cm^2. $0.5\text{ m}^2 = 5000\text{ cm}^2$
12. $800\,000\text{ cm}^3$ is larger than 0.08 m^3. $0.08\text{ m}^3 = 80\,000\text{ cm}^3$
13. 1.98 *l*
14. 20
15. 50 g
16. 60
17. 50 ml
18. 0.036 m^2
19. (a) $67\,200\text{ cm}^2$ (b) 6.72 m^2
20. $600\,000\text{ cm}^3$

Exercise 31.2 — Page 330

1. (a) kilometres (b) metres
   (c) centimetres (d) millimetres
   (e) kilograms (f) grams
   (g) litres (h) millilitres
2. 250 g
3. 30 cm
4. 200 ml
6. (a) 12 m, nearest metre, 5.9 m, nearest 100 cm
   5 *l*, nearest litre, 500 m^2, nearest 50 m^2
   (b) 1200 cm, nearest 100 cm
   5900 mm, nearest 100 mm
   5000 ml, nearest 1000 ml
   $500\,000\,000\text{ mm}^2$, nearest $50\,000\,000\text{ mm}^2$

Exercise 31.3 — Page 331

1. (a) 33 pounds (b) 35 pints
   (c) 200 inches (d) 150 mm
   (e) 20 inches (f) 180 cm
2. 20 miles
3. 7 pints
4. 170 cm
5. 62.7 kg
6. (a) 600 m (b) 4.8 km (c) 5 feet (d) 2.75 pounds (e) 45.7 litres
7. 12 gallons
8. 22 pounds
9. Yes. 4 litres is about 7 pints.
   7 pints $\div 20 = \frac{7}{20}$ pint
   which is greater than $\frac{1}{3}$ pint.
10. No. 10 kg is about 22 pounds.
11. No. 6 miles is about 9.6 km.
12. 10 pounds
13. 1500 cm^2
14. (a) 48 km/hour (b) 80 km/hour (c) 108 km/hour
15. (a) 37.5 miles per hour (b) 90 miles per hour
16. 27 metres per second
17. (a) 12.8 km/litre (b) 30 miles per gallon
18. 3.35 kg, nearest 10 g
19. 6.8 m^3, 1 d.p.

Exercise 31.4 — Page 333

1. (a) continuous (b) discrete
   (c) discrete (d) continuous
2. (a) exact (b) 4.5 mins $\leqslant t <$ 5.5 mins
   (c) 62.5 kg $\leqslant w <$ 63.5 kg
3. (a) 12.5 s $\leqslant t <$ 13.5 s
   (b) 82.55 s $\leqslant t <$ 82.65 s
4. (a) 45 50 55
   (b) 45 50 55

5. (a) nearest 0.01 m (centimetre)
   (b) $1.525\,\text{m} \leqslant h < 1.535\,\text{m}$
6. (a) 61.5 kg (b) 2.25 m (c) 12.625 s
7. (a) nearest 10 m (b) nearest 50 m
   (c) nearest 100 m
8. (a) 10 499, 9500 (b) 10 049, 9950
   (c) 10 024, 9975 (d) 549, 450
   (e) 504, 495

Exercise 31.5 — Page 334

1. 98.5°
2. $39.208\,\text{cm}^3$ (3 d.p.)
3. (a) 162 minutes (3 s.f.)
   (b) 226 minutes (3 s.f.)
4. 7.428 cm (3 d.p.)
5. 68 330.58 g (2 d.p.)
6. (a) 26.595 km, 26.485 km
   (b) 3.03 km/h (3 s.f.)
7. $5.741\,\text{cm}^2$ (3 d.p.)

Exercise 31.6 — Page 336

1. (a) area (b) length (c) length
   (d) length (e) volume (f) area
   (g) volume (h) area
2. (a) perimeter (b) area (c) volume
   (d) none (e) area (f) perimeter
   (g) perimeter (h) none (i) volume
   (j) volume (k) none (l) area
3. (a) (i) $2\pi(x + y)$ (ii) $\pi(x^2 + y^2)$, πxy
5. $\frac{1}{2}pqs$, volume
   $2\left(p + q + r + \frac{3s}{2}\right)$, edge length
   $s(p + q + r) + pq$, surface area
6. (a) correct (b) correct (c) correct
   (d) wrong (e) correct (f) wrong
7. (a), (b), (c), (e)

Review Exercise 31 — Page 337

1. (a) 3.9 cm (b) 640 km
2. Taller: Ben by about 10 cm
   Heavier: Sam by about 1 kg
3. 5 gallons
4. $15\,000\,\text{cm}^2$
5. No. $2\frac{1}{2}\,\text{ft} = 75\,\text{cm}$
6. (a) $0.04\,\text{m}^2$ (b) $400\,\text{cm}^2$
7. 44 feet per second
8. Both same speed
9. (a) $120\,\text{m}^2$ (b) 1.3 litres
10. 135 tonnes
11. 1 gallon $= \frac{4544 - 4500}{4550} \times 100 \simeq 0.977\%$
    So, rule is correct to within 1%.
12. (a) **B** (b) **B** has dimension 3
    $\frac{1}{6}\pi Hw^2 \equiv \text{L} \times \text{L}^2 = \text{L}^3$

13. (a) area (b) none (c) volume
14. (a) continuous (b) discrete
15. $165.5\,\text{cm} \leqslant$ Jean's height $< 166.5\,\text{cm}$
16. 76.85 kg
17. 0.1 s
18. 3255 g, 3245 g
19. 11.615 s, 11.605 s
20. 6.7%

CHAPTER 32

Exercise 32.1 — Page 340

1. **D**, **E**
2. (a) ΔCED (b) $CBFE$
3. ΔAXZ and ΔZYC, ΔBXZ and ΔZYB
4. **A**, **D** (SSS)
5. **A**, **D** (ASA)
6. (a) Yes, ASA (b) No (c) Yes, SAS
7. (a) Yes, SSS (b) No (c) Yes, RHS
   (d) Yes, SAS (e) No (f) Yes, ASA
8. (a) No (b) No (c) Yes, RHS (d) No
   (e) Yes, ASA (f) Yes, ASA (g) Yes, SSS
   (h) Yes, SAS (i) Yes, ASA
9. No lengths given, one triangle is larger than the other.
10. $DE = PQ$ (given) $DF = PR$ (given)
    $\angle EDF = \angle QPR$ (70°) Congruent, SAS
11. (a) $\angle BDC = 25°$
    (b) BD is common, $AB = CD$ (7 cm)
    $\angle ABD = \angle BDC$ (alt. ∠'s)
    Congruent, SAS
    (c) $\angle ADB$

Exercise 32.2 — Page 342

1. Corresponding lengths not in same ratio.
2. (a) Two circles (d) Two squares
3. $a = 4\,\text{cm}$, $b = 24\,\text{cm}$
4. (a) Scale factor $= \frac{3}{2} = 1.5$
   (b) $x = 1.8\,\text{cm}$ (c) $a = 120°$
5. 30 cm
6. (a) $x = 1.5\,\text{cm}$, $y = 2.4\,\text{cm}$, $a = 70°$
   (b) $x = 5\,\text{cm}$, $y = 1.5\,\text{cm}$, $a = 53°$
   (c) $x = 30\,\text{cm}$, $y = 17.5\,\text{cm}$, $z = 10\,\text{cm}$
7. 15 cm
8. 2.8 cm
9. 18°
10. 5 cm
11. $x = 16$, $y = 48$

Exercise 32.3 — Page 344

1. (a) $x = 10\,\text{cm}$, $y = 27\,\text{cm}$
   (b) $x = 6\,\text{cm}$, $y = 10\,\text{cm}$
   (c) $x = 12\,\text{cm}$, $y = 12\,\text{cm}$
   (d) $x = 12\,\text{cm}$, $y = 5\,\text{cm}$
2. $AB = 8\,\text{cm}$, $AE = 5\,\text{cm}$
3. (a) $AB = 2.96\,\text{cm}$, $QR = 3.25\,\text{cm}$
   (b) $XZ = 14.4\,\text{m}$, $BC = 6.25\,\text{m}$
   (c) $EG = 1.6\,\text{cm}$, $MN = 4.0\,\text{cm}$
   (d) $AC = 4.0\,\text{m}$, $XZ = 2.0\,\text{m}$

4. (a) $PR = 10\,cm$
(b) $QR = 5\,cm$, $PQ = 7.5\,cm$, $QS = 4.5\,cm$
5. (a) $AB = 6\,cm$
(b) $BC = 7.5\,cm$, $AC = 6.9\,cm$
(c) $EC = 4.6\,cm$
6. $GH = 6\,cm$, $FH = 3\,cm$

Exercise 32.4 — Page 346

1. (a) $AC = 2.4\,cm$, $AP = 2.2\,cm$
(b) $AC = 4.5\,cm$, $BP = 1\,cm$
(c) $BC = 4.2\,cm$, $BP = 1.6\,cm$
2. (a) $AQ = 2.5\,cm$, $BC = 7\,cm$
(b) $AQ = 4\frac{2}{3}\,cm$, $BQ = 16\frac{2}{3}\,cm$
3. (a) $x = 58°$ (b) $x = 56°$
4. (a) (i) 61° (ii) 29°
(b) (i) 13 cm, 5.2 cm (ii) 36 cm, 27 cm
Same
5. (b) $AB = 1.8\,cm$ (c) 12.5 cm
6. 13.5 cm²

Exercise 32.5 — Page 349

1. 30 m
2. (a) 2 : 3 (b) 9 cm
3. (a) 1 : 50 (b) 2.8 cm
4. Original area multiplied by $2^2 = 4$
5. 40 kg
6. 100 times larger
7. 2.4 cm
8. area is quartered
9. 90 cm²
10. (a) 8 cm (b) 96 cm²
11. 2.4 m
12. (a) 1.5 km (b) 5 km²
13. (a) 14 cm (b) 750 000 m² (c) 1.92 cm²
14. 8 times original volume
15. 4000 ml
16. (a) 880 cm² (b) 55 cm², 25 cm³
17. 540 litres
18. 12 kg
19. $\left(\frac{1}{2}\right)^3 = \frac{1}{8}$ of original volume
20. 25 cm³
21. 405 cm³
22. (a) 10 cm (b) 16 : 25
23. 36 cm
24. 391 ml
25. (a) 3.28 m² (b) 45 cm³
26. (a) 1 : 300 (b) 0.741 cm³ (c) 2880 m²
27. 1.41
28. (a) 33.5 cm³, 524 cm³
(b) Ratio of surface areas = 4 : 25
As a percentage, $\frac{4}{25} \times 100 = 16\%$
29. (a) 15 cm (b) 33.75π
30. 0.81 kg

Review Exercise 32 — Page 351

1. **A** and **D**, SSS
2. **A** and **E**, ASA
3. (a) $x = 70°$ (b) $y = 10\,cm$
4. $\angle PQX = 54°$, $\angle AXB = 36°$
Both triangles have same angles. $PQ = 2AB$
Triangles are similar.
5. $CD = 14.4\,cm$
6. (a) $DE = 4.5\,cm$ (b) $AC = 12\,cm$
7. (a) 6 cm (b) 8 : 27
8. (a) 3125 cm² (b) 12 000 m³
9. (a) 11.2 cm (b) 170 cm³
10. $r = 7.33\,cm$

CHAPTER 33

Exercise 33.1 — Page 354

3. $\mathbf{a} = \begin{pmatrix} 3 \\ 5 \end{pmatrix}$ $\mathbf{b} = \begin{pmatrix} 3 \\ -1 \end{pmatrix}$ $\mathbf{c} = \begin{pmatrix} -2 \\ 4 \end{pmatrix}$
$\mathbf{d} = \begin{pmatrix} -6 \\ -1 \end{pmatrix}$ $\mathbf{e} = \begin{pmatrix} 4 \\ 3 \end{pmatrix}$ $\mathbf{f} = \begin{pmatrix} 2 \\ -5 \end{pmatrix}$

Exercise 33.2 — Page 355

2. (i) $\mathbf{a} = 2\mathbf{p}$, $\mathbf{e} = -\mathbf{p}$
(ii) $\mathbf{b} = 2\mathbf{q}$, $\mathbf{c} = -\mathbf{q}$, $\mathbf{h} = -3\mathbf{q}$
(iii) $\mathbf{d} = \frac{1}{2}\mathbf{r}$, $\mathbf{f} = 2\mathbf{r}$, $\mathbf{g} = -\frac{1}{2}\mathbf{r}$

Exercise 33.3 — Page 357

2. (a) **r** (b) **s** (c) **t** (d) **q**
(e) **v** (f) **u** (g) **w** (h) **p**
3. (a)

(b) $\overrightarrow{OB} = \overrightarrow{EH} = \mathbf{y}$. OB is parallel and equal to EH, so, $OBHE$ is a parallelogram.
4. (b) (i) $\frac{1}{2}\mathbf{a} + \mathbf{b}$ (ii) $\frac{1}{2}\mathbf{a} + \mathbf{b}$ (iii) $\mathbf{a} + 2\mathbf{b}$
(c) (i) $\overrightarrow{PC} = \overrightarrow{AR}$, so, PC is parallel and equal in length to AR.
(ii) $\overrightarrow{AE} = 2\overrightarrow{PC}$, so, PC is parallel to AE and AE is $2 \times$ length of PC.

Exercise 33.4 — Page 359

1. (a) $-\mathbf{a}$ (b) $\mathbf{b} - \mathbf{a}$
(c) $\frac{1}{2}(\mathbf{b} - \mathbf{a})$ (d) $-\frac{1}{2}(\mathbf{a} + \mathbf{b})$
2. (a) (i) $\mathbf{q} - \mathbf{p}$ (ii) $2\mathbf{p}$
(iii) $2\mathbf{p} + 2\mathbf{q}$ (iv) $2\mathbf{q} - 2\mathbf{p}$
(b) parallel, $PQ = \frac{1}{2}XZ$

3. (a) (i) $\frac{1}{2}\mathbf{a}$ (ii) $\frac{1}{2}\mathbf{a} + \mathbf{b}$
(iii) $\frac{1}{2}\mathbf{a} - \mathbf{b}$ (iv) $\frac{1}{6}\mathbf{a} - \frac{1}{3}\mathbf{b}$
(v) $\frac{2}{3}\mathbf{a} + \frac{2}{3}\mathbf{b}$ (vi) $\mathbf{a} + \mathbf{b}$
(b) OYB is a straight line.

4. (a) (i) $2\mathbf{q}$ (ii) $2\mathbf{p}$ (iii) $\mathbf{q} - \mathbf{p}$
(iv) $2\mathbf{p} - 2\mathbf{q}$ (v) $\mathbf{q}$ (vi) $-\mathbf{p}$
(b) similar triangles

5. (a) (i) $\mathbf{b} - \mathbf{c}$ (ii) $-\frac{1}{2}\mathbf{b}$
(iii) $\frac{1}{2}\mathbf{a} + \frac{1}{2}\mathbf{b} - \mathbf{c}$ (iv) $\frac{1}{2}\mathbf{a} - \frac{1}{2}\mathbf{c}$
(v) $\frac{1}{2}\mathbf{a} - \frac{1}{2}\mathbf{c}$

6. (a) $4\mathbf{a} + 2\mathbf{b}$
(b) $\overrightarrow{OC} = 6\mathbf{a} + (4 + n)\mathbf{b}$ $\quad n = -1$

7. (a) $\overrightarrow{AD} = 3\mathbf{a} - 2\mathbf{b}$, $\overrightarrow{BC} = 6\mathbf{a} - 4\mathbf{b}$
(b) $ABCD$ is a trapezium.

Review Exercise 33 — Page 362

1. (a) $\mathbf{b} - \mathbf{a}$ (b) $\frac{1}{2}(\mathbf{b} - \mathbf{a})$ (c) $\frac{1}{2}(\mathbf{a} + \mathbf{b})$

2. (a) (i) E.g. $\begin{pmatrix} -8 \\ 2 \end{pmatrix}$ (ii) E.g. $\begin{pmatrix} -2 \\ 3 \end{pmatrix}$ or $\begin{pmatrix} 2 \\ -3 \end{pmatrix}$
(b) $\overrightarrow{AB} = \begin{pmatrix} 4 \\ 3 \end{pmatrix}$

3. (a) (i) $\mathbf{a} + \mathbf{c}$ (ii) $\frac{1}{2}\mathbf{a}$ (iii) $\mathbf{c} - \frac{1}{2}\mathbf{a}$
(b) (i) $\frac{2}{3}\left(\mathbf{c} - \frac{1}{2}\mathbf{a}\right) = \frac{2}{3}\mathbf{c} - \frac{1}{3}\mathbf{a}$
(ii) $\frac{2}{3}(\mathbf{a} + \mathbf{c})$
(c) P is a point on OB such that OP is $\frac{2}{3}OB$.

4. (a) (i) $\mathbf{c} - \mathbf{b}$ (ii) $\frac{1}{2}\mathbf{c}$ (iii) $\frac{1}{2}\mathbf{c}$
(b) Parallelogram, $\overrightarrow{NQ} = \overrightarrow{MP}$

5. (a) (i) $2\mathbf{a} + \mathbf{b}$ (ii) $\frac{2}{3}(\mathbf{a} - \mathbf{b})$ (iii) $\frac{1}{3}(\mathbf{b} - \mathbf{a})$
(b) (i) 2 : 1
(ii) PS and RS are parallel.
S is a common point.

CHAPTER 34

Exercise 34.1 — Page 365

1. (a) positive (b) positive (c) positive
(d) positive (e) negative (f) negative
(g) negative (h) negative (i) positive
(j) positive (k) negative (l) positive
(m) negative (n) positive (o) negative
(p) positive (q) positive (r) negative
(s) negative (t) positive (u) negative
(v) positive (w) positive (x) negative
(y) negative (z) positive

2. (a) $-\tan 80°$ (b) $-\cos 30°$ (c) $-\sin 20°$
(d) $\tan 70°$ (e) $\cos 60°$ (f) $\sin 60°$
(g) $-\tan 10°$ (h) $-\cos 30°$ (i) $-\sin 70°$
(j) $-\tan 30°$ (k) $\cos 10°$ (l) $-\sin 80°$
(m) $\tan 10°$ (n) $\sin 60°$ (o) $-\cos 60°$
(p) $-\sin 50°$ (q) $-\cos 80°$ (r) $\tan 30°$

3. (a) (i) $\frac{1}{\sqrt{2}}$ (ii) 1
(b) (i) $-\frac{1}{\sqrt{2}}$ (ii) 1 (iii) $-\frac{1}{\sqrt{2}}$
(iv) $\frac{1}{\sqrt{2}}$ (v) $-\frac{1}{\sqrt{2}}$ (vi) -1

4. (a) $\frac{1}{2}$, $\frac{\sqrt{3}}{2}$, $\frac{1}{\sqrt{3}}$
(i) $\frac{1}{2}$ (ii) $-\frac{\sqrt{3}}{2}$ (iii) $-\frac{1}{\sqrt{3}}$ (iv) $\frac{\sqrt{3}}{2}$
(b) $\frac{\sqrt{3}}{2}$, $\frac{1}{2}$, $\sqrt{3}$
(i) $-\frac{1}{2}$ (ii) $\sqrt{3}$ (iii) $-\frac{1}{2}$ (iv) $-\frac{\sqrt{3}}{2}$

Exercise 34.2 — Page 366

1. (a) 0.985 (b) −0.866 (c) 0.364
(d) −0.940 (e) 0.5 (f) −0.176
(g) 0.123 (h) −0.993 (i) −0.799
(j) 0.891 (k) −0.325 (l) 0.438
(m) −0.810 (n) −0.961 (o) −0.616
(p) 0.766 (q) −0.839 (r) 0.643
(s) 0.866 (t) −0.985

2. (a) $p = 60°, 300°$ (b) $p = 210°, 330°$
(c) $p = 45°, 225°$ (d) $p = 30°, 150°$
(e) $p = 135°, 315°$ (f) $p = 120°, 240°$
(g) $p = 230°, 310°$ (h) $p = 40°, 320°$
(i) $p = 60°, 120°$ (j) $p = 116°, 296°$
(k) $p = 10.9°, 190.9°$ (l) $p = 55.6°, 304.4°$
(m) $p = 20°, 160°$ (n) $p = 150°, 210°$
(o) $p = 35°, 215°$

3. $x = 228.6°,\ 311.4°,\ 588.6°,\ 671.4°$

4. $x = -245°,\ -115°,\ 245°$

5. $x = -307°,\ -233°,\ 53°,\ 127°$

Exercise 34.3 — Page 367

1. (a) (i) $x = 90°$
(ii) $x = 0°,\ 180°,\ 360°$
(iii) $x = 270°$
(b) (i) $x = 0°,\ 360°$
(ii) $x = 90°,\ 270°$
(iii) $x = 180°$

2. (a) $x = 45°,\ 225°$
(b) $x = 0°,\ 180°,\ 360°$
(c) $x = 135°,\ 315°$

3. (a) (i) $x = 30°,\ 150°$
(ii) $y = -2.6$

4. (b) $x = 78.5°,\ 281.5°$

Exercise 34.4 — Page 370

1. (a) $a = 5.79$ cm (b) $a = 13.1$ cm
(c) $a = 4.26$ cm

2. (a) $b = 9.92$ cm (b) $c = 8.28$ cm
(c) $b = 10.1$ cm (d) $q = 13.6$ cm
(e) $p = 1.64$ cm (f) $r = 8.48$ cm

3. (a) $\angle ACB = 55°$, $AC = 7.41$ cm, $AB = 6.54$ cm
   (b) $\angle MLN = 71.2°$, $LM = 5.84$ cm, $AB = 5.71$ cm
   (c) $\angle RPQ = 69.8°$, $QR = 15.7$ cm, $PR = 14.9$ cm
   (d) $\angle SUT = 31.8°$, $ST = 12.8$ cm, $UT = 24.1$ cm
   (e) $\angle XYZ = 31.2°$, $XY = 7.54$ cm, $ZY = 9.4$ cm
4. (a) $PR = 9.7$ cm (b) $QR = 6.8$ cm

Exercise 34.5 Page 372

1. (a) 61.7° (b) 53.7° (c) 74.7°
2. (a) 137.3° (b) 129.7° (c) 118.5°
3. 48.6°, 131.4°
4. $\angle QPR = 68.1°$, $\angle PRQ = 51.9°$, $PQ = 6.36$ cm
   $\angle QPR = 111.9°$, $\angle PRQ = 8.1°$, $PQ = 1.14$ cm
5. (a) $\angle ACB = 60.7°$, $\angle BAC = 50°$, $BC = 7.29$ cm
   (b) $\angle DFE = 25.2°$, $\angle EDF = 29.8°$, $EF = 9.91$ cm

Exercise 34.6 Page 373

1. (a) $a = 10.8$ cm (b) $a = 8.27$ cm
   (c) $a = 15.3$ cm
2. (a) $p = 5.98$ cm (b) $q = 3.45$ cm
   (c) $r = 15.3$ cm
3. 12.7 cm 5. 13.4 cm
4. 11.8 cm 6. 18 cm 7. 26.7 km

Exercise 34.7 Page 375

1. (a) $A = 130.5°$ (b) $A = 32.2°$
   (c) $A = 45.8°$
2. (a) $B = 64.5°$ (b) $C = 35.4°$
   (c) $P = 41.6°$
3. 45° 4. 130.5° 5. 102.6° 6. 119.9°

Exercise 34.8 Page 376

1. (a) 6.83 cm² (b) 14.2 cm²
   (c) 6.95 cm² (d) 16.2 cm²
2. 84.8 cm²
3. (a) 3 cm (b) 5.63 cm²
4. 65 cm²
5. 8.9 cm 7. $XY = YZ = 7.37$ cm
6. 729 cm² 8. 56.2°, 123.8°

Exercise 34.9 Page 378

1. (a) (i) 10.8 cm, 73.9°, 46.1°
   (ii) 8.44 cm, 46.4°, 23.6°
   (b) (i) 46.8 cm² (ii) 11 cm²
2. (a) 118.7°, 35.1°, 26.2°
   (b) $XP = 128.5$ m, $\angle XPY = 123°$
3. 9.24 km

4. 12.7 m
5. (a) 7.84 m (b) 29.9° (c) 14.7 m
6. (a) 5.17 cm (b) 10.8 cm²
7. (a) 28° (b) 38.2° (c) 28 cm²
8. 6.63 cm, 15.4 cm

Review Exercise 34 Page 380

1. (a) $x = 60°, 300°$ (b) $y = 210°, 330°$
2. (b) $x = 156°, -204°, -336°$
3. $x = 138.6°$, 221.4°
4. 58.1°, 121.9°
5. (a)

   (b) 9 am
6. 14.8 cm
7. (a) 484 m (b) 34 400 m² (3 s.f.)
8. 15 cm² 9. 098.6° 10. 108° (3 s.f.)

Shape, Space and Measures

Section Review

Non-calculator Paper Page 382

1. (a) $x = 118°$ (b) (i) $p = 25°$ (ii) $q = 105°$
2. (a) (b)

3. 12 cm
4. (a) Reflection in the y axis
   (b) Rotation, 90° clockwise, about (0, 0)
   (c)

5. (a) 380 m (b) 5600 m²
6. (a) 100π cm (b) 2500π cm²
7. $a = 120°$, $b = 150°$, $c = 135°$
8. (b) 116 m, 276°
9.

10. 264 cm
11. (a) 56 cm²
    (b) (i) $BC = 5$ m
    (ii) 34 m
12. (a) Enlargement, scale factor $\frac{1}{2}$, centre (1, −1)
    (b) (c)

13. (a) $2\pi a(a+b)$, $\frac{1}{2}(a+b)c$ (b) $\pi a^2 b$

14. (a) $\sin x = \frac{3}{5}$
(b) (i) $\cos y = \frac{4}{5}$ (ii) $PQ = 9$ cm

15. (a) 810 cm³ (b) 248 cm²

16. (a) (i) 160° (ii) 10° (iii) 40° (iv) 40°
(b) OA is common
$OB = OC$ (radii of circle)
$\angle OAB = \angle AOC$ (Δs OAB and OAC are isosceles, with base angles = 40°, part (a))
Δs OAB and OAC are congruent, SAS.

17. (a) 24π cm² (b) 40π cm² (c) 240π cm³

18. (a) -1 (b) $x = 225°$ (c) $y = 135°, 315°$

19. (a) (i) $6\mathbf{a}$ (ii) $9\mathbf{b} - 6\mathbf{a}$
(b) (i) $4\mathbf{a} + 3\mathbf{b}$ (ii) $8\mathbf{a} + 6\mathbf{b}$
(c) $\overrightarrow{OF} = 2\overrightarrow{OE}$
OEF is a straight line.
E is the midpoint of OF.

20. $\angle BCT = y - 0.5x$

Shape, Space and Measures — Section Review

Calculator Paper — Page 385

1. (a) 123°. $x + 57° = 180°$
(b) (i) Trapezium (ii) $y = 60°$

2. $\angle WOZ = 35°$

3. 280°

4. (a) 12 000 cm² (b) 1.2 m²

5. (b) 9.8 cm²

6. (a) 187.5 cm (b) 172 pounds

7. 72°

8. 31

9. (a) 75 cm² (b) 41.8 cm

10. (a) Reflection in $y = x$
(b) Rotation, 90° anticlockwise, about $(-1, 0)$

11. 30 788 cm³

12. 24 cm²

13.

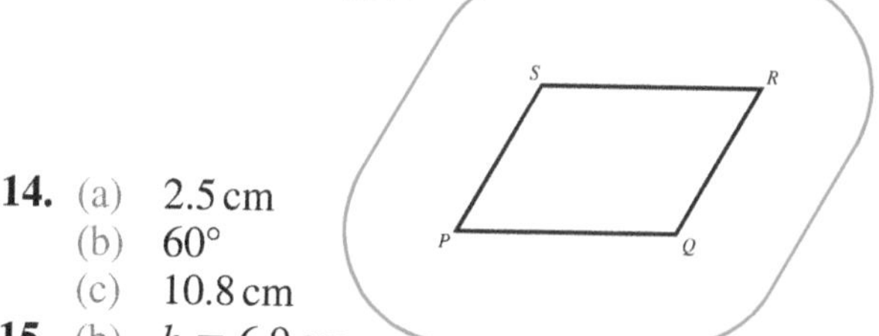

14. (a) 2.5 cm
(b) 60°
(c) 10.8 cm

15. (b) $h = 6.9$ cm
(c) $x = 41.8°$

16. $x = 38°$ $\angle DAB = \angle DCB = 52°$
$x = 180° - 90° - 52° = 38°$

17. 70.1 seconds

18. $AD = AB$ (sides of square $ABCD$)
$\angle BQA = \angle APD = 90°$ (ext. $\angle$'s square $PQRS$)
$\angle BAQ = \angle ADP = 90 - y°$
Δs ABQ and DAP congruent, ASA.

19. (a) 110 cm (b) (i) 17.5 cm (ii) 3105 cm³

20. 12.3 cm

21. (a) 15.4 cm (b) 67.7 cm² (c) $r = 7.78$ cm

22. (a)
(b) 4
(c) $x = 37°, 143°, -217°, -323°$

23. 33

CHAPTER 35

Exercise 35.1 — Page 388

1. qualitative
2. quantitative, continuous
3. quantitative, discrete
4. qualitative
5. quantitative, discrete
6. quantitative, continuous
7. quantitative, discrete
8. qualitative
9. quantitative, discrete
10. quantitative, continuous

Exercise 35.2 — Page 390

1. 4

2. (a)

Height h cm	Frequency
$145 \leqslant h < 150$	2
$150 \leqslant h < 155$	2
$155 \leqslant h < 160$	7
$160 \leqslant h < 165$	9
$165 \leqslant h < 170$	6
$170 \leqslant h < 175$	8
$175 \leqslant h < 180$	2
Total	36

(b) 5 cm (c) 7 (d) 11 (e) 32

3. (a) Francis (b) Louisa (c) Alistair

4. (a) (i)

Make	Frequency
Ford	4
Nissan	3
Vauxhall	5
Total	12

Colour	Frequency
Blue	2
Green	1
Grey	2
Red	3
White	4
Total	12

Number of doors	Frequency
2	2
3	5
4	2
5	3
Total	12

(ii)

Mileage (m)	Frequency
$0 \leqslant m < 5000$	0
$5000 \leqslant m < 10\,000$	2
$10\,000 \leqslant m < 15\,000$	0
$15\,000 \leqslant m < 20\,000$	2
$20\,000 \leqslant m < 25\,000$	1
$25\,000 \leqslant m < 30\,000$	2
$30\,000 \leqslant m < 35\,000$	4
$35\,000 \leqslant m < 40\,000$	1
Total	12

(b) (i) Vauxhall (ii) 2 (iii) 5 (iv) 5

Exercise 35.3 — Page 393

1. (a) Too personal (b) Leading
   (c) (i) Groups overlap
   (ii) ☐ less than 5 ☐ 5 to 9 ☐ 10 or more
2. (a) Too personal (b) Too open
   (c) Too open (d) Too open
5. Leading question.

Exercise 35.4 — Page 394

1. (a) 8 (b) 16 (c) 20 (d) 40 (e) 50%
2. (a) 14 (b) 20 (c) 70% (d) 80%
   (e) Disprove. 80% is greater than 70%.
3. (a) (i) 10 (ii) No. Less 8, More 12.
   (b) (i) 37 (ii) No. Girls 37, Boys 37, same.
4. (a)

	Yes	No
Men	47	11
Women	18	24

(b) Yes. Taller: $\frac{47}{58} \times 100 = 81\%$

Exercise 35.5 — Page 396

1. Small sample. One data collection time.
2. Women only. One location.
   One data collection time.
3. Advantage: confidential, wider circulation, etc.
   Disadvantage: slow, non-response, etc.
4. Fewer females than males, no-one under 18.
5. Lower school 23, Upper school 17
6. (a) Pupils from one year group only may be chosen.
   (b) 9, 11, 11, 10, 9
7. (a) May not sample all departments. (b) 49
8. 9, 14, 2
9. 1, 5, 54

Review Exercise 35 — Page 397

1. (a) **Question 1** - too open
   Question 2 - leading
   (b) Do you think that smoking is bad for your health? YES/NO
2. Only males asked. No-one under 11.
   Mainly adults surveyed.
3. Do not support.
   Women 75%, Men 75%. Same proportion.
4. (a) (i) Biased. Only asking one year group. Asks equal numbers of boys and girls.
   (ii) May pick all students from one year group.
   (b) Needs to pick 5% from each year group. Needs equal numbers of boys and girls.
5. (a) Pupils from one year group only or same sex may be chosen.
   (b) 8
6. (a) All sections may not be represented. People of one sex may not be included.
   (b) (i) 16 (ii) 10

CHAPTER 36

Exercise 36.1 — Page 399

1. Mode 1, median 2, mean 3, range 6.
2. (a) 3 (b) 5 (c) 3.5 (d) 3.9
3. (a) 10 (b) 9 (c) 9.3
4. 86.9 kg
5. 4
6. 129 cm
7. 15.4

Exercise 36.2 — Page 401

1. Mode 2, median 2.5, mean 2.7
2. (a) 5 (b) 4 (c) 5 (d) 5
3. (a) 9 (b) 9 (c) 30 (d) 8.5

Exercise 36.3 — Page 403

1. (a) $£10\,000 \leqslant s < £20\,000$, £20 350
   (b) 30 - 40 hours, 29.4 hours
2. 5.3 kg
3. £94 000
4. 27.7

Exercise 36.4 — Page 404

1. (a) (i) 0.5 minutes
   (ii) 1.95 minutes
   (b) Girls a little slower on average and more varied.

2. Women: mean 1.6, range 6
Men: mean 1.5, range 2
Women made more visits to the cinema, though the number of visits is more spread.
3. Average: Boys 6.2, Girls 7.2
Range: Boys 4, Girls $4\frac{1}{2}$
No. Girls' average greater than boys'.
Correct about variation.
4. (a) MacQuick 20 - 29, Pizza Pit 30 - 39
(b) MacQuick - mean 26 years (Pizza Pit 36.5 years)
(c) Exact ages not known.
5. Before: median 3, range 4
After: median 3, range 5
Would have been better to calculate the means.
Before 2.2, After 3.0

Exercise 36.5 — Page 405

1. Mode trainers.
Cannot calculate others.
2. Mode 15s, median 12s, mean 22.15s
Median most sensible, not affected by 200 as is mean, mode not much use.
3. Swimmer A.
Mean is lower (A 30.88s, B 31.38s)
Range less (A 1.7s, B 15s)
Median is higher (A 30.9s, B 30.0s)
4. He should use the median mark.
The median mark is the middle mark, so, half of the students will get the median mark or higher.
5. (a) Mode.
Represents the lowest cost for these data.
(b) Median.
Mean affected by **one** much higher cost.
Mode is equal to the lowest cost.

Review Exercise 36 — Page 406

1. (a) Range 6p, mode 35p
(b) Median 34p, mean 34.1p
2. (a) (i) 2 (ii) 2.28
(b) (i) 2.5 (ii) 2.63
(c) The calculation for the mean involves the number of runs scored off every ball.
3. (a) 24 cm
(b) 24.3 cm
(c) 1st group had higher mode (1st 25, 2nd 23), but lower range (1st 7, 2nd 13).
4. (a) £200 - £300
(b) £300 - £400
(c) Mean, influenced by 12 people earning £600 - £1000.
5. 16.9 minutes
6. (a) $150 \leqslant p < 155$
(b) 156 litres
(c) Estimated mean - uses all values

CHAPTER 37

Exercise 37.1 — Page 409

1. (b) £7 (c) £6.37 (d) 20%
2. (a)

Cereal	Corn flakes	Muesli	Porridge	Bran flakes
Angle	125°	100°	60°	75°

(b) Corn flakes
3. (a) 7 hours (b) 5 hours
(c) 3 (d) (i) 20 (ii) 20%
(e) E.g. boys have higher mode and larger range.
4. (a) 288 (b) 174° (c) 5 : 9

Exercise 37.2 — Page 411

1. (a) 1 | 0 means 10 litres

1	0 2 6 6 7 9
2	3 3 4 5 5 6 7 9
3	1 3 5 5
4	1 2

(b) 32 litres
2. (a) 3 | 2 means 3.2 seconds

1	5
2	4 4 5 6 7 8 8 9
3	0 1 2 2 3 5 5 6 7
4	2 2 3
5	6 6 8

(b) 3.2 seconds
3. 2 | 7 means 2.7 cm

1	8
2	0 1 4 5 6 6 7
3	1 4 5 5 6 9
4	0 2 2 5
5	4
6	0

4. (a) 17 (b) 38 pence (c) 39 pence
5. (a) Children 4 | 7 means 4.7 mins

Adults		Children
	4	7 9
9 4	5	1 3 4 9
7 5 4 1 0	6	2 3 4 5 5 6 8
9 8 7 3 3 3 0	7	1 4 6 7 9
2 2 0	8	0 2
4 2	9	
1	10	

(b) Adults have larger range.
Fastest time recorded by child, slowest time recorded by adult.

Exercise 37.3 — Page 412

1. (a) Negative (b) Positive (c) Zero
(d) Positive (e) Negative
2. (a) **B** (b) **C** (c) **D**

3. (a) 2 (b) 164 cm (c) No.
(d) Taller girls usually have larger shoe sizes than shorter girls.
4. (b) Positive correlation.
(c) Different conditions, types of road, etc.
5. (b) Negative correlation.
(c) Points are close to a straight line.

Exercise 37.4 — Page 414

1. (b) Positive correlation (d) 4.8 to 4.9 kg
2. (b) Negative correlation (d) 38 minutes
3. (b) Scatter of points suggests correlation is close to zero.
4. (b) Negative correlation.
(d) (i) 27 to 28 (ii) 91 to 92 kg
5. (b) (i) 88 - 89 (ii) 49 - 50
(c) (ii), as estimated value is within the range of known values.

Review Exercise 37 — Page 416

1. 5 | 4 means 5.4 grams

2	8						
3	5	9					
4	2	4	6	6	7	8	8
5	0	1	4	4	6	6	8
6	0	3	7				

2.

Item	TVs	VCRs	Computers	Other
Angle	140°	120°	68°	32°

3. ❶ B, ❷ C, ❸ A
4. (a) 9 (b) 50 (c) 17
(d) Highest mark scored by a boy.
Lowest mark scored by a girl.
Boys have a greater range of marks.
5. (a) 2 : 3 (b) 20%
(c) Boys: range = 3, mode = size 5.
Girls: range = 4, mode = size 4.
Girls: higher range. Boys: higher mode.
6. (a)

Fuel	Solid fuel	Electricity	Gas
Angle	24°	126°	210°

(b) (i) $\frac{5}{36}$ (ii) 72
(c) Gas. Town has larger population and proportion of gas users greater than 50%.
7. (c) (i) 80 - 85 km (ii) 11 - 13 litres
(d) (i), as estimated value is between known values.

CHAPTER 38

Exercise 38.1 — Page 419

1. (b) (i) 154 cm (ii) 15 years 4 months
2. (a) 35, 34, 35, 36, 39, 37, 40
(b) 28.8, 28.6, 28.2, 28.6, 30.6, 33
3. 18, 17.5, 17, 16, 15, 14, 13, 12, 11.25
4. 6.6
5. 3-point moving averages are:
370, 365, 375, 390, 400, 390, 400

Exercise 38.2 — Page 420

1. (a) (c)

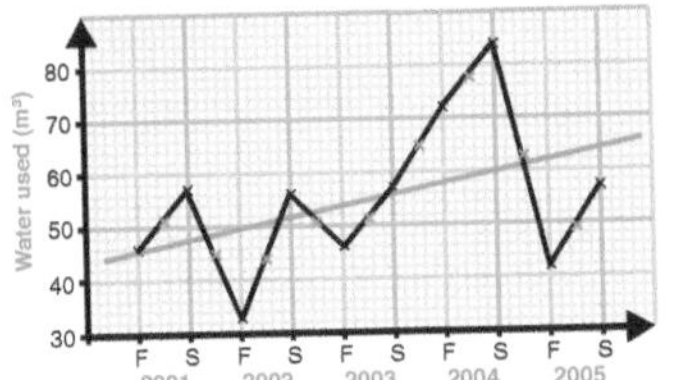

(b) 51.5, 45, 44, 50.5, 51.5, 64.5, 77.5, 62.5, 49.5
(d) There is an upward trend in the amount of water used.

2. (a) (c)

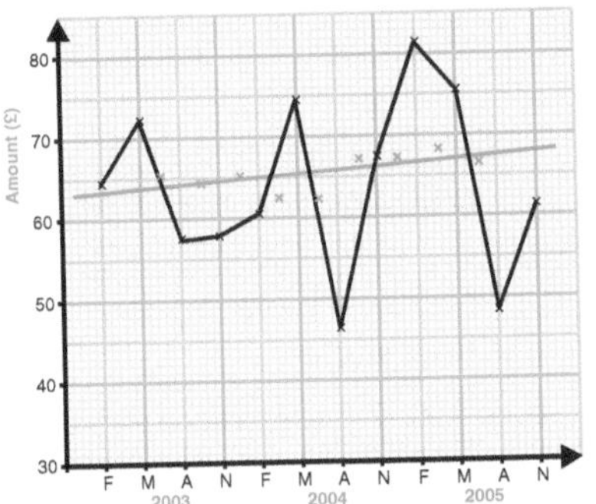

(b) £65.61, £64.55, £65.11, £62.27, £62.17, £67.39, £67.70, £68.18, £66.74
(d) There is an upward trend in the quarterly bills.

3. (a) (c)

(b) 3-point moving averages are:
35, 36, 62, 63, 65, 40, 44, 43, 41, 46
(d) Trend shows steady level of absence, greatly affected by Spring figure for Year 9.

4. (a) (c) (d)

(b) 47.8, 44.7, 43.8, 44, 43.6, 44.5, 46, 46.8, 47.3, 47.8, 49.7, 50.7, 52.2
(e) The trend shows a general increase in sales.
(f) £4800

5. (a) (c) (d)

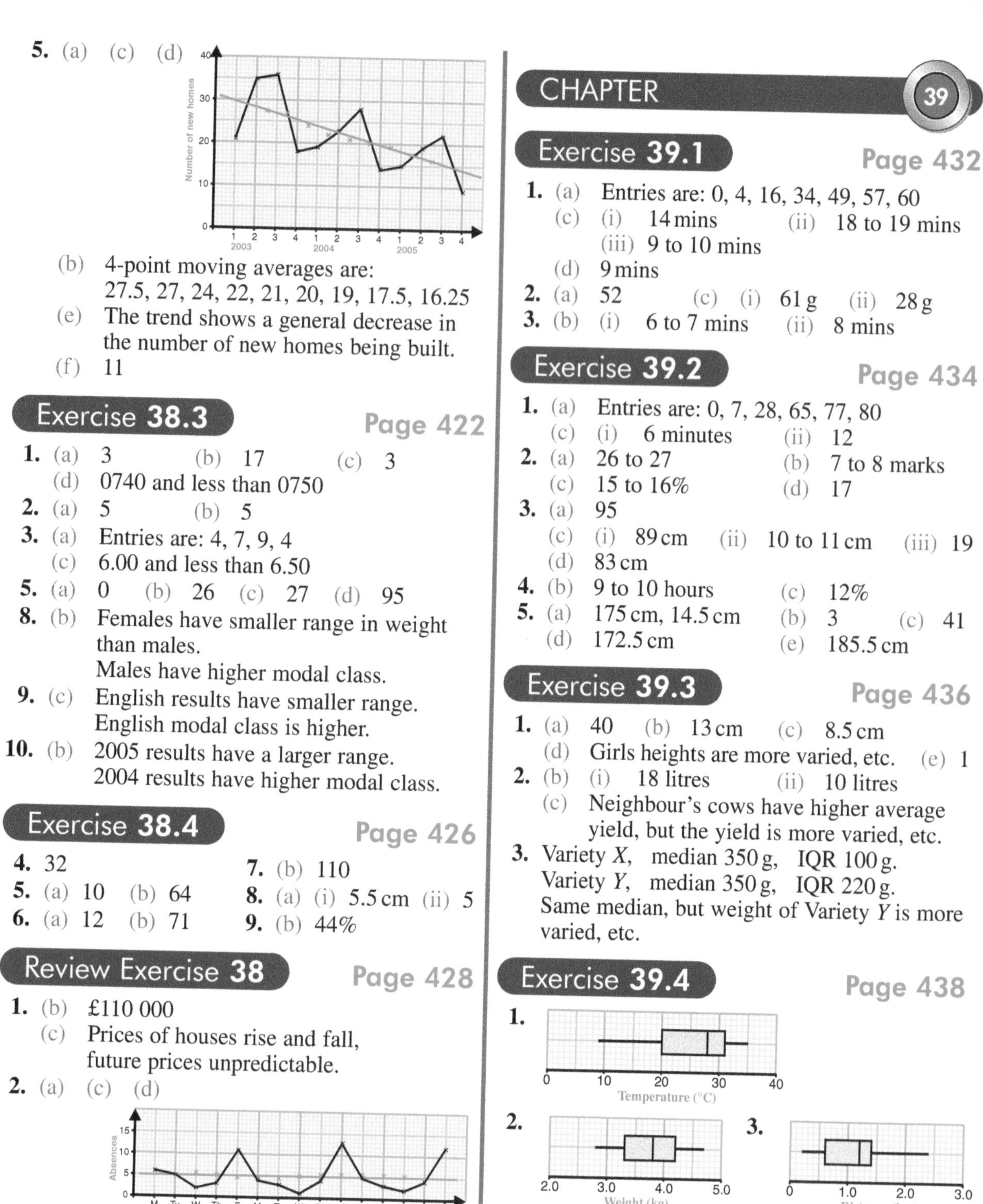

(b) 4-point moving averages are:
27.5, 27, 24, 22, 21, 20, 19, 17.5, 16.25

(e) The trend shows a general decrease in the number of new homes being built.

(f) 11

Exercise 38.3 — Page 422

1. (a) 3 (b) 17 (c) 3
   (d) 0740 and less than 0750
2. (a) 5 (b) 5
3. (a) Entries are: 4, 7, 9, 4
   (c) 6.00 and less than 6.50
5. (a) 0 (b) 26 (c) 27 (d) 95
8. (b) Females have smaller range in weight than males.
   Males have higher modal class.
9. (c) English results have smaller range.
   English modal class is higher.
10. (b) 2005 results have a larger range.
    2004 results have higher modal class.

Exercise 38.4 — Page 426

4. 32
5. (a) 10 (b) 64
6. (a) 12 (b) 71
7. (b) 110
8. (a) (i) 5.5 cm (ii) 5
9. (b) 44%

Review Exercise 38 — Page 428

1. (b) £110 000
   (c) Prices of houses rise and fall, future prices unpredictable.
2. (a) (c) (d)

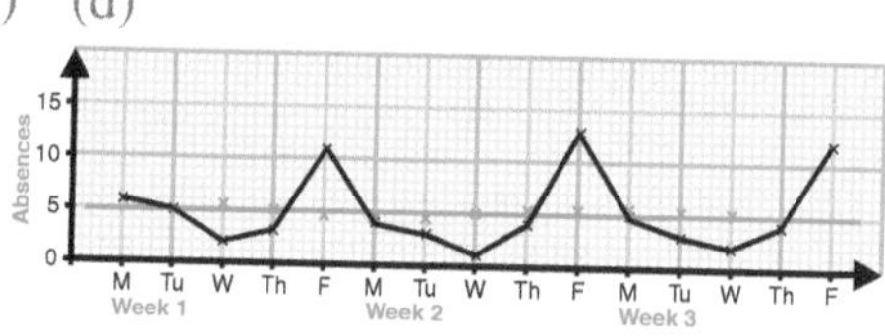

   (b) 5.4, 5, 4.6, 4.4, 4.6, 5, 5.2, 5.2, 5.4, 5.4, 5.2
   (e) Trend shows no change in the number of absences.
3. Vertical axis does not begin at zero.
   Horizontal axis is not uniform.
4. (b) Type Y
   (c) (i) Type X
   (ii) Weights spread over more weight class intervals.
5. (b) £4 (c) 35%
6. (b) 13

CHAPTER 39

Exercise 39.1 — Page 432

1. (a) Entries are: 0, 4, 16, 34, 49, 57, 60
   (c) (i) 14 mins (ii) 18 to 19 mins
   (iii) 9 to 10 mins
   (d) 9 mins
2. (a) 52 (c) (i) 61 g (ii) 28 g
3. (b) (i) 6 to 7 mins (ii) 8 mins

Exercise 39.2 — Page 434

1. (a) Entries are: 0, 7, 28, 65, 77, 80
   (c) (i) 6 minutes (ii) 12
2. (a) 26 to 27 (b) 7 to 8 marks
   (c) 15 to 16% (d) 17
3. (a) 95
   (c) (i) 89 cm (ii) 10 to 11 cm (iii) 19
   (d) 83 cm
4. (b) 9 to 10 hours (c) 12%
5. (a) 175 cm, 14.5 cm (b) 3 (c) 41
   (d) 172.5 cm (e) 185.5 cm

Exercise 39.3 — Page 436

1. (a) 40 (b) 13 cm (c) 8.5 cm
   (d) Girls heights are more varied, etc. (e) 1
2. (b) (i) 18 litres (ii) 10 litres
   (c) Neighbour's cows have higher average yield, but the yield is more varied, etc.
3. Variety X, median 350 g, IQR 100 g.
   Variety Y, median 350 g, IQR 220 g.
   Same median, but weight of Variety Y is more varied, etc.

Exercise 39.4 — Page 438

1. 2. 3. 4.

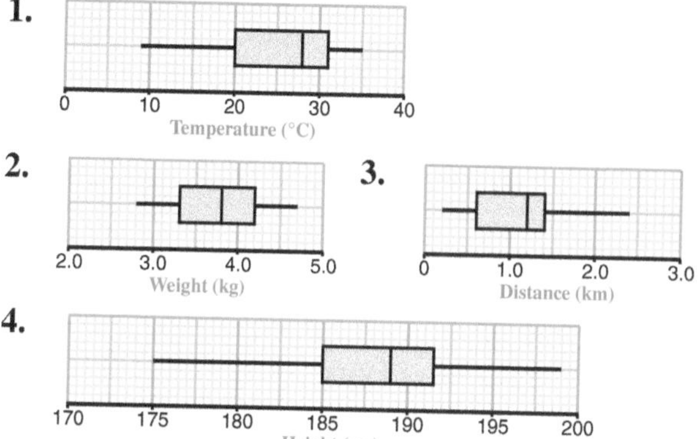

5. (a) 0.06 s (b) 0.08 s
6. (a) 72% (b) 18% (c) English (d) 21%
   (e) English mark has a higher median, Maths marks more spread.
7. (b) Males have lower median time and not spread out. Females have higher median time and greater variation.
8. (a) median = £62, lower quartile = £43, upper quartile = £75

Review Exercise 39 — Page 441

1. (a) Entries are: 2, 5, 13, 22, 28, 35, 40
   (c) 78.5g (d) 19g
   (e) The spread of the second sample is greater (IQR 38 g compared with 19 g), etc.
2. (b) (i) 57 marks (ii) 22 marks
   (c) Paper 2. Higher median, less variation.
3. (a) 13p (b) 33p (c) 5p
4.

1 2 3 4 5 6
Time (minutes)

8. (a) (i) The probabilities add to 105%
   (ii) 5%
   (b) (i) 45% (ii) 75% (iii) 80%
9. (a) 0.4 (b) 0.6
   (c) (i) Yes (ii) No
   (d) (i) 0.6 (ii) 0.6 (iii) 0.7

CHAPTER 40

Exercise 40.1 — Page 443

1. (a) Evens (b) Impossible (c) Likely
   (d) Unlikely (e) Certain (f) Unlikely
2. (a) **T** (b) **P** (c) **R** (d) **S**
3. **W** at 0, **Y** just below $\frac{1}{2}$, **V** at $\frac{1}{2}$, **X** at 1 (scale from 0 to 1)

Exercise 40.2 — Page 444

1. $\frac{2}{3}$
2. (a) $\frac{1}{4}$ (b) $\frac{1}{2}$ (c) $\frac{3}{4}$
3. (a) $\frac{3}{10}$ (b) $\frac{7}{10}$
4. (a) $\frac{1}{11}$ (b) $\frac{4}{11}$ (c) $\frac{2}{11}$
5. (a) $\frac{1}{2}$ (b) $\frac{1}{4}$ (c) $\frac{1}{52}$
6. (a) $\frac{2}{5}$ (b) $\frac{3}{5}$ (c) 1 (d) 0
7. (a) $\frac{2}{3}$ (b) $\frac{1}{3}$ (c) $\frac{3}{4}$
8. (a) $\frac{2}{5}$ (b) $\frac{3}{5}$ (c) $\frac{4}{25}$ (d) $\frac{4}{15}$ (e) $\frac{2}{5}$
9. The events are not equally likely.
10. (a) $\frac{7}{15}$ (b) $\frac{1}{2}$ (c) $\frac{1}{16}$ (d) $\frac{3}{4}$
11. (a) $\frac{1}{3}$ (b) $\frac{1}{15}$ (c) $\frac{11}{24}$ (d) $\frac{17}{50}$ (e) $\frac{21}{25}$ (f) $\frac{4}{5}$

Exercise 40.3 — Page 446

1. $\frac{7}{25}$ 2. $\frac{9}{10}$ 3. $\frac{3}{10}$
4. (a) $\frac{52}{100} = 0.52$ $\frac{102}{200} = 0.51$ $\frac{141}{300} = 0.47$
   (b) 0.47
5. $\frac{21}{30} = \frac{7}{10}$ 6. 8 7. 18 8. 25
9. (a) 300 (b) 120 (c) 150

Exercise 40.4 — Page 447

1. $\frac{3}{5}$
2. 0.4
3. 0.04
4. $\frac{47}{50}$
5. 0.6
6. (a) 0.04 (b) 0.97
7. (a) 0.5 (b) 0.3

Exercise 40.5 — Page 450

1. (a) RBG, RGB, GBR, GRB, BGR, BRG
   (b) $\frac{1}{3}$
2. (a)

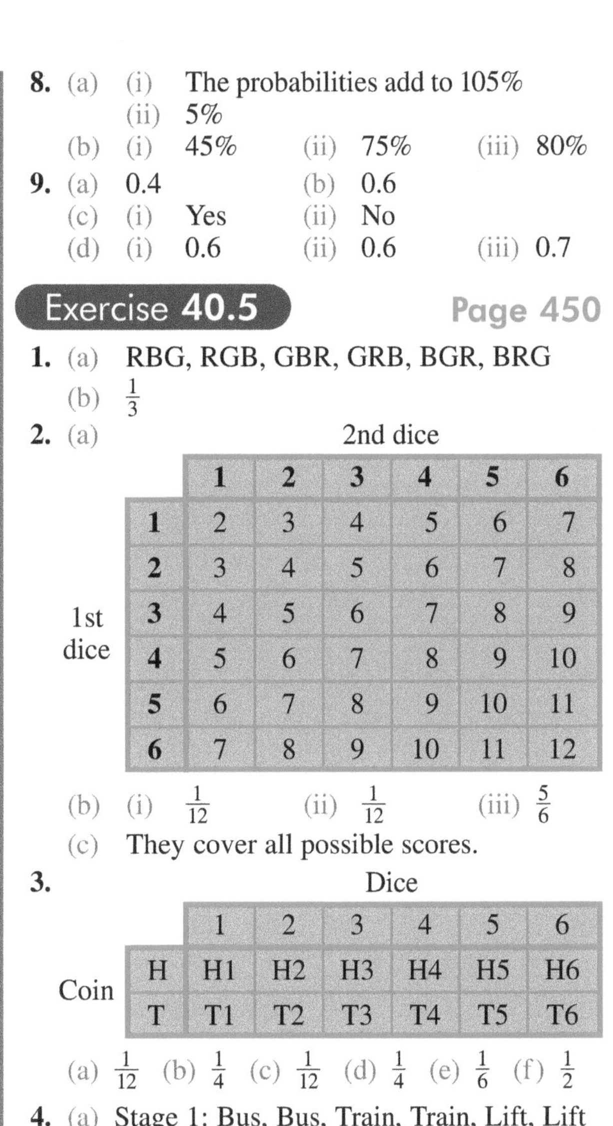

1st dice \ 2nd dice	1	2	3	4	5	6
1	2	3	4	5	6	7
2	3	4	5	6	7	8
3	4	5	6	7	8	9
4	5	6	7	8	9	10
5	6	7	8	9	10	11
6	7	8	9	10	11	12

   (b) (i) $\frac{1}{12}$ (ii) $\frac{1}{12}$ (iii) $\frac{5}{6}$
   (c) They cover all possible scores.
3.

Coin \ Dice	1	2	3	4	5	6
H	H1	H2	H3	H4	H5	H6
T	T1	T2	T3	T4	T5	T6

   (a) $\frac{1}{12}$ (b) $\frac{1}{4}$ (c) $\frac{1}{12}$ (d) $\frac{1}{4}$ (e) $\frac{1}{6}$ (f) $\frac{1}{2}$
4. (a) Stage 1: Bus, Bus, Train, Train, Lift, Lift
   Stage 2: Bus, Walk, Bus, Walk, Bus, Walk
   (b) $\frac{1}{6}$
5. (a)

1st spin \ 2nd spin	1	2	3	4
1	2	3	4	5
2	3	4	5	6
3	4	5	6	7
4	5	6	7	8

   (b) (i) $\frac{1}{16}$ (ii) $\frac{1}{8}$ (iii) $\frac{3}{16}$
6. (a) Maths, English; Maths, Science; Maths, Art; English, Science; English, Art; Science, Art
   (b) $\frac{3}{6} = \frac{1}{2}$ (c) $\frac{1}{3}$
7. (a)

Bag B \ Bag A	R	R	W
W	RW	RW	WW
W	RW	RW	WW
R	RR	RR	WR

   (c) $\frac{4}{9}$

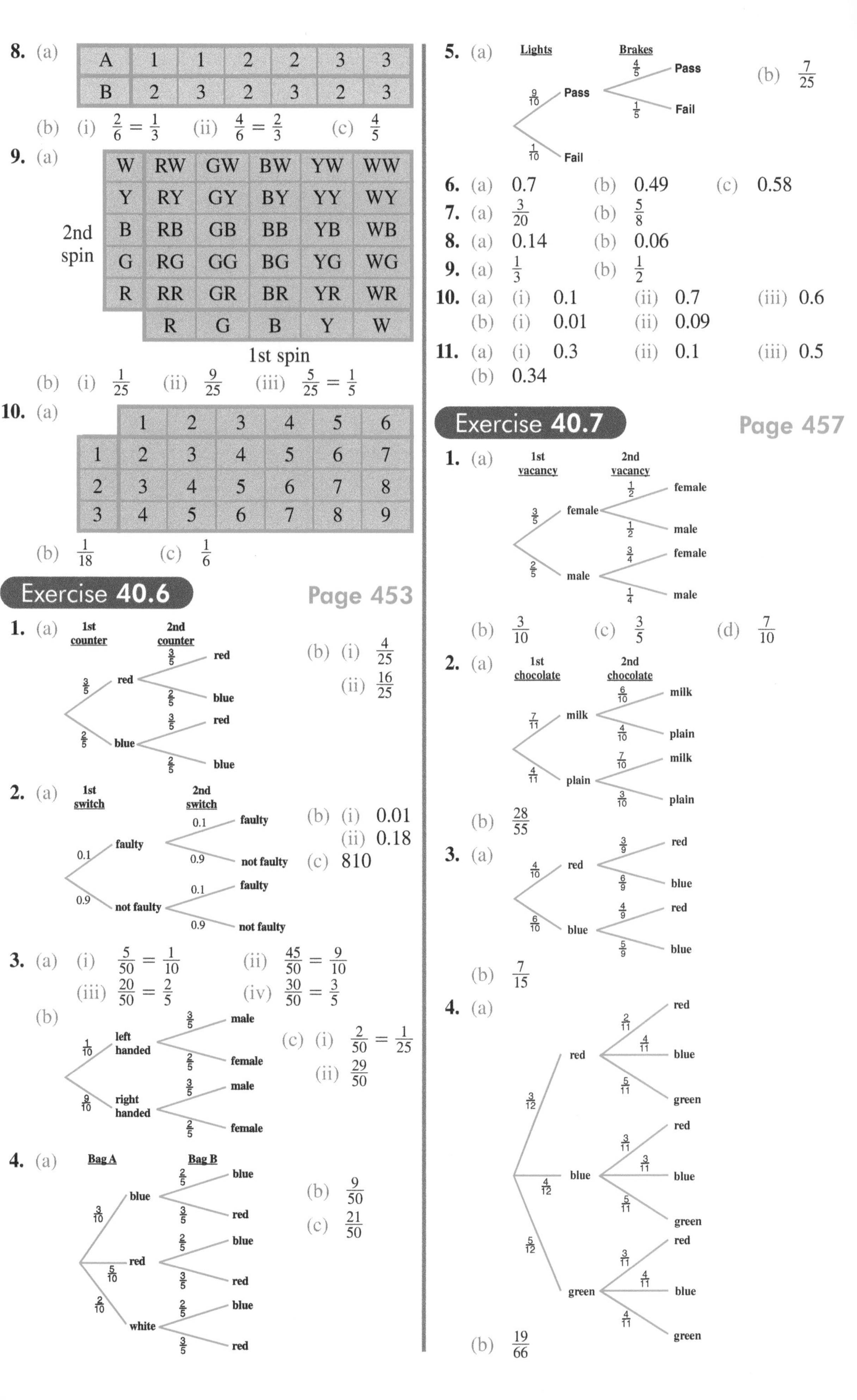

8. (a)

A	1	1	2	2	3	3
B	2	3	2	3	2	3

(b) (i) $\frac{2}{6} = \frac{1}{3}$ (ii) $\frac{4}{6} = \frac{2}{3}$ (c) $\frac{4}{5}$

9. (a) 2nd spin (rows) / 1st spin (columns)

2nd spin	R	G	B	Y	W
W	RW	GW	BW	YW	WW
Y	RY	GY	BY	YY	WY
B	RB	GB	BB	YB	WB
G	RG	GG	BG	YG	WG
R	RR	GR	BR	YR	WR

(b) (i) $\frac{1}{25}$ (ii) $\frac{9}{25}$ (iii) $\frac{5}{25} = \frac{1}{5}$

10. (a)

	1	2	3	4	5	6
1	2	3	4	5	6	7
2	3	4	5	6	7	8
3	4	5	6	7	8	9

(b) $\frac{1}{18}$ (c) $\frac{1}{6}$

Exercise 40.6 — Page 453

1. (a)

(b) (i) $\frac{4}{25}$ (ii) $\frac{16}{25}$

2. (a)

(b) (i) 0.01 (ii) 0.18 (c) 810

3. (a) (i) $\frac{5}{50} = \frac{1}{10}$ (ii) $\frac{45}{50} = \frac{9}{10}$ (iii) $\frac{20}{50} = \frac{2}{5}$ (iv) $\frac{30}{50} = \frac{3}{5}$

(b)

(c) (i) $\frac{2}{50} = \frac{1}{25}$ (ii) $\frac{29}{50}$

4. (a)

(b) $\frac{9}{50}$ (c) $\frac{21}{50}$

5. (a)

(b) $\frac{7}{25}$

6. (a) 0.7 (b) 0.49 (c) 0.58

7. (a) $\frac{3}{20}$ (b) $\frac{5}{8}$

8. (a) 0.14 (b) 0.06

9. (a) $\frac{1}{3}$ (b) $\frac{1}{2}$

10. (a) (i) 0.1 (ii) 0.7 (iii) 0.6
(b) (i) 0.01 (ii) 0.09

11. (a) (i) 0.3 (ii) 0.1 (iii) 0.5
(b) 0.34

Exercise 40.7 — Page 457

1. (a)

(b) $\frac{3}{10}$ (c) $\frac{3}{5}$ (d) $\frac{7}{10}$

2. (a)

(b) $\frac{28}{55}$

3. (a)

(b) $\frac{7}{15}$

4. (a)

(b) $\frac{19}{66}$

5. (a)

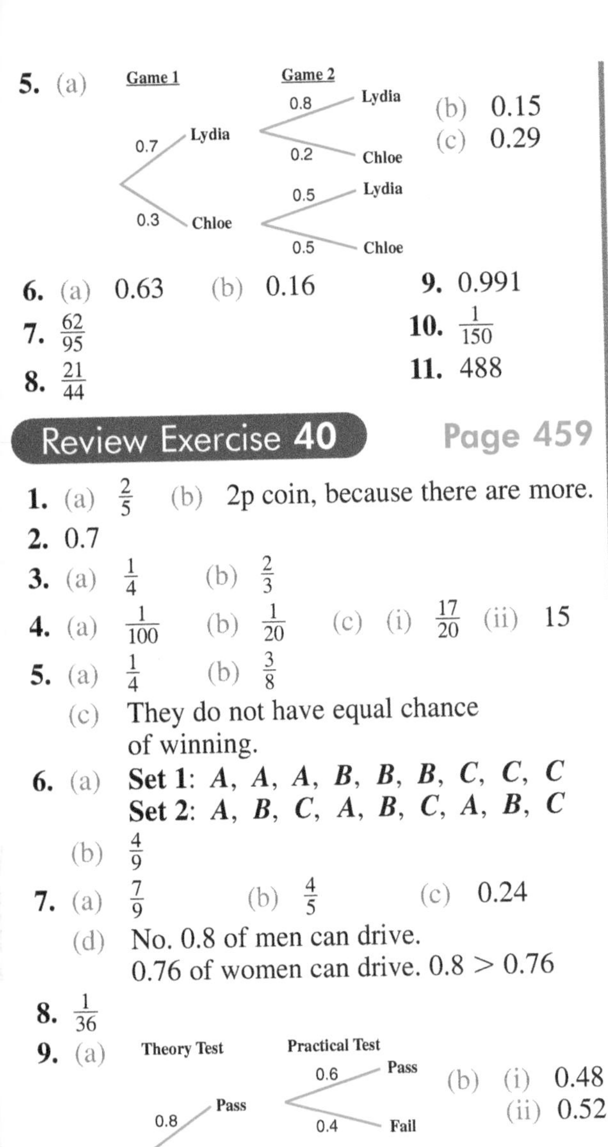

(b) 0.15
(c) 0.29

6. (a) 0.63 (b) 0.16
7. $\frac{62}{95}$
8. $\frac{21}{44}$
9. 0.991
10. $\frac{1}{150}$
11. 488

Review Exercise 40 — Page 459

1. (a) $\frac{2}{5}$ (b) 2p coin, because there are more.
2. 0.7
3. (a) $\frac{1}{4}$ (b) $\frac{2}{3}$
4. (a) $\frac{1}{100}$ (b) $\frac{1}{20}$ (c) (i) $\frac{17}{20}$ (ii) 15
5. (a) $\frac{1}{4}$ (b) $\frac{3}{8}$
   (c) They do not have equal chance of winning.
6. (a) **Set 1**: ***A, A, A, B, B, B, C, C, C***
   Set 2: ***A, B, C, A, B, C, A, B, C***
   (b) $\frac{4}{9}$
7. (a) $\frac{7}{9}$ (b) $\frac{4}{5}$ (c) 0.24
   (d) No. 0.8 of men can drive.
   0.76 of women can drive. $0.8 > 0.76$
8. $\frac{1}{36}$
9. (a)

Theory Test — Practical Test
0.8 Pass — 0.6 Pass, 0.4 Fail
0.2 Fail

   (b) (i) 0.48 (ii) 0.52
10. (a) 0.72 (b) 0.26
11. (a) $\frac{1}{25}$ (b) $\frac{1}{5}$ (c) $\frac{10}{25} = \frac{2}{5}$
12. $\frac{8}{15}$
13. (a) 0.18 (b) 0.435
14. 200
15. (a) $\frac{3}{10}$ (b) 200
16. $\frac{14}{30} = \frac{7}{15}$

Handling Data — Section Review

Non-calculator Paper — Page 462

1. (a) 3 (b) 5.1
   (c) Males have greater range, 7 compared with 3.
   Females have greater average, 5.1 compared with 3.6.
2. (a) 1H, 2H, 3H, 4H, 5H, 6H, 1T, 2T, 3T, 4T, 5T, 6T.
   (b) $\frac{1}{12}$
3. (a) Only women surveyed.
   Only one location used for survey.
   Sample size too small.
   (b) Survey equal numbers of men and women. Survey all over the country.
   Survey a much larger sample.
4. (a) $\frac{3}{20}$ (b) $\frac{7}{20}$
5. (a) $\frac{1}{6}$
   (b) Probably not, as 2 occurs twice as many times as any other number.
6. (a) $\frac{7}{9}$ (b) $\frac{6}{25}$
   (c) Can swim: girls 0.76, boys 0.8.
   Not true. $0.8 > 0.76$
7. (c) Positive correlation.
   Taller trees have wider trunks.
   (d) 13 cm is closer to the line of best fit for a tree which is 3 m tall.
8. (a) Q1: Too open, vague.
   Q2: Too long **and** leading.
   (b) How many hours per week do you read?
   ☐ less than 1 ☐ 1 to 3 ☐ more than 3
9. Estimated total weight
   $= (8 \times 40) + (10 \times 60) + (4 \times 80) = 1240$ kg
   Likely to be overloaded.
10. (a)

450 500 550 600 650 700 750
Weight (grams)

   (b) 120 g (c) (i) 0 (ii) 20
11. (a) $\frac{4}{25}$ (b) $\frac{6}{25}$ (c) $\frac{4}{25}$
12. (b) 150 days
13. (a) Choose, at random, $\frac{380}{700} \times 50 \approx 27$ men and $\frac{320}{700} \times 50 \approx 23$ women.
   (b) Both men and women are included in proportion to total population.
14. (a) (i) 39 (ii) 26
   (b) Not representative sample of TV watchers.
15. (a)

Score	0	5	6
Probability	$\frac{2}{3}$	$\frac{1}{6}$	$\frac{1}{6}$

   (b) Option A. If he chooses Option B he only has a $\frac{1}{3}$ chance of getting a higher score.

Handling Data — Section Review

Calculator Paper — Page 465

1. (a) 5.9 cm (b)

4	7	9			
5	0	3	6	8	
6	1	4	5	5	7
7	2				

4 | 7 means 4.7 cm

2. (a) 0.4 (b) 18

3. (a) 600 (b) Larger sample.

4. (a) 0.6 (b) 0.2

5. (a) E.g. Most people travel to out-of-town shopping supermarkets by car and so a high proportion of drivers will be included.

(b) 23 000

6. (a) $\frac{12}{20} = \frac{3}{5}$ (c) 500

7. (b) Positive correlation.

(d) Plotted points close to line of best fit.

8. (a) 13.4 minutes

9. (a)

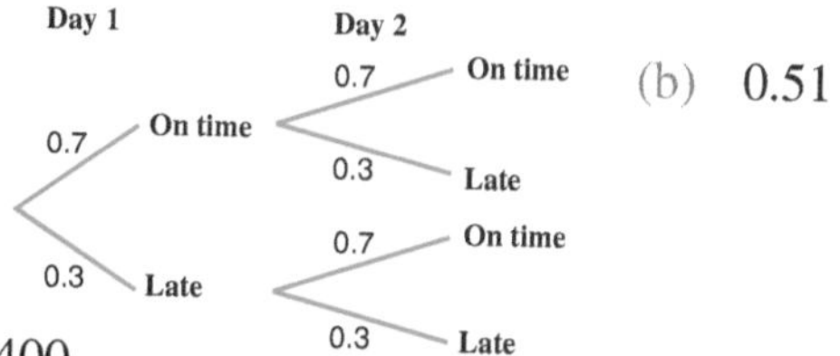

(b) 0.51

10. (a) 400

(b) There are three terms to a cycle of one year, so, a 3-point moving average will smooth out termly variations.

11. (a)

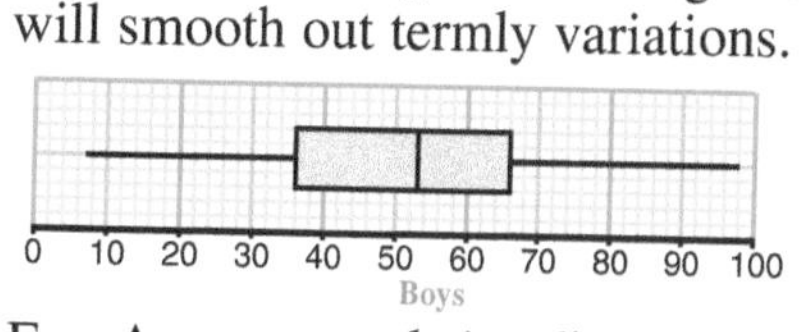

(b) E.g. Average mark (median) similar, but girls have a smaller interquartile range (marks closer together).

12. $\frac{4}{7}$

13. Ashworth 120, Brigholm 51, Corchester 29

14. (a) 8 (b) 120 (c) 51.4

15. (a)

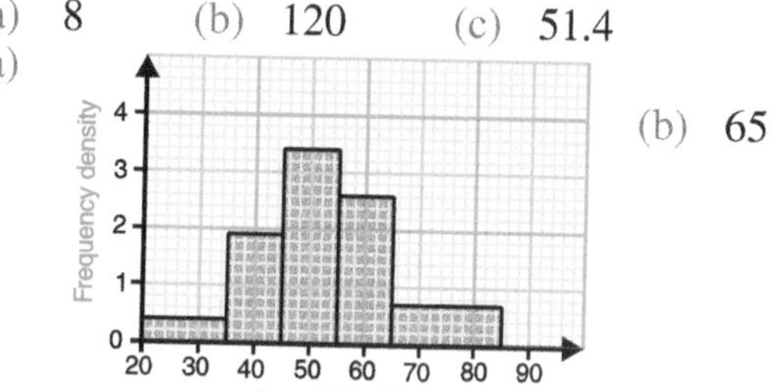

(b) 65

16. (a) 0.4875 (b) 0.3025

17. $\frac{7}{15}$

Exam Practice

Non-calculator Paper Page 468

1. (a) $x = 54°$

(b) (i) $y = 63°$ (ii) alternate angles

2. (a) 50 (b) $2t - 3$ (c) (i) $x = 7$ (ii) $t = 1\frac{1}{2}$

3. (a) 80 km/h (b) 45 minutes

(c) 30 km (d) 20 km/h

4. $5 \times 5 \times \pi = 25\pi \text{cm}^2$

5. (a) 0.55 (b) 0.1 (c) 0.6

6. (a) a^5 (b) b^4 (c) c^6

7. 32

8. $\frac{300 \times 4}{0.2} = 6000$

9. (a) $3n + 2$ (b) $3n - 1$

10. (a) $n(2n - 1)$ (b) $3 - 4x - x^2$

(c) $x = -\frac{4}{5}$

11. (a) Rotation, 90° anticlockwise, centre (0, 0).

(b) Reflection in $y = x$.

(c)

12. $LM = \sqrt{45}$ cm

13. (a) (b)

(c) 1 : 2

14. (a) $2^5 \times 3$ (b) 12

15. (a) $m = 30$

(b) (i) $3x^2y - 6y^2$ (ii) $3x^2 - x - 10$

16. $NL = 5.4$ cm

17. (a) $v = 9.2$ (b) $t = \frac{v - u}{a}$

18. (a) 2, 3 (b) $x = -2, \ y = 0.5$

(c) $x = 0$ or 5

19. (b) 50 (c) (i) 53 (ii) 29

(d) E.g. Town has lower median and ages are more closely grouped.

20. (b) $x = -1.2$ or 3.2

(c) $k = -5$

(d) $y = 1 - x, \ x = -1.8$ or 2.8

21. (a) 4×10^{11} (b) 2.5×10^{-3}

22. (a) (i) $15p^5$ (ii) $2p^3$ (iii) $\frac{x - 2}{x}$

(b) (i) $3x(x - 2y)$ (ii) $(2n + 1)(n - 3)$

23. (a) $x = 4$ cm, $y = 9$ cm (b) 90.5 mm

24. (a) $P(0, 4)$ (b) $y = 2x + 4$

25. (a) $\frac{4}{3}abc$

(b) Inconsistent dimensions.
a^3 has dimension 3, volume
bc has dimension 2, area

26. (a) $y = 1.9$ (b) $x = 2$ or 4

27. (a) $P = 0.09$ (b) $n = \frac{m}{\sqrt{P}}$

28. Coordinates of enlargement:
$P'(4, 2\frac{1}{2}), \ Q'(2, 2\frac{1}{2}), \ R'(2, 1\frac{1}{2})$

29. Population must be divided into groups. Example: By age, gender, occupation, ... Random sample chosen from each group in proportion to size of group.

30. (a) $\frac{\sqrt{3}}{6}$ (b) $\sqrt{6}$

31. (a) $A(-2, 0), B(4, 0)$ (b) $x = 3$

32. (a) $x^2 + (x + 7)^2 = 25$
$x^2 + x^2 + 14x + 49 = 25$
$2x^2 + 14x + 24 = 0$
$x^2 + 7x + 12 = 0$
(b) $A(-4, 3), B(-3, 4)$

33. $\cos PQR = \frac{4^2 + 5^2 - 7^2}{2 \times 4 \times 5} = \frac{16 + 25 - 49}{40} = -0.2$
As $\cos PQR$ is negative, angle $PQR > 90°$.

34. (a) $\frac{7}{125}$ (b) 2^{18} (c) $\frac{1}{27}$

35. $\angle QPB + \angle BSR + \angle RXQ = 180°$
$\angle QAB = 180° - \angle QPB$
$\angle BAR = 180° - \angle BSR$
$\angle QAR = 360° - \angle QAB - \angle BAR$
$= 360° - (180° - \angle QPB) - (180° - \angle BSR)$
$\angle QAR = \angle QPB + \angle BSR$, So,
$\angle QAR + \angle QXR = 180°$ and $XQAR$ is cyclic.

36. $90\pi \text{ cm}^2$

37. (a)

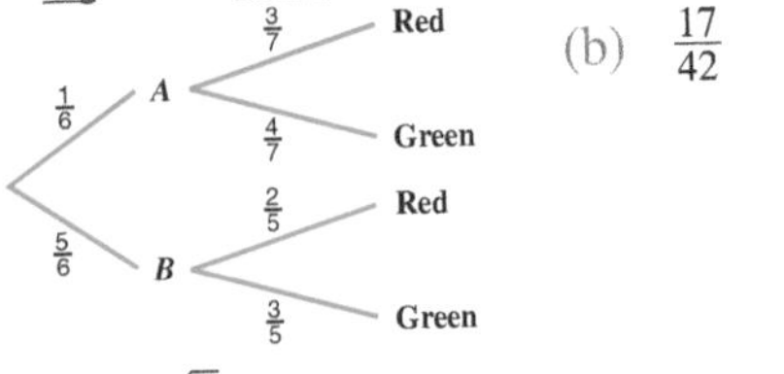
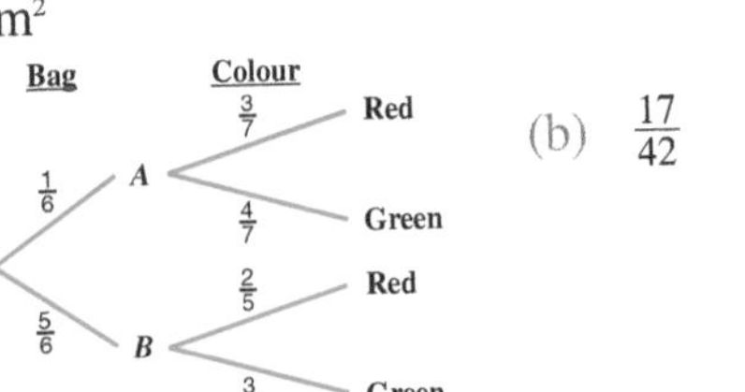
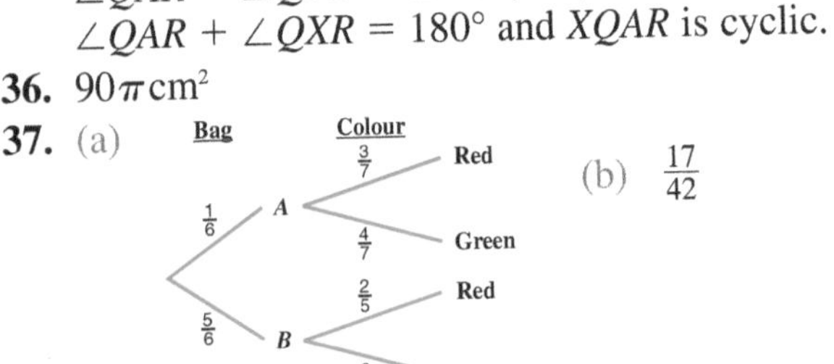
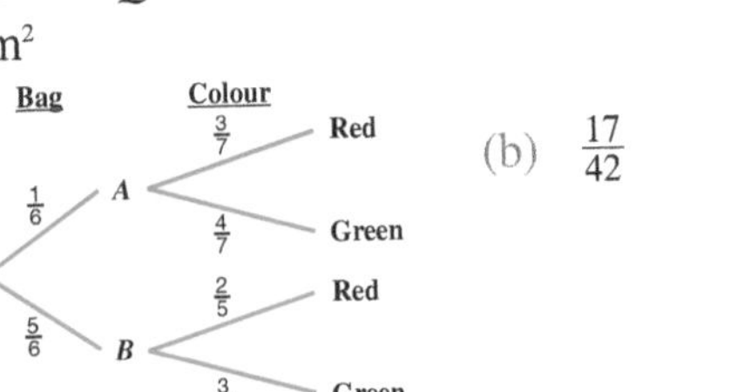

(b) $\frac{17}{42}$

38. $x = -2 \pm \sqrt{2}$

39. (a) $\overrightarrow{AB} = \mathbf{b} - \mathbf{a}$ (b) $\overrightarrow{PX} = \frac{1}{2}(\mathbf{a} + \mathbf{b})$
(c) $\overrightarrow{PY} = 2 \times \overrightarrow{PX} = 2 \times \frac{1}{2}(\mathbf{a} + \mathbf{b}) = \mathbf{a} + \mathbf{b}$
(d) $\overrightarrow{BY} = \mathbf{a}$
(e) $\overrightarrow{PA} = \overrightarrow{BY}$, so, PA is equal in length and parallel to BY.
A quadrilateral with opposite sides equal and parallel is a parallelogram.

40. (a) $y = x^3 - 5x + 6$ (b) $y = 5x - x^3$

Exam Practice

Calculator Paper

Page 472

1. 200 m^2
2. Size 1: 12.8 grams/penny
Size 2: 13.6 grams/penny
Size 2 gives more grams per penny.
3. £42.30
4. 13.4
5. (a) 503 cm (b) 3 complete turns
6. 6000 cm^2
7. 35 g
8. 45 m^2
9. (a) (i) $x = -1$ (ii) $x = 3.2$
(b) $3(a - 2b)$ (c) $3x - 6$
10. (a) $20x$ grams (b) 180 grams
11. (a) 50 mph to less than 60 mph (b) 58 mph
12. (a) £281 (b) $87\frac{1}{3}\%$
13. (a) 31.284 (b) $\frac{90 \times 10}{20 + 10} = 30$

14. 5 hours 12 minutes
15. (a) $2^2 \times 3^2$ (b) 180
16. $7 - 2n$
17. (a) If $\angle B$ is a right angle, then by Pythagoras on ΔABC.
$AC^2 = AB^2 + BC^2$ $(39^2 = 15^2 + 36^2)$
(b) $x = 105°$
18. (a) (i) $x = 1$ (ii) $x = -\frac{2}{5}$
(b) $x = 7.3$
19. 4 years
20. 170 litres
21. (a) $m > 4$ (b) $-1, 0, 1$
(c) $x = 3,\ y = -2$
22. $a = 57°,\ b = 114°,\ c = 123°$
23. (a) 333° (b) 21.7 km^2
24. (a) Missing entries: Philip 0.3, Abdul 0.4
(b) 0.46
25. £49
26. (a) $y = 3x - 4$ (b)

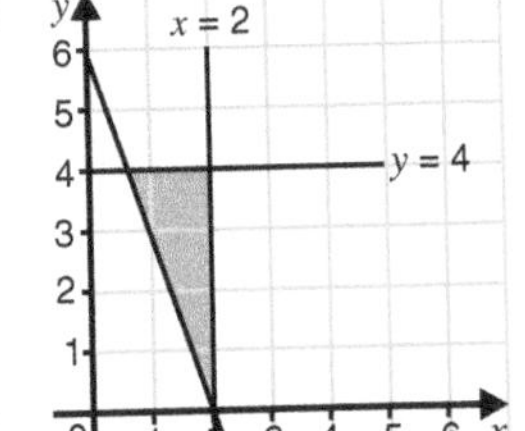

27. 22 cm^2
28. (a) $a = 25,\ b = -5$
(b) $x = 2.76$ or 7.24
(c) 5
29. (a) $W = \frac{r^3}{81}$
(b) $W = 2.67$ g
(c) $r = 11.7$ m
30. (a) $\frac{14}{33}$ (b) $\frac{49}{66}$
31. $PR = 254$ cm
32. (a) $r = \sqrt{\frac{3V}{\pi h}}$ (b) 1.6 cm^3
33. (a) (i) $(4x - 1)(x - 9)$ (ii) $x = \frac{1}{4}$ or 9
(b) $y = -\frac{1}{2},\ \frac{1}{2},\ -3$ or 3
34. (a) $x = 60°,\ 300°$ (b) $y = 210°,\ 330°$
35. (b) 6
36. $x = -0.28$ or 1.78
37. 7.7 cm
38. Distance 15.2 km, bearing 276.8°
39.

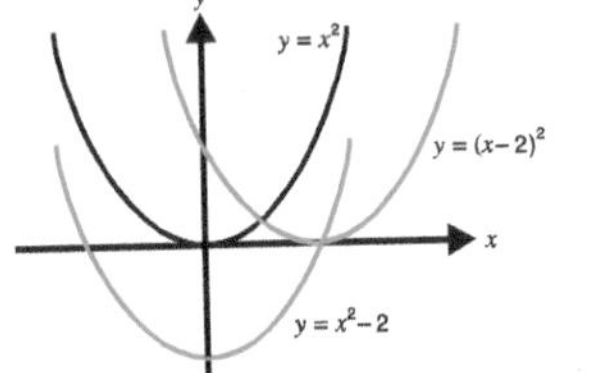

40. (a) $2x^2 + 4xy = 80$
$x^2 + 2xy = 40$
$2xy = 40 - x^2$
$y = \frac{40 - x^2}{2x}$
(b) $x = 2.2$ $(x \neq -18.2)$. Volume $= 38.7 \text{ cm}^3$
41. $\frac{55.245}{10.25} = 5.39$ (2 d.p.)

Index